ANNUAL REVIEW OF BIOCHEMISTRY

EDITORIAL COMMITTEE (1995)

Responsible for the organization of Volume 64
(Editorial Committee, 1993)

Production Editor   PAUL J. CALVI JR.
Indexing Coordinator   MARY A. GLASS
Subject Indexer   KYRA KITTS

ANNUAL REVIEW OF BIOCHEMISTRY

VOLUME 64, 1995

CHARLES C. RICHARDSON, *Editor*
Harvard Medical School

JOHN N. ABELSON, *Associate Editor*
California Institute of Technology

ALTON MEISTER, *Associate Editor*
Cornell University Medical College

CHRISTOPHER T. WALSH, *Associate Editor*
Harvard Medical School

ANNUAL REVIEWS INC.   4139 EL CAMINO WAY   P.O. BOX 10139   PALO ALTO, CALIFORNIA 94303-0139

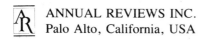

QP
501
.A7
v. 64

ANNUAL REVIEWS INC.
Palo Alto, California, USA

International Standard Serial Number: 0066-4154
International Standard Book Number: 0-8243-0864-6
Library of Congress Catalog Card Number: 32-25093

Annual Review and publication titles are registered trademarks of Annual Reviews Inc.

⊗   The paper used in this publication meets the minimum requirements of
American National Standard for Information Sciences—Permanence of Paper
for Printed Library Materials, ANSI Z39.48-1984.

Annual Reviews Inc. and the Editors of its publications assume no responsibility for the
statements expressed by the contributors to this *Review*.

Typesetting by Kachina Typesetting Inc., Tempe, Arizona; John Olson, President;
Jeannie Kaarle, Typesetting Coordinator; and by the Annual Reviews Inc. Editorial Staff

PRINTED AND BOUND IN THE UNITED STATES OF AMERICA

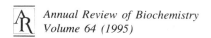

Annual Review of Biochemistry
Volume 64 (1995)

CONTENTS

(*Continued*)   v

vi   CONTENTS   *(Continued)*

SOME RELATED ARTICLES IN OTHER *ANNUAL REVIEWS*

From the *Annual Review of Pharmacology and Toxicology*, Volume 35 (1995)

> *Molecular Mechanisms and Therapeutic Approaches to the Treatment of African Trypanosomiasis*, C. C. Wang
> *The Pharmacology of the Gastric Acid Pump: The H^+,K^+ ATPase*, George Sachs, Jai Moo Shin, Carin Briving, Bjorn Wallmark, and Steve Hersey

From the *Annual Review of Neuroscience*, Volume 18 (1995)

> *Functional Interactions of Neurotrophins and Neurotrophin Receptors*, Mark Bothwell

From the *Annual Review of Genetics*, Volume 28 (1994)

> *Dynamic RNA-RNA Interactions in the Spliceosome*, Hiten D. Madhani and Christine Guthrie
> χ *and the RecBCD Enzyme of Escherichia coli*, Richard S. Myers and Franklin W. Stahl

From the *Annual Review of Immunology*, Volume 13 (1995)

> *Interleukin 12: A Proinflammatory Cytokine with Immunoregulatory Functions that Bridge Innate Resistance and Antigen-Specific Adaptive Immunity*, Giorgio Trinchieri
> *The Three-Dimensional Structure of Peptide-MHC Complexes*, Dean Madden

From the *Annual Review of Medicine*, Volume 46 (1995)

> *Class II Antigens and Disease Susceptibility*, Gerald T. Nepom
> *The Nuclear Hormone Receptor Gene Superfamily*, Ralff C. J. Ribeiro, Peter J. Kushner, and John D. Baxter

From the *Annual Review of Biophysics and Biomolecular Structure*, Volume 24 (1995)

> *Nucleic Acid Hybridization: Triplex Stability and Energetics*, G. Eric Plum, Daniel S. Pilch, Scott F. Singleton, and Kenneth J. Breslauer
> *Compact Intermediate States in Protein Folding*, Anthony L. Fink
> *DNA Analogues with Nonphosphodiester Backbones*, Peter E. Nielsen
> *Structure and Mechanism of DNA Topoisomerases*, Dale B. Wigley

Annu. Rev. Biochem. 1995. 64:1–28

TO BE THERE WHEN THE PICTURE IS PAINTED

Peter Reichard

Department of Biochemistry, Medical Nobel Institute, Karolinska Institute, 17177 Stockholm, Sweden

KEY WORDS: pyrimidine synthesis, allosteric mechanisms, radical proteins, ribonucleotide reductase, DNA evolution

CONTENTS

Roots

I have lived my whole professional life in Sweden, most of it at the Karolinska Institute, the medical school in Stockholm. However, my roots go back to another European country.

In 1925, I was born in Wiener Neustadt, a small town in Austria, about 40 km South of Vienna. In school, we learned about our town's heroic past as a bastion against invasions from the east by Magyars and Turks. By the time I arrived, that was very past history, and people made their living from small industry and the agriculture of the surrounding area, the *Steinfeld* (stone field). Most people were poor; during the depression of the 1930s they were very poor. Many of my classmates' fathers were out of work year after year. Their

1

hopeless situation makes it easy to understand their enthusiasm for the Nazi takeover in 1938; within a few weeks, everybody had a job.

My parents had a very different background. Both were born during the last years of the Habsburg Monarchy, but in different parts of the country. My mother's family was in Wiener Neustadt. The maiden name of her mother was Haiden (heathen). It was said to indicate that her ancestry went back to some of the infidels that stayed behind when the Turks were driven back from Austria in 1683. Be that as it may, the town and surrounding settlements were full of mother's relatives. Grandmother had married a manual worker, a simple man without major intellectual or political aspirations. She organized the women working in an ammunition factory during World War I, and later, after the war, became a member of the city council and then a representative for the Social Democrats in the Austrian parliament—one of its first female members. She must have been a remarkable woman since she accomplished all of this with only six years of grammar school. She died when I was one year old, and I am sorry that I never knew her.

My father was the youngest of 14 children of a Jewish family living in the ghetto of Kremsier, a small town situated in today's Czech Republic. His father was the cantor of the congregation and made part of his living by teaching the violin. Both grandparents started life somewhere in Hungary. According to family legends, my grandfather spent his early years wandering with gypsies and, while with them, learned to play the violin. Later, both grandparents came to the Austrian town of Graz, where grandfather sang in the Opera. Grandmother was a strong-willed woman who did not like the attention grandfather gave to the female singers, so the family moved to Czechoslovakia.

My parents met when my father started his career as a lawyer after World War I. Like so many bright young Jews who grew up in the outskirts of the monarchy, he had gone to Vienna to study law. He became an ardent organized socialist and was on friendly terms with many of the future leaders of the Social Democratic Party. I do not know what attracted him to the small provincial Austrian town, but once there, he could hardly avoid meeting my mother, who was secretary to the mayor. Father built a highly successful general practice, which included everything from traffic accidents to murders as well as civil and business law.

Thinking of my early childhood brings back many disparate images. I lived in a large house surrounded by a beautiful garden on the outskirts of the town. In grammar school, our teacher ruled 50 to 60 boys by the stick. Our summer vacations were in Italy, where I swam for the first time in the Adriatic and learned some Italian. This must have been the time when I fell in love with Italy and the sea. Now 60 years later, our summer place on a Swedish island is surrounded by the Baltic, and I work part time with my Italian wife at the University of Padova.

In February 1934, the economic misery in Austria exploded in a short civil war between the Social Democrats and the Catholic government that was easily won by the government. Some of the socialist leaders escaped to Czechoslovakia, but a few were executed. Father was briefly imprisoned in a small concentration camp in the vicinity of our town but was soon released.

In the meantime, I left grammar school and entered the gymnasium of the town. I received a classic education with daily Latin and Greek lessons. Chemistry and physics practically did not exist. If I ever thought of it, I presumed that I would enter law school and eventually join my father in his practice.

With the annexation of Austria by Germany in 1938, life changed completely. School became a nightmare. At home, the police searched the house in the middle of the night; I do not know for what. My parents began to talk of emigration. During the organized pogroms of the *Kristallnacht* in November, father was taken to the concentration camp of Dachau for one month. Fortunately, the former Secretary General of the Austrian Social Democratic Party, who was a close friend, had emigrated to Sweden and persuaded the Swedish party to provide an immigration visa for my father as a political refugee. Thus my Swedish life began at the age of 14 in the summer of 1939, a few months before the war. This is also one reason why, later in life, I found it difficult not to vote for the Social Democrats in Swedish elections.

Early Years in Sweden

I obtained a fellowship in a private boarding school in a small village near Stockholm. The school had a strong Quaker background and its friendly and quiet atmosphere was a tremendous relief. The school did not give grades, which was unheard of at that time. For me, it was an excellent system that provided time to adapt to new challenges and develop skills at my own pace. From Latin and Greek, I switched to modern languages and science. I was fortunate to be taught physics by a wonderful man who encouraged my growing interest in science. He recognized my limited mathematical ability and wisely directed my interest towards Chemistry. The school had a well-equipped laboratory suitable for easy analytical experiments, and I obtained considerable satisfaction from analyzing the composition of unknown mixtures of inorganic salts prepared by my teacher.

At the end of three years, I passed the examinations required to start a university education and began to think about my future life. My first choice was to become a chemical engineer. To enter the school of engineering in Stockholm, I needed some practical experience. I therefore worked for one year in a factory in the far north of Sweden. I enjoyed neither the long cold winter nor the dull work, which did not live up to my expectations. I decided no more chemistry for me.

Back in Stockholm, I was rather undecided on what to do. I had earlier considered entering medical school, but such a choice involved two problems. Medical school meant seven-and-a-half years of study followed by several years of little or no income. More important, as an alien citizen, I was not allowed to study medicine, even though I spoke Swedish fluently and had passed my examination with excellent grades. But times and rules changed, and my lack of citizenship suddenly no longer mattered. Equally important, I obtained a small monthly stipend that enabled me to contribute to the family budget. In January 1944, I was accepted as a medical student by the Karolinska Institute.

We were a small group of 40 and started with parallel classes of anatomy and general chemistry. Each subject had only one professor and one associate professor. Two senior medical students served as salaried teaching assistants (*amanuens*) and organized dissections or laboratory courses, helped by an unspecified number of younger unpaid medical students (extra *amanuenses*). These unpaid students usually also carried out a small research project while attending medical school. Graduate courses and examinations in relation to the research work did not exist. The combination of research, teaching younger students, and attending medical school meant hard work for the extra amanuenses, and most gave up after a year or two. A few remained and competed for the two salaried teaching assistantships. Successful research could finally, after considerable time, lead to a doctoral thesis.

Pehr Edman gave our first lectures in General Chemistry. This was a few years before he invented the famous Edman method for the sequential degradation of proteins. The lectures in Biochemistry were delivered by the professor, Einar Hammarsten, who clearly did not enjoy lecturing. Our ignorance during that time is difficult to comprehend today. Our German textbook, the *Lehnartz*, was descriptive, presenting chemical formulas of biological materials. It contained a few pages on nucleic acids and represented them as tetranucleotides, with stoichiometric amounts of the four bases. The general impression that we students received was that catabolic reactions could be described in chemical terms, whereas anabolism depended on cellular structure and therefore was not available for chemical experimentation. I remember the excitement with which I read *Dynamic Aspects of Biochemistry* by E. Baldwin at the end of the biochemistry course, and somewhat later, Schoenheimer's *The Dynamic State of Body Constituents*. Both books gave me a first glimpse of biochemical dynamism that was very different from the static view of Lehnartz.

Much to my delight, the introductory laboratory course involved qualitative analyses of mixtures of inorganic salts that did not cause me any problem. It was at this stage that I began to realize how much I enjoyed shaking test tubes and the outcome of simple experiments. I have done this throughout my life and still maintain contact with bench work as much as possible.

The Beginning of a Scientist

I passed the examination in Medical Chemistry with good grades, thought that maybe chemistry was not so bad after all, and began to consider a career in biological research. I was not particularly attracted to Einar Hammarsten as a teacher and decided to approach Torbjörn Caspersson, one of his former students. On my way, I met Pehr Edman by chance and told him about my intentions. He strongly discouraged me. I had looked at Caspersson as a genius who had discovered the importance of nucleic acids for protein synthesis. However, his largely technical research poorly suited my talents. In my youthful ignorance, I had not recognized this incompatibility, but followed Edman's advice and never approached Caspersson. A second incident finally brought me to Hammarsten. He invited one of my fellow students, Håkan Arvidsson, who was a distant relative of his, to start as an extra *amanuens* in the department. Håkan knew about my interests in chemistry and suggested that we both start. Thus, I was enrolled as an extra *amanuens* in Hammarsten's research.

Hammarsten was professor of Chemistry and Pharmacy, a title that started with Jöns Jacob Berzelius in 1810 when the Karolinska Institute was founded. The establishment of a new medical school had met violent opposition from the old medical faculty of the University of Uppsala. A major argument was that, with people like Berzelius as teachers, the students would not obtain the appropriate philosophical and humanistic education but be turned into a bunch of scientists. The opposition remained for a long time, and even though graduates from the Institute were allowed to practice medicine, it was not until 1908 that the Karolinska Institute could bestow upon its students the research degree of "Doctor of Medicine." No doubt, the professors of the Institute were consoled by the fact that since 1901 they selected the winners of the Nobel prize in Medicine or Physiology.

Many famous Swedish professors—among them Caspersson, Theorell, and Jorpes—started their scientific career in Einar's laboratory. He was a man who lived for his science and had the ability to instill his enthusiasm for research into his collaborators. His uncle, Olof Hammarsten, had been a well-known scientist who wrote a much-used textbook and discovered pentose in nucleic acid. Later in life, he became president of the University of Uppsala. Einar considered the decision to take this position to be an example of his uncle's poor judgment.

In his thesis, Einar Hammarsten had demonstrated that DNA from thymus was a high-molecular-weight substance when prepared without the then-usual alkaline treatment. Somewhat later, Erik Jorpes demonstrated that RNA from the pancreas did not contain equimolar amounts of the four bases and thus did not conform to the generally accepted tetranucleotide model. Their work did not receive the attention it deserved, largely due to the authority of P. A. Levene at the Rockefeller Institute, who was an advocate of the tetranucleotide concept.

At the time of my arrival, Einar was building a mass spectrometer in order to study nucleic acid metabolism with ^{15}N-labeled compounds. He was much taken by Schoenheimer's work that used ^{15}N-amino acids to measure protein turnover and planned to do similar work with nucleic acids. ^{14}C-technology was still in its infancy; moreover, the American Atomic Energy Commission had prohibited export of this radioactive isotope to Sweden, fearing that it might find its way to the Soviet Union. Thus, isotope work had to be carried out with stable isotopes or with ^{32}P.

My first assignment was to assist in the construction of the mass spectrometer. The physicist in charge soon realized my limitations. When, after one week, my screwdriver slipped and broke a major glass part of the machine, he discharged me from my duties. My friend Arvidsson had a much better hand with screwdrivers and stayed with the machine for the next two years. Not surprising, he became frustrated, left, and with time started a career in radiology.

I did not realize how lucky I was to be released from the project but thought that this was the end of my scientific career. After a week or two, I went back to the department prepared to resign. To my pleasant surprise, Einar just looked at me, with a twinkling smile in his eyes, and asked if I would prefer to crystallize deoxyribonuclease. I had never heard of the enzyme but was of course delighted. As a starting point for my work, he gave me a reprint. It was a paper by McCarty (1) describing the partial purification of deoxyribonuclease. Avery and he had used this enzyme to inactivate the pneumococcal transforming principle. I had never heard of the transforming principle and only much later did I realize the reason for Einar's interest.

Unfortunately, nobody in the laboratory, including Einar, knew how to purify an enzyme. Theorell was working next door in the building, and he would probably have been pleased to advise me. However, I did not know that he was working with enzymes, and nobody enlightened me on that point. It is difficult today to imagine the circumstances under which we worked at that time. Seminars or group meetings did not exist; everybody had to struggle by himself. The isolation and lack of guidance provided a strong selective pressure for independent students with a considerable degree of optimism. Our equipment was quite good then even though it was pitiful by present standards. I particularly remember the Köhler spectrophotometer that was used to measure the ultraviolet absorption of solutions at the wavelengths provided by a mercury lamp. The machine was placed in a small dark room and was preferentially used between midnight and four am when the streetcars outside the department were not running.

The arrival of several American visitors to the laboratory boosted my biochemical education. First came Richard Abrams. His PhD work had been in enzymology, and from him, I learned the importance of enzyme units, specific activity, and yields during purification. He was knowledgeable in many different fields, participated in the final stages of the construction of the mass spectrometer, and played a major role in getting the machine into workable shape. I am grateful for all he taught me and was happy when, 25 years later, he returned to Sweden on his sabbatical to work with me in the Nobel Institute. A little later, when I no longer worked with deoxyribonuclease, David Rittenberg and David Shemin arrived. Both were former collaborators of Rudolf Schoenheimer, my scientific hero. Shemin stayed in Stockholm for six months and determined the life span of chicken erythrocytes. He strongly influenced my development into a functioning scientist. I became his part-time assistant, took blood samples at odd hours from the wings of the chicken, and prepared heme for the ^{15}N analyses. More importantly, Shemin spent a good deal of time talking to me about how to do ^{15}N experiments, pointing to the lack of appropriate methods for the isolation of pure ^{15}N-labeled purines and pyrimidines from small amounts of nucleic acids.

At that time, chromatography in its different forms was introduced into biochemistry. Chargaff used paper chromatography to analyze purines and pyrimidines of nucleic acids. It occurred to me that partition chromatography might provide the solution to our problem. Chromatography on starch columns indeed proved a satisfactory tool, and my first publication in *Nature* described the separation of the four common ribonucleosides (2). Pehr Edman, who had conveniently just returned from a postdoctoral stint at the Rockefeller Institute, gave some advice and, more importantly, contributed to my efforts with his fraction collector. A large, old-fashioned alarm clock drove an eccentric cogwheel that made a round turntable move one step every hour and with it a series of beakers that were placed on its top. Each run was exciting because the beakers were placed loosely on the rubbermat of the turntable and sometimes moved to the wrong place before they were in their position under the column.

Now my work progressed rapidly. Starting from nucleic acids prepared by Hammarsten's method (3), I obtained ribonucleotides by alkaline hydrolysis, dephosphorylated them with prostatic phosphatase, and separated the nucleosides on my starch column. DNA was hydrolyzed with formic acid, and the free bases were again separated by using starch chromatography. This methodology was then used to analyze the ^{15}N content of separated nucleosides and bases from experiments in which labeled precursors had been administered to animals. The results of these analyses are today of little interest, except for one case involving ^{15}N-orotic acid (4). At the suggestion of Sune Bergström, we tested labeled orotic acid as the precursor of nucleic acid pyrimidines. The results were a resounding success and demonstrated a very efficient incorporation into the pyrimidines of both RNA and DNA.

These events took place within two years, and in the fall of 1949, I was ready to defend my thesis for a doctorate in medicine. A dissertation was a serious affair during which the defendant and two opponents, all dressed up in white tie, discussed the merits of the work. A faculty committee graded both the scientific content and the defense of the thesis, and on the basis of these grades, the faculty then decided if the defendant could become a *docent,* which was a prerequisite for an academic career. Olov Lindberg, professor in Cell Biology at the University of Stockholm, was one of the opponents appointed by the faculty. His kindness then and all his later encouragement were invaluable. I obtained the highest grade for both content and defense. This grade did not reflect the quality of my work alone. Hammarsten was of the opinion that all dissertations in Chemistry were much better than those in any other subject and therefore automatically deserved the highest grade.

Much had happened in Swedish academia between 1945 and 1949. I was fortunate to start my work at a time when politicians began to understand the importance of fundamental medical research. With government support the number of academic positions increased considerably. The Swedish Medical Research Council started its activity. The move of the Karolinska Institute from its downtown localities to the new campus at Solna had started. Everything pointed to a bright future for a person starting his academic career in biochemistry. Nevertheless, I hesitated. Parallel to my scientific work I had continued my medical education. I was tempted to leave halfway through medical school and do research fulltime. However, I needed only two more years to qualify as a doctor and a well-paid job in academic medicine. In addition, I married in 1949, and my wife Dagmar had postponed her budding career in law because we expected our first child. Fortunately, the Swedish system at that time permitted me to finish medical school and to continue research simultaneously, thus postponing a final decision. I am afraid that my knowledge of medical matters was rather limited when I took my final exams in 1951. Some of the research done during the previous two years had had a decisive influence on my scientific future and had eclipsed my medical studies.

The Beginnings of Ribonucleotide Reduction

The methodology I had worked out in my thesis was designed to analyze labeled pyrimidines obtained by degradation of nucleic acids. It struck me that the same methods could be used to prepare ^{15}N-nucleosides from biosynthetically labeled nucleic acids. The nucleosides could then be injected into rats as precursors of nucleic acids, similar to our previous experiments with orotic acid.

The striking outcome of experiments with labeled pyrimidine ribonucleosides (5) and deoxyribonucleosides (6) led to some bold conclusions. Ribonucleosides were incorporated into both RNA and DNA, whereas deoxyribonucleosides were used exclusively for DNA synthesis. From these results, we proposed the existence of an enzyme that transforms ribose to deoxyribose when the sugar is attached to a pyrimidine. Ribonucleotide reduction was born. Because labeled deoxynucleosides were exclusively incorporated into DNA, we hypothesized that radioactive thymidine might prove a useful tool for the visualization of DNA by radioautography. I remember with sorrow that I could not persuade any of my Swedish friends working with this technique that such experiments could be of interest.

During this time I also made the first transition from in vivo to in vitro experiments. Using slices from regenerating rat liver, I demonstrated that orotic acid was a normal intermediate in pyrimidine synthesis. Looking at these experiments today, I am struck by the scale: slices of 5 regenerating livers were incubated in a volume of 50 ml for up to 8 h. Today, we work with 50 µl samples and incubate for 20 min. Nevertheless, with this clumsy system, Ulf Lagerkvist and I later demonstrated that aspartic acid is a precursor of pyrimidines (7). Ulf had joined the laboratory a few years after myself and in his thesis devised a method for the chemical degradation of the pyrimidine ring (8). We now synthesized aspartate labeled in different positions with ^{13}C, ^{14}C, and ^{15}N and demonstrated incorporation of the whole molecule into orotic acid. By a similar technique, we also showed that carbamyl aspartate was an intermediate. Our findings were never properly recognized. We published them in *Acta Chemica Scandinavica*, aware that this was not a major biochemical journal. I suppose that we acted in the mistaken belief that our publications would transform it into one.

First Visit to the United States

After finishing medical school, I was ready for a postdoctoral period in the United States. I could count on receiving a fellowship from the Rockefeller Foundation. Support from the foundation was of enormous importance for biological research in Sweden, and Hammarsten's, as well as Caspersson's and Theorell's research at the Karolinska Institute, would not have been possible without it. I had decided to work one year at Stanford with Hubert Loring, whose work was related to my interests. One of Loring's former students, Elizabeth Anderson, was a postdoctoratal student in Sweden and described the beauty of Stanford to Dagmar and myself in glowing terms. So Stanford it had to be. Gerald Pomerat, the emissary from the Rockefeller Foundation, did not think that a visit to Loring's laboratory was a good idea and tried to persuade me to go somewhere else. I was stubborn already then and he gave in. His only condition was that we would stay at Stanford for 11 months only and would visit various American laboratories of my choice during the last month. This was not hard to accept, since travel expenses were to be paid by the Foundation.

In November 1951, Dagmar, our one-year-old son Per, and myself left Oslo by boat for New York on our first trip to the United States. My first encounter with American biochemistry made a profound impression. As Pomerat had predicted, my work in Loring's laboratory was not exciting. However, my visits to other laboratories during the final month introduced me into modern biochemical research. The period at Stanford was also the beginning of a long love story with that place. As a result, I returned many times for shorter or longer periods and worked there a total of almost four years, mostly in Arthur Kornberg's laboratory.

Traveling up and down the American continent in a ten-year-old car from Palo Alto to New York, with a two-year-old child in the backseat, was quite exciting. We must have visited almost ten towns that housed scientists active in the nucleotide and nucleic acid field and saw the best sides of my fellow scientists. I knew them through their work but had not met a single one of them. Everybody received us with open arms, me as a fellow scientist, Dagmar and Per as dear visitors to be taken care of by the family.

The discovery of my own scientific ignorance came as a shock to me and probably also to my hosts. For the first time in my life I gave seminars. The scientific content was reasonable, but I shudder when I think of the quality of my presentation. Two places stand out in my memory, Cleveland and Bethesda. In Cleveland, Harland Wood had brought together a wonderful group of scientists. We were house guests of Bob Greenberg who had begun to unravel the enzymology of purine synthesis. Before then, I did not really see a possible connection between the isotope studies that I had carried out and the enzyme work concerning nucleosides and nucleotides done by others. In Bethesda, I met Arthur Kornberg for the first time; he was to play an important role later in my life.

Our trip stopped short in New York. Our car had behaved erratically on several occasions but on the whole had served us faithfully. Now it suddenly became impossible to shift into reverse gear, which caused problems, not the least during parking. I was happy to sell it to a used-car dealer on Broadway, and after a few days extra vacation, we sailed for Gothenburg.

The Beginning of an Enzymologist

After my return to the Karolinska Institute, I continued to work on pyrimidine biosynthesis. The encounter with American science changed my approach, and I now started my second career as an enzymologist. Important for my development was a summer stay in Copenhagen during 1954, where I learned basic techniques in enzymology from Hans Klenow during efforts to prepare ribose 1,5-diphosphate. Bob Hurlbert, a postdoctoral student in Stockholm, and I had found that tissue extracts converted orotic acid to uridine nucleotides, and we suspected the involvement of ribose 1,5-diphosphate. In the meantime, Arthur

Kornberg demonstrated that 5-phosphoribosyl pyrophosphate was the true intermediate (9).

For quite some time, I studied the enzymatic formation of carbamyl aspartate from aspartate and demonstrated that two steps were required (10). The first was identical to the previously known first step in citrulline synthesis and led to the formation of an "active" carbamyl compound (Compound X) that in the second step was transcarbamylated to aspartate. Compound X had been discovered by Grisolia and required acetylglutamate for its formation. This was my first encounter with an allosteric effect, but I was not aware of it. Instead, I believed erroneously that acetylglutamate was part of the active carbamyl compound structure until it was shown, by chemical synthesis in Lipmann's laboratory, that the active compound is carbamyl phosphate (11). Purification of aspartate transcarbamylase then became simple (12). The experience was somewhat painful but emphasized the importance of outguessing an intermediate in a reaction sequence.

Aspartate transcarbamylase is famous as the first allosteric enzyme. This story began in 1956 with a report by Yates & Pardee (13) showing that the enzyme activity in extracts of *E. coli* was inhibited by cytidine and CMP. Much to my surprise neither cytidine nor CMP inhibited my pure transcarbamylase. Later experiments showed that CTP is the allosteric inhibitor. In the crude extract, CTP was formed by phosphorylation of cytidine and CMP. For once, the pure enzyme was not superior.

I now supervised my first graduate students, Nils Ringertz and Ola Sköld. Nils was officially in Caspersson's department working with sulfated nucleotides and mucopolysaccharides, somewhat outside Caspersson's own interest. I had some knowledge of nucleotide chemistry and was pleased to learn about polysaccharides in order to advise him in his work. Ola's work was on uridine kinase and phosphorylase. His experiments aroused medical interest when, together with George Klein, we showed that the development of resistance against fluorouracil could result from the loss of one or both enzymes. This was one of the first demonstrations that resistance to base analogues in mammalian cells is caused by loss of the activating enzymes.

My studies concerning pyrimidine synthesis were reasonably successful and provided me with funding from the Medical Research Council. My attempts to advance the ribonucleotide reduction story were for many years less successful. Incubation of labeled ribonucleotides with extracts from various cells never provided convincing evidence for the formation of deoxyribonucleotides. Discussions with people more knowledgeable in organic chemistry than myself were discouraging. There was no chemical precedence for this kind of reaction, and therefore the reaction did not exist. However, isotope experiments with intact cells continued to provide positive evidence. I was also greatly encouraged by Arthur Kornberg during his visit to Stockholm. When I complained about the small traces of radioactivity in dCMP after incubation of an *E. coli* extract with labeled CMP, he told me that these amounts of radioactivity were greater than what he had recovered in DNA during early efforts to demonstrate DNA replication in vitro. And then suddenly success arrived (14). I vividly recall when the sample recorded hundreds of counts instead of the usual five to ten. On that occasion, I had added just the right amount of ATP and Mg^{2+} to the bacterial extract. The function of ATP is twofold: it is an allosteric effector for the reductase, and it transforms CMP to CDP, the substrate in the reaction. With too much ATP, all CDP is transformed to inactive CTP, but I realized this much later. The first inkling of an allosteric regulation came in 1960 when I collaborated with Van and Zoe Canellakis in New Haven, Connecticut, and we observed complicated patterns of inhibition and stimulation of ribonucleotide reduction by deoxyribonucleotides with extracts from chick embryos (15). However, allostery was not yet invented, and we did not understand these phenomena for several years.

By 1960, I had become a reasonably well established scientist. At the end of my stay in New Haven, Dagmar and I again crossed America by car, this time going from New Haven to Berkeley. Many more scientists were working in my field of interest, and this time I knew them in advance. The friendly reception was the same. My seminars were better than those given in 1952 and were well received, although I was somewhat subdued by one of Arthur Kornberg's commentaries after my talk at Stanford. I recounted some complicated experiments with a crude *E. coli* extract that indirectly showed that CDP, and not CMP or CTP, is the substrate reduced by the enzyme. Arthur said that these were nice experiments but that he would have purified the enzyme first and then it would have become apparent that CDP was its substrate. Nevertheless, he was favorably impressed, and our lifelong friendship began on this occasion.

The decade from 1950 to 1960 was a period of growth. Our three children were born. We moved from a small apartment in central Stockholm to a house in the suburbs and, in 1960, bought our summer place in the Stockholm archipelago. I widened my professional competence and became a reasonably good enzymologist. Arne Tiselius, who was professor of Biochemistry with the Science faculty in Uppsala, once expressed his surprise about the number of Swedish medical doctors who were excellent biochemists. After all, none of us had any formal chemical education, and we all had to pick up knowledge while we were moving along in our science.

In 1952, after my return from the United States, I had been appointed assistant professor of Medical Chemistry. My teaching obligations were considerable. I lectured not only to medical students but also to students of dentistry and to laboratory technicians. I rather enjoyed teaching, and I believe that with time I became a good teacher. I obtained my first grants from the Medical Research Council, and later from the Swedish Cancer Society. My grants paid for two laboratory technicians. Such people, who are trained in special professional schools, constitute an important part of the Swedish research establishment. My present associate, Rolf Eliasson, comes from such a school. We started our collaboration more than 30 years ago. At the end of the decade, I obtained my first American grant from the Damon Runyon Foundation and later from the Public Health Service. These grants were tremendously important. For example, I was one of the first scientists at the Institute who could afford a scintillation counter. I could buy commercially available ^{13}H-CDP to use as a substrate for the reductase instead of ^{32}P-CDP, which we had to synthesize ourselves once a month starting from 30 mC ^{32}P$_i$.

This period also had its darker aspects. My position as assistant professor was not renewable. I applied for several professorships, one of them as associate professor at the Institute. I optimistically thought that I would get the job but was mistaken, nor did I obtain any of the other professorships. This event occurred at a time of transition in the department. Einar Hammarsten retired and Sune Bergström came from Lund to replace him. He brought a very active, large group of young people with him. Ulf Lagerkvist and myself, the leftovers from the previous era, felt rather isolated. For some time, I was torn by doubts about my future. I had an attractive offer from Philip Handler to come to Duke University, but I did not wish to leave Sweden for family reasons. There was the alternative of leaving biochemistry altogether and making use of my medical education. However, my wife Dagmar strongly advised against it. She was a pillar during these difficult times and her belief in my scientific ability provided me with the necessary staying power, and everything worked out. At the end of my seven years as assistant professor, the Medical Research Council offered me a permanent research position. Two years later, I applied for my fourth professorship and this time I succeeded. As of July 1, 1961, I was Professor of Medical Chemistry at the University of Uppsala.

Professor in Uppsala

I arrived in Uppsala at a good moment and full of energy. Compared to Stockholm, both the medical faculty and the Department of Medical Chemistry were small. My predecessor, Gunnar Blix, had discovered sialic acid but by the time I arrived had more or less retired from research. The second professor and chairman of the department, Gunnar Ågren, was also a student of Einar Hammarsten and received me with open arms. I brought all my equipment from Stockholm to Uppsala and the university provided the required funds for the remaining purchases. Within a few weeks, the three graduate students and two technicians that accompanied me started new experiments which were soon aided by new students eager to join our team.

I was also pleased that Torvard Laurent, an old friend from Stockholm, returned from several years of research in the United States and chose to come to Uppsala. He carried out pioneering work in the polysaccharide field and his expertise in physical chemistry was helpful in our work. He became my successor when I left Uppsala in 1964 and since then has built up one of the strongest Swedish departments of medical chemistry. At that time, a university professor was an important person. We were fewer than we are now. In Uppsala, every professor was formally inaugurated. On this day the church bells rang in the morning. Later, I marched in the academic procession between the president and the vice president of the university together with most of my colleagues of the faculty to the large lecture hall to give a lecture on Biochemistry and Heredity to the general public. In the evening, we had invited the whole faculty and some honoratiores from the university to a formal dinner, together with family and friends. This was the first and only time that my father, who had come from Austria was dressed in white tie. Our 11-year-old, Per, served as toastmaster. It was a memorable occasion. A few years later, this century-old tradition was abandoned: there were too many new professors each year.

I now had the means and power to organize my own department in what I considered to be the proper way. I myself had received no formal training and little instruction as a graduate student and wished my students a better fate. Soon we began to have small meetings to discuss ongoing research. In Stockholm, I had started a journal club together with Joe Bertani and his group. We now had alternating meetings in Stockholm and Uppsala once a week and were soon joined by Hans Boman from Arne Tiselius' department. Joe and Betty Bertani had arrived in Stockholm in 1958 to teach us microbial genetics, at that time an unknown subject in Sweden. During their 20 years at the Karolinska Institute, they educated the first crop of Swedish microbiological geneticists. Betty worked with me during one year to learn some biochemistry and participated in the discovery of dUTPase (16). Joe eventually became disgusted with the educational and political system of Sweden and he and Betty returned to the United States. Hans Boman was something of an anomaly in Tiselius' place. There almost everybody discovered new methodology, but Hans discovered RNA methylation (17). He later moved to the University of Umeå and founded an outstanding Department of Microbiology. Then he moved to Stockholm and became a pioneer in insect immunology. Few people have his scientific imagination. Our journal club was of high quality even though my students sometimes complained that they could not follow all the genetic lingo. I admitted I had the same problem but assured them that they would understand if they heard it often enough.

In Tiselius' Biochemistry Department were also Jerker Porath and Bo Malmström. The proximity to Jerker's group gave us an edge in our work since we could use their ingenious methods for protein fractionation before they were published. Bo Malmström became a good friend. Later, when he was professor of Biochemistry in Gothenburg and had established an outstanding group working on metalloenzymes, I often obtained their advice concerning the peculiarities of the iron center of ribonucleotide reductase. In Uppsala, I also met Lennart Philipson for the first time, then a hungry, budding molecular biologist. Lennart had a certain ability to make enemies because of his extreme outspokenness. He also had, and still has, an even greater ability to infect others with his enthusiasm and to organize research. We became good friends and together fought the battles with taxonomists and journalists to make Sweden accept molecular biology.

The first American postdoctoral students arrived. In Stockholm, American visitors had been my first teachers. Later they came to Einar Hammarsten but usually ended up working with me. A third and most distinguished group came to Hugo Theorell. Their presence contributed considerably to my growth as a scientist. Two of them, Jack Buchanan and Joe Neilands, became good friends. Jack corrected the broken English in my first paper in *Nature* (2); Joe later returned to work with me during a sabbatical. I always greatly enjoy visiting him and Juanita at their home in Berkeley, which he had built with his own hands.

And now the young Americans came to Uppsala to join our research enterprise. We began to harvest the fruits of enzyme purification. Foremost was the identification of the hydrogen donor system for the reduction. Already, after a moderate purification, we found that ribonucleotide reduction depended on reduced lipoate, but for several reasons we did not believe that lipoate was the physiological hydrogen donor. Instead, we discovered that a 12-kb dithiol-protein, named thioredoxin, could fulfill this function (18). A second protein, thioredoxin reductase, maintained the reduced status of thiols of thioredoxin with the aid of NADPH (19). Two of the new Uppsala students purified and characterized the two proteins during their thesis work. Arne Holmgren continues his elegant work on thioredoxin and in the meantime has discovered many new functions for this molecule (20). He also identified glutaredoxin as a second hydrogen donor in ribonucleotide reduction. Arne is now my successor at the Karolinska Institute. Lars Thelander developed considerable skill in physical chemistry and enzymology during his studies of the flavoprotein thioredoxin reductase (21). He later applied these skills to studies of the structure and mechanism of the *E. coli* ribonucleotide reductase and then to the mammalian enzyme. He is now professor of medical chemistry in Umeå where he can also satisfy his yearning for outdoor life and hunting. Arne, Lars, and a third student, Uno Lindberg, who was one of the few who did not work with ribonucleotide reduction, were wonderful examples of what Uppsala could offer in the way of enthusiastic future scientists.

Back at the Karolinska Institute

Even though everything was wonderful I left Uppsala after only two and a half years and returned to the Karolinska Institute to succeed Erik Jorpes as professor of Medical Chemistry. I had many friends in Stockholm with whom I hoped to be able to collaborate, foremost among them Joe Bertani and George Klein at the Karolinska Institute, and Olov Lindberg and Lars Ernster at the University of Stockholm. Dagmar and I had started our friendship with Lars Ernster, then a docent at the University, and his wife Edit, a concertmaster at the Opera. Lars was beginning to be an authority in bioenergetics, a subject he had started as a student of Lindberg and continued with such great success during his time as a professor of biochemistry at the university and into his retirement. The Ernsters remained our best friends over the years.

Also important for my decision to move was that Uppsala University started to build a new large center for biomedical research. I did not look forward to becoming heavily involved in the planning and had my doubts about how such a huge organization would function without involving a large bureaucracy. In actuality, the center functions extremely well, largely because the various groups are wisely allowed to maintain their organizational independence. My short period in Uppsala was happy and productive. It has helped me to remember that first-class biomedical research in Sweden can also be done outside the Karolinska Institute.

In Stockholm, our growing group continued the line of research started in Uppsala. All students, old and new, and several of the technicians came with me to Stockholm. The number of guest scientists increased. I found that a good way to maintain the necessary interaction within the department was to eat lunch together. One of the dishwashers who proved to be a good cook was released from her duties, and a hot lunch was served in the lunchroom every day at noon sharp. The quality of the food as well as its price sufficed to persuade almost everybody to participate. Visitors were always invited and the lunchroom also provided excellent dinners for seminar speakers. Many visitors later recalled the food better than the science. The lunches were started without authorization, but none of the top administrators of the institute protested when invited. The lunches became a resounding success that I would keep up for 25 years. Then one of the technicians in the department denounced the lunches to the central authority that oversaw all government expenditure and that was the end of our noon meals.

Ribonucleotide Reductase from E. coli

I was now back in my original department, Medical Chemistry. Enzyme purification became a major activity and eventually resulted in a pure *E. coli* ribonucleotide reductase. During purification, the reduction of CDP provided the assay. Agne Larsson, one of my early graduate students, had already found in Uppsala that the activity vis á vis ADP and GDP disappeared early during purification and had assumed that purine and pyrimidine ribonucleotides were reduced by different enzymes (22). But Agne never could purify a purine-specific enzyme. I then remembered the experiments in New Haven with the reductase from chick embryos, in which we had found that GDP reduction was stimulated by dTTP (15). To our great satisfaction, Agne found that the purified reductase preparation rapidly reduced purine ribonucleotides when supplied with dTTP. This story ended four years later with the understanding of the remarkable allosteric properties of ribonucleotide reductase (23). The enzyme consists of two proteins, at the time called B1 and B2, later renamed R1 and R2. R1 contains two sets of allosteric sites that bind nucleoside triphosphates. This chapter is not the place to describe in detail the complicated and beautiful allosteric regulation of ribonucleotide reduction that makes one enzyme produce equimolar amounts of the four building blocks for DNA synthesis. Suffice it to say that we had always used ATP in our assays, which made the protein a pyrimidine specific enzyme. With dTTP, the enzyme could reduce purine ribonucleotides.

Agne Larsson also began to study the chemical mechanism of the reaction by showing that the reduction of carbon-2′ of the ribose occurred with retention of configuration (24). Other investigators found the same stereochemistry for the reaction catalyzed by the adenosyl cobalamin—requiring reductase, discovered by Blakley and Barker (25), even though the protein structures of the two enzymes differ greatly from each other. More recently, Joanne Stubbe's elegant work has clarified the mechanism of both types of reductases in great detail and extended their mechanistic similarity (26).

It was hard work to obtain large enough amounts of the *E. coli* reductase. DNA-technology for overexpression of enzymes was not yet invented, and the methods for enzyme purification were less sophisticated than today. After a joint effort of many technicians and postdoctoral students, foremost among them Lars Thelander and Neal Brown, we arrived at a reproducible method that after two weeks of hard work gave us several milligrams of R1 and R2 (27). Structural work became possible. R2 gave a characteristic spectrum, which suggested that the presence of iron in the enzyme was linked to its activity (28). My next-door neighbor was professor Anders Ehrenberg, a former student of Hugo Theorell, and a pioneer in the application of EPR spectroscopy to biological problems. When he analyzed the R2 protein by EPR spectroscopy we were in for a major surprise. No iron signal was found, but a sharp signal with all the hallmarks of an organic free radical (29). At first, we did not believe that the protein could contain a free radical, because it had spent two weeks or more in water solution during its preparation. This was the beginning of a long and fruitful collaboration with Anders Ehrenberg, and later with Astrid Gräslund, his student and now his successor, that led to the identification of the radical and to an understanding of the interplay between the iron center and the radical. Without their expertise and enthusiasm, this work would not have been possible. Britt-Marie Sjöberg played a decisive role in this collaboration, first as a postdoctoral student and now as Professor of Molecular Biology. She localized the organic radical to tyrosine-122 of the polypeptide chain of R2 (30). The iron center consists of a pair of Fe(III) ions linked by oxygen (31). Each R2 monomer contains one iron center. This feature became evident from the three-dimensional structure of the protein (32) even though our analyses erroneously first indicated only one center per dimer. The number of tyrosyl radicals is still not settled; radicals cannot be localized by X-ray crystallography. Tyrosine-122 is deeply embedded in the protein structure in close vicinity to the iron center.

The presence of an organic free radical in R2 suggested that ribonucleotide reduction proceeds by radical chemistry and that R2 provides the required radical (33). What is the function of R1? This protein contains not only the earlier-mentioned allosteric sites but also the catalytic sites for substrates (34). In 1974, Lars Thelander did an important experiment that demonstrated the stoichiometric oxidation of thiols of R1 to disulfides during the reduction of CDP (35). The dithiol functionality of the enzyme is restored by reduced thioredoxin or glutaredoxin. Later experiments done independently by Britt-Marie Sjöberg (36) and Joanne Stubbe (37) carried this work further and identified two separate groups of dithiols. One group participated directly in the reduction of the ribose, and the other interacted with thioredoxin. On the basis of extensive, elegant experiments, Stubbe proposed a model that describes the interaction of the tyrosyl radical with the thiols at the active site during the reduction of the ribose moiety (38). Recent crystallographic results by Ulla Uhlin and Hans Eklund concerning the structure of R1 (39) strongly support this model. I look forward to an extension of the crystallography to complexes between R1 and allosteric effectors.

This brings the account of research on ribonucleotide reductase from *E. coli* up to 1994. I may have spent too many words on this subject, but the enzyme has been the most important part of my scientific life. A brash conclusion in 1950 by a 25-year-old beginner in science, in part founded on my ignorance of organic chemistry, had a decisive influence on the rest of my life in science.

The Nobel Institute and the Nobel Prize in Medicine

In 1971, I succeeded Hugo Theorell as Professor of Biochemistry. He had been the head of the Medical Nobel Institute for Biochemistry at the Karolinska Institute since 1937. The institute was originally founded with Nobel funds accumulated mainly from prize money that was not used during the war years.

In 1959, the Institute became a Department of Biochemistry within the Karolinska Institute.

According to the statutes of the Nobel foundation, a Nobel institute should assist in the work of the prize-awarding institution. The original idea may have been to check on the results of prize candidates. In medicine, the prize-awarding institution (the Nobel Faculty) originally consisted of all the professors at the Karolinska Institute. In 1982, the Swedish law concerning matters of secrecy in government affairs was liberalized drastically, and to maintain secrecy of the deliberations concerning the prize, a private Nobel assembly was put in place of the previous federal Nobel Faculty. The assembly is a self-perpetuating institution of 50 members who must be professors at the Karolinska Institute.

A first attempt to start a medical Nobel institute was made in 1918 when influential members of the Nobel Faculty suggested starting an Institute of Racial Biology. However, this was not done since it was not apparent how such an institute could help in the deliberations for the Nobel prize. A federal Institute for Racial Biology was instead started a few years later at Uppsala University. The name is disgusting, but I remember the director of the institute during the war as an outspoken enemy of the Nazi ideology.

The second attempt to start a medical Nobel institute came in 1935, when Einar Hammarsten started a campaign for an Institute in Biochemistry with Theorell as its director. This time the argument could be made that the creation of the institute would allow the Karolinska Institute to retain Theorell and that his collaboration with the Nobel Committee was of importance for the selection of worthy prize winners. The proposal stressed that the director of the institute must continue his research activities in order to maintain his usefulness for the prize work. In 1937, the faculty of the Karolinska Institute was small, and a person of Theorell's caliber must have been a great asset for the Nobel work. He could now write expert opinions on prize candidates and closely collaborate with the Nobel Committee in other ways. He won the prize himself in medicine in 1955 and was therefore excluded from prize discussions during the early 1950s.

More than just the prestige of being director of the Nobel Institute induced my move from Medical Chemistry to Biochemistry. The number of medical chemistry students increased each year and so did the size of the department and its teaching load. I was attracted by the possibility of moving to a smaller department exclusively dedicated to research and of still maintaining my professorship at the Karolinska Institute.

On my return from Uppsala in 1964, I automatically became a member of the Nobel Faculty. I was immediately elected an adjunct member of the Nobel Committee and participated heavily in its work for the next 20 years. The directorship of the Medical Nobel Institute for Biochemistry could hardly increase my obligations in this respect. My interests in Nobel matters dated back to my early days in Hammarsten's laboratory. For many years he had considerable influence on the choice of prize winners and sometimes commented rather freely on what went on, in particular when he returned from a committee meeting that did not go his way. His interest in Avery's work was why he suggested I attempt the crystallization of deoxyribonuclease. Avery's discovery in 1944 that DNA is the carrier of genetic information is certainly one of the greatest biochemical discoveries of this century and should have been rewarded with a Nobel Prize, either in medicine or chemistry. At the time of Avery's discovery, the idea that DNA and not protein is the genetic material was foreign even to scientists like Hammarsten and Caspersson who realized that nucleic acids are complicated molecules and not tetranucleotides. Avery's evidence was far from complete, and among the outspoken nonbelievers was Alfred Mirsky, a close colleague of Avery at the Rockefeller Institute. I can therefore understand why Hammarsten was hesitant in the beginning. A mistake had happened once before in 1926 when a Nobel prize in medicine was awarded for a discovery that later turned out to be incorrect. However, Avery lived until 1955. By that time the evidence for DNA was overwhelming and included Watson & Crick's discovery of the double helix. In the early 1950s, many important discoveries were backlogged because no prizes had been given during the war but it remains something of a mystery how the Medical Nobel Faculty could avoid giving a prize to Avery.

I have sometimes been asked why a certain person is not a Nobel Prize winner. No omission is as blatant as Avery's, but I can name a few other scientists who, in my opinion, would have deserved a Nobel Prize. One must remember that a Nobel Prize is given for "the most important discovery within the domain of physiology or medicine" and not for the accumulated work of a whole life. Also, Nobel's will stipulates that the prizes should be given "to those who, during the preceding year, shall have conferred the greatest benefit on mankind." For obvious reasons it is hardly ever possible to give the Prize for a discovery that was made the preceding year. The sentence is interpreted by the prize-awarding institutions to mean that the importance of the discovery should have become apparent during the previous years. In medicine, the prizes awarded to Peyton Rous and Barbara McClintock represent two extreme cases where the importance of the discoveries was appreciated only much later. However, as a rule old discoveries may find it difficult to compete with more recent ones. An incubation period of ten years is certainly not unusual.

I found my work in the Nobel Committee highly stimulating and rewarding, certainly much more so than work in any of the other committees that I attended during my professorial years. I became familiar, at least to some extent, with many areas of biomedicine outside of my own specialty. As a biochemist, I was pleased to see that an increasing number of prizes went to scientists whom

I would classify as biochemists even though their research was classified as immunology, virology, pharmacology, or something else. My colleagues usually protest when I point out that the aim of today's biomedical research is to understand life in chemical terms.

Some say that the Nobel Prizes are a disservice to science and that Prize winners stop doing important research. However, scientists usually make their most important discovery before the age of 50, and most winners are older than that when they receive the prize. There are also many examples of the Prize winner continuing his or her research successfully, two of them being Arthur Kornberg and Gobind Khorana. Once a year, the award of the Nobel Prizes focuses the interest of society on intellectual achievement. The fact that so many other Prizes, instituted more recently, often wish to relate to the Nobel Prize is also telling. There can be no doubt about the usefulness of the Nobel Prize for Swedish science. It forces the Swedish scientists participating in the prize work to maintain contact with the cutting edge of science, and it brings the most outstanding international scientists to Sweden before and after they receive their Nobel Prize.

Sabbatical Leaves

In 1966, I took my first sabbatical leave at Stanford. The medical school had been started, and Arthur Kornberg had a laboratory there in which I worked for six months. Stanford was no longer "The Farm" it had been in 1952, and on trips to Santa Cruz we no longer went through blooming orchards. But such nostalgic feelings rapidly disappeared when I viewed the scientific atmosphere. Arthur is not only an outstanding scientist but he had also created a new department with other outstanding young people who could work together. Over the years, my many visits to Stanford, first to Arthur and more recently to Roger Kornberg, gave me the opportunity of uninterrupted bench work in a congenial atmosphere. Every visit, the latest in 1994, provides stimulation and pleasure.

My work in 1966 set the pattern for later visits: I participated in ongoing work on DNA replication in Arthur's group and initiated collaborative experiments linked to my work in Stockholm. I studied the replication of *phiX* DNA with Arthur, measured the fluorescence of thioredoxin with Lubert Stryer, and did NMR experiments aimed at the stereochemistry of ribonucleotide reduction with Lois Durham in the chemistry department.

After my return to Stockholm, I began to think about experiments that would connect ribonucleotide reduction with DNA replication. Influenced by Arthur's concept of viral windows on DNA replication, I spent my next sabbatical in 1970 at the Salk Institute in La Jolla and learned from Walter Eckhart how to handle polyoma virus. This experience led to a decade of experiments in Stockholm in which we used isolated nuclei from polyoma virus infected cells to study the replication of the viral DNA. The ultimate goal was to be able to extract the enzymes from the nuclei and to achieve a complete replication of the DNA. In addition, I hoped to be able to couple DNA replication with the enzymes isolated by Lars Thelander that were involved in mammalian ribonucleotide reduction. This was utopia. We were not able to initiate new rounds of replication, mainly because the large amounts of genetically engineered T antigen required for a successful experiment were not available. Our major contribution was the discovery and description of a particular kind of RNA that initiated the synthesis of Okazaki fragments (40). This RNA was characteristically 10 nucleotides long, started with A or G, but lacked sequence specificity. Many postdoctoral and graduate students in Stockholm worked diligently on this project, foremost among them Ernst Winnacker and Göran Magnusson. Both soon became independent professors—Ernst in Munich and Göran in Uppsala—and continued to work with animal viruses.

For many years, I returned to Stanford for one month each November to do some experimental work and to interact scientifically and socially with good friends. Stanford became my second intellectual home. On returning to Sweden, I always felt that I traveled back with a portion of Stanford in my luggage. Life became poorer when, in the late 1980s, I no longer could return because of the illness of my wife Dagmar.

A Return to Intact Cells

There came the time when I wished to find out if the complicated regulation of the pure ribonucleotide reductase in the test tube was relevant for the function of the enzyme in intact mammalian cells. In his thesis, Lambert Skoog devised an enzymatic method for the measurement of the small intracellular dNTP pools (41). In S-phase cells, the pools suffice only for a few minutes of DNA replication. The sensitivity of Skoog's method is such that it could be used to measure the dATP pool in resting lymphocytes, which amounts to 1 pmol/10^6 cells. We could now study pool fluctuations caused by manipulations of the reductase. One example was the paradoxical inhibition of DNA replication by thymidine. Gunnar Bjursell found that addition of thymidine to cells not only increased the dTTP pool as expected but also depleted the dCTP pool and increased the dGTP pool, as predicted from the known allosteric regulation of the reductase (42). Also, other examples showed that intracellular events mirrored the results found with the isolated reductase.

Measurements of pool sizes represent only snapshots and tell us little about the dynamics of dNTPs in the cell. The size of a pool depends both on the rate of its synthesis and the rate of its disappearance via DNA replication and catabolism. Questions concerning the dynamics of dNTP metabolism can be answered from experiments using isotope flow kinetics. We began such experiments in the early 1980s by feeding cells in culture with labeled nucleosides and measuring the flow of isotope into deoxyribonucleosides, dNTPs, and DNA. In these experiments, measuring the specific radioactivity of dNTP pools is crucial. Unfortunately, it is all too common that investigators carrying out experiments of this type are satisfied to determine the total amount of radio-activity in the pool, assuming that the size of the pool does not change with the manipulations involved in the experiment.

We used both normal cultured cells and cells carrying specific mutations in enzymes involved in deoxyribonucleotide metabolism. We could conclude that ribonucleotide reductase is the primary target for the regulation of dNTP synthesis but that an additional type of control, involving nucleoside kinases and nucleotidases, is also important. The kinases are often called salvage enzymes, which implies that their only function is to reutilize nucleosides arising from the catabolism of nucleic acids. However, our experiments indicate that kinases also form part of a regulatory circuit, which explains why they are strictly allosterically regulated. Intracellular deoxynucleotides are continuously broken down by nucleotidases and resynthesized by kinases. This process results in substrate (or futile) cycles that help maintain the proper size of dNTP pools (43). This regulatory circuit comes into play, for example, when cells have lost the enzyme dCMP deaminase (44) or when DNA synthesis is blocked during S-phase by inhibitors (45). In both in-stances, dNTPs are continuously overproduced by ribonucleotide reductase catabolized via substrate cycles and excreted as deoxynucleosides. Substrate cycles are also of considerable medical interest since they are involved in the activation of nucleoside analogues used in the treatment of viral and malignant diseases.

Experiments on substrate cycles were initiated ten years ago in Björn Ni-cander's thesis and are continued today by Vera Bianchi, who joined our group during a sabbatical in 1984. She has later returned many times. Vera is also Professor of Cell Biology at the University of Padova and, most importantly, has been my dear wife since 1991.

In the course of these experiments, our laboratory became involved in a controversy concerning multienzyme complexes that are supposed to channel dNTPs from ribonucleotide reductase to the site of DNA replication. Chan-neling of intermediates, i.e. the movement of small molecules from one enzyme to the next without mixing with a general cytoplasmic pool, is certainly an attractive idea but must be supported by solid evidence. A multiprotein com-plex containing enzymes of dNTP synthesis and DNA replication, called replitase, was postulated to be assembled in mammalian-cell nuclei during S-phase (46). The evidence for this concept was shaky from the beginning (47), and much later work has contradicted it, not the least the finding that mammalian ribonucleotide reductase is a cytoplasmic enzyme (48). Neverthe-less, it has been difficult to bury the replitase, probably because the concept is diffuse enough to provide apparent explanations for half-baked data from experiments with intact cells.

More Ribonucleotide Reductases

Blakley & Barker had discovered that ribonucleotide reduction by extracts of *L. leichmannii* requires addition of a vitamin B_{12} derivative(25). The enzyme present in extracts from *E. coli* did not require addition of cobalamin, and when we purified the reductase from bacteria grown in the presence of ^{60}Co-labeled vitamin B_{12}, the radioactivity disappeared during purification. Clearly, *E. coli* and *L. leichmannii* contain different ribonucleotide reductases. The amino acid sequences of the two enzymes were later found to be very different. The two enzymes can be considered prototypes of two classes of ribonucleotide reductases. Class I enzymes have an R1:R2 structure and contain a stable tyrosyl radical; class II enzymes consist of a single polypeptide chain and require adenosyl cobalamin.

What about ribonucleotide reduction in higher organisms? Mammalian cell extracts did not require addition of cobalamin for the reduction of ribonucleotides, and when purified enzymes became available, we learned that reductases from higher organisms in almost all respects behave as the *E. coli* reductase. Final proof came from Lars Thelander's work that produced the first pure mammalian enzyme (49). Today one can generalize that the reductases from all higher organisms are class I enzymes.

DNA technology greatly contributed to this generalization as it has to many other insights concerning ribonucleotide reduction. Britt-Marie Sjöberg constructed the first plasmids that were used to produce large amounts of *E. coli* reductase (50) and now applies recombinant DNA technology to identify functionally important amino acids of the enzyme. Rolf Eliasson and I familiarized ourselves in the late 1980s with the new technique. Now we happily clone new genes related to new reductases and obtain the required overproducers, but we prefer to do this in collaboration with experts in the field.

I realized early on that I would not be able to learn all the technology required to study the reductase. During my first professorial years, I encouraged my students to develop the required special skills. Lars Thelander showed a good ability to master physical-chemical techniques; Arne Holmgren specialized in amino acid sequences; and Olle Karlström was our microbial geneticist. However, this approach was limited. The students became independent scientists and started to dig in their own gardens. I began to collaborate with other groups and soon relied on a network of experts who were a prerequisite for our success. I have already mentioned my indebtedness to the biophysical expertise of Anders Ehrenberg and Astrid Gräslund. It would not have been possible to apply recombinant DNA technology so successfully without Britt-Marie Sjöberg's and Elisabeth Haggård-Ljungquist's help. Outside Sweden, Fritz

Eckstein from the Max Planck Institute for Medicine in Göttingen has been a long-standing friend and collaborator whose deep knowledge of synthetic nucleotide chemistry has left many imprints on our work.

In recent years, we have had very close relations with Marc Fontecave's laboratory in Grenoble. Marc came to Stockholm as a postdoctoral student in 1985. During his graduate work on cytochrome P-450, he obtained a solid knowledge of metal biochemistry. At the time, we had just started to fractionate the components of an enzyme system from *E. coli* that generates the radical of tyrosine-122 in an inactive form of protein R2. With Marc's help, we separated the system into four proteins and identified a hitherto unknown flavin reductase as a key component that reduces the Fe(III) center of the reductase to Fe(II) (51). Molecular oxygen reoxidizes the iron center and at the same time generates the radical in Tyr122.

How does *E. coli* synthesize dNTPs in the absence of oxygen? Of the several possible answers, Nature has chosen the most interesting one which opens up new vistas into biochemistry and evolution and has provided Marc and me with a fascinating problem to solve. Not all the answers are in yet, but the collaborative efforts of our two laboratories have begun to give us a good grasp of the general picture. Collaboration with Marc has been a joy. In 1990, at the young age of 34, Marc became Professor of Inorganic Biochemistry at the University of Grenoble. There he has already built up a strong laboratory with many enthusiastic co-workers. During recent years they have spent time in Stockholm, and I visited Grenoble on several occasions.

Anaerobic *E. coli* produces a new ribonucleotide reductase with an iron-sulfur center and, in its active form, a glycyl radical (52). The amino acid sequence differs from that of a class I or II enzyme and opens a new class III of ribonucleotide reductases that can be expected to contain other anaerobic enzymes.

The glycyl radical of the new enzyme is extremely oxygen sensitive. The enzyme is isolated inside an anaerobic box, but by the time we have obtained a pure protein, most of the radical and the activity are lost. However, both can be restored when the inactive reductase is incubated with S-adenosyl methionine, NADPH, and an enzyme system of *E. coli* that consists of three defined, and by now, cloned proteins. We were greatly helped in our work by earlier elegant experiments by Joachim Knappe and his group in Heidelberg. They had shown that a similar system generates a glycyl radical in pyruvate formate lyase from *E. coli*. This enzyme catalyzes the dissimilation of pyruvate to acetyl-CoA and formate, a central route in anaerobic energy metabolism (53).

One fascinating aspect of our present work deals with the evolution of DNA (54). It is now generally believed that RNA preceded DNA during the early evolution of life and, in the beginning, provided the means for both catalysis and self-replication. This discovery raised the question, at what stage did proteins enter? DNA synthesis requires deoxyribonucleotides. It is not unreasonable to assume that these were produced from ribonucleotides. Thus, either the first ribonucleotide reductase was a protein or RNA itself could catalyze the reduction of ribonucleotides (55). Ribonucleotide reduction requires radical chemistry, and free radicals are a highly reactive species useful only if they can be contained. I find it difficult to believe that RNA can do this without being damaged itself, whereas proteins can obviously be designed for this purpose; I therefore favor the idea that the first ribonucleotide reductase was a protein. This means that proteins preceded DNA during evolution. One problem is the current existence of three apparently different reductases, but we do not know the extent of those differences. A more definite answer must await knowledge of the 3D structure of the enzymes. Currently, Joan Stubbe's data on the active cysteines of the *L. leichmanni* enzyme suggest that class I and II enzymes are structurally related. Also, the class III *E. coli* reductase shares certain features with enzymes from the other two classes, particularly with respect to allosteric regulation. In my scenario, the first protein-reductase emerged during evolution before photosynthesis and was a prerequisite for DNA synthesis. The radical chemistry required for ribotide reduction was provided by the oxygen-sensitive glycyl radical that the reductase could have usurped from the older pyruvate formate lyase. Today's class III enzymes are the closest relative of this ancient enzyme. When oxygen appeared on Earth, it required a new radical generating mechanism involving major restructuring of the protein, and this mechanism resulted in the appearance of class I and II enzymes.

Active Retirement

In the Swedish university system, a professor retires at the age of 66 to make room for a successor. For me, the date was July 1, 1991. I had prepared for retirement by not accepting new graduate students or postdoctoral fellows, but I did not cut down my research. I was in the middle of the exciting discoveries concerning the new anaerobic ribonucleotide reductase described above. The policy of the Karolinska Institute is to provide laboratory space to active professors after retirement if they can finance their research with outside funds. So far I have been able to do so and have continued working together with two long-time associates, Rolf Eliasson and Elisabet Pontis. I am also fortunate to continue collaborating with other groups, the foremost being Marc Fontecave's in Grenoble. Recently, our circle was widened by Albert Jordan from Jordi Barbé's laboratory in Barcelona. Their work has resulted in the discovery of an additional ribonucleotide reductase in *E. coli* (56). This bacterium has the potential to produce three different ribonucleotide reductases.

The most important change in my life occurred immediately after my retirement in 1991, when Vera and I married. We had collaborated in research since 1984, and after the death of my wife Dagmar, decided to extend our collaboration to the private sector. Vera teaches Cell Biology in Padova and has a research group there. As a consequence, we divide our time between Sweden and Italy and get the best of each country.

I have borrowed the title for this chapter from Francis Crick's autobiography, "The Important Thing Is to Be There when the Picture Is Painted"(57). I was there, and I hope to be there for some time to come. In looking back, I find that the citation describes a good part of my life. So much happened outside my control; so many events of my life came to me without my own doing. My own contribution was to grab the opportunities.

Most important was that I was taken to Sweden and a school that made the transition easy. My new country provided me with the opportunity to start a career in science and later gave me all the support I needed for success. Arthur Kornberg recently asked me if it would not have been better to have gone to the United States and practice biochemistry there. My answer was no. My preference for Sweden in part comes from my preference for Swedish society. But it also has to do with the satisfaction I obtained from doing what I did. I would not have been a better scientist in the United States, and given that, I believe that my achievements as a scientist and teacher left a larger imprint in Sweden than they would have in the United States.

The next most important event came in the person of Einar Hammarsten. The example of his dedication to science made a lasting impression on me, and his way of running, or rather not running, a department was what made me develop rapidly into an independent biochemist.

The third event occurred when ribonucleotide reduction came to me. Little did I know what would happen when I injected labeled ribonucleosides into rats in 1950. From there, one thing led to another. I did plan to find the enzyme for ribonucleotide reduction but did not know that when I added ATP to the extract I would get deeply involved in allosteric effects of a unique kind. Again, I planned to obtain a pure enzyme but had no plan to make the unprecedented discovery of the first stable protein radical. It may be lack of imagination that has kept my mainstream in research to a single subject for so many years, but there were so many fascinating side streams from other fields to be explored. There was of course DNA replication but also disulfide metabolism, cobalamin, allosteric effects, organic free radicals, metallo-enzymes, anaerobic metabolism, and now evolution. It was a question of keeping the eyes and mind open. To be there when the picture was painted.

Any *Annual Review* chapter, as well as any article cited in an *Annual Review* chapter, may be purchased from the Annual Reviews Preprints and Reprints service. 1-800-347-8007; 415-259-5017; email: arpr@class.org

Literature Cited

1. McCarty M. 1946. *J. Exp. Med.* 83:89–96
2. Reichard P. 1948. *Nature* 162:662–63
3. Hammarsten E. 1947. *Acta Med. Scand.* 196:634–42
4. Arvidson H, Eliasson NA, Hammarsten E, Reichard P, von Ubisch H, Bergström S. 1949. *J. Biol. Chem.* 179:169–73
5. Hammarsten E, Reichard P, Saluste E. 1950. *J. Biol. Chem.* 183:105–9
6. Reichard P, Estborn B. 1951. *J. Biol. Chem.* 188:839–46
7. Reichard P, Lagerkvist U. 1953. *Acta Chem. Scand.* 7:1207-17
8. Lagerkvist U. 1953. *Acta Chem. Scand.* 7:114–18
9. Kornberg A, Lieberman I, Simms ES. 1954. *J. Am. Chem. Soc.* 76:2027–28
10. Reichard P, Hanshoff G. 1955. *Acta Chem. Scand.* 9:519–30
11. Jones ME, Spector L, Lipmann F. 1955. *J. Am. Chem. Soc.* 77:819–20
12. Reichard P, Hanshoff G. 1956. *Acta Chem. Scand.* 10:548–66
13. Yates RA, Pardee AB. 1956. *J. Biol. Chem.* 221:757–70
14. Reichard P, Rutberg L. 1960. *Biochim. Biophys. Acta* 37: 554-55
15. Reichard P, Canellakis ZN, Canellakis ES. 1961. *J. Biol. Chem.* 236: 2514–19
16. Bertani LE, Häggmark A, Reichard P. 1961. *J. Biol. Chem.* 236:PC67-68
17. Svensson I, Boman HG, Eriksson KG, Kjellin K. 1963. *J. Mol. Biol.* 7:254–71
18. Laurent T, Moore EC, Reichard P. 1964. *J. Biol. Chem.* 239: 3436–44
19. Moore EC, Reichard P, Thelander L. 1964. *J. Biol. Chem.* 239:3445–52
20. Holmgren A. 1989. *J. Biol. Chem.* 264: 13963–66
21. Thelander L. 1967. *J. Biol. Chem.* 242: 852–59
22. Larsson A. 1963. *J. Biol. Chem.* 238: 3414–19
23. Brown NC, Reichard P. 1969. *J. Mol. Biol.* 46:39–55
24. Durham LJ, Larsson A, Reichard P. 1967. *Eur. J. Biochem.* 1:92–95
25. Blakley RL, Barker HA. 1964. *Biochim. Biophys. Res. Commun.* 16:301–97
26. Booker S, Stubbe J. 1993. *Proc. Natl. Acad. Sci. USA* 90:8352-56
27. Brown NC, Canellakis ZN, Lundin B, Reichard P, Thelander L. 1969. *Eur. J. Biochem.* 9:561–73
28. Brown NC, Eliasson R, Reichard P, Thelander L. 1969. *Euro. J. Biochem.* 9:512–18
29. Ehrenberg A, Reichard P. 1972. *J. Biol. Chem.* 247:3485–88
30. Larsson Å, Sjöberg B-M. 1986. *EMBO J.* 5:2031–36
31. Atkin CL, Thelander L, Reichard P, Lang G. 1973. *J. Biol. Chem.* 248:7464–72
32. Nordlund P, Sjöberg B-M, Eklund H. 1990. *Nature* 345:593–98
33. Reichard P, Ehrenberg A. 1983. *Science* 221:514–19
34. von Döbeln U, Reichard P. 1976. *J. Biol. Chem.* 251:2616–622
35. Thelander L. 1974. *J. Biol. Chem.* 249: 4858–62
36. Åberg A, Hahne S, Larsson A, Ormö M, Åhgren A, Sjöberg B-M. 1989. *J. Biol. Chem.* 264:12249–52
37. Mao SS, Holler TP, Yu GX, Bollinger JM, Booker S, et al. 1992. *Biochemistry* 31:9733–43
38. Stubbe J. 1989. *Annu. Rev. Biochem.* 58:257–85
39. Uhlin U, Eklund H. 1994. *Nature* 370: 533–539
40. Reichard P, Eliasson R. 1979. *Cold Spring Harbor Symp. Quant. Biol.* 43: 271–77
41. Skoog L. 1970. *Eur. J. Biochem.* 17: 202–8
42. Bjursell G, Reichard P. 1973. *J. Biol. Chem.* 248:3904–9
43. Reichard P. 1988. *Annu. Rev. Biochem.* 57:349–74
44. Bianchi V, Pontis E, Reichard P. 1987. *Mol. Cell Biol.* 7:4218–24
45. Nicander B, Reichard P. 1985. *J. Biol. Chem.* 260:9216–22
46. Reddy GPV, Pardee AB. 1982. *J. Biol. Chem.* 257:12526–31
47. Spyrou G, Reichard P. 1983. *Biochem. Biophys. Res. Commun.* 115:1022–26
48. Engström Y, Rozell B, Hansson H-A, Stemme S, Thelander L. 1984. *EMBO J.* 4:863–67
49. Thelander L, Eriksson S, Åkerman M. 1980. *J. Biol Chem.* 255:7426–32
50. Platz A, Sjöberg B-M. 1980. *J. Bacteriol.* 143:561–68
51. Fontecave M, Eliasson R, Reichard P. 1987. *J. Biol. Chem.* 262:12325–31
52. Reichard P. 1993. *J. Biol. Chem.* 268: 8383–86
53. Wagner AFV, Frey M, Neugebauer FA, Schäfer W, Knappe J. 1992. *Proc. Natl. Acad. Sci. USA* 81:996–1000
54. Reichard P. 1993. *Science* 260:1773–77
55. Benner SA, Ellington AD, Tauer A. 1989. *Proc. Natl. Acad. Sci. USA* 86: 7054–58
56. Jordan A, Gibert I, Barbé J. 1994. *J. Bacteriol.* 175:3420-27
57. Crick F. 1988. *What Mad Pursuit.* Alfred P Sloan Foundation.

Annu. Rev. Biochem. 1995. 64:29–63

THE ENVELOPE OF MYCOBACTERIA

Patrick J. Brennan

Department of Microbiology, Colorado State University, Fort Collins, Colorado
80523

Hiroshi Nikaido

Department of Molecular and Cell Biology, University of California, Berkeley,
California 94720

KEY WORDS: cell wall, arabinogalactan, mycolic acid, permeability barrier, drug resistance

CONTENTS

ABSTRACT

Mycobacteria, members of which cause tuberculosis and leprosy, produce cell
walls of unusually low permeability, which contribute to their resistance to

29

therapeutic agents. Their cell walls contain large amounts of C_{60}-C_{90} fatty acids, mycolic acids, that are covalently linked to arabinogalactan. Recent studies clarified the unusual structures of arabinogalactan as well as of extractable cell wall lipids, such as trehalose-based lipooligosaccharides, phenolic glycolipids, and glycopeptidolipids. Most of the hydrocarbon chains of these lipids assemble to produce an asymmetric bilayer of exceptional thickness. Structural considerations suggest that the fluidity is exceptionally low in the innermost part of bilayer, gradually increasing toward the outer surface. Differences in mycolic acid structure may affect the fluidity and permeability of the bilayer, and may explain the different sensitivity levels of various mycobacterial species to lipophilic inhibitors. Hydrophilic nutrients and inhibitors, in contrast, traverse the cell wall presumably through channels of recently discovered porins.

INTRODUCTION

Most mycobacterial species appear to be saprophytic inhabitants of soil, but a few are important pathogens. *Mycobacterium tuberculosis* causes tuberculosis, a disease that still kills about 3 million people every year, mostly in the developing part of the world. Although the incidence of tuberculosis had been declining steadily in industrialized countries, this trend was reversed in the United States about 10 years ago. Even more alarming is the appearance of multiple drug–resistant strains of *M. tuberculosis* (1), a problem compounded by the availability of few drugs for the treatment of tuberculosis. Another species, *M. leprae,* causes leprosy, a chronic disease afflicting about 12 million people in the world. Finally, "atypical mycobacteria" that include *M. avium* complex, *M. kansasii, M. fortuitum,* and *M. chelonae,* cause opportunistic infections in immunologically compromised patients, such as AIDS patients, although they are likely to be essentially saprophytic organisms.

Mycobacteria are problem pathogens primarily because they are resistant to most common antibiotics and chemotherapeutic agents (2). Thus *M. tuberculosis* is susceptible only to aminoglycosides (e.g. streptomycin) and rifamycins (e.g. rifampicin) among antibiotics, and to fluoroquinolones among general chemotherapeutic agents. *M. leprae* is susceptible only to rifamycins, fluoroquinolones, and dapsone (a sulfonamide). Atypical mycobacteria are especially resistant, showing occasional susceptibility only to some fluoroquinolones and to clarithromycin. Thus very few general-purpose antimicrobials can be used, although several agents active specifically against *M. tuberculosis* and a few other species—including isoniazid, ethionamide, pyrazinamide, *p*-aminosalicylic acid, and ethambutol—do exist (3). However, most of these agents have little activity against atypical species (2).

Mycobacteria are also relatively resistant to drying, alkali, and many chemical disinfectants, and this makes it difficult to prevent the transmission of *M. tuberculosis* in institutions and in urban environments in general. This general resistance and the resistance to therapeutic agents are thought to be related to the unusual structure, and the resultant low permeability, of the mycobacterial cell wall.

The characteristic component of a eubacterial cell wall is the peptidoglycan, in which polysaccharide chains composed of repeating *N*-acetyl-β-D-glucosaminyl-(1→4)-*N*-acyl-muramic acid units are crosslinked by short peptide chains linked to the acid groups of the muramic acid residues. Cell walls of Gram-negative bacteria contain outer membranes outside the peptidoglycan layer, but cell walls of the usual Gram-positive bacteria lack such membranes and are largely composed of peptidoglycan, with some additional polysaccharides or polyol phosphate polymers (teichoic acids).

Mycobacteria are a group of eubacteria that belong to a larger group of Gram-positive bacteria containing GC-rich DNA, sometimes called the actinomycete line. Within this group, mycobacteria belong to one branch, often called the *Corynebacterium-Mycobacterium-Nocardia* (CMN) branch. These bacteria produce cell walls of a unique structure, sometimes called chemotype IV cell wall, containing *meso*-diaminopimelic acid as the diamino acid in the peptidoglycan. Interestingly, the muramic acid residue is *N*-glycolylated in *Mycobacterium* and *Nocardia* (4), in contrast to the *N*-acetylation found in all other bacteria. Glycolyl groups could further tighten the peptidoglycan structure by providing additional opportunities for hydrogen bonding, but the three-dimensional structure of mycobacterial peptidoglycan does not appear to have been investigated. An important feature of the chemotype IV cell walls is the presence of a unique polysaccharide, arabinogalactan (AG), which is substituted by characteristic long-chain fatty acids, mycolic acids (5–7). Mycolic acids in *Corynebacterium* and *Nocardia* contain up to about 40 and 60 carbon atoms, respectively, but those from *Mycobacterium* usually contain 70–90 carbons. In addition, several other lipid species, many of them with unusual structures, are known to exist in mycobacterial cell wall as "free" lipids, that is as solvent-extractable lipids that are not covalently linked to the AG-peptidoglycan complex (5–9). The list of such extractable lipids is becoming longer. Recent years have witnessed the discovery and characterization of trehalose-based lipooligosaccharides (LOSs) (10) and of lipoarabinomannan (LAM) (11). The purified cell wall of *M. chelonae* was found to contain conventional glycerophospholipids (EY Rosenberg, H Nikaido, unpublished observation).

Studies of the phylogenetic relationship among mycobacterial species showed, mainly by the use of 16S rRNA sequences, that the genus *Mycobacterium* consists of two subgroups, fast growers and slow growers, and that each of these groups may be subdivided into several clusters (12–14). Some of these clusters are shown in Table 1, which additionally shows the types of mycolic

Table 1 Cell wall lipids and sensitivity to inhibitors

| | Lipid composition of cell wall | | | Susceptibility to inhibitors[a] | | | | | |
| | Mycolic acids[b] | | | Hydrophilic | | Bile | Lipophilic | | |
	% trans at proximal position of α-mycolate	Other types	Other lipids[c]	NH$_2$OH	PNBA	salts[d]	Oleate[e]	PyrB	TolB
Fast growers									
M. fortuitum	42	α', E	GPL, LOS	R		R	R(>400)	R	R
M. chelonae	51	α'	GPL	R		R	R(>400)	R	R
M. smegmatis	68	α', E	GPL, LOS	S		S	R(400)	R	R
M. phlei	75	K, W		S		S	R(400)	S	S
M. thermoresistibile	<20	α', K, M, W		S		S	S(1.6)	S	S
M. vaccae	<10	α', K, W		S		S	S(3.2)	S	S
M. aurum	<20	K, W		R/S		S	S(1.6)	S	S

Slow growers								
M. terrae	75	K, W		R	R	S	R/S	R
M. gordonae	12	M, K	SL, LOS, PDM	d	R	S	R/S	R/S
M. tuberculosis	<10	M, K	SL, PDM	S	S	S	S	S
M. bovis	<10	M, K	PGL, PDM	S	S	S	S	S
M. marinum	<10	M, K	PGL, PDM	R	d	S	S/R	S
M. kansasii	<10	M, K	GPL, LOS, PDM	d	R/S	S	S	S
M. szulgai	?	M, K	LOS	S	R			
M. malmoense	?	?	LOS	d	R			
M. avium complex	18	K, W	GPL	d[f]	R		S	R[g]
M. leprae	?	K	PGL, PDM		R			R

[a] The susceptibility data are from Reference 15, unless indicated otherwise. PNBA, p-nitrobenzoic acid; PyrB, pyronine B; and TolB, toluidine blue. R, S, and d refer to resistance, sensitivity (no growth in the presence of inhibitor), and different responses depending on the particular strains, respectively.

[b] Two lines of data on mycolic acids are shown. The left column shows the percentage fraction of α-mycolates with trans double bonds or cyclopropane groups at the proximal position ("Y" of Figure 3). These values were estimated from the analytical data of Kaneda et al (15b). The right column shows the other mycolate classes (described below in Figure 3) known to be present (7). E, M, K, and W stand for epoxy-, methoxy-, keto-, and wax-ester mycolate, respectively.

[c] Other extractable lipid species known to be present (8, 9). GPL, glycopeptidolipids; LOS, trehalose-containing lipooligosaccharides; SL, sulfolipids; PDM, phthiocerol dimycocerosate; PGL, phenolic glycolipids.

[d] Measured by growth on MacConkey agar (without crystal violet), which contains 0.075% bile salts. Data on slow growers are from ref. 16.

[e] Data on slow growers are from ref. 17. Numbers in parentheses are MIC_{40} values (in μg/ml) from ref. 18.

[f] Among M. intracellulare and M. avium, 45% and 4% respectively grew on this medium.

[g] M. intracellulare is resistant, but M. avium shows R/S phenotype.

acids present as well as the characteristic free lipids described so far. Table 1 also shows susceptibility of various species to general inhibitory agents, characteristics that suggest the degree of permeability of the cell wall (see below). It is clear that certain lipids are limited to related groups of organisms. For example, sulfolipids are essentially limited to *M. tuberculosis.* α'-Mycolates occur only among the fast growers. On the other hand, some lipids appear to be widely distributed, such as LAM and cord factors. In the following sections of this review, we describe the structures of the various lipid components and try to relate them to the probable molecular architecture of cell wall. Possible functions and biosynthesis of the various components are also discussed.

ULTRASTRUCTURAL FEATURES

M. tuberculosis is a rod-shaped bacillus of $1–4 \times 0.3–0.6$ μm (15a, b), although other mycobacterial species can occur as much shorter cocco-bacilli or curved rods. The envelope of mycobacteria consists of the plasma membrane and the wall. Since it remained difficult, until recently (19), to separate physically and study those independently, an important theme of ultrastructural investigation has been attempts to relate visible structure to chemical identity (reviewed in 20). The most interpretable images of the mycobacterial cell envelope were obtained from ultrathin sections of embedded bacteria. Recent investigation using freeze-substitution methods (21) produced images that are in accord with earlier ones (22) (Figure 1) and demonstrate that the cell wall is composed of an inner layer of moderate electron density, a wider electron-transparent layer, and an outer electron-opaque layer of variable appearance and thickness. The inner layer probably contains peptidoglycan, in that its moderate electron density is consistent with a product containing carboxyl groups that bind metal ions (23). The electron-transparent layer appears to be the hydrophobic domain of the cell wall, which is dominated by the mycolate residues covalently bound to the AG (see below). This layer has a thickness of about $9–10$ nm (20), and thus is much thicker than the $4–4.5$ nm deep layer present in the cytoplasmic membrane (Figure 1). In both cases, the transparency is caused by the failure of the water-soluble stains to penetrate into these hydrocarbon-rich regions. The nature of the outermost electron-opaque layer of the wall was debated for some time. The layer varies in thickness (from negligible to considerable), electron density, and appearance (fibrillar, granular, or homogeneous), which is attributable to differences among species, in growth conditions, and in preparation methods for microscopy. The dye ruthenium red allowed this layer to be consistently visualized (24), suggesting that a minimal structure is probably negatively charged head groups of lipids. In some species, there may be additional carbohydrate material, such as LAM, which contains phosphate and succinyl groups, and the capsular polysaccharides—glucans, mannans, arabi-

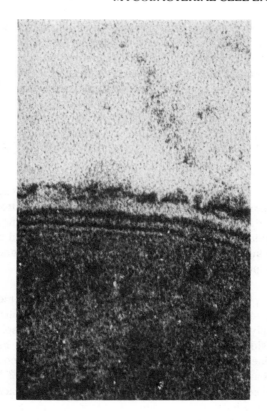

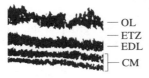

— OL
— ETZ
— EDL
⎤
⎦ CM

Figure 1 Appearance of mycobacterial envelope in thin sections. *a*. Electron micrograph of envelope and part of cell contents of *Mycobacterium phlei* 425. Cells were fixed by freeze-substitution (21) to optimize preservation of structure and to reduce extraction of lipid components by solvents used in processing. (Photograph reprinted from 20 with permission of the publisher.) Scale bar indicates 30 nm. *b*. Diagram showing interpretation of image shown in *a* in terms of layer structure described in text. Thickness of layers is enlarged about fourfold compared with the photograph. OL, outer layer; ETZ, electron-transparent zone; EDL, electron-dense layer; CM, cytoplasmic membrane. (Modified from Figure 15 of Ref. 21.)

nans, or other heteropolysaccharides. Proteins may also contribute to the binding of stains.

An important observation is that mycobacterial cell wall proper can be cleaved by the process of freeze-fracture. The presence of this fracture plane strongly supports the concept that the wall is constructed basically as a lipid bilayer (20).

CELL ENVELOPE COMPONENTS: STRUCTURE

Plasma Membrane

The appearance of the mycobacterial plasma membrane is not symmetrical in cells carefully fixed from a viable state, in that the outer, electron-dense layer is thicker than the inner layer in thin sections (20, 21) (Figure 1). There is cytochemical evidence that the "extra" thickness is associated with carbohydrate (25), and this suggests that the phosphatidylinositol mannosides (PIMs) (see below) are preferentially located in the outer leaflet.

PIMs   The phospholipids in mycobacterial envelopes are almost invariably derivatives of phosphatidic acid. The most common are phosphatidylglycerol, diphosphatidylglycerol, phosphatidylethanolamine, and phosphatidylinositol and its mannosides (PIMs). The PIMs are restricted to the bacteria of the actinomycete line (8). They are major plasma membrane components and also form the lipid base of LAM and lipomannan. The early use of ^{1}H-NMR (26) clearly established that the glycerol phosphate moiety was attached to the L-1 position of the myo-inositol ring and that the mannose residues were glycosidically linked to the 2 and 6 positions. The structures of each of the higher homologs, PIM_3, PIM_4, PIM_5, and PIM_6, have been established (27). They contain a single α-D-mannopyranosyl group at the 2 position, and a mannose oligosaccharide, for example [α-D-Man$p$-(1→2)-]$_2$-[α-D-Man$p$-(1→6)-]$_2$-α-D-Man$p$ in PIM_6, at the 6 position (27). The PIMs, notably PIM_2 and PIM_5, contain acyl functions other than those of the diacylglycerols (28, 29).

OTHER COMPONENTS   Other components associated with the plasma membrane include a number of polyterpene-based products thought to be associated with protection against photolytic damage, such as the carotenoids, and the menaquinones that are involved in electron transport (8). The carotenoids of mycobacteria are responsible for the characteristic yellow-orange color of photochromogenic mycobacteria such as M. gordonae and M. kansasii.

The known glycosylphosphopolyprenols of mycobacteria, presumably involved in cell wall biosynthesis, are also believed to be plasma-membrane-associated. The polyprenols in these are decaprenol and octahydroheptaprenol, rather than the undecaprenol found in most common bacteria. The compounds isolated thus far are β-D-mannopyranosyl phosphodecaprenol (30), β-D-mannopyranosyl phosphooctahydroheptaprenol (31), β-D-arabinofuranosyl phosphodecaprenol (32), and 6-*O*-mycolyl-β-mannopyranosyl phosphooctahydroheptaprenol (32a).

The Cell Wall Skeleton

The covalently linked skeleton of cell wall (36) may be described as a peptidoglycan to which are linked polysaccharide side chains esterified at their distal ends with mycolic acids, that is the mycolyl-AG-peptidoglycan complex, 40% of which corresponds to lipids in the form of mycolic acids (33). The peptidoglycan contains peptide side chains consisting of L-alanyl-D-isoglutaminyl-*meso*-diaminopimelyl-D-alanine, in which the diaminopimelic acids are amidated (4). This type of peptidoglycan, type Alγ (34), is one of the commonest found in bacteria. The mycobacterial peptidoglycan, however, differs in two ways from the common type, in that muramic acid is *N*-glycolylated as described above and that the crosslinks include bonds between two residues of diaminopimelic acid as well as between diaminopimelic acid and D-alanine (35).

MYCOLYL-AG   It was known already in the 1930s that the major wall polysaccharide of *M. tuberculosis* is a serologically active branched-chain AG, with the arabinose residues forming the nonreducing termini of the chains. The presence of a similar AG was shown in some seven species of mycobacteria (36). It was also demonstrated early that AG is attached to peptidoglycan through a phosphodiester link to position 6 of about 10–12% of the muramic acid residues (36). Liu & Gotschlich (37) obtained muramic acid 6-phosphate from a variety of Gram-positive bacteria, including *M. smegmatis*. There was much uncertainty, however, about the structure of the galactan component of AG.

More recent work established that the AG is rather unique in the nature of its component sugars as well as its overall structure (38) (Figure 2). Partial depolymerization of the per-*O*-alkylated polysaccharide and analyses of the generated oligomers by gas chromatography-mass spectrometry (GC-MS) and fast atom bombardment-mass spectrometry (FAB-MS) established that: (*a*) within AG, all Ara and Gal residues are in the furanose form; (*b*) the nonreducing termini of arabinan consist of a branched hexaarabinofuranosyl structure [β-D-Araf-(1→2)-α-D-Araf]$_2$-3,5-α-D-Araf-(1→5)-α-D-Araf; (*c*) the majority of the arabinan chain consists of 5-linked α-D-Araf residues with branching introduced by 3,5-α-D-Araf residues replaced at both branch positions with 5-α-D-Araf; (*d*) the arabinan chains are attached to the galactan

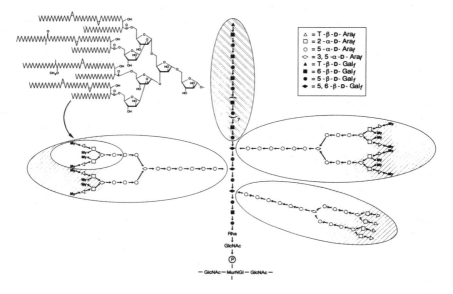

Figure 2 Chemical structure of the mycolyl-AG peptidoglycan complex of mycobacteria. The largest fragment of arabinan and galactan isolated to date are within the ellipses.

core through the C-5 of some of the 6-linked Gal*f* units; (*e*) the galactan region consists of linear alternating 5- and 6-linked β-D-Gal*f* residues; (*f*) the galactan of AG is linked to the C-6 of some muramyl residues of peptidoglycan via the diglycosylphosphoryl bridge, L-Rha*p*-(1→3)-D-GlcNAc-(1→P) (39); and (*g*) the mycolic acids are located in clusters of four on the terminal hexaarabinofuranosyl units, but only about two thirds of these arrangements are mycolated (40).

Recently, a family of arabinases and galactanases secreted by a *Cellulomonas* species was used to degrade base-solubilized AG from *M. tuberculosis* (41). The major Ara-containing degradation products were the hexaarabinofuranoside mentioned above and a linear disaccharide, α-D-Ara*f*-(1→5)-D-Ara*f*. The linear galactan backbone was degraded into cyclic oligosaccharides of the structure [5)-β-D-Gal*f*-(1→6)-β-D-Gal*f*-(1→]$_n$. More recently, oligosaccharide fragments containing up to 23 Ara residues were obtained by gentle acid hydrolysis of the per-*O*-methylated AG, and the molecular weights and alkylation patterns were determined by FAB-MS (GS Besra et al, submitted for publication). The extended nonreducing ends of the arabinan were thus shown to consist of the following unit (Figure 2):

β-D-Ara*f*(1→2)α-D-Ara*f*(1→3)

β-D-Ara*f*(1→[5)α-D-Ara*f*(1→]₃→3)

β-D-Ara*f*(1→2)α-D-Ara*f*(1→5)

β-D-Ara*f*(1→2)α-D-Ara*f*(1→3)

αDAra*f*(1→[5)αDAra*f*(1→]₆)

β-D-Ara*f*(1→[5)α-D-Ara*f*(1→]₃→5)

β-D-Ara*f*(1→2)α-D-Ara*f*(1→5)

Three such arabinans appear to be attached to the homogalactan (Figure 2). Using the same approach, an extended stretch of the galactan was also isolated consisting of 25 Gal residues, β-D-Gal*f*(1→5)-[β-D-Gal*f*(1→6)-β-D-Gal*f*(1→5) →]₁₂, devoid of any branching, demonstrating that the points of attachment of the arabinan chains to galactan are close to the reducing end of galactan, which itself is linked to peptidoglycan via the linker disaccharide-phosphate, mentioned above (Figure 2).

MYCOLIC ACIDS   Mycolic acids are high-molecular-weight α-alkyl, β-hydroxy fatty acids, present mostly as bound esters of AG, where they appear primarily as tetramycolylpentaarabinosyl clusters (Figure 2), but also in extractable lipids, mainly as trehalose 6,6′-dimycolate (cord factor). Owing to the presence of 3-hydroxy group, pyrolysis of mycolic acids releases the part of the main chain distal to the C-3, producing "meromycolic acid." Their structures can thus conveniently be discussed by separating the molecule into meromycolate moiety and the α-branch (5).

Mycobacterial mycolic acids are distinguishable from those of other genera (such as *Corynebacterium, Nocardia,* and *Rhodococcus*): (*a*) They are the largest (C_{70} to C_{90}); (*b*) they have the largest α-branch (C_{20} to C_{25}); (*c*) in the main chain (the meromycolic acid moiety), they contain one or two groups—which may be double bonds or cyclopropane rings—that are capable of producing "kinks" in the molecule; (*d*) they may contain oxygen functions additional to the β-hydroxy group; and (*e*) they may have methyl branches in the main carbon backbone. The entire structural spectrum of mycolic acids was resolved in the 1960s (5) through the use of MS, NMR, and infrared spectroscopy (IR). The α branch, except for chain length, was found to be consistently conserved among the family of mycolic acids. Structures of various types of mycolic acids are summarized in Figure 3.

The totally saturated and fairly long (typically 24-carbon) structure of the α-branch as well as the exceptional length of the meromycolic acid chain (typically almost 60-carbon) should favor strongly the regular, parallel packing of the hydrocarbon chains of mycolic acid. On the other hand, the oxygen

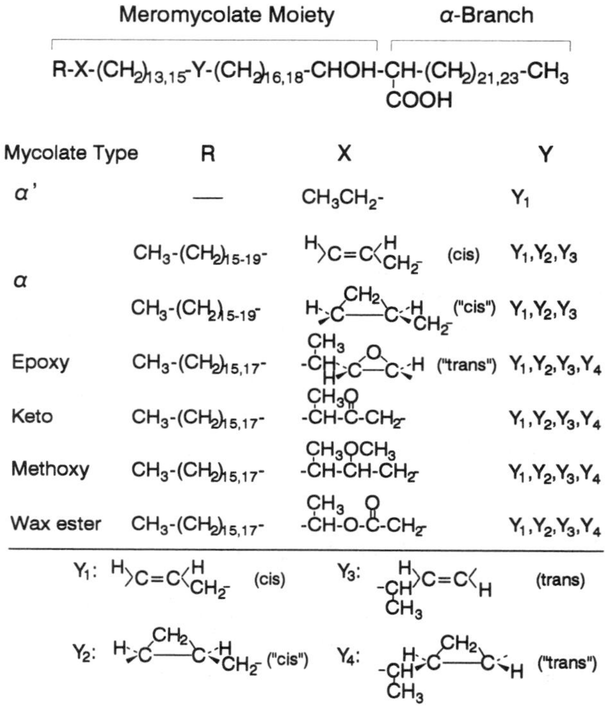

Meromycolate Moiety   α-Branch

$$R-X-(CH_2)_{13,15}-Y-(CH_2)_{16,18}-CHOH-\underset{\underset{COOH}{|}}{CH}-(CH_2)_{21,23}-CH_3$$

Mycolate Type	R	X	Y
α'	—	CH_3CH_2-	Y_1
α	CH_3-$(CH_2)_{15-19}$-	$H{>}C=C{<}^H_{CH_2}$ (cis)	Y_1,Y_2,Y_3
	CH_3-$(CH_2)_{5-19}$-	$H{\cdot}{>}C{\overset{CH_2}{\frown}}C{\cdot}{<}^H_{CH_2}$ ("cis")	Y_1,Y_2,Y_3
Epoxy	CH_3-$(CH_2)_{15,17}$-	$-\overset{CH_3}{C}H{\cdot}C{\overset{O}{\frown}}C{\cdot}H$ ("trans")	Y_1,Y_2,Y_3,Y_4
Keto	CH_3-$(CH_2)_{15,17}$-	$-CH{\overset{CH_3}{\underset{\overset{\|}{O}}{C}}}-CH_2$-	Y_1,Y_2,Y_3,Y_4
Methoxy	CH_3-$(CH_2)_{15,17}$-	$-\overset{CH_3O}{C}H-\overset{CH_3}{C}H-CH_2$-	Y_1,Y_2,Y_3,Y_4
Wax ester	CH_3-$(CH_2)_{15,17}$-	$-\overset{CH_3}{C}H-O-\overset{O}{C}-CH_2$-	Y_1,Y_2,Y_3,Y_4

Y_1: $H{>}C=C{<}^H_{CH_2}$ (cis) Y_3: $-\underset{CH_3}{C}H{>}C=C{<}^H_H$ (trans)

Y_2: $H{\cdot}{>}C{\overset{CH_2}{\frown}}C{\cdot}{<}^H_{CH_2}$ ("cis") Y_4: $-\underset{CH_3}{C}H{\cdot}C{\overset{CH_2}{\frown}}C{\cdot}{<}H$ ("trans")

Figure 3 Structures of mycolic acids from mycobacteria. Modified from Ref. 7. In some compounds, the methyl branch in Y_3 is reported to be on the proximal side of the double bond (5).

functions, as well as *cis* double bonds and *cis* cyclopropane structures, are expected to modulate this tight packing by producing kinks in the chains. Interestingly, the packing-disruptive groups are located farther away from the carboxyl end of the molecule. Thus, all of the oxygen functions (except the β-hydroxy group) are located at position X of Figure 3, usually more than 35 carbons away from the carboxyl end of the molecule. Furthermore, the *cis* structures at the proximal position Y are frequently converted into less disruptive structures such as *trans* double bonds and *trans* cyclopropane (See Table 1); studies with conventional lipids showed that the latter structures do not produce kinks in the chain and do little to prevent the tight packing of membrane lipids (42). These points are discussed below in connection with the organization of cell wall lipids.

Lipoarabinomannan (LAM)

It was known early that mycobacteria also contained "soluble," immunologically active arabinomannan (43). Misaki et al (44) showed in 1977 that this polysaccharide from *M. tuberculosis* contained an $\alpha(1\rightarrow6)$-linked D-Man*p* backbone to which were attached short side chains of $\alpha(1\rightarrow5)$-linked D-Ara*f* residues. However, their analysis was performed on alkali-treated samples, and the finding (45) in 1970 that some of the arabinomannan from *M. tuberculosis* was acylated by palmitic and tuberculostearic acids did not attract much attention until systematic study of the native arabinomannan was undertaken in 1986. Such a study (11) showed that LAM contained glycerol, inositol, and phosphate, in addition to arabinose, mannose, lactate, succinate, palmitate, and tuberculostearate, which were identified earlier. The situation was similar with lipomannan, which is essentially LAM without an arabinose-containing side chain. The phosphate was shown to occur in the form of an alkali-labile phosphatidyl*myo*inositol unit, containing the two fatty acid residues. Thus both LAM and lipomannan were the first prokaryotic versions of the membrane components anchored via phosphatidylinositol (46), a class of compounds that occur frequently in animal cells; LAM and lipomannan are multiglycosylated extensions of PIMs (47).

Detailed structural analysis of LAM has resulted in recognition of two distinct arabinan arrangements occupying the terminal end: branched hexa arabinofuranosides with the structure, $[\beta$-D-Ara*f*-$(1\rightarrow2)$ -α-D-Ara*f*$]_2$-3,5-α-D-Ara*f*-$(1\rightarrow5)$-α-D-Ara*f*, similar to that in AG, and a linear β-D-Ara*f*-$(1\rightarrow2)$-α-D-Ara*f*-$(1\rightarrow5)$-α-D-Ara*f* (48). In addition, in the case of LAM isolated from strains of *M. tuberculosis*, these two types of Ara termini are extensively capped with Man*p* residues, a product now termed ManLAM (49–51). A version of LAM, as isolated from a rapidly growing *Mycobacterium* sp. and *M. smegmatis*, is devoid of Man caps and is called AraLAM (51). It is partially capped with inositol-P residues (52). The structures of various types of LAM are schematically shown in Figure 4.

The Extractable Lipids of Cell Wall

The search for dominant antigens on the surfaces of various mycobacteria, especially "atypical" (or nontuberculous saprophytic) mycobacteria, was stimulated by the infections caused by these bacteria in immunocompromised patients (53). This led to the definition of a remarkable array of cell wall glycolipids (8, 9). We describe below the major classes of such extractable glycolipids—LOSs, phenolic glycolipids (PGLs), and glycopeptidolipids (GPLs)— as well as other classes of free lipids (See Table 1).

LOSs   Members of the LOS class of glycolipids were first found in *M. kansasii* and later in *M. malmoense, M. szulgai, M. gordonae,* and *M. butyricum* (54).

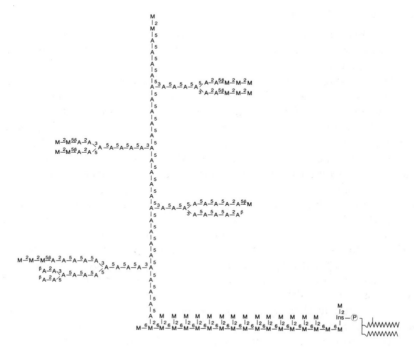

Figure 4 An attempt to present a composite structure of ManLAM from *M. tuberculosis* (Erdman strain) based on the recognition of certain small motifs and on the relative amounts of the various arabinose and mannose units in their different linkages.

Eight such glycolipids are present in smooth variants of *M. kansasii* (10). They are composed of variable residues of xylose, 3-*O*-methylrhamnose, fucose, and a novel *N*-acylamino sugar (*N*-acylkansosamine) linked to a common tetraglucose core, which itself contains an α,α′-trehalose moiety at the end (Figure 5). The terminal glucose residue of the α,α′ trehalose unit is usually acylated at positions 3, 4, and 6 by 2,4-dimethyltetradecanoic acid residues ("R" in Figure 5). The LOS from *M. gordonae* contains 6′-*O*-methyl-2,3,4,6-tetraacyltrehalose structure at one end (55).

At about the same time as the discovery of *M. kansasii* LOS, somewhat different trehalose-based LOS was discovered in *M. smegmatis* (56). The oligosaccharide unit of the *M. smegmatis* LOS contains only D-glucose residues, although two of them are pyruvylated. Furthermore, the acylation occurs on both glucose residues of the trehalose moiety, at 4′ and 6 positions (57). More recently, LOS of *M. fortuitum* biovar. *fortuitum* was found to have features of both the *M. kansasii*-type LOS and *M. smegmatis*-type LOS, in that

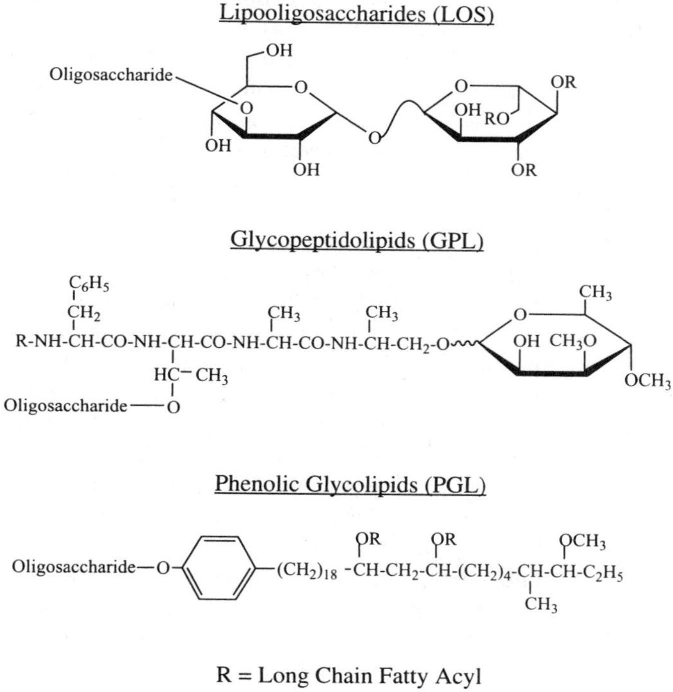

Figure 5 Generic structures of three major classes of extractable glycolipids of mycobacterial cell walls. In the case of the lipooligosaccharide class, the point of attachment of the oligosaccharide unit may also be the 4- or the 6-OH group of the acyltrehalose unit.

the acylation occurs both on the 3, 4, and 6 positions of the terminal glucose and on the 2′ position of the subterminal glucose residue of the trehalose unit (58). The chemistry and basis of the antigenicity of LOSs and their association with variable colony morphology in mycobacteria have been reviewed (54).

PGLs   The second class of glycolipids is more correctly termed glycosylphenolphthiocerol dimycocerosates, although the term PGL is generally used (8, 9, 59–61). This class includes "mycoside A" of *M. kansasii,* "mycoside G" of *M. marinum,* and "mycoside B" of *M. bovis* in the earlier literature. Their structure (Figure 5) is characterized by a very large hydrophobic moiety, containing a C_{36} phenolic diol substituted by two molecules of typically C_{34} fatty acid, mycocerosate. The oligosaccharide part contains from one to four sugar residues, and the sugars are usually not very hydrophilic, consisting often of deoxy sugars that are multiply *O*-methylated.

GPLs   Another class of mycobacterial glycolipids, GPLs ["C-mycosides" in earlier literature (62)], has been reviewed (54). As shown in Figure 5, the core head group is a short peptide, D-Phe-D-*allo*Thr-D-Ala-L-alaninol, and the alaninol is substituted by a 3,4-di-*O*-methyl-L-rhamnose. The hydroxyl group of the D-*allo*threonine residue carries an oligosaccharide substituent; its most proximal portion is usually α-L-rhamnopyranosyl-(1→2)-6-deoxy-L-talopyranose. The amino group of D-phenylalanine residue is substituted by a fatty acid residue.

GPLs are the major cell surface antigens of the *M. avium, M. intracellulare, M. scrofulaceum* group (MAIS), and they can be subdivided into 31 distinct serotypes based on the serospecific GPLs. The structures of the GPLs from 12 of these serovars have been completely defined (54). Although the most proximal sugars were common as described above, the nonreducing terminal sugars of the oligosaccharides were highly variable, containing an array of novel amido sugars, branched-chain sugars, sugar acids, and pyruvylated sugars. The use of monoclonal antibodies to the individual serovars of the *M. avium* complex in conjunction with semisynthetic neoantigens containing some of the precise terminal sugar combinations has established the antigenically dominant epitopes of individual serovars (63).

GPLs also constitute a major cell surface component in some fast growers, such as *M. chelonae* and *M. fortuitum* (Table 1). Recently, the structures of the major GPLs from *M. fortuitum* biovar. *peregrinum* have been established (64, 65). These compounds share a peptidolipid core as found in the GPLs of *M. avium,* but differ in the location and nature of the sugar residues. They do not contain 6-deoxytalose or its derivatives. Moreover, the oligosaccharide portion is linked to the alaninol residue instead of the *allo*-threonine. *M. xenopi* contained GPL of a unique structure (a "non-C-mycoside" GPL) in which the core consisted of a different lipopeptide, fatty acyl-NH-L-Ser-L-Ser-L-Phe-D-*allo*-Thr-OCH$_3$ (65, 66). The oligosaccharide hapten is linked glycosidically to the *allo*-Thr-OCH$_3$ residue and contained one or two additional fatty acyl substituents. The remaining 3-*O*-CH$_3$-6dTal*p* was glycosidically linked to the distal serine residue.

WAXES, ACYLATED TREHALOSES, SULFOLIPIDS   Several slowly growing mycobacteria contain an array of waxes, generally long-chain diols [phthiocerols A and B, phthiodiolone, phthiotriol; phthiocerol A is a mixture of 3-methoxy-4-methyl-do-(and tetra)-triacontane-9,11-diols] in which mycocerosic acids are esterified to both hydroxyl groups (see Table 1). Related to this family of waxes are the phenolphthiocerols, which form the basic core of the phenolic glycolipids. Three other families of trehalose-based lipids, cord factor (trehalose 6,6'-dimycolate), the simpler acylated trehaloses [containing a combination of saturated straight-chain C$_{16}$-C$_{19}$, C$_{21}$-C$_{25}$ mycocerosate, C$_{24}$-C$_{28}$

mycolipanolic, and C_{25}-C_{27} mycolipenic fatty acids: for example, 2, 3-di-O-acylated trehalose isolated from *M. tuberculosis* (67)], and the sulfolipids (trehalose 2'-sulfate acylated with hydroxyphthioceranic, phthioceranic, and saturated straight-chain fatty acids) have been implicated in the pathogenesis of tuberculosis. The structures and biological activities of these have been reviewed (68).

GLYCEROPHOSPHOLIPIDS   Although the cell envelope of mycobacteria has a high content of conventional glycerophospholipids, these were assumed to be the components of the plasma membrane. Recent separation of *M. chelonae* cell walls from plasma membranes, with less than 1% contamination by the latter (19), allowed the analysis of pure cell walls, which unexpectedly contained about two short-chain (C_{16-18}) fatty acid residues per bound mycolic acid residue. Some of these short-chain acids were present in PIMs (EY Rosenberg, H Nikaido, unpublished results). Interestingly, earlier studies reported the presence of glycerophospholipids, especially PIM, in crude cell wall fractions of *M. phlei* (69) and *M. tuberculosis* (70), although these reports were greeted with skepticism because of the lack of evidence for the purity of the "cell wall" fractions. Now that we know that glycerophospholipids constitute a fraction of extractable lipids at least in *M. chelonae* cell wall, we need to expand similar quantitative analysis to other mycobacterial species.

OUTER LAYERS AND CAPSULES   Some of the extractable lipids may exist outside the cell wall proper, as is discussed below. The case has been made that in *M. tuberculosis* isolates, the exocellular and surface-exposed materials are mostly polysaccharides, specifically D-glucans, D-arabino-D-mannans, and D-mannans (71).

Cell Wall Proteins

The purified cell wall of *M. chelonae* contained 30-kDa major protein (19). Similarly, *M. leprae* cell wall fraction was enriched in a 35-kDa major protein (72). The functions of these proteins are not known. Trias et al (73) isolated a pore-forming protein, porin, from the cell wall of *M. chelonae* by following its channel-forming function. The function of this 65-kDa minor protein is discussed below.

Physical Organization of Cell Wall Lipids

Knowledge of the chemistry of mycobacterial lipids unfortunately does not allow us to understand the function of the cell wall as the permeability barrier. For this purpose, we need to know the physical organization of the lipids. More than 10 years ago, Minnikin proposed a model, in which mycolic acid chains

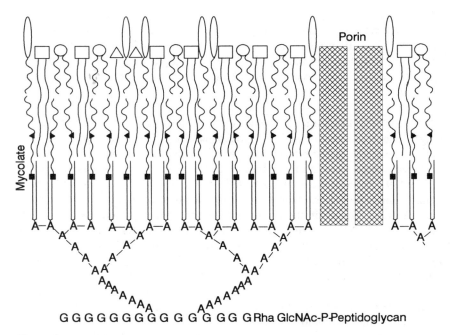

Figure 6 A model of the mycobacterial cell wall. *cis-* or *trans-*Double bond, or *cis-* or *trans-*cyclopropane group (solid squares); *cis-*double bond, *cis-*cyclopropane group, or oxygen-containing groups (solid triangles). The lipids composing the outer leaflet differ from species to species. Those containing short-chain (C_{16-18}) fatty acids include glycerophospholipids and GPL. Those containing intermediate chain–length hydrocarbons include PGL, phthiocerol dimycocerosate, and others.

are packed side by side in a direction perpendicular to the plane of the cell surface (5). It was also proposed that this mycolic acid–containing inner leaflet is covered by an outer leaflet composed of extractable lipids, the whole structure thus producing an asymmetric lipid bilayer (Figure 6).

This model has received support from several recent studies. X-ray diffraction study of purified *M. chelonae* cell wall showed that a large part of the hydrocarbon chains in the cell wall are tightly packed in a parallel array in a direction perpendicular to cell surface (19). An important feature of the X-ray data was that a large fraction of the lipid was in a nearly crystalline arrangement of presumably very low fluidity, whereas the rest showed the diffuse diffraction expected of fluid domains. The coexistence of domains of different fluidities is expected from our consideration of the mycolic acid structure. Thus the section closest to the carboxyl group contains few, if any, structures capable of producing "kinks" that increase fluidity: The α-branch is always without any double bond or cyclopropane group, and the proximal portion of the meromycolate chain also favors crystalline arrangement because the first double bond or cyclopropane group (solid squares in Figure 6) is often *trans*, which does not introduce kinks (42). [For example, the most abundant mycolic acid species in *M. smegmatis*, α-smegmamycolic acid, contains a *trans* double bond at this position (74).] The disorder increases as one approaches the outer surface, because the second double bond or cyclopropane group (solid triangles in Figure 6) has the *cis* configuration, and at the corresponding position mycolate often contains oxygen-containing substituents that would strongly disrupt the tight lateral packing (Figure 3). The outer leaflet, containing the common glycerophospholipids (see above), should have the usual high fluidity. This large change in fluidity across the thickness of the membrane is reminiscent of the Gram-negative outer membrane, which is composed of a low-fluidity lipopolysaccharide outer leaflet and a high-fluidity inner leaflet (75). The gradient of fluidity in mycobacterial cell wall seems to have an opposite orientation, with the external regions more fluid than the internal segment. This arrangement explains the observation that mycobacterial permeability barrier can be disorganized by adding surfactants from the outside (76), whereas the Gram-negative outer membrane is highly resistant to such treatment (75).

The bilayer model also fits with the electron-microscopic observations discussed already. First, there is a freeze-fracture plane within the cell wall, a finding that is consistent with the presence of bilayer. Second, the bilayer model predicts that the hydrophobic center of the cell wall, consisting of hydrocarbon chains of 70–80 carbon atoms (C_{60} meromycolate chain plus C_{16-18} lipid), would have about twice the thickness of such a layer in the plasma membrane (C_{16-18}, twice). Indeed the transparent layer of the cell wall was twice as thick as that in the plasma membrane, as described above (Figure 1).

Are there enough bound mycolic acids to cover the cell surface as a leaflet, as predicted by the model? Quantitative analytical data on *M. bovis* strain (BCG) (77) indeed suggest that this is the case (19). Finally, the model requires that the cell wall contains enough extractable lipids as components of the outer leaflet. Furthermore, because the α-branch and the meromycolate branch of mycolic acid differ in length (usually by 25 carbons or more), we need lipids containing the short-chain fatty acids with 14–18 carbons, as well as those containing fatty acid residues of about 30–40 residues. Quantitative analysis of fatty acids in *M. chelonae* cell wall showed that it contained not only glycerophospholipids and GPL but also substantial amounts of lipids containing intermediate chain–length fatty acids, although the identity of these lipids is not yet clear (EY Rosenberg, H Nikaido, unpublished results).

The tendency of mycolic acid hydrocarbons to produce tight, parallel arrays was also shown by the pioneering study of Durand et al (78). These authors demonstrated, by monolayer studies, that a cord factor containing natural, C_{80} di-*cis*-unsaturated mycolic acid residues formed a dense, presumably paracrystalline, packing at room temperature, a result indicating that the interaction between long hydrocarbon chains is strong enough to overcome the disorganizing effect of *cis* double bonds. In this study, cord factors containing corynemycolate (C_{32}) of *R* or natural configuration around the β-carbon had a much higher enthalpy of melting in comparison with a similar compound with the unnatural *S* configuration, or an analog containing β-deoxy corynemycolic acids. This finding suggests that the β-OH moiety stabilizes the mycolic acid monolayer (the inner leaflet in the bilayer of Figure 6) by intermolecular or intramolecular hydrogen bonding. This is reminiscent of the model that in the outer membrane of Gram-negative bacteria, the outer leaflet, which acts as the main permeability barrier, is stabilized by hydrogen bonding involving the β-oxygen atom of β-hydroxymyristate residues of the component lipid, lipopolysaccharide (79).

We believe that extractable lipids usually occur in the outer leaflet of the bilayer. Some lipids, however, may form an independent aggregate outside the bilayer, especially when overproduced. For example, PGLs of *M. leprae* are quite apolar and are likely to form disordered oil droplets, as do such apolar lipids as triglycerides. Indeed, PGL-I is known to occur as loosely cell-associated lipid droplets on the surface of *M. leprae* (80).

Amphiphilic lipids associate to produce organized structures in water. When the cross-section of head group is equal to that of the hydrophobic tail, the lipid tends to form bilayers, but when the cross-section of head group is larger than that of the hydrocarbon tail, the lipid forms micelles or fibrillar structures (as extension of micelles in one direction) (81). Among the free lipids, GPLs appear to fall into this category. They contain a large head group, containing 3–6 sugar units and 4 amino acids, and yet their hydrocarbon moiety consists of only one fatty acid chain (except in *M. xenopi* GPL). Indeed, *M. lepraemurium* produces fibrillar material on its surface, and this material appears to consist of GPLs (82). A similar observation was made with *M. avium* (83).

In contrast to PGL and GPL, trehalose dimycolate (cord factors), LOSs, the trehalose-containing sulfolipids, as well as more conventional glycerophospholipids such as PIMs, are ideally shaped to form bilayers, and thus are unlikely to exist outside the cell wall bilayer.

The bilayer model also fits with the unusual structure of AG. Thus there are many (usually around 12–13) Ara*f* residues between the mycolate residues and the central galactan chain (Figure 2). The presence of these many Ara residues and the exceptional flexibility of Ara*f*-(1→5)-Ara*f* linkages are expected to facilitate the lateral movement of mycolate hydrocarbons, helping their tight packing into a rather rigid structure. Such packing would otherwise be difficult, because about 16 mycolic acid residues are covalently connected to one AG molecule.

LAM and lipomannan were thought to be anchored to the plasma membrane through their PIM structures. However, now that PIM is known to exist in the cell wall bilayer, it seems equally possible that they are anchored to the outer leaflet of the cell wall. In this arrangement, the hydrophilic polysaccharide chains need not be assumed to penetrate the hydrocarbon interior of the cell wall.

MYCOBACTERIAL CELL WALL AS A PERMEATION BARRIER

Small solutes are expected to traverse the mycobacterial cell wall either through the porin channel (73) or through the lipid bilayer region, just as they traverse the Gram-negative outer membrane by using either of these mechanisms (75). Since hydrophilic solutes cannot traverse lipid bilayers, they are predicted to utilize the porin pathways in the mycobacterial cell wall as well. In contrast, lipophilic solutes are not the favored solutes for passage through porin channels (75) and thus are likely to diffuse mainly through lipid bilayers.

Permeability to Hydrophilic Solutes

The permeability of the *M. chelonae* cell wall to hydrophilic solutes cephalosporins, was experimentally determined (84) with a method introduced for measurement of outer membrane permeability in Gram-negative bacteria (85). The rate of hydrolysis of cephalosporins by intact mycobacterial cells was measured, and the cell wall permeability was calculated by assuming that drug molecules first diffuse through the cell wall (following Fick's first law of diffusion) and then are hydrolyzed by periplasmic β-lactamase (following Michaelis-Menten kinetics). The cell wall permeability measured was indeed very low: about 3 orders of magnitude lower than that of *Escherichia coli* outer membrane and 10 times lower than the permeability of the notoriously impermeable *Pseudomonas aeruginosa* outer membrane. Permeation rates had low temperature coefficients and did not increase when more lipophilic cephalosporins were used (84); these data suggest that the permeation occurred mainly through aqueous channels. Permeability to hydrophilic nutrient molecules such as glycerol and glucose was also very low (84).

When the presence of a porin was discovered in *M. chelonae* and its properties were determined (73), it became clear that the low hydrophilic permeability of mycobacteria is due to two factors. First, *M. chelonae* porin is a minor protein of the cell wall, unlike enterobacterial porins (73). Second, *M. chelonae* porin produces permeability far lower than that produced by an equal weight of *E. coli* porin (73). The presence of a pore-forming protein with similar properties was shown in *M. smegmatis* (86), and similar porins are probably distributed widely among mycobacteria.

The mycobacterial cell wall thus shows an unusually low degree of permeability to hydrophilic solutes. However, there are significant differences among various species. Recent studies showed that in *M. smegmatis* (86) and *M. tuberculosis* H37Ra (EY Rosenberg, H Nikaido, unpublished results), the cell wall permeability to β-lactams is about an order of magnitude higher than in *M. chelonae*. These results may be compared with the presumptive indicators of hydrophilic permeability shown in Table 1. Higher sensitivity to a small, hydrophilic inhibitor hydroxylamine indeed suggests that *M. tuberculosis* and *M. smegmatis* cell walls are more permeable than *M. chelonae* cell wall, as expected. It is also consistent with the earlier observation that nutrients are accumulated more rapidly by *M. smegmatis* (reviewed in 87) and by *M. tuberculosis* (88) than by *M. chelonae*. Finally, if the sensitivity to *p*-nitrobenzoate indeed gives some indication of hydrophilic permeability, we can also predict that most of the slow growers listed in Table 1 should have lower hydrophilic permeability than *M. tuberculosis*. These species are indeed more resistant to a small, hydrophilic antimycobacterial agent, ethambutol, than is *M. tuberculosis* (15a).

Permeability to Hydrophobic Solutes

A lipid bilayer is ordinarily highly permeable to lipophilic solutes. However, its permeability is inversely correlated with its fluidity (42). Fluidity decreases when the membrane lipid contains longer hydrocarbon chains with fewer *cis*-double bonds or *cis*-cyclopropane groups. The innermost part of mycobacterial cell wall is extreme in this regard, and presumably has very low fluidity. Furthermore, lipids of the type in which more than two fatty acid chains are connected to a single head group appear to reduce the permeability further (89); mycolyl-AG is an example of this type of lipid.

These principles were experimentally verified with the outer membrane of Gram-negative bacteria. There the outer leaflet consists of lipopolysaccharide, which contains 6–7 saturated fatty acid residues, all connected to a single head group (75). The permeability of the outer membrane bilayer, most probably limited by the lipopolysaccharide leaflet, was indeed up to 100-fold lower than the permeability of bilayers composed of the usual glycerophospholipids (90). Since mycobacterial cell wall is more extreme in its structure, we can expect even lower permeability through its lipid matrix. However, this pathway cannot be neglected, because it may make a more significant contribution to the permeation of hydrophobic solutes than the porin pathway, which is extremely inefficient as described above. Very lipophilic solutes cross the asymmetric bilayer of the Gram-negative outer membrane at rates similar to those with which β-lactams traverse *E. coli* porin channels (90). Since the permeability through the porin channels is about 100–1000-fold lower in the mycobacterial cell wall, the lipid bilayer pathway may remain more important for such solutes than the porin pathway, even if this bilayer is less permeable, say by a factor of 10, than that of the Gram-negative outer membrane.

The antibacterial agents of the relatively lipophilic classes, such as rifamycins, tetracyclines, macrolides, and fluoroquinolones, may thus utilize the lipid bilayer pathway in traversing the mycobacterial cell wall. If so, one can predict that, within each class, the more lipophilic derivatives would be more active against mycobacteria. This is indeed the case. For example, among fluoroquinolones, the more hydrophobic sparfloxacin is more active than the reference fluoroquinolones against many mycobacteria (91). Ciprofloxacin, when made more hydrophobic by the addition of alkyl substituents, becomes more active against *M. tuberculosis* and *M. avium* (92). With *M. leprae,* a good positive correlation was seen between the lipophilicity of fluoroquinolone and its efficacy (93; the data are analyzed in 87). Similar examples exist for other classes of agents: tetracyclines (94–96), macrolides (97–99), and rifamycins (100). In *M. avium,* which appears to have very low hydrophilic permeability, the efficacy of a small, hydrophilic agent isoniazid was improved by converting it into a hydrophobic compound, by the addition of palmitoyl substituent (101). Recently, the penetration rate of norfloxacin, a fluoroquinolone, into *M. tuberculosis* cells was shown to increase more than six times when the temperature was increased by 10°C (T Kocagoz, HF Chambers, personal communication), and this high temperature coefficient also suggests the predominant role of the lipid bilayer pathway in penetration.

We can also compare the activity of one single agent against various mycobacterial species. Table 1 shows that several fast growers, especially *M. chelonae* and *M. fortuitum,* are highly resistant to lipophilic inhibitors such as dyes and detergents. Such differences in susceptibility presumbaly reflect differences in cell wall permeability, which are also likely to affect the susceptibility of various species to lipophilic antibiotics. For example, rifampicin is active against almost all of the clinically relevant species of mycobacteria, except *M. chelonae* and *M. fortuitum* (and *M. avium* complex among the slow growers) (102). Available data indeed allow us to propose a structural explanation of such differences in permeability. Thus in resistant species (*M. chelonae,* and *M. fortuitum, M. smegmatis,* and *M. phlei* as well as a relatively resistant slow grower, *M. terrae*), substantial fractions (41-75%) of α-mycolates, which constitute a major fraction of mycolates in most organisms, contain *trans* double bonds at the proximal (inner) position (Table 1). These *trans* double bonds will decrease the fluidity, and thus permeability, of the innermost part of the bilayer (Figure 6), the part that would act as the rate-limiting permeability barrier for lipophilic solutes. In contrast, in the more susceptible species (*M. thermoresistible, M. vaccae,* and *M. aurum* among fast growers as well as most of the slow growers), fluidity and permeability of the innermost part are presumably much higher, because most of their α-mycolates contain fluidity-increasing *cis* double bonds or *cis* cyclopropane groups at the corresponding position (Table 1). Although the α-mycolates of *M. avium* group appear to contain some *trans* cyclopropane structures at the proximal position, their high reistance to lipophilic agents cannot be explained by this observation alone.

Cell Wall Barrier is a Necessary, but not a Sufficient, Factor for Resistance

Although mycobacterial cell wall is a formidable permeation barrier, production of clinically significant levels of resistance usually requires the participation of an additional resistance mechanism, such as the enzymatic inactivation or the active efflux of the agents. This is because the surface-to-volume ratio is extremely high in small bacterial cells, and thus half-equilibration across the cell wall takes place in several minutes for β-lactams, even though the permeability coefficients of the mycobacterial cell wall are exceedingly low (2). The nature of these additional resistance mechanisms for mycobacteria is discussed in Ref. 2.

MYCOBACTERIAL CELL WALL: INTERACTION WITH HOST COMPONENTS

The cell wall and its associated structures are the outermost components of a bacterial cell, and as such they play crucial roles in the interaction of pathogenic species with their host. Below we discuss some of these "cell surface" functions of the components of mycobacterial cell wall.

LAM exhibits a wide spectrum of immunoregulatory functions, but the biological implication of the in vitro data is not always clear. Earlier data using *M. leprae* LAM and AraLAM from a rapidly growing *Mycobacterium* sp. were interpreted as a suggestion that LAM suppresses immune responses, thus contributing to pathogenesis of tuberculosis and leprosy. These data include LAM-induced abrogation of T-cell activation (104), inhibition of γ-interferon-mediated activation of murine macrophages (105), scavenging of potentially cytotoxic oxygen free-radicals (106), and inhibition of protein kinase C activity (106). Although AraLAM evoked a large array of cytokines associated with macrophages, such as α-tumor necrosis factor (TNF) (107), granulocyte macrophage-colony stimulating factor, and interleukins-1a, 1b, 6, and 10 (108), this was frequently interpreted as a contributor to the disease processes: for example, the production of fever, weight loss, and tissue necrosis was emphasized in the case of α-TNF. More recently, however, ManLAM, which is present in strains of *M. tuberculosis*, was found to be much less potent in evoking α-TNF, in contrast to AraLAM found in nonvirulent species (107). Similarly, AraLAM, but not ManLAM, was found to activate the early response genes (including c-*fos* and the genes for α-TNF) in macrophages (109). Additionally, ManLAM could stimulate phagocytosis by interacting with the Man receptor (110). These results now suggest that *M. tuberculosis* strains become phagocytized efficiently but survive within the host macrophages because their ManLAM does not activate these phagocytes.

The glycolipids of PGL class are located at cell surface, and may in some cases exist even outside the bilayer structure of the cell wall proper, as described above. They have been implicated in pathogenesis. For example, PGL is thought to contribute to the intracellular survival of *M. leprae* within macrophages of individuals with lepromatous leprosy through its ability to scavenge oxygen radicals (61). The variable oligosaccharide constituents of these glycolipid antigens are usually of sufficient antigenicity as to evoke corresponding specific antibodies and thereby allow serodiagnosis of individual mycobacterioses (8) and leprosy (61). The triglycosyl phenolphthiocerol dimycocerosate of *M. tuberculosis*, PGL-Tb1 (111), is apparently not present in virulent strains of *M. tuberculosis* and therefore, unlike PGL-I from *M. leprae*, is not useful in the serodiagnosis of tuberculosis.

GPLs are also clearly present on the bacterial cell surface, as anti-GPL antibodies were repeatedly shown to react with intact cells. It has been pointed out above that some GPLs could exist as micelles or fibrils outside the cell wall bilayer proper. Species that produce large amounts of GPLs tend to produce smooth colonies, presumably because the surface of their cells is covered by hydrophilic carbohydrate moieties of GPL. In contrast, typical strains of *M. tuberculosis* are devoid of GPL, typical LOS, or PGL, and indeed produce colonies with extremely hydrophobic, rough surfaces (15). Carbohydrate layers on cell surfaces usually contribute to virulence by preventing nonspecific phagocytosis, but it is not clear whether this is also the case with mycobacteria, which survive within macrophages. Indeed both smooth, GPL-containing forms and rough, GPL-deficient forms of *M. avium* are isolated from patients (112). GPL was also often suggested to protect mycobacterial cells within the phagolysosomes, but GPL-deficient rough mutants of *M. avium* were as resistant in this environment as the smooth parent strain (113).

BIOSYNTHESIS AND ASSEMBLY

Arabinogalactan (AG)

The structure of the AG linker [α-L-Rha$p$-(1→3)- D-GlcNAc (1→P)] is strikingly similar to that of the linker involved in the attachment of teichoic acid to peptidoglycan, N-ManNAc-N-GlcNAc-1-P (114). The linkage unit plays a key role in the initiation of new teichoic acid chains (115). Thus the biosynthesis begins by the transfer of GlcNAc-1-P moiety from UDP-GlcNAc onto a polyisoprenoid lipid carrier in a reaction that is inhibited strongly by tunicamycin (116, 117). This reaction is then followed by the assembly, still on the polyisoprenoid carrier, of the rest of the linker and the teichoic acid proper. Since similar linkage units are also involved in cell wall attachment of a polysaccharide in *Micrococcus luteus* (118), and of teichoic acids in actinomycetes (119), we may assume that AG is also synthesized in a similar way. If so, the first reaction will be the transfer of GlcNAc-1-P moiety from UDP-GlcNAc to a polyprenol carrier, followed by the transfer of L-rhamnosyl moiety (presumably from dTDP-Rha). This is presumably followed by the transfer of Gal*f* units. Although the donor of Gal*f* has not been identified in mycobacteria, UDP-Gal*f* is the donor of Gal*f* in polysaccharide synthesis in *Penicillium charlesii* (120). Ara*f* is also likely to be added to the growing polysaccharide while it is still linked to the carrier. Mycobacterial cell extracts catalyze the transfer of Ara*f* from Ara*f*-P-decaprenol to AG and LAM (K Mikusova, PJ Brennan, unpublished results); the nucleotide-linked form of Ara*f*, however, is not known.

The sites of action of several antituberculosis drugs are probably located in this pathway. Ethambutol at 3.0 µg/ml inhibited the transfer of label from [^{14}C]Glc into the D-Ara residue of AG in whole cells of a drug-susceptible strain of *M. smegmatis* (121), inhibition that began almost immediately upon exposure of the cells to the drug. This dramatic decline in incorporation of label in growing *M. smegmatis* applies to the arabinan moiety of both AG and LAM (K Mikusova, PJ Brennan, unpublished results). Thus the primary mode of action of ethambutol appears to be the inhibition of synthesis of the arabinan of both AG and LAM.

Mycolic Acids

Early studies on the biosynthesis of mycolates showed that in *Corynebacterium diphtheriae*, C_{32} mycolic acids were formed by Claisen-type condensation and reduction of C_{16} fatty acids (122), and similar condensations were established for the mycolic acids of *Nocardia asteroides* and *M. smegmatis* (123). The proposed pathway for biogenesis of mycolates (Figure 7) involves four stages (124): (*a*) synthesis of C_{24}-C_{26} straight-chain saturated fatty acids to provide C-1 and C-2 atoms and the α-alkyl chain; (*b*) synthesis of C_{40}-C_{60} acids (meromycolic acids) to provide the main carbon backbone; (*c*) modification of this backbone to introduce other functional groups—one series of C_{24} to C_{32} acids are produced by a Δ-5-desaturase enzyme acting on a C_{24} acid, followed by elongation, implying that at least some of the modifications are introduced during growth of the meromycolate chains; and (*d*) the final condensation step to produce mycolic acid.

There is evidence that the point of divergence from the biosynthesis of conventional fatty acids is at the C_{24} level, with a ω-19 acid, 24:1 *cis*-5, being elongated (124a). ω-19 Acids have been detected at low concentrations in lipid extracts of mycobacteria (125), and these acids correspond structurally to the methyl terminus of the meromycolate chain. Incubation of mycobacterial cell walls with labeled acetate and extraction and oxidation of resulting mycolates to fragment the meromycolate chain at its double bonds demonstrated that the methyl terminus and the next 18 carbon atoms were virtually unlabeled (126), indicating an endogenous ω-19 precursor. Also, when wall material was extracted with hexane to remove endogenous fatty acids, synthetic 24:1 *cis*-5 significantly stimulated the incorporation of labeled acetate into mycolic acids, while a range of other acids, even some 24-carbon acids, had no effect.

The early steps in the biosynthesis of mycolic acids have not been defined (124a). Direct incorporation of appropriate labeled fatty acids such as 24:1 *cis*-5 into mycolic acids is yet to be shown. The carrier molecule must be identified. Coenzyme A (CoA) is the usual carrier, but it does not function in these cell-free systems (126). Further, it is not known how mycobacteria synthesize 24:1 *cis*-5; a 24:0 desaturase has been characterized, but its product is the ubiquitous 24:1 *cis*-15 acid, nervonic acid (127). The starting 24- and 26-carbon fatty acids are probably synthesized by fatty acid synthases (FASs) and elongases, specifically the FAS-I and -II complexes (128). In *M. tuberculosis*, these functions are linked in a multifunctional enzyme system that appears to be a de novo synthase joined to an elongase that elongates C_{16} to C_{24} or C_{26} (129). Further elongation of the C_{24} fatty acids to meromycolic acids (C_{50}-C_{56}) has been demonstrated (124, 130), but it is not clear which of these FAS/elongases are used.

α-Mycolate, which contains double bonds or cyclopropane rings as its only functional groups, is not a precursor of the more complex oxygenated mycolates. The dissociation of their biosynthesis was shown in *M. aurum*, in which radioactive label appeared in ketomycolate before α-mycolate (131). Thus, any common intermediate must be at an early stage in biosynthesis. For this, 24:1 *cis*-5 is a good candidate; judging from the structure of oxygenated mycolates, it would be expected to be oxidized and methylated to a 5-keto, 6-methyl-24-carbon precursor. Mutants and isolates impaired in mycolic acid synthesis are known, e.g. strains of actinomycetes (132), and one mutant of *M. smegmatis* that synthesizes only the unbranched meromycolates (133). These should be useful in unraveling this complex pathway.

The unique mycolyl phospholipid (Myc-PL), 6-*O*-mycolyl-β-D-mannopyranosyl-phosphooctahydroheptaprenol, has the attribute of a mycolyl carrier and may play a role in the terminal mycolyl group transfer to cell wall components. The Myc-PL is probably the product of a Claisen-type condensation reaction and contains an acyl carrier group that is equivalent to the proposed R_1 in

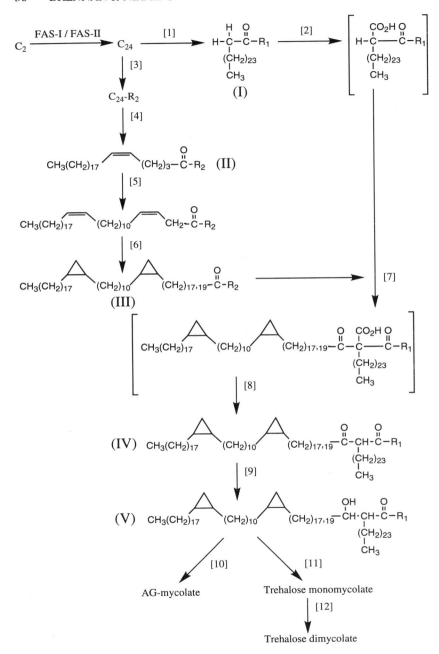

Figure 7. The other acyl carrier group R_2 could be either CoA, acyl carrier protein (ACP), or membrane-bound protein. Myc-PL may be the direct precursor of trehalose monomycolate (TM), or it may transfer the mycolate residue to trehalose (partial reaction [11] in Figure 6). Presumably, the fully formed mycolic acid is transferred to AG to form the mycolate-AG complex either from Myc-PL (partial reaction [10]) or TM (not shown). The Claisen-type condensation of the C_{26} fatty acid with the meromycolic acid to yield the ketomycolyl-R_1 intermediate (partial reactions [7, 8] in Figure 7) has not been demonstrated in a cell-free system. However, the synthesis of the short-chain C_{32}-C_{36} corynemycolic acids has been reported in vitro using C. diphtheriae (134) and C. matruchotii (135). These products were reported to appear in the form of TM. These esters were thought to be carriers of the mycolic acid to the mycobacterial cell wall, where it is transferred to the nonreducing terminal D-arabinose residue of AG via an unknown transacylation reaction.

The strongest evidence that normal mycolic acid metabolism is crucial to the survival of M. tuberculosis comes from work on the action of two well-known antituberculosis drugs, isoniazid and ethionamide, which apparently inhibit primarily mycolic acid synthesis (136). Identification of the inhA resistance gene seems to implicate the elongation stages of mycolate synthesis in isoniazid action rather than the condensation step and beyond. Ethionamide and isoniazid share similar inhibitory properties, and the missense mutations within the inhA gene confer resistance to both drugs (137). All of the features of the gene (homology to envM, NAD/NADH binding, downstream from a β-ketoacyl ACP reductase; 137; WR Jacobs, personal communication) point to an enoyl ACP reductase and therefore involvement in elongation events. Long before the genetic evidence, biochemical data suggested that isoniazid specifically inhibits the insertion of a Δ-5 double bond into a C_{24} fatty acid (138).

Extractable Lipids

The biosynthesis and genetics of only some of the glycolipids have been examined. Biosynthetic pathways for assembling the GPLs of M. avium had been proposed earlier (8), but these schemes remain to be confirmed. Recent

---

Figure 7 Proposed anabolic pathway of mycolic acids in M. tuberculosis H37Ra. This is an updated version of a previously proposed pathway (46). The key products are (I) hexacosanoate-R_1, (II) Δ-5-tetracosanoate-R_2, (III) $C_{52,54}$ meromycolate-R_2, (IV) oxomycolate-R_1, and (V) mycolate-R_2. Carrier group R_1 is believed to be the β-D-mannopyranosyl-monophosphoryl-polyisoprenol, and R_2 may be either CoA or ACP. The reactions are identified as follows: [1] elongation and introduction of carrier group R_1, [2] carboxylation, [3] introduction of carrier group R_2, [4] Δ-5-desaturation, [5] elongation and Δ-3-desaturation, [6] introduction of cyclopropane rings and elongation, [7] Claisen-type condensation, [8] decarboxylation, [9] reduction, [10] mycolate transfer to AG, [11] mycolate transfer to trehalose, and [12] trehalose mycolyltransferase. (Courtesy of Dr. G.S. Besra.)

publications (139–141) presented a novel approach in this area. The genes that confer serovar specificity to *M. avium* serovar 2, designated the *ser2* cluster, were cloned, expressed in *M. smegmatis,* and shown to encode the determinants necessary to synthesize the haptenic oligosaccharide segment of the serovar 2–specific GPL, i.e. the distal disaccharide 2,3-di-*O*-methyl-α-L-fucopyrano-syl-(1→3)-α-L-rhamnopyranose. Expression of the *ser2* genes in *M. smegmatis* produced a recombinant serovar 2–specific GPL, because the nonspecific GPL of *M. smegmatis,* singly glycosylated at the *allo*threonine residue with a 6-de-oxytalose, served as a precursor for further, serovar 2–directed glycosylation. Recently, transposon mutagenesis defined four essential loci within the *ser2* gene cluster, encoding at least the rhamnosyltransferase, the fucosyltransferase, and the methyltransferases required to methylate the fucose (142). Further, isolation of the truncated versions of the hapten induced by the transposon insertions provides genetic evidence that the GPLs of *M. avium* serovar 2 are synthesized by an initial transfer of the rhamnose unit to the 6-deoxytalose attached to the peptide core followed by fucose and finally *O*-methylation of the fucosyl unit.

There is also some limited information on the biosynthesis of PGLs and their components (143). Early work demonstrated that the carbon atoms in the methyl-branched structures in mycocerosic acids are derived from propionate (144, 145). So also are those in phthiocerol (146). The phenol-phthiocerols probably come from *p*-hydroxybenzoate, which are then elongated to yield the long-chain diol (146). The methoxyl residue in phenolphthiocerol presumably comes from methionine, by analogy with its known source in phthiocerol. Recently, *M. microti* was shown to contain two lipids, which became labeled when the cells were grown in the presence of $[2-^{14}C]$propionate (143). These were identified as phenolphthiocerol dimyco-cerosate and phenolphthiodiolone dimycocerosate, the aglycosyl derivatives of mycoside B, the phenolic glycolipid produced by *M. microti* and *M. bovis*. Cell-free extracts of the organism were able to glycosylate the lipids to form mycoside B in vitro.

Some of the enzymes for biosynthesis of the mycocerosates have been purified, and their genes cloned and sequenced (147, 148). Like some of the fatty acid synthases (133), mycocerosic acid synthase (MAS) is a multifunctional enzyme with domains that are involved in fatty acid elongation (149). Its unusual features are its substrate specificity for methylmalonyl-CoA, an acyl carrier protein–like domain that is usually associated with aggregated enzyme systems and a product mycocerosate (2,4,6,8-tetramethyloctacosan-oate) that remains bound to the enzyme. The role of the acyl carrier protein–like domain may be in binding the product, and there apparently is no thioesterase domain to release the product, perhaps to prevent uncontrolled acylation of lipids with mycocerosates. The methylmalonyl-CoA is generated by the same acyl-CoA carboxylase that generates malonyl-CoA for straight-chain fatty acid elongation in *M. tuberculosis* (150).

OUTLOOK AND CHALLENGES

Many areas remain to be explored for our better understanding of the structure and functions of the mycobacterial cell wall. In terms of structure, our knowledge of the lipids that presumably form the outer leaflet of the bilayer structure is still inadequate, and we need to define the precise arrangement of these lipids. Further, a major challenge is to understand the mechanism of assembly of this extremely complex structure, which appears to have a rather rigid interior. In terms of function, we are only beginning to get a glimpse of the possible functions of some lipid species. Mycobacterial cells must synthesize so many lipid species of unusual structures because each species performs important functions in their interactions with the nonliving or living environment. Our knowledge on the latter process, that is the role of cell wall components in pathogenesis, is unsatisfactory in spite of great efforts over many years. Perhaps a major reason for the lack of success is that we know so little about pathogenesis. Much work has been done to study the roles of cell wall components in protecting mycobacteria against oxygen radicals and in preventing the phagolysosome fusion, the two mechanisms long believed to be crucial for the intraphagocytic survival of mycobacteria. Yet recently it was reported that the major macrophage factor responsible for killing of mycobacteria is reactive nitrogen intermediates, rather than oxygen radicals (151), and that *M. tuberculosis* apparently multiplies within macrophages by escaping from the fused phagolysosomes rather than by preventing the fusion event (152). In any case, we hope that a better understanding of the functions and biosynthetic pathways of these cell wall components will be achieved, especially because recombinant DNA methods can now be used with mycobacteria (153), and that this may enable us to develop strategies that are more effective in controlling these unusually resistant pathogens.

ACKNOWLEDGMENTS

Studies in our laboratories have been supported by NIH grants AI-09644 and -33702 (to HN) and AI-18357 (to PJB).

Literature Cited

1. Collins FM. 1993. *CRC Crit. Rev. Microbiol.* 19:1–16
2. Jarlier V, Nikaido H. 1994. *FEMS Microbiol. Lett.* 123: 11–18
3. Heifets LB, ed. 1991. *Drug Susceptibility in the Chemotherapy of Mycobacterial Infections.* Boca Raton: CRC
4. Lederer E, Adam A, Ciorbaru R, Petit JF, Wietzerbin J. 1975. *Mol. Cell. Biochem.* 7:87–104
5. Minnikin DE. 1982. In *The Biology of the Mycobacteria*, ed. C Ratledge, JL Stanford, 1:95–184. London: Academic
6. Minnikin DE, Goodfellow M. 1980. In *Microbiological Classification and Identification*, ed. M Goodfellow, RG Board, pp. 189–256. London: Academic
7. Dobson G, Minnikin DE, Minnikin SM, Parlett JH, Goodfellow M, et al. 1985. In *Chemical Methods in Bacterial Systematics*, ed. M Goodfellow, DE Minnikin, pp. 237–65. London: Academic
8. Brennan PJ. 1988. In *Microbial Lipids*, ed. C Ratledge, SG Wilkinson, 1:203–98. London: Academic
9. Brennan PJ. 1989. *Rev. Infect. Dis.* 11: 420–30 (Suppl.)
10. Hunter SW, Murphy RC, Clay K, Goren MB, Brennan PJ. 1985. *J. Biol. Chem.* 258:10481–87
11. Hunter SW, Gaylord H, Brennan PJ. 1986. *J. Biol. Chem.* 261:12345–51
12. Stahl DA, Urbance JW. 1990. *J. Bacteriol.* 172:116–24
13. Rogall T, Wolters J, Flohr T, Böttger EC. 1990. *Int. J. Syst. Bacteriol.* 40:323–30
14. Pitulle C, Dorsch M, Kazda J, Wolters J, Stackerbrandt E. 1992. *Int. J. Syst. Bacteriol.* 42:337–43
15a. Wayne LG, Kubica GP. 1986. In *Bergey's Manual of Systematic Bacteriology*, ed. JG Holt, 2:1436–57. Baltimore: Willams & Wilkins
15b. Kaneda K, Imaizumi S, Mizuno S, Baba T, Tsukamura M, Yano I. 1988. *J. Gen. Microbiol.* 134:2213–29
16. Kubica GP. 1973. *Am. Rev. Respir. Dis.* 107:9–21
17. Wayne LG, Doubek JR, Russell RL. 1964. *Am. Rev. Respir. Dis.* 90:588–97
18. Saito H, Tomioka H, Yoneyama T. 1984. *Antimicrob. Agents Chemother.* 26:164–69
19. Nikaido H, Kim SH, Rosenberg EY. 1993. *Mol. Microbiol.* 8:1025–30
20. Brennan PJ, Draper P. 1994. In *Tuberculosis: Pathogenesis, Protection, and Control*, ed. B. Bloom, pp. 271–84. Washington, DC: Am. Soc. Microbiol.
21. Paul TR, Beveridge TJ. 1992. *J. Bacteriol.* 174:6508–17
22. Imaeda T, Kanetsuna F, Galindo B. 1968. *J. Ultrastruct. Res.* 25:46–63
23. Beveridge TJ, Murray RGE. 1980. *J. Bacteriol.* 141:876–87
24. Rastogi N, Frehel C, David HL. 1984. *Int. J. Syst. Bacteriol.* 34:293–99
25. Silva MT, Macedo PM. 1983. *Int. J. Lepr.* 51:225–34
26. Lee YC, Ballou CE. 1964. *J. Biol. Chem.* 239:1316–27
27. Lee YC, Ballou CE. 1965. *Biochemistry* 4:1395–404
28. Brennan PJ, Ballou CE. 1967. *J. Biol. Chem.* 242:3046–56
29. Brennan PJ, Ballou CE. 1968. *J. Biol. Chem.* 243:2975–84
30. Takayama K, Goldman DS. 1970. *J. Biol. Chem.* 245:6251–57
31. Takayama K, Schnoes HK, Semmler EJ. 1973. *Biochim. Biophys. Acta* 136: 212–21
32. Wolucka BA, McNeil MR, de Hoffmann E, Chojnacki T, Brennan PJ. 1994. *J. Biol. Chem.* 269:23328–35
32a. Besra GS, Sievert T, Lee RE, Slaydon RA, Brennan PJ, Takayama K. 1994. *Proc. Natl. Acad. Sci. USA.* In press
33. Kotani S, Kitaura T, Hirano T, Tanaka A. 1959. *Biken J.* 2:129–41
34. Schleifer KH, Kandler O. 1972. *Bacteriol. Rev.* 36:407–77
35. Ghuysen J-M. 1968. *Bacteriol. Rev.* 32: 425–64
36. Misaki A, Seto N, Azuma I. 1974. *J. Biochem.* (Tokyo) 76:15–27
37. Liu TY, Gotschlich EC. 1967. *J. Biol. Chem.* 242:471–76
38. Daffe M, Brennan PJ, McNeil M. 1990. *J. Biol. Chem.* 265:6734–43
39. McNeil MR, Daffe M, Brennan PJ. 1991. *J. Biol. Chem.* 265:18200–06
40. McNeil MR, Daffe M, Brennan PJ. 1991. *J. Biol. Chem.* 266:13217–23
41. McNeil MR, Robuck KG, Harter M, Brennan PJ. 1994. *Glycobiology* 4:165–73
42. McElhaney RN, de Gier J, van der Neut-Kok ECM. 1973. *Biochim. Biophys. Acta* 298:500–12
43. Azuma I, Ajisaka M, Yamamura Y. 1970. *Infect. Immun.* 2:347–49
44. Misaki A, Azuma I, Yamamura Y. 1977. *J. Biochem.* (Tokyo) 82:1759–70
45. Ohasi M. 1970. *Jpn. J. Exp. Med.* 40:1–14
46. Hunter SW, Brennan PJ. 1990. *J. Biol. Chem.* 265:9272–79
47. Chatterjee D, Hunter SW, McNeil MR,

Brennan PJ. 1992. *J. Biol. Chem.* 267: 6228–33

48. Chatterjee D, Bozic CM, McNeil MR, Brennan PJ. 1991. *J. Biol Chem.* 266: 9652–60

49. Chatterjee D, Lowell K, Rivoire B, McNeil MR, Brennan PJ. 1992. *J. Biol. Chem.* 267:6234–39

50. Venisse A, Berjeaud J-H, Chaurand P, Gilleron P, Puzo G. 1993. *J. Biol. Chem.* 268:12401–11

51. Prinzis S, Chatterjee D, Brennan PJ. 1993. *J. Gen. Microbiol.* 139:2649–58

52. Khoo K-H, Chatterjee D, Brennan PJ, Morris HR, Dell A. 1994. Abstr. XVII of *Int. Carbohydr. Symp. Ottawa, 1989,* pp. 2–10

53. Horsburg CR, Selik RM. 1989. *Am. Rev. Respir. Dis.* 139:4–7

54. Besra GS, Brennan PJ. 1994. In *Mass Spectrometry for the Characterization of Microorganisms,* ed. C Fenselau, pp. 203-32. Washington, DC: Amer. Chem. Soc.

55. Besra GS, McNeil MR, Khoo K-H, Dell A, Morris HR, Brennan PJ. 1993. *Biochemistry* 32:12705–14

56. Saadat S, Ballou CE. 1983. *J. Biol. Chem.* 258:1813–18

57. Kamisango K-I, Saadat S, Dell A, Ballou CE. 1985. *J. Biol. Chem.* 260:4117–21

58. Besra GS, McNeil MR, Brennan PJ. 1992. *Biochemistry* 31:6504–09.

59. Besra GS, McNeil MR, Minnikin DE, Portaels F, Riddel M, Brennan PJ. 1990. *Biochemistry* 30:7772–77

60. Dobson G, Minnikin DE, Besra GS, Mallet AI, Magnuson M. 1990. *Biochim. Biophys. Acta* 1042:176–81

61. Gaylord H, Brennan PJ. 1987. *Annu. Rev. Microbiol.* 41:645–75

62. Brennan PJ, Goren MB. 1979. *J. Biol. Chem.* 254:4205–11

63. Aspinall GO, Chatterjee C, Brennan PJ. 1994. *Adv. Carbohydr. Chem. Biochem.* In press

64. Lopez-Marin LM, Laneelle M-A, Prome D, Daffe M, Laneelle G, et al. 1991. *Biochemistry* 30:10536–42

65. Riviere M, Puzo G. 1992. *Biochemistry* 31:3575–80

66. Besra GS, McNeil MR, Rivoire B, Khoo K-H, Morris HR, et al. 1993. *Biochemistry* 32:347–55

67. Besra GS, Bolton RC, McNeil MR, Ridell M, Simpson KE, et al. 1992. *Biochemistry* 31:9832–37

68. Goren MB, Brennan PJ. 1979. In *Tuberculosis,* ed. GP Youmans, pp. 63–193. Philadelphia: Saunders

69. Akamatsu Y, Ono Y, Nojima S. 1966. *J. Biochem.* (Tokyo) 59:176–82

70. Goldman DS. 1970. *Am. Rev. Respir. Dis.* 102:543–55

71. Lemassu A, Daffe M. 1994. *Biochem. J.* 297:351–57

72. Hunter SW, Rivoire B, Mehra V, Bloom BR, Brennan PJ. 1990. *J. Biol. Chem.* 265:14065–68.

73. Trias J, Jarlier V, Benz R. 1992. *Science* 258:1479–81

74. Etemadi AH, Okuda R, Lederer E. 1964. *Bull. Soc. Chim. France* 868–70

75. Nikaido H, Vaara M. 1985. *Microbiol.Rev.* 49:1–32

76. Hui J, Godron N, Kajioka R. 1977. *Antimicrob. Agents Chemother.* 11:773–79

77. Brennan PJ, Rooney SA, Winder FG. 1970. *Irish J. Med. Sci.* 3:371–90

78. Durand E, Welby M, Laneelle G, Toccane J-F. 1979. *Eur. J. Biochem.* 93, 103–12

79. Naumann D, Schultz C, Born J, Labischinski H, Brandenburg K, et al. 1989. *J. Mol. Struct.* 214:213–46

80. Gaylord H, Brennan PJ. 1987. *Annu. Rev. Microbiol.* 41:645–75

81. Cullis PR, de Kruijf B, Hope MJ, Verkleij AJ, Nayar R, et al. 1983. In *Membrane Fluidity in Biology,* ed. AC Aloia, 1:39–81. New York: Academic

82. Draper P. 1982. *The Biology of the Mycobacteria,* ed. C Ratledge, JL Stanford, 1:9–52. London: Academic

83. Barrow WW, Ullom BP, Brennan PJ. 1980. *J. Bacteriol..* 144:814–22.

84. Jarlier V, Nikaido, H. 1990. *J. Bacteriol..* 172:1418–23

85. Zimmermann W, Rosselet, A. 1977. *Antimicrob. Agents Chemother.* 12:368–72

86. Trias J, Benz R. 1994. *Mol. Microbiol.* 14:283–90

87. Connell ND, Nikaido H. 1994. In *Tuberculosis: Pathogenesis, Protection, and Control,* ed. BR Bloom, pp. 333–52. Washington: Am. Soc. Microbiol.

88. Sundaram KS, Venkitasubramanian TA. 1978. *Antimicrob. Agents Chemother.* 13:726–30

89. Nikaido H. 1990. In *Membrane Transport and Information Storage,* ed. RC Aloia, CC Curtain, LM Gordon, pp. 165–90. New York: Wiley-Liss

90. Plesiat P, Nikaido H. 1992. *Mol. Microbiol.* 6:1323–33

91. Yajko DM, Sanders CA, Nassos PS, Hadley WK. 1990. *Antimicrob. Agents Chemother.* 34:2442–44

92. Haemers A, Leysen DC, Bollaert W, Zhang M, Pattyn SR. 1990. *Antimicrob. Agents Chemother.* 34:496–97

93. Franzblau SG, White KE. 1990. *Antimicrob. Agents Chemother.* 34:229–31

94. Wallace RJ Jr, Dalovisio JR, Pankey

GA. 1979. *Antimicrob. Agents Chemother.* 16:611–14

95. Swenson JM, Thornsberry C, Silcox VA. 1982. *Antimicrob. Agents Chemother.* 22:186–92

96. Gelber RH. 1987. *J. Infect. Dis.* 156:236–39

97. Fernandes PB, Hardy DJ, McDaniel D, Hanson CW, Swanson RN. 1982. *Antimicrob. Agents Chemother.* 33:1531–34

98. Brown BA, Wallace RJ Jr, Onyi GO, De Rosas V, Wallace RJ III. 1992. *Antimicrob. Agents Chemother.* 36:180–84

99. Gorzinski EA, Gutman SI, Allen W. 1989. *Antimicrob. Agents Chemother.* 33:591–92

100. Heifets LB, Lindhom-Levy PJ, Flory MA. 1990. *Am. Rev. Respir. Dis.* 141:626–30

101. Rastogi N, Goh KS. 1990. *Antimicrob. Agents Chemother.* 34:2061–64

102. Yates MD, Collins CH. 1981. *Tubercle* 62:117–21

103. Deleted in proof

104. Kaplan G, Gandhi RR, Weinstein DE, Levis WR, Patarroyo ME, et al. 1987. *J. Immunol.* 138:3028–34

105. Sibley LD, Hunter SW, Brennan PJ, Krahenbuhl JL. 1988. *Infect. Immun.* 56:1232–36

106. Chan J, Fan X, Hunter SW, Brennan PJ, Bloom BR. 1991. *Infect. Immun.* 59:1755–61

107. Chatterjee D, Roberts AD, Lowell K, Brennan PJ, Orme IM. 1992. *Infect. Immun.* 60:1249–53

108. Barnes PF, Chatterjee D, Abrams JS, Lu S, Wang E, et al. 1992. *J. Immunol.* 149:541–47

109. Roach TIA, Barton CH, Chatterjee D, Blackwell JM. 1993. *J. Immunol.* 150:1886–96

110. Schlesinger LS. 1993. *J. Immunol.* 150:2920–30

111. Daffe M, Lacave C, Laneelle M-A. 1987. *Eur. J. Biochem.* 167:155–60

112. Barrow WW. 1991. *Res. Microbiol.* 142:427–33

113. Rastogi N, Levy-Frebault V, Blom-Potar M, David HL. 1989. *Zbl. Bakteriol. Hyg.* A270:346–60

114. Kojima N, Araki Y, Ito E. 1985. *J. Bacteriol.* 161:299–306

115. Hancock IC, Baddiley J. 1976. *J. Bacteriol.* 125:880–86

116. Hancock IC, Wiseman G, Baddiley J. 1981. *J. Bacteriol.* 147:698–701

117. Hancock IC, Wiseman G, Baddiley J. 1976. *FEBS Lett.* 69:75–80

118. Hase S, Matsushima Y. 1979. *J. Biochem.* 81:1181–86

119. Archibald AR, Hancock IC, Harwood CR. 1993. In *Bacillus subtilis and other Gram-positive Bacteria: Biochemistry, Physiology, and Molecular Genetics.* Washington, DC: Am. Soc. Microbiol.

120. Garcia-Trejo A, Haddock LW, Chittendon GJF, Baddiley J. 1971. *Biochem. J.* 122:49

121. Takayama K, Goldman DS. 1989. *Antimicrob. Agents Chemother.* 33:1493–99

122. Gastambide-Odier M, Delaumeny JM, Lederer E. 1963. *Chem. Ind.* 1963:1285–86

123. Etemadi A-H. 1967. *Bull. Soc. Chim. Biol.* 49:695–706

124. Takayama M, Qureshi N. 1984. In *The Mycobacteria. A Sourcebook,* ed. GP Kubica, LG Wayne, pp. 315–44. New York: Dekker

124a. Wheeler PR, Ratledge C. 1994. In *Tuberculosis: Pathogenesis, Protection and Control,* ed. B Bloom, pp. 353–85. Washington, DC: Am. Soc. Microbiol.

125. Couderc F, Aurelle H, Prome D, Savagnac A, Prome JC. 1988. *Biomed. Environ. Mass Spectrom.* 16:317–21

126. Lacave C, Quemard A, Laneelle G. 1990. *Biochim. Biophys. Acta* 1045:58–65

127. Kikuchi S, Kusaka T. 1986. *J. Biochem.* 99:723–31

128. Bloch K. 1977. *Adv. Enzymol.* 45:1–84

129. Kikuchi S, Rainwater DL, Kolattukudy PE. 1992. *Arch. Biochem. Biophys.* 295:318–26

130. Qureshi N, Sathyamoorthy N, Takayama K. 1984. *J. Bacteriol.* 157:46–52

131. Lacave C, Laneelle MA, Daffe M, Montrozier H, Laneelle G. 1989. *Eur. J. Biochem.* 181:459–66

132. Embley TM, O'Donnell AG, Rostron J, Goodfellow M. 1988. *J. Gen. Microbiol.* 134:953–60

133. Kundu M, Basu J, Chakrabarti P. 1989. *FEBS Lett.* 256:207–10

134. Walker RW, Prome JC, Lacave C. 1973. *Biochim. Biophys. Acta* 326:52–62

135. Shimakata T, Iwaki M, Kusaka T. 1984. *Arch. Biochem. Biophys.* 229:329–39

136. Winder FG. 1982. See Ref. 82

137. Banerjee A, Dubnau E, Quemard V, Balasubramanian KS, Um T, et al. 1994. *Science* 263:227–30

138. Takayama K, Wang L, David HL. 1975. *J. Lipid Res.* 16:303–17

139. Belisle JT, Pascopella L, Inamine JM, Brennan PJ, Jacobs WR Jr. 1991. *J. Bacteriol.* 173:6991–97

140. Belisle JT, McNeil MR, Chatterjee D, Inamine JM, Brennan PJ. 1991. *J. Biol. Chem.* 268:10510–16

141. Belisle JT, Klaczkiewicz K, Brennan PJ, Jacobs WR Jr, Inamine JM. 1993. *J. Biol. Chem.* 268:10517–23
142. Mills JA, McNeil MR, Belisle JT, Jacobs WR Jr, Brennan PJ. 1994. *J. Bacteriol.* 176:4803–08
143. Thurman PF, Chai W, Rosankiewica JR, Rogers HJ, Lawson AM, et al. 1993. *Eur. J. Biochem.* 212:705–11
144. Gastambide-Odier M, Delaumeny J-M, Lederer E. 1963. *Biochim. Biophys. Acta* 70:670–78
145. Yano I, Kusunose M. 1966. *Biochim. Biophys. Acta* 116:593–96
146. Gastambide-Odier M, Sarda P. 1970. *Pneumonologie* 142:241–55
147. Rainwater DL, Kolattukudy PE. 1983. *J. Biol. Chem.* 258:2979–85
148. Rainwater DL, Kolattukudy PE. 1985. *J. Biol. Chem.* 260:616–23
149. Manjula M, Kolattukudy PE. 1992. *J. Biol. Chem.* 267:19388–95
150. Rainwater DL, Kolattukudy PE. 1982. *J. Bacteriol.* 151:905–11
151. Chan J, Xing Y, Magliozzo RS, Bloom BR. 1993. *J. Exp. Med.* 175:1111–22
152. McDonough KA, Kress Y, Bloom BR. 1993. *Infect. Immun.* 61:2763–73
153. Jacobs WR Jr, Kalpana GV, Cirillo JD, Pascopella L, Snapper SB, et al. 1991. *Methods Enzymol.* 204:537–55

Annu. Rev. Biochem. 1995. 64:65–95

TRIPLEX DNA STRUCTURES

Maxim D. Frank-Kamenetskii

Center for Advanced Biotechnology and Department of Biomedical Engineering, Boston University, Boston, Massachusetts 02215

Sergei M. Mirkin

Department of Genetics, University of Illinois at Chicago, Chicago, Illinois 60612

KEY WORDS: DNA triplex, H-DNA, gene-drugs, triplex-forming oligonucleotides, PNA

CONTENTS

ABSTRACT

A DNA triplex is formed when pyrimidine or purine bases occupy the major groove of the DNA double Helix forming Hoogsteen pairs with purines of the Watson-Crick basepairs. Intermolecular triplexes are formed between triplex forming oligonucleotides (TFO) and target sequences on duplex DNA. Intramolecular triplexes are the major elements of H-DNAs, unusual DNA structures, which are formed in homopurine-homopyrimidine regions of supercoiled DNAs. TFOs are promising gene-drugs, which can be used in an anti-gene strategy, that attempt to modulate gene activity in vivo. Numerous chemical modifications of TFO are known. In peptide nucleic acid (PNA), the sugar-phosphate backbone is replaced with a protein-like backbone. PNAs form

0066-4154/95/0701-0065$05.00

P-loops while interacting with duplex DNA forming triplex with one of DNA strands leaving the other strand displaced. Very unusual recombination or parallel triplexes, or R-DNA, have been assumed to form under RecA protein in the course of homologous recombination.

PERSPECTIVES AND SUMMARY

Since the pioneering work of Felsenfeld, Davies, & Rich (1), double-stranded polynucleotides containing purines in one strand and pyrimidines in the other strand [such as poly(A)/poly(U), poly(dA)/poly(dT), or poly(dAG)/ poly(dCT)] have been known to be able to undergo a stoichiometric transition forming a triple-stranded structure containing one polypurine and two polypyrimidine strands (2–4). Early on, it was assumed that the third strand was located in the major groove and associated with the duplex via non-Watson-Crick interactions now known as Hoogsteen pairing. Triple helices consisting of one pyrimidine and two purine strands were also proposed (5, 6). However, notwithstanding the fact that single-base triads in tRNA structures were well-documented (reviewed in 7), triple-helical DNA escaped wide attention before the mid-1980s.

The considerable modern interest in DNA triplexes arose due to two partially independent developments. First, homopurine-homopyrimidine stretches in supercoiled plasmids were found to adopt an unusual DNA structure, called H-DNA, which includes a triplex as the major structural element (8, 9). Secondly, several groups demonstrated that homopyrimidine and some purine-rich oligonucleotides can form stable and sequence-specific complexes with corresponding homopurine-homopyrimidine sites on duplex DNA (10–12). These complexes were shown to be triplex structures rather than D-loops, where the oligonucleotide invades the double helix and displaces one strand. A characteristic feature of all these triplexes is that the two chemically homologous strands (both pyrimidine or both purine) are antiparallel. These findings led to explosive growth in triplex studies.

During the study of intermolecular triplexes, it became clear that triplex-forming oligonucleotides (TFOs) might be universal drugs that exhibit sequence-specific recognition of duplex DNA. This is an exciting possibility because, in contrast to other DNA-binding drugs, the recognition principle of TFOs is very simple: Hoogsteen pairing rules between a purine strand of the DNA duplex and the TFO bases. However, this mode of recognition is limited in that homopurine-homopyrimidine sites are preferentially recognized. Though significant efforts have been directed toward overcoming this limitation, the problem is still unsolved in general. Nevertheless, the high specificity of TFO-DNA recognition has led to the development of an "antigene" strategy, the goal of which is to modulate gene activity in vivo using TFOs (reviewed in 13).

Although numerous obstacles must be overcome to reach the goal, none are likely to be fatal for the strategy. Even if DNA TFOs proved to be unsuitable as gene-drugs, there are already many synthetic analogs that also exhibit triplex-type recognition. Among them are oligonucleotides with non-natural bases capable of binding the duplex more strongly than can natural TFOs. Another promising modification replaces the sugar-phosphate backbone of ordinary TFO with an uncharged peptidelike backbone, called a peptide nucleic acid (PNA) (reviewed in 14). Homopyrimidine PNAs form remarkably strong and sequence-specific complexes with the DNA duplex via an unusual strand-displacement reaction: Two PNA molecules form a triplex with one of the DNA strands, leaving the other DNA strand displaced (a "P-loop") (15, 16).

The ease and sequence specificity with which duplex DNA and TFOs formed triplexes seemed to support the idea (17) that the homology search preceding homologous recombination might occur via a triplex between a single DNA strand and the DNA duplex without recourse to strand separation in the duplex. However, these proposed recombination triplexes are dramatically different from the orthodox triplexes observed experimentally. First, the recombination triplexes must be formed for arbitrary sequences and, second, the two identical strands in this triplex are parallel rather than antiparallel. Some data supported the existence of a special class of recombination triplexes, at least within the complex among duplex DNA, RecA protein, and single-stranded DNA (reviewed in Ref. 18), called R-DNA. A stereochemical model of R-DNA was published (19). However, the structure of the recombination intermediate is far from being understood, and some recent data strongly favor the traditional model of homology search via local strand separation of the duplex and D-loop formation mediated by RecA protein.

Intramolecular triplexes (H-DNA) are formed in vitro under superhelical stress in homopurine-homopyrimidine mirror repeats. The average negative supercoiling in the cell is not sufficient to induce H-DNA formation in most cases. However, H-DNA can be detected in vivo in association with an increase of DNA supercoiling driven by transcription or other factors (reviewed in 20). H-DNA may even be formed without DNA supercoiling during in vitro DNA synthesis. Peculiarly, this DNA polymerase–driven formation of H-DNA efficiently prevents further DNA synthesis (21, 22). There are preliminary indications that H-DNA may also terminate DNA replication in vivo (23). More work is required, however, to elucidate the role of H-DNA in biological systems.

STRUCTURE, STABILITY, AND SPECIFICITY OF DNA TRIPLEXES

Triplex Menagerie

Since the original discovery of oligoribonucleotide-formed triplexes, numerous studies have shown that the structure of triplexes may vary substantially. First, it was shown that triplexes may consist of two pyrimidine and one purine strand

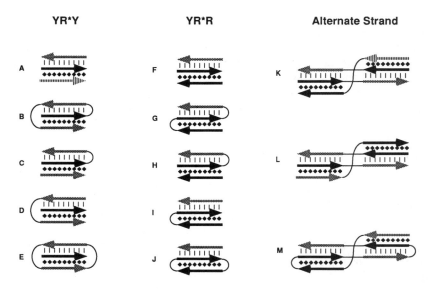

Figure 1 Triplex menagerie (see text for explanations). Solid lines, purine strands; stippled lines, pyrimidine strands; vertical lines, Watson-Crick hydrogen bonds; diamonds, Hoogsteen hydrogen bonds. Arrows indicate DNA chain polarity.

(YR*Y) or of two purine and one pyrimidine strand (YR*R). Second, triplexes can be built from RNA or DNA chains or their combinations. Third, triplexes can be formed within a single polymer molecule (intramolecular triplexes) or by different polynucleotides (intermolecular triplexes). Finally, for special DNA sequences consisting of clustered purines and pyrimidines in the same strand, triplex formation may occur by a strand-switch mechanism (alternate strand triplexes). Figure 1 summarizes numerous possible structures of triple-helical nucleic acids.

The building blocks of YR*Y triplexes are the canonical CG*C and TA*T triads shown in Figure 2. To form such triads, the third strand must be located in the major groove of the double helix that is forming Hoogsteen hydrogen bonds (24) with the purine strand of the duplex. The remarkable isomorphism of both canonical triads makes it possible to form a regular triple helix. This limits YR*Y triplexes to homopurine-homopyrimidine sequences in DNA. An important feature of the YR*Y triplexes is that formation of the CG*C triad requires the protonation of the N3 of cytosine in the third strand. Thus, such triplexes are favorable under acidic conditions (3, 4). By contrast, the YR*R triplexes usually do not require protonation (see below).

The mutual orientation of the chemically homologous strands in a triplex

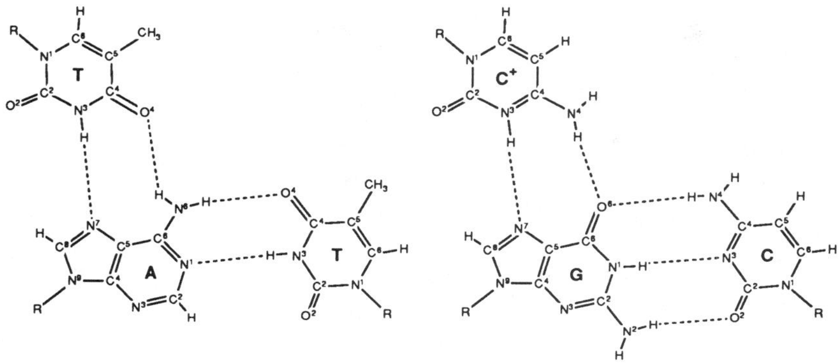

Figure 2 Canonical base triads of YR*Y triplexes: TA*T and CG*C$^+$.

(i.e. two pyrimidine strands in the YR*Y triplex or the two purine strands in the YR*R triplex), which a priori can be either parallel or antiparallel, is of paramount importance. The discovery of H- and *H-DNA (see below) indicated that both YR*Y and YR*R triplexes form as antiparallel structures (9, 25). A thorough investigation of intermolecular triplexes by different methods unambiguously demonstrated that both YR*Y and YR*R triplexes are stably formed only as antiparallel structures. The most direct data were obtained by cleaving target DNA with homopyrimidine or homopurine oligonucleotides attached to Fe•EDTA (11, 26). The observed pairing and orientation rules rigorously determine the sequence of the triplex-forming homopyrimidine strand.

YR*R triplexes are more versatile than YR*Y triplexes. Originally it was believed that they must be built from CG*G and TA*A triads (6, 25, 27). Later work showed, however, that TA*T triad may also be incorporated into the otherwise YR*R triplex. Moreover, the stability of triplexes consisting of alternating CG*G and TA*T triads is higher than that of triplexes built of CG*G and TA*A triads (26). Thus, the term YR*R triplex, though routinely used in literature, is misleading with regard to the chemical nature of the third strand. The corresponding triads are presented in Figure 3. One can see that they are not strictly isomorphous, as was the case for YR*Y triads. Another notable difference between two triplex types is that reverse Hoogsteen base-pairs are needed to form reasonable stacking interactions among CG*G, TA*A, and TA*T triads (26).

Whereas in YR*Y triplexes the sequence of the third strand is fully determined by the sequence of the duplex, the situation is different for YR*R triplexes. Here the third strand may consist of three bases, G, A, and T, where

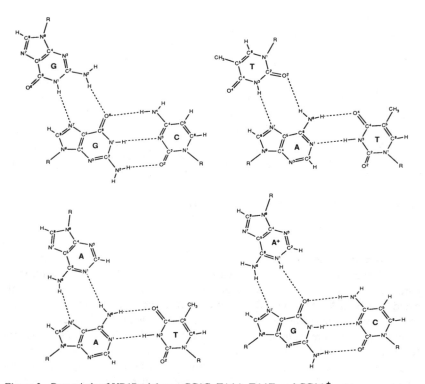

Figure 3 Base triads of YR*R triplexes: CG*G, TA*A, TA*T, and CG*A$^+$.

the guanines oppose guanines in the duplex, while adenines or thymines must oppose adenines of the duplex. A protonated CG*A$^+$ triad (Figure 3) forms so that A in the third strand may also oppose G in the duplex at acidic pH (28).

Another novel feature of YR*R triplexes is that their stability depends dramatically on the presence of bivalent metal cations (reviewed in 20). Unlike the case of YR*Y triplexes, where the requirement for H$^+$ ions has an obvious reason, the metal dependence of YR*R triplexes is an obscure function of the particular metal ion and the triplex sequence (29). Possible structural reasons for selectivity of bivalent cations in stabilization of YR*R triplexes are discussed in Ref. 30.

Despite these differences, the YR*R triplexes are similar to YR*Y triplexes in their most fundamental features: (*a*) The duplex involved in triplex formation must have a homopurine sequence in one strand, and (*b*) the orientation of the two chemically homologous strands is antiparallel.

One can easily imagine numerous "geometrical" ways to form a triplex, and those that have been studied experimentally are shown in Figure 1. The canonical intermolecular triplex consists of either three independent oligonucleotide chains (3, 4) (Figure 1A,F) or of a long DNA duplex carrying the homopurine-homopyrimidine insert and the corresponding oligonucleotide (10–12). In any case, triplex formation strongly depends on the oligonucleotide(s) concentration.

A single DNA chain may also fold into a triplex connected by two loops (Figure 1B,G). To comply with the sequence and polarity requirements for triplex formation, such a DNA strand must have a peculiar sequence: It contains a mirror repeat (homopyrimidine for YR*Y triplexes and homopurine for YR*R triplexes) flanked by a sequence complementary to one half of this repeat (31). Such DNA sequences fold into triplex configuration much more readily than do the corresponding intermolecular triplexes, because all triplex-forming segments are brought together within the same molecule (31–33).

There is also a family of triplexes built from a single strand and a hairpin. Two types of arrangements are possible for such structures (34, 35): (a) a canonical hairpin, formed by two self-complementary DNA segments, is involved in Hoogsteen hydrogen bonding with a single strand (Figure 1C,H), and (b) a "hairpin" containing a homopurine (for YR*R) or homopyrimidine (for YR*Y) mirror repeat is involved in both Watson-Crick and Hoogsteen basepair formation in a triplex (Figure 1D,I). A peculiar modification of this scheme was described in Refs. 36 and 37, where a short circular oligonucleotide could be used for triplex formation instead of a hairpin (Figure 1E,J). Such a triplex-forming oligonucleotide is of particular interest for DNA targeting in vivo (see below), since circular oligonucleotides are not substrates for degradation by exonucleases.

The structures in Figure 1 are intentionally ambiguous with regard to the 5' and 3' ends of polynucleotide chains. In fact, all these structures may exist as two chemically distinct isoforms differing in relative chain polarity. The comparative stability of the two isoforms is poorly known. Very recent data presented in Refs. 38 and 39 indicate that their free energies may differ by 1.5–2.0 kcal/mol and may depend on the loop sequence. The two isoforms may also differ topologically (see below).

So far, we have considered triplexes with their duplex part consisting of purely homopurine and homopyrimidine strands (the influence of individual mismatched triads is discussed below). It has become clear recently, however, that both sequence requirements and chain polarity rules for triplex formation can be met by DNA target sequences built of clusters of purines and pyrimidines (40–43) (see Figure 1K–M). The third strand consists of adjacent homopurine and homopyrimidine blocks forming Hoogsteen hydrogen bonds with purines on alternate strands of the target duplex, and this strand switch preserves the proper chain polarity. These structures, called alternate-strand triplexes, have been experimentally observed as both intra- (42) (Figure 1*M*) and intermolecular (41, 43) (Figure 1*K,L*) triplexes. These results increase the number of potential targets for triplex formation in natural DNAs somewhat by adding sequences composed of purine and pyrimidine clusters, although arbitrary sequences are still not targetable because strand switching is energetically unfavorable. Preliminary estimates give the minimal length of a cluster in an alternate-strand triplex as between 4 and 8 (44). A peculiar feature of alternate-strand triplexes is that two different sequences of the third strand fulfill the requirements for triplex formation for a single duplex target (Figure 1*K,L*). For a few studied targets, the efficiency of triplex formation by the two variants was quite different. Strand switching in the direction $3'$-R_n-Y_n-$5'$ along the third strand was more favorable than $3'$-Y_n-R_n-$5'$ (44).

Hybrid triplexes consisting of both DNA and RNA chain are less studied and only for YR*Y triplexes. Eight combinations of RNA and DNA chains within a triplex are possible in principle, and the relative stability of each was studied (45–47). The results from different groups differ substantially, for reasons that are yet to be understood. (Though they may be attributed to differences in sequences and/or the ambient conditions.) However, all these studies show consistently that triplexes are more stable when DNA represents the central homopurine strand than when RNA does. Affinity cleavage data also indicate that the orientation of chemically homologous chains in hybrid triplexes is antiparallel.

Fine Structure of DNA Triplexes

The structural features of DNA triplexes have been studied using such diverse techniques as chemical and enzymatic probing, affinity cleavage, and electrophoresis, all of which have provided insight into the overall structure of triplexes: (*a*) The third strand lies in the major groove of the duplex, as is deduced from the guanine N7 methylation protection (48, 49); (*b*) the orientation of the third strand is antiparallel to the chemically homologous strand of the duplex (11, 26); and (*c*) the duplex within the triplex is noticeably unwound relative to the canonical B-DNA (12). However, the fine structure of triplexes could not be elucidated at atomic resolution without more direct structural methods based on X-ray diffraction and NMR.

The first attempt to deduce the atomic structure of a poly(dT)•poly(dA)•poly(dT) triplex using X-ray fiber diffraction was performed in 1974 (50). Two important parameters of the triple helix, an axial rise equal to 3.26 Å and a helical twist of 30°, were directly determined from the fiber diffraction patterns. However, in an attempt to fit experimental data with atomic models, the

Table 1 Computed averages for various helical parameters

| | Triplexes | | | | | |
| | YR*Y | | YR*R | | | |
	X-ray	NMR	X-ray	NMR	B-DNA	A-DNA
Axial rise (Å)	3.3	3.4	—	3.6	3.4	2.6
Helical twist (°)	31	31	—	30	36	33
Axial displacement (Å)	−2.5	−1.9	—	−1.9	−0.7	−5.3
Glycosidic torsional angle	anti	anti	—	anti	anti	anti
Sugar pucker	C2' endo	C2' endo	—	C2' endo	C2' endo	C3' endo

duplex within the triplex was erroneously concluded to adopt the A conformation (50). Other studies of the atomic structure of triplexes have only recently corrected this widely accepted conclusion. NMR data (51) (see below) and infrared spectroscopy (52) convincingly demonstrated that the sugar pucker in all three strands within the triplex is of the S-type (a characteristic for B-DNA rather than A-DNA). It appeared that a B-like structure could nicely explain the original fiber diffraction data (Table 1). Moreover, this structure is stereochemically more favorable than is the original (53).

Further sophisticated NMR studies have examined inter- and intramolecular triplexes of both YR*Y and YR*R types. This data unambiguously supported the major features of the YR*Y triplexes discussed above: (a) a requirement for cytosine protonation (54–56); (b) Hoogsteen basepairing of the third strand (32, 57); and (c) antiparallel orientation of the two pyrimidine strands (32, 57). The atomic structure of the triple helix, summarized in Table 1, was also determined (51, 58). The values of all major parameters determined by NMR are very close to those determined by fiber diffraction (53). Most significantly, the deoxyribose conformation of all strands in the triplex corresponds to an S-type (C2'-endo) pucker (51). It is clear from Table 1 that the duplex within the triplex adopts a B-like configuration; the helical twist in the triplex, however, is significantly smaller than that for B-DNA.

NMR studies of the YR*R triplexes (33, 59) showed that their overall structure is similar to that of YR*Y triplexes. The important difference, however, is that reverse Hoogsteen basepairing of the third strand (as in Figure 3) was convincingly demonstrated. The helical parameters presented in Table 1 are close to those of YR*Y triplexes and suggest the formation of an unwound B-like structure. A peculiar feature of the YR*R triplexes consisting of CG*G and TA*T triads is the concerted changes in the axial rise and helical twist along the helix axis (60). This was attributed to the lack of isomorphism between CG*G and TA*T triads, discussed above.

H Form

Although the canonical Watson-Crick double helix is the most stable DNA conformation for an arbitrary sequence under usual conditions, some sequences within duplex DNA are capable of adopting structures quite different from the canonical B form under negatively superhelical stress (reviewed in 61). One of these structures, the H form, includes a triplex as its major structural element (Figure 4). Actually, there is an entire family of H-like structures (see 20 for a comprehensive review).

The term "H form" was proposed in a study of a cloned sequence from a spacer between the histone genes of sea urchin (62). It contained a $d(GA)_{16}$ stretch hypersensitive to S1 nuclease. Such S1-hypersensitive sites had been anticipated previously to adopt an unusual structure (63, 64), and numerous hypotheses had been discussed in the literature (63, 65–71). Using 2-D gel electrophoresis of DNA topoisomers (see 20), a structural transition was demonstrated without enzymatic or chemical modification (62). The pH dependency of the transition was remarkable: At acidic pH the transition occurs under low torsional tension, while at neutral pH it is almost undetectable. Because pH dependence had never been observed before for non-B-DNA conformations (cruciforms, Z-DNA, bent DNA, etc), the investigators concluded that a novel DNA conformation was formed. The structure was called the H form because it was clearly stabilized by hydrogen ions, i.e. it was a protonated structure.

The H form model proposed in Ref. 8 (Figure 4A) consists of an intramolecular triple helix formed by the pyrimidine strand and half of the purine strand, leaving the other half of the purine strand single stranded. As Figure 4A shows, this structure is topologically equivalent to unwound DNA. Two isoforms of H form are possible: one single stranded in the 5' part of the purine strand and the other single stranded in the 3' part. The existence of single-stranded purine stretches in H-DNA explains its hyperreactivity to S1 nuclease. Canonical TA*T and CG*C+ base triads stabilize the triple helix (Figure 2). The protonation of cytosines is crucial for the formation of CGC+ base triads, which explains the pH dependency of the structural transition.

The H-DNA model predicts that a homopurine-homopyrimidine sequence must be a mirror repeat to form H-DNA. It was convincingly demonstrated in Ref. 9 that mirror repeats indeed adopt the H form, while even single-base violations of the mirror symmetry significantly destabilize the structure. Chemical probing of H-DNA using conformation-specific DNA probes (reviewed in 72) provided final proof of the H-form model (25, 48, 49, 73–76). Notably, these studies revealed that different sequences preferentially adopt only one of the two possible isomeric forms of H-DNA, the one in which the 5' part of the purine strand is unstructured.

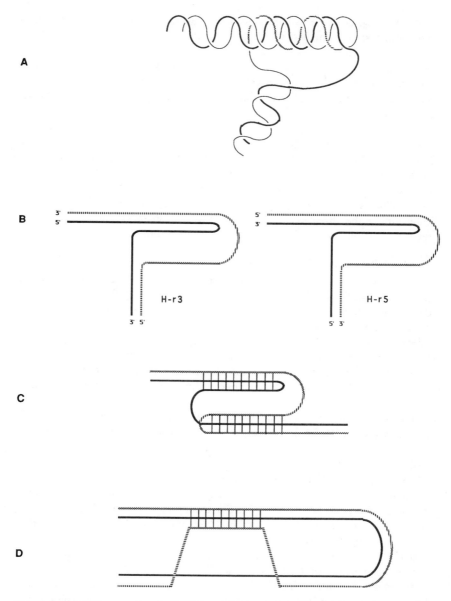

Figure 4 H-DNA menagerie. *A*. H-DNA model. Bold line, homopurine strand; thin line, homopyrimidine strand; dashed line, the half of the homopyrimidine strand donated to the triplex. *B*. Two isoforms of *H-DNA. *C*. Nodule DNA. *D*. Tethered loop. In *B–D*, solid line, homopurine strand; stippled line, homopyrimidine strand.

The structural features responsible for the difference between the two isoforms have been identified in Ref. 77. The isoform with the 3′ half of the pyrimidine strand donated to the triplex (designated H-y3) is preferentially formed at physiological superhelical densities. In this isoform, the 5′ portion of the purine strand is single stranded, and its formation is consistent with the chemical probing results described above. The other isoform (in which the 5′ half of the pyrimidine strand is donated to the triplex—designated H-y5) was only observed at low superhelical density. Topological modeling of H-DNA formation showed that the formation of the H-y3 isoform releases one extra supercoil relative to the H-y5 isoform. This explains why H-y3 is favorable at high superhelical density. Recent studies show that the mechanisms underlying preferential isomerization into the H-y3 conformation are more complex. Apparently, the presence of bivalent cations can make the H-y5 isoform preferable (78). What is more surprising, the loop sequence plays an important role in determining the direction of isomerization (79, 80). Systematic studies of factors contributing to isomerization are yet to be done.

H-DNA Menagerie

As for intermolecular triplexes, a menagerie of H-DNA-like structures exists (reviewed in 20). First, intramolecular YR*R triplex, called *H-DNA, was described in Refs. 25 and 81 (Figure 4B). This structure is also topologically equivalent to the unwound DNA and requires DNA supercoiling (82). As in intermolecular YR*R triplexes, A can be replaced with T (83) and, at acidic pH, G can be replaced with A (28) in the third strand of *H-DNA. Thus, the sequences adopting the *H form are not necessarily mirror repeated and not even necessarily homopurine-homopyrimidine (see 20 for comprehensive review).

Two isoforms of *H-DNA are possible, designated H-r3 and H-r5 according to which half of the homopurine strand is donated to the triplex (Figure 4B). Chemical probing with single-stranded, DNA-specific agents showed that H-r3 isoform is dominant.

As for all YR*R triplexes, the mechanisms of *H-DNA dependence on bivalent cations are unclear. Cation requirements are different for different sequences (20, 25, 27, 81, 84–87). For example, while *H-DNA formed by $d(G)_n \cdot d(C)_n$ sequences is stabilized by Ca^{2+}, Mg^{2+}, and Mn^{2+}, the same structure formed by $d(GA)_n \cdot d(TC)_n$ is formed in the presence of Zn^{2+}, Mn^{2+}, Cd^{2+}, and Co^{2+}. The differences in cation requirements are due to variations in neighboring triads or changes in the GC content or both. Even moderate changes in GC content (from 75% to 63%) switched cation requirement from Mg^{2+} to Zn^{2+} for a particular sequence (22). A Mg^{2+}-to-Zn^{2+} switch was reported to affect the equilibrium between H-r5 and H-r3 isoforms (86) or to substantially modify the *H-structure (87).

A hybrid of H and *H forms was described, called nodule DNA (88, 89) (Figure 4C). Nodule DNA is an analog of the intermolecular alternate-strand triplexes described above.

A peculiar H-like structure formed by two distant homopurine-homopyrimidine tracts was described in Ref. 90. It is in a way similar to an early model for S1 hypersensitivity in the human thyroglobulin gene (69). It was found that linear DNA containing both tracts at pH 4.0 and in the presence of spermidine migrates very slowly in an agarose gel. This abnormal electrophoretic mobility was attributed to the formation of a so-called tethered loop (Figure 4D). In this structure, the homopyrimidine strand of one stretch forms a triplex with a distant stretch, while its complementary homopurine strand remains single stranded. Supporting this model, it was found that the addition of excess homologous homopyrimidine, but not homopurine, single-stranded DNA prevented loop formation. Though the mechanism of tethered loop formation is not self-evident, it is allowed topologically. Chemical probing is required to prove the existence of this structure definitively.

Specificity of Triplex Formation

The specificity and stringency of triplex formation (35) has attracted serious attention for two reasons. First, the formation of triplexes is limited to the homopurine-homopyrimidine sequences or to sequences composed of adjacent oligopurine/oligopyrimidine clusters. This major limitation to the biological and theurapeutic applications of triple-helical DNAs prompted an extensive search for DNA bases that could be incorporated into the third strand of a triplex in order to recognize thymines or cytosines in the otherwise homopurine strand of the duplex. Secondly, accurate knowledge of the specificity of third-strand recognition for perfect homopurine-homopyrimidine sequences is necessary in order to target natural DNAs.

The quest for such knowledge stimulated the study of non-orthodox triads. So far most of the data have been collected for YR*Y triplexes, including all 14 noncanonical triads (other than CG*C and TA*T). One approach was to analyze the influence of mismatched triads on H-DNA formation using 2-D gel electrophoresis (91). Stability of mismatched triads in intermolecular triplexes was studied using affinity cleavage (92), melting experiments (93, 94), and NMR (95). These studies agreed that although single mismatches could be somewhat tolerated, each mismatch significantly disfavored triplex. The mismatch energies were within the range of 3–6 kcal/mol, i.e. similar to the cost of B-DNA mismatches. Thus, homopyrimidine oligonucleotides form triplexes with target sequences at a specificity comparable to that seen in Watson-Crick complementary recognition.

High sequence specificity of third-strand recognition of homopurine-homopyrimidine sequences in the duplex makes TFOs very attractive candidates for targeting genomic DNA. Supporting this conclusion, homopyrimidine TFOs equipped with Fe•EDTA have been demonstrated to cleave unique sites in yeast (96, 97) and human (98) chromosomes. They were also found to be convenient tools for affinity capture of human genomic targets (99).

However, studies widely disagreed on the relative stability of individual noncanonical triads. For example, the AT*G triplet was shown to be the most favorable in studies of intermolecular triplexes (92, 93), but it is not among the best for H-DNA (91). This contradiction could be due to the different triplex-forming sequences studied by different groups, since heterogeneity in stacking interactions within a triple helix must seriously affect its stability. This idea was recently supported by NMR studies of the AT*G triad (58, 100). It was found that guanine in this triplet is tilted out of the plane of its target AT basepair to avoid a steric clash with the thymine methyl group. This causes a favorable stacking interaction between this guanine and the thymine flanking it from the 5'-side, which is likely to be a major determinant of AT*G triplet stability. This also explains the differences between the inter- and intramolecular triplex studies: In the first case, guanine was flanked by thymine on the 5' end (92), while in the second case, it was flanked by a cytosine (91). Thus, the favorable stacking interaction was absent in the intramolecular triplex, and the AT*G triad was relatively unstable. Recently, it was shown directly that replacement of the TA*T triad on the 5' side of guanine with a CG*C triad reduces the stability of TA*G triplet (101). The clear message from these results is that the influence of nearest neighbors on triad stability must be studied to better understand the duplex-to-triplex transition. This doughty goal is not yet achieved.

Notwithstanding the difficulties discussed above, empirical rules for targeting imperfect homopurine-homopyrimidine sequences were suggested in Ref. 102. If the homopurine strand of a duplex is interrupted by a thymine or cytosine, it must be matched by a guanine or thymine, respectively, in the third strand. However, this expansion of the third-strand recognition code is premature, as was recently addressed (103). The GC*T triad, though reasonably stable, is dramatically weaker than the canonical TA*T triad. Thus, a TFO containing a thymine, intended to interact with a cytosine in the target, would bind significantly better to a different target containing adenine in the corresponding position. In the AT*G case, the triad specificity is high, but the affinity of the G for the TA pair is only modest.

Another approach to overcoming the homopurine-homopyrimidine target requirements is to incorporate artificial DNA bases within the third strand. Several studies found that non-natural bases, such as 2'-deoxynebularine or 2'-deoxyformycin A and others, may form very stable triads with cytosines and thymines intervening the homopurine strand (94, 104, 105). It is yet to be seen if the specificity and stringency of such complexes is sufficient.

Limited data are available on the mismatched triads in YR*R triplexes. By use of affinity cleavage experiments, all 13 noncanonical triplets (all combinations except CG*G, TA*A, and TA*T) were shown to disfavor triplex formation (106). The only notable exception is the CG*A triad, which is favorable under acidic pH due to the protonation of its adenine (28). Much as with homopyrimidine TFOs, purine-rich TFOs can be used specifically to target homopurine-homopyrimidine sequences in natural DNAs.

Stabilization of Triplexes

The stabilization of DNA triplexes is particularly important for any possible biological applications. As discussed above, the YR*Y triplexes are formed under acidic pH, while YR*R triplexes require millimolar concentrations of bivalent cations. Physiological pH, however, is neutral, and a high concentration of unbound bivalent cations in a cell is unlikely. Thus, numerous studies have been aimed at the stabilization of DNA triplexes at physiological conditions.

Most of the YR*Y triplexes studies have been concentrated on overcoming pH dependency. The most promising results show that polyamines, specifically spermine and spermidine, favor both inter- and intramolecular YR*Y triplexes under physiological pH (11, 107, 108). The stabilizing effect is likely due to decreased repulsion between the phosphate backbones after binding to polyamines, overcoming the relatively high density of a negative charge in triplexes. The millimolar polyamine concentrations found in the nuclei of eukaryotic cells (reviewed in 109) raise the hope for triplexes in vivo.

The requirement for cytosine protonation could be overcome by several chemical means. The incorporation of 5-methylcytosines instead of cytosines in TFOs increases the stability of YR*Y triplexes at physiological pH (110, 111), but more detailed study found that this effect is relatively small (the apparent methylation-induced ΔpK_a is only 0.5) (112). Another solution is to substitute cytosines in the third strand with non-natural bases that do not require protonation for Hoogsteen hydrogen bond formation. Indeed the substitution of cytosines with N^6-methyl-8-oxo-2-deoxyadenosines (113), pseudoisocytidines (114), 7,8 dihydro-8-oxoadenines (115), or 3-methyl-5-amino-1H pyrazolo [4.3-d] pyrimidin-7-ones (116) led to pH-independent triplex formation.

Intermolecular triplexes could be additionally stabilized if the third strand represented an oligodeoxynucleotide-intercalator conjugate. This was first demonstrated for a homopyrimidine oligonucleotide linked with an acridine derivative (10) and later shown for other oligonucleotides and intercalators (117–119). The stabilization is due to the intercalation of a ligand into DNA at the duplex-triplex junction. For reasons that are yet unclear, the most stable complex is formed when the intercalator is attached to the 5′ end of the TFO.

Particularly promising for gene targeting is an oligonucleotide-psoralen conjugate, as near-UV irradiation of a triplex formed by such a conjugate leads to crosslink formation, making the triplex irreversible (120).

An independent line of research has sought for triplex-specific ligands. One such ligand, a derivative of benzo[e]pyridoindole (BePI), has been described in Refs. 121 and 122. BePI shows preferential intercalation into a triple- rather than double-helical DNA, thus greatly stabilizing triplexes (122). Another promising triplex-binding ligand is coralyne (123).

It should be emphasized that, to be prospective drugs for gene targeting, TFOs must meet two requirements: They must bind their targets relatively strongly and not target other sequences. If a TFO has very strong affinity to its target, it can also bind to a site with one or even more mismatches. This should be especially true for non–sequence-specific stabilization of triplexes with intercalating drugs attached to TFOs. Therefore, increased stability inevitably entails decreased selectivity of the TFO. It is not at all accidental that the spectacular demonstration of sequence-selective cutting of genomic DNA with TFOs was achieved under conditions of extremely weak binding of the TFO to its target site (96–98). Systematic experimental study of sequence selectivity of all modified TFOs mentioned above is still lacking. However, it is obvious that these modified TFOs should exhibit poorer selectivity than do the original TFOs.

The stabilization of intramolecular triplexes could be achieved in several ways, the most obvious of which is to increase the negative superhelical density, since the formation of H-DNA releases torsional stress. As is discussed below, the increase of negative supercoiling does provoke triplex formation in vivo. The polyamine stabilization of H-DNA at physiological pH has already been mentioned.

A less obvious way of stabilizing H-DNA, called kinetic trapping, was described (124). It was found that oligonucleotides complementary to the single-stranded homopurine stretch in H-DNA stabilized H-DNA under neutral pH, where H-DNA alone rapidly reverts to the B conformation.

Peptide Nucleic Acid (PNA)

PNA is the prototype of an entire new class of TFO-based drugs that interact with DNA in a manner unlike that of ordinary TFOs. PNA (Figure 5A) was designed in the hope that such an oligonucleotide analog containing normal DNA bases with a polyamide (i.e. proteinlike) uncharged backbone would form triplexes with double-stranded DNA (dsDNA) much more efficiently than do the regular TFOs (14).

Instead of forming triplexes with duplex DNA, the first studied homothymine PNA oligomer, PNA-T_{10}, opened the DNA duplex in A_n/T_n tracts, forming an exceptionally strong complex with the A strand and displacing the T strand

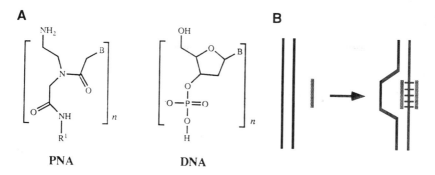

Figure 5 A. The chemical structures of PNA and DNA. *B.* P-loop formation. Bold line, DNA; stippled line, PNA.

(15, 125, 126). At the same time, model experiments with complexes formed between PNA oligomers and oligonucleotides revealed that, while PNA/DNA heteroduplexes are not much more stable under ordinary conditions than are DNA/DNA homoduplexes (127), two homopurine PNA oligomer molecules form exceptionally stable triplexes with the complementary homopurine oligonucleotide (128, 129).

These results strongly suggest an unusual mode of binding between the synthetic analog and dsDNA. Namely, two homopyrimidine PNA molecules displace the duplex DNA pyrimidine strand and form a triplex with the purine strand of DNA (15, 16, 130, 131). These complexes are called the P-loops (Figure 5B).

The P-loop is a radically different complex than that formed between duplex DNA and ordinary TFOs. Although the fact of (PNA)$_2$/DNA triplex formation during the strand-displacement reaction has been convincingly proven (16, 130, 131), the mechanism of P-loop formation remains to be elucidated. The available data indicate that the reaction most probably proceeds via a short-lived intermediate, which consists of one PNA molecule complexing with the complementary DNA strand via Watson-Crick pairing. This intermediate is formed due to thermal fluctuations (breathing) of the DNA duplex (132, 133). It is very unstable and would dissociate if it were not fixed by the second PNA oligomer in a (PNA)$_2$/DNA triplex leading to P-loop formation (see Figure 5B). This triplex is remarkably stable.

PNA forms much more stable complexes with dsDNA than do regular oligonucleotides. This makes PNA very promising as an agent for sequence-specific cutting of duplex DNA (16), for use in electron-microscopy mapping of dsDNA (15), and as a potential antigene drug (134, 135), as PNA is

remarkably stable in biological fluids in which normal peptides and oligonucleotides are quickly degraded (136).

However, serious limitations for various applications of PNA still remain.
P-loop formation proceeds through a significant kinetic barrier and strongly
depends on ionic conditions (15, 16, 125, 126). This dependency, if not by-
passed, poses significant limitations on possible sequence-specific targeting of
dsDNA by PNA under physiological conditions. Although the stringency of
$(PNA)_2$/DNA triplexes is not yet known, PNA should still target predominately
homopurine-homopyrimidine regions, just as do regular TFOs.

BIOCHEMISTRY OF TRIPLEXES

Formation and Possible Functions of H-DNA In Vivo

As is true for other unusual DNA structures, such as cruciforms, Z-DNA, and
quadruplexes, the biological role of H-DNA is yet to be established. Two
important problems must be addressed: (a) Can H-DNA be formed in cells in
principle? (b) In which biological process if any is H-DNA involved? Recently
it became clear that the answer to the first question is yes. There are currently
many hypotheses on the role of H-DNA in DNA replication, transcription, and
recombination, but more studies are needed to answer the second question.

Sequences that can form H-DNA are widespread throughout the eukaryotic
genomes (137, 138) but are uncommon among eubacteria. However, direct
detection of H-DNA in eukaryotic cells is very difficult because of the complexity of genomic DNA. Therefore, most of the studies on the detection of
H-DNA in vivo exploited *Escherichia coli* cells bearing recombinant plasmids
with triplex-forming inserts as convenient model systems. Chemical probing
of intracellular DNA proved helpful for the detection of H-DNA in vivo.
Certain chemicals, such as osmium tetroxide, chloroacetaldehyde, and psoralen, give a characteristic pattern of H- or *H-DNA modification in vitro.
Conveniently, they can also penetrate living cells. Thus, the general strategy
for detecting H-DNA in vivo was to treat *E. coli* cells with those chemicals,
isolate plasmid DNA, and locate modified DNA bases at a sequence level.
The coincidence of modification patterns in vitro and in vivo basically proved
the formation of the unusual structure in the cell.

Using this approach, the formation of both H- and *H-DNA was directly
shown (139–141). The corresponding studies were reviewed in Ref. 20, but
we briefly summarize the major findings. All these studies agreed that the level
of DNA supercoiling in vivo is the major limiting factor in the formation of
these structures. Though transient formation of H-DNA was observed in normal exponentially growing *E. coli* cells (141), formation of H-DNA was much
more pronounced when intracellular DNA supercoiling increased, due to mu-

tations in the gene for Topo I (141) or due to treatment of cells with chloramphenicol (139, 140). Environmental conditions during *E. coli* growth also significantly contributed to the appearance of triplexes. H-DNA formation was greatly enhanced when cells were growing in mildly acidic media, which somewhat decreased intracellular pH (139, 141) while *H-DNA was observed in cells growing in media with a high concentration of Mg^{2+} ions (140). Neither result is surprising, because H-DNA is stabilized by protonation while *H-DNA is stabilized by bivalent cations.

Besides the steady-state level of DNA supercoiling, determined by the balance of DNA gyrase and Topo I (reviewed in 142), the local level of supercoiling strongly depends on transcription. During the process of polymerization the RNA polymerase creates domains of high negative and positive supercoiling upstream and downstream of it, respectively (143), which may influence the formation of unusual DNA structures (144, 145). Chemical probing of intracellular DNA demonstrated transcriptionally driven formation of *H-DNA within long $d(G)_n \cdot d(C)_n$ stretches located upstream of a regulated promoter in an *E. coli* plasmid (146). Remarkably, the formation of *H-DNA stimulated homologous recombination between direct repeats flanking the structure. Thus, this work shows the formation of *H-DNA under completely physiological conditions in a cell, and implicates it in the process of recombination.

The only data on triplex DNA detection in eukaryotic cells were obtained using antibodies against triple-helical DNA (147). These antibodies were found to interact with eukaryotic chromosomes (148, 149).

Many ideas have been proposed involving H-DNA in such basic genetic processes as replication and transcription. The hypothesis regarding H-DNA in replication is based on the observation that triplex structures prevent DNA synthesis in vitro. On supercoiled templates containing *H-DNA, DNA synthesis prematurely terminates. The location of the termination site is different for different isoforms of *H-DNA, but it always coincides with the triplex boundaries as defined by chemical probing (83).

More peculiarly, H-like structures can be formed in the process of DNA polymerization and efficiently block it. Two such mechanisms were demonstrated experimentally (Figure 6A,B). It was found that $d(GA)_n$ or $d(C-T)_n$ inserts within single-stranded DNA templates cause partial termination of DNA polymerases at the center of the insert (21, 150). It was suggested that when the newly synthesized DNA chain reaches the center of the homopolymer sequence, the remaining homopolymer stretch folds back, forming a stable triplex (Figure 6A). As a result, the DNA polymerase finds itself in a trap and is unable to continue elongation.

In open circular DNA templates, H-like structures are absent due to the lack of DNA supercoiling. It was shown, however, that T7 DNA polymerase ter-

A.

B.

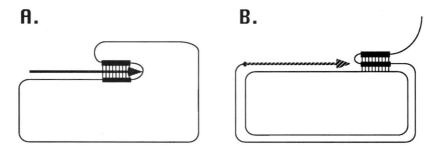

Figure 6   DNA polymerase–driven triplex formation blocks polymerization. Black boxes, the two halves of a homopurine-homopyrimidine mirror repeat involved in the formation of an intramolecular triplex; striated arrow, the newly synthesized DNA chain. *A*. Single-stranded DNA template. *B*. Double-stranded DNA template.

minated exactly at the center of *H-forming sequences. This was observed when the pyrimidine-rich but not the purine-rich strand served as a template (22). To explain this one must remember that DNA synthesis on double-stranded templates is possible due to the ability of many DNA polymerases to displace the nontemplate DNA strand (reviewed in 151). The displaced strand may fold back, promoting the formation of an intramolecular triplex downstream of the replication fork at an appropriate sequence. Conditions for DNA synthesis in vitro—i.e. neutral pH and high magnesium concentration—are optimal for the formation of YR*R triplexes. Thus, the displacement of the purine-rich (but not the pyrimidine-rich) strand provokes triplex formation which, in turn, leads to termination of DNA synthesis (Figure 6*B*).

There are only fragmentary data on the role of H motifs in the regulation of replication in vivo. Several homopurine-homopyrimidine inserts were shown to decrease the efficiency of Simian virus 40 (SV40) DNA replication (152, 153). Quite recently, the pausing of the replication fork in vivo within a $d(GA)_n \cdot d(TC)_n$ insert in SV40 DNA was demonstrated directly using a technique called two-dimensional neutral/neutral gel electrophoresis (23). Though these data make the idea of H-DNA involvement in the regulation of replication promising, it is far from proven. Future studies are crucial for the evaluation of this hypothesis.

Numerous studies concerned the possible role of H-DNA in transcription. Deletion analysis of various promoters—including *Drosophila hsp*26 (154, 155); mouse c-Ki-*ras* (156) and TGF-β3 (157); human EGFR (158), *ets*-2 (159), IR (160), and c-*myc* (161, 162); and others—showed that homopurine-homopyrimidine stretches are essential for promoter functioning.

These sequences serve as targets for nuclear proteins, presumably transcriptional activators. Several homopurine-homopyrimidine DNA-binding proteins were described, including BPG1 (163), NSEP-1 (164), MAZ (165), nm23-H2 (166), PYBP (167), Pur-1 (168), etc. Peculiarly, these proteins often bind preferentially to just one strand of the H motifs. For example, a number of mammalian proteins specifically recognize homopurine-homopyrimidine sequences in the double-helical state as well as the corresponding homopyrimidine single strands (164, 167, 169, 170). This unusual binding pattern may dramatically influence the equilibrium between different DNA conformations in the promoter in vivo.

However, the importance of the H structure for transcription was questioned in several studies. One approach is to analyze the influence of point mutations within H motifs that destroy or restore H-forming potential on the promoter's activity. No such correlation was observed for *Drosophila hsp*26 (155) and mouse c-Ki-*ras* (171) promoters. The situation with the c-*myc* promoter is more complex, since it is unclear if the canonical H-DNA or some other structure is formed even in vitro (172). Mutational analysis of the promoter gave contradictory results, with one group claiming the existence (173) and another the lack (174) of a correlation between structural potential and promoter strength. Another approach to detecting H-DNA in eukaryotic promoters is direct chemical probing followed by genomic sequencing. So far, this has only been done for the *Drosophila hsp*26 gene, and H-DNA was not observed (155).

It is hard to completely rule out the role of H-DNA in transcription based on the above results. First, it is quite possible that the structural peculiarities of promoter DNA segments may affect the interaction between promoter DNA and specific regulator proteins. The features of homopurine-homopyrimidine DNA-binding proteins described above as well as a report about the partial purification of a triplex-binding protein (175) indirectly support this idea. A study in which the influence of $d(G)_n$ stretches of varying length on the activity of a downstream minimal promoter was analyzed additionally supports this hypothesis (176). A clear reverse correlation between the ability of a stretch to form the *H configuration in vitro and its ability to activate transcription in vivo was observed. It was concluded, therefore, that short $d(G)_n$ stretches serve as binding sites for a transcriptional activator, while longer stretches adopt a triplex configuration, which prevents activator binding. Secondly, negative data on the role of H-DNA in transcription were obtained in transient assays, while it can actually work at a chromosome level. Indeed, H motif in the *Drosophila hsp*26 gene was found to affect the chromatin structure (177, 178).

Despite the wealth of data and hypotheses, there is no direct evidence that the structural features of H motifs are involved in transcriptional regulation in vivo, and further studies are required to address this issue.

Targeting Basic Genetic Processes Using TFOs

Highly sequence-specific recognition of double-helical DNAs by TFOs is the basis of an antigene strategy (reviewed in 13). The idea is that binding of a TFO to a target gene could prevent its normal functioning. Most studies of this strategy concerned the inhibition of transcription; the studies were inspired in part by the existence of functionally important homopurine-homopyrimidine stretches in many eukaryotic promoters (see the previous section), which are appropriate targets for TFOs. The antigene strategy could potentially lead to rational drug design. Very convincing data on the inhibitory effects of TFOs were obtained in various in vitro systems. There are also preliminary indications that TFOs may function in vivo as well.

The first stage that is affected by TFOs is the formation of an active promoter complex. Pioneering results were obtained for the human c-*myc* promoter, where it was found that the binding of a purine-rich TFO to the imperfect homopurine-homopyrimidine sequence 125 basepairs (bp) upstream of the P1 promoter start site blocks its transcription in vitro (179). The TFO's target is important for c-*myc* transcription, serving as a binding site for a protein(s), presumably a transcriptional activator (161, 162). At least two candidate genes coding for proteins that bind to this target have been cloned and sequenced (164, 166). Similar observations were made for the methallothionein gene promoter. In this case a homopyrimidine oligonucleotide formed a triplex with the upstream portion of the promoter, preventing the binding of the transcriptional activator Sp1 (111). This in turn drastically reduced the promoter's activity in a cell-free transcription system (179a). TFOs were also shown to prevent SP1 binding to the human DHFR (180) and *H-ras* (181) promoters. Finally, a triplex-forming oligonucleotide-intercalator conjugate was shown to act as a transcriptional repressor of the interleukin-2 receptor α gene in vitro (182), preventing the binding of the transcriptional activator NFκB. In all these cases TFOs efficiently blocked the access of the transcription factors to their binding sites.

TFOs also inhibit initiation of transcription by RNA polymerases. The pBR322 *bla*-gene contains a 13-bp homopurine-homopyrimidine target just downstream of the transcriptional start site. A 13-mer homopyrimidine oligonucleotide forming an intermolecular triplex with this target hindered initiation of transcription by *E. coli* RNA polymerase in vitro (183). Independent studies showed that this is also the case for T7 RNA polymerase (184).

Finally, eukaryotic RNA polymerase II transcription was followed in vitro from the adenovirus major late promoter (185). The transcribed portion of DNA contained a 15-bp homopurine-homopyrimidine tract that formed an intermolecular triplex with the homopyrimidine TFO. When added prior to RNA polymerase, the TFO truncated a significant portion of the transcripts.

Thus, TFOs can block transcription at different stages: promoter complex formation, initiation, and elongation. This appears to be true for both pro- and eukaryotic RNA polymerases. TFOs can be considered to be artificial repressors of transcription (186).

There is a growing number of indications that TFOs may act as repressors of transcription in cell cultures as well. The most convincing results so far were obtained for the interleukin-2 receptor α promoter (182, 187). Homopyrimidine TFOs were designed to overlap a target site and prevent binding of the transcriptional activator NFκB. They were conjugated with acridine to stabilize the triplexes, or psoralen to make triplex formation irreversible after UV irradiation. The plasmid bearing the reporter gene under the control of the IL-2Rα promoter was cotransfected with these TFOs in tissue cultures, where it was shown that TFOs block promoter activity in vivo. Particularly strong inhibition was observed after UV irradiation of cells transfected with psoralen conjugates. In the latter case, chemical probing directly demonstrated the formation of intermolecular triplex in vivo. A similar cotransfection approach was also used to target Interferon Responsive Elements in vivo (188).

A different approach was used in several studies where purine-rich TFOs were added to the growth media of cells containing target genes. To prevent oligonucleotides from degrading, their 3′ ends were protected by an amino group (189). Such oligonucleotides accumulated within cells and could be recovered in intact form. Partial transcriptional inhibition of human c-*myc* and IL2Rα genes by such TFOs has been reported (189, 190). Similar effects were observed for human immunodeficiency virus (HIV) transcriptional inhibition in chronically infected cell lines (191). Using cholesterol-substituted TFOs, the progesterone-responsive gene has also been inhibited (192). Though the inhibitory effect was never more than 50%, it is quite remarkable considering that a short oligonucleotide must find its target in an entire genome and prevent its proper interaction with cellular transcriptional machinery. Note, however, that in none of those cases was the formation of triplexes directly demonstrated. Other mechanisms of oligonucleotide-caused transcriptional inhibition must be ruled out in the future.

The use of TFOs for DNA replication inhibition is less studied. In vitro formation of putative intramolecular triplexes or H-like triplexes (see Figure 1) on single-stranded DNA templates traps many different DNA polymerases (22, 193). Purine-rich TFOs are particularly efficient even against such processive enzymes as T7 DNA polymerase and thermophilic Taq and Vent polymerases, because the conditions of DNA synthesis in vitro are favorable for YR*R triplexes. Pyrimidine-rich TFOs must be additionally crosslinked to the target to cause inhibition (194). TFOs also block DNA polymerases on double-stranded templates (195). The inhibition of DNA synthesis in vitro was observed not only when triplexes blocked the path of DNA polymerase, but also when a polymerization primer was involved in triplex formation (193). Single-stranded DNA-binding protein (SSB protein) helped DNA polymerases partially overcome the triplex barrier, but with an efficiency dramatically dependent on the triplex configuration.

Though these observations make TFOs promising candidates for trapping DNA replication in vivo, there are almost no experimental data regarding this. The only published data concern the use of an octathymidilate-acridine conjugate, which binds to a $d(A)_8$ stretch in SV40 DNA adjacent to the T antigen–binding site. In vivo it partially inhibits SV40 DNA replication, presumably by interfering with the DNA binding or with unwinding activities of the T antigen (196).

The major problem with the use of TFOs is in matching high sequence selectivity with binding that is sufficiently strong to interfere with genetic processes. Under physiological conditions, TFOs bind weakly to their targets, which by itself favors a high sequence selectivity. However, to significantly affect genetic processes, the TFO must be rather long, which limits the number of potential targets, as such long homopurine-homopyrimidine stretches are infrequent.

Three-Stranded DNA Complexes in Homologous Recombination

In this section we briefly discuss a still poorly understood three-stranded DNA complex, formed by RecA protein and, possibly, recombinant proteins from other sources. RecA protein is well known to exhibit many enzymatic activities essential for recombination (reviewed in 18, 197). The main function of RecA protein in recombination is to exchange single-stranded DNA (ssDNA) strand with its homolog in dsDNA. The sequential stages of this reaction are: (a) cooperative assembly of RecA protein molecules on the ssDNA, leading to the formation of a right-helical nucleoprotein filament called the presynaptic complex, (b) synapsis, i.e. the formation of a complex between this filament and the homologous dsDNA, and (c) the actual strand exchange, which requires ATP hydrolysis. Strand exchange proceeds in only one direction: The displacement of a linear single-stranded product starts from its 5' end.

The synapsis step requires searching for homology between the presynaptic filament and the target dsDNA. One way to do so is to use Watson-Crick complementarity rules. However, this requires a partial strand separation of the dsDNA, resulting in the formation of a so-called D-loop. In this structure, one of the DNA strands of the duplex is displaced, while the other is involved in Watson-Crick pairing with incoming ssDNA. An alternative, very attractive possibility, first postulated in Refs. 17 and 198, does not require dsDNA strand separation and invokes triplex formation. This hypothetical type of DNA triplex was later called "recombination," "parallel," or R-DNA (19,

199, 200). These names emphasize two fundamental differences between this hypothetical triplex and the well-characterized orthodox DNA triplexes described in other sections of this chapter. First, chemically homologous DNA strands are parallel in R-DNA but antiparallel in standard triplexes. Secondly, any sequence can adopt an R-DNA conformation, while homopurine-homopyrimidine stretches are strongly preferable in adopting standard triplex structure.

In important experiments on strand exchange between the partially homologous substrates (201, 202), three types of joint molecules were observed. In the case of proximal joints, the area of homology is situated at the 5′ end of the outgoing duplex strand, i.e. both synapsis and strand exchange are possible. For a distal joint (with homology at the 3′ end of the outgoing strand), RecA cannot drive strand exchange. Medial joints contain heterologous regions at both ends of the dsDNA, making strand exchange from any DNA end impossible. Since synaptic complexes were detected in all three cases, it became clear that synapsis and strand exchange are not necessarily coupled. When synaptic complexes—in particular the medial complexes—were treated with DNA crosslinking agents, crosslinks were observed between all three DNA strands involved in the complex (203), indicating a close physical proximity of the three strands.

Analysis of distal joints with very short (38–56-bp) regions of homology showed that they are remarkably stable upon the removal of RecA protein (199). In fact, joint molecules dissociated at temperatures indistinguishable from the melting temperatures of DNA duplexes of the same length and sequence. In spite of its stability, however, the complex did not form spontaneously without recombination proteins. The conclusion was that RecA and related proteins promote the formation of a novel "recombinant" DNA triplex, which otherwise cannot form, presumably due to a kinetic barrier of unknown nature. Independent studies confirmed the extreme stability of deproteinized distal joints with longer regions of homology (204). The basepairing scheme for R-DNA involving triplets for arbitrary DNA sequences was suggested in Refs. 19 and 205. The unique feature of these triplets is the interaction of the third strand with both bases of the Watson-Crick pair.

Although the above data seem to be most consistent with the idea of a "recombination" triplex formation, a careful analysis of three-stranded complexes formed under RecA protein (206) using chemical probing indicates that basepairing in the parental duplex is disrupted. The incoming ssDNA appears to form W-C pairs with the complementary strand of the duplex. It was concluded that the synapsis is accompanied by local unwinding, leading to the formation of D-loop-like structures, rather than the "recombination" triplexes (206). This conclusion was supported by the data that the N7 position of guanines, which is involved in Hoogsteen hydrogen bonding in all known triplexes in vitro (see Figure 2), is not required for the formation of three-stranded complexes by RecA protein (207).

Thus, the putative triplex between the incoming single strand and the duplex systematically avoids detection. Nevertheless, a more general question remains whether the "recombination" triplexes can be formed in principle, even if they do not play any role in recombination. This kind of triplex has recently been claimed for a postsynaptic complex formed between the outgoing single strand and the duplex yielded as a result of the strand exchange (208).

Quite recently it was suggested that a specifically designed oligonucleotide could fold back to form an intramolecular R-like structure without the assistance of any proteins (209). The main argument is that the thermal denaturation curves are biphasic, which was interpreted as subsequent triplex-to-duplex and duplex-to-single strand transitions. This is hardly a sufficient argument, and data on the chemical and enzymatic probing of such complexes provided in the same study do not support the claim.

In the absence of conclusive evidence, the existence of "recombination" triplexes, or R-DNA, remains doubtful. One of the most uncomfortable questions is the extreme thermal stability of deproteinized distal joints described in Refs. 199 and 204. None of the proposed models can satisfactorily explain this feature. It is totally unclear what is the nature of the kinetic barrier that prevents the formation of R-DNA by dsDNA and homologous oligonucleotide without any protein. It is also unclear why the medial joints, unlike the distal joints, are unstable upon deproteinization (203, 210). Additional concern is possible exonuclease contamination of the RecA protein and SSB protein preparations used for strand transfer reaction. At least in one case, such contamination was admitted to be responsible for the formation of distal junctions (211). In both original papers (199, 204), the authors claimed the lack of nuclease contamination. As shown in (211), however, exonuclease I (ExoI) is enormously activated by SSB protein. As a result, the levels of Exo I required to generate the reverse strand exchange are extremely low (1 molecule of Exo I per 20,000 molecules of RecA protein). In the light of these new findings, it seems possible that distal joints, which were as stable as duplex DNA, might actually be duplexes formed after SSB-activated trace contamination of Exo I digested the nonhomologous strand from its 3' end.

Even in the absence of a clear understanding of the structure of three-stranded joints promoted by RecA protein, they have already found interesting applications in gene targeting. The first example is called RARE, for RecA-Assisted Restriction Endonuclease cleavage (210). The rationale for this approach is that since RecA protein can form three-stranded complexes between dsDNA and oligonucleotides as short as 15 nucleotides (212), such complexes can be used to block specific methylation sites in dsDNA. After the removal of proteins and consequent dissociation of the three-stranded complexes, cleavage by methylase-sensitive restriction endonuclease is limited to the targeted site. Thus, one can cleave large DNAs at a unique site or, using pairs of oligonucleotides, separate specific DNA fragments from the genome.

ACKNOWLEDGEMENTS

We thank our colleagues for sending us valuable reprints and preprints. N Cozzarelli, J Feigon, and B Johnston for comments on the manuscript and R Cox for editorial help. Supported by grant MCB-9405794 from the National Science Foundation to S.M.M.

Any *Annual Review* chapter, as well as any article cited in an *Annual Review* chapter, may be purchased from the Annual Reviews Preprints and Reprints service.
1-800-347-8007; 415-259-5017; email: arpr@class.org

Literature Cited

1. Felsenfeld G, Davies DR, Rich A. 1957. *J. Am. Chem. Soc.* 79:2023–24
2. Riley M, Maling B, Chamberlin MJ. 1966. *J. Mol. Biol.* 20:359–89
3. Morgan AR, Wells RD. 1968. *J. Mol. Biol.* 37:63–80
4. Lee JS, Johnson DA, Morgan AR. 1979. *Nucleic Acids Res.* 6:3073–91
5. Lipsett MN. 1964. *J. Biol. Chem.* 239:1256–60
6. Broitman SL, Im DD, Fresco JY. 1987. *Proc. Natl. Acad. Sci. USA* 84:5120–24
7. Cantor CR, Schimmel PR. 1980. *Biophysical Chemistry*. San Francisco: Freeman
8. Lyamichev VI, Mirkin SM, Frank-Kamenetskii MD. 1986. *J. Biomol. Struct. Dyn.* 3:667–69
9. Mirkin SM, Lyamichev VI, Drushlyak KN, Dobrynin VN, Filippov SA, Frank-Kamenetskii MD. 1987. *Nature* 330:495–97
10. Le Doan T, Perrouault L, Praseuth D, Habhoub N, Decout JL, et al. 1987. *Nucleic Acids Res.* 15:7749–60
11. Moser HE, Dervan PB. 1987. *Science* 238:645–50
12. Lyamichev VI, Mirkin SM, Frank-Kamenetskii MD, Cantor CR. 1988. *Nucleic Acids Res.* 16:2165–78
13. Helene C. 1991. *Anticancer Drug Des.* 6:569–84
14. Nielsen PE, Egholm M, Berg RH, Buchardt O. 1993. In *Antisense Research and Applications*, ed. ST Crooke, B Lebleu, pp. 363–73. Boca Raton, FL: CRC Press
15. Cherny DY, Belotserkovskii BP, Frank-Kamenetskii MD, Egholm M, Buchardt O, et al. 1993. *Proc. Natl. Acad. Sci. USA* 90:1667–70
16. Demidov V, Frank-Kamenetskii MD, Egholm M, Buchardt O, Nielsen PE. 1993. *Nucleic Acids Res.* 21:2103–7
17. Howard-Flanders P, West SC, Stasiak A. 1984. *Nature* 309:215–20
18. West SC. 1992. *Annu. Rev. Biochem.* 61:603–40
19. Zhurkin VB, Raghunathan G, Ulyanov NB, Camerini-Otero RD, Jernigan RL. 1994. *J. Mol. Biol.* 239:181–200
20. Mirkin SM, Frank-Kamenetskii MD. 1994. *Annu. Rev. Biophys. Biomol. Struct.* 23:541–76
21. Baran N, Lapidot A, Manor H. 1991. *Proc. Natl. Acad. Sci. USA* 88:507–11
22. Samadashwily GM, Dayn A, Mirkin SM. 1993. *EMBO J.* 12:4975–83
23. Rao BS. 1994. *Gene* 140:233–37
24. Hoogsteen K. 1963. *Acta Crystallogr.* 16:907–16
25. Kohwi Y, Kohwi-Shigematsu T. 1988. *Proc. Natl. Acad. Sci. USA* 85:3781–85
26. Beal PA, Dervan PB. 1991. *Science* 251:1360–63
27. Bernues J, Beltran R, Casasnovas JM, Azorin F. 1990. *Nucleic Acids Res.* 18:4067–73
28. Malkov VA, Voloshin ON, Veselkov AG, Rostapshov VM, Jansen I, et al. 1993. *Nucleic Acids Res.* 21:105–11
29. Malkov VA, Voloshin ON, Soyfer VN, Frank-Kamenetskii MD. 1993. *Nucleic Acids Res.* 21:585–91
30. Potaman VN, Soyfer VN. 1994. *J. Biomol. Struct. Dyn.* 11:1035–40
31. Sklenar V, Feigon J. 1990. *Nature* 345:836–38

32. Haner R, Dervan PB. 1990. *Biochemistry* 29:9761–5
33. Radhakrishnan I, de los Santos C, Patel DJ. 1991. *J. Mol. Biol.* 221:1403–18
34. Giovannangeli C, Montenay-Garestier T, Rougee M, Chassignol M, Thuong NT, Helene C. 1991. *J. Am. Chem. Soc.* 113:7775–76
35. Roberts RW, Crothers DM. 1991. *Proc. Natl. Acad. Sci. USA* 88:9397–401
36. Kool E. 1991. *J. Am. Chem. Soc.* 113: 6265–66
37. Prakash G, Kool E. 1992. *J. Am. Chem. Soc.* 114:3523–28
38. Booher MA, Wang SH, Kool ET. 1994. *Biochemistry* 33:4645–51
39. Wang SH, Booher MA, Kool ET. 1994. *Biochemistry* 33:4639–44
40. Horne DA, Dervan PB. 1990. *J. Am. Chem. Soc.* 112:2435–37
41. Beal P, Dervan P. 1992. *J. Am. Chem. Soc.* 114:1470–78
42. Jayasena SD, Johnston BH. 1992. *Biochemistry* 31:320–27
43. Jayasena SD, Johnston BH. 1992. *Nucleic Acids Res.* 20:5279–88
44. Jayasena SD, Johnston BH. 1993. *Biochemistry* 32:2800–7
45. Roberts RW, Crothers DM. 1992. *Science* 258:1463–66
46. Escude C, Francois J-C, Sun J-S, Ott G, Sprinzl M, et al. 1993. *Nucleic Acids Res.* 21:5547–53
47. Han H, Dervan PB. 1993. *Proc. Natl. Acad. Sci. USA* 90:3806–10
48. Johnston BH. 1988. *Science* 241:1800–4
49. Voloshin ON, Mirkin SM, Lyamichev VI, Belotserkovskii BP, Frank-Kamenetskii MD. 1988. *Nature* 333:475–76
50. Arnott S, Selsing E. 1974. *J. Mol. Biol.* 88:509–21
51. Macaya R, Schultze P, Feigon J. 1992. *Am. Chem. Soc.* 114:781–83
52. Howard FB, Miles HT, Liu K, Frazier J, Raghunathan G, Sasisekharan V. 1992. *Biochemistry* 31:10671–77
53. Raghunathan G, Miles HT, Sasisekharan V. 1993. *Biochemistry* 32:455–62
54. de los Santos C, Rosen M, Patel D. 1989. *Biochemistry* 28:7282–89
55. Rajagopal P, Feigon J. 1989. *Nature* 339:637–40
56. Mooren MM, Pulleyblank DE, Wijmenga SS, Blommers MJ, Hilbers CW. 1990. *Nucleic Acids Res.* 18: 6523–29
57. Radhakrishnan I, Gao X, de los Santos C, Live D, Patel DJ. 1991. *Biochemistry* 30:9022–30
58. Radhakrishnan I, Patel DJ. 1994. *Structure* 2:17–32
59. Radhakrishnan I, de los Santos C, Patel DJ. 1993. *J. Mol. Biol.* 234:188–97
60. Radhakrishnan I, Patel D. 1993. *Structure* 1:135–52
61. Frank-Kamenetskii MD. 1990. In *DNA Topology and its Biological Effects*, ed. N Cozzarelli, J Wang, pp. 185–215. Cold Spring Harbor: Cold Spring Harbor Lab. Press
62. Lyamichev VI, Mirkin SM, Frank-Kamenetskii MD. 1985. *J. Biomol. Struct. Dyn.* 3:327–38
63. Hentchel CC. 1982. *Nature* 295:714–16
64. Larsen A, Weintraub H. 1982. *Cell* 29: 609–22
65. Mace HAF, Pelham HRB, Travers A. 1983. *Nature* 304:555–57
66. Shen CK. 1983. *Nucleic Acids Res.* 11: 7899–910
67. Cantor CR, Efstratiadis A. 1984. *Nucleic Acids Res.* 12:8059–72
68. McKeon C, Schmidt A, de Crombrugghe BA. 1984. *J. Biol. Chem.* 259:6636–40
69. Christophe D, Cabrer B, Bacolla A, Targovnik H, Pohl V, Vassart G. 1985. *Nucleic Acids Res.* 13:5127–44
70. Margot JB, Hardison RC. 1985. *J. Mol. Biol.* 184:195–210
71. Pulleyblank DE, Haniford DB, Morgan AR. 1985. *Cell* 42:271–80
72. Lilley DM. 1992. *Methods Enzymol.* 212:133–39
73. Hanvey JC, Klysik J, Wells RD. 1988. *J. Biol. Chem.* 263:7386–96
74. Hanvey JC, Shimizu M, Wells RD. 1988. *Proc. Natl. Acad. Sci. USA* 85: 6292–96
75. Htun H, Dahlberg JE. 1988. *Science* 241:1791–96
76. Vojtiskova M, Mirkin S, Lyamichev V, Voloshin O, Frank-Kamenetskii M, Palecek E. 1988. *FEBS Lett.* 234:295–99
77. Htun H, Dahlberg JE. 1989. *Science* 243:1571–76
78. Kang S, Wohlrab F, Wells RD. 1992. *J. Biol. Chem.* 267:1259–64
79. Kang S, Wells RD. 1992. *J. Biol. Chem.* 267:20887–91
80. Shimizu M, Kubo K, Matsumoto U, Shindo H. 1994. *J. Mol. Biol.* 235:185–97
81. Bernues J, Beltran R, Casasnovas JM, Azorin F. 1989. *EMBO J.* 8:2087–94
82. Panyutin IG, Kovalsky OI, Budowsky EI. 1989. *Nucleic Acids Res.* 17:8257–71
83. Dayn A, Samadashwily GM, Mirkin SM. 1992. *Proc. Natl. Acad. Sci. USA* 89:11406–10
84. Kohwi Y. 1989. *Nucleic Acids Res.* 17: 4493–502
85. Beltran R, Martinez-Balbas A, Bernues J, Bowater R, Azorin F. 1993. *J. Mol. Biol.* 230:966–78
86. Kohwi Y, Kohwi-Shigematsu T. 1993. *J. Mol. Biol.* 231:1090–101

87. Martinez-Balbas A, Azorin F. 1993. *Nucleic Acids Res.* 21:2557–62
88. Kohwi-Shigematsu T, Kohwi Y. 1991. *Nucleic Acids Res.* 19:4267–71
89. Panyutin IG, Wells RD. 1992. *J. Biol. Chem.* 267:5495–501
90. Hampel KJ, Ashley C, Lee JS. 1994. *Biochemistry* 33:5674–81
91. Belotserkovskii BP, Veselkov AG, Filippov SA, Dobrynin VN, Mirkin SM, Frank-Kamenetskii MD. 1990. *Nucleic Acids Res.* 18:6621–24
92. Griffin LC, Dervan PB. 1989. *Science* 245:967–71
93. Mergny JL, Sun JS, Rougee M, Montenay-Garestier T, Barcelo F, et al. 1991. *Biochemistry* 30:9791–98
94. Sun JS, Mergny JL, Lavery R, Montenay-Garestier T, Helene C. 1991. *J. Biomol. Struct. Dyn.* 9:411–24
95. Macaya RF, Gilbert DE, Malek S, Sinsheimer JS, Feigon J. 1991. *Science* 254:270–74
96. Strobel SA, Dervan PB. 1990. *Science* 249:73–75
97. Strobel SA, Dervan PB. 1991. *Nature* 350:172–74
98. Strobel SA, Doucette-Stamm LA, Riba L, Housman DE, Dervan PB. 1991. *Science* 254:1639–42
99. Ito T, Smith CL, Cantor CR. 1992. *Proc. Natl. Acad. Sci. USA* 89:495–98
100. Wang E, Malek S, Feigon J. 1992. *Biochemistry* 31:4838–46
101. Kiessling LL, Griffin LC, Dervan PB. 1992. *Biochemistry* 31:2829–34
102. Yoon K, Hobbs CA, Koch J, Sardaro M, Kutny R, Weis AL. 1992. *Proc. Natl. Acad. Sci. USA* 89:3840–44
103. Fossella JA, Kim YJ, Shih H, Richards EG, Fresco JR. 1993. *Nucleic Acids Res.* 21:4511–15
104. Stilz HU, Dervan PB. 1993. *Biochemistry* 32:2177–85
105. Rao TS, Hogan ME, Revankar GR. 1994. *Nucleosides Nucleotides* 13:95–107
106. Beal PA, Dervan PB. 1992. *Nucleic Acids Res.* 20:2773–76
107. Hampel KJ, Crosson P, Lee JS. 1991. *Biochemistry* 30:4455–59
108. Hampel KJ, Burkholder GD, Lee JS. 1993. *Biochemistry* 32:1072–77
109. Tabor CW, Tabor H. 1984. *Annu. Rev. Biochem.* 53:749–91
110. Lee JS, Woodsworth ML, Latimer LJ, Morgan AR. 1984. *Nucleic Acids Res.* 12:6603–14
111. Maher LJ, Wold B, Dervan PB. 1989. *Science* 245:725–30
112. Plum GE, Park YW, Singleton SF, Dervan PB, Breslauer KJ. 1990. *Proc. Natl. Acad. Sci. USA* 87:9436–40
113. Krawczyk SH, Milligan JF, Wadwani S, Moulds C, Froehler BC, Matteucci MD. 1992. *Proc. Natl. Acad. Sci. USA* 89:3761–64
114. Ono A, Ts'o POP, Kan L-S. 1991. *J. Am. Chem. Soc.* 113:4032–33
115. Jetter MC, Hobbs FW. 1993. *Biochemistry* 32:3249–54
116. Koh JS, Dervan PB. 1992. *J. Am. Chem. Soc.* 114:1470–78
117. Sun JS, Giovannangeli C, Francois JC, Kurfurst R, Montenay-Garestier T, et al. 1991. *Proc. Natl. Acad. Sci. USA* 88:6023–27
118. Mouscadet JF, Ketterle C, Goulaouic H, Carteau S, Subra F, et al. 1994. *Biochemistry* 33:4187–96
119. Orson FM, Kinsey BM, McShan WM. 1994. *Nucleic Acids Res.* 22:479–84
120. Takasugi M, Guendouz A, Chassignol M, Decout JL, Lhomme J, et al. 1991. *Proc. Natl. Acad. Sci. USA* 88:5602–6
121. Mergny JL, Duval-Valentin G, Nguyen CH, Perrouault L, Faucon B, et al. 1992. *Science* 256:1681–84
122. Pilch DS, Waring MJ, Sun JS, Rougee M, Nguyen CH, et al. 1993. *J. Mol. Biol.* 232:926–46
123. Lee JS, Latimer LJ, Hampel KJ. 1993. *Biochemistry* 32:5591–97
124. Belotserkovskii BP, Krasilnikova MM, Veselkov AG, Frank-Kamenetskii MD. 1992. *Nucleic Acids Res.* 20:1903–8
125. Nielsen PE, Egholm M, Berg RH, Buchardt O. 1991. *Science* 254:1497–500
126. Peffer NJ, Hanvey JC, Bisi JE, Thomson SA, Hassman CF, et al. 1993. *Proc. Natl. Acad. Sci. USA* 90:10648–52
127. Egholm M, Buchardt O, Christensen L, Behrens C, Freier SM. 1993. *Nature* 365:566–68
128. Egholm M, Nielsen PE, Buchardt O, Berg RH. 1992. *J. Am. Chem. Soc.* 114:9677–78
129. Kim SK, Nielsen PE, Egholm M, Buchardt O, Berg RH, Norden B. 1993. *J. Am. Chem. Soc.* 115:6477–81
130. Almarsson O, Bruice TC, Kerr J, Zuckermann RN. 1993. *Proc. Natl. Acad. Sci. USA* 90:7518–22
131. Nielsen PE, Egholm M, Buchardt O. 1994. *J. Mol. Recognit.* 7:165–70
132. Frank-Kamenetskii MD. 1985. In *Structure and Motion: Membranes, Nucleic Acids and Proteins*, ed. E Clementi, G Corongin, MH Sarma, RH Sarma, et al, pp. 417–43. New York: Adenine
133. Gueron M, Leroy J-L. 1992. Nucleic Acids and Molecular Biology, ed. F Eckstein, DMJ Lilley, pp1–22. Heidelberg: Springer-Verlag
134. Hanvey JC, Peffer NJ, Bisi JE, Thomson

SA, Cadilla R, et al. 1992. *Science* 258: 1481–85

135. Nielsen PE, Egholm M, Berg RH, Buchardt O. 1993. *Anti-Cancer Drug Design* 8:53–63
136. Demidov VV, Potaman VN, Frank-Kamenetskii MD, Egholm M, Buchardt O, Nielsen PE. 1994. *Biochem. Pharmacol.* 48:1310-3
137. Beasty AM, Behe MJ. 1988. *Nucleic Acids Res.* 16:1517–28
138. Manor H, Sridhara-Rao B, Martin RG. 1988. *J. Mol. Evol.* 27:96–101
139. Karlovsky P, Pecinka P, Vojtiskova M, Makaturova E, Palecek E. 1990. *FEBS Lett.* 274:39–42
140. Kohwi Y, Malkhosyan SR, Kohwi-Shigematsu T. 1992. *J. Mol. Biol.* 223: 817–22
141. Ussery DW, Sinden RR. 1993. *Biochemistry* 32:6206–13
142. Wang JC. 1985. *Annu. Rev. Biochem.* 54:665–97
143. Liu LF, Wang JC. 1987. *Proc. Natl. Acad. Sci. USA* 84:7024–27
144. Dayn A, Malkhosyan S, Mirkin SM. 1992. *Nucleic Acids Res.* 20:5991–97
145. Rahmouni AR, Wells RD. 1992. *J. Mol. Biol.* 223:131–44
146. Kohwi Y, Panchenko Y. 1993. *Genes Dev.* 7:1766–78
147. Lee JS, Burkholder GD, Latimer LJ, Haug BL, Braun RP. 1987. *Nucleic Acids Res.* 15:1047–61
148. Burkholder GD, Latimer LJ, Lee JS. 1988. *Chromosoma* 97:185–92
149. Lee JS, Latimer LJ, Haug BL, Pulleyblank DE, Skinner DM, Burkholder GD. 1989. *Gene* 82:191–99
150. Lapidot A, Baran N, Manor H. 1989. *Nucleic Acids Res.* 17:883–900
151. Kornberg A, Baker T. 1992. *DNA Replication.* New York: Freeman. 2nd ed.
152. Rao S, Manor H, Martin RG. 1988. *Nucleic Acids Res.* 16:8077–94
153. Brinton BT, Caddle MS, Heintz NH. 1991. *J. Biol. Chem.* 266:5153–61
154. Gilmour DS, Thomas GH, Elgin SCR. 1989. *Science* 245:1487–90
155. Glaser RL, Thomas GH, Siegfried E, Elgin SC, Lis JT. 1990. *J. Mol. Biol.* 211:751–61
156. Hoffman EK, Trusko SP, Murphy M, George DL. 1990. *Proc. Natl. Acad. Sci. USA* 87:2705–9
157. Lafyatis R, Denhez F, Williams T, Sporn M, Roberts A. 1991. *Nucleic Acids Res.* 19:6419–25
158. Johnson AC, Jinno Y, Merlino GT. 1988. *Mol. Cell Biol.* 8:4174–84
159. Mavrothalassitis GL, Watson DK, Papas TS. 1990. *Proc. Natl. Acad. Sci. USA* 87:1047–51

160. Tewari DS, Cook DM, Taub R. 1989. *J. Biol. Chem.* 264:16238–45
161. Davis TL, Firulli AB, Kinniburgh A. 1989. *Proc. Natl. Acad. Sci. USA* 86: 9682–86
162. Postel EH, Mango SE, Flint SJ. 1989. *Mol. Cell. Biol.* 9:5123–33
163. Clark SP, Lewis CD, Felsenfeld G. 1990. *Nucleic Acids Res.* 18:5119–26
164. Kolluri R, Torrey TA, Kinniburgh AJ. 1992. *Nucleic Acids Res.* 20:111–16
165. Bossone SA, Asselin C, Patel AJ, Marcu KB. 1992. *Proc. Natl. Acad. Sci. USA* 89:7452–56
166. Postel EH, Berberich SJ, Flint SJ, Ferrone CA. 1993. *Science* 261:478–83
167. Brunel F, Alzari PM, Ferrara P, Zakin MM. 1991. *Nucleic Acids Res.* 19: 5237–45
168. Kennedy GC, Rattner JB. 1992. *Proc. Natl. Acad. Sci. USA* 89:11498–502
169. Yee HA, Wong AKC, van de Sande JH, Rattner JB. 1991. *Nucleic Acids Res.* 19:949–53
170. Muraiso T, Nomoto S, Yamazaki H, Mishima Y, Kominari R. 1992. *Nucleic Acids Res.* 20:6631–35
171. Raghu G, Tevosian S, Anant S, Subramanian K, George DL, Mirkin SM. 1994. *Nucleic Acids Res.* 22:3271–79
172. Firulli AB, Maibenko DC, Kinniburgh AJ. 1992. *Biochem. Biophys. Res. Commun.* 185:264–70
173. Firulli AB, Maibenco DC, Kinniburgh AJ. 1994. *Arch. Biochem. Biophys.* 310: 236–42
174. DesJardins E, Hay N. 1993. *Mol. Cell. Biol.* 13:5710–24
175. Kiyama R, Camerini-Otero RD. 1991. *Proc. Natl. Acad. Sci. USA* 88:10450–54
176. Kohwi Y, Kohwi-Shigematsu T. 1991. *Genes Dev.* 5:2547–54
177. Lu Q, Wallrath LL, Allan BD, Glaser RL, Lis JT, Elgin SCR. 1992. *J. Mol. Biol.* 225:985–98
178. Lu Q, Wallrath LL, Granok H, Elgin SCR. 1993. *Mol. Cell. Biol.* 13:2802–14
179. Cooney M, Czernuszewicz G, Postel EH, Flint SJ, Hogan ME. 1988. *Science* 241:456–59
179a. Maher LJ, Oervan PB, Wold B. 1992. *Biochemistry* 31:70–81
180. Gee JE, Blume S, Snyder RC, Ray R, Miller DM. 1992. *J. Biol. Chem.* 267: 11163–67
181. Mayfield C, Ebbinghaus S, Gee J, Jones D, Rodu B, Squibb M, Miller D. 1994. *J. Biol. Chem.* 269:18232–38
182. Grigoriev M, Praseuth D, Robin P, Hemar A, Saison-Behmoaras T, et al. 1992. *J. Biol. Chem.* 267:3389–95

183. Duval-Valentin G, Thuong NT, Helene C. 1992. *Proc. Natl. Acad. Sci. USA* 89:504–8
184. Maher LJ. 1992. *Biochemistry* 31:7587–94
185. Young SL, Krawczyk SH, Matteucci MD, Toole JJ. 1991. *Proc. Natl. Acad. Sci. USA* 88:10023–26
186. Maher LJ, Wold B, Dervan PB. 1991. *Antisense Res. Dev.* 1:277–81
187. Grigoriev M, Praseuth D, Guieysse AL, Robin P, Thuong NT, et al. 1993. *Proc. Natl. Acad. Sci. USA* 90:3501–5
188. Roy C. 1993. *Nucleic Acids Res.* 21: 2845–52
189. Orson FM, Thomas DW, McShan WM, Kessler DJ, Hogan ME. 1991. *Nucleic Acids Res.* 19:3435–41
190. Postel EH, Flint SJ, Kessler DJ, Hogan ME. 1991. *Proc. Natl. Acad. Sci. USA* 88:8227–31
191. McSchan WM, Rossen RD, Laughter AH, Trial L, Kessler DJ, et al. 1992. *J. Biol. Chem.* 267:5712–21
192. Ing NH, Beekman JM, Kessler DJ, Murphy M, Jayaraman K, et al. 1993. *Nucleic Acids Res.* 21:2789–96
193. Samadashwily GM, Mirkin SM. 1994. *Gene.* 149:127–36
194. Giovannangeli C, Thuong NT, Helene C. 1993. *Proc. Natl. Acad. Sci. USA* 90:10013–17
195. Hacia JG, Dervan PB, Wold BJ. 1994. *Biochemistry* 33:6192–200
196. Birg F, Praseuth D, Zerial A, Thuong NT, Asseline U, et al. 1990. *Nucleic Acids Res.* 18:2901–8
197. Kowalczykowski SC. 1991. *Annu.*
198. Stasiak A, Stasiak AZ, Koller T. 1984. *Cold Spring Harbor Symp. Quant. Biol.* 49:561–70
199. Hsieh P, Camerini-Otero CS, Camerini-Otero RD. 1990. *Genes Dev.* 4: 1951–63
200. Camerini-Otero RD, Hsieh P. 1993. *Cell* 73:217–23
201. Wu AM, Kahn R, DasGupta C, Radding CM. 1982. *Cell* 30:37–44
202. Bianchi M, DasGupta C, Radding CM. 1983. *Cell* 34:931–39
203. Umlauf SW, Cox MM, Inman RB. 1990. *J. Biol. Chem.* 265:16898–912
204. Rao BJ, Dutreix M, Radding CM. 1991. *Proc. Natl. Acad. Sci. USA* 88: 2984–88
205. Rao BJ, Chiu SK, Radding CM. 1993. *J. Mol. Biol.* 229:328–43
206. Adzuma K. 1992. *Genes Dev.* 6:1679–94
207. Jain SK, Inman RB, Cox MM. 1992. *J. Biol. Chem.* 267:4215–22
208. Chiu SK, Rao BJ, Story RM, Radding CM. 1993. *Biochemistry* 32:13146–55
209. Shchyolkina AK, Timofeev EN, Borisova OF, Ilicheva IA, Minyat EE, et al. 1994. *FEBS Lett.* 339:113–18
210. Ferrin LJ, Camerini-Otero RD. 1991. *Science* 254:1494–97
211. Bedale WA, Inman RB, Cox MM. 1993. *J. Biol. Chem.* 268:15004–16
212. Hsieh P, Camerini-Otero CS, Camerini-Otero RD. 1992. *Proc. Natl. Acad. Sci. USA* 89:6492–96

Rev. Biochem. Biophys. Chem. 20: 539–75

Annu. Rev. Biochem. 1995. 64:97–112

SUPEROXIDE RADICAL AND SUPEROXIDE DISMUTASES

Irwin Fridovich

Department of Biochemistry, Duke University Medical Center, Durham, North Carolina 27710

KEY WORDS: hydroxyl radical, extracellular SOD, free radicals, regulation of superoxide disutase, glycation and O_2^- production, targets for O_2^-, nitric oxide

CONTENTS

ABSTRACT

O_2^- oxidizes the [4Fe-4S] clusters of dehydratases, such as aconitase, causing inactivation and release of Fe(II), which may then reduce H_2O_2 to OH^- $+OH•$. SODs inhibit such HO• production by scavenging O_2^-, but Cu, ZnSODs, by virtue of a nonspecific peroxidase activity, may peroxidize spin trapping agents and thus give the appearance of catalyzing OH• production from H_2O_2.

There is a glycosylated, tetrameric Cu, ZnSOD in the extracellular space that binds to acidic glycosamino-glycans. It minimizes the reaction of O_2^- with NO. *E. coli*, and other gram negative microorganisms, contain a periplasmic Cu, ZnSOD that may serve to protect against extracellular O_2^-. Mn(III) complexes of multidentate macrocyclic nitrogenous ligands catalyze the dismutation of O_2^- and are being explored as potential pharmaceutical agents.

SOD-null mutants have been prepared to reveal the biological effects of O_2^-. SodA, sodB *E. coli* exhibit dioxygen-dependent auxotrophies and enhanced mutagenesis, reflecting O_2^--sensitive biosynthetic pathways and DNA damage. Yeast, lacking either Cu, ZnSOD or MnSOD, are oxygen intolerant, and the double mutant was hypermutable and defective in sporulation and exhibited requirements for methionine and lysine. A Cu, ZnSOD-null *Drosophila* exhibited a shortened lifespan.

Introduction

Situations that involve opposing and well-balanced forces, or reactions, are often effectively invisible. Discovery of the true dynamic balance is further confounded when the relevant species are short lived. These statements are offered as a rationalization, or even as an apology, for how slowly we have gained an understanding of the basis of dioxygen toxicity and the nature of the counterbalancing defenses. The demonstrations that xanthine oxidase produced O_2^- (1) and that erythrocytes contained an enzyme that very efficiently and specifically catalyzed the conversion of O_2^- into $H_2O_2 + O_2$ (2) opened the door on the true state of affairs: that O_2^- is commonly produced within aerobic biological systems, and superoxide dismutases (SODs) provide an important defense against it.

Although late in getting started, we now seem to be making up for lost time. The past 25 years have witnessed a very impressive increase in our knowledge of the biology of O_2^- and of the SODs that remove it. The relevant literature has been reviewed at regular intervals (3–56). Here, we touch upon only portions of this broad field that appear to be of current interest or that promise either important insights or practical applications.

Targets

O_2^- can act as either a univalent oxidant or reductant. Investigators have exploited this dual reactivity in devising assays for the activity of SODs. Thus, in some assays, O_2^- reduces tetranitro methane (2), cytochrome *c* (2), or nitro blue tetrazolium (57), and in others it oxidizes epinephrine (2, 58), tiron (59), pyrogallol (60), or 6-hydroxydopamine (61). The ability of O_2^- to oxidize sulfite, and thus to initiate its free-radical chain oxidation, was one of the early clues to the production of O_2^- by xanthine oxidase (62).

Given the chemical diversity of biological systems, intracellular O_2^- will surely find targets it can either oxidize or reduce. For example, it can oxidize the family of dehydratases that contain [4Fe-4S] clusters at their active sites. This group of enzymes, which includes dihydroxy acid dehydratase (63), 6-phosphogluconate dehydratase (64), aconitase (65, 66), and fumarases A and B (67, 68), undergoes rapid oxidation by O_2^- with a resultant loss of Fe(II) from the cluster and concomitant inactivation (66).

Release of Fe(II) from the cluster sets the stage for the Fenton reaction in which Fe(II) reduces H_2O_2, thereby producing Fe(II)O or Fe(III) + HO•. This reaction may be the basis for the in vivo cooperation between O_2^- and H_2O_2, which produces an oxidant capable of attacking virtually any cellular target, most notably DNA (69). Much evidence indicates that this deleterious cooperative interaction occurs between O_2^- and peroxides within cells. Thus, the *sodA sodB* strain of *Escherichia coli*, which lacks both the Mn- and Fe-containing SODs, was hypersensitive not only to O_2 and to agents that increase production of O_2^- but also to H_2O_2 (70). Delivery of SODs into hepatocytes by liposomal fusion increased resistance towards an alkyl hydroperoxide (71). In an opposite approach, diethyl dithiocarbamate was used to inactivate the Cu,ZnSOD in cultured aortic endothelial cells, and this inactivation increased sensitivity towards damage by H_2O_2 (72).

Although the outcome is the same, i.e. the production of HO• or Fe(II)O from a cooperative interaction of O_2^- and H_2O_2, the in vivo and in vitro variants of this interaction differ profoundly. In vitro, O_2^- acts as a reductant towards available Fe(III) and so generates Fe(II) that can reduce H_2O_2. In contrast, O_2^- produces Fe(II) in vivo by acting as an oxidant towards susceptible [4Fe–4S] clusters. The Fe(II) released from the oxidized clusters could bind to DNA and provide a site for production of powerful oxidants immediately adjacent to this critical target. Such a mechanism could account for the enhanced dioxygen-dependent mutagenesis exhibited by *sodA sodB E. coli* (73) and for the in vivo hydroxylation of the bases of DNA (74–78).

SOD Prevents Production of HO•

As expected for a mechanism in which HO• or Fe(II)O is generated from the metal-catalyzed interaction of O_2^- with H_2O_2, the in vitro process is inhibited by SOD, or catalase, or by chelating agents that restrict the redox cycling of the catalytic metal. Such inhibitions have been observed repeatedly (79–112). The report that Cu,ZnSOD, rather than inhibit HO• production from H_2O_2, actually catalyzes it (113) contradicted all of this earlier work and demanded some explanation. Yim et al (113) used high concentrations of H_2O_2 (30 mM) and of Cu,ZnSOD (1.25 µM). Sato et al (114) concluded that Cu(II), released from the Cu,ZnSOD as it was inactivated by the H_2O_2, was the actual catalyst of HO• production. Voest et al (115) concluded that an HO•-like entity is generated at the active site of Cu,ZnSOD in the presence of H_2O_2, but that free HO• is not produced.

Another explanation for the apparent catalysis of HO• formation by Cu,ZnSOD can be derived from much earlier work on the interaction of H_2O_2 with this enzyme (116, 117). These studies showed that H_2O_2 rapidly reduced the Cu(II) at the active site and then more slowly inactivated the reduced enzyme. This inactivation could be prevented by xanthine, urate, formate, and azide, but not by alcohols. Thus the reaction of Cu(I) with H_2O_2 at the active site, apparently generated a potent oxidant [Cu(I)O, or Cu(II)–OH] that could attack an adjacent histidine residue and thus inactivate the enzyme or alternately could attack xanthine, urate, formate, or azide. These exogenous reductants would thus serve as sacrificial substrates and spare the essential histidines.

In accord with these observations, Cu,ZnSOD acts as a peroxidase towards these and other substrates (116, 117). This peroxidase activity can fully account for the results of Yim et al (113). Thus Cu,ZnSOD probably acted as a peroxidase towards DMPO, producing DMPO-OH, whichappeared to have been produced by reaction of $HO^{\bullet}$ with DMPO (Dimethyl pyolline-N-oxide. Had free $HO^{\bullet}$ actually been produced, it would have been able to convert ethanol to the hydroxyethyl radical, which is eminently trappable by DMPO. Yim et al did not observe such trapping (113), which indicates that HO• production was not being catalyzed by Cu,ZnSOD. The inability of ethanol to protect Cu,ZnSOD against inactivation by H_2O_2 (116, 117) indicates that the Cu,ZnSOD cannot catalyze the peroxidation of ethanol. Were it able to do so, Yim et al (113) would have been able to trap the hydroxyethyl radical and would have had further reason to believe, incorrectly, that free HO• was being produced through a catalytic interaction of Cu,ZnSOD with H_2O_2.

The SOD Family of Enzymes

The cyanobacteria started a gradual oxygenation of the biosphere that applied a common selection pressure to a varied anaerobic biota. At least two of the classes of SODs were among the adaptations called forth. One consists of SODs with Cu(II) plus Zn(II) at the active site, wheras the other comprises SODs with either Mn(III) or Fe(III) at the catalytic center. These enzymes have been reviewed (14, 48), and we need now only consider relatively recent developments.

Extracellular SODs (ECSODs)

O_2^- should not easily cross biological membranes, with the exception of those membranes that are richly endowed with anion channels, such as the erythrocyte stroma (118). O_2^- must consequently be detoxified in the compartment within which it is generated. This necessity explains the presence of distinct SODs in the cytosol and in the mitochondria of eukaryotic cells (119–127) and why complementation of a defect in the mitochondrial MnSOD in yeast was effective only when the leader sequence of the maize gene, which assured importation of the gene product into the mitochondrion, was present (128).

Against this background of compartmentation of SODs, the existence of extracellular SODs bespeaks the need for defense against the numerous extracellular sources of O_2^- For example, ultraviolet irradiation of water produces O_2^- (129) continually in surface waters (130). In the presence of photosensitizers and a wide variety of electron donors, irradiation with visible light will suffice (57, 131–136). Autoxidations routinely produce O$_2^-$. Moreover the collapse of cavities introduced into aerated water by ultrasonication produces O$_2^-$ (137–139), and cavities produced by turbulence presumably will also do so, albeit to a lesser degree. In living systems, the membrane-associated NADPH oxidase, which is so abundant in phagocytic leukocytes, releases O$_2^-$ into the extracellular phase (140–146).

The mammalian extracellular SOD is a Cu,ZnSOD, but unlike its dimeric cytosolic counterpart, it is tetrameric and glycosylated (147,148). This enzyme exhibits affinity for heparin and other acidic glycosamino-glycans (149, 150) because of a C-terminal heparin-binding domain that is rich in basic amino acid residues (151, 152). This heparin affinity results in binding to the endo-thelium and to other cell types (153) such that the amount of mammalian extracellular SOD actually free in the blood plasma is low but can be elevated by injection of heparin (149). The amino acid sequence of the human ECSOD is known from the sequence of the cDNA (152), which indicated the presence of an 18-residue signal peptide, characteristic of secreted proteins. Residues 96–193 show strong sequence homology to the cytosolic Cu,ZnSODs, whereas residues 1–95 do not. The residues that compose the active site are conserved in all Cu,ZnSODs, including the ECSOD.

The nucleotide sequence coding for the heparin-binding domain has been fused onto the gene coding for the cytosolic Cu,ZnSOD and the resultant fusion gene expressed in *E. coli* (154). The artificial ECSOD so produced exerted a strong antiinflammatory effect. ECSOD also protects against reperfusion injury (155–158), enhances arterial relaxation (159), and thereby diminishes hyper-tension (160). ECSOD may prove to be a useful pharmaceutical, and the production of transgenic mice that secrete the human ECSOD in milk has clearly brought us closer to realizing a convenient source (161).

Extracellular SODs have been found in severalnonmammalian sources, such as phloem sap (162), *Nocardia asteroides* (163), *Schistosoma mansoni* (164), and *Onchocerca volvulus* (165). *N. asteroides* elicits the respiratory burst of neutrophils and of monocytes, yet resists killing by these phagocytes (166). Treatments of *N. asteroides* with an antibody to the extracellular SOD in-creased its susceptibility to killing by the leukocytes, but treatment with non-specific immunoglobulin did not (167). In this organism, the surface SOD thus appears to be a pathogenicity factor.

Periplasmic SODs

Surveys of a variety of bacterial species had indicated the presence of FeSOD and/or MnSOD but not Cu,ZnSOD (168–170). Such data led to the view that the Cu,ZnSODs were characteristic of eukaryotes. Nevertheless, some studies reported finding Cu,ZnSOD in a few bacteria. The first such instance dealt with *Photobacter leiognathi*, which was found to contain both FeSOD and Cu,ZnSOD (171). Because this microorganism ordinarily lives as a symbiont in the luminescent organ of the ponyfish, a gene transfer seemed a reasonable explanation for this instance of a bacterial Cu,ZnSOD (172, 173), but this notion was subsequently dismissed (174). Furthermore, Cu,ZnSOD was found in several bacterial species including *Caulobacter crescentis* (175), *Pseudomonas diminuta* and *Pseudomonas maltophila* (176), *Brucella abortus* (177), and in several species of *Hemophilus* (178). The *C. crescentis* (179, 180) and the *B. abortus* (181) Cu,ZnSODs are periplasmic.

The much-studied *E. coli* also expresses a Cu,ZnSOD, in addition to the MnSOD and FeSOD (182). Moreover, this *E. coli* Cu,ZnSOD is selectively released by osmotic shock and thus is periplasmic. It is induced during aerobic growth, yet the net activity remains much less than the total activity. This Cu,ZnSOD seems to be important for the aerobic growth of *E. coli* mutants that cannot make the FeSOD and MnSOD (*sodA sodB*). Thus, diethyl dithiocarbamate, which inactivates Cu,ZnSOD but not FeSOD or MnSOD, inhibits the aerobic growth of the *sodA sodB* strain on a rich medium but has no effect on anaerobic growth. A true SOD-null *E. coli* i.e. a *sodA sodB sodC* strain, would probably behave like a sensitive obligate anaerobe and is a goal for future studies. Another important aim is to find out why a SOD must be targeted to the periplasm. Are there sources of O_2^- within the periplasm or does the periplasmic SOD protect against extracellular sources of O_2^-?

SOD Mimics

Because of the role of O_2^- in free-radical chain oxidations, oxygen toxicity, inflammations, reperfusion injuries, and very likely also in senescence, low-molecular weight mimics of SOD activity should be very useful. The simplicity of the O_2^- dismutation reaction, and the catalytic abilities of certain transition-metal cations, encouraged the view that such mimics could be found. To date, many reports of SOD mimics have been published—too many to adequately review here. The Mn(III) and Fe(III) complexes of substituted porphines (183, 184) are of particular interest because they are quite active and very stable. Manganese complexes are special because Mn(II), should it be liberated from the complex, does not participate in Fenton chemistry, and because in certain lactobacilli (185–187) high intracellular concentrations of Mn(II) salts have replaced SOD. This substitution indicates that low-molecular-weight Mn complexes can provide functional replacements for SOD and are well tolerated within at least certain types of cells.

Recent studies of a porphine Mn(III) complex with N-methyl pyridyl groups on the methine bridge carbons [Mn(III)TMPyP] reveal that the porphine compound eliminates the growth inhibition imposed by aerobic paraquat on a SOD-competent *E. coli*, as well as the growth inhibition imposed by oxygen on a sodA sodB strain (188). Indeed, the protective effects of this Mn(III) porphine exceeded expectations based on its ability to catalyze the dismutation of O$_2^-$ as measured in vitro. This result is at least partially explained by the large difference between the rate constants for the first and second steps of the catalytic cycle (184) and by the observation that the complex remains reduced within *E. coli* (188).

Below, the rate constant for Reaction 1 is ~2 × 10^7 M^{-1} s^{-1}, while for Reaction 2 it is ~4 × 10^9 M^{-1} s^{-1} (184).

$$\text{Por-Mn(III)} + \text{O}_2^- \leftrightarrow \text{Por-Mn(II)} + \text{O}_2 \qquad\qquad 1.$$

$$\text{Por-Mn(II)} + \text{O}_2^- + 2\text{H}^+ \leftrightarrow \text{Por-Mn(III)} + \text{H}_2\text{O}_2. \qquad\qquad 2.$$

When the dismutation is catalyzed in vitro, the slower of these steps, Reaction 1, is rate limiting. However, this step is irrelevant in vivo, because the Mn(III) porphyrin remains reduced at the expense of GSH and NADPH (188), and the rate constant of Reaction 2 limits O$_2^-$ scavenging. This Mn(III) porphine thus protects against O$_2^-$ by acting not as a superoxide dismutase but as anO$_2^-$:GSH/NADPH oxidoreductase. The concentration of this Mn(III) porphine within the cell also increases its protective effect. Thus when the complex is at 25 μM in the medium, it reaches 1 mM within *E. coli*. Mammalian cell lines are also protected against paraquat (B Day, in preparation) or against pyocyanine (PR Gardner, personal communication) by this Mn(III) porphine.

Another promising group of SOD mimics comprises the cyclic polyamine complexes of Mn(III) (190a). These compounds catalyze the dismutation of O$_2^-$ at approximately 1% of the rate exhibited by SOD. Nevertheless, they could protect endothelial cells against damage by a flux of O$_2^-$ produced by activated neutrophils or by the xanthine oxidase reaction (190b). Because catalase did not protect in this system, O$_2^-$ is apparently the damaging species. The damage may result from protonation of the O$_2^-$ in the acidic domain adjacent to the anionic cell membrane. Alternatively, the O$_2^-$ may have been converted to ONOO$^-$ by reaction with the NO produced by endothelial cells.

Oxygen Radicals from Sugars

Small sugars, such as glycolaldehyde, glyceraldehyde, or dihydroxy acetone, autoxidize by a free radical pathway in which O$_2^-$ serves as a chain propagator (191–193). Enolization precedes autoxidation, and small sugars, unable to block the carbonyl by cyclization, are consequently most readily autoxidized; in contrast, aldohexoses, which exist primarily as pyranoses, are relatively stable. Production of O$_2^-$ and H$_2$O$_2$, during autoxidation probably explains the mutagenicity of small sugars (194), their abilities to inactivate the transsulfurase called rhodanese (195), and their abilities to cause the peroxidation of polyunsaturated fatty acids (196).

The relative resistance of aldohexoses towards autoxidation is abrogated when they react with amino compounds and are converted to fructosyl amines (197). This is the situation in glycated proteins, which do autoxidize with production of O_2^- and H_2O_2 (198). Oxidative damage, subsequent to glycation, has been reported for LDL (199), collagen (200), the cytosolic Cu,ZnSOD (201), the extracellular Cu,ZnSOD (202), and serum albumin (203). The inhibition of the autoxidative degradation of such fructosylamines by SOD (197) bespeaks a role for O_2^- as a chain propagator.

SOD Mutants

An excellent way to explore the functions of an enzyme is through the phenotype of null mutants. Mutants of *E. coli* unable to produce MnSOD (*SodA*) or FeSOD (*SodB*) were first reported by Carlioz & Touati (70). This *sodA sodB* strain was indistinguishable from the parental strain under anaerobic conditions, but under aerobic conditions it exhibited dioxygen-dependent nutritional auxotrophies, hypersensitivity towards paraquat, and enhanced mutagenesis (73); these deficits were reversed by introduction of a plasmid carrying a SOD gene. Because the *sodA sodB* strain will not grow on aerobic minimal medium, owing to its multiple O_2-dependent auxotrophies, it lends itself to complementation studies. Thus *E. coli* that have reacquired a functional SOD gene can be easily selected from among many that have not by means of growth on aerobic minimal plates.

This technique has been exploited to show that any functional SOD gene will complement the *sodA sodB* strain of *E. coli*. Investigators have obtained this result with genes coding for plant FeSODs (204), human Cu,ZnSOD (204a), the *Legionella pneumophila* FeSOD (205), the *Listeria monocytogenes* MnSOD (206), the *Coxiella burnetii* FeSOD (207), and others (208, 209). This sort of complementation has been extended to other organisms. A yeast with a defect in its MnSOD was complemented by a MnSOD gene from maize (210); extra copies of a MnSOD gene protected tobacco against paraquat (211); and a gene coding for the bovine Cu,ZnSOD complemented a Cu,ZnSOD null mutant of *Drosophila melanogaster* (212).

Some of the specific deficits associated with mutations in SOD genes are instructive, but others remain to be explained. The O_2-dependent auxotrophy for branched-chain amino acids, exhibited by the *sodA sodB E. coli*, is explained by the O_2^- oxidative inactivation of [4Fe-4S]–containing dehydratases. The section on targets for O_2^- has already discussed this topic. In this instance the dihydroxy acid dehydratase, which catalyzes the penultimate step in the biosynthesis of the branched-chain amino acids, is the O_2^--sensitive dehydratase. The 41-fold increase in spontaneous mutagenesis seen with *sodA sodB E. coli* (73) reflects attack on DNA, probably by HO• or by Fe(II)O (69). We have recently observed an additional O_2-dependent deficit in *sodA sod B E. coli* that is a rapid loss of viability in stationary phase (LT Benov & I Fridovich, unpublished results). A SOD-null mutant of *Porphyromona gingivalis* exhibited a similar O_2-dependent loss of viability (214). The causes of such viability losses are not known, but damage to cell membranes is a possibility, as shown by the protective effect of osmolytes on the *sodA sodB* strain (215) and by the role of osmolytes in facilitating loss of a SOD-1–bearing plasmid from an otherwise SOD-defective yeast (216).

Increases in spontaneous mutagenesis alsooccurs in *Salmonella typhimurium* with a deletion in *oxyR*. This defect was partially complemented by plasmids bearing genes coding for either catalase or SOD (217). This observation reflects the roles of both O_2^- and hydroperoxides in generating HO• or Fe(II)O (69).

Sensitivity towards O_2 was seen in both Cu,ZnSOD (218) and MnSOD defective yeast (219). Hypersensitivity toward paraquat and increased spontaneous mutagenesis were also characteristic of Cu,ZnSOD-null yeast (220), whereas a mutant defective in both the Cu,Zn- and the Mn-SODs was O_2-sensitive, hypermutable, auxotrophic for methionine and lysine, and defective in sporulation (221). In *D. melanogaster,* a Cu,ZnSOD null mutant exhibited hypersensitivity toward paraquat, male sterility, and a shortened lifespan (222).

To date, the only humans known to bear genetic defects in the coding regions of a SOD are those with the familial variant of amyotrophic lateral sclerosis (FALS), or Lou Gehrig's Disease. This autosomal dominant condition is associated with point mutations in the Cu,ZnSOD gene (223) and with decreased cytosolic SOD activity (224, 225). Inhibition of the cytosolic Cu,ZnSOD, either with diethyl dithiocarbamate or by expression of antisense message, causes apoptosis of spinal neurons (226). Yet, whether or not moderately decreased SOD activity causes the late-onset motor-neuron degeneration that results in the progressive and ultimately fatal disease is not clear. Transgenic mice expressing a human FALS Cu,ZnSOD gene exhibited a late onset and progressive paralysis (227). The fact that these symptoms occurred against a background of normal mouse Cu,ZnSOD activity suggests that a toxic property of the mutated Cu,ZnSOD, rather than a loss of its SOD activity, is the cause of the problem.

Cu,ZnSOD catalyzes two low-level, non-SOD activities, either of which might be increased by the mutations associated with FALS. One of these is a nonspecific peroxidase activity (117), and the other is the catalysis of the nitration of tyrosyl residues by peroxynitrite (228, 229).

NO and O_2^-

The endothelium-derived relaxing factor (EDRF), which regulates smooth muscle tone and thereby blood flow and blood pressure (230), is NO (231). Even before its identification as NO, EDRF was found to react with O_2^-. Thus, the half life of EDRF is extended by SOD but not by catalase (232, 233). Furthermore O_2^-, whether produced by the endothelium or by xanthine oxidase plus xanthine, acted like a contracting factor (234). This action can now be explained on the basis of the production of peroxynitrite from $O_2^- + NO$ (235) at the diffusion-limited rate of $6.7 \times 10^9 \, M^{-1} \, s^{-1}$ (236). Peroxynitrite is a strong oxidant and reacts with thiols (237), initiates lipid peroxidation (238), and kills *E. coli* (239) and *Trypanosoma cruzi* (240). In all likelihood the reaction of O_2^- with NO significantly modulates the biological activities of both substances.

Regulation in E. coli

Two of the SODs in *E. coli*, i.e. the cytosolic MnSOD and the periplasmic Cu,ZnSOD, are induced during aerobiosis, whereas the FeSOD is expressed both aerobically and anaerobically. Why should *E. coli* make a SOD under anaerobic conditions when O_2^- production cannot occur? One answer is that a facultative organism must maintain a standby defense to ward off the toxicity of O_2^- that must be faced following the transition from anaerobic to aerobic conditions. Such abrupt transitions must, of course, be a selection pressure for enteric organisms.

Experimental evidence supports this view. For example, *E. coli* defective in the *sodB* gene, which encodes the FeSOD, exhibited a 2-h growth lag when transferred from anaerobic to aerobic media, whereas the parental strain did not (241). Induction of the MnSOD, following exposure to aerobic conditions, finally ended the growth lag. The anaerobically grown cells evidently contained an enzyme capable of the univalent reduction of oxygen. The fumarate reductase, which allows anaerobic *E. coli* to use fumarate as an electron sink, reduced O_2 to O_2^- when supplied with NADH. If the anaerobic fumarate reductase was a major source of O_2^- in the cells, after the anaerobic-to-aerobic transition, a mutational defect in the fumarate reductase should eliminate the growth lag that attended this transition. It did (241).

The biosynthesis of MnSOD within *E. coli* is transcriptionally activated as a member of the *soxRS* (242, 243) and the *soxQ* (244) regulons and is also transcriptionally repressed by the products of the *fur* (245), *arcA* (246), and *fnr* (247) genes, as well as by the integration host factor (248). All of these regulatory elements have been explored and discussed (249). Iron plays a key role in the action of two of these transcriptional repressions, i.e. that of *fur* (250) and *fnr* (251) and is moreover a component of the SoxR protein that functions as the redox sensor of the *soxRS* regulon (252). Furthermore, iron competes with manganese for insertion into the nascent MnSOD polypeptide and when inserted in place of manganese yields a catalytically inactive product (253, 254). Indeed, iron starvation, whether imposed by chelation or by depletion of the medium, has repeatedly been reported to increase MnSOD in *E. coli* even under anaerobic conditions (253, 255–259).

Is there some rationale that can be offered for this intimate involvement of iron in the regulation of the biosynthesis of MnSOD in *E. coli*? One possible scenario depends upon the great susceptibility of the [4Fe-4S]–containing dehydratases to oxidative inactivation by O$_2^-$ (63–68). Reactivation of these enzymes, which include fumarases A and B, aconitase, dihydroxy acid dehydratase, and 6-phosphogluconate dehydratase, requires Fe(II). The levels of activity of these enzymes depend upon a balance between the rates of inactivation and reactivation. Hence, when [Fe(II)] is low, and reactivation is slow, only a low rate of inactivation, achievable by elevating [MnSOD] and thereby lowering O$_2^-$, can be tolerated. Therefore a low level of Fe(II) should lead to an increase in active MnSOD, and it does so through multiple effects. Transcriptional repression by Fur and by Fnr depends upon binding of iron to these regulatory proteins; low Fe(II) will lift these repressions. Oxidation of the SoxR protein activates transcription of *soxS*, and SoxS, in turn, activates transcription of the MnSOD gene. Because SoxR is an iron-sulfur protein, we may suppose that oxidation leads to iron loss and that the active form is iron depleted. Hence, low Fe(II) will lead to activation via the *soxRS* regulon. Finally, at the level of maturation of nascent MnSOD polypeptide, low Fe(II) will favor insertion of manganese and production of the active enzyme.

Epilogue

This review does not begin to do justice to the state of knowledge of the biology of oxygen radicals. As in all aspects of biology, beauty and perceived complexity increase with increased study. We may confidently expect that this will continue in the future. There will be more.

Literature Cited

1. McCord JM, Fridovich I. 1968. *J. Biol. Chem.* 243:5753–60
2. McCord JM, Fridovich I. 1969. *J. Biol. Chem.* 244:6049–55
3. Fridovich I. 1972. *Acct. Chem. Res.* 5:321–26
4. Halliwell B. 1974. *New Phytol.* 73:1075–86
5. Fridovich I. 1974. *Adv. Enzymol.* 41:35–97
6. Fridovich I. 1975. *Annu. Rev. Biochem.* 44:147–59
7. Fridovich I. 1978. *Photochem. Photobiol.* 28:733–41
8. Halliwell B. 1978. *Cell Biol. Int. Rep.* 2:113–28

9. Fridovich I. 1978. *Science* 201:875–80
10. Hassan HM, Fridovich I. 1979. *Rev. Infect. Dis.* 1:357–69
11. Hassan HM, Fridovich I. 1980. In *Enzymatic Basis of Detoxication*, ed. WB Jacoby, 1:311–32. New York: Academic
12. Fridovich I. 1981. In *Oxygen in Living Processes*, ed. DL Gilbert, pp. 250–72. New York: Springer
13. Elstner EF. 1982. *Annu. Rev. Plant Physiol.* 33:73–96
14. Steinman HM. 1982. In *Superoxide Dismutase*, ed. LW Oberley, pp. 11–68. Boca Raton, FL: CRC
15. Freeman BA, Crapo JD. 1982. *Lab. Invest.* 47:412–26
16. Michelson AM. 1982. *Agents Actions* 11:179–201 (Suppl.)
17. Parks DA, Bulkley GB, Granger DN. 1983. *Surgery* 94:415–22
18. Karnovsky ML, Badwey JA. 1983. *J. Clin. Chem. Clin. Biochem.* 21: 545–53
19. McCord JM. 1983. *Physiologist* 26:156–58
20. Halliwell B, Gutteridge JMC. 1984. *Biochem. J.* 219:1–14
21. Marklund SL. 1984. *Med. Biol.* 62:130–34
22. DiGuiseppi J, Fridovich I. 1984. *CRC Crit. Rev. Toxicol.* 12:315–42
23. Halliwell B, Gutteridge JMC. 1985. *Mol. Aspects Med.* 8:89–93
24. McCord JM. 1986. *Free Radic. Biol. Med.* 2:307–10
25. Halliwell B. 1987. *FASEB J.* 1:358–64
26. Weiss SJ. 1986. *Acta Physiol. Scand. Suppl.* 548:9–37
27. Bannister JV, Bannister WH, Rotilio G. 1987. *CRC Crit. Rev. Biochem.* 22:111–80
28. Fridovich I. 1986. *Adv. Enzymol.* 58:61–97
29. Michelson AM. 1987. *Life Chem. Rep.* 6:1–42
30. McCord JM. 1987. *Fed. Proc. Fed. Am. Soc. Exp. Biol.* 46:2402–6
31. Touati D. 1988. *Free Radic. Biol. Med.* 5:393–402
32. Cotgreave IA, Moldeus P, Orrenius S. 1988. *Annu. Rev. Pharmacol. Toxicol.* 28:189–212
33. Bannister WH. 1988. *Free Radic. Res. Commun.* 5:35–42
34. Koppenol WH. 1988. *Prog. Clin. Biol. Res.* 274:93–109
35. Hochstein P, Atallah AS. 1988. *Mutat. Res.* 202:363–75
36. Hassett DJ, Cohen MS. 1989. *FASEB J.* 3:2574–82
37. Hassan HM. 1989. *Adv. Genet.* 26:65–97
38. Joenje H. 1989. *Mutat. Res.* 219:193–208
39. Touati D. 1989. *Free Radic. Res. Commun.* 8:1–9
40. Floyd RA. 1990. *FASEB J.* 4:2587–97
41. Grace SC. 1990. *Life Sci.* 47:1875–86
42. Leff JA, Repine JE. 1990. *Blood Cells* 16:183–91
43. Storz G, Tartaglia LA, Farr SB, Ames BN. 1990. *Trends Genet.* 6:363–68
44. Fridovich I. 1991. *Curr. Top. Plant Physiol.* 6:1–5
45. Bielski BHJ, Cabelli DE. 1991. *Int. J. Radiat. Biol.* 59:291–319
46. Fridovich I. 1989. *J. Biol. Chem.* 264: 7761–64
47. Farr SB, Kogoma T. 1991. *Microbiol. Rev.* 55:561–85
48. Beyer W, Imlay J, Fridovich I. 1991. *Prog. Nucleic Acid Res. Mol. Biol.* 40: 221–53
49. Bowler C, van Montagu M, Inzé D. 1992. *Annu. Rev. Plant Physiol. Mol. Biol.* 43:83–116
50. Harris ED. 1992. *FASEB J.* 6:2675–83
51. Asada K. 1992. In *Molecular Biology of Free Radical Scavenging Systems*, ed. JG Scandalios, pp. 173–92. Cold Spring Harbor, NY: Cold Spring Harbor
52. Taniguchi M. 1992. *Adv. Clin. Chem.* 29:1–59
53. Omar BA, Flores SC, McCord JM. 1992. *Adv. Pharmacol.* 23:109–61
54. Halliwell B, Gutteridge JMC, Cross CE. 1992. *J. Lab. Clin. Med.* 119:598–620
55. Rice-Evans CA, Diplock AT. 1993. *Free Radic. Biol. Med.* 15:77–96
56. McCord JM. 1993. *Clin. Biochem.* 26: 351–58
57. Beauchamp C, Fridovich I. 1971. *Anal. Biochem.* 44:276–87
58. Misra HP, Fridovich I. 1972. *J. Biol. Chem.* 247:3170–75
59. Greenstock CL, Miller RW. 1975. *Biochim. Biophys. Acta* 396:11–16
60. Marklund S, Marklund G. 1974. *Eur. J. Biochem.* 47:469–74
61. Cohen G, Heikkila R. 1974. *J. Biol. Chem.* 249:2447–52
62. Fridovich I, Handler P. 1958. *J. Biol. Chem.* 233: 1578–80
63. Kuo CF, Mashino T, Fridovich I. 1987. *J. Biol. Chem.* 262:4724–27
64. Gardner PR, Fridovich I. 1991. *J. Biol. Chem.* 266:1478–83
65. Gardner PR, Fridovich I. 1991. *J. Biol. Chem.* 266:19328–33
66. Flint DH, Tuminello JF, Emptage MH. 1993. *J. Biol. Chem.* 268:22369–76
67. Liochev SI, Fridovich I. 1992. *Proc. Natl. Acad. Sci. USA* 89:5892–96
68. Flint DH, Emptage MH, Guest JR. 1992. *Biochemistry* 31:10331–37

69. Liochev SI, Fridovich I. 1994. *Free Radic. Biol. Med.* 16:29–33

70. Carlioz A, Touati D. 1986. *EMBO J.* 5:623–30

71. Nakae D, Yoshiji H, Amanuma T, Kinugasa T, Farber JL, Konishi Y. 1990. *Arch. Biochem. Biophys.* 279:315–19

72. Hiraishi H, Terano A, Razandi M, Sugimoto T, Harada T, Ivey KJ. 1992. *J. Biol. Chem.* 267:14812–17

73. Farr SB, D'Ari R, Touati D. 1986. *Proc. Natl. Acad. Sci. USA* 83:8268–72

74. Cathcart R, Schwiers E, Saul RL, Ames BN. 1984. *Proc. Natl. Acad. Sci. USA* 81:5633–37

75. Adelman R, Saul RL, Ames BN. 1988. *Proc. Natl. Acad. Sci. USA* 85:2706–08

76. Richter C, Park JW, Ames BN. 1988. *Proc. Natl. Acad. Sci. USA* 85:6465–67

77. Park EM, Shigenaga M, Degan P, Korn TS, Kitzler JW, et al. 1992. *Proc. Natl. Acad. Sci. USA* 89:3375–79

78 Wagner JR, Hu CC, Ames BN. 1992. *Proc. Natl. Acad. Sci. USA* 89:3380–84

79. Beauchamp C, Fridovich I. 1970. *J. Biol. Chem.* 245:4641–46

80. Goscin SA, Fridovich I. 1972. *Arch. Biochem. Biophys.* 153:778–83

81. Cohen G, Heikkila RE. 1974. *J. Biol. Chem.* 249:2447–52

82. Elstner EF, Konz JR. 1974. *FEBS Lett.* 45:18–21

83. Kellogg EW III, Fridovich I. 1975. *J. Biol. Chem.* 250:8812–17

84. Babior B, Curnutte JT, Kipnes RS. 1975. *J. Lab. Clin. Med.* 85:235–44

85. Lown JW, Begleiter A, Johnson D, Morgan AR. 1976. *Can. J. Biochem.* 54: 110–19

86. Cone R, Hasan SK, Lown JW, Morgan AR. 1976. *Can. J. Biochem.* 54:219–23

87. Hodgson EK, Fridovich I. 1976. *Arch. Biochem. Biophys.* 172:202–05

88. Van Hemmen JJ, Meuling WJA. 1977. *Arch. Biochem. Biophys.* 182:743–48

89. Heikkila RE, Cabbat FS. 1977. *Res. Commun. Chem. Pathol. Pharmacol.* 17:649–62

90. McCord JM, Day ED Jr. 1978. *FEBS Lett.* 86:139–42

91. Halliwell B. 1978. *FEBS Lett.* 92:321–26

92. Halliwell B. 1978. *FEBS Lett.* 96:238–42

93. Winterbourn CC. 1979. *Biochem. J.* 182:625–28

94. Gutteridge JMC, Richmond R, Halliwell B. 1979. *Biochem. J.* 184: 469–72

95. Klein SM, Cohen G, Cederbaum AI. 1980. *FEBS Lett.* 116:220–22

96. DiGuiseppi J, Fridovich I. 1980. *Arch. Biochem. Biophys.* 205:323–29

97. Sagone AL, Decker MA, Wells RM, DeMocko C. 1980. *Biochim. Biophys. Acta* 628:90–97

98. Trelstad PL, Lawley KR, Holmes LB. 1981. *Nature* 289:310–12

99. Fridovich SE, Porter NA. 1981. *J. Biol. Chem.* 256:260–65

100. Lown JW, Joshua AV, Lee JS. 1982. *Biochemistry* 21: 419–28

101. Legge RL, Thompson JE, Baker JE. 1982. *Plant Cell. Physiol.* 23:171–77

102. Rowley DA, Halliwell B. 1982. *FEBS Lett.* 138:33–36

103. Gutteridge JMC. 1984. *Biochem. Pharmacol.* 33:3059–62

104. Girotti AW, Thomas JP. 1984. *Biochim. Biophys. Acta* 118:474–80

105. Gutteridge JMC. 1984. *Biochem. J.* 224: 761–67

106. Graf E, Mahoney JR, Bryant RG, Eaton JW. 1984. *J. Biol. Chem.* 259:3620–24

107. Gutteridge JM. 1985. *FEBS Lett.* 185: 19–23

108. Gutteridge JMC, Bannister JV. 1986. *Biochem. J.* 234:225–28

109. Markey BA, Phan SH, Varanii J, Ryan US, Ward PA. 1990. *Free Radic. Biol. Med.* 9:307–14

110. Gutteridge JMC, Maidt L, Poyer L. 1990. *Biochem. J.* 269:169–74

111. Tukeshelashvili LK, McBride T, Spence K, Loeb LA. 1991. *J. Biol. Chem.* 266: 6401–6

112. Egan TJ, Barthakur SR, Aisen P. 1992. *J. Inorg. Biochem.* 48:241–49

113. Yim MB, Chock PB, Stadtman ER. 1990. *Proc. Natl. Acad. Sci. USA* 87: 5006–10

114. Sato K, Akaike T, Kohno M, Ando M, Maeda H. 1992. *J. Biol. Chem.* 267: 25371–77

115. Voest E, Van Faassen E, Marx JJM. 1993. *Free Radic. Biol. Med.* 15:589–95

116. Hodgson EK, Fridovich I. 1975. *Biochemistry* 14:5294–99

117. Hodgson EK, Fridovich I. 1975. *Biochemistry* 14:5299–5303

118. Lynch RE, Fridovich I. 1978. *J. Biol. Chem.* 253:4697–99

119. Weisiger RA, Fridovich I. 1973. *J. Biol. Chem.* 248:3582–92

120. Autor A. 1982. *J. Biol. Chem.* 257: 2713–18

121. Marres CA, Van Loon AR, Oudshoorn P, Van Steeg H, Grivell LA, Slater EC. 1985. *Eur. J. Biochem.* 147:153–61

122. White JA, Scandalios JG. 1987. *Biochim. Biophys. Acta* 926:16–25

123. Ho YS, Crapo JD. 1988. *FEBS Lett.* 229:377–82

124. White JA, Scandalios JG. 1989. *Proc. Natl. Acad. Sci. USA* 86:3534–38

125. Bowler C, Alliotte T, Van Den Bulcke M, Bauw G, Venderkerckhove J, et al.

1989. *Proc. Natl. Acad. Sci. USA* 86: 3237–41

126. Wispé JR, Clark JC, Burhans MS, Kropp KE, Korfhagen TR, Whitsett JA. 1989. *Biochim. Biophys. Acta* 994:30–36

127. Church SL, Grant W, Meese EU, Trent JM. 1992. *Genomics* 14:823–25

128. Scandalios JG. 1992. In *Molecular Biology of Free Radical Scavenging Systems*, ed. JG Scandalios, pp. 117–52. Cold Spring Harbor, NY: Cold Spring Harbor Lab.

129. McCord JM, Fridovich I. 1973. *Photochem. Photobiol.* 17:115–21

130. Jardin WF, Solda MI, Gimenez SMN. 1986. *Sci. Total Environ.* 58:47–54

131. Massey V, Strickland S, Mayhew SG, Howell LG, Engel PC, et al. 1969. *Biochim. Biophys. Acta* 36:891–97

132. Ballou D, Palmer G, Massey V. 1969. *Biochim. Biophys. Acta* 36:898–904

133. Martin JP, Logsdon N. 1987. *Arch. Biochem. Biophys.* 256:39–49

134. Martin J, Colina K, Logsdon N. 1987. *J. Bacteriol.* 169:2516–22

135. Ander S. 1967. *Strahlentherapie* 132:35

136. Cunningham ML, Krinsky N, Giovanazzi SM, Peak MJ. 1985. *J. Free Radic. Biol. Med.* 1:381–85

137. Lippitt B, McCord JM, Fridovich I. 1972. *J. Biol. Chem.* 247:4688–90

138. Makino K, Mossoba MM, Riesz P. 1982. *J. Am. Chem. Soc.* 104:35–37

139. Riesz P, Kondo T. 1992. *Free Radic. Biol. Med.* 13:247–70

140. Babior BM, Kipnes RS, Curnutte JT. 1973. *J. Clin. Invest.* 52:741–44

141. Curnutte JT, Whitten DM, Babior BM. 1974. *N. Engl. J. Med.* 290:593–97

142. Cheson BD, Curnutte JT, Babior BM. 1977. *Prog. Clin. Immunol.* 3:1–65

143. Babior BM. 1978. *N. Engl. J. Med.* 298:659–68

144. Babior BM. 1984. *Blood* 64:959–66

145. Haas A, Goebel W. 1992. *Free Radic. Res. Commun.* 16:137–57

146. Bellavite P. 1988. *Free Radic. Biol. Med.* 4:225–61

147. Marklund SL, Holme E, Hellmer L. 1982. *Clin. Chim. Acta* 126:41–51

148. Marklund SL. 1982. *Proc. Natl. Acad. Sci. USA* 79:7634–38

149. Karlsson K, Marklund SL. 1987. *Biochem. J.* 242:55–59

150. Karlsson K, Lindahl U, Marklund SL. 1988. *Biochem. J.* 256:29–33

151. Adachi T, Kodera T, Ohta H, Hayashi K, Hirano K. 1992. *Arch. Biochem. Biophys.* 297:155–61

152. Hjalmarsson K, Marklund SL, Engstrom A, Edlund T. 1987. *Proc. Natl. Acad. Sci. USA* 84:6340–44

153. Karlsson K, Marklund SL. 1989. *Lab. Invest.* 60:659–66

154. Inoue M, Watanabe N, Morino Y, Tanaka Y, Amachi T, Sasaki J. 1990. *FEBS Lett.* 269:89–92

155. Johansson MH, Deinum J, Marklund SL, Sjöquist PO. 1990. *Cardiovasc. Res.* 24:500–3

156. Sjöquist PO, Carlsson L, Jonason G, Marklund SL, Abrahamsson T. 1991. *J. Cardiovasc. Pharmacol.* 17:678–83

157. Wahlung G, Marklund SL, Sjöquist PO. 1992. *Free Radic. Res. Commun.* 17:41–47

158. Hatori N, Sjöquist PO, Marklund SL, Petrsson SK, Rydén L. 1992. *Free Radic. Biol. Med.* 13:137–42

159. Abrahamsson T, Brandt U, Marklund SL, Sjöquist PO. 1992. *Circ. Res.* 70:264–71

160. Nakazono K, Watanabe N, Matsuno K, Sasaki J, Sato T, Inoue M. 1991. *Proc. Natl. Acad. Sci. USA* 88:10045–48

161. Hansson L, Edlund M, Edlund A, Johansson T, Marklund SL, et al. 1994 *J. Biol. Chem.* 269:5358–63

162. McEuen AR, Hill HAO. 1982. *Planta* 154:295–97

163. Beaman BL, Scates SM, Moring SE, Deem R, Misra HP. 1983. *J. Biol. Chem.* 258:91–96

164. Hong Z, LoVerde PT, Thakur A, Hammarskjöld ML, Rekosh D. 1993. *Exp. Parasitol.* 76:101–14

165. James ER, McLean DC Jr, Pealer F. 1994. *Infect. Immun.* 62:713–16

166. Filice GA, Beaman BL, Krick JA, Remington JS. 1980. *J. Infect. Dis.* 142:432–38

167. Beaman BL, Black CM, Doughty F, Beaman L. 1985. *Infect. Immun.* 47:135–41

168. Britton L, Malinowski DP, Fridovich I. 1978. *J. Bacteriol.* 134:229–36

169. Jurtshuck P, Lin JK, Moore ERB. 1984. *Appl. Environ. Microbiol.* 47:1185–87

170. Baumann P. 1984. *Arch. Microbiol.* 138:170–78

171. Puget K, Michelson AM. 1974. *Biochimie* 56:1255–67

172. Martin JP Jr, Fridovich I. 1981. *J. Biol. Chem.* 256:6080–89

173. Bannister JV, Parker MW. 1985. *Proc. Natl. Acad. Sci. USA* 82:149–52

174. Leunissen J, de Jong W. 1986. *J. Mol. Evol.* 23:250–58

175. Steinman HM. 1982. *J. Biol. Chem.* 257:10283–93

176. Steinman HM. 1985. *J. Bacteriol.* 162:1255–60

177. Beck BL, Tabatai LB, Mayfield JE. 1990. *Biochemistry* 29:372–76

178. Kroll JS, Langford PR, Loynds BM. 1991. *J. Bacteriol.* 173:7449–57
179. Steinman HM. 1987. *J. Biol. Chem.* 262:1882–87
180. Steinman HM, Ely B. 1990. *J. Bacteriol.* 172:2901–10
181. Stabel TJ, Sha Z, Mayfield JE. 1994. *Vet. Microbiol.* 38:307–14
182. Benov LT, Fridovich I. 1994. *J. Biol. Chem.* 269:25310–14
183. Pasternack RF, Barth A, Pasternack JM, Johnson CS. 1981. *J. Inorg. Biochem.* 15:261–67
184. Faraggi M. 1984. In *Oxygen Radicals in Chemistry and Biology*, ed. W Bors, M Saran, D Tait, pp. 419–30. Berlin: de Gruyter
185. Archibald FS, Fridovich I. 1981. *J. Bacteriol.* 145:442–51
186. Archibald FS, Fridovich I. 1981. *J. Bacteriol.* 146:928–36
187. Archibald FS, Fridovich I. 1982. *Arch. Biochem. Biophys.* 215:589–96
188. Faulkner KM, Liochev SI, Fridovich I. 1994. *J. Biol. Chem.* 269:23471–76
189. Deleted in proof
190. Deleted in proof
190a. Riley DP, Weiss RH. 1994. *J. Am. Chem. Soc.* 116:387–88
190b. Hardy MS, Flickinger AG, Riley D, Weiss RH, Ryan US. 1994. *J. Biol. Chem.* 269:18535–40
191. Robertson P, Fridovich SE, Misra HP, Fridovich I. 1981. *Arch. Biochem. Biophys.* 207:282–89
192. Mashino T, Fridovich I. 1987. *Arch. Biochem. Biophys.* 252:163–70
193. Mashino T, Fridovich I. 1987. *Arch. Biochem. Biophys.* 254:547–51
194. Garst J, Stapleton P, Johnston J. 1983. In *Oxy Radicals and Their Scavenger Systems: Proc. Int. Conf. Superoxide and Superoxide Dismutase, 1982*, ed. G Cohen, RA Greenwald, pp. 125–30. New York: Elsevier
195. Cannella C, Berni R. 1983. *FEBS Lett.* 162:180–84
196. Hicks M, Delbridge L, Yue DK, Reeve TS. 1988. *Biochem. Biophys. Res. Commun.* 151:649–55
197. Smith PR, Thornalley PJ. 1992. *Eur. J. Biochem.* 210:729–39
198. Sakurai T, Tsuchiya S. 1988. *FEBS Lett.* 236:406–10
199. Sakurai T, Kimura S, Nakano M, Kimura H. 1991. *Biochim. Biophys. Acta* 177:433–39
200. Bailey AJ, Sims TJ, Avery NC, Miles CA. 1993. *Biochem. J.* 296:489–96
201. Ookawara T, Kawamura N, Kitagawa Y, Taniguchi N. 1992. *J. Biol. Chem.* 267:18505–10
202. Adachi T, Ohta H, Hayashi K, Hirano K, Marklund SL. 1992. *Free Radic. Biol. Med.* 13:205–10
203. Hunt JV, Bottoms MA, Mitchinson MJ. 1993. *Biochem. J.* 291:529–35
204. Van Camp W, Bowler C, Villarroel R, Tsang EWT, Van Montagu M, Inzé D. 1990. *Proc. Natl. Acad. Sci. USA* 87:9903–7
204a. Natvig DO, Imlay K, Touati D, Hallewell RA. 1987. *J. Biol. Chem.* 262:14697–14701
205. Steinman HM. 1992. *Mol. Gen. Genet.* 232:427–30
206. Brehm K, Haas A, Goebel W, Kreft J. 1992. *Gene* 118:121–25
207. Heinzen RA, Frazier ME, Mallavia LP. 1992. *Infect. Immun.* 60:3814–23
208. Chambers SP, Brehm JK, Michael NP, Atkinson T, Minton NP. 1992. *FEMS Microbiol. Lett.* 70:277–84
209. Haas A, Goebel W. 1992. *Mol. Gen. Genet.* 231:313–22
210. Zhu D, Scandalios JG. 1992. *Genetics* 131:803–9
211. Bowler C, Slooten L, Vandenbranden S, de Rycke R, Butterman J, et al. 1991. *EMBO J.* 10:1723–32
212. Reveillaud I, Phillips J, Duyf B, Hilliker A, Kongpachith A, Flemin J. 1994. *Mol. Cell. Biol.* 14:1302–07
213. Deleted in proof
214. Nakayama K. 1994. *J. Bacteriol.* 176:1939–43
215. Imlay JA, Fridovich I. 1992. *J. Bacteriol.* 174:953–61
216. Rech SB, Stateva LI, Oliver SG. 1992. *Curr. Genet.* 21:339–44
217. Storz G, Christman MF, Sies H, Ames BN. 1987. *Proc. Natl. Acad. Sci. USA* 84:8917–21
218. Bilinski T, Krawiec Z, Liczmanski A, Litwinska J. 1985. *Biochem. Biophys. Res. Commun.* 130:533–39
219. van Loon APGM, Pesold-Hurt B, Schatz G. 1986. *Proc. Natl. Acad. Sci. USA* 83:3820–24
220. Gralla EB, Valentine JS. 1991. *J. Bacteriol* 173:5918–20
221. Liu XF, Elashvili I, Gralla EB, Valentine JS, Lapinskas P, Culotta VC. 1992. *J. Biol. Chem.* 267:18298–18302
222. Phillips JD, Campbell SD, Michaud D, Charbonneau M, Hilliker AJ. 1989. *Proc. Natl. Acad. Sci. USA* 86:2761–65
223. Rosen DR, Siddique T, Patterson D, Figlewicz DA, Sapp P, et al. 1993. *Nature* 362:59–62
224. Denq HX, Hentati A, Tainer JA, Iqbal Z, Cayabyab A, et al. 1993. *Science* 261:1047–51
225. Bowling AC, Schulz JB, Brown RH Jr, Beal MF. 1993. *J. Neurochem.* 61:2322–25

226. Rothstein JO, Bristol LA, Hosler B, Brown RH Jr, Kunal RW. 1994. *Proc. Natl. Acad. Sci. USA* 91:4155–59
227. Gurney ME, Pu H, Chiu AY, Dal Canto MC, Polchow CY, et al. 1994. *Science* 264:1772–74
228. Ischiropoulos H, Zhu L, Chen J, Tsai M, Martin JC, et al. 1992. *Arch. Biochem. Biophys.* 298:431–37
229. Beckman JS, Ischiropoulos H, Zhu L, van der Woerd M, Smith C, et al. 1992. *Arch. Biochem. Biophys.* 298:438–45
230. Furchgott RF, Zawadski JV. 1980. *Nature* 288:373–76
231. Palmer RMJ, Ferrige AG, Moncada S. 1987. *Nature* 327:524–26
232. Gryglewski RJ, Palmer RM, Moncada S. 1986. *Nature* 320:454–56
233. Rubanyi GM, Vanhoutte PM. 1986. *Am. J. Physiol.* 250:H822–27
234. Katusic ZS, Vanhoutte PM. 1989. *Am. J. Physiol.* 257:H33–37
235. Bielski BHJ, Arudi RL. 1985. *Inorg. Chem.* 24:3502–4
236. Huie RE, Padmaja S. 1993. *Free Radic. Biol. Med.* 18:195–99
237. Radi R, Beckman JS, Bush KM, Freeman BA. 1991. *J. Biol. Chem.* 266:4244–50
238. Radi R, Beckman JS, Bush KM, Freeman BA. 1991. *Arch. Biochem. Biophys.* 288:481–87
239. Zhu L, Gunn C, Beckman JS. 1992. *Arch. Biochem. Biophys.* 298:452–57
240. Denicola A, Rubbo H, Rodriguez D, Radi R. 1993. *Arch. Biochem. Biophys.* 304:279–86
241. Kargalioglu Y, Imlay JA. 1994. *Free Radic. Biol. Med.* 15:472 (Abstr.)
242. Greenberg JT, Monach P, Chou JH, Josephy PD, Demple B. 1990. *Proc. Natl. Acad. Sci. USA* 87:6181–85
243. Tsaneva IR, Weiss B. 1990. *J. Bacteriol.* 172:4197–4205
244. Greenberg JT, Chou JH, Monach PA, Demple B. 1991. *J. Bacteriol.* 173:4433–39
245. Privalle CT, Fridovich I. 1993. *J. Biol. Chem.* 268:5178–81
246. Tardot B, Touati D. 1993. *Mol. Microbiol.* 9:53–63
247. Beaumont MD, Hassan HM. 1993. *J. Gen. Microbiol.* 139:2677–84
248. Friedman DI. 1988. *Cell* 55:545–54
249. Compan I, Touati D. 1993. *J. Bacteriol.* 175:1687–96
250. Bagg A, Neilands JB. 1987. *Microbiol. Rev.* 51:509–18
251. Sharrocks A, Green J, Guest J. 1991. *Proc. R. Soc. London Ser. B* 245:219–26
252. Hidalgo E, Demple B. 1993. *EMBO J.* 13:138–46
253. Privalle CT, Fridovich I. 1992. *J. Biol. Chem.* 267:9140–45
254. Beyer WF Jr, Fridovich I. 1991. *J. Biol. Chem.* 266:303–8
255. Hassan HM, Moody CS. 1984. *FEMS Microbiol. Lett.* 25:233–36
256. Moody CS, Hassan HM. 1984. *J. Biol. Chem.* 259:12821–25
257. Pugh SY, Fridovich I. 1985. *J. Bacteriol.* 162:196–202
258. Touati D. 1988. *J. Bacteriol.* 170:2511–20
259. Privalle CT, Kong SE, Fridovich I. 1993. *Proc. Natl. Acad. Sci. USA* 90:2310–14

Annu. Rev. Biochem. 1995. 64:113–39

SELECTIN-CARBOHYDRATE INTERACTIONS AND THE INITIATION OF THE INFLAMMATORY RESPONSE

Laurence A. Lasky

Department of Immunology, Genentech Inc., South San Francisco, California 94080

KEY WORDS: inflammation, adhesion, leukocyte, endothelium, lectin, carbohydrate, mucin

CONTENTS

ABSTRACT

The orderly migration of various white blood cell types to inflammatory sites is a highly regulated process that involves a diversity of adhesion and signaling molecules. This cellular influx is initiated by relatively low affinity interactions that allow for leukocytes to roll along the vascular surface. This rolling phenomenon is mediated by adhesive interactions between lectin containing adhesion molecules, termed selectins, on both the vascular endothelium and leukocytes, and carbohydrate ligands immobilized on mucin-like scaffolds. This adhesion allows for a rapid recognition of various cell types under the conditions of vascular flow, with the result that inflammatory cells are specifically decelerated adjacent to sites of inflammation. This review focuses on the various biochemical aspects of the interactions between the selectins and

113

their cognate carbohydrate ligands, with an emphasis on the importance of these adhesive events to the inflammatory response.

Introduction

The highly regulated migration of white blood cells, or leukocytes, from the blood vascular system to regions of pathogenic exposure is one of the most important aspects of the immune system (1). Why this event (termed the inflammatory cascade) perplexes researchers is clarified when one ponders the physical forces the leukocyte must overcome as it emigrates from the blood vasculature into the tissue site of inflammation. For example, the rate of bulk flow of blood cells in the postcapillary venule, the most common site of leukocyte egress, is approximately 2 mmsec^{-1}. This high velocity allows for leukocytes to rapidly cover vast distances in the organism, but it also results in a rush of the cells past the inflammatory site, making the efficient migration of the white cell into the appropriate tissue locale difficult. In addition to this rapid flow rate, enormous shear forces that are engendered by both the fluid and cellular fractions of the blood are constantly bombarding the leukocyte with a degree of kinetic energy that inhibits its ability to adhere to the vessel wall and emerge at the site of pathogenic invasion.

While such problems as rapid blood-flow rate and high vascular shear forces would at first seem insurmountable, they can be overcome by various cell to cell adhesion events that allow for specific binding of the leukocyte to the endothelium of the vessel that is adjacent to the inflammatory insult (2, 3). The importance of these leukocyte-endothelium adhesion systems is highlighted by two classes of naturally occurring human mutants, one of which is described in detail below, that target different aspects of the inflammatory adhesion cascade. In both cases, these mutations can result in life threatening infections and other abnormalities of the immune system. The leukocytes of people bearing these mutants neither adhere appropriately to vessel walls nor migrate to sites of initial infection (4, 5). These unfortunate patients evidence the prodigious importance of the various leukocyte-endothelium adhesion pathways to the immunological well-being of the organism. In addition, the profoundly anti-inflammatory phenotypes of these genetic diseases suggest a mode of therapy: Inhibition of one or more of these adhesion systems by drugs might impede the deleterious migration of leukocytes that damage tissue in the many abnormal inflammatory conditions.

Researchers have demonstrated that four families of vascular adhesion molecules are involved in the migration of leukocytes during the inflammatory response (3–6). The integrin family is a group of heterodimeric proteins consisting of a large diversity of α and β chains that are expressed on the leukocyte surface in a partially cell-dependent manner. Various extracellular signaling molecules can modulate their cell-surface expression levels and adhesion/activation states. Members of this large family of heterodimeric proteins bind with high avidity to a diversity of extracellular matrix ligands, such as fibronectin or laminin, as well as to a family of immunoglobulin-like endothelial cell-surface proteins termed the immunoglobulin superfamily—the second kind of vascular adhesion proteins involved with inflammation. The functions of the leukocyte integrins and their endothelium-localized immunoglobulin superfamily ligands have been described in a number of excellent reviews and are not discussed further here. The selectin family, and specialized carbohydrates displayed by the sialomucin adhesion family, represent the other set of receptor-ligand pairs critically involved with the inflammatory cascade (7–9). The binding interactions mediated by the recognition of sialomucin-presented carbohydrates by the lectin domains of the selectins appear to be of lower avidity than those of the integrin-immunoglobulin superfamily—a finding consistent with their hypothesized roles, described in greater detail below, as adhesive initiators of the inflammatory cascade (10). This review focuses on cell biological and biochemical aspects of the selectin-sialomucin adhesion proteins, with emphasis on how lectin-carbohydrate binding interactions affect the functions of these molecules during the initiation of the inflammatory response.

Lymphocyte–High Endothelial Venule (HEV) Binding: Early Aspects of Carbohydrate-Mediated Cell Adhesion

Investigators had proposed a variety of potential physiological roles for protein-carbohydrate interactions in both plants and animals, but until the early 1980s, functional evidence for such interactions was essentially lacking. At this time, Rosen, Stoolman, and colleagues, who were studying the mechanisms of traffic between the blood vasculature and the lymph nodes, provided the first indication that the adhesive interactions between these white cells and the specialized high endothelial venules (HEV) of peripheral lymph nodes (PLN) involved protein-carbohydrate recognition. In this early work, these investigators demonstrated, using in vitro frozen-section adhesion assays (11), that the binding of lymphocytes to the HEV of PLN could be efficiently blocked by a number of monomeric, anionically charged sugars, such as mannose-6-phosphate (12). These investigators also demonstrated a degree of carbohydrate specificity, noting that other similarly charged sugars were unable to block the cell adhesive interactions. An initial criticism of these data was that this blocking might be an artefact: Only very high concentrations (~mM) of monomeric charged sugars blocked these adhesive events. The polyvalent binding interactions between the lymphocyte and the HEV may simply have required more complex carbohydrates for efficient inhibition.

These same investigators later demonstrated that a specific subset of polyvalent, anionic sugars, such as fucoidin (a polymer of fucose-4-sulfate) and yeast cell wall polyphosphomannan ester (PPME), could block these adhesive interactions at much lower concentrations (effectively ~nM) (13). In vitro and in vivo studies of the effects of the enzyme sialidase on lymphocyte-HEV binding provided further evidence for a role of carbohydrate recognition in these adhesive events (14, 15). These studies demonstrated that removal of sialic acid, an acidic (anionic) carbohydrate found on the nonreducing ends of many cell-surface glycoconjugates, resulted in a profound decrease of lymphocyte adhesion to the PLN endothelium. These data were thus consistent with the possibility that lymphocytes utilized protein-anionic carbohydrate interactions to adhere to the endothelium of PLN during their migration from blood to the lymphoid compartment.

The molecular nature of these protein-carbohydrate interactions became clearer when Rosen and colleagues demonstrated the nature of the lymphocyte cell-surface molecule that was involved with the recognition of the HEV-localized carbohydrates. Earlier work had demonstrated that a monoclonal antibody (Mab), termed MEL 14, which was directed against a ~90 kD leukocyte cell-surface glycoprotein antigen, effectively blocked the binding of these cells to the HEV in vitro (16). Because MEL 14 recognized a monomeric glycoprotein, it appeared that this leukocyte surface antigen was not an integrin. Since all integrin-like adhesion molecules require a heterodimeric complex for cell-surface expression and function. In addition, this Mab was also capable of inhibiting the migration of lymphocytes to PLN in vivo—further substantiating a role for the antigen recognized by this Mab in lymphocyte adhesion and trafficking (17). Because the antigen recognized by MEL 14 appeared to be involved with the specific homing of lymphocytes to PLN, it was termed the homing receptor. Because the lymphocyte migration to the PLN appeared to involve the homing receptor and appeared to be mediated by protein-carbohydrate interactions, it was possible that the MEL 14 antigen was a carbohydrate-binding protein. Researchers confirmed this possibility when they established that the binding of the carbohydrate PPME to the lymphocyte cell surface could be specifically inhibited by the MEL 14 Mab (18, 19). Because PPME and MEL 14 had previously been shown to block cell adhesion to the HEV of PLN, these data were consistent with the possibility that the homing receptor was a carbohydrate binding protein, or lectin, which mediated lymphocyte trafficking to the PLN by the recognition of a carbohydrate(s) specifically expressed by the HEV of these lymphoid organs. The concept that nonintegrin-type molecules such as lectins could mediate the specific adhesion of cells by the recognition of carbohydrates remained only an appealing hypothesis until the techniques of molecular biology uncovered three different lectin-like adhesion molecules.

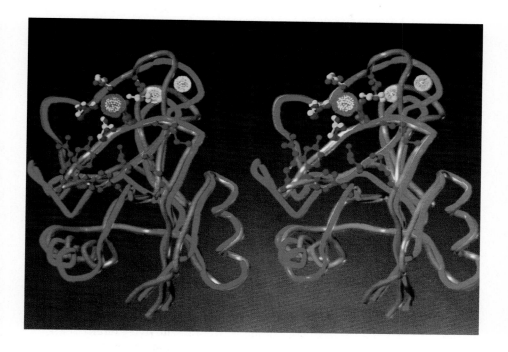

Figure 2 Stereoview of the crystal structures of the lectin domains of the mannose binding protein (MBP) and E selectin. The main chain is illustrated as magenta for the MBP and cyan for E selectin. Side chains are highlighted that have been shown to be involved with carbohydrate recognition and/or cell adhesion as well as with coordination of the single conserved calcium ion. For MBP, residues E185, N187, E193 and N205 interact with mannose and are shown in yellow. For E selectin, residues Y48, N82, N83, E92, Y94, R97, K111 and K113 have been shown to be critical for sLe[x] recognition and cell adhesion and are depicted in red. Calcium ligands that are not involved with carbohydrate recognition (E80, N105 and D106) are shown as blue side chains. The calcium ion positions in MBP are shown as green dots surrounded by orange spheres, while the conserved calcium found in both the MBP and E selectin is shown as a green sphere containing a yellow sphere.

Molecular Cloning of the Three Members of the Selectin Family of Cell Adhesion Molecules

Work performed in two other systems suggested additional cell adhesive events that do not involve the integrins. The production of Mabs directed against cell-surface molecules on cultured endothelial cells activated with inflammatory mediators such as tumor necrosis factor or IL 1 resulted in reagents capable of blocking the adhesion of granulocytic leukocytes to these activated cells (20–22). Characterization of the antigen recognized by these blocking Mabs demonstrated that it was an activation-specific ~110-kD endothelial surface glycoprotein, termed the endothelial leukocyte adhesion molecule 1 (ELAM 1) that appeared to mediate the specific adhesion of neutrophils and monocytes to inflamed endothelium. Because this glycoprotein is monomeric, these data suggested that it was not an integrin but an unfamiliar type of adhesion molecule. Concurrently, the production of Mabs against the surface molecules of thrombin-activated platelets resulted in antibodies that recognized a ~140-kD platelet glycoprotein that appeared to mediate the adhesion of neutrophils to these cells (23–26). It was also demonstrated that this glycoprotein, termed granule membrane protein 140 (gmp140) or platelet activation dependent granule external membrane protein (PADGEM), was stored in the alpha granules of platelets as well as in the Weibel-Palade bodies of endothelial cells (27, 28). In the case of the endothelial form of gmp140/PADGEM, thrombin activation of the endothelial cells resulted in the surface expression of this adhesion protein, presumably by the rapid fusion of the storage granules with the plasma membrane of the endothelial cells (29). More recently, it has also been shown that the expression of this protein, like that of ELAM 1, can also be transcriptionally activated in the endothelium by inflammatory mediators (30). Because this glycoprotein appeared to be monomeric, the molecular characterization suggested that gmp140/PADGEM was not an integrin but a novel type of cell adhesion molecule.

Molecular cloning and sequence analysis of the cDNAs encoding the MEL 14 antigen, ELAM 1, and gmp140/PADGEM revealed that these three adhesive proteins were members of a novel family of cell adhesion molecules (Figure 1) (31–35). Perhaps most exciting was the discovery that these three proteins contained domains at their N-termini that were related to a large superfamily of proteins termed the type-C lectin family (36). While the functions of many of the previously described type-C lectins were unclear, it was probable that many of these lectin domains were involved in the recognition of various carbohydrate moieties. Thus, especially in the case of the homing receptor (37–38), the cDNA sequencing data suggested that this lectin domain was likely to be involved with cell adhesion through the recognition of carbohydrates expressed on opposing cells. In addition to the N-terminal lectin

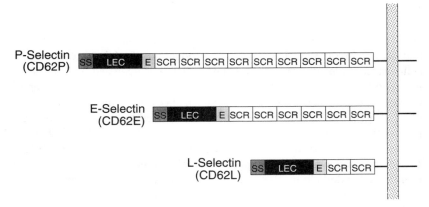

Figure 1 The selectin family of adhesion proteins. Illustrated are the three members of the selectin family, L selectin (CD62L), which is expressed on various leukocytes, E selectin (CD62E), which is expressed on endothelium activated by inflammatory mediators, and P selectin (CD62P) which is stored in alpha granules of platelets and Weibel-Palade bodies of endothelial cells and is also expressed on endothelium activated by inflammatory stimuli. The various motifs represented are: signal sequence (SS), calcium dependent (type C) lectin domain (LEC), epidermal growth factor-like domain (E), short consensus repeat (complement binding protein) domain (SCR). Not highlighted are a transmembrane anchoring motif and a short cytoplasmic tail.

domain, all three of these adhesive proteins contained an adjacent domain that was homologous to the superfamily of epidermal growth factor (egf) motifs. All three of the proteins also contained variable numbers of an extracellular motif that was homologous to those found in various complement binding (CB) proteins. In the case of the homing receptor, there were two identical copies (at the DNA level in the murine homing receptor) of these motifs. The ELAM 1 contained six motifs, and gmp140/PADGEM contained either eight or nine, presumably due to differential mRNA splicing.

More recent work has demonstrated that the number of these complement binding motifs in ELAM 1 and gmp140/PADGEM is species specific. Species specificity suggests a less than critical role for the absolute number of these motifs. Finally, all three proteins contain hydrophobic membrane anchoring domains and short cytoplasmic tails—observations consistent with the fact that they are all type-1 transmembrane proteins. In addition, a differentially spliced form of gmp140/PADGEM was described that lacked the transmembrane anchoring motif, a lack that suggested a potential secretory form of this adhesion protein. While one report in the literature suggested that there was a phosphotidyl-inositol (PI) lipid-linked form of the human homing receptor (39), analysis of the exon structure of the human (40) and murine (41) homing-receptor genes, as well as treatment of MEL 14 positive cells with PI-phospholipase C (S Stachel & L Lasky, unpublished data), suggests that this form of the homing receptor was probably due to a cloning artefact. The overall sequence homology of the lectin and egf domains among these three glycoproteins was both relatively high (~65%) and therefore consistent with their importance in carbohydrate ligand recognition (see below); the homology of the CB motifs and the transmembrane and intracytoplasmic domains was less well conserved (~40%).

These data suggested that these three adhesion molecules may have evolved from a single progenitor sequence by gene duplication; both genomic analysis and chromosomal localization studies were consistent with this possibility. Thus, all three of these genes mapped to the long arm of human and murine chromosome 1 within an approximately 200-kbp region (42). This genomic analysis also demonstrated that the selectin gene family was assembled by exon shuffling, because each protein domain was encoded on a separate exon (41–44). Similar exons, even at the level of splice-site interruption, also encode such domains when they are expressed in other protein contexts. This suggests that the lectin, egf, and CB domains were evolutionarily shuffled throughout the genome such that they assemble novel proteins (45).

This hypothesis is further strengthened by the finding that the Mel 14, ELAM 1, and gmp140/PADGEM genes also map adjacent to a genomic region that encodes a number of functional complement binding proteins that contain a region homologous to the CB region contained in the lectin-containing adhesion molecules. Thus, this position of the genome may have been rich in these types of CB exons even before the lectin and egf domains were shuffled in. In order to reflect the importance of the lectin domains and the selective nature of cell adhesion mediated by these proteins, investigators recently agreed (46) to call these lectin-containing adhesion molecules L selectin or CD62L (homing receptor, MEL 14 antigen, Leu 8, etc), E selectin or CD62E (ELAM 1), and P selectin or CD62P (gmp140/ PADGEM).

Carbohydrate Recognition by the Selectins

The pioneering work on potential carbohydrate interactions in the L selectin (homing receptor) system, together with the N-terminal lectin domains that were identified by cDNA sequencing, suggested that all three selectins might mediate cell adhesion through the recognition of carbohydrates. Various antibody studies supported this hypothesis because they demonstrated that monoclonal antibodies against the lectin domains of these adhesion molecules blocked cell adhesion (47–50). In addition, it was also established that cell adhesion mediated by these proteins was completely dependent upon calcium, a finding consistent with previous data demonstrating that carbohydrate recognition by the type C lectins was calcium dependent. Finally, work in the homing receptor system, as well as in gmp140/PADGEM, coupled with the high degree of homology that was found between the various selectins (particularly the high level of cationic lysine residues), indicated that sialic acid, a negatively charged carbohydrate, might be critical for cell adhesion mediated by all three selectins (14, 15, 51, 52). These data thus set the stage for the discovery of the nature of the carbohydrates recognized by the various selectins.

Simultaneous work from a number of laboratories demonstrated the characteristics of the minimal carbohydrate epitope recognized by the selectins. One approach demonstrated that antibodies directed against a previously described myeloid cell-surface lactosaminoglycan carbohydrate antigen, the sialyl Lewis x or sLe^x antigen (NeuNacα2-3Galβ1-4(Fuc1-3)GlcNac; Neu-Nac: sialic acid, Gal: galactose, Fuc: fucose and GlcNac: N acetyl glucosamine), were capable of blocking cell adhesion mediated by both E and P selectin (48, 53, 54). Cell adhesion to activated endothelium that was mediated by either of these two proteins could be blocked with liposomes containing a glycolipid form of the sLe^x carbohydrate. Using a very different approach, other investigators provided further support for a role of a fucosylated carbohydrate in the adhesion mediated by E selectin. These groups demonstrated that cells could bind to E selectin following the transfection of a cDNA encoding a fucosyl transferase that added fucose to a polylactosamine backbone in a specific linkage (α1-3/4) to the GlcNac residue (55–57). Finally, a third group of investigators isolated an E selectin binding glycolipid from myeloid cells and demonstrated, by mass spectrometry, that this adhesive glycolipid contained a sialylated, fucosylated compound that was related to a previously determined carbohydrate termed VIM 2 (58). While the VIM 2 carbohydrate was related to sLe^x, the fucose group was attached to the penultimate GlcNac rather than the last GlcNac, and further work demonstrated that the VIM 2 antigen was a weak, probably nonphysiological ligand for selectin-mediated adhesion (59, 60). More recently, it has been demonstrated that both the sialic acid and the fucose residues were absolutely required for carbohydrate recognition by E selectin (59). This result was consistent with previous data indicating that sialic acid was unequivocally required for adhesion mediated by both L and P selectin. It has also recently been demonstrated, using an immobilized glycolipid in a solid-phase binding assay, that all three selectins bind with differing affinities to sLe^x. This binding agrees with the hypothesis that this is the minimal oligosaccharide epitope for selectin-mediated adhesion (61). An isomeric carbohydrate with a structure related to sLe^x, termed sLe^a (NeuNacα2-3Galβ1-3(Fuc1-4)GlcNac), can also act as a selectin ligand in in vitro binding assays (59, 62). Two-dimensional NMR analysis of these two different selectin ligands suggests that these carbohydrates can adopt related conformations in solution with both the carboxylate of the sialic acid and the fucose pointing in the same general direction, a suggestion consistent with the hypothesis that both of these oligosaccharides may act in selectin-mediated carbohydrate recognition. Other ligands, such as LNF III (CD15), a nonsialylated form of sLe^x (63), sulfatide (64), an uncharacterized fraction of heparin (65); and sulfated forms of Lewis^x (sulfoLe^x) (66, 67) have also shown selectin binding activity. The physiological significance of these interactions remains to be investigated, especially in light of the sialic acid dependence demonstrated for the adhesion mediated by all selectins. In summary, these data supported the hypothesis that an sLe^x or sLe^x-like oligosaccharide could function as an adhesive ligand for selectin binding. Although, it should be emphasized that none of these studies prove definitively that sLe^x alone is sufficient to mediate high-avidity binding of leukocytes under the conditions of vascular flow.

While the sLe^x saccharide appeared to represent at least part of the ligand that was recognized by the selectins, work in the L selectin system suggested that a higher level of complexity is involved with the carbohydrate recognized by this lectin. Work using an L selectin immunoglobulin chimeric molecule (L selectin-IgG), which recognized L selectin ligands produced by the HEV of PLN, demonstrated that two proteins of molecular mass 50 kD and 90 kD could be precipitated by this chimera in an apparently carbohydrate-dependent manner (68, 69). That these proteins, which are described in greater detail below, were highly labeled by radioactive inorganic sulfate was consistent with an older observation by Ford and colleagues, who found that the HEV incorporated large quantities of inorganic sulfate and secreted a portion of the labeled material (70). Subsequent to the discovery of these sulfated molecules, researchers demonstrated that the ability of L selectin to interact with these proteins was dependent upon the sulfate residues (71). In these experiments, inhibition of the production of the high-energy sulfate precursor 3'-phosphoadenosine 5'-phosphosulfate (PAPS) by chlorate resulted in a loss of sulfate incorporation into the 50-kD and 90-kD proteins. These undersulfated glycoproteins would no longer interact in precipitation analyses with the L selectin-IgG chimera. This failure of interaction is consistent with a role for these sulfates in high-avidity recognition by the lectin domain of L selectin.

Further work demonstrated that the sulfated carbohydrates found on these two glycoprotein ligands were modified forms of sLe^x. Thus, biochemical analysis of degraded sulfated oligosaccharides (72) and comparison with various standard sugars revealed that two sulfated carbohydrates contained the following probable structures: NeuNacα2-3(SO$_4$-6)Galβ1-4(Fuc1-3)GlcNac and NeuNacα2-3Galβ1-4(Fuc1-3)(SO$_4$-6)GlcNac (73, 74). These uniquely modified forms of sLe^x, together with the requirement for the sulfate in high-affinity recognition of these glycoprotein ligands by L selectin, suggested that supplementary modifications of sLe^x-like carbohydrates could confer additional levels of binding affinity and/or specificity on the adhesive events mediated by the selectins.

Of course, the complex nature of the carbohydrate ligands for the selectins suggests that a diversity of glycosylating enzymes are involved with the production of these oligosaccharides. Because fucose and sialic acid appear to be critical functional components of the selectin oligosaccharide ligands, a great deal of work has focused on the enzymes that attach these components to the lactosaminoglycan backbone. Researchers have described many sialyl transferases and cloned three of them. It is not currently clear which, if any, of the three cloned enzymes is used for the production of the selectin carbohydrate ligands, but at least one of them, termed ST3O or ST-4, does not sialylate the appropriate precursor disaccharide and therefore cannot be involved with the production of selectin ligands (75).

The situation with the fucosyltransferases is a bit more complex. Five fucosyltransferases have so far been cloned, although until recently it was unclear which of these enzymes was responsible for fucosylation of the sLe^x determinant in myeloid cells (55, 76–81). It was originally proposed that the myeloid fucosyltransferase, Fuc-T IV or ELAM-1 ligand fucosyltransferase (ELFT), was the enzyme that produced sLe^x, but subsequent investigations failed to confirm this hypothesis (56, 76). Thus, other work demonstrated that Fuc-T IV production of an sLe^x-like carbohydrate that could act as an E-selectin ligand was cell specific and suggested that part of the confusion regarding this enzyme's role in sLe^x synthesis was due to the use of different cells or cell lines for adhesion assays (82).

The description of the fucosyltransferase Fuc-T VII (81) has recently eliminated the confusion regarding the role in myeloid sLe^x synthesis of Fuc-T IV (ELFT) versus other potential fucosylating enzymes. This enzyme has all of the characteristics expected for the catalyst that is involved in myeloid sLe^x production. These include the specific expression of this enzyme in neutrophils. (Fuc-T IV has a broader cellular distribution.) In addition, Fuc-T VII appears to produce an sLe^x that is recognized by a number of anti-sLe^x antibodies. In contrast, Fuc-T IV produced a carbohydrate that was only recognized by a subset of anti-sLe^x antibodies. Finally, Fuc-T VII appeared to endow a diversity of cells with the ability to bind to E selectin, while Fuc-T IV's ability to induce E selectin–dependent binding was highly cell specific. Taken together, these data provide strong evidence that Fuc-T VII is the myeloid-specific fucosyltransferase involved with the production of sLe^x.

However, as described below, the carbohydrate recognition situation is more complex. The various oligosaccharide ligands are presented on protein backbones that also appear to contribute to the binding avidity and specificity of the selectins. Thus, the absence of these glycoprotein scaffolds in many of the experiments described above could explain some of the apparent confusion regarding the actual enzymes involved in myeloid sLe^x synthesis. Finally, the characteristics of the very interesting enzymes involved in the sulfate modifications in the L selectin system remain to be determined (71, 73, 74).

Structural Analysis of the Selectin Lectin Domains

Because of their involvement in the inflammatory cascade, the molecular interactions between carbohydrate ligands and the lectin domains of the selectins are of considerable interest. Preliminary data indicated that several monoclonal antibodies directed against the lectin domains of all three selectins inhibited the adhesion of these molecules to either carbohydrates or cells. In the case of murine L selectin, the epitope recognized by the adhesion-blocking antibody MEL 14 was mapped to within the N-terminal 53 amino acids of the lectin domain (47). Analysis of deletion constructs of both L and E selectin revealed that removal of the adjacent egf domain abolished monoclonal antibody recognition of the lectin domain and suggested that the conformation of the lectin domain was potentially dependent upon interactions with the egf domain (47, 48). In addition, the same types of deletion analyses, together with the mapping of a cross-reacting blocking Mab on L and E selectin, supported a role for the C-terminal CB domains in high-avidity ligand recognition as well, although the mechanisms by which the CB domains support high-avidity binding are currently unknown (83–86). While much of this information suggested that an analysis of structural aspects of the lectin domain might provide information relevant to the mechanisms of carbohydrate recognition by the selectins, such insights were forthcoming only when the crystal structure of a related type C lectin, the mannose binding protein, was determined (87, 88).

E selectin was the first selectin to be analyzed on a structure-function basis. The analysis, which consisted of mutation, Mab binding, sLe^x carbohydrate recognition, and cell adhesion assays, revealed that a relatively small number of residues in the lectin domain appeared to be involved with the binding of adhesion-blocking Mabs, as well as with the recognition of carbohydrates and cell adhesion (49). A number of the residues that appeared to be responsible for the binding of blocking monoclonal antibodies and immobilized sLe^x binding were positively charged residues such as lysine. This observation was consistent with the possibility that electrostatic interactions between sialic acid and these charge side chains were involved with ligand recognition. The mutagenesis data became much more compelling when the E-selectin lectin domain was modeled based upon the crystal structure of the related type C lectin, the mannose binding protein (87, 88). Although the amino acid sequence of the mannose binding protein was only distantly related to other type C lectins, the conservation of a subset of amino acids suggested that its three dimensional structure might be representative of the structures of all type C lectin family members. Thus, the modeling of the E-selectin lectin domain, together with the mutagenesis and activity assays, was felt to be representative of the actual ligand recognition site on this lectin.

This hypothesis was reinforced when the residues that affected the function of E selectin were shown to be clustered in a small, hydrophilic pocket on the face of the lectin domain. The pocket included the single calcium binding site that was conserved between the selectins and the mannose binding protein. Further mutations that were inspired by the modeling data confirmed that this region was involved with carbohydrate recognition and cell adhesion. In addition, mutagenesis of a similar region in P selectin revealed that this same site on the lectin was involved with carbohydrate recognition and cell adhesion mediated by this selectin as well. This mutagenesis suggested that the functional site was conserved—potentially in all three selectins (50, 89–91). As expected, the amino acid residues important for carbohydrate recognition and cell adhesion were conserved in all selectin species studied. Thus, these data were consistent with the hypothesis that a defined site on one face of the E- and P-selectin lectin domains mediated carbohydrate recognition.

The validity of these data was further accentuated when the crystal structure of the lectin and egf domains of E selectin was determined by Graves and colleagues (92). Their structure revealed that the modeling data on E selectin, while not perfect, provided a surprisingly accurate depiction of the structure of this lectin domain. In addition to increasing the accuracy of the previously proposed E selectin structure, the crystal structure of this lectin, together with the crystal structure of the complex between mannose binding protein and mannose, allowed for further investigations of the nature of the carbohydrate recognition site. Analysis of the structure of the cocrystal between the mannose binding–protein lectin domain and mannose revealed that the carbohydrate interacted with a number of side chains and with the calcium at one of the two calcium-binding sites of the protein, the one that is conserved in the selectins (36, 88, 93, 94). This interaction was relatively weak—an observation consistent with the low binding affinity of mannose by the mannose binding protein. However, it also appeared to explain the mechanism of calcium dependence of these interactions, because removal of the calcium would result in a loss of both the structure of the calcium binding site and some of the direct mannose binding interactions.

These data also suggested that fucose could bind to this site in a similar manner, and the involvement of fucose in selectin-mediated carbohydrate interactions suggested that the conserved calcium binding site on the selectins might also be involved with ligand recognition. This proposal was confirmed when mutation to aspartate of two asparagine residues (82, 83) conserved between E selectin and the mannose binding protein resulted in a loss of carbohydrate binding and cell adhesion. While these asparagines are both involved with calcium ligation in E selectin, their mutation to aspartate was not expected to affect calcium binding. It did, however, remove the conserved amino groups that were previously proposed, in the mannose binding protein, to be involved with hydrogen bonding to the equatorial hydroxyls of mannose.

Thus, these data suggested that the face of E (and P) selectin that was previously characterized to be involved with carbohydrate recognition could be expanded to include the calcium site as well as the positively charged, hydrophilic region. Because the E selectin mutagenesis data, together with the crystal structure of the mannose binding protein–mannose sacharride complex, were consistent with the binding of the all-important fucose to the calcium binding site, these data also indicated that the mechanism of calcium dependence for the selectins might be due to a loss of structure and side-chain interactions between the fucose and the single calcium binding site. Finally, these data allowed for a docking model between sLe^x and E selectin that suggested that the sialic acid interacted with the cationic hydrophilic pocket while the fucose interacted with the calcium binding site (Figure 2).

While the mutagenesis, modeling, and crystal structure data on E and P selectin were consistent with the involvement of a relatively small face of the lectin domain with carbohydrate recognition and cell adhesion, other reports suggest that the interaction may be more complex. A series of peptides derived from a number of sites in the P- and E-selectin lectin domains have been shown to block cell adhesion (95, 96). These data were consistent with the involvement of a much larger area of these lectins in the recognition of carbohydrates and cell adhesion. A major conundrum with these peptides is that they cannot possibly be directly involved with carbohydrate recognition. This is because a single amino acid change in the lectin domains of either E or P selectin (for example at Lys 113 or Arg 82) completely abolishes carbohydrate recognition and cell adhesion without disrupting the overall lectin structure, a result logically inconsistent with a small peptide's retaining a carbohydrate recognition function (49, 50, 89, 92). In addition, it is difficult to understand how the calcium dependence of selectin binding can be reconciled with the nature of the inhibitory peptides, because many of these peptides are derived from sites distant (as characterized by modeling and crystal structure) from the calcium binding site. Thus, it is highly likely that the inhibitory activity of these peptides is due to secondary effects and not to the blocking of carbohydrate recognition. One such secondary effect could be the inhibition of selectin oligomerization, though there is currently no direct evidence that the selectins oligomerize on the cell surface.

An additional proposed ligand interaction site is in the egf domain (97, 98). Initial work demonstrated that deletion of the egf domain abolished lectin-specific antibody binding. This early work was consistent with the possibility that the conformation of the lectin domain was affected by the adjacent egf domain (47, 48). Unfortunately, while the crystal structure of the lectin and egf domains of E selectin revealed a number of interactions between these two motifs, the structure did not disclose the mechanism by which these interactions might affect lectin structure (92). A separate report, using the egf-specific Ly 22 antibody, suggested that the egf domain of L selectin might also be directly involved with ligand binding via protein-protein interactions (97). However, subsequent work demonstrated that the Ly 22 antibody efficiently blocked the carbohydrate-dependent recognition of the sulfated 50-kD glycoprotein ligand. Again, this blockage is consistent with the possibility that the egf domain affected the ability of the lectin domain to recognize carbohydrates (98).

Finally, a recent report demonstrates that replacement of the L selectin egf domain with the P selectin domain confers P selectin cell binding specificity on the chimeric protein (98). The authors suggest that the egf domain may be directly involved with cell adhesion and that this involvement may occur through specific protein-protein interactions. An alternative hypothesis is that the P selectin egf domain modulates the carbohydrate recognition specificity of the L selectin lectin domain such that it now recognizes myeloid cell carbohydrates. The latter conjecture is strengthened by the finding that the interactions mediated by this chimeric selectin are carbohydrate dependent and that the P selectin egf domain alone will not mediate binding. The resolution of this confusing set of data may be attained when the P selectin egf-specific ligand on the myeloid cell is isolated and characterized.

Glycoprotein Scaffolds: The Sialomucin Family of Adhesion Proteins

The elucidation of the minimal nature of the selectin carbohydrate ligands, coupled with the analysis of the structure-function aspects of the selectin lectin domains, presented the cell biologist with a conundrum. Analysis of the affinities of the interactions between the selectins and immobilized, essentially monomeric sLe^x-like ligands suggested that these interactions were relatively weak, with avidities in the high micromolar to low millimolar range (61). This low selectin-oligosaccharide avidity was matched by other type C lectins' avidities (36). In addition, the proposed interactions between the lectin domains of these proteins and their ligands suggested that these relatively weak avidities were reflected by the limited number of interactions between the amino acid side chains found in the lectin carbohydrate binding site, and the binding moieties on the saccharides (49, 50, 88, 89, 92–94). When viewed in the context of the role of these adhesion proteins in mediating leukocyte-endothelial cell adherence under the conditions of vascular shear, these relatively low avidities were puzzling. It was not clear how interactions that appeared to occur with such low avidity could be tenacious enough to mediate efficient binding of the leukocyte in the blood vessel. It thus seemed evident that other mechanisms must have evolved to insure that these adhesive interactions could occur in the presence of the high shear forces encountered in the blood vascular system.

The initial breakthrough in deciphering the mechanisms by which the avidities of the selectins for carbohydrate ligands might be increased came from investigations of the nature of the endothelial ligands recognized by L selectin. As described above, L selectin adhered to sialylated carbohydrate ligands expressed specifically on the HEV of PLN. These ligands, as characterized by the L selectin-IgG chimeric probe, appeared to be sulfated glycoproteins of approximately 50 kD and 90 kD (69). Initial work on the 50-kD sulfated glycoprotein suggested that most carbohydrate side chains on this protein were of the O-linked variety and appeared to be heterogeneous in size (69, 72). In addition, it appeared that this O-linked glycoprotein was rapidly shed from the endothelium, at least in organ cultures (69, 100). The nature of this glycoprotein was clarified by the molecular cloning of the cDNA encoding the murine and rat forms of the protein (101, 102). The sequence revealed that this glycoprotein was a small, highly O-glycosylated mucin-like molecule. Mucins are glycoproteins that contain predominately O-linked carbohydrate side chains that are attached to serine or threonine residues (103). In addition, most of the molecular weight of these mucins appears to be due to these O-linked carbohydrates. In the case of the 50-kD sulfated glycoprotein, approximately 70% of the molecular weight could be accounted for by carbohydrate. The protein sequence did not contain a transmembrane anchoring domain.

Thus, despite the fact that this mucin bound to L selectin with high avidity and in a carbohydrate-dependent manner, there did not appear to be a direct interaction between the mucin and the endothelial cell-surface. This lack of interaction suggested that either the mucin bound indirectly to the lumenal surface of the endothelium to mediate adhesion or that it functioned as a soluble molecule. Similarly, a differentially glycosylated, non-L selectin-binding form of this mucin was also expressed by mammary epithelial cells during lactation and was found in the soluble whey fraction of milk (104). Because the PLN HEV form of the 50-kD mucin appeared to be a glycosylation-dependent ligand for L selectin, it was named GlyCAM 1 (glycosylation dependent adhesion molecule 1).

The molecular nature of the 50-kD ligand was consistent with the possibility that the avidity of L selectin for carbohydrates is increased by a clustered presentation of the saccharides by a mucin-like scaffold. The characterization of the sulfated 90-kD L selectin ligand further supported this hypothesis. In contrast to the novel sequence that was found for GlyCAM 1, sequencing of the purified 90-kD L selectin ligand demonstrated that it was identical to a previously characterized mucin named CD34 (105). The structure of CD34 was similar to that of GlyCAM 1 in that it contained a large N-terminal serine-

and threonine-rich O-linked domain that was highly glycosylated (106). In contrast to GlyCAM 1, CD34 contained a classic transmembrane anchoring motif, and it was demonstrated that the protein could only be extracted from cells with non-ionic detergents. Thus it appeared likely that, unlike GlyCAM 1, CD34 encoded a membrane-immobilized L selectin ligand that presented clustered carbohydrate ligands to L selectin in order to mediate endothelial adhesion under conditions of flow. While GlyCAM 1 expression was restricted to the HEV of PLN and mesenteric lymph nodes, and to mammary epithelial cells during pregnancy and lactation, CD34 was globally expressed at a diversity of endothelial sites (107). CD34 has also been shown to be expressed by hematopoietic progenitor cells in adults as well as throughout embryonic development (107, 108). In addition, the glycosylation of CD34 differs according to the tissue in which it is expressed. For example, the CD34 expressed in PLN HEV contains the sulfated carbohydrate recognized by both L selectin and the Mab MECA 79 (69, 105). In contrast, the CD34 expressed in other, nonlymphoid sites, including hematopoietic progenitors, does not contain this unique oligosaccharide (107). Thus, it is likely that the function of this mucin is highly dependent upon the types of carbohydrate modifications it undergoes during its transit through the secretory pathway of the cell.

Analysis of P selectin–dependent adhesion suggested that a mucin-like scaffold may be involved with cell adherence mediated by this selectin as well. Initial studies demonstrated that transfection of a fucosyl transferase into CHO cells induced surface sLe^x expression and allowed for P selectin binding to occur, but they also showed that this binding was of low affinity and nonsaturable (57). The binding of P selectin to myeloid cells, by contrast, was saturable and of high affinity (~70 nM) (109). These data implied that a glycoprotein scaffold may have presented sLe^x-related carbohydrate ligands to P selectin in a unique conformation. Subsequent work demonstrated a single, ~120-KD dimerizable neutrophil surface protein that bound to P selectin with high affinity and in an apparently carbohydrate-dependent manner (110, 111). Most of the ~120-kD protein was resistant to N glycanase, an observation consistent with the possibility that it contained O-linked carbohydrates. This supposition was supported by an analysis of the ~120-kD protein with a unique protease, O-sialoglycoprotease, that specifically cleaved highly O-linked, mucin-like molecules (112, 113). The finding that the ~120-kD P selectin ligand was completely sensitive to this protease was consistent with the supposition that it was a mucin-like molecule.

Treatment of myeloid cell lines, such as HL60 cells, with the O-sialoglycoprotease completely abolished P selectin binding to the cells but had no apparent effect on the levels of sLe^x expressed on the cell surface. These data were consistent with the hypothesis that most of the sLe^x on the cell surface is neither capable of high-affinity binding to P selectin nor associated with the

~120-kD P selectin ligand. This important result suggested that the presentation of the carbohydrate in the appropriate manner by glycoprotein scaffolds is what mediates high-affinity binding of the selectins. In addition, this result suggested that the many publications that purported to demonstrate ligand activity of sLe^x-containing glycoproteins were probably reporting in vitro artefacts of binding whose avidity was to be too low to be physiologically relevant (114).

As with the L selectin mucin ligands, the molecular nature of the ~120-kD P selectin ligand was characterized by molecular cloning. Isolation of a cDNA that induced high-avidity P selectin–dependent binding on cells cotransfected with a fucosyltransferase cDNA revealed that the P selectin ligand was a large mucin that contained a transmembrane domain (115). The mucin nature of the cloned P selectin ligand supported the O-sialoglycoprotease experiments mentioned above and suggested that the mucin-like P selectin ligand was probably the major high-affinity binding site on the myeloid surface for this selectin. In contrast to the nonrepetitive mucin structure of GlyCAM 1 and CD34, the P selectin mucin ligand contained 15 tandemly repeated domains that contained high levels of threonine. Presumably, these repeated threonine-rich domains represented the O-linked sites of oligosaccharide ligand presentation by this mucin, and the tandem nature of these O-linked sites suggested that P selectin recognition of the mucin may be different in some unknown way from L selectin recognition of GlyCAM 1 and CD34. The highly specific ability of this particular mucin to present sLe^x-type carbohydrates to P selectin was underlined by the finding that transfection of another mucin-like molecule, CD43, into cells, together with the fucosyltransferase cDNA, did not induce P selectin adhesive characteristics. At least one report suggested that this same mucin could interact with E selectin as well (see below). Because of the P selectin ligand nature of this mucin, it was named the P Selectin Glycoprotein Ligand 1 (PSGL 1).

Finally, there are likely to be other physiological ligands for the selectins that may or may not have mucin-like characteristics. A recently described ligand for $\alpha_4 \beta_7$ integrin binding has been shown to be an immunoglobulin superfamily member that has been named MadCAM 1 (mucosal addressin cell adhesion molecule 1) (116, 117). This endothelial ligand contained both immunoglobulin-like domains, which appear to mediate integrin recognition, as well as a small serine- and threonine-rich site that was conjectured to have a mucin-like structure. Other work had demonstrated that the mesenteric lymph node form of MadCAM 1 contained the sulfated carbohydrate side chains that were recognizable by both L selectin and the Mab MECA 79. Such work suggested that this mucin-like domain could serve to present oligosacharides to L selectin.

This supposition was demonstrated in an in vitro system of lymphocyte rolling where the appropriately glycosylated MadCAM 1 was coated onto a capillary and shown to mediate adhesion under conditions of fluid shear that mimicked those found in the blood vasculature (118). Even though this endothelial protein is apparently sulfated, it was not detected in the L selectin-IgG precipitation analyses of Imai et al (69). Since these experiments only analyzed the highest-affinity binding interactions, it is possible that the mucin domain of MadCAM 1 presents carbohydrates in a low-affinity conformation that may only be relevant under reduced shear conditions.

The glycoprotein ligands for E selectin remain essentially uncharacterized at the molecular level. Initial work demonstrated that the binding of P selectin to myeloid cells was sensitive to a broad array of proteases, while the binding of E selectin was essentially protease resistant (119). These data suggested a difference between the glycoprotein ligands for P and E selectin, and an analysis revealed at least two novel murine myeloid cell-surface ligands recognized by E selectin. One of these ligands appeared to interact with P selectin as well (120). The commonly recognized ligand seemed to be homologous to the PSGL 1 described, while the other E selectin-specific ligand, a glycoprotein of ~150 kD appeared to present N- and not O-linked carbohydrates to this selectin. In addition, a P selectin–specific glycoprotein ligand of ~160 kD, which also appeared to present N-linked carbohydrates, has been described (121). The nature of these selectin-specific, nonmucin glycoprotein ligands will only be determined when they are molecularly cloned. Finally, there also appears to be a specific glycoprotein ligand for E selectin binding on bovine γδ T cells whose selectin adhesion is sialic acid dependent (122). Because all of the currently cloned selectin ligands are mucins that present sialylated oligosaccharide ligands to the selectins, I propose that this family of adhesion molecules be termed the sialomucin family (Figure 3).

How might the sialomucin family of adhesion glycoproteins mediate higher-avidity binding of the selectins under conditions of vascular flow? The rigid rod-like structure that has previously been proposed for other mucins (103) would likely provide for an extended, relatively inflexible matrix upon which the oligosaccharide ligands could be presented above the cellular glycocalyx to the rapidly passing cell-associated selectins. In addition to providing for this extended rod-like scaffold, it also seems likely that these mucins might present the oligosaccharide ligands to a clustered array of cell-surface selectins in a specifically arranged manner. The predominant reason for this hypothesis is that neither a single nonmucin like sLe^x-containing glycoprotein nor the majority of similarly glycosylated mucin-like proteins (such as CD43) is capable of acting as high-affinity ligands for selectin binding under the conditions of vascular flow or using soluble binding assays that probably select for the highest-avidity interactions.

While the evidence for selectin clustering is limited, the fact that deletion

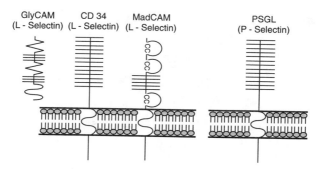

| GlyCAM | CD 34 | MadCAM | PSGL |
| (L - Selectin) | (L - Selectin) | (L - Selectin) | (P - Selectin) |

Figure 3 The sialomucin family of adhesion proteins. Illustrated are the four current members of this mucin-like family. The cross-hatched regions refer to highly O-glycosylated, mucin-like domains that present clustered arrays of carbohydrates to the lectin domains of the selectins. GlyCAM (glycosylation dependent cell adhesion molecule) is a soluble molecule, found in blood and milk, that appears to bind to L selectin. CD34 is a vascular mucin that binds to lymphocyte L selectin when the mucin is expressed on the high endothelial venules of lymph nodes. CD34 is expressed at all vascular sites in adults and embryos and is also found on hematopoietic progenitor cells, though the function of this mucin at these nonlymphoid sites is not known. MadCAM (mucosal addressin cell adhesion molecule) is a hybrid adhesion protein that contains immunoglobulin-like domains (shown as disulfide-bonded loops) and a mucin-like domain that presents carbohydrates to lymphocytes. PSGL (P selectin glycoprotein ligand) is a mucin found on the surface of myeloid cells and activated lymphocytes that present carbohydrates to endothelial P selectin.

of the cytoplasmic domain of L selectin results in a mutant adhesion lectin that is unable to mediate leukocyte rolling in vivo could be interpreted to indicate that L selectin–cytoskeletal interactions are involved in appropriate selectin clustering or cell-surface localization (123). The latter possibility is also suggested by immunoelectron-microscopic observations showing that L selectin is expressed specifically on microvillar projections on the surface of neutrophils (114). There are at least two mechanisms by which this specific clustering could be attained. In the first, each of the clustered selectins recognizes an individual oligosaccharide side chain that is expressed in an appropriately clustered manner by the sialomucin. These data would suggest a one-to-one relationship between the selectin and the appropriately presented oligosaccharide side chain. The avidity of the interaction would be greatly increased by the clustering of both interacting members. Indeed, the molecular modeling and crystallographic analyses mentioned above support a single oligosaccharide binding site on each selectin lectin domain.

A second model suggests that a unique conformational patch of oligosaccharides is formed by the novel clustering of carbohydrate side chains by the mucin scaffold (112). In this case, a number of saccharide chains would combine to construct a particular sugar face that would bind with higher affinity

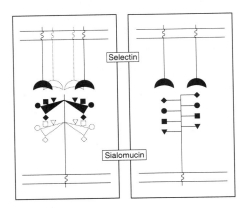

Figure 4 Two models for carbohydrate presentation by the sialomucin-like adhesion molecules. The left model illustrates the concept of homomeric clustering. Here, each O-linked carbohydrate side chain presents a complete ligand to the lectin domain of the selectin in a one-to-one manner. The increase in avidity is accomplished by clustering of the selectins in a specific manner that allows for enhanced binding. On the right is illustrated the concept of heteromeric clustering. In this model, the sialomucin presents a unique carbohydrate patch by positioning a number of different carbohydrate entities in a unique manner. In this way, a novel configuration of carbohydrates is presented to a single lectin domain in a manner that greatly increases the binding avidity of the interaction. This type of model would not necessarily require that the selectins be clustered to increase the avidity of the interaction.

to each individual selectin lectin domain. While the latter hypothesis is appealing in its novelty, it seems more likely that the appropriately clustered presentation of single oligosaccharide chains to a clustered array of selectins greatly increases the avidity of the interaction and allows it to occur under the remarkably rigorous conditions of vascular flow. This is particularly true in the case of the sulfated sLe^x-like L selectin ligand where it has been shown that a single oligosaccharide chain contains all of the apparently appropriate components for L selectin recognition (73, 74). A schematic version of these two models is illustrated in Figure 4.

Selectins In Vivo: Rolling and Drug Development

While a survey of the role of selectins during the inflammatory response is beyond the scope of this review, one should note that a great deal of information has accumulated regarding the function of selectins as initiators of inflammation. For example, a number of studies using Mabs against L, E, and P selectin have suggested that these adhesion molecules are intimately involved with a diversity of acute and chronic inflammatory conditions in the lung, gut, ear (extremities), myocardium, and pancreas (for examples, see 124–130).

In addition, immunoglobulin chimeric molecules containing the various selectins have also been used to demonstrate the involvement of these lectins in ear reperfusion, thioglycollate-induced peritonitis, and acute lung injury (131, 132). A number of experiments have demonstrated that creation of null mutants in either L, E, or P selectin in mice using homologous gene knockout technology results in animals with profound defects in leukocyte influx into the irritated peritoneum (133). An interesting ex vivo study has demonstrated a possible role for P selectin in the recruitment of lymphocytes to rheumatoid joints. The result of the study is interesting in the context of the demonstration of the expression of the P selectin ligand (possible PSGL) on T cells isolated from these inflamed areas (134).

Finally, and perhaps most compelling, there is a human disease, termed leukocyte adhesion deficiency 2 (LAD 2), that appears to illustrate in a very powerful manner the importance of selectin-carbohydrate interactions to the inflammatory response (5). In this condition, so far reported in only two cases, a metabolic defect (currently unknown) results in a complete lack of fucose production. In addition to a number of physical and mental problems these patients suffer from recurrent infections, a phenomena that was also a hallmark of the integrin deficiency disease originally termed LAD 1 (4). As expected from the lack of fucose in these patients, analysis of their neutrophils revealed a complete absence of the selectin oligosaccharide ligand sLe^x, and these neutrophils were also unable to bind to immobilized E selectin. Such defects may explain the inability of these patients to mount appropriate inflammatory responses to external pathogens. Taken together, this diversity of in vivo results underlines the importance of the adhesive interactions mediated by the selectins during the inflammatory response.

A number of elegant studies have highlighted the apparent in vivo role of selectins in the inflammatory process. Over a century ago, several investigators, using in vivo intravital microscopy of frog mesenteric venules, found that leukocytes rolled, or marginated, along the vessel wall during an inflammatory response. More recent work has demonstrated that this phenomenon is mediated by the adhesive interactions between selectins and their carbohydrate ligands. For example, an in vitro analysis of the interactions of leukocytes with immobilized P selectin in a parallel-plate laminar-flow chamber demonstrated that the immobilized selectin could mediate the rolling of the white cells under shear conditions that mirrored those proposed for the postcapillary venule (135). The leukocytes did not roll on immobilized ICAM 1, a ligand for one of the leukocyte integrins, but a combination of immobilized P selectin and ICAM 1 resulted in both rolling and firm adhesion in response to neutrophil activation by the chemotactic factor f-met-leu-phe (FMLP). These results suggested that the selectins mediated the initial low-affinity adhesion margination episode, after which, in response to chemotactic factors, the integrins mediated the high-avidity binding and extravasation events. These in vitro phenomena were recapitulated by even more compelling in vivo intravital-microscopic studies demonstrating that L selectin interaction with carbohydrates, perhaps presented by endothelial CD34, also mediated rolling along the endothelium adjacent to an inflammatory insult (136–139).

Further intravital-microscopic studies have revealed that P selectin can also mediate rolling of leukocytes along postcapillary venules. This mediation suggests a function for at least two of the selectins in the initiation of the inflammatory response via this lower-affinity interaction (140). It is likely that the deceleration events mediated by the lower-avidity rolling-type adhesion between selectins and their oligosaccharide ligands allow for the white cell to specifically sense gradients of chemotactic factors such as the chemokines, activated complement products, or bacterially derived factors. This gradient detection ultimately results in cellular activation via G-coupled receptors (141). It was also recently demonstrated that sLe^x-deficient neutrophils from LAD 2 patients are unable to roll along mesenteric venules when injected into rabbits in vivo. This observation is consistent with the deficiencies of these patients in mounting appropriate neutrophil defensive responses against pathogens (142). Because the selectins are involved with initiating inflammation, at least under the normal shear conditions found in postcapillary venules, by allowing the white cells to roll along the endothelium, it seems likely that the antagonism of the margination response would also inhibit the downstream events of chemotactic activation, integrin mediated firm adhesion, extravasation, and the deleterious effects of inflammation.

Finally, knowledge about selectin function during the inflammatory cascade and about the molecular details of the interactions between the lectin domains and the mucin-immobilized carbohydrate ligands suggests that novel classes of anti-inflammatory drugs may be derived from research on these adhesive glycoproteins. It is only a matter of time until these new drugs are clinically tested. For instance, several reports in the literature have demonstrated that oligosaccharides derived from the sLe^x structure have clear anti-inflammatory activities. Both the sialic acid-containing (sLe^x) and the sulfate (sulfo-Le^x) forms of this oligosaccharide have anti-inflammatory effects in vivo (143–146) (S Swiedler & M Matthay, personal communication).

These small molecules, which are produced by combinations of synthetic and enzymatic approaches, appear to block neutrophil influx into tissue sites in a number of inflammatory conditions. This blockage appears to occur via an inhibition of neutrophil rolling along the endothelium, although this conjecture has not been proven for all inflammatory conditions. This inhibition of rolling might prevent both the influx of white cells into the tissue sites of inflammation and the localized damage to the endothelium by the activated neutrophils. While these compounds so far appear to be extremely safe, many challenges remain before these molecules prove clinically useful—primarily because of their low affinities and short in vivo half-lives. However, there are other interpretations possible for the low in vitro affinities and apparent short half-lives in vivo of this class of compounds. For example, the low affinities in vitro may not be representative of the effectiveness of the compounds in vivo, since many of the in vitro assays are not performed under shear conditions like those found in the vasculature (61). It is likely that this shear, which constantly threatens the white cell's adherence to the endothelium, will make these compounds much more effective in vivo. In addition, the short half-lives may also be a great advantage, because they allow for a higher degree of control of the drugs serum concentrations in vivo. Thus, if the compounds have deleterious effects on the host's ability to inhibit bacterial infections, then the serum level of the drug can be decreased quickly.

Compounds that are not related to sLe^x also appear to have strong anti-inflammatory activity. For example, fucoidin, the polymer of fucose-4-sulfate that has ligand activity for P and L selectin, appears to block neutrophil rolling in vivo and acts as an anti-inflammatory compound in at least one type of animal model (147, 148). Presumably, this highly anionic carbohydrate polymer functions by competitively inhibiting the binding of sialic acid to the cationic site on the lectin domain, though this is at present a purely theoretical argument. While this result appears clinically compelling, such anionic saccharide polymers can often have severe side effects, such as undesirable anticoagulant activity.

A highly phosphorylated form of inositol phosphate also exhibits selectin antagonist activity (149). This compound has remarkably strong antiselectin effects against L and P selectin and has also shown anti-inflammatory efficacy in thioglycollate-induced peritoneal inflammation models. However, the unknown side effects of such compounds may limit the clinical usefulness of these findings. Finally, a recently reported compound whose structure is based upon fucosylated forms of glycyrrhitinic acid has shown modest antiselectin activity in vitro and in vivo (150). This compound was derived from a search of the chemical structural libraries for compounds related to sLe^x, and its activity suggests that other searches or modifications of such compounds may result in other drug leads. In summary, the initial attempts at developing small molecule antagonists based upon the structures of various selectin oligosaccharide ligands has met with some success; further work will result in even better inhibitors of the initial adherence events mediated by this class of adhesion molecules.

Summary

Discovery of the selectins, their carbohydrate ligands, the sialomucin scaffolds that present these ligands, and the in vivo biology of these novel molecules has opened up new vistas for the cell and molecular biologist, carbohydrate chemist, and the clinician interested in inflammation. The data accumulated about these compounds has resulted in new concepts about cell adhesion and the inflammatory cascade as well as the development of novel experimental anti-inflammatory compounds. While the pace of research in this field has been rapid, much of the puzzle remains to be solved. For instance, molecular pictures of the interactions between the selectin lectin domains and their cognate carbohydrate ligands, perhaps as determined by crystal structure analysis, will provide a clearer idea of how to develop better antagonists of these interactions. The physical mechanisms by which the mucin scaffolds present these saccharide ligands may also provide information that allows for the development of higher-avidity inhibitors of inflammation. Finally, analysis of the roles of the selectins in various acute and chronic inflammatory conditions in vivo will help to direct clinical development of such inflammation inhibitors.

ACKNOWLEDGMENTS

I would like to thank Dr. Bradley Graves of Roche for the use of the stereoview of the structure of E selectin, Ms. Allison Bruce for the other figures, and Dr. Olivia K. Loui-Lasky for many helpful discussions.

Any *Annual Review* chapter, as well as any article cited in an *Annual Review* chapter, may be purchased from the Annual Reviews Preprints and Reprints service.
1-800-347-8007; 415-259-5017; email: arpr@class.org

Literature Cited

1. Gallin J, Goldstein I, Snyderman R, eds. 1992. *Inflammation: Basic Principles and Clinical Correlates*. New York: Raven. 2nd ed.
2. Springer TA. 1990. *Nature* 346:425–34
3. Springer TA. 1994. *Cell* 76(2): 301–14
4. Anderson DC, Springer TA. 1987. *Annu. Rev. Med.* 38:175–94
5. Etzioni A, Frydman M, Pollack S, Avidor I, Phillips ML, et al. 1992. *N. Engl. J. Med.* 327:1789–92
6. Hemler ME. 1990. *Annu. Rev. Immunol.* 8:365–400
7. Lasky LA. 1992. *Science* 258(5084): 964–69
8. Bevilacqua MP, Nelson RM. 1993. *J. Clin. Invest.* 91:379–87
9. McEver RP. 1994. *Curr. Opin. Immunol.* 6(1):75–84
10. Butcher EC. 1992. *Adv. Exp. Med. Biol.* 323(181):181–94
11. Stamper HB, Woodruff JJ. 1976. *J. Exp. Med.* 144:828–33
12. Stoolman LM, Rosen SD. 1983. *J. Cell Biol.* 96:722–29
13. Stoolman LM, Tenforde TS, Rosen SD. 1984. *J. Cell Biol.* 99:1535–40
14. Rosen SD, Singer MS, Yednock TA, Stoolman LM. 1985. *Science* 228:1005–7
15. Rosen SD, Chi SI, True DD, Singer MS, Yednock TA. 1989. *J. Immunol.* 142:1895–1902
16. Gallatin WM, Weissman IL, Butcher EC. 1983. *Nature* 303:30–34
17. Mountz JD, Gause WC, Finkelman FD, Steinberg AD. 1988. *J. Immunol.* 140: 2943–49
18. Yednock TA, Stoolman LM, Rosen SD. 1987. *J. Cell Biol.* 104:713–23
19. Yednock TA, Butcher EC, Stoolman LM, Rosen SD. 1987. *J. Cell Biol.* 104: 725–31
20. Bevilacqua MP, Pober JS, Wheeler ME, Cotran RS, Gimbrone MA. 1985. *J. Clin. Invest.* 76:2003–11
21. Cotran RS, Gimbrone MA, Bevilacqua MP, Mendrick DL, Pober JS. 1986. *J. Exp. Med.* 164:661–66
22. Bevilacqua MP, Pober JS, Mendrick

DL, Cotran RS, Gimbrone MA. 1987. *Proc. Natl. Acad. Sci. USA* 84:9238–42

23. Hsu-Lin SC, Berman CL, Furie BC, August D, Furie B. 1984. *J. Biol. Chem.* 259:9121–26
24. McEver RP, Martin MN. 1984. *J. Biol. Chem.* 259:9799–804
25. Larsen E, Celi A, Gilbert GE, Furie BC, Erban JK, et al. 1989. *Cell* 59(2):305–12
26. Geng JG, Bevilacqua MP, Moore KL, McIntyre TM, Prescott SM, et al. 1990. *Nature* 343(6260):757–60
27. Bonfanti R, Furie BC, Furie B, Wagner DD. 1989. *Blood* 73:1109–12
28. McEver RP, Beckstead JH, Moore KL, Marshall-Carlson L, Bainton DF. 1989. *J. Clin. Invest.* 84:92–99
29. Hattori R, Hamilton KK, Fugate RD, McEver RP, Sims PJ. 1989. *J. Biol. Chem.* 264:7768–71
30. Weller A, Isenmann S, Vestweber D. 1992. *J. Biol. Chem.* 267(21):15176–83
31. Siegelman MH, van der Ijn M, Weissman IL. 1989. *Science* 243(4895):1165–72
32. Lasky LA, Singer MS, Yednock TA, Dowbenko D, Fennie C, et al. 1989. *Cell* 56(6):1045–55
33. Tedder TF, Isaacs CM, Ernst TJ, Demetri GD, Adler DA, Disteche CM. 1989. *J. Exp. Med.* 170(1):123–33
34. Bevilacqua MP, Stengelin S, Gimbrone MA, Seed B. 1989. *Science* 243(4895):1160–65
35. Johnston GI, Cook RG, McEver RP. 1989. *Cell* 56(6):1033–44
36. Drickamer K. 1993. *Biochem. Soc. Trans.* 21:456–59
37. Geoffroy JS, Rosen SD. 1989. *J. Cell Biol.* 109:2463–69
38. Imai Y, True DD, Singer MS, Rosen SD. 1990. *J. Cell Biol.* 111(3):1225–32
39. Camerini D, James SP, Stamenkovic I, Seed B. 1989. *Nature* 342(6245):78–82
40. Ord DC, Ernst TJ, Zhou LJ, Rambaldi A, Spertini O, Griffin J, Tedder TF. 1990. *J. Biol. Chem.* 265(14):7760–67
41. Dowbenko DJ, Diep A, Taylor BA, Lusis AJ, Lasky LA. 1991. *Genomics* 9(2):270–77
42. Watson ML, Kingsmore SF, Johnston GL, Siegelman MH, LeBeau MM, et al. 1990. *J. Exp. Med.* 172(1):263–72
43. Collins T, Williams A, Johnston GI, Kim J, Eddy R, et al. 1991. *J. Biol. Chem.* 266(4):2466–73
44. Johnston GI, Bliss GA, Newman PJ, McEver RP. 1990. *J. Biol. Chem.* 265(34):21381–85
45. Patthy L. 1987. *FEBS Lett.* 214(1):1–7
46. Bevilacqua M, Butcher E, Furie B, Furie B, Gallatin M, et al. 1991. *Cell* 67:233

47. Bowen BR, Fennie C, Lasky LA. 1990. *J. Cell Biol.* 110(1):147–53
48. Walz G, Aruffo A, Kolanus W, Bevilacqua M, Seed B. 1990. *Science* 250(4984):1132–35
49. Erbe DV, Wolitzky BA, Presta LG, Norton CR, Ramos RJ, et al. 1992. *J. Cell Biol.* 119(1):215–27
50. Erbe DV, Watson SR, Presta LG, Wolitzky BA, Foxall C, Brandley BK, Lasky LA. 1993. *J. Cell Biol.* 120(5):1227–35
51. True DD, Singer MS, Lasky LA, Rosen SD. 1990. *J. Cell Biol.* 111(6):2757–64
52. Corral L, Singer MS, Macher BA, Rosen SD. 1990. *Biochem. Biophys. Res. Commun.* 172(3):1349–56
53. Phillips ML, Nudelman E, Gaeta FC, Perez M, Singhal AK, Hakomori S, Paulson JC. 1990. *Science* 250(4984):1130–32
54. Polley MJ, Phillips ML, Wayner E, Nudelman E, Singhal AK, et al. 1991. *Proc. Natl. Acad. Sci. USA* 88(14):6224–28
55. Lowe JB, Stoolman LM, Nair RP, Larsen RD, Berhend TL, Marks RM. 1990. *Cell* 63(3):475–84
56. Goelz SE, Hession C, Goff D, Griffiths B, Tizard R, et al. 1990. *Cell* 63(6):1349–56
57. Zhou Q, Moore KL, Smith DF, Varki A, McEver RP, Cummings RD. 1991. *J. Cell Biol.* 115(2):557–64
58. Tiemeyer M, Swiedler S, Ishihara M, Moreland M, Schweingruber H, et al. 1991. *Proc. Natl. Acad. Sci. USA* 88:1138–42
59. Tyrrell D, James P, Rao N, Foxall C, Abbas S, et al. 1991. *Proc. Natl. Acad. Sci. USA* 88(22):10372–76
60. Macher BA, Beckstead JH. 1990. *Leuk. Res.* 14:119–30
61. Foxall C, Watson SR, Dowbenko D, Fennie C, Lasky LA, et al. 1992. *J. Cell Biol.* 117(4):895–902
62. Berg EL, Magnani J, Warnock RA, Robinson MK, Butcher EC. 1992. *Biochem. Biophys. Res. Commun.* 184(2):1048–55
63. Larsen E, Palabrica T, Sajer S, Gilbert GE, Wagner DD, et al. 1990. *Cell* 63(3):467–74
64. Aruffo A, Kolanus W, Walz G, Fredman G, Seed B. 1991. *Cell* 67(1):35–44
65. Norgard-Sumnicht KE, Varki NM, Varki A. 1993. *Science* 261:480–83
66. Yuen CT, Lawson AM, Chai W, Larkin M, Stoll MS, et al. 1992. *Biochemistry* 31(38):9126–31
67. Green PJ, Tamatani T, Watanabe T, Miyasaka M, Hasegawa A, et al. 1992.

Biochem. Biophys. Res. Commun. 188 (1):244–51

68. Watson SR, Imai Y, Fennie C, Geoffroy JS, Rosen SD, Lasky LA. 1990. *J. Cell Biol.* 110(6):2221–29
69. Imai Y, Singer MS, Fennie C, Lasky LA, Rosen SD. 1991. *J. Cell Biol.* 113 (5):1213–21
70. Andrews P, Milsom DW, Ford WL. 1982. *J. Cell Sci.* 57:277–92
71. Imai Y, Lasky LA, Rosen SD. 1993. *Nature* 361(6412):555–57
72. Imai Y, Rosen SD. 1993. *Glycoconjugate J.* 10(1):34–39
73. Hemmerich S, Bertozzi CR, Leffler H, Rosen SD. 1994. *Biochemistry* 33(16): 4820–29
74. Hemmerich S, Rosen SD. 1994. *Biochemistry* 33(16):4830–35
75. Kitagawa H, Paulson JC. 1994. *J. Biol. Chem.* 269(2):1394–1401
76. Kumar R, Potvin B, Muller WA, Stanley P. 1991. *J. Biol. Chem.* 266(32):21777–83
77. Lowe JB, Kukowska LJ, Nair RP, Larsen RD, Marks RM, et al. 1991. *J. Biol. Chem.* 266(26):17467–77
78. Weston BW, Nair RP, Larsen RD, Lowe JB. 1992. *J. Biol. Chem.* 267(6):4152–60
79. Koszdin KL, Bowen BR. 1992. *Biochem. Biophys. Res. Commun.* 187(1): 152–57
80. Easton EW, Schiphorst WE, van Drunen E, van der Schoot CE, van den Eijnden DH. 1993. *Blood* 81(11):2978–86
81. Sasaki K, Kurata K, Funayama K, Nagata M, Watanabe E, et al. 1994. *J. Biol. Chem.* 269(20):14730–37
82. Goelz S, Kumar R, Potvin B, Sundaram S, Brickelmaier M, Stanley P. 1994. *J. Biol. Chem.* 269(2):1033–40
83. Watson SR, Imai Y, Fennie C, Geoffrey J, Singer M, et al. 1991. *J. Cell Biol.* 115(1):235–43
84. Jutila MA, Watts G, Walcheck B, Kansas GS. 1992. *J. Exp. Med.* 175(6): 1565–73
85. Li SH, Burns DK, Rumberger JM, Presky DH, Wilkinson VL, et al. 1994. *J. Biol. Chem.* 269(6):4431–37
86. Bargatze RF, Kurk S, Watts G, Kishimoto TK, Speer CA, Jutila MA. 1994. *J. Immunol.* 152(12):5814–25
87. Weis WI, Kahn R, Fourme R, Drickamer K, Hendrickson WA. 1991. *Science* 254:1608–15
88. Weis WI, Drickamer K, Hendrickson WA. 1992. *Nature* 360:127–34
89. Hollenbaugh D, Bajorath J, Stenkamp R, Aruffo A. 1993. *Biochemistry* 32(12): 2960–66
90. Bajorath J, Hollenbaugh D, King G,

Harte W Jr., Eustice DC, et al. 1994. *Biochemistry* 33(6):1332–39
91. Norgard KE, Han H, Powell L, Kriegler M, Varki A, Varki NM. 1993. *Proc. Natl. Acad. Sci. USA* 90(3):1068–72
92. Graves BJ, Crowther RL, Chandran C, Rumberger JM, Li S, et al. 1994. *Nature* 367(6463):532–38
93. Drickamer K. 1992. *Nature* 360:183–86
94. Iobst S, Wormald M, Weis W, Dwek R, Drickamer K. 1994. *J. Biol. Chem.* 269(22):15505–11
95. Geng JG, Heavner GA, McEver RP. 1992. *J. Biol. Chem.* 267(28):19846–53
96. Heavner GA, Falcone M, Kruszynski M, Epps L, Mervic M, et al. 1993. *Int. J. Pept. Protein Res.* 42(5):484–89
97. Siegelman MH, Cheng IC, Weissman IL, Wakeland EK. 1990. *Cell* 61(4): 611–22
98. Kansas GS, Saunders KB, Ley K, Zakrzewicz A, Gibson RM, et al. 1994. *J. Cell Biol.* 124(4):609–18
99. Imai Y, Lasky LA, Rosen SD. 1992. *Glycobiology* 2(4):373–81
100. Brustein M, Kraal G, Mebius RE, Watson SR. 1992. *J. Exp. Med.* 176(5): 1415–19
101. Lasky LA, Singer MS, Dowbenko D, Imai Y, Henzel WJ, et al. 1992. *Cell* 69(6):927–38
102. Dowbenko D, Watson SR, Lasky LA. 1993. J. Biol. Chem. 268(19):14399–403
103. Jentoft N. 1990. *Trends Biochem. Sci.* 15:291–94
104. Dowbenko D, Kikuta A, Fennie C, Gillett N, Lasky LA. 1993. *J. Clin. Inv.* 92:952–60
105. Baumhueter S, Singer MS, Henzel W, Hemmerich S, Renz M, et al. 1993. *Science* 262:436–38
106. Brown J, Greaves MF, Molgaard HV. 1991. *Int. Immunol.* 3:175–84
107. Baumhueter S, Kyle C, Mebius R, Dybdal N, Lasky LA. 1994. *Blood.* 84: 2554–65
108. Young P, Baumhueter S, Lasky LA. 1995. *Blood.* 85:96–105
109. Ushiyama S, Laue TM, Moore KL, Erickson HP, McEver RP. 1993. *J. Biol. Chem.* 268:15229–37
110. Moore KL, Varki A, McEver RP. 1991. *J. Cell Biol.* 112(3):491–99
111. Moore KL, Stults NL, Diaz S, Smith DF, Cummings RD, et al. 1992. *J. Cell Biol.* 118(2):445–56
112. Norgard KE, Moore KL, Diaz S, Stults NL, Ushiyama S, et al. 1993. *J. Biol. Chem.* 268(17):12764–74
113. Steininger CN, Eddy CA, Leimgruber RM, Mellors A, Welply JK. 1992. *Bio-*

chem. Biophys. Res. Commun. 188(2): 760–66

114. Picker LJ, Warnock RA, Burns AR, Doerschuk CM, Berg EL, Butcher EC. 1991. *Cell* 66(5):921–33

115. Sako D, Chang X-J, Barone KM, Vachino G, White HM, et al. 1993. *Cell* 75(6):1179-86

116. Briskin MJ, McEvoy LM, Butcher EC. 1993. *Nature* 363:461–64

117. Berlin C, Berg EL, Briskin MJ, Andrew DP, Kilshaw PJ, et al. 1993. *Cell* 74(1): 185–95

118. Berg EL, McEvoy LM, Berlin C, Bargatze RF, Butcher EC. 1993. *Nature* 366:695–98

119. Larsen GR, Sako D, Ahern TJ, Shaffer M, Erban J, et al. 1992. *J. Biol. Chem.* 267(16):11104–10

120. Levinovitz A, Muhlhoff J, Isenmann S, Vestweber D. 1993. *J. Cell Biol.* 121: 449–59

121. Lenter M, Levinovitz A, Isenmann S, Vestweber D. 1994. *J. Cell Biol.* 125(2): 471–81

122. Walcheck B, Watts G, Jutila MA. 1993. *J. Exp. Med.* 178:853–63

123. Kansas GS, Ley K, Munro JM, Tedder TF. 1993. *J. Exp. Med.* 177(3):833–38

124. Seekamp A, Till GO, Mulligan MS, Paulson JC, Anderson DC, et al. 1994. *Am. J. Pathol.* 144(3):592–98

125. Mulligan MS, Miyasaka M, Tamatani T, Jones ML, Ward PA. 1994. *J. Immunol.* 152(2):832–40

126. Ma XL, Weyrich AS, Lefer DJ, Buerke M, Albertine KH, et al. 1993. *Circulation* 88(2):649–58

127. Weyrich AS, Ma XL, Lefer DJ, Albertine KH, Lefer AM. 1993. *J. Clin. Inv.* 91:2620–29

128. Winn RK, Liggitt D, Vedder NB, Paulson JC, Harlan JM. 1993. *J. Clin. Invest.* 92(4):2042–47

129. Yang X-D, Karin N, Tisch R, Steinman L, McDevitt HO. 1993. *Proc. Natl. Acad. Sci. USA* 90:10494–98

130. Lefer AM, Ma XL. 1994. *J. Appl. Physiol.* 76(1):33–38

131. Watson SR, Fennie C, Lasky LA. 1991. *Nature* 349(6305):164–67

132. Mulligan MS, Watson SR, Fennie C, Ward PA. 1993. *J. Immunol.* 151:6410–17

133. Mayadas TN, Johnson RC, Rayburn H, Hynes RO, Wagner DD. 1993. *Cell* 74(3):541–54

134. Grober JS, Bowen BL, Ebling H, Athey B, Thompson CB, et al. 1993. *J. Clin. Invest.* 91:2609–19

135. Lawrence MB, Springer TA. 1991. *Cell* 65(5):859–73

136. Ley K, Gaehtgens P, Fennie C, Singer MS, Lasky LA, Rosen SD. 1991. *Blood* 77(12):2553–55

137. Ley K, Tedder TF, Kansas GS. 1993. *Blood* 82(5):1632–38

138. von Andrian UH, Chambers JD, McEvoy LM, Bargatze RF, Arfors KE, Butcher EC. 1991. *Proc. Natl. Acad. Sci. USA* 88:7538–42

139. von Andrian UH, Chambers JDD, Berg EL, Michie SA, Brown DA, et al. 1993. *Blood* 82(1):182–91

140. Dore M, Korthuis RJ, Granger DN, Entman ML, Smith CW. 1993. *Blood* 82:1308–16

141. Tanaka Y, Adams DH, Hubscher S, Hirano H, Siebenlist U, Shaw S. 1993. *Nature* 361(6407):79–82

142. von Andrian UH, Berger EM, Ramezani L, Chambers JD, Ochs HD, et al. 1993. *J. Clin. Invest.* 91:2893–97

143. Mulligan MS, Paulson JC, de Frees S, Zheng ZL, Lowe JB, Ward PA. 1993. *Nature* 364:149–51

144. Mulligan MS, Lowe JB, Larsen RD, Paulson J, Zheng ZL, et al. 1993. *J. Exp. Med.* 178:623–31

145. Buerke M, Weyrich AS, Zheng ZL, Gaeta F, Forrest MJ, Lefer AM. 1994. *J. Clin. Invest.* 93(3):1140–48

146. Nelson RM, Dolich S, Aruffo A, Cecconi O, Bevilacqua MP. 1993. *J. Clin. Invest.* 91(3):1157–66

147. Ley K, Linnemann G, Meinen M, Stoolman LM, Gaehtgens P. 1993. *Blood* 81(1):177–85

148. Granert C, Raud J, Xie X, Lindquist L, Lindbom L. 1994. *J. Clin. Invest.* 93(3): 929–36

149. Cecconi O, Nelson RM, Roberts WG, Hanasaki K, Mannori G, et al. 1994. *J. Biol. Chem.* 269(21):15060–66

150. Rao B, Anderson M, Musser J, Foxall C, Oda Y, et al. 1994. *J. Biol. Chem.* 269(31):19663–66

Annu. Rev. Biochem. 1995. 64:141–69

DNA PROCESSING REACTIONS IN BACTERIAL CONJUGATION

Erich Lanka

Max-Planck-Institut für Molekulare Genetik, Abteilung Schuster, Ihnestrasse 73, D-14195 Berlin, Federal Republic of Germany

Brian M. Wilkins

Department of Genetics, University of Leicester, University Road, Leicester LE1 7RH, United Kingdom

KEY WORDS: conjugative plasmids, horizontal gene transfer, transfer origin, relaxosome, DNA relaxase, rolling circle replication, DNA helicase

CONTENTS

0066-4154/95/0701-0141$05.00

ABSTRACT

Bacterial conjugation is an important source of genetic plasticity. The initiation complex for conjugative transfer of transmissible plasmids—the relaxosome—is a specific DNA-protein structure that has been isolated from cells and reconstituted from purified components in vitro. Complexes containing uncleaved DNA and DNA cleaved at the *nic* site in the origin of transfer (*oriT*) coexist in equilibrium. Relaxase is usually loaded onto *oriT* by accessory DNA-binding proteins. Relaxase catalyzes cleavage of a specific phosphodiester bond at *nic* and becomes covalently linked through a tyrosyl residue to the 5′ terminus of the cleaved strand. Cleaved DNA may be unwound for transfer by a plasmid-encoded helicase. Single-strand transfer is thought to occur by a replicative rolling circle mechanism. Termination of a round of transfer is achieved by the cleaving-joining activity of the relaxase linked to the 5′ end of the transferring strand. Relationships between DNA processing reactions and conjugative interactions of cell envelopes are particularly obscure aspects of the conjugation cycle.

INTRODUCTION

Bacterial conjugation is a specialized process involving unidirectional transfer of DNA from a donor to a recipient cell by a mechanism requiring specific contact. The process is encoded by diverse plasmids and conjugative transposons, constituting the major route for their horizontal transfer. Conjugation systems are remarkable in mediating transfer between a wide range of bacterial genera and, in some cases, from bacteria to fungal and plant cells. These properties make conjugation an important source of genetic plasticity, potentiating changes of clinical, environmental and evolutionary significance (1). Conjugation also provides a cornerstone of bacterial genetics. Not surprisingly therefore, the process has been the subject of many excellent reviews to which we direct the reader.

Biochemical and molecular studies have focused on enterobacteria, reflecting the discovery of conjugation as the plasmid F-encoded process of *Escherichia coli* K-12. Plasmid F and its relatives continue to provide the paradigm. The F system is genetically complex, requiring about 40 transfer (*tra*) genes in a continuous ~33-kb DNA segment. Only four of the Tra proteins participate in the conjugative processing of the plasmid. The majority specify the mating apparatus, which includes the extracellular conjugative pilus and proteins that stabilize aggregates of mating cells (2–6). Substantial advances have come from work with other types of conjugative plasmid, in particular with RP4 and RK2 of the IncPα group (7–9). Study of naturally mobilizable plasmids, especially those of the IncQ group, has also been instructive (8, 10).

A mobilizable plasmid typically carries a limited number of *mob* genes for its own DNA processing in conjugation, but requires a coresident conjugative plasmid to supply other essential functions. The number of plasmids investigated is very extensive but can be rationalized, since members of any one incompatibility (Inc) group generally determine the same or closely related conjugation systems. A classification based on Inc groups is given in Table 1 to indicate relationships of plasmids featured in this article.

The general model of enterobacterial conjugation states that cells are brought into surface contact by retraction of the conjugative pilus on the donor cell. Resulting cellular interactions trigger the opening of a specific plasmid strand cleaved at a unique "nick" site (*nic*) in the origin of transfer (*oriT*). The open strand is then thought to be unwound by a DNA helicase and transferred with a 5′ to 3′ polarity to the recipient cell. Finally, the ends of the transferring strand are sealed to produce a circular DNA molecule. Single-stranded (ss) DNA transfer is normally associated with replication mediated by DNA polymerase III holoenzyme to generate a replacement strand in the donor cell and a complementary strand in the recipient (51).

Conjugative elements have been found in many families of Gram-positive and -negative bacteria. Examples include plasmids of agrobacteria (52), enterococci (31, 53), staphylococci (33, 54), streptococci (54), streptomyces (55), as well as conjugative transposons (56, 57). Some conjugative transposons are as small as ~16 kb, indicating that their transfer systems are considerably less complex than those of typical enterobacterial plasmids. Mobilizable insertion elements containing their own *oriT* and at least one *mob* gene constitute yet another class of transmissible unit. Examples are NBUs; these are small chromosomal segments of *Bacteroides* spp., which can be excised and mobilized by the large tetracycline-resistance conjugative transposons native to the genus (49, 50, 58).

Much remains to be learned about the conjugative elements found in organisms other than enterobacteria and it must not be assumed that the general model of conjugation pertains to all. The potential diversity of mechanisms is illustrated by the fact that while enterobacterial conjugation commonly involves a plasmid-encoded pilus (5), no such surface appendage has been detected in Gram-positive systems. However, comparisons of nucleotide and amino-acid sequences provide increasing evidence that conjugation systems are related through a series of overlapping networks. One intriguing relationship is the homology between components of the conjugation system of IncP plasmids and the Vir system of the Ti plasmid (9, 59–61). pTi is a large plasmid (~200 kb) of the plant pathogen *Agrobacterium tumefaciens,* which induces tumors through the introduction of a distinct plasmid sector, called T-DNA (20–30 kb), into the nuclear genome of the plant. The transfer process is mediated by the plasmid virulence genes (*vir*) that extend over ~30 kb (60).

Table 1 Properties of transmissible elements

Element	Inc group[a]	Type of nick region	Relaxase	Original host	Reference
Conjugative plasmids					
F	FI	F	TraI	*Escherichia coli*	11
P307	FI	F		*Escherichia coli*	12
R1	FII	F		*Salmonella paratyphi*	13
R100	FII	F	TraI	*Shigella flexneri*	14, 15
ColB4-K98	FIII	F		*Escherichia coli*	14
pED208	FIV	F		*Salmonella typhi*	16
CollB-P9	I1	P		*Shigella sonnei*	17
R64	I1	P	NikA	*Salmonella typhimurium*	18, 19
pCU1	N	F		*Salmonella sp.*	20
R46	N	F		*Salmonella typhimurium*	21
pKM101	N	F		*Salmonella typhimurium*	22
RK2	Pα	P	TraI	*Klebsiella aerogenes*	23
RP4	Pα	P	TraI	*Pseudonomas aeruginosa*	24, 25
R751	Pβ	P	TraI	*Klebsiella aerogenes*	24, 26
R388	W	F	TrwC	*Escherichia coli*	27
pTiC58		P (LB-RB)[b];	VirD2	*Agrobacterium tumefaciens*	28, 29
		Q (*oriT*)	TraA		30
pAD1				*Enterococcus faecalis*	31
pGO1		Q		*Staphylococcus aureus*	32
pSK41				*Staphylococcus aureus*	33
pIP501		Q		*Streptococcus agalactiae*	34
Mobilizable plasmids					
R1162	Q	Q	MobA	*Pseudomonas aeruginosa*	35
RSF1010	Q	Q	MobA	*Escherichia coli*	36, 37
CloDF13				*Enterobacter cloacae*	38
ColA				*Citrobacter freundii*	39
ColE1			MbeA	*Escherichia coli*	40, 41
ColK				*Escherichia coli*	42
pACYC184				*Escherichia coli*	43
pBFTM10				*Bacteroides fragilis*	44
pSC101		Q		*Salmonella panama*	45
pTF1		Q		*Thiobacillus ferrooxidans*	46
pTF-FC2		P		*Thiobacillus ferrooxidans*	47
Conjugative transposons and mobilizable insertion elements					
Tn*4399*			MocA	*Bacteroides fragilis*	48
NBU1				*Bacteroides* sp.	49
NBU2				*Bacteroides* sp.	50

[a] Only incompatibility groups defined in *E. coli* are shown. Many nonconjugative plasmids have not been assigned to a named group.
[b] LB, RB: left or right border sequence of T-DNA

In addition, pTi encodes a bacterial conjugation system called Tra, which mediates transfer of the entire plasmid between agrobacteria (52).

This review addresses recent advances in the biochemistry of DNA processing reactions fundamental to conjugation. It focuses on plasmids of the F, P, and Q groups, together with the Ti plasmid. Topics include the structure of *oriT* sites and proteins interacting with these sites in the initiation and termination of a round of DNA transfer. An essential structure is the relaxosome. This is a complex of the transfer origin in supercoiled DNA and specific plasmid-encoded proteins, which releases an open circular DNA form as the presumptive transfer intermediate. An important component of the relaxosome is relaxase, the enzyme mediating a cycle of strand cleavage and joining at *nic*. Other aspects of conjugative DNA metabolism, including systems promoting establishment of the transferred plasmid in the new host cell, have recently been discussed elsewhere (62).

ORIGIN OF TRANSFER (*oriT*) SITES

Definitions

The DNA strand transferred to the recipient cell is named the T-strand and nucleotide sequences provided here refer to this strand. The position of the phosphodiester bond cleaved at *nic* is indicated ▼ where known. The region 5′ of *nic* is referred to as upstream; it is also described as the trailing region of a transferring strand, since it is theoretically the last portion of a monomer to arrive in the recipient cell. Conversely, the region 3′ of *nic* is described as downstream and as the leading region. Here we refer to reactions at *nic* as cleaving-joining, but they are also known as nicking-ligation.

Nature of oriT Regions

The transfer origin is the only cis-acting site required for DNA transfer, and is readily defined by its ability to convert a nontransmissible vector into a mobilizable plasmid. It has been demonstrated that *oriT*s are located in the transfer gene complex, comprising intergenic portions of up to 500 bp (Table 1). Extensive nucleotide sequence similarities are found only between the *oriT* regions of closely related plasmids, such as those of the F complex, the IncP group or the IncQ group. However, significant similarities are detectable between the nick regions of different transmissible plasmids (63, 64). The nick region, defined as the recognition site for relaxase, is a short stretch of up to 10 nucleotides. Three groups of nick region are identifiable (Figure 1). In each, conserved nucleotides extend for a few nucleotides upstream or around *nic*. Although the nick regions of the three groups differ considerably by sequence, it cannot be ruled out they are structurally analogous.

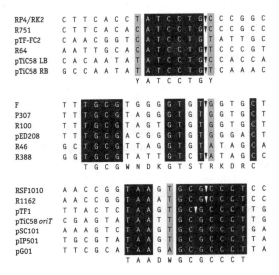

Figure 1 Alignment of nick region sequences. *nic*-sites are indicated ▼ if known. Data for R388 were provided by F de la Cruz (personal communication). The sequences shown for pGO1 and pIP501 were given by GL Archer (personal communication), and A Wang & FL Macrina (personal communication), respectively. Identical nucleotides are boxed and a consensus is shown below each sequence block. References are given in Table 1.

Structural analogies may exist between *oriT*s, plasmid vegetative replication origins and (+)-strand origins of ssDNA phages, because some have a rudimentary form of the IncP *oriT* nick region in common (YAWCYTG; ref. 9, 65). The related origin sequences are known or thought to involve the rolling circle type of replication.

Transfer origins have several features in common. They:

1. map asymmetrically with regard to the transfer (*tra*) genes such that the majority of the *tra* genes enter the recipient last;
2. have an AT content higher than flanking regions, probably facilitating strand separation in negatively supercoiled DNA;
3. have extensive secondary structure conferred by direct and indirect sequence repetitions. In F and RP4 these structural elements are known to function as recognition sites for specific DNA-binding proteins;
4. contain intrinsic bends as potential protein-binding sites to alter the *oriT* structure locally, resulting in easier access of proteins to the nick region;
5. include promoters for *tra* gene expression, which in many cases cause divergent transcription from *oriT*.

We discuss the organization of *oriT* regions on specific enterobacterial plasmids in the context of relaxosomes (see section on Relaxosomes). The Ti plasmid has two types of transfer origin. One is the *oriT* of the Tra system, which belongs to the RSF1010 group (ref. 30, Figure 1). The second is comprised of two 25-bp direct imperfect repeats, the T-DNA borders (28, 29). Any DNA between these borders is transferred to the plant cell by the Vir system. Thus, nononcogenic derivatives of the Ti plasmids, in which most of the internal sequences of the T-DNA have been replaced, are widely used as vectors for genetic transformation of plant cells. The T-DNA borders are differentially utilized since transfer has been shown to be polar in starting from the right border (66).

Information is also available on the *oriT* regions of conjugative elements of Gram-positive bacteria (Table 1). Sequence comparisons have been instructive in showing that the *oriT* portion of the staphylococcal pG01 plasmid (GL Archer, personal communication) and of the broad host range streptococcal pIP501 plasmid (A Wang & FL Macrina, personal communication) contain a nick region very similar or identical to the consensus for the RSF1010 family (Figure 1). These findings are important in showing that the transfer mechanisms of at least some Gram-positive conjugative plasmids are fundamentally similar to Gram-negative counterparts.

RELAXOSOMES

Cleavage at oriT

A common assumption in bacterial conjugation is that DNA is transferred as a specific single strand, the T-strand. It is stressed that this feature has been demonstrated directly only for conjugation directed by F (67–69), R538-1 (IncFII), and R64 (IncI1) (70). Recent evidence on the efficiency of extrachromosomal recombination of T-DNA sequences in plant cells suggests that T-DNA is likewise transferred as a single strand (71). Moreover, only for F has it been shown that the T-strand is transferred in the 5' to 3' direction (67, 68). Despite the limited range of plasmids that have been analyzed in this respect, parallels between different plasmids at the level of protein-*oriT* interactions provide convincing evidence that ssDNA transport in the 5' to 3' direction is a unifying property of the several transfer systems considered in this article.

Generation of the transmissible single strand from a plasmid DNA involves specific cleavage of the T-strand at *nic* within a DNA-protein complex called the relaxosome (24). Treatment of relaxosomes with protein-denaturing agents disrupts the complexes by releasing an open circular DNA that is an ideal substrate for sequencing (Table 1; Figure 1). A common conclusion from these studies is that one specific phosphodiester bond is cleaved between two different nucleotides in *oriT* (in most cases R/Y or Y/R; see Figure 1). This resembles the arrangement at the (+)-strand origin of replication of ssDNA phages and certain Gram-positive bacterial plasmids that replicate via asymmetric rolling circles (9). Sequencing from both sides of *nic* confirmed that in F, RP4, and RSF1010, *nic* consists of two contiguous nucleotides (26, 72–78).

Chemically the cleaved DNA specimens offer an unmodified 3' hydroxyl terminus at *nic*, which is susceptible to elongation in vitro by DNA polymerase I of *E. coli* (79). Pioneering studies on ColE1 demonstrated that a 60-kDa polypeptide is tightly associated with the 5' end at *nic* (79–81). The polypeptide is probably the ColE1 *mob3* (*mbeA*) gene product of 57,895 Da (40, 82). The 5' terminus of specifically cleaved F, RP4, and RSF1010 DNA is likewise modified. The covalently linked protein is the relaxase specified by F *traI*, RP4 *traI* and RSF1010 *mobA* (75, 76, 78). In the latter two cases, the side chain hydroxyl of tyrosine RP4 TraI Y22 (83, 84) and RSF1010 MobA Y24 (85) were identified to form a phosphodiester linkage with the 5' phosphate of the corresponding terminal nucleotide of the T-strand. Convincing evidence has also been obtained that VirD2 of the Ti plasmid is the cleavage enzyme in T-DNA processing. Y28 is apparently the active center amino acid (86).

Predictions have been made for other systems from alignments of conserved motifs in the initiator proteins of diverse replicons that replicate via the rolling circle mode. These studies suggest that in F TraI the reactive amino acid of the relaxase in also a tyrosine (Y80) located in the N-terminal region of the protein (87). Other possible tyrosines are Y16/17 and Y23/24 located very close to the N-terminus of TraI (6). The two sets of doublet tyrosines in TraI of F are conserved in TrwC of R388 but there is no equivalent for the TraI Y80 position in TrwC (88). These studies of different transfer systems support the model that the T-strand is covalently linked at its 5' terminal nucleotide at *nic* to a plasmid-encoded protein, the relaxase. The covalent complex functions as a transfer intermediate.

Multiprotein Complexes at oriT

A relaxosome is the initiation complex of the DNA transfer process present in the donor in an isolatable form. The structure was first described for ColE1 and termed relaxation complex (89). The ColE1 complex contains three proteins of 60, 16, and 11 kDa as determined by a labeling procedure and subsequent gel electrophoresis (81). The proteins probably correspond to the products of three of the ColE1 *mbe* mobilization genes, namely MbeA (57,744 Da), MbeB (19,525 Da) and MbeC (12,883 Da; ref. 82). In the ten-year period following their discovery, relaxosomes were reported for a variety of conjugative and mobilizable plasmids of enterobacteria (51). Here we focus on advances coming from the study of three such plasmids: F, RP4, and RSF1010.

F PLASMIDS   The transfer origin of F and F-like plasmids maps at one end of the *tra* gene cluster and extends over (approx. 250 bp (for reviews, see refs. 2, 4, 6). *oriT* is flanked by a single gene downstream of *nic*, called gene X or *orf169* (gene *19* in R1; ref. 90). This is the first gene of the leading region; its role is unknown. Upstream of *nic* is *traM*, marking the beginning of the *tra* gene cluster. Binding sites have been described for F-encoded proteins TraY and TraM, and host-encoded integration host factor (IHF). The latter appears to be attracted by a region of *oriT* containing an intrinsic bend. All known binding sites are clustered upstream of *nic* in the trailing region (62, 91–94).

Genetic tests of the requirements for *oriT* cleavage on F indicated a role for *traY* and the region containing *traI* (95). The two genes are at opposite ends of the Tra operon, showing that relaxation does not necessarily involve closely linked loci as is the case for IncI1, P, Q, and W plasmids (24, 36, 46, 88, 96). The *traI* locus specifies a 180-kDa polypeptide (TraI; 1756 aa) and a related 94-kDa polypeptide (TraI*; 802 aa) generated from an internal in-frame translational start (97, 98). F-like *traY* gene products are plasmid-specific *oriT* binding proteins (6, 99, 100), which are structurally similar to the Arc and Mnt repressors of phage P22 (101). TraY binding sites (*sbyC* and *sbyA*) are located from bp 40 to 100 upstream of *nic* in the trailing region (6). *sbyC* overlaps a high-affinity IHF binding site positioned from bp 28 to 54 upstream of *nic* (93).

In vitro data demonstrate that cleavage of supercoiled F *oriT* DNA takes place in the presence of TraI protein and Mg^{2+} ions (73, 74). No additional cofactors are needed for this reaction that causes cleavage at the same place in *oriT* as observed in vivo. An additional system for F and R100 *oriT* cleavage has been established, which depends on proteins TraY and IHF (SW Matson, personal communication, 101, 101a). The in vitro system requires supercoiled F *oriT* DNA, TraI, Mg^{2+} ions, but TraY and IHF must be added prior to TraI. Cleavage in this system requires less TraI protein than in the TraY/IHF-independent system. Note that proteinase K/SDS treatment is still necessary to release the DNA as a cleaved species. The reaction is also detected on linear *oriT* DNA. Complexes on linear DNA are visible by electron microscopy when all three proteins are present. These findings are consistent with data that R100 TraY is required for R100 *oriT* cleavage by the cognate TraI protein in a crude in vitro system (101, 101a).

IncQ PLASMIDS   Mobilization of RSF1010 (8684 bp; ref. 37) requires a ~1.8-kb region of the plasmid (36, 102). The minimal functional *oriT* is 38 bp long and is flanked upstream by *mobA* and *mobB*, and downstream by *mobC* (85). These genes are arranged in divergently transcribed clusters. Purified Mob proteins have been used to demonstrate in vitro strand-specific cleavage of RSF1010 *oriT* (78). In the presence of Mg^{2+} ions, MobA, MobB, and MobC-dependent cleavage was observed with linear or supercoiled DNA. Cleavage of linear DNA requires larger amounts of MobC. MobA and MobC are essential for cleavage of double-stranded DNA, whereas MobB has a stimulatory effect. The position of *nic* generated in vitro and in vivo is identical. Large protein-*oriT* complexes can be seen in the electron microscope after incubation of MobA and MobC proteins with an *oriT*-containing plasmid DNA (E Scherzinger, personal communication). MobC protein alone fails to form any observable *oriT*-specific complex with linear duplex or supercoiled DNA. Presumably MobC is brought into position through its interaction with MobA bound to *oriT*. MobC binding at *oriT* may result in a localized destabilization of the DNA helix, allowing the MobA protein to interact with its single-stranded recognition sequence that is then cleaved. The 78-kDa MobA protein, also known as RepB, possesses replicon-specific priming activity. There is no evidence implicating this activity in conjugation (103).

IncP PLASMIDS   The *oriT* site has a size of ~300 bp and is located in the Tra1 region. Flanking genes are *traJ* in the relaxase operon and *traK* in the leader operon (23, 26). These genes encode specificity determinants for *oriT* function. A Tra1 core region of ~2.2 kb is needed for heterologous mobilization by R751 and for relaxosome formation. Capturing of specifically cleaved DNA from cells requires part of the relaxase operon, namely the genes *traJ* and the 5′ three fifths of *traI*. This operon, extending to the left of RP4 *oriT*, contains a third gene *traH*, which is within *traI* but occupies a different reading frame.

These findings were confirmed by in vitro reconstitution of the RP4 relaxosome from purified components (77). Proteins TraJ and TraI in the presence of Mg^{2+} ions were sufficient for in vitro cleavage. Addition of TraH (12.9 kDa, 160 aa), an acidic oligomeric protein (calculated pI = 4.2), stabilized the protein-*oriT* DNA complex, allowing its detection on agarose gels and visualization in the electron microscope. TraH does not bind to DNA and is contained in the relaxosome through interactions with TraJ/TraI. TraH is inessential; its absence reduces the transfer frequency only marginally (104). The presence of TraK in the in vitro reactions increased the yield of cleaved *oriT* plasmids as it does in vivo (24, 64).

Negatively supercoiled *oriT* DNA is the preferred substrate for the formation of relaxosomes in vitro. This is in agreement with the basic observation that relaxosomes prepared from cells contain superhelical plasmid DNA. TraJ protein (13.5 kDa, 123 aa) binds specifically to *oriT*, recognizing a 10-bp palindrome in the right arm of the imperfect 19-bp inverted sequence repetition that is positioned upstream of *nic*(105). The TraJ binding site and *nic* are located on the same side of the DNA helix, indicating a close contact between TraJ and TraI. Relaxosome assembly is proposed to involve a cascade-like mechanism. The initial step is binding of TraJ to *oriT*, giving a complex detectable by gel electrophoresis. Next this complex is recognized by TraI, which is unable to bind to double-stranded *oriT* DNA directly. Cleavage products are identical, irrespective of whether relaxosomes were isolated from cells or reconstituted in vitro. Notably the in vitro cleavage reaction at RP4 *oriT* is independent of host chromosome-encoded components, as is the case for RSF1010. Independence of host functions may contribute to the broad transfer range of these plasmids.

The *traK* product of IncP plasmids is essential to the transfer machinery because *traK* mutants are phenotypically Tra⁻ and the gene belongs to the core, defined as the minimal region that still allows heterologous mobilization by IncPβ helper plasmids, and vice versa. The assumption that TraK protein binds to *oriT* and recognizes only cognate sequences was verified (106). TraK is a basic protein (calculated pI = 10.7) of 14.7 kDa (134 aa) that interacts exclusively with its cognate *oriT*, as shown by DNA fragment retention assay. Footprinting experiments using DNaseI or hydroxy radicals indicate that several TraK molecules interact specifically with an intrinsically bent region of *oriT* covering a range of almost 200 base pairs. The target sequence maps in the leading region adjacent to *nic*. Recognition sequences involved in relaxosome formation are close by but do not overlap the TraK binding region. TraK-*oriT* complexes were suggested to consist of DNA wrapped around a core of TraK molecules. Since *oriT*-specific cleavage requires negatively supercoiled DNA, the TraK-*oriT* complex could assist in local unwinding of the neighboring *nic* regions by introducing positive supercoiling. Despite these remarkable DNA-binding properties and the stimulatory effect on *oriT*-specific cleavage, the essential role of TraK in the transfer process is unknown.

T-DNA processing during infection of plants is initiated by site- and strand-specific incision at the T-DNA border sequences of the Ti plasmid. Two proteins—VirD1 and VirD2—are required for this reaction to occur in vitro on double-stranded DNA (107). Spontaneous relaxation of plasmid DNA does not occur, indicating that the DNA in the VirD1/VirD2 complex is topologically constrained by strong protein-DNA interactions. The characteristics of the VirD1/VirD2-mediated cleavage reaction strongly resemble those observed with relaxosomes of IncP plasmids, the analogous partners being TraJ/TraI, respectively.

Generalizations

In the systems discussed there are common observations. Cleavage on double-stranded *oriT* DNA occurs preferentially on supercoiled DNA and requires in addition to the relaxase one or more structural proteins acting as accessory factors. In the case of RSF1010 the function analogous to RP4 TraJ is probably built into MobA protein (78). It has been debated whether *oriT* cleavage is a conjugative event triggered by mating-pair formation or a reversible process occurring in the absence of recipient cells (95). The latter possibility is strongly supported by recent evidence that relaxosomes contain the cleaved species of DNA. However, only a portion of the relaxosomal DNA is found as form II DNA following treatment with proteinase K/SDS; another fraction is still supercoiled (form I) and indistinguishable from input DNA. This is interpreted to indicate that the treatment in fact freezes an equilibrium between two sorts of complex. The first consists of uncleaved form I DNA associated with relaxosomal proteins in a noncovalent manner. The second complex is comprised of specifically cleaved DNA held in the superhelical state by interactions with relaxase (77). The enzyme attached covalently to the 5′ end holds the 3′ end tightly through noncovalent interaction like a clamp conserving the superhelical status of the substrate DNA. This strong noncovalent interaction is destroyed by denaturing conditions, yielding form II DNA.

Studies with the RSF1010 in vitro system substantiate the reversibility of the relaxase reaction (78). Specifically cleaved DNA that accumulates when the reaction is carried out in low-salt/Mg^{2+} buffer rapidly disappears if, prior to SDS/proteinase treatment, the salt concentration is increased to 0.6 M NaCl or the incubation temperature is raised to 50°C. Addition of excess EDTA to the preincubated reaction mixture, however, blocks this salt- or temperature-induced reverse reaction. Following reversal, the DNA product remains supertwisted with no change in the superhelical density. These observations are most consistent with the idea that the DNA is cleaved prior to freezing of the equilibrium with SDS/proteinase and that MobA acts as a protein clamp to maintain the topological state of the DNA through covalent and noncovalent interactions. This structure resembles the transient intermediate in the cleaving-joining reaction of type-I DNA topoisomerases (108, 109).

Additional evidence for the above interpretation comes from studies with the RP4 TraI-mediated cleavage at *oriT*. Under certain conditions a variety of topoisomers is seen during the cleavage reaction with RP4 TraI mutant proteins containing serine to alanine substitutions at two different positions (S14A and S74A). This is interpreted as a decrease of binding strength of TraI in the complex to the part of the nick region upstream of *nic*. It also indicates that the joining reaction is a property of the relaxase (110). Thus, the relaxosome possesses the intrinsic capacity to break and reseal its target DNA strand in a reversible reaction without additional cofactors or a conjugative trigger.

RELAXASES

DNA Relaxase Assays

The assay for relaxases is based on an observation made in 1969 by Clewell & Helinski (89) that plasmid DNA isolated from cells in the presence of denaturing agents like SDS /protease was in form II instead of form I. A simple way to demonstrate that the conversion of form I *oriT* DNA to form II DNA has occurred by site- and strand-specific cleavage is to linearize the DNA at a unique restriction enzyme site and separate the mixture on an alkaline agarose gel into single-stranded DNA fragments. Normally three bands are visible: two single-stranded fragments, which add up to one unit length of the plasmid, and the complete single-stranded plasmid. This reaction can be applied not only to relaxosomes isolated from cells but also to complexes reconstituted in vitro. To determine the cleavage site precisely, DNA isolated from the gel is used as substrate for sequencing reactions (26, 111).

The assay on double-stranded DNA in some cases requires the presence of accessory proteins. Hence, an assay has been developed based on the property of gene A protein of phage ϕX174 (112) to cleave not only dsDNA but also ssDNA in the form of short synthetic oligodeoxynucleotides of defined sequence and length (83, 85, 86, 113). The radioactively labeled oligonucleotides used consist of the nick region extending a few nucleotides beyond *nic*. Products are separated on sequencing gels before and after proteolytic digestion and subsequently visualized by autoradiography. The assay for the joining activity of relaxases (DNA strand transfer) uses the same approach except that an acceptor oligonucleotide consisting of the sequence upstream of *nic* is also added to the reaction mixture. To allow easy detection of products of the strand transfer reaction, the length of the acceptor oligonucleotide should be different from the length of the product resulting from cleavage of the oligonucleotide containing *nic* (Figure 2).

Enzymatic Properties

In general, relaxase reactions on single-stranded oligonucleotide substrates:

1. occur when the oligonucleotide embracing the nick region is as short as ten residues;
2. require Mg^{2+}-ions;
3. neither require nor are stimulated by accessory proteins;
4. show the classical behavior of an equilibrium reaction;
5. involve covalent attachment of the protein to the T-strand via transesterification;
6. are catalyzed by the activity of the N-terminal portion of the enzyme. The active site amino acid is Y22 in RP4 TraI and is Y24 in RSF1010 MobA (83, 85).

The properties of relaxases mentioned were derived from enzymatic analyses with the purified proteins F TraI, RP4 TraI, Ti VirD2, and RSF1010 MobA (83, 85, 86, 113). Additional experimental data available for some of these proteins are summarized below.

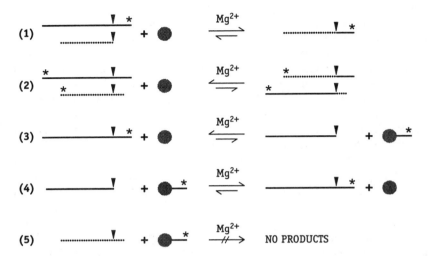

Figure 2 Dissection of DNA strand transfer reactions catalyzed by DNA relaxase in vitro. Oligonucleotide substrates containing the nick region are symbolized by solid or dotted lines that indicate different substrate lengths and demonstrate how the products were assembled. Asterisks designate the positions of the radioactive label and wedges mark the *nic* sites. DNA relaxase is drawn as a black solid circle. The relaxase-oligonucleotide adduct produced in reaction 3 was purified and used in reaction 4 and 5.

RP4TraI PROTEIN   The relaxase domain contains at least three motifs as found by sequence comparison with other potential relaxases, site-directed mutagenesis, and activities of purified mutant proteins (84, 110). Motif I of TraI contains the active site tyrosine (res. 22) and motif III is histidine-rich (63). Replacement of Y22 results in an inactive protein in vitro and in a transfer-null phenotype in vivo. Alteration of motif III indicates the importance of residues H116 and H118, which are thought to increase the nucleophilicity of the tyrosine's aromatic hydroxyl group by proton abstraction. A Ser to Ala change in res. 74 of motif II has two major effects on in vitro activities: (*a*) stable relaxosomes cannot be formed but topoisomers of the *oriT* substrate DNA are produced instead, as shown by electrophoresis of the relaxosome assembly mixture, and (*b*) cleavage of single-stranded oligonucleotide yields higher amounts of products. These observations suggest that motif II might interact noncovalently with the end of the trailing region 3′ to *nic*. This part of the nick region contains the sequence and structure that determines the cleaving and joining specificity (see above). A reduction in the strength of binding to the nick region of TraI S74A would explain the observations. The three motifs I,

II, and III catalyze the cleaving-joining reaction of TraI by concerted action. This would predict that the motifs are in a close spatial arrangement.

Ti PLASMID Tra A AND VirD2 PROTEIN   The putative relaxase of the Ti Tra system is encoded by *traA*. The gene is located next to the *oriT* region in one of the Tra2 operons that are transcribed divergently from *oriT*. TraA shares motifs with RSF1010 MobA relaxase, with the potential nucleotide binding motifs of F TraI helicase (see next section) and with the primase (DnaG) of *E. coli* (SK Farrand, personal communication). VirD2 is the relaxase of the Vir system (114). The protein possesses activities analogous to that of RP4 TraI (83, 86). The structural organization of conserved motifs in the relaxase domain of the two proteins is also very similar (84, 110). An interesting difference between the two proteins is that VirD2 cleaves T-border and RP4 nick region oligonucleotides, whereas TraI only accepts IncP nick region oligonucleotides as substrate. VirD2 is thought to pilot the T-DNA into the plant nucleus (115). Integration is thought to occur by a mode of illegitimate recombination initiated at a nick or break in the host DNA. The right T-DNA terminus is conserved around the first three nucleotides of the border repeat, possibly reflecting VirD2-conferred protection of the end from degradation or VirD2-mediated sealing to the plant DNA (116, 117).

RSF1010 MobA PROTEIN   The relaxase activity determined by plasmid RSF1010 is attributed to MobA protein (78, 118). Both the original MobA protein (708 aa) and a C-terminally truncated version, MobA* (residues 1–243), are functional in the in vitro cleavage reaction. MobA-catalyzed cleavage and joining of single-stranded DNA has been studied on viral M13 DNA and on single-stranded oligonucleotides containing RSF1010 *oriT* sequences. Cleavage of M13 DNA occurs at preferred sites and is independent of any *oriT* sequence. The MobA-mediated reaction is Mg^{2+}-ion dependent and produces numerous DNA species distinguishable on an agarose gel. After cleavage, MobA remains covalently linked to the 5′ end of the fragment generated and retains its activity for the joining reaction. Joining has been demonstrated by detection in the electron microscope of single-stranded DNA circles of different size. The circles are most likely formed by the intramolecular joining of MobA-generated fragments, each of which may contain an active enzyme intermediate tightly bound to the 5′ end. Sites of cleavage have not been determined but M13 sequences fortuitously similar to a nick region might be recognized by MobA as targets.

Mechanistic Parallels to other DNA Strand Transferases

A large number of DNA topoisomerases and DNA strand transferases are known to catalyze the breakage and joining of DNA strands via transesterification reactions. Three mechanistically distinct classes of enzyme are known. The DNA topoisomerases and many of the site-specific DNA recombinases belong to the first class, with one unique active site tyrosine or serine per protein protomer for the cleavage and joining of DNA. Because the transfer of one DNA end to another involves the cleavage and joining of a minimum of two and often four strands, a minimum of two or four polypeptides with one active site amino acid in each is needed in the DNA strand–transfer reactions catalyzed by members of this class. A second category of proteins includes phage Mu transposase and the HIV type-1 integration protein, which may not form a covalent protein-DNA intermediate (119, 120). The φX174 gene A protein (gpA) represents a third class of mechanistically distinct proteins that break and rejoin DNA strands via transesterification. The roles of gpA in the replication of φX174 DNA are well established (121). The protein cleaves the negatively supercoiled form of double-stranded φX174 DNA between nucleotides GA (pos. 4305/4306) of the viral strand and becomes covalently bound to the 5′-phosphoryl group of A via a phosphotyrosine bond. In the presence of *E. coli* DNA helicase Rep and DNA polymerase III holoenzyme, DNA chain extension initiates at the 3′-OH of G generated by gpA. Elongation of the DNA strand is accompanied by displacement of the strand linked to the gpA. After the synthesis of one full-length φX174 viral strand, gpA becomes linked to A of the newly synthesized DNA and joins the end of the displaced strand to form a covalently closed ring for packaging into a virus particle.

Among the DNA strand transferases φX174 gpA is mechanistically unique in that one of a pair of tyrosyl residues, separated by three amino acids, can function as the acceptor of a DNA 5′ phosphoryl group in the specific cleavage reaction (112, 122). Roth et al (122) have proposed a mechanism for the gpA-catalyzed DNA cleavage and joining, in which the two residues Y343 (a) and Y347 (b) are directly involved. It was postulated that (a) and (b) were equally proficient in catalyzing the initial cleavage of the form I DNA, and one of them (a) would become covalently bound to the viral or (+)strand via a phosphotyrosine linkage. After one round of (+)strand synthesis the free tyrosyl residue (b) would participate in a second cleavage at the (+)strand origin, at what is now the junction of old and newly synthesized DNA, and become linked to the 5′ phosphoryl group of the new strand. The 3′ hydroxyl group resulting from this second cleavage reaction would undergo transesterification with the existing phosphotyrosyl linkage to tyrosine (a) for the next round of transesterification. The two tyrosines would thus alternate in DNA cleavage and strand joining at the end of every replication cycle. This mechanism has been confirmed by Hanai & Wang (123). The tyrosine cluster at the N-terminal part of the F TraI and R388 TrwC protein (88), which is proposed to contain active site amino acids, might also be involved in such a two-tyrosine relay mechanism.

The replication initiation protein of filamentous phages (f1, fd, and M13) is gene II product. The protein offers two juxtaposed tyrosines, which could act as nucleophiles in the cleaving-joining reaction (112). The phages are known to replicate via rolling circles but covalent DNA-protein intermediates have not been found (124). In the case of *S. aureus* plasmid pT181 and homologues such as pC221, which also replicate via rolling circles, the RepC-like proteins have only one active site tyrosine (125, 126). These proteins exist as homodimers acting in initiation and termination of a single round of plasmid replication.

DNA UNWINDING

Generation of the T-strand after cleavage at *nic* is a poorly understood process. In principle, the following three mechanisms exist for producing a single strand from double-stranded DNA. These are:

1. nucleolytic degradation of the second strand, by analogy with DNA uptake in natural transformation (127);
2. polymerase-mediated strand displacement during de novo DNA synthesis;
3. unwinding by a DNA helicase with or without de novo DNA synthesis (128). The conjugative enzyme might be plasmid- or host-encoded.

Conjugative unwinding of DNA is best documented for the F system. The first mechanism is ruled out by the demonstration that the nontransferred strand of F is retained in the donor cell (69). The second mechanism cannot provide the sole system because F transfer was found to continue when conjugative DNA synthesis was blocked in donor cells (129, 130). Instead plasmid F is thought to specify its own conjugative unwinding enzyme as the helicase activity of the multifunctional TraI protein (1756 aa). The protein has interrelated ssDNA-dependent ATPase and helicase activities, and was first characterized as DNA helicase I of *E. coli* by Hoffmann-Berling and colleagues (131). In vitro studies with heterologous DNAs have shown that the enzyme is highly processive and proceeds unidirectionally in the 5′ to 3′ direction relative to the bound single strand (132, 133). The protein readily aggregates and was originally proposed to unwind DNA as an oligomer following the binding of ~80 molecules to a single-stranded region adjacent to duplex DNA (132). However, cooperative binding of helicase I is not essential for the unwinding function (134). Possibly the enzyme is active as a monomer, since substantial unwinding was detected in a reaction involving a 1:1 ratio of helicase I molecules to DNA substrate molecules (133).

TraI contains two potential nucleotide binding motifs (NBS) (SDKVGKTG, aa 177 to 184; VQGYAGVGKTTQ, aa 992 to 1003), both identified by sequence comparison to known conserved motifs (97, 135, 136). ATPase activity has been demonstrated for the C-terminal domain containing the second motif (98). Dash et al (137) have described two Km-values for ATP (0.27 mM and 2 μM). This finding is consistent with the presence of two potential NBS in the protein sequence, but the correlation between ATP hydrolysis and individual sites has not been tested. TrwC protein of R388 has also been shown to function as a helicase in vitro with properties similar to those of helicase I (137a). Although there are extensive sequence similarities between IncF and IncW *oriT* sequences (Figure 1), only the second NBS consensus sequence in F TraI (aa 992 to 1003, see above) seems to be conserved in TrwC [aa 493 to 504, VQGFAGTGKSHM (ref. 88)].

DNA helicase I is thought to be the primary helicase involved in unwinding the F plasmid in the donor cell (138). The idea has yet to be tested using a relaxase-proficient helicase-defective F *traI* mutant. In the proposed role the protein would migrate on the strand destined for transfer, since this has the necessary 5′ to 3′ polarity. The rate of unwinding by DNA helicase I in vitro is in the range of 1 kb per sec at 37°C (131), which is in excess of genetic estimates of the rate of chromosomal DNA transfer from Hfr donor cells (0.75 kb per sec at 37°C). The unwinding reaction could provide the motive force for DNA transport, if the enzyme is anchored with respect to the cell envelope to allow displacement of the unwound strand relative to the cell boundary. TraI protein is located in the cytoplasmic fraction of the cell (3), but one or more other *tra* gene products could function as the hypothetical anchor (139).

DNA helicase I requires a ssDNA region of about 200 nucleotides to initiate unwinding of nonspecific substrates in vitro (132). Possibly this region is generated during conjugation by another helicase, or the requirement is by-passed in vivo by a coupling of DNA unwinding to the *oriT* cleavage reaction (2, 98). The coupling might be achieved directly by TraI protein, since it is both a DNA helicase and a component of the *oriT* cleavage mechanism re-quiring a single-stranded substrate. Appreciation of these dual activities of TraI rationalizes the paradox that, whereas helicase I is active on heterologous DNA substrates in vitro, the *traI* genes of F and the closely related R100 plasmid show system-specificity in vivo (140). The specificity probably reflects char-acteristics of the *oriT* cleavage reaction rather than the unwinding process.

No conjugative helicase has been found in the IncP, IncQ, and Ti plasmid transfer systems. RSF1010 encodes a protein (RepA) with helicase and ssDNA-stimulated ATPase activities, but the protein is active in vegetative replication and inessential for RSF1010 mobilization (102, 103).

TERMINATION OF TRANSFER

Perspectives

A round of transfer terminates in the general model of conjugation with the joining of the ends of the T-strand to give a covalently closed circle. The preferred mechanism involves the processing of an intermediate that is greater than unit length, as in rolling circle replication of phage φX174. Following a cycle of unwinding coupled to replication, the covalently linked gpA molecule cleaves the regenerated origin and simultaneously joins the ends of the displaced strand to produce a monomeric circle. If unwinding is uncoupled from replication, gpA joins the ends of the displaced strand to produce the circle (121). Termination of transfer will similarly require cleavage of a reconstituted nick region if the 3'-OH terminus of the T-strand is elongated by DNA synthesis in the donor. If this terminus is protected, circularization will require reversal of the initiating *oriT* cleavage. Circularization in both models is catalyzed by relaxase linked to the 5' end of the transferring strand, with energy for the sealing reaction coming from the energy of the original phosphodiester bond conserved in the DNA-protein covalent intermediate. The two-tyrosine relay model (see section on relaxases) explains how one relaxase molecule can mediate the two DNA cleavages and single joining event in the replicative model. This combination of reactions might be catalyzed by a homodimer in the case of a relaxase with one active site.

An important question is whether or not the 3'-OH terminus of the T-strand is elongated by replication in the donor to produce a transfer intermediate greater than unit length. Many molecular investigations with F and IncI1 plasmids in the late 1960s and the 1970s addressed this question but none provided compelling evidence that a high molecular weight intermediate accumulates (51). A second approach addressed requirements for initiating replacement strand synthesis (RSS) in the donor cell. The nick site in F and RP4 DNA isolated by SDS/proteinase K is flanked by a 3'-OH terminus, but it is unknown whether the terminus is accessible as a primer in vivo (26, 72, 76, 77). Evidence for its inaccessibility is the apparent requirement for an RNA primer to initiate RSS of plasmid F and, more problematically, to initiate transfer as well as RSS of R64 (130, 141, 142). The requirement was inferred from the sensitivity of these substages of conjugation to genetic and rifampicin-induced inhibition of RNA synthesis. Experiments involving the application of such treatments to bacteria harboring F and I1-type plasmids must be interpreted with caution, because it is now recognized that rifampicin and some other procedures that inhibit RNA synthesis prevent synthesis of the short-lived antisense RNAs that regulate production of Hok killer proteins specified by these plasmids (143). The proteins are potent killing agents that damage the membrane permeability barrier from within, causing extensive degradation of stable RNA and cellular DNA under conditions used in the mating experiments (144–146). Possibly production of the proteins also deranges conjugation.

Meyer and colleagues have developed a genetic system that has proved particularly instructive for investigating termination (147–149). The system measures RecA-independent, intramolecular recombination at *oriT* sites during

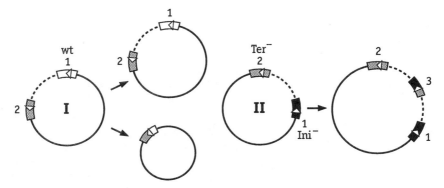

Figure 3 Structure of mobilizable plasmids used in assays of *oriT* recombination during conjugation. *oriT* sites are boxed, with arrowheads indicating *nic* sites and a clockwise direction of transfer. The continuous single line represents the plasmid replicon and the broken line indicates the genetically marked spacer. Plasmid I has a wild-type *oriT* site 1; deletant plasmids are generated when transfer is initiated at site 1 and terminated at site 2. Plasmid II has an initiation-defective site (Ini⁻) and a termination-defective site (Ter⁻); transferred plasmids commonly contain a third *oriT* (site 3) of the recombinant type. Further details are in the text (adapted from Erickson & Meyer, ref. 149).

mobilization. DNA substrates are small plasmids containing a direct repeat of *oriT*, separated in one interval by the plasmid replicon and in the other by a genetically marked spacer (Figure 3). Depending on the *oriT* used for initiation and termination of transfer, transconjugant cells can inherit the intact plasmid or a deletant plasmid with a single *oriT* and no spacer fragment. The spacer may be transferred by itself, but it cannot replicate. The frequency of termination at *oriT* can therefore be calculated from the fraction of transconjugants carrying plasmids of the deletant type. The termination frequency was significantly <1 for R1162 *oriT* (150), but approached 0.9 for F *oriT* (151). The latter value is consistent with the reported fraction of tetrameric F *oriT* plasmids reduced to monomers (152). The frequencies indicate that a mechanism exists for processing T-strands of greater than unit length, as would be generated by rolling circle replication in the donor cell.

   The *oriT* recombination system has been used to investigate sequence requirements for termination (plasmid I, Figure 3). The assay measures production of deletant plasmids following mobilization of a parental plasmid in which the *oriT* (site 2) 3' of the replicon is used for testing mutant sequences for termination efficiency. Another instructive plasmid contained an initiation-proficient and termination-defective site 2, and an initiation-deficient but termination-proficient site 1 (plasmid II, Figure 3); this gave rise to transconjugants mostly containing plasmids of greater than parental length. The novel plasmids contained an extra copy of the spacer and a third copy of the *oriT* (site 3; Figure 3). This site was recombinant, having the downstream sequence of site

2 and the upstream sequence of site 1 (151). Production of these plasmids is consistent with the prediction of the rolling circle mechanism that the 3'OH terminus at the initiating *oriT* cleavage is extended by transfer replication.

Specific recombination at R1162 *oriT* also occurs when tandem copies are cloned in a phage M13 vector and the constructs are propagated in the presence of MobA relaxase. Presumably, phage replication generates ssDNA molecules that are recognized as a transfer intermediate (153, 154). Recombination in this phage-based system implies that relaxase catalyzes in vivo cleavage and joining of *oriT* sites in ssDNA, as demonstrated in vitro with oligonucleotides corresponding to partial *oriT* sequences (see above). Taken together, the different lines of evidence are more consistent with a replicative rolling circle model of transfer than a mechanism in which the 3'-OH terminus at *nic* is protected.

Requirements for Termination

The three classes of *oriT* in Figure 1 are organized similarly in containing an upstream inverted repeat separated from *nic* by eight or nine nucleotides (Figure 4). The *oriT* recombination assay showed that the upstream or outer arm of the R1162 (RSF1010) repeat is unnecessary for initiation of transfer, but is required as part of an intact repeat for termination. The arms may form secondary structure in the trailing region of the T-strand, which traps the relaxase linked to the leading region (147, 148). It is reasonable to suppose that the *oriT* structure for specific cleavage at termination is similar to that for specific cleavage in the relaxosome. A structure similar to the hairpin envisaged as necessary for T-strand cleavage at termination would be produced in the relaxosome if the *nic*-proximal arm of the repeat is in duplex form. The possibility that the additional relaxosomal proteins destabilize the helix to allow relaxase to interact with its single-stranded recognition sequence has been suggested (see section on relaxosomes).

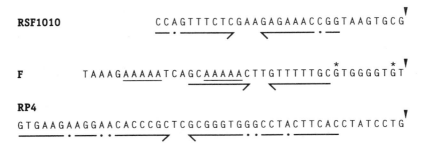

Figure 4 Sequence features in the trailing portion of the *oriT* sites of plasmids RSF1010, F, and RP4. Inverted repeats are indicated by horizontal arrows. Asterisks indicate sites on F where transition mutations inhibit *oriT*-specific cleavage and termination of transfer. Phased A-tracts are underlined. Source of information: RSF1010 (78), F (151), and RP4 (26).

Experiments with specific DNA strands indicate that R1162 MobA forms an unusually stable complex with *oriT* on the T-strand. Formation of the complex requires both arms of the repeat, and their sequences to be complementary, as well as the nucleotides separating the repeat from *nic*(155). The inverted repeat is not required for MobA-mediated cleavage of single-stranded oligonucleotides where the minimal substrate contains seven nucleotides upstream of *nic* and four downstream (see section on relaxases). Neither is the repeat required for joining of the DNA in the relaxase adduct to an acceptor oligonucleotide (85). However, competition experiments indicate that DNA with the intact repeat is a preferred substrate for specific cleavage of ssDNA (118, 148).

Efficient termination of F transfer in the *oriT* recombination assay likewise requires the inverted repeat in the trailing region as well as the overlapping phased A-tracts extending to upstream residue 35 (Figure 4). Termination also requires integrity of the sequence between the repeat and *nic*, since it was inhibited by two transition mutations in this region (151) (Figure 4). The effect of the transitions, which reduce the efficiency of *oriT* cleavage in duplex DNA and affect the consensus sequence of the nick region (Figure 1), provides further evidence that relaxase catalyzes the termination event.

INTERACTIONS BETWEEN DNA PROCESSING AND THE MATING APPARATUS

General Description of the Mating Process

There are several stages where DNA processing reactions interact with the mating apparatus. This component of the F system involves 17 or more Tra proteins predominantly located in the inner membrane, periplasm, or outer membrane of *E. coli* donors. Most of the proteins are required in pilus production (6). The conjugative contact in liquid media is thought to be initiated by interactions between the tip of the F pilus and a receptor, possibly lipopolysaccharide, in the recipient cell envelope (156, 157). Pilus retraction then brings cells into surface contact where they are stabilized by specific Tra proteins. Thin-section electron microscopy indicates that the contact area, called a conjugational junction, between the surfaces of the bacteria is unique in containing extra electron-dense material of unknown chemistry (158).

The RP4 plasmid–mating apparatus requires ten of the eleven *trb* genes in the Tra2 core region plus *traF* of the Tra1 region (159). The RP4 Tra2 region is related to the Ti *virB* operon, also consisting of 11 genes (61, 160). VirB proteins are thought to specify the T-DNA export channel in agrobacteria. A number of the proteins have a membrane localization and the finding that some are in inner and outer membrane fractions suggests that there is a single channel spanning the cell envelope (60, 161, 162).

DNA transfer is initiated by some signal generated in response to a productive mating contact. In the donor, the signal must disrupt the strong noncovalent interaction between relaxase and the 3′ end at *nic* to make the structure accessible to the transport machinery. The nature of the signal and the components involved in its transduction across the donor cell envelope to the relaxosome are unknown. The favored model, recently supported by video-enhanced light microscopy (158), is that transfer ordinarily occurs between cells in close and specific contact (6). Possibly the DNA export pore includes some of the proteins involved in pilus assembly. The route into the recipient cell is unknown; it might involve a preformed junction between outer and inner cell membranes, pilus penetration (2), or DNA import in a two-step process involving a periplasmic intermediate (158).

Coupling of DNA Transfer to the Transport Pore

Some mechanism must exist for coupling transfer of the unwound T-strand to the transport pore. Candidate proteins are members of a family including F TraD, RP4 TraG, Ti VirD4, pGO1 TrsK (pRK31 TraK), and R388 TrwB (61, 84, 159, 163). TraG contains a type A and a type B nucleotide binding motif. Both are highly conserved in the other proteins, with the exception of pGO1 TrsK (pRK31 TraK), which only contains the B motif (84). F TraD is an inner membrane protein with nonspecific DNA-binding activity (164). The protein is thought to mediate DNA transfer directly, because it is required at a stage following effective cell aggregation, *oriT*-cleavage, and initiation of unwinding (95, 130, 165). RP4 TraG is similar in being an essential transfer function that operates at a stage other than pilus formation and *oriT*-cleavage (159).

The key to understanding the possible role of this class of proteins comes from mobilization experiments. TraD is the only F Tra protein other than components of the mating apparatus known to be required for ColE1 mobilization (51, 166). TraD is not required for mobilization of the ColE1-related CloDF13 plasmid, but CloDF13 specifies a protein (MobB) with similar properties to TraD (167). Moreover, MobB contains type A and B nucleotide binding motifs (84). RP4 TraG is likewise the only RP4 Tra protein other than the mating apparatus to be essential for mobilization of RSF1010 (159). RSF1010 can be mobilized between agrobacteria by the Ti plasmid; here the process requires VirD4 together with the VirB proteins that are thought to constitute the DNA export channel (168). A unifying explanation is that RP4 TraG-like proteins couple the conjugatively activated relaxosome to the mating apparatus (169). The finding that F *traD* mutants are overpiliated (170) is compatible with this suggestion.

Cotransfer of Proteins with DNA

It is unknown whether DNA-linked relaxase remains in the donor in close proximity to the transport pore or is transferred into the recipient cell. Certain *tra* gene products are transmitted in conjugation, as shown by the abundant transfer of plasmid primase to the cytoplasm of the recipient in conjugation mediated by CoIIb and RP4. Transferred proteins, identified as polypeptides labeled in the donor, were the two in-frame translation products (210 kDa and 160 kDa) of the CoIIb *sog* primase gene and the larger (120 kDa) product of the RP4 *traC* primase gene (172–174). The plasmid enzymes bind ssDNA and function to generate primers to initiate discontinuous DNA synthesis on the transferred plasmid strand (62). Sog proteins are transferred with an estimated stoichiometry of 250 molecules of each polypeptide type per 93-kb CoIIb strand (173). Thus, the transmitted proteins may coat the T-strand. Sog proteins are multifunctional; the N-terminal region of the 210-kDa polypeptide has primase activity, whereas the C-terminal region common to both polypeptides promotes transfer of the T-strand, possibly facilitating its transport across cell membranes (172). Plasmid F does not encode a primase (62) and is transferred without detectable association with a cytoplasmic protein (174).

If the transferring strand loops into the recipient cell, some mechanism may exist to prevent jamming due to base-pairing of sequences that recur as inverted repeats. Transfer of DNA in a protein complex would obviate the problem. A number of plasmids belonging to the F and I complexes of incompatibility groups encode ssDNA-binding (SSB) proteins that are homologous to *E. coli* SSB and bind DNA cooperatively without sequence specificity (175–177). There is no evidence that these proteins escort the T-strand into the recipient cell (174), which is consistent with findings that plasmid *ssb* genes are inessential for conjugation (178, 179), are expressed transiently at elevated levels after their transfer (180), and map distantly from the Tra region in the portion of the plasmid that enters the recipient first (179, 181). The evidence therefore suggests that plasmid *ssb* genes function to promote installation of the immigrant plasmid in the new cell, rather than to facilitate DNA transport itself (62, 180).

T-DNA is transferred into the plant cell in a complex with specific Vir proteins. One such protein is VirE2 (69 kDa). This ssDNA-binding protein binds T-DNA cooperatively and nonspecifically, possibly to protect the DNA from nucleases and to enable its transfer across bacterial membranes (182–186). VirE2 contains nuclear localization signals and is thought to promote nuclear transport of the complexed T-DNA (187). A second transferred protein is VirD2 relaxase, which is covalently linked to the T-DNA. Only the N-terminal half of VirD2 is required for T-strand production. The C-terminal portion plays a role in the transit of the T-strand into the plant nucleus. The function involves a C-terminal bipartite nuclear localization signal that conforms to the motif required for numerous nuclear-localized nonplant eukaryotic proteins (115). Possibly, transmission of VirD2 into the recipient cell is atypical of relaxases, reflecting the role of the pTi-encoded protein in aiding transport of DNA into the plant nucleus.

SUMMARY AND PERSPECTIVE

Most of the biochemical understanding of bacterial conjugation has come from systems studied in *E. coli*. Significant attention is now being devoted to conjugative mechanisms in a wide range of organisms and it is anticipated that these investigations will soon be fruitful in elucidating the range of mechanistic principles. It is already clear that systems classically regarded as unrelated can share common modules and involve analogous enzymes. This point is excellently illustrated by the discovery that the Ti plasmid encodes two functionally different transfer systems, both biochemically related and evolutionary related to the transfer system of IncP plasmids.

The past five years have seen rapid progress in understanding the protein composition and *oriT* substructure of the transfer initiation complex—the relaxosome—and the biochemistry of relaxases as DNA-strand transferases. This class of enzyme, which breaks and joins ssDNA by a transesterification reaction, plays important roles in other aspects of DNA metabolism, including replication-initiation, site-specific recombination, and modulation of DNA topology. The relaxosome is remarkable in possessing the innate capacity to break and seal its target strand in the absence of any extracellular stimulus and without disrupting transcription and vegetative replication. It is now necessary to elucidate the mechanism that converts the relaxosome into a rolling circle replication machine in response to a mating contact at the cell surface, to substantiate the replicative model at the biochemical level, and to characterize enzymes that catalyze unwinding of the T-strand in systems apparently lacking a plasmid-encoded DNA helicase.

The most mysterious aspects of conjugation are structures and processes involving the bacterial cell envelope, including the composition of the DNA transport pore, the nature of the mating contact, and the mechanism that transduces the mating signal to the relaxosome. Ignorance of these aspects of the mating apparatus impedes integration of the growing knowledge of DNA-processing reactions into a unified molecular model of the conjugation cycle.

ACKNOWLEDGMENTS

We thank our many colleagues working in this subject area for contributions to this review. Work in our laboratories was supported by grants from the Deutsche Forschungsgemeinschaft Sonderforschungsbereich 344/B2 to EL and from the Medical Research Council G9321196MB to BMW. We are

grateful for the secretarial assistance of Renate Spann and Hannelore Markert.
EL thanks Heinz Schuster for generous support.

Any *Annual Review* chapter, as well as any article cited in an *Annual Review* chapter,
may be purchased from the Annual Reviews Preprints and Reprints service.
1-800-347-8007; 415-259-5017; email: arpr@class.org

Literature Cited

1. Mazodier P, Davies J. 1991. *Annu. Rev. Genet.* 25:147–71
2. Ippen-Ihler KA, Minkley EG Jr. 1986. *Annu. Rev. Genet.* 20:593–624
3. Willetts N, Skurray R. 1987. *Escherichia coli and Salmonella typhimurium: Cellular and Molecular Biology*, ed. FC Neidhardt, JL Ingraham, KB Low, B Magasanik, M Schaechter, HE Umbarger, pp. 1110–33. Washington DC: Am. Soc. Microbiol.
4. Ippen-Ihler K, Skurray RA. 1993. *Bacterial Conjugation*, ed. DB Clewell, pp. 23–52. New York: Plenum
5. Frost LS. 1993. *Bacterial Conjugation*, ed. DB Clewell, pp. 189–212. New York: Plenum
6. Frost LS, Ippen-Ihler K, Skurray RA. 1994. *Microbiol. Rev.* 58:162–210
7. Guiney DG, Lanka E. 1989. *Promiscuous Plasmids of Gram-negative Bacteria*, ed. CM Thomas, pp. 27–56. London: Academic
8. Guiney DG. 1993. *Bacterial Conjugation*, ed. DB Clewell, pp. 75–103. New York: Plenum
9. Pansegrau W, Lanka E, Barth PT, Figurski DH, Guiney DG, Haas D, Helinski DR, Schwab H, Stanisich VA, Thomas CM. 1994. *J. Mol. Biol.* 239:623–63
10. Frey J, Bagdasarian M. 1989. *Promiscuous Plasmids in Gram-negative Bacteria*, ed. CM Thomas pp. 79–94 London: Academic
11. Thompson R, Taylor L, Kelly K, Everett R, Willetts N. 1984. *EMBO J.* 3:1175–80
12. Göldner W, Graus H, Högenauer G. 1987. *Plasmid* 18:76–83
13. Ostermann E, Kricek F, Högenauer G. 1984. *EMBO J.* 3:1731–35
14. Finlay BB, Frost LS, Paranchych W, 1986. *J. Bacteriol.* 168:132–39
15. McIntire SA, Dempsey WB. 1987. *J. Bacteriol.* 169:3829–32
16. Di Laurenzio L, Frost LS, Finlay BB, Paranchych W. 1991. *Mol. Microbiol.* 5:1779–90
17. Howland CJ, Wilkins BM. 1988. *J. Bacteriol.* 170:4958–59
18. Komano T, Toyoshima A, Morita K, Nisioka T. 1988. *J. Bacteriol.* 170:4385–87
19. Furuya N, Nisioka T, Komano T. 1991. *J. Bacteriol.* 173:2231–27
20. Paterson ES, Iyer VN. 1992. *J. Bacteriol.* 174:499–507
21. Coupland GM, Brown AMC, Willetts NS. 1987. *Mol. Gen. Genet.* 208:219–25
22. Pohlman RF, Genetti HD, Winans SC. 1994. *Mol. Microbiol.* 14:655–668
23. Guiney DG, Yakobson E. 1983. *Proc. Natl. Acad. Sci. USA* 80:3595–98
24. Fürste JP, Pansegrau W, Ziegelin G, Kröger M, Lanka E. 1989. *Proc. Natl. Acad. Sci. USA* 86:1771–75
25. Ziegelin G, Pansegrau W, Strack B, Balzer D, Kröger M, Kruft V, Lanka E. 1991. *DNA Sequence* 1:303–27
26. Pansegrau W, Ziegelin G, Lanka E. 1988. *Biochim. Biophys. Acta* 951:365–74
27. Llosa M, Bolland S, de la Cruz F. 1991. *Mol. Gen. Genet.* 226:473–83
28. Yadav NS, Vanderleyden J, Bennett DR, Barnes WM, Chilton M-D. 1982. *Proc. Natl. Acad. Sci. USA* 79:6322–26
29. Zambryski P, Tempe J, Schell J. 1989. *Cell* 56:193–201
30. Cook DM, Farrand SK. 1992. *J. Bacteriol.* 174:6238–46
31. Clewell DB. 1993. *Cell* 73:9–12
32. Morton TM, Eaton DM, Johnson JL, Archer GL. 1993. *J. Bacteriol.* 175:4436–47
33. Firth N, Ridgway KP, Byrne ME, Fink PD, Johnson L, Paulsen IT, Skurray RA. 1993. *Gene* 136:13–25
34. KrahIII ER, Macrina FL. 1989. *J. Bacteriol.* 171:6005–12
35. Brasch MA, Meyer RJ. 1986. *J. Bacteriol.* 167:703–10
36. Derbyshire KM, Willetts NS. 1987. *Mol. Gen. Genet.* 206:154–60
37. Scholz P, Haring V, Wittmann-Liebold B, Ashman K, Bagdasarian M, Scherzinger E. 1989. *Gene* 75:271–88

38. Snijders A, van Putten AJ, Veltkamp E, Nijkamp HJJ. 1983. *Mol. Gen. Genet.* 192:444–51
39. Morlon J, Chartier M, Bidaud M, Lasdunski C. 1988. *Mol. Gen. Genet.* 211: 231–43
40. Chan PT, Ohmori H, Tomizawa J, Lebowitz J. 1985. *J. Biol. Chem.* 260: 8925–35
41. Bastia D. 1978. *J. Mol. Biol.* 124:601–39
42. Archer JAK. 1985. PhD thesis, Univ. Glasgow, UK
43. Rose RE. 1988. *Nucleic Acids Res.* 16: 355
44. Hecht DW, Jagielo TJ, Malamy MH. 1991. *J. Bacteriol.* 173:7471–80
45. Bernardi A, Bernardi F. 1984. *Nucleic Acids Res.* 12:9415–26
46. Drolet M, Zanga P, Lau PCK. 1990. *Mol. Microbiol.* 4:1381–91
47. Rohrer J, Rawlings DE. 1992. *J. Bacteriol.* 174:6230–37
48. Murphy CG, Malamy MH. 1993. *J. Bacteriol.* 175:5814–23
49. Li L-Y, Shoemaker NB, Salyers AA. 1993. *J. Bacteriol.* 175:6588–98
50. Li L-Y, Shoemaker NB, Salyers AA. 1995. *J. Bacteriol.* In press
51. Willetts N, Wilkins BM. 1984. *Microbiol. Rev.* 48:24–41
52. Farrand SK. 1993. *Bacterial Conjugation,* ed. DB Clewell, pp. 255–91. New York: Plenum
53. Clewell DB. 1993. *Bacterial Conjugation,* ed. DB Clewell, pp. 349–67. New York: Plenum
54. Macrina FL, Archer GL. 1993. *Bacterial Conjugation,* ed. DB Clewell, pp. 313–29. New York: Plenum
55. Hopwood DA, Kieser T. 1993. *Bacterial Conjugation,* ed. DB Clewell, pp. 293–311. New York: Plenum
56. Scott JR. 1992. *J. Bacteriol.* 174:6005–10
57. Clewell DB, Flannagan SE. 1993. *Bacterial Conjugation,* ed. DB Clewell, pp. 369–93, New York: Plenum
58. Speer BS, Shoemaker NB, Salyers AA. 1992. *Clin. Microbiol. Rev.* 5:387–99
59. Stachel SE, Zambryski PC. 1986. *Cell* 46:325–33
60. Zambryski P. 1992. *Annu. Rev. Plant Physiol. Plant Mol. Biol.* 43:465–90
61. Lessl M, Lanka E. 1994. *Cell* 77:321–24
62. Wilkins B, Lanka E. 1993. *Bacterial Conjugation,* ed. DB Clewell, pp. 105–36. New York: Plenum
63. Pansegrau W, Lanka E. 1991. *Nucleic Acids Res.* 19:3455
64. Waters VL, Hirata KH, Pansegrau W, Lanka E, Guiney DG. 1991. *Proc. Natl. Acad. Sci. USA* 88:1456–60
65. Waters VL, Guiney DG. 1993. *Mol. Microbiol.* 9:1123–30
66. Wang K, Herrera-Estrella L, van Montagu M, Zambryski P. 1984. *Cell* 38: 455–62
67. Ohki M, Tomizawa J. 1968. *Cold Spring Harbor Symp. Quant. Biol.* 33:651–57
68. Rupp WD, Ihler G. 1968. *Cold Spring Harbor Symp. Quant. Biol.* 33:647–50
69. Vapnek D, Rupp WD. 1970. *J. Mol. Biol.* 53:287–303
70. Vapnek D, Lipman MB, Rupp WD. 1971. *J. Bacteriol.* 108:508–14
71. Tinland B, Hohn B, Puchta H. 1994. *Proc. Natl. Acad. Sci. USA* 91:8000–4
72. Thompson TL, Centola MB, Deonier RC. 1989. *J. Mol. Biol.* 207:505–12
73. Reygers U, Wessel R, Müller H, Hoffmann-Berling H. 1991. *EMBO J.* 10: 2689–94
74. Matson SW, Morton BS. 1991. *J. Biol. Chem.* 266:16232–37
75. Matson SW, Nelson WC, Morton BS. 1993. *J. Bacteriol.* 175:2599–606
76. Pansegrau W, Ziegelin G, Lanka E. 1990. *J. Biol. Chem.* 265:10637–44
77. Pansegrau W, Balzer D, Kruft V, Lurz R, Lanka E. 1990. *Proc. Natl. Acad. Sci. USA* 87:6555–59
78. Scherzinger E, Lurz R, Otto S, Dobrinski B. 1992. *Nucleic Acids Res.* 20:41–48
79. Guiney DG, Helinski DR. 1975. *J. Biol. Chem.* 250:8796–803
80. Blair DG, Helinski DR. 1975. *J. Biol. Chem.* 250:8785–89
81. Lovett MA, Helinski DR. 1975. *J. Biol. Chem.* 250:8790–95
82. Boyd AC, Archer JAK, Sherratt DJ. 1989. *Mol. Gen. Genet.* 217:488–98
83. Pansegrau W, Schröder W, Lanka E. 1993. *Proc. Natl. Acad. Sci. USA* 90: 2925–29
84. Balzer D, Pansegrau W, Lanka E.1994. *J. Bacteriol.* 176:4285–95
85. Scherzinger E, Kruft V, Otto S. 1993. *Eur. J. Biochem.* 217:929–38
86. Pansegrau W, Schoumacher F, Hohn B, Lanka E. 1993. *Proc. Natl. Acad. Sci. USA* 90:11538–42
87. Ilyina T, Koonin EV. 1992. *Nucleic Acids Res.* 13:3279–85
88. Llosa M, Bolland S, de la Cruz F. 1994. *J. Mol. Biol.* 235:448–64
89. Clewell DB, Helinski DR. 1969. *Proc. Natl. Acad. Sci. USA* 62:1159–66
90. Koraimann G, Schroller C, Graus H, Angerer D, Teferle K, Högenauer G. 1993. *Mol. Microbiol.* 9:717–27
91. Di Laurenzio L, Frost LS, Paranchych W. 1992. *Mol. Microbiol.* 6:2951–59
92. Nelson WC, Morton BS, Lahue EE, Matson SW. 1993. *J. Bacteriol.* 175: 2221–28

93. Luo Y, Gao Q, Deonier RC. 1994. *Mol. Microbiol.* 11:449–69
94. Tsai M-M, Fu YHF, Deonier RC. 1990. *J. Bacteriol.* 172:4603–9
95. Everett R, Willetts N. 1980. *J. Mol. Biol.* 136:129–50
96. Furuya N, Komano T. 1991. *J. Bacteriol.* 173:6612–17
97. Bradshaw HD Jr, Traxler BA, Minkley EG Jr, Nester EW, Gordon MP. 1990. *J. Bacteriol.* 172:4127–31
98. Traxler BA, Minkley EG Jr. 1988. *J. Mol. Biol.* 204:205–9
99. Lahue EE, Matson SW. 1990. *J. Bacteriol.* 172:1385–91
100. Inamoto S, Yoshioko Y, Ohtsubo E. 1991. *J. Biol. Chem.* 266:10086–92
101. Bowie JN, Sauer RT. 1990. *J. Mol. Biol.* 211:5–6
101a. Inamoto S, Fukuda H, Abo T, Ohtsubo E.1994. *J. Biochem.* 116:838–44
102. Derbyshire KM, Hatfull G, Willetts NS. 1987. *Mol. Gen. Genet.* 206:161–68
103. Haring V, Scherzinger E. 1989. *Promiscuous Plasmids of Gram-Negative Bacteria,* ed. CM Thomas, pp. 95–124, London: Academic
104. Cole SP, Lanka E, Guiney DG. 1993. *J. Bacteriol.* 175:4911–16
105. Ziegelin G, Fürste JP, Lanka E. 1989. *J. Biol. Chem.* 264:11989–94
106. Ziegelin G, Pansegrau W, Lurz R, Lanka E. 1992. *J. Biol. Chem.* 270:1269–76
107. Scheiffele P, Pansegrau W, Lanka E. 1995. *J. Biol. Chem.* 270:1269–76
108. Champoux JJ. 1977. *Proc. Natl. Acad. Sci. USA* 74:3800–4
109. Liu LF, Wang JC. 1979. *J. Biol. Chem.* 254:11082–88
110. Pansegrau W, Schröder W, Lanka E. 1994. *J. Biol. Chem.* 269:2782–89
111. Fürste JP, Ziegelin G, Pansegrau W, Lanka E. 1987. *DNA Replication and Recombination,* ed. R McMacken, TJ Kelly, UCLA Symp. Mol. Cell Biol., New Ser. 47:553–64. New York: Liss
112. van Mansfeld ADM, van Teeffelen HAA, Baas PD, Jansz HS. 1986. *Nucleic Acids Res.* 14:4229–38
113. Sherman JA, Matson SW. 1994. *J. Biol. Chem.* 269:26220–26
114. Filichkin SA, Gelvin SB. 1993. *Mol. Microbiol.* 8:915–26
115. Howard EA, Zupan JR, Citovsky V, Zambryski PC. 1992. *Cell* 68:109–18
116. Gheysen G, Villarroel R, van Montagu M. 1991. *Genes Dev.* 5:287–97
117. Mayerhofer R, Koncz-Kalman Z, Nawrath C, Bakkeren G, Crameri A, Angelis K, Redei GP, Schell J, Hohn B, Koncz C. 1991.*EMBO J.* 10:697–704
118. Bhattacharjee MK, Meyer RJ. 1991. *Nucleic Acids Res.* 19:1129–37
119. Mizuuchi K, Adzuma K. 1991. *Cell* 66:129–40
120. Engelman A, Mizuuchi K, Craigie R. 1991. *Cell* 6:1211–21
121. Kornberg A, Baker TA. 1992. *DNA Replication.* New York: Freeman. 2nd ed.
122. Roth MJ, Brown DR, Hurwitz J. 1984. *J. Biol. Chem.* 259:10556–68
123. Hanai R, Wang JC. 1993. *J. Biol. Chem.* 268:23830–36
124. Meyer TF, Geider K. 1982. *Nature* 296:828–32
125. Thomas CD, Balson DF, Shaw WV. 1990. *J. Biol. Chem.* 265:5519–30
126. Dempsey LA, Birch P, Kahn SA. 1992. *Proc. Natl. Acad. Sci. USA* 89:3083–87
127. Stewart GJ, Carlson CA. 1986. *Annu. Rev. Microbiol.* 40:211–35
128. Matson SW, Kaiser-Rogers KA. 1990. *Annu. Rev. Biochem.* 59:289–329
129. Sarathy PV, Siddiqi O. 1973. *J. Mol. Biol.* 78:443–51
130. Kingsman A, Willetts N. 1978. *J. Mol. Biol.* 122:287–300
131. Geider K, Hoffmann-Berling H. 1981. *Annu. Rev. Biochem.* 50:233–60
132. Kuhn B, Abdel-Monem M, Krell H, Hoffmann-Berling H. 1979. *J. Biol. Chem.* 254:11343–50
133. Lahue EE, Matson SW. 1988. *J. Biol. Chem.* 263:3208–15
134. Wessell R, Müller H, Hoffmann-Berling H. 1990. *Eur. J. Biochem.* 189:277–85
135. Walker JE, Saraste M, Runswick MJ, Gay NJ. 1982. *EMBO J.* 1:945–51
136. Gorbalenya AE, Koonin EV, Donchenko AP, Blinor VM. 1989. *Nucleic Acids Res.* 17:4713–38
137. Dash PK, Traxler BA, Panicker MM, Hackney DD, Minkley EG Jr. 1992. *Mol. Microbiol.* 6:1163–72
137a. Grandoso G. Llosa M, Zabala JC, de la Cruz F. 1994. *Eur. J. Biochem.* 226:403–12
138. Abdel-Monem M, Taucher-Scholz G, Klinkert M-Q. 1983. *Proc. Natl. Acad. Sci. USA* 80:4659–63
139. Silverman PM. 1987. *Bacterial Outer Membranes as Model Systems,* ed M. Inouye, pp. 277–309. New York: Wiley
140. Willetts N, Maule J. 1985. *Genet. Res. Camb.* 47:1–11
141. Curtiss R III, Fenwick RG Jr. 1975. *Microbiology-1974,* ed. D Schlessinger, pp. 156–65. Washington, DC: Am. Soc. Microbiol.
142. Maturin LJ Sr, Curtiss R III. 1981. *J. Bacteriol.* 146:552–63
143. Nielsen AK, Thorsted P, Thisted T, Wagner EGH, Gerdes K. 1991. *Mol. Microbiol.* 5:1961–73

144. Boulnois GJ, Beddoes MJ, Wilkins BM. 1979. *J. Bacteriol.* 138:324–32
145. Ohnishi Y, Akimoto S. 1980. *J. Bacteriol.* 144:833–35
146. Sakikawa T, Akimoto S, Ohnishi Y. 1989. *Biochim. Biophys. Acta* 1007:158–66
147. Kim K, Meyer RJ. 1989. *J. Mol. Biol.* 208:501–5
148. Bhattacharjee MK, Rao X-M, Meyer RJ. 1992. *J. Bacteriol.* 174:6659–65
149. Erickson MJ, Meyer RJ. 1993. *Mol. Microbiol.* 7:289–98
150. Rao X-M, Meyer RJ. 1994. *J. Bacteriol.* 176:5958–61
151. Gao Q, Luo Y, Deonier RC. 1994. *Mol. Microbiol.* 11:449–58
152. Everett R, Willetts N. 1982. *EMBO J.* 1:747–53
153. Meyer R. 1989. *J. Bacteriol.* 171:799–806
154. Barlett MM, Erickson MJ, Meyer RJ. 1990. *Nucleic Acids Res.* 18:3579–86
155. Bhattacharjee MK, Meyer RJ. 1993. *Nucleic Acids Res.* 21:4563–68
156. Anthony KG, Sherburne C, Sherburne R, Frost LS. 1994. *Mol. Microbiol.* 13:939–53
157. Manning PA, Achtman M. 1979. *Bacterial Outer Membranes: Biogenesis and Functions*, ed. M Inouye, pp. 407–47. New York: Wiley
158. Dürrenberger MB, Villiger W, Bächi T. 1991. *J. Struct. Biol.* 107:46–156
159. Lessl M, Balzer D, Weyrauch K, Lanka E. 1993. *J. Bacteriol.* 175:6415–25
160. Lessl M, Balzer D, Pansegrau W, Lanka E. 1992. *J. Biol. Chem.* 267:20471–80
161. Citovsky V, Zambryski P. 1993. *Annu. Rev. Microbiol.* 47:167–97
162. Thorstenson YR, Kuldau GA, Zambryski PC. 1993. *J. Bacteriol.* 175:5233–41
163. Lessl M, Pansegrau W, Lanka E. 1992. *Nucleic Acids Res.* 20:6099–100
164. Panicker MM, Minkley EG Jr. 1992. *J. Biol. Chem.* 267:12761–66
165. Panicker MM, Minkley EG Jr. 1985. *J. Bacteriol.* 162:584–90
166. Willetts NS. 1980. *Mol. Gen. Genet.* 180:213–17
167. van Putten AJ, Jochems GJ, de Lang R, Nijkamp HJJ. 1987. *Gene* 51:171–78
168. Beijersbergen A, den Dulk-Ras A, Schilperoort RA, Hooykaas PJJ. 1992. *Science* 256:1324–47
169. Waters VL, Strack B, Pansegrau W, Lanka E, Guiney DG. 1992. *J. Bacteriol.* 174:6666–73
170. Armstrong GD, Frost LS, Sastry PA, Paranchych W. 1980. *J. Bacteriol.* 141:333–41
171. Deleted in proof
172. Merryweather A, Rees CED, Smith NM, Wilkins BM. 1986. *EMBO J.* 5:3007–12
173. Rees CED, Wilkins BM. 1989. *J. Bacteriol.* 171:3152–57
174. Rees CED, Wilkins BM. 1990. *Mol. Microbiol.* 4:1199–205
175. Chase JW, Merrill BM, Williams KR. 1983. *Proc. Natl. Acad. Sci. USA* 80:5480–84
176. Golub EI, Low KB. 1985. *J. Bacteriol.* 162:235–41
177. Lohman TM, Ferrari ME. 1994. *Annu. Rev. Biochem.* 63:527–70
178. Golub EI, Low KB. 1986. *Mol. Gen. Genet.* 204:410–16
179. Howland CJ, Rees CED, Barth PT, Wilkins BM. 1989. *J. Bacteriol.* 171:2466–73
180. Jones AL, Barth PT, Wilkins BM. 1992. *Mol. Microbiol.* 6:605–13
181. Kolodkin AL, Capage MA, Golub EI, Low KB. 1983. *Proc. Natl. Acad. Sci. USA* 80:4422–26
182. Citovsky V, Wong ML, Zambryski P. 1989. *Proc. Natl. Acad. Sci. USA* 86:1193–97
183. Christie PJ, Ward JE, Winans SC, Nester EW. 1988. *J. Bacteriol.* 170:2659–67
184. Das A. 1988. *Proc. Natl. Acad. Sci. USA* 85:2909–13
185. Gietl C, Koukolíková-Nicola Z, Hohn B. 1987. *Proc. Natl. Acad. Sci. USA* 84:9006–10
186. Sen P, Pazour GJ, Anderson D, Das A. 1989. *J. Bacteriol.* 171:2573–80
187. Citovsky V, Zupan J, Warnick D, Zambryski P. 1992. *Science* 256:1802–5

Annu. Rev. Biochem. 1995. 64:171–200

DNA POLYMERASE III HOLOENZYME: Structure and Function of a Chromosomal Replicating Machine

Zvi Kelman and Mike O'Donnell[1]

Microbiology Department and Hearst Research Foundation, Cornell University
Medical College, 1300 York Avenue, New York, NY 10021

KEY WORDS: DNA replication, multisubunit complexes, protein-DNA interaction,
                DNA-dependent ATPase, DNA sliding clamps

CONTENTS

[1]Howard Hughes Medical Institute

171

0066-4154/95/0701-0171$05.00

ABSTRACT

DNA polymerase III holoenzyme contains two DNA polymerases embedded in a particle with 9 other subunits. This multisubunit DNA polymerase is the *Escherichia coli* chromosomal replicase, and it has several special features that distinguish it as a replicating machine. For example, one of its subunits is a circular protein that slides along DNA while clamping the rest of the machinery to the template. Other subunits act together as a matchmaker to assemble the ring onto DNA. Overall, *E. coli* DNA polymerase III holoenzyme is very similar in both structure and function to the chromosomal replicases of eukaryotes, from yeast all the way up to humans. This review summarizes our present knowledge about the function of the 10 subunits of this replicating machine and how they coordinate their actions for smooth duplication of chromosomes.

INTRODUCTION

The main function of DNA polymerase III holoenzyme (holoenzyme) is duplication of the *E. coli* chromosome, although it acts in other areas of DNA metabolism as well (1). Holoenzyme shares special features with replicases of eukaryotes, viruses, prokaryotes, and their phages, which distinguishes holoenzyme from single-subunit polymerases such as DNA polymerase I (Pol I). Among these features are a multisubunit structure, the requirement for ATP to clamp tightly to DNA, the rapid speed of DNA synthesis, and a remarkably high processivity, such that the enzyme remains bound to DNA for thousands of polymerization events (1, 2). Replicases of most systems share amino acid sequence homology to holoenzyme. Hence, holoenzyme is likely to serve as a faithful guide to understanding the basics of replicase action in other systems.

Holoenzyme functions at the point of the replication fork with other proteins. Replication of the chromosome entails separation of the duplex DNA by helicase and topoisomerase, followed by semidiscontinuous synthesis of DNA at a speed of about 1 kilobase (kb) per second (3). The discontinuous strand (lagging) is synthesized by holoenzyme acting with a priming apparatus for repeated initiation and extension of 2000–4000 Okazaki fragments. These fragments are only 1–2 kb in length, and therefore each is completed within 1–2 s. The intracellular scarcity of holoenzyme [10–20 molecules (4)] necessitates rapid recycling upon completing one fragment and transfer to a new primer for the next fragment. Holoenzyme is clamped tightly to DNA by a sliding-clamp subunit that completely encircles the duplex (5, 6), but despite this tight grip to DNA, holoenzyme has a novel mechanism allowing it to rapidly cycle on and off DNA for action on the lagging strand (7–9).

There have been several reviews on holoenzyme in the past few years (2,

10–14), and this review is an update since the last in this series (12). The outline of how holoenzyme functions at a replication fork is presented; the reader is referred to recent reviews for more information (2, 10, 15–17, 17a). Holoenzyme also functions in repair and mutagenesis, and excellent reviews on these subjects have appeared recently (18–21).

THE HOLOENZYME PARTICLE

DNA polymerase III was first identified as the chromosomal replicase on the basis that extracts of temperature-sensitive mutants in the essential *dnaE* gene contained temperature-sensitive DNA polymerase III activity (18–24). Initial purification of DNA polymerase III utilized a template DNA that was nicked and gapped by nuclease action, and probably led to purification of the three-subunit subassembly that is now called DNA polymerase III core (core) (25). Subsequent studies utilized primed circular single-stranded (ss) DNA genomes of bacteriophages M13, G4, and ϕX174 as templates, which led to purification of holoenzyme and its subassemblies (26–33).

These early studies were hampered by the low abundance of holoenzyme in *E. coli*. There are only 10–20 copies of holoenzyme in the cell, and purification of one mg to near homogeneity requires 7400-fold enrichment from 2–3 kg of cells (29). Despite its scarcity, study of holoenzyme and its subassemblies outlined many important features of this replicating machine. For example, holoenzyme was found to be exceedingly rapid in DNA synthesis—approximately 750 nucleotides/s—consistent with the observed rate of fork movement in *E. coli* (34) and much faster than the 10–20 nucleotides/s of Pol I (35). This rapid rate results from the high processivity of holoenzyme, which extends a chain for several thousand nucleotides without dissociating from the template even once (36, 37). In contrast, Pol I dissociates rapidly from DNA, extending a primer only 10–50 nucleotides for each template-binding event (34). Holoenzyme is also distinguished from Pol I in a requirement for ATP hydrolysis (26, 30). The ATP is only needed initially by holoenzyme to clamp onto a primed template; afterward holoenzyme is rapid and processive without additional ATP (38, 39). Upon encounter with a duplex region, holoenzyme simply diffuses over the duplex, searches out the next 3' end, and reinitiates processive extension without additional ATP (40).

Identification of all the genes encoding the 10 subunits of holoenzyme has been completed recently, the proteins overproduced and purified, and the holoenzyme reconstituted from them. In Table 1 the 10 different subunits are listed in an order that explains which subunits are present in the various subassemblies of holoenzyme. The core polymerase consists of the α, ϵ, and θ subunits (41). The Pol III' subassembly contains two cores and a dimer of τ (42, 43). The presence of two polymerases in one molecular structure sup-

Table 1 DNA Polymerase III holoenzyme subunits and subassemblies

Subunit	Gene	Mass (kDa)	Function	Subassembly
α	dnaE	129.9	DNA polymerase	core ⎤
ε	dnaQ, mutD	27.5	Proofreading 3'–5' exonuclease	PolIII' ⎤
θ	holE	8.6	Stimulates ε exonuclease	PolIII* ⎤
τ	dnaX	71.1	Dimerizes core, DNA-dependent ATPase	
γ	dnaX	47.5	Binds ATP	γ-complex
δ	holA	38.7	Binds to β	
δ'	holB	36.9	Cofactor for γ ATPase and stimulates clamp loading	
χ	holC	16.6	Binds SSB	
ψ	holD	15.2	Bridge between χ and γ	
β	dnaN	40.6	Sliding clamp on DNA	

ports the hypothesis that replicative polymerases act in pairs for coordinated replication of both strands of a duplex chromosome (discussed later). The Pol III* assembly contains 9 different subunits; it lacks only β (44). The polymerase activity of each of these subassemblies can be distinguished on the basis of adding either spermidine, ssDNA-binding protein (SSB protein), ethanol, or salt to the assays (36, 45). In general the polymerases become more processive as their subunit complexity increases, but the very high speed and processivity of holoenzyme absolutely requires the β clamp (36, 45). The five-subunit γ complex is a matchmaker that couples ATP hydrolysis to load β clamps on primed DNA (7, 31, 46).

THE CORE POLYMERASE

Core contains the DNA polymerase and proofreading exonuclease activities (47). There are approximately 40 molecules of core in the cell, and therefore only half are assembled into holoenzyme (47). The three subunits of core are tightly associated and cannot be resolved short of denaturation. Individual subunits are provided through use of the genes. Study of α showed it to be the DNA polymerase (8 nucleotides/s), but it lacked exonuclease activity (47, 48). The isolated ε subunit is a potent 3′-5′ exonuclease (49), consistent with the *dnaQ/mutD* mutator phenotype (50, 51). The α and ε subunits form a tight 1:1 complex, resulting in increases in both polymerase activity (34) and exonuclease activity (52). The rate of digestion of ssDNA by ε is similar to that of core, but hydrolysis of double-stranded (ds) DNA by ε requires α for significant activity (52). Presumably the primer template-recognition site of α brings ε in contact with a basepaired 3′ end. The function of θ has yet to be identified, except for a slight stimulation of ε activity on a mismatched T-G basepair (53). The θ subunit binds ε but not α, suggesting a linear α-to-ε-to-θ arrangement in core, and structural analysis shows a single copy of each subunit (53).

Core synthesizes DNA at a rate of approximately 20 nucleotides/s and is processive for 11 nucleotides (36), similar to Pol I. However, on a singly primed ssDNA viral template, core is the weakest polymerase known. It cannot extend a unique primer full circle around a natural template no matter how much core is added or how long one waits (54). Presumably some DNA structures are absolute barriers to chain extension by core.

Ironically, core becomes the fastest polymerase in the presence of its accessory proteins (discussed below). In the absence of ε, α is stimulated by the accessory proteins, but the processivity drops to 500–1500 nucleotides, and the intrinsic speed is half that of core (34). With accessory proteins, the αε complex is as fast and processive as core (34). Hence, ε has effects on the speed and processivity of holoenzyme, not just fidelity. On the other hand, θ has no effect on the efficiency of $\alpha\varepsilon$ (53). These results are consistent with the growth defect of *dnaQ* (ε) mutants and lack thereof in *holE* (θ) null mutants (53) (described in the GENETICS section).

THE β DNA SLIDING CLAMP

The ATP-activated grip of holoenzyme to primed DNA is inherent in the accessory proteins, γ complex and β subunit. The γ complex can be resolved from the holoenzyme only by harsh treatment (29), but some γ complex exists in free form and can be purified alone (56). Presumably, γ complex was the active ingredient in elongation factor II (30). In contrast, the β subunit departs from holoenzyme easily and can be separated on a phosphocellulose column [used to be called copol III* (26, 27) and elongation factor I (30)]. The intracellular abundance of β (300 dimers per cell) made its purification possible without having to resort to resolving it from purified holoenzyme (57). Early studies using partially pure preparations indicated that γ complex coupled ATP to the assembly of β onto DNA (31). A reinvestigation of this reaction using pure proteins and primed DNA coated with SSB protein confirmed the earlier observation. One dimer of β is chaperoned to DNA in an ATP-dependent reaction catalyzed by γ complex in the absence of core to form the "preinitiation complex" (7, 46). In a second stage, the core assembles with the preinitiation complex to form the "initiation complex" in a reaction that does not require ATP (7, 31). Hence, holoenzyme has two components that recognize a primer terminus: the core polymerase and the accessory proteins themselves.

The γ complex has only weak affinity for ss and ds DNA, although it does bind to ssDNA coated with SSB protein (described later). The γ complex easily departs into solution after it places β onto primed DNA. This "β-only" preinitiation complex retains the capacity to restore highly efficient synthesis onto core (see Figure 1) (5). Following departure from the β-DNA complex, the γ complex is still active and is able to place multiple β dimers on DNA, accounting for the high specific activity of the γ complex (5, 29, 56).

The γ complex can place β onto a singly nicked plasmid (RF II), and upon linearizing the circular plasmid with a restriction enzyme, β dissociates from DNA, implying β has mobility on DNA and can slide off over ends (5). This behavior of β on DNA allowed reasoning of the nature of the β-DNA interaction. Since the affinity of β to DNA depends on the geometry of the DNA molecule, β must likewise be bound to DNA by virtue of its protein topology (i.e. by encircling the DNA like a doughnut). If the main attraction of β to DNA were through chemical forces (i.e. ionic, hydrophobic, or hydrogen bonds), as is the case with all other DNA-binding proteins before β, then upon reaching the end, β would have remained bound to DNA rather than give up its tight chemical grip.

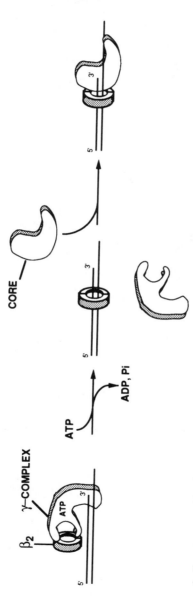

Figure 1 Two-stage assembly of a processive polymerase. The γ complex recognizes a primed template and couples hydrolysis of ATP to assemble β onto DNA. The γ complex easily dissociates from DNA and can resume its action in loading β clamps on other DNA templates. In a second step, core assembles with the β clamp to form a processive polymerase.

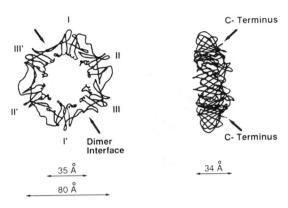

Figure 2 Structure of the β subunit. *Left*—The central cavity is lined with 12 α helices, and the outside perimeter is one continuous layer of sheet structure, which also forms the intermolecular boundaries (arrows). The sixfold appearance stems from three globular domains that compose each monomer. These domains have the same polypeptide chain–folding pattern. The six domains are labeled I, II, and III on one monomer and I′, II′, and III′ on the other monomer. *Right*—The β dimer is turned 90° relative to the view on the left. The thickness of the β ring is approximately equal to one turn of β-form DNA. The two C termini extrude from the same face of the ring (arrows). Dimensions of the inner and outer diameters of the ring and the thickness of β are shown below the diagrams.

A simple mechanism by which a β "sliding-clamp" confers processivity to core is by directly binding core, thus tethering it to DNA; the clamp would be passively pulled along with core during polymerization. Consistent with this notion, β binds to core through the α subunit even in the absence of DNA (5, 58, 59).

The β subunit as a sliding-clamp doughnut was confirmed by X-ray structure analysis (6). The β appeared as a ring-shaped head-to-tail dimer with a central cavity of sufficient diameter to accommodate duplex DNA (Figure 2). The central cavity is lined with 12 α helices, and the ring is encased by one continuous layer of antiparallel pleated sheet along the outside. The β dimer has a six-fold appearance even though it only has a true two-fold rotational axis of symmetry. The apparent six-fold symmetry derives from a three-fold repetition of a globular domain in the monomer (six domains in the dimer). The three domains have no significant amino acid homology, yet they are nearly superimposable.

The 12 α helices lining the central cavity have a common tilt and lie perpendicular to the phosphate backbone of duplex DNA. Hence, the helices may act as crossbars to prevent β from entering the grooves of DNA and facilitate the sliding motion. Further, β is quite acidic (pI = 5.2) and would be repelled by DNA, but there is a net positive surface potential inside the cavity. There is room for 1–2 layers of water molecules between the DNA and the α helices, which may insulate β from local interaction with DNA. Those that are interested in other features of the β structure are referred to several reviews (60–62).

THE γ COMPLEX MATCHMAKER

The β dimer does not assemble onto DNA by itself. The γ complex is a molecular matchmaker that hydrolyzes ATP to load β clamps onto DNA. The γ complex is composed of five different subunits in the stoichiometry $\gamma_2\delta_1\delta'_1$ $\chi_1\psi_1$ (56, 63). The δ and δ' subunits, originally thought to be related by proteolysis, are distinct proteins encoded by different genes (64–67). Interestingly, the δ' amino acid sequence shows homology to γ and τ (64, 67, 68). The γ, τ, and δ' subunits are further characterized by their appearance as doublets on an SDS polyacrylamide gel (64, 67, 69). The physical basis and the function of this microheterogeneity are not known.

The γ complex can be fragmented into a $\gamma\chi\psi$ complex (125.8 kDa) and a $\delta\delta'$ complex (75.6 kDa), and $\delta\delta'$ can be further resolved into δ and δ' (69). In early studies using partially pure fractions, the γ complex activity (elongation factor II) was subdivided into two factors, one of 125 kDa (called DnaZ protein) and one of 63 kDa (called elongation factor III) (31, 32). Presumably these factors were $\gamma\chi\psi$ and $\delta\delta'$, respectively.

The genes encoding each subunit of γ complex have now been identified (64–67). The proteins have been overproduced, purified (65, 70), and used to reconstitute the γ complex in abundance (63). No one subunit alone can assemble β onto DNA (8, 69, 74). At low ionic strength, a combination of γ and δ assembles β onto DNA, but the reaction is feeble; the δ' subunit is needed for an efficient reaction (65, 69, 75). The χ and ψ subunits are also needed at an ionic strength commensurate with that inside the cell (69).

Role of ATP

The γ complex has weak DNA-dependent ATPase activity and is stimulated by β (75). The best effector is a primed template. The only subunit of γ complex with an exact match to an ATP binding site motif is γ (76, 77), and γ binds ATP with a K_d of 2 μM (78). The γ subunit lacks significant hydrolysis activity even in the presence of DNA (75, 78). Significant DNA-dependent ATPase activity of γ requires δ and δ', implying that the $\gamma\delta\delta'$ complex recognizes the DNA template (65, 75). ATP is crosslinked to δ upon irradiation with UV light (79), and the δ sequence shows a close match to an ATP site sequence (64, 66). However, evidence against a role for ATP binding in δ action, at least in β assembly, has been gained by replacing the Lys of the putative ATP-binding site in δ with an Ala. The γ complex constituted using the mutated δ is as active as wild-type γ complex in assembly of β on DNA, and in DNA-dependent ATPase activity (H Xiao, M O'Donnell, unpublished). Mutation of the ATP binding site of γ and subsequent constitution into the γ complex destroys the ATPase activity and ability to assemble β onto DNA. Further, ATP binding site mutants of γ and τ, expressed from a plasmid, fail to complement a conditional lethal *dnaX* strain (J Walker, personal communication).

Holoenzyme hydrolyzes two molecules of ATP upon forming an initiation complex on primed DNA (39). Presumably the action here lies with the γ complex in assembling β onto DNA. That two ATP are hydrolyzed indicates that each γ protomer hydrolyzes one ATP during assembly of β onto DNA (43, 78). τ is also a DNA-dependent ATPase, however, and may contribute to the observed hydrolysis (75, 78, 80).

The K_d for interaction of Pol III* with β is approximately 1 nM in the presence of ATP; in the absence of ATP the interaction is undetectable (81). The γ complex also binds β in an ATP-dependent manner (V Naktinis, M O'Donnell, unpublished). Study of individual subunits of γ complex shows that only δ interacts with β (63). The δ-to-β interaction does not depend on ATP. A simple mechanism to explain the ATP dependence of the γ complex–β interaction and the lack of an ATP requirement for the δ-β interaction is that δ is buried within γ complex and ATP induces a conformational change that presents δ for interaction with β (Figure 3).

Addition of a large excess of β to holoenzyme circumvents the need for ATP in forming a processive polymerase (82, 83). This interesting observation implies that ATP is not needed for the β ring to open and close around DNA. However, these studies were performed using linear templates, and β may have threaded over a DNA end without opening. Indeed, ATP-independent thread-

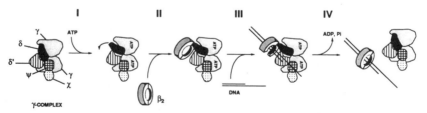

Figure 3   Putative action of γ complex in assembly of a β clamp on DNA. The diagram of γ complex is consistent with the known stoichiometry and contacts between the subunits (γ-γ, γ-ψ, γ-δ', δ-δ', and χ-ψ). In the first diagram the surface of δ that interacts with β is buried to explain its inability to bind β in the absence of ATP. Upon binding (or hydrolyzing) ATP, a conformational change exposes δ (step I) for binding β (step II). The γ complex then recognizes a primed template, thus bringing β into proximity with DNA (step III). In step IV, hydrolysis of ATP (or loss of ADP, P_i) sequesters δ back within γ complex, severing the δ-β contact and allowing β to snap shut around DNA and γ complex to dissociate. The β subunit is shown as opening at one interface and then reclosing; other possibilities exist, however, as discussed in the text.

ing of proliferating cell nuclear antigen (PCNA) (the eukaryotic homolog of β) over DNA ends has been observed (84). Further studies on the ATP-independent stimulation of holoenzyme by excess β are necessary to determine what insight the reaction provides into the clamp-loading mechanism.

Interaction of γ Complex with SSB Protein

The γ complex binds ssDNA coated with SSB protein, but not naked ssDNA (85). Study of individual subunits of γ complex showed that only χ interacts with SSB protein (Z Kelman, M O'Donnell, unpublished). The affinity of χ for SSB protein was strengthened approximately eightfold by the presence of ssDNA. The χ-to-SSB protein contact is sensitive to ionic strength and may underlie the known salt sensitivity of holoenzyme initiation complex formation (86). Holoenzyme is more resistant to potassium glutamate than to any other salt (86), consistent with potassium glutamate as the physiological osmolyte of *E. coli* (87).

A clue to further roles of the χ-to-SSB protein contact may be taken from study of mutant SSB proteins (reviewed in 88, 88a, 89). One SSB protein mutant, *SSB-113,* has a pleiotropic phenotype including defects in replication and recombination. The *SSB-113* is a missense mutant in which the penultimate amino acid at the C terminus, Pro176, is replaced with Ser. *SSB-113* binds ssDNA as tightly as wild-type SSB protein, leading to the suggestion that the C terminus of SSB protein may interact with proteins. Study of χ and *SSB-113* shows χ does not interact with *SSB-113,* implying that χ may be involved in one or more of the *SSB-113* phenotypes (Z Kelman, M O'Donnell, unpublished).

Mechanism of the γ Complex Clamp Loader

A mechanism by which γ complex may assemble β around DNA is hypothesized in Figure 3. Upon binding (or hydrolysis) of ATP by the γ complex, δ is presented for interaction with β. The γ complex recognizes a primed template, possibly aided by the χ-to-SSB protein contact. The interaction of γ complex with both DNA and β, positions β near the primer terminus where it can be assembled around the DNA.

Exactly how the β ring is opened and closed around DNA and how ATP hydrolysis is coupled to the process are unknown. Three possible mechanisms are: (*a*) Only one interface of the β dimer is opened and closed around DNA (as in Figure 3), (*b*) both interfaces are opened followed by reforming the dimer around DNA, and (*c*) the DNA is cut and rejoined after being threaded through the β ring.

A rapid monomer-dimer equilibrium for β (K_d = 35 nM) has been reported in the presence of magnesium (90, 91), suggesting the β dimer is inherently unstable and implying that β may come apart at both interfaces during assembly on DNA. In another study, however, the rate of subunit exchange between β dimers was slow, with a half-life of 3 h at 37°C, suggesting the dimer is quite stable (92). Further studies using β and γ complex are needed to define the mechanism of β assembly onto DNA.

THE τ SUBUNIT

The length of DNA needed to code for the combined mass of τ (71 kDa) and γ (47 kDa) is 3.2 kb. However, the region of DNA expressing both τ and γ is only 2.1 kb (93, 94). Further study showed γ is formed from the same gene that encodes τ (*dnaX*) by an efficient translational frameshift, which produces γ in amounts equal to those of τ (95–97). As a result, γ is the N-terminal 430 residues of τ followed by a unique C-terminal Glu. One may consider that the holoenzyme is composed of two populations: those with γ and those with τ. Examination reveals, however, that each holoenzyme molecule contains both γ and τ (98).

The τ subunit is a DNA-dependent ATPase of ill-defined function (80). From studies using pure subunits, a "τ complex" (τδδ'χψ) can be assembled and is active in loading β clamps on DNA (65, 69). Whether τ serves such a role in holoenzyme action is not known. Inability to isolate a τ complex from cell lysates suggests that τ complex is not present in vivo, and thus that τ ATPase may be put to another task.

The τ and γ subunits are the only subunits of Pol III* with oligomeric structure. The τ dimer binds two molecules of core (42, 43). The γ subunit does not bind core, and therefore the C-terminal sequence unique to τ is responsible for the τ-core interaction. Indeed, mutation of the C-terminal region of τ destroys cell viability, suggesting that the ability of τ to dimerize core is an essential function (99).

ASYMMETRIC STRUCTURE OF HOLOENZYME

Synthesis of the leading strand and synthesis of the lagging strand are quite different. The leading-strand polymerase need only remain clamped to DNA continuously, but the lagging strand is synthesized discontinuously as a series of fragments. Thus the lagging-strand polymerase must repeatedly be clamped and unclamped from DNA to cycle from one fragment to the next. The hypothesis, that replicative polymerases act in pairs for simultaneous synthesis of both strands of duplex DNA (100, 101), was extended by McHenry by suggesting that the accessory proteins may be distributed asymmetrically relative to the two polymerases to confer distinctive properties for leading and lagging strands (102).

Evidence for functional asymmetry in holoenzyme was obtained from assays using the ATP analog, ATPγS (102). In the presence of ATPγS, one-half the amount of holoenzyme is clamped onto primed DNA relative to use of ATP. After using ATP to clamp holoenzyme onto DNA, treatment with ATPγS released one half of the enzyme. It was hypothesized that of the two polymerases in the holoenzyme, one could use ATPγS to clamp onto DNA, and the other was dissociated from DNA by ATPγS.

Evidence for asymmetry in holoenzyme structure has also been obtained. The τ dimer binds two core polymerases tightly ($K_d < 17$ nM); the simplest arrangement imaginable is one core on each τ protomer (42, 43). The τ subunit also binds the γ complex (described below), leading to an organization of subunits illustrated at the bottom of Figure 4. In Figure 4, the τ dimer is assumed to be in the common isologous arrangement, in which each core-τ protomer unit is related to the other by a two-fold axis of rotation (i.e. τ is symmetric relative to the two polymerases). The γ complex is an asymmetric structure, because four of its subunits are present in a single copy (56, 63). Hence, γ complex imposes an asymmetry about the two core polymerases (as shown in Figure 4). Consistent with the holoenzyme structure in Figure 4, the composition of Pol III* showed a total of 14 polypeptides in the following composition: $\alpha_2\varepsilon_2\theta_2\tau_2\gamma_2\delta_1\delta'_1\chi_1\psi_1$ (63). This composition is consistent with Pol III* containing a Pol III′ assembly ($\alpha_2\varepsilon_2\theta_2\tau_2$) bound to one γ complex ($\gamma_2\delta_1\delta'_1\chi_1\psi_1$).

Although the subunit composition of Pol III* suggests one γ complex bound to Pol III′, the Pol III* could not be formed by mixing Pol III′ with γ complex even at high concentration (28 μM each) (63). Since a "τ complex" ($\tau\delta\delta'\chi\psi$) can be constituted (63, 65, 71), it was thought that perhaps Pol III* would assemble upon mixing "τ complex," γ complex, and core; this still did not result in Pol III*, however (63). Further study showed that the assembly of Pol III* relies on adding subunits in a defined order (63). The resulting Pol III* remains associated even when diluted to 30 nM and therefore, once formed, Pol III* does not easily fall apart (J Turner, M O'Donnell, unpublished). An essential contact needed to form Pol III* is interaction of γ with τ to form a $\gamma_2\tau_2$ tetramer (63). The γ-to-τ contact is inhibited if the $\delta\delta'\chi\psi$ subunits are added to γ or τ before mixing γ with τ (63). After the $\gamma_2\tau_2$ tetramer forms, the two core polymerases and the $\delta\delta'\chi\psi$ subunits can be assembled onto it (see Figure 4).

Study of how δ, δ′, χ, and ψ inhibit the γ-to-τ contact showed that δ and δ′, but not χ or ψ, are the culprits (63). The δ′ subunit inhibits the γ-to-τ contact if it is present on both γ and τ, but if δ′ is present on only γ or τ, the γ-to-τ contact is productive and Pol III* assembly is enabled. Hence, in Figure 4, δ′ is shown near the interface of the γ_2-τ_2 tetramer, such that binding of one δ′ occludes a second molecule of δ′. This restriction of only one δ′ on either γ or τ to enable Pol III* assembly is consistent with the stoichiometry of one δ′

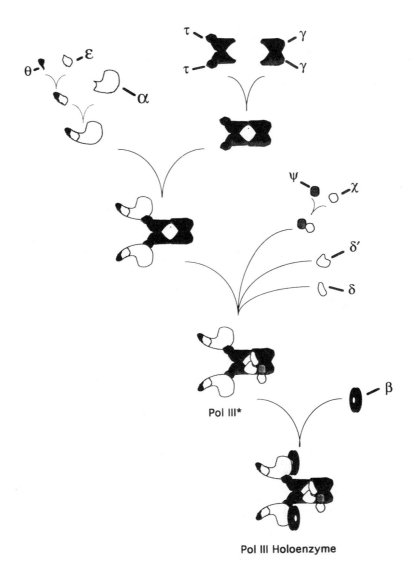

Pol III*

Pol III Holoenzyme

Figure 4 Assembly of the asymmetric holoenzyme. Organization of the 10 different subunits within the holoenzyme particle. The τ dimer and γ dimer are each shown in an isologous arrangement and the γ-τ heterotetramer is also shown as isologous. Each core polymerase is shown as a linear arrangement of α-ε-θ. The two core polymerases are attached through α to the τ dimer. The single-copy subunits, δ, δ′,χ, and ψ, assemble onto the γ-τ heterotetramer and must be added in order (see text for details). The δ′ is positioned near the τ and γ interface to explain the observation that only one δ′ is accommodated in the heterotetramer. The ability to form a $\delta_1\delta'_1$ complex is reflected in the contact of δ to δ′. The χ subunit binds ψ, which in turn binds γ (or τ). Two β dimers are shown bound to the two cores. Reflected in the final structure are the strong intersubunit contacts within holoenzyme, identified as α-ε, ε-θ, τ-α, δ-δ′, χ-ψ, γ-ψ, τ-ψ, and τ-γ (34, 43, 53, 63, 65, 69, 71).

in Pol III*, and also with the single δ in Pol III*, as it has been shown that δ-δ' forms a 1:1 complex (65). The δ subunit also inhibits the γ-to-τ contact if added to the reaction early; if δ is added after the γ-to-τ contact is established, assembly of Pol III* proceeds. This phenomenon explains why Pol III' and γ complex do not assemble to form Pol III* and may even be a useful mechanism to keep some Pol III' and γ complex as separate entities. Perhaps Pol III' and/or γ complex have separate roles in other DNA metabolic pathways, such as in repair or recombination.

Since the δ, δ', χ, and ψ subunits can be added after mixing τ with γ, their position on either γ or τ is ambiguous. The presence of core on τ decreases the association rate of these subunits with τ, and thus should bias their association toward γ (V Naktinis, M O'Donnell, unpublished). This kinetic bias may explain why Pol III', purified from *E. coli* lysates, does not contain the δδ'χψ subunits (42). It is still possible, however, that in the holoenzyme, the single-copy δδ'χψ subunits are functional with both halves of the $\gamma_2\tau_2$ tetramer. Further, the δ' subunit displays weak, but detectable, clamp-loading activity with τ, but not with γ, thereby presenting the possibility of two clamp loaders in Pol III* consisting of τδ' and γδ (65, 69).

A slightly different subunit arrangement and stoichiometry were suggested in an earlier study in which core was proposed to be dimerized by θ, and the τ dimer was proposed to bind only one core and γ the other (44). The dimerization of core by θ was indicated by a larger species of core polymerase when concentrated to 18 μM. However, later studies using reconstituted core at 73 μM showed it was only a monomer ($\alpha_1\epsilon_1\theta_1$) (53). Evidence that τ is located on one core and γ on the other lies in an observation that τ and γ complex compete for binding to core and β on primed ssDNA coated with SSB protein (44). The competition between τ and γ complex may have been, however, for sites on the template, since both τ and γ complex bind ssDNA coated with SSB protein (85, 103).

DNA footprinting studies show the holoenzyme protects approximately 30 nucleotides of the duplex portion of a primer template (J Reems, C McHenry, personal communication). Finer analysis using chemical crosslinking agents attached to specific nucleotides on the primer strand show α crosslinks to position −13, γ crosslinks to position −18, and β at position −22; no subunit crosslinks to position −27 (J Reems, C McHenry, personal communication). Fluorescence energy transfer between a fluorophore on β (Cys333) and a fluorophore on DNA (3 nucleotides back from the primer terminus) indicates a distance of 65 Å between them (104).

The arrangement of subunits within holoenzyme and how they are oriented on DNA may be learned from future work by several approaches, including crosslinking, fluorescence energy transfer, neutron scattering, 2D crystals in the electron microscope, and 3D crystals analyzed by X rays.

DNA POLYMERASE III HOLOENZYME AS A REPLICATING MACHINE

Exchange of β from γ Complex to Core

The γ complex must bind β to assemble it on a primer terminus, and core must interact with β for processivity. Since both core and γ complex recognize a primed template junction, they may interact with the same face of the β ring. Comparison of gene sequences encoding β from seven different bacteria shows that the most conserved residues lie on only one face: the face containing the two C termini (see Figure 2) (Z Kelman, M O'Donnell, unpublished). Consistent with this face as a site of interaction, point mutants in four of the five C-terminal residues of β inactivate β in replication assays and also prevent β from binding γ complex (V Naktinis, M O'Donnell, unpublished). Surprisingly, each of the C-terminal point mutations also prevented β from binding core, indicating that core and γ complex bind to β at the same place.

Why do core and γ complex have overlapping binding sites on β? The γ complex not only loads β onto DNA, but also unloads β clamps from DNA (9, 103). Hence, the competitive arrangement could ensure that while core is using β to extend DNA, it prevents γ complex from unloading β from DNA.

Studies using subassemblies (γ complex, core, and β) invoke the idea that only core and β are present on DNA during chain elongation, since γ complex acts catalytically. In fact, the overlapping binding site of γ complex and core on the β dimer is consistent with this view. Studies using the entire holoenzyme, however, show that γ complex remains with core and β on DNA (44, 105). In the holoenzyme, τ acts as a bridge between core and γ complex to hold them together (103). This arrangement may allow β to be repositioned from γ complex to core as illustrated in Figure 5. Positioning the catalytic clamp-loading activity of γ complex at a replication fork, through constant association with the holoenzyme, would be advantageous for the multiple initiation events on the lagging strand (described below).

Cycling of Holoenzyme on the Lagging Strand

The picture of a polymerase with a sliding clamp riding behind it fits nicely with continuous synthesis of the leading strand. On the lagging strand, however, the DNA is synthesized discontinuously in a series of short Okazaki fragments (1). Each fragment is only 1–2 kb, and at a speed near 1 kb/s, the polymerase will finish a fragment within a second or two and must rapidly recycle to the next RNA primer. The β clamp holding the polymerase tight to DNA would conceptually hinder rapid recycling of polymerase. One strategy to overcome this difficulty would be to produce 4000 molecules of holoenzyme, one for each Okazaki fragment. Because there are only 10–20 molecules

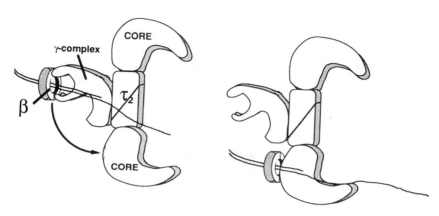

Figure 5 Core and γ complex interact with the same face of the β ring. The holoenzyme contains two core polymerases bound to a τ dimer, and one γ complex clamp loader (see Figure 4). The γ complex interacts with the C termini of the β dimer and presumably orients this face of β toward the primed site. Core interacts with some of the same C-terminal residues on β as the γ complex does. (Hence, after γ complex loads β on DNA, the core may swing into position with the β clamp.) In the holoenzyme, γ complex is held to DNA with core and β through interaction with τ.

of holoenzyme in a cell (4), however, there must be a specialized mechanism for rapid polymerase recycling.

The fact that holoenzyme is held to DNA by a ring-shaped protein suggests that holoenzyme may solve the recycling problem by sliding back along the lagging strand until it regains its position at the fork and captures the next primer. This would require holoenzyme to slide over the duplex fragment it had just finished, and over the gap of ssDNA separating it from the fork. Study of holoenzyme diffusion on DNA showed that holoenzyme slides on duplex DNA, but not on ssDNA, whether SSB protein is present or not (40). These results at first seem inconsistent with a β ring having a central cavity large enough to accommodate duplex DNA, and therefore also ssDNA (at least if SSB protein is not present). β can only slide over a short stretch of ssDNA (up to 25 nucleotides), however; β cannot slide over a 1-kb stretch of ssDNA (with or without SSB protein) (5). Presumably, ssDNA has secondary structure, such as hairpins, that block β sliding.

The mechanism of holoenzyme cycling to new primed sites has been found to lie in the ability of this highly processive enzyme to switch rapidly to a distributive mode in a novel process of partial disassembly of its multisubunit structure and then reassembly (7–9, 103). Prior to completing a template, Pol III* remains stably associated with its β clamp ($t_{1/2} \sim 5$ min), but upon completing a template, Pol III* rapidly dissociates from DNA (in less than 1 s),

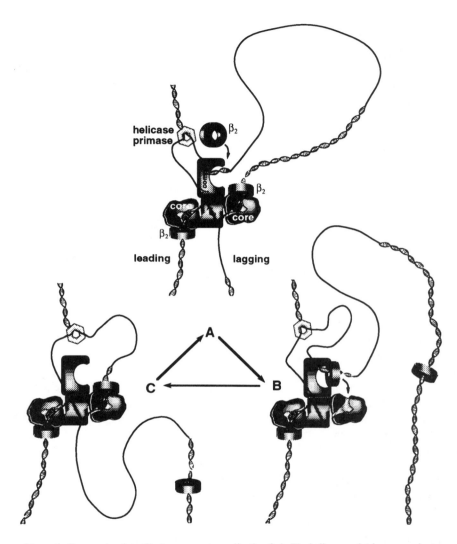

Figure 6   Proposed action of holoenzyme at a replication fork. The helicase and primase are shown as a hexamer surrounding the duplex at the forked junction. The holoenzyme structure is placed at a replication fork with one core polymerase on each strand. The γ complex is asymmetrically disposed relative to the two cores such that it points toward the lagging strand to load β clamps on primers repeatedly to initiate processive extension of Okazaki fragments. (*A*) As the lagging-strand polymerase extends an Okazaki fragment, the γ complex assembles a β clamp onto an RNA primer. (*B*) Upon completing an Okazaki fragment, the core disengages its β clamp, creating a vacancy for the new β clamp. (*C*) The new β clamp falls into place with the lagging-strand core polymerase to start the next Okazaki fragment. (Reproduced from 9)

leaving the β ring behind. Once off the completed DNA, Pol III* rapidly associates with a new β clamp at another primed site. The dynamics of these proteins on DNA imply that at the replication fork, the Okazaki fragment is extended to the very last nucleotide and then Pol III* rapidly dissociates from its β clamp and cycles to the upstream RNA primer (but only after assembly of a new β clamp on the new RNA primer).

Earlier studies concluded that Pol III* required a second β clamp on another DNA molecule to induce Pol III*'s dissociation from a completed template (8). It is now evident, however, that Pol III* does not require assistance to disengage its β clamp after completing a template (9). The earlier observations that holoenzyme remained bound to replicated DNA were likely explained by the presence of too little salt in the analysis (8, 105). At low ionic strength Pol III* binds DNA nonspecifically (5, 9).

The implication of this mechanism of polymerase recycling at a replication fork fits nicely with the overall structure of holoenzyme. In Figure 6, the holoenzyme is placed into the context of a moving replication fork and each core polymerase is shown with a β sliding clamp for processive elongation of both strands. In proceeding from diagram A to B, the γ complex assembles a β ring around a new primed site at the fork. Also in going from diagram A to B, the lagging-strand core completes an Okazaki fragment to a nick, thereby effecting its release from the β clamp and DNA. Polymerase release of the β clamp results in a vacancy in the binding site for β on the core polymerase, a logical prerequisite for association of this core with a new β clamp on the upstream RNA primer. In proceeding from diagram B to C, the lagging-strand core cycles to the new β clamp to initiate processive extension of the next Okazaki fragment.

This entire cycle of events must occur within a second or two. Can β clamps be assembled fast enough to account for a new clamp on every Okazaki fragment (i.e. 1 clamp/s)? Experiments performed at intracellular concentrations of β, DNA, γ complex, and potassium glutamate have shown that one β clamp is assembled on DNA every one-half second (9). Hence β clamp assembly appears rapid enough to account for a new β clamp for each Okazaki fragment, especially considering that the effective concentration of γ complex would be very high at a replication fork due to being held near the DNA by its presence in the holoenzyme structure.

The polymerase transfer mechanism entails stoichiometric use of β for each Okazaki fragment, consistent with the cellular abundance of β relative to holoenzyme. There are approximately 10 times more Okazaki fragments produced during chromosome replication than there are β dimers in the cell, however. Pertinent to this point is the finding that Pol III* not only loads β clamps onto DNA, but also can remove them from DNA for use at new primed sites (9).

Significant insight into the workings of holoenzyme at a replication fork have been obtained from studies using a rolling-circle system (108–112). In the rolling-circle assay the holoenzyme is present with the helicase (DnaB protein) and primase (DnaG protein) (plus or minus the other primosomal proteins, PriA–C, DnaT, and DnaC) to produce a unidirectional replication fork that peels off a long lagging strand as the fork is advanced multiple times around a circular duplex (1). This assay has been exploited to determine the processivity of proteins during fork movement and to characterize the effect on leading- and lagging-strand synthesis of different concentrations of nucleotides, salt, and proteins. Lowering the concentration of β decreased the efficiency of primer utilization on the lagging strand, a result consistent with stoichiometric consumption of one β clamp per Okazaki fragment (108, 109). Further, under some conditions, the final number of Okazaki fragments was greater than the total amount of β in the assay, consistent with eventual recycling of β clamps. Omission of τ significantly reduces replication, consistent with its structural role in dimerizing core (K Marians, personal communication). It is known that τ can replace γ in action with the δ, δ', χ, and ψ subunits in loading β clamps onto DNA (65, 69), consistent with the ability to omit γ without significant effect (K Marians, personal communication).

Another important observation in the rolling-circle system is that at a low concentration of core, Okazaki fragments are not extended to completion, suggesting that primase can induce premature release of the lagging polymerase (112). A polymerase release mechanism such as this would be a useful backup mechanism to effect the removal of a stalled holoenzyme at a site of DNA damage.

Coordination of Leading and Lagging Strands

Coordinated synthesis of the leading and lagging strands is probably necessary to survival. The issue at stake is the ability to stop one strand if the other strand is stalled, such as upon encounter with a damaged site. For example, if the leading polymerase were to continue unabated while the lagging polymerase was immobilized at a lesion, the lagging-strand template would continue to be spooled out as ssDNA. There are approximately 800 SSB protein tetramers in a cell, and therefore about 50 kb of ssDNA can be coated, after which the exposed ssDNA would be available for nuclease attack. An ssDNA scission would be difficult, if not impossible, to repair. Presumably coordinated synthesis of the two strands occurs, as DNA-damaging agents lead to cessation of replication.

It seems reasonable to expect a dimeric polymerase to be at the root of the mechanism of strand coordination. Perhaps the proximity of the two polymerases facilitates allosteric communication between them, as suggested (12). Or, since polymerases travel in spiral paths when forming a spiral duplex product (or the DNA spirals in back of the polymerase), perhaps stopping one polymerase prevents spiraling of the other. The mechanism of strand coordination is an important area for future studies.

It should be noted that a dimeric polymerase does not solve the kinetic barrier to polymerase cycling (i.e. rapid dissociation of a processive polymerase from DNA for cycling to the next primer). Although a dimeric structure would result in holding the lagging polymerase at the fork, and thereby increase its effective concentration for action on the lagging strand, dissociation reactions are independent of concentration. As discussed above, holoenzyme has a specific mechanism for rapidly dissociating from DNA upon completing a template (7–9).

COMPARISON OF HOLOENZYME TO OTHER REPLICASES

Holoenzyme can be thought of as three components: a polymerase (core), a sliding clamp (β), and a clamp loader (γ complex). At this level of resolution, the replicases of eukaryotes (Pol δ) and phage T4 are similar to holoenzyme (reviewed in 2).

The replicase of each system has these three activities of *E. coli* holoenzyme (Table 2). The clamp loader of Pol δ is the five-subunit RF-C (also called A1), and the clamp is PCNA (reviewed in 113). In T4, the clamp loader is the gene 44/62 protein complex (g44/62p) and the clamp is the gene 45 protein (g45p) (reviewed in 114). Interestingly, the sequences of all the subunits of the RF-C complex are homologous to one another (115, 116), as are the γ/τ and δ' subunits of γ complex (64, 67). The *E. coli* γ/τ and δ' subunits are also homologous to the human RF-C subunits and to T4 g44p, implying that the mechanism of clamp loading (and unloading) is common to all these systems (68).

Table 2 Comparison of the three-component structure of replicases from *E. coli*, eukaryotes, and T4 phage

	E. coli	Eukaryotes	T4 phage
Polymerase/ exonuclease	core (3 subunits)	Pol δ (2 subunits)	g43p (1 subunit)
Clamp loader (matchmaker)	γ complex (5 subunits)	RF-C complex (5 subunits)	g44/62p complex (2 subunits)
Sliding clamp	β	PCNA	g45p

The monomer mass of PCNA and of g45p is only 2/3 the mass of β, but their native mass is similar to that of β due to their trimeric aggregation state (117, 118). On the basis of the six-domain structure of the β dimer (three domains per monomer), it was hypothesized that PCNA and g45p trimers form rings of six domains, two per monomer (6). Consistent with this hypothesis, human PCNA, like β, slides on DNA and falls off over DNA ends (N Yao, Z Kelman, M O'Donnell, unpublished). Further, yeast PCNA self-loads over the ends of linear DNA, but not on circular DNA (84). In the T4 system, cryoelectron microscopy studies showed that the accessory proteins form a sliding clamp on DNA having similar dimensions as β (119). Also, studies of transcriptional activation by the T4 accessory proteins showed that they track along DNA (120–122). Recent protein-DNA crosslinking studies demonstrate that indeed all three clamps (g45p, β, and PCNA) track along DNA (123).

The crystal structure of yeast PCNA shows just how similar it is to E. coli β. The inner and outer diameters of these rings are the same, as is the six-domain structure. In fact, the topologies of the polypeptide chain–folding patterns of the two PCNA domains are the same as those of the three domains of β (123a).

A major difference between E. coli holoenzyme and eukaryotic Pol δ is that Pol δ is not organized into a twin polymerase, and the RF-C clamp loader is not physically connected to Pol δ in solution. Hence, at the current state of knowledge, the human system lacks the equivalent of the E. coli τ subunit for organizing its polymerases and clamp loader into one particle. Likewise, the T4 system lacks the equivalent of τ, and its clamp loader appears to act separately from the polymerase.

Polymerase action in cycling among Okazaki fragments during lagging-strand replication has been examined in the T4 and T7 systems. The T4 polymerase remains stably associated with its sliding clamp on a primed template, but rapidly disengages from its sliding clamp upon completing synthesis (124, 125). Hence, the T4 and E. coli systems behave similarly. Rolling-circle assays in the T4 system show that the leading and lagging strands continue even when the reaction is diluted, and therefore the lagging polymerase must be processive (126). Direct interaction between two T4 polymerase molecules suggests that the lagging polymerase binds the leading polymerase and thereby remains with the replication fork as it cycles among Okazaki fragments (127).

Studies in the T7 system also show rapid cycling of polymerase during lagging-strand replication (128). The T7 polymerase is composed of two subunits: gene 5 protein (the polymerase) and thioredoxin (the processivity factor); it lacks a clamp loader. Hence, the T7 replicase may employ a different mechanism for processivity and cycling than do the replicases of E. coli, T4, and eukaryotes. It is conceivable, however, that processivity and cycling in the T7 system share the basic principles of the other replicases. For example, the T7 polymerase may have a cavity in which a duplex fits, and thioredoxin may seal the cleft, trapping DNA inside. Polymerase cycling may possibly be achieved by partial or complete separation of the two subunits upon completing an Okazaki fragment, followed by reforming the T7 holoenzyme at the next primed site. The herpes simplex replicase is also a highly processive, two-subunit enzyme that lacks a clamp loader, like the T7 polymerase (129, 130).

ARE POLYMERASE SLIDING CLAMPS USED BY OTHER PROTEINS?

Besides the use of β by Pol III*, the β clamp also increases the processivity of DNA polymerase II (Pol II) (131, 132), an enzyme implicated in DNA repair (133, 134). The fact that β can be harnessed by two different DNA polymerases suggests that its use may generalize to yet other enzymes. For example, the β clamp may participate in recombination and repair, or in cell-cycle processes such as cell division and checkpoint control.

The hypothesis that DNA polymerase clamps may be harnessed by other enzymatic machineries is strengthened by the observation that clamps of other systems also interact with proteins besides the replicative polymerase (Table 3). PCNA is utilized by two DNA polymerases, δ and ε (135). The T4 g45p interacts with RNA polymerase (modified with g33p and g55p), specifically activating it on late gene promoters (120–122). Human PCNA forms a complex with cyclins, their associated kinases, and p21 (137, 138). Subsequent studies have shown that the p21 kinase inhibitor binds directly to PCNA and thereby inactivates Pol δ (139, 140). PCNA was also shown to interact with Gadd45, a protein that is induced upon DNA damage (140a).

HOLOENZYME IN REPAIR AND MUTAGENESIS

Holoenzyme also functions in mismatch repair and replication recovery after exposure to DNA-damaging agents (18–21). During correction of a mis-

Table 3 Multiple proteins interact with sliding clamps of prokaryotic and eukaryotic DNA polymerases

Clamp	Interacts with
E. coli β	Pol III, Pol II
T4 g45p	g43p (pol), RNA polymerase
human PCNA	Pol δ, Pol ε, p21 cell-cycle kinase inhibitor, Gadd45

match, several repair enzymes coordinate their actions to recognize the mismatch, scan the DNA to the nearest methylated site, nick the opposite strand, and excise the DNA strand all the way back to remove the mismatch. This gap is then filled in specifically by holoenzyme; no other DNA polymerase can substitute. Replication recovery occurs after cells are exposed to DNA-damaging agents; replication is stopped, but after a lag it starts up again. The predominant pathway of replication recovery is replication restart, in which it is believed that the replication machinery stops at a lesion and then synthesis is restarted past the lesion by a priming event, leaving the lesion behind for repair enzymes to act upon later. Another pathway, called targeted mutagenesis, requires RecA protein, UmuC protein, and a proteolytic form of UmuD protein (UmuD'). These proteins are hypothesized to assemble into a "mutasome" at the site of the lesion to help holoenzyme past the damaged site, resulting in an error (thus the term "targeted"). In the absence of these other factors, the holoenzyme has been shown to dissociate from DNA upon encountering a lesion, and it has been suggested that the UmuC and D' proteins may act by stabilizing the association of holoenzyme to DNA at a lesion (141–144). Further biochemical studies are needed to define these events. The recent development of an in vitro system for lesion bypass requiring RecA, UmuC, UmuD', and holoenzyme holds promise toward this end (145).

A new observation that may be pertinent to the mutagenic pathway is damage-dependent induction of a shorter version of β, called β^*. β^* comprises the C-terminal 2/3 of β, and hence each monomer contains two domains instead of three. Characterization of β^* showed it behaves as a trimer, presumably forming a six-domain ring (like PCNA and g45p), and it stimulates DNA synthesis by Pol III* (Z Livneh, personal communication). Surprisingly, β^*, in the absence of γ complex, converts core to a more salt-resistant form that is not inhibited by SSB protein. It is proposed that β^* may function in repair and mutagenesis, perhaps working specifically with core polymerase instead of Pol III*.

Another pathway for UV-induced mutagenesis is independent of replication and requires the repair genes uvrA, B, and C. An in vitro system for this pathway has been developed that depends on the UvrA, B, and C encinuclease, helicase II, and holoenzyme (146, 147). Presumably the error is caused by two closely opposed cylobutyl dimers such that only one is excised and the other is present in the repair gap, thus constraining the polymerase to cross the lesion as it fills the gap. Only holoenzyme is mutagenic in this assay; Pol I and Pol II are not, consistent with in vivo observations. The lack of a requirement for β suggests that a subassembly of the holoenzyme may perform this function.

GENETICS OF HOLOENZYME SUBUNITS

Five holoenzyme subunits are encoded by conditional lethal genes: α by *dnaE*, ε by *dnaQ*, β by *dnaN*, and γ/τ by *dnaX* (1). The remaining five subunit genes have been identified recently: the genes encoding δ, δ', χ, ψ, and θ (*holA–E*, respectively) (53, 64–67, 70–73, 148). Genetic knockout experiments of *holA* (δ) and *holB* (δ') show these genes to be essential for cell viability, consistent with the important roles of δ and δ' in assembly of β on DNA (R Krishnan, J Carter, D Berg, C McHenry, personal communication). Knockout of the *holC* gene (χ) is tolerated, but only small colonies form at 37°C and they fail to grow at 42°C (R Maurer, personal communication). Both of these phenotypes are partially corrected upon blocking induction of the SOS response. Another phenotype of *holC* cells was revealed upon study of mutations in recombination genes (*ruvA, B,* and *C,* and *recG*), which show no significant phenotype alone, but cannot tolerate interruption of *holC* (the *ruvA, holC* double mutant is suppressible by the *ruv* suppressor, *rus-1*) (R Maurer, personal communication). These results imply that χ may function in recombination as well as replication. Mutations of *holD* (ψ) have yet to be performed.

Studies of genes encoding subunits of core showed that a *dnaQ* (ε) null mutant shows not only a mutator phenotype, but also a severe growth defect (55, 149), consistent with the requirement of ε for holoenzyme to realize its full speed and processivity (34). The growth defect in the *dnaQ* null mutant is suppressible by a mutation in *dnaE*, presumably producing a more efficient α (150). A mutation in α (*dnaE173*) increases the spontaneous mutation frequency 1000-fold, and therefore α is also an important determinant of fidelity (151). Interallelic complementation of conditional lethal *dnaE* alleles is consistent with the presence of two core polymerases in the holoenzyme (152). It is tempting to speculate that one allele is defective on the lagging strand and the other is defective on the leading strand, thus explaining how the two alleles may complement. The function of θ has not been identified, other than a slight stimulation of ε in removal of a mismatch (53). Consistent with the subtle function of θ, a deletion of *holE* has no noticeable phenotype (55).

The frameshift site in the chromosomal *dnaX* gene has been mutated such that τ is produced but γ is not (99). These "γ-less" cells are viable, suggesting that γ is not essential (unless an undetectable but sufficient amount of γ is produced in these cells) (99). Presumably the τ subunit binds $\delta\delta'\chi\psi$ in γ-less cells to substitute for the γ function. Indeed a γ-less form of Pol III*, comparable to Pol III* in activity, can be reconstituted from individual subunits and appears to be present in γ-less cells, although its purification was defeated by proteolysis (99). In the same study, deletion of C-terminal residues in τ, lacking in γ, were found to be essential to cell viability. The unique property of τ, lacking in γ, is the ability to bind and dimerize core.

It seems likely that holoenzyme subunit genes will be regulated in accordance with the physiological state of the cell (12). Consistent with this notion, an element located within the coding sequence of *dnaX* has been shown to effect expression of the gene (153). The sequence of the element suggests that the binding factor may be *purO*, a regulator that binds operators involved specifically in purine synthesis, indicating that expression of holoenzyme is tied to the production of nucleotides.

THE FUTURE

The past few years have seen several significant advances in our knowledge of holoenzyme structure and function. All the genes have been identified, proving that all 10 subunits are distinct and are not proteolytic versions of larger subunits. Also, each subunit has been obtained in quantity, and binding studies show that each of them forms a complex with at least one other subunit, with consequences that can be assayed biochemically. Hence, none of these 10 proteins were spurious contaminants in holoenzyme preparations. Further, the holoenzyme particle can be reconstituted from them. The molecular basis underlying the tight grip of holoenzyme to DNA has been explained by the β sliding clamp encircling DNA; this clamp is pulled along by core while passively locking the polymerase to the template. The sliding clamp also explains how the polymerase binds tightly to DNA yet rapidly cycles off DNA upon finishing one fragment to start another. Holoenzyme demonstrates such action by recognizing the completion of the template and then hopping off its current sliding clamp and onto a new sliding clamp.

Despite this knowledge about the structure and function of holoenzyme, it is fair to say that only 3 of the 10 subunits—the polymerase, the exonuclease, and the clamp—have well-defined functions. The mechanism of the γ complex in loading the β clamp—especially the roles of ATP binding and hydrolysis, and the individual functions of the five different subunits—is still relatively obscure. The function of the ATPase activity of τ is still uncertain, and the role of θ is completely unknown. Why Pol III* releases the β clamp only upon finishing a template, and how the leftover clamps are recycled, also lack a detailed explanation. The holoenzyme is asymmetric structurally, but the extent to which this is manifested in function on leading and lagging strands remains for future study.

The imaginable responsibilities of a replicase are far more numerous than are the subunits in the holoenzyme; there is plenty for these proteins to do. Studies of how the holoenzyme interfaces with other replication proteins such as those that activate the origin, advance the replication fork, and terminate the chromosome have only just begun. Likewise, the roles of holoenzyme and its subassemblies in other processes such as recombination, repair, and mutagenesis have yet to be determined.

The availability of individual subunits in quantity, in addition to the ability to reconstitute the several subcomplexes as well as the entire holoenzyme, provides fertile ground for detailed structural studies, especially X-ray crystallography and examination of 2-D crystals in the electron microscope. It is now abundantly clear that the replicase of eukaryotes and T4 are similar in function to the *E. coli* holoenzyme. Besides the functional similarities, close to half the mass of each of these holoenzymes can be predicted to have similar three-dimensional structure from the homology in sequences among 275 kDa of the holoenzyme (γ_2, τ_2, δ'_1), 136 kDa of the T4 holoenzyme (g44p tetramer), and almost all of the 280-kDa five-subunit RF-C complex of humans. The shape of the β subunit tells a lot about its function. Perhaps proteins that work on DNA structures, rather than on specific sequences, reflect their function in their shape. It will be very interesting to see the visual appearance of the other holoenzyme subunits, especially the non-enzymatic ones.

ACKNOWLEDGMENTS

We are grateful to several people for information in advance of publication, including Drs. Bruce Alberts, Peter Geiduschek, Keven Hacker, John Kuriyan, Zvi Livneh, Ken Marians, Russell Maurer, Charles McHenry, and Jim Walker. Our work was supported by grants from the National Institutes of Health (GM38839) and the National Science Foundation (DCB9303921).

Literature Cited

1. Kornberg A, Baker TA. 1991. *DNA Replication.* New York: Freeman. 931 pp. 2nd ed.
2. Kelman Z, O'Donnell M. 1994. *Curr. Opin. Genet. Dev.* 4:185–95
3. Chandler M, Bird RE, Caro L. 1975. *J. Mol. Biol.* 94:127–32
4. Wu YH, Franden MA, Hawker JR, McHenry CS. 1984. *J. Biol. Chem.* 259: 12117–22
5. Stukenberg PT, Studwell-Vaughan PS, O'Donnell M. 1991. *J. Biol. Chem.* 266: 11328–34
6. Kong X-P, Onrust R, O'Donnell M, Kuriyan J. 1992. *Cell* 69:425–37
7. O'Donnell ME. 1987. *J. Biol. Chem.* 262:16558–65
8. Studwell PS, Stukenberg PT, Onrust R, Skangalis M, O'Donnell M. 1990. *UCLA Symp. Mol. Cell. Biol. New Ser.* 127:153–64
9. Stukenberg PT, Turner J, O'Donnell M. 1994. *Cell* 78:877–87
10. Kornberg A. 1988. *J. Biol. Chem.* 263:1–4
11. McHenry CS. 1988. *Biochim. Biophys. Acta* 951:240–48
12. McHenry CS. 1988. *Annu. Rev. Biochem.* 57:519–50
13. McHenry CS. 1991. *J. Biol. Chem.* 266: 19127–30
14. O'Donnell M. 1992. *BioEssays* 14:105–11
15. Nossal NG. 1983. *Annu. Rev. Biochem.* 53:581–615

16. Baker TA, Wickner SH. 1992. *Annu. Rev. Genet.* 26:447–77
17. Marians KJ. 1992. *Annu. Rev. Biochem.* 61:673–719
17a. Marians KJ. 1994. In *Escherichia coli and Salmonella typhimuruim.* 2nd ed. In press
18. Livneh Z, Cohen-Fix O, Skaliter R, Elizur T. 1993. *CRC Crit. Rev. Biochem. Mol. Biol.* 28:465–513
19. Echols H, Goodman MF. 1990. *Mutat. Res.* 236:301–11
20. Echols H, Goodman MF. 1991. *Annu. Rev. Biochem.* 60:477–511
21. Goodman MF, Creighton S, Bloom LB, Petruska J. 1993. *CRC Crit. Rev. Biochem. Mol. Biol.* 28:83–126
22. Kornberg T, Gefter ML. 1971. *Proc. Natl. Acad. Sci. USA* 68:761–64
23. Gefter ML, Hirota Y, Kornberg T, Wechsler JA, Barnoux C. 1971. *Proc. Natl. Acad. Sci. USA* 68:3159–53
24. Nüsslein V, Otto B, Bonhoeffer F, Schaller H. 1971. *Nature New Biol.* 324:185–86
25. Kornberg T, Gefter ML. 1972. *J. Biol. Chem.* 247:5369–75
26. Wickner W, Kornberg A. 1973. *Proc. Natl. Acad. Sci. USA* 70:3679–83
27. Wickner S, Schekman R, Geider K, Kornberg A. 1973. *Proc. Natl. Acad. Sci. USA* 70:1764–67
28. Wickner S, Kornberg A. 1974. *J. Biol. Chem.* 249:6244–49
29. McHenry C, Kornberg K. 1977. *J. Biol. Chem.* 252:6478–84
30. Hurwitz J, Wickner S. 1974. *Proc. Natl. Acad. Sci. USA* 71:6–10
31. Wickner S. 1976. *Proc. Natl. Acad. Sci. USA* 73:3511–15
32. Wickner S, Hurwitz J. 1976. *Proc. Natl. Acad. Sci. USA* 73:1053–57
33. Hurwitz J, Wickner S, Wright M. 1973. *Biochem. Biophys. Res. Commun.* 51:257–67
34. Studwell PS, O'Donnell M. 1990. *J. Biol. Chem.* 265:1171–78
35. Bryant FR, Johnson KA, Benkovic SJ. 1983. *Biochemistry* 22:3537–46
36. Fay PJ, Johanson KO, McHenry CS, Bambara RA. 1981. *J. Biol. Chem.* 256:976–83
37. Mok M, Marians KJ. 1987. *J. Biol. Chem.* 262:16644–54
38. Burgers PMJ, Kornberg A. 1982. *J. Biol. Chem.* 257:11468–73
39. Burgers PMJ, Kornberg A. 1982. *J. Biol. Chem.* 257:11474–78
40. O'Donnell ME, Kornberg A. 1985. *J. Biol. Chem.* 260:12875–83
41. McHenry CS, Crow W. 1979. *J. Biol. Chem.* 254:1748–53
42. McHenry CS. 1982. *J. Biol. Chem.* 257:2657–63
43. Studwell-Vaughan PS, O'Donnell M. 1991. *J. Biol. Chem.* 266:19833–41
44. Maki H, Maki S, Kornberg A. 1988. *J. Biol. Chem.* 263:6570–78
45. Fay PJ, Johanson KO, McHenry CS, Bambara RA. 1982. *J. Biol. Chem.* 257:5692–99
46. Maki S, Kornberg A. 1988. *J. Biol. Chem.* 263:6561–69
47. Maki H, Kornberg A. 1985. *J. Biol. Chem.* 260:12987–92
48. Maki H, Horiuchi T, Kornberg A. 1985. *J. Biol. Chem.* 260:12982–86
49. Scheuermann RH, Echols H. 1984. *Proc. Natl. Acad. Sci. USA* 81:7747–51
50. Scheuermann RH, Tam S, Burgers PMJ, Echols H. 1983. *Proc. Natl. Acad. Sci. USA* 80:7085–89
51. DiFrancesco R, Bhatnagar SK, Brown A, Bessman MJ. 1984. *J. Biol. Chem.* 259:5567–73
52. Maki H, Kornberg A. 1987. *Proc. Natl. Acad. Sci. USA* 84:4389–92
53. Studwell-Vaughan PS, O'Donnell M. 1993. *J. Biol. Chem.* 268:11785–91
54. LaDuca RJ, Fay PJ, Chuang C, McHenry CS, Bambara RA. 1983. *Biochemistry* 22:5177–88
55. Slater SC, Lifsics MR, O'Donnell M, Maurer R. 1994. *J. Bacteriol.* 176:815–21
56. Maki S, Kornberg A. 1988. *J. Biol. Chem.* 263:6555–60
57. Burgers PMJ, Kornberg A, Sakakibara Y. 1981. *Proc. Natl. Acad. Sci. USA* 78:5391–95
58. Kuwabara N, Uchida H. 1981. *Proc. Natl. Acad. Sci. USA* 78:5764–67
59. LaDuca RJ, Crute JJ, McHenry CS, Bambara RA. 1986. *J. Biol. Chem.* 261:7550–57
60. O'Donnell M, Kuriyan J, Kong X-P, Stukenberg PT, Onrust R. 1992. *Mol. Biol. Cell* 3:953–57
61. Kuriyan J, O'Donnell M. 1993. *J. Mol. Biol.* 234:915–25
62. O'Donnell M, Kuriyan J, Kong X-P, Stukenberg PT, Onrust R, Yao N. 1993. *Nucleic Acids Mol. Biol.* 8:197–216
63. Onrust R. 1993. *The structure and function of the accessory proteins of the E. coli DNA polymerase III holoenzyme.* PhD thesis. Cornell Univ. Med. Coll., New York. 204 pp.
64. Dong Z, Onrust R, Skangalis M, O'Donnell M. 1993. *J. Biol. Chem.* 268:11758–65
65. Onrust R, O'Donnell M. 1993. *J. Biol. Chem.* 268:11766–72
66. Carter JR, Franden MA, Aebersold R,

McHenry CS. 1992. *J. Bacteriol.* 174: 7013–25

67. Carter JR, Franden MA, Aebersold R, McHenry CS. 1993. *J. Bacteriol.* 175: 3812–22

68. O'Donnell M, Onrust R, Dean FB, Chen M, Hurwitz J. 1993. *Nucleic Acids Res.* 21:1–3

69. O'Donnell M, Studwell PS. 1990. *J. Biol. Chem.* 265:1179–87

70. Xiao H, Crombie R, Dong Z, Onrust R, O'Donnell M. 1993. *J. Biol. Chem.* 268: 11773–78

71. Xiao H, Dong Z, O'Donnell M. 1993. *J. Biol. Chem.* 268:11779–84

72. Carter JR, Franden MA, Lippincott JA, McHenry CS. 1993. *Mol. Gen. Genet.* 241:399–408

73. Carter JR, Franden MA, Aebersold R, McHenry CS. 1993. *J. Bacteriol.* 175: 5604–10

74. Maki S, Kornberg A. 1988. *J. Biol. Chem.* 263:6547–54

75. Onrust R, Stukenberg PT, O'Donnell M. 1991. *J. Biol. Chem.* 266:21681–86

76. Flower AM, McHenry CS. 1986. *Nucleic Acids Res.* 14:8091–101

77. Yin K-C, Blinkowa A, Walker JR. 1986. *Nucleic Acids Res.* 14:6541–49

78. Tsuchihashi Z, Kornberg A. 1989. *J. Biol. Chem.* 264:17790–95

79. Biswas SB, Kornberg A. 1984. *J. Biol. Chem.* 259:7990–93

80. Lee SH, Walker JR. 1987. *Proc. Natl. Acad. Sci. USA* 84:2713–17

81. Lasken RS, Kornberg A. 1987. *J. Biol. Chem.* 262:1720–24

82. Crute JJ, LaDuca RJ, Johanson KO, McHenry CS, Bambara RA. 1983. *J. Biol. Chem.* 258:11344–49

83. Kwon-Shin O, Bodner JB, McHenry CS, Bambara RA. 1987. *J. Biol. Chem.* 262: 2121–30

84. Burgers PMJ, Yoder BL. 1993. *J. Biol. Chem.* 268:19923–26

85. Fradkin LG, Kornberg A. 1992. *J. Biol. Chem.* 267:10318–22

86. Grip MA, McHenry CS. 1989. *J. Biol. Chem.* 264:11294–301

87. Record JT Jr., Anderson CF, Mills P, Mossing M, Roe J-H. 1985. *Adv. Biophys.* 20:109–35

88. Meyer RR, Laine PS. 1990. *Microbiol. Rev.* 54:342–80

88a. Ruvolo PP, Keating KM, Williams KR, Chase JW. 1991. *Proteins: Struct. Funct. Genet.* 9:120–34

89. Lohman TM, Ferrari ME. 1994. *Annu. Rev. Biochem.* 63:527–70

90. Grip MA, McHenry CS. 1988. *Biochemistry* 27:5210–15

91. Grip MA, McHenry CS. 1990. *J. Biol. Chem.* 265:20356–63

92. Kelman Z, Naktinis V, O'Donnell M. 1994. *Methods Enzymol.* In press

93. Kodaira M, Biswas SB, Kornberg A. 1983. *Mol. Gen. Genet.* 192:80–86

94. Mullin DA, Woldringh CL, Henson JM, Walker JR. 1983. *Mol. Gen. Genet.* 192: 73–79

95. Tsuchihashi Z, Kornberg A. 1990. *Proc. Natl. Acad. Sci. USA* 87:2516–20

96. Flower AM, McHenry CS. 1990. *Proc. Natl. Acad. Sci. USA* 87:3713–17

97. Blinkowa AL, Walker JL. 1990. *Nucleic Acids Res.* 18:1725–29

98. Hawker JR Jr, McHenry CS. 1987. *J. Biol. Chem.* 262:12722–27

99. Blinkova A, Hervas C, Stukenberg PT, Onrust R, O'Donnell ME, Walker JR. 1993. *J. Bacteriol.* 175:6018–27

100. Sinha NK, Morris CF, Alberts BM. 1980. *J. Biol. Chem.* 225:4290–93

101. Kornberg A. 1982. *1982 Supplement to DNA Replication.* New York: Freeman. 273 pp.

102. Johanson KO, McHenry CS. 1984. *J. Biol. Chem.* 259:4589–95

103. Stukenberg PT. 1993. *The dynamics of E. coli DNA polymerase III holoenzyme in an in vitro lagging strand model system.* PhD thesis. Cornell Univ. Med. Coll., New York. 192 pp.

104. Griep MA, McHenry CS. 1992. *J. Biol. Chem.* 267:3052–59

105. Burgers PMJ, Kornberg A. 1983. *J. Biol. Chem.* 258:7669–75

106. Turner J, O'Donnell M. 1994. *Methods Enzymol.* In press

107. O'Donnell ME, Kornberg A. 1985. *J. Biol. Chem.* 260:12884–89

108. Wu CA, Zechner EL, Marians KJ. 1992. *J. Biol. Chem.* 267:4030–44

109. Zechner EL, Wu CA, Marians KJ. 1992. *J. Biol. Chem.* 267:4045–53

110. Zechner EL, Wu CA, Marians KJ. 1992. *J. Biol. Chem.* 267:4054–63

111. Wu CA, Zechner EL, Hughes AJ Jr., Franden MA, McHenry CS, Marians KJ. 1992. *J. Biol. Chem.* 267:4064–73

112. Wu CA, Zechner EL, Reems JA, McHenry CS, Marians KJ. 1992. *J. Biol. Chem.* 267:4074–83

113. Downey KM, Tan C-K, So AG. 1990. *BioEssays* 12:231–36

114. Young MC, Reddy MK, von Hippel PH. 1992. *Biochemistry* 31:8675–90

115. Chen M, Pan Z-Q, Hurwitz J. 1992. *Proc. Natl. Acad. Sci. USA* 89:5211–15

116. Chen M, Pan Z-Q, Hurwitz J. 1992. *Proc. Natl. Acad. Sci. USA* 89:2516–20

117. Bauer GA, Burgers PA. 1988. *Proc. Natl. Acad. Sci. USA* 85:7506–10

118. Jarvis TC, Paul LS, von Hippel PH. 1989. *J. Biol. Chem.* 264:12709–16
119. Gogol EP, Young MC, Kubasek WL, Jarvis TC, von Hippel PH. 1992. *J. Mol. Biol.* 224:395–12
120. Herendeen DR, Kassavetis GA, Barry J, Alberts BM, Geiduschek EP. 1989. *Science* 245:952–58
121. Herendeen DR, Kassavetis GA, Geiduschek EP. 1992. *Science* 256: 1298–303
122. Tinker RL, Williams KP, Kassavetis GA, Geiduschek EP. 1994. *Cell* 77:225–37
123. Tinker RL, Kassavetis GA, Geiduschek EP. 1994. *EMBO J.* 13:5330–37
123a. Krishna TSR, Kong X-P, Gary S, Burgers P, Kuriyan J. 1994. *Cell* 79:1233–44
124. Hacker KJ, Alberts BM. 1994. *J. Biol. Chem.* 269:24203–20
125. Hacker KJ, Alberts BM. 1994. *J. Biol. Chem.* 269:24221-18
126. Selick HE, Barry J, Cha T-A, Munn M, Nakanishi M, et al. 1987. *UCLA Symp. Mol. Cell. Biol.* New Ser. 147:183–214
127. Alberts BM, Barry J, Bedinger P, Formosa T, Jongeneel CV, Kreuzer KN. 1983. *Cold Spring Harbor Symp. Quant. Biol.* 47:655–68
128. Debyser Z, Tabor S, Richardson CC. 1994. *Cell* 77:157–66
129. Hernandez TR, Lehman IR. 1990. *J. Biol. Chem.* 265:11227–32
130. Gottlieb J, Marcy AI, Coen DM, Challberg MD. 1990. *J. Virol.* 64:5976–87
131. Hughes AJ, Bryan SK, Chen H, Moses RE, McHenry CS. 1991. *J. Biol. Chem.* 266:4568–73
132. Bonner CA, Stukenberg PT, Rajagopalan M, Eritja R, O'Donnell M, et al. 1992. *J. Biol. Chem.* 267:11431–38
133. Bonner CA, Randall SK, Rayssiguier C, Radman M, Eritja R, et al. 1988. *J. Biol. Chem.* 263:18946–52
134. Bonner CA, Hays S, McEntee K, Goodman M. 1990. *Proc. Natl. Acad. Sci. USA* 87:7663–67

135. Hübscher U, Thömmes P. 1992. *Trends Biochem. Sci.* 17:55–58
136. Deleted in proof
137. Xiong Y, Zhang H, Beach D. 1992. *Cell* 71:505–14
138. Zhang H, Xiong Y, Beach D. 1993. *Mol. Biol. Cell* 4:897–906
139. Waga S, Hannon GJ, Beach D, Stillman B. 1994. *Nature* 369:574–78
140. Flores-Rozas H, Kelman Z, Dean F, Pan Z-Q, Harper JW, et al. 1994. *Proc. Natl. Acad. Sci. USA* 91:8655–59
140a. Smith ML, Chen I-T, Zhan Q, Bae I, Chen C-Y, Gilmer TM, Kastan MB, O'Conner PM, Fornace AJ Jr. 1994. *Science* 266:1376–80
141. Shavitt O, Livneh Z. 1989. *J. Biol. Chem.* 264:11275–81
142. Hevroni D, Livneh Z. 1988. *Proc. Natl. Acad. Sci. USA* 85:5046–50
143. Shwartz H, Shavitt O, Livneh Z. 1988. *J. Biol. Chem.* 263:18277–85
144. Shwartz H, Livneh Z. 1987. *J. Biol. Chem.* 262:10518–23
145. Rajagopalan M, Lu C, Woodgate R, O'Donnell M, Goodman MF, Echols H. 1992. *Proc. Natl. Acad. Sci. USA* 89: 10777–81
146. Cohen-Fix O, Livneh Z. 1992. *Proc. Natl. Acad. Sci. USA* 89:3300–4
147. Cohen-Fix O, Livneh Z. 1994. *J. Biol. Chem.* 269:4953–58
148. Carter JR, Franden MA, Aebersold R, Ryong D, McHenry CS. 1993. *Nucleic Acids Res.* 21:3281–86
149. Lancy ED, Lifsics MR, Kehres DG, Maurer R. 1989. *J. Bacteriol.* 171:5572–80
150. Lancy ED, Lifsics MR, Munson P, Maurer R. 1989. *J. Bacteriol.* 171:5581–86
151. Maki H, Mo J-Y, Sekiguchi M. 1991. *J. Biol. Chem.* 266:5055–61
152. Bryan SK, Moses RE. 1992. *J. Bacteriol.* 174:4850–52
153. Chen K-S, Saxena P, Walker JR. 1993. *J. Bacteriol.* 175:6663–70

Annu. Rev. Biochem. 1995. 64:201–33

THE ROLES OF RETINOIDS IN VERTEBRATE DEVELOPMENT[1]

Anna L. Means and Lorraine J. Gudas

Department of Pharmacology and Program in Molecular Biology, Cornell University Medical College, New York, NY 10021

KEY WORDS: homeobox, hox, limb development, anteroposterior specification, embryonic development

CONTENTS

ABSTRACT

Several lines of experimentation suggest that endogenous retinoids, metabolites of vitamin A, play a role in the anterior/posterior development of the central body axis and the limbs of vertebrates. High levels of endogenous retinoids have been detected in proximity to these developing axes in a variety of vertebrate fetuses. Teratogenesis studies suggest that both retinoid excess and deficiency are capable of disrupting the development of these axes. Finally,

[1]Abbreviations: RA, retinoic acid; AP, anterior-posterior; ZPA, zone of polarizing activity; AER, apical ectodermal ridge; pc, post-coitum; dpc, days post-coitum; CNC, cranial neural crest

0066-4154/95/0701-0201$05.00

retinoic acid receptors regulate many developmental control genes, including homeobox genes and growth factor genes.

PERSPECTIVES AND SUMMARY

Retinoids, metabolites of vitamin A, have been of interest to developmental biologists for many decades because of their teratogenic effects on fetal development. Many studies have found that retinoic acid given during pregnancy leads to many birth defects, while other studies indicate that a diet deficient in vitamin A is likewise teratogenic. Thus, normal development seems to require a careful balance of retinoids.

Within the last decade, the discovery of retinoid receptors as members of the steroid receptor superfamily of transcription factors has spurred molecular approaches to the study of retinoids. A variety of retinoid receptors, the RARs (retinoic acid receptors), RXRs (retinoid "X" receptors), and potential retinoid receptors, the orphan receptors (having no identified ligands), have been discovered (reviewed in 1). The RARs and RXRs are ligand-dependent transcriptional activators of many genes, and repressors of the expression of other genes. Some of the genes directly regulated by these receptors are genes expressed during embryogenesis. Some of these genes encode other transcription factors, and some encode peptide growth factors (reviewed in 2).

Many homeobox-containing genes are regulated by retinoic acid. Homeobox genes have been conserved in evolution as far back evolutionarily as insects. In *Drosophila*, homeobox genes were found to regulate early steps in embryonic differentiation that led to anterior/posterior specification of the different regions or compartments of the fly (reviewed in 3). Conservation of these homeobox genes in vertebrates, the fact that they encode transcription factors, and their regulation by retinoids suggest that retinoids may be one of the earliest signals in vertebrates to differentiate tissues according to their future anterior/posterior identity.

Many of the malformations resulting from retinoic acid (RA) excess occur in tissues whose origins can be traced back to the anterior/posterior development of the central body axis—that is, to the early differentiation of the neural tube and the surrounding paraxial mesenchyme. Defects in the neural tube can result not only in brain defects but in defects of the craniofacial region, eye, ear, throat, heart, and major heart arteries as well. These affected regions arise from cranial neural crest cells, which are neuroepithelial cells that migrate away from the hindbrain and undergo a transition to mesenchymal cells (as well as to gangliar cells). Cranial neural crest cells contribute to most of the facial bones, both jaws, periocular tissues, bones of the ear, the thymus, the parathyroids, and the septa of the heart and its major arteries—all tissues malformed or missing following RA treatment during development. Treatment with exogenous retinoids also leads to defects in vertebrae and ribs, structures that arise from the mesoderm lining either side of the neural tube, the somites. Just as the anterior/posterior central body axis formation seems to be affected by retinoids, the later formation of the anterior/posterior axis of the limb can be affected by excess retinoids. The regulation by retinoids of homeobox transcription factor and peptide growth factor genes has been implicated in this teratogenesis, and so perhaps in normal development.

New techniques are allowing researchers to bridge the fields of developmental and molecular biology to understand retinoid teratogenicity. Transgenic promoter analysis allows us to ask, in the embryo, what gene regulatory sites control embryonic expression; are they the sites identified in vitro as direct sites of retinoid receptor binding and activation, or perhaps the binding and activation sites of genes directly regulated by retinoid receptors? Perhaps more importantly, the function of genes in development may be assayed by either overexpressing those genes or by deleting them from the mouse genome. Overexpression may be achieved by the introduction of transgenes under the control of heterologous promoters such that the levels of the gene product are higher than in normal development or the boundaries of normal expression are extended. The technique of gene disruption by targeted mutagenesis allows us to ask what the role of a particular gene product is in development by observing development in the absence of that gene product. Such analyses have begun both for the retinoid receptors and for their target homeobox genes. As each step in the molecular cascade of retinoid-mediated gene activation is elucidated in vitro it can be assayed during fetal development, by either transgenic or gene disruption experiments.

INTRODUCTION

Structure and Bioactivity of Retinoids

Retinoids are a family of low molecular weight, hydrophobic molecules derived from vitamin A. These compounds exhibit striking effects on the growth and differentiation of many types of cells (reviewed in 2). Retinoids are also used clinically in the treatment of some types of cancer and in some dermatological diseases (4, 5).

Retinoids exert striking effects on vertebrate development. A diet either deficient in vitamin A or supplemented with the vitamin A metabolite retinoic acid (RA) can lead to teratogenicity. Much of the work on the teratogenic effects of retinoids has focused on the role of all-*trans* retinoic acid, a biologically active, teratogenic vitamin A derivative. However, a wide variety of retinoids have been found in vertebrate embryos. These retinoids include retinol (vitamin A), retinoic acid (RA), and many other metabolites. While

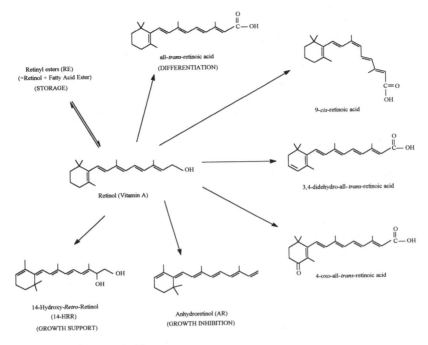

Figure 1 Structures of retinoids

some of these retinoids may be inactive storage forms, metabolic precursors, or derivatives of active forms, increasing evidence suggests that retinoids other than retinoic acid have biological activity. However, the function of each endogenous retinoid has yet to be determined. Therefore, in this review, the term "retinoid" will be used when the particular vitamin A metabolite is not clear.

The structures of vitamin A and some of its metabolites are shown in Figure 1. While most studies of teratogenicity have focused on all-*trans* retinoic acid, its stereoisomers 9-*cis* retinoic acid and 13-*cis* retinoic acid also exhibit at least some of the biological potency of all-*trans* RA. In addition, changes in the ring structure result in bioactive compounds such as 3,4-didehydroretinoic acid and 4-oxo-retinoic acid (reviewed in 6). Two retinol derivatives have also recently been shown to be important for the growth of some cells in culture. They are anhydroretinol and 14-hydroxy-4,14-retroretinol (14-HRR) (7, 8). 14-HRR is required for growth of lymphoid cells while anhydroretinol acts as an antagonist of 14-HRR and inhibits lymphoid cell growth. Most of the enzymes involved in the conversion of vitamin A (retinol) to active metabolites such as RA and 14-HRR have not yet been cloned; such cloning will certainly be important to future research.

Two types of proteins bind to retinoids to mediate and/or modulate their effects. These are known as the retinoid binding proteins (CRABP I and II, and CRBP I and II) and the retinoid receptors (RARs and RXRs). Cellular retinoid binding proteins are cytoplasmic proteins, whereas receptors are nuclear, DNA-binding proteins.

Retinoid Receptors

The two classes of retinoid receptors belong to the larger family of steroid hormone receptors. Retinoid receptors bind to all-*trans* RA and/or 9-*cis* RA, and act as sequence specific transcription factors for a variety of genes (reviewed in 1). The first-discovered and best-studied receptors are the RARs (9–14), referred to as RARα, RARβ and RARγ. Each of these genes encodes different protein isoforms as a result of alternative splicing and differential promoter usage (15, 16, 17). RXRs (retinoid "X" receptors), the second class of retinoid receptors, are closely related to RARs (18, 19, 20). Like RARs, they are sequence-specific DNA-binding proteins. RXRs have a lower affinity for all-*trans* RA than do the RARs and thus have been grouped separately. Recently, RXRs have been shown to exhibit a higher affinity for 9-*cis* RA than for all-*trans* RA(21, 22).

The steroid hormone superfamily includes proteins that have as yet no identified ligands (e.g. 23–31). Some of these may interact with retinoids other than retinoic acid and therefore may be important in the activity of retinoids.

All the members of the retinoid receptor family bind to very similar DNA sequence elements called RAREs (retinoic acid response elements) (reviewed in 1). These elements consist of tandem repeats of a 6–base pair sequence AGGTCA separated by 1–5 base pairs of variable sequence. Slight variations in these hexamer sequences and in the distance by which they are separated have profound effects on the ability of each of the receptors to bind (32). RARs and RXRs bind RAREs and regulate transcription by forming homo- or heterodimers (33, 34). RXRs can also heterodimerize with other members of the steroid hormone superfamily such as the thyroid hormone receptor, vitamin D receptor, and PPAR (peroxisome proliferator activated receptor), and thus may participate in many different signaling pathways. Although many genes are activated by RARs and RXRs, these receptors can also inhibit gene expression by several different mechanisms (for review see 1, 2).

Binding Proteins

The retinoid binding proteins represent another type of protein that can interact with retinoids. Different classes of proteins exist in this family as well, including the cellular retinol binding proteins and the retinoic acid binding proteins.

These proteins are small and are localized in the cytoplasm. They have not yet been assigned any direct function other than their ability to bind retinol or RA, respectively, but it is likely that they influence the cellular concentrations of specific retinoids.

Cellular retinol binding proteins (CRBP-I and -II) are thought to be involved in the metabolism of retinol into retinyl esters (35; for review see 36), or into retinaldehyde and then RA (37). Thus the CRBPs may facilitate the synthesis of active retinoids in the embryo.

Different functions have been ascribed to the cellular retinoic acid binding proteins (CRABP I and II). CRABPs have been hypothesized either to transport RA to the RARs (38) or to prevent RA from interacting with the RARs (39–43). Data both from cell line studies and from in vitro assays suggest that the latter is true, at least in some cell types. CRABP I overexpression reduces the ability of RA to stimulate transcription from RARE-containing promoters (39). Over-expression of CRABP I also causes an increase in metabolism of RA while expression of an antisense CRABP I, which effectively decreases the amount of CRABP I protein made, decreases the rate of RA metabolism (40). Micro-somal assays of retinoid metabolism in vitro also indicate that CRABP I may facilitate conversion of RA to more polar metabolites (37, 44). The function of CRABP II is less clear. CRABP II has a lower affinity for RA than does CRABP I (45, 46). The induction of the CRABP II gene by RA (47, 48, 49) suggests that CRABP II may function to sequester or degrade retinoids when high RA levels saturate CRABP I. In embryonal carcinoma cells, overexpression of CRABP II increases RA metabolism, but only at higher RA concentrations than those observed for CRABP I overexpression (AC Chen & LJ Gudas, submitted).

TERATOGENIC EFFECTS OF RETINOIDS

During early vertebrate development, the central body axis must be established, organizing the animal along the anterior/posterior axis (head to tail). Exogenous all-*trans* retinoic acid (RA) can have profound teratogenic effects on the establishment of this axis and the structures that arise from it. RA induces a wide variety of malformations in vertebrates, which have been well character-ized in the development of humans (50, 51, 52), hamsters (53, 54), mice, and rats (55–58). Among the most frequently described malformations is a char-acteristic pattern involving craniofacial, cardiac, thymic, and central nervous system structures, suggestive of abnormalities in one of the first anterior/pos-terior structures, the neural tube. The neural tube gives rise to both the central nervous system and the cranial neural crest cells. The latter migrate from the neural tube to form parts of the face, heart, and thymus, among other structures. Many additional malformations have been observed at lower frequencies. The spectrum of malformations depends to a great extent upon the time at which the exogenous RA is given to the embryos (53, 59). These times span periods when the central body axis is forming (60), and later periods as well.

RA may cause different types of teratogenic effects via different mechanisms, depending on its concentration and the manner in which it is delivered to the developing embryo. At lower exogenous RA concentrations, RA may activate genes that normally would not be activated at the time in development when the exogenous RA is delivered; at very high exogenous RA concentrations, RA may cause teratogenic effects primarily through cytotoxic mechanisms. For example, when RA is added exogenously to the entire embryo, both mesenchymal cells and the cells of the apical ectodermal ridge of the developing limbs exhibit pronounced cell death and thus limb reduction defects and other limb malformations result (see Figure 4 and the section below on interactions between retinoids and peptide growth factors) (61–63). However, limb duplications are seen when RA-soaked beads are implanted in the anterior portion of the developing limb bud (64). In the case of the bead implants it is likely that RA exerts its effects via a different mechanism from that seen when RA is delivered exogenously to the whole embryo.

Although much less attention has been given to the teratogenicity associated with vitamin A deficiency than to that associated with RA excess, a broad array of congenital malformations is observed in fetuses from vitamin A deficient female rats. While deficiency or excess seems to affect many of the same tissues, the frequency with which particular malformations occur varies. Following vitamin A deficiency, the most frequently observed abnormalities were in the development of the eye, including anophthalmia (absence of the eye), the genito-urinary tract, the diaphragm, the heart (such as failure of closure of the interventricular septum), and the lung (65, 66, 67). While the abnormalities associated at highest frequency with fetal vitamin A deficiency are generally different from those observed at highest frequency following exogenous RA treatment, many malformations appear to occur in the same structures in response to either RA excess or vitamin A deficiency.

Again, the complexity of the interpretation of the malformations associated with vitamin A deficiency or excess RA treatment results to some extent from the fact that the phenotype varies with the dose of RA and time of treatment. For example, it was recently shown that in zebrafish embryos treated with exogenous RA, a duplication of the retinas occurred. This duplication resulted only when RA was provided within a 2-h period during the formation of the optic primordia (59). This result contrasts markedly both with the anophthalmia associated with fetal vitamin A deficiency (65) and with the microphthalmia seen in hamsters treated with excess RA at gestational day 6.75 (53).

The complex spectrum of abnormalities associated with retinoid exposure at different times during fetal development makes it difficult to propose a unifying explanation of the molecular mechanisms of retinoid teratogenicity or its relationship to normal development. However, many of the malformations that result from excess RA occur in tissues derived from the cranial neural crest cells—neuroepithelial cells that migrate away from specific regions, called rhombomeres, of the hindbrain (50, 68). Because retinoid teratogenicity is associated with molecular changes in gene expression in these rhombomeres of the hindbrain from which the neural crest cells arise—particularly changes in the anterior/posterior boundaries of homeobox gene expression (reviewed in 69) a plausible unifying hypothesis is that many of the teratogenic effects of exogenous retinoids result from such alterations in homeobox gene expression in the neuroepithelial cells of the hindbrain. This hypothesis is developed further below.

While the role of retinoids in gene regulation may be mediated by retinoid receptors, there is no strong correlation of any one receptor with sites of teratogenesis (19, 42, 70–75). Rather, a number of correlations can be found between patterns of RAR or RXR expression and regions of retinoid susceptibility. RARα, RXRα, and RXRβ are all ubiquitously expressed and thus may mediate retinoid signaling wherever and whenever sufficiently high levels of retinoids are present, while the RARβ and RARγ genes display more limited patterns of expression (42, 76).

Recent studies have indicated that RA-induced teratogenesis may be associated with the elevation of RARβ2 transcripts (77, 78). The RARβ2 promoter (one of two promoters in the RARβ gene) contains a RARE (79, 80), and thus is directly activated by RA. It is possible that the combination of elevated levels of the RARβ gene product and exogenously added retinoids leads to the inappropriate activation of retinoic acid–responsive genes in cells. Such inappropriate gene activation could then result in the teratogenic effects observed in response to RA treatment. In zebrafish, exogenous RA induces ectopic expression of RARγ transcripts in anterior brain structures and of both RARα and RARγ mRNA in the eyes (81). Thus expression of more than one receptor type may increase in response to exogenous RA treatment and be involved in mediating the teratogenic effects of RA. Conversely, because these genes for RARs contain RARE sequence elements (reviewed in 1), and are thus directly regulated by RA, their increased expression during teratogenesis may be a symptom rather than a cause of RA teratogenesis. That is, exogenous RA can bind to receptors already present in cells and thus affect the expression of many genes in addition to the RAR genes.

Mutational analyses may give more insight into the roles of the retinoid receptors both in teratogenesis and in normal development. Recently, null mutations have been introduced into some of the retinoid receptors. RXRα null mutant mice have been generated, and these homozygous RXRα −/− mice die between embryonic day 13.5 and 16.5 as a result of heart defects (82, 83). These RXRα −/− mice also exhibit ocular malformations similar to those described above in vitamin A deficient embryos (83). In contrast to RXRα −/− mice, RARα −/− and RARγ −/− mice survive to birth, although both exhibit high postnatal lethality (84, 85, 86). The slight effects on development from loss of the RARα and RARγ genes may indicate functional redundancy. The complete functions of the RARs and RXRs may not be apparent until mice are generated carrying multiple mutations. For example, RXRα −/− and RARγ −/− mutations have recently been combined in genetic crosses. The resulting mice had a number of developmental malformations not seen in single null mutations (83). This suggests that RXR:RAR heterodimers may be important in mediating some of the effects of retinoids.

Of the retinoid binding proteins, CRBP I, CRABP I, and CRABP II are expressed in the early embryo (42, 72, 73, 87–94). CRABP II and CRBP I are expressed in some of the tissues susceptible to retinoid teratogenicity but are also expressed in other tissues that are not susceptible. However, CRABP I expression patterns strongly reflect the structures of the embryo that are susceptible to RA teratogenicity. CRABP I is detected in the hindbrain, where RA teratogenicity results in alteration, decrease, or loss of hindbrain segmentation (95–99). CRABP I is also expressed in the cranial neural crest cells that migrate from the neural tube and give rise to facial and periocular tissues, the inner ear, thymus, and the septa of the heart and its major arteries. All of these structures are teratogenized by RA. RA suppresses the outgrowth of limbs and facial structures, both of which express high levels of CRABP I in their distal, rapidly growing mesenchyme.

The two binding proteins, CRABP I and CRABP II are reported to vary as gradients across the anterior/posterior limb bud axis. However, reports of this anterior/posterior gradient of CRABP I expression vary. Maden et al (41) and Perez-Castro et al (89) reported an anterior/posterior gradient of CRABP I protein and RNA, respectively, with highest concentration of CRABP I anteriorly. Dollé et al (70) and Ruberte et al (100) reported a gradient of CRABP I RNA with highest expression posteriorly. Miyagawa-Tomita et al (101) detected no anterior/posterior gradient in CRABP I protein across the limb bud. However, Miyagawa-Tomita et al (101) did report an anterior/posterior gradient of CRABP II protein with highest expression in the posterior limb bud. No anterior/posterior gradient of CRABP II RNA was detected (100). Less equivocally, CRABP I RNA and protein are expressed in limb mesenchyme in a sharp proximodistal gradient, with high distal expression in the rapidly growing cells of the progress zone, immediately underlying the apical ectodermal ridge (70).

These expression patterns support the hypothesis that CRABP I expression suppresses the action of retinoids, because cells that are susceptible to RA express CRABP I. Exogenous RA presumably can then saturate CRABP I and free RA is then able to interact with the receptors, leading to developmental malformations. In *Xenopus*, overexpression of its only known CRABP causes many anterior/posterior defects similar to those observed after RA treatment (93). However, a homozygous mutation that decreases CRABP I expression results in mice with a normal phenotype and no increased incidence of RA-induced teratogenesis relative to wild type (102). Thus, while CRABP I expression correlates with RA susceptibility, its reduction does not alter RA susceptibility. Either CRABP I expression does not play a role in RA susceptibility or another protein, perhaps CRABP II, has redundant functions in these tissues.

RETINOIDS IN AXIAL PATTERNING DURING DEVELOPMENT

During early embryogenesis, axes are established within the embryo. First, the central body forms anterior/posterior and dorsal/ventral axes, distinguishing head from tail and back from front. Later, structures like the limb must establish these axes as well as a proximal/distal axis for proper development. As mentioned above, exogenously supplied retinoids cause malformation of many structures that arise along the anterior/posterior axis of the central body. As discussed below, retinoids can also disrupt anterior/posterior formation of limb digits. This influence of exogenous retinoids on morphological development and on regulation of developmental genes suggests that endogenous retinoids may mediate at least some of the anterior/posterior specification that occurs in normal embryogenesis. Evidence points to the involvement of retinoids in the establishment of the anterior/posterior central body axis, which forms the central nervous system and the vertebral column (Figures 2b, 2c, and 3b), and later, in the anterior/posterior specification of the limb, which forms a varied array of digits across this axis.

Early Mesodermal/Ectodermal Patterning

Axial patterning in vertebrates first becomes morphologically apparent at gastrulation. In *Xenopus*, the site of initial gastrular invagination is called the organizer, and is established by signals from a region of the more ventral vegetal hemisphere called the Nieuwkoop center (103, 104). This organizer region determines the site of invagination that comes to define the posterior end of the embryo, thus establishing the initial anterior/posterior body axis. In mammals and birds, a similar structure called the node serves this function (Figure 2a) (105, 106). In addition to having similar functions, the organizer of *Xenopus* and the node of birds and mammals share many molecular markers, including the genes for goosecoid, and members of the HNF family, pintallavis in Xenopus and HNF-3β in the mouse (107–112).

The process of gastrulation itself is inhibited by exogenously applied retinoids, although the role of endogenous retinoids in this process is not clear. Treatment with RA inhibits migration of involuting mesodermal cells and thus inhibits the formation of cranial structures, which are reliant upon this anterior migration (57, 113–117).

In *Xenopus*, retinoids may interfere with gastrulation in part by interfering with signals from the peptide growth hormone, activin, which may be one of the signaling molecules for the induction of gastrulation (43, 118). Activin can induce the expression of a homeobox gene, goosecoid, while RA can inhibit goosecoid expression (107). Goosecoid expression is capable of signaling both gastrular invagination and establishment of the anterior/posterior body axis (110, 119). Thus, by altering levels of goosecoid expression, exogenous retinoids may influence both gastrulation and axis formation.

Endogenous retinoids have been detected chromatographically in *Xenopus* embryos at gastrula and the subsequent neurula stages (when the neural plate and tube are forming) (116). The biological activity of these retinoids has been measured by extracting them from *Xenopus* embryos and applying them to tissue culture cells that carry the luciferase structural gene regulated by the retinoic acid response element (RARE) found in the RARβ gene promoter (120). These tissue culture cells respond to different concentrations of RA in a linear manner such that higher concentrations of RA yield higher amounts of luciferase activity. When measured in this way, the organizer region of *Xenopus* embryos induced high amounts of luciferase activity, with activity increasing from early to late gastrula. Similar experiments with rat and chick embryos demonstrated that the node (equivalent to the *Xenopus* organizer) also contains high levels of endogenous retinoids (120–122).

While retinoids are present in *Xenopus* organizers and mammalian or avian nodes, their activity is not apparent in the embryo until induction of the neural plate begins, immediately after mesodermal invagination (Figure 2a). This activity was measured in transgenic mouse lines that carried the lacZ reporter gene regulated by three copies of a RARE from the RARβ gene promoter (123, 124). In order for β-gal expression to be detected in embryos, RA or other retinoids capable of activating a RARE must be present. In these experiments, no β-gal activity was detected after implantation until day 7.5 pc (post-coitum), through the primitive streak or early gastrula stages. Expression was not observed until notochord and neural plate had started to form. β-Gal was then detected in all three germ layers surrounding the primitive streak, a region that becomes the posterior end of the embryo (Figure 2a). Thus, it appears that higher levels of endogenous retinoids occur in the posterior embryo at the time of anterior/posterior pattern determination (123, 124). This agrees with the aforementioned data localizing high retinoid levels to the organizer or node regions that come to lie posteriorly in the embryo (120–122).

A

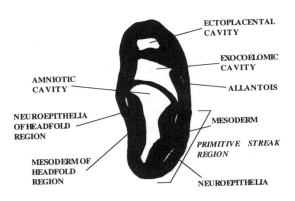

ECTOPLACENTAL CAVITY

EXOCOELOMIC CAVITY

AMNIOTIC CAVITY

ALLANTOIS

NEUROEPITHELIA OF HEADFOLD REGION

MESODERM

PRIMITIVE STREAK REGION

MESODERM OF HEADFOLD REGION

NEUROEPITHELIA

B

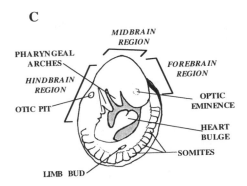

ANTERIOR NEUROPORE

NEURAL TUBE

SOMITES

OPEN NEURAL FOLDS

POSTERIOR NEUROPORE

C

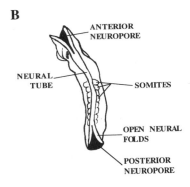

MIDBRAIN REGION

PHARYNGEAL ARCHES

FOREBRAIN REGION

HINDBRAIN REGION

OPTIC EMINENCE

OTIC PIT

HEART BULGE

SOMITES

LIMB BUD

Exogenous retinoids not only alter gastrular movements of cells but also alter subsequent processes determining anterior/posterior differentiation. In *Xenopus*, ectopic addition of RA at neurula stages (after completion of gastrulation) causes reductions or deletions in anterior ectodermal structures and in the levels of anterior molecular markers (116, 117, 125, 126). That retinoids induce posterior differentiation rather than just reduction in anterior formation is suggested by the anterior extension of the expression of genes normally confined to more posterior locales (117, 125–129).

While exogenous retinoids can alter the anterior/posterior nature of cells in *Xenopus*, the endogenous signals for this specification are not completely known but may involve endogenous retinoids. Experiments indicate that in *Xenopus* the posterior vs anterior nature of cells along the body axis is not intrinsic to these cells at the end of gastrulation but rather is dependent upon signals from the organizer region (130–132). This signal transduction and the influence of retinoids have been shown by removal of anterior, middle, or posterior sections of ectoderm and underlying mesoderm from *Xenopus* embryos immediately after gastrulation. Culture of these sections to the equivalent of the tail bud stage resulted in all three sections expressing the gene XIF3, normally only expressed in anterior structures. If these sections were removed at progressively later times they began to express genes that are appropriate to their anterior/posterior site of removal (132). Thus at the end of gastrulation, anterior/posterior cell fates have not yet been determined. However, when RA was added to the culture medium of the middle section of the embryo, anterior gene expression was repressed and a midregion gene was expressed (132). This suggests that endogenous RA supplied by specific regions of the embryo (such as the organizer) may be sufficient to posteriorize segments of the central body axis.

A role for the organizer in this signaling of anterior/posterior cell fates is supported by the results of Sive & Cheng (128). When the *Xenopus* embryo is UV irradiated, it loses its axial formation and becomes a radially symmetric, ventralized animal. However, implantation of an organizer region from an untreated embryo restores axial patterning, including formation of anterior structures such as eye and cement gland. If the organizer is taken from RA

←

Figure 2 Diagrams of murine embryos. (*a*) Lateral view of a sagittal section of 7 dpc (days post-coitum), primitive streak stage embryo. Anterior is to the left, posterior is to the right, dorsal is to the top, and ventral is to the bottom. The primitive streak is the location of gastrular involution. The node lies at the anterior tip of the primitive streak. The neuroepithelia are the cells that make up the neural plate. (*b*) Dorsal view of an 8.5–9 dpc embryo. Anterior is to the top, posterior is to the bottom. The neural tube is the precursor of the central nervous system. The somites are the precursors of vertebrae. (*c*) Lateral view of a 9–9.5 dpc embryo. The pharyngeal arches arise from cranial neural crest cells and contribute to the face, jaws, eyes, ears, thymus, and septa of the heart and major arteries.

treated embryos, posterior structures such as tails form; but anterior structures are reduced or absent (117). Exogenous RA not only affects the inductive capacity of the organizer but also affects the ability of ectodermal tissue to respond. When the organizer is coupled to undifferentiated ectodermal cells, anterior ectodermal genes are induced. However, if the ectodermal cells are taken from an RA treated embryo, they are incapable of responding normally and the induction of anterior genes is reduced.

Thus, experiments suggest that: 1. signals from the organizer region, posteriorly located in the *Xenopus* embryo, can specify anterior/posterior cell fates along the body axis; 2. RA may substitute for at least some of this signaling; 3. the organizer region contains levels of retinoids higher than in other portions of the embryo. Combined, these results suggest that endogenous retinoids supplied by the organizer may be a means of determining anterior/posterior cell fates during normal development. The molecular mechanisms underlying this determination are discussed below.

Patterning along the Anterior/Posterior Axis in the Hindbrain: Involvement of Homeotic Genes

Many of the effects of retinoid teratogenicity on ectodermal tissues have focused on the most overtly AP (anterior-posterior) segmented section of the neuroepithelium, the hindbrain. Because many structures of the face, throat, and heart are derived from the hindbrain neuroepithelium (as discussed below) its correct formation influences many tissues. The hindbrain is divided into eight morphologically distinct regions called rhombomeres (Figure 3b). Treatment with exogenous retinoids has profound effects on rhombomere segmentation and identity. At lower doses, RA alters rhombomere identity such that anterior rhombomeres have the morphology and gene expression of more posterior rhombomeres, while at higher doses of RA, segmentation is abolished (95–99). Retinoid deficiency also affects brain development (133, 134), suggesting that a carefully balanced level of endogenous retinoids regulates development of this region. This regulation may be accomplished by regulation of expression of a group of homeotic genes, the hox genes.

Homeotic genes were first described in *Drosophila* as genes that specified the identities of the different regions or compartments of the body (reviewed in 3). These genes were found to be DNA-binding transcription factors that contained high sequence homology in their DNA-binding domains, called homeoboxes. Vertebrates have conserved many of these genes throughout evolution (Figure 3a). Here we principally discuss sets of linked homeobox genes called the hox genes. These vertebrate genes are homologous to linked groups of genes in *Drosophila*, the antennapedia and bithorax complexes, which together form the HOM-C complex. The combination of HOM-C genes expressed in a given segment determines the resulting structure of that segment.

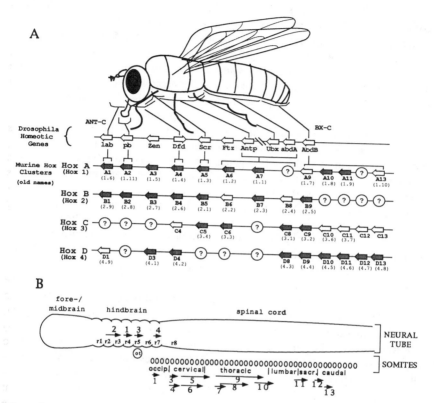

Figure 3 Hox genes are expressed colinearly in the anterior/posterior body axis. (*a*) The murine hox clusters (Hox a–d) are indicated beneath their homologous *Drosophila* counterparts. Each arrow indicates the direction of transcription for that gene. For the murine clusters, each column represents a paralog group (e.g. A3, B3, and D3). A circle containing a question mark indicates that no paralog has been identified in that cluster. Old as well as new hox nomenclatures are indicated. Filled arrows indicate genes whose expression patterns have been mapped and are diagrammed in figure section *b*. (*b*) The mouse central nervous system and somites are diagrammed. Arrows within the hindbrain indicate the anterior boundaries of expression of hox paralogous genes in the hindbrain. Arrows below indicate the anterior boundaries of hox paralogous genes in the somites. Numbers immediately above or below the arrows indicate the hox paralog group defined by the arrow. The lengths of the arrows indicate the disparities in anterior boundaries observed among different members of the same paralogous groups. Note that the arrows indicate only the anterior boundaries and that expression domains extend much farther posteriorly (reviewed in 135, 136).

For example, the most anterior compartments of the fly express only the most 3′ gene in the antennapedia complex, while more posterior compartments express this and adjacent 5′ genes. This variable combination of genes then determines the identity of each segment of the developing fly.

Similarly, in vertebrates, clusters of homeobox-containing genes (the hox complexes) tend to be expressed in a colinear fashion, with the more 3' genes expressed in more anterior compartments and the more 5' genes limited to more posterior compartments (Figure 3b) (reviewed in 135–137). The varying expression of these homeobox genes and some recent mutagenesis experiments suggest that the hox genes influence the developmental fate of many segmented structures along the anterior/posterior axis, such as the rhombomeres of the hindbrain, as well as vertebra and limb digit precursors, both of which are discussed below. As in *Drosophila*, determination of developmental fate seems to rely not necessarily on expression of a single hox gene in a given segment but on expression of multiple hox genes whose variable anterior and posterior boundaries create unique combinations in each of the segments. These combinations are known as the hox code.

In vertebrates, the HOM-C complex has been replicated such that four copies of hox complexes exist (137). Because each *Drosophila* gene has multiple homologous genes in vertebrates, each homologous group is referred to as a paralog group, with each member gene considered a paralog. The hox complexes are not exact copies of each other, so some paralog groups do not contain four paralogs (Figure 3a). Relative to the *Drosophila* HOM-C complex, each hox cluster appears to have lost some genes and replicated others. However, at least two copies exist in each paralog group, allowing for the possibility of redundant functions in vertebrate development.

Retinoids may be endogenous signaling molecules that can lead to the sequence of anterior/posterior activation of hox genes that is seen in the neural tube. Indeed, five RAREs (retinoid receptor binding sites) have been found thus far in the hox complexes (138–142; AW Langston & LJ Gudas, unpublished results).

RA can activate many of these hox genes in embryonal carcinoma cell lines, in both a temporal and a concentration dependent manner (95, 143, 144)—an observation that supports a role for retinoids in embryonic hox gene expression. In these lines, low doses of RA activate primarily 3' hox genes, and progressively higher doses of RA induce progressively more 5' hox genes. Longer RA exposure times will also lead to activation of transcription of progressively more 5' hox genes. Either of these methods of differential activation—by variation in concentration or in time of exposure—may relate to activation of these hox genes in the neural tube. As mentioned above, experiments in *Xenopus* indicate that signals for anterior/posterior specification come from the organizer region (109, 128, 130, 132). This organizer region (as well as the similar avian node) has higher retinoid concentrations than do other regions of the embryo (122, 145). Therefore, if retinoids are released from the organizer or node, posterior regions, closest to the node, would have higher concentrations of retinoids, while more distant, anterior regions would have lower concentrations. Thus, 5' (as well as 3') hox genes would be activated posteriorly, while 3' hox genes alone would be activated in the most anterior regions. Alternatively, since posterior cells migrate from the organizer or node later than do anterior cells, these posterior cells would have a longer exposure to retinoids, thus activating more 5' hox genes than the anterior cells that migrated earlier.

The comparison of hox misexpression experiments with teratogenesis experiments also suggests that retinoids influence development via hox gene regulation. Exposure of embryos to exogenous RA, effectively increasing the levels of retinoids along the body axis, extends the anterior boundaries of some of the hox genes (93, 98, 146; reviewed in 69) and can cause transformation of anterior rhombomeres into morphologically posterior rhombomeres (95–99). Similarly, transgenic overexpression of hoxa-1, achieved by placing the cDNA behind the ubiquitous β-actin promoter, transforms anterior rhombomeres 2 and 3 (not normally expressing hoxa-1) into rhombomeres that resemble rhombomere 4 (the most anterior boundary of hoxa-1 in normal development) (147). Conversely, abolishing hoxa-1 expression by introduction of a null mutation reduces some of the segmentation of the hindbrain leading to the loss of rhombomeres 4 and 5 (148–151).

The ability of RA to regulate hox gene expression in tissue culture cells and the similarities in phenotypes between retinoid teratogenicity and hox gene mutations together suggest that endogenous retinoids may establish the anterior/posterior graded expression of hox genes and thus influence anterior/posterior identities along the body axis. In this manner, retinoids may influence the identities of not only the rhombomeres of the hindbrain, but also the facial, throat, and heart tissues arising from them.

Axial Patterning of Cranial Neural Crest

The cranial neural crest (CNC) cells arise from the dorsal roof of the neural folds (in *Xenopus* and mouse) or of the recently closed neural tube (in chick) and migrate away from the neural tube (Figures 2b and 2c). Successive waves of CNC migrate laterally and ventrally from the hindbrain to form the pharyngeal (also called branchial) arches (Figure 2c). Marking of neural crest cells has indicated that the segmentation of the hindbrain is reflected in the segmentation of the CNC (Figure 3b). Rhombomeres 1 and 2 give rise to CNC of the first pharyngeal arch, rhombomere 4 to the second arch, and rhombomere 6 to the third arch. Pharyngeal arches 4, 5, and 6 arise from more posterior regions of the hindbrain (152–155).

Transplantation experiments indicate that the identity of the cranial neural crest derivatives is determined prior to migration from the neural folds (156). For example, CNC from rhombomeres 1 and 2 can be transplanted to the region of rhombomere 4 yet still give rise to pharyngeal arch 1 structures such as beaks and otic cartilage, rather than the arch 2 structures normally arising from the rhombomere 4 level (156). Thus, rhombomere identity and cranial neural crest identity appear linked. Therefore, regulation of hox genes by retinoids is implicated in neural crest differentiation as well as in rhombomere differentiation.

Hoxa-2 is expressed in the hindbrain anteriorly through rhombomere 2 (r2) and in the migrating CNC that comprise pharyngeal arches 2 and 3 but not arch 1. In hoxa-2 null mutant mice, these CNC no longer expressed the hoxa-2 gene product; thus, much of the structure of pharyngeal arch 2 was transformed to structures normally present in arch 1, which does not express the hoxa-2 gene product (157, 158). This transformation is most clearly seen in the duplication of the ossification center of the inner ear, which is derived from Meckel's cartilage in the first pharyngeal arch. Conversely, Reichert's cartilage, a second arch structure forming the styloid process of the temporal bone and the lesser horns of hyoid, is missing in the hoxa-2 null mutant mouse.

Most hox gene null mutations do not create homeotic transformations in the CNC such as that seen for the hoxa-2 gene. Rather, they result in reduction or deletion of pharyngeal arch structures, phenotypic phenomena also associated with retinoid teratogenicity. For example, a null mutation of hoxa-3 results in loss of the thymus and parathyroids, and in heart defects such as a missing carotid artery and defective pulmonary valve (159). These structures are formed from pharyngeal arch 3 CNC. Pharyngeal arch 3 CNC cells migrate from rhombomeres 5 and 6 (152, 160), which comprise the anterior boundary of hoxa-3 expression. In null mutations of hoxa-3, instead of an anterior transformation the structures from arch 3 simply do not form. Thus, the loss of hoxa-3 gene function has developmental consequences similar to those of RA teratogenicity, suggesting again that retinoids influence anterior/posterior identities via regulation of the hox genes.

Gain-of-function hox mutations also lead to defects in pharyngeal arch formation. These gain-of-function mutations usually result in expression of a hox gene into regions that do not normally express that hox gene. As RA treatment of embryos results in similar ectopic anterior expression of hox genes, it is not surprising that the overexpression of hox genes by means other than RA treatment results in similar phenotypic alterations.

For example, overexpression of hoxb-6, normally expressed into the posterior hindbrain (161, 162), affects the migration and differentiation of many pharyngeal arch 1 and 2 components. The outward phenotypes of cleft palate, microtia (abnormally small ear), and micrognathia (small jaws, especially the mandible) associated with overexpression of hoxb-6 seem to result from reduced or malformed facial and ear bones including the maxilla, mandible, zygoma, and squamosal bones (163).

While exogenous RA affects the development of many structures arising from the cranial neural crest, hox gene overexpression or deletion tends to affect subsets of CNC derived structures. Cranial neural crest cells from the neuroepithelium give rise to cranial ganglia as well as to pharyngeal arch mesenchyme. Whereas null mutations of hoxa-2 or hoxa-3 grossly affect mesenchymal differentiation, they have no apparent effect on cranial ganglia formation (157–159). Conversely, mutation of hoxa-1 affects the development of cranial ganglia but has little effect on neural crest mesenchyme (147, 149, 150). Many hox null mutations have no apparent effect on any CNC-derived structures. Thus, while an individual hox gene may be required for mesenchymal development or gangliar development, it may be required for neither if its function is redundant with other hox genes on other chromosomal clusters (paralogs). This is not surprising given the duplications of the hox complexes and similar expression domains of paralogous hox genes.

The anterior/posterior identity of a region may be achieved through a combination of a requirement for particular hox paralogs and a minimum dose requirement for all the genes within a paralogous group (Figure 3a). For example, deletion of one paralog (e.g. hoxa-1) perturbs neural development but is insufficient to perturb development of CNC mesenchymal structures, perhaps because the other paralogs (e.g. hoxb-1 and d-1) are present (148–151). This dose dependency may vary for the different structures formed from hox-expressing cells as well as for the different paralog groups. Dose dependency is illustrated in the single and double null mutations of hoxa-3 and hoxd-3 (164). Deletion of hoxd-3 had little effect on pharyngeal arch development. Mutation of hoxa-3 caused many malformations in pharyngeal arch structures, as detailed above. Double null mutations of hoxa-3 and d-3 more severely affected these structures, while the compound heterozygote, hoxa-3 +/– hoxd-3 +/–, had many of the defects seen in the hoxa-3 null mutation, even though either heterozygote alone exhibited no perceptible effects. While the two alleles of hoxd-3 are dispensable, the compound heterozygote, hoxa-3 +/– hoxd-3 +/–, and the single homozygote hoxa3 –/–, each having two missing alleles, lower the "dose" or level of that paralog group by the same amount and have similar effects. Deletion of all hoxa-3 and d-3 alleles leaves only hoxb-3 (because no paralog exists in the hox c complex), an even greater reduction in "dose" and thus a more severe phenotype. Thus, a combination of a requirement for the hoxa-3 gene and a requirement for a threshold dose of "3" paralogs seems to determine the development of these structures. The reverse was also true. Deletion of hoxd-3 caused defects in the somite-derived vertebrae, discussed below, while deletion of hoxa-3 had no effect. The defects caused by hoxd-3 deletion were more severe in the double mutant, hoxa-3 –/– hoxd3 –/–. Compound heterozygotes, hoxa-3 +/– and hoxd-3 +/–, had a phenotype very much like the hoxd-3 –/– animal. Thus, phenotype may depend on the total number of paralogous alleles expressed as well as the specific hox gene for the determination of a region's AP identity.

Retinoids can alter the expression of many of the members of the four paralogous hox complexes. Thus, the effects of exogenous RA treatment and RA deficiency are severe, often more so than are the effects of mutations of one or two of the hox genes. This observation only implicates endogenous retinoids more strongly in the regulation of anterior/posterior identity through their regulation of the many hox genes.

Axial Patterning of Paraxial Mesoderm

Anterior/posterior determination of the vertebrate central body axis is perhaps most clearly illustrated in formation of the vertebrae, which exhibit clear morphological differences along the AP axis. Vertebrae arise from pairs of somites, which are blocks of segmented, paraxial (defined as next to or along the body axis) mesoderm on either side of the neural tube (Figures 2b, c, d). The first four pairs of somites form the occipital bones at the base of the skull. The rest form the cervical, thoracic, lumbar, sacral, and caudal vertebrae in an anterior-to-posterior direction (Figure 3b).

Exogenous retinoids clearly affect vertebral formation. While large doses of RA cause deletions or gross malformations of many embryonic structures, Kessel & Gruss (165) found that lower concentrations of RA led to homeotic transformations in vertebral identity. These transformations are reminiscent of those seen in the hindbrain. As detailed below, these retinoid-associated transformations are similar to those seen in hox gene mutational analyses, suggesting that retinoid mediated regulation of hox gene expression affects development of the vertebrae as it does of the hindbrain and its resulting structures.

Following RA exposure at 7.3 dpc (i.e. early in somite determination), many posteriorizing transformations occurred at a high frequency among all distinguishable sets of vertebrae. Transformations included anterior cervical vertebrae to more posterior cervical vertebrae, the last two cervical vertebrae to rib-bearing thoracic vertebrae, thoracic to non-rib-bearing lumbar vertebrae, unfused lumbar to fused sacral vertebrae, and fused sacral to unfused caudal vertebrae. Following RA exposure at 8.5 dpc, a time when hox genes are being downregulated, anteriorizing transformations occurred, particularly in the posterior vertebrae, including posterior thoracic to anterior thoracic (eighth rib attached to sternum), lumbar to thoracic (one or two extra ribs), sacral to lumbar, and caudal to sacral (loss or presence of fused vertebrae, respectively).

RA alters segment identities in the formation of vertebrae probably through alterations in hox gene expression. In the somites, as in the neural tube, the hox genes are expressed in an anterior to posterior manner, although the anterior boundaries are slightly different from those of the neural tube (Figure 3b). Both hox expression and determination of somite identity are established early, soon after gastrular involution of this paraxial mesoderm. The 3′ members of the hox complexes are activated first and extend farthest anteriorly into the somites. More 5′ members, expressed in cells that involute later, are activated later and more posteriorly. After involution of paraxial mesoderm is complete, the hox genes are downregulated, gradually losing expression posteriorly (165). Exogenous RA may cause posterior transformations by extending hox gene expression anteriorly early in gastrulation, and may then cause anterior transformations in more posterior regions by extending the time of hox gene expression.

Deletion or misexpression of hox genes produces morphological transformations very similar to those induced by retinoid teratogenicity. Overexpression of the hoxb-6 gene, such that it is expressed throughout the embryo rather than in its normal domain posterior to the last cervical somite, caused both posteriorizing and anteriorizing transformations, as both the seventh cervical vertebra and the first lumbar vertebrae are transformed to rib-bearing vertebrae (163). Posterior transformations occurred when the hoxd-4 gene was driven by the hoxa-1 promoter so that its region of expression was extended from the cervical region anteriorly through the occipital region; in this case occipital bones were greatly reduced and seem to be partially replaced by one to four vertebra-like arches, suggesting a posterior homeotic transformation to more posterior, vertebral structures (166). Anterior transformations occurred in the posterior mouse embryo when the hoxc-8 gene expression was extended farther posteriorly in the embryo. Lumbar and even sacral vertebrae were transformed to rib-bearing vertebrae (167). The overexpression of hoxc-8 also resulted in anterior transformations of lumbar to rib-bearing vertebrae, but the extent of its expression in posterior domains was not reported (168). These transformations reflect many of those seen following RA treatment, which lead to posterior transformations owing to anterior extension of hox domains early in paraxial determination, and anterior transformations later when hox genes continue to be expressed rather than downregulated in posterior domains.

Other hox genes that were misexpressed throughout the body axis affected somitic development in such a way that deletions and malformations rather than transformations occurred. When the anterior boundaries of the hoxb-7 (169) or hoxa-7 (170), both normally restricted to somites of the thoracic region, were extended throughout the body, gross malformations or reductions in the occipital bones and the first cervical vertebra, the atlas, occurred. As with RA treatment, disruption of some hox genes appears to lead to loss of structures rather than transformations.

Many of the hox cluster genes have been targeted for loss-of-function analysis. By analogy to *Drosophila* HOM-C genes, loss of expression of a hox gene is expected to result in anteriorizing transformations. For example, the most anterior somites (the occipital somites) express the hox "1" paralog genes while the next most posterior, the cervical somites, express both the "1" and the "3" genes. Deletion of "3" genes would leave only the hox "1" genes, which is the hox code for the occipital somites. Thus, deletion of hox "3" genes is expected to result in the anterior transformations from cervical vertebrae to occipital bones. However, because of the potential for redundancy of the hox code owing to the four paralogous complexes of genes, variable results were obtained for the various hox genes mutated. For example, deletion of the most anteriorly expressed hox genes of the hoxa complex—hoxa-1, hoxa-2, and hoxa-3—had little effect on vertebrate identity, causing only minor malformations of occipital bones (148–151, 157–159).

While deletion of hoxa-3 had no obvious effect on vertebra formation (159), disruption of its paralog, hoxd-3, affected development of the occipital bones and the first and second vertebrae, the atlas and axis (171). Partial anterior transformations occurred, with the atlas often found fused to and resembling the exoccipitals and the axis resembling the atlas, its next more anterior vertebra. A double null mutation of both hoxd-3 and hoxa-3 caused similar but more severe malformations of these vertebrae (see the discussion above on dose requirements) (171). Deletion of hoxa-4 also caused anterior transformations, causing a dorsal process usually confined to the second cervical vertebra to form on the third as well (172). Deletion of hoxb-4 caused an anterior transformation, with the axis appearing morphologically more like an atlas (173). Mutation of hoxa-5 led to posterior transformations of the seventh cervical vertebra to a rib-bearing vertebra, and to anterior transformations of the first lumbar vertebra, which became rib-bearing (174). Mutation of hoxa-6 resulted in a similar posterior transformation of the seventh cervical vertebra to a rib-bearing vertebra, and also of the first thoracic to the second thoracic vertebra, which bears a distinctive dorsal process (172).

The similarity between RA teratogenic phenotypes and hox mutant phenotypes in the vertebral column, coupled to our knowledge of hox regulation by RA in tissue culture cells, is perhaps the most convincing evidence for the involvement of retinoids in AP pattern determination in vertebrates. RA treatment can cause posterior transformations throughout the vertebral column, presumably by extending hox gene expression more anteriorly. RA may also lead to anterior transformations, presumably by expanding the length of hox gene expression in posterior somites.

Patterning in the Limb: Involvement of Hox Genes

Limb development relies upon epithelial-mesenchymal interactions for growth and pattern formation along all three axes: proximal/distal, anterior/posterior, and dorsal/ventral. The teratogenic effects of retinoids on both proximal/distal regulation and anterior/posterior development indicate that endogenous retinoids may play a role in both patternings.

Two important structures governing this development are the apical ectodermal ridge (AER) (175) and an area of mesenchyme referred to as the zone of polarizing activity (Figure 4). The AER is an epithelial structure that functions in limb development via expression of growth factors important for limb outgrowth and patterning (discussed below).

The zone of polarizing activity, or ZPA, is defined functionally (Figure 4). It is a posterior-distal region of limb bud mesenchyme that, when transplanted to the anterior side of a host limb, induces a mirror image duplication of digits, with posterior digits now next to the implanted ZPA as well as next to the host's own ZPA (reviewed in 176). Two hypotheses account for this polarizing activity. One is that the ZPA releases a signal or "morphogen" through the limb bud that results in a high signal concentration near the ZPA and progressively lower concentrations at greater distances from the ZPA. A high concentration of the signal would specify the formation of posterior digits, and lower concentrations would result in anterior digits (177, 178). The other hypothesis is that the ZPA establishes anterior/posterior digit specificity by a series of cell–cell contacts rather than by graded action over a distance. In this model, the ZPA acts locally on adjacent cells, which then act upon cells adjacent to them, and so on such that the ZPA "signal" is passed throughout the limb field (179).

One group of chemical compounds, the retinoids, has been found to mimic the polarizing effects of the ZPA (64, 180, 181). Both RA and 3,4-didehydro-RA caused mirror image digit duplications when beads soaked in these compounds were grafted on the anterior edge of the limb bud (Figure 4). No effect of either compound was observed when grafted to the posterior margin (unless high doses were used, as discussed below). Thaller & Eichele (182) found that when many limb buds were divided into posterior and anterior halves and analyzed by HPLC, the posterior halves had two-fold higher levels of RA. When limb bud tissue was applied to cell lines expressing a reporter gene driven by an RARE, high levels of active retinoids were detected, but the relative anterior vs posterior levels were not quantified (121). These results suggest that retinoids are strong candidates for signaling molecules in the developing limb. Recently, application of the newly discovered peptide hormone, sonic hedgehog, was found to induce mirror image duplications similar to those induced by the ZPA and RA. Sonic hedgehog is expressed in the ZPA and thus may be part of the signaling pathway. However, because RA can induce sonic hedgehog gene expression, retinoids may act before sonic hedgehog in the generation of ZPA activity (183).

Whatever the signal produced by the ZPA, one of its consequences is the spatial and temporal activation of hox genes (184, 185). Genes from both the hoxa and hoxd complexes are expressed as gradients across the limb bud. The hoxa complex genes are expressed in a more proximal/distal manner (186), while the hoxd complex genes are expressed in an anterior/posterior manner (70, 187,

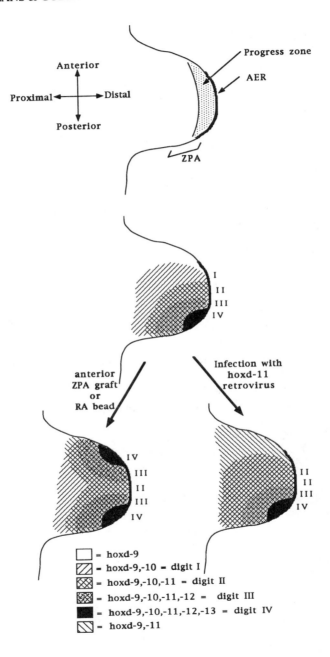

176). The specific genes of the complexes that are expressed in an axial manner in the early limb bud are located at the 5′ ends of the hoxa and hoxd complexes. These genes are duplications of a gene homologous to the *Drosophila* AbdB gene, part of the bithorax complex (Figure 3a). The hoxb complex has only one copy of this gene. It remains to be seen if the hoxc complex AbdB-like genes will have graded patterns of expression in the limb bud.

The most 3′ of the AbdB-like genes in the hoxd complex, hoxd-9, is expressed first in the distal posterior margin of the limb bud. As its expression pattern gradually extends to cross the entire limb bud, the more 5′ genes are gradually expressed from the posterior-distal margin. The digits formed from the resulting hox patterns vary according to species because the number of digits formed varies. For the chick hindlimb, discussed here, the pattern is as follows. Eventually hoxd-9 fills the limb bud, hoxd-10 extends anteriorly to the region of digit 1 formation, hoxd-11 extends to the region of digit 2 formation, hoxd-12 extends to digit 3, and hoxd-13 remains only posteriorly expressed in the region of digit 4 expression (Figure 4).

When either ZPA cells or retinoids are applied to the anterior limb bud margin, these hox AbdB-like genes are activated from the anterior margin in a mirror image of the pattern of posterior expression (Figure 4). Duplication of hoxd gene expression precedes and thus may determine digit formation and duplication. Supporting this causal role for the hoxd genes are the following data: When hoxd-11, which is normally expressed anteriorly to digit 2, is expressed throughout the limb, no digit field expresses exclusively hoxd-9 and d-10, the "code" for the first digit. Instead, the second most anterior digit field expresses hoxd-9, d-10, and d-11. As a result, two digit 2s are formed, one replacing digit 1 (188). Because RA can regulate hox genes and these genes in turn control AP identity of digits, the involvement of retinoids in anterior/posterior patterning is indicated.

Figure 4 Alteration of hox gene expression alters AP axis formation of digits. The chick hindlimb bud, which forms four digits, is diagrammed. Anterior/posterior and proximal/distal axes are indicated. The progress zone is the region of rapidly proliferating cells in the distal mesenchyme. The AER (apical ectodermal ridge) is the layer of ectoderm immediately overlying the progress zone. The ZPA (zone of polarizing activity) is a region of posterior distal mesenchyme responsible for anterior/posterior specification. Hox genes d-9 through d-13 are expressed in overlapping patterns in the limb bud, initiating at the posterior edge and expanding progressively anteriorly as indicated. Transplantation of a ZPA from one limb bud to the anterior border of another can respecify both hox gene expression and digit formation, causing a mirror image duplication of both. Likewise, implantation of a retinoic acid–containing bead can cause mirror image duplications of hox expression and digit formation. That alteration in digit pattern can result from altered hox expression is demonstrated by expansion of hoxd-11 expression throughout the limb via retrovirus infection. The limb subsequently contains no region of solely hoxd-12 and d-13, but has an expanded region of hoxd-11, d-12, and d-13 expression. This respecifies digit formation such that two digit II and no digit I form. See text for references.

Proximal/distal limb development is also affected by RA, although loss or malformation rather than homeotic transformations generally results from RA teratogenicity. Application of high doses of RA to the limb bud shortly after evagination has begun causes truncation of the limb (51, 62, 188). Members of the hoxa complex may be involved in this proximal/distal regulation. Genes at the 5' end of this complex are initially expressed from the posterior-distal margin (the region of the ZPA). However, genes of the hoxa cluster are then expressed throughout the anterior/posterior axis such that an AP gradient is no longer observed. Rather, a proximal/distal gradient of hoxa cluster genes is seen, with progressively overlapping domains from the distal end of the limb bud. The expression of the 3' most AbdB-like gene, hoxa-10, extends most proximally, hoxa-11 extends intermediately, and hoxa-13 remains confined to the most distal limb mesenchyme. A null mutation of hoxa-11, normally expressed midway between hoxa-10 and a-13, results in fusion and other malformations of the forelimb bones (189). These results suggest that hoxa-11 may regulate development of this region, whereas hoxa-10 and a-13 may regulate more proximal and more distal structures, respectively, such as pha-langes and upper limb bones. As with retinoid teratogenicity, alteration of this "hox code" results in loss or malformation rather than homeotic transformation.

Another hox gene involved in digit formation is the hoxb-8 gene, which is not a member of the AbdB-like paralogs. Hoxb-8 is normally expressed along the posterior margin of the limb bud, including the ZPA region. Ectopic expression of the hoxb-8 gene under the control of the RARβ2 promoter, which resulted in its expression throughout the anterior/posterior axis of the limb bud, caused mirror image duplications in the forelimb similar to those seen when the ZPA or RA is transplanted to the anterior margin of the limb bud (190).

Other homeobox-containing genes, not in the hox clusters but more distantly related to the HOM-C genes of *Drosophila*, are also expressed in the limb, but their roles in patterning have yet to be elucidated. Evx-1, a homeobox-con-taining gene similar to the *Drosophila* even-skipped gene, is expressed at highest levels in the posterior, most distal part of the limb bud; expression gradually decreases in an anterior-proximal direction (191). Hox7.1 is ex-pressed more highly in the distal mesenchyme, in the rapidly proliferating "progress zone" cells (192, 193). Hox8.1 is expressed both in the AER ecto-derm overlying the progress zone and in the anterior mesenchyme (186, 194, 195). Hox7.1 and 8.1 may have roles in maintaining growth and in signal transduction between epithelia and underlying mesenchyme (196, 197).

Patterning in the Limb: Interactions Between Retinoids and Peptide Growth Factors

Up to this point we have considered the effects of retinoids on embryonic development, concentrating on both the functions of endogenous retinoids during development and the effects of exogenously added retinoids on the developing fetus. It is important to note that retinoids act in vivo in concert with a variety of other hormones and growth factors, some of which are discussed here. A more detailed discussion of the relationships between growth factors and retinoids can be found in the review by Gudas et al (2).

Fibroblast growth factors are small peptides that influence the growth and differentiation of a variety of cell types. There are currently nine known members of the FGF peptide family (198, 199). In *Xenopus*, FGFs are important mediators of mesoderm induction (200), and they influence limb development in the mouse and chick (e.g. 201). As discussed above, the ZPA, a small group of mesenchyme cells located at the posterior margin of the developing limb bud, can specify position across the anterior/posterior axis of the limb. The cells of the apical ectodermal ridge (AER) are also important in signaling by the ZPA, because when the apical ectodermal ridge is removed from the developing limb there is a dramatic decrease in ZPA activity. Thus, the AER supplies signals to the underlying mesenchyme that are necessary for proper mesenchymal cell growth and differentiation. Remarkably, this decrease in ZPA activity of the posterior mesenchymal cells after removal of the AER was prevented by implanting a FGF-4-soaked bead in the developing bud (202). Similarly, Niswander & Martin (203) demonstrated that FGF-4 could stimulate the growth of murine limb bud mesenchyme, an observation suggesting that FGF-4, normally expressed in the AER, mediates this function of the apical ectodermal ridge. When limb bud cells were cultured in vitro, FGF-2 (bFGF) was also able to maintain ZPA signaling activity (204). Moreover, in the absence of the apical ectodermal ridge FGF-4 can interact with RA to generate ZPA activity at the anterior margin of the limb bud; as a result, mirror image digit duplications occur along the anteroposterior axis (205). Although the mechanisms by which FGF-4 and possibly other members of the FGF family interact with RA to generate and maintain ZPA activity are not generally understood, sonic hedgehog gene expression can be activated in limb mesenchyme by RA plus FGF-4; and once induced, sonic hedgehog expression can be maintained by FGF-4 alone (205). These data indicate that both RA and FGF-4 may exert their effects via upregulation of the peptide hormone, sonic hedgehog.

The transforming growth factor β (TGF-β) family consists of a large number of related peptides that can regulate both proliferation and differentiation in a wide variety of normal and neoplastic cells. This large family of peptide factors includes inhibins, activins, the bone morphogenetic proteins, and the protein products of the *Drosophila* decapentaplegic gene complex (for review see 206). One member of the TGF-β superfamily, the gene nodal, is detected in the primitive streak of mouse embryos at about the time of gastrulation (207, 208). Several members of the TGF-β superfamily also affect the developing limb bud. For instance, local application of TGF-β1 to distal regions of the chick limb bud causes specific skeletal elements in the limb to be reduced or missing (209). The related peptide BMP-2, which is expressed in the apical ectodermal ridge, can inhibit limb growth when it is applied to forelimb buds in a culture system (203). The BMP-2 gene is also expressed in a domain of mesenchymal cells in the limb that co-localizes with the ZPA. The application of RA to the anterior part of the limb bud, which generates a new ZPA, activates BMP-2 gene expression in the anterior limb bud cells (210). However, because BMP-2 itself has no detectable ZPA activity, the activation of BMP-2 expression in response to RA may be part of the ZPA signaling pathway (i.e. BMP-2 may not be the signaling molecule) (210).

That RA can regulate the expression of some genes in the TGF-β superfamily has been shown in a number of cell culture systems and in developing mouse embryos. For example, the level of endogenous BMP-2 transcripts increased approximately 11-fold after treatment of F9 embryonal carcinoma cells with retinoic acid and dibutyryl cyclic AMP, whereas the message for the closely related BMP-4 gene decreased approximately 12-fold after RA addition (211). RARγ is presumably involved in the upregulation of the BMP-2 gene in this model cell culture system, because the disruption of both copies of the RARγ gene in an F9 cell line (an RARγ −/− F9 line) resulted in the loss of the RA-inducible expression of the BMP-2 gene (212, 213).

In conclusion, the interactions between retinoids and the FGF and TGF-β growth factor families are complex and depend on the cell type and particular family member. It is not now possible to predict whether retinoids and particular members of these growth factor families will synergize, enhance, or antagonize each other.

FUTURE PROSPECTS

Here we have discussed many recent studies linking retinoid teratogenicity to molecular changes in gene expression both during the formation of the central body axis and during the later development of the limbs. However, much experimental work has been performed using exogenously added RA, and there is no proof to date that gradients of endogenously synthesized RA play a role in the establishment of the anterior/posterior body axis during development. Vital information on the roles of retinoids in normal development should come from 1. further research into the intracellular metabolism of vitamin A (retinol) and its metabolic derivatives (such as RA) in the embryo, including the identification and cloning of the enzymes involved in its metabolism; 2. research concerning the mechanisms by which the homeobox gene clusters are regulated by retinoids (because the homeobox genes are crucial intermediates in the RA signaling pathway); and 3. research concerning the ligands for and functions of other potential retinoid receptors.

ACKNOWLEDGMENTS

We thank Drs. Lap Ho, John Wagner, Rosemary Bachvarova, Jane Love, Teresa Faria, and Alex Langston for critically reading this manuscript; Alex Langston and Charles Achkar for assistance with figures; and Ms. Taryn Resnick for editorial assistance. This work was supported by National Institutes of Health grants R01CA39036 to LJG and 1F32CA09336 to ALM.

Any *Annual Review* chapter, as well as any article cited in an *Annual Review* chapter,
may be purchased from the Annual Reviews Preprints and Reprints service.
1-800-347-8007; 415-259-5017; email: arpr@class.org

Literature Cited

1. Mangelsdorf DJ, Umesono K, Evans RM. 1994. See Ref. 6, pp. 319–50
2. Gudas LJ, Sporn MB, Roberts AB. 1994. See Ref. 6, pp. 443–520
3. Lewis EB. 1992. *J. Am. Med. Assoc.* 267:1524–31
4. Hong WK, Itri LM. 1994. See Ref. 6, pp. 597–630
5. Peck GL, DiGiovanna JJ. 1994. See Ref. 6, pp. 631–58
6. Sporn MB, Roberts AB, Goodman DS, eds. 1994. *The Retinoids: Biology, Chemistry, and Medicine.* New York: Raven. 679 pp. 2nd ed.
7. Eppinger TM, Buck J, Hammerling U. 1993. *J. Exp. Med.* 178:1995–2005
8. Buck J, Grun F, Derguini F, Chen Y, Kimura S, et al. 1993. *J. Exp. Med.* 178:675–80
9. Benbrook D, Lernhardt E, Pfahl M. 1988. *Nature* 333:669–72
10. Brand N, Petkovich M, Krust A, Chambon P, de Thé H, et al. 1988. *Nature* 332:850–53
11. Giguere V, Ong ES, Segui P, Evans RM. 1987. *Nature* 330:624–29
12. Krust A, Kastner P, Petkovich M, Zelent A, Chambon P. 1989. *Proc. Natl. Acad. Sci. USA* 86:5310–14
13. Petkovich M, Brand NJ, Krust A, Chambon P. 1987. *Nature* 330:444–50
14. Zelent A, Krust A, Petkovich M, Kastner P, Chambon P. 1989. *Nature* 339:714–17
15. Kastner P, Krust A, Mendelsohn C, Garnier JM, Zelent A, et al. 1990. *Proc. Natl. Acad. Sci. USA* 87:2700–4
16. Leroy P, Krust A, Zelent A, Mendelsohn C, Garnier JM, et al. 1991. *EMBO J.* 10:59–69
17. Zelent A, Mendelsohn C, Kastner P, Krust A, Garnier JM, et al. 1991. *EMBO J.* 10:71–81
18. Mangelsdorf DJ, Ong ES, Dyck JA, Evans RM. 1990. *Nature* 345:224–29
19. Mangelsdorf DJ, Borgmeyer U, Heyman RA, Zhou JY, Ong ES, et al. 1992. *Genes Dev.* 6:329–44
20. Blumberg B, Mangelsdorf DJ, Dyck JA, Bittner DA, Evans RM, De Robertis EM. 1992. *Proc. Natl. Acad. Sci. USA* 89:2321–25
21. Heyman RA, Mangelsdorf DJ, Dyck JA, Stein RB, Eichele G, et al. 1992. *Cell* 68:397–406
22. Levin AA, Sturzenbecker LJ, Kazmer S, Bosakowski T, Huselton C, et al. 1992. *Nature* 355:359–61
23. Nakshatri H, Chambon P. 1994. *J. Biol. Chem.* 269:890–902
24. Carlsberg C, van Huijsduijnen RH, Staple JK, DeLamarter JF, Becker-André M. 1994. *Mol. Endocrinol.* 8:757–70
25. Jonk LJC, de Jonge MEJ, Pals CEGM, Wissink S, Vervaart JMA et al. 1994. *Mech. Dev.* 47:81–97
26. Cooney AJ, Tsai SY, O'Malley BW, Tsai M-J. 1992. *Mol. Cell. Biol.* 12:4153–63
27. Tran P, Zhang X-K, Salbert G, Hermann T, Lehmann JM, Pfahl M. 1992. *Mol. Cell. Biol.* 12:4666–76
28. Harding HP, Lazar MA. 1993. *Mol. Cell. Biol.* 13:3113–21
29. Luo X, Ikeda Y, Parker KL. 1994. *Cell* 77:481–90
30. Duncan SA, Manova K, Chen WS,

Hoodless P, Weinstein DW, et al. 1994. *Proc. Natl. Acad. Sci. USA* 91:7598–602

31. Baes M, Gulick T, Choi H-S, Martinoli MG, Simha D, Moore DD. 1994. *Mol. Cell. Biol.* 14:1544–52
32. Nagpal S, Saunders M, Kastner P, Durand B, Nakshatri H, Chambon P. 1992. *Cell* 70:1007–19
33. Zhang XK, Hoffmann B, Tran PB, Graupner G, Pfahl M. 1992. *Nature* 355:441–46
34. Kliewer SA, Umesono K, Mangelsdorf DJ, Evans RM. 1992. *Nature* 355:446–49
35. Levin MS. 1993. *J. Biol. Chem.* 268:8267–76
36. Ong DE, Newcomer ME, Chytil F. 1994. See Ref. 6, pp. 283–317
37. Napoli JL, Posch KP, Fiorella PD, Boerman MH. 1991. *Biomed. Pharmacother.* 45:131–43
38. Takase S, Ong DE, Chytil F. 1986. *Arch. Biochem. Biophys.* 247:328–34
39. Boylan JF, Gudas LJ. 1991. *J. Cell Biol.* 112:965–79
40. Boylan JF, Gudas LJ. 1992. *J. Biol. Chem.* 267:21486–91
41. Maden M, Ong DE, Summerbell D, Chytil F. 1988. *Nature* 335:733–35
42. Ruberte E, Dollé P, Chambon P, Morriss-Kay G. 1991. *Development* 111:45–60
43. Fiorella PD, Napoli JL. 1991. *J. Biol. Chem.* 266:16572–79
44. Fiorella PD, Napoli JL. 1994. *J. Biol. Chem.* 269:10538–44
45. Ong DE, Chytil F. 1978. *J. Biol. Chem.* 253:4551–54
46. Bailey JS, Siu CH. 1988. *J. Biol. Chem.* 263:9326–32
47. Giguere V, Lyn S, Yip P, Siu CH, Amin S. 1990. *Proc. Natl. Acad. Sci. USA* 87:6233–37
48. Durand B, Saunders M, Leroy P, Leid M, Chambon P. 1992. *Cell* 71:73–85
49. Astrom A, Pettersson U, Chambon P, Voorhees JJ. 1994. *J. Biol. Chem.* 269:22334–39
50. Lammer EJ, Chen DT, Hoar RM, Agnish ND, Benke PJ, et al. 1985. *N. Engl. J. Med.* 313:837–41
51. Rizzo R, Lammer EJ, Parano E, Pavone L, Argyle JC. 1991. *Teratology* 44:599–604
52. Dai WS, LaBraico JM, Stern RS. 1992. *J. Am. Acad. Dermatol.* 26:599–606
53. Shenefelt R. 1972. *Teratology* 5:103–18
54. Wiley MJ. 1983. *Teratology* 28:341–53
55. Kochhar DM, Penner JD. 1987. *Teratology* 36:67–75
56. Rutledge JC, Shourbaji AG, Hughes LA,

Polifka JE, Cruz YP, et al. 1994. *Proc. Natl. Acad. Sci. USA* 91:5436–40
57. Morriss GM. 1972. *Am. J. Anat.* 113:241–50
58. Kraft JC, Löfberg B, Chahoud I, Bochert G, Nau H. 1989. *Toxicol. Appl. Pharmacol.* 100:162–76
59. Hyatt GA, Schmitt EA, Marsh-Armstrong NR, Dowling JE. 1992. *Proc. Natl. Acad. Sci. USA* 89:8293–97
60. Morriss-Kay G. 1991. *Semin. Dev. Biol.* 2:211–18
61. Jiang H, Kochhar DM. 1992. *Teratology* 46:333–40
62. Alles AJ, Sulik KK. 1989. *Teratology* 40:163–71
63. Sulik KK, Dehart DB. 1988. *Teratology* 37:527–37
64. Tickle C, Lee J, Eichele G. 1985. *Dev. Biol.* 109:82–95
65. Wilson JG, Roth CB, Warkany J. 1953. *Am. J. Anat.* 92:189–217
66. Warkany J, Schraffenberger E. 1946. *Arch. Ophthalmol.* 35:150–69
67. Wilson JG, Warkany J. 1949. *Am. J. Anat.* 85:113–55
68. Webster WS, Johnston MC, Lammer EJ, Sulik KK. 1986. *J. Craniofac. Genet. Dev. Biol.* 6:211–22
69. Langston AW, Gudas LJ. 1994. *Curr. Opin. Genet. Dev.* 4:550–55
70. Dollé P, Ruberte E, Kastner P, Petkovich M, Stoner CM, et al. 1989. *Nature* 342:702–5
71. Dollé P, Ruberte E, Leroy P, Morriss-Kay G, Chambon P. 1990. *Development* 110:1133–51
72. Ruberte E, Dolle P, Krust A, Zelent A, Morriss-Kay G, Chambon P. 1990. *Development* 108:213–22
73. Ruberte E, Friederich V, Chambon P, Morriss-Kay G. 1993. *Development* 118:267–82
74. Ellinger-Ziegelbauer H, Dreyer C. 1993. *Mech. Dev.* 41:33–46
75. Pfeffer PL, De Robertis EM. 1994. *Mech. Dev.* 45:147–53
76. Mendelsohn C, Ruberte E, Chambon P. 1992. *Dev. Biol.* 152:50–61
77. Soprano DR, Gyda M III, Jiang H, Harnish DC, Ugen K, et al. 1994. *Mech. Dev.* 45:243–53
78. Rowe A, Richman JM, Brickell PM. 1991. *Development* 111:1007–16
79. Sucov HM, Murakami KK, Evans RM. 1990. *Proc. Natl. Acad. Sci. USA* 87:5392–96
80. de Thé H, Vivanco-Ruiz M, Tiollais P, Stunnenberg H, Dejean A. 1990. *Nature* 343:177–80
81. Joore J, van der Lans GBLJ, Lanser PH, Vervaart JMA, Zivkovic D, et al. 1994. *Mech. Dev.* 46:137–50

82. Sucov HM, Dyson E, Gumeringer CL, Price J, Chien KR, Evans RM. 1994. *Genes Dev.* 8:1007–18
83. Kastner P, Grondona JM, Mark M, Gansmuller A, LeMeur M, et al. 1994. *Cell* 78:987–1003
84. Lohnes D, Kastner P, Dierich A, Mark M, LeMeur M, Chambon P. 1993. *Cell* 73:643–58
85. Lufkin T, Lohnes D, Mark M, Dierich A, Gorry P, et al. 1993. *Proc. Natl. Acad. Sci. USA* 90:7225–29
86. Li E, Sucov HM, Lee K-F, Evans RM, Jaenisch R. 1993. *Proc. Natl. Acad. Sci. USA* 90:1590–94
87. Vaessen M-J, Kootwijk E, Mummery C, Hilkens J, Bootsma D, Van Kessel AG. 1989. *Differenta* 40:99–105
88. Dencker L, Annerwall E, Busch C, Eriksson U. 1990. *Development* 110:343–52
89. Perez-Castro AV, Toth-Rogler LE, Wei L-N, Nguyen-Huu MC. 1989. *Proc. Natl. Acad. Sci. USA* 86:8813–17
90. Gustafson A-L, Dencker L, Eriksson U. 1993. *Development* 117:451–60
91. Harnish DC, Soprano KJ, Soprano DR. 1992. *Teratology* 46:137–46
92. Maden M, Horton C, Graham A, Leonard L, Pizzey J, et al. 1992. *Genes Dev.* 37:13–23
93. Dekker E-J, Vaessen M-J, van den Berg C, Timmermans A, Godsave S, et al. 1994. *Development* 120:973–85
94. Ho L, Mercola M, Gudas LJ. 1994. *Mech. Dev.* 47:53–64
95. Papalopulu N, Lovell-Badge R, Krumlauf R. 1991. *Nucleic. Acids Res.* 19:5497–506
96. Marshall H, Nonchev S, Sham MH, Muchamore I, Lumsden A, Krumlauf R. 1992. *Nature* 360:737–41
97. Morriss-Kay GM, Murphy P, Hill RE, Davidson DR. 1991. *EMBO J.* 10:2985–95
98. Kessel M. 1993. *Neuron* 10:379–93
99. Wood H, Pall G, Morriss-Kay G. 1994. *Development* 120:2279–85
100. Ruberte E, Friederich V, Morriss-Kay G, Chambon P. 1992. *Development* 115:973–87
101. Miyagawa-Tomita S, Kitamoto T, Momma K, Takao A, Momoi T. 1992. *Biochem. Biophys. Res. Commun.* 185:217–23
102. Gorry P, Lufkin T, Dierich A, Rochette-Egly C, Décimo D, et al. 1994. *Proc. Natl. Acad. Sci. USA* 91:9032–36
103. Nieuwkoop PD. 1973. *Adv. Morphog.* 10:1–39
104. Spemann H, Mangold H. 1924. *Roux's Arch. Dev. Biol.* 100:599–638
105. Waddington CH. 1932. *Proc. Trans. R. Soc. London* 211:179–230
106. Beddington RS. 1994. *Development* 120:613–20
107. Cho KWY, Blumberg B, Steinbeisser H, De Robertis EM. 1991. *Cell* 67:1111–20
108. Dirksen ML, Jamrich M. 1992. *Genes Dev.* 6:599–608
109. Altaba ARI, Jessell TM. 1992. *Development* 116:81–93
110. Blum M, Gaunt SJ, Cho KW, Steinbeisser H, Blumberg B, et al. 1992. *Cell* 69:1097–106
111. Ang S-L, Rossant J. 1994. *Cell* 78:561–74
112. Altaba ARI, Prezioso VR, Darnell JE, Jessell TM. 1993. *Mech. Dev.* 44:91–108
113. Morriss GM. 1975. In *New Approaches to the Evaluation of Abnormal Embryonic Development,* ed. D Neubert, HJ Merker, pp. 678–87. Stuttgart: Thieme
114. Morriss GM, Steele CE. 1974. *J. Embryol. Exp. Morphol.* 32:505–14
115. Morriss GM, Steele CE. 1977. *Teratology* 15:109–19
116. Durston AJ, Timmermans JP, Hage WJ, Hendriks HF, de Vries NJ, et al. 1989. *Nature* 340:140–44
117. Sive HL, Draper BW, Harland RM, Weintraub H. 1990. *Genes Dev.* 4:932–42
118. Green JBA, Smith JC. 1990. *Nature* 347:391–94
119. Niehrs C, Steinbeisser H, De Robertis EM. 1994. *Science* 263:817–20
120. Chen Y, Huang L, Solursh M. 1994. *Dev. Biol.* 161:70–76
121. Wagner M, Han B, Jessell TM. 1992. *Development* 116:55–66
122. Chen Y, Huang L, Russo AF, Solursh M. 1992. *Proc. Natl. Acad. Sci USA* 89:10056–59
123. Rossant J, Zirngibl R, Cado D, Shago M, Giguére V. 1991. *Genes Dev.* 5:1333–44
124. Balkan W, Colbert M, Bock C, Linney E. 1992. *Proc. Natl. Acad. Sci. USA* 89:3347–51
125. Altaba ARI, Jessell TM. 1991. *Development* 112:945–58
126. Altaba ARI, Jessell T. 1991. *Genes Dev.* 5:175–87
127. Cho KW, De Robertis EM. 1990. *Genes Dev.* 4:1910–16
128. Sive HL, Cheng PF. 1991. *Genes Dev.* 5:1321–32
129. Taira M, Otani H, Jamrich M, Dawid IB. 1994. *Development* 120:1525–36
130. Dixon JE, Kintner CR. 1989. *Development* 106:749–57
131. Altaba ARI. 1992. *Development* 116:67–80

132. Sharpe CR. 1991. *Neuron* 7:239–47
133. Wolbach SB, Bessey OA. 1941. *Arch. Pathol.* 32:689–722
134. Mellanby E. 1944. *Proc. R. Soc.* 132:28–46
135. Kessel M, Gruss P. 1990. *Science* 249:374–79
136. Krumlauf R, Marshall H, Studer M, Nonchev S, Sham MH, Lumsden A. 1993. *J. Neurobiol.* 24:1328–40
137. Krumlauf R. 1994. *Cell* 78:191–201
138. Langston AW, Gudas LJ. 1992. *Mech. Dev.* 38:217–28
139. Popperl H, Featherstone MS. 1993. *Mol. Cell. Biol.* 13:257–65
140. Moroni MC, Vigano MA, Mavilio F. 1993. *Mech. Dev.* 44:139–54
141. Marshall H, Studer M, Popperl H, Aparicio S, Kuroiwa A, et al. 1994. *Nature* 370:567–71
142. Studer M, Popperl H, Marshall H, Kuroiwa A, Krumlauf R. 1994. *Science* 265:1728–32
143. Simeone A, Acampora D, Nigro V, Faiella A, D'Esposito M, et al. 1991. *Mech. Dev.* 33:215–27
144. Faiella A, Zappavigna V, Mavilio F, Boncinelli E. 1994. *Proc. Natl. Acad. Sci. USA* 91:5335–39
145. Hogan BLM, Thaller C, Eichele G. 1992. *Nature* 359:237–41
146. Conlon RA, Rossant J. 1992. *Development* 116:357–68
147. Zhang M, Kim H-J, Marshall H, Gendron-Maguire M, Lucas DA, et al. 1994. *Development* 120:2431–42
148. Dollé P, Lufkin T, Krumlauf R, Mark M, Duboule D, Chambon P. 1993. *Proc. Natl. Acad. Sci. USA* 90:7666–70
149. Mark M, Lufkin T, Vonesch JL, Ruberte E, Olivo JC, et al. 1993. *Development* 119:319–38
150. Carpenter EM, Goddard JM, Chisaka O, Manley NR, Capecchi MR. 1993. *Development* 118:1063–75
151. Lufkin T, Dierich A, Le Meur M, Mark M, Chambon P. 1991. *Cell* 66:1105–19
152. Lumsden A, Sprawson N, Graham A. 1991. *Development* 113:1281–91
153. Le Lievre CS, Le Douarin NM. 1975. *J. Embryol. Exp. Morphol.* 34:125–54
154. Kirby ML, Gale TF, Stewart DE. 1983. *Science* 220:1059–61
155. Bockman DE, Kirby ML. 1984. *Science* 223:498–500
156. Noden DM. 1983. *Dev. Biol.* 96:144–65
157. Gendron-Maguire M, Mallo M, Zhang M, Gridley T. 1993. *Cell* 75:1317–31
158. Rijli FM, Mark M, Lakkaraju S, Dierich A, Dollé P, Chambon P. 1993. *Cell* 75:1333–49
159. Chisaka O, Capecchi MR. 1991. *Nature* 350:473–79
160. Sechrist J, Serbedzija GN, Scherson T, Fraser SE, Bronner-Fraser M. 1993. *Development* 118:691–703
161. Schughart K, Kappen C, Ruddle FH. 1988. *Br. J. Cancer (58 Suppl. 9):9–13*
162. Graham A, Papalopulu N, Krumlauf R. 1989. *Cell* 57:367–78
163. Kaur S, Singh G, Stock JL, Schreiner CM, Kier AB, et al. 1992. *J. Exp. Zool.* 264:323–36
164. Condie BG, Capecchi MR. 1994. *Nature* 370:304–7
165. Kessel M, Gruss P. 1991. *Cell* 67:89–104
166. Lufkin T, Mark M, Hart C, Dollé P, Le Meur M, Chambon P. 1992. *Nature* 359:835–41
167. Jegalian BG, De Robertis EM. 1992. *Cell* 71:901–10
168. Pollock RA, Jay G, Bieberich CJ. 1992. *Cell* 71:911–23
169. McLain K, Schreiner C, Yager KL, Stock JL, Potter SS. 1992. *Mech. Dev.* 39:3–16
170. Kessel M, Balling R, Gruss P. 1990. *Cell* 61:301–8
171. Condie BG, Capecchi MR. 1993. *Development* 119:579–95
172. Kostic D, Capecchi MR. 1994. *Mech. Dev.* 46:231–47
173. Ramirez-Solis R, Zheng H, Whiting J, Krumlauf R, Bradley A. 1993. *Cell* 73:279–94
174. Jeannotte L, Lemieux M, Charron J, Poirier F, Robertson EJ. 1993. *Genes Dev.* 7:2085–96
175. Summerbell D, Lewis JH, Wolpert L. 1973. *Nature* 244:492–96
176. Tabin CJ. 1992. *Development* 116:289–96
177. Tickle C, Summerbell D, Wolpert L. 1975. *Nature* 254:199–202
178. Wolpert L. 1975. *Ann. Biomed. Eng.* 3:401–5
179. Bryant SV, Muneoka K. 1986. *Trends Genet.* 2:153–56
180. Summerbell D. 1983. *J. Embryol. Exp. Morphol.* 78:269–89
181. Tickle C, Alberts B, Wolpert L, Lee J. 1982. *Nature* 296:564–66
182. Thaller C, Eichele G. 1987. *Nature* 327:625–28
183. Riddle RD, Johnson RL, Laufer E, Tabin C. 1993. *Cell* 75:1401–16
184. Nohno T, Noji S, Koyama E, Ohyama K, Myokai F, et al. 1991. *Cell* 64:1197–205
185. Izpisua-Belmonte JC, Tickle C, Dollé P, Wolpert L, Duboule D. 1991. *Nature* 350:585–89
186. Yokouchi Y, Ohsugi K, Sasaki H, Kuroiwa A. 1991. *Development* 113:431–44

187. Izpisua-Belmonte JC, Falkenstein H, Dollé P, Renucci A, Duboule D. 1991. *EMBO J.* 10:2279–89
188. Morgan BA, Izpisua-Belmonte JC, Duboule D, Tabin CJ. 1992. *Nature* 358: 236–39
189. Small KM, Potter SS. 1993. *Genes Dev.* 7:2318–28
190. Charité J, de Graaff W, Shen S, Deschamps J. 1994. *Cell* 78:589–601
191. Niswander L, Martin GR. 1993. *Development* 119:287–94
192. Robert B, Sassoon D, Jacq B, Gehring WJ, Buckingham M. 1989. *EMBO J.* 8:91–100
193. Hill RE, Jones PF, Rees AR, Sime CM, Justice MJ, Copeland NJ, Jenkins NA, Graham E, Davidson DR. 1989. *Genes Dev.* 3:26–37
194. Robert B, Lyons G, Simandl B, Kuroiwa A, Buckingham M. 1991. *Genes Dev.* 5:2363–74
195. Coelho CND, Sumoy L, Rodgers BJ, Davidson DR, Hill R, et al. 1991. *Mech. Dev.* 34:143–54
196. Ros MA, Lyons G, Kosher RA, Upholt WB, Coelho CND, Fallon JF. 1992. *Development* 116:811–18
197. Izpisua-Belmonte J-C, Duboule D. 1992. *Dev. Biol.* 152:26–36
198. Basilico C, Moscatelli D. 1992. *Adv. Cancer Res.* 59:115–65
199. Mason IJ. 1994. *Cell* 78:547–52
200. Kimelman D, Maas A. 1992. *Development* 114:261–69
201. Riley BB, Savage MP, Simandl BK, Olwin BB, Fallon JF. 1993. *Development* 118:95–104
202. Vogel A, Tickle C. 1993. *Development* 119:199–206
203. Niswander L, Martin GR. 1993. *Nature* 361:68–71
204. Anderson R, Landry M, Muneoka K. 1993. *Development* 117:1421–33
205. Niswander L, Jeffrey S, Martin GR, Tickle C. 1994. *Nature* 371:609–12
206. Roberts AB, Sporn MB. 1992. In *Cancer Surveys: Growth Regulation by Nuclear Hormone Receptors*, ed. pp. 205–20. London: Imperial Cancer Res. Fund
207. Zhou X, Sasaki H, Lowe L, Hogan BLM, Kuehn MR. 1993. *Nature* 361: 543–47
208. Conlon FL, Lyons KM, Takaesu N, Barth K, Kispert A, et al. 1994. *Development* 10:1919–28
209. Hayamizu TF, Sessions SK, Wanek N, Bryant SV. 1991. *Dev. Biol.* 145:164–73
210. Francis PH, Richardson MK, Brickell PM, Tickle C. 1994. *Development* 120: 209–18
211. Rogers MB, Rosen V, Wozney JM, Gudas LJ. 1992. *Mol. Biol. Cell* 3:189–96
212. Boylan JF, Lohnes D, Taneja R, Chambon P, Gudas LJ. 1993. *Proc. Natl. Acad. Sci. USA* 90:9601–5
213. Boylan JF, Lufkin T, Achkar CC, Taneja R, Chambon P, Gudas LJ. 1995. *Mol. Cell. Biol.* In press

Annu. Rev. Biochem. 1995. 64:235–57

PLASMA LIPID TRANSFER PROTEINS

Alan Tall

Department of Medicine, Columbia University, New York, New York 10032

KEY WORDS: lipid transfer protein, cholesteryl ester transfer protein, phospholipid transfer protein, high density lipoproteins, atherosclerosis, mutation, lipopolysaccharide binding protein, transgenic mouse

CONTENTS

ABSTRACT

The plasma lipid transfer proteins mediate the transfer and exchange of phospholipids and neutral lipids between the plasma lipoproteins. The cholesteryl ester transfer protein (CETP) and the phospholipid transfer protein (PLTP) are members of the lipid transfer/lipopolysaccharide binding gene family. The CETP contains binding sites for cholesteryl ester and triglycerides and probably acts by a carrier-mediated mechanism. The CETP mediates catabolism of HDL cholesteryl esters, with secondary decreases in HDL size and protein

235

0066-4154/95/0701-0235$05.00

content. The CETP plays a central role in reverse cholesterol transport i.e. the centripetal movement of cholesterol from the periphery back to the liver. CETP gene expression is upregulated in response to increased dietary cholesterol or endogenous hypercholesterolemia. Although CETP reduces HDL levels, its role in reverse cholesterol transport suggests a dominant anti-atherogenic action in vivo.

Perspectives and Summary

The plasma lipoproteins are continuously remodeled during their transit through the plasma compartment, owing to the action of lipid metabolizing enzymes and lipid transfer proteins. These activities have a major effect on the composition, size, and concentration of the lipoproteins. Lipid transfer processes involving neutral lipids and phospholipids are mediated by specialized plasma proteins, called lipid transfer proteins. The plasma lipid transfer proteins include the cholesteryl ester transfer protein (CETP, also called lipid transfer protein-1) and the phospholipid transfer protein (PLTP, also called lipid transfer protein-2). The CETP and PLTP belong to a gene family including lipopolysaccharide binding proteins, which we refer to as the lipid transfer protein/lipopolysaccharide binding protein (LTP/LBP) gene family. Even though structurally unrelated, the plasma lipid transfer proteins are functionally analogous to intracellular lipid transfer proteins that are involved in both the transfer of lipids between organelles (1) and in lipoprotein assembly (2). Lipid transfer proteins appear to have binding sites for transferred lipids and to operate by carrier-mediated mechanisms (1, 3). Recently, major insights into the function of lipid transfer proteins have been gained from genetic deficiency states with remarkable phenotypes (2, 4, 5). Genetic deficiency of CETP, highly prevalent in the Japanese, results in marked increases in high density lipoprotein (HDL) levels (4). HDL is generally thought to have an antiatherogenic role. However, the CETP also plays a central role in reverse cholesterol transport, i.e. the movement of cholesterol from peripheral tissues to the liver. The overall relationship of CETP to atherogenesis is unclear.

Molecular Characterization of Plasma Cholesteryl Ester Transfer Protein

The plasma cholesteryl ester transfer protein (CETP) is a hydrophobic glycoprotein that appears as a broad band or doublet of M_r 66,000 to 74,000 in SDS polyacrylamide gels (6). At high concentrations, purified CETP self-associates, forming dimers and higher order-multimers (6, 7). In isoelectric focusing gels, the CETP shows nine bands with isoelectric points ranging from 4.6 to 5.4 (8). With neuraminidase treatment these collapse to two bands around pI 5.6 (8). The cloning of the CETP cDNA revealed a 476 amino acid sequence with a high content of hydrophobic residues scattered throughout and four potential

N-linked glycosylation sites (9). The difference between the deduced M_r of the CETP polypeptide (M_r 53,000) and that of the mature CETP protein (M_r 66,000–74,000) is largely due to posttranslational modification by addition of asparagine (N)-linked carbohydrate (10). Active, purified, recombinant CETP contains about 1 mole of phosphatidylcholine per mole CETP but no bound neutral lipid (P Kussie & A Tall, unpublished data). The purification of CETP from plasma is tedious. However, recombinant CETP derived by expression in CHO cells can be purified to homogeneity in a simple two-step procedure: hydrophobic chromatography followed by anion exchange chromatography (7). Recombinant CETP, which has specific activity identical to that of plasma CETP, is suitable for use in metabolic and structural studies.

The CETP gene consists of 16 exons, encompassing about 25 Kbp of genomic DNA (11), residing in the 16q12-21 region of the long arm of chromosome 16 near the LCAT locus (12). In humans, the organs with the highest content of the CETP mRNA are the liver, spleen, and adipose tissue, with lower levels of expression in the small intestine, adrenal, kidney, and heart (9, 13). In various species, the tissue showing the most conserved expression of the CETP mRNA is probably adipose (13).

Molecular Characterization of Plasma Phospholipid Transfer Protein

Human plasma contains a phospholipid transfer protein (PLTP) that is biochemically and genetically distinct from the plasma CETP (14, 15). The purified PLTP has a M_r of about 81,000 (16, 17). The cDNA encodes a polypeptide of 476 amino acids (M_r 55,000) with a high content of hydrophobic residues and six potential N-linked glycosylation sites (18). The PLTP gene has been mapped to chromosome 20 where other members of the LTP/LBP gene family are found. Of the major tissues containing the human PLTP mRNA, placenta and pancreas contain equivalent levels, which in turn are greater than those in the lung, kidney, heart, liver, skeletal muscle, and brain (listed in descending order). The PLTP cDNA was cloned from a human umbilical vein endothelial cell library, and the PLTP mRNA was found to be expressed in microvasculature, artery, and vein-derived endothelium.

Lipid Transfer Specificity of CETP and PLTP

The CETP has the ability to facilitate the exchange of neutral lipids (triglycerides, cholesteryl esters, and retinyl esters) and phospholipids between the plasma lipoproteins. In plasma, the CETP mediates an approximately equimolar exchange of HDL cholesteryl esters with very low density lipoprotein (VLDL) triglycerides (19). This hetero-exchange process results in the net transfer of cholesteryl esters from HDL to VLDL and reciprocal net transfer of triglycerides from VLDL to HDL (20). In human plasma, the CETP also stimulates the hetero-exchange of VLDL triglycerides with LDL cholesteryl esters (19). The CETP is responsible for all of the neutral lipid exchange activity in human plasma (19, 21). Investigation of the preference of CETP for different molecular species of cholesteryl esters reveals a moderate specificity (22, 23). The highest transfer rate was observed for cholesteryl oleate (C18:1); saturation or further unsaturation of the C18 chain decreased the transfer efficiency as did decreasing the fatty acid chain length of the cholesteryl ester. These differences probably reflect differential binding of cholesteryl esters to CETP rather than altered availability in the lipoprotein surface (23). Cholesteryl ester (CE) specificity is somewhat different from that of the molecular species of CE generated in plasma by the LCAT reaction that preferentially employs polyunsaturated fatty acids (24). This difference suggests that CETP may play a role in providing 18:1 fatty acids to certain tissues.

The PLTP facilitates the net transfer of phospholipids from unilamellar vesicles into HDL and also the exchange of phospholipids between lipoproteins; the PLTP has no neutral lipid transfer activity (14, 15, 25, 26). The net transfer of phospholipid into HDL results in the formation of larger, less dense species (14). In subjects with complete deficiency of CETP due to a splicing defect of the CETP gene, an investigation of residual lipid transfer activities shows that CETP contributes all of the neutral lipid transfer activity in plasma and about half of the activity mediating phospholipid exchange between lipoproteins (15).

Lipid Transfer Proteins as Members of the Lipid Transfer/Lipopolysaccharide Binding Protein Gene Family

The CETP cDNA has been cloned from human, monkey, rabbit, and hamster, with about 80-95% homology between the species (9, 13, 27, 28). Although the CETP cDNA shows no general homology to the apolipoprotein cDNAs, there is a conserved pentapeptide (VLTLA) within the signal sequences of human CETP, lipoprotein lipase, apoA-I, and apoA-IV, suggesting a common conserved function perhaps related to translational or posttranslational regulation of expression (11). The PLTP and CETP genes appear to have evolved from an ancient family of lipopolysaccharide binding proteins. The human PLTP and CETP cDNAs are homologous to each other and to members of the LPS binding protein gene family including the plasma LPS binding protein (LBP) and the neutrophil bactericidal permeability increasing protein (BPI) (29). Sequence alignments of CETP and PLTP indicate that about 20% of their amino acids are identical (18). In CETP and LBP 23% of the amino acids residues are identical and a further 23% represent conservative substitutions. The homology is seen throughout the sequences. Furthermore, the exon/intron organizations of the CETP and PLTP genes are conserved (X-C Jiang & AR Tall, unpublished data).

Lipid Transfer Proteins In Various Species

Plasma cholesteryl ester transfer activity in various species ranges from unde-tectable to levels several times that in humans (30). Cholesteryl ester transfer activity was significantly correlated with VLDL cholesteryl ester concentration but not with HDL or LDL cholesteryl ester concentrations. However, only species deficient in cholesteryl ester transfer activity accumulate appreciable amounts of large CE and apoE-enriched HDL in response to a high cholesterol diet.

Expression of CETP In Various Cells and Tissues

A wide variety of cell types secrete cholesteryl ester transfer activity, secrete CETP, or contain the CETP mRNA (31–35). However, in each case, the levels of secreted activity or the masses of CETP are extremely low. HepG2 cells and CaCo-2 cells contain authentic CETP mRNA, but the abundance is less than one tenth that of the native tissue (3). The low levels of secreted activity and mRNA suggest that transformed cell lines have lost their ability to express CETP. Similarly, primary cell cultures (of hepatocytes, lymphocytes, and adipocytes) rapidly lose their CETP mRNA. These observations are explained, in part, by the observation that the CETP gene promoter contains a binding site for the transcription factor C/EBP-alpha (a CAAT box binding factor); C/EBP-alpha binds to and *trans*-activates the CETP promoter. C/EBP-alpha is expressed in fully differentiated hepatocytes, and its mRNA disappears rapidly from hepatocytes in primary culture in parallel with CETP mRNA (36). In the liver, both parenchymal and nonparenchymal cell fractions contain the CETP mRNA; differences are evident between species (3, 36, 37). The studies in HepG2 cells and human-CETP transgenic mice suggest that, in humans, hepatocytes are likely to be an important source of CETP.

Hepatic synthesis appears to be the major source of plasma CETP in primates (38). In monkeys receiving a high fat, high cholesterol diet, there was a strong correlation between plasma CETP concentration and the abundance of hepatic CETP mRNA, and the output of CETP in liver perfusates. Adipose-tissue CETP mRNA abundance shows a weaker but independent correlation with plasma CETP levels, suggesting that adipose tissue synthesis of CETP may contribute to plasma levels.

In addition to the expression of CETP by a wide variety of cell types, CETP activity is found in interstitial fluids. Appreciable levels of CETP activity and/or protein are present in cerebrospinal (39) and follicular fluid (40). CETP is present within the female reproductive tract, where it may enhance sperm capacitation, a complex process involving membrane lipid changes that in-crease the sperm's ability to penetrate the egg (40). Purified CETP can enhance sperm capacitation in vitro (40). The mechanisms are obscure but presumably involve membrane lipid changes secondary to CETP activity. It seems unlikely that CETP is essential for sperm capacitation, because homozygotes that lack plasma CETP are not infertile. However, the possibility of decreased fertility of female subjects with homozygous CETP deficiency has not been investigated.

Regulation of CETP Gene Expression by Cholesterol and Other Factors

In a variety of species, plasma CETP activity, plasma CETP mass, and hepatic and peripheral CETP mRNA levels increase in response to a high cholesterol, high fat diet (13, 41–44). Dietary cholesterol appears to be the major factor responsible for these increases; however, high saturated-fat diets causing hypercholesterolemia may also increase plasma CETP levels (45). Cholesterol loading of cultured macrophages leads to an increase in CETP secretion (35). The low expression of CETP mRNA in cultured cells has limited investigation into the mechanisms of these effects.

The regulation of CETP gene expression has been investigated in human-CETP transgenic mice (43). A CETP transgene containing the flanking sequences of the human CETP transgene was expressed in a pattern similar to that in humans. Moreover, in response to a high cholesterol diet there was a marked (4–10-fold) induction of hepatic CETP mRNA due to an increase in gene transcription and an increase in plasma CETP mass and activity. In contrast, mice expressing a similar CETP transgene under control of the mouse metallothionein promoter (instead of the natural promoter) showed no change in hepatic CETP mRNA on a high cholesterol diet and no alteration of plasma CETP mass or activity. The results suggest that the increase in hepatic CETP mRNA and plasma CETP induced by the high cholesterol diet is entirely mediated by increased transcription of the CETP gene. This effect requires the natural flanking regions of the human CETP gene.

In other studies (L Masucci-Magoulas, X Jiang, A Plump, J Breslow, A Tall, unpublished data) the natural flanking region of the CETP transgene was bred into homozygous apoE knock-out mice. This resulted in a profound induction of plasma CETP levels associated with a marked increase in hepatic CETP mRNA; a similar response was not observed for the metallothionein promoter CETP transgene. Thus endogenous hyperlipidemia may result in increased CETP gene expression by a mechanism similar to that of dietary cholesterol.

CETP gene expression is influenced by corticosteroids and lipopolysaccharide (LPS) (46). Plasma CETP levels in humans and CETP transgenic mice are decreased by corticosteroid treatment. In mice, this effect is mediated by a decrease in hepatic CETP mRNA levels and requires the natural flanking regions of the CETP gene (46). In response to LPS, mice show a marked reduction in hepatic CETP mRNA. This reduction is prevented by adrenalectomy suggesting that it is mediated by adrenal corticosteroid release. In humans, plasma CETP levels are increased during the second and third trimesters of pregnancy, and women have higher levels of plasma CETP than men (47, 48). This difference suggests that sex hormones have an effect on CETP expression. However, because rabbits are hypolipidemic during pregnancy and have decreased plasma CETP levels (49), the increase in CETP during human pregnancy may be secondary to the hyperlipidemia of pregnancy.

Alternative Splicing of CETP mRNA

All human tissues containing the CETP mRNA contain an alternatively spliced variant in which exon 8–derived sequences are directly linked to exon 10–derived sequences, without alteration of the translational reading frame (50). Exon 9 encodes 60 amino acids in the central portion of the CETP molecule. The exon 9-deleted form varies from about 20% of the total CETP mRNA in the liver to about 40–60% in the spleen. Cellular expression of the exon 9-deleted cDNA leads to synthesis of a smaller protein that is poorly secreted. Recent experiments in transgenic mice show that expression of the exon 9-deleted variant of CETP does not lead to changes in plasma lipoprotein levels (T Ping-Yang, J Breslow & A Tall, unpublished data). These results are consistent with the suggestion that this exon 9-deleted variant represents a form of CETP that does not have a major effect on plasma lipoprotein levels (51). Thus, alternative splicing could represent a regulatory device operating on a posttranscriptional level to decrease the expression of CETP.

Mechanisms of Lipid Transfer by CETP

CETP-facilitated lipid transfer is probably a carrier mediated process (52). Upon binding the donor (HDL), CE in the surface of HDL exchanges with a neutral lipid molecule in a binding site on CETP. The CETP then dissociates from the HDL, diffuses and binds the acceptor lipoprotein, and exchanges its bound neutral lipid with a neutral lipid molecule in the surface of the acceptor. Phospholipid exchange probably occurs by similar mechanisms but involves a different binding site, or sites, on CETP. It is also possible that after exchange of its neutral lipid molecule, CETP remains bound to the donor lipoprotein and diffuses with it until it collides with the acceptor (VLDL or LDL). CETP could then dissociate from the donor and bind the acceptor, or the exchange event could occur in a ternary complex. Kinetic experiments performed with partially purified preparations of CETP have been interpreted to support both carrier-mediated (53) and ternary mechanisms (54) of neutral lipid transfer. Experimental evidence indicates that the lipid transfer mechanism involves several necessary events, each of which may influence the efficiency of lipid transfer. These events include the binding of CETP to the donor and/or acceptor lipoproteins, the availability of neutral lipid in the lipoprotein surface, and the entry of neutral lipid into binding sites on CETP.

Pattnaik & Zilversmit (55) showed that the phosphorylcholine head-groups of phospholipids in the lipoprotein surface play an important role in the binding of CETP. They concluded that the phospholipid phosphate groups are the primary sites of lipoprotein interaction with CETP, and that the stability of the complexes increases with the negative charge of the lipoproteins. Alterations of HDL, LDL, or VLDL that increased the binding of CETP to the lipoprotein resulted in enhanced efficiency of lipid transfer (56). The inhibition of binding of CETP to HDL inhibited lipid transfer (57). Based on the effects of different salts on the binding of CETP to lipoproteins, Nishida and coworkers (58) concluded that the binding of CETP to lipoproteins is determined by a delicate balance of hydrophobic and electrostatic interactions and that optimal degrees of interaction with both donor and acceptor particles are required for the maximal degree of CE transfer.

The concentration of neutral lipid in the lipoprotein surface is an important factor regulating lipid transfer efficiency. Morton & Steinbrunner (59) evaluated the dependence of CETP activity on the concentrations of cholesteryl ester and/or triglyceride in the phospholipid bilayer of substrate particles using liposomes with variable neutral lipid content and LDL. When liposomes lacked neutral lipid there was no facilitated neutral lipid transfer from LDL to the liposomes. PC transfer, which depended on the same CETP-donor particle-binding interactions as those required for neutral lipid transfer, was not altered by changes in neutral lipid content of the liposomes. That lipid transfer in either direction followed the same kinetics showed that transfer between the two particles is tightly coupled and bidirectional. Similarly, with LDL as neutral lipid donor and discoidal HDL particles as acceptor, CETP only mediated neutral lipid transfer from LDL to the discoidal HDL when the latter contained neutral lipid (1 or 2 mole% cholesteryl oleate) (60). These studies suggest that cholesteryl ester (CE) and triglyceride (TG) transfer activities are determined by the concentrations of these lipids in the phospholipid surface of donor and acceptor particles (59, 60). At low TG and CE concentrations, CETP binds to a liposome or discoidal HDL particle, but the interaction does not result in a neutral lipid transfer event. The overall rate of neutral lipid transfer and the competition between TG and CE for transfer depend on the concentrations of these lipids in the phospholipid bilayer or monolayer.

The CETP has binding sites for neutral lipids and phospholipids that are involved in the lipid transfer process. These sites were initially demonstrated by incubating PC vesicles containing small amounts of radiolabeled neutral lipids (CE or TG) with CETP followed by separation of the vesicles from CETP by gel filtration (52). The CETP with bound lipid was eluted at approximately its monomeric molecular weight. Each mole of CETP bound up to 0.9

mole CE or 0.2 mole TG and about 11 mole of PC. The CETP with bound lipid was isolated and incubated with LDL, which resulted in the transfer of lipids to LDL. Under various conditions, the CETP was isolated either as an apparent monomer with bound lipid or in complexes with vesicles. These results indicated that CETP has phospholipid and neutral lipid binding sites that readily equilibrate with lipoprotein lipids. Such observations suggest that the CETP can act as a carrier of lipids between the lipoproteins.

Further evidence implicates the neutral lipid binding site in the mechanism of neutral lipid transfer. In particular, a neutralizing mAb was found to inhibit the binding of neutral lipid by CETP (61). The epitope of the neutralizing antibody was mapped to the C-terminus of the CETP molecule. Deletion and point mutants of the C-terminus were found to have decreased neutral lipid transfer activity with generally parallel impairment of CE and TG transfer but with no effect on PC transfer activity (62). Furthermore, a C-terminal deletion mutant with severely impaired CE and TG transfer activity was shown to have parallel defects in its ability to bind CE and TG molecules (63). These observations suggest that a C-terminal neutral lipid binding site plays an essential role in the lipid transfer mechanism. Most likely, this site can be occupied by a CE or TG molecule but a different site (or sites) is involved in phospholipid exchange mediated by CETP.

Recently, phospholipid/cholesteryl ester monolayers were used as a model to study the surface binding and desorption of CETP (64). CETP bound to the monolayers with a maximum surface pressure of 23 mN/m; the K_d was 1 nM. Thus, at plasma concentrations of CETP (23 nM in normolipidemic subjects), lipoprotein surfaces are likely to be saturated with CETP (i.e. with little non-lipoprotein-bound CETP). The binding of CETP to the lipid interface decreased linearly with increasing initial surface pressure; CETP was excluded at pressure greater than 31 mN/m. The exclusion pressure is slightly lower than that of apoA-I and the calculated pressure of the HDL surface. This lower pressure implies that the binding of CETP to lipoproteins may be tenuous and that changes in surface pressure may regulate its interaction with donor and acceptor particles.

CETP catalyzed the desorption of cholesteryl oleate from mixed lipid monolayers in a concentration-dependent fashion (64). Similar studies with apoA-I and apoA-IV established that cholesteryl ester desorption was not caused by changes in surface pressure or cholesteryl ester solubility. The desorption rate was proportional to CETP concentration, but at all concentrations, surface radioactivity remained constant until surface pressure reached a plateau. The observation that surface radioactivity remains constant until surface pressure approaches equilibrium is probably a consequence of the effect of interfacial pressure on the rate of CETP dissociation, i.e. the desorption rate increases as the surface pressure increases. Hence, the rate of dissociation will be maximal only after CETP binding reaches equilibrium. The observed dependence of the rate constant for CE desorption on CETP subphase concentration implies a second-order process in which the rate of desorption is a function of the interfacial concentrations of both CE and CETP.

The results indicate that CETP binds to the phospholipid monolayer with an affinity equivalent to that of the plasma apolipoproteins and effects desorption of CE molecules from phospholipid molecules by a carrier mechanism only (64). The relatively low equilibrium surface pressure of CETP suggests that when bound to lipid the entire CETP molecule may not penetrate the surface. This supposition is consistent with 1. the fact that unlike the plasma apolipoproteins, which unfold upon binding to interfaces (65), the CETP has a very stable tertiary structure; and 2. the apparent lack of change in conformation upon binding to the lipid-water interface (66).

The binding of CETP to emulsions and the air-water interface has also been studied (66). Using centrifugal separation, the CETP bound to the surface of a lipid microemulsion with a K_d of 10 nM and a maximum saturation binding level of 8 protein molecules per particle regardless of the presence of apoA-I. Circular dichroism measurements indicated that the protein in solution is predominantly in the beta-sheet/beta-turn conformation with some alpha-helix, and this profile did not undergo drastic change by its binding to the lipid surface. When examined at the air-water interface, CETP molecules occupied the same area per amino acid as apolipoproteins in the monolayer but had a higher collapse pressure of its monolayer (18 dyn/cm) and stayed at the interface even after overcompressing the monolayer far beyond the collapse pressure. These observations were thought to suggest that CETP is an exceptionally active surface protein.

Genetic data suggested dominant inhibitory effects of CETP mutations (7); CETP forms dimers or higher-order self-associated forms under certain conditions. For these reasons, it was suggested that the active molecular species of CETP might be a dimer (7). However, experiments carried out using cross-linking and fluorescence energy transfer measurements indicate that CETP binds to HDL particles as a monomer and is monomeric in solution at physiological concentrations (60). Thus, it is unlikely that the mechanism of lipid transfer involves formation of a dimer.

Structure-Function Relationships of the CETP Molecule

An informative approach to the investigation of structure-function relationships in CETP has involved the production of CETP mAbs and epitope analysis by site-directed mutagenesis. The epitope of a neutralizing CETP mAb was localized to the C-terminal 26 amino acids of CETP (61). Several other neutralizing mAbs have been produced with C-terminal epitopes, while a variety of nonneutralizing mAbs have been found to have distinct, non-C-terminal epitopes. Deletional and site-directed mutagenesis of the C-terminal region of CETP (476 amino acids) shows that the epitope of TP2 is located between amino acids 463 and 475 and also involves aspartate 460 (67). Deletion of amino acids 465–475 gives rise to well-secreted mutant proteins with low CE and TG transfer activities (0–20% of wild-type CETP) but well-preserved phospholipid transfer activity (62). Kinetic analysis of the C-terminal deletion mutant (i.e. deletion of residues 470–475) by varying the concentrations of donor and acceptor lipoproteins, indicated a major catalytic defect, as shown by markedly decreased maximum CE transfer activity (apparent V_{max}) and only a slightly increased concentration of the donor and acceptor at 50% V_{max} (apparent K_m) (63). Direct HDL binding assays of the mutant gave similar results for the mutant and wild-type CETP molecules. By contrast, the binding of both CE and TG to the mutant CETP was found to be markedly impaired. These results indicate that amino acid residues 470–475 are involved in neutral lipid binding by CETP but are not required for the association of CETP with lipoproteins. Thus, the C-terminus forms part of, or is required for entry into, a neutral lipid binding site directly involved in the lipid transfer mechanism.

Point mutagenesis of single amino acids at the C-terminus (between amino acids 468 and 475) shows that the amino acids binding TP2 are different from those required for neutral lipid transfer activity (67). Organization of the C-terminus as a helix can rationalize the results. The amino acids necessary for binding the neutralizing mAb are clustered on the polar face, while those needed for CE transfer are on the nonpolar face. The assignment of the C-terminus as a helix is also supported by the fact that circular dichroism spectroscopy of the C-terminal 26-amino-acid peptide reveals a high content of alpha-helix. Thus, the mAb probably binds to the hydrophilic face of this helical segment of CETP and therefore neutralizes neutral lipid transfer activity by local steric hindrance. The antibody may immobilize the putative flexible helical tail of CETP.

Distribution of CETP in Plasma; Binding to Lipoprotein

Analysis of plasma by gel filtration or apoA-I affinity chromatography indicates that the large majority of CETP is associated with HDL (68). Analysis of normolipidemic plasma by immunoaffinity chromatography indicates that most of the CETP is found in the LpA-I fraction. However, a significant proportion of the CETP is also found in LpA-I/A-II, particularly in males and hyperlipidemic subjects who have lower levels of LpA-I. In fact, in hyperlipidemic males, the major fraction of CETP is found in LpA-I/A-II. These data indicate that LpA-I has a higher CETP binding affinity than LpA-I/A-II. Analysis of plasma by agarose gel electrophoresis indicates that the major portion of CETP is associated with pre-beta migrating HDL (69, 70). Using two-dimensional gel electrophoresis separation, Francone et al (70) observed

CETP signals in particles with a size range as wide as that of apoA-I and found nearly a third of plasma CETP comigrating with a fraction containing apoA-I, apoD, and LCAT (identified as LpA-I pre-beta 3) and of a size larger than the major apoA-I lipoproteins. A similar size distribution of CETP-containing particles was found for immunoaffinity isolated HDL particles (68).

The CETP is able to bind to all lipoprotein classes in vitro. As determined by rapid gel filtration, the K_d values for VLDL, LDL, and HDL were, respectively, 130 nM, 60 nM, and 39 nM (58). These K_d values predict that, in plasma, essentially all of the CETP will be bound to lipoproteins, and the distribution will be 74%, 24%, and 1% associated with HDL, LDL, and VLDL, respectively, with only about 1% of CETP in free form. Only the K_d for VLDL is within the range of the physiological concentrations. Thus, physiological factors that increase the affinity of CETP for VLDL, such as lipolysis (see below), or changes in VLDL concentration (for example during alimentary lipemia or in dyslipidemia), will likely increase cholesteryl ester transfer into VLDL.

Recently, the binding of CETP to recombinant HDL of defined composition and structure has been studied (60). The binding of CETP to discoidal complexes of phospholipid and apoA-I containing small amounts of cholesterol, cholesteryl ester, or phosphatidylinositol occurred with about an order of magnitude higher affinity than the binding of CETP to spherical plasma HDL-3. This greater affinity suggests that nascent HDL particles, such as those thought to be formed at sites of lipolysis (71), may be the preferred targets for CETP activity. These model discoidal complexes may also resemble the pre-beta HDL containing a major portion of CETP in plasma (70).

Modulation of Lipid Transfer Activity by Changes in Substrate Lipoproteins

Lipolysis of VLDL by lipoprotein lipase stimulates the CETP-mediated transfer of CE from HDL to VLDL (72). This phenomenon is related to accumulation of fatty acids in the surface of VLDL, which greatly enhances the binding of CETP to these particles (56). Although it is uncertain if fatty acids build up sufficiently in lipoproteins in vivo to mediate effects similar to those observed in vitro, such a buildup is likely within tissues (e.g. in adipose tissue at the site of lipolysis) or in pathological conditions associated with increased fatty acid/albumin ratios such as uncontrolled diabetes, nephrosis, sepsis, or prolonged fasting (73). The distribution of CETP and LPL mRNAs in similar tissues also suggests a coordinate action of CETP and LPL at the site of lipolysis (13).

The enrichment of lipoproteins with unesterified cholesterol results in enhanced net transfer of CE from HDL to VLDL (74). This process is thought to result from decreased back-transfer of triglyceride from VLDL to HDL; this back-transfer is related to a decrease in the surface concentration of CE in VLDL due to increased free cholesterol. As noted above, the concentration of neutral lipid in the phospholipid surface is thought to play an important part in the modulation of CE transfer. These findings may have relevance to pathophysiological states such as dysbetalipoproteinemia where there is a buildup of free cholesterol in lipoproteins and (in some reports; 75) an accelerated transfer of CE from HDL to triglyceride-rich lipoproteins.

Although there do not appear to be specific protein cofactors influencing CETP activity, various apolipoproteins may modulate the lipid transfer reaction. Thus, apoA-I, apoA-II, apoE, and apoCIII were all found to stimulate the transfer of neutral lipid by CETP when added to lipid microemulsions (76). The activation of lipid transfer by these apolipoproteins was proportional to their calculated surface concentrations. The mechanism has not been elucidated. ApoA-IV, however, may play an important role in modulating the surface pressure of HDL. ApoA-IV moves onto HDL that have been acted on by LCAT; this action may help to maintain the surface pressure and thereby enable the desorption of CETP with bound cholesteryl ester (64).

Protein Inhibitors of Lipid Transfer

Son and Zilversmit (77) first purified a protein inhibitor of plasma neutral lipid transfer activity. The inhibitor was a sialoglycoprotein of M_r 32,000, pI 3.9–4.3; it suppressed cholesteryl ester and triglyceride transfer activities to similar extents. The percent reduction of lipid transfer between lipoproteins by the inhibitor was independent of the concentration of the transfer protein but was decreased at higher lipoprotein concentrations. This result suggests that the inhibitor interacts with substrates rather than with the transfer protein. Even though various apolipoproteins (including apoA-I) can display similar inhibitory properties at high concentrations, it appears that the inhibitory activity in lipoprotein-deficient plasma is due primarily to the inhibitor protein rather than apoA-I (78). The inhibitor has been reported to be associated primarily with the HDL (79) or LDL fractions (78) of human plasma. The predominant association with LDL has been shown by several methods (e.g. gel filtration), and it has been suggested that the previously reported association with HDL may have arisen during the isolation procedure (79). In a comparison of the effects of the inhibitor in transfer assays using different lipoprotein pairs, it was found that inhibition was greatest for VLDL-LDL transfers and least for HDL-VLDL transfers. When added to incubated plasma, the inhibitor changed the observed dominant transfer of CE from LDL to VLDL to transfer from HDL to VLDL.

Role of Lipid Transfer Proteins in Mediating Size Changes of HDL Subclasses

A factor that converts HDL into populations of larger and smaller particles in the absence of other lipoproteins has been demonstrated in plasma of several species. This so-called conversion activity can also change discoidal complexes of phospholipid, cholesterol, and apoA-I into smaller particles. Lagrost et al (80) showed that the HDL conversion process could be mediated by partially purified CETP and blocked by the addition of CETP mAb (TP2). More recently, purified phospholipid transfer protein was shown to have HDL conversion activity (16). Moreover, antibodies to PLTP, which inhibited net phospholipid transfer activity, blocked the increase in size of HDL observed in incubated plasma (17). Thus, it appears that both CETP and PLTP have HDL conversion activity. PLTP may play a dominant role in human plasma.

In addition to promoting formation of HDL particles of various sizes, CETP promotes the formation of pre-beta HDL, an effect requiring the presence of LDL or VLDL (81). The pre-beta HDL may be similar to the small HDL formed by the conversion reaction. Whereas CETP promotes the movement of apoA-I from alpha-migrating HDL to pre-beta- migrating HDL, LCAT induces the formation of alpha-migrating HDL from pre-beta-migrating HDL. The pre-beta-migrating HDL may be the optimal acceptor for cellular cholesterol (82). Thus, the conversion activity of CETP or PLTP may be important in the formation of small HDL particles that are initial cholesterol acceptors in the reverse cholesterol transport pathway.

CETP Transgenic Mice

CETP transgenic mice have provided insights into the role of CETP in lipoprotein metabolism and atherogenesis. Mice normally lack significant plasma CE transfer activity. Expression of a human CETP transgene under the control of the mouse metallothionein promoter resulted in human-like levels of plasma CETP activity and mass and 20–30% decreases in HDL cholesterol levels (83). There was no significant change in the total cholesterol content of VLDL and LDL. However, there was a decrease in the C/CE ratio in plasma and in all lipoprotein fractions of transgenic mouse plasma. This decrease suggested stimulation of plasma cholesterol esterification. Despite the reduction in HDL cholesterol levels, there were minimal changes in HDL size or apoA-I content in the CETP transgenic mice. Much more dramatic reductions in HDL cholesterol (60–70%) and apoA-I (40%) resulted when CETP transgenic mice were crossed with human apoA-I transgenic mice (84). Human apoA-I expression in mice results in human-like speciation of the HDL resulting in the formation of distinctive HDL-2 and HDL-3 subclasses (85). This humanized HDL shows stronger binding of CETP than mouse HDL and apparently acts as a better substrate for human CETP.

The effects of hypertriglyceridemia on CETP expression were examined in transgenic mice (86). Human apoCIII overexpression in mice results in hypertriglyceridemia (87). When the effects of hypertriglyceridemia and CETP activity were combined, there were additive reductions in HDL cholesterol levels and marked synergistic reductions of apoA-I levels and HDL size. In apoCIII/apoA-I transgenic mice, CETP expression produced a 70% decrease in apoA-I levels and a dramatic reduction of HDL peak particle size from about 10.2 and 8.8 nm to a single class of small particles of diameter 7.6 nm. Turnover studies indicated that the reduction in HDL-cholesterol was primarily due to an increase in fractional catabolic rate, while the decrease in apoA-I reflected both an increase in fractional catabolic rate and a decrease in transport rate. The phenotype of these mice (i.e. with high triglycerides, low HDL cholesterol, and small HDL size) resembles the common hypertriglyceridemia, low HDL phenotype in humans; it provides direct proof of the important role of CETP in the strong inverse relationship between plasma triglyceride and HDL cholesterol levels observed in human population samples.

However, a preliminary study of atherogenesis in mice containing different combinations of apoCIII, apoA-I, and CETP transgenes indicates that animals expressing CETP have fewer lesions than those without CETP (T Hayek, E Rubin, J Breslow, X-C Jiang, and AR Tall, unpublished data). If confirmed, these results would suggest that, even though HDL levels are reduced as a result of CETP expression, the dynamics of reverse cholesterol transport are improved by the presence of CETP and the result is less atherosclerosis.

Some in vitro studies have suggested that apoA-II can act as an inhibitor of CETP activity (88). In order to assess a possible modulation of the effects of CETP by apoA-II, human CETP transgenic mice were crossbred with transgenic mice expressing human apoA-II (89). CETP expression resulted in reductions of HDL CE and increases in VLDL CE in mice expressing human apoA-II alone or in combination with apoA-I and apoCIII. These changes indicate that apoA-II does not inhibit the CE transfer activity of CETP. However, the presence of the apoA-II transgene did result in different consequences of CETP expression. In the presence of apoA-II, CETP caused more prominent increases in HDL triglyceride. CETP expression caused dramatic reductions in HDL size and apoA-I content in apoA-I-apoCIII transgenic mice but not in apoA-II/apoA-I/apoCIII transgenic mice. These results, and some in vitro studies, suggested that apoA-II might inhibit the action of hepatic lipase on HDL (90). This was confirmed by studies of the transgenic mouse HDL. Thus, human apoA-II–enriched particles are resistant to reductions in size and apoA-I content, reflecting inhibition of hepatic lipase by apoA-II. These findings may help to explain the apparent pro-atherogenic properties of apoA-II (91, 92). Hepatic lipase may stimulate reverse cholesterol transport by promoting formation of small HDL particles that are optimal mediators of cholesterol efflux and also by providing a driving force for cholesterol transfer from HDL into hepatocytes (93).

Higher levels of CETP expression in transgenic mice also result in increased VLDL and LDL cholesterol and increased apoB levels (94). The increase in plasma apoB levels and in VLDL and LDL cholesterol levels in high express-ing CETP transgenic mice results in part from down-regulation of hepatic LDL receptors. CETP expression also resulted in lower levels of HMGCoA reduc-tase mRNA, suggesting a general effect of CETP expression on cholesterol homeostasis in the liver. Liver cholesterol content increased as a result of CETP expression. This increase suggested an increased return of cholesterol from plasma to the liver. Assuming a steady state of cholesterol balance in these mice on chow diets, the decrease in HMGCoA reductase in the liver implies increased cholesterol synthesis in the periphery. Such an increase could reflect increased reverse cholesterol transport as a result of CETP expression due to increased plasma cholesterol esterification that in turn reflects the presence of CETP. An overall increase in reverse cholesterol transport could explain the decrease in atherosclerosis in apoCIII transgenic mice expressing CETP.

High-level expression of simian CETP in transgenic mice of the C57BL6 background resulted in the accelerated development of focal atherosclerotic lesions in the proximal aorta (95). In mice with various levels of CETP expression, the extent of atherosclerosis was proportional to plasma CETP levels and to the ratio of VLDL + LDL cholesterol to HDL cholesterol. Although these findings suggest an atherogenic potential of CETP expression, they are inconsistent with other transgenic studies as noted above.

Human Genetic Deficiency of Lipid Transfer Proteins

Mutations of the CETP gene result in decreased plasma CETP levels and increased HDL. Several mutations have been described in Japanese (4, 7, 96a, b) and European (97a) populations. The first described mutation was an intron 14 splicing defect caused by a point mutation (G:A) in the first nucleotide of intron 14 (96a). Subsequently, a missense mutation (D442:G) (7), a nonsense mutation (Q309:stop) (96b), and another intron 14 splicing defect (T insertion at position +3 of intron 14) (96a) were found in Japanese subjects. The intron 14 splicing defects are null alleles displaying codominant expression. The D442G mutant allele appears to be partially expressed and genetically domi-nant. Recently, several novel CETP gene mutations, including two missense mutations (A373P, R451:Q) and a cytosine deletion of codon 38, were discov-ered in German subjects; in addition, homozygosity for a polymorphic variant (I405V) was thought to be overrepresented in subjects with high HDL (97a). Genetic CETP deficiency is highly prevalent in the Japanese population. Screening of subjects with high HDL cholesterol by PCR-single strand con-formational polymorphism analysis of the CETP gene showed that the intron

14A and D442:G mutations (but not other mutations) are common in subjects with elevated HDL cholesterol (97b). Furthermore, the prevalences of the D442G and Int 14A mutations were extremely high in a general population sample of Japanese men ($n = 236$) with heterozygote frequencies of 7% and 2% respectively. Because three mutations were found in a population sample of 35 German subjects with HDL cholesterol above the 90th percentile, CETP gene mutations are likely not uncommon among subjects with elevated HDL cholesterol in Caucasian populations (97a).

Both homozygous and heterozygous genetic deficiencies of CETP result in marked alterations in the plasma lipoproteins (7, 96a). The most prominent feature is elevated HDL cholesterol and apoA-I levels. Subjects with homozygous CETP deficiency commonly have HDL cholesterol values over 100 mg/dl and apoA-I levels more than twice normal. The increase is primarily seen in larger HDL particles (HDL-2 and HDL-1). Large HDL species enriched in apoE accumulate in the HDL and LDL density ranges; these enriched species resemble the HDL particles that accumulate in cholesterol-fed species lacking cholesteryl ester transfer activity in plasma (98). Heterozygotes also have significantly higher HDL cholesterol and apoA-I levels, and higher ratios of HDL-2 to HDL-3 than unaffected family members and control populations (96a).

The effect of heterozygous CETP deficiency on HDL-cholesterol levels has been studied in a small general population sample of Japanese men (97b). Mutants displayed a wide range of HDL-cholesterol values, with a median 60–69 mg/dl, compared to the population median of 40–49 mg/dl. In this population sample, genetic CETP deficiency accounted for 10% of the overall variance in HDL-cholesterol levels and thus has a larger effect than environmental factors such as obesity and cigarette smoking. In addition, the mutation accounted for a major part of the skewing of HDL-cholesterol values towards higher values, a phenomenon observed in Japanese and other populations.

In addition to its effects on HDL, heterozygous CETP deficiency alters other lipoprotein fractions. In particular, there is a decreased content of CE in VLDL and IDL, which reflects the role of these lipoproteins as acceptors of CE transferred from HDL (99). This effect is seen as a decreased ratio of CE to TG in VLDL and intermediate density lipoprotein (IDL), with a graded reduction comparing homozygotes, heterozygotes, and unaffected family members. The effects on LDL are variable. Homozygotes for null alleles, such as the intron 14 splicing defects, have no detectable CETP and appear to have a reduction of mean LDL cholesterol and apoB levels. In addition, the LDL is triglyceride enriched, contains CE apparently derived from ACAT activity, and shows unusually sharp speciation into four or five distinctly sized subspecies (100, 101). However, changes in LDL cholesterol and apoB are seen neither in heterozygotes nor in homozygotes with the D442G mutation who have low but detectable CETP activity in plasma (97b). This suggests that the changes in LDL are only seen in the absence of CETP.

Genetic CETP deficiency serves as a paradigm illustrating the physiology of the cholesteryl ester transfer process and the catabolic pathways of HDL in humans. The phenotype shows that CETP plays a major role in the catabolism of HDL cholesteryl esters and suggests that the initial acceptors of HDL CE are VLDL and IDL. The secondary elevations of apoA-I show that apoA-I levels may be regulated by lipid transfer processes and suggest that apoA-I is catabolized more slowly from larger HDL particles. Metabolic turnover studies in homozygotes with the Int 14A defect show markedly delayed catabolism of apoA-I (fractional catabolic rate 0.135 pools/day) compared to normal subjects (FCR 0.196 pools/day) (102).

An intriguing question is the relationship of genetic CETP deficiency to coronary artery disease. Although the lipoprotein phenotype resulting from CETP deficiency, namely increased HDL and decreased CE content of VLDL and IDL, is one usually associated with decreased incidence of coronary heart disease, studies of the physiology of CETP indicate its central role in the process of reverse cholesterol transport. The anecdotal evidence of longevity in some families with CETP deficiency (96a) is tempered by the discovery of CETP-deficient subjects with premature coronary heart disease. Thus, one proband with the D442G mutation is a 47-year-old male with an HDL cholesterol of 84 mg/dl who presented with bilateral xanthelasma and vasospastic angina with normal resting coronary angiogram (97b). Because genetic CETP deficiency is common, many subjects will be found with coincidental defects. The elucidation of the relationship of CETP deficiency and coronary heart disease will require systematic, population-based clinical studies.

Variation in Plasma CETP Concentrations in Human Populations

Plasma CETP levels are about 2 g/ml in normolipidemic males and slightly higher in females (69). Correlational studies fail to show any consistent, strong relationships between plasma CETP and HDL levels in cross-sectional population studies (69, 103). By contrast, in genetic CETP deficiency states CETP levels are inversely related to HDL cholesterol. Plasma CETP levels are markedly increased in dysbetalipoproteinemia (Type III hyperlipidemia) and are normalized with lipid lowering (103). Plasma CETP levels are moderately increased in subjects with combined hyperlipidemia or increased LDL levels and show fairly consistent correlations with LDL cholesterol levels and , in some cases, with VLDL cholesterol levels (103, 104). It is uncertain to what extent elevated CETP levels are caused by or contribute to dyslipidemia. As indicated above, CETP gene expression is highly sensitive to both exogenous and endogenous hypercholesterolemia. This sensitivity suggests that CETP expression may be driven by a plasma cholesterol sensor; perhaps this sensor is the cholesterol content of cell membranes. Thus the correlations between hypercholesterolemia and CETP levels likely represent a response to, rather than a cause of, hyperlipidemia.

Role of CETP in Reverse Cholesterol Transport

The plasma CETP plays a central role in the process of reverse cholesterol transport, i.e. the centripetal movement of cholesterol from peripheral tissues to the liver via the plasma compartment. There is direct evidence that CETP in plasma increases the rate at which plasma cholesteryl esters return to the liver (94, 105). Whether CETP influences the rate of efflux of cholesterol from peripheral tissues into plasma is more controversial. The finding that HMGCoA reductase mRNA is down-regulated in the livers of CETP transgenic mice implies a lower rate of cholesterol biosynthesis in the liver compared to non-transgenic mice (106). This finding may indicate an increased rate of cholesterol biosynthesis in the periphery and thus suggest increased cholesterol removal from peripheral tissues and increased return to the liver (see above). The observation that CETP gene expression is sensitive to endogenous or exogenous hypercholesterolemia also suggests that CETP is a key regulatory molecule in the process of reverse cholesterol transport.

CETP enhances the removal of CE from plasma by stimulating the movement of CE from its sites of synthesis in HDL to chylomicron and VLDL remnants and to LDL. The latter are cleared by receptor-mediated processes mediated by the LDL receptor-related protein (LRP) and the LDL receptor in the liver. There are several mechanisms by means of which CETP may stimulate the efflux of cholesterol from peripheral tissues into HDL. These mechanisms include stimulation of the LCAT reaction (107), formation of HDL particles that are optimal mediators of cellular cholesterol efflux (81), and direct removal of cholesterol or cholesteryl esters from peripheral cells (108, 109). Both CETP and PLTP may stimulate the LCAT reaction by providing substrate phospholipid, and CETP may also stimulate LCAT by enhancing CE removal (25, 107). The activity of CETP does not appear to be rate limiting for the LCAT reaction in incubated normolipidemic plasma (19). However, changes in lipoprotein C/CE ratio in CETP transgenic mice suggest that the presence of CETP leads to stimulation of the LCAT reaction in vivo (83). Furthermore, acute inhibition of CETP in hamsters using a CETP mAb (TP2) results in a marked increase in C/CE in plasma and in lipoproteins. This increase suggests an acute state of LCAT deficiency (R Schnizter-Polokoff, AR Tall, unpublished data). Thus, even though CETP is not rate limiting for the LCAT reaction in normolipidemic plasma, it may be in vivo. The optimal substrates for LCAT in vivo may be nascent HDL, and formation of CE in such particles may be accelerated by removal of CE.

Relationship of CETP to Atherosclerosis

The above analysis indicates a central role for CETP in the reverse cholesterol transport pathway. Recent evidence suggests that the reverse cholesterol transport pathway has a direct relationship to development of atherosclerosis. Over-expression of human apoA-I results in increased HDL levels in transgenic mice and resistance to atherosclerosis (110, 111). Isolated apoA-I or LCAT deficiencies are associated with premature atherosclerosis (112, 113). Hepatic lipase overexpression in transgenic mice results in reduced HDL levels and decreased cholesterol content in aorta (114). If the reverse cholesterol transport pathway is antiatherogenic as suggested, CETP may have an antiatherogenic function. Paradoxically, however, CETP deficiency results in increased HDL levels, and the vast majority of observational studies in human populations indicate that high HDL levels are associated with a lower rate of coronary heart disease (115).

The experimental evidence regarding CETP and atherosclerosis is mixed. CETP expression may be positively correlated with the development of atherosclerosis: 1. The lipoprotein changes of CETP deficiency (increased HDL and decreased CE in VLDL and IDL) are usually associated with resistance to atherosclerosis. 2. In a study of dietary atherosclerosis in monkeys, plasma CETP levels were positively correlated with the extent of coronary atherosclerosis (38). However, this relationship reflected the correlation of plasma CETP levels with LDL cholesterol and HDL cholesterol. In multiple regression analysis only LDL cholesterol concentration appeared as an independent variable significantly correlated with atherosclerosis. 3. In many human dyslipidemias associated with accelerated atherosclerosis, there is increased plasma CETP concentration and/or an increased rate of net transfer of CE from HDL to apoB-containing lipoproteins (3). Alcohol intake, typically associated with increased HDL and decreased atherosclerosis, leads to lower CETP levels (116). Exercise condition, also resulting in beneficial lipoprotein effects is also associated with lower CETP levels possibly as a result of weight loss and improved insulin action (117). However, these associations could indicate that CETP is increased when there is a need for reverse cholesterol transport. 4. CETP transgenic mice may have an increase in early atherosclerotic lesions (95). However, in other transgenesis models, CETP expression does not increase, and may in fact decrease, atherosclerosis (see above).

Other observations suggest that CETP has a beneficial effect on atherogenesis: 1. CETP is a key regulated molecule of the reverse cholesterol transport pathway. The weight of evidence indicates that stimulation of reverse cholesterol transport leads to removal of cholesterol from cells in atheromata. 2. CETP may be able to directly facilitate removal of cholesteryl esters from macrophage foam cells (108), the vascular interstitium, or dying smooth muscle cells (109). 3. Probucol therapy increases CETP levels and decreases HDL (118, 119). Although the evidence in humans is equivocal, most animal studies indicate that probucol has an antiatherogenic action. However, probucol also lowers LDL cholesterol and has antioxidant effects. Thus, it is unclear to what extent the effects of probucol on CETP are related to its antiatherogenic properties.

The overall effects of CETP on atherogenesis may thus vary depending on metabolic context. However, if the other components of the reverse cholesterol transport pathway are functioning normally, CETP predominant action is probably antiatherogenic. Further progress in understanding the relationship of CETP to atherogenesis will come from investigations of the relationship of genetic CETP deficiency to coronary heart disease, from further studies in CETP transgenic mice, and from studies of the effects of CETP inhibition or expression on the vascular biology of the arterial wall and the rate of atheroma formation.

Any *Annual Review* chapter, as well as any article cited in an *Annual Review* chapter, may be purchased from the Annual Reviews Preprints and Reprints service. 1-800-347-8007; 415-259-5017; email: arpr@class.org

Literature Cited

1. Wirtz KWA. 1991. *Annu. Rev. Biochem.* 60:73–99
2. Wetterau JR, Aggerbeck LP, Bouma M-E, Eisenberg C, Munck A, et al. 1992. *Science* 258:999–1001
3. Tall AR. 1993. *J. Lipid Res.* 34:1255–74
4. Brown ML, Inazu A, Hesler CB, Agellon LB, Mann C, et al. 1989. *Nature* 342:448–51
5. Bankaitis VA, Aitken JR, Cleves AE, Dowhan W. 1990. *Nature* 347:561–62
6. Hesler CB, Swenson TL, Tall AR. 1987. *J. Biol. Chem.* 262:2275–82
7. Takahashi K, Jiang X-C, Sakai N, Yamashita S, Hirano K, et al. 1993. *J. Clin. Invest.* 92:2060–64
8. Kato H, Nakanishi T, Arai H, Nishida HI, Nishida T. 1989. *J. Biol. Chem.* 264:4082–87
9. Drayna D, Jarnagin AS, McLean J, Henzel W, Kohr W, et al. 1987. *Nature* 327:631–34
10. Stevenson SC, Wang S, Deng LP, Tall AR. 1993. *Biochemistry* 32:5121–26
11. Agellon L, Quinet E, Gillette T, Drayna D, Brown M, Tall AR. 1990. *Biochemistry* 29:1372–76
12. Lusis AJ, Zollman S, Sparkes RS, Klisak I, Mohandas T, et al. 1987. *Genomics* 1:232–35
13. Jiang X-C, Moulin P, Quinet E, Goldberg IJ, Yacoub LK, et al. 1991. *J. Biol. Chem.* 266:4631–39
14. Tall AR, Abreu E, Shuman JS. 1983. *J. Biol. Chem.* 258:2174–86
15. Brown ML, Hesler C, Tall AR. 1990. *Curr. Opin. Lipidol.* 1:122–27
16. Jauhiainen M, Metso J, Pahlman R, Blomqvist S, van Tol A, Ehnholm C. 1993. *J. Biol. Chem.* 268:4032–36
17. Tu AY, Nishida HI, Nishida T. 1993. *J. Biol. Chem.* 268:23098–105
18. Day JR, Albers JJ, Lofton-Day CE, Gilberg TL, Ching AFT, et al. 1994. *J. Biol. Chem.* 269:9388–91
19. Yen FY, Deckelbaum RJ, Mann CJ, Marcel YL, Milne RW, Tall AR. 1989. *J. Clin. Invest.* 83:2018–24
20. Morton RE, Zilversmit DB. 1983. *J. Biol. Chem.* 258:11751–57
21. Hesler CB, Milne RW, Swenson TL, Weech PK, Marcel YL, Tall AR. 1988. *J. Biol. Chem.* 263:5020–23
22. Green SR, Pittman RC. 1991. *J. Lipid Res.* 32:457–67
23. Morton RE. 1986. *J. Lipid Res.* 27:523–29
24. Glomset JA. 1968. *J. Lipid Res.* 9:155–67

25. Tollefson JH, Ravnik S, Albers JJ. 1988. *J. Lipid Res.* 29:1593–602
26. Lagrost L, Athias A, Gambert P, Lallemant C. 1994. *J. Lipid Res.* 35:825–35
27. Nagashima M, McLean JW, Lawn RM. 1988. *J. Lipid Res.* 29:1643–49
28. Pape ME, Rehberg EF, Marotti KR, Melchior GW. 1991. *J. Clin. Invest.* 85:357–63
29. Schumann RR, Leong RSR, Flaggs GW, Gray PW, Wright SD. 1990. *Science* 249:1429–31
30. Ha YC, Barter PJ. 1982. *Comp. Biochem. Physiol. B* 71:265–69
31. Faust RA, Albers JJ. 1987. *Arteriosclerosis* 7:267–75
32. Swenson TL, Simmons JS, Hesler CB, Bisgaier C, Tall AR. 1987. *J. Biol. Chem.* 262:16271–74
33. Faust RA, Albers JJ. 1988. *J. Biol. Chem.* 263:8786–89
34. Faust RA, Cheung MC, Albers JJ. 1989. *Atherosclerosis* 77:77–82
35. Faust RA, Tollefson JH, Chait A, Albers JJ. 1990. *Biochim. Biophys. Acta* 1042:404–9
36. Agellon LB, Zhang PQ, Jiang X-C, Mendelsohn L, Tall AR. 1992. *J. Biol. Chem.* 267:22336–39
37. Pape ME, Ulrich RG, Rea TJ, Marotti KR, Melchior GW. 1991. *J. Biol. Chem.* 266:12819–31
38. Quinet E, Tall AR, Ramakrishnan R, Rudel L. 1991. *J. Clin. Invest.* 87:1559–66
39. Albers JJ, Tollefson JH, Wolfbauer G, Albright RE, Jr. 1992. *Int. J. Clin. Lab. Res.* 21:264–66
40. Ravnik SE, Albers JJ, Muller CH. 1993. *Fertil. Steril.* 59:629–38
41. Son Y-SC, Zilversmit DB. 1986. *Arteriosclerosis* 6:345–51
42. Quinet E, Agellon L, Marcel Y, Lee YC, Whitlock ME, Tall AR. 1990. *J. Clin. Invest.* 85:357–63.
43. Jiang X-C, Agellon LB, Walsh A, Breslow JL, Tall AR. 1992. *J. Clin. Invest.* 90:1290–95
44. Martin LJ, Connelly PW, Nancoo D, Wood N, Zhang ZJ, et al. 1993. *J. Lipid Res.* 34:437–46
45. Quig DW, Zilversmit DB. 1990. *Annu. Rev. Nutr.* 10:169–93
46. Masucci-Magoulas L, Moulin P, Jiang X-C, Richardson H, Walsh AM, et al. 1994. *J. Clin. Invest.* In press
47. Silliman K, Tall AR, Kretchmer N, Forte TM. 1993. *Metabolism* 42:1592–99
48. Iglesias A, Montelongo A, Herrera E, Lasuncion MA. 1994. *Clin. Biochem.* 27:63–68
49. Quig DW, Zilversmit DB. 1986. *Proc. Soc. Exp. Biol. Med.* 182:386–92
50. Inazu A, Quinet EM, Brown ML, Moulin P, Tall AR. 1991. *Biochemistry* 31:2352–58
51. Quinet E, Yang TP, Marinos C, Tall AR. 1993. *J. Biol. Chem.* 268:16891–94
52. Swenson TL, Brocia RW, Tall AR. 1988. *J. Biol. Chem.* 263:5150–57
53. Barter PJ, Jones ME. 1980. *J. Lipid Res.* 23:1328–41
54. Ihm J, Quinn DM, Busch SJ, Chataing B, Harmony JAK. 1982. *J. Lipid Res.* 23:1328–41
55. Pattnaik NM, Zilversmit DB. 1979. *J. Biol. Chem.* 254:2782–86
56. Sammett D, Tall AR. 1985. *J. Biol. Chem.* 260:6687–97
57. Morton RE. 1985. *J. Biol. Chem.* 260:12593–99
58. Nishida HI, Arai H, Nishida T. 1993. *J. Biol. Chem.* 268:16352–60
59. Morton RE, Steinbrunner JV. 1990. *J. Lipid Res.* 31:1559–67
60. Bruce C, Davidson WS, Lund-Katz S, Kussie P, Phillips M, Tall AR. 1994. *Circulation* 90:1–31
61. Swenson TL, Hesler CB, Brown ML, Quinet E, Trotta PP. 1989. *J. Biol. Chem.* 264:14318–26
62. Wang S, Deng LP, Milne RW, Tall AR. 1992. *J. Biol. Chem.* 267:17487–90
63. Wang S, Kussie P, Deng L, Tall A. 1994. *J. Biol. Chem.* In press
64. Weinberg R, Cook VR, Jones JB, Kussie P, Tall AR. 1994. *J. Biol. Chem.* 269:29588–29591
65. Hesler CB, Brown ML, Feuer DS, Marcel YL, Milne RW, Tall AR. 1989. *J. Biol. Chem.* 264:11317–25
66. Ohnishi T, Hicks LD, Oikawa K, Kay CM, Yokoyama S. 1994. *Biochemistry* 33:6093–99
67. Wang SK, Wang X-B, Deng LP, Rassart E, Milne RW, Tall AR. 1993. *J. Biol. Chem.* 268:1955–59
68. Moulin P, Cheung MC, Bruce C, Zhong S, Cocke T, et al. 1994. *J. Lipid Res.* 35:793–802
69. Marcel YL, McPherson R, Hogue M, Czarnecka H, Zawadzlo Z, et al. 1990. *J. Clin. Invest.* 85:10–17
70. Francone OL, Gurakar A, Fielding C. 1989. *J. Biol. Chem.* 264:7066–72
71. Tall AR. 1990. *J. Clin. Invest.* 86:379–84
72. Tall AR, Sammett D, Vita G, Deckelbaum RJ, Olivecrona T. 1984. *J. Biol. Chem.* 259:9587–94
73. Cistola DP, Small DM. 1991. *J. Clin. Invest.* 87:1431–41
74. Morton RE. 1988. *J. Biol. Chem.* 263:12235–41
75. Tall AR, Granot E, Tabas I, Williams KJ, Brocia R, et al. 1987. *J. Clin. Invest.* 79:1217–25

76. Ohnishi T, Yokoyama S. 1993. *Biochemistry* 32:5029–35
77. Son Y-SC, Zilversmit DB. 1984. *Biochim. Biophys. Acta* 795:473–80
78. Morton RE, Greene DJ. 1994. *J. Lipid Res.* 35:836–47
79. Nishide T, Tollefson JH, Albers JJ. 1989. *J. Lipid Res.* 30:149–58
80. Lagrost L, Gambert P, Dangremont V, Athias A, Lallemant C. 1990. *J. Lipid Res.* 31:1569–75
81. Kunitake ST, Mendel CM, Hennessy LK. 1992. *J. Lipid Res.* 33:1807–16
82. Castro GR, Fielding CJ. 1988. *Biochemistry* 27:25–29
83. Agellon LB, Walsh A, Hayek T, Moulin P, Jiang X-C, et al. 1991. *J. Biol. Chem.* 266:10796–801
84. Hayek T, Chajek-Shaul T, Walsh A, Agellon LB, Moulin P, et al. 1992. *J. Clin. Invest.* 90:505–10
85. Rubin EM, Ishida BY, Clift SM, Krauss RM. 1991. *Proc. Natl. Acad. Sci. USA* 88:434–38
86. Hayek T, Azrolan N, Verdery RB, Walsh A, Chajek-Shaul T, et al. 1993. *J. Clin. Invest.* 92:1143–52
87. Breslow JL. 1993. *Proc. Natl. Acad. Sci. USA* 90:8314–18
88. Guyard-Dangremont V, Lagrost L, Gambert P. 1994. *J. Lipid Res.* 35:982–92
89. Zhong S, Goldberg IJ, Bruce C, Rubin E, Breslow J, Tall AR. 1994. *J. Clin. Invest.* 94:2457–67
90. Thuren T, Wilcox RW, Sisson P, Waite P. 1991. *J. Biol. Chem.* 266:4853–61
91. Warden CH, Hedrick CC, Qiao J-H, Castellani LW, Lusis AJ. 1993. *Science* 261:469–72
92. Schultz JR, Verstuyft JG, Gong EL, Nichols AV, Rubin EM. 1993. *Nature* 365:762–64
93. Bamberger M, Lund-Katz S, Phillips MC, Rothblat GH. 1985. *Biochemistry* 24:3693–701
94. Jiang X-C, Masucci-Magoulas L, Mar J, Lin M, Walsh A, et al. 1993. *J. Biol. Chem.* 268:27406–12
95. Marotti KR, Castle CK, Boyle TP, Lin AH, Murray RW, Melchior GW. 1993. *Nature* 364:73–75
96a. Inazu A, Brown ML, Hesler CB, Agellon LB, Koizumi J, et al. 1990. *N. Engl. J. Med.* 323:1234–38
96b. Gotoda T, Kinoshita M, Shimano H, Harada K, Shimada M. 1993. *Biochem. Biophys. Res. Commun.* 194:519–24
97a. Funke H, Wiebusch H, Fuer L, Assmann G. 1994. *Circulation* 90:1–241
97b. Inazu A, Jiang X-C, Haraki T, Kamon N, Koizumi J, et al. 1994. *J. Clin. Invest.* In press
98. Yamashita S, Sprecher DL, Sakai N, Matsuzawa Y, Tarui S, Hui DY. 1991. *J. Clin. Invest.* 86:688–95
99. Koizumi J, Inazu A, Koizumi I, Uno Y, Kajinami K, et al. 1991. *Atherosclerosis* 90:189–96
100. Bisgaier CL, Siebenkas MV, Brown ML, Inazu A, Koizumi J, et al. 1991. *J. Lipid Res.* 32:21–33
101. Sakai N, Matsuzawa Y, Hirano K, Yamashita S, Nozaki S, et al. 1991. *Arterioscler. Thromb.* 11:71–79
102. Ikewaki K, Rader DJ, Sakamoto T, Nishiwaki M, Wakimoto N, et al. 1993. *J. Clin. Invest.* 92:1650–58
103. McPherson R, Mann CJ, Tall AR, Hogue M, Martin L, et al. 1991. *Arterioscler. Thromb.* 11:797–804
104. Groener JE, van Gent T, van Tol A. 1989. *Biochim. Biophys. Acta* 1002:93–100
105. Whitlock ME, Swenson TL, Ramakrishnan R, Leonard MT, Marcel YL, et al. 1989. *J. Clin. Invest.* 84:129–37
106. Dietschy JM, Turley SD, Spady DK. 1993. *J. Lipid Res.* 34:1637–57
107. Chajek T, Aron L, Fielding CJ. 1980. *Biochemistry* 19:3673–77
108. Morton RE. 1988. *J. Lipid Res.* 29:1367–77
109. Stein Y, Stein O, Olivecrona T, Halperin G. 1985. *Biochim. Biophys. Acta* 834:336–45
110. Rubin EM, Krauss RM, Spangler EA, Verstuyft JG, Clift SM. 1991. *Nature* 353:265–67
111. Paszty C, Maeda N, Verstuyft J, Rubin EM. 1994. *J. Clin. Invest.* 94:899–903
112. Breslow JL. 1989. In *The Metabolic Basis of Inherited Disease*, ed. CR Scrivner, AL Beaudet, WS Sly, D Valle, p. 1251. New York: McGraw-Hill
113. Ng DS, Leiter LA, Vezina C, Connelly PW, Hegele RA. 1994. *J. Clin. Invest.* 93:223–29
114. Busch SJ, Barnhart RL, Martin GA, Fitzgerald MC, Yates MT, et al. 1994. *J. Biol. Chem.* 269:16376–82
115. Gordon DJ, Rifkind BM. 1989. *N. Engl. J. Med.* 321:1311–16
116. Hannuksela M, Marcel YL, Kesaeniemi YA, Savolainen MJ. 1992. *J. Lipid Res.* 33:737–44
117. Seip RL, Moulin P, Cocke T, Tall A, Kohrt WM, et al. 1993. *Arterioscler. Thromb.* 13:1359–67
118. Franceschini G, Sirtori M, Vaccarino V, Gianfranceschi G, Rezzonico L, et al. 1989. *Arteriosclerosis* 9:462–69
119. McPherson R, Hogue M, Milne RW, Tall AR, Marcel YL. 1991. *Arterioscler. Thromb.* 11:476–81

Annu. Rev. Biochem. 1995. 64:259–86

THE MOLECULAR BIOLOGY OF HEPATITIS DELTA VIRUS

Michael M. C. Lai

Howard Hughes Medical Institute and Department of Microbiology, University of Southern California School of Medicine, Los Angeles, California 90033

KEY WORDS: RNA-dependent RNA synthesis, RNA-protein interaction, RNA editing, ribozymes

CONTENTS

ABSTRACT

Hepatitis delta virus (HDV) contains a circular, viroid-like RNA genome, the only animal viral RNA of its kind. It possesses a ribozyme activity, which can autocatalytically cleave and ligate itself. The ribozyme has a unique structural requirement different from other known ribozymes. HDV RNA undergoes RNA-dependent RNA replication via a double rolling circle mechanism, which

259

0066-4154/95/0701-0259$05.00

is probably mediated by cellular RNA polymerase II, utilizing modified cellular transcription machineries. HDV RNA encodes a single protein, hepatitis delta antigen, which is a nuclear, RNA-binding phosphoprotein and required for viral RNA replication. During replication, HDV RNA undergoes a specific RNA editing event to extend its open reading frame and produce a longer, isoprenylated delta antigen, which suppresses RNA replication and initiates viral particle assembly. Ribozyme, cell-mediated RNA-dependent RNA replication, and RNA editing are some of the unique properties and unresolved issues of the molecular biology of HDV.

PERSPECTIVES AND SUMMARY

RNA-dependent RNA synthesis is the mechanism by which all RNA viruses, except retroviruses, replicate their genomes. This process is normally carried out by RNA-dependent RNA polymerases encoded by the viral genomes. The polymerase usually can replicate only its own RNA; thus, each RNA virus must encode its own polymerase. No comparable RNA-dependent RNA polymerase has been detected in normal cells, except for a few reports of RNA-dependent RNA synthesis in eukaryotic cells (1, 2). This limitation poses a quandary for a special class of RNA viruses, including viroids and hepatitis delta virus (HDV) that do not encode any polymerase and yet replicate autonomously by RNA-dependent RNA synthesis in the cells. Therefore, eukaryotic cells must be able to replicate these RNAs. Both viroid and HDV RNA are small, circular RNAs (300-400 nucleotides for viroids and 1700 nucleotides for HDV) with a large number of intramolecular complementary sequences. Some viroids and HDV RNA contain a ribozyme with a self-cleavage activity. They cause diseases either on their own or together with a helper virus. Thus, these RNAs are exceptionally active and versatile and offer insights into the unusual functional capacities of normal cells, since both RNAs rely on host cell functions for their replication.

In this review, I concentrate on the molecular biology of HDV, particularly the properties and functions of HDV RNA and its gene product, which in the past few years, have revealed surprising insights into several issues of interest, such as RNA-dependent RNA synthesis, ribozymes, RNA-protein interaction, RNA editing, and protein isoprenylation.

INTRODUCTION—HISTORY AND CLASSIFICATION

Dr. M Rizzetto of Turin, Italy, discovered in 1977 a new nuclear antigen in the hepatocytes of patients infected with hepatitis B virus (HBV) (3). The antigen, termed delta antigen (HDAg), initially was thought to represent a new HBV-encoded antigen. Subsequently, successful transmission of infectious hepatitis to chimpanzees and the identification of a new type of transmissible virus particles in chimpanzee sera established that HDAg is derived from a new virus, HDV (4). HDV particles contain an envelope consisting of HBV surface antigen (HBsAg) and an internal nucleocapsid consisting of HDAg and viral RNA; thus, the production and transmission of HDV relies on the presence of HBV to provide HBsAg(4). HDV infection has been reported worldwide, but is particularly prevalent in the Mediterranean basin, Middle East, South America, West Africa, and certain South Pacific islands (5, 6). Although HDV transmission relies on HBV, geographical variations in the prevalence rate of HDV do not correspond to those of HBV. HDV infection is often associated with a severe form of hepatitis with a propensity to lead to chronic infections; however, the disease outcome appears to vary among virus isolates (7) and is probably influenced by the nature of the coinfecting HBV as well.

HDV is a subviral particle (8). It forms virus particles only in the presence of a helper virus, HBV; thus, HDV can be considered a satellite virus of HBV. Unlike a classical satellite virus, HDV does not share sequence similarity with the helper virus, and it can replicate independently of HBV. HDV is the only animal virus that contains a circular RNA genome. Thus, taxonomically, HDV constitutes a unique entity. Although HDV resembles plant pathogens, viroids, and virusoids (8) structurally and in mode of replication, it is sufficiently different to be assigned a separate floating genus, *Deltaviridae*. The final taxonomic status of HDV awaits future studies.

VIRION STRUCTURE

The virus particles isolated from sera of infected humans and chimpanzees are enveloped, spherical particles of homogeneous size with a diameter of approximately 36 nm (4, 9). They have a buoyant density in cesium chloride of 1.24–1.25 g/cm^3 and a sedimentation velocity intermediate between that of HBV and the 22-nm empty HBsAg particles (4, 9). The virus particles produced from tissue-culture cells after cotransfection of HDV cDNA and HBsAg also have a similar morphology (10), but larger and more heterogeneous HDV particles produced from a slightly different culture system have been reported (11). After the virus particle is disrupted with nonionic detergents, an internal nucleocapsid is released and HDAg becomes detectable (12, 13). The nucleocapsid is a spherical particle with a diameter of approximately 19 nm (13), which consists of HDAg and HDV RNA. HDAg consists of two protein species (27 and 24 kDa) of variable ratios (12, 14, 15). The number of HDAg molecules in nucleocapsids has been estimated to be 70 per RNA molecule (13). If this estimate is precise, it is incompatible with a regular icosahedral structure. Furthermore, because of the variable ratio of the two HDAg species in the virus particles, the HDV nucleocapsid is not expected to have a precisely symmetrical structure. Its exact structure has not been resolved. The envelope of HDV particles consists of lipid and three protein components (S, M, and L) of HBsAg (12), which are supplied by the coinfecting HBV. The relative ratio of the three HBsAgs in HDV is 95:5:1 (S:M:L), which is more similar to that of empty HBsAg particles than to their ratio in the infectious HBV particles (12). This difference in the ratio of the three HBsAg components between HDV and HBV may account for the slight differences in the host range and target cell specificity between these two viruses (see below).

GENOME STRUCTURE

The HDV particle contains a RNA genome in complex with HDAg. This genome is approximately 1700 nucleotides long and is single stranded and circular, with a very high G+C content (60%) and a high degree (almost 70% of nucleotides) of intramolecular base pairing (16–18). The electron microscopic examination of HDV RNA under denaturing conditions showed a circular RNA (19).When examined under nondenaturing conditions, HDV RNA appeared as a double-stranded, unbranched, rod-shaped structure, half the length of the denatured RNA (19). This property reflects the high degree of intramolecular base pairing. RNA sequence analyses confirmed that HDV RNA is a nearly perfectly arranged, internally complementary molecule (16, 18). So far, at least 14 different HDV isolates from different parts of the world have been sequenced, and all range from 1670 to 1685 nucleotides in length. Based on the extent of sequence similarity, HDV isolates can be classified into three genotypes (20). Genotype I comprises the bulk of the isolates, including isolates from every part of the world. They have a sequence similarity higher than 80% and show a wide range of pathogenicity. Genotype II is represented by a single isolate from Japan, which differs in sequence from genotype I by nearly 25% (21). Genotype III comprises three isolates from South America (20) and is associated with a fulminant form of hepatitis (7). This genotype has only a 60–65% sequence similarity with the other two genotypes. Whether sequence heterogeneity of HDV RNA is responsible for the different clinical pictures seen in delta hepatitis patients is unclear. Besides sequence heterogeneity of different HDV isolates, the viral RNA genome also exhibits another level of sequence heterogeneity, i.e. HDV RNA in any patient represents a collection of RNA molecules (quasispecies) with slight sequence heterogeneity (16, 22, 23). Furthermore, HDV RNA undergoes continuous evolution throughout the infection, with an evolution rate of approximately 3×10^{-2}–3×10^{-3} substitutions/nucleotide per year (24, 24a). The sequence evolution rate varies with the clinical stage of the disease (24).

For the purpose of genome organization, a *Hind*III restriction enzyme site

VIROID DOMAIN   HDAg-CODING DOMAIN

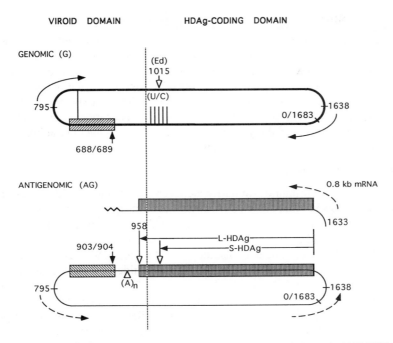

Figure 1 Schematic diagrams of the structure of HDV RNA. The antigenomic (AG) RNA and mRNA are detected only in the cells. The nucleotide numbers are according to Reference 17 and represented in genomic orientation even on the antigenomic strand. The genomic RNA is represented in clockwise orientation, while antigenomic RNA is counterclockwise. Nucleotides 688/689 and 903/904 are ribozyme cleavage sites for genomic and antigenomic RNAs respectively. The hatched boxes represent the ribozyme domain. Nucleotide 1015 (Ed) denotes RNA editing site. (A)n represents polyadenylation signal. The UV cross-linking site is indicated by a vertical line in the viroid domain.

present in the cDNA of the prototype HDV RNA was designated nucleotide 1 (16). Thus, the ends of the rod-like structure of HDV RNA correspond to nucleotides 795 and 1638, respectively (Figure 1; The numbering of nucleotides in this review is according to reference 17). With these two points connected as the dividing line, the HDV RNA sequences can be divided into two halves that are nearly complementary to each other (16, 18). The sequence heterogeneity among different HDV isolates appears to be clustered between nucleotides 1 and 615 (20, 22, 23); yet several highly conserved regions correspond to important functional domains (see below). HDV RNA shares no extensive sequence similarity with other known viral or cellular RNAs. Limited sequence similarity between part of HDV RNA (from nucleotide 615 to 950) and viroids and virusoid RNAs has led to the suggestion that HDV is phylogenetically related to the plant pathogens (25). Also, a tertiary RNA structure detectable by UV cross-linking is demonstrable in both HDV RNA and some viroid RNAs (26). However, HDV RNA is at least four to five times the size of the typical viroid RNA and can encode a protein, whereas viroid RNAs do not. In addition, nucleotides 680–725 (the ribozyme domain) have been noted to be partially homologous to the 7S RNA of signal recognition particles (27, 28) that are involved in protein translocation. The functional significance of this homology is not clear. HDV RNA is the only animal virus known to have a circular RNA structure.

Ribozyme Structure and Properties

Because of the structural similarity between HDV and viroid RNAs, researchers suspected that HDV RNA might have ribozyme activities similar to those associated with some viroid and virusoid RNAs. Such an activity has been demonstrated in HDV RNA and its complementary strand (29–32). The ribozyme activity could be demonstrated in RNA fragments encompassing the HDV RNA sequence from approximately nucleotide 650 to nucleotide 800 (Figure 1), which self-cleaved in the presence of magnesium ions at physiological pH and temperature. The cleavage site was mapped between the U and the G at nucleotides 688/689 (30, 31). A similar ribozyme activity occurs at the complementary region on the antigenomic RNA (29, 30). The self-cleavage reaction is a *trans*-esterification reaction, yielding a 5′-OH and a 2′–3′-cyclic monophosphate terminus (30, 31). The minimum RNA fragment containing the ribozyme activity consists of approximately 85 nucleotides, which are almost entirely derived from the 3′ side of the cleavage site (33, 34). Only one nucleotide at the 5′ side of the cleavage site is required for cleavage, and this can be any nucleotide (34). This unique attribute of ribozyme has been exploited to generate RNA species of a uniform 3′ end when the HDV ribozyme is added to the 3′ end of the desired RNA sequence (35). The minimum ribozyme self-cleaves at a very high efficiency. The presence of additional sequence at both ends often affects the self-cleavage activities, probably because of the alteration of the RNA conformation required for self-cleavage (30, 31, 34, 36).

The sequence and structural requirement of the minimum ribozyme domain has been extensively studied, and several structural models have been proposed (37–39). For a detailed discussion of mutagenesis studies and the biochemical properties of various ribozymes, the readers should refer to the review by Been (38). These models propose that four helical structures, one of which may be formed by a pseudoknot (40), are required for ribozyme activity. (Figure 3 includes a simplified ribozyme model). Besides the overall RNA conformation, specific nucleotides at various positions are also required to either form a catalytic center or maintain overall RNA conformations. Some HDV ri-

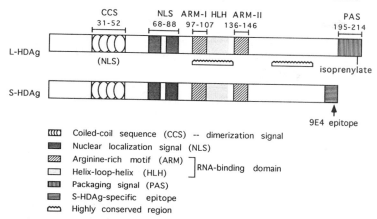

Figure 2 Structural domains of hepatitis delta antigen (HDAg). The domains are described in the text. 9E4 epitope is defined by an S-HDAg–specific monoclonal antibody (9E4) (78). The cryptic NLS (65) overlaps the coiled-coil sequence.

bozymes were active even at very high concentrations of denaturants (41, 42), and some ribozyme activities were enhanced by repeated denaturation and renaturation (43), indicating that some RNA helices are very stable and that the RNAs exist as various conformers. The HDV ribozyme can be divided into two separate molecules, which undergo a cleavage reaction in trans (33, 44–46). These *trans*-acting ribozymes have allowed the studies of the ribozyme kinetics. HDV RNA represents a novel type of ribozyme in that the structure and sequence requirement of HDV ribozyme is different from other known ribozymes, such as hammerhead (47) or hairpin ribozymes (48). The antigenomic HDV ribozyme is very similar to the genomic ribozyme in its sequence and overall RNA conformation (49), because it is localized at the complementary region on the complementary RNA strand (Figure 1).

In addition to the self-cleavage reaction, the HDV ribozyme can also undergo self-ligation, which has been demonstrated in vitro by using a complementary cDNA fragment to bring the two ribozyme-mediated cleavage products together (50). This ligation reaction is a reversal of the cleavage reaction, but in contrast to the cleavage reaction, the in vitro ligation reaction is too slow to account for the RNA ligation involved in HDV RNA replication (see below). A faster self-ligation reaction has been reported (32); however, it was later shown to result from incomplete self-cleavage rather than from true ligation (45). Thus, cellular factors are likely required to facilitate RNA ligation in vivo. The sequence and structural requirement for the self-ligation of HDV RNA has not been studied.

The HDV ribozyme activities are required for HDV RNA replication (51). This requirement was demonstrated by site-specific mutagenesis of the sequence at the ribozyme cleavage site of HDV RNA. The in vitro self-cleavage activities of the mutant RNAs (52) paralleled their abilities to replicate in vivo (51), suggesting that the ribozyme-mediated cleavage and/or ligation reactions are important processes of HDV RNA replication. The full-length, native HDV RNA has an unbranched, rod-shaped structure (19), but does not have the Helical structures of HDV ribozymes. Therefore, the circular or linear monomeric or multimeric HDV RNAs are unlikely to have a ribozyme activity. Thus, it is most likely that only the nascent RNA made during HDV RNA replication, before it folds into a rod-like structure, will contain ribozyme activities. Conceivably, specific or nonspecific cellular factors may alter the conformation of HDV RNA and either enhance or reduce the ribozyme activities. Preliminary data indeed have identified such a factor (A Daniel & MMC Lai, unpublished observation). Similar factors have been shown to alter the activities of hammerhead ribozymes (53, 54). However, HDAg does not appear to regulate the ribozyme activity (55) despite its strong RNA-binding activity (see below).

Open Reading Frames (ORFs)

Multiple open reading frames of different lengths are present on both the genomic and antigenomic strands of HDV RNA. Some of these ORFs can encode proteins of more than 100 amino acids (16, 17). Most of them are not preserved in all HDV isolates, and thus, they are not essential for, but may play regulatory roles in, HDV replication. Most of these potential gene products have not been detected. The only ORF preserved in all HDV isolates and whose product has been detected is the one present on the antigenomic strand (ORF 5 in Reference 16 or ORF −2 in Reference 17) that depending on the RNA sequence, encodes a protein of either 214 or 195 amino acids. This protein is discussed in detail below.

Based on HDV RNA structure, biochemical properties, and sequence-relatedness to other RNAs, it has been proposed that HDV RNA can be divided into two structural domains (26) (Figure 1): (a) Nucleotide 600 to approximately 980 is a viroid-like domain. This domain contains both the genomic and antigenomic ribozyme activities and contains a UV-induced RNA cross-linking site similar to that seen in some viroids (26). (b) The remaining three-quarters of the RNA genome include the ORF for HDAg present on the antigenomic strand. Thus, HDV RNA can be considered a recombinant RNA molecule composed of viroid RNA and, possibly, cellular RNA sequences encoding HDAg.

HEPATITIS DELTA ANTIGEN (HDAg)

HDAg, the signature protein detected in the nuclei of hepatocytes of delta hepatitis patients (3), is the only HDV-encoded protein detected so far. It is a phosphoprotein (56), a component of the viral nucleocapsid, and it usually consists of two protein species: a 27-kDa (214 amino acids) large HDAg (L-HDAg) and a 24-kDa (195 amino acids) small HDAg (S-HDAg) (12, 14, 15, 57). These two HDAgs are identical in sequence, except that the L-HDAg contains an additional 19 amino acids at the C-terminus. The relative ratios of these two species vary from patient to patient. Although an earlier study suggested that L-HDAg could be translated from the S-HDAg-encoding ORF by using a suppressor tRNA (58), current results indicate that two separate ORFs on different RNAs encode L-HDAg and S-HDAg respectively. Indeed, in every patient examined, HDV RNA contains two RNA species, one encoding L-HDAg and the other encoding S-HDAg (59). A single nucleotide at the amber termination codon for S-HDAg is altered in some RNAs, so that the ORF extends for 19 additional amino acids. The N-terminal two-thirds of the protein are highly basic, while the C-terminal one-third is relatively uncharged. The two proteins contain several structural domains (Figure 2):

1. A coiled-coil structure (amino acids 31–52 from the N-terminus), consists of four leucines or isoleucines, each separated by six amino acids. Therefore, it has the classical features of the leucine-zipper sequence (60–62), although one of the potential helical loops is interrupted by a proline residue. This coiled-coil sequence enables HDAg to form dimers and oligomers either with itself or between S-HDAg and L-HDAg (61–63). The leucine residues within this coiled-coil sequence are important for HDAg dimerization because site-specific mutations of these leucines disrupted protein dimerization (61). Mutations of one of the leucines to glycine diminished protein dimerization, whereas mutations of two leucines completely eliminated dimerization. A helix-loop-helix (HLH) domain in the middle of HDAg (63; CZ Lee & MMC Lai, unpublished observation) may also contribute to protein-protein interactions.

2. The nuclear localization signal (NLS) is located between amino acids 68 and 88 from the N-terminus. The NLS consists of a bipartite, basic amino acid–rich region (64). Both components are required for nuclear localization of the HDAg. The HDAg can enter nuclei as a protein complex; therefore, even an HDAg without an NLS can be transported into nuclei as long as it forms a complex with another HDAg molecule containing an NLS (64, 65). Another cryptic NLS located in amino acids 35–50 has no activity unless a downstream hydrophobic domain is removed (65).

3. The RNA-binding domain (amino acids 95–146 from the N-terminus) consists of two stretches of arginine-rich sequences (66), that are similar to the arginine-rich motifs (ARMs) present in several other RNA-binding proteins, such as *rev* and *tat* of human immunodeficiency virus (67). Both ARMs are required for the binding of HDAg to HDV RNAs; deletion or mutation of either one diminished or completely eliminated RNA binding (66). The spacer se-

quence between the ARMs, which may assume a helix-loop-helix structure, is also required, although the sequence specificity is not yet clear (63, 66). Another stretch of sequence (amino acid 2–27 from the N-terminus) may also contain a cryptic RNA-binding activity (68). RNA-binding of HDAg appears to be specific for HDV RNAs (69). It binds to several regions of HDV RNA, and the binding to one region could be competed away by a different region of HDV RNA (69). Furthermore, HDAg binds to the rod-like structure of HDV RNA (70). However, a double-stranded RNA structure consisting of RNA sequences other than that of HDV did not bind to HDAg (69, 70). Both the genomic and antigenomic strands of HDV RNA can bind HDAg (69), and both the large and small HDAgs bind with equal affinity (71). HDAg complexes with HDV RNA in the virion (13, 69) and in the nuclei of infected cells (13).

4. The C-terminal 19-amino-acid extension of L-HDAg is unique to L-HDAg. The fourth amino acid from the C-terminus is a cysteine residue, which constitutes an isoprenylation motif (72) and is prenylated with a geranyl-geranyl prenoid (73, 74). The addition of the geranyl-geranyl prenoid instead of a farnesyl group contradicts the previously reported sequence specificity for isoprenylation (72, 74). The prenylation probably allows the L-HDAg to acquire the ability to interact with membrane. L-HDAg has been shown to interact directly with the envelope proteins (HBsAg) of HBV rather than with the lipid (75, 76). This interaction requires isoprenylation (76). Besides the cysteine residue, the remaining residues within this 19–amino acid extension also play some significant roles in the biochemical functions of L-HDAg. Mutations or deletions of these amino acids affected isoprenylation and also the interaction of L-HDAg with HBsAg (74, 77). In genotype II and III HDVs, these 19 amino acids are almost completely diverged, except for the last 4 residues (20, 21); thus, the significance of these amino acids probably lies in their contribution to the overall conformation of the protein rather than the primary amino acid sequence. The presence of isoprenylates results in the alteration of the conformation of HDAg, particularly near its C-terminus (77). For instance, S-HDAg contains a unique conformation at the C-terminus. This conformation is detectable with a monoclonal antibody (9E4), which is specific for S-HDAg and does not react with L-HDAg (78) (Figure 2). When isoprenylation was inhibited by site-specific mutagenesis of the cysteine residue, this epitope became exposed in L-HDAg (77). Thus, this is a conformational epitope that is masked by the prenylate residue (77).

5. HDAg has been shown to be phosphorylated at serine residues (56). The sites of phosphorylation have not been determined. L-HDAg is phosphorylated approximately six times more heavily than S-HDAg (71). The functional significance of phosphorylation has not been studied.

6. the C-terminal domain (amino acids 146–195) of S-HDAg has not been assigned any function. This is a highly conserved region (20) (Figure 2). This

Table 1   Comparison of biochemical properties and biological activities of the large and small HDAg

	Large	Small	References
Amino acid residues	214	195	16, 17, 20
Nuclear localization	+	+	64, 65
Dimerization	+	+	60–63
RNA-binding	+	+	56, 63, 66, 68–70
Phosphorylation	+++ (6x)	+	56, 71
Isoprenylation	+	–	73, 74, 78
C-terminus–specific epitope (9E4)	–	+	77, 78
Trans-acting function for RNA synthesis	–	+	80
Trans-dominant suppression of RNA synthesis	+	–	82, 83
Interaction with HBsAg	+	–	75, 76
Suppression of polyadenylation of HDV mRNA	+	+	85, 86
Stabilization of HDV RNA	+	+	87

domain may contain a *trans*-acting function for transcription similar to that of DNA-dependent transcription factors. This possibility has been tested by fusing this domain to the GAL4 DNA-binding domain and then testing it on a GAL4-reporter system (79). However, no *trans*-acting activity was detected by such an assay (79a).

Despite the structural similarities between the two antigen species, they have very distinct biological activities (Table 1). S-HDAg is required for HDV RNA replication (80). This activity works in *trans*, i.e. an HDV RNA encoding a defective S-HDAg cannot replicate. However, the replication activity is restored when an S-HDAg is supplied in *trans* (80). This function apparently requires all the known functional domains of S-HDAg because most of the S-HDAg mutants defective in various domains lost this activity (55, 61, 63, 66, 77, 81). Thus, S-HDAg is apparently transported to the nuclei, forms a multimer complex, and binds to HDV RNA to carry out the *trans*-acting function. However, the *trans*-activating functions of S-HDAg for HDV RNA replication could not be demonstrated in HDV RNA replication in vitro (see below).

In contrast, L-HDAg inhibits HDV RNA replication by a *trans*-dominant negative suppression mechanism (82, 83). The presence of as little as 5% of L-HDAg in the HDAg pool can inhibit HDV RNA replication almost totally (82). Furthermore, the *trans*-dominant inhibitory activity of L-HDAg depends on its ability to form a complex with S-HDAg (55, 61) and probably results from the conformational difference between L-HDAg and S-HDAg (77, 78).

This interpretation is consistent with the finding that no specific amino acid sequences are required for this *trans*-dominant inhibitory activity, e.g. even a truncated form of S-HDAg without the last 19 amino acids (61), or an S-HDAg with irrelevant sequences at the C-terminus (63) can inhibit HDV RNA replication. Furthermore, the L-HDAg without the RNA-binding domain can still inhibit RNA replication (55), indicating that its *trans*-dominant negative effects are indirect. Also, the presence of isoprenylates accentuates this inhibitory function (77). Thus, researchers have hypothesized that S-HDAg forms an oligomer complex whose conformation is critical for its activity (61). The presence of a L-HDAg protein with a different conformation in this complex will alter the overall conformation of this complex, resulting in the loss of its biological functions (61). Another biological function of L-HDAg is its involvement in HDV particle assembly. The presence of L-HDAg is required for virion assembly (10, 11, 84). Apparently, L-HDAg provides the initiation points for the interactions with HBsAg (76). S-HDAg can be packaged into virus particles, but only when L-HDAg is present, probably as a result of interactions between S-HDAg and L-HDAg (11, 81, 84). In addition, both S-HDAg and L-HDAg have been reported to suppress the polyadenylation of the HDV mRNA (85, 86) (see below) and stabilize HDV RNA (87) (Table 1). The biological significance of these activities is not clear.

VIRAL REPLICATION CYCLE

In Vitro and In Vivo Models for Studying HDV Replication

HDV infection of cultured cell lines so far has not been successful. This failure is probably related to the fact that the envelope of HDV is made up of HBsAg of HBV, which cannot infect cultured cell lines. The only cell culture susceptible to HDV infection is primary hepatocytes from woodchucks or chimpanzees (88–90). It has been reported that, using HDV stocks that are essentially free of the helper woodchuck hepatitis virus, the virus can successfully infect these primary cultures (90). The finding led to the argument that virus infection can occur in the absence of a helper virus. However, the low efficiency of infection and the difficulty of obtaining primary hepatocytes make this system ineffectual for molecular biology studies.

The most successful system for studying HDV replication has been the transfection of HDV cDNA into cultured cell lines, which bypasses the processes of virus attachment and penetration. Typically, a dimer or trimer cDNA copy of HDV RNA under the control of a foreign promoter was transfected into cell lines in the absence of any helper virus. The production of HDV RNA was usually noticeable within two or three days after DNA transfection. The detected HDV RNAs represent RNA replication because both genomic and antigenomic RNAs and longer-than-full-length RNA intermediates could be detected (51, 80). This approach has proven so useful that the majority of our knowledge on HDV replication has come from studies using this approach. Even an HDV cDNA without a foreign promoter, such as a recircularized monomer HDV cDNA, is as effective a source for replicating HDV RNA (91, 92). Thus, HDV cDNA has a strong endogenous promoter. This system is also useful for studying the function of HDAg, because HDAg can be supplied in trans to complement the defect of HDAg-coding capacity of HDV RNA (66, 80). This approach has been successful with most mammalian cell lines; however no success has been reported for amphibian, insect, or yeast cells.

Another approach is the use of in vitro-transcribed HDV RNA for transfection into cultured cells (93, 94). This approach has proven to be more difficult than DNA transfection. It has worked only when RNA transfection was performed in cells stably expressing HDAg. In contrast to cDNA transfection, even when HDV RNA was cotransfected with a plasmid encoding HDAg, no RNA replication was detected (79a, 93). Thus, a stably expressed HDAg provides a unique helper function for HDV RNA replication; the reason why is not clear. Successful RNA replication has been reported using monomer HDV RNA of both genomic and antigenomic senses and even HDV RNA containing some irrelevant sequences (79a, 93). More recently, HDV particles and ribonucleoprotein complex have also been successfully transfected by liposomal agents into cultured cell lines (95).

For studies of HDV replication in animals, chimpanzees and woodchucks are the most commonly used animal systems (3, 96, 97). The experiments are usually performed on animals that have been infected with hepadnaviruses. These infections closely mimic HDV infections in humans, leading to hepatitis and liver damage, as seen in natural infections in humans (3, 96, 97). Experimental HDV infection of helper-virus-free animals has also been reported (97a). HDV cDNA as described above can also be used for intrahepatic injection into chimpanzees, leading to HDV replication and hepatitis (98). This appears to be a useful system for studying not only the molecular biology of HDV replication in vivo but also HDV pathogenesis. A small animal model system (mice) was recently reported (99). When woodchuck-derived HDV was injected intraperitoneally or intravenously, HDV infections of hepatocytes occurred in both wild-type and SCID mice (99). Unfortunately, the efficiency of infection was low, infecting no more than 0.6% of hepatocytes, and no hepatitis developed. The virus did not spread between cells, and both kinds of mice were cleared of virus within two to three weeks.

An unusual approach is the use of transgenic mice expressing HDV RNA (JM Polo & MMC Lai, unpublished observation). When an HDV dimer cDNA placed under either a liver-specific promoter (albumin) or a universal promoter (β-actin promoter) was integrated into transgenic mice, the expressed RNA replicated very well in several tissues, particularly in the skeletal muscles and several other organs, including brain, testes, and kidney. Surprisingly, HDV replication in the liver was relatively poor. This pattern of extrahepatic HDV replication is distinctly different from the in vivo infection of humans or animals. Although these transgenic mice do not provide an understanding of natural HDV infection, they offer an opportunity to study HDV replication in different host environments. An extension of this approach is the intramuscular injection of HDV cDNA into mice, which leads to the replication of HDV RNA in the skeletal muscles (JM Polo & MMC Lai, unpublished observation). Mice then develop humoral antibodies against HDAg. This approach provides a novel system for studying HDV replication.

Finally, HDV RNA replication has been studied in vitro using isolated nuclei or nuclear extracts from infected or uninfected cells (94, 100). Nuclear extracts from HDV RNA-replicating cells have been used for studying the metabolic requirements of HDV RNA run-on synthesis, e.g. the sensitivity of HDV RNA replication to α-amanitin was demonstrated in such a system (94). Nuclear extracts from normal cells also could support HDV RNA replication using an exogenous HDV RNA template (94, 100). This approach established unequivocally that HDV RNA replication can be carried out by cellular transcription machineries. S-HDAg was not required and had no effects on HDV RNA replication in these systems. Thus, these in vitro replication systems may not reflect accurately the mechanism of HDV RNA replication in vivo. An additional step forward is the use of the reconstituted in vitro RNA replication system, which is made by inclusion of various purified transcription factors, using exogenous HDV RNA as a template (JM Taylor, personal communication). This approach will likely yield further information on the mechanism of HDV RNA replication.

Host Range and Early Events of HDV Replication

Since the envelope of HDV consists of HBsAg, the host range of HDV is similar to that of HBV; however, there is a slight difference: HBV infects some extrahepatic tissues, such as pancreas or peripheral blood cells (101, 102); in contrast, extrahepatic replication has never been demonstrated for HDV. Also, HDV can infect woodchucks, even when HBV was used as a helper (90, 96), whereas HBV cannot infect woodchucks. This slight difference in host range could be the result of the different ratios of the three components of HBsAg in HDV and HBV (12). Nevertheless, the hepatotropism of HDV in vivo is mainly determined by the interactions between HBsAg and cellular receptors because the transfection of HDV cDNA into hepatic and nonhepatic cell lines, thus bypassing the receptors, resulted in HDV replication. The receptors for HDV have not been identified, and thus, the mechanism of virus adsorption and penetration is unclear.

Once the virus enters the cells, HDV RNA must be transported to the nucleus, where RNA replication occurs (103). Whether RNA enters the nucleus as a ribonucleoprotein (RNP) complex or uncoated RNA is unkown. HDAg is likely responsible for the nuclear transport of HDV RNA (94), because HDAg contains a nuclear localization signal (64). However, since viral RNP contains L-HDAg, which inhibits HDV RNA replication (82), the nuclear transport of RNP could potentially interfere with the initiation of HDV RNA synthesis. Both the naked HDV RNA (93, 94) or RNP (95), when transfected into various cell lines, led to HDV replication. The mechanism of their nuclear transport is unclear. Another unresolved issue is whether there is an additional level of restriction on tissue tropism of HDV once the virus gets into the cells. Liver-specific differentiation factors apparently are not required for HDV replication; however, the findings that HDV RNA cannot replicate in insect cells or amphibian oocytes, and that HDV replication occurs preferentially in certain tissues, particularly the skeletal muscles, of transgenic mice (JM Polo & MMC Lai, unpublished observation), suggest that some cellular factors are involved in HDV replication.

Metabolic Requirements for HDV RNA Replication

HDV RNA replication takes place in the nuclei, since both HDAg and HDV RNA are detected in the nuclei (103). Although S-HDAg is required for HDV RNA replication in vivo (80), it is not required for in vitro replication (94, 100). Thus, the sole function of HDAg in HDV RNA replication may be the nuclear transport of the incoming viral RNA (94). However, this argument is inconsistent with the finding that both S-HDAg and L-HDAg have a nuclear localization signal (64) and yet only S-HDAg facilitates HDV RNA replication. HDV RNA replication is insensitive to actinomycin D, suggesting that it does not go through a DNA-dependent RNA synthesis (104); however, it is sensitive to α-amanitin at 1 μg/ml, which is the concentration believed to inhibit RNA polymerase II (94). Furthermore, in in vitro replication systems, HDV RNA replication is sensitive to antibodies against pol II (100). Therefore, although HDV RNA replication does not go through a DNA intermediate, DNA-dependent RNA transcription machineries, particularly those associated with pol II, are probably involved in HDV RNA replication. It is noteworthy that HDV RNA replication takes place even in the absence of HBV, as demonstrated in HDV cDNA transfection experiments. Thus, although HBV provides a helper function for the formation of HDV particles, it does not provide functions for HDV RNA replication. HDV RNA is capable of autonomous RNA replication, utilizing exclusively the transcription machineries supplied by the host cells. Since HDV undergoes RNA-dependent RNA replication without a DNA intermediate, the cellular transcription machineries must be able to support

RNA-dependent RNA replication. The studies of HDV RNA replication thus revealed a novel capacity of cellular transcription machineries.

Double Rolling-Circle Replication

Because of its circular RNA structure, the mechanism of HDV RNA replication was inferred initially from the studies of viroid replication, which undergoes a rolling-circle replication process (105). Available evidence confirms that HDV RNA indeed replicates by a double rolling-circle mechanism. In this model, the circular genomic RNA serves as a template for the initial round of RNA replication to generate an antigenomic RNA intermediate, which is longer than the genomic RNA. This RNA intermediate is self-cleaved by HDV ribozyme into a monomer RNA that is then ligated, probably also by the ribozyme, into an antigenomic, monomeric, circular RNA. This circular antigenomic RNA subsequently serves as a template for a second round of rolling circle replication to generate a circular genomic HDV RNA.

Evidence in support of this double rolling circle model includes the following:

1. Multimer HDV RNAs of both genomic and antigenomic senses have been detected in infected livers (106) or in cells transfected with HDV DNA (80, 91, 92). Since the transfected DNA used was either a dimer or monomer, and the RNA intermediates of dimers or even trimers were detected, this is most consistent with a rolling circle replication. Furthermore, the monomeric RNAs of both genomic and antigenomic sense occur in both linear and circular forms (106), again consistent with a rolling-circle replication model.

2. Site-specific mutations of the HDV ribozyme cleavage sites resulted in a concomitant reduction or loss of the ribozyme activity in vitro and RNA replication in vivo (51). Among all of the 3'-cleavage-site mutants, only the G→A mutations retained partial self-cleavage activity in vitro (52). Correspondingly, this was the only mutant RNA that retained the replication activity in vivo (51). This result clearly indicated that ribozyme activity is required for HDV RNA replication. Furthermore, when the mutation was on the antigenomic ribozyme and the transfected DNA expressed multimeric antigenomic HDV RNA, RNA replication still did not occur, which indicates that the unprocessed multimeric RNA cannot be used as a template for replication (51). This interpretation is most consistent with a double rolling circle mechanism. Thus, both the genomic and antigenomic ribozymes are required for HDV RNA replication.

3. The self-ligation activity has been demonstrated for HDV RNA, although the ligation reaction in vitro is very slow (50). This activity will allow the monomeric HDV RNA to be ligated into a circular RNA. However, the biological role of the self-ligation activity in HDV RNA replication has not been experimentally established.

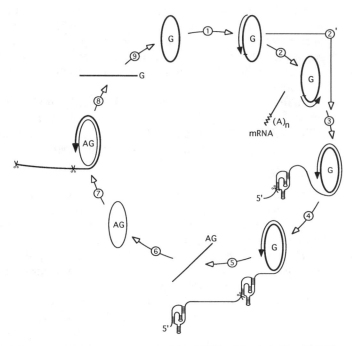

Figure 3   The double rolling-circle model of HDV RNA replication. (Step 1) RNA replication starts from near the top of the rod structure, generating antigenomic transcript. (Step 2) The nascent RNA is processed and polyadenylated, generating the 0.8 kb mRNA. This step is skipped in the later round of RNA replication. (Step 3, 4) Polyadenylation is inhibited by HDAg, allowing RNA replication to extend into the larger RNA product. The ribozyme domain (38, 39) is formed on the nascent RNA, allowing self-cleavage to occur. A simplified ribozyme model is shown. (Step 5) Two self-cleaving events generate a linear monomeric antigenomic RNA. (Step 6) Self-ligation results in the formation of a circular RNA. (Steps 7–9) Rolling circle replication of the genomic strand, mimicking steps 3–6. (Step 2′) step 2 is skipped in most rounds of replication, particularly when HDAg is present (Reference 85).

4. Most of the RNA intermediates detected in the cells were dimeric and trimeric HDV RNAs of distinct sizes; no heterogeneous RNA species in between these multimeric RNAs were detected. These results suggest that HDV RNA intermediates are precisely processed into distinct monomeric and multimeric forms, which is consistent with the processing of HDV RNA at precise sites.

One difficulty with this replication model is that the multimeric RNA likely forms the rod-shaped structure, which will not have ribozyme activities. Also, how can the multimer RNA be synthesized without being cleaved by an active ribozyme during RNA synthesis? Thus, whether the multimeric RNAs are indeed the intermediate of RNA replication is questionable. I would propose an alternative model of double rolling circle replication (Figure 3) in which the HDV monomer RNA is generated from the nascent RNA by the ribozyme as soon as the nascent RNA includes the ribozyme domain. Therefore, the dimeric and trimeric HDV RNA should represent the dead-end products of incomplete ribozyme cleavage.

The relative ratio of genomic and antigenomic RNA is approximately 15:1 (106). Researchers have estimated that each infected cell contains approximately 300,000 copies of HDV genomic RNA (106). The initiation and termination points of RNA replication have not been determined.

Subgenomic mRNA Species

In addition to the monomeric and multimeric HDV RNAs of genomic and antigenomic sense, which are present in the nuclei, an HDV-specific, 0.8-kb, poly(A)-containing RNA species has been detected in the cytoplasm of the infected liver or HDV cDNA-transfected cells (106, 107). This RNA is of antigenomic sense and encompasses the ORF for HDAg. The 5′-end of this RNA starts near one end (nucleotide 1633) of the rod structure, upstream from the HDAg ORF, and ends near a eukaryotic polyadenylation signal (AAUAAA) located approximately 60 nucleotides downstream of the ORF (107) (Figure 1). It was hypothesized that this mRNA may be generated during the first round of RNA replication from the incoming HDV RNA (107), and that its start site corresponds to the initiation point of HDV RNA replication (Figure 3, step 1). The nascent RNA is processed and polyadenylated at the polyadenylation signal, generating the 0.8-kb, poly(A)-containing RNA, which serves as an mRNA for translation of HDAg (Step 2). Once HDAg is made, it suppresses the polyadenylation signal and allows subsequent rounds of RNA replication to continue past the polyadenylation signal and generate monomeric and multimeric HDV RNAs (85) (steps 3, 4). The inhibition by HDAg of polyadenylation requires the presence of the complementary sequence opposite to the polyadenylation signal, and both L-HDAg and S-HDAg can inhibit polyadenylation (85, 86). Furthermore, the mutations of the polyadenylation signal inhibited HDV RNA replication, which could be restored by a *trans*-acting S-HDAg (85), supporting the role of the polyadenylation signal in the synthesis of HDAg. This model predicts that the 0.8-kb RNA is synthesized only early in the infection (85), but this prediction has not been experimentally proven, and the existence of the 0.8-kb mRNA is controversial. The amount of this mRNA, if detectable at all, is extremely small in most of the experimental systems studied, including various cultured cell lines transfected with HDV cDNA, in contrast to the large amount of HDAg detectable. Furthermore, in the transgenic mouse system, which constitutively expresses a large amount of HDAg, no 0.8-kb mRNA was detected (JM Polo & MMC Lai, unpublished observation). Thus, the functional significance of 0.8-kb mRNA species remains to be studied.

Enzymology and Mechanism of RNA-Dependent RNA Replication

The α-amanitin sensitivity of HDV RNA replication in nuclear lysates (94) suggested that pol II itself or pol II–transcribed cellular factors are involved in HDV RNA replication. Its sensitivity to antibodies against pol II (100) further directly implicates pol II. These data suggest that HDV RNA replication utilizes a mechanism similar to that of viroid RNA replication, which may also rely on pol II of the host cells (108). Whether pol II is directly involved in HDV RNA replication, and if so, how it is converted from a DNA-dependent to an RNA-dependent RNA polymerase are areas of active research. Preliminary studies using an in vitro replication system consisting of reconstituted transcription factors suggested that only a few basal pol II–associated transcription factors are required for HDV RNA replication (JM Taylor, personal communication). No HDAg is required. The mechanism by which the basal transcription factors participate in the replication of HDV RNA is not yet clear. A basal transcription factor TATA-binding protein (TBP) may bind specifically to HDV RNA, and several other transcription factors, including SP1 and AP-1, may bind to HDAg, which in turn, binds specifically to HDV RNA (M Chao & MMC Lai, unpublished observation). These interactions have recently been demonstrated both in vitro and in vivo. Furthermore, antibodies against TBP, SP1, and most significantly, pol II, can also precipitate HDV RNA, indicating that they are in the same ribonucleoprotein complex (M Chao & MMC Lai, unpublished observation). Thus, pol II and some of its associated transcription factors can interact directly or indirectly with HDV RNA. Conceivably, this is how pol II transcription machineries are converted to utilize the RNA template. The functional roles of these transcription factors in HDV RNA replication have not yet been directly demonstrated. However, HDV RNA cannot replicate in cells deficient in AP-1 transcription factor, but replication is restored when cells are transfected with AP-1 (M Chao, KS Jeng & MMC Lai, unpublished observation). These preliminary studies did not establish that AP-1 is a necessary factor for HDV RNA replication but suggest that AP-1 can stimulate HDV RNA replication. Therefore, the ability of HDV RNA to interact with certain transcription factors is probably the reason that HDV RNA can be replicated by pol II. This is a novel property for viral or cellular RNAs and a previously unrecognized property of pol II transcription factors, and it raises the interesting possibility that pol II transcription machineries may have the ability to replicate some cellular RNAs as well.

According to this model, HDV RNA replication can take place even in the absence of HDAg, because some of the basal pol II–transcription factors can directly interact with HDV RNA and thus are sufficient to recruit pol II to carry out HDV RNA replication. This RNA replication in the absence of HDAg has been demonstrated in in vitro replication systems (94, 100). However, the lack of functional roles for S-HDAg in vitro contrasts with the in vivo data, which indicate that S-HDAg is absolutely required for HDV RNA replication (66, 80). The RNP in the nucleus contains mostly S-HDAg but very little L-HDAg (13), consistent with the role of S-HDAg in HDV RNA replication in vivo. So far, there is no in vitro replication system in which the effect of HDAg can be demonstrated; thus, the precise role of HDAg in RNA replication remains unclear. One possibility is that S-HDAg is required for the activated level of HDV RNA replication because of its ability to recruit additional transcription factors to the RNA replication complex. The in vitro replication systems may be able to support only a basal level of RNA replication. Another possibility that HDAg is required only for the nuclear transport of HDV RNA (94) has not been ruled out. However, the finding that S-HDAg mutants with a defective coiled-coil sequence could not *trans*-activate HDV RNA synthesis (55, 61), although they still could be transported to the nuclei, suggests a more direct role for S-HDAg in RNA replication. Still another possible role for HDAg is to stabilize HDV RNA (87) (Table 1). This possibility could not explain the functional difference between L-HDAg and S-HDAg.

Translation

Although HDV RNA contains several ORFs on both the genomic and antigenomic RNA (16, 17), only one of them, the ORF that encodes HDAg and is present on the antigenomic strand, is preserved among all the isolates. This is the only viral protein detected in all HDVs. An additional ORF encoding a 27-amino acid peptide sharing partial sequence similarity to HBV polymerase protein has also been recognized (109). Researchers proposed that this ORF represents a different C-terminal extension product of HDAg, since it partially overlaps the C-terminus of L-HDAg (110). The antibody to this putative protein has been detected in some delta hepatitis patients, suggesting that this protein can be synthesized under some conditions (110). However, the functional significance of this small ORF remains unclear, since the protein product has not been directly identified.

The mechanism of biogenesis of HDAg is controversial. HDAg is possibly translated from the 0.8-kb mRNA (107). However, this mRNA is present in very small quantities in contrast to the large amount of HDAg synthesized. Furthermore, this mRNA could not be detected in some cells. The ORF for HDAg does not have the perfect context for translational initiation (111); yet, the ORF is efficiently translated. In contrast, some other nonconserved ORFs have a better context for translational initiation, but they are poorly translated both in vitro and in vivo (MMC Lai, unpublished observation). It is not clear how HDAg is synthesized in cells that do not transcribe the 0.8-kb mRNA. An intriguing possibility is that the circular HDV RNA could serve as the mRNA for HDAg. This would be consistent with the finding that mutations of the autocleavage site of HDV RNA, which inhibited RNA replication, still allowed HDAg to be synthesized (51), suggesting that RNA replication and autocleavage activity are not required for the translation of HDAg. Thus, HDAg conceivably could be translated from a multimer or circular RNA. This unconventional translation mechanism would negate the requirement for a conventional context for translational initiation. Two separate HDV RNA molecules containing ORFs of different lengths are used for the translation of L-HDAg and S-HDAg respectively. The origin of these two RNAs is discussed in the following section. Because HDAg is required for RNA replication in vivo, HDAg is presumably continuously synthesized, but the effects of inhibitors of protein synthesis on HDV RNA replication have not been studied.

RNA Editing

The two HDAg species, S-HDAg and L-HDAg, have opposing biological activities: S-HDAg is required for HDV RNA replication (80), whereas L-HDAg inhibits it (82) and is required for virus assembly (84). These properties suggest that S-HDAg is required early in the HDV life cycle when HDV RNA replication is active; in contrast, L-HDAg should not be present until later in the infection when sufficient viral RNA has been made, and virus particles are being assembled. This temporal regulation of HDAg expression has been demonstrated in cells transfected with HDV cDNA—initially only S-HDAg was synthesized after a HDV dimeric cDNA with a functional ORF for S-HDAg was transfected into cells. L-HDAg appeared three to four days later in the cell culture (112). A similar observation was made during serial passages of a cell line stably transformed with HDV cDNA (113). Correspondingly, an HDV RNA species with a larger ORF encoding L-HDAg emerged (112). Sequence analysis demonstrated that a specific mutation had occurred at the amber termination codon of the ORF for S-HDAg (nucleotide 1015 on the genomic RNA) (112). Site-specific mutagenesis studies indicate that the substrate for this specific RNA-editing event is not the antigenomic strand, which contains the ORF for HDAg, but rather the genomic-sense RNA, on which a U is converted to a C, resulting in an A→G conversion on the antigenomic RNA (114). As a result, the termination codon of the ORF for S-HDAg is eliminated, and the ORF is extended for 19 additional amino acids. This editing event requires both specific sequences and a double-stranded RNA structure surrounding the editing site; thus, mutations that disrupted the base pairing around the editing site eliminated RNA editing (114, 115). Whether this editing requires RNA replication is controversial (112). In one scenario RNA editing occurs in the absence of HDV RNA replication and does not require either form of HDAg (115). RNA editing can occur very efficiently when the genomic RNA is incubated with nuclear extracts from a variety of mammalian and insect cells (115). These findings would appear to be inconsistent with the temporal patterns of the regulation of RNA editing in HDV-infected cells. However, these findings have not been reproduced. The RNA sequence and the structure that had been shown to be required for RNA editing (114) were not conserved in HDV of genotypes II and III (20). Also, a Central African isolate of HDV has a different nucleotide mutation at the editing site, i.e. the termination codon UAG is converted to A̲AG instead of the more common UG̲G (116). Moreover, in transgenic mice stably expressing HDV RNA, RNA editing did not occur (JM Polo & MMC Lai, unpublished observation). Thus, the biochemical requirement and biological significance of this editing event remain unclear. As discussed above, S-HDAg should be present early in infection, while L-HDAg appears later. However, in clinical situations, the relative ratio of S-HDAg to L-HDAg did not correlate with the clinical stages of disease or the severity of the infection (117). How this RNA-editing event is regulated and the biological consequences, particularly in those virus isolates that do not undergo efficient RNA editing, are currently unresolved.

Virus Assembly

Because HDV particles consist of HBsAg, HDAg (both L-HDAg and S-HDAg), and RNA, they are assembled only in the presence of the helper virus, HBV. Cotransfection experiments indicated that HBsAg and L-HDAg are necessary and sufficient for the assembly of virus particles, whereas HDV RNA or S-HDAg are not required (10, 11, 60, 84). Even though S-HDAg is present in HDV particles, it could not initiate HDV particle assembly, suggesting that the primary initiation event for HDV assembly is the interaction between L-HDAg and HBsAg. Researchers have shown that L-HDAg binds to HBsAg in in vitro binding assays (75, 76). This interaction, which presumably triggers the formation of HDV particles, requires isoprenylation and specific amino acid sequences at the C-terminus of L-HDAg (77). Mutations of either isoprenylation signal or sequences within this 19–amino acid stretch affected virus assembly (73, 74, 81).

One puzzling feature of this interaction is that HBsAg is present in the cytoplasm, while L-HDAg is present in the nuclei. How the two proteins interact remains a mystery. A study showed that L-HDAg, but not S-HDAg, is translocated gradually from nucleolus to nucleoplasm, while HBsAg is around the perinuclear region (117a), suggesting the possible site of interaction between L-HDAg and HBsAg. A mutant HDAg with a defective nuclear localization signal still can be packaged (81), suggesting that the site of interaction is in the cytoplasm. S-HDAg does not interact with HBsAg (76); however, in the presence of L-HDAg, S-HDAg is copackaged into virus particles (13, 60, 84), presumably because of the interaction between S-HDAg and L-HDAg (61). An L-HDAg mutant with a deleted leucine-zipper sequence could not help in copackaging of S-HDAg, although the mutant L-HDAg itself could be packaged (60). The relative ratio of these two HDAgs in HDV particles varies. Since S-HDAg and L-HDAg have different conformations (78), the variable ratio between the two may preclude the formation of a regular, symmetrical nucleocapsid structure. However, the possibility that L-HDAg and S-HDAg are present in different virus particles cannot be ruled out.

HDV RNA is not required for virus particle formation. However, when HDV RNA is present, it can be incorporated into virus particles (10, 11, 84), probably because of its interaction with HDAg. Although HDAg can interact with both genomic- and antigenomic-sense HDV RNA (69, 71), only genomic-sense RNA is found in the virus particles. The basis of the selectivity of RNA packaging in vivo is not yet clear. Besides the structural domains required for RNA-HDAg binding, there may be a specific packaging signal on the HDV RNA. A nonreplicating, 348-nucleotide, viroid-like domain of HDV RNA can be packaged into virus particles (87). However, one cannot infer from this study that this domain contains a specific packaging signal, because this transfection system lacked the specificity for distinguishing genomic from antigenomic RNA. Recent studies showed that, although L-HDAg alone is sufficient to package HDV RNA into virus particles, efficient RNA packaging requires S-HDAg (63a). Thus, RNA packaging involves another step of regulation of specificity.

HDAg binds to HDV RNA, forming an RNP structure in the virus particles (13, 69). The RNP structures isolated from virus particles and from the nuclei of infected cells show some structural differences (13). Virion RNP contains approximately 70 HDAg molecules per RNA molecule, while the number for nuclear RNP is 30 (13). These numbers have yet to be confirmed. The antigenomic RNA can also form RNP with HDAg in the nuclei (13). Most RNPs in the nuclei contain mainly S-HDAg (13). Some of these probably represent the replication complex for HDV RNA replication. How this is differentiated from RNP used for virus assembly is not clear.

Only the small form (major form) of HBsAg is required for virus particle formation (10). However, for the assembled HDV particles to be infectious, the large form of HBsAg must be present (118, 119), probably because the large form of HBsAg is required for the viral interaction with the cellular receptor (120). The middle form of HBsAg, which contains the binding sites for polymerized human albumin, is not necessary for the infectivity of the assembled virus particles (121).

MOLECULAR BASIS OF HDV PATHOGENESIS

HDV infection is generally associated with hepatitis that is more severe than that caused by HBV alone, but the degree of severity appears to vary with geography. In South America, particularly virulent delta hepatitis epidemics have occurred(7) that may be associated with a particular genotype of HDV isolate in the region (20). Which genetic element of HDV is responsible for the variation in the severity of hepatitis is unknown. HDV pathogenesis can be attributed to the following mechanisms: First, HDAg or HDV RNA may be directly cytotoxic to hepatocytes. S-HDAg, but not L-HDAg, is directly cytotoxic to the cells when expressed in large quantities (104, 113, 122). However, there are many stably transformed cell lines expressing either S-HDAg or L-HDAg without apparent cytotoxicity (123). Moreover, transgenic mice that express S-HDAg or L-HDAg at a level similar to that seen in HDV infections in humans and animals do not show any pathology (124). Nevertheless, this mechanism cannot be completely ruled out because of the finding that HDAg can interact with certain transcription factors and other cellular proteins (M Chao & MMC Lai, unpublished observation) and thus potentially interfere with transcription or other functions of host cells. HDV RNA replication may also cause cytotoxicity. This is an attractive hypothesis because of the possibility that HDV RNA replication usurps the pol II transcription machineries (94, 100), thereby interfering with gene expressions of the host cells. However, in many cell lines in which HDV RNA undergoes active replication, no apparent cytotoxicity was detected. Furthermore, transgenic mice that express HDV RNA and undergo active RNA replication, show no signs of hepatitis (JM Polo & MMC Lai, unpublished observation). HDV RNA replication is very active in the skeletal muscles in these transgenic animals. In some mice, some of the skeletal muscles expressing HDAg were slightly atrophied (JM Polo & MMC Lai, unpublished observation). Thus, direct cytotoxicity of HDV RNA replication can not be rigorously ruled out. An additional potential mechanism of HDV cytotoxicity is that HDV RNA shares some sequence homology with the 7S RNA of signal recognition particles (27, 28)that are involved in protein translocation. Thus, HDV RNA may interfere with protein synthesis or processing of host cells. These potential mechanisms have not been experimentally confirmed.

The second type of mechanisms are immune-mediated.Inflammatory cells surround infected hepatocytes in chronic delta hepatitis (125). Various autoantibodies have also been detected in the serum of delta hepatitis patients (126), suggesting the possible involvement of immune responses in HDV pathogenesis. However, corticosteroid therapy does not have beneficial effects on delta hepatitis (127). Thus, the significance of immune responses in this disease has also not been established.

Third, delta hepatitis may be the outcome of interactions between HDV and HBV and thus can be modulated by underlying or accompanying HBV infection. This possibility has been supported by studies of patients who received liver transplants (128, 129). In these patients, the underlying HDV infection in the transplanted livers usually did not cause any noticeable histopathology or clinical diseases until HBV superinfection occurred, and then massive hepatitis developed. Delta hepatitis patients with active HBV replication, generally, have more severe liver diseases and clinical outcomes than patients without active HBV replication (130). Possibly, HBV replication helps the spread of HDV, or the synergism between HBV and HDV replication accentuates the liver damage. Thus, the nature of HBV infection potentially has profound effects on HDV pathogenesis.

RELATIONSHIP BETWEEN HDV AND HBV INFECTIONS

HDV can be considered a satellite virus of HBV. HDV infections rely on HBV because HBV provides HBsAg for HDV particle formation. Besides this helper function from HBV, HDV replication is completely independent of HBV. However, many clinical studies have indicated that HDV infection suppressed the serological and molecular markers of HBV infections; for example, HBsAg and HBV DNA are inhibited or reduced during the course of HDV infections (5, 98, 131). Therefore, HDV replication can likely interfere with the HBV gene expression. A previous study showed that the expression of HDAg alone indeed inhibited the synthesis of the HBV-specific mRNAs (132). This inhibition could be the result of interaction between HDAg and cellular factors required for the transcription of HBV genes, since HDV and HBV likely share the same transcription machineries of the host cells. However, this study (132) has not been reproduced. Nevertheless, the relationship between HDV and HBV is similar to that between a defective-interfering RNA and its helper virus RNA, although there is no sequence relationship between HDV and HBV. On one hand, HBV provides a helper function for HDV, while on the other, HDV inhibits HBV gene expression and replication. The molecular mechanism for this interference should be of sufficient interest to the future development of the therapy of chronic B hepatitis.

CONCLUSIONS

HDV represents a novel class of animal virus. It has a unique RNA structure and a novel mechanism of RNA-dependent RNA replication. These unique features are expanding the horizon of our knowledge of viruses and cells. They reveal interesting properties of ribozymes and show the previously unrecognized capability of the cellular transcription machineries to replicate RNA.

Many unanswered questions remain: What is the mechanism of RNA-dependent RNA replication? What is the mechanism of the regulation of RNA editing? How is HDAg translated from RNA? What is the property of this new ribozyme? What is the function of HDAg? What is the mechanism of HDV-induced hepatitis? The answers to these questions will undoubtedly reveal novel insights into the molecular biology of cells and viruses.

Literature Cited

1. Volloch V, Schweitzer B, Xun Z, Rits S. 1991. *Proc. Natl. Acad. Sci. USA* 88:10671–75
2. Schiebel W, Haas B, Marinkovic S, Klanner A, Sanger HL. 1993. *J. Biol. Chem.* 268:11851–57
3. Rizzetto M, Canese MG, Arico S, Crivelli O, Trepo C, et al. 1977. *Gut* 18:997–1003
4. Rizzetto M, Hoyer B, Canese MG, Shih JWK, Purcell RH, Gerin JL. 1980. *Proc. Natl. Acad. Sci. USA* 77:6124–28
5. Rizzetto M, Purcell RH, Gerin JL. 1980. *Lancet* 1:1215–18
6. Ponzetto A, Forzani B, Parravicini PP, Hele C, Zanetti A, Rizzetto M. 1985. *Eur. J. Epidemiol.* 1:257–63
7. Hadler SC, de Monzon MA, Rivero D, Perez M, Bracho A, Fields H. 1992. *Am. J. Epidemiol.* 136:1507–16
8. Diener TO, Pruisner SB. 1985. *Subviral Pathogens of Plants and Animals. Viroids and Prions,* ed. K Maramorosch, JJ McKelvey Jr, pp. 3–18. Orlando: Academic
9. Bonino F, Hoyer B, Shih JWK, Rizzetto M, Purcell RH, Gerin JL. 1984. *Infect. Immunity* 43:1000–5
10. Wang CJ, Chen PJ, Wu JC, Patel D, Chen DS. 1991. *J. Virol.* 65:6630–36
11. Ryu WS, Bayer M, Taylor J. 1992. *J. Virol.* 66:2310–15
12. Bonino F, Heermann KH, Rizzetto M, Gerlich WH. 1986. *J. Virol.* 58:945–50
13. Ryu WS, Netter HJ, Bayer M, Taylor J. 1993. *J. Virol.* 67:3281–87
14. Bergmann KF, Gerin JL. 1986. *J. Infect. Dis.* 154:702–6
15. Pohl C, Baroudy BM, Bergmann KF, Cote PJ, Purcell RH, et al. 1987. *J. Infect. Dis.* 156:622–29
16. Wang KS, Choo QL, Weiner AJ, Ou JH, Najarian RC, et al. 1986. *Nature* 323:508–14
17. Makino S, Chang MF, Shieh CK, Kamahora T, Vannier DM, et al. 1987. *Nature* 329:343–46
18. Kuo MYP, Goldberg J, Coates L, Mason W, Gerin J, Taylor J. 1988. *J. Virol.* 62:1855–61
19. Kos A, Dijkema R, Arnberg AC, van der Merde PH, Schellekens H. 1986. *Nature* 323:558–60
20. Casey JL, Brown TL, Colan EJ, Wignall FS, Gerin JL. 1993. *Proc. Natl. Acad. Sci. USA* 90:9016–20
21. Imazeki F, Omata M, Ohto M. 1991. *Nucleic Acids Res.* 19:5439
22. Chao YC, Chang MF, Gust I, Lai MMC. 1990. *Virology* 178:384–92
23. Chao YC, Lee CM, Tang HS, Govindarajan S, Lai MMC. 1991. *Hepatology* 13:345–52
24. Lee CM, Bih FY, Chao YC, Govindarajan S, Lai MMC. 1992. *Virology* 188:265–73
24a. Imazeki F, Omata M. Ohto M. 1990. *J. Virol.* 64:5594–99
25. Elena SF, Dopazo J, Flores R, Diener TO, Moya A. 1991. *Proc. Natl. Acad. Sci. USA* 88:5631–34
26. Branch AD, Benenfeld BJ, Baroudy BM, Wells FV, Gerin JL, Robertson HD. 1989. *Science* 243:649–52
27. Negro F, Gerin JL, Purcell RH, Miller RH. 1989. *Nature* 341:111
28. Young B, Hicke B. 1990. *Nature* 343:28
29. Sharmeen L, Kuo MYP, Dinter-Gottlieb G, Taylor J. 1988. *J. Virol.* 62:2674–79
30. Kuo MYP, Sharmeen L, Dinter-Gottlieb G, Taylor J. 1988. *J. Virol.* 62:4439–44
31. Wu HN, Lin YJ, Lin FP, Makino S, Chang MF, Lai MMC. 1989. *Proc. Natl. Acad. Sci. USA* 86:1831–35
32. Wu HN, Lai MMC. 1989. *Science* 243:652–54
33. Wu HN, Wang YJ, Hung CF, Lee HJ, Lai MMC. 1992. *J. Mol. Biol.* 223:233–45
34. Perrotta AT, Been MD. 1990. *Nucleic Acids Res.* 18:6821–27
35. Pattnaik AK, Ball LA, LeGrone AW, Wertz GW. 1992. *Cell* 69:1011–20
36. Belinsky MG, Dinter-Gottlieb G. 1991. *Nucleic Acids Res.* 19:559–64

37. Perrotta AT, Been MD. 1993. *Nucleic Acids Res.* 21:3959–65
38. Been MD. 1994. *Trends Biochem. Sci.* 19:251–56
39. Tanner NK, Schaff S, Thill G, Petit-Koskas E, Crain AM. 1994. *Curr. Biol.* 4:488–98
40. Perrotta AT, Been MD. 1991. *Nature* 350:434–36
41. Rosenstein SP, Been MD. 1990. *Biochemistry* 29:8011–16
42. Smith JB, Dinter-Gottlieb G. 1991. *Nucleic Acids Res.* 19:1285–89
43. Wu HN, Lai MMC. 1990. *Mol. Cell Biol.* 10:5575–79
44. Branch AD, Robertson HD. 1991. *Proc. Natl. Acad. Sci. USA* 88:10163–67
45. Perrotta AT, Been MD. 1992. *Biochemistry* 31:16–21
46. Puttarajan M, Perrotta AT, Been MD. 1993. *Nucleic Acids Res.* 21:4253–58
47. Forster AC, Symons RH. 1987. *Cell* 50:9–16
48. Hampel A, Tritz R. 1989. *Biochemistry* 28:4929–33
49. Rosenstein SP, Been MD. 1991. *Nucleic Acids Res.* 19:5409–16
50. Sharmeen L, Kuo MYP, Taylor J. 1989. *J. Virol.* 63:1428–30
51. Macnaughton TB, Wang YJ, Lai MMC. 1993. *J. Virol.* 67:2228–34
52. Wu HN, Hwang ZS. 1992. *Nucleic Acids Res.* 20:5937–41
53. Bertrand EL, Rossi JJ. 1994. *EMBO J.* 13:2904–12
54. Herschlag D, Khosla M, Tsuchihashi Z, Karpel RL. 1994. *EMBO J.* 13:2913–24
55. Lazinski DW, Taylor JM. 1993. *J. Virol.* 67:2672–80
56. Chang MF, Baker SC, Soe LH, Kamahora T, Keck JG, et al. 1988. *J. Virol.* 62:2403–10
57. Roggendorf M, Pahlke C, Bohm B, Rasshofer R. 1987. *J. Gen. Virol.* 68:2953–59
58. Weiner AJ, Choo QL, Wang KS, Govindarajan S, Redeker AG, et al. 1988. *J. Virol.* 62:594–99
59. Xia YP, Chang MF, Wei D, Govindarajan S, Lai MMC. 1990. *Virology* 178:331–36
60. Chen PJ, Chang FL, Wang CJ, Lin CJ, Sung SY, Chen DS. 1992. *J. Virol.* 66:2853–59
61. Xia YP, Lai MMC. 1992. *J. Virol.* 66:6641–48
62. Wang JG, Lemon SM. 1993. *J. Virol.* 67:446–54
63. Chang MF, Sun CY, Chen CJ, Chang SC. 1993. *J. Virol.* 67:2529–36
63a. Wang HW, Chen PJ, Lee CZ, Wu HL, Chen DS. 1994. *J. Virol.* 68:6363–71

64. Xia YP, Yeh CT, Ou JH, Lai MMC. 1992. *J. Virol.* 66:914–21
65. Chang MF, Chang SC, Chang CI, Wu K, Kang HY. 1992. *J. Virol.* 66:6019–27
66. Lee CZ, Lin JH, Chao M, McKnight K, Lai MMC. 1993. *J. Virol.* 67:2221–27
67. Lazinski D, Grzadzielska E, Das A. 1989. *Cell* 59:207–18
68. Poisson F, Roingeard P, Baillou A, Dubois F, Bonelli R, et al. 1993. *J. Gen. Virol.* 74:2473–77
69. Lin JH, Chang MF, Baker SC, Govindarajan S, Lai MMC. 1990. *J. Virol.* 64:4051–58
70. Chao M, Hsieh SY, Taylor J. 1991. *J. Virol.* 65:4057–62
71. Hwang SB, Lee CZ, Lai MMC. 1992. *Virology* 190:413–22
72. Moores SL, Schaber MD, Mosser SD, Rands E, O'Hara MB, et al. 1991. *J. Biol. Chem.* 266:14603–10
73. Glenn JS, Watson JA, Havel CM, White JM. 1992. *Science* 256:1331–33
74. Lee CZ, Chen PJ, Lai MMC, Chen DS. 1994. *Virology* 199:169–75
75. de Bruin W, Leenders W, Kos T, Yap SH. 1994. *Virus Res.* 31:27–37
76. Hwang SB, Lai MMC. 1993. *J. Virol.* 67:7659–62
77. Hwang SB, Lai MMC. 1994. *J. Virol.* 68:2958–64
78. Hwang SB, Lai MMC. 1993. *Virology* 193:924–31
79. Sadowski I, Ma J, Triezenberg S, Ptashne M. 1988. *Nature* 335:563–64
79a. Hwang SB, Jeng KS, Lai MMC. 1994. *The Unique Hepatitis Delta Virus.* ed. G. Dinter-Gottlieb. In press
80. Kuo MYP, Chao M, Taylor J. 1989. *J. Virol.* 63:1945–50
81. Chang MF, Chen CJ, Chang SC. 1994. *J. Virol.* 68:646–53
82. Chao M, Hsieh SY, Taylor J. 1990. *J. Virol.* 64:5066–69
83. Glenn JS, White JM. 1991. *J. Virol.* 65:2357–61
84. Chang FL, Chen PJ, Tu SJ, Wang CJ, Chen DS. 1991. *Proc. Natl. Acad. Sci. USA* 88:8490–94
85. Hsieh SY, Taylor JM. 1991. *J. Virol.* 65:6438–46
86. Hsieh SY, Yang PY, Ou JT, Chu CM, Liaw YF. 1994. *Nucleic Acids Res.* 22:391–96
87. Lazinski DW, Taylor JM. 1994. *J. Virol.* 68:2879–88
88. Choi SS, Rasshofer R, Roggendorf M. 1988. *Virology* 167:451–57
89. Sureau C, Jacob JR, Eichberg JW, Lanford RE. 1991. *J. Virol.* 65:3443–50
90. Taylor J, Mason W, Summers J, Gold-

berg J, Aldrich C, et al. 1987. *J. Virol.* 61:2891–95

91. Macnaughton TB, Beard MR, Chao M, Gowans EJ, Lai MMC. 1993. *Virology* 196:629–36

92. Tai FP, Chen PJ, Chang FL, Chen DS. 1993. *Virology* 197:137–42

93. Glenn JS, Taylor JM, White JM. 1990. *J. Virol.* 64:3104–7

94. Macnaughton TB, Gowans EJ, McNamara SP, Burrell CJ. 1991. *Virology* 184:387–90

95. Bichko V, Netter HJ, Taylor J. 1994. *J. Virol.* 68:5247–52

96. Ponzetto A, Cote PJ, Popper H, Hoyer BH, London WT, et al. 1984. *Proc. Natl. Acad. Sci. USA* 81:2208–12

97. Ponzetto A, Negro F, Popper H, Bonino F, Engle R, et al. 1988. *Hepatology* 8:1655–61

97a. Netter HJ, Gerin JL, Tennant BC, Taylor JM. 1994. *J. Virol.* 68:5344–50

98. Sureau C, Taylor J, Chao M, Eichberg JE, Lanford RE. 1989. *J. Virol.* 63:4292–97

99. Netter HJ, Kajino K, Taylor JM. 1993. *J. Virol.* 67:3357–62

100. Fu TB, Taylor J. 1993. *J. Virol.* 67:6965–72

101. Halpern MS, England JM, Deery DT, Petcu DJ, Mason WS, Molnar-Kimber KL. 1983. *Proc. Natl. Acad. Sci. USA* 80:4865–69

102. Pasquinelli C, Laure F, Chatenoud L, Beaurin G, Gazengel C, et al. 1986. *J. Hepatol.* 3:95–103

103. Gowans EJ, Baroudy BM, Negro F, Ponzetto A, Purcell RH, Gerin JL. 1988. *Virology* 167:274–78

104. Macnaughton TB, Gowans EJ, Jilbert AR, Burrell CJ. 1990. *Virology* 177:692–98

105. Branch AD, Robertson HD. 1984. *Science* 223:450–55

106. Chen PJ, Kalpana G, Goldberg J, Mason W, Werner B, et al. 1986. *Proc. Natl. Acad. Sci. USA* 83:8774–78

107. Hsieh SY, Chao M, Coates L, Taylor J. 1990. *J. Virol.* 64:3192–98

108. Rackwitz HR, Rohde W, Sanger HL. 1981. *Nature* 291:297–301

109. Khudyakov YE, Makhov AM. 1990. *FEBS Lett.* 262:345–48

110. Khudyakov YE, Favorov MO, Fields HA. 1993. *Virus Res.* 27:13–24

111. Kozak M. 1989. *J. Cell Biol.* 108:229–41

112. Luo G, Chao M, Hsieh SY, Sureau C, Nishikura K, Taylor J. 1990. *J. Virol.* 64:1021–27

113. Macnaughton TB, Gowans EJ, Reinboth B, Jilbert AR, Burrell CJ. 1990. *J. Gen. Virol.* 71:1339–45

114. Casey JL, Bergmann KF, Brown TL, Gerin JL. 1992. *Proc. Natl. Acad. Sci. USA* 89:7149–53

115. Zheng H, Fu TB, Lazinski D, Taylor J. 1992. *J. Virol.* 66:4693–97

116. Tang JR, Hantz O, Vitvitski L, Lamelin JP, Parana R, et al. 1993. *J. Gen. Virol.* 74:1827–35

117. Govindarajan S, Hwang SB, Lai MMC. 1993. *Hepatitis Delta Virus,* ed. SJ Hadziyannis, JM Taylor, F Bonino, pp. 139–43. Wiley

117a. Wu JC, Chen CL, Lee SD, Sheen IJ, Ting LP. 1992. *Hepatology* 16:1120–27

118. Sureau C, Moriarty AM, Thornton GB, Lanford RE. 1992. *J. Virol.* 66:1241–45

119. Sureau C, Guerra B, Lanford RE. 1993. *J. Virol.* 67:366–72

120. Neurath AR, Kent SBH, Strick N, Parker K. 1986. *Cell* 46:429–36

121. Sureau C, Guerra B, Lee H. 1994. *J. Virol.* 68:4063–66

122. Cole SM, Gowans EJ, Macnaughton TB, Hall PDM, Burrell CJ. 1991. *Hepatology* 13:845–51

123. Chen PJ, Kuo MYP, Chen ML, Tu SJ, Chiu MN, et al. 1990. *Proc. Natl. Acad. Sci. USA* 87:5253–57

124. Guilhot S, Huang SN, Xia YP, La Monica N, Lai MMC, Chisari FV. 1994. *J. Virol.* 68:1052–58

125. Negro F, Baldi M, Bonino F, Rocca G, Demartini A, et al. 1988. *J. Hepatol.* 6:8–14

126. Bonino F, Brunetto MR, Negro F. 1991. *The Hepatitis Delta Virus,* ed. JL Gerin, RH Purcell, M Rizzetto, pp. 137–46. New York: Wiley-Liss

127. Rizzetto M, Verme G, Recchia S, Bonino F, Farci P, et al. 1983. *Ann. Intern. Med.* 98:437–41

128. Ottobrelli A, Marzano A, Smedile A, Recchia S, Salizzoni M, et al. 1991. *Gastroenterology* 101:1649–55

129. Davies SE, Lau JYN, O'Grady JG, Portmann BC, Alexander GJM, et al. 1992. *Am. J. Clin. Pathol.* 98:554–58

130. Smedile A, Rosina F, Saracco G, Chiaberge E, Lattore V, et al. 1991. *Hepatology* 13:413–16

131. Arico S, Aragona M, Rizzetto M, Caredda F, Zanetti A, et al. 1985. *Lancet* 2:356–58

132. Wu JC, Chen PJ, Kuo MYP, Lee SD, Chen DS, Ting LP. 1991. *J. Virol.* 65:1099–104

Annu. Rev. Biochem. 1995. 64:287–314

THE MULTIPLICITY OF DOMAINS IN PROTEINS

Russell F. Doolittle

Center for Molecular Genetics University of California, San Diego, La Jolla, California, 92093-0634

KEY WORDS: protein domains, domainal evolution, modules, exon shuffling, evolution, combinatorial advantage

CONTENTS

ABSTRACT

The domainal nature of proteins is well established. What is less certain is how many domains are evolutionarily mobile in that they occur in otherwise

287

0066-4154/95/0701-0287$05.00

nonhomologous proteins or in different sequential locations in homologous proteins. The combinatorial advantage of shuffling domains around into diverse settings is obvious. Those domains that have been shuffled about in recent evolutionary times, within the last half billion years or so, can usually be identified on the basis of sequence resemblances alone. Contrarily, domains that were rearranged in ancient times may only be apparent after three-dimensional analysis, their sequence resemblances having been eroded over time. The shuffling of domains in recently evolved proteins has been greatly promoted by introns, but this does not imply that all domainal rearrangements involve introns. Only a small fraction of known exons show evidence of having been shuffled. Taken in aggregate, the available data best fit a scenario whereby a relative small number of genes encoding domain-sized polypeptides has been expanded by duplication and modification with a burst of exceptional genomic rearrangement.

PERSPECTIVES AND SUMMARY

It is generally accepted that there is a level of organization in proteins that overlaps the classical categories of tertiary and quaternary, i.e. sequentially consecutive residues in polypeptide chains tend to fold into more or less compact modules called domains (1). In this regard, a domain can be defined as that part of a protein that can fold up independently of neighboring sequences, although the term is often used less precisely. Crystallographers often interchange the term *domain* with the word *fold* (used as a noun). Sequence-oriented investigators often use the term domain when they mean sequence motif or signature sequence, although such sequences are often characteristic of domains.

The underlying basis of domain organization is the cohesion between sidechains that stabilizes unique structures. As such, there is a need for a critical number of residues in a sequence, typically 50–100, before a domain can be realized, although smaller segments can be further stabilized by folding around metal centers, or by forming appropriate disulfide bonds, or in the case of short repetitive units, packing against each other. Contrarily, some domains are much larger (2). As it happens, the average polypeptide chain length for proteins or their constituent subunits is about 350 residues (3), and most proteins contain two or more domains, although, naturally, many proteins are single domains.

The principal questions before us are, How many kinds of domains exist, and what was their origin? An extreme view is based on the idea that a large number of polypeptides was concocted early in the history of living systems and that all the world's proteins, past and present, are the result of shuffling those primordial structural units (4). An opposing view is that there were only a few small polypeptides in the early stages of life and that most contemporary proteins are descended from them as a result of genetic duplication and descent with modification (5). Perhaps the truth will turn out to be somewhere between these two extremes, a modest number of starter types giving rise to the first genuinely modular proteins. We can attempt to evaluate these propositions retrospectively by considering the inventory of contemporary domains.

At the same time we can address some subsidiary questions, including: Are new domains still being evolved, and if so, how frequently? Also, does a similar fold always imply common ancestry? If we can conduct the census along phylogenetic grounds, perhaps we can obtain some idea of the general multiplicity. Thus, some domains are clearly ubiquitous and occur in all kinds of organisms. Accordingly, they must predate the last common ancestor of all contemporary life forms. Others appear to be narrowly distributed, occurring only in animals, for example. In some cases, this may be an illusion resulting from extensive sequence decay, and three-dimensional studies will be needed to settle the point.

The emphasis in this review is on domains that have been shuffled about during evolution, whether they were moved recently or in the very ancient past. The evolutionarily mobile domain is often called a module (as in *modular* furniture). Some authors prefer to limit the use of this term to domains that are genetically bordered by introns (6). In my view, this is too restrictive. By our criterion here, a domain need only occur in two different settings in otherwise nonhomologous proteins, or at clearly different locations in homologous proteins, to be considered evolutionarily mobile. Indeed, some domains have been so successful in their structural and functional adaptations that they occur in numerous proteins in very different settings.

IDENTIFYING DOMAINS

The domains in a given protein may be packed together tightly or loosely. If it is the latter, they can often be dissected by limited proteolysis, providing suitable target peptide bonds are available in the connecting segments. Sometimes, also, domains can be distinguished by scanning calorimetry because of differences in thermostability (7).

The ideal way of distinguishing domains is by direct observation of the three-dimensional structures obtained by X-ray diffraction. The number of known three-dimensional structures is still small relative to the number of available sequences, however, and most of the domains we are addressing here were uncovered on the basis of sequence analysis alone. The problem is complicated by the fact that amino acid sequences are eroded during evolution at a much faster rate than are three-dimensional structures (Figure 1). The number of known three-dimensional structures is growing rapidly, however, and in the future this will doubtless be the method of choice for establishing

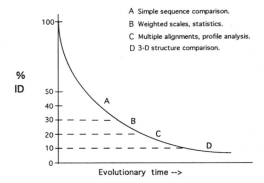

Figure 1   Divergence of two sequences from a common ancestor occurs as a negative exponential. If two (sufficiently long) sequences are more than 30% identical (*A*), simple visual inspection is enough to warrant a judgment about common ancestry. As the resemblance decays (*B–D*), increasingly sophisticated means are required to validate the relationship.

domain relationships. In the meantime, sequence searching methods have been extremely successful in identifying members of the major families.

Sequence Searching Programs

There are numerous computer programs available for searching an amino acid sequence against a database in an effort to find related sequences, but because the subject is relentlessly reviewed (8,9), only a brief comment is needed here. The most popular program currently in use is called Blast (10), having replaced FastA (11) as the most frequently used routine. If a sequence belongs to a known family, then a variety of pattern searches is available (12–14); in general, these will be more sensitive than the search of a single sequence. In any case, it always pays to conduct the operation iteratively, which is to say, the search should be reexecuted with the inclusion of whatever other sequences are initially uncovered.

Three-Dimensional Comparison Programs

A priori, we expect that there will be limits to the effectiveness of relationships based on amino acid sequences alone. Sequences change more or less stochastically, and eventually a point will be reached where not enough information remains to distinguish genuine homology from chance similarity (Figure 1). This is especially true in the case of smaller domains composed of fewer than a hundred residues, which includes many of the most common domains. Family memberships will in general be undercounted, and distant relationships may have to be established exclusively by three-dimensional structures.

A number of programs are available for searching three-dimensional structures against databases of other three-dimensional structures [recently reviewed by Orengo (15)]. Occasionally it is possible to find an unexpected relationship without a formal computer search. An interesting case in point has to do with the domain ("fold") now known as the cystine knot dimerization domain. The fold was originally observed in nerve growth factor (16). Subsequently, a structure appeared for the growth factor known as TGF-beta (17); it, too, was reported as a new fold. A sharp-eyed reader (18) saw a similarity between the two structures, however, and the two proteins have in fact turned out to be members of the same family (19). The structure of platelet-derived growth factor showed it to be related (20), and computer searching has now extended the family to include several connective tissue growth factors (21). A recent modelling study has revealed that the Norrie disease protein is also in this family, as are parts of the proteins mucin and von Willebrand factor (22). In all of these cases, the sequence resemblances are marginal at best.

Computer analysis is rapidly replacing the casual inspection of new three-dimensional structures. Actin is a protein so far known only in eukaryotic cells, and yet when its crystal structure was determined, to everyone's surprise, it was found to resemble the ATPase domain of heat shock 70, a ubiquitous chaperone protein (23). When the structure was scrutinized by careful computer searching, the family was extended to include not only hexokinase (23) but a number of other kinases as well (24). Moreover, phycocyanins turn out to have structures that are clearly similar to globins (25), and now the relationship has been extended to bacterial toxins (26). In another case, cold shock protein has been found to have the exact same fold as an oligonucleotide-binding protein (27), which in turn has the same structure as an oligosaccharide-binding protein (28, 29). Whether it will be possible to root today's multiplicity of domains back to their starter types on the basis of three-dimensional comparisons remains to be seen, but significant progress along these lines is already evident.

Classifying Domains

To a degree, how one classifies domains depends on one's notion of how they came into being in the first place. If, for example, all domains are descended from a single common ancestor, then a structure-based classification derived from a phylogenetic tree is an ideal representation. At the other extreme, if all domains have been concocted independently from some kind of random "big bang" (4, 30), then phylogenetic trees will be useless in this regard, and some kind of arbitrary classification will have to be imposed. A simple structure-based classification is widely used when three-dimensional information is available (31); assignments to classes can be made automatically by computer (15, 32). But three-dimensional data are often lacking, and schemes are needed that follow functional, cytogeographic, or when appropriate, phylogenetic lines.

So far, no single means of classification has proven altogether satisfactory. Although there is some utility in discussing domains as being either extracellular or intracellular, or, in the case of eukaryotes, nuclear or otherwise compartmentalized, we now know of many exceptions to such efforts at cellular delimitization. Thus, the immunoglobulin domain was initially thought to be restricted to an extracellular existence, but it has now been found in intracellular proteins, too (33). Similarly, phylogenetic restrictions have turned out not to be as rigid as first believed (34); again, the immunoglobulin domain has been found in several bacterial proteins and may prove to be ubiquitous. Finally, in the past, domains have been roughly categorized into those that contain disulfide bonds and those that do not. Now we realize that there are several domain families in which some members have disulfides and some do not. Again, the immunoglobulin domain is a good example.

Nomenclature

The names of the individual domains present a problem; clearly, we are at a point where a systematic nomenclature needs to be imposed. Many domains have been named on the basis of the first protein in which they were found, whether or not that protein was a good representative. Others have been named arbitrarily by function, others by undecipherable acronyms that strain the imagination; several domains have more than one name. For the moment, the reader will have to bear with a certain amount of anagrammatic confusion. But the discovery of new domains is slowing, and the time has come for workers to agree on a simple comprehensive system (35).

EXONS, INTRONS, AND MODULES

There is a widespread but mistaken belief that all exons encode domains. This lore has been encouraged by the notion that primitive proteins were encoded by "minigenes" that were spliced together, the genome eventually maturing into a state where coding regions were separated by introns (36, 37). The popular "introns early" scenario (38) holds that the prokaryotic lineage has lost all (coding region) introns over the course of time as a secondary adaptation concordant with streamlining the genome or quickening replication times (39). The "introns late" doctrine, on the other hand, holds that the class of introns that interrupt protein coding regions are a latter-day invention, making their impact after the occasion of the major endosymbiotic events that led to the appearance of mitochondria in eukaryotes (40)

Structural Aspects

Many studies have been conducted that attempt to show that intron positions in contemporary protein coding regions correlate with rudimentary structures (41–45). Although some of the earliest of these studies were vague about what constituted basic structural elements, two recent efforts have tried to settle the point with a more rigorous approach in which observed intron distributions are measured against what would be expected by chance (46, 47). Both come down in favor of the null hypothesis: To wit, in ancient proteins there is little or no correlation of present-day intron positions with recognized elements of protein structure.

This is not to say that introns do not aid and abet the movement of defined domains about the genome. In many instances, particularly evident in animal proteins that have been fashioned within the last billion years, it is clear that introns have been greatly involved (48, 49). It may even prove to be the case that the enormous success of the metazoan radiation is the direct result of the introduction of introns into coding regions. All the same, it must be emphasized that most exons do not encode domains, and that many movable domains are in fact themselves interrupted by introns. In these latter cases, structural integrity demands that the full complement of exons be shuffled as a unit.

We must ask why the genes of some animal proteins fit so well with the concept of exon shuffling while others do not (Table 1). As a general rule, those that have well-delineated domains set off by introns have been fashioned more recently. Beyond that, the discordance in other instances may be attributable to two causes. First, some proteins may have been assembled by exon shuffling, but subsequent loss and gain of introns has obscured the relationship. Several of the proteins in Table 1 likely fall into this category. Second, some proteins may have been assembled in a modular fashion, but without the involvement of introns. Consider a typical zymogen protease such as trypsinogen: It has two well-defined and similarly folded domains, but none of its four introns occur at the obvious joining point (59). Clearly, a simple contiguous duplication gave rise to the structure, and at some later time the gene was invaded by introns.

Genetic Aspects of Domain Rearrangement

DNA replication is given to a variety of iterative errors, and chance duplications can range from a few base pairs to entire chromosomes and even genomes. Various kinds of recombination, homologous (legitimate) or illegitimate, occur as the result of random or nonrandom breakage and reunion of DNA. Homologous recombination is the result of mistaken mismatching of similar DNA sequences, and as a result, there is a tendency for duplication to beget more duplication (67). The more recombination occurs at a given locus,

Table 1 Correlations (or not) of animal protein domains with exons

Good correlation	Poor or no correlation
tissue plasminogen activator (50)	trypsinogen (59)
LDL-receptor (51)	serpins (60)
vonWillebrand factor (52)	complement C6 (61)
protein S (vitronectin) (53)	complement C3–C5 (62)
link protein (54)	α spectrin (63)
fibronectin (55)	preadipocyte EGF-like protein-1 (64)
factor XIII b chain (56)	notch (65)
homing receptor (57)	lin12 (66)
lipoprotein-associated coagulation inhibitor (58)	

the more opportunities there will be for mismatch. In compensation, intron sequences drift away much more rapidly than coding sequences, thereby diminishing opportunities for homologous recombination. On the other hand, introns often provide safe havens for transposable elements or highly repetitive sequences like the Alu family, and these in turn can become involved in mismatches that give rise to rearrangements (68). Additionally, there is a good deal of rearrangement by more obscure mechanisms that usually fall under the rubric "illegitimate recombination."

There is an additional genetic constraint, as first pointed out by Patthy (69), imposed by the necessity for shuffled exons to have compatible introns so that neighboring domains will not find their coding to be out of register. Interestingly, the vast majority of introns flanking movable domains in animal proteins are of the phase I variety, the terminal codons of the separated exons being interrupted after the first nucleotide position (70).

THE PHYLOGENETIC DISTRIBUTION OF DOMAINS

Some Ancient Domains

This review is organized as though one could classify domains arbitrarily along phylogenetic and cytogeographical lines. In line with the author's bias that the number of primordial domains will ultimately prove to be small, we begin with the ancient proteins, which have been defined as those proteins that are common to all or most creatures on the Earth today (34). It is presumed that such proteins were in existence before the last common ancestor, which is to say about two billion years ago, give or take a half a billion. In this regard, we must take note of the modular structure of ubiquitous mainstream enzymes that occur in the cytoplasm, mentioning, if only in passing, ligand-binding domains such as the nucleotide-binding fold (71) or the flavin-binding (72) or heme-binding (73) modules. Most of these domains are now locked into their positions in proteins at the core of metabolism, and the evidence of their ancient mobility is apparent only from three-dimensional comparisions (74), all sequence resemblance having been eroded long ago (although, as happens in the case of the nucleotide-binding fold, occasionally a weak sequence motif, or signature sequence, remains that can be diagnostic). In contrast, many non-catalytic domains still move around in various positions peripheral to catalytic domains, in prokaryotes and eukaryotes alike, as made clear by a number of recent examples (75).

During this discussion, we must be careful not to exaggerate the evolutionary power of shuffling modules about in proteins. Although the combinatorial advantage is significant, it is not unlimited, and it has sometimes proved more expedient to reinvent a feature than to depend on genetic shuffling. Nucleotide-binding folds, for example, appear to have been reinvented on several occasions (76). Furthermore, we must distinguish between a system in which only structural components are exchanged (4), on the one hand, and one in which already functional domains are swapped, on the other. In the latter case, one would expect only a small fraction of combinations ever to be useful, although there are some settings where the utilization of an already existing domain seems made to order. For example, the regulation of gene expression seems a perfect realm for combinatorics, a small number of DNA-binding domains being paired with many different ligands or metabolites.

In this regard, the modular nature of numerous DNA binding proteins from bacteria has been firmly established both by X-ray crystallographic studies and primary structure comparisons. In many bacterial repressor proteins, for example, there are DNA-binding domains associated with various ligand-binding domains (77). A compelling example is the catabolite gene activator protein (CAP), which binds to DNA in a sequence-dependent manner after it has been complexed with cAMP, and which has a distinctly separate cAMP-binding domain in tandem with a helix-turn-helix domain that is clearly homologous to many other DNA-binding proteins (78). Modular proteins are also at the heart of bacterial signaling systems, the transmitter-receiver basis of which is largely combinatorial (79). Prokaryotic enhancer proteins also have a eukaryote-like modularity (80).

In retrospect, it should not have been surprising that many of these domains are common to prokaryotes and eukaryotes. Thus, the helix-turn-helix structural motif, first observed in bacteria (81), has its counterpart in the eukaryotic homeodomain (82–84). The most graphic illusrtation of how structurally similar prokaryotic and eukaryotic DNA-binding domains can be is afforded by the virtual superimposability of the eukaryotic POU-specific domain (Pit-Oct-Unc) on the lambda repressor (85, 86)

Still, prokaryotes and eukaryotes may have some domains that are unique to themselves. Indeed, each may have major lineages that have domains that were fashioned along the way; a number of schemes have been put forth for how new structures might be conceived, including frame-shifts (87) and intron capture (88). Certainly fungi, plants, and animals, respectively, have many sequences that appear not to have counterparts elsewhere (see below). Either new domains have been concocted, or old ones have been selectively lost by some lineages, or, in many cases, relationships have been blurred by excessive sequence change, and the related structures have yet to be studied by three-dimensional approaches (Figure 1). For example, steroid-binding receptors, modular DNA-binding proteins with shuffled domains (89), are so far a strictly animal phenomenon (90, 91), although the possibility of their existence in fungi has been suggested (92).

Intracellular Modular Proteins Apparently Unique to Eukaryotes

One module with a characteristic sequence motif that has so far been found only in eukaryotes is the widespread WD-40 domain, first identified in beta-transducins from animals and cell-division-cycle proteins from yeast (93), but now known to occur in upwards of 50 different settings (94). This domain, which is only about 40 amino acids long, has a definable sequence pattern (X..GH..X..WD) that has been conserved over the billion years since fungi, plants, and animals last had a common ancestor; it has not been found in bacteria. The domain, which is usually found as a set of 4–8 repeats, is associated with a staggering number of functions and interactions with other proteins. Its proposed structure, modeled from similar sequences in parts of known structures (94), is a small five-stranded antiparallel sheet. Its evolutionary origin remains mysterious.

There are other apparently critical modules that are known to be common to fungi, plants, and animals, but which have not yet been traced to a prokaryotic origin. For example, the dimerization domain that came to be known as the leucine zipper (95) was first identified by sequence searching when it was found to be common to the carboxyl-terminal segment of the yeast GCN4 protein and vertebrate oncogenes like *jun* and *myc* (96); databases now contain scores of related sequences, but all of them are from eukaryotes.

Another set of widely distributed domains that is so far restricted to eukaryotes is involved in recognizing and binding phosphorylated proteins. The src homology region 2 (SH2) is known to bind phosphorylated tyrosines in proteins (97, 98). It frequently occurs in association with another well-defined domain, the src homology region 3 (SH3), the function of which is not yet clear (97). Interestingly, its three-dimensional structure (99) has now been reported to resemble that of a photosystem protein from a cyanobacterium (100).

Eukaryotic cells also have a commonly shuffled domain called the PH domain (Pleckstrin Homology), which may be involved in the recognition and binding of phosphorylated serines and threonines (101). There is also a 42-amino acid repeat that occurs in a wide variety of fungal and animal cells that apparently participates in multiple interactions within the nucleus, including the maintenance of the nucleolus. In mammals it was first called plakoglobin (102); the Drosophila homologue corresponds to a mutant called *armadillo* (103), and the first observation in yeast involved a suppressor of temperature-sensitive mutations (104). It is also found in beta-catenin, an adhesive junction protein in animals (105). The number of repeats within any given setting may be as high as eight.

Finally, there is the 33-residue repeat known as the ankyrin repeat. Computer searching reveals that this domain is spread through a wide variety of fungal and animal proteins, and also in a few scattered bacteria (106). The distribution is unusual enough that the possibility of horizontal gene transfer has been raised. As in the case of other relatively short segments, the domains presumably gain stability by packing against each other, and as a result, the sequence motif never occurs as an isolated unit.

EXTRACELLULAR MOSAIC PROTEINS IN ANIMALS

It was the explosion of movable domains that occurred with the evolution of metazoans that offers us our richest demonstration of the construction of new proteins by shuffling modules. Here we have two great advantages. First, the intense research focus on animal proteins, combined with large-scale genome projects on humans, mice, fruit flies, and worms, has provided an enormous database. Second, and more importantly, the shuffling has occurred much more recently and is therefore more evident. Amino acid sequences alone are usually sufficient for following the dispersal of units. It appears, also, that the shuffling has been enormously more prevalent in animals and, indeed, has been part and parcel of the evolution of metazoan development, giving rise to the many unique proteins needed in cell-to-cell signaling and to the systems of defense and repair that are so essential to a multicellular existence (Figure 2).

Animal tissues are a marvel of structural complexity and intercellular communication. Many different cells exist in an extracellular matrix (ECM) that is itself an enormous assemblage of macromolecules, many of the components of which appear to be present in all animals but virtually nowhere else (113–115). The most abundant of these proteins is collagen, the highly repetitive structure of which was likely generated de novo as a part of animal emergence. The unexpected structural diversity of this protein, forming cables in some

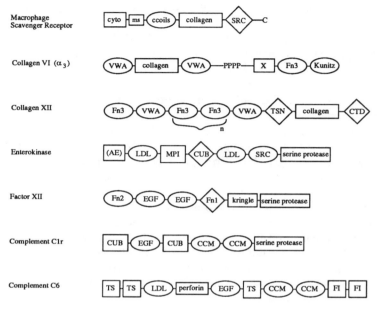

Figure 2 Some representative mosaic proteins. All the examples are from animal extracellular proteins (61, 107–112). Codes: CCM, complement control module; LDL, low-density lipoprotein receptor; Fn1, Fn2, and Fn3, fibronectin type I, II, and III repeats, respectively; EGF, epidermal growth factor–like; TS, thrombospondin type 1 repeat; TSN, thrombospondin amino-terminal domain; VWA, von Willebrand type A repeat; SRC, macrophage scavenger receptor domain; CUB, complement subcomponent; FI, factor I cryptic protease; MPI, metalloproteinase inhibitor; CTD, collagen carboxyl-terminal domain; AE, alternative exon; ms, membrane-spanner. Signal peptides are not shown; PPPP denotes proline-rich region. (Unfortunately, this figure was prepared before the recommendation on standardized abbreviations noted in Table 2.)

instances and sheets in others, is partly attributable to its being associated with a cadre of different noncollagen domains (Figure 2). In addition to collagen, there is a potpourri of syndecans and perlecans, laminins and mucins, all contributing to what used to be called "the ground substance," the nonprotein portion of which has long been known to be rich in sulfated mucopolysaccharides. It turns out that many of the mobile modules are themselves attachment sites for glyco-sulfates (116–119), while many others bind to these heparinic materials.

Approximately 50 different domains have been identified that are shuffled about in various animal extracellular proteins (120, 121). Of these, a few are much more abundant than all the others combined. These include the so-called EGF-like domain and the immunoglobulin (Ig) and fibronectin type III (Fn3) domains (it should be noted that Ig and Fn3 are not restricted to animals or to

Table 2   Some modules occurring in extracellular proteins of animals that do not depend on disulfide bonds for their integrity

Code [a]	Module	Length (res.)	Reference [b]
FA58A (FA)	coagulation factor V/VIII, type A	300	122
VWFA (VA)	von Willebrand factor, type A	200	123
LAMG (LG)	laminin G-like (A-type module)	190	124
FA58C (FC)	coagulation factor V/VIII, type C	150	122
ClQ (CQ)	collagen/complement Clq	140	125
CADH (CA)	cadherin-like	130	126
IGSF (IG)	immunoglobulin	100	127
FN3 (F3)	fibronectin, type III	90	55
HEMOP (HX)	hemopexin-like	60	53
LDLY (LY)	"YWTP" repeat, LDL-receptor	50	51
LRP (LR)	leucine-rich (tolloid)	30	12

[a] An international workshop on extracellular protein modules was held in Angelholm, Sweden, September 24–28, 1994 (P Bork, S Forsen, A Pastore & M Sunnerhagen, organizers), during which the participants discussed and endorsed a standardized scheme for referring to some 50 different modules found in extracellular proteins. The common abbreviations use 3–5 characters, but a two-letter designation (in parentheses above) is recommended for use in cartoons. The complete list, which was mainly compiled by P Bork and A Bairoch, appears in Reference 128.

[b] An effort has been made to cite the original observation of evolutionary mobility, but in some cases only the original sequence report is given, whether or not a module was described.

an extracellular existence). Other frequently shuffled modules include the calcium-dependent carbohydrate-recognition domain (C-lectins), complement control protein modules, cadherins, collagen segments, some domains originally observed in the LDL receptor, and about a dozen others (Tables 2 and 3). Most of these domains are found in each other's company, a mosaicism attributable to the genetic and structural compatability discussed above.

Some modules occur in remarkably long tandem repeats. The complement control protein module (also known as the beta-2 glycoprotein and sushi domain) occurs in six complement proteins, at least four complement regulatory proteins, and assorted noncomplement proteins as well. The complement control protein CR1 (not to be confused with C1r) is composed of 31 such domains strung together, and coagulation factor XIII b chain is wholly composed of 10 such repeats. Interestingly, a number of different proteins that contain this domain are found at the same locus on chromosome 1 in humans (140), suggesting that much of the duplication and shuffling is relatively recent. This chromosomal region may be likened to an active volcano from which further eruptions can be expected.

Many of these evolutionarily mobile domains are versatile in their binding properties. They form good packing units that bind to themselves as well as to other modules; many bind ligands. Their attachments to ligands may be loose and temporary or tight and permanent. They are seldom involved directly

Table 3 Some animal protein modules whose structure is dependent on disulfide bonds

Code[a]	Module	Length (res.)	Cysteines	Reference[b]
VWFB (VB)	von Willebrand factor, type B	30	8	124
SOMAB (SO)	somatomedin (vitronectin)	40	8	128
LDLRA (LA)	LDL receptor, type A	40	6	51
FN1 (F1)	fibronectin, type I	40	4	55
EGF (EG)	epidermal growth factor–like	40	6	129
FOLL1 (FS)	follistatin (ovomucoid)	50	6–10	130
PDOM (PD)	P domain (trefoil)	60	6	132
FN2 (F2)	fibronectin, type II	60	4	55
TSP1 (T1)	thrombospondin, type I	60	4–6	133
CCP (CP)	complement control protein (β_2GP, sushi, SCR)	65	4	134
FIMAC (FM)	complement factor I, MAC proteins	70	8–10	135
CTCK (CK)	C-terminal cystine knot	90	6–11	18
KUNIT (KU)	Kunitz-type inhibitor	90	4–6	130
LINK (LK)	link protein/proteoglycan	100	4	54
SRCR (SR)	scavenger receptor	110	6	136
VWFC (VC)	von Willebrand factor, type C	110	10	124
CUB (CU)	complement subcomponent module	110	2–4	137
KRING (KR)	kringle	120	6	138
CLECT (CL)	C-type lectin	130	4–6	139

[a] See footnote to Table 2.
[b] See footnote to Table 2.

in catalysis. Many are calcium dependent, consistent with the millimolar concentrations of calcium that occur in animal extracellular fluids. The majority depend on disulfide bonds for their structural integrity (Table 3), another advantage of an existence in the extracellular milieu.

Functionally, many are involved in cell recruitment, others in protein gathering or the assembling of other extracellular aggregates. The protein-protein interactions are often casual associations more than firm attachments. The specificity may be narrow or loose, the latter often reflecting a kind of cross-talk between systems or pathways. In animal bloods and fluids, mosaic proteins containing these various interacting domains are often fellow travelers rather than conjugal partners. In defense systems, these mosaics exhibit multiple interactions that keep a variety of substances in the vicinity of points of vulnerability such as vessel walls (141). Other aggregates seem to operate as a kind of macromolecular wolf pack, waiting in swarms for the inevitable microbial attack.

Many of these domains are genetically joined to proteases; others are themselves protease inhibitors (58, 131, 142). This is in keeping with the principal mode of extracellular regulation involving limited proteolysis. The vertebrate blood coagulation and fibrinolysis schemes have been assembled primarily by shuffling approriate modules into position at the amino-termini of serine proteases. On the other hand, many other mobile domains are involved in transmembrane signaling and constitute the extracellular portions of membrane receptors that are kinases or phosphatases, in keeping with intracellular regulation being mainly dependent on phosphorylation and dephosphorylation. Regulation emerges as a dominant role.

The EGF-Like Domain

Epidermal growth factor is a small (53 residues in human), stand-alone extracellular polypeptide stabilized by three disulfide bonds, which when bound to suitable cell surface receptors stimulates processes leading to mitosis and growth (143). The cloning of the unexpectedly large precursor of this protein (144, 145) revealed that the growth factor was one of a number of repetitive homologous units. Computer analysis further revealed that some of the homologues in the precursor were significantly related to doubly repeated segments in several proteins involved in blood coagulation (146).

Since that time, related sequences have been identified in more than 70 different vertebrate and invertebrate proteins (35), often occurring in long strings, and never in the cytoplasm, a natural consequence of their integrity depending on disulfide bonds. NMR studies have indicated that the structure is a compact bundle of five short beta strands (147). The family can be divided into two subclasses, one of which is ordinarily modified posttranslationally and binds calcium (148).

Recently, the domain has been found in a plant protein kinase (149). Even further afield, the domain has been found in the malaria-causing protozoan, *Plasmodium falciparum*, as a string of four units in one instance (150) and as a block of two in another (151). The sequence resemblance is sufficiently strong that there can be no doubt of homology, to the point where the question arises as to whether the parasite may have acquired the domain from some animal host (see below). Coincidentally, another protozoan, *Euplotes raikus*, makes a pheromone growth factor, the size of which is about the same as EGF and which contains six cysteines also (152), but the disulfide pairing is different (153) and NMR studies show it to be composed of several alpha helices (154).

The Immunoglobulin Domain

Some three decades ago it was proposed, on the basis of preliminary amino acid sequence data, that the heavy and light chains of immunoglobulins were not only homologous, but that these proteins had been assembled during the course of evolution by the extensive duplication of a primitive gene corresponding to a polypeptide chain of about 100 amino acids (155, 156). At the time there was no indication that these structures would turn out to be among the most common of rudimentary protein modules, or even that similar structures would be involved in other aspects of the immune response. Today it is clear that the superfamily not only includes antibodies and T-cell receptors, but also hundreds of other proteins involved in animal cell-cell attachment and communication (127, 157). The original family relationship was based very much on the existence of paired cysteines for what was thought to be an essential disulfide bond. As it happens, many members of the family lack the disulfide, showing that the characteristic sandwich structure can be sustained without auxiliary stabilization (157).

The diversity of the proteins in which these domains are found is simply staggering. Many of them are membrane bound and are involved in all sorts of cell adhesion events. Many others serve as the extracellular portions of membrane-bound receptors, the cytoplasmic portions of which are often kinases or phosphatases (158). A remarkable number are also associated with the nervous system and are referred to by a bewildering array of acronyms (e.g. N-CAMs, MAG, L1, TAG-1) and fascinating cognates (e.g. contactin, fasciculin, neuroglian, amalgam) (159). Moreover, the same kinds of domains are components of intracellular proteins, such as the skeletal muscle C-protein (33) and a smooth muscle light chain kinase, as well as the long connection proteins titin (160), twitchin (161), and projectin (162).

For a long time, it was thought that the immunoglobulin domain was confined to animals. Indeed, extension of the family to invertebrates remained tentative for a considerable period (127). In 1989, however, an X-ray structure of a bacterial chaperonin protein revealed a structure that was exceedingly similar to the immunoglobulin fold (163). There is no apparent sequence resemblance between this protein and animal immunoglobulins, although the point is disputed. As we shall see, however, the bulk of the evidence favors common ancestry.

The Fibronectin Type III Domain

The fibronectin type III (Fn3) domain has a distinctive motif (164), the structure of which was predicted on the basis of a multiple alignment of many cytokines to be a seven-stranded beta sandwich similar to that of immunoglobulins (165). The prediction has subsequently been borne out by experiment (166, 167). Although the structure is very similar to the immunoglobulin fold, it is distinguished in that its strand arrangement is a three-on-four sandwich (EBA:GFCD) instead of the four-on-three (DEBA:GFC) found in the class typified by the immunoglobulin constant region. Fn3 units do not have disulfide bonds (Table 2).

The domain occurs in a wide variety of animal proteins, both extracellular and intracellular. It is especially common in the extracellular portions of membrane-bound tyrosine kinase and tyrosine phosphatase receptors, where it is often found in the company of immunoglobulin domains. Similarly, the two of these domains together comprise the intracellular proteins titin, twitchin, and projectin (160–162). The domain has been estimated to occur in 2% of all animal proteins (168, 169); as we shall see, this may be a significant underestimate.

The domain has yet to be identified in plant or fungal proteins, but, remarkably, it has been readily identified by sequence analysis in a family of extracellular enzymes in bacteria, and the spectre of horizontal gene transfer has been raised to account for its existence in those enzymes (see below). The apparent absence from fungi and plants may be an artifact of the limits of sequence searching, however, even when a relatively sophisticated pattern search is employed. That many Fn3 domains were missed by sequence searching is apparent from the recent report of an Fn3 domain revealed by the X-ray structure of coagulation factor XIII (170). Similarly, the recently reported X-ray structure of beta galactosidase from *E. coli* showed that it has two domains of this type (171), and a previously reported cellulase from bacteria has one also (172). These structures will likely turn out to be ubiquitous, and it will be surprising indeed if three-dimensional studies of plant and fungal proteins do not turn them up eventually.

Although the Fn3 domain can be distinguished from the immunoglobulin domain both by sequence and three-dimensional criteria, the structural correspondence is so close that the two may be derived from a common ancestor. Indeed, a compelling study (173) based on a careful analysis of three-dimensional structures leads to the conclusion that all immunoglobulin and Fn3 domains arose from an ancestral form that most probably resembled the bacterial domains such as PapD.

Some Other Common Domains

A few other domains have been extensively shuffled about in animal extracellular proteins (Tables 2 and 3). Recently, there have been some excellent reviews focused on these modules, and only a few brief comments are made here about some of them. The nomenclature for many of these modules remains a problem. For example, the complement control protein modules were originally observed in a blood plasma protein of unknown function called the beta-2 glycoprotein and were initially referred to as B2GP repeats (134). They were subsequently found in enormous numbers in various complement proteins and have been rechristened as complement control modules. At about the same time, the discovery by Japanese workers of their presence in the hemolymph clotting system of the horseshoe crab led to them being dubbed "sushi domains" (174).

Lectin domains also have a dizzying nomenclature (175). The calcium-dependent C-lectins, which occur either singly or in sets of two or three in widely diverse settings, have been reviewed recently (176), as have the collagen-associated collectins (177). The cadherins are another widely spread group that binds to a great diversity of substances in a calcium-dependent manner (178).

Two aspects about the large number of modern mosaic proteins need emphasizing: timeliness and structural compatability. That so many of them appear at about the same time in a phylogenetic sense and that so many of them are associated with each other is not a mere coincidence. Let me try to illustrate the connection with a few more comments about collagen domains and where they are found.

For a protein whose structure was thought to have been determined with finality a generation ago, collagen has turned out to exist in a bewildering variety of forms and in association with a cadre of other domains, including the Fn3 domain and various von Willebrand factor repeats (Figure 2). The Gly-Xxx-Pro repeat that typifies collagen is well known to form a tightly woven structure composed of three helices. Collagen occurs in two quite different settings, now called fibrillar and nonfibrillar. The fibrillar forms mature to very long stretches of the collagen motif flanked by globular procollagen segments that are removed by limited proteolysis. Remarkably, the fibrillar collagen gene in sponges has virtually the same exon arrangement as is found in mammals, indicating that the system has been in place since the very dawn of animals (179). The nonfibrillar forms are sometimes called "interrupted" and are found as mosaics in association with a variety of noncollagen domains (180–184).

Any noncollagen domains that find themselves situated immediately adjacent to collagenous segments, whether at their amino or carboxyl ends, must be able to accomodate a threefold symmetrical association. Casual examination reveals that this group includes the thrombospondin and von Willebrand C domains, both of which have cysteine spacings in common with the "pro-segment" of procollagen and are clearly homologous with it. Additionally, the headpieces of C1q obviously accomodate threefold packing ("pawnshop model"), as must the headpieces of the macrophage scavenger receptor family (185, 186). The 200-residue von Willebrand A domain, which lacks cysteine, is often found with collagenous segments on either side and must pack with a threefold axis. Moreover, members of the collectin family (177) have variably long collagenous segments with short coiled coil "necks" that separate them from carboxyl-terminal lectin domains; these, too, must obey the threefold rule. By extension, Fn3 units and kringles, which are also found in chains containing collagenous segments, must be able to pack similarly.

If we may continue with the theme of structural compatability, the number of reported mosaic proteins containing three-stranded alpha-helical coiled coils is also surprising. There was a time when only vertebrate fibrinogen was thought to contain such a structure, but now the list includes various laminins (187, 188), thrombospondin (133), the collectins (177), the macrophage scavenger receptor (185) mentioned above, and a bevy of others. Scavenger receptor proteins were originally observed on the surfaces of macrophages (185). They are remarkable mosaics, combining compact globular domains with extended structures once thought to typify fibrous proteins (Figure 2). The MSCR module has now been found in several dozen different secreted or membrane-associated proteins, mostly involved in host defense (186).

Having just mentioned the three-stranded coiled coils that occur in fibrinogen, I should add that these are associated with large (250 residues) carboxylterminal domains previously referred to as FReDs, for fibrinogen-related domains (189). These are found in a number of extracellular proteins in both invertebrate and vertebrate animals. Most of these occurrences have been reviewed recently (189), but additional reports continue to pour in (190, 191), the most interesting of which is for the protein ficolin, which is a hybrid of collagen and fibrinogen (192).

Fibrinogen, of course, is the precursor of the fibrin blood clot, and I would be remiss if I did not say a few words about the interaction of fibrin with other substances by way of selected domains. Fibronectin (55) received its name because of this interaction, which is now mostly attributable to its Fn1 domains. But another moveable domain has been implicated for some other proteins. Thus, kringles were first identified in prothrombin and subsequently in several other proteins involved in blood clotting and fibrinolysis. In several of these proteins, the ability to bind to fibrin has been attributed to one or more of these domains. The most astonishing occurrence involves apolipoprotein a, which is a plasminogen homologue that contains 30 or more kringles at its amino terminus. In humans, apolipoprotein a isoforms with excessive numbers of kringles may correlate with thrombotic events (193). Beyond that, there have been some interesting attempts to follow the diaspora of kringle structures by sequence alignment and computer analysis (194, 195).

Most of the domains mentioned in the preceding paragraphs share a number of features. They all appear to exist in both vertebrate and invertebrate animals, and most of them appear to be able to coexist with each other. They dominate certain physiological systems peculiar to animals, including blood clotting and the immune system. Clearly, the advent of metazoans was a boom time for mixing and matching domains.

MODULES UNIQUE TO PLANT OR FUNGAL PROTEINS

Plants, animals, and fungi last shared a common ancestor about a billion years ago, and, as we noted earlier, they share many modular proteins that are involved in gene expression and intracellular regulation. But the three kingdoms have adapted quite different strategies for many of their extracellular involvements. Accordingly, it is perhaps not so surprising that most of the commonly shuffled domains that occur in extracellular proteins in animals have not yet been identified in plants or fungi. An interesting exception was the discovery of a somatomedin domain and four obvious hemopexin-type sequence repeats in the cytosolic plant protein PA2, a major storage protein component of pea seeds, leading to the suggestion that these two domains existed in the common ancestor of plants and animals (196).

Most of the other common domains (Tables 2 and 3), including anything resembling an immunoglobulin domain, remain among the missing. This may be partly attributable to the fact that the number of reported plant and fungal sequences is still relatively small, or it may be that the sequences have changed to the extent where they will only be recognized by three-dimensional studies, as has proven to be the case for some bacterial homologues. There is no indication that the somatomedin and hemopexin domains are changing any less rapidly than the others, however.

On the other hand, highly repetitive proteins that appear to be unique to plants are well known (197), and it seems reasonable that through one route or another some new modules have evolved during the time since plants last shared ancestry with animals. As for the fungi, the size and breadth of the ongoing yeast megasequencing projects tends to negate the notion that not enough data have been accumulated to find the missing domains. Rather, the surprise so far is that anywhere from a third to a half of all open reading frames (orfs) appear to be unidentifiable and unique to fungi (198). Again, this may reflect enormous sequence change that exceeds the limits of sequence searching. It is curious, though, what a large fraction of the yeast unidentified orfs resemble other unidentified open frames in yeast (198). These findings tend to confirm an earlier observation of an extensive family of proteins typified by a 34-residue repeat that also appeared limited to fungi (199).

EVOLUTIONARILY MOBILE DOMAINS IN BACTERIA

We have already commented that numerous ubiquitous proteins, principally those involved in mainstream metabolism, were constructed in a modular fashion in ancient times, as shown by obviously rearranged domains revealed by X-ray crystallography. That this mode of invention has continued throughout the evolutionary history of prokaryotes, including very recent happenings, is evidenced by the many examples of module shuffling observed on the basis of amino acid sequence similarity alone.

Thus, there are numerous instances of repetitive or transposed parts of proteins that occur in different settings. For example, the ribosomal protein S1, as originally sequenced in *E. coli,* was found to be composed, in its entirety, of six homologous segments about 90 residues long (200). This was an unexpected situation, because functionally the protein was thought to be composed of an mRNA-binding domain, on the one hand, and a ribosomal-binding domain, on the other. Somehow the same primitive structure had adapted to both of these functions (both involve RNA binding, of course). Given the nature of the repeat, it was not surprising, later, when the enzyme polynucleotide phosphorylase was found to have a carboxyl-terminal domain that is clearly homologous with the S1 repeats (201). A simple phylogenetic tree suggests that the enzyme acquired it from the ribosomal protein and not the other way around, in that the enzyme domain is actually more similar to two of the S1 repeats than those repeats are to the others.

Another example of domain swapping uncovered by sequence analysis involves the *E. coli* DNA repair protein UvrA. This large protein (940 residues) is composed of a series of repeated segments that are clearly homologous with a family of periplasmic transport proteins (202). More recent computer surveys have expanded the superfamily greatly (203).

The recency of some of these exchange events can be gauged by examining how similar repeated segments are in different genomes. The problem is complicated in that many of the recently shuffled domains also seem to be involved in horizontal gene transfers. In many soil bacteria, carbohydrate-binding domains are found at widely differing locations in a diversity of extracellular glycosidases (204–206). Some of these domains have actually been observed in an occasional eukaryote (207). Furthermore, fibronectin type III (Fn3) domains are strewn about many of these same enzymes (168, 208), and they are simply too similar in sequence to animal Fn3 units to have been passed in a conventional manner from a common ancestor (168, 209). These proteins bear all the earmarks of "exon shuffling without introns."

A number of prokaryotic structural proteins are also built up in an obvious domain fashion, including wall-attachment proteins in various bacteria. Some of these are long, multidomained proteins that contain coiled coil segments (210, 211). Others are highly repetitive and exhibit sequence motifs observed in other microbial proteins (212).

A fascinating example of module shuffling in the fashioning of bacterial mosaic proteins is offered by a set of membrane proteins that bind animal blood plasma proteins. A number of these are known that bind specific animal proteins with great specificity, including binders for fibrinogen, fibronectin, immunoglobulins, and plasma albumin. The staphylococcal protein A that binds to IgG, for example, is well known to most molecular biologists because of its great experimental utility. Less well known is that there are streptococcal proteins that not only bind to IgG, but also bind to plasma albumin. Protein G, for example, binds IgG with its amino-terminal domain and albumin with its carboxyl-domain (213). Moreover, certain strains of *Peptostreptococcus* are able to bind immunoglobulin light chains with a protein denoted protein L (214). Other strains of this bacterium have a mosaic protein (PAB) that is clearly the result of very recent shuffling together of domains from proteins G and L and that binds albumin (215). The mechanisms for moving these domains about is presumed to involve homologous or nonhomologous recombination, or perhaps a conjugative plasmid (215).

Many other examples of rearranged modules in bacterial proteins could be cited, including the components of various nitrogen fixation pathways (216) or the prokaryotic phosphotransferase system (217), but the point should have been made by now that domain shuffling has been a persistent activity among bacteria.

SOME UNEXPECTED OCCURRENCES

Earlier, it was mentioned that tandem arrays of EGF-like domains are found in the malarial parasite, *Plasmodium falciparum* (150, 151). The question arises, Does this indicate the existence of this domain in common ancestors of protists and the later diverging eukaryotes? If that were the case, we would expect to find similar arrays in plants and fungi, where they are rarely observed. Did the parasite acquire the gene for such a domain from some animal host? Or is the remarkable sequence resemblance the result of sequence convergence? Arguments have been made against such unlikely occurrences (218).

Several other evolutionarily mobile domains have been found in this same protozoan, including a well-defined thrombospondin domain with a remarkable resemblance to one that occurs in the complement protein properdin (219, 220) and a domain found in coagulation factor VIII (120). Again, some kind of horizontal acquisition may have occurred. It should be mentioned also that the trypanosome *T. cruzi* has been reported to have a repeat similar to one that occurs in the LDL receptor, and also an Fn3 module (221).

An EGF-like domain has also been identified in vaccinia virus (222). Interestingly, vaccinia also contains another of the commonly swapped domains that appears in Table 3, the complement control protein module (223). Poxviruses, of which vaccinia is one, are well known to acquire genes and pieces of genes from their mammalian hosts, and several other structures cited in this review have been found in their genomes also (75, 106)

HOW MANY DOMAINS?

Each month a popular biochemical journal contains a section devoted to claims and counter-claims of new domainal relationships based on sequence resemblances (224–232). Where will it end? Recent estimates about the total number of protein structures we can expect, based on the number of unique folds already in the structure data bank, are in the range of 500–1000 (233, 234). There are a few hundred folds known at present (15), and many new reported structures are being assigned to already known families. Furthermore, as in the case of sequence analyses, it is now being recognized that what were thought to be separate groups may indeed share common ancestry (235). The question is not how many unique groups can one point to today, but rather how many primordial groups were there in the beginning? As an example, immunoglobulin domains can be distinguished from Fn3 domains, but the very close structural resemblance between the two families indicates strongly that they share common ancestry (173).

What is the minimum number of starter types that can account for today's inventory of domains? The number may be much smaller than the estimates based on current inventories, perhaps less than 20 and maybe as few as 10. These numbers emerge from a consideration of fundamental domain structures, which largely reduces to about 20 major types (31). These classes can be reduced further to two or three all-alpha, a few all-beta, and a small number of alpha-beta structures.

The dilemma we face is distinguishing structural convergence from evolutionary divergence (218). Are there so many parallel alpha-beta barrels because this is a fundamentally sturdy structure? Or, is it because it evolved once, was found to be sturdy and functional, and has been propagated by duplication and modular exchange ever since (236)? In the absence of data to the contrary, the latter explanation seems more reasonable (237).

Furthermore, some contemporary folds must have been fashioned since those protean times. In particular, disulfide-stabilized domains, which are so well represented in the extracellular mosaic proteins of animals (Table 3), probably could not have been concocted until after the Earth's atmosphere became oxidizing. Witness the absence of most of these domains from prokaryotes and cytoplasmic environments in general. Almost certainly, also, many of them will prove to have descended from common ancestors.

SUMMARY AND CONCLUSIONS

Domainal shuffling has occurred throughout the course of protein evolution. There are certain natural realms where the combinatorial advantage of mixing and matching domains can be very advantageous, but there are limits to its usefulness. Because amino acid sequences relentlessly diverge, the history of genetic rearrangements affecting the structure of ancient proteins is often revealed only by three-dimensional studies. In more recently evolved proteins, shuffled modules are often readily apparent on the basis of sequence considerations alone. In many modern animal proteins, it is clear that the rearrangement process has been greatly accelerated by the participation of introns. This is not the same as saying that all exons encode domains, however. Certainly the majority of exons in eukaryotic genes do not. Moreover, many evolutionarily mobile domains, while genetically flanked by introns, also contain internal introns that do not participate in the rearranging. Several examples of recently rearranged modules in bacterial proteins are now known where domainal shuffling obviously occurred without introns, and there is no reason to think the same prosaic processes were not at work in ancient times.

ACKNOWLEDGEMENTS

I thank Peer Bork for numerous discussions about protein modules and for sending me materials in advance of their publication. I am grateful to Karen Anderson and Chad Manning for help in the preparation of the manuscript.

Literature Cited

1. Wetlaufer DB. 1973. *Proc. Natl. Acad. Sci. USA* 70:697–701
2. Murzin AG. 1994. *Curr. Opin. Struct. Biol.* 4:441–49
3. Sommer SS, Cohen JE. 1980. *J. Mol. Evol.* 15:37–57
4. Dorit RL, Schoenbach L, Gilbert W. 1990. *Science* 250:1377–82
5. Doolittle RF. 1981. *Science* 214:149–59
6. Patthy L. 1991. *Curr. Opin. Struct. Biol.* 1:351–61
7. Donavan JW, Mihalyi E. 1974. *Proc. Natl. Acad. Sci. USA* 71:4125–28
8. Doolittle RF. 1994. *Curr. Opin. Biotechnol.* 5:24–28
9. Altschul SF, Boguski MS, Gish W, Wooton JC. 1994. *Nat. Genet.* 6:119–29
10. Altschul SF, Gish W, Miller W, Myers EW, Lipman DJ. 1990. *J. Mol. Biol.* 215:403–10
11. Pearson WR, Lipman DJ. 1988. *Science* 85:2444–48
12. Patthy L. 1987. *J. Mol. Biol.* 198:567–77
13. Gribskov M, McLachlan AD, Eisenberg D. 1987. *Proc. Natl. Acad. Sci. USA* 84:4355–58
14. Rohde K, Bork P. 1993. *CABIOS* 9:183–89
15. Orengo C. 1994. *Curr. Opin. Struct. Biol.* 4:429–40
16. McDonald NQ, Lapatto R, Murray-Rust J, Gunning J, Wlodawer A, Blundell TL. 1991. *Nature* 354:411–14
17. Daopin S, Piez KA, Ogawa Y, Davies DR. 1992. *Science* 257:369–73
18. Swindells M. 1992. *Science* 258:1160–61
19. Daopin S, Cohen GH, Davies D. 1992. *Science* 258:1161–62
20. Oefner C, D'Arcy A, Winkler FK, Eggiman B, Hosang M. 1992. *EMBO J.* 11:3921–26
21. Bork P. 1993. *FEBS Lett.* 327:125–30
22. Meitinger T, Meindl A, Bork P, Rost B, Sander C, et al. 1993. *Nat. Genet.* 5:376–80
23. Flaherty KM, McKay DB, Kabsch W, Holmes KC. 1991. *Proc. Natl. Acad. Sci. USA* 88:5401–45
24. Bork P, Sander C, Valencia A. 1992. *Proc. Natl. Acad. Sci. USA* 82:7290–94
25. Pastore A, Lesk AM. 1990. *Proteins: Struct. Funct. Genet.* 8:133–55
26. Holm L, Sander C. 1991. *FEBS Lett.* 315:301–6
27. Schnuchel A, Wiltscheck R, Czisch M, Herrler M, Willimsky G, et al. 1993. *Nature* 364:169–71
28. Schindelin H, Marahiel MA, Heinemann U. 1993. *Nature* 364:164–68
29. Murzin AG. 1993. *EMBO J.* 12:861–67
30. White SK. 1994. *Annu. Rev. Biophys. Biomol. Struct.* 23:407–39
31. Richardson JS. 1981. *Adv. Protein Chem.* 34:167–339
32. Holm L, Ouzounis C, Sander C, Tuparev G, Vriend G. 1992. *Protein Sci.* 1:1691–98
33. Einheber S, Fischman DA. 1990. *Proc. Natl. Acad. Sci. USA* 87:2157–61

34. Doolittle RF. 1992. *Protein Sci.* 1:191–200
35. Campbell ID, Bork P. 1993. *Curr. Opin. Struct. Biol.* 3:385–92
36. Seidel HM, Pompliano DL, Knowles JR. 1992. *Science* 257:1489–90
37. Dorit RL, Gilbert W. 1991. *Curr. Opin. Struct. Biol.* 1:973–77
38. Gilbert W, Glynias M. 1993. *Gene* 135:137–44
39. Doolittle WF. 1978. *Nature* 272:581–82
40. Palmer JD, Logsdon JM. 1991. *Curr. Opin. Gen. Dev.* 1:470–77
41. Lonberg N, Gilbert W. 1985. *Cell* 40:81–90
42. Michaelson AM, Blake CCF, Evans ST, Orkin SH. 1985. *Proc. Natl. Acad. Sci. USA* 82:6965–69
43. Straus D, Gilbert W. 1985. Mol. Cell. Biol. 5:3497–506
44. Burke J, Hwang P, Anderson L, Lebo R, Gorin F, Fletterick R. 1987. *Proteins: Struct. Funct. Genet.* 2:177–87
45. Gilbert W, Marchionni M, McKnight G. 1986. *Cell* 46:151–54
46. Weber K, Kabsch W. 1994. *EMBO J.* 13:1280–86
47. Stolzfus A, Spencer DF, Zuker M, Logsdon JM Jr, Doolittle WF. 1994. *Science* 265:202–7
48. Gilbert W. 1978. *Nature* 271:501
49. Doolittle RF. 1985. *Trends Biochem. Sci.* 10:233–37
50. Ny T, Elgh F, Lund B. 1984. *Biochemistry* 81:5355–59
51. Sudhof TC, Russell DW, Goldstein JL, Brown MS, Sanchez-Pescador R, Bell GI. 1985. *Science* 228:893–95
52. Mancuso DJ, Tuley EA, Westfield LA, Worrall NK, Shelton-Inloes BB, et al. 1989. *J. Biol. Chem.* 264:19514–27
53. Jenne D, Stanley KK. 1987. *Biochemistry* 26:6735–42
54. Kiss I, Deak F, Mestric S, Delius H, Soos J, et al. 1987. *Biochemistry* 84:6399–403
55. Patel RS, Odermatt E, Schwarzbauer JE, Hynes RO. 1987. *EMBO J.* 6:2565–72
56. Bottenus RE, Ichinose A, Davie EW. 1990. *Biochemistry* 29:11195–209
57. Dowbenko DJ, Diep A, Taylor BA, Lusis AJ, Lasky LA. 1991. *Genomics* 9:270–77
58. van der Logt CPE, Reitsma PH, Bertina RM. 1991. *Biochemistry* 30:1571–77
59. Craik CS, Choo Q-L, Swift GH, Quinto C, Rutter WJ. 1984. *J. Biol. Chem.* 259:14255–64
60. Wright HT. 1993. *J. Mol. Evol.* 36:136–43
61. Hobart MJ, Fernie B, DiScipio RG. 1993. *Biochemistry* 32:6198–205
62. Vik DP, Amiguet P, Moffat GJ, Fey M, Amiguet-Barras F, et al. 1991. *Biochemistry* 30:1080–85
63. Kotula L, Laury-Kleintop LD, Showe L, Sahr K, Linnenbach AJ, et al. 1991. *Genomics* 9:131–40
64. Smas CM, Green D, Sul HS. 1994. *Biochemistry* 33:9257–65
65. Wharton KA, Johansen KM, Xu T, Artavanis-Tsakonas S. 1985. *Cell* 43:567–81
66. Greenwald I. 1985. *Cell* 43:583–90
67. Doolittle RF. 1979. Protein evolution. In *The Proteins*, ed. H Neurath, RL Hill, Vol. 4:1–118. New York: Academic. 2nd ed.
68. Stoppa-Lyonnet D, Carter PE, Meo T, Tosi M. 1990. *Proc. Natl. Acad. Sci. USA* 87:1551–55
69. Patthy L. 1987. *FEBS Lett.* 214:1–7
70. Patthy L. 1994. *Curr. Opin. Struct. Biol.* 4:383–92
71. Rossmann MG, Moras D, Olsen KW. 1974. *Nature* 250:194–99
72. Correll CC, Ludwig ML, Bruns CM, Karplus PA. 1993. *Prot. Sci.* 2:2112–33
73. Vasudevan SG, Armarego WLF, Shaw DC, Lilley PE, Dixon NE, Poole RK. 1991. *Mol. Gen. Genet.* 226:49–58
74. Smith GCM, Tew DG, Wolf CR. 1994. *Proc. Natl. Acad. Sci. USA* 91:8710–14
75. Bork P, Doolittle RF. 1994. *J. Mol. Biol.* 236:1277–82
76. Schulz GE. 1992. *Curr. Opin. Struct. Biol.* 2:61–67
77. Bustos SA, Schleif RF. 1993. *Proc. Natl. Acad. Sci. USA* 90:5638–42
78. Steitz TA, Ohlendorf DH, McKay DB, Anderson WF, Matthews BW. 1982. *Proc. Natl. Acad. Sci. USA* 79:3097–100
79. Parkinson JS, Kofoid EC. 1992. *Annu. Rev. Genet.* 26:71–112
80. North AK, Klose KE, Stedman KM, Kustu S. 1993. *J. Bacteriol.* 175:4257–73
81. Sauer RT, Yocum RR, Doolittle RF, Lewis M, Pabo CO. 1982. *Nature* 298:447–51
82. Laughon A, Scott MP. 1984. *Nature* 310:25–31
83. Quian YQ, Billeter M, Otting G, Muller M, Gehring, Wuthrich K. 1989. *Cell* 59:573–80
84. Affolter M, Percival-Smith A, Muller M, Billeter M, Qian YQ, et al. 1991. *Cell* 64:879–80
85. Assa-Munt N, Mortishire-Smith RJ, Aurora R, Herr W, Wright P. 1993. *Cell* 73:193–205
86. Holm L, Sander C. 1993. *J. Mol. Biol.* 233:123–38
87. Keese PK, Gibbs A. 1992. *Proc. Natl. Acad. Sci. USA* 89:9489–93

88. Golding GB, Tsao N, Pearlman RE. 1994. *Proc. Natl. Acad. Sci. USA* 91: 7506–09
89. Laudet V, Hanni C, Coll J, Catzeflis F, Stehelin D. 1992. *EMBO J.* 11:1003–13
90. Evans RM. 1988. *Science* 240:889–95
91. Green S, Chambon P. 1988. *Trends Genet.* 4:309–14
92. Moore DD. 1990. *New Biol.* 2:100–5
93. Fong HKW, Hurley JB, Hopkins RS, Miake-Lye R, Johnson MS, et al. 1986. *Proc. Natl. Acad. Sci. USA* 83:2162–2166
94. Neer EJ, Schmidt CJ, Manbudripad R, Smith TF. 1994. *Nature* 371:297–300
95. Landschulz WH, Johnson PF, McKnight SL. 1988. *Science* 240:1759–64
96. Vogt PK, Bos TJ, Doolittle RF. 1987. *Proc. Natl. Acad. Sci. USA* 84:3316–19
97. Koch CA, Anderson D, Moran MF, Ellis C, Pawson T. 1991. *Science* 252:668–74
98. Waksman G, Kominos D, Robertson SC, Pant N, Baltimore D, et al. 1992. *Nature* 358:646–53
99. Fry MJ, Panayotou G, Booker GW, Waterfield MD. 1993. *Protein Sci.* 2: 1785–97
100. Falzone CJ, Kao YH, Zhao J, Bryant DA, Lecomte JTJ. 1994. *Biochemistry* 33:6052–62
101. Musacchio A, Gibson T, Rice P, Thompson J, Saraste M. 1993. *Trends Biochem. Sci.* 18:343–48
102. Franke WW, Goldschmidt MD, Zimbelmann R, Mueller HM, Schiller DL, Cowin P. 1989. *Proc. Natl. Acad. Sci. USA* 86:4027–31
103. Peifer M, Wieschaus E. 1990. *Cell* 63: 1167–78
104. Yano R, Oakes ML, Tabb MM, Nomura M. 1994. *Proc. Natl. Acad. Sci. USA* 91:6880–84
105. Peifer M, McCrea PD, Green KJ, Wieschaus E, Gumbiner BM. 1992. *J. Cell. Biol.* 118:681–91
106. Bork P. 1993. *Proteins: Struct. Funct. Genet.* 17:363–74
107. Kodama T, Freeman M, Rohrer L, Zabrecky J, Matsudaira P, Krieger M. 1990. *Nature* 343:531–35
108. Bonaldo P, Colombatti A. 1989. *J. Biol. Chem.* 264:20235–39
109. Yamagata M, Yamada KM, Yamada SS, Shinomura T, Tanaka H, et al. 1991. *J. Cell Biol.* 115:209–21
110. Kitamoto Y, Yuan X, Wu Q, McCourt DW, Sadler JE. 1994. *Proc. Natl. Acad. Sci. USA* 91:7588–92
111. McMullen BA, Fujikawa K. 1985. *J. Biol. Chem.* 260:5328–41
112. Leytus SP, Kurachi K, Sakariassen KS, Davie EW. 1986. *Biochemistry* 25: 4855–63
113. Schmid V, Bally A, Beck K, Haller M, Schlage WK, Weber C. 1991. *Hydrobiologia* 216/217:3–10
114. Irwin DH. 1991. *TREE* 6:131–34
115. Miklos GLG, Campbell HD. 1992. *Curr. Opin. Genet. Dev.* 2:902–6
116. Goldstein LA, Zhou DFH, Picker LJ, Minty CN, Bargatze RF, et al. 1989. *Cell* 56:1063–72
117. Fisher LW, Termine JD, Young MF. 1989. *J. Biol. Chem.* 264:4571–76
118. Bernfield M, Kokenyesi R, Kato M, Hinkes MT, Spring J, et al. 1992. *Annu. Rev. Cell Biol.* 8:365–93
119. Hinkes MT, Goldberger OA, Neumann PE, Kokenyesi R, Bernfield M. 1993. *J. Biol. Chem.* 268:11440–48
120. Bork P. 1991. *FEBS Lett.* 286:47–54
121. Bork P. 1992. *Curr. Opin. Struct. Biol.* 2:413–21
122. Vehar GA, Keyt B, Eaton D, Rodriguez H, O'Brien DP, et al. 1984. *Nature* 312:337–42
123. Shelton-Inloes BB, Titani K, Sadler JE. 1986. *Biochemistry* 25:3164–71
124. Beck K, Hunter I, Engel J. 1990. *FASEB J.* 4:148–60
125. Reid KBM. 1974. *Biochem. J.* 141:189–203
126. Shirayoshi Y, Hatta K, Hosoda M, Tsunasawa S, Sakiyama F, Takeichi M. 1986. *EMBO J.* 5:2485–88
127. Williams A. 1987. *Immunol. Today* 8: 298–303
128. Bork P, Bairoch A. 1995. *Trends Biochem. Sci.* In Press
129. Jenne D, Stanley KK. 1985. *EMBO J.* 4:3153–57
130. Doolittle RF, Feng DF, Johnson MS. 1984. *Nature* 307:558–60
131. Laskowski M Jr, Kato I. 1980. *Annu. Rev. Biochem.* 49:593–626
132. Thim L. 1989. *FEBS Lett.* 250:85–90
133. Lawler J, Hynes RO. 1986. *J. Cell. Biol.* 103:1635–48
134. Lozier J, Takahashi N, Putnam FW. 1984. *Proc. Natl. Acad. Sci. USA* 81: 3640–44
135. Larson RS, Corbi AL, Berman L, Springer T. 1989. *J. Cell. Biol.* 108:703–12
136. Freeman M, Ashkenas J, Rees DJG, Kingsley DM, Copeland NG, et al. 1990. *Proc. Natl. Acad. Sci. USA* 87: 8810–14
137. Bork P, Beckmann G. 1993. *J. Mol. Biol.* 231:539–45
138. Sottrup-Jensen L, Claeys H, Zajdel M, Petersen TE, Magnusson S. 1978. *Progr. Chem. Fibrinolysis Thrombolysis* 3:191–209
139. Drickamer K. 1988. *J. Biol. Chem.* 263: 9557–60
140. Carroll MC, Alicot EM, Katzman PJ,

Klickstein LB, Smith JA, Fearon DT. 1988. *J. Exp. Med.* 167:1271–80
141. Preissner KT, Jenne D. 1991. *Thrombosis Haemost.* 66:123–32
142. Bolander ME, Young ME, Fisher LW, Yamada Y, Termine FD. 1988. *Proc. Natl. Acad. Sci. USA* 85:2919–23
143. Carpenter G, Cohen S. 1990. *J. Biol. Chem.* 265:7709–12
144. Gray A, Dull TJ, Ullrich A. 1983. *Nature* 303:722–25
145. Scott J, Urdea M, Quiroga M, Sanchez-Pescador R, Fong N, et al. 1983. *Science* 221:236–40
146. Doolittle RF, Feng DF, Johnson MS. 1984. *Nature* 307:558–60
147. Montelione G, Wuthrich K, Burgess AW, Nice EC, Wagner G, et al. 1992. *Biochemistry* 31:236–49
148. Rees DJ, Jones IM, Handford PA, Esnouf MP, Smith KJ, Brownlee GG. 1988 *EMBO J.* 7:2053–61
149. Kohurn BD, Lane L, Smith TA. 1992. *Proc. Natl. Acad. Sci. USA* 89:10989–92
150. Kaslow DC, Quakyi IA, Syin C, Raum MG, Keister DB, et al. 1988. *Nature* 333:74–76
151. Blackman MJ, Ling IT, Nicholls SC, Holder AA. 1991. *Mol. Biochem. Parasitol.* 49:29–34
152. Raffioni S, Miceli C, Vallesi A, Chowdhury SK, Chait BT, et al. 1992. *Proc. Natl. Acad. Sci. USA* 89:2071–75
153. Stewart AE, Raffioni S, Chaudhary T, Chait BT, Luporini P, Bradshaw RA. 1992. *Protein Sci.* 1:777–85
154. Brown LR, Mronga S, Bradshaw RA, Ortenzi C, Luporinin P, Wuthrich K. 1993. *J. Mol. Biol.* 231:800–16
155. Singer SJ, Doolittle RF. 1966. *Science* 153:13–25
156. Hill RL, Delaney R, Fellows RE Jr, Lebowitz HE. 1966. *Proc. Natl. Acad. Sci. USA* 56:1762–69
157. Williams AF, Barclay AN. 1988. *Annu. Rev. Immunol.* 6:381–405
158. Yoshihara Y, Oka S, Ikeda J, Mori K. 1991. *Neurosci. Res.* 10:83–105
159. Krueger NX, Saito H. 1992. *Proc. Natl. Acad. Sci. USA* 89:7417–21
160. Labeit S, Barlow DP, Gautel M, Gibson T, Holt J, et al. 1990. *Nature* 345:273–76
161. Ayme-Southgate A, Vigoreaux J, Benian G, Pardue ML. 1991. *Proc. Natl. Acad. Sci. USA* 88:7973–77
162. Higgins DG, Labeit S, Gautel M, Gibson TJ. 1993. *J. Mol. Evol.* 38:395–404
163. Holmgren A, Branden C-I. 1989. *Nature* 342:248–51
164. Patthy L. 1990. *Cell* 61:13–14
165. Bazan JF. 1990. *Proc. Natl. Acad. Sci. USA* 87:6934–69
166. de Vos AM, Ultsch M, Kossiakoff AA. 1992. *Science* 255:306–12
167. Leahy DJ, Hendrickson WA, Aukhil I, Erickson HP. 1992. *Science* 258:987–91
168. Bork P, Doolittle RF. 1992. *Proc. Natl. Acad. Sci. USA* 89:8990–94
169. Bork P, Doolittle RF. 1993. *Protein Sci.* 2:1185–87
170. Yee VC, Pedersen LC, Le Trong I, Bishop PD, Stenkamp RE, Teller DC. 1994. *Proc. Natl. Acad. Sci. USA* 91:7296–300
171. Jacobson RH, Zhang X-J, DuBose RF, Matthews BW. 1994. *Nature* 369:761–66
172. Juy M, Amit AG, Alzari PM, Poljak RJ, Claeyssens M, et al. 1992. *Nature* 357:89–91
173. Bork P, Holm L, Sander C. 1994. *J. Mol. Biol.* 242:309–20
174. Iwanaga S, Miyata T, Tokunaga F, Muta T. 1992. *Thrombosis Res.* 68:1–32
175. Bevilacqua M, Butcher E, Furie B, Furie B, Gallatin M, et al. 1991. *Cell* 67:233
176. Drickamer K. 1993. *Curr. Opin. Struct. Biol.* 3:393–400
177. Hoppe HJ, Reid KBM. 1994. *Protein Sci.* 3:1143–58
178. Klambt C, Muller S, Lutzelschwab R, Rossa R, Totzke F, Schmidt O. 1989. *Dev. Biol.* 133:425–36
179. Exposito JY, van der Rest M, Garrone R. 1993. *J. Mol. Evol.* 37:254–59
180. Chu ML, Zhang RZ, Pan TC, Stokes D, Conway D, et al. 1990. *EMBO J.* 9:385–93
181. Rehn M, Pihlajaniemi T. 1994. *Proc. Natl. Acad. Sci. USA* 91:4234–38
182. Oh SP, Kamagata Y, Muragaki Y, Timmons S, Ooshima A, Olsen BR. 1994. *Proc. Natl. Acad. Sci. USA* 91:4229–33
183. Myers JC, Kivirikko S, Gordon MK, Dion AS, Pihlajaniemi T. 1992. *Proc. Natl. Acad. Sci. USA* 89:10144–48
184. Koller E, Winterhalter KH, Trueb B. 1989. *EMBO J.* 8:1073–77
185. Krieger M, Acton S, Ashkenas J, Pearson A, Penman M, Resnick D. 1993. *J. Biol. Chem.* 268:4569–72
186. Resnick D, Pearson A, Krieger M. 1994. *Trends Biochem. Sci.* 19:5–8
187. Engel J. 1992. *Biochemistry* 31:10643–51
188. Hunter I, Schulthess T, Engel J. 1992. *J. Biol. Chem.* 267:6006–11
189. Doolittle RF. 1992. *Protein Sci.* 1:1563–77
190. Norenberg U, Wille H, Wolff JM, Frank R, Rathjen FG. 1992. *Neuron* 8:849–63
191. Yamamoto T, Gotoh M, Sasaki H, Terada M, Kitajima M, Hirohashi S. 1993. *Biochem. Biophys. Res. Commun.* 193:681–87

192. Ichijo H, Hellman U, Wernstedt C, Gonez LJ, Claesson-Welsh L, et al. 1993. *J. Biol. Chem.* 268:14505–13
193. Koschinsky ML, Beisiegel U, Henne-Bruns D, Eaton DL, Lawn RM. 1990. *Biochemistry* 29:640–44
194. Ikeo K, Takahashi K, Gojobori T. 1991. *FEBS Lett.* 287:146–48
195. Pesole G, Gerardi A, di Jeso F, Saccone C. 1994. *Genetics* 136:255–60
196. Jenne D. 1991. *Biochem. Biophys. Res. Commun.* 176:1000–6
197. Kreis M, Forde BG, Rahman S, Miflin BJ, Shewry PR. 1985. *J. Mol. Biol.* 183:499–502
198. Koonin EV, Bork P, Sander C. 1994. *EMBO J.* 13:493–503
199. Sikorski RS, Boguski MS, Goebl M, Hieter P. 1990. *Cell* 60:307–17
200. Doolittle RF, Woodbury NW, Jue RA. 1982. *Biosci. Rep.* 2:405–12
201. Regnier P, Grunberg-Manago M, Portier C. 1987. *J. Biol. Chem.* 262:63–68
202. Doolittle RF, Johnson MS, Husain I, Van Houten B, Thomas DC, Sancar A. 1986. *Nature* 323:451–53
203. Gorbalenya AE, Koonin EV. 1990. *J. Mol. Biol.* 213:583–91
204. Meinke A, Gilkes NR, Kilburn DG, Miller RC, Warren RAJ. 1991. *J. Bacteriol.* 173:7126–35
205. Fujii T, Miyashita K. 1993. *J. Gen. Microbiol.* 139:677–86
206. Gilkes NR, Henrissat B, Kilburn DG, Miller RC, Warren RAJ. 1991. *Microbiol. Rev.* 55:303–15
207. Ramalingam R, Blume JE, Ennis HL. 1992. *J. Bacteriol.* 174:7834–38
208. Hansen CK. 1992. *FEBS Lett.* 305:91–96
209. Little E, Bork P, Doolittle RF. 1994. *J. Mol. Evol.* 39:631–43
210. Lupas A, Engelhardt H, Peters J, Santarius U, Volker S, Baumeister W. 1994. *J. Bacteriol.* 176:1124–233
211. Engel AM, Cejka Z, Lupas A, Lottspeich F, Baumeister W. 1992. *EMBO J.* 11:4369–78
212. Foster SJ. 1993. *Mol. Microbiol.* 8:299–310
213. Guss B, Eliasson M, Olsson A, Uhlen M, Frej AK, et al. 1986. *EMBO J.* 5:1567–75
214. Kastern W, Sjobring U, Bjorck L. 1992. *J. Biol. Chem.* 267:12820–25
215. deChateau M, Bjorck L. 1994. *J. Biol. Chem.* 269:12147–51
216. Ouzounis C, Bork P, Sander C. 1994. *Trends Biochem. Sci.* 19:199–200
217. Saier MH, Reizer J. 1990. *Res. Microbiol.* 141:1033–38
218. Doolittle RF. 1994. *Trends Biochem. Sci.* 19:15–18
219. Goundis D, Reid KBM. 1988. *Nature* 335:82–85
220. Robson KJH, Hall JRS, Jennings MW, Harris TJR, Marsh K, et al. 1988. *Nature* 335:79–82
221. Pereira MEA, Mejia JS, Ortega-Barria E, Matzilevich D, Prioli RP. 1991. *J. Exp. Med.* 179–91
222. Blomquist MC, Hunt LT, Barker WC. 1984. *Proc. Natl. Acad. Sci. USA* 81:7363–67
223. Kotwal GJ, Moss B. 1988. *Nature* 335:176–78
224. Beckmann G, Bork P. 1993. *Trends Biochem. Sci.* 18:40–41
225. Jank B, Habermann B, Schweyen RJ. 1993. *Trends Biochem. Sci.* 18:427–28
226. Reizer J, Reizer A, Saier MH. 1993. *Trends Biochem. Sci.* 18:247–48
227. Kwong PD, McDonald NQ. 1993. *Trends Biochem. Sci.* 18:425–26
228. Creighton TE, Kemmink J. 1993. *Trends Biochem. Sci.* 18:424–25
229. McDonald NQ, Kwong PD. 1993. *Trends Biochem. Sci.* 18:208–9
230. Petrou S, Ordway RW, Singer JJ, Walsh JV. 1993. *Trends Biochem. Sci.* 18:41–42
231. Grandori R, Carey J. 1994. *Trends Biochem. Sci.* 19:72
232. Koonin EV, Bork P. 1994. *Trends Biochem. Sci.* 19:234–35
233. Chothia C. 1992. *Nature* 357:543–44
234. Blundell TL, Johnson MJ. 1993. *Protein Sci.* 2:877–83
235. Murzin AG, Chothia C. 1992. *Curr. Opin. Struct. Biol.* 2:895–903
236. Farber GK. 1993. *Curr. Opin. Struct. Biol.* 3:409–12
237. Doolittle RF, Blundell TL. 1993. *Curr. Opin. Struct. Biol.* 3:377–78

Annu. Rev. Biochem. 1995. 64:315–43

EUKARYOTIC PHOSPHOLIPID BIOSYNTHESIS

Claudia Kent

Department of Biological Chemistry, University of Michigan Medical School, Ann Arbor, Michigan 48109-0606

KEY WORDS: phosphatidylcholine, phosphatidylserine, phosphatidylethanolamine, phosphoinositides, phosphatidate

CONTENTS

ABSTRACT

The current status of the biochemistry of phospholipid biosynthesis is presented. The review focuses on the identification and characterization of mo-

315

lecular tools such as purified enzymes and cloned genes and cDNAs for those enzymes. The enzymes discussed are those involved in the biosynthesis of the major phospholipid classes, namely, phosphatidate, phosphatidylserine, phosphatidylethanolamine, phosphatidylcholine, sphingomyelin, phosphatidylinositol and its phosphorylated derivatives, and cardiolipin. The review centers on the pathways in mammals and yeast. Novel genetic approaches used to delineate pathways and clone cDNAs are discussed. The regulatory roles played by some of the enzymes involved in controlling the biosynthetic pathways are presented.

PERSPECTIVE

Studies on eukaryotic phospholipid biosynthesis have a long history, dating back to the 1950s when most of the pathways were delineated, largely through work in Eugene Kennedy's laboratory. Subsequent studies over the next 25 years shed some light on the importance and regulation of individual pathways, but progress in the biochemical and molecular analysis was slow. Many enzymes in these pathways are membrane-associated and difficult to purify, and kinetic analysis of reactions with insoluble substrates and products was not straightforward. Developments over the last ten years have helped to clean up many areas that were previously murky, and to provide pristine enzymes and genes or cDNAs. This review will focus on these enzymes and molecular tools; space limitations necessarily limit discussion of many intriguing papers on regulation, especially in the areas of phosphatidylcholine (PC) and phosphatidylinositol (PI) metabolism, which are being pursued rapidly and enthusiastically. The reader interested in more background or more details in specific areas will find a number of excellent reviews helpful (1–5).

MAKING THE BACKBONE: PHOSPHATIDATE

The goal of the first enzymes of glycerophospholipid biosynthesis is to make phosphatidate (PA), the diacylglycerophosphatide that serves as precursor for all glycerophospholipids (Figure 1).

Glycerol-3-phosphate Acyltransferase

Glycerol-3-phosphate acyltransferase is the first committed step in glycerolipid biosynthesis. As such, it would seem that this step would be highly regulated. Glycerol-3-phosphate acyltransferase is clearly a lipogenic enzyme because it is found in high levels in lipogenic tissues (6), and its activity shows the expected changes upon adipocyte differentiation (7, 8) as well as starvation and refeeding (9). On the other hand, it has not been satisfactorily demonstrated

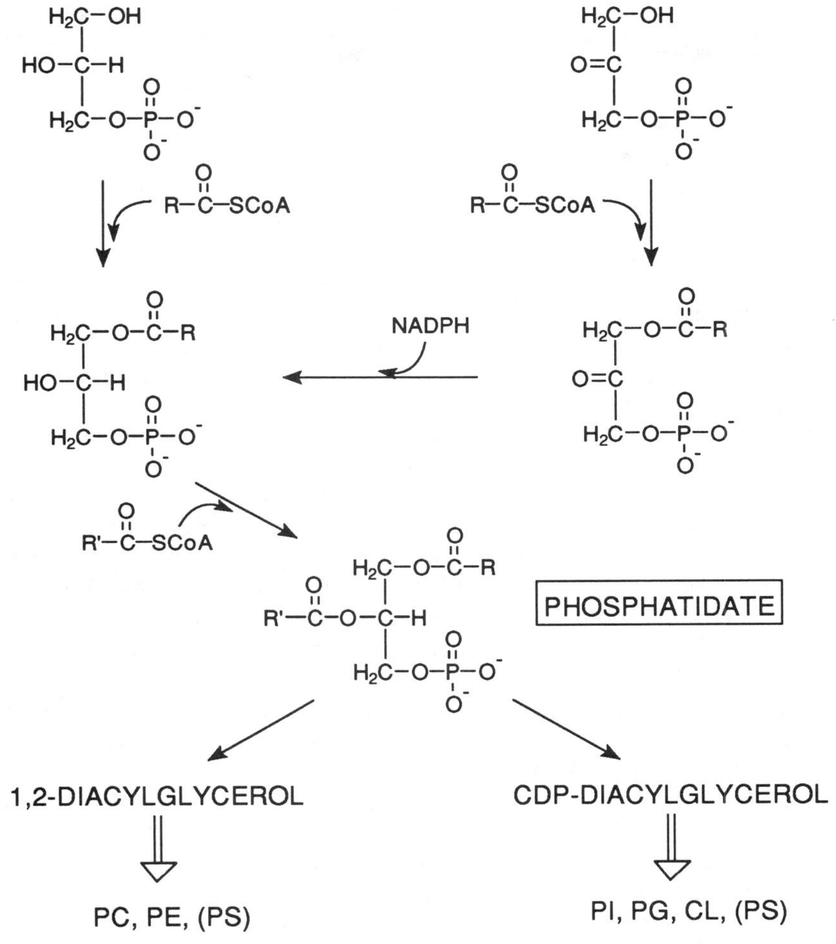

Figure 1 The central role of phosphatidate in phospholipid biosynthesis. Conversion of phosphatidate to diacylglycerol is required for biosynthesis of PC, PE, and, in higher eukaryotes, PS. Conversion to CDP-diacylglycerol is used for the biosynthesis of PI, PG, CL, and, in yeast, PS.

that glycerol-3-phosphate acyltransferase is regulatory for glycerolipid biosynthesis.

In mammals, two glycerol-3-phosphate acyltransferases have been identified, one mitochondrial and one microsomal. They can be distinguished by their sensitivities to N-ethylmaleimide (NEM) (10). The microsomal enzyme is inhibited by NEM while the mitochondrial enzyme is not sensitive to NEM.

The mitochondrial enzyme prefers saturated over unsaturated fatty acyl groups, while the microsomal enzyme does not show such a substrate specificity. This has led to the speculation that the mitochondrial enzyme is responsible for the observed preference for saturated fatty acyl groups in position 1 of phospholipids synthesized in vivo, although this remains to be proven.

Obtaining molecular tools for the yeast and mammalian glycerol-3-phosphate acyltransferases has been slow. Despite a report in 1986 on the isolation of a yeast mutant defective in glycerol-3-phosphate acyltransferase (11), this gene has not yet been isolated and characterized. In the mammalian field, however, a cDNA that is likely to encode the mitochondrial glycerol-3-phosphate acyltransferase has been isolated. Sul's laboratory first reported the isolation of this cDNA as one that is hormonally and nutritionally regulated (12). Sequencing of the cDNA revealed an open reading frame encoding a protein with a molecular weight of 90,000, referred to as p90 (6). The amino acid sequence of p90 is about 25% identical to the sequence of *Escherichia coli* glycerol-3-phosphate acyltransferase (13). The 6.8-kb mRNA for p90 increases with fasting and refeeding or with insulin treatment of streptozotocin-treated rats, and these changes are consistent with those that have been observed for the mitochondrial glycerol-3-phosphate acyltransferase. Furthermore, levels of p90, as detected by Western blots, show changes in fasted/refed mice as well as 3T3-L1 adipocytes that correspond with changes in mitochondrial glycerol-3 phosphate acyltransferase activity (9). Expression of the cDNA for p90 in CHO cells results in about an eightfold increase in mitochondrial glycerol-3-phosphate acyltransferase activity, consistent with the identity of p90 as this enzyme (9). It is still possible that p90 is a regulatory protein that modulates glycerol-3-phosphate acyltransferase activity. In light of the conservation of sequence between p90 and the *E. coli* enzyme, however, p90 is highly likely to be the mitochondrial glycerol-3-phosphate acyltransferase.

Considerable attention has been paid to plant glycerol-3-phosphate acyltransferase because of the presumed importance of this enzyme in determining the positional distribution of fatty acids, which in turn plays an important role in the sensitivity of plants to chilling (14). Glycerol-3-phosphate acyltransferase has been purified to apparent homogeneity from the postmicrosomal supernate from cocoa seeds (15) as well as from the soluble fraction from chloroplasts from squash (16) and pea (17). The molecular weight of the chloroplast enzymes in both the native and denatured states is about 42,000 (17, 18). The cytosolic enzyme, on the other hand, is reported to have a small (20 kDa) subunit, which oligomerizes to form the native enzyme with a molecular weight of 200,000 (15). cDNAs for the chloroplast enzymes from squash (16), pea (17), and *Arabidopsis* (19) have been isolated and sequenced. The identity of the cDNA clones as encoding glycerol-3-phosphate acyltransferase is indicated by the fact that the deduced amino acid sequences of the proteins contain sequences of peptides obtained from the pure proteins (16, 17) Furthermore, expression of the cDNA in *E. coli* results in production of large amounts of soluble glycerol-3-phosphate acyltransferase activity (19), whereas the enzyme from *E. coli* is membrane associated. The deduced protein sequences of the chloroplast enzymes are about 60–70% identical to each other, but surprisingly, there is no significant sequence similarity to the enzyme from *E. coli* in either the overall sequence or in isolated regions. Perhaps the membrane-associated enzymes of *E. coli* and rat mitochondria evolved separately from the soluble enzymes of plant chloroplasts.

Dihydroxyacetone Phosphate Acyltransferase

Both the mitochondrial and microsomal glycerol-3-phosphate acyltransferases catalyze the acylation of dihydroxyacetone phosphate, but there is a separate, peroxisomal dihydroxyacetone phosphate acyltransferase that does not acylate glycerol-3-phosphate. As discussed previously in depth (2), the peroxisomal enzyme appears to be specific for the synthesis of ether lipids, while the microsomal and mitochondrial enzymes function only in glycerolipid synthesis. In yeast, glycerol-3-phosphate acyltransferase and dihydroxyacetonephosphate acyltransferase have been reported to be the same enzyme since both activities are reduced in a single mutant (11). The activity of dihydroxyacetone phosphate acyltransferase, however, is reduced to a much greater extent than that of glycerol-3-phosphate acyltransferase (20), suggesting that the two activities do not reside in the same protein but that they may be coordinately regulated.

CONVERSION TO DIACYLGLYCEROL: PHOSPHATIDATE PHOSPHATASE

After the synthesis of PA, the pathways of glycerolipid biosynthesis separate into two branches in which the next intermediate to be formed is either diacylglycerol or CDP-diacylglycerol (Figure 1). Diacylglycerol is the precursor for PC, phosphatidylethanolamine (PE), and triacylglycerols, and CDP-diacylglycerol is used for biosynthesis of phosphatidylglycerol (PG), PI and, in yeast, phosphatidylserine (PS). In yeast, the diacylglycerol branch may be less important in rapidly growing cells, in which the CDP-diacylglycerol branch is the principal pathway for PC biosynthesis. In mammals, PC is made only through the diacylglycerol branch. The two enzymes that lead off from PA, CDP-diacylglycerol synthetase and PA phosphohydrolase, are of interest because of their expected roles in controlling the flux into the two branches. In addition, there has been a great deal of interest in PA phosphohydrolase because of its implied role in signal transduction pathways, in which PA, produced by the action of phospholipase D, is converted to *sn*-1,2-diacylglycerol (diacylglycerol). Because both PA and diacylglycerol can act as second messengers, the conversion of one to the other by PA phosphohydrolase is likely to be a site of control.

There are several forms of PA phosphohydrolase in eukaryotic cells. Activity can be found in soluble as well as membrane-associated forms in both mammals (21) and yeast (22). Membrane-associated PA phosphohydrolase activity in animals has been reported to be associated with microsomes (23) and with plasma membranes (24, 25). The cytosolic and microsomal enzymes are stimulated by Mg^{2+} and inhibited by sulfhydryl reagents such as NEM, whereas the plasma membrane enzyme is neither stimulated by Mg^{2+} nor sensitive to sulfhydryl reagents (25, 26). It is possible that the cytosolic and microsomal enzymes are interconvertible forms of the same enzyme, since fatty acids cause translocation of the cytosolic form to microsomes in cell-free extracts (26–28) and in intact cells (29). A plasma membrane–associated PA phosphohydrolase activity that faces the extracellular environment has been observed in neutrophils and Chinese hamster ovary cells (30). Because this ecto-PA phosphohydrolase is NEM-insensitive, it is likely to be the same as that reported in isolated plasma membrane fractions (25, 26).

The first form of PA phosphohydrolase to be purified to apparent homogeneity was a membrane-associated form from yeast (31, 32), the molecular weight of which is 104,000. The enzyme appears to function as a monomer. This form of PA phosphatase is sensitive to sulfhydryl reagents such as NEM. An antibody to the enzyme also detects a 45-kDa form, but peptide mapping and a pulse-chase study shows that the 45-kDa form is not a degradation product of the 104-kDa form. The 45-kDa form is also sensitive to sulfhydryl reagents. The two forms have distinct subcellular locations and regulatory properties (32, 33).

A membrane-associated form of phosphatidate phosphohydrolase has been purified to apparent homogeneity from porcine thymus membranes (34). The molecular weight of the enzyme by SDS-PAGE is 83,000; it is not clear if the native enzyme is a monomer or oligomer. Because the purified enzyme is insensitive to Mg^{2+} and sulfhydryl reagents, it is likely to be the plasma membrane–associated enzyme. Hopefully, we can anticipate isolation of antibodies and cDNAs for this isoform in the near future.

CONVERSION TO CDP-DIACYLGLYCEROL

The synthesis of CDP-diacylglycerol might also be expected to be a regulated step, since this is a branch point in glycerophospholipid synthesis. In yeast as well as mammals the activity is predominately microsomal, with some activity also in the mitochondrial inner membrane (35–38). The enzyme from yeast mitochondria has been purified to near homogeneity (39). The molecular weight of the native enzyme, determined by radiation inactivation, is 114,000. Two bands are apparent on SDS gels with molecular weights of 54,000 and 56,000. It is not clear if both "subunits" are needed for activity, nor is it known if one is a proteolytic degradation product of the other.

In yeast, CDP-diacylglycerol synthase is regulated in response to water-soluble precursors, in that the presence of inositol and choline in the growth medium causes a 55% decrease in enzyme activity in cell extracts. An interesting yeast mutant, *cdg1*, was found to be defective in CDP-diacylglycerol synthase activity (40). This mutant exhibits pleiotropic effects on phospholipid metabolism, and CDP-diacylglycerol synthase partially purified from the mutant appears normal, so *CDG1* is likely to be a regulatory gene rather than a structural gene. The fact that the level of CDP-diacylglycerol synthetase is controlled by this gene suggests that regulation of this enzymatic step is important in yeast. Regulation of this enzyme has not been demonstrated in mammals, although increased accumulation of CDP-diacylglycerol in response to hormonal stimulation has been reported (41–43), suggesting that the levels of this important intermediate are regulated in vivo.

PHOSPHATIDYLSERINE

PS can be considered both an end product and a biosynthetic intermediate, since it is both found in eukaryotic cell membranes and serves as a precursor of PE. The role of PS in the membrane may not be as important as its biosynthetic role, however, because cells defective in the synthesis of PS can grow in the presence of ethanolamine (44), PE (45), or, in yeast, choline (46–48). PS is synthesized by phosphatidyl transfer from CDP-diacylglycerol in yeast or from PE or PC in mammals. The latter process is usually referred to as base exchange.

Base Exchange

Serine has long been known to be incorporated into PS by Ca^{2+}-dependent base exchange (49–51). The role of base-exchange enzymes in PS biosynthesis in vivo has been elucidated by elegant genetic studies, primarily from the work of Kuge, Nishijima, and Akamatsu. These workers have developed several protocols for the isolation of mutants of Chinese hamster ovary cells defective in phospholipid biosynthesis. One protocol (52) developed to detect mutants in choline base-exchange yielded a temperature-sensitive mutant, referred to as mutant 64, in which choline base-exchange activity is decreased by over 90% at 40°C. The rate of incorporation of $^{32}P_i$ into PS is similarly decreased in intact cells at 40°C. Several revertants were isolated in which both PS synthesis and choline exchange activity are normal, indicating that both defects result from one mutation. These observations thus implicate a choline-serine exchange enzyme as the principal means by which PS is synthesized in these cells.

Another mutant with defective choline exchange activity was isolated as a PS auxotroph (53). In this mutant, PSA-3, levels of PS and PE are about 30% and 50% of wild-type values, respectively. Ethanolamine labeling of PE via the CDP-ethanolamine pathway is normal. Choline base-exchange is virtually absent in the mutant, and rates of serine and ethanolamine exchange are reduced. The authors propose that there are two base-exchange enzymes, both using serine and ethanolamine as substrates, but only one using choline. They further propose that the choline base-exchange enzyme (later termed PS synthase I) is defective in the PS auxotroph, but the other base-exchange enzyme, or PS synthase II, is normal. In support of this, choline inhibits serine base-exchange activity in extracts from the wild-type, but not mutant, cells. In a test of the ability of these cells to take up and incorporate exogenous ^{32}P-labeled phospholipids, wild-type CHO cells can incorporate label from [^{32}P]PC into PS and PE, but mutant PSA-3 cannot (45). Mutant PSA-3 is not defective in utilization of exogenous [^{32}P]PE to make PS.

In a related study, Voelker & Frazier (44) isolated a CHO mutant that requires ethanolamine for growth, and PS can replace ethanolamine as the growth requirement. This mutant was also found to be markedly defective in choline, but not ethanolamine, base-exchange activity.

These observations support the scheme for PS biosynthesis shown in Figure 2, in which the primary route of entry of serine into PS is through base exchange with PC, catalyzed by PS synthase I. The PS is then decarboxylated and the resulting PE can be a substrate for base exchange through PS synthase II. The ethanolamine produced from the second base-exchange reaction is recycled into PE through the CDP-ethanolamine pathway. In fact, there may be more

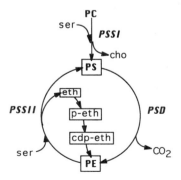

Figure 2 Phosphatidylserine biosynthesis in mammals. PSSI and PSSII refer to phosphatidylserine synthases I and II; PSD is phosphatidylserine decarboxylase; ser, serine; cho, choline; eth, ethanolamine; p-eth, phosphoethanolamine; cdp-eth, CDP-ethanolamine.

than two base-exchange enzymes in cells, with enzymes in different tissues and subcellular locations exhibiting distinct substrate specificities and abilities to be stimulated by agonists (54).

Mutant PSA-3 has been used to clone cDNAs that convert this PS auxotroph to prototrophy. Two distinct cDNAs have been isolated. One of these, *pssC*, is discussed later in the section on PE biosynthesis. The other, *pssA*, possibly encodes PS synthase I (55). Expression of this cDNA in either PSA-3 or mutant 64 restores PS levels to normal and results in very high PS synthase I activities. The protein encoded by the cDNA is 471 residues in length and is predicted to contain five membrane-spanning segments. Whether or not this is truly PS synthase I or a regulatory protein required for PS synthase I activity awaits expression of this cDNA in an appropriate heterologous system.

Phosphatidylserine Synthase

This enzyme catalyzes the transfer of the phosphatidyl moiety from CDP-diacylglycerol to serine; this membrane-bound enzyme and its gene have been well studied (1). This enzyme is considered the first committed step in the biosynthesis of PC in yeast, and, as with other enzymes in this pathway, is subject to regulation by the combination of inositol and choline (56). This regulation is dependent on the presence of the octomer, 5'-TTCACATG, in the 5'-upstream region of the PS synthase gene (57). This is the inversion of the consensus sequence for control by water-soluble precursors, CATRTGAA found for the phospholipid *N*-methyltransferase genes (58). PS synthase is also competitively inhibited by inositol (59). Since PS synthase and PI synthase compete for CDP-diacylglycerol, the control by water-soluble precursors plus the competitive inhibition by inositol would favor the synthesis of PI. Furthermore, PS synthase is quite sensitive to the phospholipid environment of the membrane (60).

The activity of PS synthase is inhibited by phosphorylation by cAMP-dependent protein kinase (61). That this is important in vivo is suggested by the fact that the ratio of PI to PS decreases in an adenylate cyclase–defective mutant, *cyrl* (62). When cAMP is added to the mutant cells, the wild-type ratio of PI to PS is restored. This is accompanied by a decrease in PS synthase activity, consistent with phosphorylation of the enzyme by cAMP-dependent protein kinase. PI synthase activity is not affected by cAMP.

PHOSPHATIDYLETHANOLAMINE

In eukaryotes, PE can be made either by decarboxylation of PS or through incorporation of ethanolamine into PE via the CDP-ethanolamine pathway.

Decarboxylation of Phosphatidylserine

As with other enzymes in the pathway for PC synthesis in yeast, PS decarboxylase is controlled by the levels of inositol and choline. In the presence of both these precursors, PS decarboxylase activity is repressed by 70–80% (63, 64). A gene encoding PS decarboxylase (*PSD1*) has been cloned from yeast by complementation of either *E. coli* (65) or yeast (66) mutants. The yeast *PSD1* gene encodes a protein with a molecular weight of 56,558 that is 35% identical to that from *E. coli*. PS decarboxylase from CHO cells has been cloned by complementation of the phosphatidylserine auxotroph, PSA-3, discussed earlier (53). The complementing cDNA, termed *pssC*, converts PSA-3 to prototrophy (67). PS synthase I activity remains low in cells expressing *pssC*, but PS decarboxylase activity is twice as high as normal; PS levels are normal and PE levels are 40% higher than normal. The *pssC* cDNA encodes a protein of about 370 residues (the N-terminal portion appears to be missing) that is 44% identical to PS decarboxylase of yeast (68). Expression of *pssC* in yeast under the GAL1 promoter results in a galactose-dependent increase in PS decarboxylase activity, indicating that *pssC* is the structural gene for this enzyme. The fact that increased decarboxylase activity can overcome a PS synthase I defect suggests that the residual PS synthase I activity in PSA-3 is inhibited by cellular PS, which is consistent with inhibition of this enzyme by PS in assays in vitro (69) and in vivo (70).

Surprisingly, disruption of the yeast *PSD1* gene for PS decarboxylase has little if any effect on phospholipid composition or growth rates although PS decarboxylase activity is decreased by at least 95% (65, 66). The residual PS decarboxylase activity, however, is hydroxylamine sensitive (66), consistent with the concept that the residual activity is truly due to a second PS decarboxylase. Considering the unusual location of PS decarboxylase in mitochondria (see below), it will be of considerable interest to learn the location of the putative second PS decarboxylase.

The CDP-Ethanolamine Pathway

This pathway is analogous to the major pathway for synthesis of PC, discussed below. In fact, considerable evidence indicates that the first enzymes of the pathways, ethanolamine kinase and choline kinase, are one and the same enzyme in yeast and mammals. For example, the two enzymatic activities from rat liver copurify to homogeneity and the activities from several rat tissues are quantitatively immunoprecipitated by the same titer of an antibody prepared against the purified enzyme (71). In yeast, expression of choline kinase in *E. coli* results in concomitant expression of ethanolamine kinase activity, indicating the same protein catalyzes the two activities (72). In addition, a mutation in the choline kinase gene results in reduced ethanolamine kinase activity (72).

On the other hand, choline kinase activity in the mutant is reduced to a greater extent than is ethanolamine kinase activity, leaving open the possibility that an additional protein catalyzes ethanolamine kinase activity. Ethanolamine kinase and choline kinase appear to be separate enzymes in some plants (73), insects (74), and protozoa (75).

CTP:phosphoethanolamine cytidylyltransferase has been purified to homogeneity from rat liver cytosol (76). The enzyme has a subunit molecular weight of 50,000 and fractionates as a soluble enzyme that is independent of lipids for activity. Although this enzyme appears soluble upon fractionation, it has a high content of hydrophobic amino acids (76), and ultrastructural studies indicate that the enzyme is enriched in the regions of the rough endoplasmic reticulum, although some of it also appears cytosolic (77). This suggests that the enzyme might have a weak affinity for membranes. Indeed, the phosphoethanolamine cytidylyltransferase in castor bean endosperm is membrane associated, primarily mitochondrial (78).

The last enzyme in the pathway, ethanolaminephosphotransferase, is predominantly located in the endoplasmic reticulum (35). The phosphotransferase has not been purified, but the enzyme from yeast has been cloned (79) and sequenced (80). The deduced gene product has a molecular weight of 44,525 and is predicted to contain seven transmembrane-spanning segments.

The Role of the CDP-Ethanolamine Pathway

In yeast the primary path of PE biosynthesis is through PS decarboxylation, and the CDP-ethanolamine pathway is auxiliary. In mammals, however, the contribution of the CDP-ethanolamine pathway has been called into question by the realization that the principal pathway for PE biosynthesis in several mammalian cell lines is via decarboxylation of PS (81, 82).

It would seem, then, that mammalian cells could get along well without the CDP-ethanolamine pathway. On the contrary, some mammalian cells require ethanolamine for growth (83, 84). What, then, is the role of the CDP-ethanolamine pathway in these cells? In human retinoblastoma cells, labeled ethanolamine and phosphate are incorporated into PE and ethanolamine plasmalogen at similar rates, while the rate of incorporation of labeled serine into ethanolamine plasmalogen is considerably lower than its rate of incorporation into PE (85). This suggests that the CDP-ethanolamine pathway may be required for biosynthesis of ethanolamine plasmalogen. In support of this, labeling of rat tissues in vivo showed that [³H]serine is incorporated into diacyl PS and diacyl PE but not ethanolamine plasmalogen, while [³H]ethanolamine is incorporated into PE as well as ethanolamine plasmalogen (86). Two rat mammary carcinoma cell lines, which differ from each other in whether or not they require ethanolamine for growth, were examined for possible differences in PE biosynthesis (84). Cells that require ethanolamine were found to be defective in their ability to incorporate exogenous ^{32}P-labeled PE, but not PC, into PS, suggesting that PS synthase II may be defective. Unfortunately, base-exchange activities were not measured in cell-free extracts. It appears, though, that the CDP-ethanolamine pathway may be used for the synthesis of distinct pools of phospholipid in some cells, and in others it may serve to augment PE synthesis if PS synthase II activities are low.

Transport of Phosphatidylserine and Phosphatidylethanolamine

A fascinating aspect of PE biosynthesis by decarboxylation is that the two participating enzymes have different subcellular locations. PS synthase is largely microsomal, presumably in the endoplasmic reticulum, whereas PS decarboxylase is predominantly in the inner mitochondrial membrane (88, 89). This presents questions as to how PS is transported to the mitochondrion, how the newly made PE is transported to its final cellular destination, and how these processes are regulated.

Systems used to study coupled PS transport and decarboxylation include intact cells (89), permeabilized cells (90, 91), and combinations of purified subcellular fractions (92–97). In none of the systems studied is there a requirement for a cytosolic protein or factor; thus, there appears to be no role of soluble phospholipid transfer proteins in PS transport to mitochondria. There is a requirement for energy for some unidentified steps: PS synthesis and transport are disrupted in intact cells by energy poisons (89), and permeabilized cells require ATP for both PS synthesis and translocation (90, 91). ATP does not play a direct role in PS synthesis and translocation but appears to be needed for sequestration of Ca^{2+} by the endoplasmic reticulum (91, 92, 94).

Once PS is in the mitochondrial outer membrane, transfer to the inner membrane does not require an energized membrane, a functioning respiratory chain, or exogenous ATP (96). It appears likely that transfer from the outer to inner membrane occurs through zones of adhesion: Purified zones of adhesion are quite efficient at synthesizing PE from PS (97). In addition, mixtures of mitoplasts from *chol* yeast cells, deficient in PS synthase, plus microsomes from wild-type cells are much less efficient at synthesizing PE. In this case, transfer through zones of adhesion is not possible; the only mechanism for transfer of PS from the wild-type microsomes to the mutant inner membranes is through collision.

Another exciting aspect of studies of PS transport is the discovery that mammalian cells have a membrane fraction, specialized in PS transport, that may be a unique organelle. Jean Vance discovered this membrane fraction while studying PS transport and decarboxylation in reconstituted systems from rat liver (93). She found that crude mitochondria are quite capable of incorporating [^{3}H]serine into PS, then PE, and even PC. (Incorporation of label into PC presumably occurs through transport of PE from mitochondria to the endoplasmic reticulum, the site of PE N-methyltransferase activity.) When the mitochondria are further purified, they lose all ability to incorporate [^{3}H]serine into lipid, consistent with the removal of membranes containing PS synthase. A fraction of membranes, called fraction X, that was removed from the mitochondria turns out to have higher specific activities of PS synthase and glucose-6-phosphatase than purified microsomes, and a similar level of PE N-methyltransferase. This implies that fraction X may be a specialized, mitochondrial-associated, endoplasmic reticulum–like organelle. In related work, a membrane fraction prepared from rat liver mitochondria is capable of incorporating [^{3}H]serine into PS and PE (95). If the fraction is prepared by mild sonication to remove the endoplasmic reticulum–like membrane, the resulting fraction is deficient at phospholipid synthesis. PE synthesis is restored by adding back endoplasmic reticulum, suggesting that the formation of the connection between mitochondria and the endoplasmic reticulum–like fraction may be reversible.

In a remarkable coincidence, Dennis Vance and his coworkers cloned a cDNA for a PE N-methyltransferase that appears to be localized to fraction X (98). An antibody to the methyltransferase binds to a protein with a subunit molecular weight of 20 kDa, the expected size calculated from the sequence. This protein is highly enriched in membrane fraction X but not found in the endoplasmic reticulum. This is the first protein known to be unique to this fraction.

PHOSPHATIDYLCHOLINE

Pathways

There are two pathways for PC biosynthesis (Figure 3), one by methylation of PE and the other by transfer of phosphocholine from CDP-choline to diacylglycerol. The former pathway predominates in yeast, with the CDP-choline pathway serving an auxiliary function. In mammals the CDP-choline pathway is the predominant pathway for PC biosynthesis in all tissues. Liver is the only tissue in which a significant amount of PC is made by methylation of PE (99). [Although phospholipid methyltransferase activity has been reported in nonhepatic tissues, the activity is extremely low relative to the activity of the CDP-choline pathway (100, 101).]

The CDP-Choline Pathway

Regulation of the CDP-choline pathway occurs through both of the first two steps in this pathway, choline kinase and CTP:phosphocholine cytidylyltransferase (3).

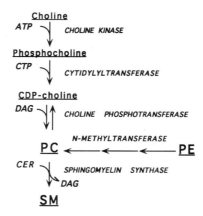

Figure 3 Pathways for biosynthesis of phosphatidylcholine and sphingomyelin. DAG, *sn*-1,2-diacylglycerol; CER, ceramide.

CHOLINE KINASE   This soluble enzyme has been purified to homogeneity from rat kidney (102), liver (71), and brain (103). The subunit molecular weights of the pure enzymes are reported to be 42,000 for kidney (102), 47,000 for liver (71), and 47,000 for brain (103). From gel filtration the brain enzyme appears to be a dimer (103) and the rat liver enzyme, a tetramer (71). Western blots with antibody to the rat liver enzyme detect both 47-kDa and 42-kDa forms in kidney, suggesting that multiple sizes of choline kinase exist in this tissue. Considerable additional evidence supports the concept that there are multiple isoforms of choline kinase: For example, multiple forms are seen upon isoelectric focusing and chromatofocusing (104, 71) and immunoprecipitation (104).

Yeast choline kinase was cloned by complementation of a choline kinase mutant (72). The deduced molecular weight of the protein encoded by this gene, called *CKI*, is 66,316. An isoform of rat liver choline kinase was cloned by screening an expression library with an antibody to the rat brain enzyme (105). The deduced molecular weight of this isoform is 49,743. Expression of both the yeast gene (72) and the rat cDNA (105) in *E. coli* results in production of choline kinase activity, demonstrating that these genes truly encode choline kinase. A second mammalian choline kinase has been cloned from human glioblastoma cells (106) and is highly similar to the rat liver form. The deduced molecular weight of the human form is 52,065 and contains two additional sequences not found in the rat. One of the extra peptides is 3 residues long and the other, 18 residues. The rest of the human and rat sequences are 90% identical. The two mammalian enzymes are not strikingly similar to the yeast enzyme, but show significant identity within three isolated regions (105).

Choline kinase activity appears to be correlated with mitogenesis. The first convincing demonstration that this enzyme is regulatory made use of mitogen-treated Swiss 3T3 cells: The twofold increased rate of phosphatidylcholine biosynthesis in the mitogen-treated cells is due to a similar increase in choline kinase activity (107, 108). A correlation of increased choline kinase activity with expression of the *ras* oncogene has been noted in murine fibroblasts (109, 110) and *Xenopus* oocytes (111), and this increase could be responsible for the elevated phosphocholine levels seen in cells expressing activated *ras* (112). Mitogenesis in response to PDGF is inhibited by hemicholinium-3, an inhibitor of choline kinase, suggesting that the elevation in choline kinase activity is important for PDGF-mediated signal transduction (113). In fact, phospho-choline itself has been reported to be mitogenic, although very high extracel-lular levels are required for the effect (113). It is hoped that the cloning of mammalian isoforms for choline kinase will lead to molecular elucidation of other isoforms, and further studies detailing which isoform may be important for mitogenesis.

CTP:PHOSPHOCHOLINE CYTIDYLYLTRANSFERASE   With respect to regulation of the pathway, the cytidylyltransferase (CT) has received much more attention than choline kinase since CT was recognized to be rate limiting (114) and regulatory (115) about 15 years ago. The first and only CT to be purified to homogeneity was that from rat liver (116, 117). CT from rat is a homodimer (118, 119), the subunit molecular weight of which is 42,000. The gene for CT from yeast (120) was cloned by complementation of a yeast mutant (121). A cDNA for rat liver CT was then cloned (122). The two sequences are 65% identical within a central region of 160 residues, and this segment was proposed to be a catalytic core (122) (Figure 4). Further support for the catalytic nature of this segment is the demonstration that a bacterial CTP:glycerol-3-phosphate cytidylyltransferase is significantly similar to this region (123, 124).

In addition to the sequences from yeast and rat liver, cDNAs for CT from Chinese hamster ovary cells (125), mouse testis (126), and HeLa cells (127) have been cloned and sequenced. The coding regions for mammalian CTs are highly conserved; the deduced protein sequence for the rat is 96–99% identical to those of the other species. No evidence for isoforms of CT has been observed.

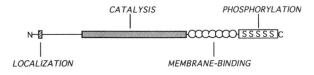

Figure 4   Functional regions of phosphocholine cytidylyltransferase.

CT has long been known to reside in two distinct intracellular pools: soluble and membrane-associated. The soluble enzyme is activated by certain lipids in vitro, while the membrane-associated enzyme is active in the absence of added lipids. In many cells, activation of CT occurs concomitantly with translocation of the soluble to the membrane form (3, 4). This shift suggests that the soluble pool is a reservoir of inactive enzyme, ready to become rapidly activated. Activation of CT does occur without appreciable translocation to membranes, however (128, 129), suggesting that multiple mechanisms regulate CT activity.

In order to reach a full understanding of the regulatory mechanisms that control the activity of CT, it is necessary to identify the cellular locations of the soluble and membrane-associated enzymes. The soluble enzyme has long been assumed to be cytosolic, because it is readily released from cells by permeabilization. The membrane-associated enzyme has been localized by gradient sedimentation studies to the endoplasmic reticulum (130) and Golgi apparatus (131, 132). The concept that CT is cytoplasmic, however, has been called into question by a series of experiments that indicate the enzyme is nuclear in several cell lines. That the membrane-associated enzyme is associated with the nuclear envelope was assessed by immunofluorescence microscopy (133, 134) and Percoll gradient sedimentation (135). The soluble enzyme was demonstrated to be nuclear by immunofluorescence microscopy and enucleation (136).

The nuclear localization signal in mammalian CT has been localized to residues 8–28 (137). This contains a stretch of five basic residues, ^{12}RKRRK, which is similar to a canonical nuclear targeting sequence (138) but which alone cannot function as a targeting signal. Deletion of these five residues from the entire targeting signal, however, renders β-galactosidase cytoplasmic. Yeast CT is membrane associated (139), but the organelle with which it is associated has not been reported.

The way in which CT interacts with cellular membrane in vivo and lipids in vitro is of considerable interest. There are no obvious transmembrane segments in the CT sequence, but then a truly transmembrane protein would not be expected to reversibly associate with membranes as does CT. Cornell and coworkers noted that mammalian CT contains an 80-residue stretch, just C-terminal to the catalytic domain (Figure 4), that is predicted to form two amphipathic helices interrupted by a turn, and they proposed that this is the site of membrane binding (122). Evidence to support this concept comes from studies employing limited proteolysis of CT, in which progressive digestion of CT from the C-terminus does not interfere with binding of CT to lipid vesicles until the putative amphipathic helix is removed (140). Although the nature of the binding site on the membrane has not yet been revealed, it has been proposed that CT binds to membranes enriched in nonbilayer-forming lipids (141). Whether lipid alone or a specific protein-protein interaction, e.g. to a CT receptor, is involved in binding of CT to membranes remains to be determined.

A major goal of research in this area is to know how the activity of CT is modulated. The possibility that CT is regulated by phosphorylation was suggested by indirect evidence, including the ability of cAMP analogues to inhibit (142) and phorbol esters to stimulate (143, 144) PC synthesis. CT was first demonstrated to be a phosphoprotein in vivo by immunoprecipitation from ^{32}P-labeled HeLa cell extracts (128). Surprisingly, however, no change in the phosphorylation state of CT occurs in response to cAMP (145, 146) and phorbol esters (128). On the other hand, treatment of Chinese hamster ovary cells with phospholipase C (147) and HeLa cells with fatty acids (134), both of which cause nearly complete translocation of the enzyme, promote extensive dephosphorylation. Furthermore, choline depletion, which causes translocation of CT in HepG2 cells, converts highly phosphorylated CT to a form that is less phosphorylated (148). Activation and translocation of CT by phospholipase C is inhibited by okadaic acid, suggesting that dephosphorylation is important for the change in enzyme activity and location (147). In fact, okadaic acid itself reduces the amount of membrane-associated CT in hepatocytes (149) and increases the level of phosphorylated enzyme (150), consistent with the importance of the phosphorylation state for membrane association. On the other hand, the changes in activity and location in response to fatty acids in the HeLa system are not inhibited by okadaic acid (134), suggesting that different mechanisms may be responsible for dephosphorylating CT in the different systems.

The sites phosphorylated in soluble CT have been determined by expressing CT in a baculovirus system (151, 152), purifying ^{32}P-labeled CT from this system, and subjecting radiolabeled phosphopeptides to chemical sequencing (153). CT is exclusively phosphorylated on Ser residues (128, 153) and all 16 Ser residues from S315 to the C-terminus are phosphorylated, with the possible exception of S339. There are seven Ser-Pro sequences in this region, suggesting phosphorylation by a proline-directed kinase such as MAP kinase or cdc2 kinase. The latter possibility is especially interesting because CT activity has recently been shown to change with the cell cycle, consistent with modulation by a cyclin-dependent kinase (154).

The question as to how a cell would handle an excess of CT is important for understanding the mechanisms by which it regulates the CDP-choline pathway. In yeast, overexpression of yeast CT results in highly elevated levels of CDP-choline but no increase in the rate of PC synthesis (120). It is possible that CT is not rate limiting in yeast, or that the supply of diacylglycerol is also limiting, as has been observed in some mammalian cells (146, 155). Two mammalian overexpression systems have been studied with rather different results. In one system, rat liver CT is expressed in a Chinese hamster ovary cell mutant that has low levels of a temperature-sensitive CT (125, 156). When overexpressed by about eightfold over wild-type levels, the extra CT does not result in increased CDP-choline levels or rates of PC synthesis (125). This is consistent with the concept that most of CT in normal cells is in an inactive reservoir and that the level of active CT is tightly controlled. Overexpression of rat liver CT in COS cells, however, does result in increased PC synthesis as well as markedly increased levels of glycerophosphocholine, indicating an increased rate of PC degradation occurs to maintain constant PC levels (157). Thus the tightness of control of CT activity appears to vary considerably from cell to cell. The situation in CHO cells reflects that most often found when enzymes of lipid metabolism are overexpressed: Changes, if any, in rates of lipid synthesis are far lower than anticipated from changes in levels of enzymes (67, 120, 158–160)

DIACYLGLYCEROL:CHOLINEPHOSPHOTRANSFERASE This tightly membrane-bound enzyme is in the endoplasmic reticulum (35, 161). It has not yet been purified to homogeneity from any source. A genetic approach was taken to clone the gene for the yeast enzyme: A mutant strain was isolated by screening for clones defective in cholinephosphotransferase activity, and the gene, termed *CPT1*, was cloned by complementation of the mutant (162). The gene encodes a 407-residue protein that is highly hydrophobic (163) and is 55% identical to the sequence of yeast ethanolaminephosphotransferase (80). Hydropathy analysis predicts seven transmembrane domains for both these proteins. The high degree of similarity between ethanolamine- and cholinephosphotransferase allowed for construction of chimeric proteins, analysis of which resulted in mapping certain functional regions of the proteins (164). The region responsible for CDP-aminoalcohol specificity is about 110 residues in the cytoplasmic segment between the first and second transmembrane segments. The first three transmembrane segments control the diacylglycerol acyl chain specificity.

Methylation of Phosphatidylethanolamine

In yeast, the major pathway of PC biosynthesis is the methylation of PE. Two methyltransferases have been cloned by genetic complementation (165). The *PEM1* gene encodes a protein with a subunit molecular weight of 101,202. Its gene product is membrane associated and catalyzes the methylation of PE, but not the further methylation of PME nor PDME. The other gene, *PEM2*, encodes a smaller protein (M_r = 23,150). The *PEM2* gene product, also membrane associated, catalyzes the methylation of PE, PME, and PDME. Because of these different substrate specificities, the authors refer to the gene product of *PEM1* as a PE methyltransferase and the product of *PEM2* as a phospholipid methyltransferase. The sequence of the *PEM2* gene product has a low but significant similarity (24% identity) to a portion near the amino terminus of the *PEM1* gene product. The *pem1pem2* double mutant requires choline for growth, as does a mutant missing *PEM2*. Cells missing the *PEM1* gene can grow without choline, however, indicating that the *PEM2* methyltransferase can carry out all three methylations in vivo. The role of *PEM1* in the cell is not clear, except that the specific activity of this enzyme is much higher than the activity of *PEM2* with PE.

There also appear to be at least two mammalian methyltransferases. Two such enzymes could be distinguished in rat pituitary extracts, one enzyme preferring PE and the other more active with PME and PDME (166). A 25-kDa form (167) and an 18-kDa form (168) have been purified from rat liver microsomes. These forms are similar to the smaller yeast enzyme in that they catalyze the methylation of all three substrates.

A cDNA for a phospholipid methyltransferase has been cloned from rat liver (98). The molecular weight of the encoded protein is 22,300, and the expressed protein product catalyzes the conversion of PE to PC, so it appears similar to yeast *PEM2* in function and size. This cDNA is 44% identical to yeast *PEM2* in deduced amino acid sequence. This methyltransferase is found only in liver and does not appear to be the major PE *N*-methyltransferase in rat liver, but it is specific for the so-called fraction X discussed previously. The cloning of the other mammalian methyltransferase(s) will, therefore, be of considerable interest.

SPHINGOMYELIN

Biosynthesis of this lipid is closely tied to that of PC, since the phosphocholine moiety of sphingomyelin (SM) is donated directly from PC by the SM synthase reaction (Figure 4). For the past 13 years the location of SM synthase has been the subject of some dispute, with a number of researchers reporting that it is in the plasma membrane and others finding it in the Golgi apparatus (see citations in 169). Papers still appear that seem to demonstrate convincingly that the enzyme is nearly exclusively in either the Golgi apparatus (169) or plasma membrane (170) or, even, both organelles (171). It is reasonable to propose that there are at least two isozymes of SM synthase, one in the Golgi apparatus and another in the plasma membrane, and that their relative levels vary considerably from one cell to the next.

PHOSPHATIDYLINOSITOL

There has been an enormous amount of interest recently in the pathways of phosphoinositide metabolism, in light of the importance of these lipids in signal transduction. Phosphatidylinositol-4,5-bisphosphate (PI-4,5-P_2) is the substrate for phospholipase C that is activated in response to a wide variety of hormonal stimuli. The diacylglycerol produced acts as a second messenger to activate protein kinase C while the other product, inositol-1,4,5-trisphosphate, causes release of Ca^{2+} from internal stores. Furthermore, 3-phosphorylated derivatives of phosphoinositides are of great interest because the enzyme responsible for their synthesis, PI 3-kinase, is associated with a number of tyrosine-phosphorylated receptors (172) and with the *ras* oncogene (173). It would take several reviews to cover all the interesting recent developments in phosphoinositide metabolism; this review focuses only on the biosynthesis of these lipids.

Pathways

PI-4,5-P_2 is formed by the sequential phosphorylation of PI, first by a PI 4-kinase, then by PI-4-P 5-kinase (Figure 5). The pathway of synthesis of 3-phosphorylated derivatives of PI is not yet clear. PI 3-kinase can use PI, PI-4-P, or PI-4,5-P_2 as substrates (174), so the possibility exists that the phosphate at position 3 is added either before or after PI is phosphorylated at position 4. Measurement of the specific radioactivities of the individual phosphate moieties of PI-3,4,5-P_3 in NIH 3T3 cells stimulated with platelet-derived growth factor in the presence of ^{32}Pi supports a pathway in which PI is first phosphorylated by PI 3-kinase before being further phosphorylated at positions 4 and 5 (175). The same approach in another laboratory using Swiss 3T3 cells stimulated by the same growth factor, however, supports the addition of the phosphate to position 3 after one or both of the other positions have been phosphorylated (176). On the other hand, the identification of PI-3-P 4-kinase activity in human erythrocytes (177) and platelets (178) further supports the pathway in which the 3-phosphate is added first.

It is of interest that the PI-specific phospholipases C that cleave PI-4,5-P_2 do not hydrolyze 3-phosphorylated derivatives of PI (179). In fact, it is not yet known which of the 3-phosphorylated derivatives of PI are important for signal transduction.

Phosphatidylinositol Synthase

The first enzyme of phosphoinositide synthesis is PI synthase (Figure 5). PI synthase of yeast has been reported to be in both microsomes and the outer

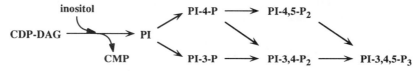

Figure 5 Possible pathways of polyphosphoinositide biosynthesis. CDP-DAG, CDP-diacylglycerol.

mitochondrial membrane (38). Considerable enzyme activity has also been found in a fraction of secretory vesicles (180), suggesting the presence of the enzyme in plasma membranes. In mammals, PI synthase exists in both endoplasmic reticulum and plasma membranes in pituitary tumor cells (181) and in kidney (182). Enzymatic activities in the two membrane fractions differ markedly in their activation by Mn^{2+} (181) and extractability by high salt (183), suggesting that the enzymes in the different fractions are dissimilar.

PI synthase was first purified to homogeneity from yeast (184). The molecular weight of the major protein band on SDS gels is 34,000. A gene for yeast PI synthase, *PIS*, was cloned by complementing a mutant defective in this activity; the deduced molecular weight of the enzyme is 24,823 (185). Disruption of the *PIS* gene is lethal. That this gene actually encodes PI synthase was confirmed by its expression in *E. coli* under an inducible promoter (186), which resulted in inducible PI synthase activity as measured in crude extracts. The difference in subunit molecular weights between the purified and cloned enzymes suggests that these two are different isozymes.

A mammalian PI synthase has been purified from human placenta to a very high specific activity of 19–35 μmol/min/mg (187). The purified fraction shows a predominant band on SDS gels of 24,000, but this band has not yet been convincingly demonstrated to be PI synthase. PI synthase from rat liver, previously reported to be highly purified, appeared to have only one predominant band with a molecular weight of 60,000 (188), but the specific activity of this purified fraction was less than 0.1 % of that purified from human placenta.

A cDNA for a human PI 4-kinase has been obtained by searching DNA banks for sequences similar to PI 3-kinase, then cloning and expressing such a sequence (205a). This deduced protein sequence of this kinase, called PI4Kα, is 854 residues long and is 44% identical to the yeast *SST4* kinase. The human PI 4-kinase also shares similarity to mammalian and yeast PI 3-Kinases.

Phosphatidylinositol 4-Kinase

The next enzyme in the conversion of PI to PI-4,5,-P_2 is PI 4-kinase. A number of PI 4-kinases have been identified, purified, and cloned. In yeast, PI 4-kinase activity can be detected in microsomes (189), plasma membranes (190), and cytosol (191). Two forms of the microsomal kinase have been purified. The first has a subunit molecular weight of 45,000 (189, 192) and the second has a subunit molecular weight of 55,000 (193). The two forms are distinct in several properties, including amino acid compositions, isoelectric points, peptide maps, and reactivity with an antibody to the 45-kDa form. The two forms also appear to be differentially regulated in the cell (190, 194).

A soluble PI 4-kinase has been highly purified from yeast; enzyme activity is associated with a protein whose subunit molecular weight is 125,000 (195).

Activity is stimulated by Triton X-100, as are the activities of several mammalian PI 4-kinases. The *PIK1* gene encoding this enzyme has been cloned (196). The deduced molecular weight of the encoded subunit is 119,929. A mutant in which the *PIK1* gene is disrupted is not viable, indicating that this soluble kinase plays a vital role in this organism. The *PIK1* gene is highly similar in certain regions to P110, a PI 3-kinase from mammals (197), and Vps34p, a PI 3-kinase from yeast (198). The PI 4-kinase encoded by the *PIK1* gene is a nuclear protein (199), suggesting that this kinase participates in a nuclear PI cycle.

An additional PI 4-kinase was found to be the product of the *SST4* gene, a defect in which results in resistance to the protein kinase inhibitor, staurosporine (200). Cell homogenates of the *sst4* mutant have greatly decreased PI 4-kinase activity. The deduced molecular weight of the protein encoded by the *SST4* gene is 214,605, considerably larger than any of the other phosphoinositide kinases sequenced so far. Within a 400 residue stretch the *SST4* kinase is about 27% identical to the PI 3-kinases, yeast Vps34p and mammalian p110. The sequences of the two cloned yeast PI 4-kinases are dissimilar except for about 180 residues at the C-terminal, in which they are 44% identical, suggesting that this C-terminal portion is the catalytic domain. It will be of considerable interest to determine the function of the rest of this large protein.

Mammalian PI 4-kinase has been purified to apparent homogeneity from an acetone powder of bovine uterus (201), A431 cell membranes (202), and porcine liver microsomes (203). In all cases the subunit molecular weight is 55,000. The purified enzyme is a monomer associated with a micelle of Triton X-100 (202). Another isoform of PI 4-kinase has been identified in bovine uterus (204) and bovine brain (205). This isoform is distinguished from the 55-kDa form by its larger size upon gel filtration and its decreased sensitivity to adenosine. PI-3-P 4-kinases have been identified in human erythrocytes (177) and platelets (178) but have not yet been purified to homogeneity.

Phosphatidylinositol 3-Kinase

PI 3-kinase has been purified to apparent homogeneity from rat liver cytosol (174). The molecular weight of the undenatured enzyme is 190,000, and two bands (M_r = 110,000 and 85,000) are seen on SDS gels. This PI 3-kinase can use PI, PI-4-P, or PI-4,5-P as substrates. The 190-kDa form of PI 3-kinase has also been purified from bovine thymus cytosol, as well as a form whose native and denatured molecular weight is 110,000 (206). This smaller form is identical to the 110-kDa subunit of the 190-kDa form by peptide mapping, indicating that the 110-kDa band is the catalytic subunit. It is interesting that the specific activity of the smaller form is five times greater than that of the larger form, suggesting that the 85-kDa subunit may have an inhibitory function.

That the 110-kDa band is the catalytic unit of PI 3-kinase was confirmed by the cloning of this subunit from bovine brain and its expression as an active enzyme in a baculovirus system (197). The protein sequence is very similar to that of the yeast protein, Vps34p, a 100-kDa protein involved in protein sorting (198). Vps34p was also demonstrated to have PI 3-kinase activity (207). A cDNA for an additional human 110-kDa isoform has been identified as being only 42% identical to the bovine isoform (208), so there may be multiple PI 3-kinase isoforms that interact with an 85-kDa subunit. The 85-kDa subunit of PI 3-kinase functions in linking the catalytic subunit to receptors and related proteins (209–211). The catalytic 110-kDa subunit, however, interacts directly with the *ras* oncogene protein (173).

Phosphatidylinositol Phosphate 5-Kinase

Human red blood cells have both soluble and membrane-associated PI-4-P 5-kinase activity. A membrane-associated form was isolated by salt extraction of membranes and highly purified (212). SDS gel analysis of the purified preparation shows a prominent band at 53 kDa. This band copurifies with PI-4-P 5-kinase activity, and an antibody prepared to the 53-kDa band isolated from gels immunoprecipitates kinase activity. An additional form of PI-4-P 5-kinase in human blood cells can be distinguished from the 53-kDa form immunologically (213). This may be the same as a form purified from bovine erythrocytes membranes, for which the subunit molecular weight appears to be 68,000 (214). Clearly the multiplicity of isoforms in this pathway provides opportunities for diverse modes of regulation.

PHOSPHATIDYLGLYCEROL AND CARDIOLIPIN

Cardiolipin or diphosphatidylglycerol in eukaryotes is found exclusively in the inner mitochondrial membrane, and its biosynthesis is also mitochondrial. The first unique step in the synthesis of this lipid is the conversion of CDP-diacylglycerol and *sn*-glycerol-3-phosphate to phosphatidylglycerolphosphate, followed by hydrolysis of the phosphomonoester to form PG. In eukaryotes PG reacts with an additional molecule of CDP-diacylglycerol to form cardiolipin, while in *E. coli*, cardiolipin is formed by the condensation of two molecules of PG. In yeast, phosphatidylglycerolphosphate synthase, the first step in this pathway, is repressed by growing cells in the presence of inositol (215), but phosphatidylglycerolphosphate phosphatase (216) and cardiolipin synthase (217) are not repressed by inositol.

While the enzymes of cardiolipin synthesis have neither been purified nor cloned from any eukaryotic source, a genetic approach to the pathway has been initiated in CHO cells. A mutant with defective phosphatidylglycerolphosphate synthase activity was isolated by screening for colonies with defective activity of this enzyme (218). Enzymatic activity is more thermolabile in the mutant than in wild-type cells, and the mutant cells grow more slowly at 40°C. A revertant was isolated in which both phosphatidylglycerolphosphate synthase activity and growth at 40°C are restored, indicating that both defects are due to a single mutation. The rate of synthesis of PG is extremely low in the mutant, and the rate of synthesis of cardiolipin is also reduced. In mutant cells grown at 40°C, PG levels are reduced to less than 10% of normal, and cardiolipin levels are about 30% of normal.

The effects of the mutation in phosphatidylglycerolphosphate synthase on mitochondrial structure and function have been analyzed (219). Ultrastructural studies show that cells grown at 40°C, containing low levels of cardiolipin, have mitochondria that are swollen and much less electron-dense than normal. ATP production and oxygen consumption are reduced by about 40% at 40°C, and the activity of the respiratory chain, as indicated by rotenone-sensitive NADH oxidase activity, is reduced by 60%. When individual components of the respiratory chain are analyzed, there is an 80% reduction in activity of Complex I, in which electrons are transferred from NADH to ubiquinone. The other components of the respiratory chain appear normal, so cardiolipin appears to be needed specifically by complex I. This mutant is unable to grow at 40°C when galactose is substituted for glucose, making the cell more dependent on respiration. Such a conditional-lethal phenotype should allow the cloning of the mammalian phosphatidylglycerol phosphate synthase gene.

CONCLUDING COMMENTS

Rapid progress has been made recently in purifying and cloning enzymes of phospholipid biosynthesis, and this progress will continue. The next few years are expected to be rich in the use of molecular techniques to determine how these enzymes function in the cell. The use of genetic approaches with yeast and mammalian cells promises to yield insights into the roles of these enzymes and pathways that we currently do not even imagine. The work of Bankaitis and coworkers (220) on the interactions of the pathways of PC synthesis and secretion is an example. These approaches as well as immunological studies (221) promise to reveal regulatory proteins that interact with and control the enzymes of phospholipid biosynthesis. Furthermore, we can expect that the functions of phospholipids in signal transduction pathways will continue to expand and that the roles played by distinct isozymes in those pathways will become clear.

ACKNOWLEDGMENTS

Work in this laboratory is supported by grants from the National Cancer Institute and the American Cancer Society. I would like to thank George Carman for helpful input regarding yeast phospholipid pathways.

Literature cited

1. Carman GM, Henry SA. 1989. *Annu. Rev. Biochem.* 58:635–69
2. Kennedy EP. 1986. In *Lipids and Membranes: Past, Present, and Future*, ed. JAF Op den Kamp, B Roelofsen, KWA Wirtz, pp. 171–206. Amsterdam: Elsevier
3. Kent C. 1990. *Progr. Lipid Res.* 29:87–105
4. Tijburg LBM, Geelen MJH, van Golde LMG. 1989. *Biochim. Biophys. Acta* 1004:1–19
5. Voelker DR. 1991. *Microbiol. Rev.* 55:543–60
6. Shin D-H, Paulauskis JD, Moustaid N, Sul HS. 1991. *J. Biol. Chem.* 266:23834–39
7. Coleman RA, Reed BC, Mackall JC, Student AK, Lane MD, Bell RM. 1978. *J. Biol. Chem.* 253:7256–61
8. Kuri-Harcuch W, Green H. 1977. *J. Biol. Chem.* 252:2158–60
9. Yet SF, Lee S, Hahm YT, Sul HS. 1993. *Biochemistry* 32:9486–91
10. Saggerson ED, Carpenter CA, Cheng CH, Sooranna SR. 1980. *Biochem J.* 190:183–89
11. Tillman TS, Bell RM. 1986. *J. Biol. Chem.* 261:9144–49
12. Paulauskis JD, Sul HS. 1988. *J. Biol. Chem.* 263:7049–54
13. Lightner VA, Bell RM, Modrich P. 1983. *J. Biol. Chem.* 258:10856–61
14. Wolter FP, Schmidt R, Heinz E. 1992. *EMBO J.* 111:4685–92
15. Fritz PJ, Kauffman JM, Robertson CA, Wilson MR. 1986. *J. Biol. Chem.* 261:194–99
16. Ishizaki O, Nishida I, Agata K, Eguchi G, Murata N. 1988. *FEBS Lett.* 238:424–30
17. Weber S, Wolter F-P, Buck F, Frentzen M, Heinz E. 1991. *Plant Mol. Biol.* 17:1067–76
18. Bertrams N, Heinz E. 1981. *Plant Physiol.* 68:653–57
19. Nishida I, Tasaka Y, Shiraishi H, Murata N. 1993. *Plant Mol. Biol.* 21:267–77
20. Minskoff SA, Racenis PV, Granger J, Larkins L, Hajra AK, Greenberg ML. 1994. *J. Lipid Res.* In press
21. Bell RM, Coleman RA. 1980. *Annu. Rev. Biochem.* 49:459–87
22. Hosaka K, Yamashita S. 1984. *Biochim. Biophys. Acta* 796:102–9
23. Brindley DN, Hubscher G. 1965. *Biochem. Biophys. Acta* 106:495–509
24. Kent C, Vagelos PR. 1976. *Biochim. Biophys. Acta* 436:377–86
25. Jamal Z, Martin A, Gomez-Munoz A, Brindley DN. 1991. *J. Biol. Chem.* 266:2988–96
26. Day CP, Yeaman SJ. 1992. *Biochim. Biophys. Acta* 1127:87–94
27. Martin-Sanz P, Hopewell R, Brindley DN. 1984. *FEBS Lett.* 175:284–88
28. Hopewell R, Martin-Sanz P, Martin A, Saxton J, Brindley DN. 1985. *Biochem. J.* 232:485–91
29. Aridor-Piterman O, Lavie Y, Liscovitch M. 1992. *Eur. J. Biochem.* 204:561–68
30. Perry DK, Stevens VL, Widlanski TS, Lambeth JD. 1993. *J. Biol. Chem.* 268:25302–10
31. Lin YP, Carman GM. 1989. *J. Biol. Chem.* 264:8641–45
32. Morlock KR, McLaughlin JJ, Lin YP, Carman GM. 1991. *J. Biol. Chem.* 266:3586–93
33. Quinlan JJ, Nickels JT Jr, Wu WI, Lin YP, Broach JR, Carman GM. 1992. *J. Biol. Chem.* 267:18013–20
34. Kanoh H, Imai S, Yamada K, Sakane F. 1992. *J. Biol. Chem.* 267:25309–14
35. Jelsema CL, Morré DJ. 1978. *J. Biol. Chem.* 253:7960–71
36. Hostetler KY, Van den Bosch H. 1972. *Biochim. Biophys. Acta* 260:380–86
37. Mok AY, McDougall GE, McMurray WC. 1992. *FEBS Lett.* 312:236–40
38. Kuchler K, Daum G, Paltauf F. 1986. *J. Bacteriol.* 165:901–10
39. Kelley MJ, Carman GM. 1987. *J. Biol. Chem.* 262:14563–70
40. Klig LS, Homann MJ, Kohlwein SD, Kelley MJ, Henry SA, Carman GM. 1988. *J. Bacteriol.* 170:1878–86
41. Heacock AM, Seguin EB, Agranoff BW. 1993. *J. Neurochem.* 60:1087–92
42. Stubbs EB, Agranoff BW. 1993. *J. Neurochem.* 60:1292–99
43. Stubbs EB Jr, Walker BA, Owens CA, Ward PA, Agranoff BW. 1992. *J. Immunol.* 148:2242–47
44. Voelker DR, Frazier JL. 1986. *J. Biol. Chem.* 261:1002–8

45. Kuge O, Nishijima M, Akamatsu Y. 1986. *J. Biol. Chem.* 261:5795–98
46. Atkinson KD, Jensen B, Kolat AI, Storm EM, Henry SA, Fogel S. 1980. *J. Bacteriol.* 141:558–64
47. Kovac L, Gbelska I, Poliachova V, Subik J, Kovacova V. 1980. *Eur. J. Biochem.* 111:491–501
48. Nikawa J-I, Yamashita S. 1981. *Biochim. Biophys. Acta* 665:420–26
49. Hubscher G, Dils RR, Pover WFR. 1959. *J. Biol. Chem.* 36:518–28
50. Borkenhagen LF, Kennedy EP, Fielding L. 1961. *J. Biol. Chem.* 236:PC28–29
51. Suzuki TT, Kanfer JN. 1985. *J. Biol. Chem.* 260:1394–99
52. Kuge O, Nishijima M, Akamatsu Y. 1985. *Proc. Natl. Acad. Sci. USA* 82:1926–30
53. Kuge O, Nishijima M, Akamatsu Y. 1986. *J. Biol. Chem.* 261:5790–94
54. Siddiqui RA, Exton JH. 1992. *J. Biol. Chem.* 267:5755–61
55. Kuge O, Nishijima M, Akamatsu Y. 1991. *J. Biol. Chem.* 266:24184–89
56. Poole MA, Homann MJ, Bae-Lee MS, Carman GM. 1986. *J. Bacteriol.* 168:668–72
57. Kodaki T, Nikawa J-I, Hosaka K, Yamashita S. 1991. *J. Bacteriol.* 173:792–95
58. Kodaki T, Hosaka K, Nikawa J-I, Yamashita S. 1991. *J. Biochem.* 109:276–87
59. Kelley MJ, Bailis AM, Henry SA, Carman GM. 1988. *J. Biol. Chem.* 263:18078–85
60. Bae-Lee M, Carman GM. 1990. *J. Biol. Chem.* 265:7221–26
61. Kinney AJ, Carman GM. 1988. *Proc. Natl. Acad. Sci. USA* 85:7962–66
62. Kinney AJ, Bae-Lee M, Panghaal SS, Kelley MJ, Gaynor PM, Carman GM. 1990. *J. Bacteriol.* 172:1133–36
63. Overmeier JH, Waechter CJ. 1991. *Arch. Biochem. Biophys.* 290:5511–16
64. Lamping E, Kohlwein SD, Henry SA, Paltauf F. 1991. *J. Bacteriol.* 173:6432–37
65. Clancey CJ, Chang SC, Dowhan W. 1993. *J. Biol. Chem.* 268:24580–90
66. Trotter PJ, Pedretti J, Voelker DR. 1993. *J. Biol. Chem.* 268:21416–24
67. Kuge O, Nishijima M, Akamatsu Y. 1991. *J. Biol. Chem.* 266:6370–76
68. Li Q, Dowhan W. 1988. *J. Biol. Chem.* 263:11516–22
69. Hasegawa K, Kuge O, Nishijima M, Akamatsu Y. 1989. *J. Biol. Chem.* 264:19887–92
70. Nishijima M, Kuge O, Akamatsu Y. 1986. *J. Biol. Chem.* 263:5784–88
71. Porter TJ, Kent C. 1990. *J. Biol. Chem.* 265:414–22
72. Hosaka K, Kodaki T, Yamashita S. 1989. *J. Biol. Chem.* 264:2053–59
73. Macher BA, Mudd JB. 1976. *Arch. Biochem. Biophys.* 177:24–30
74. Ramabrahmam P, Subrahmanyam D. 1981. *Arch. Biochem. Biophys.* 207:55–62
75. Ancelin ML, Vial HJ. 1986. *FEBS Lett.* 202:217–23
76. Vermeulen PS, Tijburg LBM, Geelen MJH, van Golde LMG. 1993. *J. Biol. Chem.* 268:7458–64
77. van Hellemond J, Slot JW, Geelen MJH, van Golde LMG, Vermeulen PS. 1994. *J. Biol. Chem.* 269:15415–18
78. Wang XM, Morres TS Jr. 1991. *J. Biol. Chem.* 266:19981–87
79. Hjelmstad RH, Bell RM. 1988. *J. Biol. Chem.* 263:19748–57
80. Hjelmstad RH, Bell RM. 1991. *J. Biol. Chem.* 266:5094–103
81. Voelker DR. 1984. *Proc. Natl. Acad. Sci. USA* 81:2669–73
82. Miller MA, Kent C. 1986. *J. Biol. Chem.* 261:9753–61
83. Murakami H, Masuei H, Sato GH, Sueoka N, Chow TP, Kano-Sueoka T. 1982. *Proc. Natl. Acad. Sci. USA* 79:1158–62
84. Kano-Sueoka T, King DM. 1987. *J. Biol. Chem.* 262:6074–81
85. Yorek MA, Rosario RT, Dudley DT, Spector AA. 1985. *J. Biol. Chem.* 269:2930–36
86. Arthur G, Page L. 1991. *Biochem. J.* 273:121–25
87. Baranska J. 1982. *Adv. Lipid. Res.* 19:163–84
88. Dennis EA, Kennedy EP. 1972. *J. Lipid Res.* 13:263–67
89. Voelker DR. 1985. *J. Biol. Chem.* 260:14671–76
90. Voelker DR. 1989. *Proc. Natl. Acad. Sci. USA* 86:9921–25
91. Voelker DR. 1990. *J. Biol. Chem.* 265:14340–46
92. Voelker DR. 1989. *J. Biol. Chem.* 264:8019–25
93. Vance JE. 1990. *J. Biol. Chem.* 265:7248-56
94. Vance JE. 1991. *J. Biol. Chem.* 266:89–97
95. Ardail D, Gasnier F, Lermé F, Simonot D, Louisot P, Gateau-Roesch O. 1993. *J. Biol. Chem.* 268:25985–92
96. Simbeni R, Paltauf F, Daum G. 1990. *J. Biol. Chem.* 265:281–85
97. Simbeni R, Pon L, Zinser E, Paltauf F, Daum G. 1991. *J. Biol. Chem.* 266:10047–49
98. Cui Z, Vance JE, Chen MH, Voelker

DR, Vance DE. 1993. *J. Biol. Chem.* 268:16655–63
99. Ridgeway ND. 1989. In *Phosphatidylcholine Metabolism*, ed. DE Vance, pp. 103–20. Boca Raton, FL: CRC
100. Blusztajn JK, Wurtman RJ. 1981. *Nature* 290:417–18
101. Yang EK, Blusztajn JK, Pomfret EA, Zeisel SH. 1988. *Biochem. J.* 256:821–28
102. Ishidate K, Nakagomi K, Nakazawa Y. 1985. *J. Biol. Chem.* 259:14706–10
103. Uchida T, Yamashita S. 1990. *Biochim. Biophys. Acta* 1043:281–88
104. Tadokoro K, Ishidate K, Nakazawa Y. 1985. *Biochim. Biophys. Acta* 835:501–13
105. Uchida T, Yamashita S. 1992. *J. Biol. Chem.* 267:10156–62
106. Hosaka K, Tanaka S, Nikawa J-I, Yamashita S. 1992. *FEBS Lett.* 304:209–32
107. Warden CH, Friedkin M. 1984. *Biochim. Biophys. Acta* 792:270–80
108. Warden CH, Friedkin M. 1985. *J. Biol. Chem.* 260:6006–11
109. Macara IG. 1989. *Mol. Cell. Biol.* 9:325–28
110. Teegarden D, Taparowsky EJ, Kent C. 1990. *J. Biol. Chem.* 265:6042–47
111. Lacal JC. 1990. *Mol. Cell. Biol.* 10:333–40
112. Kent C. 1994. In *Lipobiology*, ed. RW Gross. Greenwich, CT: JAI. In press
113. Cuadrado A, Carnero A, Dolfi F, Jimenez B, Lacal JC. 1993. *Oncogene* 8:2959–68
114. Vance DE, Trip EM, Paddon HB. 1980. *J. Biol. Chem.* 255:1064–69
115. Sleight RG, Kent C. 1980. *J. Biol. Chem.* 255:10644–50
116. Weinhold PA, Rounsifer ME, Feldman DA. 1986. *J. Biol. Chem.* 261:5104–10
117. Feldman DA, Weinhold PA. 1987. *J. Biol. Chem.* 262:9075–81
118. Weinhold PA, Rounsifer ME, Charles L, Feldman DA. 1989. *Biochim. Biophys. Acta* 1006:299–310
119. Cornell R. 1989. *J. Biol. Chem.* 264:9077–82
120. Tsukagoshi Y, Nikawa J, Yamashita S. 1987. *Eur. J. Biochem.* 169:477–85
121. Nikawa J, Yonemura K, Yamashita S. 1983. *Eur. J. Biochem.* 131:223–29
122. Kalmar GB, Kay RJ, Lachance A, Aebersold R, Cornell RB. 1990. *Proc. Natl. Acad. Sci. USA* 87:6029–33
123. Mauel C, Young M, Karamata D. 1991. *J. Gen. Microbiol.* 137:929–41
124. Park YS, Sweitzer TD, Dixon JE, Kent C. 1993. *J. Biol. Chem.* 268:16648–54
125. Sweitzer TD, Kent C. 1994. *Arch. Biochem. Biophys.* 311:107–16

126. Rutherford MS, Rock CO, Jenkins NA, Gilbert DJ, Tessner TG, et al. 1993. *Genomics* 18:698–701
127. Kalmar GB, Kay RJ, LaChance A, Cornell RB. 1994. *Biochim. Biophys. Acta*. In press
128. Watkins JD, Kent C. 1990. *J. Biol. Chem.* 265:2190–97
129. Weinhold PA, Charles L, Rounsifer ME, Feldman D. 1991. *J. Biol. Chem.* 266:6093–100
130. Terce F, Record M, Ribbes G, Chap H, Douste-Blazy L. 1988. *J. Biol. Chem.* 263:3142–49
131. Higgins JA, Fieldsend JK. 1987. *J. Lipid Res.* 28:268–78
132. Vance JE, Vance DE. 1988. *J. Biol. Chem.* 263:5898–909
133. Watkins JD, Kent C. 1992. *J. Biol. Chem.* 267:5686–92
134. Wang Y, MacDonald JIS, Kent C. 1993. *J. Biol. Chem.* 268:5512–18
135. Morand JN, Kent C. 1989. *J. Biol. Chem.* 264:13785–92
136. Wang Y, Sweitzer TD, Weinhold PA, Kent C. 1993. *J. Biol. Chem.* 268:5899–904
137. Wang Y, Kent JC. 1995. *J. Biol. Chem.* 270:354–60
138. Garcia-Bustos J, Heitman J, Hall MN. 1991. *Biochim. Biophys. Acta* 1071:83–101
139. Tsukagoshi Y, Nikawa J-I, Hosada K, Yamashita S. 1991. *J. Bacteriol.* 173:2134–36
140. Craig L, Johnson JE, Cornell RB. 1994. *J. Biol. Chem.* 269:3311–17
141. Jamil H, Hatch GM, Vance DE. 1993. *Biochem. J.* 291:419–27
142. Pelech SL, Pritchard PH, Vance DE. 1981. *J. Biol. Chem.* 256:8283–86
143. Guy GR, Murray AW. 1982. *Cancer Res.* 42:1980–85
144. Kinzel V, Kreibich G, Hecker E, Suss R. 1979. *Cancer Res.* 39:2743–50
145. Watkins JD, Wang Y, Kent C. 1992. *Arch. Biochem. Biophys.* 292:360–67
146. Jamil H, Utal AK, Vance DE. 1992. *J. Biol. Chem.* 267:1752–60
147. Watkins JD, Kent C. 1991. *J. Biol. Chem.* 266:21113–17
148. Weinhold PA, Charles L, Feldman DA. 1994. *Biochim. Biophys. Acta* 1210:335–47
149. Hatch GM, Tsukitani Y, Vance DE. 1991. *Biochim. Biophys. Acta* 1081:25–32
150. Hatch GM, Jamil H, Utal Ak, Vance DE. 1992. *J. Biol. Chem.* 267:15751–58
151. MacDonald JIS, Kent C. 1993. *Protein Expr. Purif.* 4:1–7
152. Luche MM, Rock CO, Jackowski S.

1993. *Arch. Biochem. Biophys.* 301: 114–18
153. MacDonald JIS, Kent C. 1994. *J. Biol. Chem.* 269:10529–37
154. Jackowski S. 1994. *J. Biol. Chem.* 269: 3858–67
155. Lim P, Cornell R, Vance DE. 1986. *Biochim. Biophys. Acta* 64:692–98
156. Esko JD, Wermuth MM, Raetz CRH. 1981. *J. Biol. Chem.* 256:7388–93
157. Walkey CJ, Kalmar GB, Cornell RB. 1994. *J. Biol. Chem.* 269:5742–49
158. Ohta A, Waggoner K, Louie K, Dowhan W. 1981. *J. Biol. Chem.* 256:2219–25
159. Ohta A, Waggoner K, Radomiska-Pyrek S, Dowhan W. 1981. *J. Bacteriol.* 147: 552–62
160. Hiraoka S, Nukui K, Uetake N, Ohta A, Shibuya I. 1991. *J. Biochem.* 110: 443–49
161. Braell WA. 1988. *Anal. Biochem.* 170: 328–34
162. Hjelmstad RH, Bell RM. 1987. *J. Biol. Chem.* 262:3909–17
163. Hjelmstad RH, Bell RM. 1990. *J. Biol. Chem.* 265:1755–64
164. Hjelmstad RH, Morash SC, McMaster CR, Bell RM. 1994. *J. Biol. Chem.* 269:20995–1002
165. Kodaki T, Yamashita S. 1987. *J. Biol. Chem.* 262:15428–35
166. Prasad C, Edwards RM. 1981. *J. Biol. Chem.* 256:13000–3
167. Pajares MA, Villalba M, Mato JM. 1986. *Biochem. J.* 237:699–705
168. Ridgeway ND, Vance DE. 1987. *J. Biol. Chem.* 262:17231–39
169. Futerman AH, Stieger B, Hubbard AL, Pagano RE. 1990. *J. Biol. Chem.* 265: 8650–57
170. van Helvoort A, van't Hof W, Ritsema T, Sandra A, van Meer G. 1994. *J. Biol. Chem.* 269:1763–69
171. Kallen KJ, Quinn P, Allan D. 1994. *Biochim. Biophys. Acta* 1166:305–8
172. Cantley LC, Auger KR, Carpenter C, Duckworth B, Graziani A, et al. 1991. *Cell* 64:281–302
173. Rodriguez-Viciana P, Warne PH, Dhand R, Vanhaesebroeck B, Gout I, et al. 1994. *Nature* 370:527–32
174. Carpenter CL, Duckworth BC, Auger KR, Cohen B, Schaffhausen BS, Cantley LC. 1990. *J. Biol. Chem.* 265:19704–11
175. Cunningham TW, Majerus PW. 1991. *Biochem. Biophys. Res. Commun.* 175: 568–76
176. Hawkins PT, Jackson TR, Stephens LR. 1992. *Nature* 358:157–59
177. Graziani A, Ling LE, Endemann G, Carpenter CL, Cantley LC. 1992. *Biochem. J.* 284:39–45
178. Yamamoto K, Graziani A, Carpenter C,

Cantley LC, Lapetina EG. 1990. *J. Biol. Chem.* 265:22086–89
179. Serunian LA, Haber MT, Fukui T, Kim JW, Rhee SG, et al. 1989. *J. Biol. Chem.* 264:17809–15
180. Kinney AJ, Carman GM. 1990. *J. Bacteriol.* 172:4115–17
181. Imai A, Gershengorn MC. 1987. *Nature* 325:726–28
182. Galvao C, Shayman JS. 1990. *Biochim. Biophys. Acta* 1044:34–42
183. Cubitt AB, Gershengorn MC. 1989. *Biochem. J.* 257:639–44
184. Fischl AS, Carman GM. 1983. *J. Bacteriol.* 154:304–11
185. Nikawa J-I, Kodaki T, Yamashita Y. 1987. *J. Biol. Chem.* 262:4876–81
186. Nikawa J-I, Kodaki T, Yamashita S. 1988. *J. Bacteriol.* 170:4727–31
187. Antonsson BE. 1994. *Biochem. J.* 297: 517–22
188. Takenawa T, Egawa K. 1977. *J. Biol. Chem.* 252:5419–23
189. Belunis CJ, Bae-Lee M, Kelley MJ, Carman GM. 1988. *J. Biol. Chem.* 263: 18897–903
190. Nickels JT Jr, Buxeda RJ, Carman GM. 1994. *J. Biol. Chem.* 269:11018–24
191. Talwalkar RT, Lester RL. 1974. *Biochim. Biophys. Acta* 360:306–11
192. Buxeda RJ, Nickels JT Jr, Belunis CJ, Carman GM. 1991. *J. Biol. Chem.* 266: 13859–65
193. Nickels JT Jr, Buxeda RJ, Carman GM. 1992. *J. Biol. Chem.* 267:16297–303
194. Buxeda RJ, Nickels JT Jr, Carman GM. 1993. *J. Biol. Chem.* 268:6248–55
195. Flanagan CA, Thorner J. 1992. *J. Biol. Chem.* 267:24117–25
196. Flanagan CA, Schnieders EA, Emerick AW, Kunisawa R, Admon A, Thorner J. 1993. *Science* 262:1444–48
197. Hiles ID, Otsu M, Volinia S, Fry MJ, Gout I, et al. 1992. *Cell* 70:419–29
198. Herman PK, Emr SD. 1990. *Mol. Cell. Biol.* 10:6742–54
199. Garcia-Bustos JF, Marini F, Stevenson I, Frei C, Hall MN. 1994. *EMBO J.* 13:2352–61
200. Yoshida S, Ohya Y, Goebl M, Nakano A, Anraku Y. 1994. *J. Biol. Chem.* 269:1166–71
201. Porter FD, Li YS, Deuel TF. 1988. *J. Biol. Chem.* 263:8989–95
202. Walker DH, Dougherty N, Pike LJ. 1988. *Biochemistry* 27(17):6504–11
203. Hou WM, Zhang ZL, Tai HH. 1988. *Biochim. Biophys. Acta* 959:67–75
204. Li YS, Porter FD, Hoffman RM, Deuel TF. 1989. *Biochem. Biophys. Res. Commun.* 160:202–9
205. Endemann G, Dunn SN, Cantley LC. 1987. *Biochemistry* 26:6845–52

205a. Wong K, Cantley LC. 1994. *J. Biol. Chem.* 269:28878–84
206. Shibasaki F, Homma Y, Takenawa T. 1991. *J. Biol. Chem.* 266:8108–14
207. Schu PV, Takegawa K, Fry MJ, Stack JH, Waterfield MD, Emr SD. 1993. *Science* 260:88–91
208. Hu P, Mondino A, Skolnik EY, Schlessinger J. 1993. *Mol. Cell Biol.* 13:7677–88
209. Escobedo JA, Navankasattusas S, Kavanaugh WM, Milfay D, Fried VA, Williams LT. 1991. *Cell* 65:75–82
210. Skolnik EY, Margolis B, Mohammadi M, Lowenstein E, Fischer R, et al. 1991. *Cell* 65:83–90
211. Otsu M, Hiles I, Gout I, Fry MJ, Ruiz-Larrea F, et al. 1991. Cell 65:91–104
212. Ling LE, Schulz JT, Cantley I.C. 1989. *J. Biol. Chem.* 264:5080–88
213. Bazenet CE, Ruano AR, Brockman JL, Anderson RA. 1990. *J. Biol. Chem.* 265:18012–22
214. Jenkins GH, Fisette PL, Anderson RA. 1994. *J. Biol. Chem.* 269:11547–54
215. Greenberg ML, Hubbell S, Lam C. 1988. *Mol. Cell Biol.* 8:4773–79
216. Kelly BL, Greenberg ML. 1990. *Biochim. Biophys. Acta* 1046:144–50
217. Tamai KT, Greenberg ML. 1990. *Biochim. Biophys. Acta* 1046:214–22
218. Ohtsuka T, Nishijima M, Akamatsu Y. 1993. *J. Biol. Chem.* 268:22908–13
219. Ohtsuka T, Nishijima M, Suzuki K, Akamatsu Y. 1993. *J. Biol. Chem.* 268:22914–19
220. McGee TP, Skinner HB, Whitters EA, Henry SA, Bankaitis VA. 1994. *J. Cell Biol.* 124:273–87
221. Feldman DA, Weinhold PA. 1993. *J. Biol. Chem.* 268:3127–35

Annu. Rev. Biochem. 1995. 64:345–73

TRANSCRIPTIONAL REGULATION OF GENE EXPRESSION DURING ADIPOCYTE DIFFERENTIATION[1]

Ormond A. MacDougald and M. Daniel Lane

Department of Biological Chemistry, The Johns Hopkins University School of Medicine, Baltimore, Maryland 21205

KEY WORDS: differentiation, preadipocyte, C/EBP, adipose tissue

CONTENTS

[1]ABBREVIATIONS: Abbreviations used in this review include: SCD1, stearoyl-CoA desaturase I; GLUT4, insulin-sensitive glucose transporter; MIX, methylisobutylxanthine; LPL, lipoprotein lipase; $PGF_{2\alpha}$, prostaglandin $F_{2\alpha}$; PGI_2, prostaglandin I2 (or prostacyclin); IGF, insulin-like growth factor; FGF, fibroblast growth factor; PDGF, platelet-derived growth factor; G3PDH, glycerol-3-phosphate dehydrogenase; C/EBP, CCAAT/enhancer binding protein; TNF, tumor necrosis factor; ARE, adipose regulatory element; ARF, adipose regulatory factor; PRE, preadipocyte repressor element; CUP, C/EBP undifferentiated protein; ADD, adipose determination differentiation factor; DBP, D-site binding protein; Ig/EBP, immunoglobulin/enhancer binding protein; LAP, liver activator protein; LIP, liver inhibitory protein; PPAR, peroxisome proliferator-activated receptor; FAAR, fatty acid-activated receptor; RAR, retinoicacid receptor; RXR, retinoid x receptor.

345

0066-4154/95/0701-0345$05.00

ABSTRACT

Cell culture models (e.g. 3T3-L1 cells) have been developed for studying the process of adipocyte differentiation. Differentiation can be induced by adding insulin-like growth factor I, glucocorticoid, fatty acids, and an agent that increases intracellular cAMP level. The adipocyte differentiation program is regulated by transcriptional activators such as CCAAT/enhancer binding proteinα (C/EBPα), peroxisomal proliferator activated receptor γ2 (PPARγ2), fatty acid activated receptor (FAAR), and transcriptional repressors such as preadipocyte repressor element binding protein (PRE) and C/EBP undifferentiated protein (CUP). These transcription factors coordinate the expression of genes involved in creating and maintaining the adipocyte phenotype including the insulin-responsive glucose transporter (GLUT4), stearoyl CoA desaturase 1 (SCD1), and the fatty acid binding protein (422/aP2).

PERSPECTIVES AND SUMMARY

Adipocytes serve an important function in the energy economy of vertebrate organisms by providing a massive energy reserve that can be mobilized upon demand. Thus, when caloric intake exceeds expenditure, metabolic flux is directed into pathways leading to triacylglycerol synthesis for storage in the adipocyte. Conversely, when caloric expenditure exceeds intake, this energy reserve is mobilized to meet the deficiency and to provide needed physiological fuel, i.e. free fatty acids, for other cell types. These functions are the raison d'être of the adipocyte.

During development the adipocyte acquires the full complement of enzymes and accessory proteins with which to carry out and regulate these lipogenic and lipolytic functions (1). The regulatory proteins include the receptors and second messenger systems necessary to confer responsiveness to the hormones that regulate these processes, e.g. insulin, which promotes lipogenesis, and counter-regulatory hormones such as glucagon, ACTH, epinephrine, and glucocorticoids, which promote lipolysis. Although adipose tissue appears late in embryonic development, major expansion of the adipocyte population occurs only after birth. This expansion coincides with the need for an energy reserve enabling the newborn to survive in a new environment in which the intervals between nutrient intake may be lengthy.

The adipose lineage arises relatively late in the developmental pathway from a multipotent stem cell of mesodermal origin that also gives rise to the muscle and cartilage lineages. The availability of immortalized cell lines (notably the 10T1/2 stem cell line and the 3T3 and Ob1771 preadipocyte lines), which represent different stages of adipocyte development, has greatly facilitated characterization of the differentiation program and the identification of many cis-regulatory elements and trans-acting factors involved in the coordinate expression of adipocyte genes during differentiation. The first recognizable, though poorly understood, step in the program is commitment of the multipotent stem cell to the adipocyte lineage. Although it seems likely that activation of a specific gene or genes underlies commitment to the adipocyte lineage, this gene has been elusive. Following commitment to the adipose lineage, quiescent preadipocytes become susceptible to the complement of exogenous inducers, i.e. IGF-1, glucocorticoid, and an agent that increases the intracellular level of cAMP, which trigger subsequent differentiation events. Upon induction, the cells undergo several rounds of mitotic clonal expansion and then again become quiescent as the coordinate transcriptional activation of adipocyte genes is initiated. Expression of these genes is accompanied by dramatic biochemical and morphological changes that give rise to the adipocyte phenotype. If analogous to myogenic differentiation, a single (or a limited number of) differentially expressed transcription factor(s) would be sufficient to coordinately activate adipocyte genes. CCAAT/enhancer binding protein α (C/EBPα) possesses many of the characteristics required of such a "master regulator," e.g. in cell culture models, C/EBPα coordinately activates transcription of many adipocyte genes. Indeed, evidence shows that C/EBPα is not only required but is sufficient to trigger differentiation of preadipocytes without use of exogenous hormonal inducers (2, 3). A critical piece of evidence that removes all doubt concerning the role of C/EBPα in adipocyte development in vivo was recently provided by gene knockout experiments in the mouse (4, 5). Thus, disruption of the C/EBPα gene by homologous recombination gave rise to mice that failed to develop adipose tissue shortly after birth.

C/EBPα appears to play other roles in adipocyte differentiation as well. C/EBPα is believed to terminate the mitotic clonal expansion preceding entry into the terminally differentiated state (6) and appears to maintain the fully differentiated state by autoactivating transcription of its own gene (7). Finally, C/EBPα appears to be part of a system that senses and integrates nutrient/hormonal signals in response to changes in the energy status of the animal. It was recently found, for example, that the expression of C/EBPα, and thereby certain adipocyte genes, is regulated by glucocorticoid and insulin in the fully differentiated adipocyte (8, 9).

Members of the steroid/thyroid receptor gene family of transcription factors, e.g. PPARγ2, FAAR, and the RXRs, also appear to play an important role in preadipocyte differentiation (10–12). PPARγ2 binds to several adipose response elements (ARE) within the adipose-specific enhancer of the 422/aP2 gene (10). Of particular interest, a long-chain fatty acid, linoleic acid, induces reporter gene activity driven by promoters containing multiple AREs in cells transfected with PPARγ2. The fatty acid-ARE interaction may have broader implications, since fatty acids have been identified as inducers of preadipocyte gene expression and differentiation in certain preadipose cell contexts. The fact that both the 422/aP2 and PEPCK genes possess functional C/EBP and PPARγ2 cis-acting elements raises the question: Do these transcription factors function dependently or independently to activate and sustain coordinate adipocyte-specific gene transcription during adipocyte differentiation?

Important inroads have been made in our attempts to understand the terminal stages of adipocyte differentiation; however, little progress has been made in understanding the process of determination, which is a fertile area for future research. This review covers our present knowledge of the factors and second messenger pathways involved in inducing adipocyte differentiation, the sequence of events in the differentiation program, and the transcription factors that have been implicated as regulators of these processes. The reader is referred to several previous reviews (1, 13–15) that deal with the broader aspects of adipocyte differentiation and development.

CELL CULTURE MODELS

Two types of cell lines, representative of cells arrested at different stages of adipocyte development, have served as useful models: 1. multipotent stem cell lines (such as the 10T1/2 and Balbc/3T3 lines) that have not undergone commitment to the adipocyte lineage, but have the capacity to do so, and 2. preadipocyte cell lines (such as the 3T3-L1, 3T3-F442A and Ob1771 lines) that have undergone commitment and when appropriately induced, differentiate into adipocytes. It is likely that further subdivision of these two types of cell lines will be possible as markers for other checkpoints in the developmental pathways are identified.

Studies with the 10T1/2 stem cell line have proven particularly instructive for understanding certain aspects of the determination process. When exposed to 5-azacytidine, an inhibitor of DNA methylation, these multipotent cells undergo determination (commitment) giving rise to stable preadipocyte, pre-myocyte, and prechondrocyte cell lines (16). Transfection of 10T1/2 cells with large segments of hypomethylated genomic DNA from 5-azacytidine-treated cells leads to myogenic conversion at a frequency consistent with expression of a single gene as the cause of commitment to the myogenic lineage (17).

These findings led to the isolation of the MyoD gene and later to other transcription factors in the MyoD gene family that when transfected into 10T1/2 cells activate the expression of muscle genes (reviewed in 18). While commitment to the adipocyte lineage might be expected to follow this precedent, this approach has not as yet proven successful in the search for the gene(s) that controls commitment to this lineage.

By definition, preadipocytes have passed through the commitment checkpoint in development and, when appropriately stimulated, differentiate into cells possessing the biochemical and morphological phenotype of adipocytes. The most extensively characterized preadipocyte cell lines are the 3T3-L1, 3T3-F442A, and Ob1771 lines. The 3T3-L1 and 3T3-F442A cell lines were derived from disaggregated mouse embryo cells and were selected for their propensity to accumulate triacylglycerol lipid droplets (19–21). Ob17 cells were derived from the stroma of epididymal fat pads from adult *ob/ob* genetically obese mice (22), and several sublines, notably the Ob1771 line, were selected. While these and other preadipocyte cell lines exhibit many similar characteristics, their responsiveness to agents that induce/ repress differentiation differs considerably. These differences are most likely due to differences in the stage at which preadipocyte development was arrested during cloning, presumably representing early, intermediate, and late stages of preadipocyte differentiation.

The most compelling evidence that preadipocyte cell lines represent faithful models of preadipocyte differentiation in vivo comes from transplantation studies. When 3T3-F442A preadipocytes were injected subcutaneously into Balb-C athymic mice, normal fat pads developed at the site of injection within 5 weeks (23). In addition, a considerable body of evidence shows that fully differentiated "adipocytes" derived from the above-mentioned preadipocyte lines faithfully mimic the metabolism of adipocytes isolated from adipose tissue. Extensive biochemical analysis has revealed that the accumulation of cytoplasmic triacylglycerol is closely correlated with the coordinate expression of virtually every enzyme of the pathways of de novo fatty acid and triacylglycerol biosynthesis (see Table 1 in Ref. 1). Preadipocytes in culture also acquire the proteins necessary for lipolysis of triacylglycerol, uptake and intracellular translocation of fatty acids, and regulation of these processes by lipogenic and lipolytic hormones (24–31). Finally, detailed electron micrographic studies have verified that mature 3T3-L1 adipocytes possess the ultrastructural features of adipocytes in situ (32).

INDUCERS AND SECOND MESSENGER PATHWAYS

External inducers are required for the induction of differentiation of preadipocytes, although the complement of inducers needed differs somewhat

with different cell lines. Conditions have been developed for inducing differentiation of 3T3-L1, 3T3-F442A, 1246, Ob1771, and primary stromal preadipocytes in the absence of serum (33–38). This advance has made it possible
to assess the requirements for differentiation using chemically defined cell
culture media. Three modulators, i.e. IGF-1, glucocorticoid, and cAMP (or
agents that can serve as substitutes), are generally considered necessary for
the induction of preadipocyte differentiation either in serum-containing or
serum-free medium. In addition, recent evidence indicates that long-chain fatty
acids function as inducers of preadipocyte differentiation, particularly for
certain preadipocyte lines (39).

These and other findings have implicated four second-messenger signaling
pathways for the induction of adipocyte differentiation (Figure 1). These include: 1. the IGF-1-activated tyrosine kinase pathway, 2. the glucocorticoid/
prostacyclin pathway, 3. the cAMP-dependent protein kinase pathway, and 4.
the fatty acid activated receptor pathway. It should be emphasized that the
requirements and activation patterns vary somewhat among cell lines, presumably because of differences in the stage of development at which the different
lines are arrested. Other agents have been reported to influence adipocyte
differentiation (reviewed in 1).

IGF-1

Early studies identified insulin (at high concentration) and growth hormone as
inducers of preadipocyte differentiation (25, 40, 41). However, it was later
discovered that both hormones exerted their effects on differentiation through
the IGF-1 receptor (42) and thus could be replaced by physiological concentrations of IGF-1. Preadipocytes possess a large number of IGF-1 receptors,
but few insulin receptors. Moreover, insulin can bind to the IGF-1 receptor,
but only at nonphysiologically high concentrations, and therefore can mimic

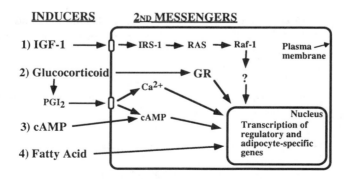

Figure 1   Inducers of preadipocyte differentiation and their second messenger pathways.

IGF-1. On the other hand growth hormone functions through an autocrine or paracrine mechanism in preadipocytes by activating expression and secretion of IGF-1 (43–45). Recognition that IGF-1 is an authentic inducer of pre-adipocyte differentiation indicates the involvement of a tyrosine kinase–mediated signaling pathway in the differentiation process, since the IGF-1 receptor is a ligand-dependent tyrosine kinase.

Several other lines of evidence have implicated tyrosine kinase–mediated signaling in preadipocyte differentiation. Overexpression of phosphotyrosine phosphatase (PTPase) 1B by 3T3-L1 preadipocytes blocks differentiation into adipocytes (46; and K Liao & MD Lane, unpublished data). This blockade can be reversed and the capacity to differentiate restored by exposing the cells to vanadate (a potent inhibitor of PTPase action) at the time they are treated with differentiation inducers, including an IGF-1 agonist. The tyrosine kinase in-volved may be the IGF-1 receptor tyrosine kinase. Evidence from Santos' laboratory (47, 48) has implicated the IGF-1 receptor tyrosine kinase (i.e. insulin at high concentration acting through the IGF-1 receptor; see above) in the activation of the Ras signaling pathway (Figure 1), and thereby in differ-entiation of 3T3-L1 preadipocytes. Expression of a transfected Ras oncogene (constitutively active) was sufficient to induce preadipocyte differentiation in the absence of IGF-1 or insulin (47). Presumably, phosphorylation of IRS-1 (insulin receptor substrate-1), a proximal target of the IGF-1 receptor (49, 50), triggers the activation of Ras by recruitment of a GTP exchange promoting protein, such as SOS, via SH_2/SH_3 adapters (51). Additional evidence indicates that the differentiation signal from activated Ras is mediated by the cytosolic serine/threonine protein kinase, Raf-1, since expression of a Raf-1 oncogene (constitutively active) is sufficient to induce adipocyte differentiation, albeit to a lesser extent than with external differentiation inducers (52). Although expression of a dominant negative raf-1 markedly reduced expression of an adipocyte gene (i.e. the *422/aP2* gene), the lack of a complete blockade suggests that a parallel pathway may also be required for maximal stimulation of preadipocyte differentiation (52). While this work implicates Ras and Raf-1 as signal mediators in the differentiation pathway initiated by IGF-1, the events beyond Raf-1 have yet to be elucidated.

Glucocorticoid and cAMP

Glucocorticoid has long been recognized as an inducer of differentiation of the 3T3 preadipocyte lines in serum-containing media (25). More recent studies show that a chemically defined (serum-free) medium containing transferrin, fetuin, growth hormone, triiodothyronine, and high concentrations of insulin (or physiological concentrations of insulin and IGF-1) is sufficient to support expression of early adipocyte markers, such as lipoprotein lipase and the α2 chain of collagen type VI (53, 54), as Ob1771 preadipocytes achieve growth arrest at confluence. However, an additional mitogenic/adipogenic agent such as glucocorticoid, arachidonic acid, or PGI_2 (prostacyclin) is required for the progression of these preadipocytes through clonal expansion and terminal differentiation (55–58). It has since been determined that glucocorticoids activate phospholipase A2, which causes the release of arachidonic acid for prostaglandin (specifically PGI_2 and $PGF_{2\alpha}$) synthesis. An increase in PGI_2 leads to increases in the intracellular levels of cAMP and Ca^{2+}, both of which appear to be required for differentiation of Ob1771 preadipocytes (59). Although glucocorticoid gives rise to an increased cAMP concentration sufficient to induce differentiation of Ob1771 cells, this is not the case for certain other preadipocyte lines. For example, 3T3-L1 preadipocytes require both glucocorticoid and cAMP (i.e. dibutyryl-cAMP, forskolin or MIX, a cAMP phosphodiesterase inhibitor) for maximal differentiation.

Fatty Acids

Fatty acids were first found to induce the expression of the 422/aP2 gene in Ob1771 preadipocytes (60–62). More recent evidence suggests that long-chain saturated fatty acids can induce differentiation of these cells (39). Thus, exposure of growth-arrested cells to palmitic acid for 3 days promotes postconfluent mitosis, accumulation of triacylglycerol, and the induction of several adipocyte markers including glycerol-3-phosphate dehydrogenase and acyl-CoA synthase. Induction of differentiation by fatty acid does not require its further metabolism since 2-bromopalmitate, a nonmetabolizable derivative, also stimulates the adipogenic program (39). Like most differentiation inducers, fatty acids function synergistically with other agents, such as growth hormone or methylisobutylxanthine. The ability of saturated fatty acids to induce both the mitotic clonal expansion and differentiation of preadipocytes has also been investigated in vivo. Shillabeer & Lau (63) found that feeding rats a diet containing highly saturated fat caused a threefold increase in the number of retroperitoneal adipocytes compared to the number of adipocytes in animals fed a diet high in polyunsaturated fat.

Recent evidence suggests that activation of expression of the 422(aP2) gene and of preadipocyte differentiation may be mediated by receptors of the PPAR (peroxisome proliferation activator receptor) class which are members of the steroid/thyroid hormone receptor gene family (10–12). Both naturally occurring long-chain polyunsaturated fatty acids, such as 5,8,11,14-eicosatetraynoic and arachidonic acids, as well as synthetic hydrophobic peroxisome proliferating agents, e.g. clofibrate, bezafibrate and WY-14643, activate the PPARs. Of interest in this connection, fatty acids and peroxisome proliferators activate PPARγ2 which binds to and regulates enhancer sequences in the 422(aP2) gene that confer adipose-specific expression (10).

Another fatty acid activated receptor (FAAR), related to the PPARs, was recently sequenced (11). Preliminary results indicate that FAAR is an early adipocyte marker as its expression is induced at the time Ob1771 preadipocytes achieve confluence. Furthermore, when the FAAR gene is transfected into 3T3-C2 fibroblasts, the cells acquire the capacity to undergo fatty acid-induced expression of the endogenous 422/aP2 gene. While further work is needed to understand the role and mechanism of fatty acids in preadipocyte differentiation, there is a compelling teleological basis for their involvement. An abundant supply of fatty acids might be expected to stimulate clonal expansion and differentiation of preadipocytes to accommodate their storage in adipose depots. A large body of evidence indicates that adipocyte cell hyperplasia and hypertrophy are associated with obesity.

SEQUENCE OF EVENTS IN THE DIFFERENTIATION PROGRAM

Through analysis of adipocyte development with multipotent stem cell and preadipocyte model systems, as well as in developing adipose tissue, it has been possible to identify key events in the development process. The sequence of events in the program of adipocyte development is described in this section, the key features of which are illustrated in Figure 2.

Determination

Treatment of multipotent 10T1/2 stem cells with 5-azacytidine causes hypomethylation of genomic DNA, which appears to activate regulatory genes that commit the cells to one of several lineages, i.e. the adipocyte, myocyte, or chondrocyte lineage (16). Through analysis of this process, it was discovered that conversion of 10T1/2 cells to the myocyte lineage is associated with the activation of the genes encoding muscle-specific transcription factors such as MyoD, myogenin, myf5, and mrf4. While it was originally concluded that activation of the MyoD gene (or its family members) by hypomethylation committed 10T1/2 cells to the myocyte lineage, there is now some uncertainty as to whether hypomethylation of this gene(s) per se was responsible for this commitment (reviewed in 18, 64). It has been suggested that members of the MyoD gene family may instead represent muscle-specific transcription factors that participate at various stages of the differentiation process, and that hypomethylation may have activated a determination gene that lies upstream of the MyoD gene family in the pathway of muscle development.

Although transcription factors that commit stem cells to the adipocyte lineage have not been identified, recent work in Freytag's laboratory (3) suggests that "commitment" of the multipotent Balb/c 3T3 cells to the preadipocyte lineage can be effected by the enforced expression of C/EBPα. Unlike most

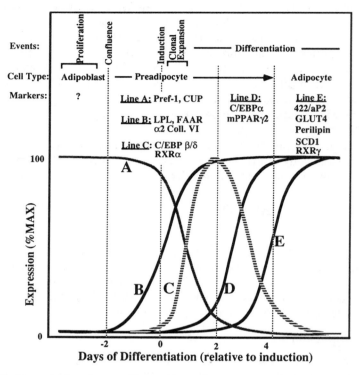

Figure 2 Stages in the adipocyte development program illustrating the events and temporal expression of various preadipocyte and adipocyte markers.

cell lines, Balb/c 3T3 cells can tolerate high levels of C/EBPα during cell proliferation, perhaps due to the concommitant expression of a dominant negative factor (e.g. C/EBPβ/LIP, CHOP/gadd153), which could block the antimitogenic action of C/EBPα. By analogy, elevated levels of MyoD, which is also antimitogenic, are also tolerated by some cell lines due to the presence of Id, a dominant negative helix-loop-helix factor (18, 64). Although Balb/c 3T3 cells expressing C/EBPα do not differentiate spontaneously, ectopic expression of C/EBPα appears to create a permissive state from which the cells can undergo differentiation when treated with exogenous inducing agents. As C/EBPα is expressed in a number of other tissues, e.g. liver, intestine, and lung, it seems unlikely that expression of this transcription factor alone can be responsible for adipose-specific determination.

Cell/Cell Contact at Confluence

Cell/cell contact appears to be a prerequisite for subsequent preadipocyte differentiation. In cell culture as preadipocytes arrest at the G_0/G_1 cell cycle boundary at confluence, they begin to express the "early" markers of adipocyte differentiation, including lipoprotein lipase (65), the mouse equivalent of the human $\alpha 2$ chain of Type VI collagen (65), and FAAR (fatty acid-activated receptor) (11; see below). Induction of these early markers is most likely mediated through an autocrine/paracrine mechanism(s) that is activated by cell/cell contact. Cell/cell contact prior to mitotic clonal expansion, the next step in the program, appears to be essential for preadipocyte differentiation. The importance of cell surface protein-protein and/or protein-carbohydrate interactions between cells or with the extracellular matrix is emphasized by other findings (66). For example, the use of fibronectin as a substratum for 3T3-L1 preadipocytes interferes with their differentiation (67).

Preadipocyte Factor 1 (Pref-1) is a cell-surface protein that appears to play a role in maintaining preadipocytes in the undifferentiated state (68, 69). It is a highly glycosylated transmembrane protein, which possesses six tandem epidermal growth factor–like repeats. Expression of Pref-1 is repressed when 3T3-L1 preadipocytes are induced to differentiate. Moreover, the constitutive expression of a transfected Pref-1 gene prevents differentiation (68). Proteins containing EGF-like repeats, such as Notch and Delta in *Drosophila,* have been observed across broad phylogenetic boundaries as important for directing cell fate decisions. Cells expressing Dorsal or Notch communicate with each other through protein-protein interactions between their extracellular EGF-like domains. Pref-1 may act similarly to promote cell/cell communication and thereby arrest cells at the preadipocyte stage until expression of Pref-1 ceases. Alternatively, the extracellular domain of Pref-1, which contains the EGF-like repeats, may be cleaved by a protease, thereby releasing the EGF-like domain for autocrine or paracrine function (68).

Mitotic Clonal Expansion

Following cell/cell interaction and growth arrest at confluence, preadipocytes must receive the appropriate combination of mitogenic and adipogenic signals (provided by external modulators; see Inducers and Second Messenger Pathways) to proceed through the required mitotic clonal expansion and subsequent differentiation. During clonal expansion ex vivo (in cultured cells; 30) and in vivo (70), preadipocytes undergo several rounds of mitosis. Certain mitogens, e.g. FGF and PDGF, stimulate postconfluent mitosis but do not induce differentiation. Thus, DNA replication per se appears to be necessary but not sufficient for induction of differentiation. Presumably, DNA replication and the accompanying changes in chromatin structure increase the accessibility of cis-elements to trans-acting factors which activate (or derepress) transcription of the gene(s) that gives rise to the adipocyte phenotype.

Coordinate Gene Expression

As clonal expansion ceases, expression of C/EBPα, a transcription factor implicated in the coordinate activation of adipose-specific genes, is induced. Expression of C/EBPα is followed closely by the transcriptional activation of many adipocyte genes including the 422/aP2, GLUT4, and SCD1 genes, which are transactivated by C/EBPα. Not surprisingly, the coordinate changes in the cellular levels of adipocyte proteins during differentiation are due almost entirely to changes in the transcription rates of the corresponding genes (30, 71).

Based on two-dimensional electrophoretic analysis of cell extracts before, during, and after adipose conversion, it has been estimated that the expression of at least 100 proteins is altered within 5 hours of initiating differentiation (during clonal expansion), and that at least 200 additional proteins are differentially expressed by the time terminal differentiation is achieved (72). Many of these proteins whose genes are differentially expressed during preadipocyte differentiation have been identified. A comprehensive list of these proteins has been compiled (1). Proteins regulated during differentiation fall into several classes, including those involved in lipogenesis and lipolysis (e.g. acetyl-CoA carboxylase, SCD1, hormone sensitive lipase, ATP:citrate lyase), hormone action and signaling (e.g. insulin receptor, β-adrenergic receptors, glucocorticoid receptor, insulin receptor substrate-1 and G-proteins), cytoskeleton and extracellular matrix structure/function (e.g. actin, tubulin, collagens, Pref-1), as well as secretory proteins with extra-adipocyte functions (e.g. angiotensinogen, adipsin, apolipoprotein E, and IGF-1). Additions to the list of differentially expressed proteins previously compiled by Cornelius et al. (see Table 1 in Ref. 1) include phosphatidylinositol-3-kinase (73), perilipin (74, 75), mitochondrial acyl transferase (76), adenosine receptors A_1 and A_2 (77), complement components C3, Ba, and Bb (14), as well as the transcription factors PPARα (12), RXRα (12), RXRγ (12), mPPARγ2 (10), FAAR (11), CUP (78), PRE (79), and C/EBPα (80), C/EBPβ (81), and C/EBPδ (81).

Terminal Differentiation

Even after differentiation has been initiated and preadipocytes begin to acquire some adipocyte characteristics, the cells retain the ability to "dedifferentiate" and reenter mitosis. Loss of these characteristics can be initiated by disrupting cell/cell contact, or by exposing cells to agents such as retinoic acid (82) or cytokines such as TNFα (83). However, once the cells have gone beyond a specific stage, they can no longer undergo mitosis or dedifferentiation, and thus they have become "terminally differentiated" (84, 85).

Maintenance of the terminally differentiated state appears to be facilitated by the sustained expression of C/EBPα. Two functions of C/EBPα—its capacity to block mitosis when expressed at high levels (6) and its ability to transactivate expression of a number of adipocyte genes (86, 87)—contribute to maintenance of the terminally differentiated state. The continued expression of C/EBPα has been attributed to a C/EBP binding site within its proximal promoter that allows autoactivation of its own expression (7, 80, 88, 89; see below). Agents such as TNFα and other cytokines found in monocyte-conditioned media appear to disrupt this autoactivation cycle leading to a loss of C/EBPα and subsequent regression of the adipocyte phenotype (83, 90, 91).

Changes in Morphology

Prior to differentiation, the preadipocyte cell lines described above are morphologically similar to fibroblastic "preadipose" cells in the stroma of adipose tissue. When induced to differentiate, 3T3 preadipocytes lose their fibroblastic character, assume a "rounded-up" appearance, and acquire the morphological and biochemical characteristics of adipocytes. The rounding up of the differentiating preadipocyte may be due to changes in expression of cytoskeletal and/or extracellular proteins involved in matrix formation (65, 92, 93). Consistent with this view, the cellular content of the cytoskeletal proteins actin and tubulin decreases, and a switch in the expression of collagen types occurs, concomitant with the transition from fibroblast to adipocyte morphology. Soon after the induction of differentiation (3–4 days), many cytoplasmic triacylglycerol-containing vacuoles appear and, after an extended period in culture, coalesce to form unilocular fat droplets, thus giving rise to the typical "signet ring" appearance of mature white adipocytes. Electron micrographic analysis has shown that, ultrastructurally, 3T3-L1 adipocytes are essentially identical to adipocytes observed in vivo (32).

TRANSCRIPTIONAL CONTROL

The dramatic changes in cell phenotype that accompany preadipocyte differentiation are largely the result of coordinate transcriptional activation of adipocyte-specific genes (30, 71, 86, 87) and concomitant silencing of preadipocyte genes not required for adipocyte functions (30, 93). Efforts launched nearly 10 years ago have led to the identification of a group of cis-acting elements and their cognate trans-acting factors that serve to coordinate expression of these genes during differentiation. Presumably, the second messenger systems triggered by the external differentiation inducers described above (see Figure 1 and above) activate (or repress) expression of these trans-acting factors. For convenience, these factors can be grouped into two classes: 1. those (e.g. c-fos, c-myc, fra-1, and c-jun; 94) associated with mitotic clonal expansion, and 2. those which serve as "master regulators" of adipocyte gene transcription (e.g. C/EBPα, PPAR/FAAR, and possibly others including HNF/ *forkhead*, homeodomain/POU transcription factors, and other members of the C/EBP gene family), since they control the transcription of sets of genes and in some cases (e.g. C/EBPα) have additional functions during differentiation. Most of these putative master regulators are differentially expressed during differentiation of preadipocytes into adipocytes, lending credence to their involvement in the process. The time windows during which each of these factors is expressed must be rigidly controlled to facilitate the orderly progression of the differentiation program. Presumably, this is accomplished through a combination of controls including the timely regulated expression of positive-acting (C/EBPα, PPARγ2, FAAR) and repressive (CHOP/gadd153, C/EBPβ/LIP, CUP, PRE) transcription factors, as well as the activation or inhibition of these factors by covalent modification, notably phosphorylation/dephosphorylation. In the discussion that follows we describe the characteristics of the transcription factors that have been implicated in adipocyte development and attempt to show where and how these factors function in the differentiation program.

The CCAAT/Enhancer Binding Protein (C/EBP) Family

C/EBPα, the first member of this family to be identified was purified and cloned as a nuclear factor that binds to the CCAAT motif and "core homology" sequences found in certain viral promoters/enhancers, as well as the albumin gene promoter (see review by McKnight—95). Following the cloning of the C/EBPα gene (96), other members of this gene family, including C/EBPβ (81, 97–100), C/EBPδ (81, 98), C/EBPγ (Ig/EBP; 101), CHOP/gadd153 (102, 103), DBP (104), CRP1 (98), and a1/EBP (105) were cloned and sequenced. While members of this family share considerable amino acid sequence identity within the C-terminal domain, the sequences within the much larger N-terminal section are divergent.

The C-terminal basic region/leucine zipper (bZIP) domain, which consists of 55-65 amino acids, confers both the ability to bind DNA (basic region) and to form dimers (leucine zipper) with themselves or with other family members (Figure 3). The N-terminal 50–75% of these proteins house transactivation and transrepression domains. The leucine zipper domain consists of five heptad repeats, which form dimers through association of the amphipathic α-helices in a coiled-coil structure reminiscent of the interactions between keratins or tropomyosin (106–108). Hydrophobic interactions are mediated, in part, by highly conserved leucine residues and also by other hydrophobic amino acids, which interdigitate as "knobs in holes." In addition, electrostatic interactions between the two polypeptides contribute to the strength of interaction between homo- and heterodimers and, indeed, are critical for determining which classes of bZIP transcription factors form heterodimers (109). It appears that dimerization, and thus DNA binding, can be modulated by posttranslational modifications such as phosphorylation (110). In addition to mediating homodimer and heterodimer formation, the leucine zipper region has also been implicated in cell type–specific control of reporter gene activity (111), perhaps through protein-protein interactions with other types of transcription factors [e.g. glucocorticoid receptor (112) or NF-κB/Rel (113, 114)].

Several other families of bZIP transcription factors have been described (e.g. the fos/jun and the CREB/ATF families), whose ability to heterodimerize is generally restricted to members of that particular class (reviewed in 115). Some exceptions to this rule, such as C/ATF binding to C/EBPβ, could have important consequences for adipocyte gene expression, since heterodimerization directs C/EBPβ to cAMP response elements to which it does not normally bind (e.g. in the proenkephalin promoter), or to which it binds with relatively low affinity (e.g. the somatostatin gene promoter; 116).

Located just amino-terminal of the leucine zipper is the region rich in basic amino acids that interacts specifically with C/EBP binding sites to form high-affinity protein-DNA interactions. Although unstructured in the absence of DNA, the basic region assumes an α-helical conformation upon interaction with a C/EBP-binding site (107, 117). Specific amino acids within the basic region are involved in site-specific binding, and these determine the binding site preferences of the different classes of bZIP transcription factors (118). While the consensus binding site for C/EBP transcription factors has been reported to be a dyad symmetrical sequence, ATTGCGCAAT, a broad array of substitutions are tolerated. The structure of the dimerized bZIP transcription factor, first modeled as a Y-shaped complex (119), has been confirmed by the cocrystal structure of a bZIP protein bound to an oligonucleotide (120). The leucine zipper can be visualized as the stem of a Y and the bifurcating arms as the two α-helical basic regions that bind within the major groove of target DNA. Specific bZIP-DNA interactions appear to be due to 1. hydrogen bonding/VanderWaals forces between nucleotides and amino acid side chains, 2. the sequence-dependent ability of DNA to distort upon binding by protein, and 3. the preorganization of the DNA binding site into a distorted shape recognized only by specific bZIP proteins (121). Phosphorylation within the basic region influences binding to DNA and may be an important regulational device (122).

The region N-terminal of the bZIP domain (comprising 50 to 75% of the molecule) contains the transactivation domains presumed to interact directly (or indirectly) with components of the transcriptional preinitiation complex (111, 123, 124). The N-terminal section of the molecule can be visualized as large extensions of the bifurcating arms that lie beyond the DNA binding domain of the Y-shaped dimer. Mutational analysis of the N-terminal region of C/EBPα suggests that transactivation is complex and involves interactions between three activator regions (any two of which are necessary for transactivation) and a modulatory/inhibitory region (111, 123, 124). The activity of these regions is affected by posttranslational modification, e.g. transactivation by C/EBPβ is increased when serine 105 is phosphorylated by protein kinase C (125).

C/EBPα and C/EBPβ have alternative translation products that lack the N-terminal portion of the molecules and have altered functional properties. Initiating translation at the third AUG of the C/EBPβ mRNA eliminates the transactivation domain and produces a dominant-negative isoform, i.e. p18$^{C/EBPβ}$ (LIP; 126). In contrast, initiating translation at the third AUG of C/EBPα mRNA gives rise to a 30-kDa isoform that lacks adipogenic and antimitotic activity, but which still transactivates certain adipocyte promoters (7). The truncated isoforms of C/EBPα and C/EBPβ are expressed in vivo and are developmentally regulated; the ratio of truncated to full-length isoforms changes during liver and adipocyte development.

C/EBPα   A large body of evidence indicates that C/EBPα has multiple roles in adipocyte differentiation including: coordinate transcriptional activation of a group of adipocyte genes (86, 87); termination of mitotic clonal expansion as preadipocytes enter the quiescent terminally differentiated state; and transcriptional autoactivation of the C/EBPα gene to sustain expression in the terminally differentiated state (7, 89). The first indication that C/EBPα might be involved in adipogenesis was its high level of expression in adipose tissue and the induction of its expression during differentiation of 3T3-L1 preadipocytes (80, 86, 127). A connection to adipocyte gene expression was established by the discovery that C/EBPα is the differentiation-induced nuclear factor that binds specifically to sites (with nucleotide sequence similarity) within the promoters of three adipocyte genes (the 422/aP2, SCD1, and GLUT4 genes) that are coordinately expressed during differentiation. Furthermore, it was shown both with intact 3T3-L1 preadipocytes (86, 87) and with a cell-free transcription system (128) that C/EBPα transactivates reporter gene transcription from the promoters of these genes. Mutation of the C/EBP binding sites in these promoters obliterates transactivation. Definitive proof that expression of C/EBPα is required for preadipocyte differentiation was obtained using antisense C/EBPα RNA (88, 129). Constitutive expression of antisense C/EBPα RNA not only blocked expression of C/EBPα, but also expression of 422(aP2), SCD1 and GLUT4 and accumulation of cytoplasmic triacylglycerol (88). Furthermore, rescue of the adipocyte phenotype was accomplished by expressing a complementary sense C/EBPα RNA(88).

Attempts to directly test the effect of C/EBPα on differentiation by constitutive expression of C/EBPα in preadipocytes have been hampered by the fact that C/EBPα is antimitogenic; hence, stable cell lines expressing this factor cannot be propagated (130; RJ Christy, F-T Lin, MD Lane, unpublished results). However, Umek et al (6) obtained a transfected 3T3-L1 cell line that constitutively expressed a conditionally active p42 C/EBPα-estrogen receptor fusion protein. It was observed that activation of the fusion protein by exposure of the cells to estrogen blocked mitosis but did not induce adipocyte differentiation. When induced using standard differentiation inducers supplemented with estrogen, however, expression of 422(aP2) mRNA and the accumulation of cytoplasmic triglyceride were accelerated. Using a different approach, Freytag & Geddes (130) found that following transfection of 3T3-L1 preadipocytes with a C/EBPα expression vector, foci could be selected that expressed C/EBPα. These foci could not be propagated, however, presumably because of the antimitotic activity of C/EBPα; nevertheless, 12% underwent spontaneous differentiation in the absence of inducing agents, suggesting that C/EBPα might be sufficient to induce differentiation. In view of these findings, it is surprising that expression of the C/EBPα-estrogen receptor fusion protein did not induce differentiation (6). The estrogen receptor domain may interfere with the adipogenic function of C/EBPα, much as mutations within MyoD can dissociate the myogenic and growth inhibitory properties of MyoD (131).

Two recent reports (2, 3), provide proof that C/EBPα can initiate adipocyte differentiation. Using a LacSwitch inducible p42$^{C/EBP\alpha}$ expression system, it was shown that exposure to IPTG induced p42$^{C/EBP\alpha}$ and led to expression of adipocyte marker genes and acquisition of the adipocyte phenotype at high frequency (50–60% of the cells differentiated) without exogenous inducers of preadipocyte differentiation (2). Furthermore, ectopic expression of C/EBPα using a retroviral expression system induced adipogenesis in preadipocyte cell lines and fibroblastic cell lines that do not normally develop into adipocytes (3). Thus, it appears that C/EBPα is not only required but is sufficient to induce adipocyte differentiation. The essential role of C/EBPα in differentiation in vivo was recently verified by disruption of the C/EBPα gene in mice (4, 5). Although newborn C/EBPα "knock-out" mice die shortly after birth due to hypoglycemia, survival is extended by glucose administration. Importantly, adipose tissue in the scapular region, which is the first to develop, failed to appear in the knock-out mice during this period of extended survival.

As indicated above, the single mRNA encoding C/EBPα gives rise to both 42-kDa and 30-kDa alternative translation products (7, 132). Mutational analysis has shown that translation leading to p30$^{C/EBP\alpha}$ is initiated at the third in-frame methionine codon (7). The first two AUGs in the message possess unfavorable translation initiation sequences (reviewed in Kozak; 133), and are bypassed due to leaky ribosomal scanning during translation. The p30$^{C/EBP\alpha}$ and p42$^{C/EBP\alpha}$ isoforms are both capable of transactivating certain adipocyte gene promoters (7), whereas only the p42 isoform transactivates the albumen gene promoter, indicating that transactivation by p30$^{C/EBP\alpha}$ may be restricted to certain promoters or cell types (132). In contrast to p42$^{C/EBP\alpha}$, p30$^{C/EBP\alpha}$ is not antimitogenic and thus can be constitutively expressed when transfected into preadipocyte cell lines (7). Constitutive expression of p30$^{C/EBP\alpha}$ by transfected preadipocytes is insufficient to trigger differentiation; however, when transfected cells are treated with external hormone inducers, expression of p30$^{C/EBP\alpha}$ accelerates the differentiation program.

The antimitotic activity of p42$^{C/EBP\alpha}$ is of particular interest because C/EBPα may serve to inhibit mitotic clonal expansion of preadipocytes as they enter the terminally differentiated state (reviewed in 13). Consistent with this view, expression of C/EBPα occurs at the point in the differentiation program when clonal expansion ceases (Figure 2; 80). It should be noted that during differentiation of 3T3-L1 preadipocytes and during hepatocyte development in the rat, the cellular p42/p30 C/EBPα ratio decreases. The relative level of p42 reaches a maximum as proliferation ceases. In view of the differences in the antimitotic and adipogenic capacities of p42 and p30$^{C/EBP\alpha}$, it appears that the 12-kDa N-terminal domain may serve different functions. Although not tested with preadipocyte cell lines, several studies (111, 123, 124) indicate that this N-terminal segment contains transactivation elements between amino acid residues 1 to 70 and 71 to 90. Another transactivation domain is located within p30$^{C/EBP\alpha}$ (amino acids 126 to 200), as is an internal modulatory/inhibitor domain (residues 126 to 155). Further work will be required to determine how the N-terminal 12-kDa domain interacts with the transcriptional apparatus and functions in mitotic growth control.

The C/EBPα gene promoter   One approach to understanding the events that precede expression of C/EBPα in the pathway of adipocyte development is through the identification of *cis*-regulatory elements within the C/EBPα promoter and the transcription factors that interact with these elements. DNA sequence analysis coupled with DNase I footprinting of the 5′-flanking region of the mouse gene has identified several binding sites for known and for as yet uncharacterized *trans*-acting factors within the proximal promoter (78, 80, 89). These include binding sites for C/EBP, CUP (C/EBP undifferentiated protein), Sp1, Zif268/Krox20, NF1, USF, c-Myc, and the nuclear protein(s) that bind to the basal transcription element. The C/EBP and CUP binding sites are differentially footprinted by preadipocyte versus adipocyte nuclear extracts and have been extensively investigated (78, 80). Several lines of evidence suggest that C/EBPα autoactivates transcription of its own gene by interacting with the C/EBP binding site, thereby maintaining expression of C/EBPα as preadipocytes enter the fully differentiated state. It has been shown that: 1. C/EBPα antisense RNA not only blocks expression of C/EBPα protein, but also blocks transcription of the C/EBPα gene (88), 2. the C/EBPα promoter is transactivated by C/EBPα (7, 89), and 3. inducible (2) or ectopic (3) expression of C/EBPα is sufficient to activate expression of the endogenous C/EBPα gene.

Located 30–40 bp 5′ of the C/EBP binding site is a bipartite *cis*-element with binding sites for CUP and an Sp1-like GT box binding protein (78, 80). Binding of CUP to this element is greatly enhanced by its interaction with the Sp1-like protein (78). During differentiation of 3T3-L1 preadipocytes into adipocytes, the DNA-binding activity of CUP decreases (probably due to decreased expression of CUP) concomitant with a markedly increased expression of C/EBPα. Preliminary evidence (78) suggests that the C/EBP binding site participates with the adjacent CUP regulatory element (see below) to maintain the C/EBPα gene in a repressed state prior to differentiation. Based on this and other evidence (13, 78), a model was proposed in which CUP, an Sp1-like protein, and Sp1 form a complex that bridges the CUP/GT box and C/EBP binding elements, thereby blocking access to the C/EBPα binding site. Thus, when differentiation is induced and CUP expression ceases, the bridging complex would collapse allowing C/EBP homologues (initially C/EBPβ and/or C/EBPδ and eventually C/EBPα), to access the C/EBP binding site and activate C/EBPα gene expression.

C/EBPβ AND C/EBPδ   In addition to C/EBPα, two other C/EBP homologues, C/EBPβ and C/EBPδ, are believed to function in preadipocyte differentiation (81). Although the expression of each of these homologues is high in proliferating preadipocytes (P Cornelius, MD Lane, unpublished data), the levels decline upon growth arrest at confluence, and remain suppressed until differentiation is induced with external modulators. Expression of both isoforms then increases immediately, reaching a maximum by 24 h and then decreases, C/EBPδ declining precipitously and C/EBPβ at a much slower rate. The temporal expression of C/EBPβ and C/EBPδ during differentiation, and the presence of a C/EBP binding site within the C/EBPα promoter have led to the hypothesis that C/EBPβ and/or C/EBPδ may, in part, be responsible for activating transcription of the C/EBPα gene. It is not likely that either or both isoforms can activate the C/EBPα gene since both C/EBPβ and C/EBPδ are expressed at high levels by proliferating preadipocytes, when C/EBPα expression is suppressed, and since constitutive expression of C/EBPβ in a variety of cell types, including the 3T3-L1 preadipocyte, is insufficient to induce the adipogenic program (3). It is possible that binding of C/EBPβ or C/EBPδ to the C/EBP binding site in the C/EBPα gene promoter must await the release of that site from repression by the CUP-containing protein complex (see above).

C/EBPβ has several alternative translation products, the smallest of which, p18[C/EBPβ] (LIP; liver inhibitory protein; 126) lacks the transactivation domain in the N-terminal region of p41$^{C/EBP\beta}$ (LAP; liver activator protein; 134), and thus acts as a dominant-negative regulator of C/EBP-induced transcription. Since LIP and LAP as well as C/EBPδ can homodimerize and heterodimerize, these homologues have the potential to modulate the activities of other members of the C/EBP family and thereby regulate the differentiation process in a manner not interpretable with our current state of knowledge.

CHOP/GADD 153   CHOP was cloned from a 3T3-L1 adipocyte library based on its ability to interact with the leucine zipper domain of C/EBPβ (102), while the cDNA for its hamster homologue, gadd153, was cloned from a Chinese hamster ovary cell library while screening for genes activated upon growth arrest or DNA damage (103). This bZIP transcription factor is a member of the C/EBP family and is unique in that the amino acid sequence corresponding to the basic-DNA binding region of other C/EBPs contains two α-helix breaking prolines. Since CHOP can heterodimerize with other C/EBP homologues (α and β) and block their ability to bind to C/EBP binding sites, CHOP acts as a dominant-negative isoform (102). It was recently reported (135), however, that although CHOP/C/EBP heterodimers do not bind to classical C/EBP binding sites, they can bind to other sites in a manner similar to the redirection of binding of C/EBPβ to cAMP response elements when heterodimerized to C/ATF (116). In another context, it has been observed that expression of CHOP/gadd153 in 3T3-L1 and HeLa cells is inversely related to the concentration of glucose in cell culture media (136). By maintaining glucose at a high level throughout the course of differentiation of 3T3-L1 preadipocytes, the expression of CHOP/gadd153 is held at a low level, thus clearly dissociating CHOP/gadd153 as a critical regulator in the adipogenic program.

ROLE OF C/EBP TRANSCRIPTION FACTORS IN TERMINALLY DIFFERENTIATED CELLS   In addition to its role in differentiation, C/EBP$^\alpha$ may also have important functions in terminally differentiated cells. It has been suggested that C/EBP$^\alpha$ may serve as a central regulator of energy metabolism (137). Several facts are consistent with this hypothesis: First, C/EBP$^\alpha$ is expressed in tissues (7, 127, 138, 139), i.e. white and brown adipose tissue, liver and small intestine, that play important roles in the global energy economy (notably lipid and carbohydrate metabolism) of the animal. Secondly, it was recently found (4, 5) that mice lacking a functional C/EBP$^\alpha$ gene (due to a C/EBP$^\alpha$ gene knockout) survive fetal development, but die shortly after birth due to hypoglycemia, most likely due to a lack of liver glycogen and possibly an inability to carry out gluconeogenesis. Third, the genes encoding several key enzymes of energy metabolism—PEPCK (140), which is essential for both gluconeogenesis and glyceroneogenesis; acetyl CoA carboxylase (141), the rate limiting enzyme for de novo fatty acid biosynthesis; GLUT4 (87, 142), the insulin-responsive glucose transporter; the insulin receptor (143), and SCD1 (86), which is necessary for the desaturation of saturated fatty acids—all possess functional C/EBP binding sites in their proximal promoters. Fourth, C/EBPα (and certain of its other homologues) appears to function in the second messenger signaling pathways of glucocorticoids and insulin, key regulators of energy metabolism, by coordinating the effects of these hormones on adipocyte gene expression (8, 9). Thus, treatment of fully differentiated 3T3-L1 adipocytes with glucocorticoid causes a rapid and transient decrease in the expression of C/EBPα, a rapid and transient increase in the expression of C/EBPδ, and no change in the expression of C/EBPβ (8). Insulin, on the other hand, causes a persistent decrease in the expression of C/EBPα while transiently inducing the expression of both C/EBPβ (LAP and LIP) and C/EBPδ (9). In addition to its suppressive effect on transcription of the C/EBPα gene, insulin also causes rapid dephosphorylation of C/EBPα protein, apparently through activation of a nuclear phosphatase. Hence, it appears that insulin regulates both the level and activity of C/EBPα through multiple mechanisms (9). The suppression of C/EBPα by insulin may be involved in mediating the insulin-induced downregulation of GLUT4 and SCD1 (144). Taken together, these findings suggest that C/EBPα is a central regulator of energy metabolism.

The 422/aP2 Adipose-Specific Enhancer and the Role of mPPARγ2

422/aP2 is a small cytosolic fatty acid-binding protein that is expressed exclusively in white and brown adipose tissue. Early studies with 3T3- preadipocyte cell lines suggested that cis-elements within the proximal promoter region of the 422/aP2 gene [i.e. binding sites for C/EBP, AP1, and the glucocorticoid receptor (86, 145–147)] are required for expression in adipocytes. However, subsequent studies utilizing transgenic mice revealed that a 518 bp enhancer, located ~5.4 kb 5' to the transcription start site, was necessary to direct reporter gene expression to adipose tissue (148, 149). Analysis of enhancer-nuclear protein interactions revealed that five binding sites (adipose regulatory elements; ARE), located within a 183 bp segment near the 5' end of the enhancer, accounted for most of the enhancer activity (150, 151). NF-1 binds to ARE-1 while adipose regulatory factor-6 (ARF6) binds to ARE6 and ARE7, which appear to be necessary and sufficient for adipocyte-specific gene expression, since multiple copies of either element are capable of driving reporter gene expression in differentiating adipocytes (but not preadipocytes), and since mutation of the ARE6 markedly reduces adipocyte-specific expression (150, 151). A cDNA encoding a nuclear hormone receptor, mPPARγ2 (peroxisome proliferator activator receptor γ2), which binds to ARE7 as a heterodimer with RXRα, appears to be an important component of the ARF6 complex (10). mPPARγ2 is a member of the peroxisome proliferator-activated receptor family and is identical to mPPARγl (152–154) except for an additional N-terminal 30 amino acids and a different 5′ untranslated sequence, presumably due to differential promoter usage (10). While functional differences between the two isoforms have not been assessed, mPPARγ2 is expressed in adipose tissue, is activated by fatty acids, and is sufficient to activate the adipocyte-specific enhancer in nonadipocyte cell lines .

Fatty Acid Activated Receptor (FAAR)

The gene encoding a PPAR-related transcription factor, designated fatty acid activated receptor (FAAR), has recently been cloned and sequenced (11). FAAR shares many properties with mPPARγ2 in that it is expressed in adipose tissue (and other fat metabolizing tissues excluding liver) and binds, as a heterodimer with RXR, to peroxisome proliferator-response elements. Moreover, the expression of FAAR is induced in Ob1771 preadipocytes as the cells achieve confluence just prior to undergoing terminal differentiation. Transfection of a FAAR expression vector into 3T3-C2 fibroblasts is sufficient to confer fatty acid responsivity to the endogenous 422/aP2 and fatty acid transporter (FAT; 155) genes, thus demonstrating a functional role for this receptor in the regulation of gene expression by fatty acids.

While mPPARγ2 (10), FAAR (11), and other members of the family (PPARα; 12) have not been directly linked to the regulation of adipocyte differentiation, they most likely play a role, since both peroxisome proliferators and fatty acids have been reported to induce, or potentiate, differentiation of several preadipocyte cell lines (12, 39, 156). In addition to their inferred role in preadipocyte differentiation, PPAR and FAAR appear to function in regulating lipid homeostasis. For example, these receptors confer fatty acid-responsivity to adipocytes (10, 11) by inducing the proteins involved in fatty acid transport (FAT) and binding (422/aP2). These receptors may also regulate fatty acid metabolism in adipocytes, since in other systems PPARs are known to regulate peroxisomal β-oxidation by controlling the transcription of the gene encoding the rate-limiting enzyme, i.e. acyl-CoA oxidase, as well as other enzymes in this pathway (157, 158). The fact that peroxisome proliferators and 9-cis retinoic acid cooperatively activate the acyl-CoA oxidase gene promoter in CV-1 cells, and the fact that mPPARγ2, FAAR, and PPARα must heterodimerize with RXR to bind to DNA suggest an interesting convergence of signal transduction pathways, particularly in light of the inhibitory effect that high concentrations of retinoic acid have on preadipocyte differentiation (12, 159).

Differentiation-Specific Element

Adipose tissue is the major nonhepatic source of angiotensinogen, the glycoprotein precursor of the vasopressor angiotensin II. A cis-regulatory element has been identified in the angiotensinogen gene that appears to regulate induction of angiotensinogen expression during differentiation of 3T3-L1 preadipocytes into adipocytes (160). This 14-bp *cis*-acting element, referred to as differentiation-specific element (DSE), is located about 1000 bp 5' of the start-site of transcription and mediates a high level of reporter gene expression in the absence of exogenous inducers of differentiation (160). The DSE possesses nucleotide sequence similarity to the homeodomain/POU class of transcription factors; thus, binding to the DSE by a nuclear factor(s) present in preadipocyte nuclear extracts is competed by homeo/POU binding site oligonucleotides (160). As homeo/POU transcription factors play important roles in other development systems, it will be important to ascertain whether the DSE and its cognate *trans*-acting factors are involved in the regulation of other adipocyte genes.

ADD1 (Adipocyte Determination and Differentiation Factor 1)

A gene encoding a novel member of the basic helix-loop-helix-leucine zipper transcription factor family (ADD1) was cloned from a rat adipocyte cDNA library based on its ability to bind to an E-box in the fatty acid synthase promoter (161). Since expression of ADD1 is regulated at various stages in the adipocyte development program, it was proposed that ADD1 functions in both adipocyte determination and differentiation. ADD1 mRNA is expressed at low levels in several rat tissues including white and brown adipose tissue (161). 3T3-F442A preadipocytes, as well as multipotent 10T1/2 stem cells express low levels of ADD1 mRNA. Upon differentiation of both 3T3-F442A and determined 10T1/2 preadipocytes into adipocytes, expression of the ADD1 mRNA increases. Although a target gene for ADD1 has not been identified, an E-box-containing promoter element from the fatty acid synthase gene can mediate transactivation by ADD1 (161). In view of the fact that sequence analysis shows that ADD1 is a truncated form of the sterol response element–binding protein, its postulated role in adipose determination and differentiation remains uncertain.

HNF3/forkhead

Transcriptional activation of the LPL gene occurs as adipoblasts contact neighboring cells and cease dividing at confluent cell density, hence, LPL serves as one of the early markers of the adipoblast-to-preadipocyte transition (65). Enerback et al (162) have identified two potential regulatory regions, designated LP-a and LP-b, within the promoter of the LPL gene, which confer differentiation-specific regulation. DNA sequence analysis of these regions revealed considerable identity with the consensus binding sites for the HNF3/*forkhead* family of transcription factors. Gel retardation studies with LP-a and

LP-b-containing oligonucleotides verified the presence of HNF3/*forkhead*-like factors in nuclear extracts from 3T3-F442A cells and showed that the binding activity of these factors increased during differentiation (162). Furthermore, multimers of the LP-a or LP-b elements confer differentiation-dependent expression to heterologous promoter-reporter gene constructs transfected into 3T3-F442A cells (162). Whether this class of transcription factor is involved in the regulation of other adipocyte genes is as yet unknown.

Preadipocyte Repressor Element (PRE)

A *cis*-acting element identified within the promoter of the stearoyl-CoA desaturase-2 (SCD2) gene appears to repress the gene in preadipocytes prior to differentiation (79). This preadipocyte repressor element (PRE) binds a 58-kDa nuclear protein present in 3T3-L1 preadipocytes but not present (or inactive) in adipocytes. A single copy of the PRE confers differentiation-specific control to the SV40 enhancer/promoter in that the reporter gene is strongly repressed in preadipocytes and HeLa cells, but not in adipocytes (79). These findings suggest that the PRE and its *trans*-acting factor(s) maintain the SCD2 gene, and possibly other genes, in a repressed state until adipocyte differentiation is initiated.

STAGE-SPECIFIC GENE EXPRESSION DURING ADIPOCYTE DEVELOMENT

The appropriate timing of gene expression is critical for the orderly progression of events in adipocyte development. Illustrated in Figure 2 are the temporal patterns of expression of selected genes that either regulate or contribute to acquisition of the adipocyte phenotype. To date, the gene(s) that commits progression from the mesodermal stem cell to the adipoblast stage of development has not been identified. Nevertheless, numerous gene markers for the preadipocyte and adipocyte stages are now known. As adipoblasts contact neighboring cells and enter the growth-arrested state (at the G_0/G_1 boundary of the cell cycle), early markers such as LPL and the $\alpha 2$ chain of type VI collagen are expressed (65), perhaps through transactivation of their respective genes by members of the HNF3/*forkhead* family of *trans*-acting factors (162). Growth arrest at confluence also leads to induction of the FAAR (11) gene (and somewhat later, the PPARγ2 gene; 10), which has been proposed as a pleiotropic transcriptional activator of a group of genes involved in lipid metabolism/homeostasis.

Following induction of differentiation (with external modulators) of growth-arrested preadipocytes, there is an immediate transient expression of cell di-

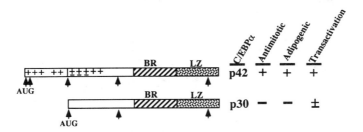

Figure 3 Schematic diagrams of p42$^{C/EBP\alpha}$ and p30$^{C/EBP\alpha}$ illustrating the C-terminal location of the basic region (BR)/leucine zipper (LZ) domains, as well as regions implicated in activation (+) and inhibition (-) of transcription. The arrow heads represent potential start sites of translation. p42$^{C/EBP\alpha}$ and p30$^{C/EBP\alpha}$ are initiated at the first and third methionine, respectively. The antimitotic, adipogenic, and transactivational properties of each isoform are also shown.

vision-associated transcription factors (e.g. c-Fos, c-Jun, and c-Myc; 94, 163) as the cells initiate several rounds of clonal expansion. A number of events then occur in rapid succession: The expression of Pref-1 (68), CUP (78), and PRE (79) decreases, and the expression of C/EBPβ, C/EBPδ, and RXRα (12) increases (and then declines) followed by expression of C/EBPα and cessation of clonal expansion. Several events are correlated with and are believed to be involved in the transcriptional activation of the C/EBPα gene: A decrease in the level (or activity) of CUP may serve to derepress the gene (13, 78), and an increased expression of C/EBPβ and C/EBPδ (81) may serve to activate the gene through interaction with its C/EBP binding site. Induction of C/EBPα is thought to have a three-fold effect: termination of clonal expansion (13); transcriptional activation of a large group of adipocyte genes [e.g. 422/aP2, GLUT4, SCD1, PEPCK, FSP27, and the insulin receptor (30, 86, 140, 143, 164)]; and autoactivation of the C/EBPα gene through its C/EBP binding site (7, 89).

ACKNOWLEDGMENTS

Research from the authors' laboratory was supported by reseach grants from the NIH and a national service research award to OA MacDougald from the NIH.

Literature Cited

1. Cornelius P, MacDougald OA, Lane MD. 1994. *Annu. Rev. Nutr.* 14:99–129
2. Lin F-T, Lane MD. 1994. *Proc. Natl. Acad. Sci. USA* 91:8757–8761
3. Freytag SO, Paielli DL, Gilbert JD. 1994. *Genes Dev.* 8:1654–63
4. Wang N, Bradley A, Finegold M, Matzuk M, Darlington GJ. 1994. *Am. Soc. Hum. Genet.* (Abstr)
5. Wang N, Bradley A, Matzuk MM, Darlington GJ. 1994. *Mol. Biol. Cell* 5(suppl):336a (Abstr)
6. Umek RM, Friedman AD, McKnight SL. 1991. *Science* 251:288–92
7. Lin F-T, MacDougald OA, Diehl AM, Lane MD. 1993. *Proc. Natl. Acad. Sci. USA* 90:9606–10
8. MacDougald OA, Cornelius P, Lin F-T, Chen SS, Lane MD. 1994. *J. Biol. Chem.* 269:19,041–47
9. MacDougald OA, Cornelius P, Liu R, Lane MD. 1994. *J. Biol. Chem.* 270:647–54
10. Tontonoz P, Hu E, Graves RA, Budavari AI, Spiegelman BM. 1994. *Genes Dev.* 8:1224–34
11. Amri E-Z, Bonino F, Ailhaud G, Abumrad NA, Grimaldi PA. 1995. *J. Biol. Chem.* Vol. 270:2367–71
12. Chawla A, Lazar MA. 1994. *Proc. Natl. Acad. Sci. USA* 91:1786–90
13. Vasseur-Cognet M, Lane MD. 1993. *Curr. Opin. Gen. Devel.* 3:238–45
14. Spiegelman BM, Choy L, Hotamisligil GS, Graves RA, Tontonoz P. 1993. *J. Biol. Chem.* 268:6823–26
15. Ailhaud G, Grimaldi P, Negrel R. 1992. *Annu. Rev. Nutr.* 12:207–33
16. Taylor SM, Jones PA. 1979. *Cell* 17:771–79
17. Lassar AB, Paterson BM, Weintraub H. 1986. *Cell* 47:649–56
18. Olson EN, Klein WH. 1994. *Genes Dev.* 8:1–8
19. Green H, Kehinde O. 1974. *Cell* 1:113–16
20. Green H, Kehinde O. 1975. *Cell* 5:19–27
21. Green H, Kehinde O. 1976. *Cell* 7:105–13
22. Negrel R, Grimaldi P, Ailhaud G. 1978. *Proc. Natl. Acad. Sci. USA* 75:6054–58
23. Green H, Kehinde O. 1979. *J. Cell Phys.* 101:169–72
24. Mackall JC, Student AK, Polakis SE, Lane MD. 1976. *J. Biol. Chem.* 251:6462–64
25. Student AK, Hsu RY, Lane MD. 1980. *J. Biol. Chem.* 255:4745–50
26. Ntambi JM, Buhrow SA, Kaestner KH, Christy RJ, Sibley E, et al. 1988. *J. Biol. Chem.* 263:17291–300
27. Kaestner KH, Ntambi JM, Kelly TJ, Lane MD. 1989. *J. Biol. Chem.* 264:14,755–61
28. Reed BC, Lane MD. 1980. *Proc. Natl. Acad. Sci. USA.* 77:285–89
29. Rubin CS, Hirsch A, Fung C, Rosen OM. 1978. *J. Biol. Chem.* 253:7570–78
30. Bernlohr DA, Bolanowski MA, Kelly TJ, Lane MD. 1985. *J. Biol. Chem.* 260:5563–67
31. Coleman RA, Reed BC, Mackall JC, Student AK, Lane MD, Bell RM. 1978. *J. Biol. Chem.* 253:7256–61
32. Novikoff AB, Novikoff PM, Rosen OM, Rubin CS. 1980. *J. Cell Biol.* 87:180–96
33. Guller S, Corin RE, Mynarcik DC, London BM, Sonenberg M. 1988. *Endocrinology* 122:2084–89
34. Hauner H. 1990. *Endocrinology* 127:865–72
35. Hausman GJ. 1992. *J. Anim. Sci.* 70:106–14
36. Schmidt W, Poll-Jordan G, Loffler G. 1990. *J. Biol. Chem.* 265:15489–95
37. Catalioto R-M, Gaillard D, Ailhaud G, Negrel R. 1992. *Growth Factors* 6:255–64
38. Serrero G, Lepak NM, Goodrich SP. 1992. *Endocrinology* 131:2545–51
39. Amri E-Z, Ailhaud G, Grimaldi P-A. 1994. *J. Lipid Res.* 35:930–37
40. Zezulak KM, Green H. 1986. *Science* 233:551–53
41. Nixon R, Green H. 1984. *Endocrinology* 114:527–32
42. Smith PJ, Wise LS, Berkowitz R, Wan C, Rubin CS. 1988. *J. Biol. Chem.* 263:9402–8
43. Doglio A, Dani C, Fredrikson G, Grimaldi P, Ailhaud G. 1987. *EMBO J.* 6:4011–16
44. Gaskins HR, Kim JW, Wright JT, Rund LA, Hausman GJ. 1990. *Endocrinology* 126:622–30
45. Nougues J, Reyne Y, Barenton B, Chery T, Garandel V, Soriano J. 1993. *Int. J. Obesity* 17:159–67
46. Liao K, Lane MD. 1994. *FASEB J.* S82: A1235 (abstr)
47. Benito M, Porras A, Nebreda AR, Santos E. 1991. *Science* 253:565–68
48. Porras A, Nebreda AR, Benito M, Santos E. 1992. *J. Biol. Chem.* 267:21124–31
49. Izumi T, White MF, Kadowaki T, Takaku F, Akanuma Y, Kasuga M. 1987. *J. Biol. Chem.* 262:1282–87

50. Myers MG, Sun XJ, Cheatham B, Jachna BR, Glasheen EM, et al. 1993. *Endocrinology* 132:1421–30
51. Skolnik EY, Batzer A, Li N, Lee C-H, Lowenstein E, et al. 1993. *Science* 260: 1953–55
52. Porras A, Muszynski K, Rapp UR, Santos E. 1994. *J. Biol. Chem.* 269:12741–48
53. Amri E-Z, Dani C, Doglio A, Etienne J, Grimaldi P, Ailhaud G. 1986. *Biochem. J.* 238:115–22
54. Dani C, Doglio A, Amri E-Z, Bardon S, Fort P, et al. 1989. *J. Biol. Chem.* 264:10,119–25
55. Gaillard D, Negrel R, Lagarde M, Ailhaud G. 1989. *Biochem. J.* 257:389–97
56. Gaillard D, Wabitsch M, Pipy B, Negrel R. 1991. *J. Lipid Res.* 32:569–79
57. Catalioto R-M, Gaillard D, Maclouf J, Ailhaud G, Negrel R. 1991. *Biochim. Biophys. Acta* 1091:364–69
58. Negrel R, Gaillard D, Ailhaud G. 1989. *Biochem. J.* 257:399–405
59. Vassaux G, Gaillard D, Ailhaud G, Negrel R. 1992. *J. Biol. Chem.* 267: 11092–97
60. Amri E-Z, Bertrand B, Ailhaud G, Grimaldi P. 1991. *J. Lipid Res.* 32:1449–56
61. Amri E-Z, Ailhaud G, Grimaldi P. 1991. *J. Lipid Res.* 32:1457–63
62. Grimaldi PA, Knobel SM, Whitesell RR, Abumrad NA. 1992. *Proc. Natl. Acad. Sci. USA* 89:10930–34
63. Shillabeer G, Lau DC. 1994. *J. Lipid Res.* 35:592–600
64. Weintraub H. 1993. *Cell* 75:1241–44
65. Dani C, Amri E-Z, Bertrand B, Enerback S, Bjursell G, et al. 1990. *J. Cell. Biochem.* 43:103–10
66. Ibrahimi A, Bonino F, Bardon S, Ailhaud G, Dani C. 1992. *Biochem. Biophys. Res. Commun.* 187:1314–22
67. Spiegelman BM, Ginty CA. 1983. *Cell* 35:657–66
68. Smas CM, Sul HS. 1993. *Cell* 73:725–34
69. Smas CM, Green D, Sul HS. 1994. *Biochemistry* 33:9257–65
70. Cook JR, Kozak LP. 1982. *Dev. Biol.* 92:440–48
71. Cook KS, Hunt CR, Spiegelman BM. 1985. *J. Cell Biol.* 100:514–20
72. Sadowski HB, Wheeler TT, Young DA. 1992. *J. Biol. Chem.* 266:4722–31
73. Saad MJA, Folli F, Araki E, Hashimoto N, Csermely P, Kahn CR. 1994. *Mol. Endocrinol.* 8:545–57
74. Greenberg AS, Egan JJ, Wek SA, Garty NB, Blanchette-Mackie EJ, Londos C. 1991. *J. Biol. Chem.* 266:11341–46
75. Greenberg AS, Egan JJ, Wek SA, Moos MC, Londos C, Kimmel AR. 1993. *Proc. Natl. Acad. Sci. USA.* 90:12035–39
76. Yet S-F, Lee S, Hahm YT, Sul HS. 1993. *Biochemistry* 32:9486–91
77. Vassaux G, Gaillard D, Mari B, Ailhaud G, Negrel R. 1993. *Biochem. Biophys. Res. Commun.* 193:1123–30
78. Vasseur-Cognet M, Lane MD. 1993. *Proc. Natl. Acad. Sci. USA* 90:7312–16
79. Swick AG, Lane MD. 1992. *Proc. Natl. Acad. Sci. USA* 89:7895–99
80. Christy RJ, Kaestner KH, Geiman DE, Lane MD. 1991. *Proc. Natl. Acad. Sci. USA.* 88:2593–97
81. Cao Z, Umek RM, McKnight SL. 1991. *Genes Dev.* 5:1538–52
82. Hoerl BJ, Wier ML, Scott RE. 1984. *Exp. Cell Res.* 155:422–34
83. Ron D, Brasier AR, McGehee RE, Habener JF. 1992. *J. Clin. Invest.* 89: 223–33
84. Wier ML, Scott RE. 1986. *J. Cell Biol.* 102:1955–64
85. Wang H, Scott RE. 1993. *Cell Prolif.* 26:55–66
86. Christy RJ, Yang VW, Ntambi JM, Geiman DE, Landschulz WH, et al. 1989. *Genes & Dev.* 3:1323–35
87. Kaestner KH, Christy RJ, Lane MD. 1990. *Proc. Natl. Acad. Sci. USA.* 87: 251–55
88. Lin F-T, Lane MD. 1992. *Genes & Dev.* 6:533–44
89. Legraverend C, Antonson P, Flodby P, Xanthopoulos KG. 1993. *Nucleic Acids Res.* 21:1735–42
90. Stephens JM, Pekala PH. 1991. *J. Biol. Chem.* 266:21839–45
91. Stephens JM, Pekala PH. 1992. *J. Biol. Chem.* 267:13,580–84
92. Bernlohr DA, Angus CW, Lane MD, Bolanowski MA, Kelly TJ. 1984. *Proc. Natl. Acad. Sci. USA.* 81:5468–72
93. Spiegelman BM, Farmer SR. 1982. *Cell* 29:53–60
94. Stephens JM, Butts MD, Pekala PH. 1992. *J. Mol. Endocrinol.* 9:61–72
95. McKnight SL. 1992. CCAAT/enhancer binding protein. In *Transcriptional Regulation* ed. SL McKnight, KR Yamamoto, pp. 771–796. Cold Spring Harbor, New York: Cold Spring Harbor Laboratory
96. Landschulz WH, Johnson PF, Adashi EY, Graves BJ, McKnight SL. 1988. *Genes & Dev.* 2:786–800
97. Akira S, Isshiki H, Sugita T, Tanabe O, Kinoshita S, et al. 1990. *EMBO J.* 6: 1897–906
98. Williams SC, Cantwell CA, Johnson PF. 1991. *Genes & Dev.* 5:1553–67

99. Poli V, Mancini FP, Cortese R. 1990. *Cell* 63:643–53
100. Chang C-J, Chen T-T, Lei H-Y, Chen D-S, Lee S-C. 1990. *Mol. Cell. Biol.* 10:6642–53
101. Roman C, Platero JS, Shuman J, Calame K. 1990. *Genes Devel.* 4:1404–15
102. Ron D, Habener JF. 1992. *Genes & Dev.* 6:439–53
103. Fornace AJ, Nebert DW, Hollander MC, Luethy JD, Papathanasiou M, et al. 1989. *Mol. Cell. Biol.* 9:4196–203
104. Mueller CR, Maire P, Schibler U. 1990. *Cell* 61:279–91
105. Bowers WJ, Ruddell A. 1992. *J. Virol.* 66:6578–86
106. Landschulz WH, Johnson PF, McKnight SL. 1988. *Science* 240:1759–64
107. O'Shea EK, Rutkowski R, Kim PS. 1989. *Science* 243:538–42
108. Landschulz WH, Johnson PF, McKnight SL. 1989. *Science* 243:1681–88
109. Vinson CR, Hai T, Boyd SM. 1993. *Genes Dev.* 7:1047–58
110. Wegner M, Cao Z, Rosenfeld MG. 1992. *Science* 256:370–73
111. Nerlov C, Ziff EB. 1994. *Genes Dev* 8:350–62
112. Nishio Y, Isshiki H, Kishimoto T, Akira S. 1993. *Mol. Cell. Biol.* 13:1854–62
113. LeClair KP, Blanar MA, Sharp PA. 1992. *Proc. Natl. Acad. Sci. USA.* 89: 8145–49
114. Stein B, Cogswell PC, Baldwin AS. 1993. *Mol. Cell. Biol.* 13:3964–74
115. Lamb P, McKnight SL. 1991. *TIBS* 16: 417–22
116. Vallejo M, Ron D, Miller CP, Habener JF. 1993. *Proc. Natl. Acad. Sci. USA.* 90:4679–83
117. O'Neil KT, Shuman JD, Ampe C, DeGrado WF. 1991. *Biochemistry* 30: 9030–34
118. Johnson PF. 1993. *Mol. Cell. Biol.* 13: 6919–30
119. Vinson CR, Sigler PB, McKnight SL. 1989. *Science* 246:911–16
120. Ellenberger TE, Brandl CJ, Struhl K, Harrison SC. 1992. *Cell* 71:1223–37
121. Paolella DN, Palmer CR, Schepartz A. 1994. *Science* 264:1130–33
122. Mahoney CW, Shuman J, McKnight SL, Chen H-C, Huang KP. 1992. *J. Biol. Chem.* 267:19396–403
123. Friedman AD, McKnight SL. 1990. *Genes Dev.* 4:1416–26
124. Pei D, Shih C. 1991. *Mol. Cell Biol.* 11:1480–87
125. Trautwein C, Caelles C, van der Geer P, Hunter T, Karin M, Chojkier M. 1993. *Nature* 364:544–47
126. Descombes P, Schibler U. 1991. *Cell* 67:569–79
127. Birkenmeier EH, Gwynn B, Howard S, Jerry J, Gordon JI, et al. 1989. *Genes Dev.* 3:1146–56
128. Cheneval D, Christy RJ, Geiman D, Cornelius P, Lane MD. 1991. *Proc. Natl. Acad. Sci. USA* 88:8465–69
129. Samuelsson L, Stromberg K, Vikman K, Bjursell G, Enerback S. 1991. *EMBO J.* 10:3787–93
130. Freytag SO, Geddes TJ. 1992. *Science* 256:379–82
131. Weintraub H, Davis R, Tapscott S, Thayer M, Krause M, et al. 1991. *Science* 251:761–66
132. Ossipow V, Descombes P, Schibler U. 1993. *Proc. Natl. Acad. Sci. USA.* 90: 8219–23
133. Kozak M. 1991. *J. Biol. Chem.* 266: 19,867–70
134. Descombes P, Chojkier M, Lichtsteiner S, Falvey E, Schibler U. 1990. *Genes Dev.* 4:1541–51
135. Barone MV, Crozat A, Tabaee A, Philipson L, Ron D. 1994. *Genes Dev.* 8:453–64
136. Carlson SG, Fawcett TW, Bartlett JD, Bernier M, Holbrook NJ. 1993. *Mol. Cell. Biol.* 13:4736–44
137. McKnight SL, Lane MD, Gluecksohn-Waelsch S. 1989. *Genes Dev.* 3:2021–24
138. Rehnmark S, Antonson P, Xanthopoulos KG, Jacobsson A. 1993. *FEBS Lett.* 318:235–41
139. Chandrasekaran C, Gordon JI. 1993. *Proc. Natl. Acad. Sci. USA* 90:8871–75
140. Park EA, Gurney AL, Nizielski SE, Hakimi P, Cao Z, et al. 1993. *J. Biol. Chem.* 268:613–19
141. Tae H-J, Luo X, Kim K-H. 1994. *J. Biol. Chem.* 269:10475–84
142. Kaestner KH, Christy RJ, McLenithan JC, Braiterman LT, Cornelius P, et al. 1989. *Proc. Natl Acad. Sci. USA.* 86: 3150–54
143. McKeon C, Pham T. 1991. *Biochem. Biophys. Res. Commun.* 174:721–28
144. Flores-Riveros JR, McLenithan JC, Ezaki O, Lane MD. 1993. *Proc. Natl. Acad. Sci. USA.* 90:512–16
145. Cook JS, Lucas JJ, Sibley E, Bolanowski MA, Christy RJ, et al. 1988. *Proc. Natl. Acad. Sci. USA.* 85:2949–53
146. Herrera R, Ro HS, Robinson GS, Xanthopoulos KG, Spiegelman BM. 1989. *Mol. Cell. Biol.* 9:5331–39
147. Yang VW, Christy RJ, Cook JS, Kelly TJ, Lane MD. 1989. *Proc. Natl. Acad. Sci. USA.* 86:3629–33
148. Ross SR, Graves RA, Greenstein A, Platt KA, Shyu H-L, et al. 1990. *Proc. Natl. Acad. Sci. USA.* 87:9590–94

149. Graves RA, Tontonoz P, Ross SR, Spiegelman BM. 1991. *Genes Dev.* 5: 428–37
150. Graves RA, Tontonoz P, Platt KA, Ross SR, Spiegelman BM. 1992. *J. Cell. Biochem.* 49:219–24
151. Graves RA, Tontonoz P, Spiegelman BM. 1992. *Mol. Cell. Biol.* 12:1202–8
152. Zhu Y, Alvares K, Huang Q, Rao MS, Reddy JK. 1993. *J. Biol. Chem.* 268: 26817–20
153. Chen F, Law SW, O'Malley BW. 1993. *Biochem. Biophys. Res. Commun.* 196: 671–77
154. Kliewer SA, Forman BM, Blumberg B, Ong ES, Borgmeyer U, et al. 1994. *Proc. Natl. Acad. Sci. USA.* 91:7355–59
155. Abumrad NA, El-Maghrabi MR, Amri E-Z, Lopez E, Grimaldi PA. 1993. *J. Biol. Chem.* 268:17665–68
156. Brandes R, Arad R, Benvenisty N, Weil S, Bar-Tana J. 1990. *Biochim. Biophys. Acta* 1054:219–24
157. Kliewer SA, Umesono K, Noonan DJ, Heyman RA, Evans RM. 1992. *Nature* 358:771–74
158. Keller H, Dreyer C, Medin J, Mahfoudi A, Ozato K, Wahli W. 1993. *Proc. Natl. Acad. Sci. USA.* 90:2160–64
159. Stone RL, Bernlohr DA. 1990. *Differentiation* 45:119–27
160. McGehee RE, Ron D, Brasier AR, Habener JF. 1993. *Mol. Endocrinology* 7:551–60
161. Tontonoz P, Kim JB, Graves RA, Spiegelman BM. 1993. *Mol. Cell. Biol.* 13:4753–59
162. Enerback S, Ohlsson BG, Samuelsson L, Bjursell G. 1992. *Mol. Cell. Biol.* 12:4622–33
163. Freytag SO. 1988. *Mol. Cell. Biol.* 8: 1614–24
164. Danesch U, Hoeck W, Ringold GM. 1992. *J. Biol. Chem.* 267:7185–93

Annu. Rev. Biochem. 1995. 64:375–401

HUMAN CARBONIC ANHYDRASES AND CARBONIC ANHYDRASE DEFICIENCIES

William S. Sly and Peiyi Y. Hu

Edward A. Doisy Department of Biochemistry and Molecular Biology, St. Louis University School of Medicine, 1402 South Grand Boulevard, St. Louis, Missouri 63104

KEY WORDS   bone resorption, renal tubular acidosis, mental retardation, CO_2/HCO_3^- transport, gene family, isozyme

CONTENTS

ABSTRACT

Carbonic anhydrases (CAs I–VII) are products of a gene family that encodes seven isozymes and several homologous, CA- related proteins. All seven

375

0066-4154/95/0701-0375$05.00

isozymes have been cloned, sequenced, and mapped, and the intron-exon organization of five genes established. They differ in subcellular localizations, being cytoplasmic (CA I, II, III, and VII), GPI-anchored to plasma membranes of specialized epithelial and endothelial cells (CA IV), in mitochondria (CA V), or in salivary secretions (CA VI). They also differ in kinetic properties, susceptibility to inhibitors, and tissue-specific distribution. Structural and kinetic studies of recombinant natural and mutant CAs have greatly increased our understanding of the structural requirements for catalysis. Studies of the effects of CA inhibitors over many years have implicated CAs in a variety of physiological processes. Analyses of human and animal CA deficiencies provide unique opportunities to understand the individual contributions of different isozymes to these processes.

SUMMARY AND PERSPECTIVE

Carbonic anhydrases (CA I–CA VII) are the products of a gene family that encodes seven distinct isozymes and several additional CA-related proteins with sequence homology but no CA activity. The seven isozymes differ widely in their kinetics, in susceptibility to different inhibitors, in subcellular localization, and in tissue-specific distribution. They participate in a variety of physiological processes that involve pH regulation, CO_2 and HCO_3^- transport, ion transport, and water and electrolyte balance. Functions that depend on CAs, directly or indirectly, include H^+ secretion, HCO_3^- reabsorption, HCO_3^- secretion, bone resorption, and production of aqueous humor, cerebrospinal fluid, gastric acidity, and pancreatic juice. Metabolic roles include important steps in ureagenesis, gluconeogenesis, and lipogenesis.

The past decade has seen a great renewal of interest in CAs as the genes for each of the family members were cloned, their products expressed in large amounts in bacteria and purified, and their structures modified by site-directed mutagenesis and explored by teams of structural biologists and kineticists. Decades of accumulated hypotheses could be tested directly. In addition, many interesting new questions were raised by the discovery of new members of the family.

Adding greatly to interest in this enzyme family was the discovery 11 years ago that CA II deficiency is the basic defect underlying the inherited human syndrome of osteopetrosis with renal tubular acidosis and brain calcification. Studies of patients with this disorder clarified the role of CA II in bone resorption and in renal acidification, and suggested its importance in normal brain development. In addition, these studies illuminated roles for other CAs that were not affected by the mutations producing CA II deficiency. Not surprisingly, the extraordinary lessons provided from this one example stimulated interest in identifying other human CA deficiencies, and in producing animal models for these deficiencies.

References 1–14 provide a partial list of reviews and books on CA research from the last decade. Future research is likely to focus on finding hereditary disorders resulting from deficiencies in the other CAs, developing murine models for each of these enzyme deficiencies using gene-targeted inactivation of the respective gene, defining the regulatory elements that govern the differences in tissue-specific expression of the different isozymes, and defining structure/function correlations that explain their different catalytic properties and susceptibilities to inhibitors. Hopefully, these studies will suggest specific inhibitors for individual isozymes that can be exploited in clinical medicine.

THE CARBONIC ANHYDRASE ISOZYMES—A BRIEF FAMILY HISTORY

Discovery of the first enzyme with CA activity in bovine erythrocytes took place only 60 years ago (15, 16), and recognition of more than one CA in human erythrocytes only 30 years ago (20); characterization of the last four members of the enzyme family took place only in the last five years.

CA activity was studied extensively in various tissues by physiologists and pharmacologists for nearly three decades following its discovery (reviewed in 17, 18) before a CA was first purified in 1960. The enzyme source was bovine erythrocytes, which contain a single, high-activity enzyme (19). Purification from human erythrocytes in 1961 (20) disclosed two forms, one in larger amounts than the other and having less activity, and a second form, present in smaller amounts, that has high activity similar to the bovine enzyme. These high and low activity isozymes, initially designated CA C and CA B, came to be known as CA II and CA I. The amino acid sequences of both CA I (21, 22) and CA II (23, 24) were reported in the early 1970s, and the x-ray crystal structures for CA II (25) and CA I (26) in 1972 and 1975, respectively (27).

Subsequently discovered isozymes were assigned numerical names in the order of their discovery. A sulfonamide-resistant CA was found in homogenates of male but not female rat livers in 1974 (28, 29), and a sulfonamide-resistant CA was isolated from chick muscle two years later (30). Subsequently, it became clear that the same 29-kDa protein had been purified from rabbit skeletal muscle in 1972 (31) and named BMP for basic muscle protein. Its lack of reactivity to antisera to CA I and II, and its insensitivity to sulfonamides, led it to be recognized as a distinct isozyme called CA III (32).

Although membrane-associated enzyme had been reported earlier, its purification from bovine lung in 1982 allowed CA IV to be identified as a distinct isozyme. The lung enzyme required strong detergent to solubilize it from membranes, was 52 kDa, much larger than the 29-kDa isozymes previously described, was resistant to solubilization in 5% sodium dodecylsulfate, and contained carbohydrate. The human enzyme was isolated subsequently from brush border membranes of kidney (34, 35) and human lung (35) and has been studied intensively over the past five years.

CA V is a nuclear-encoded, mitochondrial isozyme. A CA activity in mitochondria was suspected since 1959 (36) and supported by observed effects of CA inhibitors on mitochondrial metabolism (39). It was demonstrated by O^{18} exchange studies by 1980 (37, 38). Recently, a partial N-terminal sequence was obtained from enzyme isolated from guinea pig mitochondria (40), and a cDNA was isolated from mouse liver that was homologous with the partial sequence from guinea pig and appeared to encode a CA with a mitochondrial leader peptide (41). Expression of this cDNA (42) and the homologous human cDNA (43) allowed the mitochondrial isozyme to be characterized.

Isozyme CA VI is a secretory glycoprotein found in saliva. Although CA activity in saliva was recognized in 1946, it was not until 1979 that salivary CA from sheep was isolated and found to be distinct from other CAs (44). The rat enzyme was characterized in 1984 (45), the human enzyme in 1987 (46), and the complete sequence for the sheep enzyme reported in 1989 (47).

CA VII is unique in that, to this date, it is only a "virtual enzyme." Unlike the other isozymes that were recognized first as proteins for which genes were subsequently identified, the gene for CA VII was identified by homology to other CAs before the protein was recognized (48).

STRUCTURE OF THE CATALYTIC SITE AND REACTION MECHANISM

Structure of the Catalytic Site

X-ray crystallographic structures are available for three of the CA isozymes (49). The structure of CA II, the high activity form from human erythrocytes, has been determined and refined to 1.54Å resolution (50–52). The enzyme has a roughly spherical structure with the active site comprising a conical cleft about 15Å deep. One side of the cavity is formed by hydrophobic residues (Figure 1). The other side contains hydrophilic residues including Thr199 and Glu106. The zinc ion is located at the bottom of this cleft, and tetrahedrally coordinated to the imidazoles of the three histidine residues (His94, His96, His119) and to a water molecule (called the "zinc water") that ionizes to a hydroxide ion with a pK about 7 (53). This zinc coordination polyhedron is a conserved feature among CAs.

Amino acid residues Thr199-Glu106 contribute to a hydrogen bonding network with the Zn-OH⁻ that maintains the catalytically competent structure optimal for nucleophilic attack by the zinc-bound hydroxide on the CO_2 sub-

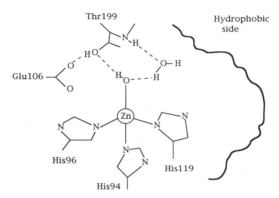

Figure 1 Diagrammatic representation of the active site of CA II. The zinc ion is tetrahedrally coordinated to the imidazole groups of three His residues and to a hydroxide or water (the zinc water) molecule. Thr199 hydrogen binds to the hydroxide ion or water and to Glu106 through its Oγl. Its peptide nitrogen binds to the "deep water" and may be the site of CO_2 binding (adapted from 10 and 86).

strate (54–59). Thr199 forms two hydrogen bonds from Oγl, donating a hydrogen bond to the carboxyl group of Glu106 and accepting a hydrogen bond from the zinc hydroxide. The peptide nitrogen of Thr199 also forms a hydrogen bond to a second water molecule in the deepest end of the cavity (referred to as the "deep water"). The Zn-OH⁻/Thr199/Glu106 hydrogen bond network, which is absolutely conserved in all nonplant CA isozymes (4, 8), restricts the orientation of the zinc-bound OH⁻, so that one of its lone electron pairs is directed toward the CO_2 molecule located in a hydrophobic pocket (55, 58).

Structural studies of inhibitors (10, 60–70) and HCO_3^- (71) binding to the metal ion show that the Zn-OH⁻/Thr199/Glu106 network is also important for binding bicarbonate, sulfonamide inhibitors, and many anionic inhibitors (see Ref. 10 for a comprehensive review of these studies). Almost all inhibitors appear to displace the deep water situated between the NH of Thr199, the zinc ion, and the hydrophobic portion of the active site (Figure 1). Some coordinate with the metal ion, and many contribute a hydrogen bond to the hydroxyl of Thr199. In fact, the position of the Thr199 and its ability to offer a lone pair of electrons and act as a hydrogen bond acceptor has led to the suggestion (10, 52, 59) that Thr199 plays a gatekeeper function, limiting access of the metal ion to anions and inhibitors that can serve as hydrogen bond donors. However, some inhibitors like azide can bind the metal ion without being able to contribute a hydrogen bond (66). In this case, the zinc-bound azide nitrogen makes a nonhydrogen bonded van der Waals contact with the hydroxyl group of Thr199 (66). Cyanide and cyanate displace the deep water and bind in the hydrophobic cavity without binding zinc. Instead, they hydrogen bond to the

NH of Thr199 in the binding site, which was suggested to be the position of the substrate CO_2 when attacked by the zinc bound hydroxide (57, 71).

The Catalytic Mechanism

The CAs efficiently catalyze the hydration of CO_2 to bicarbonate and a proton. One isozyme, CA II, is one of the fastest enzymes known, with a maximum turnover rate for the CO_2 hydration reaction of more than 10^6 s^{-1}. CAs also catalyze the hydrolysis of aromatic and aliphatic esters. The kinetics of enzyme catalysis and inhibition have been studied extensively and recently reviewed (9, 53, 72). Considerable evidence suggests that the reaction involves two steps (53, 57), conversion of CO_2 to HCO_3^-, leaving water as a ligand on the zinc (equation 1), and transfer of proton to solvent buffer through a proton shuttle group, His64 (equation 2).

$$+H_2O$$
$$EZn\text{-}OH^- + CO_2 \leftrightarrow EZn\text{-}HCO3- \leftrightarrow EZn\text{-}H_2O + HCO_3^- \qquad 1.$$
$$B$$
$$His64\text{-}EZn\text{-}H_2O \leftrightarrow H^+\text{-}His64\text{-}EZn\text{-}OH^- \leftrightarrow His64\text{-}EZn\text{-}OH^- \qquad 2.$$
$$BH^+$$

In the first step shown in Equation 1, the zinc-bound OH^- adds to CO_2 to yield a bicarbonate with a hydroxyl coordinated to the zinc (Figure 2a). In the second step, displacement of the zinc-bound bicarbonate by a water molecule releases bicarbonate and restores zinc-bound water. The proton release reaction (Equation 2), which is facilitated by His64, is the rate-limiting step for the high activity isozymes like CA II (53, 73) (Figure 2b). Inhibition of CA II by heavy metals like copper and mercury is explained by structural studies showing that they bind to His64 (69). The dependence on buffer to accept the proton shuttled by His64 explains why a minimum buffer concentration is required to achieve the maximal rate of turnover (53, 73, 75). Replacement of His64 in CA II with Ala by site-directed mutagenesis results in dramatic decreases in k_{cat} in the CO_2 hydration reaction, with the actual magnitude of the decreases depending on the pH, the buffer, and the buffer concentration (73–75). In fact, the activity of His64Ala CA II is nearly restored to that of wild type in the presence of sufficient imidazole buffer (and several other buffers) which apparently can bypass the rate-limiting proton shuttle and directly accept protons from the zinc water (73, 74). When additional single mutations were made on the background of the His64Ala mutant enzyme to introduce a histidine at position 62, 67, or 200, the k_{cat} of the enzyme increased to nearly 5% of that of the wild type enzyme (76). These results show, on the one hand, that histidine groups at certain other locations can participate in proton transfer.

Figure 2 a. Diagrammatic representation of CO_2/HCO_3^- interconversion. *b.* Diagrammatic representation of the proton transfer step in which a proton from the zinc water transfers to the His64 proton shuttle from which it transfers to solvent buffer.

On the other hand, they show that the location of the proton shuttle at His64 is critically important for optimum proton transfer.

Other combinations of kinetic and structural studies on wild type and mutant enzymes produced by site-directed mutagenesis have clarified a number of other important structure-function relationships. These include the role of His200 in explaining the higher affinity of CA I than CA II (Thr200) for HCO_3^- (71, 77, 78), the zinc binding coordination and structure of several enzyme-inhibitor complexes (10), the mobility of the protein shuttle residue His64 (79), the key roles of Glu106 and Thr199 in the catalytic function of CA II (57, 58), and the dispensability of Tyr7 for catalytic function of CA II (58). Studies of the hydrophobic pocket adjacent to the zinc-bound hydroxide (80-83) have implicated this region as the site for CO_2 association, provided evidence that this region modulates the catalysis of CO_2 hydration, and defined the structural requirements for the hydrophobic residues in this region.

Differences in properties between the naturally occurring isozymes CA II and CA III have been explored by site-directed mutagenesis. Changing Phe198 in CA III to Leu198 to make it correspond to CA II at this position did increase the k_{cat}/K_m 25-fold and made it more CA II–like (84). However, replacing four active site residues of CA II (His64, Asn67, Leu198, and Val207) with Lys64, Arg67, Phe198, and Ile207, as found in the low-activity muscle-specific CA III, did not reduce the level of activity of the modified CA II nearly to that of

CA III (85). Additional structural features must contribute to the activity differences between CA II and CA III.

As was the case with His64Ala, several individual replacements have been particularly instructive in testing hypotheses regarding structural requirements for catalysis. Among these are the substitutions for Thr199 and Glu106 that alter the hydrogen bonding network of the active site (57–59). Replacement of Thr200 in CA II with His, the residue normally found in this position in CA I–type enzymes (77, 78), also dramatically affects catalysis, in large part by increasing the affinity for HCO_3^- and making HCO_3^- dissociation a rate-contributing if not rate-limiting step for CO_2 hydration by Thr200His CA II, as it is for CA I (77, 78). It was this substitution that stabilized the bicarbonate complex with Thr200His CA II sufficiently to allow structural studies of the enzyme-bicarbonate complex (71). The dramatic progress in this area has been summarized in recent reviews (10, 86).

CONSERVATION OF PRIMARY STRUCTURE AMONG THE HUMAN CAS

Figure 3 (p. 388) presents the alignment of the amino acids in the seven human isozymes, as determined directly or deduced from their cDNAs (13, 87–90). Numbering is based on that of CA I. The 40 residues boxed (15% of the total aligned sequence) are identical in all seven CAs. Comparisons of sequences of each CA with those of the others, and those of CAs from other amniotes, shark, and algae, have been used to deduce the evolutionary patterns of descent from an ancient common precursor (4, 8, 13). It was estimated that CA IV and CA VI are the oldest mammalian CAs evolutionarily, that CA I, II, and III diverged most recently (i.e. between 300 and 400 million years ago), and that CA V and CA VII occupy intermediate positions. The percent identities in the human CAs are consistent with this interpretation. Thus, the putative latecomers, CAs I, II, and III, show 58–60% identity with each other in amino acids at similar positions (46). The percent similarities between the more ancient CA IV and other CAs are: CA I, 31%; CA II, 36%; CA III, 33%; CA VI, 33%; and CA VII, 32% (87). Intermediate is the sequence of CA V, the most recent human CA sequence reported, which reveals the following percent identities with other CAs: CA I, 47%; CA II, 49%; CA III, 44%; CA VII, 48%; CA IV, 30%; and CA VI, 35% (43).

Note that CAs IV, V, and VI have N-terminal and C-terminal sequences that extend beyond the regions of alignment with the cytoplasmic CAs, CA I, II, III, and VII. The N-terminal sequences of CAs IV, V, and VI were deduced from their cDNAs but are not present in the mature proteins. The 18 and 17 amino acid hydrophobic signal peptides at the N-termini of CA IV and CA VI are presumed to be removed cotranslationally as the enzymes are synthesized and translocated into the endoplasmic reticulum of the cells that express them. Likewise, the precursor for CA V contains a 38–amino acid N-terminal mitochondrial leader peptide that is not present in the mature protein isolated from mitochondria (43, 88). The N-terminal leader peptide is assumed to be removed following import into mitochondria.

Both CA IV and CA VI also have additional C-terminal amino acids that extend beyond the aligned sequences. In the case of CA IV, C-terminal processing removes the C-terminal 28 amino acids, and the C-terminus of the 266 amino acid cleaved amino terminal portion of the CA IV precursor is transferred to a glycosylphosphatidylinositol (GPI) anchor (87). Thus, the C-terminal 28 amino acids in CA IV (underlined in Figure 3) do not appear in the mature protein. By contrast, the 30–amino acid C-terminal extension of CA VI is not removed from the secretory glycoprotein, based on the 36–kDa size of the human salivary glycoprotein following removal of carbohydrates (46).

PROPERTIES OF INDIVIDUAL HUMAN ISOZYMES

The Cytosolic Isozymes CA I, II, III, and VII

CARBONIC ANHYDRASE II   CA II is a high-activity isozyme with a maximum turnover rate for CO_2 hydration of 1×10^6 s^{-1} (91), and has the widest distribution, being expressed in the cytosol of some cell types in virtually every tissue or organ (13). Cell types expressing CA II include osteoclasts in bone, oligodendrocytes in brain, epithelium of the choroid plexus (brain) and the ciliary body (eye), lens, Müller cells in retina, liver (mainly perivenous hepatocytes), kidney (proximal tubule, distal tubule, and intercalated cells of the cortical collecting ducts), acinar cells in salivary glands, pancreatic duct cells, gastric parietal cells, endometrium of the uterus, endothelial cells, epithelial cells of seminal vesicle and ductus deferens (92), spermatozoa (93), erythrocytes, and platelets. It has also been reported recently in neutrophils (94), type II epithelial cells of lung (95), endothelial cells and epithelial cells of duodenum, intestine, and colon (96), and zona glomerulosa cells of the adrenal (97).

The physiological roles of CA II in these cell types are diverse (6, 13). In some cells, CA II plays a major role in contributing to acid base homeostasis. It contributes to H$^+$ secretion by gastric parietal cells, by renal tubular cells that secrete H$^+$ to produce urinary acidification, and by osteoclasts that secrete H$^+$ to acidify the bone-resorbing compartment. CA II promotes HCO$_3^-$ secretion by pancreatic duct cells that contribute HCO$_3^-$ to pancreatic juice, by ciliary body epithelium (which produces aqueous humor), by choroid plexus (which produces cerebrospinal fluid), by salivary gland acinar cells (which produce saliva), and by distal colonic epithelium, where H$^+$ and HCO$_3^-$ secretion are coupled to Cl$^-$ and Na$^+$ reabsorption and contribute to electrolyte and water balance (98). CA II also promotes CO_2 exchange in proximal tubules in the kidney, in the erythrocytes, and in lung. It has been suggested that it contributes to fatty acid and amino acid synthesis (13).

The transcriptional regulation that appears to account for the different tissue-specific expression of CA II was recently reviewed (13), as was the evidence for hormone regulation of CA II in uterus, bone, and prostate. Recent evidence suggests that there is up-regulation of CA II in kidney in metabolic acidosis (99).

CARBONIC ANHYDRASE I   CA I is five to six times as abundant as CA II in human erythrocytes, but has only about 15% of the activity ($k_{cat} = 2 \times 10^5$ s^{-1}) (91). Thus, in intact adult erythrocytes, CA I contributes about 50% of the total CA activity (100). CA I is much more sensitive to inhibition by Cl^- and other halide ions than is CA II, and less sensitive to sulfonamide inhibitors. It is expressed in epithelium of the large intestine, corneal epithelium, lens, ciliary body epithelium, sweat glands, adipose tissue, and myoepithelial cells (13). Recent reports add neutrophils (94) and zona glomerulosa cells of the adrenal gland (97) to this list.

CA I is not detectable in human fetal erythrocytes but is switched on at 40 weeks gestation at the time of expected normal birth (101). It rises to adult levels over the first year of life. Transcriptional and developmental regulation of the CA I gene have been reviewed recently (13). CA I is also strongly expressed in colon, but from a colon-specific promoter that is distinct from the erythrocyte promoter (5, 102, 103).

Although CA I is the most abundant nonhemoglobin protein in red blood cells, no hematologic abnormalities result from its absence (13). In addition, since neither felids nor ruminants express erythrocyte CA I (4), its physiological importance is unclear. Presumably, other CAs or other mechanisms can compensate for its absence in isolated CA I deficiency. However, one might expect CA I to be essential for survival in the absence of CA II.

CARBONIC ANHYDRASE III   CA III is a very low-activity isozyme, its activity for CO_2 hydration being only 1% that of CA II. Part of its low activity is attributable to absence of the His64 to carry out the proton shuttle function (73, 104). Part is attributable to the presence of Phe198 in place of Leu198 (as in CA II) (105, 106). Changing Phe198 to Leu198 increases catalytic activity by 25-fold (105). CA III is quite insensitive to inhibition by concentrations of sulfonamides that readily inhibit the other isozymes. Phe198 probably also contributes to sulfonamide resistance by sterically inhibiting access to the cavity (107). Because of its sulfonamide resistance, the CO_2 hydration activity observed in tissues in the presence of 1–5 µm acetazolamide is generally assumed to be CA III (108).

Its major site of expression is skeletal muscle, where it can represent 8% of the soluble protein of slow-twitch (type I fiber) red skeletal muscle. It is expressed in other fiber types in lower amounts (109) and in high levels in adipose cells, at least in the rat (110). CA III is expressed at lower levels in other tissues, including salivary glands, smooth muscle cells in uterus, red cells, prostate, lung, kidney, colon, and testis (4).

Developmental studies show that CA III is expressed at low levels in human muscle during early fetal development but increases rapidly during the last trimester to reach 50–60% of adult levels at birth (111, 112). Developmental studies have also been done in the rat, mouse, and in myogenic cell lines (reviewed in 13). Of interest is the expression of CA III in notochord at even higher levels than in developing muscle in the embryonic mouse (113).

Although CA III was first purified in 1972, and we have since learned a great deal about its structure, kinetics, hormonal regulation by thyroid (reviewed in 4), and developmental regulation, its functional role is still a mystery (81).

CARBONIC ANHYDRASE VII   As mentioned earlier, CA VII is only a virtual enzyme which we know conceptually from its gene, its cDNA, and its mRNA detected by in situ hybridization (13, 48). The mRNA is expressed primarily in the cytosol of the salivary glands. The cDNA has not yet been expressed, nor has the protein been isolated from salivary glands. Thus, its activity, properties, and functional role remain speculative.

The Membrane-Associated CA, CA IV

CARBONIC ANHYDRASE IV   CA IV is a fascinating isozyme whose importance has been appreciated only recently. It was originally purified from bovine lung and found to be a membrane-associated glycoprotein of 52 kDa that was resistant to SDS, apparently stabilized by disulfide bonds (33). The purified enzyme was a high-activity enzyme like CA II and was even more resistant to halide ions, although somewhat less resistant to sulfonamide inhibitors. The human enzyme was found to have many similar properties, but it is smaller (35 kDa) and contains no carbohydrate (34, 35). In lacking carbohydrate, the human enzyme is unique among the nine mammalian CA IVs so far studied. The CA IVs vary in molecular mass from 39–52 kDa, and contain 1–5 N-linked oligosaccharide chains. All of the other mammalian CA IVs are reduced to 35–36 kDa when the enzymes are treated with endoglycosidase F to remove N-linked oligosaccharide chains (114).

CA IV, the only membrane-anchored CA, is anchored to membranes by a glycosylphosphatidylinositol anchor and can be released from membranes by treatment with phosphatidylinositol-specific phospholipase C (35, 87, 114).

The cDNA for the human enzyme has been cloned, sequenced, and expressed, and the biosynthesis and turnover of the enzyme studied (87). The genomic sequences have also been characterized (115). Recent kinetic studies (116) confirmed its high-activity character, its resistance to salt inhibition, and its somewhat decreased sensitivity to sulfonamide inhibitors (averaging a 17-fold decrease in sensitivity compared to CA II with a battery of different sulfonamide inhibitors).

CA IV is expressed on the apical surfaces of epithelial cells of some segments of the nephron, the apical plasma membrane in the lower gastrointestinal tract, and the plasma face of endothelial cells of certain capillary beds. In rat kidney, it is highly expressed on the apical brush border membranes (and somewhat less expressed on the basolateral membranes) of proximal tubular cells and cells of the thick ascending limb (117). In the lung, it is expressed on the plasma face of the pulmonary microvasculature (118). Developmental studies show that the CA IV mRNA is not expressed in fetal lung (though it is expressed in fetal kidney). Comparisons of mRNA levels in the adult rat by northern blot analysis suggested that the highest levels of expression are in colon, followed by brain, lung, kidney, and heart. In brain, CA IV has been localized to the plasma face of the cortical capillaries and has been proposed as a marker for the blood-brain barrier (119). CA IV is also expressed on the plasma face of endothelial cells of the choriocapillaris of human eye (120), the microcapillaries of skeletal and cardiac muscle (121), and the microvasculature in rat and human colon (S Parkkila, WS Sly, unpublished observations). It has also been demonstrated on specific epithelial cells of the human reproductive tract (122) and abundantly expressed on the apical plasma membrane of the colon (S Parkkila, WS Sly, unpublished observations).

The physiological role of CA IV in kidney is to facilitate HCO_3^- reabsorption by catalyzing its dehydration to CO_2 (11, 12). In fact, CA IV is the luminal CA in the proximal tubule of the kidney that was inferred from micropuncture studies to mediate 85% of bicarbonate reabsorption in kidney. In other tissues (like brain, skeletal muscle, and heart muscle), it promotes CO_2 flux by facilitating CO_2 hydration to form bicarbonate and accelerating CO_2 removal by the microvasculature from tissues generating CO_2 from metabolism. In lung, it acts to promote CO_2 exchange from the blood to the alveoli by facilitating HCO_3^- dehydration to CO_2 which diffuses across the alveolar membrane. In colon, its role is less clear. Since Na^+ and Cl^- reabsorption are coupled to HCO_3^- secretion that is inhibited by CA inhibitors, a CA is known to be involved (98). The relative contributions of cytoplasmic and membrane CAs to this process are not yet clear. CA IV may facilitate recycling of secreted HCO_3^- and H^+, converting them to CO_2 which can be reabsorbed to promote another round of Na^+ and Cl^- uptake.

The Secretory Form of CA in Saliva, CA VI

CARBONIC ANHYDRASE VI   Human CA VI is a 42-kDa glycoprotein that has been purified and characterized from human saliva (46). Its cDNA was cloned from human parotid. The deduced amino acid sequence presented in Figure 3 is 72% identical to that of sheep CA VI (89), and the two cysteines, which were shown to be linked by a disulfide bridge in the sheep enzyme, are conserved. The human cDNA predicts three potential glycosylation sites. Complete and partial digestion products of 36 kDa and 39 kDa, respectively, following treatment with endoglycosidase, suggest that two of these sites are used in the human enzyme (46).

CA VI has also been purified from rat saliva and characterized (45). Both the rat CA VI and the sheep CA VI resemble CA II in being high-activity isozymes (42, 45). The purified human isozyme had considerably lower activity (only 2–3% of that of CA II) (46). Although it resembles CA II in some of its inhibitory properties, it is distinct from CA II in its sensitivity to chloride, acetazolamide, and methazolamide (46).

Human CA VI was found to be expressed only in salivary glands, being prominent in the serous acinar cells of the submandibular gland and parotid gland (124). The mean enzyme content in saliva is $6.8 + 4.3$ µg/ml (125), and the estimated daily output in saliva is 10–14 mg per day. Although the cytoplasmic CA II in submandibular and parotid glands mediates HCO_3^- secretion into saliva, the secreted CA VI probably plays a role of pH regulation in saliva, using the buffer provided by CA II, and may have a protective effect in esophagus and stomach (126).

Mitochondrial CA, CA V

CARBONIC ANHYDRASE V   The cDNA for human mitochondrial CA V was recently cloned from a human liver cDNA library (43). See Figure 3 for the deduced amino acid sequence. Expression of the cDNA in COS cells produced proteins of the masses expected for the precursor (34 kDa) and mature (30 kDa) mitochondrial enzyme. Only the 30-kDa protein was detected in mitochondria isolated from adult human liver. Processing from precursor to mature CA V in COS cells involves removal of a 38-amino acid mitochondrial leader sequence. Because the amount of activity expressed was only 0.06% of the activity expressed from the same vector containing the cDNA for human CA II, the human CA V was thought to be a "low-activity isozyme." However, Heck et al (42) recently reported purification of mature murine CA V from bacteria expressing the murine cDNA. They found that the k_{cat} for hydration increased dramatically with pH, with an apparent pKa of 9.2. The k_{cat} at pH 7 was only 2.2×10^3 s^{-1}, but it increased with pH to 3×10^5 s^{-1} at pHs above 9.2. From these results, they concluded that the activity at alkaline pH for

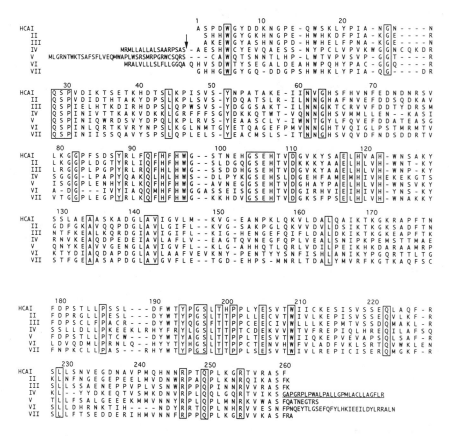

Figure 3 Comparison of amino acid sequences of human CAs I–VII. Homologous residues in all seven isozymes are boxed. Gaps are introduced to optimize alignment. The arrow (upper left) indicates the sites of cleavage of the leader sequences of CA IV, V, and VI. Amino acids to the left of the arrow are deduced from the cDNA sequences and are not found in the mature CA IV, V, or VI. The underlined hydrophobic C-terminal sequence of CA IV is cleaved off during GPI anchoring and not found in the mature GPI-anchored enzyme. The numbering system used is based on that of human CA I. Sequence data for CA I, II, III, and VII are taken from Ref. 13. Sequences for CA IV, V, and VI are from references 87, 88, and 89, respectively.

mouse CA V is quite appreciable. Although the enzyme resembled CA I in its catalytic properties, its sensitivity to inhibitors most resembled those of CA II. Thus, the suggestion that human CA V is a low-activity isozyme, based on the activity expressed in crude extracts of transfected COS cells (43), should be viewed with caution. Also, an earlier estimate of the turnover number of

$24,000 \text{ s}^{-1}$ for CA V purified from rat liver (127) refers to studies at pH near 7. Possibly, the maximal k_{cat} of the rat enzyme is considerably higher (42).

There are two obvious differences in amino acids near the active site between human CA V and CA II. Tyr7 is replaced in human CA V by Thr, and His64 is replaced by Tyr. Heck et al (42) replaced Tyr64 with His64 in the murine enzyme by site-directed mutagenesis and concluded that the unique kinetic properties of murine CA V are not explained by the Tyr64. The replacement of Tyr7 in CA V with Thr is probably not a critical difference either because Liang et al reported that Tyr7 is not essential for CA II (58).

The distribution of CA V in human tissues has not been established. The mouse mRNA for CA V was detected only in liver on northern blots of mRNA from seven tissues examined (41). Rat CA V has been detected in many more tissues (128, 129), but there is some question whether the antibody used in these studies may have recognized malate dehydrogenase instead of CA V (130). In a recent study using antibodies to synthetic peptides corresponding to the C-termini of mouse and rat CA Vs (131), the mouse antibody detected CA V only in liver of nine tissues examined. However, CA V was detected in six of nine rat tissues (130). The signal was most intense in rat liver, followed by heart, lung, kidney, spleen, and intestine. No signal was detected in brain, testis, or muscle.

CA V has been suggested to be important for two metabolic pathways that depend in part on mitochondrial enzymes (39). One is gluconeogenesis, where CA V may be required to supply HCO_3^- to pyruvate carboxylase in mitochondria. The second is ureagenesis, where CA V may be required to supply HCO_3^- to carbamyl phosphate synthetase in mitochondria. Involvement of CA V in these processes has been suggested to explain the effects of CA inhibitors on these pathways (reviewed in 132).

Other CA-Related Proteins and Unrelated CAs

CA-RELATED PROTEINS   There are several examples of proteins with sequence homology to CAs that do not have CA activity. One is a gene product found in vaccinia virus encoded by the D8 gene (133). This intronless gene encodes a 304–amino acid, transmembrane protein with over 30% similarities to mammalian CAs. Its extracellular domain is CA-like, but it has no CA activity. It is a nonessential viral protein, the function of which is unclear. Another viral example is the erb-A gene of avian erythroblastosis virus (AEV) (134). The 3' end of this gene (domain 2) encodes a CA-like domain with similarity to residues 10–183 of CA I. Presumably, both of these viral examples reflect resourceful acquisition of CA sequences for some structural purpose by organisms that did not need to conserve the active site residues required for CA activity.

Three mammalian examples have also been discovered recently. A CA-related mouse brain cDNA was isolated, the deduced sequence of which shows 33–41% homology with CA I, II, III, and VI (135). Two active site substitutions, His94Arg and Gln92Glu, explain its lack of CA activity. This protein, named CARP for CA-related protein, is expressed in Purkinje cells. The human homologue is 98% identical (136), reflecting a high degree of conservation for this Purkinje cell protein with a yet unknown function.

Another highly conserved CA-related protein isolated from a cDNA library from brain is a member of the protein tyrosine phosphatase called PTP-zeta (137). This transmembrane protein has a cytoplasmic domain that encodes two PTPase domains, only one of which is active, and has a large 1616–amino acid extracellular receptor domain whose N-terminal 266 amino acids are homologous to CA. The CA-like structure was proposed to bind a small molecule ligand. The same gene was cloned by another group (138), who mapped it to human chromosome 7 and found its expression limited to brain.

Another CA-related tyrosine phosphatase, RPTP gamma, is encoded by a candidate tumor suppressor gene for renal cell carcinoma that maps to human chromosome 3p14.2-p21 (139, 140). It also has two tandem PTPase domains in the cytoplasmic tail, and its large extracellular domain contains a 266–amino acid stretch of homology to CAs. Although 11 of 19 active site residues are conserved, only 1 of 3 His residues that coordinate with zinc is conserved, making it unlikely that it has any CA activity. The murine homologue of this gene is expressed in developing brain, and both alleles are found defective in mouse L cells. Inactivating the CA-like domain, the putative receptor for some small molecule ligand, is thought to inactivate the tumor suppressor gene.

UNRELATED CARBONIC ANHYDRASES   The periplasmic CA from green algae *Chlamydomonas reinhardtii* (141) is homologous to the animal CAs, with conservation of all of the active site residues thought to be critical for catalysis including 92, 94, 96, 106, 107, 117, 119, and 199 (see Figure 3). Thus, it appears to be related to the animal CAs and to be derived from the same common ancient ancestral gene.

For this reason, it came as a surprise when the recently determined sequences deduced from the cDNAs from spinach (142, 143) and from pea chloroplasts (144) were completely unrelated to CAs in the animal kingdom. This was even more surprising since the spinach enzyme has been characterized as a zinc-containing, high-activity enzyme with nearly 50% the activity of CA II and as having some affinity for sulfonamide inhibitors. It would appear that the plant kingdom achieved a solution to the need for CA activity by a completely different evolutionary strategy than that used by Chlamydomonas and all the members of the animal kingdom so far studied.

GENE ORGANIZATION AND CHROMOSOME LOCALIZATION

The structures of human genes for the CA I, II, III, IV, and VII (Figure 4) are similar in possessing six introns that separate exons 1–7. However, the human CA I gene has two additional noncoding exons called exon 1a and 1b, at the 5' end of the gene (145), with 36 kb between exon 1a and the first coding exon, 1c.

The human CA IV gene also has an additional exon (115) (exon 1a) that encodes the signal sequence. Exons 1b through 7 encode the remaining coding sequences and the 3' end untranslated region. The positions of introns 3, 4, 5, and 6 are identical with the corresponding positions of introns in the genes for the soluble CA isozymes (CA I, II, III, and VII). However, the positions of introns 1b and 2 in CA IV differ from the positions of the corresponding introns in genes for the soluble isozymes.

The genes for human CAs I-VII have been assigned to chromosomes 1, 8, 16, and 17. The CA I, II, and III genes (CA1, CA2, CA3) are clustered in a stretch of about 180 kb on chromosome 8q22 in the order of CA1, CA3, CA2. The CA II and CA III genes are transcribed in the same direction and opposite to that of CA I (146). CA IV was assigned to 17q23 (115), and CA VI was assigned to chromosome 1p36.22–33 (147). CA V was recently assigned to chromosome 16 (43), and CA VII was previously mapped on 16q22 (48).

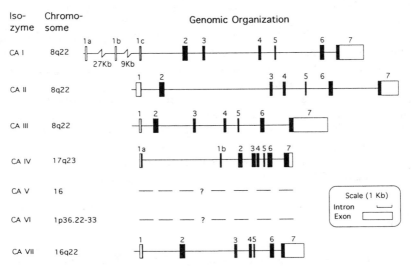

Figure 4 Structures and chromosomal localizations of human carbonic anhydrase isozyme genes. Numbered exons are indicated by boxes in which the coding regions are filled in black. The scale for exons is double that for introns. Sources: Human CA I (145); CA II (155); CA III (112); CA IV (115); CA V (88); CA VI (89); CA VII (48).

CARBONIC ANHYDRASE DEFICIENCY DISEASES

CA II Deficiency Syndrome

DISCOVERY OF CA II DEFICIENCY AND ITS IMPACT   To date, the only known disease attributable to a deficiency of CA is the CA II deficiency syndrome (11), which is associated with osteopetrosis with renal tubular acidosis and cerebral calcification (McKusick catalog #259730) (148–151). In 1983, Sly et al (152) reported that the three affected members of the original American family with this disorder (150) lacked CA II in their erythrocytes, and that their normal-appearing parents and many of their first degree relatives had half-normal levels of CA II. CA I, the other CA isozyme normally present in erythrocytes, was present in normal amounts. These observations led them to propose that CA II deficiency was the primary defect in this newly recognized metabolic disorder. Since then, more than 50 patients with this syndrome have been reported, all of whom were shown to have CA II deficiency. Detailed clinical manifestations and pathogenesis are reviewed in 11, 12, 153, and 154.

This discovery attracted great interest for several reasons. First, it provided a biochemical explanation for one of several inherited forms of osteopetrosis, a method to distinguish patients with this form of osteopetrosis from those with other forms, and to screen for heterozygous carriers of this recessive form of osteopetrosis. Second, at a more fundamental level, it provided clear genetic evidence for a role of CA in bone resorption, and specifically implicated isozyme II in the process. Third, it allowed the distinct individual roles of CA II and CA IV in HCO_3^- reabsorption and acidification by the proximal renal tubules in the kidney to be delineated. Fourth, it clearly demonstrated an essential role for CA II in supporting H^+ secretion in the distal parts of the nephron that mediate distal urinary acidification. Fifth, the mental retardation and brain calcification seen in over 90% of affected patients demonstrated the importance of CA II (already known to be expressed in, and limited to, oligodendrocytes in brain) in normal brain development. Finally, the initial report appeared in April, 1983 (152), five months before the symposium (1) celebrating the 50th anniversary of the discovery of CAs, and the five decades of research that followed from that discovery. Naturally, the report that CA II deficiency is the basis for a newly recognized metabolic disorder affecting bone, brain, and kidney added considerable excitement to the anniversary celebration (1).

MOLECULAR GENETICS OF CA II DEFICIENCY   Seven different mutations in the CA II structural gene have been identified by PCR amplification of genomic DNA from patients with this disorder (Figure 5). The first mutation was identified in a Belgian patient homozygous for a C-to-T transition in exon 3,

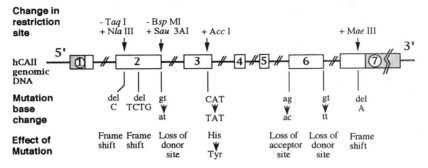

Figure 5 Seven structural mutations found to date in the human CA II gene. The + or − signs indicate the restriction sites introduced (+) or removed (−) by the mutation. Capital letters and the lower cases represent base pairs in exons and introns respectively. A gt→at change in the 5' end of intron 2 destroyed a splice junction donor site. This is the commonest mutation in Arabic patients (159). A C→G transition in exon 3 results in a replacement of the conserved histidine at position 107 with tyrosine (His107Tyr). This mutation was identified in a homozygous Belgian patient (155), a homozygous Italian patient (PY Hu, P Strisciuglio, J Ciccolella, and WS Sly, unpublished data), and also as one of two mutations in the three American sisters who were compound heterozygotes. Their second mutation is an A→C transversion at the 3' end of intron 5, which destroys a splice junction acceptor site (156). A single-base deletion in the coding region of exon 7 results in a frameshift at codon 227, which changes the next 12 amino acids and introduces a UGA stop codon 22 amino acids earlier than in the normal enzyme. This mutation was found to be common in Hispanic patients from the Caribbean islands (160, 161). Two different deletions in exon 2 and a gt→tt change in the 5' end of intron 6 were discovered recently in two Italian families and one American family (PY Hu, P Strisciuglio, J Ciccolella, and WS Sly, unpublished data).

which results in replacement of the conserved histidine at position 107 with tyrosine (His107Tyr) (155). The three affected sisters in the American family in which CA II deficiency was first reported were also found to have this mutation. However, they were compound heterozygotes, having inherited the His107Tyr mutation from their mother and a splice acceptor mutation in the 3' end of intron 5 from their father (156). Neither the Belgian patient nor the American patients were mentally retarded, as were most subsequently reported patients. Frequent skeletal fractures were the most disabling manifestation of this disease (157). When the CA II cDNA containing the His107Tyr mutation was expressed in *E. coli*, some CA activity was detected. These experiments led to the suggestion that a small amount of residual CA II activity in patients with the His107Tyr mutation may allow them to escape mental retardation (156). Detailed kinetic analysis of the mutant His107Tyr CA II isolated from overexpressing bacteria was reported by Tu et al (158).

The third structural gene mutation identified is a splice junction mutation at the 5' end of intron 2, which was found in patients from Kuwait, Saudi Arabia, Algeria, and Tunisia (159), and all of the patients of Arabic descent so far studied. More than 75% of the CA II-deficient patients so far recognized have been Arabic (157) and have been severely affected. In these patients with the "Arabic mutation," mental retardation and metabolic acidosis were prominent, while bone fractures were less frequent (157, 159).

A frameshift mutation resulting from a single-base deletion in the coding region of exon 7 was found in a mildly affected Hispanic girl, who is the only patient reported so far without renal tubular acidosis (160, 161). This single-base deletion results in a frameshift at codon 227 that changes the next 12 amino acids and introduces a UGA stop codon at codon 239. The truncated enzyme resulting from the mutation is 22 amino acids shorter than the 260 amino acids in normal CA II and is inactive. However, when expressed in bacteria, the mutant allele produced 0.05% of the activity expressed by the normal allele. This mutant enzyme activity resides in a small fraction of near-normal size enzyme (29 kDa), which had about 10% of normal specific activity. Protein sequencing showed that the first 11 amino acids following the frameshift mutation were abnormal in the 29–kDa mutant protein, as predicted, after which the reading frame was restored to normal. The last 23 amino acids of the 29–kDa mutant protein were the same as in normal CA II. These results can be explained by a ribosomal −1 translational frameshift that restores the reading frame 11 codons after the original mutation and allows completion of full-length CA II. Subsequently referred patients from seven independent Hispanic families, all derived from Caribbean ancestors, were found by sequencing or restriction site analysis to be homozygous for the same mutation. However, some of these Hispanic patients had severe clinical manifestations, including severe renal tubular acidosis, anemia, and hepatosplenomegaly. The basis for the wide clinical variability in these patients is not clear. However, these findings raise the interesting possibility that individual variation in efficiency of frameshift suppression could contribute to clinical heterogeneity among patients with identical frameshift mutations.

Recently, three new mutations were discovered in the CA II structural gene of CA II–deficient patients (PY Hu, P Strisciuglio, J Ciccolella, and WS Sly, unpublished data). Two deletion mutations were found in exon 2. A four base pair deletion was identified in an American family, and a single base deletion in an Italian family. Two other Italian families were studied recently, in one of which the CA II–deficient patient was homozygous for the already described His107Tyr mutation. In the other family, the CA II–deficient patient was homozyous for a splice junction mutation at the 5′ end of intron 6 (Figure 5).

CA I Deficiency Causes No Disease

The first report of human CA deficiency was a deficiency of CA I in a family originating from the Greek island of Icaria (163). However, except for the reduction in red cell CA I to trace levels, the individuals homozygous for the deficiency gene in this family had no hematological or other abnormalities. Recently, it was reported that CA I–deficient members of this family had a missense mutation (Arg246His) in exon 7 of their CA I gene (164). This mutation is the probable cause of the CA I deficiency. Arg246 is not only conserved among all seven human CA isozymes (Figure 3) but is also present in all animal CAs and even in the CA-related proteins (13). This conservation led to the suggestion that Arg246 plays a critical role in the intramolecular structures of all CA molecules, and the substitution of His at this position results in an unstable molecule (164).

Even though CA I normally contributes to CA activity in vivo in humans, unless some low level activity of the Arg246His CA I is present in nonerythroid tissues, other CA isozymes or alternative processes can apparently substitute for the function of CA I when it is absent. The failure to find any clinical abnormalities in CA I deficiency led to the conclusion that the earlier report of a partial deficiency of CA I in a patient with distal renal tubular acidosis and deafness was merely a coincidental association (152).

A survey of red cell CA I concentrations in 3376 individuals in a Japanese population revealed that 25 had about half the normal level of CA I (165).

Other CA Genes in Search of a Disease

Will other CA deficiencies be found, or will the loss of other CA genes be harmless, as is the case for CA I? We hypothesize that loss of certain other CA genes would be quite deleterious. For example, CA IV is responsible for most of the bicarbonate reclamation by kidney. Thus, the phenotype of CA IV deficiency should include proximal renal tubular acidosis, at the minimum. How serious the loss would be of CA IV from brain, lung, eye, heart, and skeletal muscle would determine whether CA IV deficiency is compatible with life, and if so, what other handicaps such patients might have.

CA V deficiency may also be deleterious. Loss of CA V would probably lead to impairment of gluconeogenesis and ureagenesis. The defect in gluconeogenesis might produce fasting hypoglycemia, and the defect in ureagenesis, hyperammonemia. Many patients with unexplained fasting hypoglycemia or hyperammonemia or both are considered potential candidates for CA V deficiency. The availability of the cDNA and genomic sequences for both CA IV and CA V make it possible to examine DNA for mutations in patients suspected of having these deficiencies.

It seems less likely that loss of CA VI would produce a debilitating disease than loss of CA IV or CA V. However, if it does indeed have an important role in pH regulation in the upper alimentary tract as suggested (123), its loss could be significant.

Another difficult phenotype to predict is that of CA III deficiency. Despite the abundance of CA III in red skeletal muscle, its precise role in, and importance to, muscle is still unclear. Is it in muscle for CO_2 hydration—or does it have another more important role? Also intriguing is the role of CA III in adipose tissue, and its apparent decline with the onset of obesity (108). It is difficult to anticipate what might happen in adipose tissue with loss of CA III. It seems likely that targeted mutagenesis in embryonic stem cells will provide murine models for individual deficiencies of each of the CAs in the next few years. Although these models will be important, the murine model does not always have the same features as the human disease. This is obvious from the CA II–deficient mouse (discussed below) that lacks the bone and brain findings of human CA II deficiency.

ANIMAL MODELS OF CA DEFICIENCY

CA I Deficiency in the Pigtail Macaque

The pigtail macaque has a common mutation that produces CA I deficiency in homozygous animals (166, 167). The mutation ($-85C \rightarrow G$) creates an AUG codon 6 bp 5' of the end of exon 1a and creates an upstream open reading frame (ORF) terminating 5 bp from the normal AUG (13). This mutation is of unusual interest as an example of an upstream ORF inhibiting translation from a downstream AUG, as an example of a nondeleterious mutation that occurred in a founder population prior to the Pleistocene glaciation, and because it has a peculiar *cis* negative regulatory effect on CA II expression, even though the CA II promoter is 100 kb away from the CA I mutation in exon 1a of the CA I gene. However, this is not a disease model, as CA I deficiency produces no disease in the pigtail macaque, as is also true in humans.

Mouse CA II Deficiency

A CA II–deficiency allele was produced in the mouse by chemical mutagenesis using ethylnitrosourea (169). Surprisingly, the homozygous CA II–deficient mice showed neither osteopetrosis nor cerebral calcification. These mice did exhibit growth impairment and a defect in renal acidification, and they were later found to develop calcifications in blood vessels with advanced age (170). The mutation was identified as a C-to-T substitution introducing a stop codon in place of Gln155 (Gln155X) (JS Platero, WS Sly, unpublished observations).

It is unclear why the CA II–deficient mouse lacks some important features of the human disease. Nonetheless, the absence of CA II made possible precise histochemical localization studies that clarify the distribution of membrane-associated CA IV in the murine kidney (171) and the eye (172). Brechue et al (173) used cell fractionation techniques to define the subcellular distribution of CA IV in the CA II–deficient mouse and also characterized its renal acidification abnormalities. Others (174) have used the CA II–deficient mouse in seizure studies and found that CA II deficiency confers increased resistance to seizure-inducing stimuli. Biesecker et al (175) studied the effects of transplantation of CA II–deficient mice with normal bone marrow, and of normal mice with marrow from CA II–deficient animals. The latter suggested that the renal acidification defect, and the markedly increased sensitivity of these animals to acid loading, result from both the erythrocyte and the renal CA II deficiency.

Two groups have reached different conclusions regarding the oligodendrocyte findings in brain and the neuropathology. Ghandour et al (176) made detailed morphological and histochemical analyses of oligodendrocytes and astrocytes and concluded that there were no abnormalities in one-year-old CA II–deficient mice. By contrast, Cammer et al (177) found evidence of shrunken oligodendrocytes in white matter and gray matter, and increased abundance of astrocytes in the white matter. Possibly these differences are due to differences in age or diet. Another possibility is uncontrolled genetic differences between CA II–deficient mice in different laboratories. In many cases, the CA II mutation has not been removed by outbreeding from the heavily mutagenized strains in which it was induced. Genetic differences between strains developed from the original CA II–deficient mice by different laboratories could be very significant.

Literature Cited

1. Tashian RE, Hewett-Emmett D, eds. 1984. *Biology and Chemistry of the Carbonic Anhydrases. Ann. NY Acad. Sci.* 429:1–640
2. Deutsch HF. 1987. *Int. J. Biochem.* 19:101–13
3. Fernley RT. 1988. *Trends Biochem.* 13:356–59
4. Tashian RE. 1989. *BioEssays* 10:186–92
5. Edwards Y. 1990. *Biochem. Soc. Trans.* 18:171–75
6. Dodgson SJ, Tashian RE, Gros G, Carter ND, eds. 1991. *The Carbonic Anhydrases: Cellular Physiology and Molecular Genetics.* New York: Plenum
7. Botre F, Gros G, Storey BT, eds. 1991. *Carbonic Anhydrase: From Biochemistry and Genetics to Physiology and Medicine.* Weinheim, Germany: VCH
8. Hewett-Emmett D, Tashian RE. 1991. See Ref. 6, pp. 15–32
9. Silverman DN. 1991. *Can. J. Bot.* 69:1070–78
10. Liljas A, Håkansson K, Jonsson BH, Xue Y. 1994. *Eur. J. Biochem.* 219:1–10
11. Sly WS, Hu PY. 1995. In *The Molecular and Metabolic Bases of Inherited Disease,* ed. CR Scriver, AL Beaudet, WS Sly, D Valle, pp. 4113–4124. New York: McGraw-Hill. 7th ed.
12. Sly WS, Sato S, Zhu XL. 1991. *Clin. Biochem.* 24:311–18
13. Tashian RE. 1992. *Adv. Genet.* 30:321–56
14. Hewett-Emmett D, Hopkins PJ, Tashian R, Czelusniak J. 1984. *Ann. NY Acad. Sci.* 429:338–57
15. Meldrum NU, Roughton FJW. 1933. *J. Physiol.* 80:113–42
16. Stadie WC, O'Brien H. 1933. *J. Biochem.* 103:521–29
17. Maren TH. 1967. *Physiol. Rev.* 47:595–81
18. Maren TH. 1988. *Annu. Rev. Physiol.* 50:695–717

19. Lindskog S. 1960. *Biochim. Biophys. Acta* 39:218–26
20. Nyman PO. 1961. *Biochim. Biophys. Acta* 52:1–12
21. Andersson B, Nyman PO, Strid L. 1972. *Biochem. Biophys. Res. Commun.* 48:670–77
22. Lin KTD, Deutsch HF. 1973. *J. Biol. Chem.* 248:1885–93
23. Henderson LED, Henriksson D, Nyman PO. 1973. *Biochem. Biophys. Res. Commun.* 52:1388–94
24. Lin KTD, Deutsch HF. 1974. *J. Biol. Chem.* 249:2329–37
25. Liljas A, Kannan KK, Bergsten PC, Waara I, Fridborg K, et al. 1972. *Nature* 235:131–37
26. Kannan KK, Notstrand B, Fridborg K, Lövgren S, Ohlsson A, Petef M. 1975. *Proc. Natl. Acad. Sci. USA* 72:51–55
27. Tashian RE, Carter ND. 1976. In *Advances in Human Genetics,* ed. H Harris, K Hirschhorn, pp. 1–56. New York: Plenum
28. Garg LC. 1974. *J. Pharmacol. Exp. Ther.* 189:557–62
29. Garg LC. 1974. *Biochem. Pharmacol.* 23:3153–61
30. Holmes RS. 1976. *J. Exp. Zool.* 197:289–95
31. Blackburn MN, Chirgwin JM, James GT, Kempe TD, Parsons TF, et al. 1972. *J. Biol. Chem.* 247:1170–79
32. Koester MK, Register AM, Notman EA. 1977. *Biochem. Biophys. Res. Commun.* 76:196–204
33. Whitney PL, Briggle TV. 1982. *J. Biol. Chem.* 257:12056–59
34. Wistrand PJ, Knuuttila KG. 1989. *Kidney Int.* 35:851–59
35. Zhu XL, Sly WS. 1990. *J. Biol. Chem.* 265:8795–801
36. Datta PK, Shepherd TH. 1959. *Arch. Biochem. Biophys.* 81:124–29
37. Dodgson SJ, Forster RE, Storey BT, Mela L. 1980. *Proc. Natl. Acad. Sci. USA* 77:5562–68
38. Vincent SH, Silverman DN. 1982. *J. Biol. Chem.* 275:6850–55
39. Dodgson SJ. 1991. See Ref. 6, pp. 267–305
40. Hewett-Emmett D, Cook RG, Dodgson SJ. 1986. *Isozyme Bull.* 19:13
41. Amor-Gueret M, Levi-Strauss M. 1990. *Nucleic Acids Res.* 18:1646
42. Heck RW, Tanhauser SM, Manda R, Tu CK, Laipis PJ, Silverman DN. 1994. *J. Biol. Chem.* 269:24742–46
43. Nagao Y, Platero JS, Waheed A, Sly WS. 1993. *Proc. Natl. Acad. Sci. USA* 90:7623–27
44. Fernley RT, Wright RD, Coghlan JP. 1979. *FEBS Lett.* 105:299–302
45. Feldstein JB, Silverman DN. 1984. *J. Biol. Chem.* 259:91–96
46. Murakami H, Sly WS. 1987. *J. Biol. Chem.* 262:1382–88
47. Fernley RT, Wright RD, Coghlan JP. 1989. *Biochemistry* 27:2815–20
48. Montgomery JC, Venta PJ, Eddy RL, Fukushima YS, Shows TB, Tashian RE. 1991. *Genomics* 11:835–48
49. Eriksson AE, Liljas A. 1991. See Ref. 6, pp. 33–48
50. Eriksson AE. 1988. *Acta Univ. Ups. Compr. Summ. Upps. Diss. Fac. Sci.* 164:1–36
51. Eriksson AE, Jones TA, Liljas A. 1988. *Proteins Struct. Funct. Genet.* 4:274–82
52. Håkansson K, Carlsson M, Svensson LA, Liljas A. 1992. *J. Mol. Biol.* 227:1192–204
53. Silverman DN, Lindskog S. 1988. *Acc. Chem. Res.* 21:30–36
54. Coleman JE. 1967. *Nature* 214:193–94
55. Merz KM Jr. 1990. *J. Mol. Biol.* 214:799–802
56. Merz KM Jr. 1991. *J. Am. Chem. Soc.* 113:406–11
57. Krebs JF, Ippolito JA, Christianson DW, Fierke CA. 1993. *J. Biol. Chem.* 268:27458–66
58. Liang ZW, Xue YF, Behravan G, Jonsson BH, Lindskog S. 1993. *Eur. J. Biochem.* 211:821–27
59. Xue YF, Liljas A, Jonsson BH, Lindskog S. 1993. *Proteins Struct. Funct. Genet.* 17:93–106
60. Eriksson AE, Kylsten PM, Jones TA, Liljas A. 1988. *Proteins Struct. Funct. Genet.* 4:283–93
61. Vidgren J, Liljas A, Walker NPC. 1990. *Int. J. Macromol.* 12:342–44
62. Vidgren J, Svensson LA, Liljas A. 1993. *Int. J. Macromol.* 15:97–100
63. Lindahl M, Svensson LA, Liljas A. 1993. *Proteins Struct. Funct. Genet.* 15:117–82
64. Lindahl M, Liljas A, Habash J, Harrop S, Helliwell JR. 1992. *Acta Crystallogr. B* 48:281–85
65. Jonsson BH, Håkansson K, Liljas A. 1993. *FEBS Lett.* 322:186–90
66. Nair SK, Christianson DW. 1993. *Eur. J. Biochem.* 213:507–15
67. Mangani S, Liljas A. 1993. *J. Mol. Biol.* 232:9–14
68. Håkansson K, Briand C, Zaitsev V, Xue YF, Liljas A. 1994. *Acta Crystallogr. D* 50:101–4
69. Håkansson K, Wehnert A, Liljas A. 1994. *Acta Crystallogr. D* 50:93–100
70. Mangani S, Håkansson K. 1992. *Eur. J. Biochem.* 210:867–71
71. Xue YF, Vidgren J, Svensson LA, Liljas

A, Jonsson BH, Lindskog S. 1993. *Proteins Struct. Funct. Genet.* 15:80–87
72. Christianson DW. 1991. *Adv. Protein Chem.* 42:281–85
73. Tu CK, Silverman DN, Forsman C, Jonsson BH, Lindskog S. 1989. *Biochemistry* 28:7913–18
74. Engstrand C, Forsman C, Liang ZW, Lindskog S. 1992. *Biochim. Biophys. Acta* 1122:321–26
75. Ren XL, Lindskog S. 1992. *Biochem. Biophys. Acta* 1120:81–86
76. Liang ZW, Jonsson BH, Lindskog S. 1993. *Biochem. Biophys. Acta* 1203: 142–46
77. Behravan G, Jonsson BH, Lindskog S. 1990. *Eur. J. Biochem.* 190:351–57
78. Behravan G, Jonasson P, Jonsson BH, Lindskog S. 1991. *Eur. J. Biochem.* 198:589–92
79. Nair SK, Christianson DW. 1991. *J. Am. Chem. Soc.* 113:9455–58
80. Krebs JF, Fierke CA. 1993. *J. Biol. Chem.* 268:948–54
81. Alexander RS, Nair SK, Christianson DW. 1991. *Biochemistry* 30:11064–72
82. Fierke CA, Calderone TL, Krebs JF. 1991. *Biochemistry* 30:11054–63
83. Nair SK, Calderone TL, Christianson DW, Fierke CA. 1991. *J. Biol. Chem.* 266:17320–25
84. Chen X, Tu CK, LoGrasso PV, Laipis PJ, Silverman DN. 1993. *Biochemistry* 32:7861–65
85. Ren X, Jonsson BH, Lindskog S. 1991. *Eur. J. Biochem.* 201:417–20
86. Lindskog S, Liljas A. 1993. *Curr. Opin. Struct. Biol.* 3:915–20
87. Okuyama T, Sato S, Zhu XL, Waheed A, Sly WS. 1992. *Proc. Natl. Acad. Sci. USA* 89:1315–19
88. Nagao Y, Platero JS, Waheed A, Sly WS. 1993. *Am. J. Hum. Genet.* 53 (Suppl.):932 (Abstr.)
89. Aldred P, Fu P, Barrett G, Penschow JD, Wright RD, et al. 1991. *Biochemistry* 30:569–75
90. Murakami H, Marelich GP, Grubb JH, Kyle JW, Sly WS. 1987. *Genomics* 1: 159–66
91. Khalifah RG. 1971. *J. Biol. Chem.* 246: 2561–73
92. Kaunisto K, Parkkila S, Tammelu T, Rönnberg L, Rajaniemi H. 1990. *Histochemistry* 94:381–86
93. Parkkila S, Kaunisto K, Kellokumpu S, Rajaniemi H. 1991. *Histochemistry* 95: 477–82
94. Campbell AR, Andress DL, Swenson ER. 1994. *J. Leukocyte Biol.* 55:343–48
95. Fleming RE, Moxley MA, Waheed A, Crouch EC, Sly WS, Longmore WJ.

1994. *Am. J. Respir. Cell Mol. Biol.* 10:499–505
96. Lonnerholm G, Selking O, Wistrand PJ. 1985. *Gastroenterology* 88:1151–61
97. Parkkila AK, Parkkila S, Juvonen T, Rajaniemi H. 1993. *Histochemistry* 99: 37–41
98. Dagher PC, Egnor RW, Charney AN. 1993. *Am. J. Physiol.* 264:G569–75
99. Schwartz GJ, Winkler CA, Zavilowitz BJ, Bargiello T. 1993. *Am. J. Physiol.* 265:F764–72
100. Dodgson SJ, Forster RE II, Sly WS, Tashian RE. 1988. *J. Appl. Physiol.* 65:1472–80
101. Brady HJM, Sowden JC, Edwards M, Lowe N, Butterworth PHW. 1989. *FEBS Lett.* 257:451–56
102. Brady HJW, Edwards M, Linch DC, Knott L, Barlow JH, Butterworth PHW. 1990. *Br. J. Haematol.* 76:135–42
103. Brady HJM, Lowe N, Sowden JC, Edwards M, Butterworth PHW. 1991. *Biochem. J.* 277:903–5
104. Tu CK, Paranawithana SR, Jewell DA, Tanhauser SM, LoGrasso PV, et al. 1990. *Biochemistry* 29:6400–5
105. Chen X, Tu CK, LoGrasso PV, Laipis PJ, Silverman DN. 1993. *Biochemistry* 32:7861–65
106. LoGrasso PV, Tu CK, Chen X, Taoka SC, Laipis PJ, Silverman DN. 1993. *Biochemistry* 32:5786–91
107. Eriksson AE, Liljas A. 1991. See Ref. 6, pp. 33–48
108. Lynch CJ, Brennan WA Jr, Vary PC, Carter N, Dodgson SJ. 1993. *Am. J. Physiol.* 264:E621–30
109. Zheng A, Rahkila P, Vuori J, Rasi S, Takala T, Vaananen HK. 1992. *Histochemistry* 97:77–81
110. Lynch CJ, Hazen SA, Horetsky RL, Carter ND, Dodgson SJ. 1993. *Am. J. Physiol.* 265:C234–43
111. Jeffery S, Edwards Y, Carter N. 1980. *Biochem. Genet.* 18:843–49
112. Lloyd J, Brownson C, Tweedie S, Charlton J, Edwards YH. 1987. *Genes Dev.* 1:594–602
113. Lyons GE, Buckingham ME, Tweedie S, Edwards YH. 1991. *Development* 111:233–44
114. Waheed A, Zhu XL, Sly WS, Wetzel P, Gros G. 1992. *Arch. Biochem. Biophys.* 294:550–56
115. Okuyama T, Batanian JR, Sly WS. 1993. *Genomics* 16:678–84
116. Maren TH, Wynns GC, Wistrand PJ. 1993. *Mol. Pharmacol.* 44:901–5
117. Brown D, Zhu XL, Sly WS. 1990. *Proc. Natl. Acad. Sci. USA* 87:7457–61
118. Fleming RE, Crouch EC, Ruzicka CA,

Sly WS. 1993. *Am. J. Physiol.* 265: L627–35

119. Ghandour MS, Langley OK, Zhu XL, Waheed A, Sly WS. 1992. *Proc. Natl. Acad. Sci. USA* 89:6823–27

120. Hageman GS, Zhu XL, Waheed A, Sly WS. 1991. *Proc. Natl. Acad. Sci. USA* 88:2716–20

121. Sender S, Gros G, Waheed A, Hageman GS, Sly WS. 1994. *J. Histochem. Cytochem.* 42:1229–36

122. Parkkila S, Parkkila AK, Kaunisto K, Waheed A, Sly WS, Rajaniemi H. 1993. *J. Histochem. Cytochem.* 41:751–57

123. Amthauer R, Kodukula K, Gerber L, Udenfriend S. 1993. *Proc. Natl. Acad. Sci. USA* 90:3973–77

124. Parkkila S, Kaunisto K, Rajaniemi L, Kumpulainen T, Jokinen K, Rajaniemi H. 1990. *J. Histochem. Cytochem.* 38: 941–47

125. Parkkila S, Parkkila AK, Vierjoki T, Stahlberg T, Rajaniemi H. 1993. *Clin. Chem.* 39:2154–57

126. Parkkila S, Parkkila AK, Juvonen T, Rajaniemi H. 1994. *Gut* 35:646–50

127. Dodgson SJ, Cherian K. 1989. *Am. J. Physiol.* 257:E791–96

128. Carter ND, Dodgson SJ, Quant PA. 1990. *Biochim. Biophys. Acta* 1036: 237–41

129. Carter ND. 1991. See Ref. 7, pp. 168–77

130. Ohlinger DE, Lynch CE, Forster RE II, Dodgson SJ. 1993. *FASEB J.* 7:A676 (Abstr.)

131. Nagao Y, Srinivasan M, Platero JS, Svendrowski M, Waheed A, Sly WS. 1994. *Proc. Natl. Acad. Sci. USA* 91: 10330–34

132. Dodgson SJ. 1991. See Ref. 6, pp. 297–306

133. Niles EG, Seto J. 1988. *J. Virol.* 62: 3772–78

134. Debuire B, Henry C, Benaissa M, Biserte G, Claverie JM, et al. 1984. *Science* 224:1456–59

135. Kato K. 1990. *FEBS Lett.* 271:137–40

136. Skaggs LA, Bergenhem NC, Venta PJ, Tashian RE. 1993. *Gene* 126:291–92

137. Krueger NX, Saito H. 1992. *Proc. Natl. Acad. Sci. USA* 89:7417–21

138. Levy JB, Canoll PD, Silvennoinen O, Barnea G, Morse B, et al 1993. *J. Biol. Chem.* 268:10573–81

139. Wary KK, Lou ZW, Buchberg AM, Siracusa LD, Druck T, et al. 1993. *Cancer Res.* 53:1498–502

140. Barnea G, Silvennoinen O, Shaanan B, Honegger AM, Canoll PD, et al. 1993. *Mol. Cell. Biol.* 13:1497–506

141. Fukuzawa H, Fujiwara S, Tachiki A, Miyachi S. 1990. *Nucleic Acids Res.* 18:6441–42

142. Burnell JN, Gibbs MJ, Mason JG. 1989. *Plant Physiol.* 92:37–40

143. Fawcett TW, Browse JA, Nokolita M, Bartlett SC. 1990. *Nucleic Acids Res.* 18:6441–42

144. Roeske CA, Ogren WL. 1990. *Nucleic Acids Res.* 18:3413

145. Lowe N, Brady HJM, Barlow JH, Sowden JC, Edwards M, Butterworth PHW. 1990. *Gene* 93:277–83

146. Lowe N, Edwards YH, Edwards M, Butterworth PHW. 1991. *Genomics* 10: 882–88

147. Sutherland GR, Baker E, Fernandez KW, Callen W, Aldred P, et al. 1989. *Cytogenet. Cell Genet.* 50:149–50

148. Guibaud P, Larbre F, Freycon M-T, Genoud J. 1972. *Arch. Fr. Pediatr.* 29: 269–86

149. Vainsel M, Fondu P, Cadranel S, Rocmans C, Gepts W. 1972. *Acta Paediatr. Scand.* 61:429–34

150. Sly WS, Lang R, Avioli L, Haddad J, Lubowitz H, McAlister W. 1972. *Am. J. Hum. Genet.* 24(Suppl.):A34

151. Ohlsson A, Stark G, Sakati N. 1980. *Dev. Med. Child. Neurol.* 22:72–84

152. Sly WS, Hewett-Emmett D, Whyte MP, Yu Y-SL, Tashian RE. 1983. *Proc. Natl. Acad. Sci. USA* 80:2752–56

153. Cochat P, Loras-Duclaux I, Guibaud P. 1987. *Pediatrie* 42:121–28

154. Whyte MP. 1993. *Clin. Orthop. Relat. Res.* 294:52–63

155. Venta PJ, Welty RJ, Johnson TM, Sly WS, Tashian RE. 1991. *Am. J. Hum. Genet.* 49:1082–90

156. Roth DE, Venta PJ, Tashian RE, Sly WS. 1992. *Proc. Natl. Acad. Sci. USA* 89:1804–08

157. Strisciuglio P, Sartorio R, Pecoraro C, Lotito F, Sly WS. 1990. *Eur. J. Pediatr.* 149:337–40

158. Tu CK, Couton JM, Van Heeke G, Richards NG, Silverman DN. 1993. *J. Biol. Chem.* 268:4775–79

159. Hu PY, Roth DE, Skaggs LA, Venta PJ, Tashian RE, et al. 1992. *Hum. Mutat.* 1:288–92

160. Hu PY, Ernst AR, Sly WS. 1992. *Am. J. Hum. Genet.* 51(Suppl.):A29

161. Hu PY, Ernst AR, Sly WS, Venta PJ, Skaggs LA, Tashian RE. 1994. *Am. J. Hum. Genet.* 54:602–8

162. Venta PJ, Tashian RE. 1991. *Nucleic Acids Res.* 19:4795

163. Kendall AG, Tashian RE. 1977. *Science* 197:471–72

164. Wagner LE, Venta PJ, Tashian RE. 1991. *Isozyme Bull.* 24:35

165. Goriki K, Hazama R, Yamakido M. 1984. *Ann. NY Acad. Sci.* 429:276

166. Tashian RE, Goodman M, Headings

VE, DeSimone J, Ward RE. 1971. *Biochem. Genet.* 5:183–200

167. DeSimone J, Magid E, Tashian RE. 1973. *Biochem. Genet.* 8:165–74
168. Maren TH, Raybury CS, Lidell NE. 1976. *Science* 191:469–72
169. Lewis SE, Erickson RP, Barnett LB, Venta PJ, Tashian RE. 1988. *Proc. Natl. Acad. Sci. USA* 85:1962–66
170. Spicer SS, Lewis SE, Tashian RT, Schulte BA. 1989. *Am. J. Pathol.* 134: 947–54
171. Ridderstråle Y, Wistrand PJ, Tashian R. 1992. *J. Histochem. Cytochem.* 40: 1665–73
172. Ridderstråle Y, Wistrand PJ, Brechue WF. 1994. *Invest. Ophthalmol. Vis. Sci.* 35:2577–84
173. Brechue WF, Saffran EK, Kinne RHK, Maren TH. 1991. *Biochem. Biophys. Acta* 1066:201–7
174. Velisek L, Moshe SL, Cammer W. 1993. *Dev. Brain Res.* 72:321–24
175. Biesecker L, Erickson RP, Tashian RE. 1994. *Biochem. Med. Metab. Biol.* 51: 61–65
176. Ghandour MS, Skoff RP, Venta PJ, Tashian RE. 1989. *J. Neurosci. Res.* 23:180–90
177. Cammer W, Zhang H, Cammer M. 1993. *J. Neurol. Sci.* 118:1–9

Annu. Rev. Biochem. 1995. 64:403–34

COLLAGENS: Molecular Biology, Diseases, and Potentials for Therapy

Darwin J. Prockop

Department of Biochemistry and Molecular Biology, Jefferson Institute of Molecular Medicine, Jefferson Medical College of Thomas Jefferson University, Philadelphia, Pennsylvania 19107

Kari I. Kivirikko

Collagen Research Unit, Biocenter and Department of Medical Biochemistry, University of Oulu, 90220 Oulu, Finland

KEY WORDS: fibrosis, osteogenesis imperfecta, transgenic mice, gene therapy, antisense oligonucleotides, antisense gene

CONTENTS

ABSTRACT

The collagen superfamily of proteins now contains at least 19 proteins formally defined as collagens and an additional ten proteins that have collagen-like domains. The most abundant collagens form extracellular fibrils or network-like structures, but the others fulfill a variety of biological functions. Some of

403

0066-4154/95/0701-0403$05.00

the eight highly specific post-translational enzymes involved in collagen bio-synthesis have recently been cloned. Over 400 mutations in 6 different colla-gens cause a variety of human diseases that include osteogenesis imperfecta, chondrodysplasias, some forms of osteoporosis, some forms of osteoarthritis, and the renal disease known as the Alport syndrome. Many of the disease phenotypes have been produced in transgenic mice with mutated collagen genes. There has been increasing interest in the possibility that the unique post-translational enzymes involved in collagen biosynthesis offer attractive targets for specifically inhibiting excessive fibrotic reactions in a number of diseases. A number of experiments suggest it may be possible to inhibit collagen synthesis with oligo- nucleotides or antisense genes.

THE COLLAGEN FAMILY OF PROTEINS AND GENES

At least 19 proteins are now known as collagens, and at least an additional 10 proteins have collagen-like domains. Initially, collagens were defined as pro-teins of the extracellular matrix that contained large domains comprised of repeating -Gly-X-Y- sequences and that folded into a unique triple-helical structure. Screening of cDNA and genomic DNA libraries with probes for collagens revealed a large number of related proteins with varying lengths of repeating -Gly-X-Y- sequences. Because they were discovered in searches for collagens, the proteins encoded were defined as collagens, even though in some cases the triple-helical domains were small and most of the protein structure was globular. Also, a few proteins studied in other contexts were found to contain triple-helical domains but were not defined as collagens. The variety in the structures of different collagens and related proteins implies that they have vastly different biological functions.

Because of the extensive literature on collagens, in this chapter we concen-trate primarily on recent advances in the field. For more detailed data, the reader is referred to several previous reviews on the structures and functions of collagens (1–11), on the structures of collagen genes (12–14), on the bio-synthesis of the proteins (1, 6, 7, 15, 16), and on mutations in patients and mice (2, 17–26). A vast literature is also available on *cis*-regulatory elements for collagen genes, and several transcription factors have been characterized. The topic, however, is not reviewed here, in part because there are conflicting data from similar experiments with different gene constructs in cell transfection assays. Also, data from cell transfection assays are not always consistent with data from experiments in transgenic mice.

Structure and Functions of the Collagen Triple Helix

The collagen triple helix is formed from three polypeptide chains that are each coiled into a left-handed helix. The three chains are then wrapped around each other into a right-handed super-helix so that the final structure is a rope-like rod (1, 2, 4, 6, 7). The presence of glycine as every third amino acid in the repeating -Gly-X-Y- sequence of each chain is essential, because a larger amino acid will not fit in the center of the triple helix where the three chains come together. Proline is frequently in the X-position of the -Gly-X-Y- sequence and 4-hydroxyproline is frequently in the Y-position. These two amino acids limit rotation of the polypeptide chains. The triple helix is further stabilized by hydrogen bonds and water bridges, many of which require the presence of 4-hydroxyproline. The conformation of the triple helix places the side chains of amino acids in the X- and Y-positions on the surface of the molecule. This arrangement explains the ability of many collagens to polymerize, since the multiple clusters of hydrophobic and charged side chains direct self-assembly into precisely ordered structures. The triple helix is relatively rigid. In some contexts, the resistance of the molecule to extension or compression is important for the biological function of the protein. In many collagens, the triple helix is interrupted by globular sequences that make the molecule more flexible, but the precise functions of the globular sequences are unknown.

Types of Collagen

For simplicity, the superfamily of collagens (6) can be divided into several classes on the basis of the polymeric structures they form or related structural features (see Figure 1): (a) collagens that form fibrils (types I, II, III, V, and XI), (b) collagens that form network-like structures (the type IV family, and types VIII and X), (c) collagens that are found on the surface of collagen fibrils and are known as fibril-associated collagens with interrupted triple helices (FACITs that include types IX, XII, XIV, XVI, and XIX), (d) the collagen that forms beaded filaments (type VI), (e) the collagen that forms anchoring fibrils for basement membranes (type VII), (f) collagens with a transmembrane domain (types XIII and XVII), and (g) the newly discovered types XV and XVIII collagens that have been only partially characterized. An additional group (h) consists of proteins containing triple-helical domains that have not been defined as collagens.

FIBRIL-FORMING COLLAGENS   All these collagens (types I–III, V, and XI) are similar in size and in that they contain large triple-helical domains with about 1000 amino acids or 330 -Gly-X-Y- repeats per chain. In addition, they are also first synthesized as larger precursors, and the precursors need to be processed to collagens by cleavage of N-propeptides and C-propeptides by specific proteinases. Finally, they are similar in that they all assemble into cross-striated fibrils in which each molecule is displaced about one-quarter of its length relative to its nearest neighbor along the axis of the fibril (Figure 1). Type I is the most abundant collagen and is found in a variety of tissues. Many

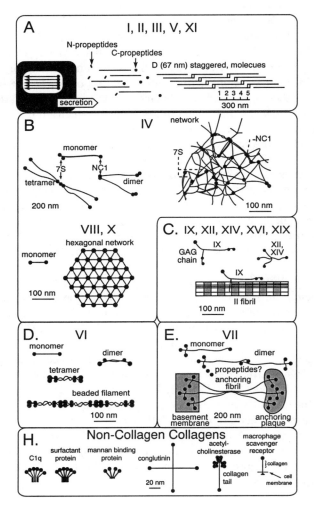

Figure 1 Schematic for the structure of various collagens. The figure is modified after the figure presented by Hulmes (6) and reproduced here with permission. The letters refer to the classifications used in the text. Because the protein structures are still unknown, the scheme does not present collagens with a transmembrane domain (types XIII and XVII) and the family of two newly discovered collagens (types XV and XVIII).

Table 1   Collagen types and the location of their genes on human chromosomes[a]

Type	Gene	Chromosome	Expression
I	COL1A1	17q21.3-q22	Most connective tissues
	COL1A2	7q21.3-q22	
II	COL2A1	12q13-q14	Cartilage, vitreous humor
III	COL3A1	2q24.3-q31	Extensible connective tissues, e.g. skin, lung, vascular system
IV	COL4A1	13q34	Basement membranes
	COL4A2	13q34	
	COL4A3	2q35-q37	
	COL4A4	2q35-q37	
	COL4A5	Xq22	
	COL4A6	Xq22	
V	COL5A1	9q34.2-q34.3	Tissues containing collagen I, quantitatively minor component
	COL5A2	2q24.3-q31	
	COL5A3		
VI	COL6A1	21q22.3	Most connective tissues
	COL6A2	21q22.3	
	COL6A3	2q37	
VII	COL7A1	3p21	Anchoring fibrils
VIII	COL8A1	3q12-q13.1	Many tissues, especially endothelium
	COL8A2	1p32.3-p34.3	
IX	COL9A1	6q12-q14	Tissues containing collagen II
	COL9A2	1p32	
	COL9A3		
X	COL10A1	6q21-q22	Hypertrophic cartilage
XI	COL11A1	1p21	Tissues containing collagen II
	COL11A2	6p21.2	
	COL2A1[b]	12q13-q14	
XII	COL12A1	6	Tissues containing collagen I
XIII	COL13A1	10q22	Many tissues
XIV	COL14A1		Tissues containing collagen I
XV	COL15A1	9q21-22	Many tissues
XVI	COL16A1	1p34-35	Many tissues
XVII	COL17A1	10q24.3	Skin hemidesmosomes
XVIII	COL18A1	21q22.3	Many tissues, especially liver and kidney
XIX	COL19A1	6q12-q14	Rhabdomyosarcoma cells

[a] For chromosome locations see References 21, 44, 75, 102, 214.
[b] The $\alpha3(XI)$ chain of type XI collagen is encoded by the same gene as the $\alpha1(II)$ chain of type II.

of the other fibril-forming collagens have a more selective tissue distribution (Table 1).

Among the new developments concerning fibril-forming collagens is the discovery of alternative splicing of exons in the N-terminal propeptides of type II (27–29) and type XI collagens (30–32). Because of alternative splicing, the coding sequences of an additional exon (27–29) are present in the type II procollagen formed in noncartilaginous tissues early in embryonic development (33–37). Another new development concerning fibril-forming collagens is the finding that many fibrils in vivo are composed of two or more different collagen types (4, 6, 9). An additional discovery is that hybrid molecules containing chains of both type V and type XI collagens are present in some tissues such as a hybrid molecule containing the $\alpha 2(V)$ chain and the $\alpha 1(XI)$ chain (4, 6, 9, 38–40). Therefore, type V and type XI collagens can probably be considered as a single kind of collagen comprised of five different chains (9), i.e. $\alpha 1(V)$, $\alpha 2(V)$, $\alpha 3(V)$, $\alpha 1(XI)$ and $\alpha 2(XI)$.

The gene structures of the fibril-forming collagens show a great deal of similarity (12–14). One common feature of the genes is that the major triple-helical domain of each chain is coded for by 42 exons. Most of the exons are 54 bp and the others are either twice 54, three times 54, or combinations of 45 and 54 bp exons. Also, each exon begins with a complete codon for glycine, and therefore the exon codes for a discrete number of -Gly-X-Y- tripeptide units. In addition, the pattern of exon sizes is similar in all the genes and has been highly conserved throughout evolution.

The genes for the $\alpha 2(I)$ chain of type I collagen and $\alpha 1(III)$ chain of type III collagen contain alternative promoters that code for different polypeptides (41, 42). The alternative promoter of the COL1A2 gene is located within intron 2, and the transcript contains a short open reading frame that is out of frame with the collagen coding sequence (41). Thus, this RNA cannot encode a collagen but may encode a noncollagen polypeptide. The transcript appears early in embryogenesis in tissues derived from neuroectoderm, but at later stages of development, it is found almost exclusively in hyaline cartilage (41). The alternative promoter of the COL3A1 gene is located in intron 23, and the transcript may encode either a noncollagen polypeptide or a truncated collagen (42). This transcript appears transiently in limb mesenchyme and then decreases to low levels in intact cartilage (42). The functions of these alternative transcripts are currently unknown.

NETWORK-FORMING COLLAGENS   These collagens include the family of type IV collagens found in basement membranes and type VIII and X collagens (Figure 1). The collagenous domain of a type IV collagen molecule is longer than in the fibril-forming collagens and consists of about 1400 amino acids in -Gly-X-Y- repeats that are frequently interrupted by short noncollagenous sequences. The N-terminus of a molecule contains a small noncollagenous domain and the C-terminus a major noncollagenous domain of about 230 amino acids (Figure 1). The molecules self-assemble to form net-like structures in which monomers associate at the C-termini to form dimers and at the

N-termini to form tetramers. In addition to these end-to-end interactions, the triple-helical domains intertwine to form supercoiled structures (43–45).

Although most of the type IV collagen in basement membranes consist of a combination of $\alpha1(IV)$ and $\alpha2(IV)$ chains, some basement membranes contain smaller amounts of molecules of $\alpha3(IV)$ and $\alpha4(IV)$ (46–49) or of $\alpha5(IV)$ and $\alpha6(IV)$ chains (50–56) that are similar but not identical. Further variation in the structure of type IV collagens is caused by alternative splicing of RNA transcripts for the $\alpha3(IV)$ chain (57, 58). Also of interest is that the genes of type IV collagens are found in pairs with head-to-head orientations on different chromosomes so that the promoter regions overlap (Table 1). The $\alpha1(IV)$ and $\alpha2(IV)$ chain genes are head-to-head on chromosome 13 (43, 44), the $\alpha3(IV)$ and $\alpha4(IV)$ chain genes are head-to-head on chromosome 2 (44, 47, 48, 59); and the $\alpha5(IV)$ and $\alpha6(IV)$ chain genes are head-to-head on the X chromosome (44, 50, 51, 54, 55, 60). The structures of these genes differ distinctly from those of the fibrillar collagens. Only a few exons are 54 or 45 bp, and many exons coding for the triple-helical domain begin with a split codon for glycine in which the first G of the codon is in the preceding exon (12–14, 44, 60).

The two other network-forming collagens, types VIII and X, are very different in structure from type IV but similar to each other (4–7, 9, 61). The $\alpha1(VIII)$, $\alpha2(VIII)$, and $\alpha1(X)$ chains all contain a collagenous sequence of almost the same size and with eight imperfections in similar positions in the -Gly-X-Y- sequences. The genes for these two collagens all contain only three exons, and almost all of the coding sequences are found in the large third exon (4, 12–14, 61). Descemet's membrane, which separates the corneal endothelial cells from the stroma, consists of stacks of hexagonal lattices made of type VIII collagen (4). Type X collagen is among the most specialized of the collagens and is synthesized primarily by hypertrophic chondrocytes in the deep-calcifying zone of cartilage (4–7, 62). The assembled form of type X collagen resembles the hexagonal lattice of type VIII in Descemet's membrane (62).

FACIT COLLAGENS   These collagens (types IX, XII, XIV, XVI, and XIX) do not form fibrils themselves but are found attached to the surfaces of preexisting fibrils of the fibril-forming collagens (3–11). All these collagens are characterized by short triple-helical domains interrupted by short noncollagenous sequences.

The type IX collagen molecule consists of three triple-helical domains and four noncollagenous domains. The protein is commonly found on the surface of fibrils of type II collagen covalently bound to molecules of type II collagen in antiparallel orientation (Figure 1). One unusual feature of collagen IX is that it often occurs as a proteoglycan in which a single glycosaminoglycan side chain is covalently attached to the second noncollagenous domain of the α2(IX) chain. In ocular and embryonic tissues, type IX collagen occurs in a form with a short α1(IX) chain lacking nearly all of the N-terminal globular domain. This short α1(IX) chain is transcribed from an alternative promoter located between exons 6 and 7 of the α1(IX) gene (63). The expression patterns of the long and short forms seem to be both temporally and spatially regulated. During avian development, the switch in expression from the short form to the long form occurs at the beginning of chondrogenesis during the early development of the vertebral column (64, 65).

Type XII and XIV collagens show several structural similarities to type IX collagen, particularly in the C-terminal collagenous domains (3, 6, 8, 9, 66–70). These two collagens also contain glycosaminoglycan side chains attached to the large N-terminal globular domain. The RNA transcripts for type XII and XIV collagens undergo alternative splicing that varies the structures of the N-terminal globular domain. In the longest form of type XII collagen, the N-terminal globular domain contains 18 fibronectin type III repeats and four repeats homologous to the von Willebrand factor A domain (8, 9). In the longest form of type XIV collagen, the N-terminal domain contains eight and two of these repeats respectively (69). Type XVI collagen (71, 72) and the recently discovered type XIX collagen (73, 74) also show similarities in structure to the FACIT collagens and are therefore classified into this subgroup.

BEADED FILAMENT–FORMING COLLAGEN   The only collagen known to form beaded filaments is type VI (75) (Figure 1). Each of the three different chains of the protein contains a very short triple-helical domain, and the remainder consists of large N-terminal and C-terminal globular domains (75). Recently, researchers found that the N-terminal globular region of the α3(VI) chain is much larger than the same region in the other two chains (4–7, 75). The N-terminal and C-terminal globular domains of all three chains contain 200-residue repeats with significant similarities to the A domains of von Willebrand factor. The C-terminal region of the α3(VI) chain also contains three additional domains that show similarities to salivary proteins, to fibronectin type III repeats, and to Kunitz-type protease inhibitors (4–7, 9, 75). Several α2(VI) and α3(VI) chain variants result from alternative splicing of the repetitive noncollagenous subdomains (4–7, 9, 75, 76).

COLLAGEN OF ANCHORING FIBRILS   Type VII collagen forms anchoring fibrils (Figure 1) that link basement membranes to anchoring plaques of type IV collagen and laminin in the underlying extracellular matrix (77–82). The triple-helical domain of type VII collagen, which is longer than the triple helix of any other collagen, contains 1530 amino acids in -Gly-X-Y- repeats that are interrupted at 19 separate sites (82). The large N-terminal globular domain contains a segment homologous to cartilage matrix protein. The segment is followed by nine fibronectin type III repeats, one segment homologous to von Willebrand factor A domain, and a segment that is cysteine and proline rich. The smaller C-terminal globular domain contains a segment homologous to Kunitz-type protease inhibitors (82). The protein is first assembled into antiparallel dimers formed by a small overlap at the C-terminal globular ends. The C-terminal globular domains appear to be cleaved during the assembly of dimers and the dimers are stabilized by disulfide bonds. The dimers then associate laterally and in register to become the main constituents of anchoring fibrils. The gene for type VII collagen is about 31 kb and has 118 exons (83). It therefore has more exons than any other known gene.

COLLAGENS WITH A TRANSMEMBRANE DOMAIN   Two recently discovered collagens contain a transmembrane domain and, therefore, are probably not secreted into the extracellular matrix. Type XIII collagen (11, 84–87) is found in many tissues. In contrast, type XVII collagen (80, 88–91) is found primarily in the hemidesmosomes of the skin and is one of the two antigens that produce the autoimmune disease known as bullous pemphigoid. These two collagens are not homologous in structure, but they both contain a single transmembrane N-terminal domain that is apparently cytoplasmic. The remainder of the molecule is extracellular (11, 87, 90, 91). One of the most remarkable features of type XIII collagen is that it undergoes extensive alternative splicing that can generate several hundred forms of the protein (84, 92, 93). The alternative splicing is unique among collagens in that it involves -Gly-X-Y- sequences. How this alternative splicing alters the potential of the protein for folding into a triple helix in which all three chains must be the same length is not clear.

FAMILY OF TYPES XV AND XVIII   The newly discovered type XV (94–98) and XVIII (97, 99–103) collagens have a large N-terminal globular domain, a highly interrupted triple helix, and a large C-terminal globular domain. Both collagens also contain several potential attachment sites for serine-linked glycosaminoglycans and asparagine-linked oligosaccharides, observations that suggest these collagens may be extensively glycosylated. Both type XV and type XVIII collagens are found in many tissues, but type XVIII collagen is expressed at much higher levels in the liver. Type XVIII collagen is transcribed from two alternative promoters, and the RNA transcript from one of the promoters is further modified by alternative splicing within sequences coding for the N-terminal globular domain (103). The longest variant of type XVIII collagen has an N-terminal cysteine-rich sequence that is homologous to three noncollagenous proteins in the family of G–protein coupled receptors (103).

"NONCOLLAGEN" COLLAGENS   The group of proteins containing collagenous sequences but not defined as collagens includes the subcomponent C1q of

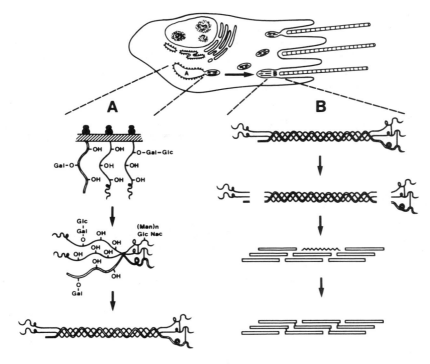

Figure 2 Schematic for the biosynthesis of a fibril-forming collagen. (*A*) Intracellular events that involve post-translational hydroxylation and glycosylation, association of polypeptide chains, and folding of the triple helix. (*B*) Extracellular events that involve cleavage of the N- and C-propeptides, self-assembly of collagen into fibrils, and cross-linking of the fibrils. Reproduced with permission from Reference 108.

complement, the tail structure of acetylcholinesterase, pulmonary surfactant proteins SP-A and SP-D, mannan-binding protein, conglutinin, collectin-43, the bacterial enzyme pullanase, and type I and type II macrophage scavenger receptors (6, 11, 104–107). The type I and II macrophage scavenger receptors resemble type XIII and XVII collagens in that they contain a single transmembrane domain preceded by a short cytoplasmic N-terminal domain. The rest of the molecule is extracellular (11, 105).

BIOSYNTHESIS

General Features

The fibril-forming collagens are first synthesized as larger precursor molecules known as procollagens. The intracellular steps in the assembly of a procollagen are (Figure 2) cleavage of signal peptides, hydroxylation of Y-position proline and lysine residues to 4-hydroxyproline and hydroxylysine; hydroxylation of a few X-position proline residues to 3-hydroxyproline, addition of galactose or both galactose and glucose to some of the hydroxylysine residues, addition of a mannose-rich oligosaccharide to one or both of the propeptides, association of the C-terminal propeptides through a process directed by the structure of these domains, and formation of both intrachain and interchain disulfide bonds. After the C-propeptides have associated and each chain has acquired about 100 4-hydroxyproline residues, a nucleus of triple helix forms in the C-terminal region, and the triple helical conformation is then propagated to the N-terminus in a zipper-like manner (6, 7, 16, 21, 108).

After secretion of procollagen from fibroblasts, the N-propeptides are cleaved by a procollagen N-proteinase and the C-propeptides by a separate procollagen C-proteinase. The collagen then self-assembles into fibrils. Finally, lysyl oxidase converts some lysine and hydroxylysine residues to aldehyde derivatives that form a complex series of cross-links.

The assembly and processing steps of many nonfibrillar collagens are the same, but there are notable exceptions. Many collagens contain N- and/or C-terminal noncollagenous domains that are not removed and therefore not called propeptides. Several collagens undergo N-glycosylation. Three collagens (types IX, XII, and XIV) are modified by addition of glycosaminoglycan side chains, and two additional collagens (types XV and XVIII) have potential attachment sites for such chains. The triple helices of a few collagens that lack large C-terminal globular domains (e.g. type XII) may fold by means of a mechanism that does not involve formation of a nucleus in the C-terminus (109).

Intracellular Processing

Recently, analyses of cDNAs provided the complete amino acid sequences for the α subunit of prolyl 4-hydroxylase from human (110), chick (111), nematode *Caenorhabditis elegans* (112), and for the β subunit from several organisms (113). In addition, complete amino acid sequences have been reported for human (114) and chick (115) lysyl hydroxylase. Prolyl 4-hydroxylase from vertebrates is an $\alpha_2\beta_2$ tetramer and lysyl hydroxylase an α_2 dimer, but the subunit structure of prolyl 3-hydroxylase is currently unknown (116, 117). No significant homology is found between the primary structures of lysyl hydroxylase and the two types of subunits of prolyl 4-hydroxylase in spite of the marked similarities in the catalytic properties between these two enzymes (116, 117).

The two catalytic sites in the tetramer of prolyl 4-hydroxylase are located in the α subunits (113, 116, 117). Even though the enzyme from all the vertebrate sources studied is an $\alpha_2\beta_2$ tetramer, cloning and expression of the α subunit from *C. elegans* revealed that its prolyl 4-hydroxylase is an $\alpha\beta$ dimer

(112). Also, an isoform of the α subunit of the vertebrate enzyme, defined as $\alpha(II)$ subunit, was recently discovered (118). Expression studies have demonstrated that the recombinant enzyme from vertebrates can form both $[\alpha(I)]_2\beta_2$ and $[\alpha(II)]_2\beta_2$ tetramers (118, 119), but whether the recombinant enzyme can also form tetramers that contain both α subunits is currently unknown. There appear to be no major differences in the tissue distribution of the two α subunits (118). Most of the catalytic properties of the $[\alpha(I)]_2\beta_2$ and $[\alpha(II)]_2\beta_2$ tetramers are highly similar, but a surprising difference is that poly(L-proline) is a very poor inhibitor of the $[\alpha(II)]_2\beta_2$ enzyme (118), whereas it is a highly effective competitive inhibitor of the $[\alpha(I)]_2\beta_2$ enzyme (117).

Nucleotide sequencing of the cDNA for the β subunit of prolyl 4-hydroxylase indicated that this polypeptide is identical to the enzyme protein disulfide isomerase (PDI) (120). Moreover, the β subunit has PDI activity even when present in the native prolyl 4-hydroxylase tetramer (121). PDI catalyzes thiol: disulfide interchange in vitro, leading to net protein disulfide formation, reduction, or isomerization depending on the reaction conditions. Researchers regard it as the in vivo catalyst of disulfide bond formation in the biosynthesis of a large number of secreted and cell surface proteins, including collagens (122, 123).

The PDI activity of the PDI/β subunit is not directly involved in the hydroxylation reaction of prolyl 4-hydroxylase. This finding is based on recent data obtained by expression of a recombinant prolyl 4-hydroxylase tetramer in insect cells (119, 124). The PDI/β polypeptide has two -Cys-Gly-His-Cys-sequences that represent two independently acting catalytic sites for the isomerase activity (124, 125). When both these sequences were modified to -Ser-Gly-His-Cys-, the polypeptide had no PDI activity but still associated with the α subunits to form the $\alpha_2\beta_2$ tetramer, and this tetramer proved to be fully active prolyl 4-hydroxylase (124). Expression studies have further demonstrated that in the absence of the PDI/β subunit, the α subunit forms highly insoluble aggregates (118, 119, 126). Therefore, one function of the PDI/β subunits in the prolyl 4-hydroxylase tetramer is to keep the α subunits in a catalytically active, nonaggregated conformation.

Recent reports indicate that the cellular PDI/β polypeptide may have several additional functions (117, 122, 123, 127–129). One is to serve as a major cellular thyroid hormone–binding protein in the endoplasmic reticulum. A second function is to act as a chaperone-like polypeptide that nonspecifically binds peptides in the lumen of the rough endoplasmic reticulum. A third function is to serve as the smaller subunit of the microsomal triglyceride transfer protein. Further suggested functions are to serve as a dehydroascorbate reductase (130) and to act as a developmentally regulated retinal protein termed r-cognin (131). The PDI/β subunit thus appears to be an unusually versatile polypeptide that has many biological functions.

The two hydroxylysyl glycosyltransferases involved in the biosynthesis of collagens have been extensively characterized (6, 7, 16), but the genes have not yet been cloned.

Recent reports also suggest that chain association and folding of type I and IV collagens may involve a specific molecular chaperone protein called Hsp47 or colligin (132–134). Hsp47 binds specifically to type I procollagen and to types I and IV collagens in vitro. Cellular levels of the protein parallel the rates of synthesis of type I or type IV collagen in many experimental situations. Cross-linking studies in intact cells demonstrated association of type I procollagen with Hsp47, and this association was increased when cells were heat shocked or treated with the iron chelator α,α'-dipyridyl that effectively inhibits the hydroxylation of proline residues (132). Treatment of cells with antisense oligonucleotides to Hsp47 decreased the rate of synthesis of type I procollagen (134). However, Hsp47 does not bind to type III procollagen, and researchers have not clearly established that it has an essential role in procollagen biosynthesis.

Extracellular Events

Extracellular collagen fibrils are formed by secretion of a soluble procollagen that is then enzymatically processed to an insoluble collagen. The mechanisms by which other collagens are incorporated into an insoluble extracellular matrix are more obscure, since presumably they must be soluble during intracellular assembly. One possible mechanism is that such collagens are secreted as soluble proteins that bind to other macromolecules after secretion to form insoluble heteromolecules.

Both the N- and C-propeptides of procollagens must be cleaved by specific proteinases for the proteins to self-assemble into fibrils under physiological conditions. The N-propeptides of both types I and II procollagens are cleaved by the same specific procollagen N-proteinase (6, 7, 16). The N-propeptide of type III procollagen is probably cleaved by a different type III N-proteinase (6, 7, 16). Whether other specific N-proteinases are required to cleave other procollagens such as types V and XI is unclear. Contrary to earlier reports, type I N-proteinase extracted from bovine tissues was recently shown to be the same protein as the better-characterized enzyme from chick embryos (135). Also, researchers recently demonstrated that if type I procollagen is aggregated by addition of polyethylene glycol (136), the rate of cleavage by the C-proteinase is increased 10- to 15-fold. The rate of cleavage by the N-proteinase is increased about fourfold. Because the turnover numbers with monomeric type I procollagen are low, it may be that the enzymes in vivo may act on secreted aggregates of procollagen (137).

The self-assembly of fibril-forming collagens has been studied for many years by warming and neutralizing solutions of the collagen extracted from tissues with cold acidic buffers (137, 138). More recently, the process was studied in a system in which pCcollagen, a soluble and partially processed precursor lacking the N-propeptide, is cleaved to collagen by purified procollagen C-proteinase in a physiological buffer and at physiological temperature ranges (137). Cleavage of pCcollagen to collagen reduces the solubility of the protein by about 1000-fold. The resulting collagen reproducibly self-assembles into tightly packed fibrils (137, 139).

One series of experiments was carried out by isolating type I procollagen and cleaving it with C-proteinase to generate type I pNcollagen (137). The pNcollagen assembled into thin, sheet-like structures that were cross-striated in longitudinal sections and of a uniform thickness. Mixtures of type I collagen and pNcollagen copolymerized to form a variety of pleomorphic fibrils. The results were consistent with the hypothesis that under some circumstances type I pNcollagen has a biological role in altering the morphology of type I collagen fibrils (137, 140).

Type III pNcollagen also formed true copolymers with type I collagen, and the copolymerization generated fibrils that were thinner than fibrils generated from type I collagen alone (141). The results were consistent with a model in which type III pNcollagen can regulate the diameter of type I collagen fibrils by coating their surface. However, the effects on fibril diameter required at least a 1:1 ratio of type III pNcollagen to type I collagen (141).

In related experiments, recombinant type II pCcollagen (142, 143) was used for fibril assembly by incubation with C-proteinase (144). The kinetics for the assembly of type II collagen fibrils differed markedly from those for the assembly of type I collagen, and the critical concentration at 37°C was about 50-fold greater. Also, the type II collagen fibrils were relatively thin and formed three-dimensional networks. The results indicated that the differences in primary structure between type II and I collagens are sufficient to explain many of the characteristic differences in morphology between these two kinds of fibrils seen in tissues (144).

The system for generating type I collagen fibrils by enzymic cleavage of type I pCcollagen made it possible to follow the growth of fibrils from the intermediate stages. The first fibrils detected had a blunt end and a pointed tip (145). Initial growth of the fibril was exclusively from the pointed tip. Later, tips appeared on the blunt ends and the fibrils grew in both directions. Both the initial tips and the later tips were nearly paraboloidal (145). Based on the observations, a model of fibril growth was developed (146), the essential features of which were a distinctive structural nucleus that formed at each end of a growing fibril. The growth of the fibril then occurred by propagation of the two structural nuclei. The structural nuclei had similar spiral helical conformations, and assembly and propagation of each structural nucleus required just two kinds of specific binding steps (146). Similar studies on fibrils of recombinant type II collagen demonstrated that the tips were again nearly paraboloidal (147). However, the monomers in the tips of the two types were oriented differently. In tips of type I collagen fibrils, all the monomers were oriented so that the N-termini pointed toward the end of the tip (145). In fibrils of type II collagen, all the monomers were oriented so that the C-termini pointed toward the growing tip (147).

Recently, Fourier transformed infrared spectroscopy (FTIR) was used to study the lag period of fibrils assembled by neutralizing and warming solutions of collagen extracted with cold acidic buffers (138). The results are consistent with the conclusion that as the temperature is raised, the triple helix tightens or stiffens but then relaxes again as fibrils are formed (138, 148).

The lysyl oxidase that forms cross-links in collagen fibrils is a highly insoluble copper-containing protein (149). Complete cDNA-derived amino acid sequences have now been reported for the enzyme from several sources (150–153). The enzyme was identical to a tumor suppressor protein known as *rrg* (154, 155). Earlier work had demonstrated that lysyl oxidase activity is markedly low in the culture medium of many malignantly transformed human cell lines (156), and recently, these cells were also found to have very low levels of lysyl oxidase mRNA (157).

Potentials for Inhibiting Fibrosis

Normal wound healing involves the formation of scars and fibrous tissue that largely consist of collagen fibrils. Although moderate degrees of fibrous tissue are beneficial in wound repair, fibrous material often accumulates in excessive amounts and impairs the normal function of the affected tissue. Such excessive accumulation of collagen becomes an important event in scarring of the skin following burns or traumatic injury and in fibrosis of the liver, lungs, and kidneys following injury to these organs. Therefore, there has been considerable interest in agents that can inhibit or modulate collagen synthesis in fibrotic diseases. Potential target sites for inhibiting collagen synthesis include transcription of the genes, translation of the mRNAs, and some of the unique post-translational enzymes involved in the biosynthesis of the protein.

Recent studies have demonstrated that synthesis of type I collagen can be specifically inhibited in cell culture by the use of antisense oligonucleotides (158–160). However, the degree of inhibition obtained is highly variable and rarely exceeds 50% (158, 159). In related experiments, an antisense gene to human type I collagen in which only the 3'-half was inverted was shown to be highly effective in inhibiting collagen synthesis in transgenic mice expressing an internally deleted human COL1A1 minigene (161). The results raised the possibility that chimeric gene constructs that contain intron sequences and in which only part of the gene is inverted may be particularly effective as antisense genes that can inhibit collagen synthesis in fibrotic conditions. However, both the antisense-oligonucleotide strategy and the antisense-gene strategy appear to present considerable problems in the delivery of the agents in ways that will be effective in inhibiting fibrosis in vivo.

Several of the post-translational enzymes appear to be attractive targets for specific inhibition because they are unique to collagen biosynthesis. These include prolyl 4-hydroxylase, procollagen C-proteinase, and perhaps also lysyl hydroxylase and lysyl oxidase. Numerous compounds are now known that inhibit prolyl 4-hydroxylase competitively with respect to some of its cosubstrates or the peptide substrate (116, 117). For example, pyridine 2,4-dicarboxylate inhibited prolyl 4-hydroxylase with a K_i of 2 µM. The problem of cell membrane permeability was in part overcome by the design of lipophilic proinhibitors that were converted to the active inhibitors intracellularly (21, 117, 162). One such derivative inhibits hepatic collagen accumulation in two models of liver fibrosis in rats (21, 162). The recent success in expression of an active recombinant human prolyl 4-hydroxylase in insect cells (119) should make it possible to define the critical structural features of the enzyme by site-directed mutagenesis (119, 163) and to produce adequate amounts of the enzyme for crystallization so that more effective inhibitors can be designed.

Procollagen C-proteinase is another attractive target for inhibition of fibrosis. Most of the available evidence suggests that procollagen cannot participate in fibril assembly unless the C-propeptide is specifically cleaved from the precursor (6, 7, 16, 137). The only challenge to this proposal comes from experiments in which recombinant procollagen was synthesized with a mutation at the cleavage site so that the protein was not cleaved by C-proteinase. Some of the protein synthesized in cell culture was cleaved by nonspecific proteases (164; J Bateman, personal communication). However, such nonspecific cleavage is unlikely to generate collagen that is assembled into normal fibrils. Initial studies suggested that basic amino acids and peptides may specifically inhibit C-proteinase (see 108), but the development of more effective agents such as peptidomimetics is still in the early stages.

Several attempts have been made to develop inhibitors for lysyl hydroxylase. Minoxidil and many of its derivatives have the surprising effect of reducing both lysyl hydroxylase activity (165) and the mRNA (166, 167) in cultured cells. Their mechanism of action is unknown. Also, whether inhibition of lysine hydroxylation will in itself be effective in inhibiting fibrosis is unclear.

One of the first targets explored for inhibition of fibrosis was lysyl oxidase. β-Aminopropionitrile has long been known to be a suicide inhibitor of the enzyme (149). Recently, several derivatives were developed that are even more effective (149). However, whether an inhibition of lysyl oxidase will prevent fibrosis is unclear, because the cross-linking of collagens occurs long after fibril assembly.

MUTATIONS IN MEN AND MICE

Mutations in Patients

TYPE I COLLAGEN   Almost 200 different mutations have now been characterized in the COL1A1 and COL1A2 genes that code for proα1(I) and proα2(I) chains of type I procollagen (2, 17–22, 25; for details, see 18). Most of these mutations have been identified in patients with osteogenesis imperfecta (OI), but they are also found in patients with related disorders (Table 2).

OI is characterized by brittle bones but also involves other tissues rich in type I collagen so as to produce blue sclerae, abnormal teeth, thin skin, weak tendons, and hearing loss. In the most severe forms, bones and other tissues are so fragile that death occurs in utero or shortly after birth. In more moderate forms, the disease is not lethal, but the patients have repeated fractures after minor trauma that may lead to permanent deformities of limbs and other bony structures.

Table 2   Diseases caused by mutations in collagen genes or deficiencies in the activities of post-translational enzymes of collagen synthesis

Gene or enzyme	Disease
COL1A1; COL1A2	Osteogenesis imperfecta
	Osteoporosis[a]
	Ehlers-Danlos syndrome type VIIA, VIIB
COL2A1	Several chondrodysplasias
	Osteoarthritis[a]
COL3A1	Ehlers-Danlos syndrome type IV
	Aortic aneurysms[a]
COL4A3; COL4A4	Alport syndrome, autosomally inherited forms
COL4A5	Alport syndrome, X-linked form
COL4A5 and COL4A6	Alport syndrome with diffuse esophageal leiomyomatosis, X-linked
COL7A1	Epidermolysis bullosa, dystrophic forms
COL9A1	Osteoarthritis[b]
COL9A2	Multiple epiphyseal dysplasia[c]
COL10A1	Schmid metaphyseal chondrodysplasia
COL11A2	Stickler syndrome, nonocular form[c]
Lysyl hydroxylase	Ehlers-Danlos syndrome type VI
Type I N-proteinase	Ehlers-Danlos syndrome type VIIC
Lysyl oxidase	Occipital horn syndrome[d]
	Menkes syndrome[d]

[a] In a subset of patients.
[b] Demonstrated only in transgenic mice.
[c] Demonstrated only by genetic linkage.
[d] Secondary to an abnormality in copper metabolism.

Essentially all patients with OI have mutations in type I collagen. Most of the mutations are single base substitutions that convert a codon for an obligate glycine in the repeating -Gly-X-Y- sequence of the triple helix to a codon for an amino acid with a bulkier side chain (2, 17–22). Other mutations include deletions, insertions, RNA splicing defects, and null alleles. The null alleles primarily cause the mild type I variant of OI (168). The mutations inactivating the alleles have been difficult to define, but four patients were recently found to have a premature translation termination codon that decreased the cytoplasmic level of the mRNA (J Körkkö, P Paassilta, J Zhuang, H Kuivaniemi, L Ala-Kokko, et al, in preparation).

Mutations that cause synthesis of structurally altered proα chains of type I procollagen generally cause more severe phenotypes than null alleles (2, 17–22). One of two molecular mechanisms are usually involved. Some substitutions for obligate glycines interrupt the zipper-like folding of the triple helix and generate unfolded procollagen that first accumulates in fibroblasts and is then degraded. The effects of the mutations are amplified because both the normal and mutated chains present in the same molecule are degraded in a process referred to as procollagen suicide or a dominant negative effect (17, 18, 108). The effects of other glycine substitutions are explained by their consequences on the nucleated growth of collagen fibrils. One such mutation was a heterozygous substitution of cysteine for glycine at position 748 of the α1(I) chain that introduced a flexible kink into the triple helix (2, 18, 169). Studies on fibril formation in vitro (see biosynthesis) demonstrated that molecules with the cysteine kink copolymerized into fibrils with the normal molecules, but the presence of the kinked molecules delayed fibril formation, reduced the total amount of collagen incorporated into the fibrils, and drastically altered the morphology of the fibrils (18, 138, 170) (Figure 3). Other glycine substitutions that do not affect folding have similar effects on fibril assembly (171–173).

In OI, bones are fragile in part because of a marked reduction in bone mass (osteopenia). Mutations in type I collagen have also been found in a few patients who have little evidence of OI but who have osteopenia and fractures characteristic of osteoporosis (18, 20, 21, 25, 174, 175). A recent survey suggested that 1–3% of patients with osteoporosis have a mutation in either the COL1A1 or COL1A2 gene (175).

Some mutations in the type I collagen genes produce a disease known as the type VII variant of EDS (Table 2), a syndrome characterized by joint hypermobility and skin abnormalities (18, 21, 23, 25). The disease is caused by a failure to cleave the N-propeptide from type I procollagen. The persistence of the N-propeptide on the molecule drastically alters fibril formation so that the fibrils become thin and highly irregular in cross-section. The mutations causing EDS VII are either RNA splicing mutations that eliminate the amino

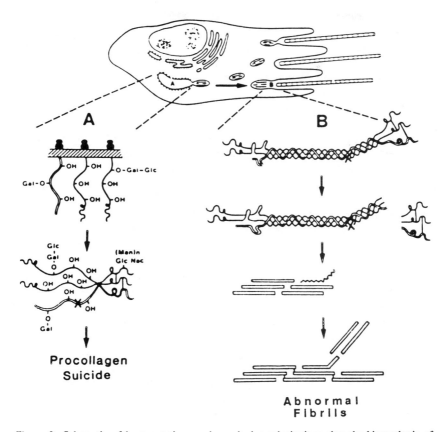

Figure 3 Schematic of how mutations such as glycine substitutions alter the biosynthesis of collagen. (*A*) Illustration of a glycine substitution that prevents the zipper-like folding of the triple helix and leads to degradation of both normal and abnormal proα chains by a procollagen suicide or dominant negative mechanism. (*B*) Illustration of a glycine substitution that does not interfere with the folding of the triple helix but produces a conformational change such as a kink in the protein. (Modified and reproduced with permission from Reference 108.)

acids of the cleavage site for the N-propeptide (subtypes VIIA and VIIB) or mutations that decrease the activity of the cleaving enzyme, procollagen N-proteinase (subtype VIIC).

TYPE II COLLAGEN   Over 50 different mutations in the COL2A1 gene have been shown to cause a heterogeneous group of disorders of cartilage that are known as chondrodysplasias and are characterized by short-limbed dwarfism and skeletal deformities (18, 21, 24, 176–182). The mutations include amino acid substitutions, deletions, insertions, RNA splicing defects, and stop codons for premature termination of translation. All five stop codons so-far characterized have been identified in the single phenotype of the Stickler syndrome that involves vitreous degeneration and retinal detachment in addition to degeneration of joint cartilage (21, 183–187).

Mutations in type II collagen (Table 2) are also found in about 2% of patients with early-onset familial osteoarthritis (177). One mutation is a substitution of cysteine for arginine at amino acid position 519 of the $\alpha1(II)$ chain, that has now been found in four apparently unrelated families with severe, early-onset osteoarthritis and mild chondrodysplasia (179, 188, 189). Two additional mutations in patients with a similar disease phenotype are a serine-for-glycine substitution at position $\alpha1$-274 (190) and at $\alpha1$-976 (179). In addition, mutations in other genes for collagens such as types IX and XI (Table 1) may be found as predisposing factors for osteoarthritis in some families (see Mutations in Transgenic Mice, below).

TYPE III COLLAGEN   About 50 different mutations in the COL3A1 gene have been found in patients with EDS IV (18, 21, 23, 191, 192), the most severe form of EDS that can cause sudden death from rupture of large arteries and other hollow organs in addition to skin and joint changes (23, 25). The mutations include glycine substitutions, deletions, RNA splicing defects and null alleles (18, 21, 191, 192). Mutations in the COL3A1 gene have also been found in a subset of patients who have arterial aneurysms but who exhibit little (193) or no evidence (18, 21, 194–196) of other connective tissue manifestations. Mutations in the COL3A1 gene may also be a predisposing factor for intracranial aneurysms (18, 21, 197), but they appear to be a rare cause of this disease (198).

TYPE IV COLLAGEN   No mutations have been identified in the genes COL4A1 and COL4A2 that encode the two major α chains of type IV collagen, but mutations have been found in the genes coding for the minor type IV collagen polypeptides (Table 2). The Alport syndrome is a progressive heritable kidney disease characterized by hematuria caused by structural changes in the glomerular basement membrane. This disease is also associated with hearing loss and ocular lesions. The gene coding for the $\alpha5(IV)$ chain was mapped to the locus for the X-linked form of the Alport syndrome (50, 51), and subsequently, more than 50 different mutations in the COL4A5 gene were found in families with this disorder (21, 44, 199, 200). The mutations include amino acid substitutions, large deletions, and gene rearrangements such as inversions, insertions, and duplications. Although the Alport syndrome is primarily X-linked, autosomally inherited forms also exist, and recently, heterozygous mutations in both the COL4A3 and COL4A4 genes were characterized in autosomally inherited forms of the disease (44, 201).

Deletions involving the head-to-head 5′-ends of both the COL4A5 and COL4A6 genes have been found in several patients (54, 202) who have the Alport syndrome together with diffuse esophageal leiomyomatosis, a rare syndrome characterized by proliferation of smooth muscle cells in the esophagus, tracheobronchial tree, and the female genital tract (44).

TYPE VII COLLAGEN   About 20 mutations in the COL7A1 gene have been found in patients with the dystrophic form of epidermolysis bullosa, a disease characterized by severe blistering and scarring of the skin from minor trauma (78–81, 203–206). As a consequence of these mutations, the anchoring fibrils that link the basement membrane to the anchoring plaques in the skin are either reduced in amount or completely absent (203). Mutations in the COL7A1 gene were found in both the dominantly and recessively inherited forms of the disease (78–81). The mutations include amino acid substitutions, an insertion-deletion, and premature translation termination codons (81, 203–206).

TYPE X COLLAGEN   More than 10 different mutations have now been characterized in type X collagen (207–213) in patients with Schmidt metaphyseal chondrodysplasia, a disease that is characterized by shortening of limbs and bowing of legs aggravated by walking. The mutations include amino acid substitutions, deletions, and premature translation termination codons. All the mutations so far characterized alter the structure of C-terminal noncollagenous domain of the polypeptide, an observation suggesting that the mutant chains are unable to associate to form triple-helical molecules.

OTHER COLLAGENS   Genetic linkage was found between the COL9A2 gene locus (214) and multiple epiphyseal dysplasia (EDM2). Also, genetic linkage was found between the COL11A2 gene locus and a non-ocular form of the Stickler syndrome (215).

Since correct expression of collagen genes appears to be essential for the structural integrity of many tissues, mutations in more than 30 different collagen genes (Table 1) can probably produce disease phenotypes. Therefore, research in this area is still at a very early stage (20). Also, similar disease phenotypes are probably produced by mutations in genes for other matrix proteins, since several diseases of cartilage have been linked to loci that do not contain any known collagen genes (18, 26).

POST-TRANSLATIONAL ENZYMES   A deficiency of lysyl hydroxylase is found in most but not all families with EDS VI (21, 23, 108), a disease characterized by hyperextensible skin and joints, scoliosis, and ocular fragility. One mutation in the gene for the enzyme was a homozygous duplication of seven exons that appears to be caused by a recombination of Alu sequences (216, 217). The same mutation was found in several apparently unrelated families (216–218). Additional mutations include a homozygous translation termination codon (219) and several amino acid substitutions (220).

EDS VIIC is caused by a deficiency in type I procollagen N-proteinase (221, 222), but no mutations in the genes have been characterized so far.

Deficiencies of lysyl oxidase, a copper-containing protein, are seen in two rare and severe X-linked recessive diseases, the occipital horn syndrome and Menkes syndrome (21, 108, 223). The diseases are caused by defects in copper metabolism and lead to secondary defects in the cross-linking of collagen, but the mechanisms producing a deficiency of lysyl oxidase are unclear. Skin fibroblasts from patients with these diseases contain and secrete reduced amounts of the lysyl oxidase protein (21, 224), and recently, two (157, 224) out of three (157, 224, 225) studies reported that these cells also contain reduced amounts of lysyl oxidase mRNA. These observations suggest that the abnormality in copper metabolism somehow influences the synthesis or stability of the mRNA for lysyl oxidase.

Mutations in Transgenic Mice

Transgenic mice are particularly useful for studying matrix proteins because most of the proteins are large, insoluble, and difficult to test for function. They are also useful for studying the consequences of disease-causing mutations in matrix genes, since the mutations affect many tissues that cannot be examined fully in patients. Experiments have been carried out in which mutated collagen genes were randomly inserted in transgenic mice to produce dominant negative effects, and more recently, a few experiments were carried out involving knock-out of collagen genes. In some instances, the results from these two types of experiments have been complementary. In others they are apparently contradictory.

The first experiments with transgenic mice used a retrovirus infection of mouse embryos. By chance, the retrovirus was inserted in the first intron of the COL1A1 gene and prevented expression of the gene in most tissues (see226). Homozygous mice died in utero with liver necrosis and bleeding. Heterozygous mice survived and had decreased collagen content, decreased mechanical strength of bones, and hearing loss (226). No fractures were detected. Subsequently, transgenic mice were prepared that expressed a mutated COL1A1 gene in which a cysteine codon was substituted for an obligate glycine (227). Four of seven founder mice had some of the phenotypic features of OI such as poor mineralization in most bony structures. Again, however, no fractures were demonstrated and no breeding lines of the transgenic mice could be developed. In contrast, extensive fractures were observed in several lines of transgenic mice expressing a mini-gene version of the human COL1A1 gene in which exons 6 to 46 were deleted in-frame to cause synthesis of shortened proα1(I) chains (228). Mice expressing high levels of the mini-gene protein developed a lethal phenotype of extensive fractures because the shortened proα chains produced procollagen suicide or a dominant negative effect through protein depletion (228). Mice expressing lower levels developed a milder phenotype resembling human osteoporosis (229, 230). Extensive breeding of the transgenic mice in the inbred line demonstrated marked phenotypic variability and incomplete penetrance that apparently is an inherent property of expression of mutated collagen genes (230). In related experiments, a naturally occurring recessive mutation that produced bowed and brittle bones in mice was shown to be a single base deletion in the C-propeptide that caused synthesis of proα2 chains that could not associate with proα1 chains (231). As a result, the only type I procollagen synthesized was homotrimers of proα1(I) chains.

In experiments with the COL2A1 gene, transgenic mice expressing an internally deleted version of the human gene developed a phenotype similar to some human chondrodysplasias with dwarfism, a short snout, a cranial bulge, a cleft palate, and delayed mineralization of bone (232). At the same time, similar phenotypes of severe chondrodysplasias were seen in transgenic mice expressing mouse COL2A1 genes mutated either by substitution of a cysteine codon for glycine at amino acid position 85 in the α1(II) chain (233) or by a deletion that eliminated amino acids 4–18 of the α1(II) chain (234). In older mice from some of the same lines, the evidence of chondrodysplasias was less marked, and the most striking features were degenerative changes of articular cartilage similar to osteoarthritis (235, 236). A surprising finding was that over-expression of a normal mouse COL2A1 gene in transgenic mice produced abnormally thick collagen fibrils in cartilage, apparently because of an imbalance in the amounts of different collagens being synthesized in the tissue (237). Transgenic mice over-expressing the normal gene developed a chondrodysplasia.

Transgenic mice were prepared that expressed a COL3A1 gene in which a methionine codon was substituted for lysine at amino acid 939, the cross-linking site in the triple-helical region of the protein (238, 239). The phenotype was mild, but pregnant females apparently developed uterine dysfunction (238). The same mice had morphologic changes in healing dermal wounds, but no drastic consequences occurred on wound repair (239).

Transgenic mice expressing a mutated COL9A1 gene (240) and transgenic mice in which the gene was inactivated (241) developed early onset osteoarthritis. The results suggested, therefore, that a subset of patients with osteoarthritis may have mutations in type IX collagen.

In transgenic mouse experiments with the COL10A1 gene, the results with dominant negative experiments and gene inactivation experiments were different. Transgenic mice expressing constructs of the chicken gene with an in-frame deletion in the central triple-helical domain of the collagen (242) developed skeletal abnormalities and a deficiency of leukocytes similar to the Schmidt type of human metaphyseal chondrodysplasia. The observations in transgenic mice (242) largely provided the basis for defining mutations in the same gene that cause Schmidt phenotypes in patients (see above). However, mice with homozygous inactivation of the COL10A1 gene did not develop any apparent phenotype (243). The initial explanation of these conflicting data was that mutations producing a dominant negative effect were more severe in terms of phenotype than mutations that inactivated the gene. Subsequently, researchers found that mice harboring a 68–amino acid deletion from the triple-helical domain of the type X collagen gene did not develop any gross abnormalities of the skeleton (244). At the moment, there is no simple way to resolve these conflicting observations.

Transgenic mice expressing an internally deleted COL11A2 gene developed a mild phenotype with a short snout, prominent forehead, shortened limbs, and shortened tail (245). Also, DNA linkage to the locus for COL11A1 was found in cho mice that have a naturally occurring autosomal recessive chondrodysplasia (246). The $\alpha 1(XI)$ chain was absent from tissues, and the collagen fibrils in cartilage from the mice were unusually thick.

Potentials for Gene Therapy

Researchers have conducted preliminary experiments to test the potentials of gene therapy for either controlling collagen deposition in fibrotic conditions or rescuing the phenotypes produced by mutated genes. As indicated above, oligonucleotides partially inhibited expression of collagen genes in cell culture experiments (158–160). A related approach showed that an antisense gene to the human COL1A1 gene is effective in transgenic mice (161). The experiments were carried out with mice that developed a lethal phenotype of fragile bones because they expressed an internally deleted mini-gene for the $pro\alpha 1(I)$ chain of human type I procollagen. The mice were bred to transgenic mice that expressed an antisense gene that was similar to the internally deleted human COL1A1 mini-gene, but the 3′-half of the gene was inverted so as to code for hybrid sense and antisense RNA. In mice that inherited both genes, the incidence of lethal fragile bone phenotype was reduced from 92 to 27% (161), and there was a corresponding decrease in the tissue levels of the human mini-$pro\alpha 1(I)$ chains. The experiment may have succeeded where similar experiments with antisense cDNAs failed in the past, because the transcript from the antisense gene contained inverted intron sequences and, therefore, interacted with the sense transcript in the nucleus.

In related experiments, normal mice were subjected to lethal body irradiation and then received stromal cells of bone marrow from a transgenic mouse containing a DNA marker consisting of a human COL1A1 mini-gene. After five months, the donor cells were found to account for 3–5% of the cells in bone marrow, spleen, bone, cartilage, and lung (247). Also, the human COL1A1 mini-gene was expressed in bone. Because stromal cells from bone marrow are readily expanded in culture, they may be a useful source of long-lasting precursor cells for gene therapy of bone and cartilage diseases.

Techniques were also developed for site-directed insertion of an exogenous collagen gene into an endogenous gene locus (248). In stable transfection experiments with the human tumor cell line known as HT-1080, a construct containing a short 5'-fragment from the COL1A1 gene linked to 30 kb of the COL2A1 gene was found to be inserted at a high frequency into both alleles of the endogenous COL1A1 gene (248). Transfected cells with the targeted insertion expressed relatively high levels of the exogenous gene. The results suggested that it may be possible to target insertion of exogenous genes into predetermined loci of collagen genes. The targeted insertion has the advantage that it provides controlled expression of the exogenous genes and avoids the potential dangers of random insertions.

SUMMARY

Collagens are defined as proteins that: (a) contain several repeats of the amino acid sequence -Gly-X-Y- in which the X-position is frequently proline and the Y-position is frequently 4-hydroxyproline and (b) have the potential for three chains with such repeated sequences to fold into a characteristic triple helix. At least 19 proteins and more than 30 gene products are now formally defined as collagens. An additional 10 proteins have collagen-like domains. Therefore, the number of proteins classified within this super-family is rapidly expanding. The most abundant collagens form extracellular fibrils, but others form network-like structures. Still others bind to the surfaces of collagen fibrils, form beaded-filaments, serve as anchoring fibrils for the skin, and are transmembrane proteins. The exon-intron patterns of most of the collagen genes are complex. Most of the genes are widely distributed in the genome, but three pairs of genes are located in a unique head-to-head arrangement with overlapping promoters.

Recently, several of the eight highly specific posttranslational enzymes involved in collagen biosynthesis were cloned. The β subunit of prolyl 4-hydroxylase is identical to the enzyme protein disulfide isomerase and appears to have several other distinct functions. Experiments with recombinant procollagens that are the soluble precursors of fibril-forming collagens have confirmed previous indications that collagens self-assemble into fibrils by the classical mechanism of nucleation and propagation. The results also indicated that the differences in primary structure between the two most abundant fibril-forming collagens (types I and II) are sufficient to explain many of the characteristic differences in morphology between these two kinds of fibrils.

There has been increasing interest in the possibility that the unique post-translational enzymes involved in collagen biosynthesis offer attractive targets for specifically inhibiting the excessive deposition of collagen fibrils that occurs in scars and in the fibrotic response of injury seen in most tissues. Inhibition of prolyl 4-hydroxylase and procollagen C-proteinase appear to be particularly attractive strategies.

The important roles of collagens in biology have been illustrated by the recent demonstrations that over 400 mutations in 6 different collagen types cause a variety of human diseases. Mutations in the type I collagen genes can produce defects of bones and related tissues that range from lethal osteogenesis imperfecta to osteoporosis. Mutations in the gene for type II collagen can produce cartilage disorders ranging from lethal chondrodysplasias to early-onset osteoarthritis. Mutations in the gene for the type X collagen, which is expressed in hypertrophic chondrocytes, also produce chondrodysplasias, and genes for two additional collagens of cartilage have been linked to similar phenotypes. Mutations in type III collagen produce defects of blood vessels and other tissues that range from a severe form of the Ehlers-Danlos syndrome to familial aortic aneurysms. Mutations in several polypeptide chains of the type IV collagens of basement membranes produce the renal disease known as the Alport syndrome that may be associated with diffuse esophageal leiomyomatosis. Mutations in the type VII collagen that forms anchoring fibrils for basement membranes cause severe blistering and scarring of the skin. Mutations in other collagens are also likely to produce human diseases, but some diseases with similar phenotypes are clearly linked to noncollagen genes.

Experiments with transgenic mice have been particularly useful for defining the structure and function of collagens because the proteins are large, insoluble, and difficult to test for function. In addition, transgenic mice have been useful in demonstrating the consequences of disease-causing mutations in collagen genes, since the mutations affect many tissues that cannot be examined fully in patients. Experiments with mutated genes for type I collagen reproduced the phenotypes of osteogenesis imperfecta and osteoporosis. Experiments with mutated genes for type II collagen reproduced the phenotypes of severe chondrodysplasias and osteoarthritis. Similar experiments with the type III collagen gene produced only a mild phenotype. Experiments with the type IX gene produced early onset osteoarthritis. Experiments with a dominant negative version of the type X gene produced a phenotype of a specific chondrodysplasia, but a knock-out experiment with the same gene produced no apparent phenotype.

In exploring potentials for gene therapy, oligonucleotides were shown to partially inhibit expression of collagen genes in cell culture. In stable cell-transfection experiments, a hybrid gene containing a 5' fragment of the type I collagen gene was found to be inserted into both alleles of the normal locus of the type I gene at a high frequency. The results suggested it may be possible to target insertion of exogenous genes into predetermined loci of collagen genes. A phenotype of severe osteogenesis imperfecta in transgenic mice expressing a mutated human type I collagen gene was partially rescued by breeding these mice to other transgenic mice expressing an antisense version of the same human gene.

Literature Cited

1. Fleischmajer R, Olsen BR, Kühn K. 1990. *Structure, Molecular Biology, and Pathology of Collagen. Ann. NY Acad. Sci.* 580:1–592
2. Engel J, Prockop DJ. 1991. *Annu. Rev. Biophys. Biophys. Chem.* 20:137–52
3. Shaw LM, Olsen BR. 1991. *Trends Biochem. Sci.* 16:191–94
4. van der Rest M, Garrone R. 1991. *FASEB J.* 5:2814–23
5. Burgeson RE, Nimni ME. 1992. *Clin. Orthop.* 282:250–72
6. Hulmes DJS. 1992. *Essays Biochem.* 27:49–67
7. Kielty CM, Hopkinson I, Grant ME. 1993. In *Connective Tissue and Its Heritable Disorders: Molecular, Genetic and Medical Aspects*, ed. PM Royce, B Steinmann, pp. 103–47. New York: Wiley-Liss
8. Mayne R, Brewton RG. 1993. *Curr. Opin. Cell Biol.* 5:883–90
9. van der Rest M, Garrone R, Herbage D. 1993. *Adv. Mol. Cell Biol* 6:1–67
10. Brewton RG, Mayne R. 1994. In *Extracellular Matrix Assembly and Structure*, eds. PD Yurchenco, DE Birk, RP Mecham, pp. 129–70. San Diego: Academic
11. Pihlajaniemi T, Rehn M. 1995. *Prog. Nucleic Acid Res. Mol. Biol.* 50:225–62
12. Sandell LJ, Boyd CD. 1990. In *Extracellular Matrix Genes*, eds. LJ Sandell, CD Boyd, pp. 1–56. New York: Academic
13. Vuorio E, de Crombrugghe B. 1990. *Annu. Rev. Biochem.* 59:837–72
14. Chu M-L, Prockop DJ. 1993. See Ref. 7, pp. 149–65
15. Olsen BR. 1991. In *Cell Biology of*
Extracellular Matrix*, ed. ED Hay, pp. 177–220. New York: Plenum. 2nd ed.
16. Kivirikko KI. 1995. In *Principles of Medical Biology*, ed. EE Bittar, N Bittar. Greenwich, CT: JAI. In press
17. Prockop DJ. 1990. *J. Biol. Chem.* 265: 15349–52
18. Kuivaniemi H, Tromp G, Prockop DJ. 1991. *FASEB J.* 5:2052–60
19. Byers PH, Steiner RD. 1992. *Annu. Rev. Med.* 43:269–82
20. Prockop DJ. 1992. *N. Engl. J. Med.* 326:540–46
21. Kivirikko KI. 1993. *Ann. Med.* 25:113–26
22. Byers PH. 1993. See Ref. 7, pp. 317–50
23. Steinmann B, Royce PM, Superti-Furga A. 1993. See Ref. 7, pp. 351–407
24. Horton WA, Hecht JT. 1993. See Ref. 7, pp. 641–75
25. Prockop DJ, Kuivaniemi H, Tromp G. 1994. In *Harrison's Principles of Internal Medicine*, ed. KJ Isselbacher, E Braunwald, JD Willson, JB Martin, AS Fauci, DL Kasper, pp. 2105–17. New York: McGraw-Hill. 13th ed.
26. Jacenko O, Olsen BR, Warman ML. 1994. *Am. J. Hum. Genet.* 54:163–68
27. Ryan MC, Sandell LJ. 1990. *J. Biol.Chem.* 265:10334–39
28. Sandell LJ, Morris N, Robbins JR, Goldbring MB. 1991. *J. Cell Biol.* 114:1307–19
29. Nah HD, Upholt WB. 1991. *J. Biol. Chem.* 266:23446–52
30. Morris N, Oxford JT, Doege K. 1994. *Matrix Biol.* 14:361 (Abstr.)
31. Zhidkova NI, Justice SK, Mayne R. 1994. *Matrix Biol.* 14:365 (Abstr.)

32. Tsumaki N, Kimura T. 1994. *Matrix Biol.* 14:364 (Abstr.)
33. Thorogood P, Bee J, von der Mark K. 1986. *Devl. Biol.* 116:497–509
34. Hayashi M, Ninomiya Y, Hayashi K, Linsenmayer TF, Olsen BR. 1988. *Development* 103:27–36
35. Fitch JM, Menzer A, Mayne R, Linsenmayer TF. 1989. *Development* 105:85–89
36. Kosher RA, Solursh M. 1989. *Devl. Biol.* 131:558–66
37. Bernard M, Yoshioka H, Rodriguez E, van der Rest M, Kimura T, Ninomiya Y, et al. 1988. *J. Biol. Chem.* 263:17159–66
38. Niybizi C, Eyre DR. 1989. *FEBS Lett.* 242:314–18
39. Kleman J-P, Hartmann DJ, Ramirez F, van der Rest M. 1992. *Eur. J. Biochem.* 210:329–35
40. Mayne R, Brewton RG, Mayne PM, Baker JA. 1993. *J. Biol. Chem.* 268:9381–86
41. Bennett VD, Adams SL. 1990. *J. Biol. Chem.* 265:2223–30
42. Nah H-D, Niu Z, Adams SL. 1994. *J. Biol. Chem.* 269:16443–48
43. Timpl R. 1989. *Eur. J. Biochem.* 180:487–502
44. Hudson BG, Reeders ST, Tryggvason K. 1993. *J. Biol. Chem.* 268:26033–36
45. Yurchenco PD. 1994. See Ref. 10, pp. 351–88
46. Gunwar S, Saus J, Noelken ME, Hudson BG. 1990. *J. Biol. Chem.* 265:5466–69
47. Morrison KE, Germino GG, Reeders ST. 1991. *J. Biol. Chem.* 266:34–39
48. Kamagata Y, Mattei MG, Ninomiya Y. 1992. *J. Biol. Chem.* 267: 23753–58
49. Mariyama M, Kalluri R, Hudson BG, Reeders S. 1992. *J. Biol. Chem.* 267: 1253–58
50. Myers JC, Jones TA, Pohjalainen E-R, Kadri AS, Goddard AD, et al. 1990. *Am. J. Hum. Genet.* 46:1024–33
51. Hostikka SL, Eddy RL, Byers MG, Höyhtyä M, Shows TB, Tryggvason K. 1990. *Proc. Natl. Acad. Sci. USA* 87: 1606–10
52. Pihlajaniemi T, Pohjalainen E-R, Myers JC. 1990. *J. Biol. Chem.* 265:13758–66
53. Zhou J, Hertz JM, Leinonen A, Tryggvason K. 1992. *J. Biol. Chem.* 267: 12475–81
54. Zhou J, Mochizuki T, Smeets H, Antignac C, Lauriala L, et al. 1993. *Science* 261:1167–69
55. Oohashi T, Sugimoto M, Mattei M-G, Ninomiya Y. 1994. *J. Biol. Chem.* 269: 7520–26
56. Zhou J, Ding M, Zhao Z, Reeders ST. 1994. *J. Biol. Chem.* 269:13193–99

57. Bernal D, Quinones S, Saus J. 1993. *J. Biol. Chem.* 268:12090–94
58. Feng L, Xia Y, Wilson CB. 1994. *J. Biol. Chem.* 269:2342–48
59. Mariyama M, Zheng K, Yang-Feng TL, Reeders ST. 1992. *Genomics* 13:809–13
60. Zhou J, Leinonen A, Tryggvason K. 1994. *J. Biol. Chem.* 269:6608–14
61. Muragaki Y, Jacenko O, Apte S, Mattei MG, Ninomiya Y, Olsen BR. 1991. *J. Biol. Chem.* 266:7721–27
62. Schmid TM, Cole AA, Chen Q, Bonen DK, Luchene L, Linsenmayer TF. 1994. See Ref. 10, pp. 171–206
63. Nishimura I, Muragaki Y, Olsen BR. 1989. *J. Biol. Chem.* 264:20033–41
64. Hayashi M, Hayashi K, Iyama K-I, Trelstad RL, Linsenmayer TF, Mayne R. 1992. *Dev. Dyn.* 194:169–76
65. Swiderski RE, Solursh M. 1992. *Dev. Dyn.* 194:118–27
66. Watt LS, Lundstrum GP, McDonough AM, Keene DR, Burgeson RE, Morris NP. 1992. *J. Biol. Chem.* 267:20093–99
67. Gerecke DR, Foley JW, Castagnola P, Gennari M, Dublet B, Cancedda R, et al. 1993. *J. Biol. Chem.* 268:12177–84
68. Oh SP, Griffith M, Hay ED, Olsen BR. 1993. *Dev. Dynam.* 196:37–46
69. Wälchli C, Trueb J, Kessler B, Winterhalter H, Trueb B. 1993. *Eur. J. Biochem.* 212:483–90
70. Wälchli C, Koch M, Chiquet M, Odermatt BF, Trueb B. 1994. *J. Cell Sci.* 107:669–81
71. Pan TC, Zhang R-Z, Mattei M-G, Timpl R, Chu M-L. 1992. *Proc. Natl. Acad. Sci. USA* 89:6565–69
72. Yamaguchi N, Kimura S, McBride OW, Hori H, Yamada Y, et al. 1992. *J. Biochem.* 112:856–63
73. Yoshioka H, Zhang H, Ramirez F, Mattei M-G, Moradi-Ameli M, et al. 1992. *Genomics* 13:884–86
74. Myers JC, Yang H, D'Ippolito JA, Presente A, Miller MK, Dion AS. 1994. *J. Biol. Chem.* 269:18549–57
75. Timpl R, Chu ML. 1994. See Ref. 10, pp. 207–42
76. Zhang R-Z, Pan T-C, Timpl R, Chu M-L. 1993. *Biochem. J.* 291:787–92
77. Parente MG, Chung LC, Ryynänen J, Woodley DT, Wynn KC, et al. 1991. *Proc. Natl. Acad. Sci. USA* 88:6931–35
78. Uitto J, Christiano AM. 1992. *J. Clin. Invest.* 90:687–92
79. Burgeson RE. 1993. *J. Invest. Dermatol.* 101:252–55
80. Gerecke DR, Gordon MK, Wagman DF, Champliaud MF, Burgeson RE. 1994. See Ref. 10, pp. 417–39
81. Bruckner-Tuderman L. 1994. *Ann. Med.* 26:165–71

82. Christiano AM, Greenspan DS, Lee S, Uitto J. 1994. *J. Biol. Chem.* 269:20256–62
83. Christiano AM, Hoffman GG, Chung-Honet LC, Lee S, Chung W, et al. 1994. *Genomics* 21:169–79
84. Pihlajaniemi T, Tamminen M. 1990. *J. Biol. Chem.* 265:16922–28
85. Tikka L, Elomaa O, Pihlajaniemi T, Tryggvason K. 1991. *J. Biol. Chem.* 266:17713–19
86. Juvonen M, Pihlajaniemi T, Autio-Harmainen H. 1993. *Lab. Invest.* 69:541–51
87. Hägg P, Jaakkola S, Pihlajaniemi T. 1995. *Matrix Biol.* 14:351 (Abstr.)
88. Li K, Sawamura D, Giudice GJ, Diaz LA, Mattei M-G, et al. 1991. *J. Biol. Chem.* 266:24064–69
89. Giudice GJ, Emery DJ, Diaz LA. 1992. *J. Invest. Dermatol.* 99:243–50
90. Hopkinson SB, Riddelle KS, Jones JCR. 1992. *J. Invest. Dermatol.* 99:264–70
91. Li K, Tanai K, Tan EML, Uitto J. 1993. *J. Biol. Chem.* 268:8825–34
92. Juvonen M, Pihlajaniemi T. 1992. *J. Biol. Chem.* 267:24693–99
93. Juvonen M, Sandberg M, Pihlajaniemi T. 1992. *J. Biol. Chem.* 267:24700–7
94. Myers JC, Kivirikko S, Gordon MK, Dion AS, Pihlajaniemi T. 1992. *Proc. Natl. Acad. Sci. USA* 89:10144–48
95. Huebner K, Cannizzaro LA, Jabs EW, Kivirikko S, Manzone H, et al. 1992. *Genomics* 14:220–24
96. Kivirikko S, Heinämäki P, Rehn M, Honkanen N, Myers JC, Pihlajaniemi T. 1994. *J. Biol. Chem.* 269:4773–79
97. Muragaki Y, Abe N, Ninomiya Y, Olsen BR, Ooshima A. 1994. *J. Biol. Chem.* 269:4042–46
98. Kivirikko S, Saarela J, Myers JC, Autio-Harmainen H, Pihlajaniemi T. 1995. Submitted
99. Rehn M, Pihlajaniemi T. 1994. *Proc. Natl. Acad. Sci. USA* 91:4234–38
100. Oh SP, Kamagata Y, Muragaki Y, Timmons S, Ooshima A, Olsen BR. 1994. *Proc. Natl. Acad. Sci. USA* 91:4229–33
101. Rehn M, Hintikka E, Pihlajaniemi T. 1994. *J. Biol. Chem.* 269:13929–35
102. Oh SP, Warman ML, Seldin, MF, Cheng S-D, Knoll JHM, et al. 1994. *Genomics* 19:494–99
103. Reh M, Pihlajaniemi T. 1995. *J. Biol. Chem.* 270. In press
104. Reid KBM. 1993. *Biochem. Soc. Trans.* 21:464–68
105. Krieger M, Acton S, Ashkenas J, Pearson A, Penman M, Resnick D. 1993. *J. Biol. Chem.* 268:4569–72
106. Jensenius JC, Laursen SB, Zheng Y,
107. Holmskov U. 1994. *Biochem. Soc. Trans.* 22:95–100
107. Haagsman HP. 1994. *Biochem. Soc. Trans.* 22:100–6
108. Prockop DJ, Kivirikko KI. 1984. *N. Engl. J. Med.* 311:376–86
109. Mazzorana M, Gruffat H, Sergeant A, van der Rest M. 1993. *J. Biol. Chem.* 268:3029–32
110. Helaakoski T, Vuori K, Myllylä R, Kivirikko KI, Pihlajaniemi T. 1989. *Proc. Natl. Acad. Sci. USA* 86:4392–96
111. Bassuk JA, Kao WW-Y, Herzer P, Kedersha NL, Seyer JA, et al. 1989. *Proc. Natl. Acad. Sci. USA* 86:7382–86
112. Veijola J, Koivunen P, Annunen P, Pihlajaniemi T, Kivirikko KI. 1994. *J. Biol. Chem.* 269:26746–53
113. Kivirikko KI, Helaakoski T, Tasanen K, Vuori K, Myllylä R, et al. 1990. *Ann. NY Acad. Sci.* 580:132–42
114. Hautala T, Byers MG, Eddy RL, Shows TB, Kivirikko KI, Myllylä R. 1992. *Genomics* 13:62–69
115. Myllylä R, Pihlajaniemi T, Pajunen L, Turpeenniemi-Hujanen T, Kivirikko KI. 1991. *J. Biol. Chem.* 266:2805–10
116. Kivirikko KI, Myllylä R, Pihlajaniemi T. 1989. *FASEB J.* 3:1609–17
117. Kivirikko KI, Myllylä R, Pihlajaniemi T. 1992. In *Post-Translational Modifications of Proteins*, ed. JJ Harding, MJC Crabbe, pp. 1–51. Boca Raton: CRC
118. Helaakoski T, Annunen P, Vuori K, MacNeil IA, Pihlajaniemi T, Kivirikko KI. 1995. *Proc. Natl. Acad. Sci. USA.* In press
119. Vuori K, Pihlajaniemi T, Marttila M, Kivirikko KI. 1992. *Proc. Natl. Acad. Sci. USA* 89:7467–70
120. Pihlajaniemi T, Helaakoski T, Tasanen K, Myllylä R, Huhtala M-L, et al. 1987. *EMBO J.* 6:643–49
121. Koivu J, Myllylä R, Helaakoski T, Pihlajaniemi T, Tasanen K, Kivirikko KI. 1987. *J. Biol. Chem.* 262:6447–49
122. Noiva R, Lennarz WJ. 1992. *J. Biol. Chem.* 267:3553–56
123. Freedman RB, Hirst TR, Tuite MF. 1994. *Trends Biochem. Sci.* 19:331–36
124. Vuori K, Pihlajaniemi T, Myllylä, R, Kivirikko KI. 1992. *EMBO J.* 11:4213–17
125. Vuori K, Myllylä R, Pihlajaniemi T, Kivirikko KI. 1992. *J. Biol. Chem.* 267:7211–14
126. John DCA, Grant ME, Bulleid NJ. 1993. *EMBO J.* 12:1587–95
127. Wetterau JR, Combs KA, McLean LR, Spinner SN, Aggerbeck LP. 1991. *Biochemistry* 30:9728–35
128. Noiva R, Freedman RB, Lennarz, WJ. 1993. *J. Biol. Chem.* 268:19210–17

129. Otsu M, Omura F, Yoshimori T, Kikuchi M. 1994. *J. Biol. Chem.* 269: 6874–77
130. Wells WW, Xu DP, Yang Y, Rocque PA. 1990. *J. Biol. Chem.* 265:15361–64
131. Krishna Rao ASM, Hausman RE. 1993. *Proc. Natl. Acad. Sci. USA* 90: 2950–54
132. Nakai A, Satoh M, Hirayashi K, Nagata K. 1992. *J. Cell Biol.* 117:903–14
133. Clark EP, Jain N, Brickenden A, Lorimer IA, Sanwal BD. 1993. *J. Cell Biol.* 121:193–99
134. Sauk JJ, Smith T, Norris K, Ferreira L. 1994. *J. Biol. Chem.* 269:3941–46
135. Hojima Y, Mörgelin MM, Engel J, Boutillon M-M, van der Rest M, McKenzie J, et al. 1994. *J. Biol. Chem.* 269:11381–90
136. Hojima Y, Behta B, Romanic AM, Prockop DJ. 1994. *Anal. Biochem.* 223: 173–80
137. Prockop DJ, Hulmes DJS. 1994. See Ref. 10, pp. 47–90
138. Veis A, George A. 1994. See Ref. 10, pp. 15–45
139. Kadler KE, Hojima Y, Prockop DJ. 1987. *J. Biol. Chem.* 262:15696–701
140. Romanic AM, Adachi E, Hojima Y, Prockop DJ. 1992. *J. Biol. Chem.* 267: 22265–71
141. Romanic AM, Adachi E, Kadler KE, Hojima Y, Prockop DJ. 1991. *J. Biol. Chem.* 266:12703–9
142. AlaKokko L, Hyland J, Smith C, Kivirikko KI, Jimenez SA, Prockop DJ. 1991. *J. Biol. Chem.* 266:14175–58
143. Fertala A, Sieron A, Ganguly A, Li S-W, Ala-Kokko L, et al. 1994. *Biochem. J.* 298:31–37
144. Fertala A, Sieron AL, Hojima Y, Ganguly A, Prockop DJ. 1994. *J. Biol. Chem.* 269:11584–89
145. Holmes DF, Chapman JA, Prockop DJ, Kadler KE. 1992. *Proc. Natl. Acad. Sci. USA* 89:9855–59
146. Silver D, Miller J, Harrison R, Prockop DJ. 1992. *Proc. Natl. Acad. Sci. USA* 89:9860–64
147. Fertala A, Holmes DF, Kadler KE, Prockop DJ. 1994. *Matrix Biol.* 14:359 (Abstr.)
148. George A, Veis A. 1991. *Biochemistry* 30:2372–77
149. Kagan HM, Trackman PC. 1991. *Am. J. Respir. Cell Mol. Biol.* 5:206–10
150. Trackman PC, Pratt AM, Wolanski A, Tang SS, Offner GD, et al. 1991. *Biochemistry* 30:8282
151. Hämäläinen E-R, Jones TA, Sheer D, Taskinen K, Pihlajaniemi T, Kivirikko KI. 1991. *Genomics* 11:508–16
152. Mariani TJ, Trackman PC, Kagan HM, Eddy RL, Shows TB, et al. 1992. *Matrix* 12:242–48
153. Wu Y, Rich CB, Lincecum J, Trackman PC, Kagan HM, Foster JA. 1992. *J. Biol. Chem.* 267:24199–206
154. Contente S, Kenyon K, Rimoldi D, Friedman RM. 1990. *Science* 249:796–98
155. Kenyon K, Contente S, Trackman PC, Tang J, Kagan HM, Friedman RM. 1991. *Science* 253:802
156. Kuivaniemi H, Korhonen R-M, Vaheri A, Kivirikko KI. 1986. *FEBS Lett.* 195: 261–64
157. Hämäläinen E-R, Kemppainen R, Kuivaniemi H, Tromp G, Vaheri A, et al. 1994. *Matrix Biol.* 14:419 (Abstr.)
158. Colige A, Sokolov BP, Nugent P, Baserga R, Prockop DJ. 1991. *Biochemistry* 32:7–11
159. Laptev AV, Lu Z, Colige A, Prockop DJ. 1994. *Biochemistry* 33:11033–39
160. Marini JC, Wang Q. 1994. *Matrix Biol.* 14:405
161. Khillan JS, Li S-W, Prockop DJ. 1994. *Proc. Natl. Acad. Sci USA* 91:6298–7302
162. Hanauske-Abel HM. 1991. *J. Hepatol.* 13:S8–16
163. Lamberg A, Pihlajaniemi T, Kivirikko KI. 1995. *J. Biol. Chem.* 270:In press
164. Lee S-T, Kessler E, Greenspan DS. 1990. *J. Biol. Chem.* 265:21992–6
165. Murad S, Tennant MC, Pinnell SR. 1992. *Arch. Biochem. Biophys.* 292: 234–38
166. Hautala T, Heikkinen J, Kivirikko KI, Myllylä R. 1992. *Biochem. J.* 283:51–54
167. Yeowell HN, Ha V, Walker LC, Murad S, Pinnell SR. 1992. *J. Invest. Dermatol.* 99:864–69
168. Willing MC, Pruchno CJ, Atkinson M, Byers PH. 1992. *Am. J. Hum. Genet.* 51:508–15
169. Vogel BE, Doelz R, Kadler KE, Hojima Y, Engel J, Prockop DJ. 1988. *J. Biol. Chem.* 263:19249–55
170. Kadler KE, Torre-Blanco A, Adachi E, Vogel BE, Hojima Y, Prockop DJ. 1991. *Biochemistry* 30:5081–88
171. Torre-Blanco A, Adachi E, Romanic AM, Prockop DJ. 1992. *J. Biol. Chem.* 267:4968–73
172. Lightwood SJ, Holmes DF, Brass A, Grant ME, Byers PH, Kadler KE. 1992. *J. Biol. Chem.* 267:25521–28
173. Romanic AM, Spotila L, Adachi E, Engel J, Hojima Y, Prockop DJ. 1994. *J. Biol. Chem.* 269:11614–49
174. Spotila LD, Constantinou CD, Sereda L, Ganguly A, Riggs BL, Prockop DJ. 1991. *Proc. Natl. Acad. Sci. USA* 88: 5423–27

175. Spotila LD, Colige A, Sereda L, Constantinou-Deltas CD, Whyte MP, et al. 1994. *J. Bone Miner. Res.* 9:923–32

176. Vikkula M, Metsäranta M, Ala-Kokko L. 1994. *Ann. Med.* 26:107–14

177. Ritvaniemi P, Körkkö J, Bonaventure J, Vikkula M, Hyland J, et al. 1995. *Arthritis Rheum.* In press.

178. Winterpacht A, Hilbert M, Schwarze U, Mundlos S, Spranger J, Zabel BU. 1993. *Nat. Genet.* 3:323–26

179. Williams CJ, Rock M, Harland L, Considine E, McCarron S, et al. 1994. *Matrix Biol.* 14:391 (Abstr.)

180. Winterpacht A, Schwarze U, Menger H, Mundlos S, Spranger J, Zabel B. 1994. *Matrix Biol.* 14:392 (Abstr.)

181. Tiller GE, Weis MA, Polumbo PA, Cohn DH, Rimoin DL, Eyre DR. 1994. *Matrix Biol.* 14:390 (Abstr.)

182. Wilkin DJ, Bogaert R, Wilcox WR, Rimoin DL, Eyre DR, Cohn DH. 1994. *Matrix Biol.* 14:390 (Abstr.)

183. Ahmad NN, Ala-Kokko L, Knowlton RG, Jimenez SA, Weaver EJ, et al. 1991. *Proc. Natl. Acad. Sci. USA* 88:6624–27

184. Brown DM, Nichols BE, Weingeist TA, Sheffield VC, Kimura AE, Stone EM. 1992. *Arch. Ophthalmol.* 110:1589–93

185. Ahmad NN, McDonald-McGinn DM, Zackai EH, Knowlton RG, LaRossa D, et al. 1993. *Am. J. Hum. Genet.* 52:39–45

186. Ritvaniemi P, Hyland J, Ignatius J, Kivirikko KI, Prockop DJ, Ala-Kokko L. 1993. *Genomics* 17:218–21

187. Williams CJ, Ganguly A, McCarron S, Considine E, Michels V, Prockop DJ. 1994. *Matrix Biol.* 14:390 (Abstr.)

188. Ala-Kokko L, Baldwin CT, Moskowitz RW, Prockop DJ. 1990. *Proc. Natl. Acad. Sci. USA* 87:6565–68

189. Pun YL, Moskowitz RW, Lie S, Sundstrom WR, Block SR, et al. 1994. *Arthritis. Rheum.* 37:264–69

190. Spranger J, Winterpacht A, Zabel B. 1994. *Eur. J. Pediatr.* 153:56–65

191. Pal G, Keek JMF, Muijs MA. 1994. *Matrix Biol.* 14:387 (Abstr.)

192. Goldstein JA, Schwarze U, Witz A, Byers PH. 1994. *Matrix Biol.* 14:392 (Abstr.)

193. Kontusaari S, Tromp G, Kuivaniemi H, Ladda RL, Prockop DJ. 1990. *Am. J. Hum. Genet.* 47:112–20

194. Kontusaari S, Tromp G, Kuivaniemi H, Romanic AM, Prockop DJ. 1990. *J. Clin. Invest.* 86:1465–73

195. Tromp G, Wu YL, Prockop DJ, Madhatheri SL, Kleinert C, et al. 1993 *J. Clin. Invest.* 91:2539–45

196. Anderson DW, Tromp G, Kuivaniemi H, Ricketts M, Pope FM, et al. 1994. *Matrix Biol.* 14:392 (Abstr.)

197. Majamaa K, Savolainen E-R, Myllylä V. 1992. *Biochim. Biophys. Acta* 1138:191–96

198. Kuivaniemi H, Prockop DJ, Wu Y, Madhatheri SL, Kleinert C, et al. 1993. *Neurol.* 43:2652–58

199. Barker DF, Hostikka SL, Zhou J, Chow LT, Oliphant AR, et al. 1990. *Science* 248:1224–27

200. Tryggvason K, Zhou J, Hostikka SL, Shows TB. 1993. *Kidney Int.* 43:38–44

201. Mochizuki T, Lemmink HH, Mariyama M, Antignac C, Gubler M-C, et al. 1994. *Nat. Genet.* 8:77–82

202. Antignac C, Zhow J, Sanak M, Cochat P, Roussel B, et al. 1992. *Kidney Int.* 42:1178–83

203. Christiano AM, Greenspan DS, Hoffman GG, Zhang X, Tanai Y, et al. 1993. *Nat. Genet.* 4:62–66

204. Hilal L, Rochat A, Duquesnoy P, Blanchet-Bardon C, Wechsler J, et al. 1993. *Nat. Genet.* 5:287–93

205. Christiano AM, Ryynänen M, Uitto J. 1994. *Proc. Natl. Acad. Sci. USA* 91:3549–53

206. Christiano AM, Anhalt G, Gibbons S, Bauer EA, Uitto J. 1994. *Genomics* 21:160–68

207. Warman M, Abbott M, Apte SS, Hefferon T, McIntosh I, et al. 1993. *Nat. Genet.* 5:79–82

208. McIntosh I, Abbott MH, Warman ML, Olsen BR, Francomano CA. 1994. *Hum. Mol. Genet.* 3:303–7

209. Wallis GA, Rash B, Sweetman WA, Thomas JT, Super M, et al. 1994. *Am. J. Hum. Genet.* 54:169–78

210. Dharmavaram RM, Elberson MA, Peng M, Kirson LA, Kelley TE, Jimenez SA. 1994. *Hum. Mol. Genet.* 3:507–9

211. McIntosh I, Abbott MH, Francomano CA. 1994. *Matrix Biol.* 14:397 (Abstr.)

212. Chan D, Cole WG, Rogers J, Bateman JF. 1994. *Matrix Biol.* 14:396 (Abstr.)

213. Dharmavaram RM, Elberson MA, Peng M, Kirson LA, Kelly TE, Jimenez SA. 1994. *Matrix Biol.* 14:377 (Abstr.)

214. Briggs MD, Choi HC, Warman ML, Loughlin JA, Wordsworth P, et al. 1994. *Am. J. Hum. Genet.* 55:678–84

215. Brunner HG, van Beersum SEC, Warman ML, Olsen BR, Ropers H-H, Mariman ECM. 1994. *Hum. Mol. Genet.* 3:1561–64

216. Hautala T, Heikkinen J, Kivirikko KI, Myllylä R 1993. *Genomics* 15:399–404

217. Pousi B, Hautala T, Heikkinen J, Pajunen L, Kivirikko KI, Myllylä R. 1994. *Am. J. Hum. Genet.* 55:899–906

218. Marshall M, Walker L, Murad S, Pinnell

S, Yeowell H. 1994. *Matrix Biol.* 14:398 (Abstr.)

219. Hyland J, Ala-Kokko L, Royce P, Steinmann B, Kivirikko KI, Myllylä R. 1992 *Nat. Genet.* 2:228–31

220. Ha VT, Marshall MK, Elsas LJ, Pinnell SR, Yeowell HN. 1994. *J. Clin. Invest.* 93:1716–21

221. Smith LT, Wertelecki W, Milstone LM, Petty EM, Seashore MR, et al. 1992. *Am. J. Hum. Genet.* 51:235–44

222. Nusgens BV, Verellen-Dumoulin C, Hermanns-Le T, De Paepe A, Nuytinck L, et al. 1992. *Nat. Genet.* 1:214–17

223. Danks DM. 1993. See Ref. 7, pp. 487–506

224. Gacheru S, McGee C, Uriu-Hare JY, Kosunen T, Packman S, et al. 1993. *Arch. Biochem. Biophys.* 301:325–29

225. Yeowell HN, Marshall MK, Walker LC, Ha V, Pinnell SR. 1994. *Arch. Biochem. Biophys.* 308:299–305

226. Bonadio J, Saunders TL, Tsai E, Goldstein SA, Morris-Winman J, et al. 1990. *Proc. Natl. Acad. Sci. USA* 87:7145–49

227. Stacey A, Bateman J, Choi T, Mascara T, Cole W, Jaenisch R. 1988. *Nature* 332:131–36

228. Khillan JS, Olsen AS, Kontusaari S, Sokolov B, Prockop DJ. 1991. *J. Biol. Chem.* 266:23373–79

229. Pereira R, Khillan JS, Helminen H, Hume EL, Prockop DJ. 1993. *J. Clin. Invest.* 91:709–16

230. Pereira R, Halford K, Sokolov B, Khillan JS, Prockop DJ. 1994. *J. Clin. Invest.* 93:1765–69

231. Chipman SD, Sweet HO, McBride DJ, Davisson MT, Marks SC, et al. 1993. *Proc. Natl. Acad. Sci. USA* 90:1701–5

232. Vandenberg P, Khillan JS, Prockop DJ, Helminen H, Kontusaari S, Ala-Kokko L. 1991. *Proc. Natl. Acad. Sci. USA* 88:7640–44

233. Garofalo S, Vuorio E, Metsäranta M, Rosati R, Toman D, et al. 1991. *Proc. Natl. Acad. Sci. USA* 88:9648–52

234. Metsäranta M, Garofalo S, Decker G, Rintala M, de Crombrugghe B, Vuorio E. 1992. *J. Cell Biol.* 118:203–12

235. Helminen HJ, Kiraly K, Pelttari A, Tammi MI, Vandenberg P, et al. 1993. *J. Clin. Invest.* 92:582–95

236. Metsäranta M, Säämänen A-M, Vuorio E. 1994. *Matrix Biol.* 14:411 (Abstr.)

237. Garofalo S, Metsäranta M, Ellard J, Smith C, Horton W, et al. 1993. *Proc. Natl. Acad. Sci. USA* 90:3825–29

238. Toman D, Starcher B, Mascara T, Robberson D, Smith C, et al. 1994. *Matrix Biol.* 14:413 (Abstr.)

239. Quaglino D, Toman D, de Crombrugghe B, Davidson JM. 1994. *Matrix Biol.* 14:412 (Abstr.)

240. Nakata K, Ono K, Miyazaki J-I, Olsen BR, Muragaki Y, et al. 1993. *Proc. Natl. Acad. Sci. USA* 90:2870–74

241. Fässler R, Schnegelsberg PNJ, Dausman J, Shinya T, Muragaki Y, et al. 1994. *Proc. Natl. Acad. Sci. USA* 91:5070–74

242. Jacenko O, LuValle P, Olsen BR. 1993. *Nature* 365:56–61

243. Rosati R, Horan GSB, Pinero GJ, Garofalo S, Keene DR, et al. 1994. *Nat. Genet.* 8:129–35

244. Elima K, Eerola I, Markkula M, Kananen K, Rosati R, et al. 1994. *Matrix Biol.* 14:414 (Abstr.)

245. Vandenberg P, Khillan J, Prockop DJ. 1994. *Matrix Biol.* 14:415 (Abstr.)

246. Li Y, Lacerda DA, Warman ML, Beier DR, Yoshioka H, et al. 1995. *Cell.* In press

247. Pereira RF, Halford KW, O'Hara M, Leeper D, Pollard M, et al. 1995. *Proc. Natl. Acad. Sci. USA* 92:In press

248. Ganguly A, Smelt S, Mewar R, Fertala A, Sieron AL, et al. 1994. *Proc. Natl. Acad. Sci. USA* 91:7365–69

Annu. Rev. Biochem. 1995. 64:435–61

STRUCTURE AND ACTIVITIES OF GROUP II INTRONS

François Michel and Jean-Luc Ferat

Centre de Génétique Moléculaire du C.N.R.S., Avenue de la Terrasse, 91190 Gif-sur-Yvette, France

KEY WORDS: group II ribozymes, self-splicing, intron-encoded proteins, intron mobility, intron evolution

CONTENTS

ABSTRACT

Group II introns are found in eubacteria and eubacterial-derived, organellar genomes. They have ribozymic activities, by which they direct and catalyze the splicing of the exons flanking them. This chapter reviews the secondary structure and known tertiary interactions of the ribozymic component of group II introns in relation to the problems of specifying splice sites and building a catalytic core. We pay special attention to the relationship between the transesterification and hydrolytic modes of initiating splicing and the stereospecificities of these reactions. A number of group II introns encode proteins of the

435

reverse transcriptase family; the activity of these proteins enables the host introns to change genomic locations by mechanisms that are only beginning to be deciphered. Finally, we briefly discuss multipartite and post-transcriptionally edited group II introns, together with the intron microcosm of *Euglena gracilis* chloroplasts and the possible relationships between group II and spliceosome-catalyzed splicing processes.

SUMMARY AND PERSPECTIVES

A typical group II intron contains both a multidomain RNA endowed with catalytic (ribozymic) activity and the coding sequence of a polyprotein, one domain of which is related to the polymerase domain of reverse transcriptases (see Figure 1, below). From this duality stem the contrasting views of group II intervening sequences as either active participants in splicing reactions or mobile genetic elements. These past years have seen a number of reviews that have dealt with at least one of these aspects (1–6).

As introns, group II intervening sequences interrupt primary transcripts, from which they are subsequently removed during the splicing process. Still, group II introns need not be continuous molecules. Some are bipartite, resulting from the noncovalent assembly of two separate transcripts, and at least one group II intron appears to be tripartite. The corresponding splicing reaction is called *trans*-splicing (reviewed in 7), as opposed to *cis*-splicing, in which the exons to be ligated belong to the same polynucleotide chain.

Group II introns have been reported from organellar and bacterial genomes. While they have never been found in functional form in the nucleus of eukaryotes, their mode of excision closely resembles that of nuclear premessenger introns, because the freed intron molecule is in lariat (branched) form. In both group II- and spliceosome-catalyzed splicing, branching results from attack of the phosphodiester bond at the 5′ intron-exon junction by the 2′OH group of an intron A residue that is bulging out of a helix. Such similarity is suggestive of a common origin (8–10), but could also result from chemical necessities (11).

A major attraction of group II introns is that they have ribozyme abilities (just like group I introns, to which, nevertheless, they are not related). However, even those group II introns that can direct a complete self-splicing reaction in the absence of protein helpers do so only under nonphysiological salt and temperature conditions and depend on proteins for in vivo activity. Because only a minority of group II introns are proven ribozymes, we must rely in most cases on structural similarity to allocate an intron to group II. Secondary structure models of group II introns comprise a central wheel from which radiate six spokes that define the six major ribozyme domains (see Figure 2, below) [the secondary structure of group II introns was reviewed in detail by

Michel et al (1)]. This core structure contains disappointingly few conserved nucleotides and only the first few bases of the intron and some of those constituting the small domain V may be regarded as characteristic of group II. Even such a restrictive definition may not apply to the chloroplast genomes of euglenoid protists (reviewed in 12), where bona fide, although somewhat defective, group II members coexist with numerous smaller introns that share the same lariat-forming mode of excision but lack any recognizable correlate of even domain V. Just like nuclear premessenger introns, which are excised by spliceosomes, these group III introns must rely for their excision on an external machinery, presumably made of RNA and proteins. Whether this machinery, or the spliceosome, bear any relation to the group II ribozyme core, can probably not be decided until more is learned about tertiary structure in each of these systems. Unfortunately, the elucidation of tertiary interactions has been slow in coming, and the study of group II introns is lagging behind that of other ribozymes of comparable size (group I introns, RNase P) for which approximate three-dimensional models have been proposed (13–15).

The biology of group II introns should prove no less fascinating than their ribozymic feats. Just like many group I introns, several protein-encoding group II introns are proven mobile elements, capable of homing (i.e. of converting intron-less alleles to the intron-plus state) as well as transposing to some new insertion sites (reviewed in 2). Contrary, however, to the situation in group I introns, whose ribozyme and protein-encoding components are engaged into a merely symbiotic relationship that allows them to liberally exchange partners, the association between group II–encoded proteins and group II ribozymes is an intimate one. Evolutionary trees of group II ribozymes are largely congruent with those of intron-encoded proteins and, as we might expect from the involvement of reverse transcription in the mobility process (16), at least some mutations in the ribozymic sections of the intron prevent transposition. Nevertheless, the relationship between splicing and mobility of group II introns remains unclear, and deciphering the biochemical pathways that allow group II introns to move remains a primary goal.

DISTRIBUTION OF GROUP II INTRONS

Most known group II introns were discovered serendipitously during the sequencing of organelle genomes, where they interrupt genes encoding proteins, tRNAs, and ribosomal RNAs. They exist in relatively small numbers in fungal mitochondria, are predominant in plant mitochondria (17, 18) and chloroplasts (19, 20), are present in algae (21–24), and are overwhelmingly frequent in *Euglena gracilis* chloroplasts (25). Once homologous introns inserted at the same location in related organisms are discarded, roughly 100 known group II introns in noneuglenoid organisms are left. Some 25 of these introns contain

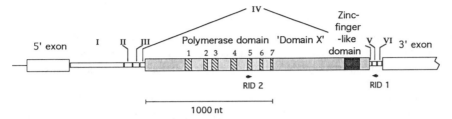

Figure 1 Organization of a typical protein-encoding bacterial group II intron (redrawn from 28). Small boxes labeled with Roman numbers symbolize structural domains of the group II ribozyme (see Figure 2). The ORF within the intron is shown as a large gray box; numbers refer to the seven amino acid motifs conserved in the polymerase domains of reverse transcriptases (26, 27). RID-1 (complementary to the 3′ half of ribozyme domain V) and RID-2 (derived from the consensus coding sequence of polymerase motif 5) are degenerate primers for amplification of group II introns from genomic DNA (28) (see text).

long open reading frames (ORFs) in addition to their structurally conserved ribozyme core, and almost all of the proteins that can be translated from these ORFs belong to a single family clearly related to reverse transcriptases (26, 27).

Recently, two strategies were developed to search for additional group II introns. One is experimental and uses two degenerate primers to tentatively amplify group II intron fragments from DNA extracts by means of the polymerase chain reaction (PCR). Using one primer that matched part of ribozyme domain V and another that covered one of the conserved motifs of the presumptive polymerase-encoding domain of group II–coding sequences (Figure 1), Ferat and colleagues (28, 29) were able to detect group II introns in both cyanobacteria (*Calothrix, Anabaena, Nostoc*) and proteobacteria (*Azotobacter, Escherichia*). Interestingly, these two bacterial groups include the presumptive ancestors of chloroplasts and mitochondria, respectively. Ironically, even though entire introns and surrounding DNA stretches were subsequently cloned and sequenced, the exact nature of the interrupted genes has often remained a mystery. One exception, however, is *Escherichia coli,* which has four introns so far identified in natural isolates: All of these introns interrupt either proven or putative mobile DNA elements (29).

In this era of genome sequencing, searching for additional group II introns by looking into databases with consensus sequences matching domain V was recently shown (30) to be a valid alternative to probing of actual DNA molecules with degenerate oligonucleotides. Among group II members recovered by this nonexperimental approach was one of the four *E. coli* introns identified by PCR, a defective version of which had been lying unrecognized in DNA databases since 1991. However, neither the computer-based nor the experimental search strategies are exhaustive. Any intron whose domain V sequence is too divergent from the consensus (or that lacks an open reading frame with a matching motif, in the case of the experimental search as currently implemented) will not be picked.

The exact breadth of the ecological niche of group II introns remains to be determined. They have never been reported from the nucleus of eukaryotes, except in the form of remnants recently transferred from the mitochondrion (31).

GROUP II INTRONS AS RIBOZYMES

In Vitro Activity

The original demonstration of group II in vitro self-splicing (32–34), consisting of the two-step transesterification pathway depicted in Figure 3 (see below), involved two yeast mitochondrial introns. These are the last intron of the gene encoding subunit I of cytochrome oxidase and intron 1 of the cytochrome *b* gene. Those two introns, known as *ai5γ* and *bi1*, belong to the same subgroup within subdivision IIB (1) and appear to be similar enough in structure and activity to be largely interchangeable. They have remained experimentalists' favorites for reasons beyond the historical: Very few of the other known group II introns are proven ribozymes in the sense that they can be shown to have some catalytic activity in vitro. Although failures are seldom publicized (see 35 for one exception), it is commonly known that many chloroplast introns have been investigated for in vitro activity in vain (36). Even some of the reported catalysts, such as two other fungal (37, 38) and one algal (39) mitochondrial introns as well as two of the recently discovered bacterial members of group II (28, 29), are mostly inactive, except under rather extreme conditions. As for introns *ai5γ* and *bi1*, they themselves require very unphysiological conditions [e.g. 100 mM magnesium, 500 mM $(NH4)_2SO_4$ and 45°C (40)] for optimal in vitro activity.

Not surprisingly in this context, group II introns depend on proteins for in vivo activity. This dependence has been established in particular for the well-studied yeast mitochondrial group II introns, the excision of all four of which requires expression of the nuclear-encoded *MRS2* gene (41); the two introns that include ORFs also depend on the maturase function of the proteins they encode (see next section), and one of them is also under control of a helicase-related protein (42). Interestingly, both the latter and the protein encoded by the *MRS2* locus have other, unknown functions in mitochondrial biogenesis, which is consistent with the possibility that their involvement in splicing is a relatively recent evolutionary development [note in this respect that there is both genetic (43) and biochemical (44) evidence that proteins not specifically involved in splicing can nevertheless suppress splicing-defective mutations or facilitate in vitro self-splicing].

Secondary Structure and Tertiary Interactions

SECONDARY STRUCTURE   As is true of most natural RNAs with a conserved structure, the secondary structure of group II introns, defined as the set of all contiguous classical (Watson-Crick and wobble) base pairs that can be drawn in. planar representation, was established by comparative sequence analysis (45–47). Secondary structure models (Figure 2) were reviewed in detail in 1989 (1), together with the tertiary interactions known at that time. Relatively

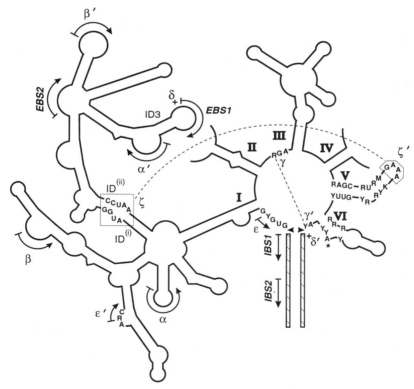

Figure 2   Structure and consensus nucleotides of the ribozyme component of group II introns. The secondary structure model is that of intron *ai5γ* (48, 49). Residues shown are consensus ones for both subgroup IIA and IIB introns (R stands for purine, Y for pyrimidine, M for A or C, K for G or U). Tertiary interactions are indicated by dashed lines, curved arrows, and/or Greek lettering (EBS and IBS are exon-binding and intron-binding sites, respectively). Plus signs indicate the nucleotides that participate in the δ-δ′ interaction. Roman numbers refer to the six major structural domains. Arrowheads indicate intron-exon junctions. The asterisk shows the lariat branchpoint.

little has been learned since about group II structure, and this review focuses instead on the functional significance of structural components and interactions.

Two major subdivisions (IIA and IIB) of group II introns were proposed (1), based on rather subtle structural and sequence differences. Although the majority of introns sequenced since 1989 still fall into either of these two major subgroups, a few have proved somewhat intermediate (e.g. 29), and additional structural diversity has become apparent with the sequencing of numerous plant mitochondrial introns (17, 18), only some of which belong to previously recognized minor subtypes. Not only universally conserved components, but in fact virtually all elements of group II structure are subject to strong selective pressures, which manifest themselves by the existence of numerous compensatory base changes between closely related sequences (e.g. Figure 2 in 48).

In contrast with the striking conservation of group II structure, the at least 600 or so residues that make up the ribozyme component of a typical group II intron include few generally conserved nucleotides. In fact, virtually any position admits exceptions, especially when the *Euglena gracilis* introns are taken into account. As seen in Figure 2, the only strongly conserved sequences are intron boundaries (with consensus sequences GUGYG and AY), a few nucleotides in the central wheel, domain V, and the few sites within domain I that have been shown to interact with these components.

TERTIARY INTERACTIONS   Tertiary structure should ideally be defined based on sequential folding and/or energetic considerations (50–52). In the absence of relevant data, any nonclassical base-base interaction or any contact involving presumably preformed secondary structure components (terminal and internal loops, helices) located at distant sites in secondary-structure models, is commonly designated as *tertiary*. According to the latter criterion, the two proven intron-exon pairings, EBS1-IBS1 and EBS2-IBS2 (49) (EBS and IBS stand for exon- and intron-binding sites, respectively), fall into the tertiary category, even though they consist of contiguous classical base pairs. Additional tertiary, though classical, interactions should include the following: (*a*) The extended pairing α-α′ (proposed in Reference 48) was experimentally demonstrated in *ai5γ* (53) and is potentially present in many other introns. (*b*) Another extended pairing, β-β′, is less common, but is still identifiable in several self-splicers and a number of related introns. (*c*) The pairing ε-ε′ (54) involves intron bases 3 and 4 and two consecutive bases in the IC1 internal loop. As indicated by the pattern of in vitro compensations, the ε-ε′ pairing seemingly consists of two consecutive Watson-Crick pairs, but additional interactions involving these bases must exist, because little variation is allowed in nature (Figure 2) and the wild-type combinations perform best in vitro. (*d*) The pairing γ-γ′ (54) is an isolated Watson-Crick base pair residing between the last intron nucleotide and one of the residues in the central group II wheel. (*e*) Finally, δ-δ′ is another isolated, presumably Watson-Crick base pair, between the first base of the 3′ exon and the nucleotide immediately 5′ to the EBS1 sequence (1, 48, 55).

An interaction between the first (G) and penultimate (A) nucleotides of intron *ai5*γ was suggested from the U:G and C:G combinations providing some measure of compensatory effects (56). Finally, another nonclassical contact, between the terminal loop of hairpin V and the internal loop at the junction of helices D$^{(i)}$ and D$^{(ii)}$ (ζ-ζ′ in Figure 2) was recently uncovered (56a) (see below).

Those interactions not involved in selection of splice sites are likely to stabilize the correct folding of the catalytic core and disrupting them should result in a reduced fraction of active molecules, as seen with α-α′ (53). What is not yet known is whether group II ribozymes will turn out to be as compact (57), stable (51, 52), and static (52) structures as group I introns, which seemingly undergo only minor rearrangements in between the two steps of the splicing reaction. The issue is of importance not only for modeling undertakings, but also in view of current discussions regarding the magnitude of rearrangements at the catalytic site(s) during the splicing process (see next subsections).

Defining Targets for Activity

5′ SPLICE SITE   In the vast majority of group II introns, including all the ones that have been shown to function in vitro, part of the terminal loop of the ID3 helix (the EBS1 sequence; Figure 2) can base pair with a contiguous stretch of residues at the 3′ end of the 5′ exon (the IBS1 sequence). All reactions catalyzed in vitro by group II introns (Figure 3) involve the 3′ extremity of IBS1 as either an attacking or leaving group.

Integrity of the EBS1-IBS1 pairing is essential both for lariat formation and stability of the intermediate complex in the normal transesterification pathway (49). The EBS1-IBS1 pairing is also both necessary and sufficient for reversal of exon ligation by attack of the terminal 3′OH group of excised intron molecules either in lariat (58, 59) or linear (60) form: Not only the authentic 3′ exon (58), but a variety of segments of foreign RNAs and even a single phosphate group at the end of an IBS1 sequence can thus be charged on the 3′ end of the intron, albeit with varied efficiencies (59, 61, 62). Nor does there seem to be any additional requirement for complete intron integration (by debranching of the lariat intermediate), which was observed to occur downstream of IBS1 motifs in noncognate RNAs (59). Predictably, chimeric molecules with an RNA 5′ exon ending in the IBS1 sequence and a DNA 3′ exon are also cleaved, resulting in addition of the DNA piece at the end of the intron

a) Transesterification   reactions

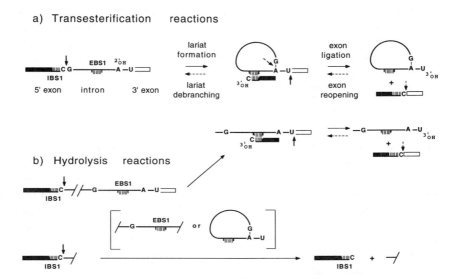

b) Hydrolysis   reactions

Figure 3 Reactions catalyzed by the ribozyme components of group II introns. Residues shown correspond to intron *ai5γ*. Vertical and slanted arrows point to reactive phosphodiester bonds (dashed lines refer to reverse reactions) and 2'OH and 3'OH indicate attacking groups in trans-esterification reactions.

(61). However, to what extent a DNA 5' exon constitutes an acceptable substrate is not clear, because both success and failure were reported using essentially the same molecules (contrast 60 and 61).

What is true of second-step reversal is also true of the hydrolysis reaction catalyzed by group II introns, which, under some conditions, requires only an IBS1 motif for activity. Both linear and lariat versions of the intron can cleave the ligated exons (40) as well as a 5' exon with a single G at its 3' end (63) and various IBS1 surrogate sequences in a molecule lacking the IBS2 motif at the authentic splice site (64). The linear or lariat intron then acts as a true catalyst, since it is left unchanged in the process (Figure 3). In a detailed study of the specificity of this cleavage reaction for intron *ai5γ* (55), hydrolysis was shown to occur opposite to a fixed site in the ID3 terminal loop, this site being the one facing the intron-exon junction in a wild-type situation. In contrast, integration of intron *bi1* into foreign RNAs is often imprecise, with a fraction of molecules being cleaved one nucleotide upstream of the expected site (61, 62), even though pairing of the IBS1 and EBS1 sequences into other registers seems unlikely in that system (among possibly significant differences between these studies is the absence of the EBS2-IBS2 pairing in the latter case).

BRANCHPOINT AND 3′ SPLICE SITE   Most group II introns have a bulging A on the 3′ side of the basal helix of domain VI at either seven or eight nucleotides from the 3′ splice site. This A constitutes the lariat branch point in all cases investigated so far (32, 34, 65, 66). However, a few exceptional group II introns lack such a bulging A (reviewed in 1). Whether these introns are nonetheless excised as lariats remains to be determined.

At least for intron *ai5γ*, lariat formation is rate-limiting for the normal transesterification reaction course, so that the intron–3′ exon branched reaction intermediate is barely detectable. An independent assay for second-step efficiency may nevertheless be set up by partly disrupting the EBS2-IBS2 pairing, thus weakening the interaction between the 5′ exon and the intron, so that exon ligation now competes with dissociation of the intermediate complex. It could thus be demonstrated (49, 54) that all three interactions surrounding the 5′ splice site (IBS1-EBS1, IBS2-EBS2 and ε-ε′; see Figure 4) are important both before and after 5′ cleavage. On the other hand, evidence does not support the existence of pairings that involve nucleotides surrounding the 3′ splice site prior to the second step of the transesterification pathway, because changing the bases involved either had no effect on cleavage at the 5′ splice site (54) or if defects were observed in the 5′ cleavage reaction they could not be compensated by additional substitutions (56).

In contrast to the situation at the 5′ splice site, there appear to be multiple determinants of 3′ splice site selection. Predictably enough, deletion of domain VI precludes lariat formation, but it does not completely prevent the formation of ligated exons through the use of either the authentic or surrogate 3′ splice sites (55, 62, 67). A detailed study of 3′–splice site selection by intron *ai5γ* in the absence of domain VI (55) led to the conclusion that favored sites were the ones preceded by a base that could engage into the γ-γ′ pairing (Figure 4), and followed by one that could form a Watson-Crick pair with the base immediately upstream of the EBS1 sequence (δ-δ′ interaction in Figure 4). Complementarity between the latter two sites is the rule in subgroup IIA introns (1) and had already been proposed to contribute to the proper alignment of the exons for the ligation step by the same sort of guiding mechanism that exists in group I introns (68–70). However, not all δ-δ′ Watson-Crick combinations support ligation, which suggests that there is yet another interaction involved in recognition of the first base of the 3′ exon. Still another interaction that appears to be confined to the second stage of the splicing process is the one between the first and penultimate intron residues (56): Base changes at the penultimate position offer no compensation for the reduced rate of branch formation (71) resulting from substitution of the first intron nucleotide. Interestingly, substitution of the penultimate nucleotide of intron *bi1* not only impairs exon ligation but also results in the balance shifting between lariat

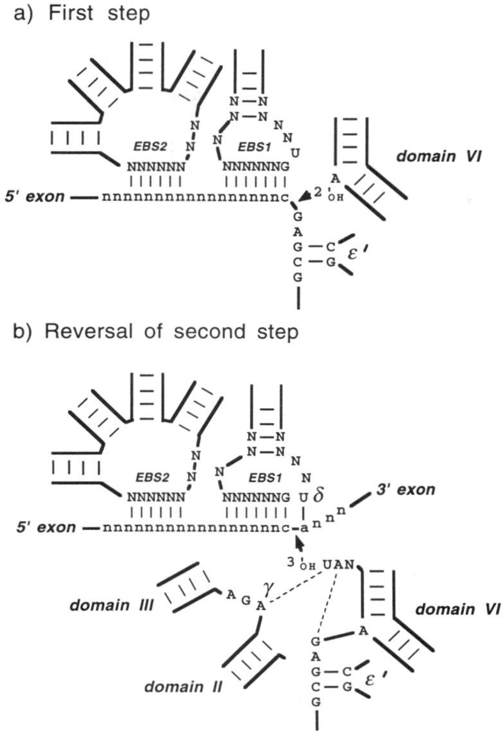

a) First step

b) Reversal of second step

Figure 4   Interactions involved in the first and second steps of the splicing process. Exon and intron bases are in lower and upper case, respectively. Bases specified are the ones surrounding intron-exon junctions as well as those known to interact with them in intron *ai5γ*. Phosphodiester bonds are shown as heavy lines, classical base pairs as thin lines and isolated tertiary interactions as dashed lines.

formation and debranching in favor of the latter process (58; M Mueller, personal communication).

Surprisingly, intron *bi1* molecules with a short 5′ extension in addition to a complete domain VI and an authentic 3′ exon have been reported to support an alternative ligation reaction in which the terminal 3′OH of free 5′ exon transcripts attacks the 5′ terminal triphosphate of intron transcripts or, alternatively, one of the internal phosphodiester bonds upstream of the intron (60). This reaction seems rather inefficient; there appears to be little specificity for one internal bond over the next and it is unknown whether domain VI is required or not.

Transesterification vs Hydrolysis

Transesterification, in which the 2'OH of the bulging A of domain VI attacks the 5' intron-exon junction, is the prevalent way to initiate excision of complete introns in vitro (except under extreme ionic conditions), and possibly the only one in vivo. Nevertheless, in an alternative group II splicing pathway, cleavage at the 5' splice site results rather from hydrolysis (Figure 3). Hydrolysis is of course the only possible reaction mode when the branch site is missing or inaccessible (72), but transesterification is also prevented when the 5' end of the intron does not coincide with the 3' end of the IBS1 sequence (which then defines the cleavage site). This is the case in *trans*-splicing setups (63), when the IBS1 motif is not followed by a sufficient length of intron residues for hybrid intron molecules to form (73), and also applies to constructs in which a spacer segment is inserted 5' of the first intron residue (55). In fact, insertion of a single A at the 5' end of intron *bi1* is enough to greatly reduce transesterification at the expense of hydrolysis, with concomitant shift of the cleavage site by one nucleotide in the latter reaction mode (62). Finally, some experimental conditions, such as the use of high concentrations of KCl, promote hydrolysis at the expense of branch formation (40). 5' Splice-site hydrolysis is not necessarily less efficient than transesterification (55). Rather, hydrolysis seems to be repressed in normal, complete precursor molecules.

Recent mechanistic studies indicate that most, if not all, cases of 5' splice-site hydrolysis should be regarded as related to the reversal of exon ligation, rather than to the first step of the transesterification pathway. On one hand, WJ Michels & AM Pyle (73a) took advantage of their ability to monitor the chemical step and established that phosphorothioate linkages in Rp conformation are preferentially cleaved by a ribozyme consisting of domains I and V of intron *ai5*γ. On the other, RA Padgett, M Podar, SC Boulanger & PS Perlman (73b) have shown that both the first and second step of the normal, transesterification-initiated, splicing pathway of the same intron are blocked when Rp, but not Sp, phosphorothioates are introduced at the 5' and 3' splice sites, respectively. Because both steps proceed with inversion of stereochemistry, as expected from an SN2 mechanism, the bond connecting the 5' and 3' exons in the final product is in Rp conformation. Thus, it is the ligated exons, but not the precursor molecules, that are in the right phosphorothioate conformation for the hydrolysis reaction observed by Pyle and collaborators.

That the two steps of a splicing reaction should have the same stereospecificity in spite of the fact that each one inverts the conformation of the reactive bond has been interpreted as evidence of distinct catalytic sites (74). However, as pointed out by Steitz & Steitz (75), these data can be reconciled with a single active site by assuming that the substrates for the first and second step (the first nucleotides of the intron and 3' exon, respectively) are bound with two distinct geometries that differ by a rotation of 120° around the reactive phosphorus. Existence of two possible binding modes for the nucleotide 3′ of the reactive bond is precisely what the available evidence suggests for group II–catalyzed reactions. On one hand, the target for attack by the terminal 3′OH group of the intron in reverse splicing is likely to be an ordinary phosphodiester bond connecting two consecutive Watson-Crick base pairs in a double-stranded helix, since the base immediately 5′ to the EBS1 sequence of the vast majority of subgroup IIA introns (the ones in which the EBS1 sequence is located asymmetrically within the ID3 loop) is complementary to the first base of the 3′ exon (1) [this guiding mechanism also operates in intron *ai5γ* (55), even though it is a member of subgroup IIB]. On the other hand, evidence from comparative sequence analysis (1) indicates that extension of the IBS1-EBS1 pairing into the intron could interfere with branch formation, because it is prevented in those group II introns in which ligation appears to be guided by Watson-Crick pairing: Whenever the exon 3′ of a subgroup IIA intron with an asymmetrical D3 loop happens to begin with a G, resulting in the base 5′ to EBS1 being a C, the first nucleotide of the intron is changed into an U, instead of the overwhelmingly preferred G. This observation strongly suggests that the appropriate conformation for transesterification is one in which interactions between the ribozyme and first few intron residues prevent the first nucleotide of the intron from being in helical continuity with the 5′ exon. What then of the situation in 5′ splice-site hydrolysis? As shown by Jacquier & Jacquesson-Breuleux (55), extension of the *ai5γ* EBS1-IBS1 pairing to the base immediately 3′ of the cleavage site is not only tolerated, but contributes to selection of the proper pairing register: This is again evidence that hydrolysis and reversal of exon ligation have the same substrate preferences (note, however, that 5′ splice sites in the right conformation for transesterification might also be susceptible to hydrolysis, in which case two hydrolytic modes of reaction would need to be distinguished).

Possible Role of Domain V in Catalysis

Ribozyme domain I includes the binding sites for bases surrounding the 5′ intron-exon junction, and domain VI contains the attacking group for initiation of splicing by transesterification but is not required for 5′ splice-site hydrolysis under high salt conditions (67, 76) or when the IBS1 target sequence is separated from the intron (55). At the opposite end of the functional spectrum, domain IV is highly variable and often quite short, and even its role in bringing together the upstream and downstream parts of introns whose ribozyme components are interrupted by an insertion can be dispensed with in vitro, as first shown by Jarrell et al (76). Relatively little is known of the exact role in group II ribozymic activity of the remaining structural domains. Molecules with a partial or complete deletion of domain II still function under high salt conditions (67, 77, 78) but tend to accumulate the intron–3' exon lariat intermediate, which suggests that this domain is mainly important for the second transesterification reaction. Consistent with the fact that some nucleotide substitutions within domain III severely affect splicing both in vivo and in vitro (34), its removal is only partly compensated for by very high salt concentrations (67). Nevertheless, domain III is dispensable for 5' splice site hydrolysis (67; 73a). In spite of its relatively short size, and just as could have been surmised from the striking conservation in evolution of both its structure and primary sequence (Figure 2), domain V is thus the only component other than domain I that is absolutely essential for group II ribozyme activity (67, 73a, 76). Interestingly, while domain V is readily recognized by the upstream ribozyme domains when supplied as a separate molecule (76), domain VI and the 3' exon must be covalently linked to domain V for a 3' splice junction to be efficiently utilized (79).

Binding of domain V by the rest of the ribozyme was quantitatively investigated in several recent studies that either used an indirect, kinetic approach (73a, 80, 81) or estimated K_d values from gel-filtration assays (80). It could thus be confirmed that domain I constitutes the primary binding site for domain V and that the affinity of these two RNA fragments for one another is rather high (with K_m and K_d values ranging from 0.3 to 1 μM), given the fact that there is no evidence of classical base pairing being involved in their association.

Even though domain V cannot be dispensed with, even at the highest practical concentrations of magnesium and monovalent ions, it could serve merely as a small scaffold to stabilize the active form of domain I. Alternatively, it could provide some of the component groups of the group II catalytic center; similarly, magnesium is essential both to the folding of, and catalysis by, known ribozymes (reviewed in 82). To distinguish between these possibilities, Chanfreau & Jacquier (83) investigated the ability of phosphorothioate-substituted or chemically modified transcripts of domains V and VI not only to participate in transesterification reactions (lariat formation and exon ligation), but to bind the rest of the ribozyme. Thus, while modification of the bases at the distal tip of domain V is detrimental to reaction, it also strongly interferes with binding, suggesting that their primary role may be merely structural; in fact, the strongly conserved GAAA terminal loop was recently shown to bind the internal loop and neighboring nucleotides at the junction of domain I helices $D^{(i)}$ and $D^{(ii)}$ (ζ-ζ' interaction in Figure 2) (56a). In contrast, modification of any of three consecutive bases (A2, G3, and C4—numbering is from the 5' extremity of domain V) blocked both transesterification steps without interfering with binding of domain V; these bases are among the best conserved ones in group II introns (1), and mutational analyses have confirmed their importance for catalysis (P Perlman, personal communication). Also, at least two Rp oxygens are required for reaction, but not for binding (these are the oxygens carried by the phosphates 5' to G3 and C25—the latter is part of the two-nucleotide bulge on the 3' side of domain V and could lie close to G3 in space).

Interestingly, all modifications within domain V that have detectable effects interfere with both transesterification reactions. Thus, even though lariat formation and exon ligation involve different types of phosphodiester bonds, there is no evidence so far that the nucleotides directly or indirectly involved in catalysis differ, or that the ribozyme undergoes major rearrangements during the course of the splicing process.

GROUP II–ENCODED PROTEINS

Sequence Analyses

About one-fourth of the known group II introns potentially encode proteins, and there are examples of protein-encoding introns within both the IIA and IIB ribozyme-based subdivisions (see e.g. 28). In contrast to group I, where ORFs occupy many different locations, the bulk of group II protein-coding sequences are always inserted in domain IV, the least conserved of the six ribozyme domains (Figure 1). Interestingly, in most mitochondrial introns that interrupt protein-coding genes, the open reading frame extends in the 5' direction to the 5' splice site, where it fuses with the exonic reading frame. The primary translation product of these introns is probably therefore a chimeric protein encoded by both exon and intron sequences. However, only shorter products, which presumably result from subsequent processing events, have so far been detected (84–86).

Nearly all group II–encoded proteins are related and most of them include a domain with obvious homology to the polymerase domain of reverse transcriptases (26); only in the proteins encoded by the lysine tRNA introns of plant chloroplasts is much of this domain either missing or modified beyond recognition. Within the reverse transcriptase family, the presumptive group II polymerase domains form a monophyletic assemblage whose sister groups are the ones constituted by the proven or putative polymerases of those retroelements that are encoded by prokaryotic genomes (27, 87, 88). These include bacterial retrons (89), a mitochondrial plasmid of *Neurospora crassa* (90, 91), and a chloroplast-encoded sequence (92). Further phylogenetic affinities of these subdivisions are with retrotransposons devoid of long terminal repeats rather than with retroviruses and their allies.

It was recognized early on that large sections of group II–encoded proteins other than the polymerase domain have been conserved in evolution (e.g. 93, 94), and claims have been made that these proteins share additional similarities with retroelement-encoded polypeptides (e.g. 95, 96). In fact, in the COOH-terminal section of several group II–encoded proteins, a domain with zinc finger–like sequences (Figure 1) has substantial similarity to a section of the *McrA* endonuclease of *E. coli* (28). Weaker similarities exist with three proven or putative DNA endonucleases encoded by group I introns and the DNase domain of bacteriocins (97, 98).

The current consensus view regarding both group I– and group II–encoded proteins is that while some of them became secondarily involved in splicing, their primary evolutionary role is to ensure mobility of the entire intron. Whatever their exact function, proteins related to reverse transcriptases are expected to recognize RNA and, therefore, coevolve with the ribozyme components of the introns that encode them. In fact, comparison of polymerase domains on one hand, and ribozyme sequences and structures on the other, yields largely congruent evolutionary trees (29; J-M Fontaine & S de Goër, personal communication). This is in complete contrast with the situation in group I introns, which encode proteins belonging to several distinct families: The evolutionary trees of these molecules reveal extensive reshuffling of ribozyme and protein components (see e.g. 99).

Function in Splicing

There is genetic evidence that, just as for some fungal group I introns (100), the proteins encoded by the first two introns (*ai1* and *ai2*) of the cytochrome oxidase subunit I gene of *Saccharomyces cerevisiae* are involved in the excision of these introns. Some of the splicing-deficient mutations in introns *ai1* and *ai2* map in protein-coding sequences and can be complemented in *trans* by genomes expressing a seemingly normal intron polypeptide (84, 86, 101). Once irrelevant nucleotide changes have been dealt with (86), it appears that the one available *trans*-recessive mutation in intron *ai2* (a missense codon) and all four ones in intron *ai1* happen to map somewhere in between the polymerase and zinc finger–like domains. This section of the group II proteins, which has no equivalent in other retroelements, has been baptized domain X (102) (Figure 1) and proposed to be endowed with maturase activity (86, 102), whatever that may be (just like in group I, nothing is yet known of the biochemical basis of maturase function). The original raison d'être for maturase activity was autoregulation (100): By promoting splicing, the protein should contribute to the destruction of its own message (recall that most group I and group II fungal mitochondrial proteins appear to be translated by readthrough from the upstream exon). Yet, part at least of domain X is present in all group II intron ORFs, quite a number of which (in particular all the ones in bacterial and chloroplast introns) are translated autonomously. It will be especially interesting in this context to determine whether or not the proteins encoded by bacterial introns play any part in the splicing process.

Function in Mobility

Although one group I intervening sequence was recognized as a mobile DNA element before splicing had been discovered (103, 104), evidence that group II introns are mobile has long been circumstantial, consisting of the fact that related introns were often found at different sites in different organisms [reviewed in Reference 1; also, the presence of extremely similar group II introns in the cytochrome oxidase subunit I genes of the yeasts *S. cerevisiae* and *Kluyveromyces lactis* has been interpreted as reflecting some horizontal transmission event (105)]. However, group II introns have now been caught in the act of moving. When *S. cerevisiae* strains that possess introns *ai1* and/or *ai2* are crossed to strains that lack either one or both of these introns, most (often 90% or more) of the progeny is found to have received the intron(s) (106) [the close relative of *ai2* in *K. lactis* behaves in the same way (107)]. This phenomenon, first observed with group I introns, is called intron homing (reviewed in 5, 108, 109) and is accompanied by preferential transmission of markers flanking the intron in a polar fashion. However, unlike group I introns, group II introns would seem to induce mostly or exclusively asymmetrical conversion, in a 3' to 5' direction from the intron (110) (some uncertainty remains, since crosses were heterospecific and the segment 3' of the donor intron was interrupted by three group I introns in receptor molecules).

In addition to homing, some of the protein-encoding fungal group II introns have been shown by PCR amplification to transpose to other genomic locations at low frequencies (111, 112). Remarkably, all transposition events occur immediately 3' of an IBS1-like motif, i.e. a stretch of nucleotides that can form base pairs with the EBS1 intron sequence; this includes the authentic IBS1 site, where insertion of a second copy of the intron ultimately generates by recombination an extrachromosomic circular version of the intervening sequence (113). Interestingly, in both mitochondrial genomes and *E. coli*, introns with closely related sequences tend to lie in the genomic vicinity of each other, which has been interpreted as evidence that transposition tends to occur preferentially in *cis* along the same molecule (29).

Evidence that the proteins encoded by group II introns are actually responsible for the phenomenon of mobility remains indirect, consisting of the fact that a splicing-deficient, *trans*-recessive mutation in the reading frame of intron *ai1* blocks both homing and transposition (106, 111). Nevertheless, all proposed models for group II mobility involve reverse transcription, and convincing evidence of reverse transcriptase activity associated with proteins encoded by group II introns was recently obtained. Fusion of the reading frame in the first intron of the cytochrome oxidase subunit I gene of *Podospora anserina* with the *TYA* ORF of the yeast retrotransposon Ty resulted in recovery of a reverse transcriptase activity from virus-like particles (114). Even more promising (16), a reverse transcriptase activity associated with yeast mitochondrial ribonucleoprotein particles was shown to depend on the presence and integrity of the reading frames in introns *ai1* and *ai2*. Reverse transcription is clearly highly specific for the two introns and their flanking exons, but the exact nature of the templates, initiation sites, and extent of cDNA synthesis remains somewhat uncertain: Of two major probable initiation sites for cDNAs covering yeast intron *ai2*, one was in domain VI of the intron and the other was in the 3' exon, immediately downstream of the intron, with product molecules extending past the 5' end of the intron (16). Actually, the *ai1* and *ai2* reverse transcriptases should not be completely specific for the introns that encode them, since the presence of introns *ai1* and/or *ai2* is responsible for the loss of both group I and group II introns from yeast mitochondrial DNA (115). Intron deletion occurs in response to selection for reversion of (somewhat leaky) splicing-deficient mutations and may involve not only the affected intron, but one or several of the neighboring intervening sequences (116, 117), exactly as expected from occasional reverse transcription of mature mRNA molecules and subsequent conversion of genomic DNA by the resulting cDNAs.

Synthesis of abundant cDNA molecules from a site immediately downstream of an intron in unspliced precursor RNA could certainly contribute to the preferential transmission of that intron in crosses and, if the synthesis extends into the 5' exon, to the observed dissymmetrical conversion of flanking sequences during homing (110). Reverse transcription followed by recombination with genomic DNA could similarly account for the occasional transposition of group II introns to novel sites on the condition that the template used was one in which an intron had inserted itself by reverse splicing (111). One strong argument in favor of such a mechanism is that reverse splicing and transposition both occur immediately 3' of sequences that mimic the IBS1 motif (see above). Also, the fact that a *cis*-dominant splicing-deficient mutation in intron *ai1* blocks both homing (106) and transposition (111) has been widely interpreted as evidence that group II introns need to be splicing competent to be mobile. However, this mutation (G2166 in 84) is located close to the 3' end of intron *ai1* and could act merely by preventing priming of reverse transcription. Furthermore, one difficulty with a reverse splicing mechanism is that both the sense and antisense strands of the same gene can be targets for insertion of a group II intron (113) [recall, however, that not merely RNA, but also single-stranded DNA can become ligated to the 3' end of group II introns (60)].

Homing of group I introns rests on efficient cleavage of the two DNA strands of intron-less alleles at, or close to, the site of insertion of the intron, followed by repair and recombination using the intron-containing DNA as template (reviewed in 5, 108). The same mechanism could account for efficient swamping of intron-less alleles by cDNA copies of an unspliced precursor molecule in group II homing [alternatively, the cleaved DNA molecules could be used as primers for reverse transcription, as is the case for at least one nonretroviral retrotransposon (118)]. The facts are that, as already mentioned, many group II–encoded proteins include a domain with putative DNA endonuclease activity, but nothing is known of possible targets. Only recently was biochemical evidence of DNA cleavage in connection with a group II intron reported (119): In *Podospora anserina,* a prominent double-stranded cut was detected in the cytochrome oxidase subunit I gene, close to the 5′ splice site of the first intron, which has long been known to generate abundant extrachromosomal circular DNA copies of itself (120). The cleavage site has been precisely determined for the upper strand and found to lie within the intron, some four to six nucleotides from the splice junction. No information exists concerning the macromolecule responsible for the endonuclease activity and its ability to cut intron-less DNA.

To summarize, there is currently no clear consensus regarding the exact mechanisms of group II intron mobility. Nor is it clear to what extent homing, which converts entire populations of molecules to the intron-plus state, and transposition, which is a rare event, are related in terms of detailed molecular processes. Finally, even though all working models sensibly assume that the reverse transcriptase activity of intron-encoded proteins ought to play an essential role in mobility, there is no reason why some introns could not instead rely for homing on a group I–type mechanism in which RNA is not involved [the ribozyme-encoding sections of group I introns do not seem to play any part in the homing process (121, 122)]. In fact, not all proteins encoded by group II introns include a polymerase domain, and the ribozyme domain IV of one intron [in the coding sequence for the small subunit mitochondrial rRNA of the fungus *Trimorphomyces papilionaceus* (123)] is interrupted not by a group II–related ORF, but rather, by the coding sequence of a protein belonging to the so-called LAGLI...DADG family of group I DNA endonucleases!

DIVERSIFYING EVOLUTION OF GROUP II INTRONS

Evolutionary strategies are dictated by the environment. Because opportunities for genetic exchanges between distantly related organisms abound in the prokaryotic world (124, 125), bacterial members of group II may have been selected for traits that would make them good colonists (self-splicing, opportunism in the choice of insertion sites), rather than for their ability to endure by making themselves indispensable to their host. Consistent with this idea, all four *E. coli* introns are inserted into proven or putative mobile elements (29). Just the opposite should be true of those members of group II that became trapped in the ever-shrinking universe of ancestral organelles. As might have been expected, many of these introns have become highly dependent on host factors for splicing (see above); some share some of their hosts' most bizarre idiosyncrasies, such as the need for editing after transcription; and yet others have adopted a split structure that makes them essentially unfit not only for transmission between species, but for any mode of transposition. In some cases, even entire populations, just like island birds turned flightless, have gone so far down the domestication line to renounce altogether the means for mobility. For example, plant chloroplast introns do not seem to have changed locations since the divergence of bryophytes from the line leading to angiosperms, which is consistent with the fact that none of them has retained the capacity to encode a protein with a complete polymerase domain (19, 20); in contrast, only one of the mitochondrial group II introns has remained at the same site (17, 18). This section reviews some aspects of these ongoing diversification processes.

Trans-Splicing and Editing

Two secondary developments that appear so far confined to plant organellar genomes are fragmentation and editing of group II introns (reviewed in 7, 18, 126). Fragmentation results from DNA rearrangements that split genes within the coding sequence of introns, so that sections 5' and 3' of the breakpoint become part of separate transcription units, often located far apart on the genome. This phenomenon has so far been found to affect group II, but not group I, introns. There are three known cases of split introns from chloroplasts—two in *Chlamydomonas* (127) and one in land plants (128)—and at least six from the mitochondria of angiosperms, where fragmentation is an ongoing process, as shown by the fact that introns that are *cis*-splicing (uninterrupted) in some organisms may be *trans*-splicing in others. In eight out of nine cases, the (or one) breakpoint is within domain IV, consistent with the fact that this domain is the one that is most often interrupted by large insertions. At least one intron appears to be synthesized in (at least) three pieces: As indicated by analyses of sequences, transcripts, and transformed strains, the central section (including domains II and III, part of domain I, and the 5' part of domain IV) of the first intron of the *Chlamydomonas psaA* gene is encoded by a separate locus (127).

How is the efficiency and specificity of splicing ensured in split introns? While tertiary interactions between domains I and V may be numerous enough to ensure efficient bimolecular recognition in in vitro settings (67, 76, 80), additional factors are likely to be involved in vivo. This is clearly the case in *Chlamydomonas,* where mutants affected in the maturation of the *psaA* transcript belong to no less than 13 nuclear complementation groups (127).

Post-transcriptional editing, consisting mostly, but not exclusively, of C-to-U substitutions, appears to affect virtually all transcripts in higher plant mitochondria and also exists in chloroplasts (reviewed in 129). Both the ribozyme (130–132) and protein-coding (130) sections of group II introns are liable to include sites of editing. It is not yet known whether editing is essential for splicing, but this appears likely, because lack of correction of some normally edited sites would leave A:C mismatches in otherwise highly conserved pairings. An understanding of the specificity of plant mitochondrial editing remains completely lacking, and in this respect, it is interesting that even closely related intron sequences within the same genome may differ in patterns of editing (133).

The Euglena Microcosm

About 40% of the chloroplast genome of *Euglena gracilis* (25) is devoted to introns, which number at least 155 (reviewed in 12); there are even introns inserted into other introns (1, 134). About half of these introns form a separate class called group III. Group III introns are characterized by their short and remarkably uniform size (91–119 nucleotides). Their conserved features are rather elusive ones, whether in terms of sequence or structure, but nonetheless suggest a relationship with group II: Nucleotides 2 and 5 are normally U and G, and the 3' terminal sequence of most group III introns has the potential for forming a hairpin structure with a bulging A on its 3' side. The suggestion (see 12) of a connection between group III and group II introns has been considerably reinforced by the finding that at least some group III introns appear to be excised as lariats, using the bulging A as branchpoint (135).

The other introns in the *Euglena gracilis* chloroplast genome are classified as group II because they include recognizable counterparts of domains V and VI (12, 136). Although some of these introns appear to belong to previously recognized group II subtypes (1), most of them are unusually short for group II membership (the smallest one is only 270 nucleotides). Secondary structure models have been proposed for a number of sequences, but whether these models should be regarded as credible or not remains a subjective matter (compare discussions in 1 and 12). The problem is one of methodology. Secondary structure modeling by comparative analysis should rest on sequence alignment and a subsequent search for compensatory base changes, rather than on an attempt to maximize similarity to a preexisting consensus model. Unfortunately, the sequences of *Euglena gracilis* group II introns are too divergent from one another to be reliably aligned, and even recognizable secondary-structure elements, such as domain V, tolerate surprisingly many mismatches. Nor can putative pairings be verified experimentally, because none of the introns tested self-splices in vitro and it has not yet been possible to put DNA back into *Euglena gracilis* chloroplasts. One way to resolve this problem, and assess proposed models, is to sequence the same introns in related genomes. Comparison (12) of homologous sequences from *Euglena gracilis* and *Astasia longa* (a colorless flagellate) seems to confirm that some at least of the euglenoid introns have retained the group II central wheel, although perhaps little more information [another intron that would seem to fit this description is in the *rpeB* gene of chloroplasts of *Rhodella violacea,* a red alga (137)].

The situation in the *Euglena gracilis* chloroplast genome is intellectually attractive because it provides us with an idea of the evolutionary potential of group II–catalyzed splicing (see 12). In particular, group III introns should turn out to be excised by a complex ribonucleoprotein machinery, at the heart of which could lie fragments of a group II intron. And since the *Euglena gracilis* introns seem to constitute an evolutionary continuum, with all degrees of defectiveness, it might even be possible to reconstitute some of the events that transformed self-splicing molecules into mere substrates of an externally encoded splicing machine.

Did Spliceosome-Catalyzed Splicing Evolve from Group II Self-Splicing?

The splicing of introns in the nuclear-encoded premessenger RNAs of eukaryotes requires small ribonucleoparticles (U1, U2, U4, U5, and U6 in fungi, metazoans, and plants) assembled into spliceosomes. Much has already been written on the possible common evolutionary origins of spliceosomes and group II introns [see recent reviews by Nilsen (138), Weiner (11), and Wise (139) and references therein], and especially in view of our still very fragmentary understanding of the two systems. Mechanistic similarities between the splicing reactions catalyzed by group II introns and eukaryotic spliceosomes are truly impressive. Not only do both systems make use of the 2′ OH group of a bulging A residue to attack the phosphodiester bond at the 5′ splice site, but as has recently become apparent, they both tolerate Sp, but not Rp, phosphorothioate stereoisomers at both stages of the splicing process (73b, 74, 140). Nor is there anything unreasonable about assuming (e.g. 141) that progressive fragmentation of a self-splicing group II intron by the same processes that operate nowadays in plant organelles could have generated the small nuclear RNAs (snRNAs) of the spliceosome, with the intron retaining only substrate sequences (at the splice and branch sites). Unfortunately, while much has now been deciphered of the interactions of snRNAs both between themselves and with the pre-mRNA molecule, attempts to relate this knowledge with what is known of the structure of group II introns have been just as unconvincing as those of linguists struggling to identify common elements in idioms that diverged too long ago.

Current views of spliceosomal–group II relationships (reviewed in 138, 139) would equate the U2-U6 pairings with domain V, the base pairs between U6 and intron nucleotides 4–6 with ε-ε′, and the ill-defined interactions between U5 and the exons to be ligated with the EBS1-IBS1 and δ-δ′ pairings (no one would question that the pairing between U2 and the nucleotides surrounding the branch site can be regarded as equivalent to domain VI). Leaving aside the fact that U4 and U1 are left with no plausible counterparts, one problem with this scheme is that, in contrast with the situation in group II domain V (83), several of the residues that participate in the U2-U6 pairing are specifically involved in exon ligation (142, 143). Also, components assumed to be homologous have completely different sequences in group II introns and snRNAs [in fact, the very existence of consensus sequences in the spliceosomal system was challenged by the recent realization that nuclear premessenger introns have at least one subclass with highly divergent boundary sequences (144)].

Hence, rather than attempting to compare sequences, one should identify the tertiary interactions responsible for evolutionary conservation of base pairs. But when this was done for bases at the intron termini, the answer has been a rather sobering one. Although the last nucleotide of group II introns forms a classical base pair with one of the residues in the segment connecting domains II and III (the γ-γ' pairing in Figure 2), no such interaction has been uncovered in spliceosomal introns, whose last nucleotide instead seems to interact with the one at the 5′ extremity by nonclassical base pairing (145, 146) (the type of base pair proposed would exclude an additional Watson-Crick interaction). What then of the first nucleotide of group II introns? It does appear to interact with the 3′ end of the molecule, but the base contacted is the penultimate one (56). Such data might have been regarded as proof that spliceosomes and group II introns are unrelated after all, but given the dominant tone in the current literature (see 11 for an exception), they are more likely to be interpreted as evidence that the two systems have diverged to such an extent that tertiary structure became somewhat altered.

CONCLUDING REMARKS

Group II introns have obviously suffered from competition with the fast-moving group I field. Yet, as a distinct type of ribozymes endowed with the ability to enlist the protein they encode in order to move from one genomic location to another, group II introns are fascinating biological entities in their own right. Progress in understanding the biology of group II introns has long been hampered by the fact that known examples were confined to organellar genomes, which do not lend themselves well to reverse genetics. However, now that bacterial members of group II have been discovered, we should not need wait very long for a detailed understanding of the molecular mechanisms underlying their mobility. But even after that, much will remain to be done before we fully understand the intricate RNA structures responsible for group II ribozymic activity or the variety of evolutionary strategies that make it possible for group II introns to thrive in contemporary organisms.

ACKNOWLEDGMENTS

We are indebted to many of our colleagues, who sent us preprints, manuscripts, and unpublished information. Special thanks are due to Maria Costa, for many helpful discussions; to her, Alain Jacquier, and Guillaume Chanfreau, for carefully reading the manuscript; and to Raymond Samaha, Otello Stampacchia, Renée Schroeder, and Fernanda de Carvalho, for their brave help in conveying the documentation for this review into port.

Literature Cited

1. Michel F, Umesono K, Ozeki H. 1989. *Gene* 82:5–30
2. Lambowitz AM, Perlman PS. 1990. *Trends Biochem. Sci.* 15:440–44
3. Woolford JL Jr, Peebles CL. 1992. *Curr. Opin. Genet. Dev.* 2:712–19
4. Saldanha R, Mohr G, Belfort M, Lambowitz AM. 1993. *FASEB J* 7:15–24
5. Lambowitz AM, Belfort M. 1993. *Annu. Rev. Biochem.* 62:587–622
6. Cech TR. 1993. See Ref. 147, pp. 239–69
7. Bonen L. 1993. *FASEB J.* 7:40–46
8. Sharp PA. 1985. *Cell* 42:397–400
9. Cech TR. 1986. *Cell* 44:207–10
10. Jacquier A. 1990. *Trends Biochem. Sci.* 15:351–54
11. Weiner AM. 1993. *Cell* 72:161–64
12. Copertino DW, Hallick RB. 1993. *Trends Biochem. Sci.* 18:467–71
13. Michel F, Westhof E. 1990. *J. Mol. Biol.* 216:585–610
14. Westhof E, Altman S. 1994. *Proc. Natl. Acad. Sci. USA* 91:5133–37
15. Harris ME, Nolan JM, Malhotra A, Brown JW, Harvey SC, Pace NR. 1994. *EMBO J.* 13:3953–63
16. Kennell JC, Moran JV, Perlman PS, Butow RA, Lambowitz AM. 1993. *Cell* 73:133–46
17. Oda K, Yamato K, Ohta E, Nakamura Y, Takamura M, et al. 1992. *J. Mol. Biol.* 223:1–7
18. Knoop V, Brennicke A. 1993. In *Plant Mitochondria*, ed. A Brennicke, U Kück, pp. 221–32. Weinheim: VCH Chemie
19. Ohyama K, Fukuzawa H, Kohchi T, Shirai H, Sano T, et al. 1986. *Nature* 332:572–74
20. Shinozaki K, Ohme M, Tanaka M, Wakasugi T, Hayashida N, et al. 1986. *EMBO J.* 5:2043–49
21. Kück U, Choquet Y, Schneider M, Dron M, Bennoun P. 1987. *EMBO J.* 6:2185–95
22. Kück U. 1989. *Mol. Gen. Genet.* 218:257–65
23. Kono M, Satoh H, Okabe Y, Abe Y, Nakayama K, Okada M. 1991. *Plant Mol. Biol.* 17:505–8
24. Manhart JR, von der Haar RA. 1991. *J. Phycol.* 27:613–17
25. Hallick RB, Hong L, Drager RG, Favreau MR, Monfort A, et al. 1993. *Nucleic Acids Res.* 21:3537–44
26. Michel F, Lang BF. 1985. *Nature* 316:641–43
27. Xiong Y, Eickbush TH. 1990. *EMBO J.* 9:3353–62
28. Ferat J-L, Michel F. 1993. *Nature* 264:358–61
29. Ferat J-L, Le Gouar M, Michel F. 1994. *C.R. Acad. Sci.* 317:141–48
30. Knoop V, Brennicke A. 1994. *Nucleic Acids Res.* 22:1167–71
31. Knoop V, Brennicke A. 1994. *J. Mol. Evol.* 39:144–50
32. Peebles CL, Perlman PS, Mecklenburg KL, Petrillo ML, Tabor JH, et al. 1986. *Cell* 44:213–23
33. Van der Veen R, Arnberg AC, Van der Horst G, Bonen L, Tabak HF, Grivell LA. 1986. *Cell* 44:225–34
34. Schmelzer C, Schweyen RJ. 1986. *Cell* 46: 557–65
35. Schäfer B, Merlos-Lange AM, Anderl C, Welser F, Zimmer M, Wolf K. 1991. *Mol. Gen. Genet.* 225:158–67
36. Rochaix J-D. 1992. *Annu. Rev. Cell Biol.* 8:1–28
37. Schmidt U, Riederer B, Mörl M,

Schmelzer C, Stahl U. 1990. *EMBO J.*
9:2289–98
38. Hebbar SK, Belcher SM, Perlman PS.
1992. *Nucleic Acids Res.* 20:1747–54
39. Kück U, Godehardt I, Schmidt U. 1990.
Nucleic Acids Res. 18:2691–97
40. Jarrell KA, Peebles CL, Dietrich RC,
Romiti SL, Perlman PS. 1988. *J. Biol.
Chem.* 263:3432–39
41. Wiesenberger G, Waldherr M, Schweyen RJ. 1992. *J. Biol. Chem.* 267: 6963–69
42. Séraphin B, Simon M, Boulet A, Faye
G. 1989. *Nature* 337:84–87
43. Wiesenberger G, Link TA, von Ahsen
U, Waldherr M, Schweyen RJ. 1990. *J.
Mol. Biol.* 217:23–37
44. Coetzee T, Herschlag D, Belfort M.
1994. *Genes Dev.* 8:1575–88
45. Michel F, Jacquier A, Dujon B. 1982.
Biochimie 64:867–81
46. Michel F, Dujon B. 1983. *EMBO J.*
2:33–38
47. Schmelzer C, Schmidt C, May K,
Schweyen RJ. 1983. *EMBO J.* 2:2047–52
48. Michel F, Jacquier A. 1987. *Cold Spring
Harbor Symp. Quant. Biol.* 52:201–12
49. Jacquier A, Michel F. 1987. *Cell* 50:17–29
50. Cole PE, Yang SK, Crothers DM. 1972.
Biochemistry 11:4358–68
51. Banerjee AR, Jaeger JA, Turner DH.
1993. *Biochemistry* 32:153–63
52. Jaeger L, Westhof E, Michel F. 1993.
J. Mol. Biol. 234:331–46
53. Harris-Kerr CL, Zhang M, Peebles CL.
1993. *Proc. Natl. Acad. Sci. USA* 90:
10658–62
54. Jacquier A, Michel F. 1990. *J. Mol.
Biol.* 213:437–47
55. Jacquier A, Jacquesson-Breuleux N.
1991. *J. Mol. Biol.* 219:415–28
56. Chanfreau G, Jacquier A. 1993. *EMBO
J.* 12:5173–80
56a. Costa M, Michel F. 1995. *EMBO J.* In
press
57. Latham JA, Cech TR. 1989. *Science*
245:276–82
58. Augustin S, Müller MW, Schweyen RJ.
1990. *Nature* 343:383–86
59. Mörl M, Schmelzer C. 1990. *Cell* 60:
629–36
60. Mörl M, Niemer I, Schmelzer C. 1992.
Cell 70:803–10
61. Mueller MW, Stocker P, Hetzer M,
Schweyen RJ. 1991. *J. Mol. Biol.* 222:
145–54
62. Wallasch C, Mörl M, Niemer I, Schmelzer C. 1991. *Nucleic Acids Res.* 19:
3307–14
63. Jacquier A, Rosbash M. 1986. *Science*
234:1099–104
64. Mueller MW, Schweyen RJ, Schmelzer
C. 1988. *Nucleic Acids Res.* 16:7383–95
65. Copertino DW, Hallick RB. 1991.
EMBO J. 10:433–42
66. Kim J-K, Hollingsworth MJ. 1993.
Curr. Genet. 23:175–80
67. Koch JL, Boulanger SC, Dib-Hajj SD,
Hebbar SK, Perlman PS. 1992. *Mol.
Cell Biol.* 12:1950–58
68. Davies RW, Waring RB, Ray JA, Brown
TA, Scazzocchio C. 1982. *Nature* 300:
719–24
69. Michel F, Hanna M, Green R, Bartel
DP, Szostak JW. 1989. *Nature* 342:391–95
70. Suh ER, Waring RB. 1990. *Mol. Cell.
Biol.* 10:2960–65
71. Peebles CL, Belcher SM, Zhang M,
Dietrich RC, Perlman PS. 1993. *J. Biol.
Chem.* 268:11929–38
72. Van der Veen R, Kwakman JHJM, Grivell LA. 1987. *EMBO J.* 12:3827–31
73. Altura R, Rymond B, Séraphin B, Rosbash M. 1989. *Nucleic Acids Res.* 17:
335–54
73a. Michels WJ, Pyle AM. 1995. *Biochemistry.* In press
73b. Padgett RA, Podar M, Boulanger SC,
Perlman PS. 1994. *Science* 266:1685–88
74. Moore MJ, Sharp PA. 1993. *Nature*
365:364–68
75. Steitz JA, Steitz TA. 1993. *Proc. Natl.
Acad. Sci. USA* 90:6498–502
76. Jarrell KA, Dietrich RC, Perlman PS.
1988. *Mol. Cell. Biol.* 8:2361–66
77. Kwakman JHJM, Konings D, Pel HJ,
Grivell LA. 1989. *Nucleic Acids Res.*
17:4205–16
78. Bachl J, Schmelzer C. 1990. *J. Mol.
Biol.* 212:113–25
79. Dib-Hajj SD, Boulanger SC, Hebbar
SK, Peebles CL, Franzen JS, Perlman
PS. 1993. *Nucleic Acids Res.* 21:1797–804
80. Pyle AM, Green JB. 1994. *Biochemistry*
33:2716–25
81. Franzen JS, Zhang M, Peebles CL. 1993.
Nucleic Acids Res. 21:627–34
82. Pyle AM. 1993. *Science* 261:709–14
83. Chanfreau G, Jacquier A. 1994. *Science*
266:1383–87
84. Carignani G, Groudinski O, Frezza D,
Schiavon E, Bergantino E, Slonimski
PP. 1983. *Cell* 35:733–42
85. Bergantino E, Carignani G. 1989. *Mol.
Gen. Genet.* 223:249–57
86. Moran JV, Mecklenburg KL, Sass P,
Belcher SM, Mahnke D et al. 1994.
Nucleic Acids Res. 22:2057–64
87. Xiong Y, Eickbush TH. 1988. *Mol. Biol.
Evol.* 5:675–90
88. Eickbush TH. 1994. In *The Evolutionary*

Biology of Viruses, ed. SS Morse, pp. 121–57. New York: Raven

89. Inouye S, Inouye M. 1991. *Annu. Rev. Microbiol.* 45:163–86
90. Nargang FE, Bell JB, Stohl LL, Lambowitz AM. 1984. *Cell* 38:441–53
91. Wang H, Kennell JC, Kuiper MTR, Sabourin JR, Saldanha R, Lambowitz AM. 1992. *Mol. Cell. Biol.* 12:5131–44
92. Boer PH, Gray MW. 1988. *EMBO J.* 7:3501–8
93. Lang BF, Ahne F, Bonen L. 1985. *J. Mol. Biol.* 184:353–66
94. Neuhaus H, Link G. 1987. *Curr. Genet.* 11:251–57
95. Doolittle RF, Feng DF, Johnson MS, McClure MA. 1989. *Q. Rev. Biol.* 64:1–30
96. McClure MA. 1991. *Mol. Biol. Evol.* 8:835–56
97. Gorbalenya AE. 1994. *Protein Sci.* 3: 1117–20
98. Shub DA, Goodrich-Blair H, Eddy SR. 1994. *Trends Biochem. Sci.* 19:402–4
99. Cummings DF, Michel F, McNally KL. 1989. *Curr. Genet.* 16:407–18
100. Lazowska J, Jacq C, Slonimski PP. 1980. *Cell* 22:333–48
101. Carignani G, Netter P, Bergantino E, Robineau S. 1986. *Curr. Genet.* 11:55–63
102. Mohr G, Perlman PS, Lambowitz AM. 1993. *Nucleic Acids Res.* 21:4991–97
103. Borst P, Bos JL, Grivell LA, Groot GSP, Heyting C, et al. 1977. See Ref. 148, pp. 213–54
104. Jacq C, Kujawa C, Grandchamp C, Netter P. 1977. See Ref. 148, pp. 255–70
105. Hardy CM, Clark-Walker GD. 1991. *Curr. Genet.* 20:99–114
106. Meunier B, Tian GL, Macadre C, Slonimski PP, Lazowska J. 1990. In *Structure, Function and Biogenesis of Energy Transfer Systems,* ed. E Quagliariello, S Papa, F Palmieri, C Saccone, pp. 169–74. Amsterdam: Elsevier
107. Skelly PJ, Hardy CM, Clark-Walker GD. 1991. *Curr. Genet.* 20:115–20
108. Dujon B. 1989. *Gene* 82:91–114
109. Perlman PS, Butow RA. 1989. *Science* 246:1106–9
110. Lazowska J, Meunier B, Macadre C. 1994. *EMBO J.* 13:4963–72
111. Mueller MW, Allmaier M, Eskes R, Schweyen RJ. 1993. *Nature* 366:174–76
112. Sellem C, Lecellier G, Belcour L. 1993. *Nature* 366:176–78
113. Schmidt WM, Schweyen RJ, Wolf K, Mueller MW. 1994. *J. Mol. Biol.* 243: 157–66
114. Fassbender S, Brühl K-H, Ciriacy M, Kück U. 1994. *EMBO J.* 13:2075–83
115. Levra-Juillet E, Boulet A, Séraphin B,

Simon M, Faye G. 1989. *Mol. Gen. Genet.* 217:168–71
116. Gargouri A, Lazowska J, Slonimski PP. 1983. In *Mitochondria 1983: Nucleo-Mitochondrial Interactions,* ed. RJ Schweyen, K Wolf, F Kaudewitz, pp. 259–68. Berlin: de Gruyter
117. Merlos-Lange AM, Kanbay F, Zimmer M, Wolf K. 1987. *Mol. Gen. Genet.* 206:273–78
118. Luan DD, Korman MH, Jakubczak JJ, Eickbush TH. 1993. *Cell* 72:595–605
119. Sainsard-Chanet A, Begel O, Belcour L. 1994. *J. Mol. Biol.* 242:630–43
120. Osiewacz HD, Esser K. 1984. *Curr. Genet.* 8:299–305
121. Bell-Pedersen D, Quirk S, Clyman J, Belfort M. 1990. *Nucleic Acids Res.* 18:3763–70
122. Eddy SR, Gold L. 1992. *Proc. Natl. Acad. Sci. USA* 89:1544–47
123. Hong SG, Young-Won K, Hack-Sung J. 1993. *Korean J. Microbiol.* 31:471–77
124. Heinemann JA. 1991. *Trends Genet.* 7:181–85
125. Amabile-Cuevas CF, Chicurel ME. 1992. *Cell* 70:189–99
126. Wissinger B, Brennicke A, Schuster W. 1992. *Trends Genet.* 8:322–28
127. Goldschmidt-Clermont M, Choquet Y, Girard-Bascou J, Michel F, Schirmer-Rahire M, Rochaix J-D. 1991. *Cell* 65: 135–43
128. Kohchi T, Umesono K, Ogura Y, Komine Y, Nakahigashi K, et al. 1988. *Nucleic Acids Res.* 16:10025–36
129. Bass BL. 1993. See Ref. 147, pp. 383–418
130. Wissinger B, Schuster W, Brennicke A. 1991. *Cell* 65:473–82
131. Knoop V, Schuster W, Wissinger B, Brennicke A. 1991. *EMBO J.* 10:3483–93
132. Binder S, Marchfelder A, Brennicke A, Wissinger B. 1992. *J. Biol. Chem.* 267: 7615–23
133. Lippok B, Brennicke A, Wissinger B. 1994. *Mol. Gen. Genet.* 243:39–46
134. Copertino DW, Hallick RB. 1991. *EMBO J.* 10:433–42
135. Copertino DW, Hall ET, Van Hook FW, Jenkins KP, Hallick RB. 1994. *Nucleic Acids Res.* 22:1029–36
136. Keller M, Michel F. 1985. *FEBS Lett.* 179:69–73
137. Bernard C, Thomas JC, Mazel D, Mousseau A, Castets AM, et al. 1992. *Proc. Natl. Acad. Sci. USA* 89:9564–68
138. Nilsen TW. 1994. *Cell* 78:1–4
139. Wise JA. 1993. *Science* 262:1978–79
140. Maschhoff KL, Padgett RA. 1993. *Nucleic Acids Res.* 21:5456–62
141. Sharp PA. 1991. *Science* 254:663

142. Madhani HT, Guthrie C. 1992. *Cell* 71:803–17
143. McPheeters DS, Abelson J. 1992. *Cell* 71:819–31
144. Hall SL, Padgett RA. 1994. *J. Mol. Biol.* 239:357–65
145. Parker R, Siliciano PG. 1993. *Nature* 361:660–62
146. Carothers AM, Urlaub G, Grunberger D, Chasin LA. 1993. *Mol. Cell. Biol.* 13:5085–98
147. Gesteland RF, Atkins JF, eds. 1993. *The RNA World.* Cold Spring Harbor, NY: Cold Spring Harbor Lab.
148. Bandlow W, Schweyen RJ, Wolf K, Kaudewitz F, eds. 1977. *Mitochondria 1977.* Berlin: de Gruyter

Annu. Rev. Biochem. 1995. 64:463–91
Copyright © 1995 by Annual Reviews Inc. All rights reserved

GENERATION, TRANSLOCATION, AND PRESENTATION OF MHC CLASS I–RESTRICTED PEPTIDES

Marie-Thérèse Heemels and Hidde Ploegh

Center for Cancer Research and Department of Biology, Massachusetts Institute of Technology, Cambridge Massachusetts 02139

KEY WORDS: proteasome, LMP, TAP peptide transporter, chaperones, calnexin

CONTENTS

463

ABSTRACT

The T lymphocytes of the vertebrate immune system look for changes that take place within the organism by examining a display of peptides at the cell surface. These peptides are presented by the products of the major histocompatibility complex (MHC). MHC class I products present peptides derived by proteolysis of cytosolic proteins by the multicatalytic protease, the proteasome. These peptides are translocated from the cytosol into the endoplasmic reticulum by a dedicated peptide transporter, the transporter associated with antigen presentation (TAP). TAP consists of two subunits, and translocates peptides that are approximately 8–12 residues in length. The COOH terminal residue of the peptide is a major determinant in the specificity of translocation. Following translocation, peptides bind to MHC class I molecules, which depend on the peptide ligand as well as on interactions with chaperonins for proper folding. These complexes then egress from the ER and are transported to their final destination, the cell surfacce.

INTRODUCTION

The vertebrate immune system serves to purge pathogens and infectious agents from the body: Antigen-specific antibody and cellular responses are essential for elimination of many invaders. While the antibody response can deal with extracellular antigens, such as bacteria, it does not always eradicate the virally infected cell that sustains production of new viral particles. Cytotoxic T lymphocytes are responsible for removal of these virus factories. In addition, these T cells are thought to be involved in destruction of cells on their way to malignant transformation (immune surveillance).

Whereas immunoglobulins can recognize antigens without the involvement of any additional molecules, T cells are incapable of doing so. The antigen-specific receptors on T cells (TCR) recognize foreign protein fragments (8–14 residues) bound to proteins encoded by the major histocompatibility complex (MHC) (1, 2). These peptide-binding MHC molecules are cell surface glycoproteins and are commonly categorized in two classes: class I MHC and class II MHC products. These two categories of MHC proteins, while homologous and structurally quite similar, have a related but distinct function. Class I MHC products present peptides to T cells that express the accessory molecule CD8, whereas class II MHC molecules present peptides to CD4 positive T cells. The antigen-specific receptor on T cells (TCR) and the appropriate accessory molecules must interact with an MHC molecule occupied by an antigenic peptide to be activated and participate in an immune response. Activation of CD4+ T cells generally leads to the production of cytokines (e.g. interferons and interleukins) that have profound immunoregulatory effects. In contrast, CD8+ T cells kill the virally infected or transformed cell and are therefore referred to as cytotoxic T lymphocytes (CTL) (3).

MHC class I molecules consist of a transmembrane glycoprotein subunit, the heavy chain, in association with a soluble light chain, β2-microglobulin, whereas class II molecules consist of two MHC-encoded transmembrane glycoprotein subunits, the α and the β chain. The subunits of class I and II molecules are cotranslationally translocated into the secretory pathway. The short C-terminal tails of the transmembrane proteins are exposed to the cytoplasm, and their peptide-binding portions are positioned in the lumen of the endoplasmic reticulum (ER), the topological equivalent of the extracellular space.

From the cell biological point of view, the key differences between class I and class II restricted antigen presentation are the source of the proteins from which antigenic peptides are derived, and the compartment in which the peptides are generated (4, 5). MHC class II molecules have evolved to accommodate peptides derived from exogenous proteins: Class II molecules are delivered to the endocytic pathway by transient association of the class II heterodimer with a third polypeptide, the invariant chain (Ii). The removal of the invariant chain by proteolysis liberates the peptide binding pocket of class II molecules and allows the binding of peptides generated in the endocytic pathway by the activity of proteases that reside there (6, 7). In contrast, most of the events that lead to the formation of a peptide-loaded class I molecule take place in the ER (8, 9), and therefore, endocytosed proteins do not usually form substrates for presentation by MHC class I molecules. Although the binding of peptides to MHC class I molecules takes place in the ER, peptides presented by class I molecules are largely derived from cytosolic proteins synthesized by the antigen-presenting cell. The presentation of such peptides relies on protein degradation by cytosolic proteases (antigen processing); the generated fragments must therefore cross the ER membrane prior to their binding to class I (3). A dedicated peptide transporter is responsible for shuttling peptides from the cytosol into the lumen of the secretory pathway (10, 11).

Most nucleated cells express MHC class I molecules; the process of proteolysis, peptide translocation, binding of peptides, and class I expression is constitutive. In the absence of viral infection, the peptide content of class I molecules is derived entirely from the cell's own proteins. Every class I–positive cell will display at the surface a complex family of peptide-MHC complexes whose peptide content reports on the types of proteins expressed by that cell (12). Viral infection, mutation, and malignant transformation all result in altered patterns of protein expression, both qualitatively and quantitatively. This process changes association of the relevant peptides with class I molecules so that the corresponding T cells may be signaled to mount a cytolytic attack. Under normal circumstances, mature T cells do not respond to self-peptides presented to them in the context of "self" MHC products. During the development of T cells, only those with low affinity for MHC and self-peptides expressed in the thymus are allowed to mature and enter the circulation (13). In addition, T cells that have left the thymus are rendered tolerant to tissue-specific antigens in the periphery (14). How exactly this tolerance is established, and whether special types of peptides are required in the course of T cell development, are hotly debated issues outside the scope of the present review. This review describes biochemical aspects of antigen processing and presentation by class I MHC molecules. This field has recently been expertly reviewed (3); we therefore describe the more recent developments in antigen processing and presentation, focus on events that occur prior to peptide loading of class I molecules, and indicate areas that have yet to yield to detailed biochemical analysis.

MHC CLASS I PROTEINS

The Structure of Class I Molecules

MHC class I molecules are composed of a membrane-anchored heavy chain (approximately 45 kDa); a soluble, noncovalently associated light chain, β_2-microglobulin (β_2m, 12 kDa,); and a peptide, usually 8-10 residues long. The transmembrane segment anchors the protein in the membrane and may mediate interactions with accessory proteins (15) required for proper assembly and folding of class I molecules. The extracellular portion is responsible for peptide display to T cells and consists of four domains. The α1 and α2 domains together comprise the peptide-binding site, which is composed of two α helices supported by an eight-stranded β-pleated sheet (16). This part of the molecule is supported by a structure that consists of the α3 domain and β_2m. Both of these domains show sequence and structural similarity to immunoglobulin constant-region domains. The peptide binding portion of the class I molecule is contacted by the antigen-specific T-cell receptor (TCR) (17), whereas the α3 domain of the class I heavy chain interacts with the T cell's accessory molecule CD8 (18, 19).

Both peptides and β_2m are required for the formation of a stable class I complex. The absence of either peptides (20, 21) or β_2m (22–25) leads to retention of the incompletely assembled complex in the ER and results in reduced cell surface expression. The requirement for peptide or β_2m is not absolute—so-called empty class I heterodimers can be transported to the cell surface (20, 21), and free heavy chains are expressed at the cell surface in the absence of β_2m (26) but in a far less efficient manner. Those class I molecules that find themselves at the cell surface in the absence of peptide are usually quite unstable; they unfold and are degraded. These properties may have been selected to decrease the likelihood that class I molecules are inadvertently loaded at the cell surface by peptides generated at a distance, which would pose the potential risk of a cytolytic T cell attack on an innocent bystander. The most remarkable feature of the MHC class I products is the extensive allelic variation concentrated in the $\alpha 1$ and $\alpha 2$ domains that form the peptide binding groove. Consequently, each allelic product presents a unique set of peptides for scrutiny by cytotoxic T cells (12).

The prevalent MHC class I heavy chains are encoded within the MHC by three different loci, designated H-2K, D, and L (in mice), and HLA-A, B, and C (in humans). The products of the class I MHC genes are codominantly expressed, hence individuals can express a maximum of six different class I molecules at the cell surface. In addition to these so-called classical class I loci, numerous other class I loci specify the nonclassical class I products. The function of the latter is less well established, but given their structural similarity to the classical class I loci, their modus operandi may be similar (27).

Peptide Binding

The elucidation of the crystal structure of MHC class I molecules revealed the architecture of its peptide-binding groove (16, 17, 28). All class I structures solved to date have a closed peptide-binding groove and conserved features that hold onto the peptide termini (29). As a consequence, peptide binding by classical class I products usually requires free NH_2^- and COOH-termini (30, 31), and in general, the bound peptides display a rather narrow size distribution (8–10 residues) (32, 33). Peptides that bind to class I molecules are tightly bound primarily by virtue of contacts of the peptide's amino acid side chains with the class I molecule (28). Pockets along the groove (designated A through F) may accommodate predominant amino acid side chains of the peptide (NH_2^- through COOH terminal side chains respectively) and thereby anchor the peptide onto the class I molecule (34, 35). While the A and F pockets are fairly conserved, B through E have distinct sizes and characters in different allelic variants of class I and thereby impose different sequence constraints on the bound peptides. As a result, class I-binding peptides contain allele-specific sequence motifs defined by the position and the identity of—usually two—anchoring residues [one of which is the C-terminus (12)]. Alignment of all peptides presently known to bind to class I molecules illustrates the limited variation in carboxy termini of presentable peptides: Ile, Leu, Val, Arg, Lys, Tyr, or Phe are found at the carboxy terminus in 95% of all class I restricted epitopes (36). Broken down along species lines, this conservation is even more pronounced: For peptides bound to mouse class I molecules, only hydrophobic and aliphatic residue side chains have been observed. This phenomenon may not reflect the constraints imposed by the nature of the F-pocket only. Early features of the antigen-processing machinery, such as the specificity of the proteases or peptide transporters, may contribute as well (discussed below).

ANTIGEN PROCESSING

The Cytosol as the Primary Site for Antigen Processing

The protein from which the epitope is to be presented must be partially degraded and processed, before peptides can be presented by class I molecules. Although proteases are present in the ER (37), the available evidence indicates that the cytosol, not the ER, is the primary source of presentable peptides.

Many proteins that contribute the presented peptides, such as viral nucleoproteins, lack the signal sequences necessary to enter the ER. Nevertheless, their epitopes are presented via class I molecules. Moreover, deletion of the ER-signal sequence from the surface glycoprotein influenza hemagglutinin precludes entrance of this protein into the secretory pathway but does not affect the presentation of the epitope contained (38). Lysosomal proteases are also unlikely candidates for the generation of the majority of class I restricted peptides, even though cytosolic proteins can reach lysosomes by a process referred to as autophagy (39). Release of breakdown fragments from lysosomes back into the cytosol could in theory make peptides available for class I restricted presentation. However, the function of lysosomal proteases depends on organelle acidification. Neutralization of otherwise acidic lysosomal compartments, using ammonium chloride or chloroquine, inhibits peptide presentation via MHC class II but not class I molecules (40). A small amount of class I molecules present on phagocytic macrophages can apparently sample peptides derived from endocytosed matter, similar to the action of class II molecules (41–44). The invariant chain (Ii), a polypeptide normally found associated with class II molecules and widely believed to target class II complexes to the endosomes, has been found associated with a subset of class I molecules (45; M Sugita & MB Brenner, submitted). Class I molecules in a complex with Ii could thus intersect the endocytic route by exploiting the targeting signal borne by the Ii and hence may be loaded with exogenously derived peptides for class I restricted presentation or processed into peptides themselves for binding to (46), and presentation by MHC class II molecules (47).

The majority of antigens can enter the class I restricted antigen pathway only via the cytosol. Normally the pool of proteins actively synthesized in the cytosol comprises the source of these antigens. Native intact protein is also properly processed for class I restricted presentation when artificially introduced into the cytosol— by means of osmotic lysis of pinosomes, membrane fusion, or electroporation, for example (48–50). Townsend and colleagues showed that targeting a protein for rapid cytosolic degradation by constitutive fusion to ubiquitin enhanced the presentation of its epitope (51). Circumventing processing altogether by cytosolic expression of signal sequence-less, presentable peptides on minigenes also leads to sensitization of the appropriate CTL (52, 53). These observations demonstrate that cytosolic processing renders epitopes suitable for class I restricted presentation and that they can enter the class I presentation pathway in the absence of any identifiable signals that would transfer such peptides to the lumen of the ER. A most persuasive piece of evidence for the involvement of cytosolic proteases is the identification of a dedicated peptide transporter (see below). This transporter translocates peptides from the cytosol to the ER, and in its absence, antigen presentation is largely impaired.

The Proteasomes

The majority of proteolytic activity in the cytosol is attributed to the activity of proteasomes. The involvement of these multicatalytic protease complexes has often been suggested in antigen processing. The proteasome (54–56) is a multisubunit complex with an M_r of 700,000. Based on the sedimentation coefficient, this protease complex is referred to as 20S proteasome. Its cylindrical structure is formed by at least 13–15 distinct but related subunits of approximately 21–31 kDa; the protease complex displays minimally three distinct proteolytic activities in vitro. Characterization of its substrate specificity, using small peptides with a fluorophore or chromophore reporter group, has revealed that the proteasome cleaves COOH-terminally of hydrophobic, basic, and acidic amino acid residues (chymotryptic, tryptic, and peptidylglutamyl peptide hydrolyzing activities respectively). The 20S proteasome is thought to form the catalytic core of a higher molecular structure, the 26S proteasome complex (~1500 kDa). Although occasionally the 26S complex has been shown to degrade proteins independently of ubiquitin conjugation (57), it is widely held responsible for proteolysis of ubiquitin-modified proteins (54–56, 58).

The Low-Molecular-Mass Polypeptides LMP2 and LMP7

Chromosome walking and jumping in the MHC led to the identification of two novel proteasome subunits: the low molecular mass polypeptides LMP2 and LMP7 (59–63). The location of the genes encoding the LMPs within the MHC, their limited polymorphic nature, and their γIFN-inducible expression (63, 64) are properties shared by most components involved in antigen presentation and were taken as evidence implicating the LMPs in antigen processing. The MHC-encoded proteasome subunits might physically tie the proteasome to the transport machinery (62), thereby increasing the efficiency of peptide presentation, but no data in support of this suggestion have been obtained. Alternatively, the LMPs could alter the proteolytic activities of the proteasome and so favor the generation of peptides most commonly found in class I molecules (65).

In the absence of hard evidence, these hypotheses were challenged immediately. Mutant cell lines (66), with a large deletion in the MHC encompassing both LMP genes and peptide transporter genes, display grossly impaired class I assembly and are incapable of presenting endogenously synthesized antigens (67, 68). Antigen presentation was restored upon transfection of these mutant cells with the cDNAs encoding peptide transporter subunits and selection for fully restored class I expression (69–72). Moreover, peptides extracted from class I molecules appeared very similar, as judged by HPLC analysis, regardless of whether the cells they originated from expressed LMPs or not (69). Although subtle changes in efficacy of proteolysis could not be excluded, these data raised doubts about the role of LMPs in antigen processing and presentation, and suggested that their presence is not essential to this process.

The Role of the Proteasome, LMP2, and LMP7 in Antigen Processing

Subsequent studies put the LMP2 and LMP7 back on the map. Upon treatment of cells with γIFN, LMP2 and LMP7 are incorporated in the proteasome at the expense of other subunits (64, 73, 74). Three groups tested purified proteasomes from cells lacking or expressing LMP subunits for their capacity to cleave fluorogenic oligopeptides as model substrates (75–78). Although the presence of LMPs clearly induces alterations in the specificity of the proteasome, no consensus has been reached as to the nature of these changes. Driscoll et al (75) and Gaczynska et al (76) both observed an increased rate of cleavages C-terminally of hydrophobic and positively charged residues in the presence of LMPs, and in addition, the latter group detected reduced rates for cleavage C-terminal of acidic residues. This shift in specificity would favor peptides with carboxy termini most commonly found bound to class I molecules. In contrast to these observations, Boes et al (78) observed a reduced efficiency in the generation of hydrophobic carboxy termini when the proteasomes contain the LMP subunits. Nevertheless, antigen presentation might also benefit from these latter alterations in specificity, as the authors showed that an authentic class I epitope buried within a 25-residue peptide was wasted by proteasomes lacking LMPs but spared in their presence. Further experimentation will determine whether these discrepancies result from differences in experimental procedures or variations in composition of the isolated proteasomes (e.g. tissue- or species-specific differences).

The cleavage preference of the proteasome has been determined using low-molecular-weight substrates. The use of these substrates is not adequate to account for the possible influence of the larger sequence context of the peptide bond to be hydrolyzed. The flanking amino acids or the positioning of an epitope within the protein can (79) but need not (80, 81), affect its intracellular processing and presentation. In vitro proteolysis of larger polypeptides shows that known epitopes can be generated by the proteasome but, in addition, reveals that the cleavage of a limited assortment of short peptides may not accurately reflect the proteasome's activity on intact proteins (56, 78, 82). There is a need for more experiments that explore breakdown of intact proteins by proteasomes under physiological conditions of proteasome activation, preferably supplemented with translocation studies of the fragments generated.

In an attempt to clarify the role of LMPs in antigen presentation, mice with disrupted LMP2 or LMP7 genes were generated by means of homologous recombination (83, 84). Mice lacking the LMP7 gene (83) express reduced levels of class I on the cell surface (25–45% reduction), thereby firmly establishing the role of the LMP7 in antigen presentation by controlling availability of class I restricted peptides. Class I levels were restored by incubating LMP7-deficient cells in the presence of peptide, indicating that peptide supply is indeed the limiting factor in class I assembly. Although cells derived from the mutant mice are still capable of processing and presenting male specific H-Y antigen to an H-Y specific T-cell, they do so with reduced (50–70%) efficiency. Further research will soon resolve whether this lowered efficiency results from a defect in processing of the H-Y antigen or from the decreased class I expression. A more detailed biochemical analysis, and a comparison of presentation of full-length and mini-gene encoded epitopes, will be required to appreciate the magnitude of the processing defect. There is a simple explanation why the alterations in class I expression seen in the LMP7 knock-out animals have not been observed in earlier studies using mutant cell lines (69, 70). The deletion mutants lacking LMPs and TAPs were transfected with the latter, and transfectants were selected for high class I expression, thereby favoring cells expressing either multiple copies of the peptide transporters or displaying another compensatory trait.

The deletion of the LMP2 gene does not result in as pronounced a phenotype as disruption of the LMP7 gene, because these mice have no reduction in class I expression. However, careful analysis of the LMP2 mutant mice reveals features subtly distinct from wild type litter mates (84). Proteasomes isolated from mutant spleen and liver cleave at a lower rate C-terminally of hydrophobic and basic residues, and increased cleavage is seen after acidic residues. The processing and presentation of an epitope from influenza nucleoprotein to a specific CTL seems impaired (2.5– 5– fold reduction), but presentation of cytosolically delivered ovalbumin is not impaired. Upon infection of mice with Influenza, but not Sendai virus, the LMP2-deficient mice manifest a 5– 6– fold reduction in CTL precursor frequency. Disruption of the LMP2 gene differentially affects the production of distinct epitopes. Differences in the amount of protein available for processing, or differential effects on the generation of functional epitopes could both explain these results.

Rock and colleagues searched for possible inhibitors of the proteasome (85) to discover if the proteasome is responsible for all class I restricted epitopes generated in the cytosol. Three peptide aldehydes, formally inhibitors of chymotryptic activity, were assayed for their effects on the activity of the proteasome. Use of fluorogenic substrates shows that all three peptide aldehydes inhibit the chymotryptic, peptidylglutamyl, and to a lesser extent, tryptic activities of purified 20S and the 26S proteasome.

Based on their inhibitory capacities, these peptide aldehydes could be ranked in a specific order. Although the potency of these inhibitors was comparable for both 20S and 26S proteasomes, this order changed when the inhibitors were tested against other proteases. The drugs prevented hydrolysis of unmodified and ubiquitin-tagged proteins by 26S proteasomes as well as the turnover of intracellularly delivered ovalbumin. Cytosolic deposition of an antigenic peptide, not requiring any further processing, could bypass the block imposed by the drug and sensitize the cells for cytolytic attack. Cells cultured in the presence of these drugs rapidly loose their capacity to properly assemble class I molecules. This loss is a property reminiscent of defective class I complex formation in peptide transporter-deficient cells and is most likely due to the absence of a suitable pool of peptides. In agreement with this hypothesis, rapid dissociation of unstable class I complexes formed in the presence of the inhibitors (85), or in the absence of TAP (86, 87), could be prevented by inclusion of appropriate peptides in the cell lysates.

While these results are encouraging, the unknown effects of the inhibitors on other, yet to be identified, cytosolic proteases may complicate the interpretation of the results. A temperature sensitive (*ts*) mutant defective in ubiquitin conjugation was found less efficient in presenting cytosolically disposed ovalbumin to class I restricted T cells at the nonpermissive temperature (88), providing further evidence in support of ubiquitin-dependent protein degradation in class I restricted antigen presentation. However, conflicting results have been obtained with additional Ts ubiquitin-conjugating enzyme (E1) defective cell lines (89). Combined, most of these results support the notion that generation of most class I presentable peptides critically depends on the proteolytic activity of the proteasome.

Are Additional Proteases Involved in Antigen Processing?

Is the proteasome the sole protease (complex) involved in the generation of presentable peptides? If so, the residue in the protein sequence preceding the N-terminal amino acid of a class I restricted epitope displays the same preference as that seen for the C-termini generated by the proteasome. Elliott and colleagues determined the nature of the amino acid preceding the N-terminal residue in the native protein for 64 class I epitopes (36). When they compared the expected amino acid frequencies—based on the natural abundance of each amino acid—with the observed frequencies, the N-terminus seemed to result from random cleavage. This result suggests that other proteases, either of cytosolic origin or localized in the lumen of the ER, are involved in the processing of peptides (aminopeptidase or other endopeptidase activities for example).

In at least one documented case, proteolysis by an ER-resident protease leads to the generation of presentable epitopes. Peptides derived from ER signal sequences have been isolated from HLA-A2 class I molecules (90, 91). The signal sequence targets a protein for cotranslational translocation into the secretory pathway. During or after translocation of these proteins into the ER, the enzyme signal peptidase cleaves off the signal peptide, which is now free to bind to class I as such, or does so following further proteolysis. Binding of these peptides to HLA-A2 is independent of the expression of the TAP peptide transporter, confirming that these peptides are generated within the lumen of the ER itself.

It is also conceivable that the peptide transporter translocates peptides that are only approximately the size preferred by class I products, and further trimming is required to allow presentation. To explore this possibility, Eisenlohr et al analyzed the efficiency of intracellularly synthesized peptides that were too long to fit the groove of a class I molecule without further processing to sensitize cells for lysis by T cells (92). The presentation of a recombinant vaccinia virus-encoded peptide of 12 residues—the inappropriate size—was enhanced by simultaneous expression of a dipeptidyl-carboxy-peptidase (93) targeted to the ER. Since the expression of the protease was exclusively directed to the ER, the peptide must have been processed within the secretory pathway following its translocation. Thus, antigenic peptides can be cut to a size favored by MHC class I molecules provided that the right proteases are present within the ER.

To characterize the contribution of ER-resident proteases in antigen processing, Link et al constructed a recombinant vaccinia virus encompassing two MHC class I presentable epitopes in tandem (94). The resulting 17-residue peptide was preceded by a ER signal sequence to facilitate entry into the secretory pathway. Either epitope was efficiently presented in TAP-deficient cell lines but only when located at the COOH-terminal end of the tandem construct. While imprecision in signal sequence cleavage by signal peptidase (an unlikely occurrence) could lead to destruction of the N-terminal epitope, these data are consistent with the presence of amino peptidase activity within the ER. Whether such activity accounts for the seemingly random generation of the N-terminus of a presentable peptide (36) is not known. Using the same strategy, Elliott and colleagues obtained similar results, but in addition, they targeted large fragments of influenza nucleoprotein to the ER (residues 328–498) in which the presentable epitope (residues 366–374) was embedded somewhere in the middle. The 150-residue polypeptide, but not the full-length protein, was processed in the ER to the appropriate 9-residue epitope as recognized by influenza-specific CTLs. These observations not only indicate the presence of aminopeptidase activity but also suggest that endopeptidase and/or carboxypeptidase activity resides within the ER as well (T Elliott, A Willis, V Cerundolo, A Townsend, submitted).

Additional peptide trimming within the ER may take place prior to or after binding to class I molecules. The direct involvement in peptide generation of the MHC molecule itself was suggested by Rammensee and colleagues to explain their observation that specific T-cell epitopes could be isolated only from cells that expressed the appropriate MHC class I restriction element (12, 95, 96). However, once the peptide is firmly anchored in the peptide binding groove, it is likely to be protected from further degradation. Because peptides, having entered the ER, will be released back into the cytosol unless specifically retained (97–99), further trimming could equally well take place in the cytosol followed by TAP dependent uptake of the processed peptide and binding to class I molecules (100). This possibility does not apply for the observations of Link and Elliott, and their respective coworkers, because their experiments were conducted in TAP deficient cells.

THE TRANSPORTER ASSOCIATED WITH ANTIGEN PROCESSING

Structure of the Peptide Transporter

Peptides generated in the cytosol are translocated across the ER by a dedicated peptide transporter, which is known as the transporter associated with antigen presentation (TAP) (101–103) and is localized to membranes of the ER and *cis* golgi (104). The existence of such a transporter became apparent from studies of mutant cell lines in which class I assembly was impaired, seemingly owing to lack of suitable peptides (20, 21, 67, 105; for review, see 10). The mutation and deletions in these mutant cell lines mapped within the MHC, where two additional genes were soon identified: *TAP1* and *TAP2* (105–108). Each gene encoded a transporter-like protein, and upon transfection of *TAP1* and *TAP2* cDNAs, class I assembly was restored and the mutant cells were able to present antigens normally (109–111). These experiments established that both gene products were required for proper assembly of MHC class I molecules.

A most interesting discovery is the occurrence of a TAP mutation in humans, observed in two severely immunodeficient patients. As in cell lines, this TAP defect results in a marked reduction in class I expression (112). *TAP1* and *TAP2* are members of the ATP–binding cassette(ABC)–transporters (113). They are more homologous to one another than to other transporters in this family and show strong sequence conservation among mammals (e.g. mouse, rat, and human). Both protein products, like other ABC transporters, contain the so-called Walker A and Walker B motifs (GX$_4$GKS/T and RX$_{6\text{-}8}$(hydrophobic)$_4$D respectively) typical for a nucleotide binding site (114). Hydrophobicity plots indicate the presence of a hydrophilic domain containing the ATP-binding site and multiple membrane spanning segments (6–10). Although the exact number of the transmembrane regions and the topology of TAP have not yet been experimentally verified, the nucleotide binding site is most likely located in the cytosol; the hydrophobic regions are embedded in the ER membrane. The presumed transmembrane segments contain polar residues possibly indicating that these domains could form a hydrophilic pore through which peptides can be transported. Antisera raised against the domain that contains the nucleotide binding site inhibit peptide translocation across microsomal membranes (see below) in agreement with a cytosolic disposition of this segment (99).

Based on the structure of other members of the ABC-transporter proteins, which contain 2 ATP binding sites and 12–16 transmembrane segments [e.g.the multidrug resistance transporter (MDR) and Sterile 6 (STE6)] (113), a TAP1/TAP2 heterodimer seemed a plausible functional complex. Antibodies raised against synthetic peptides of the COOH-terminal part of TAP1 did indeed co-immuneprecipitate TAP2, establishing at least a physical association between the two proteins (115, 116). Until the stochiometry of this complex is established, the transporters may function in multimeric complexes containing multiple copies of TAP1 and TAP2. Similarly, the existence of additional, yet to be identified, subunits essential for TAP function remains a possibility.

Polymorphism of the Peptide Transporter

The very location of the *TAP* genes within the MHC has generated considerable discussion of, and interest in, *TAP* polymorphism. Several diseases are associated with the MHC (117). As most of the HLA genes are in linkage disequilibrium, the assignment of disease susceptibility to individual genes in the HLA region is complex. Many of these diseases are thought to have an autoimmune component to their etiology and could result from an inappropriate response against MHC molecules complexed with self peptides. The identification and characterization of new genes in the MHC (e.g. *TAP*, *LMPs*) could conceivably result in disease associations that are stronger with these novel genes than with their classical neighbors. The notion that *TAP* can affect class I restricted antigen presentation puts it high on the list of potential disease susceptibility genes. Moreover, polymorphism in *TAP1* and *TAP2* may alter the function of their gene products, influencing the set of peptides that are made available for class I restricted antigen presentation in an allele-specific manner, with obvious implications for HLA-associated diseases. At present, the reported sequence variations for human and mouse *TAP1* and *TAP2* are limited and there is no persuasive evidence for their involvement in predisposition towards an HLA-linked disease.

For the mouse, the identification of restriction fragment length variants led to the assignment of four *TAP1* and seven *TAP2* alleles (118). The extent of sequence variation for these alleles is limited (J. Monaco, personal communication). Single-strand conformation analysis (SSCP) and DNA sequencing identified four possible human *TAP1* alleles and eight possible *TAP2* alleles. PCR-based screens of an extended panel of homozygous typing cell lines and Caucasoid controls revealed the occurrence of three *TAP1* and five *TAP2* alleles (119–122). One of the allelic variations found in the hydrophobic cores of both TAP1 and TAP2 (position 333 and 379 respectively: Ile → Val) alters the substrate specificity of the multi-drug resistance–associated transporter, P-glycoprotein (123), but at present, functional polymorphism of the transporter has been observed only for the rat (97, 124) and, more recently, the Syrian hamster (M Lobigs, RV Blanden & A Müllbacher, submitted).

Long before the molecular identification of peptide transporters, an unusual observation was made in the rat. A polymorphic locus in the MHC class II region was able to alter the antigenic and biochemical properties of a rat class I molecule, RT1A[a] (125–127). Two variants of this so called class I modifier locus (*cim*-locus) were identified, *cim*[a] and *cim*[b]. In the presence of the *cim*[a] allele, the RT1A[a] class I molecule is rapidly transported to the cell surface. In the presence of the *cim*[b] allele, transport of RT1A[a] is impaired and recognition by certain T cells is abolished. After the identification of the *TAP1* and *TAP2* genes within the *cim* locus (106), an elegant series of experiments by Powis et al established that allelic variation in rat *TAP2* caused the altered behavior of the RT1A[a] class I molecule (127). Four rat TAP2 alleles identified by sequence analysis can unambiguously be assigned as *cim*[a] or *cim*[b] because 25 out of 29 amino acid substitutions correlate with *cim*-type. Peptides associated with the RT1A[a] class I molecules synthesized by cells of *cim*[a] or *cim*[b]-type show different elution profiles on HPLC in that the peptides provided by the *cim*[a] background appear more hydrophilic. These results established that allelic variation in the transporter genes could alter the repertoire of peptides bound by class I. Eighteen of the 25 *cim*-correlated amino acid substitutions were nonconservative and all but two were located in the hydrophobic, presumed transmembrane-spanning domains (120). These domains may form a channel in which sequence alterations can alter the amino acid sequence specificity.

Peptide Translocation Assay Systems

Essentially two types of assay have been used to study TAP-dependent transport across the ER membrane. In one assay system, lymphocytes are permeabilized with the bacterial toxin streptolysin O to allow delivery of radiolabeled peptide substrates to the cytosol and from there into the lumen of the ER. The alternative assay exploits translocation of radio-labeled peptide across microsomal membranes prepared from mouse or rat livers. The key tool in establishing peptide transport assays was the TAP-negative control. Transporter-deficient cell lines (66, 128) have been indispensable to an understanding of the TAP dependency of the translocation process: The difference in peptide transport observed in TAP-transfectants and their TAP-deficient parent represents the extent of peptide transport mediated by TAP. Likewise, the availability of TAP1-deficient animals, generated by homologous recombination (129), was essential for establishing the conditions in which TAP-dependent peptide translocation into liver microsomes could be measured. For mouse and rat, data from the two systems can be readily compared, because the TAP-deficient cell line has been transfected with cDNAs encoding the human, rat, and murine transporters (130). Attempts at using microsomes from lymphoid cells to explore TAP-dependent transport have met with results that do not easily fit those obtained in the other systems (131). In lymphocyte-derived microsomes, peptide gained access to class I binding sites in the absence of added ATP. Proteolytic treatment of the microsomes supposedly destroyed most if not all extra-luminal protein domains, yet the treatment did not affect the entry of peptide into the lumen of the vesicles. Microsomes from TAP-deficient cells did not behave any differently from control microsomes. Peptide translocation has also been explored in canine pancreatic microsomes (132, 133), again with results that do not readily agree with those obtained in liver microsomes or permeabilized cells, in particular where the requirement for added ATP is concerned. Peptide translocation was observed in the absence of added ATP. Some, but not all, minigene-encoded peptides expressed in peptide transporter deficient cells can be presented to specific T cells (134). We will focus on the major route of peptide translocation across the ER membrane and discuss those systems in which consistent TAP-dependent results have been obtained.

During the course of experimentation, it became apparent that, for translocation to be readily measured, a retention device for translocated peptides must be present in the lumen of the ER to prevent efflux of the translocated peptides (97–99). This function can either be a class I molecule capable of interacting with the translocated peptide or the attachment of an N-linked oligosaccharide to the peptide on the luminal side of the ER. Retention of radiolabeled peptides on class I limits the set of substrates that can be tested in a translocation assay to those capable of interacting with class I. Because, in these assays, peptide translocation is determined by two factors, substrate specificity of TAP and affinity of the peptide for class I molecules, a meaningful comparison of different peptide substrates is precluded. The same argument applies to competition assays for uptake of a radiolabeled class I epitope as a means to establish the specificity of TAP. While this type of assay is perhaps closest to the physiological situation, its obligate linkage to the binding specificity of class I molecules is a limitation.

The second peptide trap relies on the observation that short peptides with a glycosylation consensus sequence (Asn-X-Thr/Ser and X≠Pro) can be glycosylated in the lumen of microsomal vesicles (135, 136). A lumenally disposed lipid-linked oligosaccharide with the composition (GlcNAc)2(Man)9(Glc)3 is transferred *en-bloc* onto the peptide. The bulk of this oligosaccharide allows peptides to be retained in the lumen of the microsomes (97, 132, 133). Although the necessity for a glycosylation site imposes some sequence constraints on the reporter peptides, numerous peptides can be tested in direct translocation assays. Provided that competitor peptides do not contain a glycosylation site, competition studies can be used for comparison of substrates and to determine the affinity of the transporter for naturally occurring epitopes. Again, competition studies should be interpreted with caution: Peptides need not be translocated across the membrane for competition to be apparent.

One formal issue deserves to be mentioned. The evidence that the available transport systems truly measure TAP-dependent peptide translocation is indirect and based largely on genetic arguments. TAP is involved, but the absence of a reconstituted system containing no other components except the TAP subunits leaves open the formal possibility that the TAP subunits, rather than translocating peptides themselves, facilitate transport mediated by other gene products. In view of the structural attributes of TAP and the results obtained for peptide transport by allelic variants of the transporter (see below), we consider this possibility unlikely.

Peptide Translocation Is ATP Dependent and Substrate Specific

Three groups almost simultaneously demonstrated ATP-dependent peptide transport mediated by the TAP peptide translocator. Shepherd and colleagues showed that in the presence of ATP, peptides could accumulate in microsomal vesicles derived from wild-type but not TAP knock-out mice (103). Both Androlewicz et al (101) and Neefjes et al (102) showed ATP-dependent transport in semi-intact cells. Whereas the former used class I binding of iodinated substrates (101), the latter utilized glycosylation acceptor peptides (102) as reporter substrates. Hydrolysis of ATP was required for peptide translocation: In the presence of a non-hydrolyzable analog no translocation was observed

(102, 103). The processes of peptide binding and peptide translocation could be distinguished by using a crosslinkable peptide. In the absence of ATP no translocation was observed, but under similar conditions the peptide could still be crosslinked to TAP (137). Thus, ATP hydrolysis is required for the translocation process itself rather than for peptide binding to the translocator. Elements of both TAP1 and TAP2 contribute to formation of the peptide binding site (138, 139), and accordingly crosslinkable peptides were found complexed with both subunits (138).

The structure of a class I molecule, and of its binding groove in particular, does not allow all possible peptides to bind and to be presented. Class I binding peptides have one common denominator: They all anchor in the peptide binding cleft with their COOH-terminal amino acid side chain in the F-pocket. If the peptide transporter can select a peptide by virtue of its C-terminal amino acid side chain, the efficacy of antigen presentation may be enhanced by a coordinated specificity of the proteins involved in antigen presentation (class I, the proteasome, and the peptide transporter).

Substitution analysis of the COOH-terminal amino acid of a peptide transporter substrate revealed that the identity of the COOH-terminal amino side chain is an important element not only for class I binding but also in the selectivity of the peptide transport process. Whereas the F-pockets of human class I alleles accommodate basic and hydrophobic amino acid side chains, the mouse class I molecules prefer hydrophobic residues in the F-pocket. Accordingly, the human transporter appears to be more relaxed in its C-terminal preference and can accommodate a variety of C-terminal residues (130, 137); the mouse TAP complex prefers Ile, Leu, Val, Met, Cys, and the aromatic residues Trp, Phe, and Tyr as C-termini (99, 130). High-affinity peptide binding to class I molecules requires free NH_2- and COOH-peptide termini. Peptide transport is reduced by acetylation or amidation of the peptide N- or C-terminus respectively (99, 130). The exception is H-2 M3[a], a nonclassical class I molecule that binds N-formylated peptides of mitochondrial and bacterial origin (140) and is recognized by CTLs in a TAP-dependent fashion (111).

Polymorphism of rat TAP complexes changes the set of peptides translocated by the peptide transporter (97, 130). The *cim*[a] variant is least restrictive in terms of COOH-terminal preference and has a specificity similar to the human transporter. The *cim*[b] transporter allele transports only a subset of peptides translocated by *cim*[a], preferring those with an aromatic or hydrophobic COOH-terminus; *cim*[b] thereby matches the specificity of the mouse translocator. Moreover, *cim*[a] selects from a complex peptide library (2304 peptides) with diverse COOH-termini, a more hydrophilic set of peptides than does *cim*[b]. Thus, as a consequence of a differential preference of *cim*[a] and *cim*[b] for COOH-terminal residues of peptides (97), the rat class I molecule RT1A[a] binds a different set of peptides when served by one or the other transporter allele. Using a similar approach, no functional polymorphism could be demonstrated for the mouse transporter in six different H-2 haplotypes, including that of the NOD mouse (141).

There is an emerging consensus that a transport-compatible COOH-terminus is necessary but not sufficient for a peptide to be translocated (97, 130; MTh Heemels & HL Ploegh, manuscript submitted). Other residues contribute to the specificity of the translocation process but apparently not in a species-specific manner (130). The residues less favored as the N-terminus are equally poor for all species examined. Nevertheless, the cim^a transporter selects a population of peptides from a library of 2304 peptides, closely resembling the starting material (97). The contribution of other residues, at least for the rat transporters, is possibly secondary to the identity of the COOH-terminus.

Length Preference of the Peptide Transporter

Class I molecules can accommodate peptides longer than nine residues (90, 91, 142, 143). A subset of HLA-A2 class I molecules associates with fragments of signal sequences. Cleavage of the fragments occurs in the lumen of the ER. These longer peptides are also associated with HLA-A2 molecules in TAP-deficient cells (90, 91, 144). However, a subset of HLA-B27 molecules, identified by their preferential reactivity with a monoclonal antibody MARB4, accommodates peptides heterogeneous in size, ranging from 8 to 33 amino acids, in what appears to be a TAP-dependent manner (143). Does the peptide transporter translocate peptides over 30 residues in length?

Competition studies (97, 99, 137) and direct translocation assays (98) have shown that the transporter prefers peptides of approximately the size usually found associated with class I molecules. The efficiency of translocation drops significantly for substrates shorter than 8 and longer than approximately 14 amino acid residues. Both permeabilized cells and microsomal preparations contain proteases, which readily cut the longer input peptides to size. Although longer peptides compete for uptake of a reporter substrate, the peptide transporter selectively translocates degradation products rather than the intact peptide (98; MTh Heemels & HL Ploegh, submitted). The simplest explanation for the discrepancy between the results obtained for HLA-B27 and the length preference of TAP is that longer peptides are occasionally transported. However, whether the HLA-B27 molecules that accommodate the longer peptides are expressed at the cell surface is unclear. The HLA-B27 molecules at the cell surface may represent the molecules in stable association with short peptides only, whereas the molecules carrying polypeptides up to 33 amino acids may be retained within the cell. The TAP-dependency of the presentation of longer peptides was assessed solely by lack of surface expression of the MARB4 reactive subset of HLA-B27 molecules.

Polymorphism in TAP can also affect the length preference of a transporter as was shown for the two allelic variants of the rat TAP. The cim^b transporter, most limited in the COOH-terminal residues it can accommodate, appears to be more relaxed in terms of peptide length than cim^a (MTh Heemels & HL Ploegh, submitted). This difference in length selectivity becomes apparent for peptides of around 12–14 residues in length, where the cim^b transporter is almost twice as effective as cim^a. From a mechanistic point of view, these observations raise the possibility of a compensatory trait. Less specific interactions with peptide C-terminal residues might put more constraints on other recognition elements, one of which might be proper positioning of the N-terminus.

The peptide-generation machinery, the transporter and the structure of the class I peptide-binding groove itself, impose sequence constraints on the epitopes presented to cytotoxic T cells. Substrate selectivity apparently has a hierarchy. The proteases generate a more diverse set of peptides than is actually transported, and a more diverse array of peptides is transported than will bind to class I. Even though the mouse TAP-complex appears to be highly selective as compared to the human TAP-complex, it does not seem to drastically limit peptide binding to the mouse class I molecules it serves, because the specificities of mouse TAP and mouse class I are matched. The situation is different for the recombinant rat strains where transporter specificity and class I specificity may not always be matched. The cim^b transporter allele clearly restricts antigen presentation via the RT1A^a class I molecule. However, wild rats that express this particular class I allele invariably carry the least restrictive cim^a allele. Laboratory-bred rats in which the class I locus was dissociated from its usual transporter by recombination are the exception. Thus, although the TAP complex limits the spectrum of peptides available in the ER, natural selection may have resulted in a situation that avoids major restriction by the transporter by matching the specificity of the class I molecules and the transporters that serve them.

A LINK BETWEEN TAP AND CLASS I

The TAP complex interacts physically with β_2m-heavy chain complexes (145, 146). Dissociation of the transporter from the class I molecules coincides with the exit of the complexes from the ER and, at least in vitro, this dissociation can be induced by the addition of peptides. The transporter binds to preassembled complexes only; no association between the transporter and free heavy chains could be detected. This interaction appears to be mediated by the TAP1 subunit (138). These observations suggest the possibility of metabolic channeling for peptide loading of class I molecules. By putting the newly synthesized class I molecule in close proximity of TAP, the deposition of peptide in the peptide binding groove of class I may be facilitated and dilution, efflux, and degradation of translocated peptides counteracted. The TAP-class I interaction may somehow trigger folding step(s) of class I complexes, yet several observations indicate that the presence of TAP is not a prerequisite for folding. Although class I molecules predominantly reside in the ER in the absence of TAP-dependent peptides, properly folded complexes can accumulate at the cell surface, provided that their structure is stabilized. Stabilization of class I by the addition of appropriate peptides (20), anti-class I antibodies to the culture medium (147), or cell culture at reduced temperature (21) enhances the cell surface expression of properly conformed class I molecules. A peptide preceded by a signal sequence crosses the membrane without the need of a peptide transporter and is efficiently presented by TAP-deficient cells to specific T-cells (148, 149). Moreover, peptides generated at the site of class I assembly (signal sequence-derived peptides) in TAP-deficient cells can associate with HLA-A2 molecules in a stable manner (90, 91). D^b molecules equipped with a glycosyl-phosphatidyl-inositol (GPI) membrane anchor cannot efficiently present endogenously synthesized antigens, whereas these molecules can present exogenously added peptides. Although folding is not impaired for the GPI-anchored form of D^b, peptide loading is impaired (G Waneck, personal communication). Should class I molecules link to TAP via their transmembrane segment, the association is likely to be perturbed when the transmembrane segment is replaced by a GPI anchor. However, peptide presentation of the nonclassical, GPI-anchored class I molecule Qa-2 depends on peptides supplied by TAP (150). Under normal circumstances, Qa-2 molecules may either be capable of associating with TAP or can be loaded with peptide even when not in close proximity of the peptide transporter. Clearly, more work is required to establish a physiological role and the molecular details for TAP–class I interactions. An alternative explanation for the presentation defect of the GPI-linked D^b molecules is a disturbed interaction with its chaperones; consequently, the class I molecules would not be retained long enough in the ER to allow peptide binding to occur.

CHAPERONES IN MHC CLASS I ASSEMBLY

Chaperones in the Endoplasmic Reticulum

Although peptides can bind to free heavy chains (151–153), most newly synthesized class I heavy chains fold, associate with β_2m, and then bind peptide (8, 154, 155). Although assembly can be reconstituted in vitro from purified subunits, newly synthesized polypeptides must complete folding and assembly before they progress to the cell surface. This process is thought to be mediated by transient interactions with molecular chaperones that reside in the lumen of the ER, such as BiP, GRP94, and calnexin (156–158). The best characterized chaperone implicated in the early assembly events of class I is calnexin, an ER-resident membrane protein of 64 kDa with an Mr 88,000 in mouse (p88) and 90kD in human (IP90) (159, 160). Calnexin has been shown to interact with a variety of membrane-bound and soluble glycoproteins and is thought to play a role in the retention of incompletely assembled oligomeric structures within the ER (161–165).

A Role for Calnexin in MHC Class I Assembly

Calnexin associates with newly synthesized class I heavy chains during, or soon after, its translocation into the ER (161, 166). In the absence of β_2m, assembly of class I complexes cannot be completed, and consequently, the heavy chains remain tightly bound to calnexin until they are degraded. In the presence of all class I subunits heavy chains, β_2m, and peptides, the interaction with calnexin is transient. The sequence of events between the initial association of calnexin with class I heavy chains and the final stages, in which the properly folded complexes egress from the ER, is not entirely clear. What induces dissociation of class I from calnexin?

In the mouse cell line RMA-S that expresses a mutant (frame shift) *TAP2*, calnexin remains associated with the empty class I heavy chain-β_2m dimer; calnexin dissociates at a rate similar to the slow rate of transport of class I molecules to the cell surface observed in these cells (167). Empty class I molecules, under certain conditions, can reach the cell surface, which indicates that the retention of peptide-deficient class I molecules by calnexin is not absolute. *Drosophila melanogaster* cells transfected with mouse class I subunits in the presence and absence of canine calnexin provided a clear answer (168). Because *D. melanogaster* lacks the machinery to process and transport peptides (K Früh, MR Jackson, P Sempé, et al, submitted) these cells produce large quantities of empty class I molecules. These peptide-deficient class I molecules are transported much more rapidly than their counterparts in mouse TAP-deficient cells, suggesting that the retention mechanism(s), present in mammalian cells is not operational in *Drosophila* (169). Coexpression of canine calnexin with the mouse heavy chain and β_2m not only accelerated the assembly of the class I complex, but its presence also drastically reduced the transport rate of these complexes to the cell surface. Comparison of the rate at which the MHC class I complexes dissociated from calnexin and their rate of ER-to-golgi transport revealed that the two processes occurred with comparable kinetics and resembled the kinetics seen in TAP-deficient mouse cells. Thus, canine calnexin aids assembly and retains peptide-deficient mouse heavy chain–β_2m complexes within the cell. Such a retention system largely prevents expression of empty class I molecules at the cell surface, thereby avoiding an undesirable situation in which antigenic peptides could bind at the cell surface and sensitize an innocent bystander cell for lysis. Heavy chain–β_2m complexes associate not only with calnexin but also with the peptide transporter; both appear to dissociate upon peptide binding (145, 146, 167). Although the simultaneous association of TAP and calnexin with class I molecules has not yet been observed, the two proteins could certainly work in conjunction (146).

The results obtained for class I molecules in cell lines of human origin do not easily agree with those obtained for mouse subunits expressed in mouse, or *D. melanogaster,* cells. In human cell lines, the dissociation of calnexin from the heavy chain seems independent of the expression of any peptide transporter (170). The association of β_2m with the heavy chain, rather than with peptide binding, seems to trigger the release of the chaperone. In the scheme proposed by Sugita & Brenner (170) the heavy chain dissociates from calnexin upon β_2m binding, giving rise to an instable folding intermediate that can be stabilized subsequently by peptide. Human cells express reduced levels of class I molecules at the cell surface if peptides supplied by the TAP are lacking. If β_2m, rather than peptides, induces dissociation of the heavy chain from calnexin, an additional retention molecule (e.g. BiP or another chaperone) may prevent expression of peptide-deficient class I molecules. Whereas in mouse cells the chaperone BiP is thought to play a minor role, in human cells both calnexin and BiP can appear in a complex with class I molecules (167).

Unlike their human counterparts, mouse class I molecules expressed in TAP-deficient human cells are readily transported to the cell surface in stable association with human β_2m (171, 172). These molecules, although seemingly devoid of peptide, are not known to be retained by any chaperone. Domain swapping experiments have indicated that this species-specific retention is dictated by the α1 and α2 domains of the human class I molecules (173). Either the association of mouse class I heavy chain with human β_2m imposes a proper conformation to pass quality control or mouse heavy chains are intrinsically less capable of interacting with human chaperones. In contrast, human heavy chain and β_2m expressed in TAP-deficient mouse cells are largely retained within the ER (172, 174). Therefore, human and mouse class I molecules and their interaction with chaperones may have species-specific differences. The expression of the appropriate combinations of calnexins and heavy chain-β_2m complexes in *D. melanogaster* cells should clarify these issues.

Calnexin and Its Substrates: How Do They Interact?

One hypothesis proposes a role for N-linked oligosaccharide trimming as a means for calnexin to monitor the protein's folding state. First, calnexin binds to many, if not all, soluble and membrane glycoproteins (164). With the exception of the CD3 ε T-cell receptor subunit expressed in COS cells (165), calnexin has not been found attached to nonglycosylated proteins. Serum albumin is not glycosylated, but the closely related α-fetoprotein is glycosylated; whereas the former does not associate with calnexin, the latter protein does. Second, upon treatment with tunicamycin, cells lose their capacity to synthesize and transfer core oligosaccharides, and concurrently the binding of any newly synthezised protein to calnexin is blocked (164). Also, glucosidase inhibitors that prevent the cleavage of glucose from the N-linked oligosaccharides interfere with the binding of calnexin, suggesting that the glucoses must be trimmed for recognition by calnexin to occur. Complete deglucosylation abolishes protein interaction with the chaperone (175). Thus, calnexin interacts with proteins whose oligosaccharides contain more than one but not all glucose residues.

Although at present no evidence supports a direct interaction of calnexin with the glucose-containing N-linked oligosaccharides, Hammond et al (176) have proposed that, following glycosylation, the newly synthesized proteins enter a deglucosylation-reglucosylation cycle until folding is completed and the polypeptide is released from calnexin. Unless one assumes that calnexin has multiple ways of interacting with molecules in the process of folding, this model does not easily agree with observations in support of a role for the transmembrane region of the protein substrate in its interaction with calnexin (15). Deletion of the transmembrane region of the class I heavy chain or replacement of this domain by a GPI anchor abolished a crosslinkable interaction with calnexin. Furthermore, many glycoproteins readily fold in the presence of either tunicamycin treatment or glucosidase inhibitors (177). Mutant class I molecules from which the glycosylation sites have been deleted are normally transported to the cell surface in the presence of TAP and retained in its absence (173). The latter observations suggest that glucose modifications are neither essential for proper folding nor for retention of class I molecules within the ER. Calnexin can interact with a large variety of seemingly unrelated polypeptides—both secretory and transmembrane proteins. The changes in conformation that are likely to occur in class I molecules upon peptide binding must be rather subtle, and the relationship of such conformational change to maturation of a secretory glycoprotein such as α1-antitrypsin is not immediately apparent. The precise role of calnexin in class I folding and assembly, and its mode of interactions, remain unclear; involvement of other proteins in the early stages of protein folding in the ER must be considered.

Other Chaperones Involved in Antigen Presentation

In addition to calnexin and class I-TAP interactions, other accessory proteins aid in the assembly of peptide loaded class I molecules. Townsend and coworkers observed a 105 kD molecule that could occasionally be recovered in complex with mouse class I molecules (87). The identity and the significance of this protein were never established, but it may be similar to one of the two proteins subsequently described by Feuerbach & Burgert (178). Two proteins with a molecular weight of 100 and 110 kD could easily be coprecipitated with class I complexes in association with a third, adenovirus-derived, protein E3/19K (178). Upon close examination, the presence of p100 and p110 could also be revealed in the absence of E3/19K, and both proteins could be displaced by the addition of peptides. To date, no further progress has been reported.

A more provocative suggestion has been put forward for grp94/gp96, an ER-resident stress protein (179). grp94 isolated from several tumor cell lines can be used as an immunogen that will protect animals from a subsequent challenge with the tumor from which the grp94 preparation was derived but not from a challenge with an antigenically different tumor. These observations suggested that grp94 would be aberrant in tumors, yet sequencing did not reveal any tumor-specific mutations (For review see 180). Sequence analysis and biochemical data suggested that grp94 is an ATP binding and hydrolyzing entity capable of binding peptides (181). Since CD8$^+$ T cells appeared to be responsible for the immunity elicited by grp94, peptides complexed with grp94 might somehow be transferred onto class I molecules in the responding animal. A protein that could facilitate peptide loading onto class I has been proposed (87, 131, 171, 181), and Li & Srivastava recently suggested that grp94 might be that protein (182). Although no direct evidence in favor of this suggestion has yet been presented, the peptide-binding properties of grp94 and its role in antigen presentation are presently being explored in several laboratories.

CONCLUDING REMARKS AND FUTURE DIRECTIONS

The generation of antigenic peptides, assembly of MHC products, and transport of class I molecules to the cell surface will continue to attract attention; many questions remain. Presentable peptides are likely to originate from proteins that are degraded by the proteasome. However, the rules that underlie proteolysis in the cytosol and the factors that regulate this process are still poorly defined. The derivation of the epitope's N-terminus is puzzling and may be a result of additional trimming by (yet-to-be-identified) proteases located in the cytosol, the ER, or both. The mechanism of action of the peptide transporter is not understood at all. Although (some of) the component parts have been identified and elements of its specificity are now understood, there is no plausible mechanistic model for translocation. Exchange or deletion of (parts of) domains and site-directed mutagenesis coupled with functional studies could help elucidate the correlation between structure and function. The significance of the interaction between TAP with MHC class I molecules is not understood, and our knowledge of the role of accessory proteins—new members of which continue to be identified—in folding and peptide loading of MHC class I molecules is limited. MHC class I molecules will continue to provide useful information for many of the questions that pertain to folding and the involvement of molecular chaperones. The structure of class I products is well established and numerous reagents are available to monitor their distinct conformational states.

The significance of certain steps in antigen presentation may be best highlighted by the way in which viruses try to evade it. Interference with MHC class I cell surface expression can abolish detection by the immune system of virally infected or transformed cells and thus benefit viral replication or uncontrolled cell growth (183). A priori, all aspects of class I assembly and trafficking could be acted upon to evade immune surveillance. Down-regulation of expression of MHC class I (184, 185) or TAP (186, 187), interference with folding through effects on chaperones, intracellular retention of MHC class I complexes (188–192), inhibition of peptide transport (A Hill, in preparation), or antigen processing, all provide ways to avoid destruction by a T-cell attack. Furthermore, mutations in the region of the viral protein that provides an antigenic peptide could decrease its recognition by T cells, ablate binding to class I molecules altogether, reduce their ability to be recognized by the transporter, or alter proteolytic cleavage and hence availability of the epitope. The selective pressures exerted by the host defense during millions of years of evolution have driven many viral pathogens to evolve inventive immune evasion strategies. Studying their behavior is likely to further our knowledge on class I restricted antigen presentation and may identify the chinks in their armor.

ACKNOWLEDGMENTS

We acknowledge helpful discussions with members of the laboratory and thank colleagues for sending preprints. Supported by NIH grant AI 33456.

Literature Cited

1. Buus S, Sette A, Colon SM, Jenis DM, Grey HM. 1986. *Cell* 47:1071–77
2. Townsend ARM, Rothbard J, Gotch FM, Bahadur G, Wraith D, McMichael AJ. 1986. *Cell* 44:959–68
3. Yewdell JW, Bennink JR. 1992. *Adv. Immunol.* 52:1–123
4. Braciale TJ, Morrison LA, Sweetser MT, Sambrook J, Gething MJ, Braciale VL. 1987. *Immunol. Rev.* 98:95–114
5. Neefjes JJ, Schumacher TN, Ploegh HL. 1991. *Curr. Opin. Cell. Biol.* 3:601–9
6. Unanue ER. 1992. *Curr. Opin. Immunol.* 4:63–69
7. Cresswell P. 1994. *Annu. Rev. Immunol.* 12:259–93
8. Bijlmakers M-J, Ploegh HL. 1993. *Curr. Opin. Immunol.* 5:21–26
9. Jackson MR, Peterson PA. 1993. *Annu. Rev. Cell Biol.* 9:207–35
10. DeMars R, Spies T. 1992. *Trends Cell. Biol.* 2:81–86
11. Römisch K. 1994. *Trends Cell Biol.* 4:311–14

12. Rammensee HG, Falk K, Rötzschke O. 1993. *Annu. Rev. Immunol.* 11:213–44
13. von Boehmer H. 1994. *Cell.* 76:219–28
14. Möller G, ed. 1993. *Immunol. Rev.* 133: 1–240
15. Margolese L, Waneck GL, Suzuki CK, Degen E, Flavell RA, Williams DB. 1993. *J. Biol. Chem.* 268:17959–66
16. Bjorkman PJ, Saper MA, Samraoui B, Bennett WS, Strominger JL, Wiley DC. 1987. *Nature* 329:506–12
17. Bjorkman PJ, Saper MA, Samraoui B, Bennett WS, Strominger JL, Wiley DC. 1987. *Nature* 329:512–18
18. Blue ML, Craig KA, Anderson P, Branton KR, Schlossman SF. 1988. *Cell* 54:413–21
19. Salter RD, Norment AM, Chen BP, Clayberger C, Krensky AM,, et al. 1989. *Nature* 338:345–47
20. Townsend A, Öhlen C, Bastin J, Ljunggren HG, Foster L, Kärre K. 1989. *Nature* 340:443–48
21. Ljunggren HG, Stam NS, Öhlen C, Neefjes JJ, Höglund P, et al. 1990. *Nature* 346:476–80
22. Gomez B, Jones EA, Barnstable CJ, Solomon E, Bodmer WF. 1978. *Tissue Antigens* 11:96–112
23. Rein RS, Seemann GHAS, Neefjes JJ, Hochstenbach FMH, Stam NJ, Ploegh HL. 1987. *J. Immunol.* 138:1178–83
24. Koller BH, Marrack P, Kappler JW, Smithies O. 1990. *Science* 248:1227–30
25. Zijlstra M, Bix M, Simister NE, Loring JM, Raulet DH, Jaenisch R. 1990. *Nature* 344:742–46
26. Allen H, Fraser J, Flyer D, Calvin S, Flavell R. 1986. *Proc. Natl. Acad. Sci. USA* 83:7447–51
27. Shawar SM, Vyas JM, Rodgers JR, Rich RR. 1994. *Annu. Rev. Immunol.* 12:839–80
28. Garrett T, Saper M, Bjorkman P, Strominger J, Wiley D. 1989. *Nature* 342:692–96
29. Stern LJ, Wiley DC. 1994. *Structure* 2:245–51
30. Schumacher TNM, De Bruin MLH, Vernie LN, Kast WM, Melief CJM, et al. 1990. *Nature* 350:703–6
31. Bouvier M, Wiley DC. 1994. *Science* 265:398–402
32. Fremont DH, Matsumura M, Stura EA, Peterson PA, Wilson IA. 1992. *Science* 257:919–27
33. Zhang W, Young ACM, Imarai M, Nathenson SG, Sacchettini JC. 1992. *Proc. Natl. Acad. Sci. USA* 89:8403–7
34. Madden DR, Gorga JC, Strominger JL, Wiley DC. 1991. *Nature* 353:321–25
35. Matsumura M, Fremont D, Peterson PA, Wilson IA. 1992. *Science* 257:927–34

36. Elliott T, Smith M, Driscoll P, McMichael A. 1993. *Curr. Biol.* 3:854–65
37. Klausner RD, Sitia R. 1990. *Cell* 62:611–14
38. Townsend ARM, Bastin J, Gould K, Brownlee GG. 1986. *Nature* 42:575–77
39. Seglen PO, Bohley P. 1992. *Experientia* 48:158–72
40. Morrison LA, Lukacher AE, Braciale VL, Fan D, Braciale TJ. 1986. *J. Exp. Med.* 163:903–21
41. Carbone FR, Bevan MJ. 1990. *J. Exp. Med.* 171:377–87
42. Rock KL, Gamble S, Rothstein L. 1990. *Science* 249:918–21
43. Kovacsovics-Bankowski M, Clark K, Benacerraf B, Rock KL. 1993. *Proc. Natl. Acad. Sci. USA* 90:4992–5046
44. Pfeifer JD, Wick MJ, Roberts RL, Findlay K, Normark SJ, Harding CV. 1993. *Nature* 361:359–62
45. Cerundolo V, Elliott T, Elvin J, Bastin J, Townsend A. 1992. *Eur. J. Immunol.* 22:2243–48
46. Chicz RM, Urban RG, Lane WS, Gorga JC, Stern LJ, et al. 1992. *Nature* 358:764–68
47. Chen BP, Madrigal A, Parham P. 1990. *J. Exp. Med.* 172:779–88
48. Moore MW, Carbone FR, Bevan MJ. 1988. *Cell* 54:777–85
49. Yewdell JW, Bennink JR, Hosaka Y. 1988. *Science* 240:637–40
50. Harding CV. 1992. *Eur. J. Immunol.* 22:1865–69
51. Townsend A, Bastin J, Gould K, Brownlee G, Andrew M, et al. 1988. *J. Exp. Med.* 168:1211–24
52. Sweetser MT, Morrison LA, Braciale VL, Braciale TJ. 1989. *Nature* 342:180–82
53. Whitton JL, Oldstone MBA. 1989. *J. Exp. Med.* 170:1033–38
54. Driscoll J, Finley D. 1992. *Cell* 68:823–25
55. Goldberg AL, Rock KL. 1992. *Nature* 357:375–79
56. Rivett J. 1993. *Biochem. J.* 291:1–10
57. Murakami Y, Matsufuji S, Kameji T, Hayashi S, Igarashi K, et al. 1992. *Nature* 360:597–99
58. Seufert W, Jentsch S. 1992. *EMBO J.* 11:3077–80
59. Brown MG, Driscoll J, Monaco JJ. 1991. *Nature* 353:355–57
60. Glynne R, Powis SH, Beck S, Kelly A, Kerr L-A, Trowsdale J. 1991. *Nature* 353:357–60
61. Kelly A, Powis SH, Glynne R, Radley E, Beck S, Trowsdale J. 1991. *Nature* 353:667–68
62. Martinez CK, Monaco JJ. 1991. *Nature* 353:664–67

63. Ortiz-Navarrete V, Seelig A, Gernold M, Frentzel S, Kloetzel PM, Hämmerling GJ. 1991. *Nature* 353:662–64
64. Yang Y, Waters JB, Früh K, Peterson PA. 1992. *Proc. Natl. Acad. Sci. USA* 89:4928–32
65. Parham P. 1990. *Nature* 348:674–75
66. DeMars R, Rudersdorf R, Chang C, Petersen J, Strandtmann J, et al. 1985. *Proc. Natl. Acad. Sci. USA* 82:8183–87
67. Cerundolo V, Alexander J, Anderson K, Lamb C, Cresswell P, et al. 1990. *Nature* 345:449–52
68. Hosken NA, Bevan MJ. 1990. *Science* 248:367–69
69. Arnold D, Driscoll J, Androlewicz M, Hughes E, Cresswell P, Spies T. 1992. *Nature* 360:171–74
70. Momburg F, Ortiz-Navarrete V, Neefjes J, Goulmy E, van de Wal Y, et al. 1992. *Nature* 360:174–77
71. Yewdell J, Lapham C, Bacik I, Spies T, Bennink J. 1994. *J. Immunol.* 152: 1163–70
72. Zhou X, Momburg F, Liu T, Abdel Motal UM, Jondal M, et al. 1994. *Eur. J. Immunol.* 24:1863–68
73. Akiyama K-Y, Yokota K-Y, Kagawa S, Shimbara N, Tamura T, et al. 1994. *Science* 265:1231–34
74. Belich MP, Glynne RJ, Senger G, Sheer D, Trowsdale J. 1994. *Curr. Biol.* 4: 769–76
75. Driscoll J, Brown M, Finley D, Monaco J. 1993. *Nature* 365:262–64
76. Gaczynska M, Rock KL, Goldberg AL. 1993. *Nature* 365:264–67
77. Howard JC. 1993. *Nature* 365:211–12
78. Boes B, Hengel H, Ruppert T, Multhaup G, Koszinowski UH, Kloetzel P-M. 1994. *J. Exp. Med.* 179:901–9
79. del Val M, Schlicht H-J, Ruppert T, Reddehase J, Koszinowski UH. 1991. *Cell* 66:1145–53
80. Chimini G, Pala P, Sire J, Jordan BR, Maryanski JL. 1989. *J. Exp. Med.* 169: 297–302
81. Hahn YS, Braciale VL, Braciale TJ. 1991. *J. Exp. Med.* 174:733–36
82. Dick LR, Aldrich C, Jameson SC, Moomaw CR, Pramanik BC, et al. 1994. *J. Immunol.* 152:3884–94
83. Fehling HJ, Swat W, Laplace C, Kühn R, Rajewsky K, et al. 1994. *Science* 265:1234–37
84. Van Kaer L, Ashton-Rickardt PG, Eichelberger M, Gaczynska M, Nagashima K, et al. 1994. *Immunity* 1:533–541
85. Rock KL, Gramm C, Rothstein L, Clark K, Stein R, et al. 1994. *Cell* 78:761–71
86. Schumacher TNM, Heemels M-T, Neefjes JJ, Kast WM, Melief CJM, Ploegh HL. 1990. *Cell* 62:563–67
87. Townsend A, Elliott T, Cerundolo V, Foster L, Barber B, Tse A. 1990. *Cell* 62:285–95
88. Michalek ME, Grant E, Gramm C, Goldberg A, Rock K. 1993. *Nature* 363: 552–54
89. Cox JH, Bennink JR, Yewdell JW. 1994. *FASEB J.* 8:A4334
90. Henderson RA, Michel H, Sakaguchi K, Shabanowitz J, Appella E, et al. 1992. *Science* 255:1264–66
91. Wei ML, Cresswell P. 1992. *Nature* 356:443–46
92. Eisenlohr LC, Bacik I, Bennink JR, Bernstein K, Yewdell JW. 1992. *Cell* 71:963–72
93. Sherman LA, Burke TA, Biggs JA. 1992. *J. Exp. Med.* 175:1221–26
94. Snyder LH, Bennink J, Yewdell J. 1994. *J. Exp. Med.* 180:2389–2394
95. Falk K, Rötzschke O, Rammensee H-G. 1990. *Nature* 351:248–51
96. Rotzschke O, Falk K, Deres K, Schild H, Norda M, et al. 1990. *Nature* 348: 252–54
97. Heemels M-T, Schumacher TNM, Wonigeit K, Ploegh HL. 1993. *Science* 262:2059–63
98. Momburg F, Roelse J, Hämmerling GJ, Neefjes JJ. 1994. *J. Exp. Med.* 179: 1613–23
99. Schumacher TNM, Kantesaria DV, Heemels M-T, Ashton-Rickardt PG, Shepherd JC, et al. 1994. *J. Exp. Med.* 179:533–40
100. Roelse J, Grommé M, Momburg F, Hämmerling G, Neefjes J. 1994. *J. Exp. Med.* 180:1591–1598
101. Androlewicz MJ, Anderson KS, Cresswell P. 1993. *Proc. Natl. Acad. Sci. USA* 90:9130–34
102. Neefjes JJ, Momburg F, Hämmerling G. 1993. *Science* 261:769–71
103. Shepherd JC, Schumacher TNM, Ashton-Rickardt PG, Imaeda S, Ploegh HL, et al. 1993. *Cell* 74:577–84
104. Kleijmeer M, Kelly A, Geuze HJ, Slot JW, Townsend A, Trowsdale J. 1992. *Nature* 357:342–44
105. Spies T, Bresnahan M, Bahram S, Arnold D, Blanck G, et al. 1990. *Nature* 348:744–47
106. Deverson EV, Gow IR, Coadwell WJ, Monaco JJ, Butcher GW, Howard JC. 1990. *Nature* 348:738–41
107. Monaco JJ, Cho S, Attaya M. 1990. *Science* 250:1723–26
108. Trowsdale J, Hanson I, Mockridge I, Beck S, Townsend A, Kelly A. 1990. *Nature* 348:741–44
109. Powis SJ, Townsend A, Deverson EV,

Bastin J, Butcher GW, Howard JC. 1991. *Nature* 354:528–31

110. Spies T, DeMars R. 1991. *Nature* 351: 323–24

111. Attaya M, Jameson S, Martinez CK, Hermel E, Aldrich C, et al. 1992. *Nature* 355:647–49

112. de la Salle H, Hanau D, Fricker D, Urlacher A, Kelly A, et al. 1994. *Science* 265:237–40

113. Higgins CF. 1992. *Ann. Rev. Cell Biol.* 8:67–113

114. Walker JE, Saraste M, Runswick MJ, Gay NJ. 1982. *EMBO J.* 1:945–51

115. Kelly A, Powis SH, Kerr L-A, Mockridge I, Elliott T, et al. 1992. *Nature* 355:641–44

116. Spies T, Bresnahan M, Bahram S, Arnold D, Blanck G, et al. 1990. *Nature* 348:744–42

117. Carpenter CB. 1991. In *Harrison's Principles of Internal Medicine*, J. D. Wilson, E. Braunwald, K. J. Isselbacher, R. G. Petersdorf, ed. JB Martin, AS Fauci, RK Root, 1:86–92. New York: McGraw-Hill

118. Gaskins HR, Monaco JJ, Leiter EH. 1992. *Science* 256:1826–28

119. Colonna M, Bresnahan M, Bahram S, Strominger JL, Spies T. 1992. *Proc. Natl. Acad. Sci. USA* 89:3932–36

120. Powis SH, Mockridge I, Kelly A, Kerr L-A, Glynne R, et al. 1992. *Proc. Natl. Acad. Sci. USA* 89:1463–67

121. Carrington M, Colonna M, Spies T, Stephens JC, Mann DL. 1993. *Immunogenetics* 37:266–73

122. Powis SH, Tonks S, Mockridge I, Kelly AP, Bodmer JG, Trowsdale J. 1993. *Immunogenetics* 37:373–80

123. Gros P, Dhir R, Croop J, Talbot F. 1991. *Proc. Natl. Acad. Sci. USA* 88:7289–93

124. Powis SJ, Deverson EV, Coadwell WJ, Ciruela A, Huskisson NS, et al. 1992. *Nature* 357:211–15

125. Livingstone AM, Powis SJ, Diamond AG, Butcher GW, Howard JC. 1989. *J. Exp. Med.* 170:777–95

126. Livingstone AM, Powis SJ, Günther D, Cramer DV, Howard JC, Butcher GW. 1991. *Immunogenetics* 34:157–63

127. Powis SJ, Howard JC, Butcher GW. 1991. *J. Exp. Med.* 173:913–21

128. Ljunggren HG, Kärre K. 1985. *J. Exp. Med.* 142:1745–59

129. Van Kaer L, Ashton-Rickardt PG, Ploegh HL, Tonegawa S. 1992. *Cell* 71:1205–14

130. Momburg F, Roelse J, Howard JC, Butcher GW, Hämmerling G, Neefjes JJ. 1994. *Nature* 367:648–51

131. Levy F, Gabathuler R, Larsson R, Kvist S. 1991. *Cell* 67:265–74

132. Koppelman B, Zimmerman DI, Walter P, Brodsky FM. 1992. *Proc. Natl. Acad. Sci. USA* 89:3908–12

133. Bijlmakers MJE, Neefjes JJ, Wojcik-Jacobs EHM, Ploegh HL. 1993. *Eur. J. Immunol* 23:1305–13

134. Zweerink HJ, Gammon MC, Utz U, Sauma SY, Harrer T, et al. 1993. *J. Immunol.* 150:1763–71

135. Bause E, Hettkamp H. 1979. *FEBS lett.* 108:341–44

136. Geetha-Habib M, Park HR, Lennarz WJ. 1990. *J. Biol. Chem.* 256:13655–60

137. Androlewicz MJ, Cresswell P. 1994. *Immunity* 1:7–14

138. Androlewicz MJ, Ortmann B, Endert PM, Spies T, Cresswell P. 1994. *Proc. Natl. Acad. Sci. USA.* 91:12716–12720

139. van Endert PM, Tampé R, Meyer TH, Tisch R, Bach J-F, McDevitt HO. 1994. *Immunity* 1:491–500

140. Shawar SM, Rodgers JR, Cook RG, Rich R. 1991. *Immunol. Res.* 10:365–75

141. Schumacher TNM, Kantesaria DV, Serreze DV, Roopenian DC, Ploegh HL. 1994. *Proc. Natl. Acad. Sci. USA.* 91: 13004–13008

142. Joyce S, Kuzushima K, Kepecs G, Angeletti RH, Nathenson SG. 1994. *Proc. Acad. Natl. Sci. USA* 91:4145–49

143. Urban RG, Roman MC, Lane WS, Strominger JL, Rehm A, et al. 1994. *Proc. Natl. Acad. Sci. USA* 91:1534–38

144. Heemels M-T, Ploegh HL. 1993. *Curr. Biol.* 3:380–83

145. Ortmann B, Androlewicz MJ, Cresswell P. 1994. *Nature* 368:864–67

146. Suh W-K, Cohen-Doyle MF, Fruh K, Wang K, Peterson PA, Williams DB. 1994. *Science* 264:1322–26

147. Ortiz-Navarrrete V, Hämmerling GJ. 1991. *Proc. Natl. Acad. Sci. USA* 88: 3594–97

148. Anderson K, Cresswell P, Gammon M, Hermes J, Williamson A, Zweerink H. 1991. *J. Exp. Med.* 174:489–92

149. Bacik I, Cox JH, Anderson R, Yewdell JW, Bennink JR. 1994. *J. Immunol.* 152:381–87

150. Tabaczewski P, Stroynowski I. 1994. *J. Immunol.* 152:5268–74

151. Elliott T, Cerundolo V, Elvin J, Townsend A. 1991. *Nature* 351:402–6

152. Elliott T, Cerundolo V, Townsend A. 1992. *Eur. J. Immunol.* 22:3121–25

153. Machold RP, Andrée S, Van Kaer L, Ljunggren H-G, Ploegh HL. 1995. *J. Exp. Med.* In press

154. Burshtyn DB, Barber BH. 1993. *J. Immunol.* 151:3082–93

155. Neefjes JJ, Hämmerling GJ, Momburg F. 1993. *J. Exp. Med.* 178:1971–80

156. Hurtley SM, Helenius A. 1989. *Annu. Rev. Cell Biol.* 5:277–307
157. Gething MJ. 1991. *Curr. Opin. Cell. Biol.* 3:610–14
158. Gething MJ, Sambrook J. 1992. *Nature* 355:33–45
159. Ahluwalia N, Bergeron JJM, Degen E, Williams DB. 1992. *J. Biol. Chem.* 267: 10914–18
160. Galvin K, Krishna S, Ponchel F, Frohlich M, Cummings DE, et al. 1992. *Proc. Natl. Acad. Sci. USA* 89:8452–56
161. Degen E, Williams DB. 1991. *J. Cell. Biol.* 112:1099–115
162. David V, Hochstenbach F, Rajagopalan S, Brenner MB. 1992. *J. Biol. Chem.* 268:9585–92
163. Hochstenbach F, David V, Watkins S, Brenner MB. 1992. *Proc. Natl. Acad. Sci. USA* 89:4734–38
164. Ou W-J, Cameron PH, Thomas DY, Bergeron JJM. 1993. *Nature* 364:771–76
165. Rajagopalan S, Xu Y, Brenner MB. 1994. *Science* 263:387–89
166. Rajagopalan S, Brenner MB. 1994. *J. Exp. Med.* 180:407–12
167. Degen E, Cohen-Doyle M, Williams D. 1992. *J. Exp. Med.* 175:1653–61
168. Jackson MR, Song ES, Yang Y, Peterson PA. 1992. *Proc. Natl. Acad. Sci. USA* 89:12117–21
169. Jackson MR, Cohen-Doyle MF, Peterson PA, Williams DB. 1994. *Science* 263:384–87
170. Sugita M, Brenner MB. 1994. *J. Exp. Med.* 180:2163–2171
171. Alexander J, Payne JA, Murray R, Frelinger JA, Cresswell P. 1989. *Immunogenetics* 29:380–88
172. Anderson KS, Alexander J, Wei M, Cresswell P. 1993. *J. Immunol.* 151: 3407–19
173. Alexander J, Payne A, Shigekawa B, Frelinger JA, Cresswell P. 1990. *Immunogenetics* 31:169–78
174. van Santen HM, Woolsey A, Ashton Rickardt PG, Van Kaer L, Baas EJ, et al. 1995. *J. Exp. Med.* In press
175. Hammond C, Helenius A. 1993. *Curr. Biol.* 3:884–86
176. Hammond C, Braakman I, Helenius A. 1994. *Proc. Natl. Acad. Sci. USA* 91: 913–17
177. Elbein AD. 1991. *FASEB J.* 5:3055–63
178. Feuerbach D, Burgert H-G. 1993. *EMBO J.* 12:3153–61
179. Booth C, Koch GLE. 1989. *Cell* 59:729–37
180. Srivastava PK, Maki RG. 1991. *Curr. Top. Microbiol. Immunol.* 167:109–23
181. Rothman JE. 1989. *Cell* 59:591–601
182. Li Z, Srivastava PK. 1993. *EMBO J.* 12:3143–51
183. Maudsley DJ, Pound JD. 1991. *Immunol. Today* 12:429–31
184. Schrier PL, Bernards R, Vaessen RTMJ, Houweling A, van der Eb AJ. 1983. *Nature* 305:771–75
185. Scheppler JA, Nicholson JKA, Swan DC, Ahmed-Ansari A, McDougal JS. 1989. *J. Immunol.* 143:2858–66
186. Cromme FV, Airey J, Heemels M-T, Ploegh HL, Keating PJ, et al. 1994. *J. Exp. Med.* 179:335–40
187. Rotem-Yehudar R, Winograd S, Sela S, Coligan JE, Ehrlich R. 1994. *J. Exp. Med.* 180:477–88
188. Burgert HG, Kvist S. 1985. *Cell* 41:987–97
189. del Val M, Hengel H, Hacker H, Hartlaub U, Ruppert T, et al. 1992. *J. Exp. Med.* 176:729–38
190. Beersma MFC, Bijlmakers MJE, Ploegh HL. 1993. *J. Immunol.* 151:4455–64
191. Warren AP, Ducroq RJ, Lehner PJ, Borysiewicz LK. 1994. *J. Virol.* 68:2822–29
192. York IA, Roop C, Andrews DW, Riddell SR, Graham FL, Johnson DC. 1994. *Cell* 77:525–35

Annu. Rev. Biochem. 1995. 64:493–531

STRUCTURE AND FUNCTION OF VOLTAGE-GATED ION CHANNELS

William A. Catterall

Department of Pharmacology, SJ-30, University of Washington, Seattle, Washington 98195

KEY WORDS: Ion channel, action potential, electrical excitability, ion transport, membrane proteins

CONTENTS

493

0066-4154/95/0701-0493$00.00

ABSTRACT

Voltage-gated ion channels are responsible for generation of electrical signals in cell membranes. Their principal subunits are members of a gene family and can function as voltage-gated ion channels by themselves. They are expressed in association with one or more auxiliary subunits which increase functional expression and modify the functional properties of the principal subunits. Structural elements that are required for voltage-dependent activation, selective ion conductance, and inactivation have been identified, and their mechanisms of action are being explored through mutagenesis, expression in heterologous cells, and functional analysis. These experiments reveal that this family of channels is built upon a common structural theme with variations appropriate for functional specialization of each channel type.

INTRODUCTION

The voltage-gated sodium, calcium, and potassium channels are responsible for the generation of conducted electrical signals in neurons and other excitable cells. The permeability increase resulting from activation of these channels is biphasic. Upon depolarization, permeability to sodium, calcium, or potassium increases dramatically over a period of 0.5 to hundreds of msec and then decreases to the baseline level over a period of 2 msec to a few seconds. This biphasic behavior results from two experimentally separable gating processes that control ion channel function: activation, which controls the rate and voltage dependence of the permeability increase following depolarization, and inactivation, which controls the rate and voltage dependence of the subsequent return of the ion permeability to the resting level during a maintained depolarization. The voltage-gated ion channels can therefore exist in three functionally distinct states or groups of states: resting, active, and inactivated. Both resting and inactivated states are nonconducting, but channels that have been inactivated by prolonged depolarization are refractory unless the cell is repolarized to allow them to return to the resting state. The ion conductance of the activated ion channels is both highly selective and remarkably efficient. Selectivity among the physiological ions ranges from 12-fold for sodium channels to over 1000-fold for calcium channels, and all three classes of ion channels conduct ions across biological membranes at rates approaching their rates of free diffusion through solution. Understanding the molecular bases for voltage-dependent activation, rapid inactivation, and selective and efficient ion conductance is a major goal of current research on these critical signaling proteins.

STRUCTURE AND FUNCTION OF SODIUM CHANNEL SUBUNITS

Purification and Characterization

The initial determination of the subunit structure of the rat brain Na^+ channel took advantage of neurotoxins that bind with high affinity and specificity to the channel complex and thus could be used as molecular probes to identify its protein components. Five groups of neurotoxins that act at different receptor sites on the Na^+ channel have been described (1–3). Briefly site 1 binds tetrodotoxin, saxitoxin and μ-conotoxin which block ion conductance. Site 2 binds the toxins, batrachotoxin, veratridine, grayanotoxin, and aconitine, resulting in persistent activation of the Na^+ channel. Site 3 binds the polypeptide α-scorpion toxins and sea anemone toxins which slow or block inactivation. Agents which bind at this site also enhance the persistent activation of the Na^+ channel caused by toxins acting at neurotoxin receptor site 2. Receptor site 4 binds a second class of scorpion toxins (β-scorpion toxins) that shift the voltage dependence of activation to more negative membrane potentials without modifying Na^+ channel inactivation. Finally, receptor site 5 binds the brevetoxins and ciguatoxins, agents that cause repetitive neuronal firing, shift the voltage dependence of Na^+ channel activation, and block Na^+ channel inactivation.

Direct chemical identification of the 260 kD α subunit and the 36 kD β1 subunit of the rat brain sodium channel in situ was accomplished by specific covalent labeling of neurotoxin receptor sites 3 and 4 on the Na^+ channel complex with photoreactive derivatives of α- and β-scorpion toxins, respectively (4–6). Separation of two photoreactive derivatives of an α-scorpion toxin by ion-exchange chromatography allowed selective labeling of each subunit (7). Radiation inactivation studies were also used to probe the molecular sizes of the subunits of the Na^+ channel. Measurements of the target size for inactivation of either tetrodotoxin or α-scorpion toxin binding to the Na^+ channel from rat brain or eel electroplax revealed a structure of 230 kD to 266 kD (8, 9). In contrast, radiation inactivation studies of β-scorpion toxin binding activity of rat brain sodium channels implicated two polypeptides of 266 kD and 45 kD, consistent with a role of both α and β1 subunits in formation of neurotoxin receptor site 4 (10).

The sodium channel from electric eel electroplax was purified using the binding of radiolabeled tetrodotoxin as a specific assay (11). It consisted of a single polypeptide of 280 kD, similar in size to the α subunit of the rat brain sodium channel (12), with a high level of N-linked carbohydrate (13).

Purification of the intact sodium channel from rat brain using high affinity binding of saxitoxin as an assay revealed a complex of one α subunit of 260 kD and two distinct β subunits: β1 with an apparent molecular mass of 36 kD, and β2 with an apparent molecular mass of 33 kD (14–17). The β2 subunit is covalently attached to the α subunit by disulfide linkage while the β1 subunit is noncovalently associated. The subunit stoichiometry of 1:1:1 yielded a molecular weight (329 kD) in close agreement with that of the solubilized oligomeric channel (316 kD) (18). Partial proteolytic maps showed that the β1 and β2 subunits were distinct and probably unrelated (17). Both the β1 and β2 subunits were covalently labeled by a hydrophobic probe specific for transmembrane segments of proteins in mixed micelles of Triton X-100 and phosphatidylcholine or in reconstituted phosphatidylcholine vesicles, and both subunits were preferentially extracted into the nonionic detergent Triton X-114 in a phase separation procedure, as expected for intrinsic membrane proteins (19). The α subunits and both β subunits are heavily glycosylated (17, 20). The apparent molecular weights of the deglycosylated subunits were determined to be 220 kD, 23 kD, and 21 kD for α, β1 and β2, respectively, which suggests that a substantial fraction of the mass of the native subunits is carbohydrate and that much of the protein mass is exposed on the extracellular surface (17). These experiments led to a heterotrimeric model for the subunit structure of the brain sodium channel as illustrated in Figure 1A.

The Na^+ channel from rat and rabbit skeletal muscle sarcolemma contains a large α subunit of 260 kD and a small β subunit of approximately 38 kD (21–23). Enzymatic deglycosylation of the purified β subunit yielded a core peptide of 26.5 kD. Although the 38 kD subunit of the Na^+ channel could be resolved into a doublet of 37 kD and 39 kD, stoichiometric analysis suggested that there is only a single β subunit associated noncovalently with each large α subunit. The noncovalent association implies that the β subunit in purified skeletal muscle sodium channels is β1-like. Studies using affinity-purified polyclonal antibodies to the purified rat brain β1 subunit identified immunoreactive β1-like subunits in rat skeletal muscle that appeared as a closely spaced doublet of 41 and 38 kD on SDS-PAGE, consistent with the results obtained using purified sodium channels (24).

Cloning and Primary Structure of the Sodium Channel α Subunit

Cloning of the α subunit of the sodium channel from eel electroplax by Noda, Numa, and colleagues (25) gave the initial insight into the primary structure of a voltage-gated ion channel. Using oligonucleotides encoding short segments of the electric eel electroplax sodium channel and antibodies directed

→

Figure 1   Subunit structures of the voltage-gated ion channels. The arrangement and biochemical properties of subunits of the voltage-gated ion channels are illustrated. Ψ, site of probable N-linked glycosylation; P, site of cAMP-dependent protein phosphorylation; -S-S-, inter-subunit disulfide bond.

A.

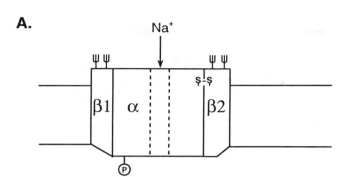

B.

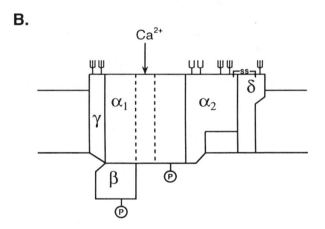

C.

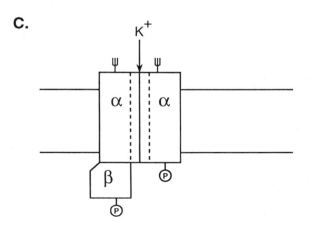

against it, Noda et al (25) isolated cDNAs encoding the entire polypeptide from expression libraries of electroplax mRNA. The deduced amino acid sequence revealed a protein with four internally homologous domains, each containing multiple potential alpha-helical transmembrane segments (Figure 2A). The wealth of information contained in this deduced primary structure has revolutionized research on the voltage-gated ion channels.

The cDNAs encoding the electroplax sodium channel were used to isolate cDNAs encoding three distinct, but highly homologous, rat brain sodium channels (types I, II, and III) (26, 27). cDNAs encoding the alternatively spliced type IIA sodium channel were isolated independently by screening expression libraries with antibodies against the rat brain sodium channel α subunit (28, 29). The type II gene contains two adjacent exons encoding segment IS3 (30) that are alternatively spliced into mature mRNA in a developmentally regulated manner. The type II is most prominent in embryonic and neonatal brain while the type IIA form is most prominent in the adult brain (30, 31). cDNAs encoding the type II/IIA sodium channel were used as probes to isolate cDNAs encoding sodium channel α subunits expressed in skeletal muscle and heart by low-stringency hybridization (32–34). The $\mu1$ sodium channel α subunit is expressed primarily in adult skeletal muscle (32); the h1 sodium channel α subunit is expressed primarily in heart and also in uninnervated or denervated skeletal muscle (33, 34). These sodium channels have a close structural relationship to the three brain sodium channel α subunits. In general, the similarity in amino acid sequence is greatest in the homologous domains from transmembrane segment S1 through S6, while the intracellular connecting loops are not highly conserved.

Complementary DNAs encoding α subunits of three distinct sodium channels from *Drosophila* have been cloned by cross-hybridization, and most of the primary structures of the corresponding sodium channels have been deduced (35–37). Thus, it appears that *Drosophila* also has multiple sodium-channel genes as observed in rat. Presumably these distinct genes have distinct roles in electrical excitability in both species.

More recently, new putative sodium channel α subunits have been cloned from glial and heart cDNA libraries (38, 39). The amino acid sequences retain the four-domain structure of the other α subunits and many of their other conserved features, but are distinctly more divergent than the other α subunits previously characterized. It has been suggested that these new glial/heart sodium channels define a new sub-family of sodium channels.

Cloning and Primary Structure of the β1 and β2 Subunits

cDNA clones encoding the β1 subunit of the rat brain Na$^+$ channel were isolated using a combination of polymerase chain reaction and library-screening techniques based on the amino acid sequence of the amino terminus of the

A. Na⁺ Channel

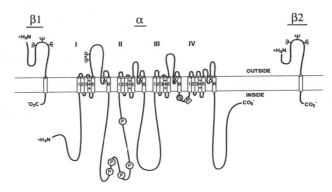

B. Ca²⁺ Channel

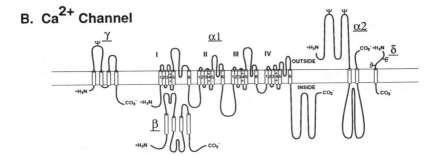

C. K⁺ Channel

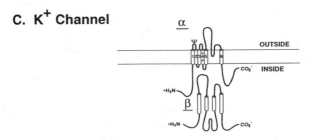

Figure 2   Transmembrane organization of the ion channel subunits. The primary structures of the subunits of the voltage-gated ion channels are illustrated. Cylinders represent probable alpha helical segments. *Bold lines* represent the polypeptide chains of each subunit with length approximately proportional to the number of amino acid residues. Ψ, sites of probable N-linked glycosylation; P, sites of demonstrated protein phosphorylation.

purified protein (40). The deduced primary structure indicates that the β1 subunit is a 22,851-dalton protein that contains a small cytoplasmic domain, a single putative transmembrane segment, and a large extracellular domain with four potential N-linked glycosylation sites (Figure 2A), consistent with previous biochemical data (17, 19). Northern blot analysis revealed a 1400-nucleotide mRNA in rat brain, heart, and spinal cord, and at low levels in rat skeletal muscle.

A similar approach was taken to cloning the β2 subunit. It also has a single transmembrane segment, a small intracellular carboxyl terminal domain, and a large glycosylated extracellular amino-terminal domain (unpublished results).

Functional Role of Na⁺ Channel Subunits

Early biochemical experiments pointed to a dominant functional role for the α subunits. Covalent labeling and radiation-inactivation studies implicated α subunits in formation of the receptor sites for tetrodotoxin and saxitoxin (site 1) (8) and both α- and β-scorpion toxins (sites 3 and 4) (4, 5, 7). In addition, Na⁺ channels purified from eel electroplax and chicken heart contained only α subunits but retained high affinity for tetrodotoxin and, in the case of eel electroplax, ion-conductance activity (12, 41, 42). Because tetrodotoxin and saxitoxin are thought to block the pore of the Na⁺ channels and the scorpion toxins affect activation and inactivation gating, these biochemical results suggested that α subunits were involved in both ion conductance and gating.

Potential functional roles of the β1 and β2 subunits were analyzed in purified and reconstituted sodium channel preparations. Selective removal of the β1 subunit from the αβ2 complex of detergent-solubilized or reconstituted rat brain Na⁺ channels resulted in the complete loss of [³H]saxitoxin-binding activity, veratridine-activated ²²Na⁺ influx, α-scorpion toxin binding activity, and voltage-activated ion conductance (43, 44). Tetrodotoxin quantitatively stabilized the solubilized complex against the loss of the β1 subunit and loss of functional activities. In contrast, removal of the β2 subunit by reduction of disulfide bonds yielded a preparation of αβ1 that retained Na⁺ channel functions. These studies showed that a complex of α and β1 subunits is both necessary and sufficient for channel function in the purified state and suggested that β1 subunits, but not β2 subunits, are required to stabilize the functional state of purified brain sodium channels in detergent solution.

The role of the α and β1 subunits in Na⁺ channel function in an intact cell was tested directly through expression of the α and β1 subunits in *Xenopus* oocytes. RNA encoding α subunits from brain, skeletal muscle, and heart alone is sufficient to encode functional Na⁺ channels in *Xenopus* oocytes (28, 29, 32, 45–48). However, their inactivation is slow relative to that observed in neurons or skeletal muscle, and their voltage dependence of inactivation is shifted to more positive membrane potentials (29, 49–51). Co-expression of low-molecular-weight RNA from rat brain or skeletal muscle accelerates inactivation, shifts the voltage dependence of inactivation to more negative membrane potentials, and increases the peak amplitude of Na^+ current expressed from cloned α subunits (29, 32, 50, 51). These results suggested a possible role for the low-molecular-weight $\beta1$ and/or $\beta2$ subunits in Na^+ channel function. Co-expression of type IIA α subunits and $\beta1$ subunits in *Xenopus* oocytes resulted in a 2.5-fold increase in the amplitude of the peak Na^+ current, an increase in the rate of activation, a 5-fold acceleration in the rate of inactivation, and a 19 mV shift in the hyperpolarizing direction of steady state fast inactivation (40, 52). Co-expression of rat or human $\beta1$ subunits with the rat skeletal muscle $\mu1$ α subunit or rat brain type III α subunits in *Xenopus* oocytes gave similar results (52–54).

Na^+ channel α subunits have also been expressed both stably and transiently in mammalian cells in culture (55–57). Stable lines of Chinese hamster ovary cells expressing only the type IIA rat brain Na^+ channel α subunit generate Na^+ currents with a normal time course (55, 56), in spite of the lack of $\beta1$ subunits detectable by Northern blot, Western blot, and activity assays in *Xenopus* oocytes (58). However, the level of functional Na^+ channel expression in these cell lines is low relative to mRNA levels. Co-expression of $\beta1$ subunits with type IIA α subunits in mammalian cell lines results in an increase in the peak Na^+ current and the total number of Na^+ channels detected by saxitoxin binding, as well as a shift in the voltage dependence of activation and inactivation (58). Thus, $\beta1$ subunits in mammalian cells may stabilize the Na^+ channel complex in the plasma membrane, which results in increased functional expression, and also may alter the voltage dependence of channel gating. Additional experiments are needed to address whether association of different $\beta1$ subunits with α subunits can modify the function of the α subunits.

Mechanism of Action of β1 Subunits

$\beta1$ subunits have multiple effects on sodium channel function—increased peak current, accelerated activation and inactivation, and altered voltage dependence of inactivation. These multiple effects may result from distinct molecular interactions between α and $\beta1$ subunits or from a single molecular mechanism that alters multiple aspects of sodium-channel function. Cloned skeletal muscle sodium channel α subunits and type IIA and III brain sodium channel α subunits expressed in *Xenopus* oocytes exhibit two prominent gating modes: a rapid mode in which inactivation is complete in a few milliseconds, and a slow gating mode in which inactivation is slow and single sodium channels open repeatedly during long depolarizations (51, 59, 60). The effects of $\beta1$ subunits on the time courses of activation and inactivation, and possibly effects on the voltage dependence of inactivation, appear to result from a shift from the slow to the rapid gating mode induced by co-expression of the $\beta1$ subunit (52, 54). Thus, a single molecular interaction with the $\beta1$ subunit may be sufficient to change the energetic relationship between these two gating modes of the α subunits and cause a shift of sodium channels to the rapid gating mode. This shift in gating mode would affect the multiple aspects of sodium channel function that differ between the two functionally distinct modes of gating.

STRUCTURE AND FUNCTION OF CALCIUM CHANNEL SUBUNITS

Molecular Properties of the Subunits of Skeletal-Muscle Calcium Channels

The calcium channels in the transverse tubule membranes of skeletal muscle have served as a primary biochemical and molecular model for studies of calcium channels because of their abundance. These channels serve two critical physiological roles. Like other calcium channels, they mediate calcium entry in response to depolarization. However, the voltage-gated calcium channels in skeletal muscle activate very slowly, and the calcium entering vertebrate skeletal muscle through voltage-gated calcium channels is not required for muscle contraction. It appears to serve to replenish cellular calcium during periods of rapid activity and to increase intracellular calcium in response to tetanic stimulation, which leads to increased contractile force. The primary physiological role for the skeletal muscle calcium channel is to serve as a voltage sensor in excitation-contraction coupling. Voltage-gated calcium channels in the transverse tubule membranes are thought to interact physically with the calcium-release channels located in the sarcoplasmic reticulum membrane. Voltage-driven conformational changes in the voltage-gated calcium channels then activate the calcium release from the sarcoplasmic reticulum via protein-protein interactions (61–63).

The voltage-gated calcium channels in skeletal muscle are L-type (64–66). They mediate long-lasting currents with slow voltage-dependent inactivation, they have a large single-channel conductance (about 25 pS) and a high voltage of activation, and they are specifically inhibited by dihydropyridine calcium-channel antagonists. The initial purification of the L-type calcium channels from skeletal muscle took advantage of their high density in transverse tubule membranes to provide an enriched starting material and employed the specific, high-affinity binding of dihydropyridine calcium channel antagonists to identify the calcium channel protein (67, 68). A heterogeous α subunit band (67, 68) and associated β subunits of 50 kD and γ subunits of 33 kD (67) were identified as components of the calcium channel in the initial purification studies as assessed by comigration during column chromatography and sucrose-gradient sedimentation. Subsequent experiments demonstrated that the heterogenous α subunit band contained not only the principal $\alpha 1$ subunits with an apparent molecular mass of 175 kD but also a disulfide-linked dimer of $\alpha 2$ and δ subunits with apparent molecular masses of 143 kD and 27 kD, respectively (69–73). These results, together with analysis of the biochemical properties of the subunits, led to a model of calcium channel subunit structure illustrated in Figure 1B (2, 70). The specific association of these proteins as a multisubunit complex was supported by the copurification of each subunit with the dihydropyridine-binding activity and calcium-conductance activity of the calcium channel (67, 70, 74, 75), by co-immunoprecipitation of all these proteins by antibodies directed against the $\alpha 1$ subunits (69, 70, 76), and by co-immunoprecipitation of the calcium channel complex by antibodies against each auxiliary subunit (77–79). Estimates of stoichiometry indicated that each mol of calcium channel complex contains approximately 1 mol of each of the five subunits. The biochemical and molecular properties of each of the subunits of skeletal muscle calcium channels are considered below.

The $\alpha 1$ subunit of skeletal muscle calcium channels was cloned by library screening based on amino acid sequence (80). The cDNA predicts a protein of 1873 amino acids with a molecular weight of 212 kD, considerably larger than the estimate of 175 kD for the $\alpha 1$ subunits of purified calcium channels. Analysis of the $\alpha 1$ subunits of purified calcium channels and calcium channels in transverse tubule membranes using sequence-directed antibodies showed that most (more than 90%) were truncated in their carboxyl terminal domain between residues 1685 and 1699, which resulted in a 190-kD form that runs anomalously in SDS gels at 175 kD (81, 82). Only a small fraction (less than 10%) contained the full-length $\alpha 1$ subunit encoded by the cDNA. Both forms are detected in rat skeletal muscle cells in culture suggesting that both may be present in vivo (83). Since no mRNA that encodes the more abundant, truncated form has been identified, the truncated form may be produced by specific proteolytic processing in vivo.

The $\alpha 2$ subunit of skeletal muscle calcium channels is a hydrophobic glycoprotein with an apparent molecular mass of 143 kD before deglycosylation and 105 kD after deglycosylation (70, 84). It contains both high-mannose and complex carbohydrate chains. Cloning and sequencing cDNAs that encode the $\alpha 2$ subunit defined a protein of 1106 amino acids with a molecular mass of 125 kD, multiple potential transmembrane segments, and multiple consensus sites for N-linked glycosylation (85) (Figure 2B). The predicted $\alpha 2$ protein was 20 kD larger than the apparent molecular mass of the deglycosylated $\alpha 2$ subunit, which suggested that a portion of the protein encoded by the $\alpha 2$ cDNAs may not be present in the mature $\alpha 2$ subunit that had been characterized biochemically. Subsequent studies have shown that both $\alpha 2$ and δ subunits are encoded by the same mRNA (see below).

The β subunits are hydrophilic proteins that are not glycosylated and therefore are likely to be located on the intracellular side of the membrane (70). They are phosphorylated by multiple protein kinases, including protein kinase C and cAMP-dependent protein kinase, which regulate the function of many calcium channels (70, 75, 86–88). cDNA cloning and sequencing revealed a protein of 524 amino acids with a predicted molecular mass of 58 kD (89). In agreement with biochemical data, the primary structure does not include any potential transmembrane segments but contains multiple consensus sites for phosphorylation by different protein kinases (Fig. 2B).

The γ subunit of skeletal muscle calcium channels is a hydrophobic glycoprotein with an apparent molecular mass of 30 kD without deglycosylation and 20 kD following deglycosylation (70). Cloning and sequencing cDNAs encoding γ subunits revealed a protein of 222 amino acid residues with a molecular mass of 25 kD (90, 91). The deduced primary structure contained three predicted hydrophobic transmembrane segments and multiple sites for N-linked glycosylation.

The δ subunit appears on SDS gels as a doublet of 24 and 27 kD proteins, which are both hydrophobic and glycosylated (70, 92). Determination of the amino acid sequences of peptides derived from the δ subunit showed that it was encoded by the same mRNA as the α2 subunit (93, 94). The mature α2 subunit is truncated at alanine 934 of the α2δ precursor protein; residues 935–1106 constitute the disulfide-linked δ subunit. This sequence encodes a protein of 16 kD and contains a single transmembrane segment and three consensus sequences for N-linked glycosylation. The doublet on SDS gels represents two differently glycosylated forms of the δ subunit.

Subunits of Purified Cardiac Calcium Channels

Like the skeletal muscle calcium channels, the principal cardiac calcium channels are L-type (95, 96). Antibodies against α2δ subunits of skeletal muscle calcium channels detect corresponding subunits in cardiac preparations (97, 98). Partially purified cardiac calcium channels appear to contain α1, α2δ, and β subunits (99–103), but the relatively low abundance of calcium channels in cardiac tissue and the difficulty of controlling proteolysis during lengthy purification procedures have frustrated attempts at complete purification of an intact cardiac calcium channel complex. The α1 subunit with an apparent mass of 165 kD to 190 kD has been directly identified by photoaffinity labeling with photoreactive dihydropyridines (99, 100, 104).

Subunits of Purified Neuronal Calcium Channels

Multiple types of calcium channels, which differ in physiological and pharmacological properties, are expressed in neurons. At least three types of high-voltage-activated calcium channels have been distinguished in addition to L-type

(105–108). N-type, P-type, and Q-type channels all have intermediate single-channel conductances (about 15 pS) and can mediate calcium currents with varying rates of voltage-dependent inactivation, depending on their subunit composition (see below) and on other factors. They are best distinguished by their pharmacological properties: N-type are specifically inhibited by ω-conotoxin GVIA, P-type are most sensitive to ω-agatoxin IVA, and Q-type are most sensitive to ω-conotoxin MVIIC. Because the concentration of calcium channels in skeletal muscle transverse tubules is much higher than in neuronal membranes, the biochemical properties of these channels in neurons are not as well established. Immunoprecipitation with specific antibodies gainst $\alpha2\delta$ subunits revealed a complex of polypeptides with sizes corresponding to $\alpha1$, $\alpha2\delta$, and β subunits of dihydropyridine-sensitive L-type calcium channels in the brain (79, 109, 110). A novel 100 kD protein was also identified as a specifically associated component of the L-type calcium channel complex from brain (79). The ω-conotoxin-sensitive N-type calcium channels purified from rat brain contain an $\alpha1$ subunit, a 140 kD $\alpha2$-like subunit, and β subunits of 60 kDa to 70 kD, as identified by antibodies against the skeletal muscle forms of these subunits (111–115). Both L-type and N-type calcium channels from brain appear to lack a γ subunit, but proteins of approximately 100 kD are specifically associated with N-type calcium channels as well as L-type calcium channels from brain and may be additional, brain-specific subunits or associated proteins (79, 112, 114, 115).

Multiple Isoforms of the Subunits of Calcium Channels

Five additional genes encoding the $\alpha1$ subunits of calcium channels have been identified by cDNA cloning and sequencing using the skeletal muscle $\alpha1$ subunit as a probe, and these are thought to encode the major types of high-voltage-activated calcium channels defined by physiological and pharmacological properties (108, 116, 117). The $\alpha1$ subunits fall into two groups based on amino acid sequence similarity. The class C and D genes encode L-type calcium channels in which the sequences are greater than 75% identical to skeletal muscle L-type $\alpha1$ subunits. The class C gene is the primary calcium channel in the heart and is widely expressed in other tissues. The class D gene is expressed in neuroendocrine cells and neurons. The class A, B, and E genes encode non-L-type calcium channels, expressed primarily in neurons, in which the amino acid sequences are only 25 to 40% identical to the skeletal muscle α subunits. In general, the level of amino acid sequence identity among the $\alpha1$ subunits is greatest in the transmembrane regions and least in the large intracellular loops connecting domains I, II, and III and in the intracellular amino-terminal and carboxy terminal domains. Most of the $\alpha1$ subunit genes also encode alternatively spliced segments that increase their molecular diversity. The functional significance of alternative splicing has not yet been defined for any of these isoforms.

Four genes encoding calcium channel β subunits ($\beta1-\beta4$) have been characterized, and three of these have multiple splice products that have been identified (89, 118–123). In general, the amino acid sequences of the β subunits have two conserved segments in a central core region and are divergent in the carboxy and amino-terminal segments. All of the isoforms have consensus sequences for physphorylation by multiple protein kinases. Similar isoforms of the β subunit have been described in human tissues (124).

$\alpha2\delta$ mRNAs that are recognized by hybridization at high stringency with cDNA probes encoding the skeletal muscle isoform are detected in total RNA from a wide range of tissues (85, 125). These $\alpha2\delta$ primary transcripts are apparently the products of the same gene, but are differentially processed in human tissues to yield at least three isoforms which were designated by Williams et al (124) as $\alpha2_a$ (expressed in skeletal muscle), $\alpha2_b$ (expressed in neuronal tissues), and $\alpha2_c$ (expressed in aorta) (124, 125).

In contrast to the $\alpha2\delta$ and β subunits, there is no evidence to-date for multiple isoforms of the γ subunits. cDNA probes derived from the coding sequence of the γ subunit hybridize to a single mRNA species from skeletal muscle, but show weak or no specific hybridization in brain and several other tissues (90, 91, 126). Thus, the γ subunit may be encoded by a single gene which is expressed primarily or exclusively in skeletal muscle.

Functional Roles of Calcium Channel Subunits

Like the α subunits of Na^+ channels, the $\alpha1$ subunits of some calcium channels can serve as voltage-gated ion channels when expressed alone. These include the L-type calcium channels from heart ($\alpha1_{C1}$) (127) and skeletal muscle (128) and the class A and B $\alpha1$ subunits from rat brain (129–131). However, the auxiliary subunits of calcium channels have substantial effects on expression and gating of the $\alpha1$ subunits.

The $\alpha1$ subunit of the skeletal muscle L-type calcium channel alone directs the synthesis of a low density of functional calcium channels in only a small fraction of the transfected cells in a mammalian L cell line (128). The currents expressed in L cells activated at least 10 times slower than calcium currents in skeletal muscle. However, the time courses of activation and inactivation of calcium currents in these cells were dramatically accelerated by co-expression of the β_{1a} subunit (132, 133). Co-expression of $\alpha2\delta$ and γ subunits had little or no effect on the amplitude and kinetics of calcium currents of the skeletal muscle $\alpha1$ subunits expressed in L cells (132). However, $\alpha2$ subunits increase the activity of purified calcium channels reconstituted in lipid bilayers (134).

Expression of the smooth-muscle splice variant of the Class C $\alpha1$ subunit in CHO cells also revealed substantial effects of auxiliary subunits (135). Peak calcium channel currents were increased, the kinetics of activation and inactivation were accelerated, and the voltage dependence of activation and inactivation were shifted slightly to more negative membrane potentials. The $\alpha2\delta$ subunit had further effects on the kinetics of activation.

Expression of mRNA that encodes the class C cardiac L-type calcium channel $\alpha1$ subunit ($\alpha1_{C1}$) in *Xenopus* oocytes alone is sufficient to direct efficient synthesis of a functional calcium channel that is modulated by dihydropyridines (127). Similarly, mRNAs that encode the alternatively spliced smooth muscle and brain isoforms of the class C $\alpha1$ subunit direct the synthesis of functional channels in this system (136–138). The $\alpha2\delta$ subunit substantially increased the amplitude of the calcium current when co-expressed with the cardiac $\alpha1$ subunit in *Xenopus* oocytes, accelerated its activation and inactivation, and shifted the voltage dependence of inactivation to more negative membrane potentials (127, 139). Co-expression of the cardiac $\alpha1$ subunit with the β_{1a} subunit in *Xenopus* oocytes increased the peak current, accelerated activation, and shifted the voltage dependence of activation to more negative membrane potentials (139, 140). Similar increases in peak current were observed when the β_{1a} subunit was expressed with the vascular smooth muscle isoform of the cardiac $\alpha1$ subunit in *Xenopus* oocytes (137) or when the cardiac $\alpha1$ subunit was expressed with cardiac β_2 subunits (121, 141). Although the β_{1a} subunit did not alter the kinetics of the smooth muscle isoform of the cardiac $\alpha1$ subunit (137), the cardiac β_2 subunits accelerated activation of the calcium current in oocytes expressing the cardiac $\alpha1$ subunit and shifted the voltage dependence of activation to more negative membrane potentials (121, 141). β_3 was more effective at increasing current than either β_{2a} or β_{2b}. Thus, the proteins encoded by the different β subunit genes display varied effects on the calcium currents mediated by cardiac $\alpha1$ subunits.

Co-expression of both $\alpha2\delta$ and β_{1a} subunits with the cardiac $\alpha1$ subunit in oocytes resulted in a greater increase in current amplitude, activation at more negative membrane potentials, a steeper voltage dependence of activation and inactivation, and a more rapid current time course compared to expression of the $\alpha1$ subunit with either of these subunits alone (139, 141). These observations suggest a synergistic interaction between the actions of the $\alpha2\delta$ and β subunits on the functional properties of the $\alpha1$ subunit.

The amplitudes of the calcium currents of the $\alpha1$ subunits of class A, class B N-type, class D L-type, and class E brain calcium channels are greatly increased by co-expression of $\alpha2\delta$ and β subunits (124, 129, 130, 142–145). The effects of the β subunit are dominant, with less-prominent effects of $\alpha2\delta$ in most experiments. γ subunits do not have a major effect. In many cases, significant expression of calcium channel function is only observed when auxiliary subunits are co-expressed. However, nuclear injection of expression vectors in *Xenopus* oocytes gives enhanced expression of the $\alpha1$ subunits alone (129, 138). Under these conditions, the effects of auxiliary subunits on the function of α1 subunits are observed most clearly, since substantial expression of the α1 subunit alone is achieved for comparison. With the class B, N-type calcium channel, the β_{1b} subunit increased expression of calcium current, increased the rates of activation and inactivation, and shifted the voltage dependence of inactivation toward more negative membrane potentials (129). In addition, co-expression of β_1, β_2, or β_3 subunits with a class A calcium channel α1 subunit causes a progressive increase in the rate of inactivation, which suggests that assembly of α1 subunits with different β subunits during biosynthesis of calcium channels or exchange of β subunits on pre-existing calcium channels may alter their functional properties (143). Like the L-type calcium channels, in which the α1 subunits are functionally autonomous but are modulated by their associated auxiliary subunits, the α1 subunits of neuronal non-L-type calcium channels can function as voltage-gated calcium channels, but they appear to require the presence of associated subunits for efficient functional expression, and expression of different auxiliary subunits alters their functional properties.

Mechanism of Regulation by β Subunits

The importance and multiplicity of the effects of the β subunits on calcium channel function are striking. In general, both the number of high-affinity sites for drug and toxin binding and the amplitude of peak calcium currents are increased, the time course of the calcium current is substantially accelerated, and the voltage dependence of activation and inactivation is altered. What molecular mechanisms can account for these multiple effects? Experiments to date provide some initial clues to possible mechanisms underlying these actions. As noted above, the β_{1a} subunit is a substrate for phosphorylation by numerous protein kinases, and the other β subunits have multiple consensus sites for phosphorylation. Moreover, the β_{1a} subunit is selectively dephosphorylated by phosphoprotein phosphatases (146). Thus, co-expression of different β subunits may induce different regulation by protein phosphorylation/dephosphorylation mechanisms. Since dephosphorylation can influence both the number of active calcium channels and the rate and extent of inactivation of calcium channels (147, 148), the effects of β subunits on these aspects of calcium channel expression and function may be due in part to phosphorylation/dephosphorylation mechanisms. Alternatively, recent experiments show that β subunits also influence the coupling of gating charge movement to channel opening (149). Like other voltage-gated ion channels, depolarization of voltage-gated calcium channels causes a series of voltage-driven transmembrane movements of charged amino acid residues in the α1 subunits that serve as gating charges. Following this movement of gating charges, the calcium channels can open but the probability of opening is much less than 1.0. For the class C α1 subunits expressed in *Xenopus* oocytes, co-expression of β_{2a} increased the efficiency of coupling of gating charge movement to channel opening by 5-fold. This provocative result suggests that the β subunit can lower the energy barrier for opening the channel pore and can thereby increase peak calcium currents without increasing the number of calcium channels expressed. Since the binding of dihydropyridine drugs to these L-type calcium channels is also influenced by channel state, it is conceivable that the number of high-affinity binding sites observed in expressed channels is also changed by this mechanism without a change in the actual number of calcium channel α1 subunit proteins expressed. This mechanism of action of the β subunits of the calcium channels would be similar to the mechanism of action proposed for the sodium channel β1 subunits that reduce the energy barrier for shift of channel state to a fast gating mode for sodium channels expressed in *Xenopus* oocytes.

A site of interaction of β subunits with α1 subunits has been identified by a novel library-screening method (150). cDNAs that encode segments of the intracellular loop between domains I and II of the α1 subunit were found to express fusion proteins that bind β subunits specifically. Co-expression of the corresponding peptides prevents the effects of β subunits on expression and function of α1 subunits. This segment of the α1 subunit interacts with a conserved motif in the β subunit (151). This interaction domain may mediate the change in energy of activation that increases the efficiency of calcium current expression upon co-expression of β subunits.

INTERACTIONS OF CALCIUM CHANNELS WITH INTRACELLULAR EFFECTOR PROTEINS

Calcium-Release Channels

In skeletal muscle, the L-type voltage-gated calcium channels in the transverse tubule membrane are thought to physically contact the ryanodine-sensitive calcium-release channels in the sarcoplasmic reticulum (SR) membrane. Voltage-driven conformational changes in the voltage-gated calcium channels serve to activate the SR calcium-release channel through protein-protein interactions. Thus, the SR calcium-release channel serves as an effector of the voltage-gated calcium channel in the process of excitation-contraction coupling. Mice with the *muscular dysgenesis* mutation have defective excitation-contraction coupling due to a mutation in the α1 subunit of the skeletal muscle calcium channel (152). Injection of an expression vector encoding the α1 subunit restores excitation-contraction coupling. The site of interaction with the SR calcium-release channel is in large intracellular loop-connecting domains II and III of the α1 subunit (153). Evidently, this loop can transmit a signal from voltage-dependent activation of the transverse tubule calcium channels that can activate the SR calcium-release channel.

Synaptic Membrane Proteins

Neuronal calcium channels in presynaptic nerve terminals initiate the process of neurotransmitter release by allowing the rapid entry of calcium in response to depolarization. The increase in calcium concentration which initiates neurotransmitter release is highly localized, and presynaptic calcium channels and synaptic vesicles are thought to be closely associated in active zones. Recent evidence indicates that this association may be mediated in part through binding of presynaptic calcium channels to synaptic membrane proteins. N-type calcium channels, which are involved in release of neurotransmitters at many synapses, bind to the synaptic plasma protein syntaxin and possibly also to the synaptic vesicle protein synaptotagmin (154–156). Recent studies with fusion proteins representing different segments of the N-type calcium channel α1 subunit show that the site of interaction is in the intracellular loop between homologous domains II and III (157). Thus, the intracellular loop between domains II and III may serve as an effector interaction domain for neurotransmitter release in presynaptic terminals as well as for excitation-contraction coupling in skeletal muscle. Additional interactions for this domain of calcium channel α1 subunits may participate in other cellular signaling processes that are initiated by depolarization or by calcium influx.

STRUCTURE AND FUNCTION OF K+ CHANNEL SUBUNITS

Principal Subunits of K+ Channels

Voltage-gated K+ channels are functionally diverse (158–160). They can be classified into two major groups based on physiological properties: delayed rectifiers which activate after a delay following membrane depolarization and either inactivate slowly or do not inactivate at all, and A-type K+ channels which are fast-activating and inactivate. The molecular structure of the voltage-gated K+ channels was first revealed by molecular cloning of the gene encoding the *Shaker* mutation in *Drosophila* (160, 161). A-type K+ channels and delayed rectifier K+ channels in *Drosophila* and vertebrates have principal subunits whose polypeptide backbones are 60 to 80 kD and are homologous in structure to a single domain of the α or α1 subunits of Na+ or Ca2+ channels (Figure 2C). They form homotetramers and heterotetramers that are fully functional as voltage-gated ion channels, and they are alternatively spliced adding additional diversity (160, 161). However, several lines of evidence now indicate that, like Na+ and Ca2+ channels, they contain one or more auxiliary subunits as components of their oligomeric structure in situ.

Subunit Composition of Purified Neuronal K+ Channels

Some brain K+ channels contain a receptor for dendrotoxins (DTX), a family of neurotoxins isolated from the venom of the black mamba snake,

Dendroaspis polyepsis (162). These basic polypeptide toxins inhibit A-type K⁺ channels, resulting in the facilitation of neurotransmitter release and epileptiform activity. The DTX binding site is the receptor for three additional peptide ligands: mast cell degranulating peptide (MCD), β-bungarotoxin (β-BTX), and charybdotoxin (CTX) (162). These toxins have been successfully used as molecular probes to identify and purify components of A-type and delayed rectifier K⁺ channels from mammalian brain.

Photo-affinity labeling of potassium channels with β-bungarotoxin reveals peptide components of approximately 75 kDa and 28 kDa in chick brain (163). Purification of DTX-binding proteins from detergent-solubilized mammalian brain membranes revealed a noncovalently associated glycoprotein complex containing polypeptides of 74–80 kD in association with smaller polypeptides of 42 kD, 38 kD, and 35 kD when analyzed by SDS-PAGE (164–166). The 35 kD polypeptide was observed in variable amounts and may be a proteolytic fragment of one of the other polypeptides (164). Binding of $[^{125}I]$DTX and $[^{125}I]$ MCD to the purified receptor was inhibited by β-BTX, demonstrating that K⁺ channels with binding sites for all three ligands were co-purified. The 80 kD DTX-binding protein from rat brain represents a family of pharmacologically and structurally related glycoproteins (164, 167, 168). Neuraminidase treatment reduced its apparent molecular weight to 70 kD, and subsequent treatment with endoglycosidase F further reduced this to 65 kD, indicating that the 80 kD subunit is a sialylated membrane glycoprotein that is exposed to the extracellular surface. In contrast, treatment of the 38 kD polypeptide with neuraminidase or endoglycosidase F had no effect on its mobility on SDS-PAGE, indicating that this component most likely did not contain N-linked sugars (168, 169). The N-terminal amino acid sequence of the 80 kD subunit from bovine brain was virtually identical to that deduced from the cDNA of the mouse/rat homologue of the *Shaker* family of K⁺ channels, RCK 5 (MK2/RBK2), a K⁺ channel protein from rat brain known to be a DTX-sensitive, delayed rectifier (166, 168). This conclusively demonstrated that the DTX receptor protein is a K⁺ channel.

Antibodies against a fusion protein generated from a cDNA that encodes a delayed rectifier K⁺ (drk) channel polypeptide from rat brain (170) were used to identify its protein components. Similar to K⁺ channels identified via their sensitivity to DTX, this putative delayed rectifier K⁺ channel also contained a low molecular weight subunit. Anti-drk antibodies specifically immunoprecipitated a complex of 130 kD and 38 kD polypeptides. Because the antibodies recognized the 130 kD polypeptide exclusively and co-precipitated the 38 kD subunit, it was concluded that these two polypeptides were immunologically distinct proteins in a heterooligomeric complex with a 1:1 stoichiometry. Thus, both delayed rectifier and A-type neuronal K⁺ channels may have β subunits (Figure 1C).

Molecular Properties of an Auxiliary Subunit of a K⁺ Channel

Recent work by Scott et al. (171) has resulted in the cloning and primary structure determination of the first potassium channel auxiliary subunit. The β subunit of the dendrotoxin-sensitive potassium channel from bovine brain is a protein of 367 amino acids (41 kD) which has no hydrophobic segments and is not glycosylated. Its amino acid sequence contains several consensus sites for protein phosphorylation, and the protein is rapidly phosphorylated by cAMP-dependent protein kinase. Analysis of probable secondary structure reveals four major probable alpha helical segments. In all of these respects, the potassium channel β subunit closely resembles the calcium channel β subunit. However, no extended amino acid sequence homology is detectable, indicating that the evolutionary relationship between the two proteins is distant.

Like the calcium channel β subunits, potassium channel β subunits have important functional effects when co-expressed with α subunits. The time courses of both activation and inactivation are accelerated (172). Moreover, multiple potassium channel β subunits have been identified and they have differential effects on function of the α subunits. Thus, the potassium channel β subunits have a functional analogy with calcium channel β subunits as well as a general structural similarity. It will be of interest to determine whether the mechanisms of action of these two classes of auxiliary subunits are similar.

STRUCTURAL BASIS FOR ION CHANNEL FUNCTION

Purification, molecular cloning, and determination of the primary structures of the principal subunits of sodium, calcium, and potassium channels have provided a molecular template for probing the relationship between their structure and function [reviewed in (2, 173, 174)]. The structure of the principal subunits of each of these channels is based on the same motif (Figure 2; see Figure 4 also): four homologous transmembrane domains which contain six probable transmembrane alpha helices and surround a central ion pore. In potassium channels, each domain is a separate gene product and the functional channel is a homo- or hetero-tetramer. In sodium and calcium channels, four domains are contained within a single functional α or α1 subunit. Although each channel contains associated auxiliary subunits, the principal subunits can carry out the basic functions of the voltage-gated ion channels by themselves and therefore must contain the necessary structural elements for ion channel function within them. This section reviews the current status of research on the molecular basis of the three major elements of voltage-gated ion channel function: voltage-dependent activation, ion conductance, and inactivation.

VOLTAGE-DEPENDENT ACTIVATION

Activation of the voltage-gated ion channels is thought to result from a voltage-driven conformational change which opens a transmembrane pore through the protein. Depolarization of the membrane exerts an electrical force on voltage sensors that contain the gating charges of the channel located within the transmembrane electrical field. These gating charges are likely to be charged amino acid residues located in transmembrane or membrane-associated segments of the protein. The movement of the gating charges of the sodium channel through the membrane driven by depolarization has been directly measured as an outward gating current (175). The gating current for sodium channels was estimated to correspond to the movement of approximately six charges all the way across the membrane; movement of a larger number of charges across a fraction of the membrane electric field would be equivalent. More detailed analyses of potassium channel gating currents suggest even larger gating charge movements. Identification of the voltage-sensors and gating charges of the voltage-gated ion channels is the first critical step toward understanding the molecular basis of voltage-dependent activation.

Inspection and analysis of the primary structure of the sodium channel α subunit, the first member of the voltage-gated ion channel gene family, led to the prediction that the fourth transmembrane segment in each domain (the S4 segment) might serve as the voltage sensor (2, 173, 176). These segments contain repeated motifs of a positively charged residue followed by two hydrophobic residues. For comparison among different channels, these positive charges have been numbered sequentially from the extracellular end of the segment (Figure 3). The prediction that these positive charges serve as gating charges has been tested by mutagenesis and expression studies of both sodium and potassium channels. In each case, the gating charge has been inferred from measurements of the steepness of voltage-dependence of channel activation at low levels of activation where this provides an indirect estimate of gating charge.

Neutralization of the four positively charged residues in the S4 segment of domain I of the sodium channel by site-directed mutagenesis has major effects on the voltage dependence of activation (177). Neutralization of the arginine residue in position 1 (R1) had little effect on the steepness of sodium channel activation, but neutralization of the positively charged residues in positions 2 through 4 in this S4 segment reduced the apparent gating charge by 0.9 to 1.8 charges. Combined neutralization of multiple charged residues and mutations of positive charges to negative charges causes progressively increasing reduction of gating charge, but the reduction of apparent gating charge is less than proportional to the expected reduction in total charge of the S4 segment. In addition, most of the mutations also caused shifts of the voltage dependence of activation to more positive or more negative membrane potentials.

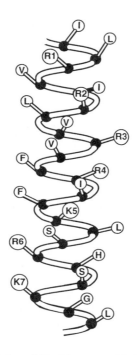

Figure 3   An S4 segment of a K⁺ channel. The S4 segment of the Shaker K⁺ channel of *Drosophila* is illustrated in a ball-and-stick alpha helical model. Amino acids are illustrated in single-letter code and positively charged amino acids in every third position are numbered from the extracellular end of the helix.

Neutralization of the positively charged residues in the S4 segments of potassium channels also caused reduction in apparent gating charge and shifts of the voltage dependence of activation (178–180). For the *Shaker* potassium channel of *Drosophila,* neutralization of R1 (Figure 1B) causes an unexpectedly large reduction in apparent gating charge while charge reversal has little additional effect (179, 180). Most mutations at this site caused a positive shift in the voltage dependence of gating (179, 180). In contrast, neutralization and charge reversal at R2 and K7 caused changes in apparent gating charge that were closely correlated with the changes in charge caused by the mutations (179, 180), and neutralization of R3 causes a substantial reduction in one component of gating current (181). Analysis of the shifts in voltage dependence caused by mutation of each of the positively charged residues in an S4 helix showed an alternating pattern of voltage shifts for most residues: positive shifts for neutralization in positions 1, 3, 5, and 7; and negative shifts for positions

2 and 4 (182). Only the arginine in position 6 deviates from this pattern. These results are consistent with a model in which all of the gating charges in odd-numbered positions make salt bridges that stabilize the activated state while those in positions 2 and 4 make salt bridges that stabilize the closed state (180, 182). Most mutations that reduce positive charge at R2 caused a negative shift in the voltage dependence of activation while those at K7 caused a positive shift. Overall, the studies of S4 segments in potassium channel gating support the conclusion that the positively charged residues in these segments are indeed gating charges involved in the voltage sensors of the ion channels. The individual residues appear to contribute differentially to the overall apparent gating charge, and their size and chemical properties (other than charge) also have important influences on gating.

If the S4 segments must move through the protein structure during the process of activation, the size and shape of the hydrophobic residues in these segments should also have an important influence on voltage-dependent activation. Mutation of a single leucine residue to phenylalanine in an S4 segment of a sodium channel shifts the voltage dependence of gating 20 mV (183). Similarly, mutation of several hydrophobic residues in the S4 segments of potassium channels also causes dramatic shifts (up to 80 mV) in the voltage dependence of activation (184, 185). In contrast, mutations of several hydrophobic residues in other transmembrane segments did not have major effects on activation (185). These results, together with the effects of charge neutralization mutations, provide strong support for identification of the S4 segments as the voltage sensors of the voltage-gated ion channels and for identification of the positively charged residues within them as the gating charges of the channel.

The mechanism by which the S4 segments serve as voltage sensors is not known. The "sliding helix" or "helical screw" models (2, 176) proposed that the entire S4 helix moves across the membrane along a spiral path exchanging ion pair partners between its positively charged residues and fixed negative charges in surrounding transmembrane segments. This model implies a large (but unknown) energy barrier for breaking and re-making numerous ion pairs within the protein structure, and suggests an approximate equivalence of gating charge movement among the different charged residues in the S4 helices. Because neutralization of individual charged residues has very different effects on the voltage dependence of channel activation, it is unlikely that this simple model can be correct in detail. A more complex "propagating helix" model proposes that the S4 transmembrane segments undergo an alpha helix-beta sheet transition which propagates outward to move charged residues across the membrane (176). This model also has no direct experimental support, but it has the potential to accommodate at least some of the differences observed among individual residues because not all charged residues in the S4 segment are proposed to move the same distance across the membrane. Thus, although the S4 segments are clearly implicated as the voltage sensors of the voltage-gated ion channels, the mechanism through which they initiate activation of the channels remains unknown.

ION CONDUCTANCE

Essentially all models for the structure of the voltage-gated ion channels include a transmembrane pore in the center of a square array of homologous transmembrane domains. Each domain in these models would contribute one fourth of the wall of the pore. Identification of the segments which line the transmembrane pore and define the conductance and ion selectivity of the channels is of great interest and importance. A number of toxins, drugs, and inorganic cations are blockers of the voltage-gated ion channels. In several cases, detailed biophysical analysis of their mechanism of action indicates that these molecules enter and bind within the transmembrane pores of the channels and compete with permeant ions for occupancy of the pore (186). These channel blockers therefore serve as molecular markers and specific probes of the pore region of the ion channels. Amino acid residues that form the extra-cellular and intracellular mouths of the transmembrane pores have been iden-tified by their interaction with pore-blocking drugs and toxins.

The Extracellular Mouth of the Pore

Tetrodotoxin and saxitoxin are thought to block sodium channels by binding with high affinity to the extracellular mouth of the pore (186). Their binding is so specific that they were used as a marker in the initial purification of sodium channels from excitable cell membranes. Block of their binding by protonation or covalent modification of carboxyl residues led to the model that these cationic toxins bind to a ring of carboxyl residues at the extracellular mouth of the pore (186). These residues have now been identified by site-di-rected mutagenesis. Noda et al (187) neutralized glu387 in rat brain sodium channel II by mutagenesis to glutamine and expressed and analyzed the func-tional properties of the mutant channel. The affinity for tetrodotoxin was reduced over 10,000-fold. This amino acid residue is located in segment SS2 (Figures 1A and 4) on the extracellular side of the S6 transmembrane segment in domain I of the sodium channel. Subsequent extension of their analysis identified acidic amino acid residues in the same position as glu387 in each domain that were all required for high-affinity tetrodotoxin binding (188). These residues are therefore likely to surround the extracellular opening of the pore and contribute to a receptor site for tetrodotoxin. In addition to the ring of carboxyl residues, a second ring of amino acids located three residues on the amino-terminal side of these is also required for tetrodotoxin binding

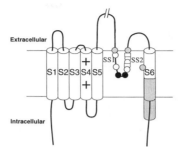

Figure 4 Pore-forming region of the homologous domains of voltage-gated ion channels. The transmembrane folding pattern of a single homologous domain of a voltage-gated ion channel is illustrated as in Figure 1. The short segments SS1 and SS2 and the positions of amino acid residues and segments which have been implicated in pore formation are illustrated. *Open circles,* residues required for ion conductance and selectivity of K$^+$ channels; *open squares,* residues required for ion conductance and selectivity of Na$^+$ and Ca^{2+} channels; *shaded circles,* residues required for high affinity binding of pore blockers of K$^+$ channels; *filled circles,* residues required for both ion conductance and binding of pore blockers of K$^+$ channels; *shaded bar,* segment of Ca^{2+} channels binding phenylalkylamine pore blockers.

(Figures 1A and 4). These are acidic amino acids in domains I and II, basic in domain III, and neutral in domain IV. If this region is in alpha helical conformation, this second set of residues required for tetrodotoxin binding would fall on the same side of the helix and form a second inner ring of residues at the opening of the pore.

Cardiac sodium channels bind tetrodotoxin with 200-fold lower affinity than brain or skeletal muscle sodium channels but retain all of the eight residues described above that are required for high-affinity binding. However, at position 385, two residues toward the amino-terminal from glu387 in the brain sodium channel II sequence, there is a change of a tyrosine or phenylalanine in the brain and skeletal muscle channels to cysteine in the cardiac sodium channel. Mutation of this residue from cys to phe or tyr causes an increase of 200-fold in the affinity of the cardiac sodium channel and the converse mutation causes a loss of affinity of 200-fold in the brain or skeletal muscle channel (189–191). Thus, it is likely that this residue also contributes in an essential way to the tetrodotoxin receptor site. Cadmium is a high-affinity blocker of cardiac sodium channels but not of brain or skeletal muscle sodium channels. Substitution of this critical cysteine in the skeletal muscle or brain sodium channel confers high-affinity block by cadmium on these channels (189, 190). Analysis of the voltage dependence of cadmium block suggests that this ion passes 20% of the way through the membrane electrical field in reaching its binding site formed by this cysteine residue (190). Thus, this residue may be approximately 20% of the way through the electrical field within the pore of the sodium channel.

The outer mouth of the potassium channel has been mapped in a similar way. The polypeptide charybdotoxin is an extracellular blocker of potassium channels which binds in the outer mouth of the pore. Identification of amino acids which contribute to binding of charybdotoxin reveals glutamic acid, aspartic acid, and threonine residues which are required for high-affinity binding (192, 193). The required residues cluster on both sides of the SS1-SS2 region as illustrated in Figure 4, and the residues closest to these short segments are most important for charybdotoxin binding. Conversion of any of these residues to a positively charged amino acid increases the K_d for charybdotoxin binding more than 300-fold, suggesting that positively charged amino acids in the toxin may normally interact with the negatively charged and hydroxylic residues in these positions in the wild-type channel.

Tetraethylammonium ions also block potassium channels from the extracellular side. Analysis of their affinity for block of the same family of potassium channel mutants reveals that residues on both sides of SS1 and SS2 are required, with the amino acid residue in position 449 on the carboxyl terminal side of SS2 being dominant (194). Tyrosine or phenylalanine in this position confers high-affinity block. Carboxyl residues in this position give intermediate affinity, and positively charged residues prevent tetraethylammonium binding completely (194). Phenylalanine residues in this position in all four subunits of a potassium channel can participate in binding tetraethylammonium ion, which suggests that the four phenyl rings coordinate a single tetraethylammonium molecule through cation-π orbital interactions (195, 196). These residues are similar in position in the amino acid sequence to those that are required for tetrodotoxin binding to sodium channels. Thus, it seems likely that tetraethylammonium ions in potassium channels and tetrodotoxin in sodium channels occupy similar receptor regions at the extracellular mouth of the pore when they block the channels.

The Intracellular Mouth of the Pore

Local anesthetics and related antiarrhythmic drugs are thought to bind to a receptor site on the sodium channel that is accessible only from the intracellular side of the membrane and is more accessible when the sodium channel is open (186). Similarly, the phenylalkylamine class of calcium channel antagonists are characterized as intracellular open-channel blockers, and tetraethylammonium and related monoalkyltrimethylammonium derivatives can block potassium channels from the intracellular side of the membrane when the channel is open (186). Both biochemical and molecular biological approaches have been used to probe the peptide segments of the principal subunits of the voltage-gated ion channels that interact with these intracellular pore blockers.

Verapamil and related phenylalkylamine calcium channel antagonist drugs are the highest affinity ligands among the diverse intracellular pore blockers. Desmethoxyverapamil and its photoreactive azido derivative ludopamil have K_ds for equilibrium binding to purified calcium channels in the range of 30 nM, and therefore can be used as highly specific binding probes of their receptor site in the intracellular mouth of the calcium channel. Covalent labeling of purified calcium channels with ludopamil results in incorporation into the $\alpha 1$ subunit only (197). The site of covalent labeling was located by extensive proteolytic cleavage of the labeled $\alpha 1$ subunit followed by identification of the photolabeled fragments by immunoprecipitation with site-directed antipeptide antibodies (197). All of the covalent label recovered was incorporated into a peptide fragment containing the S6 segment of domain IV of the $\alpha 1$ subunit and several amino acid residues at the intracellular end of this transmembrane segment. Since phenylalkylamines act only from the inside of the cell, it was concluded that the intracellular end of transmembrane segment IVS6 and the adjacent intracellular residues form part of the receptor site for phenylalkylamines and therefore part of the intracellular mouth of the calcium channel (197, 198).

Analysis of mutations that alter block of potassium channels from the intracellular side by tetraalkylammonium ions point to both the amino acid residues between SS1 and SS2 and the intracellular end of the S6 segment as components of the intracellular mouth of the pore. Mutation of a critical threonine residue (position 441 in the *Shaker* potassium channel) in the sequence between SS1 and SS2 to the closely related amino acid serine is sufficient to increase the K_d for block of potassium channels by intracellular tetraethylammonium ion 10-fold (199). Tetraethylammonium must traverse only 15% of the membrane electrical field in reaching this binding site from the intracellular solution. Thus, it is likely that this threonine residue forms part of a binding site for tetraethylammonium ions at the intracellular mouth of the potassium channel. Alkyltriethylammonium ions with long carbon chains in the alkyl group also require this threonine residue for high-affinity binding and block. In addition, mutations of a threonine residue near the middle of the S6 segment (position 469 in the *Shaker* channel) to a hydrophobic residue increase the affinity for C8 and C10 alkyltriethylammonium ions (200). A more hydrophobic residue in this position gives higher affinity for the alkyltriethylammonium ions, and the effect is greater for larger alkyl substituents consistent with hydrophobic interactions between residues at this position and the alkyl group of the substituted tetraalkylammonium ion. Thus, in potassium channels as well as calcium channels (197, 198), the S6 segments contribute to formation of the intracellular mouth of the pore and to binding of hydrophobic pore-blocking drugs.

Local anesthetics which block Na^+ channels also interact with amino acid residues in S6 segments. Analysis of the S6 segment in domain IV of the Na^+ channel α subunit by alanine-scanning mutagenesis revealed two aromatic amino acid residues in positions 1764 and 1771 which were required for high-affinity binding of a local anesthetic (201). These two residues are located approximately 11 Å apart on the same face of the proposed S6 α helix and may interact with the aromatic and positively charged amino groups of the drug molecule which are spaced 10 to 15 Å apart. It is likely that these pore-blocking drugs also interact with amino acid residues in the S6 segment of other domains of the Na^+ channel as they bind within the pore surrounded by all four domains.

Ion Conductance and Selectivity

Consistent with the idea that the amino acid residues that are required for binding of pore blockers are also required for interaction with permeant ions, changes in these residues have dramatic effects on ion conductance and selectivity. A clear demonstration of the close relationship between the amino acid residues that determine pore blocking properties and those that determine ion conductance and selectivity came from studies of chimeric potassium channels in which the SS1/SS2 region is transferred between channel types that differ in both single-channel conductance and affinity for tetraethylammonium ion at intracellular and extracellular sites (202). Such chimeric channels have the ion conductance, affinity for extracellular tetraethylammonium ion, and affinity for intracellular tetraethylammonium ion specified by the SS1/SS2 region with little effect of the remainder of the channel structure. Small changes in individual amino acids within this region also have dramatic effects on ion selectivity (Figure 4). Changes of threonine to serine or phenylalanine to serine increase the conductance of Rb^+ and NH_4^+ (203). Coordinate changes of leucine to valine and valine to isoleucine in two potentially interacting positions in the deep pore region are responsible for the differences in conductance and binding affinity for intracellular tetraethylammonium ion between chimeric channels which differ in the SS1/SS2 region (204). Deletion of two residues (tyr445, gly446) from the SS2 region of the *Shaker* voltage-gated potassium channel to yield a sequence similar in length to the corresponding region of the distantly related cyclic nucleotide-gated ion channels causes loss of potassium selectivity and increased channel block by divalent cations which are characteristic of the cyclic nucleotide-gated ion channels (205). These results indicate that even small alterations in the amino acid residues in the SS1/SS2 region have crucial effects on ion conductance and selectivity, supporting the conclusion that these residues form part of the lining of the transmembrane pore and interact directly with permeant ions. Moreover, changes in amino acid sequence in this region can also determine ion selectivity properties of different families of potassium channels. It is surprising that so many of the amino acid residues in this region are hydrophobic and that hydrophobic residues are critical determinants of ion selectivity. It remains to be determined how these residues interact with permeant ions to allow rapid and selective ion conductance.

As for potassium channels, changes in the amino acid residues in the SS1/SS2 region that are important for binding of the pore blocker tetrodotoxin are also critical determinants of sodium channel ion conductance and selectivity. Single-channel conductance values for mutations which neutralize single charges among the six negatively charged amino acid residues that are important for tetrodotoxin binding also reduce single-channel conductance, in some cases to as little as 10% of wild-type levels (188). Sodium and calcium channels have similar overall structures, but strikingly different ion selectivity. Sodium ions are essentially impermeant through calcium channels in the presence of calcium, but are rapidly permeant in the absence of divalent cations (186). This property is thought to arise from high-affinity binding of calcium ions to two sites in the ion conductance pathway that blocks sodium ion entry and allows rapid calcium conductance. Calcium ions are less than 10% as permeable as sodium ions through the sodium channel. Remarkably, mutation of only two amino acid residues in the sodium channel is sufficient to confer calcium channel-like permeability properties (206). Mutation of lysine 1422 and alanine 1714 to negatively charged glutamate residues not only altered tetrodotoxin binding but also caused a dramatic change in the ion selectivity of the sodium channel from sodium-selective to calcium-selective. In addition, these changes created a high-affinity site for calcium binding and block of monovalent ion conductance through the sodium channel, as has been previously described for calcium channels. Thus, in the mutant sodium channel with two additional negative charges near the extracellular mouth of the putative pore region, monovalent cation conductance is high in the absence of calcium. At calcium concentrations in the 10 µM range, monovalent cation conductance is strongly inhibited by high affinity calcium binding. As calcium concentrations are increased, calcium conductance is preferred over sodium conductance. These results mirror the ion conductance properties of calcium channels and indicate that a key structural determinant of the ion selectivity difference between calcium and sodium channels is specified by the negatively charged amino acid residues at the mouth of the putative pore-forming region.

These results on Na^+ channels imply that the corresponding amino acid residues in Ca^{2+} channels are also important determinants of their ion selectivity. Consistent with this expectation, mutation of the corresponding amino acid residues in calcium channels does cause marked alterations in ion selectivity, ion binding, and block of the pore by divalent cations (207, 208). These results demonstrate the importance of amino acid residues in analogous positions within the pore region in defining the selectivity of Na$^+$ and Ca^{2+} channels.

INACTIVATION

Fast Inactivation of Sodium Channels

Fast inactivation of the sodium channel acquires most of its voltage dependence from coupling to voltage-dependent activation, and the inactivation process can be specifically prevented by treatment of the intracellular surface of the sodium channel with proteolytic enzymes (175). These results led to the proposal of an autoinhibitory, "ball-and-chain" model for sodium channel inactivation in which an inactivation particle tethered on the intracellular surface of the sodium channel (the ball) diffuses to a receptor site in the intracellular mouth of the pore, binds, and blocks the pore during the process of inactivation (175). This model predicts that an inactivation gate on the intracellular surface of the sodium channel may be responsible for its rapid inactivation.

The sodium channel segments that are required for fast inactivation have been identified by use of a panel of site-directed anti-peptide antibodies against peptides corresponding to short (15 to 20 residue) segments of the α subunit. These anti-peptide antibodies were applied to the intracellular surface of the sodium channel from the recording pipet in whole-cell voltage clamp experiments or from the bathing solution in single-channel recording experiments in excised, inside-out membrane patches (209, 210). In both cases, only one antibody, directed against the short intracellular segment connecting homologous domains III and IV (Figures 1 and 5), inhibited sodium channel inactivation. Inhibition of fast inactivation of antibody-modified sodium channels in membrane patches was complete. The binding and effect of the antibody were voltage-dependent. At negative membrane potentials at which sodium channels are not inactivated, the antibody bound rapidly and inhibited channel inactivation; at more positive membrane potentials at which the sodium channel is inactivated, antibody binding and action were greatly slowed or prevented. Based on these results, it was proposed that the segment that this antibody recognizes is directly involved in the conformational change leading to channel inactivation. During this conformation change, this inactivation gating segment was proposed to fold into the channel structure, serve as the inactivation gate by occluding the transmembrane pore, and become inaccessible to antibody binding (209, 210) (Figure 5).

A similar model is supported by site-directed mutagenesis experiments (177). Expression of the sodium channel α subunit in *Xenopus* oocytes as two pieces corresponding to the first three domains and the fourth domain results in channels that activate normally but inactivate 20-fold more slowly than

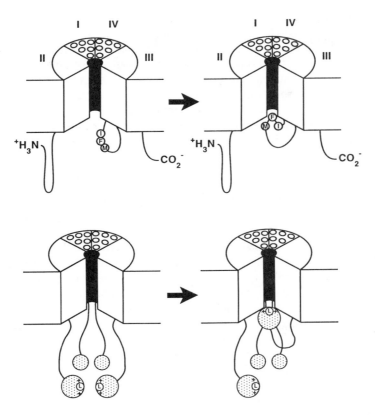

Figure 5   Mechanisms of inactivation of Na$^+$ and K$^+$ channels. The hinged-lid mechanism of Na$^+$ channel inactivation (A) and the ball-and-chain mechanism of K$^+$ channel inactivation (B) are illustrated. The intracellular loop connecting domains III and IV of the Na$^+$ channel is depicted as forming a hinged lid. The critical residues leu7 (L) and phe1489 (F) are shown as occluding the intracellular mouth of the pores in the K$^+$ channel and the Na$^+$ channel, respectively.

normal. The physiological characteristics of these cut channels are similar to those of sodium channels with inactivation blocked by the site-directed antibody. In contrast, sodium channel α subunits cut between domains II and III have normal functional properties. These two independent approaches using site-directed antibodies and cut mutations provide strong support for identification of the short intracellular segment connecting domains III and IV as an inactivation gating loop.

The inactivation gating loop contains highly conserved clusters of positively charged and hydrophobic amino acid residues. Neutralization of the positively charged amino acid residues in the inactivation gating loop of the sodium channel by site-directed mutagenesis does not have a profound effect on channel inactivation (211, 212), although neutralization of the cluster of positively charged residues at the amino-terminal end of the loop does slow inactivation and shift the voltage dependence of both activation and inactivation (212). In contrast, deletion of the 10-amino acid segment at the amino-terminal end of the loop completely blocks fast sodium channel inactivation (212). Scans of the hydrophobic amino acid residues in this segment of the inactivation gating loop by mutation to the hydrophilic but uncharged residue glutamine show that mutation of the three-residue cluster IFM to glutamine completely blocks fast sodium channel inactivation (213). The single phenylalanine in the center of this cluster at position 1489 in sodium channel II is the critical residue (Figure 5). Conversion of this residue to glutamine is sufficient by itself to almost completely prevent fast channel inactivation. Mutation of the adjacent isoleucine and methionine residues to glutamine also has substantial effects. Substitution of glutamine for isoleucine slows inactivation 2-fold and makes inactivation incomplete, leaving 10% sustained current at the end of long depolarizations. Substitution of glutamine for methionine slows inactivation 3-fold. The interaction of phe1489 with the receptor of the inactivation gating particle is likely to be hydrophobic, because there is a close correlation between the hydrophobicity of the residue at that position and the extent of fast sodium channel inactivation (214). On the basis of these results, it has been proposed that these residues serve as the inactivation gating particle entering the intracellular mouth of the transmembrane pore of the sodium channel and blocking it during channel inactivation. The intracellular loop between domains III and IV therefore serves as an inactivation gate and closes the transmembrane pore of the sodium channel from the intracellular side of the membrane.

This model for inactivation implies that the intracellular mouth of the pore has a receptor site that binds the inactivation particle. In agreement with this, a pentapeptide containing the IFM motif restores inactivation to mutant Na$^+$ channels with phe1489 replaced by glutamine (215). The freely diffusible peptide can restore inactivation with the rapid kinetics and steep voltage dependence of the native Na$^+$ channel. Alteration of the IFM motif by substitution of glutamine or alanine residues prevents restoration of inactivation, consistent with the conclusion that the pentapeptide does indeed bind to a receptor that requires the same structure as the inactivation particle itself.

A candidate for the inactivation gate receptor region has been identified by alanine-scanning mutagenesis of the S6 segment in domain IV (216). Mutation of three adjacent hydrophobic residues at the intracellular end of this transmembrane segment to alanine produces Na$^+$ channels that inactivate incompletely, having greater than 85% of their Na$^+$ current remaining at the end of 15 ms depolarizing pulses. This mutation also causes prolonged single-channel openings and frequent reopenings that are characteristic of Na$^+$ channels in which the fast inactivation process has been disrupted. This site is one helical turn on the intracellular side of the amino acid residues that are required for binding of local anesthetic drugs. This may provide a molecular explanation for the finding that local anesthetic drugs can be trapped in the channel by closure of the inactivation gate (186).

A Hinged-lid Model of Sodium Channel Fast Inactivation

The proposed inactivation gate is a short structured loop of the channel which places the inactivation particle (IFM) close to the mouth of the transmembrane pore. The inactivation gate loop resembles the "hinged lids" of allosteric enzymes that are rigid peptide loops that fold over enzyme active sites and control substrate access. Conformational changes induced by the binding of allosteric ligands move the hinged lid away from the active site and allow substrate access and catalytic activity. In analogy to the allosteric enzymes, it has been proposed that the sodium channel inactivation gate functions as a hinged lid that pivots to place the inactivation particle in a position to bind to the intracellular mouth of the transmembrane pore of the sodium channel (213) (Figure 5). The three-dimensional structures of some hinged lids of allosteric enzymes that are known from x-ray crystallographic and two-dimensional NMR studies provide a valuable structural model for design of further experiments to define the mechanism of sodium channel inactivation in more detail.

Inactivation of Potassium Channels

In contrast to the results obtained with sodium channels, the inactivation particle responsible for rapid inactivation of potassium channels by occlusion of the intracellular mouth of the pore is located at the amino-terminus of the polypeptide (217). Removal of the amino-terminus, by deletion mutagenesis, prevents fast inactivation. Mutations of single positively charged amino acids in that region slow inactivation, and mutation of leu7 to a hydrophilic amino acid prevents fast inactivation almost completely (217). Free peptides with amino acid sequences modeled on the inactivation particle restore fast, voltage-dependent inactivation to mutant potassium channels whose amino-termini have been deleted (218). These results fit the expectations of the "ball-and-chain" model of ion channel inactivation originally proposed for sodium channels by Armstrong and Bezanilla (reviewed in 175). The amino-terminal segment of the potassium channels is envisioned as an inactivation particle tethered on the end of a chain of approximately 200 amino acids (Figure 5). The positively charged and hydrophobic residues in the ball are thought to interact with an inactivation receptor at the intracellular mouth of the channel through a process of restricted diffusion and binding. This mechanistic model predicts that shortening the chain of amino acids connecting the ball to the rest of the channel should accelerate inactivation, while elongating the chain should slow inactivation. The model also predicts that extracellular permeant ions should oppose inactivation as they diffuse through the pore and compete with the inactivation particle for its binding site. These effects were in fact observed for expressed potassium channels. Fifty residue segments deleted or inserted in the "chain" between the N-terminal inactivation particle and the membrane accelerate or slow inactivation (217). External potassium ions also slow inactivation and speed recovery from inactivation as expected if inactivation cannot occur while the pore is occupied by permeant ions (219). These results support the "ball-and-chain" mechanism as a valid model of potassium channel inactivation.

The "receptor" occupied by the N-terminal inactivation particle as it occludes the mouth of the potassium channel may include amino acid residues in the short intracellular loop connecting transmembrane segments S4 and S5. Mutagenesis of charged residues in this loop to neutral, hydrophilic ones and mutations of hydrophobic residues to alanine markedly reduce fast potassium channel inactivation (220). Mutations in this loop also reduce conductance and alter ion selectivity of the potassium channels (221). Thus, this short intracellular loop may contribute to formation of the intracellular mouth of the pore, along with sequences in the intracellular end of the S6 segment and the SS1/SS2 region as described above, and may serve to form a receptor for the amino-terminal inactivation particle.

Literature Cited

1. Catterall WA. 1980. *Annu. Rev. Pharmacol. Toxicol.* 20:15–43
2. Catterall WA. 1988. *Science* 242:50–61
3. Strichartz G, Rando T, Wang GK. 1987. *Annu. Rev. Neurosci.* 10:237–67
4. Beneski DA, Catterall WA. 1980. *Proc. Natl. Acad. Sci. USA* 77:639–43
5. Darbon H, Jover E, Couraud F, Rochat H. 1983. *Biochem. Biophys. Res. Commun.* 115:415–22
6. Jover E, Massacrier A, Cau P, Martin MF, Couraud F. 1988. *J. Biol. Chem.* 263:1542–48

7. Sharkey RG, Beneski DA, Catterall WA. 1984. *Biochemistry* 23:6078–86
8. Levinson SR, Ellory JC. 1973. *Nature New Biol.* 245:122–23
9. Barhanin J, Schmid A, Lombet A, Wheeler KP, Lazdunski M. 1983. *J. Biol. Chem.* 258:700–2
10. Angelides KJ, Nutter TJ, Elmer LW, Kempner ES. 1985. *J. Biol. Chem.* 260:3431–39
11. Agnew WS, Levinson SR, Brabson JS, Raftery MA. 1978. *Proc. Natl. Acad. Sci. USA* 75:2606–10
12. Agnew WS, Moore AC, Levinson SR,

Raftery MA. 1980. *Biochem. Biophys. Res. Commun.* 92:860–66
13. Miller JA, Agnew WS, Levinson SR. 1983. *Biochemistry* 22:462–70
14. Hartshorne RP, Catterall WA. 1981. *Proc. Natl. Acad. Sci. USA* 78:4620–24
15. Hartshorne RP, Messner DJ, Coppersmith JC, Catterall WA. 1982. *J. Biol. Chem.* 257:13888–91
16. Hartshorne RP, Catterall WA. 1984. *J. Biol. Chem.* 259:1667–75
17. Messner DJ, Catterall WA. 1985. *J. Biol. Chem.* 260:10597–604
18. Hartshorne RP, Coppersmith J, Catterall WA. 1980. *J. Biol. Chem.* 255:10572–75
19. Reber BF, Catterall WA. 1987. *J. Biol. Chem.* 262:11369–74
20. Grishin EV, Kovalenki VA, Pashikov VN, Shamotienko G. 1984. *Biol. Membr.* 1:858–66
21. Kraner SD, Tanaka JC, Barchi RL. 1985. *J. Biol. Chem.* 260:6341–47
22. Casadei JM, Gordon RD, Barchi RL. 1986. *J. Biol. Chem.* 261:4318–23
23. Roberts RH, Barchi RL. 1987. *J. Biol. Chem.* 262:2298–2303
24. Sutkowski EM, Catterall WA. 1990. *J. Biol. Chem.* 265:12393–99
25. Noda M, Shimizu S, Tanabe T, Takai T, Kayano T. 1984. *Nature* 312:121–27
26. Noda M, Ikeda T, Kayano T, Suzuki H, Takeshima H, et al. 1986. *Nature* 320:188–92
27. Kayano T, Noda M, Flockerzi V, Takahashi H, Numa S. 1988. *FEBS Lett.* 228:187–94
28. Goldin AL, Snutch T, Lubbert H, Dowsett A, Marshall J, et al. 1986. *Proc. Natl. Acad. Sci. USA* 83:7503–7
29. Auld VJ, Goldin AL, Krafte DS, Marshall J, Dunn JM, et al. 1988. *Neuron* 1:449–61
30. Sarao R, Gupta SK, Auld VJ, Dunn RJ. 1991. *Nucleic Acids Res.* 19:5673–79
31. Yarowski PJ, Krueger BK, Olson CE, Clevinger EC, Koos RD. 1991. *Proc. Natl. Acad. Sci. USA* 88:9453–57
32. Trimmer JS, Cooperman SS, Tomiko SA, Zhou JY, Crean SM, et al. 1989. *Neuron* 3:33–49
33. Kallen RG, Sheng ZH, Yang J, Chen LQ, Rogart RB, Barchi RL. 1990. *Neuron* 4:233–42
34. Rogart RB, Cribbs LL, Muglia LK, Kephart DD, Kaiser MW. 1989. *Proc. Natl. Acad. Sci. USA* 86:8170–74
35. Salkoff L, Butler A, Wei A, Scavarda N, Giffen K. 1987. *Science* 237:744–49
36. Ramaswami M, Tanouye MA. 1989. *Proc. Natl. Acad. Sci. USA* 86:2079–82
37. Loughney K, Kreber R, Ganetzky B. 1989. *Cell* 58:1143–54

38. Gautron S, Dos Santos G, Pinto-Henrique D, Koulakoff A, Gros F, Berwald-Netter Y. 1992. *Proc. Natl. Acad. Sci. USA* 89:7272–76
39. George AL Jr, Knittle TJ, Tamkun MM. 1992. *Proc. Natl. Acad. Sci. USA* 89:4893–97
40. Isom LL, De Jongh KS, Patton DE, Reber BFX, Offord J, et al. 1992. *Science* 256:839–42
41. Lombet A, Lazdunski M. 1984. *Eur. J. Biochem.* 141:651–60
42. Correa AM, Bezanilla F, Agnew WS. 1990. *Biochemistry* 29:6230–40
43. Messner DJ, Catterall WA. 1986. *J. Biol. Chem.* 261:211–15
44. Messner DJ, Feller DJ, Scheuer T, Catterall WA. 1986. *J. Biol. Chem.* 261:14882–90
45. Noda M, Ikeda T, Suzuki T, Takeshima H, Takahashi T, et al. 1986. *Nature* 322:826–28
46. Suzuki H, Beckh S, Kubo H, Yahagi N, Ishida H, et al. 1988. *FEBS Lett.* 228:195–200
47. Cribbs LL, Satin J, Fozzard HA, Rogart RB. 1990. *FEBS Lett.* 275:195–200
48. White MM, Chen L, Kleinfield R, Kallen RG, Barchi RL. 1991. *Mol Pharmacol.* 39:604–8
49. Joho RH, Moorman JR, VanDongen AMJ, Kirsch GE, Silberberg H, et al. 1990. *Mol. Brain Res.* 7:105–13
50. Krafte DS, Snutch TP, Leonard JP, Davidson N, Lester HA. 1988. *J. Neurosci.* 8:2859–68
51. Krafte DS, Goldin AL, Auld VJ, Dunn RJ, Davidson N, Lester HA. 1990. *J. Gen. Physiol.* 96:689–706
52. Patton DE, Isom LL, Catterall WA, Goldin AL. 1994. *J. Biol. Chem.* 269:17649–55
53. Cannon SC, McClatchey AI, Gusella JF. 1993. *Pfluegers Arch.* 423:155–57
54. Bennett PB Jr, Makita N, George AL Jr. 1993. *FEBS Lett.* 326:21–24
55. Scheuer T, Auld VJ, Boyd S, Offord J, Dunn R, Catterall WA. 1990. *Science* 247:854–58
56. West JW, Scheuer T, Maechler L, Catterall WA. 1992. *Neuron* 8:59–70
57. Ukomadu C, Zhou J, Sigworth FJ, Agnew WS. 1992. *Neuron* 8:663–76
58. Isom LL, Scheuer T, Brownstein AB, Ragsdale DS, Murphy BJ, Catterall WA. 1994. *J. Biol. Chem.* In press
59. Moorman JR, Kirsch GE, Vandongen AM, Joho RH, Brown AM. 1990. *Neuron* 4:243–52
60. Zhou J, Potts JF, Trimmer JS, Agnew WS, Sigworth FJ. 1991. *Neuron* 7:755–85
61. Catterall WA. 1991. *Cell* 64:871–74

528   CATTERALL

62. Adams BA, Beam KG. 1990. *FASEB J.* 4:2809–16
63. Rios E, Brum G. 1987. *Nature* 325:717–20
64. Sanchez JA, Stefani E. 1978. *J. Physiol.* 283:197–209
65. Almers W, Fink R, Palade PT. 1981. *J. Physiol.* 312:177–217
66. Rosenberg RL, Hess P, Reeves JP, Smilowitz H, Tsien RW. 1986. *Science* 231:1564–66
67. Curtis BM, Catterall WA. 1984. *Biochemistry* 23:2113–18
68. Borsotto M, Barhanin J, Fosset M, Lazdunski M. 1985. *J. Biol. Chem.* 260: 14255–63
69. Leung AT, Imagawa T, Campbell KP. 1987. *J. Biol. Chem.* 262:7943–46
70. Takahashi M, Seagar MJ, Jones JF, Reber BF, Catterall WA. 1987. *Proc. Natl. Acad. Sci. USA* 84:5478–82
71. Vaghy PL, Striessnig J, Miwa K, Knaus H.-G, Itagaki K, et al. 1987. *J. Biol. Chem.* 262:14337–42
72. Hosey MM, Barhanin J, Schmid A, Vandaele S, Ptasienski J, O'Callahan C., Cooper C, Lazdunski M. 1987. *Biochem. Biophys. Res. Commun.* 147: 1137–45
73. Sieber M, Nastainczyk W, Zubor V, Wernet W, Hofmann F. 1987. *Eur. J. Biochem.* 167:117–22
74. Curtis BM, Catterall WA. 1986. *Biochemistry* 25:3077–83
75. Flockerzi V, Oeken H-J, Hofmann F, Pelzer D, Cavalie A, Trautwein W. 1986. *Nature* 323:66–8
76. Morton ME, Froehner SC. 1987. *J. Biol. Chem.* 262:11904–7
77. Leung AT, Imagawa T, Block B, Franzini-Armstrong C, Campbell KP. 1988. *J. Biol. Chem.* 263:944–1001
78. Sharp AH, Campbell KP. 1989. *J. Biol. Chem.* 264:2816–25
79. Ahlijanian MK, Westenbroek RE, Catterall WA. 1990. *Neuron* 4:819–32
80. Tanabe T, Takeshima H, Mikami A, Flockerzi V, Takahashi H, et al. 1987. *Nature* 328:313–18
81. De Jongh KS, Merrick DK, Catterall WA. 1989. *Proc. Natl. Acad. Sci. USA* 86:8585–89
82. De Jongh KS, Warner C, Colvin AA, Catterall WA. 1991. *Proc. Natl. Acad. Sci. USA* 88:10778–82
83. Lai Y, Seagar MJ, Takahashi M, Catterall WA. 1990. *J. Biol. Chem.* 265: 20839–48
84. Burgess AJ, Norman RI. 1988. *Eur. J. Biochem.* 178:527–33
85. Ellis SB, Williams ME, Ways NR, Brenner R, Sharp AH, et al. 1988. *Science* 241:1661–64
86. Curtis BM, Catterall WA. 1985. *Proc. Natl. Acad. Sci. USA* 82:2528–32
87. Jahn H, Nastainczyk W, Röhrkasten A, Schneider T, Hofmann F. 1988. *Eur. J. Biochem.* 178:535–42
88. O'Callahan CM, Ptasienski J, Hosey MM. 1988. *J. Biol. Chem.* 263:17342–49
89. Ruth P, Röhrkasten A, Biel M, Bosse E, Regulla S, Meyer HE, Flockerzi V, Hofmann F. 1989. *Science* 245:1115–8
90. Jay SD, Ellis SB, McCue AF, Williams ME, Vedvick TS, Harpold MM, Campbell KP. 1990. *Science* 248:490–92
91. Bosse E, Regulla S, Biel M, Ruth P, Meyer HE, Flockerzi V, Hofmann F. 1990. *FEBS Lett.* 267:153–56
92. Vandaele S, Fosset M, Galizzi J-P, Lazdunski M. 1987. *Biochemistry* 26:5–9
93. De Jongh KS, Warner C, Catterall WA. 1990. *J. Biol. Chem.* 265:14738–41
94. Jay SD, Sharp AH, Kahl SD, Vedvick TS, Harpold MM, Campbell KP. 1991. *J. Biol. Chem.* 266:3287–93
95. Reuter H. 1983. *Nature* 301:569–74
96. Tsien RW. 1983. *Annu. Rev. Physiol.* 45:341–58
97. Schmid A, Barhanin J, Coppola T, Borsotto M, Lazdunski M. 1986. *Biochemistry* 25:3492–95
98. Takahashi M, Catterall WA. 1987. *Biochemistry* 26:5518–26
99. Schneider T, Hofmann F. 1988. *Eur. J. Biochem.* 174:369–75
100. Chang FC, Hosey MM. 1988. *J. Biol. Chem.* 263:18929–37
101. Tokumaru H, Anzai K, Abe T, Kirino Y. 1992. *Eur. J. Pharmacol. Mol. Pharmacol.* 227:363–70
102. Kuniyasu A, Oka K, Ide-Yamada T, Hatanaka Y, Abe T, et al. 1992. *J. Biochem.* 112:235–42
103. Tuana BS, Murphy BJ, Yi Q. 1987. *Mol. Cell. Biochem.* 76:173–84
104. Ferry DR, Goll A, Glossmann H. 1987. *Biochem. J.* 243:127–35
105. Tsien RW, Lipscombe D, Madison DV, Bley KR, Fox AP. 1988. *Trends Neurosci.* 11:431–38
106. Bean BP. 1989. *Annu. Rev. Physiol.* 51:367–84
107. Llinas R, Sugimori M, Hillman DE, Cherksey B. 1992. *Trends Neurosci.* 15: 351–55
108. Zhang J-F, Randall AD, Ellinor PT, Horne WA, Sather WA, et al. 1993. *Neuropharmacology* 32:1075–88
109. Takahashi M, Catterall WA. 1987. *Science* 236:88–91
110. Takahashi M, Fujimoto Y. 1989. *Biochem. Biophys. Res. Commun.* 163: 1182–88

111. Ahlijanian MK, Striessnig J, Catterall WA. 1991. *J. Biol. Chem.* 266:20192–97
112. McEnery MW, Snowman AM, Sharp AH, Adams ME, Snyder SH. 1991. *Proc Natl. Acad. Sci. USA* 88:11095–99
113. Sakamoto J, Campbell KP. 1991. *J. Biol. Chem.* 266:18914–19
114. Witcher DR, De Waard M, Sakamoto J, Franzini-Armstrong C, Pragnell M, et al. 1993. *Science* 261:486–89
115. Leveque C, El Far O, Martin-Moutot N, Sato K, Kato R. et al. 1994. *J. Biol. Chem.* 269:6306–12
116. Snutch TP, Reiner PB. 1992. *Curr. Opin. Neurobiol.* 2:247–53
117. Soong TW, Stea A, Hodson CD, Dubel SJ, Vincent SR, Snutch TP. 1994. *Science* 260:1133–36
118. Powers PA, Liu S, Hogan K, Gregg RG. 1992. *J. Biol. Chem.* 267:22967–72
119. Pragnell M, Sakamoto J, Campbell KP. 1991. *FEBS Lett.* 291:253–58
120. Hullin R, Singer-Lahat D, Freichel M, Biel M, Dascal N, et al. 1992. *EMBO J.* 11:885–90
121. Perez-Reyes E, Castellano A, Kim HS, Bertrand P, Baggstrom E, et al. 1992. *J. Biol. Chem.* 267:1792–97
122. Castellano A, Wei X, Birnbaumer L, Perez-Reyes E. 1993. *J. Biol. Chem.* 268:3450–55
123. Castellano A, Wei X, Birnbaumer L, Perez-Reyes E. 1993. *J. Biol. Chem.* 268:12359–66
124. Williams ME, Feldman DH, McCue AF, Brenner R, Velicelebi G, et al. 1992. *Neuron* 8:71–84
125. Kim H-L, Kim H, Lee P, King RG, Chin H. 1992. *Proc. Natl. Acad. Sci. USA* 89:3251–55
126. Biel M, Hullin R, Freundner S, Singer D, Dascal N, et al. 1991. *Eur. J. Biochem.* 200:81–88
127. Mikami A, Imoto K, Tanabe T, Niidome T, Mori Y, et al. 1989. *Nature* 340:230–33
128. Perez-Reyes E, Kim HS, Lacerda AE, Horne W, Wei XY, et al. 1989. *Nature* 340:233–36
129. Stea A, Dubel SJ, Pragnell M, Leonard JP, Campbell KP, Snutch TP. 1993. *Neuropharmacology* 32:1103–16
130. Mori Y, Friedrich T, Kim M-S, Mikami A, Nakai J, et al. 1991. *Nature* 350:398–402
131. Stea A, Tomlinson WJ, Soong TW, Bourinet E, Dubel SJ, et al. 1994. *Proc. Natl. Acad. Sci. USA.* 91:10576–80
132. Varadi G, Lory P, Schultz D, Varadi M, Schwartz A. 1991. *Nature* 352:159–62
133. Lacerda AE, Kim HS, Ruth P. Perez-Reyes E, Flockerzi V, et al. 1991. *Nature* 352:527–30
134. Gutierrez LM, Brawley RM, Hosey MM. 1991. *J. Biol. Chem.* 266:16387–94
135. Welling A, Bosse E, Cavalié A, Bottlender R, Ludwig A, et al. 1993. *J. Physiol.* 471:749–65
136. Biel M, Ruth P, Bosse E, Hullin R, Stühmer W, et al. 1990. *FEBS Lett.* 269:409–12
137. Itagaki K, Koch WJ, Bodi I, Klöckner U, Slish DF, Schwartz A. 1992. *FEBS Lett.* 297:221–25
138. Tomlinson WJ, Stea A, Bourinet E, Charnet P, Nargeot J, Snutch TP. 1993. *Neuropharmacology* 32:1117–26
139. Singer D, Biel M, Lotan I, Flockerzi V, Hofmann F, Dascal N. 1991. *Science* 253:1553–57
140. Wei X, Perez-Reyes E, Lacerda AE, Schuster G, Brown AM, Birnbaumer L. 1991. *J. Biol. Chem.* 266:21943–47
141. Hullin R, Singer-Lahat D, Freichel M, Biel M, Dascal N, et al. 1992. *EMBO J.* 11:885–90
142. Williams ME, Brust PF, Feldman DH, Patthi S, Simerson S, et al. 1992. *Science* 257:389–95
143. Sather WA, Tanabe T, Zhang J-F, Mori Y, Adams ME, Tsien RW. 1993. *Neuron* 11:291–303
144. Ellinor PT, Zhang J-F, Randall AD, Zhou M, Schwarz TL, et al. 1993. *Nature* 363:455–58
145. Brust PF, Simerson S, McCue AF, Deal CR, Schoonmaker S, et al. 1993. *Neuropharmacology* 32:1089–1102
146. Lai Y, Peterson BZ, Catterall WA. 1993. *Neurochem.* 61:1333–9
147. Pelzer D, Pelzer S, McDonald TF. 1990. *Rev. Physiol. Biochem. Pharmacol.* 114:108–206
148. Armstrong DL. 1989. *Trends Neurosci.* 12:117–22
149. Neely A, Wei X, Olcese R, Birnbaumer L, Stefani E. 1993. *Science* 262:575–78
150. Pragnell M, De Waard M, Mori Y, Tanabe T, Snutch TP, Campbell KP. 1994. *Nature* 368:67–70
151. De Waard M, Pragnell M, Campbell KP. 1994. *Neuron* 13:495–503
152. Tanabe T, Beam KG, Powell JA, Numa S. 1988. *Nature* 336:134–39
153. Tanabe T, Beam KG, Adams BA, Niidome T, Numa S. 1990. *Nature* 346:567–69
154. Leveque C, Hoshino T, David P, Shoji-Kasai Y, Leys K, et al. 1992. *Proc. Natl. Acad. Sci. USA* 89:3625–29
155. Bennett MK, Calakos N, Scheller RH. 1992. *Science* 257:255–59
156. Yoshida A, Oho C, Omori A, Kuwahara

R, Ito T, Takahashi M. 1992. *J. Biol. Chem.* 267:24925–28
157. Sheng Z-H, Rettig J, Takahashi M, Catterall WA. 1994. *Neuron.* 13:1303–13
158. Rudy B. 1988. *Neuroscience* 25:729–49
159. Rehm H, Tempel BL. 1991. *FASEB J.* 5:164–70
160. Pongs O. 1992. *Physiol. Rev.* 72:69–88
161. Jan LY, Jan Y-N. 1992. *Annu. Rev. Physiol.* 54:537–55
162. Dreyer F. 1990. *Rev. Physiol. Biochem. Pharmacol.* 115:94–128
163. Schmidt RR, Betz H. 1989. *Biochemistry* 28:8346–50
164. Rehm H, Lazdunski M. 1988. *Proc. Natl. Acad. Sci. USA* 85:4919–23
165. Parcej DN, Dolly JO. 1989. *Biochem. J.* 257:899–903
166. Newitt RA, Houamed KM, Rehm H, Tempel B. 1991. *Epilepsy Res.* 4:263–73
167. Rehm H, Newitt RA, Tempel BL. 1989. *FEBS Lett.* 249:224–28
168. Scott VES, Parcej DN, Keen JN, Findlay JBC, Dolly JO. 1990. *J. Biol. Chem.* 265:20094–97
169. Rehm H, Pelzer S, Cochet C, Chambaz E, Tempel BL, et al. 1989. *Biochemistry* 28:6455–60
170. Trimmer JS. 1991. *Proc. Natl. Acad. Sci. USA* 88:10764–68
171. Scott VES, Rettig J, Parcej DN, Keen JN, Findlay JBC, et al. 1993. *Proc. Natl. Acad. Sci. USA* 91:1637–41
172. Rettig J, Heinemann SH, Wunder F, Lorra C, Parcej DN, et al. *Nature* 369:289–94
173. Numa S. 1989. *Harvey Lect.* 83:121–65
174. Jan LY, Jan YN. 1989. *Cell* 56:13–25
175. Armstrong CM. 1981. *Physiol. Rev.* 61:644–82
176. Guy HR, Conti F. 1990. *Trends. Neurosci.* 13:201–6
177. Stühmer W, Conti F, Suzuki H, Wang X, Noda M. 1989. *Nature* 339:597–603
178. Papazian DM, Timpe LC, Jan YN, Jan LY. 1991. *Nature* 349:305–10
179. Liman ER, Hess P. 1991. *Nature* 353:752–56
180. Logothetis DE, Movahedi S, Satler C, Lindpaintner K, Nadal-Ginard B. 1992. *Neuron* 8:531–40
181. Perozo E, Santacruz-Toloza L, Stefani E, Bezanilla F, Papazian DM. 1994. *Biophys. J.* 66:345–54
182. Tytgat J, Nakazawa K, Gross A, Hess P. 1993. *J. Biol. Chem.* 268:23777–79
183. Auld VJ, Goldin AL, Krafte DS, Catterall WA, Lester HA, et al. 1990. *Proc. Natl. Acad. Sci. USA* 87:323–27
184. McCormack K, Tanouye MA, Iverson LE, Lin J-W, Ramaswami M, et al. 1991. *Proc. Natl. Acad. Sci. USA* 88:2931–35

185. Lopez GA, Jan YN, Jan LY. 1991. *Neuron* 7:327–36
186. Hille B. 1992. *Ionic Channels of Excitable Membranes*, Sunderland, MA: Sinauer
187. Noda M, Suzuki H, Numa S, Stühmer W. 1990. *FEBS Lett.* 259:213–16
188. Terlau H, Heinemann SH, Stühmer W. Pusch M, Conti F, et al. 1991. *FEBS Lett.* 293:93–96
189. Satin J, Kyle JW, Chen M, Bell P, Cribbs LL, et al. 1992. *Science* 256:1202–5
190. Backx PH, Yue DT, Lawrence JH, Marban E, Tomaselli GF. 1992. *Science* 257:248–51
191. Heinemann SH, Terlau H, Imoto K. 1992. *Pfluegers Arch.* 422:90–92
192. MacKinnon R, Miller C. 1989. *Science* 245:1382–85
193. MacKinnon R, Heginbotham L, Abramson T. 1990. *Neuron* 5:767–71
194. MacKinnon R, Yellen G. 1990. *Science* 250:276–79
195. Heginbotham L, MacKinnon R. 1992. *Neuron* 8:483–91
196. Kavanaugh MP, Varnum MD, Osborne PB, Christie MJ, et al. 1991. *J. Biol. Chem.* 266:7583–87
197. Striessnig J, Glossmann H, Catterall WA. 1990. *Proc. Natl. Acad. Sci. USA* 87:9108–9112
198. Catterall WA, Striessnig J. 1992. *Trends Pharmacol. Sci.* 13:256–62
199. Yellen G, Jurman ME, Abramson T, MacKinnon R. 1991. *Science* 251:939–42
200. Choi KL, Mossman C, Aubé J, Yellen G. 1993. *Neuron* 10:533–41
201. Ragsdale DS, McPhee JC, Scheuer T, Catterall WA. 1994. *Science* 265:1724–28
202. Hartmann HA, Kirsch GE, Drewe JA, Taglialatela M, Joho RH, Brown AM. 1991. *Science* 251:942–44
203. Yool AJ, Schwarz TL. 1991. *Nature* 349:700–4
204. Kirsch GE, Drewe JA, Hartmann HA, Taglialatela M, de Biasi M, et al. 1992. *Neuron* 8:499–505
205. Heginbotham L, Abramson A, MacKinnon R. 1992. *Science* 258:1152–55
206. Heinemann SH, Terlau H, Stühmer W, Imoto K, Numa S. 1992. *Nature* 356:441–43
207. Yang J, Ellinor PT, Sather WA, Zhang J.-F, Tsien RW. 1993. *Nature* 356:158–61
208. Tang S, Mikala G, Bahinski A, Yatani A, Varadi G, Schwartz A. 1993. *J. Biol. Chem.* 268:13026–29
209. Vassilev PM, Scheuer T, Catterall WA. 1988. *Science* 241:1658–61

210. Vassilev PM, Scheuer T, Catterall WA. 1989. *Proc. Natl. Acad. Sci. USA* 86: 8147–51
211. Moorman JR, Kirsch GE, Brown AM, Joho RH. 1990. *Science* 250: 688–91
212. Patton DE, West JW, Catterall WA, Goldin AL. 1992. *Proc. Natl. Acad. Sci. USA* 89:10905–9
213. West JW, Patton DE, Scheuer T, Wang Y, Goldin AL, Catterall WA. 1992. *Proc. Natl. Acad. Sci. USA* 89:10910–14
214. Scheuer T, West JW, Wang YL, Catterall WA. 1993. *Biophys. J.* 64:A88
215. Eaholtz G, Scheuer T, Catterall WA. 1994. *Neuron* 12:1041–48
216. McPhee JC, Ragsdale DS, Scheuer T, Catterall WA. 1994. *Proc. Natl. Acad. Sci. USA.* In press
217. Hoshi T, Zagotta WN, Aldrich RW. 1990. *Science* 250:533–38
218. Zagotta WN, Hoshi T, Aldrich RW. 1990. *Science* 250:568–71
219. Demo SD, Yellen G. 1991. *Neuron* 7: 743–53
220. Isacoff EY, Jan YN, Jan LY. 1991. *Nature* 353:86–90
221. Slesinger PA, Jan YN, Jan LY. 1993. *Neuron* 11:739–49

NOTE ADDED IN PROOF

Remarkable progress has been made in the past several years toward definition of the structural elements that are responsible for the basic functions of the voltage-gated ion channels. Beyond the new insights into the structure and function of the individual channels that have been gained, a striking commonality of functional design has emerged that allows a range of channel gating and permeability properties to be derived from subtle variations on a common structural theme. This commonality of function is observed in the common structures for voltage sensing in the S4 segments, the related structures of the pore–forming regions of the channels and the subtle variations in their sequence which dramatically alter ion selectivity, the similar locations for receptor sites for extracellular and intracellular pore-blocking drugs among different channel types, and in the common mechanistic basis for inactivation of sodium and potassium channels, even though the structural basis for inactivation of the two channels is different. Future investigations using the presently available methods of protein chemistry, mutagenesis, functional expression, and molecular modeling should delineate the primary structural basis for the essential functions of the ion channels and for the functional differences between different channel types at ever higher resolution. However, a mechanistic understanding of these processes in terms of protein structure will ultimately require determination of the three-dimensional structure of a member of the ion channel family.

Annu. Rev. Biochem. 1995. 64:533–61
Copyright © 1995 by Annual Reviews Inc. All rights reserved

COMMON THEMES IN ASSEMBLY AND FUNCTION OF EUKARYOTIC TRANSCRIPTION COMPLEXES

Leigh Zawel and Danny Reinberg

Howard Hughes Medical Institute, Department of Biochemistry, Robert Wood Johnson Medical School, University of Medicine and Dentistry of New Jersey, Piscataway, New Jersey 08854-5635

KEY WORDS: RNA polymerase II, activation of RNAPII transcription, general transcription factors, TFIID, TFIIB, TFIIE, TFIIH

CONTENTS

ABSTRACT

Eukaryotes contain three distinct RNA polymerase enzymes, each responsible for the transcription of a subclass of nuclear genes. Despite this division of labor, each RNA polymerase system follows a common blueprint to execute the loading of the polymerase onto the relevant promoter region. The RNA polymerase II system appears unique in that after RNA polymerase II has loaded onto the DNA, two auxiliary factors, TFIIE and TFIIH, are necessary for its escape from the promoter region.

The complexity of the RNA polymerase II initiation pathway provides a multitude of potential targets for transcriptional activators. Tight control over transcription initiation levels is afforded by multiple cofactors that both enhance and repress.

533

Introduction

A major evolutionary distinction separating prokaryotes from eukaryotes was the move from one enzyme that can faithfully transcribe DNA into RNA, RNA polymerase, to three. All three eukaryotic polymerases maintain considerable sequence similarity in their largest subunit (1) and actually have five subunits in common (2–5). Despite these similarities, each interacts with distinct sets of transcription factors to mediate accurate transcription initiation of a subclass of nuclear genes. RNA polymerase I (RNAPI) transcribes only genes encoding large ribosomal RNAs. While this may seem a narrow function, the product of these multicopy genes represents up to four fifths of the RNA being synthesized at any instant in a rapidly growing cell. RNAPII transcribes all of the cell's protein-coding messenger RNAs and other small RNAs (snRNAs) with the exception of the U6 RNA, which is transcribed by RNAPIII. RNAPIII also synthesizes tRNAs and the 5S RNA component of ribosomes.

Transcription of the above gene families requires that the relevant RNA polymerase associate with the promoter region and form a stable initiation complex. This is accomplished with the help of a set of auxiliary factors distinct for each transcription system. Despite the fact that the promoters regulating the expression of different gene families contain a diverse array of DNA sequence elements and configurations, each RNA polymerase follows a common blueprint to build an initiation complex.

Regardless of the gene, all promoters have a core sequence element, which is recognized by a factor that binds DNA specifically and provides a nucleation site for complex formation (Figure 1). This DNA-protein complex is not sufficient to mediate loading of the polymerase onto the DNA. In all three transcription systems, a second component is required to link the polymerase with the nucleating factor, thus forming a preinitiation complex. As discussed below in some detail, RNAPII transcription is unique in that after the polymerase has become stably associated with promoter sequences, not one but two additional factors are necessary to modulate transcription initiation (6, 7). This necessity is perhaps related to an unusual structure known as the CTD (C terminal domain), which occurs at the C terminus of the largest subunit of RNAPII but not of RNAPs I and III (2, 3).

DNA Elements and Protein Factors Nucleating Initiation Complex Assembly

The promoter structure of nuclear genes varies considerably with each gene family. A common denominator exists, however. All promoters have a core sequence that is recognized specifically by a DNA-binding transcription factor. By binding their respective element(s), each of the factors described below provides the foundation for the assembly of the initiation complex (Figure 1*I*).

I. Nucleation - a core promoter element is
          recognized specifically by a
          DNA binding protein or protein complex

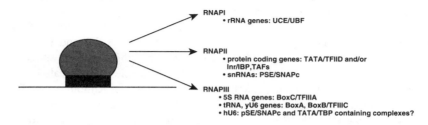

RNAPI
• rRNA genes: UCE/UBF

RNAPII
• protein coding genes: TATA/TFIID and/or
  Inr/IBP,TAFs
• snRNAs: PSE/SNAPc

RNAPIII
• 5S RNA genes: BoxC/TFIIIA
• tRNA, yU6 genes: BoxA, BoxB/TFIIIC
• hU6: pSE/SNAPc and TATA/TBP containing complexes?

II. Factors which link RNAP with the CORE promoter

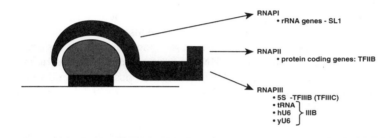

RNAPI
• rRNA genes - SL1

RNAPII
• protein coding genes: TFIIB

RNAPIII
• 5S -TFIIIB (TFIIIC)
• tRNA ⎤
• hU6 ⎬ IIIB
• yU6 ⎦

III. RNAP entry

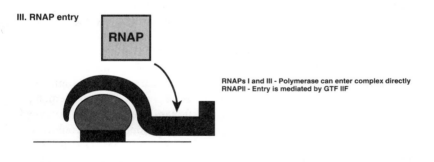

RNAP

RNAPs I and III - Polymerase can enter complex directly
RNAPII - Entry is mediated by GTF IIF

IV. Complex maturation

RNAPs I and III - no further steps required following Polymerase entry
RNAPII - Requires two more GTFs, TFIIE and TFIIH

Figure 1 *I*. Nucleation. Listed are core promoter elements and the factors that recognize them,
mediating nucleation on RNAPI, II, and III promoters. *II*. Following nucleation, each RNA
polymerase requires an intermediary component that links the polymerase with the core. These
components are listed, as in *I*. *III*. RNAPs I and III enter the initiation complex unassisted, whereas
RNAPII is delivered by the general transcription factor (GTF) TFIIF (see text for details). *IV*.
Complex maturation. Only the RNAPII system requires the assembly of initiation factors subsequent
to the association of polymerase.

RNAPI promoters have at least two elements, a GC-rich upstream control element (UCE) and a core region that overlaps the transcription start site. Both UCE and the core region are recognized by the upstream binding factor (UBF) (8–10). In the human system, UBF is a dimer of 94- (UBF2) and 97-kDa (UBF1) polypeptides encoded by the same gene but differentially spliced. Isolation of cDNAs encoding UBF from numerous species has revealed this factor to be highly conserved throughout evolution (10, 11). Although RNAPI transcription is highly species specific (see below), specificity is not intrinsic to UBF; human UBF (hUBF) can bind to the core of the *Xenopus* as well as mouse rRNA promoters. hUBF contains five DNA-binding motifs known as HMG (high mobility group) domains (11–13). HMG domains were first described in the nonhistone proteins HMG-1 and 2, which interact with DNA nonspecifically (11). The fact that the UBF-DNA interaction is sequence specific suggests that one or more UBF-HMG domains evolved from a primordial nonspecific DNA-binding structure. In fact, mutational analyses demonstrated that the most amino-terminal HMG box is necessary and sufficient for specific DNA binding (14). The hUBF C terminus has been found to be critical for an interaction with the RNAPI transcription factor SL1 (see below) (11, 14). A single HMG domain can distort the DNA by as much as 130° (13). Because a UBF dimer contains 10 such domains, it was not surprising to find that UBF binding resulted in negative supercoiling of the DNA (15). The UBF-mediated constraint of the core promoter region into a precise higher-order structure may facilitate presentation of the DNA surface to the RNAPI transcription complex.

RNAPII promoters have two possible core elements that can mediate nucleation either independently or collectively: the TATA box, which in mammals is typically located 30 nucleotides upstream of the transcription start site, and the initiator which, like the RNA Polymerase I (Pol I) core element, encompasses the transcription start site (16). All class two promoters contain one or both of these core elements (17). On TATA-containing promoters, initiation commences when TFIID binds the TATA motif. TFIID is a large multiprotein complex whose TATA-binding specificity is intrinsic to a single 38-kDa subunit, the TATA binding protein (TBP) (18). Like UBF, analysis of cDNAs encoding TBP from numerous species indicates considerable conservation of structure (19–22). This conservation is only in the carboxy-terminal 180 amino acids of the protein, which contain two imperfect direct repeats of around 90 amino acids. The three-dimensional structure of *Arabidopsis* TBP was determined by X-ray crystallography at 2.6 Å resolution (23). Later, the cocrystal structure of yeast (24) and *Arabidopsis* (25) TBP bound to the DNA was solved. These studies predicted that TBP resembled a molecular saddle in which the C-terminal direct repeats straddled the DNA. Significantly, this portion alone is capable of recognizing the TATA motif and modulating basal transcription (20). The function of the amino-terminal portion of TBP, which is evolutionarily divergent, is unclear. TBP binds to DNA in the minor groove (26, 27). TBP binding to the TATA box results in the DNA becoming severely bent to accommodate the convex undersurface of the TBP saddle. A consequence of this orientation is that the entire upper surface of TBP is exposed and available for protein–protein interactions. In fact, an incredibly large number of diverse proteins have been reported to interact specifically with TBP (28–40).

In humans, TBP is found tightly associated with polypeptides called TAFs (TBP-associated factors), which can only be dissociated under denaturing conditions (41–43). In the past few years we have learned that many different TAFs occur in the cell; moreover, different combinations of TAFs bestow distinct biochemical properties upon the TBP-containing complex. For example, the association of TBP with one set of TAFs results in the reconstitution of the RNAPII transcription factor, TFIID. Other TAFs reconstitute an RNAPI-specific complex, SL1, and still other TAFs reconstitute an RNAPIII-specific complex, TFIIIB (reviewed in 44–46). The observation that mutations to TBP compromised transcription by all three RNAP enzymes (47, 48) suggested that each of these complexes plays a critical role in cellular transcription (see below). Substantial progress has been made in the characterization of the RNAPII TAFs (TAFIIs). cDNAs have been isolated for a multitude of human and *Drosophila* TAFIIs (49–58). One or more RNAPII TAFs are required for the response of the RNAPII machinery to transcriptional activators (41–43, 46). Specific interactions between TAFIIs and certain activators have been demonstrated: dTAF40 with VP16 (50), dTAF150 with NTF-1 (59), dTAF110 with Sp1 (53), hTAF55 with CTF (R Roeder, unpublished observations). These studies culminated in the reconstitution of the entire *Drosophila* TFIID complex using recombinant polypeptides (59). Recombinant dTFIID supported activation by an assortment of activators. Consistent with the ability of some activators to interact specifically with certain TAFs, TAF subsets were able to mediate the response to specific activators. For example, a complex containing only TBP, TAFII150, and TAFII250 mediated a response to NTF-1 (59).

The initiator (Inr) motif can mediate nucleation on class two promoters that lack a TATA motif. This is not to imply that TATA-mediated and Inr-mediated complex assembly pathways are necessarily exclusive. There is evidence indicating that the Inr can direct complex formation even in the presence of a TATA motif and also that the Inr and the TATA motif function synergistically (17, 60, 61). Studies from our laboratory using the Ad-MLP indicate that while the TATA motif is the dominant element in vitro, in vivo, the Inr is dominant, i.e. in vivo, mutations in the TATA resulted in decreased levels of transcription from aberrant start sites while mutations in the Inr abolished transcription entirely (62). Three models, none of which are mutually exclusive, have been invoked to explain nucleation, as mediated by the Inr. (*a*) There is evidence indicating that RNAPII has an intrinsic ability to recognize and associate with the Inr in a manner stimulated by GTFs (17, 62). (*b*) In some TATA-containing promoters, the DNase I footprint of TFIID extends from the TATA motif to downstream of +30, suggesting that one or more TFIID TAFs bind the Inr and/or surrounding sequences. Direct evidence for this mechanism comes from the Tjian laboratory, who found that recombinant *Drosophila* TAF150 can independently and specifically recognize DNA containing an Inr motif (58). Thus, in the absence of a TATA motif, TAF–Inr interactions may anchor the TFIID complex to the promoter (63–65). (*c*) The assembly of initiation complexes may be nucleated by Inr-binding proteins (IBPs) such as TFII-I (66, 67) or YY1 (68), which interact with TBP and/or other components of the transcription apparatus (17, 69).

Though these scenarios appear quite different, they are similar in two respects: In all cases, the requirement for TFIID is maintained and TBP is brought to the initiation complex through protein–protein interactions. One notable exception is the adeno-associated virus (AAV) P5 promoter, on which transcription has been reconstituted in the absence of TBP or any TBP-containing complexes, using only TFIIB, RNAPII, and YY1 (see below). This represents the only report of RNAPII transcription in the absence of TBP (69). Complex assembly in the absence of TBP is conceivable in this case since YY1, in addition to interacting specifically with the Inr element, interacts with TFIIB and RNAPII (69).

The snRNA genes are transcribed by Pol II and do not fall into either of the two above categories. On these TATA-less promoters, nucleation is thought to be mediated through the binding of a novel multiprotein complex to an element located around 43 nucleotides upstream of the transcription start site, the proximal sequence element (PSE). Hernandez and colleagues have termed this complex the SNAPc complex, and report that it consists of TBP and several TAFs (71). In contrast, the Pugh and Roeder laboratories have independently purified a complex that recognizes the PSE yet appears to be devoid of TBP (BF Pugh, unpublished observations; RG Roeder, unpublished observations). In any case, it seems clear that nucleation on snRNA gene promoters occurs through the recognition of the PSE.

RNAPIII-transcribed genes are more complex to sum up, as at least three promoter classes have been described (reviewed in 72). Predominantly tRNA genes make up one class of Pol III–transcribed genes. tRNA gene promoters have two separated and variably spaced 10-bp elements, box A and box B. In striking contrast to Pol I and Pol II promoter elements, the A/B boxes are located downstream of the transcription start site, within the transcribed region (72). The A and B box regions have been localized to the D and TψCG stem-loop structures, respectively, in the tRNA (73). Nucleation on tRNA promoters occurs when TFIIIC recognizes the distal box, Box B, which is a high-affinity binding site. Box A orients TFIIIC in the 5' end of the coding region and sets the transcription start site, normally around 20 bp upstream of the Box A 5' border (74, 75). TFIIIC is a multisubunit factor which, in yeast, is composed of polypeptides of 132, 120, 90, 74, 62, and 55 kDa (72). cDNAs encoding the 132-, 120-, and 74-kDa subunits have been isolated (76, and references therein).

Human TFIIIC has been separated into two components, C1 and C2 (77). C1, the composition of which is unknown, cannot bind DNA independently but is induced by C2 to footprint over the A box. C2 may be analogous to yTFIIIC as this component consists of five subunits and binds the tRNA B box promoter element (78). Recently, a human cDNA was isolated encoding a B box–interacting subunit of C2 with a molecular weight of around 243,000 (78, 79). Despite the fact that the B box DNA sequence is highly conserved from yeast to humans, human p243 did not share significant homology (10% similarity) with the yeast TFIIIC B box–interacting subunit. Thus, this factor may have undergone a substantial degree of evolutionary divergence.

The U6 snRNA gene represents another class of RNAPIII-transcribed genes (reviewed in 45). Paradoxically, the human U6 gene promoter is devoid of gene internal elements but contains a typical RNAPII promoter element, a TATA motif, at around −30. hU6 promoters also contain a PSE characteristic of RNAPII-transcribed snRNA genes, which is recognized by a PSE-binding complex (71). This recognition event is instrumental in initiation of the RNAPII-transcribed snRNA genes (U2) and in fact, if the TATA motif in the hU6 promoter is mutated, the U6 PSE directs RNAPII transcription (80). Since the U6 TATA motif is functional with respect to its ability to bind TBP (81), and the PSE is recognized by SNAPc, a complex involved in RNAPII recruitment, an obvious question is how does the human U6 promoter modulate initiation for RNAPIII but not for RNAPII?

Apparently, Pol III specificity on the human U6 gene results from the combined effects of the TATA and the PSE. The binding of the SNAPc complex to the PSE apparently prevents TFIID from binding to the TATA and directing RNAPII transcription. Recent studies by Hernandez and coworkers indicate that the spacing between the TATA and PSE is critical to promoter viability and specificity (N Hernandez, personal communication). Perhaps spacing is crucial in this context because the SNAPc complex requires anchorage at both the PSE (via a TAF) and the TATA motif (via TBP). In any case, the combination of the SNAPc complex and the TATA motif on the U6 promoter results in a surface that is specific for RNAPIII; nucleation on the hU6 promoter is apparently jointly mediated by two elements, the TATA and the PSE.

The yeast U6 promoter contains an intragenic A and B box in addition to a TATA motif, but no PSE. Whereas in most RNAPIII genes, the A and B boxes are separated by around 30–90 bp, in the yU6 gene the spacing is around 120 bp (75). In this context, the TATA motif is not essential for transcription (82, 83). This is likely a result of the ability of TFIIIC to bind the A and B boxes, and therefore nucleate a RNAPIII-specific complex. The TFIIIC-binding site (box B) apparently defines RNAPIII specificity in the yU6 promoter. Removal of the A/B boxes converts the yU6 promoter to an RNAPII promoter in vivo; however, reintroduction of the B box alone restores RNAPIII specificity. Thus, it appears that as a result of the suboptimal spacing between the A and B boxes on the yU6 promoter (120 bp), RNAPIII recruitment may be facilitated by the TATA box–TFIIIB interaction (S Roberts, S Hahn, unpublished results).

The final class of Pol III genes is exemplified by the 5S RNA gene. 5S RNA promoters contain no A and B boxes and hence, no TFIIIC-binding site. Nucleation is mediated by another intragenic control region, the box C element. Historically, this was the first intragenic core promoter element to be defined (84, 85). Subsequently, it was shown that this element was recognized by TFIIIA, a 40-kDa polypeptide (86–88) that contains multiple zinc finger domains (89). A DNA-bound TFIIIA promotes the subsequent association of TFIIIC. Alternatively, preformed TFIIIA-IIIC complexes may bind to DNA in a single step (RG Roeder, unpublished observations). Once TFIIIC is bound, complex assembly is similar to what occurs in tRNA promoters.

In summary, all RNA polymerase classes use a common strategy to select the site for initiation complex assembly: A core element in the promoter is specifically recognized by an auxiliary transcription factor (Figure 1*I*). This DNA-protein complex is insufficient for the association of RNA polymerase. RNAPs I, II, and III each require an intermediary factor that recognizes the nucleation site and that can interact directly with the polymerase.

Integration of RNA Polymerase into the Initiation Complex Requires a Mediator

In the RNAPI system, bridging between the UBF-DNA complex and polymerase is provided by SL1, which is a multisubunit complex consisting of TBP and three TAFs of 110, 63, and 48 kDa (90). Like the human TFIID complex, the conserved C-terminal core of TBP is sufficient to mediate interaction with the RNAPI TAFs (91). cDNAs encoding each of the SL1 TAFs have been isolated, and each SL1 TAF has been found capable of interacting independently with TBP (R Tjian, unpublished observations). Interestingly, the association of any of the three SL1 TAFs with TBP precludes the interaction of TBP with the TFIID-TAF250 and most likely inhibits the formation of the TFIID complex (R Tjian, unpublished observations). DNase I footprinting analyses suggest that human SL1 has no DNA-binding ability and relies on

UBF to interact with the core promoter (92). Curiously, mouse SL1 can recognize the mouse core region independently (93). The association of SL1 with UBF results in an extended area of protection upstream on the rRNA promoter. SL1 bound at the core region is recognized by RNAPI.

One unique aspect of the RNAPI transcription system is the species-specific nature of the promoter recognition process. While frog and human UBF can recognize and bind to rRNA promoter sequences from either species, neither can support transcription initiation in a heterologous system. Species specificity apparently results from numerous specific interactions including, but perhaps not limited to, the UBF-SL1 interaction and the interaction between SL1 and the promoter (93–96). This latter specificity may result from the Pol I TAFs. Core hTBP expressed in a mouse cell line can associate with mouse Pol I TAFs; this hybrid SL1 complex recognizes only mouse rDNA (96). In comparing the mouse and human SL1 TAFs, it was found that only the largest TAF is different (110 kDa in mouse, 95 kDa in humans) and thus represents a good candidate for an effector of species specificity.

In the case of RNAPII, the IID-DNA complex is recognized by a single polypeptide, TFIIB (97, 98). TFIIB interacts with both TBP and RNAPII and can be thought of as a molecular bridge. TFIIB also interacts with TFIIF, the factor that delivers RNAPII to the initiation complex (28). Multiple lines of evidence indicate not only that TFIIB interacts with RNAPII, but also that this interaction is critical in the determination of the transcription start site. As expected, the powerful tool of yeast genetics provided the first inkling of such a relationship. Hampsey and coworkers isolated the yeast homolog of TFIIB, SUA7, as a supressor of a start-site alteration in the cyc1 promoter (99). Subsequently, a mutation with the same phenotype was found in the gene encoding the largest subunit of RNAPII. Yeast strains harboring both mutations were not viable (100). These findings were in good agreement with the studies of Ha et al, which demonstrated a direct interaction between TFIIB and RNAPII in vitro (28). Definitive biochemical evidence of this interaction and its importance in positioning RNAPII at the transcription start site was obtained by the Kornberg group in studies employing fractionated *Sacharomyces cerevisiae* and *Schizosacharomyces pombe* transcription systems. The transcription start site in *S. cerevisiae* is normally 120 bp downstream of the TATA, while in *S. pombe* it is only 40 bp downstream. When substituted individually, neither *S. pombe* TFIIB nor RNAPII was capable of functionally replacing the *S. cerevisiae* counterpart. When *S. pombe* TFIIB and RNAPII were substituted as a pair, not only was transcription restored, but the transcription start site was also effectively converted to the *S. pombe* location. These studies categorically established that contact between TFIIB and RNAPII is instrumental in positioning the transcription start site (101).

Using highly purified factors, we have found that the stable association of

RNAPII with the DAB or DB complex requires TFIIF (102, 103). Others have suggested that RNAPII associates with the promoter in a complex with TFIIB and TFIIF (104). Indeed, we have found that TFIIF and TFIIB interact in solution (28), and the ability of TFIIF to interact with RNAPII in solution is well documented (105). Thus, TFIIB and TFIIF may collectively recruit RNAPII in a manner analogous to that of SL1 and TFIIIB in the RNAPI and RNAPIII systems, respectively (see below).

Strategies for integrating RNAPIII into initiation complexes vary substantially with different RNAPIII promoters. In all cases, however, it is clear that the association of the multisubunit TFIIIB complex with the promoter is the pivotal step for RNAPIII recruitment. In a series of elegant papers, Geiduschek and coworkers demonstrated that yTFIIIB alone correctly positions RNAPIII at the transcription start site (106). yTFIIIB is positioned just upstream of the transcription start site on the 5S and tRNA genes (72). Once TFIIIB has been recruited, TFIIIC (and TFIIIA, in the case of the 5S RNA gene) can be stripped off the DNA by high salt or heparin. TFIIIB is resistant to these treatments and is sufficient to position RNAPIII correctly through multiple rounds of reinitiation. The TFIIIB-DNA complex is also stabile to challenge by DNA templates (106). In summary, TFIIIC is necessary to recruit TFIIIB to the DNA, and TFIIIB alone is necessary and sufficient to bring in RNAPIII.

The makeup of TFIIIB is understood in yeast but not in humans. In both species, TBP is an essential component (107–110). yTFIIIB consists of at least two additional polypeptides of 70 and 90 kDa (111, 112). The gene encoding the 70-kDa subunit, otherwise known as BRF1 (also as PCF4 and TDS4) was isolated as an extragenic suppressor of a temperature-sensitive TBP mutation (113–115). Interestingly, amino acid sequence analysis of the BRF1 gene product revealed remarkable similarity with the RNAPII GTF, TFIIB. Sharing an overall similarity of 44%, the amino-terminal half of BRF1 contains almost the entire structure of TFIIB, including a potential zinc finger domain and a pair of direct repeats. Both TFIIB and BRF1 interact with TBP as well as with their respective RNA polymerases (28, 116–118). Despite these similarities, BRF1 cannot substitute for TFIIB in a reconstituted RNAPII transcription system (115). The C terminus of BRF1 is not similar to that of TFIIB but is important for RNAPIII transcription and may preclude BRF1 from functioning in the RNAPII system.

How does TFIIIB enter into the RNAPIII initiation complex? On the 5S RNA promoter, the TFIIIA-core complex is recognized by TFIIIC, which allows the binding of TFIIIB. There is some evidence indicating that a preformed TFIIIA-TFIIIC complex exists in the cell, which can bind the promoter, perhaps expediting RNAPIII initiation (RG Roeder, unpublished observations). On tRNA promoters, TFIIIA is not required because TFIIIC can bind on its own to the A/B boxes. A bound TFIIIC is sufficient to recruit TFIIIB. Since the 120-kDa subunit of yTFIIIC crosslinks in the immediate vicinity of where TFIIIB binds, this subunit is thought to play a role in TFIIIB recruitment (72). Furthermore, the BRF component of TFIIIB is known to interact with the TFIIIC-DNA complex, and this interaction is stabilized by TBP (107).

On the U6 promoter, TFIIIB entry is understood reasonably well in yeast but poorly in humans. As alluded to above, the yeast U6 promoter contains intragenic A and B boxes in addition to a TATA motif. Complex formation may perhaps simulate what occurs in tRNA promoters; i.e. TFIIIC binds the B box and recruits TFIIIB. The observation that TFIIIC is essential for U6 transcription in vivo (83), but is not required in a highly purified reconstituted system (119), suggests that the role of TFIIIC may not be as simple in the U6 promoter as in tRNA promoters. It is conceivable that in vitro, in the absence of TFIIIC, yTFIIIB associates with the yeast U6 promoter by virtue of the TATA–TBP interaction (120). Alternatively, there is some evidence indicating that TFIIIC may be involved in alleviating chromatin-induced repression of the U6 gene (121).

On the hU6 promoter, there are no intragenic control elements and thus, TFIIIC appears to have no role. The hU6 gene promoter contains a TATA motif and further upstream, a PSE. TBP or a TBP-containing complex bound at the TATA, and the SNAPc complex at the PSE, are thought to recruit TFIIIB collectively (71). In this context, a TFIIIB component devoid of TBP is alone sufficient to fulfill the IIIB requirement as long as free TBP is provided (45).

Promoter Clearance

In the RNAPI and III systems, the association of the polymerase with the promoter signals the completion of the complex assembly process (114, 122) (Figure 1*IV*). Addition of ribonucleoside triphosphates to these complexes triggers the onset of transcription. In two regards, the RNAPII system is quite distinct.

The first is that even after the polymerase has been loaded onto the promoter, the complex is not competent to initiate transcription. This requires the association of two more GTFs, TFIIE and TFIIH (Figure 1*IV*). TFIIE is a heterodimer of 34- and 56-kDa subunits (6, 123). TFIIH is a multisubunit factor, the precise polypeptide composition of which is unclear. TFIIH activity copurifies with polypeptides of 34, 38, 41, 44, 50, 62, 80, and 89 kDa (124). cDNAs have been isolated for several of these subunits (125–128). A possible explanation as to why the RNAPII system requires a pair of GTFs following polymerase entry is that this pair somehow modifies the polymerase and/or the initiation complex. One might expect such a modification to involve the CTD, since this unusual motif is unique to RNAPII. CTD refers to a series of tandem repeats (52 in human, 43 in *Drosophila,* and 26 in yeast) of the hepatapeptide sequence YSPTSPS, occurring in the C terminus of the largest subunit of RNAPII. Phosphorylation of the serine, threonine, and/or tyrosine residues within the CTD generates at least two RNAPII isoforms, one with the CTD hyperphosphorylated, the other nonphosphorylated (129–131). TFIIH contains a kinase activity capable of phosphorylating the CTD of RNAPII as well as an ATPase and a DNA helicase activity (reviewed in 132).

The second major distinction is that initiation by RNAPII, but not by RNAPs I and III, requires the hydrolysis of β-γ bond of ATP prior to the onset of transcription (122, 133, 134). The nature of this energy requirement is discussed below.

Clues as to the functional significance of CTD phosphorylation were suggested by the following observations: The nonphosphorylated form of RNAPII (RNAPIIA) was found to participate in complex assembly and transcription of sequences proximal to the promoter (135–137). In contrast, the phosphorylated form (RNAPIIO) is the species of RNAPII that catalyzes RNA chain elongation (130, 137, 138). Together, these observations have fueled speculation that conversion of RNAPII from its underphosphorylated to its highly phosphorylated form is a determinant signaling the transition from transcription initiation to elongation. The observation that the nonphosphorylated CTD, but not the phosphorylated form, interacted with TBP (29) as well as with TFIIE (139) further suggested that CTD phosphorylation triggered a conformational change in the initiation complex (103, 140).

Several groups independently determined that phosphorylation of the CTD was catalyzed by TFIIH (141–143). TFIIE apparently stimulates this reaction (141, 144, 145). To the dismay of many, the TFIIH subunit responsible for kinase activity remains unknown. No TFIIH subunits for which cDNAs have been isolated contain kinase consensus motifs. What is encouraging is that TFIIH homologs from human, rat, and yeast all contain such an activity (141–143). Recent studies suggest that the kinase activity associated with TFIIH is cyclin H–dependent and is identical to the previously described cdk7 (R Sheiketthar, F Mermelstein, D Reinberg, unpublished observations).

Sequence analyses of the cDNAs encoding TFIIH subunits have resulted in the discovery that several TFIIH subunits are essential components of cellular pathways of nucleotide excision repair (NER) (124, 127, 146–149). As at least two of these subunits, ERCC2 and ERCC3, possess ATPase and helicase activities, it is not surprising that TFIIH also contains such activities (146, 150, 151). Like the CTD kinase, TFIIE regulates the TFIIH-ATPase/helicase activities. TFIIE was found to repress native TFIIH as well as recombinant ERCC3 helicase activity (146) but to stimulate TFIIH ATPase activity (144). TFIIE was without effect on the ERCC2 helicase activity (146). The fact that TFIIE inhibits both TFIIH and ERCC3 helicase activity completely, but is without effect on the ERCC2 helicase activity, suggests that ERCC2 helicase is silent within the context of the TFIIH complex.

The discovery that TFIIH contained polypeptides involved in NER begged the question: Could TFIIH participate in this cellular process? This question was especially poignant in light of the observation that actively transcribed genes are more efficiently repaired than are silent loci; i.e. these two cellular processes are likely coupled (reviewed in 152, 153). Purified TFIIH was capable of restoring DNA excision repair activity to extracts defective for either ERCC3 (XPB) or ERCC2 (XPD) activity (146, 154). This effect was specific, as extracts deficient for other NER proteins such as XPA were not complemented. Similar observations have been reported in yeast (148, 149). The discovery that TFIIH participates in NER does not necessarily implicate TFIIH as the factor that couples NER and transcription (155). TFIIH may facilitate transcription–NER coupling; however, genetics have clearly demonstrated the requirement for other factors such as CS-A and CS-B (ERCC6) in the coupling process (156).

As alluded to above, transcription initiation by RNAPII is unique in that it requires a hydrolyzable source of ATP (133, 134). Phosphorylation of the CTD by TFIIH cannot account for this requirement as GTP was also found capable of donating phosphate in this reaction (141, 142, 151), and since reactions containing a CTD-less form of RNAPII maintain a requirement for ATP hydrolysis (157). The discovery that TFIIH contains two helicase activities and ATPase activity suggested that TFIIH might nonetheless be involved in the requirement for ATP hydrolysis.

Systematic mutational analyses of the yeast homologs of ERCC2 and ERCC3—RAD3 and RAD25, respectively—have provided enormous insight into this issue. While both ERCC2 and ERCC3 are essential genes in yeast (158, 159), the establishment of yeast harboring temperature-sensitive mutations allowed a functional analysis of each protein in vivo. At nonpermissive temperatures, RNAPII transcription was severely reduced in both ERCC2 and ERCC3 mutants (158–160). To determine if the ATPase/helicase of either protein was involved in RNAPII transcription, point mutations were generated in the nucleotide-binding domain of ERCC2 and ERCC3. In ERCC2, such mutations were not lethal, and resulted only in defects to NER (148, 161). In contrast, similar mutations to ERCC3 were lethal and left extracts transcriptionally dead (160). Thus, while ERCC2 is essential for transcription, its function is not tied to its helicase/ATPase activity. The helicase/ATPase activity of ERCC3 is essential for RNAPII transcription and is an excellent candidate for mediating the ATP hydrolysis requirement (162).

TFIIE inhibits ERCC3, but not ERCC2 helicase activity (146). Thus, if the ERCC3 helicase is required for initiation, open complex formation, and/or promoter clearance, the TFIIE-imposed inhibition must be overcome. We propose that following CTD phosphorylation, an event that TFIIE stimulates, the initiation complex undergoes a conformational change resulting in the departure of TFIIE from the complex. This notion is supported by the observations that both TBP (29) and TFIIE (139) interact with the nonphosphorylated but not the phosphorylated form of RNAPII. The removal of TFIIE functionally enables the ERCC3 ATPase/helicase so that the hydrolysis of ATP and subsequent initiation events can proceed.

Further insight as to the transcriptional role(s) of TFIIE and TFIIH came when the reconstitution of RNAPII transcription on the IgE promoter was reported in the absence of TFIIE and TFIIH. These studies established that the requirement for TFIIE and TFIIH was closely related to the degree of supercoiling in the template DNA (163, 164). This relaxed GTF requirement was initially thought to be related to the high AT content of the IgE promoter. Subsequently, however, these observations were reproduced on the Ad-ML, HSP70, and numerous other promoters (165–168). While it is clear that transcription can occur with subsets of GTFs, on the AAV P5 promoter this is most dramatic. Transcription was reconstituted on the AAV P5 promoter with just two components (in addition to RNAPII): YY1, and the GTF TFIIB (69). This represents the first promoter to be transcribed in vitro in the absence of TBP, an especially surprising result in view of the fact that this promoter contains a TATA motif. Presumably—since YY1 interacts with the Inr, TFIIB, and RNAPII—TBP is not essential for loading the polymerase onto this promoter in vitro. Whether this occurs in vivo is uncertain. It is possible that transcription originating from such "minicomplexes" is an in vitro phenomenon only possible with abnormally high concentrations of factors. Moreover, it is unclear whether this putative initiation complex (YY1, TFIIB, RNAPII) represents a bona fide stable transcription initiation complex. We suspect that initiation complexes containing all of the GTFs are more stabile under physiological conditions and hence, more viable.

The observation that relaxed (linear) DNA templates require TFIIE/H, whereas highly supercoiled DNAs do not, suggested that TFIIE/H might be involved in mediating an energy requirement specific to relaxed or linear templates. Highly supercoiled templates have been found to escape the requirement for a hydrolyzable ATP source (165–168), presumably because of the energy stored within supercoiled DNA. The expenditure of this energy may be necessary to mediate unwinding of the DNA around the transcription start site, i.e. open complex formation, since templates that contain a stretch of noncomplementary sequence around the start site do not require hydrolyzable ATP for initiation (169). Alternatively, Goodrich and Tjian demonstrated that ATP hydrolysis (or DNA supercoiling) was required, not during the formation of the first few phosphodiester bonds in the RNA, but subsequently, for promoter clearance (165). Most likely, these two processes are tightly coupled and hence, the hydrolysis of ATP provides energy to accommodate not one step but rather a transition from an early phase of initiation to a later one.

Regulation of RNAPII Transcription

RNAPII is distinct from RNAPs I and III in yet one other regard. As a result of the immense number of different genes transcribed by RNAPII, each according to specific programs governed by various environmental and developmental cues, the regulatory burden faced by RNAPII is unique. This regulation is multitiered and is accomplished largely through the action of a large family of regulatory proteins that have the ability to recognize specific sequences located upstream or downstream of the core promoter region. These sequence elements are distinct from core promoter elements because they have no apparent role in basal transcription. Regulatory proteins contain at least two functional domains, one that tethers the factor to the DNA, i.e. a sequence-specific DNA-binding domain, and another that activates or in some cases represses transcription (170–172). The mechanisms through which regulatory proteins exert their influence over the RNAPII machinery are the focus of the closing section.

Activators and Their Targets

In discussing initiation pathways as above, there is an unfortunate tendency to lose sight of the cell as a living biological system. One must keep in mind that in the context of the cell, the network of protein–DNA and protein–protein interactions that drive initiation complex assembly is limiting for several reasons: (a) there is competition among thousands of genes for a common GTF pool; (b) the access of factors to promoter elements is limited due to the packaging of the DNA into chromatin; (c) as discussed below, there are protein factors that specifically interfere with transcription complex formation; (d) core promoter elements often deviate from the consensus and are consequently weakened in their affinity for specific DNA-binding proteins. One can certainly bypass these limitations in vitro, as most of the GTFs have now been obtained in recombinant form and can thus be provided in enormous excess to naked DNA templates. In fact, operationally, transcription directed by core promoter elements in the absence of regulatory components has been defined as basal transcription. Whether this indeed occurs in higher eukaryotic systems is unlikely. The default expression state of any given gene is probably close to silent as a result of chromatin structure and nonspecific and specific repressors like those discussed below. Activation is what happens when the RNA output from a given gene increases in response to physiological cues. If the formation of a transcription-competent complex involves multiple steps, it is likely that an activator may target some or all of these discrete steps. Accordingly, it has been observed that activators can interact with numerous components of the basal transcription machinery (33–39, 173–176). For reasons discussed below, we suspect that the activator targeting of the TBP component of TFIID, of TFIIB, and of TFIIH is especially significant (Figure 2).

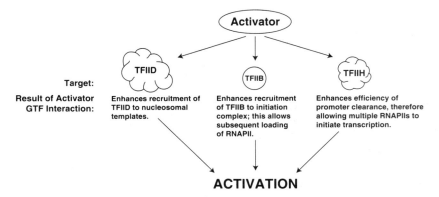

Figure 2   Many activators interact with TFIID, TFIIB, and TFIIH. Listed are the mechanisms through which activation might occur via each of these targets.

Almost 10 years ago, the observation was made that the assembly of naked template DNA into nucleosomes interferes with the ability of the DNA to be transcribed (177–180). It is now clear that nucleosomes repress transcription by blocking the access of transcription factors to the DNA (reviewed in 181). Originally, it was noticed that this inhibition could be precluded by preincubating the template DNA with TFIID prior to reconstituting the DNA into minichromosomes (178). It has since become evident that the association of some transcriptional activators, but not others, with nucleosomal DNA can induce a change in the nucleosome structure that facilitates the binding of the GTFs to the DNA (182–185). This process is facilitated by the SWI/SNF complex (see below). Thus, activators function in part by gaining greater access of GTFs to promoter sequences occluded by nucleosomes, a phenomenon termed antirepression (181).

An important feature of antirepression is that in the presence of nucleosomes, no basal transcription occurs. Since activators can overcome this repression, the net fold stimulation that an activator potentiates is increased on nucleosomal templates as compared to on naked DNA. In addition to increasing activator potency, there is also evidence suggesting that packaging of DNA into nucleosomes is essential for activator function in promoters where the activator binding site is located far upstream (greater than 1 kbp) of the core region (186). These results are not undisputed, however, as others claimed to have reconstituted long-range activation in vitro from naked DNA templates using nuclear extracts (187, 187a). Irrespective of the nature of chromatin, antirepression alone cannot account for the role(s) activators play in gene regulation, since activators are capable of stimulating transcription from naked DNA templates.

Yeast genetics have provided further insight into the relationship between gene expression and chromatin. In *S. cerevisiae,* the loss of histone components (H3, H4) resulted in the activation of numerous otherwise inactive gene promoters (188–190). A family of yeast proteins were identified, called SWI proteins, that seemed to be necessary for the transcription of tightly regulated genes (191). Interestingly, mutations that altered chromatin components such as H3 or HMG-1-like proteins alleviated the necessity for SWI proteins for the transcription of SWI-dependent genes (192). Based on these observations, it was theorized that the function of SWI proteins was to modulate chromatin structure so as to facilitate transcription. SWI1, 2, 3 proteins were found to exist in the cell as a large multiprotein complex (193, 194) and subsequently, as predicted, it was proven biochemically that this multisubunit complex stimulated the binding of GAL4 to nucleosomal templates. This SWI/SNF complex interacts directly with nucleosomal DNA, and this interaction is apparently dependent on an ATPase activity that is intrinsic to the complex (195). A homologous complex has been described in humans, and in these studies, the SWI/SNF complex was also found to stimulate the binding of TFIID with TFIIA to nucleosomal DNA (196, 197). The SWI/SNF complex apparently does not remove nucleosomes but somehow alters their structure to facilitate transcription factor access to the DNA.

In summary, the activation of a given gene requires the removal or restructuring of the architecture of chromatin and concomitantly or immediately thereafter, the establishment of transcription complexes at the promoter region. When the expression of a gene is no longer necessary, how is the "silent state" re-established? Very little is known about this phenomenon. Our studies and others indicate that once formed, the TFIID-DNA complex is extremely stabile, so initiation complexes probably do not simply fall apart. The ADI/MOT1 factor reported by Auble & Hahn might possibly be involved in this process. ADI can specifically displace TBP bound to the TATA motif in an ATP-dependent fashion (198, 199). ADI represses transcription in vitro and thus is most likely a tightly regulated activity in vivo. ADI/MOT1 is essential for yeast viability and interestingly, mutation of ADI/MOT1 increases the basal level of numerous otherwise inducible genes (200).

As stated above, activators function not only by alleviating chromatin-mediated repression, but also by stimulating the activity of the basal transcription apparatus. Although the mechanism of this latter function remains unclear, direct physical interactions have been demonstrated between acidic activators and TBP (32, 32a), TFIIB (173, 201), and TFIIH (176). Importantly, VP16 mutants with reduced activation potential are compromised in their ability to interact with each of these GTFs. Multiple lines of evidence suggest the targeting of TFIIB by activators is especially relevant (202). The acidic activator GAL4-AH was found to facilitate the association of TFIIB, but not TBP or TFIID, with initiation complexes formed on naked DNA templates (173, 203). Importantly, double point mutations in TFIIB have been identified that behave as does the wild-type protein in basal transcription but are defective for activated transcription and for interaction with VP16 (201). The interaction between TFIIB and a *Drosophila* activator has been demonstrated in vivo (204).

The association of TFIIB is a rate-limiting step in initiation complex assembly (173). This makes sense in light of the discovery that TFIIB is capable of interacting with numerous components of the initiation complex, including TBP, RNAPII, and TFIIF (28). The two models that follow predict that activators stimulate transcription by manipulating these interactions. Green and coworkers have determined that regions within the amino and carboxy termini of TFIIB interact with one another, creating a compacted structure in which the TFIIF and RNAPII interaction surfaces are occluded. The binding of VP16 to TFIIB appears to induce a conformational change in the structure of TFIIB (205). This structural alteration is thought to open the molecule and expose the surfaces within TFIIB that are critical to initiation complex architecture.

In addition, the VP16-TFIIB interaction may stimulate transcription by increasing the number of initiation events that occur. The presence of TFIIB at the promoter is necessary for the loading of RNAPII. Our studies and others have shown that after RNAPII exits the promoter, TFIIB is released from the complex (206; L Zawel, P Kumar, D Reinberg, manuscript in preparation). A result of this release is that the complex that remains at the promoter is unable to receive another molecule of RNAPII for subsequent rounds of initiation. We suspect that by physically interacting with TFIIB, some activators increase the rate of TFIIB reassociation or retain TFIIB at the promoter, following the promoter clearance step. In this manner, the activator maintains the promoter in a state accessible to the polymerase and capable of multiple initiation events.

A recent report indicates that TFIIH is directly contacted by the activation domain of different activators, including VP16, p53 (176), and EBNA2 (206a). Both interactions have been mapped to the 62-kDa subunit of TFIIH. These observations are particularly intriguing in light of the enzymatic activities ascribed to TFIIH, i.e. a kinase with specificity for the CTD and an ATPase/helicase essential for transcription initiation. It will be interesting to analyze whether the interaction of TFIIH with an activator has any effect on either of the TFIIH enzymatic activities.

Preliminary studies indicate that activators may influence the role TFIIH plays during promoter clearance. Kumar et al have analyzed abortive vs productive RNA synthesis and found that, in agreement with Goodrich & Tjian (165), initiation complexes lacking IIE/IIH were capable of initiating transcription. Initiation in the absence of IIE/IIH resulted in an increased rate of abortive RNA synthesis. A major distinction, however, was observed as the elongation complex reached around +16 with respect to the transcription start site. Complexes lacking IIE/IIH stalled between +12 and +18. The subsequent addition of IIE/IIH to these reactions did not result in these complexes returning to elongation phase, suggesting that IIE/IIH must be incorporated into the initiation complex prior to or at the moment of initiation for complete elongation competence. Complexes containing IIE/IIH paused between +12 and +18, after which a large fraction of them resumed elongation. Interestingly, the addition of GAL4-VP16 to these reactions decreased the extent of abortive transcription and increased the proportion of complexes that passed +12 to +18, suggesting yet another link between activator function and TFIIH (P Kumar, D Reinberg, unpublished observations). In a related report, Yankulov et al have found that activators can substantially affect the efficiency of the elongation process (207).

Cofactors, Mediators, and Adaptors

Despite the plethora of activator–GTF and activator–TAF interactions that have been described (see above), the addition of transcriptional activators to highly purified reconstituted transcription reactions results only in low-level enhancement of transcription. Activation seems to require additional components, which help transduce the activation signal from the activator to the basal machinery, resulting in an increased fold stimulation. Biochemical fractionation of HeLa cell nuclear extracts has resulted in the isolation of cofactors, which both enhance (PC1, PC2, PC3, PC4, ACF) and repress (NC1, NC2/Dr1, Dr2/PC3/Topoisomerase I, MOT1/ADI, NOT1-4, TUP1) transcription (30, 31, 199, 208–213). All of these cofactors appear to be general effectors (as opposed to being gene specific). Positively acting components appear not to stimulate basal transcription but enhance the response to an activator. Some negatively acting components (Dr1/NC2, Dr2) repress basal transcription, but this repression is overcome by activators so that the net fold activation is greater in their presence than in their absence. Of all these cofactors, the most seems to be known about Dr1, Dr2, and PC4. cDNAs encoding each of these have been isolated and the activities well characterized.

While searching for activities that enhanced the response of the basal machinery to activators, two groups independently isolated DNA topoisomerase I (Topo I) (30, 209). Topo I interacts with the TBP component of TFIID and can be found in association with the TFIID complex. Through this interaction, Topo I interferes with the ability of TFIID to bind to the TATA motif (but not to interact with other GTFs) and ultimately represses basal transcription. Accordingly, Topo I repression was far more severe on TATA-containing promoters, where TBP-TATA recognition is a nucleation event, than on TATA-less promoters. Oddly enough, Merino et al isolated Topo I by following an activity that enhanced the response to activators (30). Similarly, Roeder and coworkers isolated Topo I as a positive cofactor, PC3 (209). In both cases Topo I was isolated because, although it represses basal transcription, this repression can be overcome by numerous activators.

We recently learned that although alleviation of Topo I–mediated repression is activator dependent, it is mediated by TFIIA (A Merino, D Reinberg, unpublished observations). While it had long been thought that TFIIA had only a marginal effect on basal transcription, our observations regarding Topo I, and a series of reports from several other groups, now clearly indicate that TFIIA plays an important role in the activation process (214–218).

Preliminary studies indicate that Topo I is contained within the elongation complex and furthermore, that Topo I can increase the efficiency of RNAPII elongation (L Zawel, P Kumar, D Reinberg, unpublished observations). We thus suspect that upon removal from the TFIID complex, Topo I translocates to the elongation complex, where it facilitates elongation. Interestingly, the DNA relaxation activity of Topo I is not required for the repression/activation/elongation functions (30; L Zawel, D Reinberg, unpublished observations). We suspect that in the case of elongation, Topo I may act by altering the conformation of RNAPII and, in fact, we have detected a specific interaction between Topo I and RNAPII (L Zawel, P Kumar, D Reinberg, unpublished observations).

PC4 was recently purified to near homogeneity and its cDNA isolated (212, 212a). PC4 is a single polypeptide of approximately 19 kDa that binds single-stranded DNA. PC4 appears to be a genuine coactivator, or mediator, as it enhances the response to multiple activator types but does not affect basal transcription. Consistent with this function, recombinant PC4 was found capable of interacting with both a DA-DNA complex and a DNA-bound activator.

We have recently isolated a related activity, CofA, which together with PC4, is required for a high-level response to acidic-type activators. CofA is distinct from PC4 since CofA is only required with acidic activators while PC4 functions with glutamine-rich, acidic, and proline-rich activators (212). The requirement for CofA with acidic activators may not have been obvious in previous studies (212, 212a) because a crude TFIIA fraction that likely contained CofA was employed. CofA also stimulates basal transcription; this stimulation requires the CTD of RNAPII, however. Whether CofA interacts with the CTD is currently under investigation. In the presence of other components, perhaps PC4, the effect of CofA on basal transcription was not observed, suggesting that PC4 may suppress the effect of CofA on basal transcription (J Inostroza, M Sheldon, D Reinberg, unpublished observations).

Dr1 is a 19-kDa phosphoprotein that represses basal transcription in vitro and in vivo by virtue of its ability to interact with TBP and inhibit the subsequent assembly of TFIIB (31). This repression can be overcome by some, but not all, transcriptional activators. Dr1 is found tightly associated with a number of polypeptides, the characterization of which is ongoing (F Mermelstein, D Reinberg, unpublished observations). Dr1 is a general repressor of transcription and differs from previously described specific repressors such as Id, IkB, Kruppel (Kr), evenskipped (eve), and engrailed (en) in that Dr1 does not function when tethered to the DNA (211). Dr1 does, however, share sequence homology in its repression domain with several gene-specific and sequence-specific DNA-binding transcriptional repressors (31, 211).

The factors required for activation in the yeast system are presently less clear than in the human system. For example, it is now well established that in humans, TFIID TAFs are required for transcriptional activation. For many years it was thought that TAFs did not exist in the yeast cell and that TBP existed as a single polypeptide. As a result, many believed TAFs were not required for activation in yeast. In fact, a report surfaced claiming activation in a reconstituted yeast system that contained recombinant TBP (219). The fact that one of the reaction components used in this study was a crude fraction, however, left open the possibility that TAFs were involved.

Weil and coworkers were the first to describe TAFs in yeast, although it was unclear from this work whether these proteins were required for RNAPI, II, or III transcription (220). The Green laboratory took steps towards unifying opposing factions when, in a recent study, a functional RNAPII activation assay was used to search for yeast TAFs. These studies have provided clear evidence that TAFs exist in yeast in a large, multisubunit complex. Some of these TAFs are homologous to TAFs identified in humans and *Drosophila*; they are essential to yeast viability; and most importantly, they are involved in mediating activation (221).

Do cofactors, such as those identified in humans, also occur in yeast? The phenomenon known as squelching argues that they do. Squelching, or activator interference, is the inhibition of transcription observed when an activator is present at artificially high concentrations (172). Presumably, this inhibition reflects the sequestration by the activator (unbound to the DNA) of either (*a*) an adaptor component necessary to bridge the activator with the basal machinery and/or (*b*) a basal component (172, 222). Two potential yeast mediators, ADA2 and ADA3, were identified as a result of their ability to reduce the toxicity of GAL4-VP16 in vivo. While ADA2 and ADA3 are not essential genes, mutations to these proteins compromised the ability of some activators to stimulate transcription in vivo (223, 224). The behavior of purified ADA2 and ADA3 in the context of a defined reconstituted transcription system has not yet been investigated.

Kelleher et al partially purified a yeast activity based on its ability to reverse squelching between GAL4-Vp16 and GCN4 (225). This mediator was required for activation from both activators but did not affect basal transcription (226). Recently, Kornberg and colleagues claim to have obtained highly purified mediator. This mediator is a large complex containing some 20 polypeptides including RNAPII, TFIIF, GAL11, multiple SRB proteins (see below), and several unidentified polypeptides. It is unclear whether this mediator represents the same activity reported earlier, as the current activity is highly stimulatory towards basal transcription. The current mediator also appears to stimulate CTD phosphorylation, although it remains unclear whether this feature is important for enhancement of activation (227).

The mediator described by the Kornberg group appears to be identical to another multisubunit complex reported by Young and coworkers. In an effort to investigate the function of the CTD, Young and coworkers sought to isolate extragenic suppressors of the cold-sensitive phenotype indicative of yeast harboring CTD truncations. Nine different yeast genes capable of suppressing CTD truncations have been isolated: SRB2 and SRB4–SRB11 (228–230; C Thompson, R Young, manuscript in preparation). All nine SRB proteins copurify stoichiometrically in a high-molecular-weight complex that contains RNAPII and several initiation factors including TFIIB and TFIIH. Only TBP and TFIIE are required for this holoenzyme complex to acquire transcriptional competence (230). Transcription reactions reconstituted in this manner are responsive to transcriptional activators; however, the activation levels obtained are low, suggesting the absence of one or more mediator components from the holoenzyme.

It is possible that these holoenzyme complexes mediate transcription initiation in vivo in a much abbreviated fashion as compared with the stepwise assembly of each initiation factor. The idea that preformed assemblies of factors exist in vivo is not surprising. In the course of developing purification schemes for the GTFs, we and others have put considerable effort into finding methods that could separate the initiation factors from one another. It is clear that TFIIE interacts with TFIIH and RNAP II, TFIIF interacts with RNAPII and TFIIB, TFIIB interacts with TBP and RNAPII, etc. What is novel about the holoenzyme is that these preformed complexes appear to be functional. Significantly, neither Kornberg's nor Young's holoenzyme contains the full set of basal transcription factors. A consequence of this is that even in the presence of the holoenzyme, initiation complex assembly will be a multistep process, offering several potential targets for transcriptional activators, i.e. TFIID binding, TFIIB assembly, TFIIE assembly. Although no TFIIE–activator interactions have been reported, TFIIE remains an attractive activator target, since it appears to regulate TFIIH activity (see above).

Perspectives and Summary

The transcription workload in higher eukaryotes is split up such that three distinct RNA polymerase enzymes are responsible for the synthesis of a subset of cellular RNAs. Although the promoters that direct transcription of each gene class have diverged considerably, each RNA polymerase system follows a common program to execute the loading of the polymerase onto the DNA. The initiation complex in all three systems is nucleated by a component that specifically recognizes a core promoter element. Next there enters a second component, which has the ability to bridge the polymerase with the nucleating factor. Finally, the polymerase enters the complex, and in the RNAPI and III systems, this complex is competent to initiate transcription. While the auxiliary components for each system appear at first glance to be unique, it is now clear that TBP plays an important role in each system. TBP interacts with multiple sets of proteins called TAFs. TBP-containing complexes participate in transcription initiation of RNAP I, II, or III, depending on the set of associated TAFs.

Perhaps as a result of the awesome regulatory burden placed upon RNAPII, initiation in this system is a bit more complex. Initiation requires the participation of two auxiliary factors, TFIIE and TFIIH, after the polymerase has associated with the promoter. The participation of these two factors appears to be a consequence of a unique structure present at the C terminus of the largest subunit of RNAPII (CTD). It has been suggested that TFIIE regulates the activity of TFIIH, a significant task since TFIIH contains two enzymatic activities, an ATPase/DNA helicase and a kinase capable of phosphorylating the CTD.

The complexity of the initiation pathway in the RNAPII system provides a multitude of steps, which could potentially be subject to regulation. Among these steps are (a) the recognition of promoter sequences by TFIID, (b) the recruitment of TFIIB, a factor essential for RNAPII entry, and (c) the escape of RNAPII from the promoter, a process that appears to be controlled by TFIIH (Figure 2). The control of these processes is accomplished, in part, through specific interactions between transcriptional activators and each of these general transcription factors (GTFs). These interactions alone are not sufficient to provide the necessary control. The TAFs composing the TFIID complex (TAFIIs) are required for activation. Recent studies suggest that TAFs may mediate the response to some activators via direct TAF–activator interactions. In addition, some TAFs directly interact with the basal machinery. While the interactions between activators and TAFs, activators and GTFs, and TAFs and GTFs clearly contribute to activation, they alone may not be sufficient. Other molecules—mediators or adaptors—also appear necessary.

The net contribution of transcriptional activators is that they enhance the efficiency with which the basal transcription complex performs. One activator, VP16, has been shown to have the ability to interact with TFIID, TFIIB, and TFIIH. As a result, VP16 targets not only complex nucleation and RNAPII recruitment, but also RNAPII escape, ensuring that multiple RNAPII molecules can enter and exit a single promoter highly efficiently. VP16, however, is a paradigm for a eukaryotic transcriptional activator. Cellular transcriptional activators do not necessarily contact all three of these GTFs. Cellular promoters, however, generally contain recognition elements for multiple activator types. Some of these activators may engage a subset of GTFs; i.e. activator A may contact TFIID and TFIIB while activator B contacts only TFIIH, or an activator C may contact yet another GTF, for example SRF with TFIIF (175).

Recently, the dogma of step-by-step initiation complex assembly has been challenged by the discovery in yeast of multiprotein complexes containing RNAPII, multiple basal components, and other polypeptides. We envision that initiation complex formation mediated by these preformed complexes remains susceptible to the influence of activators through several potential mechanisms. So far, none of the preformed complexes has been found to contain the entire complement of basal components. It remains unclear whether GTFs that are present are contained in stoichiometric proportions. Thus, activators may help to facilitate recruitment of initiation factors not present within the complex. Activators may also facilitate recruitment of the entire holoenzyme complex. Moreover, after the complex is recruited, the activator may influence the efficiency of promoter clearance.

While in recent years we have made enormous strides in our understanding of the factors required for RNAPII initiation, our understanding of how this process is regulated is still in the early developmental stages. This is certainly the new frontier in transcription research.

ACKNOWLEDGMENTS

We thank all in the scientific community who responded to our inquiries by sending reprints and communicating unpublished material. We apologize for failing to cite work that may have been relevant but was not included due to the enormous amount of material that was covered and to space limitations. We are extremely grateful to Dr. Steve Hahn for reading the manuscript and providing helpful comments. D.R. is supported by grants from the NIH and by the Howard Hughes Medical Institute. D.R. was a recipient of an American Cancer Society Faculty Research Award. L.Z. was supported by an NIH training grant (GM08360).

Literature Cited

1. Allison LA, Moyle M, Shales M, Ingles CJ. 1985. *Cell* 42:599–610
2. Young RA. 1991. *Annu. Rev. Biochem.* 60:689–715
3. Thuriaux P, Sentenac A. 1992. In *The Molecular Biology of the Yeast Saccha-romyces cerevisiae: Gene Expression,* ed. JR Broach, JR Pringle, EW Jones, pp. 1–48. Cold Spring Harbor, NY: Cold Spring Harbor Lab.
4. Woychik NA, Liao S-M, Kolodziej PA, Young RA. 1990. *Genes Dev.* 4:313–23

5. McKune K, Woychik N. 1994. *Mol. Cell. Biol.* 14:4155–59
6. Inostroza J, Flores O, Reinberg D. 1991. *J. Biol. Chem.* 266:9304–8
7. Flores O, Lu H, Reinberg D. 1992. *J. Biol. Chem.* 267:2786–93
8. Smale ST, Tjian R. 1985. *Mol. Cell. Biol.* 5:352–62
9. Haltiner MM, Smale ST, Tjian R. 1986. *Mol. Cell. Biol.* 6:227–35
10. Paule MR. 1994. In *Transcription: Mechanisms and Regulation*, ed. RC Conaway, JW Conaway, pp. 83–106. New York: Raven
11. Jantzen H-M, Admon A, Bell SP, Tjian R. 1990. *Nature* 344:830–36
12. Leblanc B, Read C, Moss T. 1993. *EMBO J.* 12:513–25
13. Wolffe AP. 1994. *Science* 264:1100–1
14. Jantzen H, Chow AM, King DS, Tjian R. 1992. *Genes Dev.* 6:1950–63
15. Bazett-Jones DP, Leblanc B, Herfort M, Moss T. 1994. *Science* 264:1134–37
16. Smale ST, Baltimore D. 1989. *Cell* 57: 103–13
17. Weis L, Reinberg D. 1992. *FASEB J.* 6:3300–10
18. Lewin B. 1990. *Cell* 61:1161–64
19. Hoffmann A, Sinn E, Yamamoto, T, Wang J, Roy A, et al. 1990. *Nature* 346:387–90
20. Peterson MG, Tanese N, Pugh BF, Tjian R. 1990. *Science* 248:1625–30
21. Kao CC, Lieberman P, Schmidt MC, Zhou Q, Pei R, Berk AJ. 1990. *Science* 248:1646–50
22. Gasch A, Hoffmann A, Horikoshi M, Roeder RG, Chua N-H. 1990. *Nature* 346:390–94
23. Nikolov DB, Hu S-H, Lin J, Gasch A, Hoffman A, et al. 1992. *Nature* 360:40–46
24. Kim Y, Geiger J, Hahn S, Sigler PB. 1993. *Nature* 365:512–20
25. Kim Y, Nikolov DB, Burley SK. 1993. *Nature* 365:520–27
26. Starr DB, Hawley DK. 1991. *Cell* 67: 1231–40
27. Lee DK, Horikoshi M, Roeder RG. 1991. *Cell* 67:1241–50
28. Ha I, Roberts S, Maldonado E, Sun X, Kim L, et al. 1993. *Genes Dev.* 7:1021–32
29. Usheva A, Maldonado E, Goldring A, Lu H, Houbavi C, et al. 1992. *Cell* 69:871–81
30. Merino A, Madden K, Lane WS, Champoux J, Reinberg D. 1993. *Nature* 365: 227–32
31. Inostroza JA, Mermelstein F, Ha I, Lane WS, Reinberg D. 1992. *Cell* 70:477–89
32. Stringer KF, Ingles CJ, Greenblatt J. 1990. *Nature* 345:783–86
32a. Ingles CJ, Shales M, Cress WD, Triezenberg S, Greenblatt J. 1991. *Nature* 351:588–90
33. Emili A, Greenblatt J, Ingles CJ. 1994. *Mol. Cell. Biol.* 14:1582–93
34. Lieberman P, Berk AJ. 1991. *Genes Dev.* 5:2441–54
35. Lee WS, Kao CC, Bryant GO, Liu X, Berk AJ. 1991. *Cell* 67:365–76
36. Horikoshi N, Maguire K, Kralli A, Maldonado E, Reinberg D, Weinmann R. 1991. *Proc. Natl. Acad. Sci. USA* 88:5124–28
37. Kerr LD, Ransone LJ, Wamsley P, Schmitt MJ, Boyer TG, et al. 1993. *Nature* 365:412–19
38. Xu X, Prorock C, Ishikawa H, Maldonado E, Ho Y, Gelinas C. 1993. *Mol. Cell. Biol.* 13:6733–41
39. Seto E, Usheva A, Zambetti GP, Momand J, Horikoshi N, et al. 1992. *Proc. Natl. Acad. Sci. USA* 89:12028–32
40. Cortes P, Flores O, Reinberg D. 1992. *Mol. Cell. Biol.* 12:413–21
41. Dynlacht BD, Hoey T, Tjian R. 1991. *Cell* 66:563–76
42. Zhou Q, Lieberman PM, Boyer TG, Berk AJ. 1992. *Genes Dev.* 6:1964–74
43. Tanese N, Pugh BF, Tjian R. 1991. *Genes Dev.* 5:2212–24
44. Sharp PA. 1992. *Cell* 68:1–3
45. Hernandez N. 1993. *Genes Dev.* 7: 1291–308
46. Tjian R, Maniatis T. 1994. *Cell* 77:5–8
47. Cormack BP, Struhl K. 1992. *Cell* 69: 685–96
48. Schultz MC, Reeder RH, Hahn S. 1992. *Cell* 69:697–702
49. Ruppert S, Wang EH, Tjian R. 1993. *Nature* 362:175–81
50. Goodrich JA, Hoey T, Thut C, Admon A, Tjian R. 1993. *Cell* 75:519–30
51. Dynlacht BD, Weinzierl R, Admon A, Tjian R. 1993. *Nature* 363:176–79
52. Weinzierl R, Dynlacht BD, Tjian R. 1993. *Nature* 362:511–17
53. Hoey T, Weinzierl R, Gill G, Chen J-L, Dynlacht BD, Tjian R. 1993. *Cell* 72: 247–60
54. Yokomori K, Chen J-L, Admon A, Zhou S, Tjian R. 1993. *Genes Dev.* 7:2587–99
55. Kokubo T, Gong D-W, Yamashita S, Takada R, Roeder RG, et al. 1993. *Mol. Cell Biol.* 13:7859–63
56. Hisatake K, Hasegawa S, Takada R, Nakatani Y, Horikoshi M, et al. 1993. *Nature* 362:179–81
57. Kokubo T, Gong D-W, Yamashita S, Horikoshi M, Roeder RG, et al. 1993. *Genes Dev.* 7:1033–46
58. Verrijzer CP, Yokomori K, Chen J-L, Tjian R. 1994. *Science* 264:933–41
59. Chen J-L, Attardi LD, Verrijzer CP,

Yokomori K, Tjian R. 1994. *Cell* 79:93–105

60. Roeder RG. 1991. *Trends Biol. Sci.* 16:402–8
61. Concino MF, Lee RF, Merryweather JP, Weinmann R. 1984. *Nucleic Acids Res.* 12:7423–33
62. Carcamo J, Buckbinder L, Reinberg D. 1991. *Proc. Natl. Acad. Sci. USA* 88:8052–56
63. Kaufmann J, Smale S. 1994. *Genes Dev.* 8:821–29
64. Purnell BA, Emanuel PA, Gilmour DS. 1994. *Genes Dev.* 8:830–42
65. Martinez E, Chiang C-M, Ge H, Roeder RG. 1994. *EMBO J.* 13:3115–26
66. Roy AL, Meisterernst M, Pognonec P, Roeder RG. 1991. *Nature* 354:245–48
67. Roy AL, Malik S, Meisterernst M, Roeder RG. 1993. *Nature* 365:355–59
68. Seto E, Shi Y, Shenk T. 1991. *Nature* 354:241–44
69. Usheva A, Shenk T. 1994. *Cell* 76:1115–21
70. Deleted in proof
71. Sadowski CL, Henry RW, Lobo SM, Hernandez N. 1993. *Genes Dev.* 7:1535–48
72. Kassavetis GA, Bardeleben C, Bartholomew B, Braun BR, Joazeiro C, et al. 1994. See Ref. 10, pp. 107–26
73. Darnell J, Lodish H, Baltimore D. 1986. In *Molecular Cell Biology*, 9:305–71. New York: Scientific American Books
74. Geiduschek EP, Tocchini-Valentini GP. 1988. *Annu. Rev. Biochem.* 57:873–914
75. Brow DA, Guthrie C. 1990. *Genes Dev.* 4:1345–56
76. Marck C, Lefebvre O, Carles C, Riva M, Chaussivert N, et al. 1993. *Proc. Natl. Acad. Sci. USA* 90:4027–31
77. Yoshinaga SK, Boulanger PA, Berk A. 1987. *Proc. Natl. Acad. Sci. USA* 84:3585–89
78. L'Etoile ND, Fahnestock ML, Shen Y, Aebersold R, Berk A. 1994. *Proc. Natl. Acad. Sci. USA* 91:1652–56
79. Lagna G, Kovelman R, Sukegawa J, Roeder RG. 1994. *Mol. Cell. Biol.* 14:3053–64
80. Lobo SM, Hernandez N. 1989. *Cell* 58:55–67
81. Lobo SM, Lister J, Sullivan ML, Hernandez N. 1991. *Genes Dev.* 5:1477–89
82. Burnol A-F, Margottin F, Schultz P, Marsolier M-C, Oudet P, et al. 1993. *J. Mol. Biol.* 233:644–56
83. Eschenlauer JB, Kaiser MW, Gerlach VL, Brow DA. 1993. *Mol. Cell. Biol.* 13:3015–26

84. Sakonju S, Bogenhagen DF, Brown DD. 1980. *Cell* 19:13–25
85. Bogenhagen DF, Sakonju S, Brown DD. 1980. *Cell* 19:27–35
86. Shastry BS, Ng S-Y, Roeder RG. 1982. *J. Biol. Chem.* 257:12979–86
87. Bieker JJ, Roeder RG. 1984. *J. Biol. Chem.* 259:6158–64
88. Moorfield B, Roeder RG. 1994. *J. Biol. Chem.* 269:20857–65
89. Miller J, McLachlan AD, Klug A. 1985. *EMBO J.* 4:1609–14
90. Comai L, Tanese N, Tjian R. 1992. *Cell* 68:965–76
91. Rudloff U, Eberhard D, Grummt I. 1994. *Proc. Natl. Acad. Sci. USA* 91:8229–33
92. Learned RM, Learned TK, Haltiner MM, Tjian R. 1986. *Cell* 45:847–57
93. Bell SP, Jantzen H-M, Tjian R. 1990. *Genes Dev.* 4:943–54
94. Bell SP, Pikaard CS, Reeder RH, Tjian R. 1989. *Cell* 59:489–97
95. Learned RM, Cordes S, Tjian R. 1985. *Mol. Cell. Biol.* 5:1358–59
96. Rudloff U, Eberhard D, Tora L, Stunnenberg H, Grummt I. 1994. *EMBO J.* 13:2611–16
97. Buratowski S, Hahn S, Guarente L, Sharp PA. 1989. *Cell* 56:549–61
98. Maldonado E, Ha I, Cortes P, Weis L, Reinberg D. 1990. *Mol. Cell. Biol.* 10:6335–47
99. Pinto I, Ware DE, Hampsey M. 1992. *Cell* 68:977–88
100. Berroteran RW, Ware DE, Hampsey M. 1994. *Mol. Cell. Biol.* 14:226–37
101. Li Y, Flanagan PM, Tschochner H, Kornberg RD. 1994. *Science* 263:805–7
102. Flores O, Lu H, Killeen M, Greenblatt J, Burton Z, Reinberg D. 1991. *Proc. Natl. Acad. Sci. USA* 88:9999–10003
103. Zawel L, Lu H, Cisek LJ, Corden JL, Reinberg D. 1993. *Cold Spring Harbor Symp. Quant. Biol.* 58:187–98
104. Conaway RC, Garrett KP, Hanley JP, Conaway JW. 1991. *Proc. Natl. Acad. Sci. USA* 88:6205–9
105. Killeen MT, Greenblatt JF. 1992. *Mol. Cell. Biol.* 12:30–37
106. Kassavetis GA, Braun BR, Nguyen LH, Geiduschek EP. 1990. *Cell* 60:235–45
107. Kassavetis GA, Joazeiro CA, Pisano M, Geiduschek EP, Colbert T, et al. 1992. *Cell* 71:1055–64
108. Lobo SM, Tanaka M, Sullivan ML, Hernandez N. 1992. *Cell* 71:1029–40
109. Taggart A, Fisher TS, Pugh BF. 1992. *Cell* 71:1015–28
110. White RJ, Jackson SP. 1992. *Cell* 71:1041–53
111. Bartholomew B, Kassavetis GA, Geiduschek EP. 1991. *Mol. Cell. Biol.* 11:5181–89

112. Kassavetis GA, Bartholomew B, Blanco JA, Johnson TE, Geiduschek EP. 1991. *Proc. Natl. Acad. Sci. USA* 88:7308–12
113. Lopez de Leon A, Librizzi M, Puglia K, Willis IM. 1992. *Cell* 71:211–20
114. Buratowski S, Zhou H. 1992. *Cell* 71:221–30
115. Colbert T, Hahn S. 1992. *Genes Dev.* 6:1940–49
116. Huet J, Conesa C, Manaud N, Chaussivert N, Sentenac A. 1994. *Nucleic Acids Res.* 22:3433–39
117. Werner M, Chaussivert N, Willis IM, Sentenac A. 1993. *J. Biol. Chem.* 268:20721–24
118. Cormack BP, Struhl K. 1993. *Science* 262:244–48
119. Moenne A, Camier S, Anderson F, Margottin F, Beggs J, Sentenac A. 1990. *EMBO J.* 9:271–77
120. Joazeiro C, Kassavetis GA, Geiduschek EP. 1994. *Mol. Cell. Biol.* 14:2798–2808
121. Burnol A-F, Margottin F, Huet J, Almouzni G, Prioleau M-N, et al. 1993. *Nature* 362:475–77
122. Lofquist AK, Li H, Imboden MA, Paule MR. 1993. *Nucleic Acids Res.* 21:3233–38
123. Okhuma Y, Sumimoto H, Horikoshi M, Roeder RG. 1990. *Proc. Natl. Acad. Sci. USA* 87:9163–67
124. Schaeffer L, Moncollin V, Roy R, Staub A, Mezzina M, et al. 1994. *EMBO J.* 13:2388–92
125. Gileadi O, Feaver WJ, Kornberg RD. 1992. *Science* 257:1389–91
126. Fischer L, Gerard M, Chalut C, Lutz Y, Humbert S, et al. 1992. *Science* 257:1392–95
127. Schaeffer L, Roy R, Humbert S, Moncollin V, Vermeulen W, et al. 1993. *Science* 260:58–63
128. Humbert S, van Vuuren H, Lutz Y, Hoeijmakers JHJ, Egly J-M, et al. 1994. *EMBO J.* 13:2393–98
129. Dahmus ME. 1981. *J. Biol. Chem.* 256:3332–39
130. Cadena DL, Dahmus ME. 1987. *J. Biol. Chem.* 262:12468–74
131. Baskaran R, Dahmus ME, Wang J. 1993. *Proc. Natl. Acad. Sci. USA* 90:11167–71
132. Drapkin R, Reinberg D. 1994. *Trends Biol. Sci.* 19:504–8
133. Bunick D, Zandomeni R, Ackerman S, Weinmann R. 1982. *Cell* 29:877–86
134. Sawadogo M, Roeder RG. 1984. *J. Biol. Chem.* 259:5321–26
135. Lu H, Flores O, Weinmann R, Reinberg D. 1991. *Proc. Natl. Acad. Sci. USA* 88:10004–8
136. Chestnut JD, Stephens JH, Dahmus ME. 1992. *J. Biol. Chem.* 267:10500–6
137. O'Brien T, Hardin S, Greenleaf A, Lis JT. 1994. *Nature* 370:75–77
138. Weeks JR, Hardin SE, Shen J, Lee JM, Greenleaf A. 1993. *Genes Dev.* 7:2329–44
139. Maxon ME, Goodrich JA, Tjian R. 1994. *Genes Dev.* 8:515–24
140. Zawel L, Reinberg D. 1992. *Curr. Opin. Cell Biol.* 4:488–95
141. Lu H, Zawel L, Fischer L, Egly JM, Reinberg D. 1992. *Nature* 358:641–45
142. Serizawa H, Conaway RC, Conaway JW. 1992. *Proc. Natl. Acad. Sci. USA* 89:7476–80
143. Feaver WJ, Gileadi O, Li Y, Kornberg RD. 1991. *Cell* 67:1223–30
144. Ohkuma Y, Roeder RG. 1994. *Nature* 368:160–63
145. Serizawa H, Conaway JW, Conaway RC. 1994. *J. Biol. Chem.* 269:20750–56
146. Drapkin R, Reardon JT, Ansari A, Huang JC, Zawel L, et al. 1994. *Nature* 368:769–72
147. Weeda G, Ham RC, Masurel R, Westerveld A, Odijk H, et al. 1990. *Mol. Cell. Biol.* 10:2570–81
148. Feaver WJ, Svejstrup JQ, Bardwell AJ, Buratowski S, Gulyas KD, et al. 1993. *Cell* 75:1379–87
149. Wang Z, Svejstrup JQ, Feaver WJ, Wu X, Kornberg RD, Friedberg EC. 1994. *Nature* 368:74–75
150. Serizawa H, Conaway RC, Conaway JW. 1993. *J. Biol. Chem.* 268:17300–8
151. Roy R, Schaeffer L, Humbert S, Vermeulen W, Weeda G, Egly J-M. 1994. *J. Biol. Chem.* 269:9826–32
152. Hanawalt P, Mellon I. 1993. *Curr. Biol.* 3:67–69
153. Drapkin R, Sancar A, Reinberg D. 1994. *Cell* 77:9–12
154. van Vuuren AJ, Vermeulen W, Ma L, Weeda G, Appeldorn E, et al. 1994. *EMBO J.* 13:1645–53
155. Buratowski S. 1993. *Science* 260:37–38
156. Troelstra C, Gool A, Wit J, Vermeulen W, Bootsma D, et al. 1992. *Cell* 71:939–53
157. Laybourn PJ, Dahmus ME. 1990. *J. Biol. Chem.* 265:13165–73
158. Guzder SN, Qiu H, Sommers CH, Sung P, Prakash L, et al. 1994. *Nature* 367:91–94
159. Qiu H, Park E, Prakash L, Prakash S. 1993. *EMBO J.* 7:2161–71
160. Guzder SN, Sung P, Bailly V, Prakash L, Prakash S. 1994. *Nature* 369:578–81
161. Sung P, Higgins D, Prakash L, Prakash S. 1988. *EMBO J.* 7:3263–69
162. Drapkin R, Reinberg D. 1994. *Nature* 369:523–24
163. Parvin JD, Timmers M, Sharp PA. 1992. *Cell* 68:1135–44

164. Parvin JD, Sharp PA. 1993. *Cell* 73: 533–40
165. Goodrich JA, Tjian R. 1994. *Cell* 77: 145–56
166. Parvin JD, Shykind BM, Meyers RE, Kim J, Sharp PA. 1994. *J. Biol. Chem.* 269:18414–21
167. Tyree CM, George CP, Lira-DeVito LM, Wampler SL, Dahmus ME, et al. 1993. *Genes Dev.* 7:1254–65
168. Timmers HT. 1994. *EMBO J.* 13:391–99
169. Tantin D, Carey M. 1994. *J. Biol. Chem.* 269:17397–400
170. Mitchell PJ, Tjian R. 1989. *Science* 245: 371–78
171. Struhl K. 1989. *Trends Biol. Sci.* 14: 137–40
172. Ptashne M, Gann A. 1990. *Nature* 346: 329–31
173. Lin Y, Green MR. 1991. *Cell* 64:971–81
174. Baniahmad A, Ha I, Reinberg D, Tsai S, Tsai M-J, O'Malley BW. 1993. *Proc. Natl. Acad. Sci. USA* 90:8832–36
175. Zhu H, Joliot V, Prywes R. 1994. *J. Biol. Chem.* 269:3489–97
176. Xiao H, Pearson A, Coulombe B, Truant R, Zhang S, et al. 1994. *Mol. Cell. Biol.* 14:7013–24
177. Tsuda M, Hirose S, Suzuki Y. 1986. *Mol. Cell. Biol.* 6:3928–33
178. Workman JL, Roeder RG. 1987. *Cell* 51:613–22
179. Lorch Y, LaPointe JW, Kornberg RD. 1987. *Cell* 49:203–10
180. Knezetic JA, Luse DS. 1986. *Cell* 45: 95–104
181. Paranjape SM, Kamakaka RT, Kadonaga JT. 1994. *Annu. Rev. Biochem.* 63:265–97
182. Workman JL, Abamyr SM, Cromlish WA, Roeder RG. 1988. *Cell* 55:211–19
183. Workman JL, Roeder RG, Kingston RE. 1990. *EMBO J.* 9:1299–308
184. Taylor I, Workman JL, Schuertz TJ, Kingston RE. 1991. *Genes Dev.* 5:1285–98
185. Workman JL, Taylor I, Kingston RE. 1991. *Cell* 64:533–44
186. Laybourn PJ, Kadonaga JT. 1992. *Science* 257:1682–85
187. Carey M, Leatherwood J, Ptashne M. 1990. *Science* 247:710–12
187a. Miralles VJ, Cortes P, Stone N, Reinberg D. 1989. *J. Biol. Chem.* 264: 10763–772
188. Han M, Grunstein M. 1988. *Cell* 55: 1137–45
189. Durrin LK, Mann RK, Grunstein M. 1992. *Mol. Cell. Biol.* 12:1621–29
190. Mann RK, Grunstein M. 1992. *EMBO J.* 11:3297–306
191. Peterson CL, Herskowitz I. 1992. *Cell* 68:573–83
192. Kruger W, Herskowitz I. 1991. *Mol. Cell. Biol.* 11:4135–46
193. Peterson CL, Dingwall A, Scott MP. 1994. *Proc. Natl. Acad. Sci. USA* 91: 2905–8
194. Cairns BR, Kim Y-J, Sayre M, Laurent BC, Kornberg RD. 1994. *Proc. Natl. Acad. Sci. USA* 91:1950–54
195. Cote J, Quinn J, Workman JL, Peterson CL. 1994. *Science* 265:53–60
196. Kwon H, Imbalzano AN, Khavari PA, Kingston RE, Green MR. 1994. *Nature* 370:477–81
197. Imbalzano AN, Kwon H, Green MR, Kingston RE. 1994. *Nature* 370:481–85
198. Auble DT, Hahn S. 1994. *Genes Dev.* 7:844–56
199. Auble DT, Hansen KE, Mueller C, Lane WS, Thorner J, Hahn S. 1994. *Genes Dev.* 8:1920–34
200. Davis JL, Kunisawa R, Thorner J. 1992. *Mol. Cell. Biol.* 12:1879–92
201. Roberts SGE, Ha I, Maldonado E, Reinberg D, Green MR. 1993. *Nature* 363: 741–44
202. Hahn S. 1993. *Nature* 363:672–73
203. Choy B, Green MR. 1993. *Nature* 366: 531–36
204. Colgan J, Wampler S, Manley JL. 1993. *Nature* 362:549–53
205. Roberts S, Green MR. 1994. *Nature* 371:717–20
206. Reines D. 1991. *J. Biol. Chem.* 266: 10510–17
206a. Tong X, Drapkin R, Reinberg D, Kieff E. 1995. *Proc. Natl. Acad. Sci. USA.* In press
207. Yankulov K, Blau J, Purton T, Roberts S, Bentley DL. 1994. *Cell* 77:749–59
208. Collart MA, Struhl K. 1994. *Genes Dev.* 8:525–37
209. Kretzschmar M, Meisterernst M, Roeder RG. 1993. *Proc. Natl. Acad. Sci. USA* 90:11508–12
210. Kretzschmar M, Stelzer G, Roeder RG, Meisterernst M. 1994. *Mol. Cell. Biol.* 14:3927–37
211. Yeung KC, Inostroza JA, Mermelstein FH, Kannabiran C, Reinberg D. 1994. *Genes Dev.* 8:2097–109
212. Ge H, Roeder RG. 1994. *Cell* 78:513–23
212a. Kretzschmar M, Kaiser K, Lottspeich F, Meisterernst M. 1994. *Cell* 78:525–34
213. Kretzschmar M, Kaiser K, Lottspeich F, Meisterernst M. 1994. *Cell* 78:525–34
214. Ma D, Watanabe H, Mermelstein F, Admon A, Oguri K, et al. 1993. *Genes Dev.* 7:2246–57
215. Sun X, Ma D, Sheldon M, Yeung K, Reinberg D. 1994. *Genes Dev.* 8:2336–48

216. Yokomori K, Zeidler MP, Chen J-L, Verrijzer CP, Mlodzik M, et al. 1994. *Genes Dev.* 8:2313–23
217. Lieberman PM, Berk AJ. 1994. *Genes Dev.* 8:995–1006
218. Ozer J, Moore PA, Bolden AH, Lee A, Rosen CA, et al. 1994. *Genes Dev.* 8:2324–35
219. Kelleher RJ III, Flanagan PM, Chasman DI, Ponticelli AS, Struhl K, et al. 1992. *Genes Dev.* 6:296–303
220. Poon D, Weil PA. 1993. *J. Biol. Chem.* 268:15325–28
221. Reese JC, Apone L, Walker SS, Griffin LA, Green M. 1994. *Nature* 371:523–27
222. Berger SL, Cress WD, Cress A, Triezenberg SJ, Guarente L. 1990. *Cell* 61: 1199–208
223. Berger SL, Pina B, Silverman N, Marcus GA, Agapite J, et al. 1992. *Cell* 70:251–65
224. Pina B, Berger S, Marcus GA, Silverman N, Agapite J, et al. 1993. *Mol. Cell. Biol.* 13:5981–89
225. Kelleher RJ III, Flanagan PM, Kornberg RD. 1990. *Cell* 61:1209–15
226. Flanagan PM, Kelleher RJ III, Sayre MH, Tschochner H, Kornberg RD. 1991. *Nature* 350:436–38
227. Kim Y-J, Bjorklund S, Li, Y, Sayre M, Kornberg RD. 1994. *Cell* 77:599–608
228. Koleske AJ, Buratowski S, Nonet M, Young RA. 1992. *Cell* 69:883–94
229. Thompson CM, Koleske AJ, Chao DM, Young RA. 1993. *Cell* 73:1361–75
230. Koleske AJ, Young RA. 1994. *Nature* 368:466–69

Annu. Rev. Biochem. 1995. 64:563–91

HOW GLYCOSYL-PHOSPHATIDYLINOSITOL-ANCHORED MEMBRANE PROTEINS ARE MADE

Sidney Udenfriend

Roche Institute of Molecular Biology, Roche Research Center, Nutley, New Jersey 07110

Krishna Kodukula

Bristol-Myers Squibb Pharmaceutical Research Institute, Wallingford, Connecticut 06492

KEY WORDS: membrane anchor, signal peptides, COOH-terminal processing, transamidase

CONTENTS

ABSTRACT

Glycosylphosphatidylinositol (GPI) linkage is a fairly common means of anchoring membrane proteins to eukaryotic cells, although the exact function of the GPI linkage is not clear. The nascent form of a typical GPI protein contains a hydrophobic NH_2-terminal signal peptide that directs it to the ER. There the signal peptide is removed by NH_2-terminal signal peptidase. Nascent forms of GPI-linked proteins contain a second hydrophobic peptide at their COOH terminus. The COOH-terminal peptide is also removed during processing and

563

the GPI moiety is ultimately linked to what had been an internal sequence in the nascent protein. Two independent pathways are involved in the biosynthesis of GPI proteins, GPI formation, and processing of the nascent protein with attachment of the GPI moiety. Studies in whole cells and in cell-free systems indicate that structural requirements around the COOH-terminal cleavage site of nascent proteins are similar to those at the cleavage site of NH_2-terminal signal peptidase. However, COOH-terminal processing requires a transmidase for which evidence is presented as well as a proposed mechanism of its action.

INTRODUCTION

Cell surface proteins are required for many important physiologic functions, among them cell-cell recognition, cell adhesion, receptors, and nutrient and ion transporters. Until recently researchers believed that attachment of surface proteins to the membrane always occured via transmembrane peptide sequences. A recently recognized alternative means of anchoring surface proteins is via some type of lipid that is attached to the protein posttranslationally. Lipid anchors are not rare. The most ubiquitous of them is glycosylphosphatidylinositol (GPI), which is now recognized as being a major means of anchoring membrane proteins to eukaryotic cells (Figure 1). Many proteins in the central

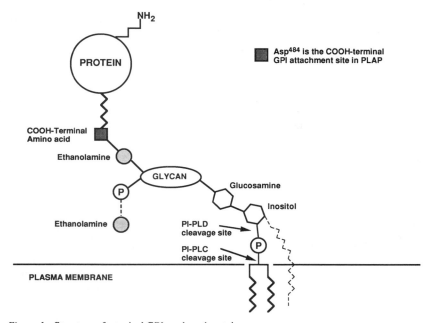

Figure 1   Structure of a typical GPI-anchored protein.

nervous system are GPI-anchored, among them Thy-1 (1, 2), ciliary neurotrophic factor receptor (3); cell adhesion molecule, N-CAM (4), and oligodendrocyte myelin protein (5). Prion proteins that are responsible for some neurodegenerative diseases, such as Scrapie and Creutzfeld-Jacob diseases, are also GPI-anchored (6, 7).

The discovery of GPI-linked proteins started with the isolation of bacterial phospholipases by Slein and Logan in the 1960s (8–11). In the late 1970s Ikezawa (12) and Low (13) independently showed that the enzyme alkaline phosphatase (AP) could be released from the plasma membranes of intact cells by treatment with bacterial PI-PLC (phosphotidylinositol specific phospholipase C). From this experiment, they concluded that AP was bound to the membrane by some type of GPI anchor. In addition to his own studies, Low, by making PI-PLC available to many investigators, soon proved that AP was not unique and that many other proteins i.e. acetylcholinesterase (14), Thy-1 (15), and trypanosomal variant surface glycoproteins (VSG) (16), could be released from cells by PI-PLC and subsequently shown to be linked to the plasma membrane in the same manner as AP. The structure of the GPI moiety of VSG was elucidated by Ferguson et al (17). Subsequently, researchers ascertained the structure(s) of many parasitic and mammalian GPIs (for reviews, see 18–25). Today, it is recognized that GPI linkage is a fairly common means of anchoring membrane proteins to mammalian, yeast, and parasitic cells. The exact function of the GPI linkage is not clear. However, one of its functions is apparently to direct proteins to the apical side of polarized cells (26), where the GPI proteins have increased mobility in the membrane as compared to proteins linked by a transmembrane peptide sequence. GPI linkage is not limited to a specific class or type of protein but can involve enzymes, receptors, cell-specific antigens, adhesion molecules, transporters, etc. For example, GPI proteins act as receptors [for urokinase (27)] and transporters [for folate (28)] even though they do not possess a transmembrane or cytoplasmic domain. Abnormalities in GPI synthesis have also been implicated in disease processes such as paroxysmal nocturnal hemoglobinuria (PNH) (29, 30). The GPI moiety itself, or in its protein-linked form from the malarial parasites *Plasmodium falciparum* and *Plasmodium berghei,* has been shown to mediate host signal transduction (31). The inositol glycan, when cleaved from the protein, has been implicated in signaling with respect to insulin and nerve growth factor (32, 33). More recently, two unusual GPI-anchored proteins or peptides were reported (34, 35). The CD-24 signal transducing molecule, which is comprised of 31–35 amino acids, is a major surface antigen on small cell lung carcinomas (34). The CAMPATH-1 antigen is even smaller, only 12 amino acids in its mature form, and antibodies against it are effective for complement-mediated cell lysis (35). Both peptides are present in abundance on all human lymphocytes. The recent review of GPI proteins by Field

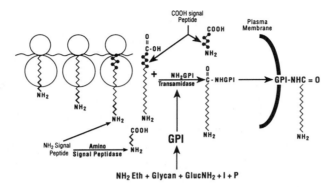

Figure 2 Scheme for the biosynthesis of GPI-anchored membrane proteins. NH2Eth, ethanolamine; GlucNH2, glucosamine; I, inositol; P, phosphate. [Reprinted from Figure 1 of Reference 74)

and Menon (23) contains a comprehensive list of many recognized GPI proteins and discusses the possible functions of the GPI-anchor.

Although some still consider them unusual entities, GPI proteins are widespread in nature. They share a common and unique pathway of biosynthesis that starts in the endoplasmic reticulum (ER) and is most probably completed in the Golgi. The nascent form of a typical GPI protein (Figure 2) contains a hydrophobic amino terminal signal peptide that is typical of proteins that are processed in the ER. During processing, the NH_2-terminal signal peptide is removed by the NH_2-terminal signal peptidase. In the absence of any other signals, the protein would then be glycosylated and secreted from the cells. However, as shown in Figure 2, nascent proteins destined to be GPI-linked contain a second hydrophobic peptide at their COOH terminus. Like the NH_2-terminal signal peptide, the COOH-terminal peptide is removed during processing, and in the final product, the GPI moiety is linked to what had been an internal sequence in the nascent protein.

BIOSYNTHESIS OF THE GPI MOIETY

Two independent metabolic pathways are involved in the formation of a GPI linked moiety. One is the biosynthesis of the GPI moiety; the other involves processing of the nascent protein and covalent attachment of the completed GPI moiety to yield the mature GPI protein. The chemical structure and biosynthesis of the GPI moiety have been under investigation in many laboratories, and intermediates in the process have been identified and characterized in parasites such as *Trypanosoma* and *Leishmania* species as well as in mammalian cells (see 22, 23, 25, and references therein). Several GPI anchor–deficient mutants have been identified or prepared by complementation cloning (36–40), and some of the enzymes involved in anchor biosynthesis have been cloned (41, 42). Biosynthesis takes place on the cytoplasmic face of the endoplasmic reticulum (43, 44), and GPI is then translocated into the lumen with the phosphatidyl group associated with the inner leaflet of the lipid bilayer (45).

CRITERIA INDICATING THAT A PROTEIN CONTAINS A GPI ANCHOR

Whereas numerous investigators have been involved in elucidating the structure and biosynthetic pathway of the GPI moiety, relatively few have studied the process of GPI addition to a protein. Most of our knowledge concerning the processing of nascent proteins to their GPI-linked forms comes mainly from two groups, our own (46–62) and that of Caras and coworkers (63–73). Whereas all earlier reviews emphasized the GPI moiety, here we concentrate on the protein component of GPI proteins from the first demonstration that a membrane protein is GPI-anchored, to the signal sequences, enzymes, and cofactors that are involved in its processing from the nascent protein. Short reviews of the protein events involved in formation of GPI proteins appeared recently (74–76).

At the cDNA level, that a protein may be processed to a GPI form can be deduced from a hydropathy plot. Proteins destined to be GPI-anchored contain hydrophobic signal sequences both at their NH_2 and COOH termini, with the remainder of the cDNA deduced sequence being essentially hydrophilic. At the protein level, GPI anchoring was first detected by demonstrating release from the cell surface by PI-PLC. This is still the simplest method, but it is not always reliable because in some forms of GPI, such as human erythrocyte acetylcholinesterase (77, 78), an additional palmitoyl group on the inositol results in resistance to PI-PLC. Similarly, the procyclic acidic repetitive protein (PARP) of *Trypanosoma brucei* also contains an additional fatty acid (79). Metabolic labeling of a protein by radioactive elements of GPI such as ethanolamine, inositol, or fatty acids is another frequently used method of establishing that a protein is GPI-anchored. Incorporation of at least two of the above components should be used because there are other mechanisms whereby either ethanolamine or fatty acids can be incorporated into proteins (48, 50, 80–82). Furthermore, one must demonstrate that the radioactivity incorporated is not a metabolite of the radioactive precursor.

If all the above procedures are carried out, one can be fairly certain that one is dealing with a GPI protein. However, the residue in the nascent protein to which the GPI is attached must still be determined. As shown in Figure 2, a

COOH-terminal signal peptide is lost during the process of GPI addition, and the GPI moiety is linked to the new COOH-terminal amino acid residue. For convenience, the amino acid in the nascent protein destined to become the new COOH terminus and accept the GPI residue has been designated as the ω site (60, 61); residues following this site are then ω+1, ω+2, etc. The value of introducing the term ω in this way becomes apparent when comparing the cDNA deduced amino acid sequences of nascent proteins that have been fully characterized, which vary greatly in size, from 532 residues for human placental alkaline phosphatase (PLAP) to about 162 residues for Thy-1. Nevertheless, all can be aligned at their known or putative ω sites. The ω notation serves other useful purposes, as discussed later. Relatively few of the many known GPI proteins have had their ω sites determined experimentally because this process involves isolation and purification of the protein and enzymatic fragmentation, isolation, and sequencing of the peptide(s) containing a GPI moiety. As with NH_2-terminal processing, the cleavage sites of COOH-terminally processed nascent proteins are now generally deduced, as described below.

In those few instances where the ω site had been determined experimentally, GPI-containing peptides released after protease treatment were detected in the lysates by one of two methods. Where the protein could possibly be labeled metabolically, either with radioactive ethanolamine, inositol, or fatty acid, the peptide was detected and purified by monitoring radioactivity in the protease digest (50, 68, 81, 82). Production of a site-directed antibody upstream and close to the expected ω site was used to detect and isolate the GPI peptide in digests of placental alkaline phosphatase (PLAP) (50). Whatever method is used, one must clone the protein to determine the appropriate protease to use so that the GPI peptide generated will be small enough to permit sequencing it entirely. Amino acid analysis of the GPI-linked peptide must be used to corroborate the presence of the ω amino acid.

BIOSYNTHESIS OF GPI PROTEINS

Most eukaryotic cells express an array of GPI proteins on their surface and, therefore, contain GPI, the putative transamidase, and accessory factors needed to convert a nascent protein to its GPI-linked form. Investigations on the biosynthesis of GPI proteins may be divided into two categories, studies with intact cells and studies in cell-free systems. By designing specific deletion and site mutations and transfecting the corresponding cDNAs into appropriate cells to generate the mutant proteins, investigators determined some characteristics of the COOH-terminal signal peptide, i.e. optimal length, hydrophobicity, and sequence requirements at and near the ω site. To investigate the mechanism involved in the formation of GPI proteins one must turn to cell-free systems.

The presence of NH_2-terminal signal peptides on nascent proteins that also contain a COOH-terminal signal peptide indicated that processing to the GPI form must take place in the ER. Investigators demonstrated this processing by coupling rough microsomal membranes (RM), as a source of ER, to a translation system containing an appropriate mRNA (52, 56, 83).

Studies with Intact Cells

Most of the studies reported thus far were carried out in Cos-7 or CHO cells, transiently or permanently transfected with the cDNA of a GPI-linked protein or with mutated forms of the cDNA designed to investigate specific properties of the protein that influence GPI linkage. Variables studied include 1. the size and degree of hydrophobicity of the COOH-terminal signal peptide, 2. the presence or absence of known cytoplasmic domains following the signal peptide, 3. the interchangeability of COOH-terminal signal peptides from one GPI protein to another and, 4. mutation of amino acid substituents at or near the ω site. The presence of an NH_2-terminal signal peptide is universal among nascent forms of GPI proteins, and this peptide directs them into the ER. In most cases, the NH_2-terminal signal peptide is cleaved during processing. However, Howell et al (84) recently showed that cleavage of the NH_2-terminal peptide is not mandatory for processing to the GPI-linked form to occur. That study was carried out with a chimeric protein linking neutral endopeptidase, which is a type II membrane protein with an NH_2-terminal signal peptide that was not cleavable, and the COOH-terminal signal sequence of DAF. Dual linked proteins, containing transmembrane peptide sequences as well as a GPI-anchor, occur naturally in cells, e.g. ponticulin in *Dictyostelium* (85) and Sm23 in *Schistosoma mansoni* (86).

By analogy with the NH_2-terminal signal peptide (87–92) the COOH-terminal hydrophobic peptide of nascent GPI proteins serves as a recognition site for binding to the appropriate enzyme, in this case the putative transamidase. The hydrophobic sequence in a COOH-terminal signal peptide from GPI proteins that have been fully characterized varies from 17–31 residues, and the exact amino acid composition differs considerably from protein to protein. In fact, hybrids containing the protein portion of one and the COOH-terminal signal peptide of another protein have been produced that are GPI anchored (67). Researchers also have prepared mutants in which the COOH-signal peptide sequence of a GPI protein is coupled to an otherwise secreted protein and thereby convert it to a GPI-anchored form (69, 70).

The GPI proteins that have been most widely used for studies in intact cells are placental alkaline phosphatase (PLAP) (47, 51, 53–55, 61, 62, 80), decay accelerating factor (DAF) (67–73), and hybrid forms of the latter with other proteins. Antibodies to PLAP and DAF proteins were used for analytical purposes. However, PLAP has an important advantage in that it is an enzyme whose activity can also be used to monitor experiments. Berger et al (51) carried out a systematic study with deletion and substitution mutants of PLAP in which they varied the size of the COOH-terminal signal peptide, either shortening or lengthening it and also changing its amino acid composition. They found that for GPI-linking to occur a newly synthesized protein must contain an uncharged, predominantly hydrophobic amino acid sequence of a certain minimal length at its COOH terminus. Although an elongated form of preproPLAP with 17 consecutive hydrophobic residues in the COOH-terminal sequence yielded a GPI-anchored membrane product, mutants with 13 or fewer such hydrophobic residues yielded proteins that were no longer GPI-anchored but were secreted into the culture medium. Cassette mutants were used to demonstrate that the precise amino acid sequence of the COOH-terminal region could be varied considerably as long as minimal hydrophobicity and length was maintained. Caras et al (65, 66) produced truncated and random mutations in the COOH-terminal peptide of DAF and came to similar conclusions.

Berger et al (53) found that elongating PLAP, by adding the COOH-terminal cytoplasmic domain of the membrane-associated G protein of the vesicular stomatitis virus (VSV) to it, converted it from a GPI-anchored form to an integral transmembrane protein. The fusion protein had a 29-residue hydrophilic sequence following a 19-residue hydrophobic sequence that was designed to act as a COOH-terminal signal peptide, including a typical ω site. This structure would indicate that the ω site and the hydrophobic portion of a signal peptide cannot be located far from the COOH terminus of a nascent protein, as is true of all the GPI proteins characterized thus far (60). However, Caras (67) reported that attachment of a GPI anchor took place at a site located 100 plus residues from the COOH terminus of a nascent protein. Those studies were carried out with a fairly complex fusion protein designed to contain the entire coding sequence including the signal recognition site (ω site plus COOH-terminal signal peptide) of DAF, followed by 191 amino acids of human growth hormone (hGH), without DAF's NH_2-terminal signal peptide. Although a GPI protein was produced upon transfection of the corresponding chimeric cDNA into Cos-7 cells, proof of the exact sequence of events leading to the formation of a mature GPI protein will have to be obtained in a cell-free system (see below).

Micanovic et al (54) carried out an essentially site-specific mutagenesis of the ω residue of PLAP which is normally aspartic acid. They found that mutants with glycine, alanine, cysteine, serine, or asparagine at the ω site became GPI-anchored, appeared in the plasma membrane, and were enzymatically active when expressed in Cos-7 cells; whereas the leucine and valine mutants yielded relatively small amounts of activity. Although mutants with glutamic acid, glutamine, proline, tryptophan, phenylalanine, threonine, methionine, and tyrosine were expressed equally well by the cells, there was little processing to

GPI-anchored forms as evidenced by the mere traces of PLAP enzyme activity that appeared on the plasma membrane. The bulk of the inactive mutant proteins remaining in the cells was not GPI-anchored and had little PLAP enzyme activity. Presumably these proteins were retained in the ER in the proform and subsequently degraded. Current evidence indicates that proteins with uncleaved COOH-terminal signal sequences are hung up in the ER (72, 93) and eventually get degraded (72, 93, 94). The same six amino acids (with small substituents on the β carbon atom) are also the only ones that have been found at ω sites of those GPI proteins that have been fully characterized (60). The findings by Micanovic et al (54, 55) in intact cells were subsequently confirmed in studies carried out in cell-free systems (see below). Moran et al (68) identified serine-319 as the ω site of DAF by metabolically labeling a human growth hormone–DAF fusion protein with ethanolamine followed by isolating and sequencing an ethanolamine-containing tryptic peptide derived from the fusion protein. They showed that amino acids immediately NH$_2$-terminal to the ω site (i.e. ω-1, ω-3, ω-5, ω-7, ω-9) could be deleted without affecting GPI anchoring of the resulting protein to the plasma membrane. However, deletion of the serine at the ω site led to loss of this activity. They also carried out saturation mutagenesis at the ω site and found, as did Micanovic et al (54, 55), that glycine, alanine, aspartate, and asparagine could substitute efficiently for serine for successful GPI anchoring. In the studies of Moran et al, cysteine at the ω site proved to be inactive, contrary to findings with a known GPI protein (1, 2) and to the studies of Micanovic et al on PLAP (54). In the PLAP studies, cysteine was weakly but definitely active. Because PLAP could also be measured by its enzyme activity, independently of immunoprecipitation, small amounts of activity could be readily confirmed. In both the PLAP and DAF studies, valine was much less active than cysteine and one or two other amino acids exhibited small amounts of activity.

Although there appears to be no selectivity with respect to amino acids that are NH$_2$-terminal to the ω site (68), this is not the case for those residues that are in the immediate vicinity of and COOH-terminal to the ω site, i.e. ω+1 and ω+2. In all but one of the characterized GPI-proteins (60), only glycine, alanine, and serine are found at the ω+2 site. By contrast, although mainly small amino acids appear at the ω+1 sites of known proteins, there are many exceptions. Nevertheless ω, ω+1, and ω+2 seem to represent a small amino acid domain at the NH$_2$-terminal end of the COOH-terminal signal peptide. Kodukula et al (61) investigated the selectivity of amino acids at the ω+1 and ω+2 sites of PLAP by using site mutagenesis. At the ω+1 site, processing to varying degrees was observed with all the amino acids except proline, when they were substituted for alanine, the normal constituent. By contrast, the only substituents at the ω+2 site that led to processing were glycine and alanine and only trace activity with serine and cysteine. In native proteins serine has thus far been found only at the $\omega+2$ sites of VSG type proteins of trypanosomes (18, 20). The presence of threonine at the $\omega+2$ site of DAF (68) is another exception to the experimentally determined allowable substitutions reported by Gerber et al (60) and Kodukula et al (61). In all other mammalian proteins that have been characterized, only glycine or alanine has been found at the $\omega+2$ site.

Studies in Cell-Free Systems

Observations on GPI addition in intact cells revealed that nascent proteins destined to be processed to a GPI form must have both NH_2- and COOH-terminal signal peptides, the first for translocation into the ER and the second for recognition by a putative GPI transamidase. The latter is apparently present in the ER. Whole-cell studies provided important information about primary structural requirements for a functional COOH-terminal signal peptide and of the permissible amino acids at the site of cleavage and GPI addition, the ω site. However, no amount of studies in intact cells can reveal the details of the mechanism of the enzyme-catalyzed reaction, which purportedly involves a transamidase. For that purpose, cell-free systems and, eventually, purified enzyme and substrates must be employed.

$$\text{proprotein} + \text{GPI} \xrightarrow[\text{transamidase}]{\text{putative}} \text{GPI-protein} + \text{COOH-terminal signal peptide}$$

Because these events apparently take place in the ER, initial attempts to demonstrate the overall reaction were carried out with such a system. By coupling crude microsomal membranes (RM), prepared from CHO or WISH cells to a rabbit reticulocyte lysate translation system, Bailey et al (52) and Micanovic et al (55) demonstrated the sequential conversion of the nascent protein, preproPLAP, first to proPLAP then to mature GPI-linked PLAP. They identified the three products of cotranslational processing not only by their relative mobility on gels but also by their interaction with three site-specific, precipitating antibodies. The antibodies were prepared against synthetic peptides representing sites within the cDNA deduced sequence of PLAP (49). The sites were selected to distinguish intermediates and products during the sequential processing of the nascent protein (Figure 3). Fasel et al (83) demonstrated translation and processing of both Thy-1 mRNA and DAF mRNA to mature GPI-linked forms in a rabbit reticulocyte lysate system coupled to RM from dog pancreas.

The studies by Bailey et al (52) and Micanovic et al (55) on PLAP clearly demonstrated the release of the NH_2-terminal signal peptide from the prepro form of PLAP to yield the pro form followed by the release of the COOH-terminal signal peptide and formation of mature GPI-linked PLAP. Fasel et al

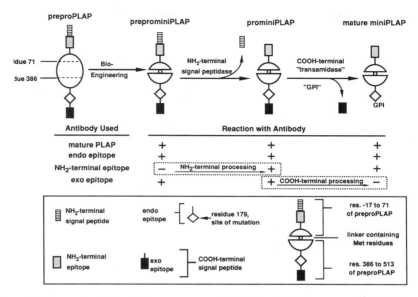

Figure 3 Structural relationship between preprominiPLAP and wild-type preproPLAP. Also shown is the stepwise processing of prepro-miniPLAP first to prominiPLAP and then to mature GPI and the immunologic determinants that were used to help elucidate the steps in processing to the pro and mature forms (Adapted from Figure 8 in Reference 56).

(83) in their experiments on Thy-1 and DAF did not identify any intermediate pro form. However, PLAP, Thy-1, and DAF, like most nascent proteins, are not ideal substrates for investigating the cell-free biosynthesis of a GPI protein. Proteins such as PLAP and DAF are in the range of 40–70 kDa. Thus, changes of 2–3 kDa between precursors and products during processing, resulting from the loss of a signal peptide, are not easily discernible. Furthermore, even smaller proteins such as Thy-1 are highly glycosylated, hence changes in mass due to the loss of signal peptides during processing are obscured. These problems make it difficult to monitor processing of the nascent forms of GPI proteins in cell-free systems. When Blöbel faced comparable problems during studies on NH_2-terminal processing of secreted proteins in the ER, he solved them by utilizing as a substrate preprolactin, a small protein (~ 20 kDa) that is devoid of glycosylation sites (95, 96). Unfortunately, a comparable native protein, small and devoid of glycosylation sites, was not available for monitoring both NH_2- and COOH-terminal processing. The problem was resolved by the design of a smaller and simpler substrate for GPI processing derived by deletion of about two-thirds of the internal sequence of preproPLAP 513 (56). The engineered protein, preprominiPLAP 208, retains both signal pep-

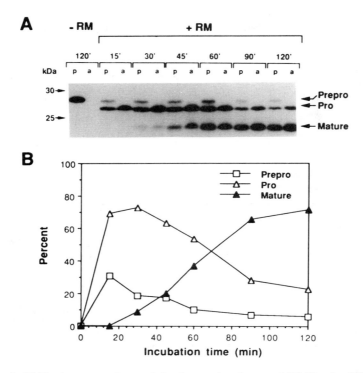

Figure 4 (A) The time course of cotranslational processing of prepprominiPLAP serine-179. At the end of each time period samples were immunoprecipitated with either polyclonal-Ab (p) or amino-Ab (a). The sample in the absence of RM was incubated for 120 min. (B) The absorbance of each band shown in A was plotted at the times indicated. (Adapted from Figure 8 in Reference 56).

tides and all the epitopes for site-directed antibodies to PLAP (49) but is devoid of glycosylation sites, the catalytic site, and most of the cysteine residues. PrepprominiPLAP markedly simplified studies in cell-free systems and made it possible to obtain direct evidence for the step by step conversion to the pro form and then to the mature GPI-linked form, with the concomitant loss of the two signal peptides (Figure 4).

Proof that the final product of processing contained the GPI moiety was still needed. Such evidence is not difficult to obtain in intact cells, because the amounts formed (nanomoles) are adequate for demonstrating the incorporation of the elements of GPI (i.e. ethanolamine, inositol, fatty acids) or release by PI-PLC. However, the small amounts of putative GPI linked miniPLAP formed under cell-free conditions, about 20–100 fmoles per experiment, rule out such direct approaches. Incorporation of the labeled components of GPI into VSG in trypanosomal lysates has been demonstrated (43). However, the trypanosomal system apparently does not lend itself to the cotranslational types of studies that have been carried out with miniPLAP from RM of mammalian cells. Thus far, identification of the mature miniPLAP product as being GPI-linked has been based on indirect observations, which are based on comparisons with the deletion mutant miniPLAP 179, which is miniPLAP that terminates at aspartic acid-179 (the ω site) but without the GPI moiety attached. On SDS gels, the Mr of the putative GPI-linked miniPLAP formed by processing prepromini-PLAP indicates that it is about 1.5 kDa greater in mass than miniPLAP 179. The difference is approximately the mass of the GPI moiety. In addition, when subjected to partitioning between Triton X-114 and water, at least 50% of the mature product obtained through processing of preprominiPLAP partitions into the detergent phase, a characteristic of GPI proteins, while miniPLAP 179 stays almost entirely in the aqueous phase (57). On treatment with PI-PLC or PI-PLD, which specifically cleave the lipid moiety from GPI-linked proteins, most of the mature product appears in the aqueous phase. GPI-anchored proteins are characteristically hydrophobic and become hydrophilic on treatment with inositol specific phospholipases.

The availability of mutant T cells that are defective in GPI biosynthesis provided another approach to investigating incorporation of GPI-anchor into mature, COOH-terminally processed proteins. Rough microsomes prepared from several such mutant T cells had little or no COOH-terminal processing activity but retained their NH_2-terminal signal peptidase activity. A dolichyl-phosphate-mannose synthase gene from yeast, which rescues dolichyl-phosphate-mannose-deficient mutant cells with respect to producing GPI-anchored Thy-1 (97), restored the ability of isolated RM to carry out COOH-terminal processing of prominiPLAP. GPI enriched lipid extracts from CHO, HeLa, and wild-type T cells, when reconstituted in the RM from GPI deficient mutant cells, also restored COOH-terminal processing but to varying degrees (58). The putative mature miniPLAP, formed by GPI-deficient RM and reconstituted with the lipid extracts, comigrated with GPI-linked miniPLAP from CHO cells. Furthermore, specific site-directed antibodies (49) interacted with this protein in exactly the same manner as with mature miniPLAP formed by RM from HeLa or CHO cells. Phosphatidylethanolamine and other known lipids could not replace the lipid extracts.

Cotranslational processing of prominiPLAP corroborated data obtained in intact cells regarding the structural requirements at and near the ω site for GPI addition. As before, one of the six small amino acids, glycine, alanine, serine, cysteine, aspartic acid, and asparagine, was required at the ω site, although some activity was also observed with valine and leucine (56). Relative activities of the various $\omega+1$ and $\omega+2$ mutants (60) were also the same as in intact cells (61). As with intact cells, substitution at the $\omega+2$ site was essentially limited to glycine and alanine; by contrast, except for proline and tryptophan, almost any amino acid could serve at the $\omega+1$ site (60).

The cotranslational system also permitted localization of COOH-terminal GPI processing to the ER. This was accomplished in several ways. The direct approach was to fractionate the RM by sedimentation and to demonstrate that all the cotranslational activity resided in the ER fraction. Specific enzyme markers were used to identify the fractions containing ER and golgi (SE Maxwell, R Amthauer, K Kodukula & S Udenfriend, unpublished results). Experiments with Brefeldin also localized GPI processing activity to the ER. Brefeldin is known to disaggregate the Golgi (98), and Takami et al (99) showed that in Brefeldin treated JEG-3 cells, GPI-anchored proteins such as alkaline phosphatase did not appear on the cell surface. Those findings in intact cells were corroborated and extended to show that RM prepared from those same Brefeldin-treated cells processed preprominiPLAP to mature GPI-anchored miniPLAP just as well as RM from control cells (62). Thus, COOH-terminal processing and linkage to GPI is complete prior to the protein's entrance into the Golgi. Additional evidence that processing of proproteins to GPI proteins takes place in the ER was obtained in translation independent studies (see below) where hydrolysis of ATP was found to be a requirement (59). One of the ATP-requiring entities involved in processing of GPI-anchored proteins in the ER is the immunoglobulin heavy chain binding protein (BiP or GRP 78) (100, 101). BiP, being a resident ER protein, binds to metabolic intermediates (apparently the pro forms) formed during the biosynthesis of three different GPI proteins, PLAP, carbonic anhydrase, and miniPLAP (62). More direct evidence that BiP binds to the pro form of a GPI protein was obtained by transfecting Cos-7 cells with a mutant form of PLAP containing proline at the ω site. This mutant undergoes processing at the NH_2-terminus but is apparently not recognized by the transamidase and, as a result, accumulates as the pro form in the ER. After a 2-h chase, during which all preproPLAP and other intermediates were converted to the proPLAP form, treatment with antibodies to PLAP precipitated BiP along with the mutant proPLAP. Treatment of the cell lysate with ATP prior to immunoprecipitation released the BiP. BiP is associated with nascent proteins that, for various reasons, cannot exit from the ER (102).

Kodukula et al (56) and Amthauer et al (59) investigated the kinetics of cotranslational processing and observed that NH_2-terminal processing occurred so rapidly that the pro form appeared almost instantaneously, whereas the mature (GPI form) appeared only after a lag of about 10–20 min (Figure 4). Membranes separated from the translation mixture after 20 min and reisolated (preloaded RM) contained mainly prepro- and prominiPLAP with only small amounts of mature miniPLAP. When preloaded RM were further incubated in translation buffer alone, there was a small and consistent increase in mature

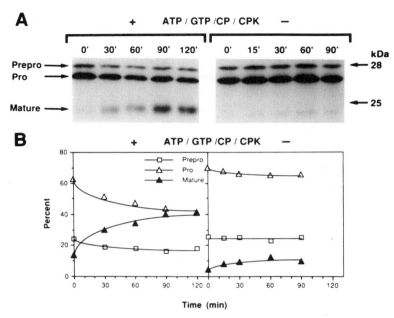

Figure 5 Time course of translation-independent COOH-terminal processing of prominiPLAP. (*A*) RM preloaded with prominiPLAP and incubated in the presence (+) or absence (−) of ATP/GTP/ creatine phosphate (CP)/creatine kinase (CPK). Immunoprecipitations were carried out with polyclonal antibody. (*B*) The relative amounts of each form of miniPLAP in *A*, obtained by densitometry. (Reprinted from Figure 3 in Reference 59).

miniPLAP (59). Addition of ATP, GTP, Mg^{2+}, and an ATP-generating system, consisting of creatine phosphate and phosphokinase, stimulated conversion to mature miniPLAP to levels comparable to those obtained cotranslationally (Figure 5). In preloaded RM, translation is arrested, and thus mature GPI-linked miniPLAP increases exclusively at the expense of prominiPLAP. The increased activity was abolished by addition of hexokinase/glucose or by apyrase, both of which deplete ATP. Stimulation by GTP was less than that with ATP, but its effect was not influenced by the ATP-generating system or by hexokinase/glucose, showing that its actions did not result from conversion to ATP. GTP elicited a consistent additional increase of about 25–35% in activity over ATP alone. Nonhydrolyzable ATP analogues, such as the γ-thio, β,γ-imido, and β,γ-methylene forms of ATP, could not substitute for ATP, indicating that ATP hydrolysis is a prerequisite. Similar experiments with nonhydrolyzable GTP analogues showed that GTP hydrolysis is also essential for processing of prominiPLAP to the GPI-linked form.

At first glance, the requirement of an energy source as well as the hydrolysis of GTP seems to rule out a transamidase type of mechanism for GPI processing. The experiments by Amthauer et al (59) also appear to contradict those of Mayor et al (43), who had earlier demonstrated that the incorporation of the labeled components of GPI into VSG in lysates of *T. brucei* did not require, and was not stimulated by, ATP or GTP and was not affected by hexokinase/glucose. These apparently contradictory findings are, however, reconcilable. In the translation-independent system of Amthauer et al (59) there was a small but appreciable conversion of prominiPLAP to the GPI form, even in the absence of ATP and GTP (Figure 5), which indicates the presence of a pool of prominiPLAP that does not require an external source of energy for cleavage and GPI addition (as in Figure 6, *top, middle*). A comparable pool of the pro form of VSG in trypanosomal lysates would explain the findings of Mayor et al (43). Conceivably, ATP and/or GTP may be required for two steps in COOH-terminal processing: generation of GPI itself and translocation of prominiPLAP to the transamidase site in the ER. The observed lag in cell-free processing and the stimulation by ATP and GTP may indicate that GPI formation is rate limiting in the RM system. However, addition of mannosamine or phenylmethylsulfonyl fluoride, inhibitors of GPI synthesis (103, 104), had no effect on translation-independent conversion of prominiPLAP to the GPI form. Furthermore, the GPI-anchor is now known to be synthesized on the cytoplasmic leaflet of the ER (45), in which case some of the synthetic enzymes may be lost during isolation of RM. If GPI synthesis is not rate limiting then proper folding of the pro form of the protein and/or translocation in the ER remain to be addressed. In the translation independent system (59), membrane insertion of the nascent protein (105) as well as SRP-dependent (106) and independent transport (107) have already been completed so that ATP and GTP are no longer required for those reactions.

The ATP requirement may be for proper folding of the pro protein and mediated by a chaperonin such as BiP (100, 101), and that of GTP may be for translocation to the transamidase site (Figure 6, *top*). The bulk of prominiPLAP in preloaded membranes appears to be newly formed, representing a pool present in the lumen of the ER with its COOH-terminal signal peptide within the membrane but not in contact with the putative transamidase (Figure 6, *top, left*). The observed lag of about 20 min between the appearance of prominiPLAP and that of the mature GPI-anchored protein during cotranslational processing is the strongest evidence for such an intermediate prior to COOH-terminal signal cleavage and GPI addition. Such large lag times were not observed in pulse-chase experiments in intact cells during the processing of VSG (108–110), Thy-1 (111), N-CAM (4), or PLAP (112). However, the kinetics of protein processing observed in intact cells is bound to be different from that in cell-free systems. In the cell-free systems many of the components can become rate limiting, which reveals their essential role in the process. A

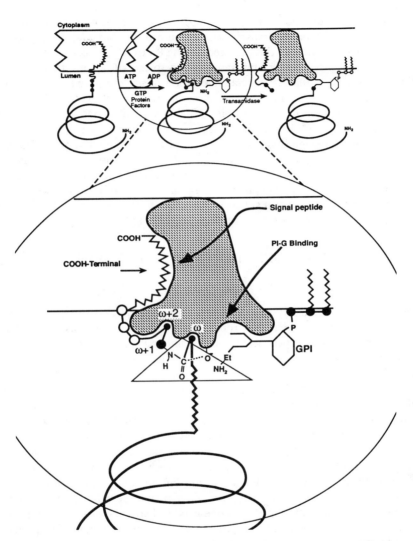

Figure 6 Proposed mechanism of processing of prominiPLAP in the ER. (*Top*) Intermediates and products (reprinted from Figure 7 in Reference 59). (*Bottom*) Detailed view of an intermediate "transamidase-substrate-GPI" complex. (adapted from Figure 5 in Reference 60 and Figure 1 in Reference 87).

second and smaller pool of proprotein appears to be present in the ER that is already at the site of the transamidase (Figure 6, *top, middle*). This pool of promimiPLAP obviously does not require ATP or GTP for conversion to the GPI form because folding and translocation to the site of the transamidase have already taken place. The available experimental evidence favors a model in which the putative transamidase is in the membrane of the ER with its recognition site for the hydrophbic COOH-terminal signal peptide on the lumenal side of the bilayer; the catalytic site of the enzyme and the cleavage/attachment recognition sites on the proprotein are in the lumen. The recognition site for the reactive portion of GPI is also in the lumen of the ER with its terminal ethanolamine in close proximity to the catalytic site of the transamidase. Once signal peptide cleavage and GPI addition have taken place, the latter appears to act as a signal for targeting the protein, most likely to the apical portion of the plasma membrane (26, 113, 114).

MECHANISM OF GPI ADDITION TO PROTEINS

Prediction of the ω Site in a cDNA Deduced Protein

Investigators customarily determine the primary sequence of a protein from its cDNA. Because the cDNA represents the unprocessed nascent protein, additional laborious experiments must be undertaken to determine modifications such as NH_2-terminal or COOH-terminal peptide cleavage and GPI substitution. In the case of NH_2-terminal processing, so much data were obtained on experimentally determined sites of cleavage that researchers realized that the signal peptides must be of a certain size and hydrophobicity. Most important however, was that certain small amino acids were found at the −1, −3 sites of the fully characterized proteins (87, 90, 92). By compiling such data, von Heijne (115) proposed a −1, −3 rule for predicting the cleavage site of a hydrophobic NH_2-terminal peptide on a nascent protein that has an accuracy of about 80% (87, 115). As a result, one rarely comes upon reports of experimentally determined NH_2 termini of mature secreted proteins in the literature. In fact, only rarely does one need to know the exact NH_2-terminal amino acid of a cDNA deduced protein. The number of mature GPI proteins that have had their ω sites determined experimentally is very small by comparison. However, data from those proteins and from site mutational studies (54–56, 60, 61, 68) now permit prediction of the ω site at the COOH terminus with about the same degree of accuracy as for the cleavage sites at the NH_2-terminus. Among the 20 characterized GPI proteins (60), the number of residues within the COOH-terminal signal peptide varies from about 10 to 30 mainly hydrophobic amino acids. Only glycine, alanine, cysteine, serine, aspartic acid, or asparagine is found at the ω site. These same small amino acids are mainly present at the $\omega+1$ sites, but one of the proteins has a glutamate and another a threonine. At the $\omega+2$ sites, except for one protein with threonine, all mammalian proteins contain either alanine or glycine; most trypanosomal proteins contain serine. These observations indicate that the ω site (cleavage site) is part of a small amino acid domain at the amino end of a hydrophobic COOH-terminal signal peptide, with cleavage occurring between the ω and $\omega+1$ sites.

More direct approaches to determine the amino acid requirements at and near the ω site were carried out by site mutational studies in which mutated cDNAs were transfected into cells, and the appearance of the GPI protein on the cell surface was monitored. Micanovic et al (54) and Kodukula et al (61) transiently transfected Cos-7 cells with mutant cDNAs and monitored the appearance of cell surface PLAP by immunocytochemistry as well as by enzyme activity released into the medium before and after treatment with PI-PLC. In these studies essentially saturation mutagenesis was carried out at one of the three sites (ω, $\omega+1$, or $\omega+2$) while the other two were kept constant. The relative extent of cleavage and GPI addition in cells transfected with each of the mutants was; at the ω site, serine $\approx$ asparagine > alanine $\approx$ glycine $\approx$ aspartate > cysteine >>> valine $\approx$ leucine. Seven other amino acids exhibited trace activity that was only detectable because of the sensitivity of the alkaline phosphatase assay. Of the amino acids tested at the $\omega+1$ site, most yielded relatively large amounts of GPI-linked product; only the proline mutant was totally inactive. At the $\omega+2$ site, the activity profile was: alanine $\approx$ glycine > serine > threonine > aspartate > valine; glutamate, histidine, and proline were totally inactive. Proline was the only amino acid that exhibited zero activity when present at any of the ω, $\omega+1$, or $\omega+2$ sites. Transfections with all mutant cDNAs resulted in the same degree of processing to the pro form, indicating that their only effect was on COOH-terminal processing.

Moran et al (68) carried out mutations on an hGH-DAF fusion protein to investigate allowable substitutions in the region of the cleavage/attachment site. In this construct, serine-319 was the normal ω site of DAF. They found that the ω site could not be deleted but that deletions could be made NH_2-terminal to it at $\omega-1$, $\omega-3$, $\omega-5$, $\omega-7$, and $\omega-9$ without affecting processing to a GPI form. They also carried out saturation site mutagenesis at the ω site and found, as did Micanovic et al (54), that glycine alanine, aspartate, and asparagine substituted best for serine and that the valine, glutamate, and cysteine ω mutants exhibited low activity. In a subsequent report, Moran and Caras (69) prepared a more complex fusion protein composed of the amino terminus of hGH linked at its COOH-terminus to a 15-residue fragment of the LDL receptor (LDLR) that bore a slight homology to the area around the cleavage site of DAF. At the COOH terminus of LDLR, Moran and Caras attached the 17-residue hydrophobic COOH-terminal signal peptide of DAF. The fusion product itself (HLD) did not lead to GPI linking. By mutating a valine-glutamate sequence in the LDLR residue to serine-glycine, they succeeded in obtaining processing to a GPI form. They concluded from this result that only two amino acids are necessary for signaling and that the serine represented a new ω site and the glycine a new ω+1 site. The residue at the ω+2 sites in both these fusion proteins was isoleucine, whereas in DAF, threonine is normally present at the ω+2 position. In this construct, as well as in some of the other fusion proteins used by these authors (65–67), optimal GPI processing required that the ω site be separated from the hydrophobic sequence by a hydrophilic spacer of about 10–12 residues and that the exact composition of the spacer was not important. Moran and Caras (70) showed that by increasing the length of the LDLR fragment in the LDL fusion protein and substituting other hydrophobic sequences for the DAF signal sequence, they could obtain new ω sites. However, the locations of these new sites were only inferred. Moran and Caras (73) also transfected cells with VSG cDNA (the GPI-linked coat protein of *T. brucei*) as well as with fusion products of VSG with hGH. They found that the normal ω, ω+1, ω+2 (aspartic acid–serine–serine) region of VSG serves poorly for GPI processing in Cos-7 cells but that this problem could be corrected by mutating ω+2 from a serine to an alanine.

Another parasitic protein, the circumsporozoite (CS) protein of *Plasmodium berghei*, which from its cDNA deduced sequence should be GPI linked, contains a cysteine-serine-serine motif at its ω, ω+1, and ω+2 sites (116). Like VSG, when transfected into Cos-7 cells, processing to a GPI form was poor. Apparently, requirements for GPI processing in parasitic protozoa are similar but not identical to those of mammalian cells (73). These findings, however, lend further support to the importance of the limited substitutions permissible (glycine and alanine) at the ω+2 site in mammalian cells.

Whereas previous investigators varied only one of the three sites (ω, ω+1, ω+2) in their studies, Coyne et al (117) varied all three sites simultaneously. They did this by adding complete synthetic putative COOH-terminal signal peptides (up to 25 residues) to the membrane protein, CD46, which is itself not GPI-anchored. They found that the cleavage/attachment site favored small amino acids, such as serine, alanine, and glycine at both the ω and ω+2 sites; a spacer domain of 8–12 residues and a hydrophobic domain of at least 11 residues.

Nuoffer et al (118) performed saturation mutagenesis at the ω site of the Gas1 protein of *Saccharomyces cerevisiae* and again found that the same small amino acids—asparagine, serine, glycine, alanine, aspartic acid, and cysteine—permitted GPI addition essentially in the same ranked order as reported by Micanovic et al (54) and Moran et al (68). They also concluded that fairly large amino acids are tolerated at the ω+1 site, and smaller residues are preferred at the ω+2 position. However, the efficiency of GPI anchoring varied depending on the size of amino acid, either at the ω or ω+2 position, with one almost compensating for the other. Mutations also carried out at the $\omega+3$ and $\omega+4$ sites led Nuoffer et al to conclude, as did Coyne et al (117), that the region between the small amino acid domain and the hydrophobic region of the COOH-terminal signal peptide is not just a space filler.

The requisites for GPI processing of a nascent protein determined on intact cells reflect many variables other than the interaction of the substrates, including 1. synthesis of the proprotein, 2. formation of GPI, which is catalyzed by an array of enzymes, 3. transfer of GPI to a site near the transamidase, and 4. production of the putative transamidase. Although none of the above components of the reaction have yet been purified, functional GPI has been obtained in crude lipid extracts of cells, but quantities are very limiting (58, 119, 120). The overall reaction can be carried out by isolated ER coupled to a translation system; this is much simpler than the use of intact cells. Gerber et al (60) determined the amino acid requirements at the $\omega+1$ and $\omega+2$ sites, utilizing preprominiPLAP as substrate, generated by a rabbit reticulocyte translation system, and RM from HeLa cells for processing. The prepro, pro, and GPI-linked forms were identified and assayed by an array of site-specific precipitating antibodies, followed by SDS-PAGE. At the $\omega+1$ site, of the 10 amino acids investigated, all yielded activity except proline. Substitution at the $\omega+2$ site was extremely limited. Alanine and glycine were about equally active, and traces of activity were observed with cysteine and serine, whereas aspartate, threonine, glutamate, methionine, tryptophan, valine, histidine, and proline were totally inactive. In comparing the studies by Kodukula et al (56) and Gerber et al (60) with those by Micanovic et al (54) and Kodukula et al (61), one should note that different substrates were used as well as different cells. One system was based on in vitro translated and processed protein, i.e. preprominiPLAP in RM from HeLa or CHO cells. In the other, preproPLAP was the starting point following transfection of PLAP cDNA into Cos-7 cells. Nevertheless, both systems generated essentially the same results, the cell-free system being much more stringent in its requirements.

Kodukula et al (61) accumulated their data for allowable amino acid substitutions at the ω, $\omega+1$, and $\omega+2$ sites and incorporated them into the hierarchical values shown in Table 1, which can be used to predict the ω site in a cDNA-deduced sequence of a putative GPI protein. Such predictions can be fairly reliable, at least in mammalian proteins, when a few simple rules are followed: 1. Substitutions at the ω position are restricted to one of the six small amino acids (54); 2. at the $\omega+1$ position, except for proline and tryptophan, all other substitutions are possible; 3. Glyane and alanine are highly favored at the $\omega+2$ position in mammalian cells, but serine apparently serves well at the $\omega+2$ site in pathogenic protozoa (60, 61, 73); 4. the $\omega+2$ site is usually followed by a spacer, or in the vernacular of the -1, -3 rule a hinge region (87, 90, 92, 115), of five to seven amino acids that is frequently rich in charged

Table 1 Experimentally determined hierarchy of amino acids allowable at $\omega+1$ and $\omega+2$ sites[a]

Mutant	ω	$\omega+1$	$\omega+2$
Ala	0.4+	1.0+*	1.0+*
Arg	ND	0.5	ND
Asn	0.8+	+	ND
Asp	0.4+*	0.4+	0.1
Cys	0.2+	0.3	0
Gln	0	ND	ND
Glu	0	+	0
Gly	0.4+	+	0.7+
His	ND	ND	0
Leu	0.1	ND	ND
Lys	0	ND	ND
Met	0	0.3	ND
Pro	0	0	0
Ser	1.0+	0.6+	0.3+
Thr	0	0.3+	0.1+
Trp	0	>0.1	(0)
Tyr	0	ND	ND
Val	0.1	ND	0.1

[a] The data were obtained from Cos-7 cells transfected with wild-type and site-specific mutants of PLAP [54 and 61]. Each value for ω, $\omega+1$ and $\omega+2$ represents the average of two or more experiments. The mutant yielding the highest activity at each site was arbitrarily given a value 1.0; all other values are relative to it. The parentheses around the zero for tryptophan indicate that the experiment was carried out in the cell-free system. + indicates that this amino acid is present at the corresponding site of a characterized protein. * indicates the wild-type amino acid that is present at the respective position in PLAP (adapted from Kodukula et al [61], Table-II).

amino acids and proline; and 5. a stretch of hydrophobic amino acids follows the hinge region, residing at the terminus of the signal peptide. The ω, $\omega+2$ rule, the presence of a hinge region, and the size range of the hydrophobic sequence generally accord with observations by Gerber et al (60) and Kodukula et al (61) as well as with those of Moran and Caras (66–68), Coyne et al (117), and Nuoffer et al (118).

Using the hierarchical values in Table 1, the probability of a specific amino acid being the actual ω site in a nascent protein can be arrived at by multiplying the individual probabilities of putative ω and $\omega+2$ sites. Thus, serine at an ω site with a probability of 1.0 and alanine at $\omega+2$ with a probability of 1.0, would give an overall probability of $1.0 \times 1.0 = 1.0$ (most probable). With cysteine at an ω site with a probability of 0.2 and alanine at an $\omega+2$ site with a probability of 1.0, the overall probability would be 0.2 (less probable).

However, with cysteine at an ω site and serine (0.2) at an ω+2 site (0.3), the overall probability would be 0.06 (even less probable). The ω, ω+2 rule was applied to the 20 characterized GPI-anchored proteins (60) (Table 1). In three of the proteins, the ω site was the only site with any degree of probability. In twelve others, more than one site was possible, but the verified site had by far the highest probability. In two proteins the verified ω site was one of two with about equal probabilities. In three proteins the actual cleavage site had a lower probability than one at a nearby site(s). However, one probable site was in a hinge region. Thus, an ω, ω+2 rule, excluding probable sites within a hinge region, would have predicted the experimentally determined cleavage/GPI addition site in approximately 80% of the nascent proteins. This reliability is comparable to that of the −1, −3 rule that is now almost universally used to predict the amino terminus of a mature protein whose cDNA-deduced sequence contains a typical NH$_2$-terminal signal peptide (115). We (121) have presented a more detailed review on prediction of the ω site.

More recently, Antony and Miller (122) have reported a statistical interpretation of essentially the same criteria discussed above. These authors employed χ^2 analysis to compute artificial endoproteolysis and arrive at a statistical minimum. Because the data set employed in that study was small, the reliability of the method is difficult to assess. But, as more and more sequences are analyzed, as in the case of NH$_2$-terminal signal peptide cleavage, one may be able to replace the experimental methodology with purely theoretical calculations in predicting actual sites of cleavage in a cDNA-deduced nascent protein.

The Putative Transamidase

As shown in Figure 7, substrate recognition sites for COOH-terminal signal peptidase are remarkably like those for NH$_2$-terminal signal peptidase. Figure

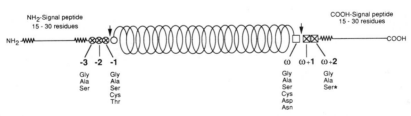

Figure 7 The structure of a typical nascent protein destined to be GPI anchored. Arrows point to the sites of cleavage of the two signal peptides from the mature protein (helical structure). The solid lines in the signal peptides represent hydrophobic domains; jagged areas represent the hydrophilic domains, those nearest the cleavage sites being putative hinge regions. Amino acids shown at the −1 and −3 sites are those reported in most nascent proteins (87). Those amino acids at the ω and ω+2 positions also represent observations on characterized proteins but were also verified experimentally. The intervening sites (−2 at the NH$_2$ terminus and ω+2 at the COOH end) are rather promiscuous. (Reprinted from Figure 4 in Reference 60).

A Peptidase   **B** Transamidase

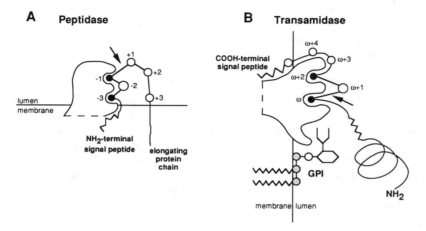

Figure 8 Comparison of the binding of NH₂- and COOH-terminal signal peptides to their respective cleavage enzymes. (*A*) The representation of the lumenal portion of the NH₂-terminal signal peptidase with a signal peptide bound at the −1 and −3 sites is adapted from the model proposed by von Heijne (87). The arrowhead indicates the site of cleavage between the −1 site of the signal peptide and the +1 site, the latter representing the NH₂ terminus of the mature protein. (*B*) Representation of the lumenal portion of the putative COOH-terminal transamidase with a signal peptide bound at the ω and ω+2 sites. The arrowhead indicates the site of cleavage between the ω+1 site of the signal peptide and the ω site, the latter representing the COOH terminus of the mature protein that, concomitant with cleavage, combines with the NH₂ group of the GPI moiety. The transamidase is depicted vertically in the lumen to emphasize similarities between the binding of the two substrates with their respective cleavage enzymes. (Reprinted from Figure 6 in Reference 61).

8 represents a comparison of the mechanism for NH₂-terminal signal processing proposed by von Heijne (115) based on the −1, −3 rule with the proposed mechanism for COOH-terminal signal processing based on an ω, ω+2 rule (61). Although there is no direct evidence yet that the latter is indeed a transamidase, there is much reason to believe so and some indirect evidence that it is not a peptidase. First of all, NH₂-terminal processing merely involves cleavage of a peptide, whereas COOH-terminal processing involves cleavage within a fully translated protein as well as addition of the GPI moiety. If GPI processing involved a peptidase, then an energy source and a second specific enzyme would be needed to add the GPI moiety. Studies in intact cells indicate that GPI addition proceeds so rapidly as to rule out a two step reaction (108–112). Trypanosome lysates add the GPI moiety to VSG in the absence of ATP or GTP (43), whereas isolated RM require ATP and GTP for optimal GPI addition to miniPLAP (59). However, appreciable processing does occur

in their absence and the energy requirement is for translocation and/or proper folding of the proprotein in the ER rather than for GPI addition (59).

The schematic drawing representing a transamidase-proprotein-GPI complex in the ER (Figure 6, *bottom*) presented by Gerber et al (60) takes into account the features of the COOH-terminal signal peptide, including the hydrophobic sequence, the hinge region, and the small amino acid domain with selective binding sites for the ω and $\omega+2$ residues. The carbonyl group of the ω residue of the proprotein is shown as being activated by a nearby hydrophilic oxygen (or SH group) on the transamidase. This aspect of the model is based on known mechanisms of transamidation and transpeptidation (123–125). The binding site for the free GPI moiety on the transamidase is shown with the NH_2 group of its ethanolamine residue in proximity to the activated carbonyl group of the proprotein, which enables nucleophilic attack (as shown in Figure 9) to form the ethanolamide linkage accompanied by cleavage of the signal peptide between the ω and the $\omega+1$ residues. Further evidence for an activated carbonyl intermediate at the ω site comes from the finding of small amounts of cleaved, but not GPI-linked, miniPLAP, formed during cotranslational processing (58, 59). By analogy with other transamidases and transpeptidases, such intermediates would be expected to react to some extent with water, resulting in an "apparent peptidase" action. In the studies by Kodukula et al (58) utilizing a GPI deficient mutant, both GPI addition and hydrolysis were totally dependent on the presence of added GPI. Only when some GPI linkage occurred did the hydrolytic reaction become evident. An explanation for this absolute requirement for GPI for both linkage and hydrolysis is that GPI may also serve in an allosteric manner to maintain the transamidase in an appropriate state to react with the pro form of a GPI-linked protein.

Transamidase activity, or GPI processing, has been studied mainly in isolated RM. However, purified ER, isolated by differential sedimentation, can be repeatedly washed with high salt and EDTA with no loss in activity (126). They can also be treated with proteinase K and still retain processing activity (SE Maxwell, R Amthauer, K Kodukula, S Udenfriend, unpublished results). RM, as well as purified and washed ER, can also be stored for long periods of time at −70°C with essentially no loss in transamidase activity. Thus, the enzyme, while present in the ER, is quite stable. The stability of NH_2-terminal signal peptidase is well known (95, 96). It can be solubilized from the RM (96) and be functionally reconstituted into proteoliposomes (127). Although the stability of the transamidase in the ER argues well for the feasibility of purification, serious problems must be overcome. For instance, transamidases, unlike peptidases, require two substrates, in this case both a proprotein and GPI. To measure GPI transamidase activity during purification would require either a pure proprotein or a synthetic propeptide that would contain a COOH-terminal signal peptide at its COOH terminus. Successful experiments have been carried out on purified NH_2-termi-

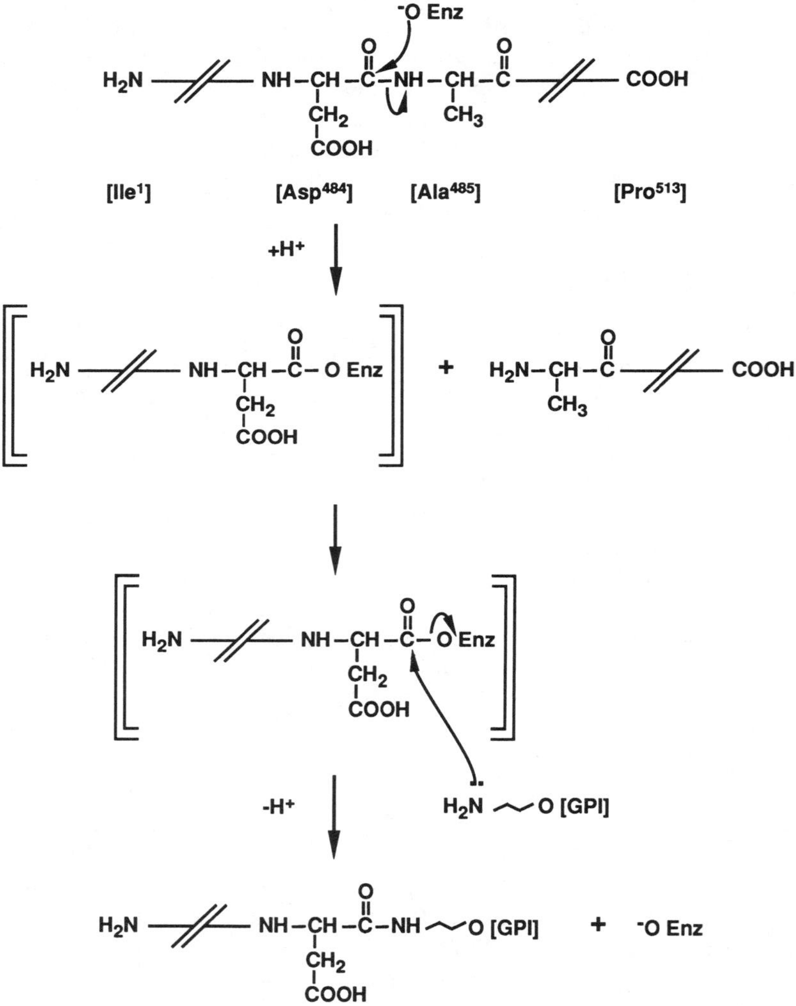

Figure 9 Proposed intermediate steps in the transamidation of proPLAP to GPI-linked PLAP.

nal signal peptidase utilizing synthetic peptides as substrates (128). Despite attempts to use peptide substrates for monitoring transamidase activity (126, 129), getting peptides into the ER still poses a problem, hence their use awaits purification of the transamidase and the ability to reconstitute its activity. A major problem in attaining a purified reconstituted system is the GPI moiety. All we know about the GPI structure has come from studies with material cleaved from GPI proteins and from in vitro biosynthesis of extremely small amounts (18-25). Furthermore free GPI has thus far been available only as trace material in crude extracts, and its presence has been detectable only in metabolic studies on its biosynthesis (43, 119, 120). A source of GPI, or a substitute for it, is required before purification of the transamidase can proceed and the properties of the enzyme be investigated.

Literature Cited

1. Seki T, Spurr N, Obata F, Goyert S, Goodfellow P, Silver J. 1985. *Proc. Natl. Acad. Sci. USA* 82:6657–61
2. Tse AGD, Barclay N, Watts A, Williams AF. 1985. *Science* 230:1003–8
3. Davis S, Aldrich TH, Valenzuela DM, Wong V, Furth ME, et al 1991. *Science* 253:59–63
4. He H-T, Finne J, Goridis C. 1987. *J. Cell Biol.* 105:2489–500
5. Mikol DD, Stefansson K. 1988. *J. Cell Biol.* 106:1273–79
6. Stahl N, Borchelt DR, Hsiao K, Prusiner SB. 1987. *Cell* 51:229–40
7. Cashman NR, Loertscher R, Nalbantoglu J, Shaw I, Kascsak RJ, et al. 1990. *Cell* 61:185–92
8. Slein MW, Logan GF. 1960. *J. Bacteriol.* 80:77–85
9. Slein MW, Logan GF. 1962. *J. Bacteriol.* 83:359–69
10. Slein MW, Logan GF. 1963. *J. Bacteriol.* 85:369–81
11. Slein MW, Logan GF. 1965. *J. Bacteriol.* 90:69–81
12. Ikezawa H, Yamanegi M, Taguchi R, Miyashita T, Ohyabu T. 1976. *Biochim. Biophys. Acta* 450:154–64
13. Low DE, Finean J. 1977. *Biochem. J.* 167:281–84
14. Low MG, Finean J. 1977. *FEBS Lett.* 82:143–46
15. Low MG, Kincade PW. 1985. *Nature* 318:62–64
16. Cross GAM. 1984. *J. Cell. Biochem.* 24:79–90
17. Ferguson MAJ, Low MG, Cross GAM. 1985. *J. Biol. Chem.* 260:14547–55
18. Ferguson MAJ, Williams AF. 1988. *Annu. Rev. Biochem.* 57:285–320
19. Low MG. 1989. *Biochim. Biophys. Acta* 988:427–54
20. Cross GAM. 1990. *Annu. Rev. Cell Biol.* 6:1–39
21. Lublin DM. 1992. *Curr. Top. Microbiol. Immunol.* 178:141–62
22. Englund PT. 1993. *Annu. Rev. Biochem.* 62:121–38
23. Field MC, Menon AK. 1993. *CRC Critical Reviews in Biochemistry,* ed. MJ Schlesinger, pp. 83–134. Boca Raton: CRC
24. McConville MJ, Ferguson MAJ. 1993. *Biochem. J.* 294:305–24
25. Menon AK. 1994. *Methods Enzymol.* 230:418–42
26. Lisanti MP, Rodriguez-Boulan E, Saltiel AR. 1990. *J. Membr. Biol.* 117:1–10
27. Ploug M, Ronne E, Behrendt N, Jesen AL, Blasi F. 1991. *J. Biol. Chem.* 266: 1926–33
28. Verma RS, Gullapalli S, Antony AC. 1992. *J. Biol. Chem.* 267:4119–27
29. Rosse WF. 1990. *Blood* 75:1595–1601
30. Schubert J, Ostendorf T, Schmidt RE. 1994. *Immunol. Today* 15:299–301
31. Schofield L, Hackett F. 1993. *J. Exp. Med.* 177:145–53
32. Low MG, Saltiel AR. 1988. *Science* 239:268–75
33. Misek DE, Saltiel AR. 1992. *J. Biol. Chem.* 267:16266–73
34. Jackson D, Waibel R, Weber E, Bell J, Stahel RA. 1992. *Cancer Res.* 52:5264–70
35. Xia MQ, Hale G, Lifely MR, Ferguson MAJ, Campbell D, et al 1993. *Biochem. J.* 293:633–40
36. Conzelmann A, Spiazzi A, Bron C, Hyman R. 1988. *Mol. Cell Biol.* 8:674–78
37. Fatemi SH, Tartakoff AM. 1988. *J. Biol. Chem.* 263:1288–94
38. Gupta D, Tartakoff A, Tisdale E. 1988. *Science* 242:1446–49

39. Sugiyama E, DeGasperi R, Urakaze M, Chang H-M, Thomas LJ, et al. 1991. *J. Biol. Chem.* 266:12119–22
40. Urakaze M, Kamitani T, DeGasperi R, Sugiyama E, Chang H-M, et al. 1992. *J. Biol. Chem.* 267:6459–62
41. Miyata T, Takeda J, Lida Y, Yamada N, Inoue N, et al. 1993. *Science* 259:1318–20
42. Inoue N, Kinoshita T, Orii T, Takeda J. 1993. *J. Biol. Chem.* 268:6882–85
43. Mayor S, Menon AK, Cross GAM. 1991. *J. Cell Biol.* 114:61–71
44. Vidugiriene J, Menon AK. 1993. *J. Cell Biol.* 121:987–96
45. Vidugiriene J, Menon AK. 1994. *J. Cell Biol.* 127:333–41
46. Garattini E, Margolis J, Heimer E, Felix A, Udenfriend S. 1985. *Proc. Natl. Acad. Sci. USA* 82:6080–84
47. Berger J, Howard AD, Gerber L, Cullen BR, Udenfriend S. 1987. *Proc. Natl. Acad. Sci. USA* 84:4885–89
48. Howard AD, Berger J, Gerber L, Familletti P, Udenfriend S. 1987. *Proc. Natl. Acad. Sci. USA* 84:6055–59
49. Bailey CA, Howard A, Micanovic R, Berger J, Heimer E, et al. 1988. *Anal. Biochem.* 170:532–41
50. Micanovic R, Bailey CA, Brink L, Gerber L, Pan Y-CE, et al. 1988. *Proc. Natl. Acad. Sci. USA* 85:1398–1402
51. Berger J, Howard AD, Brink L, Gerber L, Hauber J, et al. 1988. *J. Biol. Chem.* 263:10016–21
52. Bailey CA, Gerber L, Howard AD, Udenfriend S. 1989. *Proc. Natl. Acad. Sci. USA* 86:22–26
53. Berger J, Micanovic R, Greenspan R, Udenfriend S. 1989. *Proc. Natl. Acad. Sci. USA* 86:1457–60
54. Micanovic R, Gerber LD, Berger J, Kodukula K, Udenfriend S. 1990. *Proc. Natl. Acad. Sci. USA* 87:157–61
55. Micanovic R, Kodukula K, Gerber L, Udenfriend S. 1990. *Proc. Natl. Acad. Sci. USA* 87:7939–43
56. Kodukula K, Micanovic R, Gerber L, Tamburrini M, Brink L, Udenfriend S. 1991. *J. Biol. Chem.* 266:4464–70
57. Kodukula K, Cines D, Amthauer R, Gerber L, Udenfriend S. 1992. *Proc. Natl. Acad. Sci. USA* 89:1350–53
58. Kodukula K, Amthauer R, Cines D, Yeh ETH, Brink L, et al. 1992. *Proc. Natl. Acad. Sci. USA* 89:4982–85
59. Amthauer R, Kodukula K, Brink L, Udenfriend S. 1992. *Proc. Natl. Acad. Sci. USA* 89:6124–28
60. Gerber L, Kodukula K, Udenfriend S. 1992. *J. Biol. Chem.* 267:12168–73
61. Kodukula K, Gerber LD, Amthauer R, Brink L, Udenfriend S. 1993. *J. Cell Biol.* 120:657–64
62. Amthauer R, Kodukula K, Gerber LD, Udenfriend S. 1993. *Proc. Natl. Acad. Sci. USA* 90:3973–77
63. Caras IW, Weddell GN, Davitz MA, Nussenzweig DV, Martin DWJ. 1987. *Science* 238:1280–83
64. Caras IW, Davitz MA, Rhee L, Weddell G, Martin DWJ, Nussenzweig V. 1987. *Nature* 325:545–49
65. Caras IW, Weddell GN, Williams SR. 1989. *J. Cell Biol.* 108:1387–96
66. Caras IW, Weddell GN. 1989. *Science* 243:1196–98
67. Caras IW. 1991. *J. Biol. Chem.* 113:77–85
68. Moran P, Raab H, Kohr WJ, Caras IW. 1991. *J. Biol. Chem.* 266:1250–57
69. Moran P, Caras IW. 1991. *J. Cell Biol.* 115:329–36
70. Moran P, Caras IW. 1991. *J. Cell Biol.* 115:1595–1600
71. Keller G-A, Siegel MW, Caras IW. 1992. *EMBO J.* 11:863–74
72. Field MC, Moran P, Li W, Keller G-A, Caras IW. 1994. *J. Biol. Chem.* 269:10830–37
73. Moran P, Caras IW. 1994. *J. Cell Biol.* 125:333–43
74. Udenfriend S, Micanovic R, Kodukula K. 1991. *Cell Biol. Int. Rep.* 15:739–59
75. Udenfriend S, Kodukula K, Amthauer R. 1992. *Cell. Mol. Biol.* 38:11–16
76. Amthauer R, Kodukula K, Udenfriend S. 1992. *Clin. Chem.* 38:2510–16
77. Roberts WL, Myher JJ, Kuksis A, Low MG, Rosenberry TL. 1988. *J. Biol. Chem.* 263:18766–75
78. Roberts WL, Santikarn S, Reinhold VN, Rosenberry TL. 1988. *J. Biol. Chem.* 263:18776–84
79. Field MC, Menon AK, Cross GAM. 1991. *EMBO J.* 10:2731–39
80. Takami N, Ogata S, Oda K, Misumi Y, Ikehara Y. 1988. *J. Biol. Chem.* 263:3016–21
81. Ogata S, Hayashi Y, Takami N, Ikehara Y. 1988. *J. Biol. Chem.* 263:10489–94
82. Ogata S, Hayashi Y, Misumi Y, Ikehara Y. 1990. *Biochemistry* 29:7923–27
83. Fasel N, Rousseaux M, Schaerer E, Medof ME, Tykocinski ML, Bron C. 1989. *Proc. Natl. Acad. Sci. USA* 86:6858–62
84. Howell S, Lanctot C, Boileau G, Crine P. 1994. *J. Biol. Chem.* 269:16993–96
85. Hitt AL, Lu TH, Luna EJ. 1994. *J. Cell Biol.* 126:1421–31
86. Koster B, Strand M. 1994. *Arch. Biochem. Biophys.* 310:108–17
87. von Heijne G. 1983. *Eur. J. Biochem.* 133:17–21

88. Blobel G. 1983. *Methods Enzymol.* 96: 663–82
89. Rapoport TA. 1990. *Trends Biochem. Sci.* 15:355–58
90. Nothwehr SF, Gordon JI. 1990. *J. Biol. Chem.* 265:17202–8
91. Gilmore R. 1993. *Cell* 75:589–92
92. Jain RG, Rusch SL, Kendall DA. 1994. *J. Biol. Chem.* 269:16305–10
93. Delahunty MD, Stafford FJ, Yuan LC, Shaz D, Bonifacino JS. 1993. *J. Biol. Chem.* 268:12017–27
94. Lippincott-Schwartz J, Bonifacino JS, Yuan LC, Klausner RD. 1988. *Cell* 54: 209–20
95. Lingappa VR, Devillers-Thiery A, Blobel G. 1977. *Proc. Natl. Acad. Sci. USA* 74:2432–36
96. Jackson RC, Blobel G. 1977. *Proc. Natl. Acad. Sci. USA* 74:5598–602
97. DeGasperi R, Thomas LJ, Sugiyama E, Chang HM, Beck PJ, et al. 1990. *Science* 250:988–89
98. Klausner RD, Donaldson JG, Lippincott-Schwartz J. 1992. *J. Cell Biol.* 116: 1071–80
99. Takami N, Oda K, Fujiwara T, Ikehara Y. 1990. *Eur. J. Biochem.* 194:805–10
100. Munro S, Pelham HRB. 1986. *Cell* 46: 291–300
101. Rothman JE. 1989. *Cell* 59:591–601
102. Bole DG, Hendershot LM, Kearny JF. 1986. *J. Cell Biol.* 102:1558–66
103. Lisanti MP, Field MC, Caras IW, Menon AK, Rodriguez-Boulan E. 1991. *EMBO J.* 10:1969–77
104. Masterson WJ, Ferguson MAJ. 1991. *EMBO J.* 10:2041–45
105. Waters MG, Blobel G. 1986. *J. Cell Biol.* 102:1543–50
106. Connolly T, Gilmore R. 1989. *Cell* 57: 599–610
107. Klappa P, Mayinger P, Pipkorn R, Zimmermann M, Zimmermann R. 1991. *EMBO J.* 10:2795–803
108. Bangs JD, Hereld D, Krakow JL, Hart GW, Englund PT. 1985. *Proc. Natl. Acad. Sci. USA* 82:3207–11
109. Bangs JD, Andrews NW, Hart GW,

Englund PT. 1986. *J. Cell Biol.* 103: 255–63
110. Ferguson MAJ, Duszenko M, Lamont GS, Overath P, Cross GAM. 1986. *J. Biol. Chem.* 261:356–62
111. Conzelmann A, Spiazzi A, Bron C. 1987. *Biochem. J.* 246:605–10
112. Takami N, Oda K, Ikehara Y. 1992. *J. Biol. Chem.* 267:1042–47
113. Lisanti MP, Caras IW, Davitz MA, Rodriguez-Boulan E. 1989. *J. Cell Biol.* 109:2145–56
114. Lisanti MP, Bivic AL, Saltiel AR, Rodriguez-Boulan E. 1990. *J. Membr. Biol.* 113:155–67
115. von Heijne G. 1986. *Nucleic Acids Res.* 14:4683–91
116. Eichinger DJ, Arnot DE, Tam JP, Nussenzweig V, Enea V. 1986. *Mol. Cell Biol.* 6:3965–72
117. Coyne KE, Crisci A, Lublin DM. 1993. *J. Biol. Chem.* 268:6689–93
118. Nuoffer C, Horvath A, Riezman H. 1993. *J. Biol. Chem.* 268:10558–63
119. Doering TL, Raper J, Buxbaum LU, Hart GW, Englund PT. 1990. *Methods* 1:288–96
120. Mayor S, Menon AK. 1990. *Methods* 1:297–305
121. Udenfriend S, Kodukula K. 1995. *Methods Enzymol.* In press
122. Antony AC, Miller ME. 1994. *Biochem. J.* 298:9–16
123. Tipper DJ, Strominger JL. 1965. *Proc. Natl. Acad. Sci. USA* 54:1133–41
124. Berne P, Schmitter J, Blanquet S. 1990. *J. Biol. Chem.* 265:19551–59
125. Tate SS, Meister A. 1981. *Mol. Cell. Biochem.* 39:357–68
126. Kodukula K, Maxwell SE, Udenfriend S. 1995. *Methods Enzymol.* In press
127. Nicchitta CV, Blobel G. 1990. *Cell* 60: 259–69
128. Caulfield MP, Duong LT, Baker RK, Rosenblatt M, Lively MO. 1989. *J. Biol. Chem.* 264:15813–17
129. Broomfield SJ, Hooper NM. 1993. *FASEB J.* 7:A1044

Annu. Rev. Biochem. 1995. 64:593–620

PROTEIN-RNA RECOGNITION

David E. Draper

Department of Chemistry, Johns Hopkins University, Baltimore, Maryland 21218

KEY WORDS: RNA structure, transfer RNA, ribosomal RNA, ribonucleoprotein particles

CONTENTS

ABSTRACT

Specific interactions between RNAs and proteins are fundamental to many
cellular processes, including the assembly and function of ribonucleoprotein
particles (RNPs), such as ribosomes and spliceosomes and the post-transcriptional regulation of gene expression. Among the complexes studied to date are
small RNAs bound to individual amino acids, tRNAs and tRNA fragments
bound to their cognate aminoacyl-tRNA synthetases, and a variety of proteins

bound to RNA single strands, hairpins, irregular helices, and tertiary structures stabilized by bound cations. Several proteins use a β-sheet surface to bind RNAs, and others insert an α-helix into the widened major groove of a non-canonical RNA helix. Distortion or rearrangement of the RNA structure by bound protein is a common theme. The structural details of protein-RNA complexes are being resolved by nuclear magnetic resonance (NMR) and X-ray crystallography, but thorough thermodynamic analyses of recognition mechanisms have yet to be performed.

PERSPECTIVES AND SUMMARY

RNAs participate in many fundamental cellular processes. Their roles as informational molecules and as structural components of essential ribonucleoprotein particles (RNPs) have long been acknowledged. In the past decade a great deal of excitement has surrounded the findings that RNAs are capable of enzymatic activity (1) and make direct contributions to substrate recognition and perhaps catalysis in RNPs such as the ribosome (2) and the spliceosome (3). There has also been a growing realization that much of gene regulation occurs at the posttranscriptional level; processing, turnover, and translation of mRNAs are all important targets (4). Nearly all of the functions discovered for RNAs entail reversible or irreversible binding of proteins. The questions of how a protein recognizes a specific RNA site, what effect it has on the RNA structure and dynamics, and how it promotes a specific RNA function are thus central to a number of basic problems in molecular biology.

The problem of protein-RNA recognition first arose in relation to transfer RNAs: How do aminoacyl-tRNA synthetases achieve the specificity needed to insure faithful translation of the genetic code? A large number of elegant in vitro and in vivo studies of this problem (5–8) have been complemented in the last few years by several crystal structures of tRNAs bound to their cognate synthetases (9–11); these systems are presently the richest available sources of information about protein-RNA recognition. Work on other protein-RNA complexes has been slower to develop, in large part because almost all other RNAs are either difficult to purify from cells or very large in size. Early work on ribosomal protein-rRNA interactions, for instance, did not generally achieve the sophistication of concurrent tRNA studies largely because it was very tedious to prepare and study RNA fragments containing the protein binding sites (12). The development of enzymatic and chemical methods for large-scale synthesis of virtually any RNA sequence has alleviated these experimental difficulties (13, 14) and paved the way for application of X-ray crystallography and NMR to a wider range of RNA structural problems (15, 16).

Current work on protein-RNA complexes can be roughly divided into three areas: characterization of RNA target sites, identification of protein structures recognizing RNA, and studies of recognition mechanisms. The first question asked in relation to many protein-RNA complexes is what RNA features are contacted by the protein and used to discriminate against other potential binding sites. "Footprint" and mutagenesis approaches have given a low-resolution answer for many complexes, but it has been difficult to extract a detailed map of protein-RNA contacts from these kinds of experiments. One may identify a particular nucleotide or base as critical for protein recognition because its deletion or modification affects protein binding, but the vexing question of whether the feature is really contacted by the protein or required for correct folding of the RNA remains. RNAs may adopt complex and irregular structures, but only for tRNA is the tertiary structure known in detail. Currently the only well-defined sets of RNA-protein contacts are derived from NMR studies of small amino acid or peptide-RNA complexes and from X-ray studies of tRNA-synthetase cocrystals.

To define protein structures recognizing RNA, investigators have searched for sequence homologies among RNA-binding proteins. These exercises have revealed several "motifs." The structure of the best known of these, the RNP motif, has been solved by crystallography (17) and shows a β-sheet surface that can be adapted for recognition of a variety of RNA structures. Of the hundreds of RNA-binding protein sequences known, many do not fall into any "motif" category, at least at the sequence level; the structural diversity of RNA may require a corresponding variety of protein structures.

It has become evident in the last few years that simple maps of protein and RNA contacts will not provide a complete solution to the problem of binding specificity. In most cases in which the free and protein-bound structures of an RNA are known, the RNA structure has been substantially altered by the protein, even to the extent that it alters hydrogen bonding. The energetics of RNA structural rearrangements is therefore an important factor in determining the strength and specificity of a protein-RNA complex. Sorting out the energetic balance between RNA folding and protein-RNA contacts will be experimentally difficult but is an essential step in interpreting the structural information that is now being accumulated.

Many protein-RNA complexes with diverse functions have been identified, and this review cannot catalog even a fraction of them. Only in relatively few systems have attempts been made to determine the structural and thermodynamic origins of recognition. I concentrate on those few that illustrate the themes and directions of research in this area. A recent book contains a number of excellent reviews that discuss specific topics in more detail (18).

GENERAL CONSIDERATIONS: CONTRASTS WITH DNA-PROTEIN RECOGNITION

It is useful to consider briefly the major findings from protein-DNA studies as a prelude to the more complex problem of protein-RNA recognition. The

DNA helix is a regular structure in which the main variations are the hydro-gen-bond acceptors and donors presented by the base pairs in the two grooves. Selection of a specific DNA site by a protein was therefore predicted to rely on a network of hydrogen bonds to bases, most likely in the major groove because of the greater sequence discrimination possible there (19). This "direct readout" mechanism has proved to be basically correct for many DNA binding proteins. The potential importance of "indirect readout" was also recognized early on. That is, some DNA sequences might exhibit subtle variations in backbone conformation or be easily deformed into unusual structures. Protein interactions with the backbone could then be a source of binding specificity (20). Protein-induced distortions of DNA structure are common and range from the "kinks" seen in the *Eco* RI nuclease site (21) to the large helix bending and base pair unstacking in a complex with TATA-binding protein (22).

The direct readout mechanism presumes that protein-DNA contacts contrib-uting to binding specificity are primarily hydrogen bonds to bases and hydro-phobic contacts with thymine methyls, whereas nonspecific binding free energy is contributed by ionic and hydrogen bonds with the sugar-phosphate backbone. Other kinds of interactions may come into play in the indirect readout mechanism. For example, the distorted DNA structure complexed with the TATA-binding protein contacts an unusually hydrophobic protein surface, and aromatic amino acids intercalate into the helix (22).

RNAs differ from helical DNA in several ways that affect the possibilities for protein recognition. First, segments of Watson-Crick helix are interspersed with bulge, internal, and hairpin loops. NMR studies of frequently found "tetraloop" hairpins (23, 24) and conserved ribosomal RNA internal and hair-pin loops (25, 26) have revealed specific but irregular structures containing noncanonical base pairs and unstacked bases. Second, helices are A-form and typically less than a full turn in length. A-form helices have a very deep and narrow major groove, and it has been supposed that this steric restriction dictates protein recognition in the shallow minor groove only. However, the major groove next to loops or distortions introduced by noncanonical pairs may be quite accessible (27). The consequence of these differences is that proteins are confronted with a much richer diversity of RNA hydrogen bonding and stacking configurations than is possible in a standard DNA helix. An additional aspect of RNA to consider is that "tertiary" interactions may link different parts of an RNA and create complex shapes, as do the interactions between tRNA hairpin loops that constrain the RNA to an "L" shape. The fact that the three-dimensional structure of an RNA may be unique raises the possibility that specific protein recognition could take place entirely by con-tacts with the sugar-phosphate backbone. This situation would be an extreme case of "indirect readout" in that the bases would drive protein recognition only by specifying the overall shape of the RNA, not by directly contacting the protein.

A final consideration is that deformability of the nucleic acid structure is likely to be more important for RNA recognition than for DNA recognition. The DNA helix is a rather stable structure, whereas RNA mismatches and loops have less favorable free energies of formation. The combination of a richer variety of irregular structures in RNA and the ease with which some of these can be deformed suggests that RNA binding proteins will use a greater range of binding strategies than DNA binding proteins and that "indirect readout" will be widespread.

AMINO ACID AND PEPTIDE-RNA INTERACTIONS

Specific Recognition of Amino Acids

The method of in vitro selection allows one to extract from pools of up to $\sim 10^{14}$ different RNA sequences only those molecules binding a specific ligand of choice (28, 29). A number of small molecules (e.g. ATP or protein cofactors) covalently coupled to a column matrix have been used as ligands, and RNA binding affinities approaching 10^9 M^{-1} have been reported. The relevance to protein recognition is that RNA structures specifically binding individual amino acid side chains have been discovered. Sequences discriminating D-tryptophan from L-tryptophan and other aromatic molecules have been found, and bind the ligand with affinities of $\sim 10^5$ M^{-1} (30). Three different sequences recognizing arginine have been selected, with affinities near 10^4 M^{-1} (31). The critical sites in these RNAs are probably small internal loops. More surprising was the discovery of an RNA specific for L-valine side chains (32). Although the affinity is not large ($\sim 10^2$ M^{-1}), competition with other aliphatic amino acids showed that this RNA can extract ~ 0.9 kcal/mol of binding free energy from a single methylene. This interaction is not as efficient as that of some proteins, which can obtain ~ 1.5 kcal/mol, but it is enough for a significant contribution to protein recognition. Although hydrophobic regions of sugars and bases represent a substantial fraction of the surface area in an RNA helix, there are no large patches of contiguous hydrophobic groups (33), which has made it difficult to imagine extensive hydrophobic contacts with proteins. The internal loop in this selected RNA evidently creates a sufficiently hydrophobic pocket for efficient interactions with hydrocarbons.

Although only three amino acids have been used for in vitro selection, it seems likely that an RNA specifically recognizing any of the amino acid side chains is possible. Only a few of these specific interactions are necessary to achieve the binding affinity of a typical RNA binding protein (10^7–10^9 M^{-1}, or $-\Delta G \approx 10$–13 kcal/mol), and perhaps fewer are needed to achieve a sufficient level of discrimination against other sites. In addition to the amino acid side chains, the peptide backbone itself can make hydrogen bonds to the RNA; an example is the selection of G nucleotides by ribonuclease T_1 (34).

Arginine-TAR Recognition

The Tat protein from human immunodeficiency virus (HIV) binds to a small RNA hairpin, the *trans*-activation responsive element (TAR, Figure 1) (35). Peptides as small as 14 residues from the C-terminus of the 86-residue protein bound TAR specifically (36). The peptides are rich in arginine, and it was found that the peptide (arg)$_9$ bound TAR tightly in a gel shift assay, whereas (lys)$_9$ did not; substitutions of arginine at position 5 or 6 within the (lys)$_9$ tract were active (37). Recognition of TAR by arginine alone (38) or even by guanidinium (39) was then demonstrated. Although the arginine binding affinity is only $\sim 10^3$ M^{-1}, discrimination against a single base change in the bulge loop was seen (38). The RNA features needed for either arginine or peptide binding were deduced from mutagenesis and chemical modification experiments (Figure 1): The bulge loop must be ≥ 2 nt and contain U at the 5' position; the G26/C39 and A27/U38 base pairs cannot vary; and the 5' phosphates of A22 and U23 are protected from ethylation (38, 40, 41). The RNA must form a specific structure accommodating the arginine side chain, perhaps with the guanidinium group hydrogen bonded to the two protected phosphates.

The nature of the recognized structure was clarified by NMR studies of the RNA either free or complexed with argininamide. The three bulged pyrimidines are loosely stacked into the helix in the free RNA. Upon addition of ligand, the bulge unstacks and U23 moves near A27 of the required A-U pair (42). A U•A-U triple base interaction was suggested, and was supported by the functional substitution of a C$^+$•G-C triple and observation of expected nuclear Overhauser effects (NOEs) from either triple (43). Formation of this base triple rearranges the RNA bulge such that arginine in the major groove is next to U23 and is hydrogen bonded to the A22 and U23 phosphates and to G26. Circular dichroism (CD) experiments also showed that there is a significant rearrangement of RNA structure upon arginine binding (39). This small, specific complex illustrates what is becoming a major theme in protein-RNA recognition: An RNA may substantially rearrange its structure to accommodate a bound protein.

Is the arginine-TAR complex relevant to RNA recognition by the intact Tat protein? The sequence requirements for transactivation, Tat binding, and arginine binding are identical (summarized in 43). A more detailed analysis of the TAR RNA specificity determinants for Tat was carried out using chemically synthesized RNAs containing modifications or deletions of single hydrogen bond acceptors and donors (44). Among the bases of the putative base triple, modification of either the U23 O4 or the A27 N7 reduces binding

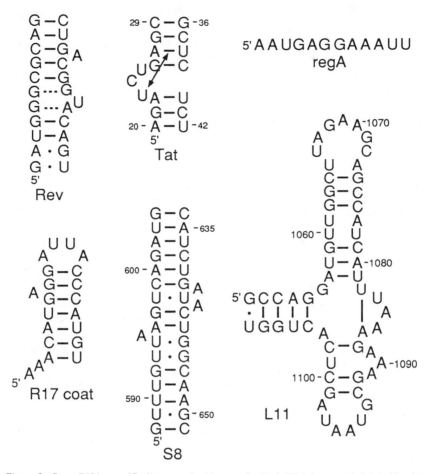

Figure 1 Some RNAs specifically recognized by protein. Each RNA fragment is labeled by the binding protein; origins of the RNAs and specific recognition features are discussed in the text. The double-headed arrow in "Tat" indicates a putative triple base interaction. Ribosomal RNA fragments (bound by S8 and L11 proteins) are numbered according to the *Escherichia coli* small and large subunit rRNAs, respectively.

substantially, whereas methylation of A27 N6 has only a very weak effect. The latter has been interpreted to mean that the intact Tat protein does not induce formation of the base triple (44), but the effect of this modification on base triple stability and arginine binding has not been tested. Because Tat binds TAR with ~10^6 greater affinity than arginine, additional protein-RNA interactions beyond those encountered in an arginine-TAR complex are to be expected. The contribution of additional protein contacts to binding specificity is a matter for discussion (45, 46).

Five RNA structures are now known to recognize arginine: The three selected from random pools and the TAR RNA have already been mentioned, and the group I intron binding site for guanine is also selective for arginine (47). Whether there is any commonality to these sites at the structural level will be interesting to see; RNA may be quite versatile at creating small arrays of specific hydrogen bond acceptors and donors.

Rev Peptide–RRE Recognition

A second protein-RNA complex from HIV-1 has been studied intensively and has also yielded a small peptide-RNA complex. The protein Rev (116 amino acids) regulates the levels of HIV-1 mRNAs and specifically binds the Rev responsive element (RRE), a several-hundred nucleotide mRNA structure (48, 49). Several Rev proteins bind each RRE (49–51). A limited region of the RRE was identified by several methods as important for high-affinity Rev binding, and a small hairpin from this region (called IIB, Figure 1) binds Rev with 1:1 stoichiometry (52, 53). In vitro selections from pools of either highly mutagenized RRE sequences or random RNA sequences yielded sets of related RNAs retaining recognition (54, 55), and comparison of these shows that an internal loop (GG opposed to AUG) flanked by conserved base pairs is important. The G-G mismatch within this loop can be substituted by A-A, which suggests that a noncanonical pair forms. The potential G-A pair is invariant. Chemical modification experiments and substitutions of modified bases are also consistent with a structure containing G-G and G-A noncanonical pairs (52, 56).

Selection of high-affinity Rev binding sites from random pools also revealed a second motif, a helix with a UU bulge (57). The strongest binding RNAs contain both the UU bulge and the internal loop separated by three Watson-Crick pairs (55), raising the possibility that the protein makes two distinct sets of contacts with the RNA. The wild-type sequence has a bulged A one base pair closer to the internal loop than the selected UU bulge. Although the A itself is probably not a recognition feature, the bases surrounding it are important. Perhaps the two different bulges can induce similar helix distortions needed by the protein.

The Rev protein has an arginine-rich basic region (10 of 17 residues at positions 34–50) similar to Tat. This 17mer is active in an in vitro splicing assay (58), binds the RRE IIB hairpin specifically, and gives essentially the same footprint and dependence on base substitutions as intact protein (56). In contrast to the situation with TAR RNA, oligo-arginine binds tightly but not specifically to the RRE fragment; a threonine, asparagine, and four arginines within the sequence are important for specificity (59). Circular dichroism spectra show that the peptide adopts an α-helical conformation when bound, and modifications of the peptide termini that stabilize the α-helix also enhance specific binding (59). The important residues are distributed around 260° of the helix. Based on protection studies, it has been proposed that the α-helix lies in the major groove of the IIB stem and contacts approximately 10 base pairs (59).

The small RNA-peptide complex defined by the above studies has been examined by two groups using NMR spectroscopy (60, 61). The results are as expected: The internal loop contains G-G and G-A pairs and a looped-out U. The extra A two base pairs from the loop is also bulged out. In one study, the internal loop is unstructured in the absence of peptide, and four new base pairs are formed upon peptide binding (60). Using slightly different solution conditions, the other NMR group argues that the internal loop is structured in the absence of ligand, although in exchange between different conformations. When the peptide is added, new imino resonances appear and the loop structure is stabilized (61). The two groups also disagree on one detail of the noncanonical pairs; one group has both the G-G and the G-A pairs having all *anti* base conformations (60), whereas the other group finds the Gs of the G-G pair *syn* (61). These are minor issues at present; higher-resolution studies that resolve the backbone structure can be expected in the near future. The remarkable result from physical studies conducted to date is that a peptide α-helix and a distorted RNA helix structure are mutually stabilized in a specific complex.

tRNA-SYNTHETASE RECOGNITION

The tRNA Identity Problem

The question of how aminoacyl-tRNA synthetases achieve their specificity was perhaps the first protein-RNA recognition problem to be seriously investigated. A great deal of progress has been made in recent years, and a number of reviews on the topic have appeared (5–8). All tRNAs must have sufficiently similar three-dimensional structures in order to fit interchangably into the translation apparatus. In this context, each of the 20 synthetases must be able to select its cognate isoacceptor tRNAs from the tRNAs present in a cell. This constraint seemingly limits the recognition strategies to discrimination between bases at specific positions within the tRNA framework (but see the discussion of the tRNA[Ser]-synthetase crystal structure below). The main experimental approach to the problem has been to determine the minimum number of base changes needed to cause one tRNA species to be selectively aminoacylated by a noncognate synthetase in a so-called "identity swap".

Identity-swap experiments have now been performed for a number of tRNAs using both in vitro and in vivo methods. In the former, variant tRNAs are synthesized by transcription with T7 RNA polymerase and assayed for aminoacylation with purified synthetases. These tRNAs lack modified bases, of course, but for tRNA^{Phe}, the unmodified RNA folds correctly (at least when Mg^{2+} is present; 62) and is aminoacylated efficiently (63). [In only two known cases is a modified base essential for synthetase recognition. Lysidine in the anticodon of tRNA is necessary for proper discrimination by the isoleucine- and methionine-tRNA synthetases (64, 65), and a 2-thiouridine in the anticodon of tRNA^{Glu} is also a positive determinant (66).] The in vivo approach involves expressing variants of a suppressor tRNA and measuring the incorporation of correct or incorrect amino acids at a specific position within a reporter gene (67). The anticodon cannot be altered in these experiments, which represents a disadvantage for studies of those tRNAs in which it is an important identity element. In vivo experiments are advantageous in that they look for what determines specificity in the context of competition between all synthetases and tRNAs, and can more easily determine whether a variant loads incorrect amino acids as well as the correct one.

Identity elements used by 19 of the 20 synthetases have been identified (see tables in 7, 8), and a distinctive pattern emerges. The majority of the synthetases recognize one or more of three features: (*a*) at least one base of the anticodon, (*b*) one or more of the last three base pairs in the acceptor stem, and (*c*) the so-called discriminator base between the acceptor stem and the CCA terminus (position 73). The latter was noted some time ago as a base that was always invariant among the isoacceptors of an amino acid (68). At least eight synthetases rely on other features as well. The three tRNA-synthetase cocrystal structures that have been solved (see below) reveal the kinds of contacts synthetases make with these different tRNA features. It is interesting to note that discrimination based on acceptor stem base pairs involves entirely different strategies in the tRNA^{Asp} and tRNA^{Gln} cases.

An identity element may affect the specificity of tRNA aminoacylation because it is recognized by the correct synthetase (positive element) or because it prevents interactions with incorrect synthetases (negative element). A negative element can mask the presence of positive elements (reviewed in 6). A particularly striking illustration is a set of three identity elements at the CCA terminus of tRNAs specific for alanine, histidine, and glycine. Each determinant is a positive element for one or two of the cognate synthetases, and a negative element for the others (69).

Studies of tRNA^{Ala} identity are of note for their general relevance to mechanisms of protein-RNA recognition. A major determinant of specificity is a single wobble pair, G3-U70, in the acceptor stem; introduction of this pair into tRNA^{Phe} or tRNA^{Cys} is sufficient to cause these tRNAs to be aminoacylated with alanine (70). This extreme reliance of the synthetase on a single feature

allows a simple hairpin duplicating the acceptor stem and 3' terminal nucleotides to be specifically recognized and aminoacylated by the synthetase (71).
A number of synthetic variants of the recognized "microhelix" have been
tested, and it is clear from substitutions of base analogues that an unpaired
exocyclic amine at G3 is essential (72). Also contributing, but to a lesser
degree, are three 2' hydroxyls within 5 Å of the G3 amine, and the neighboring
purine N2 of G2-C71 (73). Thus specific synthetase contacts in the acceptor
stem minor groove are used to select tRNA^{Ala}. Curiously, substitution of
G3-C70 for the wobble pair and the selection of pseudorevertants in vivo
yielded a mutant tRNA^{Ala} with a U•U mismatch in place of a G-C pair in the
anticodon stem (74). Altering contacts of the protein with the anticodon stem
may also alter the contacts in the acceptor stem.

tRNA-Synthetase Cocrystal Structures

Three different tRNAs have been crystalized with their cognate synthetases,
and the structures have been solved by X-ray diffraction. These structures
provide instructive correlations with the extensive identity studies described
above.

In the tRNA^{Gln}-synthetase complex, the protein interacts with the anticodon
loop, with the acceptor stem and CCA terminus, and with the inside of the
tRNA "L" (9, 75). Contacts with the acceptor stem are made in the minor
groove, and the backbone carbonyl of pro181 and the carboxyl of asp235
hydrogen bond to the N2s of G2 and G3, respectively. Mutation of asp235 to
glycine relaxes the specificity of the enzyme (76), and mutation of the G2/C71
or G3/C70 pairs lowers the aminoacylation rate (77). A surprising rearrangement of the 3' terminal nucleotides apparently takes place in the complex: The
last base pair (U1/A72) is entirely disrupted, and a sharp bend in the backbone
hydrogen bonds the N2 of G73 with the phosphate between C71 and A72. The
bend and the disruption of U1/A72 cause an unstacking of bases, which must
unfavorably contribute to the protein-binding free energy. The increased cost
of unstacking a C1/G72 pair, and the inability of an A73 to form a correct
hydrogen bond in the rearranged structure, probably account for the deleterious
effects of these mutations (77).

Within the anticodon loop, each of the three anticodon bases inserts into a
separate pocket of the synthetase and forms several hydrogen bonds to the
protein, whereas the remaining four bases of the loop are stacked and paired
in a continuation of the anticodon stem (75). The synthetase must be able to
accommodate both C and U at position 34 of the anticodon in order to ami-
noacylate both isoacceptors of the tRNA. C34 in the crystal structure is hydrogen bonded to protein at all three Watson-Crick pairing positions, and U
cannot occupy the same site. However, if U34 shifted slightly within the
pocket, a hydrogen bond between O4 and the guanidinium of an arginine could

be formed. Whether this is indeed the case remains to be tested. The extensive structure formed between the protein and anticodon loop suggests that the entire loop contributes to the identity of the tRNA.

Synthetases are divided into two groups, based on whether the ATP binding site is a Rossman fold (group I) and on whether the 2′ (group I) or 3′ (group II) terminal hydroxyl is charged. Gln-tRNA synthetase is a group I enzyme; the crystal structure of the group II asp-tRNA synthetase with tRNA^{Asp} has also been solved (10). Although the primary interactions between tRNA^{Asp} and its synthetase occur in the anticodon loop and CCA terminus, as is the case with gln-tRNA synthetase, binding takes place from the major groove side of the anticodon and acceptor stems rather than from the minor groove side. The single-stranded GCCA 3′ terminus remains in an approximately helical conformation, and protein contacts extend two base pairs into the acceptor stem. The anticodon bases have been unstacked and pulled into the protein structure, as in the tRNA^{Gln}-synthetase structure. Four of the five identity elements for tRNA^{Asp}—the three anticodon bases and G73 near the 3′ terminus—are extensively contacted by the protein (78). The fifth element, a rare G10/U25 pair in the D stem, is probably needed to insure correct contact of G25 with protein at the corner of the tRNA "L."

The most recently reported tRNA-synthetase cocrystal structure is another group II enzyme, seryl-tRNA synthetase, that has two novel features (11). First, the anticodon is not an identity element, since serine anticodons come from two distinct groups. Second, the tRNA has a large (7 bp) variable arm. The contacts made by the tRNA with the protein are therefore quite different from those made by the tRNA^{Gln}- and tRNA^{Asp}-synthetase complexes. Approximately one third of the tRNA is not visible in the crystal; the anticodon stem and loop and the variable loop project beyond the edge of the protein, and the CCA terminus and the last few pairs of the acceptor stem are also disordered, although near protein. Remarkably, the contacts made with the remaining structure are almost all to the backbone rather than to the bases, and the two base-specific contacts probably do not contribute to the binding specificity. One protein contact is made to G19 in the D loop, which is almost universally conserved among tRNAs. The other contact is two hydrogen bonds between a glutamate carboxyl and a G-C pair in the variable arm; whereas the G-C is conserved, the glutamate is not. Extensive contacts, probably both direct and water mediated, are made along the backbone of the variable stem, at the 3′ base of the acceptor stem, and in the T loop. Recognition must therefore be largely determined by shape, especially that of the variable arm. Eight base changes in the acceptor stem and D stem are needed to change tRNA^{Leu} (which has a long variable arm) to tRNA^{Ser} (79). The structure does not provide an obvious rationale for how these eight bases specify tRNA^{Ser} identity, probably because the base substitutions at these positions subtly affect RNA structure rather than direct protein contacts. A more recent in vitro study of the conversion of tRNA^{Leu} into tRNA^{Ser} concluded that no base-specific recognition occurs with the synthetase (80). Moreover, the tRNA^{Ser} D loop contains two extra nucleotides and a distinctive set of triple base interactions different from the known structures of tRNA^{Phe} and tRNA^{Asp}. To what degree the tRNA structure has been affected by protein binding will be determined by structural studies of tRNA^{Ser} alone.

RNA STRUCTURES RECOGNIZED BY PROTEINS

As yet, there is no apparent limit to the variety of RNA secondary structures recognized by specific proteins; all hairpin, bulge, internal, and junction loops are potential targets, as are the helices themselves. Shown in Figure 1 and discussed below are four of the best-studied protein-RNA complexes, presented in order of increasing complexity of the RNA structure. An indicator of how much progress remains to be made in the field is the fact that a detailed model of the protein-RNA contact surface is available for only one of these interactions (but see also the discussion of the RNP motif below).

T4 regA Protein

The regA protein of bacteriophage T4 is a translational repressor that regulates the synthesis of many early genes by binding directly to the mRNA. Two studies have used RNase protection to define the target site as a short sequence overlapping the initiation codon (81, 82). Neither yielded any evidence for stable RNA secondary structure in the binding site. This finding would seem to be a simple case of a protein recognizing a specific single-stranded sequence, but the mechanism used by the protein to discriminate between RNAs has remained elusive. A 12mer from the regulated gene 44 mRNA binds the protein with ~10^7 M^{-1} affinity (Figure 1), and single base changes at 6 of the 12 positions each reduce the affinity by 10–100 fold (83). However, this gene 44 recognition sequence is entirely upstream of the initiation codon, whereas in other genes the AUG contributes to binding (84). No consensus sequence has emerged from comparisons of the regulated genes. NMR studies of the 12mer and one of the weakly binding variants revealed similar single-stranded structure for the two (85), ruling out any preexisting unusual structure as a recognition feature. We are left with the hypothesis that the protein stabilizes some specific structure that can be adopted only by certain sequences.

R17 Coat Protein

One of the first proteins found to bind a specific site on a messenger RNA is the R17 coat protein, which recognizes a hairpin containing the ribosome binding site for the phage replicase (86). This was also the first protein (excluding aminoacyl-tRNA synthetases) whose RNA recognition site was intensely scrutinized at the nucleotide level. Uhlenbeck and his colleagues have measured the protein binding affinity for more than 100 variants of a 21-nucleotide hairpin (Figure 1) synthesized by ligation of short oligomers (87) or by transcription of DNA with T7 RNA polymerase. [RNAs as short as 16 nucleotides can bind tightly (88).] Variants may bind the protein several orders of magnitude more weakly than does the wild-type sequence, and the results have been reviewed in detail (89). Later studies that used the R17 coat protein to select high affinity binding sites from pools of random RNA sequences confirmed the results obtained with synthesized variants (89a). Essentially three features are examined by the protein. First, a hairpin is required. Most sequence variants that substitute Watson-Crick pairs but preserve the hairpin structure bind as well as the wild-type sequence. Second, a four-base hairpin loop of sequence RNYA is preferred (R = purine, Y = pyrimidine). A specific role has been hypothesized only for the pyrimidine, which may form a transient Michael adduct (addition across the 5,6 double bond) with the protein (90). Some enzymes that modify pyrimidine nucleotides are known to use this mechanism (91). Whether such an adduct forms in the coat protein case remains uncertain. Substitution of 5-BrU for the pyrimidine gives the expected increase in the coat protein binding affinity, as the bromine should stabilize a Michael adduct (92), but the complex is unaffected by mutation of the putative cysteine donor (93).

The third recognition feature is a single bulged purine at a specific location in the helix; deletion of the bulge, substitution of a pyrimidine, or a shift in the bulge site by one base pair reduces the protein affinity by at least three orders of magnitude (94). The possibility that the coat protein might form a hydrogen bond with the purine was tested by substitution of modified bases at the required position. The results were inconclusive in that elimination of any one hydrogen bond acceptor or donor did not have a major effect; it is not clear at this point whether the protein recognizes some helix distortion caused by a purine bulge or whether a direct contact (such as purine stacking with an aromatic amino acid) is important. The homologous coat proteins of related RNA phages also recognize hairpin loops with single base bulges in their mRNAs, but GA phage recognition is unaffected if the bulge is moved by one base pair, and in Qβ phage the bulge can be deleted entirely (95). These coat proteins must have evolved slightly different RNA recognition mechanisms.

The structure of the MS2 coat protein (closely related to the R17 coat protein) is known from X-ray crystallography of both the intact phage particle (96) and a mutant protein that binds the RNA hairpin but does not assemble phage (96a). The protein dimer exposes a large β-sheet surface to the solvent, and mutagenesis of residues in the middle six strands (three from each monomer) confirm the importance of this surface in determining RNA binding affinity and specificity (97, 98). Recently the structure of a hairpin bound to the coat protein dimer has been solved at 3.0 Å resolution (98a). The RNA stem and loop structure forms a crescent contacting the β-sheet surface of the protein dimer. The buldged A and the loop A occupy similar positions in the two monomers, inserted into pockets between lysine and valine side chains. However, hydrogen bonding to serine and threonine hydroxyls is different in the two cases. The essential loop pyrimidine is stacked against a tyrosine and (in the case of C) hydrogen bonded to asparagine, as suggested by studies with mutant proteins (98). The RNA recognition features deduced from binding studies are thus rationalized by the observed structure of the protein-RNA complex.

Ribosomal Protein S8

A number of ribosomal proteins that bind independently to specific sites on ribosomal RNAs were identified in the early days of ribosome structure studies (12). These proteins present a rich variety of protein-RNA interaction mechanisms (99). Some of these sites have been reduced to fairly simple structures. An example is S8 protein, which protects an irregular helix from base- and phosphate-specific chemical reagents (Figure 1) (100). A small duplex RNA formed from two strands of 18–20 nucleotides each bind S8 as well as intact 16S rRNA (101, 101a). A set of three bulged A residues are protected by S8 from chemical reagents; substitutions at two of the As are not tolerated, and the third A may be replaced by U but not deleted. At least three of the helix bases near the bulges (including U598 of the U•U mismatch) are highly conserved and cannot be substituted, even with compensatory changes to preserve pairing (101). Several helix phosphates on either side of the region containing the bulges are protected from an alkylating reagent by S8 (100). The protein-RNA contacts probably extend over a little more than one full turn of helix, and the bulged As undoubtedly distort the helix approximately in the middle of the bound region (100, 101b). It is tempting to think that the bulges are specific contacts for S8, but they may also provide a uniquely distorted backbone for protein recognition.

A striking confirmation of the essential S8 recognition-site structure came from studies of *spc* mRNA, which encodes a number of ribosomal proteins, including S8. Translation is regulated by binding of S8 to a target site that was identified by RNase protection, mutagenesis, and comparisons of regulated mRNAs from related organisms (101a, 102, 103). The same bulged A residues and helix pairs shown to be important in the rRNA fragment are also found in the mRNA site. Ribosomal proteins recognize very similar secondary structures in both mRNA and rRNA in several other instances as well (104).

Ribosomal Protein L11

The large subunit protein L11 recognizes a larger rRNA domain of ~58 nt. The site was originally defined by isolation of an RNase-protected fragment able to rebind the protein (105) and by chemical and enzymatic "footprints" (106). The domain, which is highly conserved, is a junction of three helices with an unusual "lone" base pair contained within the junction loop (Figure 1) (106, 107). Protein binding affinities for a number of mutations that collectively change every base within the domain have been measured (108, 109). Only a few of the mutations decrease the affinity by even fivefold, and these are clustered around the junction. Bases protected from chemical reagents by L11 are not among the sites sensitive to mutagenesis. A recent hydroxyl radical footprinting study noted strong protections along the A-U-rich stem adjacent to the junction (110), although the protein is not particularly affected by sequence changes in this helix.

In view of these results, it is possible that L11 contacts primarily the RNA backbone, rather than bases, and that recognition is at the level of three-dimensional shape. However, this proposal is difficult to prove in the absence of a high-resolution structure of the complex. Thermodynamic studies have identified an extensive and stable tertiary structure within this RNA (111, 112). The structure is stabilized by Mg^{2+} and NH_4^+ or K^+, which are bound in preference to other ions and strongly promote the binding of L11 (113, 114; Y Xing & DE Draper, unpublished observations). L11 likely requires a particular RNA tertiary folding, which is stabilized by the bound ions. Structural studies are needed to determine whether the protein recognizes only the shape of the RNA or whether specific base contacts are also important.

Thermodynamics of Protein-RNA Complexes

Quantitative studies of protein-RNA binding thermodynamics and kinetics have been performed for relatively few systems. The properties of the four proteins discussed above are representative and are briefly summarized here (83, 116–118). The dependence of the binding constant on added salt should be high if a protein relies primarily on electrostatic interactions with backbone phosphates. For all four proteins, the salt dependences of the binding constants are modest: $-\delta \log K / \delta \log[M^+] = 1–2$ for L11 and regA proteins, or 4–5 for S8 and R17 coat proteins. Interpretation of salt dependence in terms of the numbers of electrostatic interactions is variously complicated by the release of anions, by specific uptake of Mg^{2+} and monovalent ions, and by the irregular shape of the RNA. One can only conclude that the nonelectrostatic contribution to the binding free energy in these cases is substantial.

The temperature dependences of S8 and L11 binding to RNA are quite weak, suggesting $\Delta H \approx 0$, whereas the apparent enthalpy of R17 coat protein–RNA association is −19 kcal/mol. However, the temperature dependence of binding will have contributions from both the intrinsic enthalpy of the protein-RNA contacts and the enthalpy of any RNA unfolding in that temperature range. In the case of L11, the negative enthalpy of RNA unfolding apparently compensates for a positive enthalpy of protein binding on the order of 10 kcal/mol (Y Xing & DE Draper, unpublished observations).

The kinetics of R17 coat protein binding to RNA are slightly slower than expected for a diffusion-controlled reaction, whereas the S8-rRNA association rate is perhaps an order of magnitude slower than expected. The reactions may be slow because the protein must pick one of a number of RNA conformations present in solution, or because an RNA conformational rearrangement limits the binding rate. Detailed kinetics have not been pursued for either of these systems, and whether the association reaction proceeds in more than one step is unknown. As a greater level of structural sophistication is reached in studies of several small protein-RNA complexes, a more detailed look at the kinetics and thermodynamics of the binding mechanisms may be very revealing.

ASSEMBLY OF RNPs

Many cellular RNAs bind several different proteins to form a large ribonucleoprotein particle (RNP) with a specific function. The ribosome is a classic example, containing ~55 proteins and 4500 nt of RNA in *E. coli*. The spliceosome is an even more complex RNP; the assembly of individual snRNPs and their subsequent incorporation into the complete particle is still under investigation (120). A number of smaller RNPs also perform specific functions in the cell. How do protein-RNA and protein-protein contacts direct the assembly and function of these oligomeric protein-RNA complexes? In the best-studied RNPs, discussed below, a set of primary binding proteins individually bind to the RNA and potentiate binding of secondary proteins that have no independent affinity for RNA. An obvious explanation for this kind of ordered addition is that secondary and primary proteins interact directly. Alternatively, primary binding proteins may alter the RNA conformation and indirectly promote secondary protein binding to otherwise cryptic RNA sites. Both mechanisms may operate for any one protein, of course, and distinguishing them can be difficult. The following discussion summarizes the assembly of two RNPs that illustrate the principles involved.

The signal recognition particle (SRP) binds to ribosomes and promotes translocation of certain proteins across the endoplasmic reticulum. It consists of six proteins and a single, ~300-nt RNA. The proteins are easily dissociated, purified, and reconstituted with the RNA in vitro (121). Two pairs of proteins, SRP9/SRP14 and SRP68/SRP72, form heterodimers. These dimers bind RNA readily, although the individual polypeptides do not. SRP19 by itself binds directly to the RNA, where it stimulates binding of SRP54. Although an attempt was made to measure binding cooperativity between SRP9/SRP14 and SRP68/SRP72, which interact at opposite ends of the extended RNA secondary structure (122, 123), none was detected (124). Each pair of proteins appears to bind the RNA as an independent unit, and the RNA is a passive participant in the assembly of this RNP.

The *E. coli* 30S ribosomal subunit was the first RNP whose spontaneous assembly was reproduced in vitro from individually purified components (21 proteins and a single, ~1500-nt RNA) (125). By omitting proteins from the assembly reaction, Nomura and his colleagues were able to work out a detailed map of the cooperative interactions that took place between proteins during assembly (126). This assembly map was later reproduced in an entirely different way by Noller and colleagues, who examined the RNA "footprint" afforded by each protein as it took its place in the subunit (127). The ordered assembly of proteins can proceed in three or four layers. For example, the primary binding protein S4 recognizes the entire 5' third of the rRNA (128, 129). S16 binding is dependent on the prior addition of S4, and these two proteins in turn promote S5 and S12 interactions. S20 and S17 also bind independently within the same 5' rRNA domain and promote S16, S5, and S12 binding to some degree. Considerable evidence indicates that much of the cooperativity among these proteins is RNA mediated. First, all six proteins strongly protect some portion of the RNA from chemical reagents; thus, no protein binds to the particle without either contacting the RNA or changing the RNA conformation. Second, 30S subunit assembly stops after the binding of a first subset of the proteins, and a heating step is required before the remaining proteins can assemble (130). The unimolecular activation is most likely an RNA conformational change. Third, the effects of different protein combinations on the chemical reactivity of the RNA suggest protein-induced RNA conformational changes (summarized in 127). For example, S20 and S17 individually protect the identical set of nucleotides in one hairpin, as if both proteins induced the same conformational change in this part of the structure. Moreover, the reactivities of some bases are first enhanced by S4 and subsequently reduced by S16, suggesting that S4 exposes these nucleotides for S16 recognition. Examination of nucleotide reactivities as a function of time reveals a number of similar situations in which nucleotide reactivity initially increases and then is suppressed as assembly proceeds (131). These studies contrast strongly with the SRP assembly discussed above and give the impression that ribosomal proteins that bind early in assembly manipulate the rRNA to induce additional protein-RNA interactions.

The binding of proteins S4 and S20 to many mutants of the 5' domain rRNA has been measured and suggests that the proteins organize the RNA structure in a specific way. Small deletions or base substitutions in six loops and an

A•A mismatch weaken S4 binding by 5–10 fold; the sites are distributed throughout the 5' domain (132). This is in contrast to the S4 "footprint", which is confined to the vicinity of the mismatch (133). It is unlikely that the 24-kDa S4 directly contacts all seven of the mutation-sensitive sites; some of these must be important for proper folding of the rRNA when bound to S4. A second set of mutations that alter loop sequences or substitute a G•A mismatch affect S20 binding (134). The sites important for S20 binding are located in three hairpins interspersed between the sites contributing to S4 recognition. These two small proteins appear to promote a specific structure through much of the ~160-kDa 5' domain, perhaps by binding a few key RNA features.

PROTEIN MOTIFS RECOGNIZING RNA

RNP Motif

An eight–amino acid sequence was found twice in heterogeneous nuclear RNP (hnRNP) protein A1 and four times in a yeast poly(A) binding protein (135, 136), suggesting a repeated domain structure for these RNA binding proteins. The domain has now been found in a large number of eukaryotic proteins associated with mRNA and the splicing apparatus (reviewed in 137–139). It has been variously termed RNP consensus sequence (RNP-CS), RNA recognition motif (RRM), and RNP motif; here, we refer to it by the latter term. The domain is now known to consist of two sequences, six and eight amino acids long, separated by ~30 amino acids and located in an 80–90 residue context. Both of the short sequences have conserved tyrosine and phenylalanine residues. The structures of two of these domains have been solved, notably in a 2.8-Å resolution crystal structure of the snRNP U1A protein (17) and also at lower resolution by NMR studies of the same U1A protein (140) and a human hnRNP protein (141). The two middle strands of a four-stranded antiparallel β-sheet contain the short conserved sequences. Two of the conserved aromatic residues point into the solution, whereas other conserved hydrophobic and aromatic residues form part of the interior core.

The first suggestion as to how the RNP motif contacts RNA came from studies of the two RNP motifs in the A1 hnRNP. The motifs were prepared as 92–amino acid polypeptides and were found to bind single-stranded DNA quite well. p(dT)$_8$ was cross-linked by UV to two phenylalanine residues in each polypeptide (142); these residues correspond to the two very conserved phenylalanines that point into solution in the U1A crystal structure. Although these cross-links are probably relevant to the A1-RNA binding mechanism, p(dT)$_8$ is certainly not the natural substrate for the protein, and recent in vitro selection experiments suggest that it has considerable specificity. The intact protein selects RNAs containing one or two copies of UAGGGA/U, which resembles the vertebrate splice site consensus sequences, from a random pool (143). The "winner" sequence binds with a nanomolar dissociation constant approximately 300-fold more tightly than an unrelated RNA. Which other parts of the protein are responsible for the additional binding affinity toward specific sequences is not yet known.

It is significant that the specificity of the intact A1 hnRNP protein is not the sum of the individual RNP motif specificities, since polypeptides containing one of the two individual RNP motifs tend to select different sequences (143). A similar situation occurs in the poly(A) binding protein of yeast; poly(A) binding activity develops only when at least two of the four N-terminal RNP motif domains are present (144). How RNP motif domains interact with each other in a larger protein to confer new specificity remains to be determined.

A number of studies have attempted to identify RNA features recognized by U1A protein, a hairpin in the U1 snRNP RNA. The protein protects a number of phosphates from alkylation, both in the hairpin loop and adjacent stem (145), and the binding has a rather steep salt dependence (146). Extensive contact with RNA phosphates is implied. Mutation of one of several bases in the loop reduces binding affinity by as much as six orders of magnitude (145–147). U1A is homologous to U2B", which recognizes a hairpin in U2 RNA (148). The hairpin loops are similar in sequence, and a single base change combined with a single base deletion or insertion interconverts the protein binding specificities of the loops.

To identify U1A amino acids contacting the RNA, swaps of peptides sequences between the U1A and U2B" proteins were made. A few amino acids adjacent to one of the conserved β-sheet strands were initially identified as crucial determinants of RNA binding specificity (148), but further experiments showed that a number of amino acids scattered throughout the protein are important in some way (149). Once the crystal structure of U1A was available (17), mutations could be directed to residues with solvent-exposed side chains. Several amino acids within the consensus sequence that were located in the β-sheet structure could not be mutated without destroying specific binding (145). Taken together, the RNA and protein data suggested extensive contacts between the solvent-exposed surface of the β-sheet and the hairpin loop and stem (145).

NMR studies of the U1A-hairpin complex have now helped to resolve the positioning of the U1 hairpin on the protein surface (150), and specific contacts between the protein and RNA have been determined by a 1.92 Å resolution x-ray structure of the complex (150a). The entire solvent face of the β-sheet contributes to recognition, and a C-terminal α-helix not seen in the crystal structure of the protein alone is also important. [NMR studies of the interaction between U_8 and the hnRNP C RNP motif have also come to the conclusion that the β-sheet face of the protein interacts with RNA (151).] Lysine and arginine residues position the backbone of the hairpin on the surface of the protein. A protein loop between two β-sheet strands inserts into the RNA loop at its base. Seven of the loop bases (AUUGCAC) lie on the β-sheet and contact the protein extensively. The remaining three 3′ loop bases (UCC) do not interact with the protein. The protein essentially recognizes the single-stranded AUUGCAC sequence, which explains its affinity for the same sequence in the context of an internal loop (150a). The protein backbone (amide and carbonyl groups) and amino acid side chains hydrogen bond extensively with the recognized bases, and two aromatic residues stack with bases. An NMR study of the RNA bound to U1A suggests that the loop structure changes significantly upon binding to the protein (147).

One of the two conserved β-sheet sequences of the RNP motif is also found in the cold-shock domain (CSD) or Y-box family of DNA binding proteins (152). These proteins were originally described as highly conserved transcription factors distributed from bacteria to mammals, but some members of the class are now known to mask mRNA translation in oocytes and bind both DNA and RNA (153). The bacterial CSD protein is a five-strand β-barrel, with the RNP sequence located on one of the strands (154, 155). A similar five-strand β-barrel is found in staphylococcal nuclease and the in anticodon recognition domain of some aminoacyl-tRNA synthetases (156). The β-sheet structure of the RNP motif proteins is thus partly conserved in a number of nucleic acid binding proteins and may be a particularly advantageous framework for designing nucleic acid recognition surfaces.

Zinc Fingers

The "zinc-finger" protein folding motif was first noticed in the TFIIIA protein found associated with 5S rRNA in *Xenopus* oocytes. Nine repeats of a $CX_4CX_{12}HX_{3-4}H$ sequence and a corresponding amount of associated zinc ion suggested a small protein fold stabilized by zinc coordination to the four C and H residues (157, 158). The motif has since been found to be very common in regulatory proteins that bind DNA. In these cases, recognition takes place by wrapping of several fingers around the DNA so that a short α-helix in each finger makes sequence-specific contacts in the major groove (159). TFIIIA also binds DNA, where it regulates expression of the 5S rRNA gene, but the detailed mechanism by which it recognizes 5S rRNA is not yet clear. An early proposal was that the DNA regulatory site and the 5S RNA both contain a similar A-form or distorted helix structure, each of which contacts the protein in the same way. However, the fact that base pair substitutions at similar positions in the two nucleic acids have quite different effects on protein binding rules out this simple mechanism (160). In fact, it has been impossible to identify any RNA sequence feature that is essential for recognition; only disruption of the overall RNA structure, especially at the three-helix junction, significantly reduces TFIIIA binding affinity (161). Fingers 4–7 were recently found to be sufficient for RNA recognition, whereas the first three fingers contribute more strongly to DNA binding (162, 163). Perhaps the protein has devised two recognition surfaces, one that selects 5S rRNA on the basis of backbone shape and one that recognizes a specific DNA sequence.

Despite the prototype of TFIIIA as an RNA binding protein, the majority of zinc-finger proteins are thought to target DNA, not RNA. Several other protein folds containing zinc and binding RNA have been described. Among these are the phage mu *com* protein, a 62–amino acid protein with seven cysteine and histidine residues and one bound zinc ion. This protein regulates *mom* translation by specific binding to the mRNA (164). Another phage translational regulator, the T4 gene 32 protein, also contains a bound zinc ion (165). Retroviral nucleocapsid proteins each have two $CX_2CX_4HX_4C$ motifs that have been shown to bind zinc ion tightly. NMR-derived structures show the expected organization of the sequence motifs around the two bound ions, but the rest of the protein (55–56 amino acids total) is largely disordered (166, 167). The functional role of this protein motif remains to be resolved. Mutations in the protein implicate zinc coordination as essential for viral RNA packaging in vivo (168), but the only nucleic acid binding properties detected in vitro are nonspecific affinities for single-stranded RNAs and are independent of the zinc content of the protein (169). Perhaps the functionally important RNA ligands have not yet been identified. The same sequence motif has been found in nonviral proteins, including a factor that influences splice-site selection in yeast mRNA processing (170).

Double-Stranded RNA Recognition and Other Motifs

A number of proteins thought to bind double-stranded RNA contain a 65–68 amino acid consensus sequence with approximately 30% amino acid identity. These proteins include activities as diverse as the bacterial RNA processing nuclease RNase III (171), the human dsRNA-dependent protein kinase DAI (172), and a eukaryotic double-stranded RNA adenosine deaminase involved in RNA editing, as well as a number of double-stranded RNA binding proteins of unknown function (173, 174, and references therein). The consensus sequence itself is responsible for RNA binding activity (173). Specificity for double-stranded RNA and DNA-RNA hybrids has been demonstrated for some proteins (174), whereas RNase III is most active with helices containing certain internal loop structures (171). Further details of the protein fold or of its RNA recognition mechanism are not yet available.

Several other protein sequence motifs associated with RNA binding proteins have been identified (175, 176), such as an unusual arginine-glycine repeat ("RGG box", 177) and the "KH domain" (178). How these sequences are involved in RNA recognition has not yet been established but will certainly be investigated in the next few years.

THEMES AND PROSPECTS

Protein-RNA complexes have diverse cellular functions; the associated target RNA sizes, interaction mechanisms, and protein characteristics are widely varied as well. Generalizations about a "typical" protein-RNA complex are not possible, but it is useful to note several themes that have appeared.

A theme of major importance is that protein-induced RNA conformational changes are common and substantial. The idea of "indirect readout," discussed in relation to DNA binding proteins, is present in two senses in these recognition mechanisms: Both irregularity in the RNA structure and the ability of the RNA to adopt a new or altered structure are important determinants of specificity. A critical thermodynamic question is, how much free energy is expended by the protein in driving the RNA into the recognized conformation? The total free energy of interaction can be divided into a favorable term for the contacts between the protein and RNA, and an unfavorable free energy for isomerization of the RNA from the free to bound structures. (Conformational adaptations may take place in the protein as well and thus contribute a third term to the interaction energy.) When two RNA sequences are recognized by a protein, the one that more easily folds into the correct structure will bind more tightly, even if all the contacts are identical in the final complex. Separation of binding free energy into RNA isomerization and protein contact terms is not easy to accomplish experimentally. The general problem of how to dissect the thermodynamics of protein-DNA complexes when coupled conformational changes are present was discussed recently (179), and progress has been made in resolving the thermodynamics of *Eco* RI-DNA recognition (180). The more complex problem of protein-RNA recognition has yet to be approached in this way.

Generalizations about what kinds of RNA structures are most advantageous for protein recognition do not seem possible because virtually any RNA secondary structure is a potential recognition site. It is true that loops and mismatches are the most common targets for proteins, but these kinds of irregularities are the rule in RNA. Protein contacts in a distorted major groove within an internal or hairpin loop are common. From the protein perspective, a β-sheet surface seems to be a useful framework for designing RNA recognition, since it has already been identified as the RNA contact region in the RNP motif, in some aminoacyl-tRNA synthetases, and in phage coat proteins. The TATA-box DNA binding protein uses the same strategy to create a large nucleic acid recognition surface (22). However, α-helices are also important for RNA recognition, as illustrated by the Rev-derived peptide (59).

Comparison of homologous RNA sequences has been an enormously powerful tool for deducing RNA secondary structure (107), but this approach will probably not be as useful for analysis of contacts within conserved protein-RNA complexes. Metzenberg et al have argued that compensatory changes occur between a ribosomal protein sequence and its RNA target (181). However, in many cases homologous complexes achieve specificity by somewhat different mechanisms. Different requirements for a purine bulge in related phage coat protein complexes with hairpins have already been mentioned. The retroviral Tat protein utilizes a different hairpin structure in the bovine and human immunodeficiency viruses (182), and the ribosomal protein S4 is conserved in function between yeast and bacteria (183) but for the most part requires nonconserved features of the rRNA (132). It will be interesting to see to what extent complexes with homologous functions differ as higher-resolution structures become available.

Much work remains to be done in the area of protein-RNA recognition, and progress on several fronts can be expected in the next few years. At present, structural information at the atomic level is available only for the tRNA-synthetase complexes and for two protein-hairpin complexes. High-resolution structures of a wide range of protein and RNA complexes are clearly needed. The current situation should improve as more proteins and their RNA binding sites are biochemically defined and prepared in quantity, and as crystallization and nuclear magnetic resonance (NMR) methods are improved. As I emphasized above, more detailed thermodynamic studies are also essential. Only the barest outlines of the energetics of protein-RNA specificity are currently available, and virtually nothing is known about kinetic mechanisms. The tendency of proteins to trap altered RNA structures means that structural information alone will not be sufficient to resolve questions of protein-RNA binding specificity in any quantitative way; thermodynamic and kinetic studies are therefore essential as well. Well-defined systems are now available for this kind of work, and the next few years should see considerable progress.

ACKNOWLEDGMENTS

I thank all the people who kindly sent me manuscripts prior to publication. Work on protein-RNA interactions in my laboratory and preparation of this review have been supported by NIH grant GM29048.

Literature Cited

1. Cech TR, Bass BL. 1986. *Annu. Rev. Biochem.* 55:599–630
2. Noller HF. 1991. *Annu. Rev. Biochem.* 60:191–227
3. Guthrie C. 1991. *Science* 253:157–63
4. McCarthy JEG, Tuite MF, eds. 1990. *Post-Transcriptional Control of Gene Expression.* Berlin: Springer-Verlag
5. Normanly J, Abelson J. 1989. *Annu. Rev. Biochem.* 58:1029–49
6. Schimmel P. 1989. *Biochemistry* 28:2747–59
7. Giegé R, Puglisi JD, Florentz C. 1993. *Prog. Nucleic Acid Res. Mol. Biol.* 45:129–205
8. McClain WH. 1993. *J. Mol. Biol.* 234:257–80
9. Rould MA, Perona JJ, Söll D, Steitz TA. 1989. *Science* 246:1135–42
10. Ruff M, Krishaswamy S, Boeglin M, Poterszman A, Mitschler A, et al. 1991. *Science* 252:1682–89
11. Biou V, Yaremchuk A, Tukalo M, Cusack S. 1994. *Nature* 263:1404–10
12. Zimmermann RA. 1980. In *Ribosomes: Structure, Function, and Genetics,* ed. G Chambliss, GR Craven, J Davies, K Davis, L Kahan, M Nomura, pp. 135–69. Baltimore: Univ. Park Press
13. Usman N, Cedergren R. 1992. *Trends Biochem. Sci.* 17:334–39
14. Milligan JF, Uhlenbeck OC. 1989. *Methods Enzymol.* 180:51–62
15. Doudna J, Grosshans C, Gooding A, Kundrot CE. 1993. *Proc. Natl. Acad. Sci. USA* 90:7829–33
16. Nikonowicz EP, Pardi A. 1993. *J. Mol. Biol.* 232:1141–56
17. Nagai K, Oubridge C, Jessen TH, Li J, Evans PR. 1990. *Nature* 348:515–20
18. Nagai K, Mattaj I, eds. 1994. *RNA-Protein Interactions.* Oxford: Oxford Univ. Press
19. Seeman N, Rosenberg JM, Rich A. 1976. *Proc. Natl. Acad. Sci. USA* 73:804–8
20. von Hippel PH, McGhee JD. 1972. *Annu. Rev. Biochem.* 41:231–98
21. McClarin JA, Frederick CA, Wang B-C, Greene P, Boyer HW, et al. 1986. *Science* 234:1526–41
22. Kim JL, Nikolov DB, Burley SK. 1993. *Nature* 365:520–27
23. Varani G, Cheong C, Tinoco I. 1991. *Biochemistry* 30:3280–89
24. Heus HA, Pardi A. 1991. *Science* 252:191–94
25. Wimberly B, Varani Gl, Tinoco I. 1993. *Biochemistry* 32:1078–87
26. Szewczak AA, Moore PB, Chan Y-L, Wool IG. 1993. *Proc. Natl. Acad. Sci. USA* 90:9581–85
27. Weeks KM, Crothers DM. 1993. *Science* 261:1574–77
28. Ellington AD, Szostak JW. 1990. *Nature* 346:818–22
29. Tuerk C, Gold L. 1990. *Science* 249:505–10
30. Famulok M, Szostak JW. 1992. *J. Am. Chem. Soc.* 114:3990–91
31. Connell GJ, Illangesekare M, Yarus M. 1993. *Biochemistry* 32:5497–502
32. Majerfeld I, Yarus M. 1994. *Nature Struct. Biol.* 1:287–92
33. Alden CJ, Kim S-H. 1979. *J. Mol. Biol.* 132:411–34
34. Heinemann U, Saenger W. 1981. *Nature* 299:27–31
35. Dingwall C, Ernberg I, Gait MJ, Green SM, Heaphy S, et al. 1989. *Proc. Natl. Acad. Sci. USA* 86:6925–29
36. Weeks KM, Ampe C, Schultz SC, Steitz TA, Crothers DM. 1990. *Science* 249:1281–85
37. Calnan BJ, Tidor B, Biancalana S, Hudson D, Frankel AD. 1991. *Science* 252:1167–71
38. Tan JS, Frankel AD. 1992. *Proc. Natl. Acad. Sci. USA* 89:2723–26
39. Tan RY, Frankel AD. 1992. *Biochemistry* 31:10288–94
40. Weeks KM, Crothers DM. 1991. *Cell* 66:577–88
41. Delling U, Reid LS, Barnett RW, Ma MYX, Climie S, et al. 1992. *J. Virol.* 66:3018–25
42. Puglisi JD, Tan RY, Calnan BJ, Frankel AD, Williamson JR. 1992. *Science* 257:76–80
43. Puglisi JD, Chen L, Frankel AD, Williamson JR. 1993. *Proc. Natl. Acad. Sci. USA* 90:3680–84
44. Hamy F, Asseline U, Grasby J, Iwai S, Pritchard C, et al. 1993. *J. Mol. Biol.* 230:111–23
45. Churcher MJ, Lamont C, Hamy F, Dingwall C, Green SM, et al. 1993. *J. Mol. Biol.* 230:90–110
46. Weeks KM, Crothers DM. 1992. *Biochemistry* 31:10281–87
47. Yarus M. 1989. *Biochemistry* 28:980–88
48. Zapp M, Green MR. 1989. *Nature* 342:714–16
49. Daly TJ, Cook KS, Gray GS, Maione TE, Rusche JR. 1989. *Nature* 342:816–19
50. Cook KS, Fisk GJ, Hauber J, Usman N, Daly TJ, Rusche JR. 1991. *Nucleic Acids Res.* 19:1577–83
51. Mann DA, Mikaélian I, Zemmel RW,

Green SM, Lowe AD, et al. 1994. *J. Mol. Biol.* 241:193–207

52. Iwai S, Pritchard C, Mann DA, Karn J, Gait MJ. 1992. *Nucleic Acids Res.* 20:6465–72

53. Tiley LS, Malim MH, Tewary HK, Stockley PG, Cullen BR. 1992. *Proc. Natl. Acad. Sci. USA* 89:758–62

54. Bartel DP, Zapp ML, Green MR, Szostak JW. 1991. *Cell* 67:529–36

55. Jensen KB, Green L, MacDougal-Waugh S, Tuerk C. 1994. *J. Mol. Biol.* 235:237–47

56. Kjems J, Calnan BJ, Frankel AD, Sharp PA. 1992. *EMBO J.* 11:1119–29

57. Tuerk C, MacDougal-Waugh S. 1993. *Gene* 137:33–39

58. Kjems J, Frankel AD, Sharp PA. 1991. *Cell* 67:169–78

59. Tan RY, Chen L, Buettner JA, Hudson D, Frankel AD. 1993. *Cell* 73:1031–40

60. Battiste JL, Tan RY, Frankel AD, Williamson JR. 1994. *Biochemistry* 33:2741–47

61. Peterson RD, Bartel DP, Szostak JW, Horvath SJ, Feigon J. 1994. *Biochemistry* 33:5357–66

62. Hall KB, Sampson JR, Uhlenbeck OC, Redfield AG. 1989. *Biochemistry* 28:5794–801

63. Sampson JR, Uhlenbeck OC. 1988. *Proc. Natl. Acad. Sci. USA* 85:1033–37

64. Muramatsu T, Yokoyama S, Horie N, Matsuda A, Ueda T, et al. 1988. *J. Biol. Chem.* 263:9261–67

65. Muramatsu T, Nishikawa K, Nemoto F, Kuchino Y, Nishimura S, et al. 1988. *Nature* 336:179–81

66. Sylvers LA, Rogers KC, Shimizu M, Ohtsuka E, Söll D. 1993. *Biochemistry* 32:3836–41

67. Normanly J, Ogden RC, Horvath SJ, Abelson J. 1986. *Nature* 321:213–19

68. Crothers DM, Seno T, Söll DG. 1972. *Proc. Natl. Acad. Sci. USA* 69:3063–67

69. Francklyn C, Shi J-P, Schimmel P. 1992. *Science* 255:1121–25

70. Hou Y-M, Schimmel P. 1988. *Nature* 333:140–45

71. Francklyn C, Schimmel P. 1989. *Nature* 337:478–81

72. Musier-Forsyth K, Usman N, Scaringe S, Doudna J, Green R, Schimmel P. 1991. *Science* 253:784–86

73. Musier-Forsyth K, Schimmel P. 1992. *Nature* 357:513–15

74. Hou Y-M, Schimmel P. 1992. *Biochemistry* 31:10310–14

75. Rould MA, Perona JJ, Steitz TA. 1991. *Nature* 352:213–18

76. Perona JJ, Swanson RN, Rould MA, Steitz TA, Söll D. 1989. *Science* 246:1152–54

77. Jahn M, Rogers MJ, Söll D. 1991. *Nature* 352:258–60

78. Pütz J, Puglisi JD, Florentz C, Giegé R. 1991. *Science* 252:1696–99

79. Normanly J, Ollick T, Abelson J. 1992. *Proc. Natl. Acad. Sci. USA* 89:5680–84

80. Asahara H, Himeno H, Tamura K, Nameki N, Hasegawa T, Shimizu M. 1994. *J. Mol. Biol.* 236:738–48

81. Winter RB, Morrissey L, Gauss P, Gold L, Hsu T, Karam J. 1987. *Proc. Natl. Acad. Sci. USA* 84:7822–26

82. Webster KR, Adari HY, Spicer EK. 1989. *Nucleic Acids Res.* 17:10047–68

83. Webster KR, Spicer EK. 1990. *J. Biol. Chem.* 265:19007–14

84. Unnithan S, Green L, Morrissey L, Binkley J, Singer B, et al. 1990. *Nucleic Acids Res.* 18:7083–92

85. Szewczak AA, Webster KR, Spicer EK, Moore PB. 1991. *J. Biol. Chem.* 266:17832–37

86. Bernardi A, Spahr PF. 1972. *Proc. Natl. Acad. Sci. USA* 69:3033–37

87. Carey J, Cameron V, de Haseth PL, Uhlenbeck OC. 1983. *Biochemistry* 22:2001–10

88. Gott JM, Pan T, LeCuyer KA, Uhlenbeck OC. 1993. *Biochemistry* 32:13399–404

89. Witherell GW, Gott JM, Uhlenbeck OC. 1990. *Prog. Nucleic Acid Res. Mol. Biol.* 40:185–220

89a. Schneider D, Tuerk C, Gold L. 1992. *J. Mol. Biol.* 228:862–69

90. Romaniuk PJ, Uhlenbeck OC. 1985. *Biochemistry* 24:4239–44

91. Santi DV, Hardy LW. 1987. *Biochemistry* 26:8599–606

92. Talbot SJ, Goodman S, Bates SRE, Fishwick CWG, Stockley PG. 1990. *Nucleic Acids Res.* 18:3521–28

93. Peabody DS. 1990. *Nucleic Acids Res.* 17:6017–27

94. Wu H-N, Uhlenbeck OC. 1987. *Biochemistry* 26:8221–27

95. Witherell GW, Uhlenbeck OC. 1989. *Biochemistry* 28:71–76

96. Golmohammadi R, Valegård K, Fridborg K, Liljas L. 1993. *J. Mol. Biol.* 234:620–39

96a. Ni C-Z, Syed R, Kodandapuni R, Wickersham J, Peabody DS, Ely KR. 1995. *Structure.* In press

97. Peabody DS. 1993. *EMBO J.* 12:595–600

98. Lim F, Spingola M, Peabody DS. 1994. *J. Biol. Chem.* 269:9006–10

98a. Valegård K, Murray JB, Stockley PG, Stonehouse NJ, Liljas L. 1994. *Nature* 371:623–26

99. Draper DE. 1994. In *RNA-Protein In-*

teractions, ed. K Nagai, I Mattaj. Oxford: Oxford Univ. Press. In press
100. Mougel M, Eyermann F, Westhof E, Romby P, Expert-Bezançon A, et al. 1987. J. Mol. Biol. 198:91–107
101. Mougel M, Allmang C, Eyermann F, Cachia C, Ehresmann B, Ehresmann C. 1993. Eur. J. Biochem. 215:787–92
101a. Wu H, Jiang L, Zimmermann RA. 1994. Nucleic Acids Res. 22:1687–95
101b. Allmang C, Mougel M, Westhof E, Ehresmann B, Ehresmann C. 1994. Nucleic Acids Res. 22:3708–14
102. Cerretti DP, Mattheakis LC, Kearney KR, Vu L, Nomura M. 1988. J. Mol. Biol. 204:309–29
103. Gregory RJ, Cahill PBF, Thurlow DL, Zimmermann RA. 1988. J. Mol. Biol. 204:295–307
104. Draper DE. 1989. Trends Biochem. Sci. 14:335–38
105. Schmidt FJ, Thompson J, Lee K, Dijk J, Cundliffe E. 1981. J. Biol. Chem. 256:12301–5
106. Egebjerg J, Douthwaite SR, Liljas A, Garrett RA. 1990. J. Mol. Biol. 213:275–88
107. Gutell RR, Larsen N, Woese CR. 1994. Microbiol. Rev. 58:10–26
108. Ryan PC, Draper DE. 1991. Proc. Natl. Acad. Sci. USA 88:6308–12
109. Ryan PC, Lu M, Draper DE. 1991. J. Mol. Biol. 221:1257–68
110. Rosendahl G, Douthwaite S. 1993. J. Mol. Biol. 234:1013–20
111. Laing LG, Draper DE. 1994. J. Mol. Biol. 237:560–76
112. Lu M, Draper DE. 1994. J. Mol. Biol. 244:572–85
113. Laing LG, Gluick TC, Draper DE. 1994. J. Mol. Biol. 237:577–87
114. Wang Y-X, Lu M, Draper DE. 1993. Biochemistry 32:12279–82
115. Deleted in proof
116. Carey J, Uhlenbeck OC. 1983. Biochemistry 22:2610–15
117. Ryan PC, Draper DE. 1989. Biochemistry 28:9949–56
118. Mougel M, Ehresmann B, Ehresmann C. 1986. Biochemistry 25:2756–65
119. Deleted in proof
120. Lamond AI. 1993. BioEssays 15:595–603
121. Siegel V, Walter P. 1985. J. Cell. Biol. 100:1913–21
122. Strub K, Moss J, Walter P. 1991. Mol. Cell. Biol. 11:3949–59
123. Siegel V, Walter P. 1988. Proc. Natl. Acad. Sci. USA 85:1801–5
124. Janiak F, Walter P, Johnson AE. 1992. Biochemistry 31:5830–40
125. Held WA, Mizushima S, Nomura M. 1973. J. Biol. Chem. 248:5720–30

126. Nomura M, Held WA. 1974. In Ribosomes, ed. M Nomura, A Tissières, P Lengyel, pp. 193–223. Cold Spring Harbor, NY: Cold Spring Harbor Lab.
127. Stern S, Powers T, Changchien L-M, Noller HF. 1989. Science 244:783–90
128. Ehresmann C, Stiegler P, Carbon P, Ungewickell E, Garrett RA. 1980. Eur. J. Biochem. 103:439–46
129. Vartikar JV, Draper DE. 1989. J. Mol. Biol. 209:221–34
130. Held WA, Nomura M. 1973. Biochemistry 12:3273–81
131. Powers T, Daubresse G, Noller HF. 1993. J. Mol. Biol. 231:362–74
132. Sapag A, Vartikar JV, Draper DE. 1990. Biochem. Biophys. Acta 1050:34–37
133. Stern S, Wilson RC, Noller HF. 1986. J. Mol. Biol. 192:101–10
134. Cormack RS, Mackie GA. 1991. J. Biol. Chem. 266:18525–29
135. Adam SA, Nakagawa T, Swanson MS, Woodruff TK, Dreyfuss G. 1986. Mol. Cell. Biol. 6:2932–43
136. Sachs AB, Bond MW, Kornberg RD. 1986. Cell 45:827–35
137. Dreyfuss G, Swanson MS, Piñol-Roma S. 1988. Trends Biochem. Sci. 13:86–91
138. Kenan DJ, Query CC, Keene JD. 1991. Trends Biochem. Sci. 16:214–20
139. Dreyfuss G, Matunis MJ, Piñol-Roma S, Burd CG. 1993. Annu. Rev. Biochem. 62:289–321
140. Hoffman DW, Query CC, Golden BL, White SW, Keene JD. 1991. Proc. Natl. Acad. Sci. USA 88:2495–99
141. Wittekind M, Görlach M, Friedrichs M, Dreyfuss G, Mueller L. 1992. Biochemistry 31:6254–65
142. Merrill BM, Stone KL, Cobianchi F, Wilson SH, Williams KR. 1988. J. Biol. Chem. 263:3307–13
143. Burd CG, Dreyfuss G. 1994. EMBO J. 13:1197–1204
144. Burd CG, Matunis EL, Dreyfuss G. 1991. Mol. Cell. Biol. 11:3419–24
145. Jessen T-H, Oubridge C, Teo CH, Pritchard C, Nagai K. 1991. EMBO J. 10:3447–56
146. Hall KB, Stump WT. 1992. Nucleic Acids Res. 20:4283–90
147. Hall KB. 1994. Biochemistry 33:10076–88
148. Scherly D, Boelens W, Dathan NA, van Venrooij WJ, Mattaj IW. 1990. Nature 345:502–6
149. Scherly D, Kambach C, Boelens W, van Venrooij WJ, Mattaj IW. 1991. J. Mol. Biol. 219:577–84
150. Howe PWA, Nagai K, Neuhaus D, Varani G. 1994. EMBO J. 16:3873–81
150a. Oubridge C, Ito N, Evans PR, Teo C-H, Nagai K. 1994. Nature 372:432–38

151. Görlach M, Wittekind M, Beckman RA, Mueller L, Dreyfus SG. 1992. *EMBO J.* 11:3289–95
152. Landsman D. 1992. *Nucleic Acids Res.* 20:2861–64
153. Wolffe AP. 1994. *BioEssays* 16:245–51
154. Schindelin H, Marahiel MA, Heinemann U. 1993. *Nature* 364:164–68
155. Schnuchel A, Wiltscheck R, Czisch M, Herrler M, Willimsky G, et al. 1993. *Nature* 364:169–71
156. Cavarelli J, Rees B, Ruff M, Thierry J-C, Moras D. 1993. *Nature* 362:181–84
157. Miller J, McLachlan AD, Klug A. 1985. *EMBO J.* 4:1609–14
158. Brown RS, Sander C, Argos P. 1985. *FEBS Lett.* 186:271–74
159. Pavletich NP, Pabo CO. 1993. *Science* 261:1701–7
160. You QM, Baudin F, Romaniuk PJ. 1991. *Biochemistry* 30:2495–500
161. Baudin F, Romaniuk PJ, Romby P, Brunel C, Westhof E, et al. 1991. *J. Mol. Biol.* 218:69–81
162. Pieler T, Theunissen O. 1993. *Trends Biochem. Sci.* 18:226–29
163. Clemens KR, Wolf V, McBryant SJ, Zhang PH, Liao XB, et al. 1993. *Science* 260:530–33
164. Hattman S, Newman L, Murthy HMK, Nagaraja V. 1991. *Proc. Natl. Acad. Sci. USA* 88:10027–31
165. Giedroc DP, Qui H, Khan R, King GC, Chen K. 1992. *Biochemistry* 31:765–74
166. Summers MF, Henderson LE, Chance MR, Bess JW, South TL, et al. 1992. *Protein Sci.* 1:563–74
167. Déméné H, Jullian N, Morellet N, de Rocquigny H, Cornille F, et al. 1994. *J. Biomol. NMR* 4:153–70
168. Gorelick RJ, Henderson LE, Hanser JP, Rein A. 1988. *Proc. Natl. Acad. Sci. USA* 85:8420–24
169. Khan R, Giedroc DP. 1992. *J. Biol. Chem.* 267:6689–95
170. Frank D, Guthrie C. 1992. *Genes Dev.* 6:2112–21
171. Schweisguth DC, Chelladurai BH, Nicholson AW, Moore PB. 1994. *Nucleic Acids Res.* 22:604–12
172. Green SR, Mathews MB. 1992. *Genes Dev.* 6:2478–90
173. St. Johnston D, Brown NH, Gall JG, Jantsch M. 1992. *Proc. Natl. Acad. Sci. USA* 89:10979–83
174. Bass BL, Hurst SR, Singer JD. 1994. *Curr. Biol.* 4:301–14
175. Mattaj I. 1993. *Cell* 73:837–40
176. Burd CG, Dreyfuss G. 1994. *Science* 265:615–21
177. Kiledjian M, Dreyfuss SG. 1992. *EMBO J.* 11:2655–64
178. Siomi H, Choi M, Sioli MC, Nussbaum RL, Dreyfuss G. 1994. *Cell* 77:33–39
179. Spolar RS, Record MT. 1994. *Science* 263:777–84
180. Lesser DR, Kurpiewski MR, Waters T, Connolly BA, Jen-Jacobsen L. 1993. *Proc. Natl. Acad. Sci. USA* 90:7548–52
181. Metzenberg S, Joblet C, Verspieren P, Agabian N. 1993. *Nucleic Acids Res.* 21:4936–40
182. Chen L, Frankel AD. 1994. *Biochemistry* 33:2708–15
183. Alksne LE, Anthony RA, Liebman SW, Warner JR. 1993. *Proc. Natl. Acad. Sci. USA* 90:9538–41

Annu. Rev. Biochem. 1995 64:621–51

TRANSCRIPTIONAL RESPONSES TO POLYPEPTIDE LIGANDS: The JAK-STAT Pathway

C. Schindler

Department of Medicine, Columbia University Medical Center, New York, N.Y.

J.E. Darnell, Jr.

Laboratory of Molecular Cell Biology, The Rockefeller University, New York, N.Y.

KEY WORDS: signal transduction/JAK and STAT proteins; JAKs; STATs; transcriptional
             control

CONTENTS

0066-4154/95/0701–0621$05.00

ABSTRACT

Cytokines and growth factors regulate multiple aspects of cell growth through their interactions with specific receptors. These receptors initiate signals directed at both the cytoplasmic and the nuclear compartments. Many of the nuclear signals culminate in the induction of new genes. Characterization of the ability of IFN-α to rapidly induce new genes has led to the identification of a new signaling paradigm, the JAK-STAT (Signal Transducers and Activators of Transcription) pathway. In the IFN-α pathway, two receptor associated tyrosine kinases from the JAK family, Jak1 and Tyk2, mediate the activation of two latent cytoplasmic transcription factors, Stat1 and Stat2. More recent studies have not only determined that this pathway is used extensively, but have led to the identification of additional components (e.g., Jak2, Jak3, Stat3, Stat4, Stat5, and Stat6). This review will examine how these components mediate the transduction of signal directly from receptor to nucleus.

INTRODUCTION

Initially uncovered by experiments aimed at understanding IFN-α and IFN-γ-induced transcriptional activation, a new pathway of signal transduction from the cell surface to genes in the nucleus has been recently recognized (1). The pathway is called the JAK-STAT pathway: First, one or more members of the JAK family (2–4) of tyrosine kinases associated with a transmembrane receptor is activated after ligand-receptor attachment, which leads to the phosphorylation on tyrosine of one or more of a family of latent cytoplasmic transcription factors called STATs, for signal transducers and activators of transcription (5). These latter proteins perform a dual role, first as signal transducers by acting as substrates of the JAKs, and after phosphorylation, homo- or heterodimerization, and nuclear translocation, by acting as transcriptional activators (5, 6). While IFN-α and IFN-γ were the first polypeptide ligands described that trigger this pathway, it is now known that many other ligands also activate proteins in the pathway. The details of the early experiments with the IFNs have been summarized in a number of recent reviews (1, 7–9) and will be only briefly summarized here. We will mainly discuss work from the past two years during which time more has been learned about the IFN pathway, and recent discoveries involving the many other polypeptides that act through the pathway.

THE DISCOVERY OF THE JAK-STAT PATHWAY

When cultured cells are treated with IFN-α or with IFN-γ, transcription of specific previously quiescent genes begins within a few minutes (10–14). The cDNAs that correspond to IFN-α and IFN-γ-induced mRNAs were cloned, and the two different sequence response elements in the genomic DNA responsible for the ligand-dependent activated transcriptional response were identified (15–18). A protein complex called ISGF-3 that was activated by IFN-α (19–21) and one called GAF, the IFN-γ activated factor (17, 18), each of which bound specifically to the identified IFN-α or IFN-γ response element, were identified. The three proteins comprising the IFN-α induced transcription factor, ISGF-3 were characterized, and their genes cloned (22–26). Two of these proteins were found to be phosphorylated on a single tyrosine residue after IFN-α treatment, then to become associated and translocated to the nucleus where they could participate in DNA binding to drive IFN-α specific transcription (27). The first of the new proteins, Stat1, was found to exist in two forms, Stat1α (nominally 91 kD) and Stat1β (nominally 84 kD), due to differential splicing [(25); R Yan, S Qureshi, Z Zhong, Z Wen and JE Darnell, submitted]; Stat2 (nominally 113 kD) had significant homology to Stat1 but was encoded by a separate gene (23). IFN-α led to phosphorylation of both Stat1 (α and β) (27) and Stat2, while IFN-γ led to the phosphorylation of Stat1 (α and β) but not Stat2 (5, 28).

Contemporaneously with the description of the STAT proteins, the JAK family of kinases (2–4) was demonstrated to be required for the ligand-dependent tyrosine phosphorylation of the STAT proteins (29–31). A number of cell lines mutant in different proteins in the IFN-α and IFN-γ pathway (diagnosed by dividing the non-responding cell lines into complementing groups of cells) have been developed (1, 32, 33). One of these cell lines, U1, could be restored to IFN-α responsiveness by transfection of a cDNA clone encoding Tyk2, a previously identified tyrosine kinase of then-unknown function (29). Soon thereafter, additional mutant cell lines pinpointed the requirement of Jak1 and Jak2, members of the same family of tyrosine kinases, in the response to IFN-α (Jak1 and Tyk2) and IFN-γ (Jak1 and Jak2) (30, 31).

These two sets of observations—the characterization of the STAT proteins that functioned after tyrosine phosphorylation in the IFN-α and IFN-γ pathway, and the discovery of a requirement for tyrosine kinases in the IFN pathways— established the outline of the JAK-STAT pathway (1).

It soon became evident that many additional ligands activate these STAT proteins and employ JAK kinases (34–39). More recently, new STAT and JAK proteins have been discovered (6, 40–45a). We begin by summarizing the information about the pathway, starting with the ligand-receptor interactions and progressing to the nucleus.

RECEPTORS AND KINASES

A large number of single transmembrane-spanning receptors has been described. Two families of these receptors, the cytokine family and the receptor tyrosine kinase family, have been shown to activate the JAK-STAT pathway. Receptors and the JAKs they activate are listed in Table 1.

The Cytokine Receptor Family

Receptors that bind cytokines constitute a large heterogenous family that can be divided into two groups (46–48). Class I includes the receptors for IL-2, IL-3, IL-4, IL-5, IL6, IL-7, IL-9, IL-11, IL-12, IL-15, Epo, PRL, GH, G-CSF, GM-CSF, LIF, CNTF, and thrombopoietin (49–52). These receptors share a cytokine-binding domain with a conserved cysteine motif (a set of four conserved cysteines and one tryptophan) and WSXWS motif (a membrane proxial region encoding Trp-Ser-Xxx-Trp-Ser). The class II receptors share overall structural features, but are more divergent. They share a conserved tryptophan and one cysteine pair with the class I receptors. Class II receptors, which include the receptors for IFN-α, IFN-γ and IL-10, (46, 53–56), encode an additional conserved cysteine pair, and several conserved prolines and tyrosines. Another notable feature of both class I and II families of receptors is their multicomponent nature. This suggests that some receptors could share components, providing a potential explanation for some of the remarkable functional redundancy observed among several cytokines. This speculation has in fact been validated for several subsets of cytokines. The majority of cytokine receptors consist of at least two receptor components, a ligand binding chain and a signal transducing chain. (Some authors refer to the ligand binding chain as α and the signaling chain as β, but this nomenclature has not been followed in all cases.) Several sets of cytokines with overlapping functions have been shown to bind receptors consisting of a unique ligand-binding chain and a shared signaling chain (57). In some cases, a signal chain appears to encode both ligand-binding and signal-transducing functions (e.g. GH receptor). Additionally, the ability of cytokines to activate cytoplasmic tyrosine kinases, which appear to play a crucial role in signal transduction, has been shown to be mediated by the β-chain for many cytokine receptors.

The Janus Kinase Family

The Janus (or JAK) kinases represent a distinct family of soluble tyrosine kinases that have been strongly implicated in the signal transduction of many members of the cytokine family (1, 3, 4, 8, 29–31, 43, 58, 59). The four members of the family, which include Tyk2, Jak1, Jak2, and Jak3, have molecular weights ranging from 125–135 kDa. These kinases share an overall structural pattern with seven conserved domains and a sequence identity of 35–45%. Notable features include the absence of SH2 and SH3 domains, and the presence of two tandem tyrosine-kinase domains. Only the most carboxyl-terminal domain is believed to be functional. In general, these kinases are catalytically inactive in resting cells but are associated with the cytoplasmic

Table 1 Activation of JAK kinases by various ligands

	JAKs			
Ligands	tyk2	Jak1	Jak2	Jak3
IFN family				
IFN-α/β	+	+	−	−
IFN-γ		+	+	−
IL-10	+	?	?	−
gp130 family*				
IL-6	+	+	+	?
IL-11	?	+	?	?
OnM	?	+	+	?
LIF	?	+	+	?
CNTF	−/+	+	+	?
G-CSF	?	+	?	?
IL-12***	+	−	+	−
γ-C family				
IL-2	−	+	−	+
IL-4	−	+	−	+
IL-7***	−	+	−	+
IL-9***	−	+	−	+
IL-13***	−	+	?	?
IL-15***	?	+	?	+
gp140 family				
IL-3***	−	−	+	−
IL-5***	−	−	+	−
GM-CSF***	−	−	+	−
Growth hormone family				
EPO	?	−	+	−
GH	?	−	+	−
PRL	?	+/−	+	−
Receptor tyrosine kinases				
EGF	?	+	+	?
PDGF	?	+	+	?
CSF-1	?	+	+	?

*The gp130 family is listed together because this group of ligands utilize the gp130 chain or a closely related molecule in their receptors.

***Unpublished observations are personal communications from J Ihle and J O'Shea.

domains of receptor chains. They are rapidly activated by a ligand-stimulated phosphorylation on a tyrosine within the kinase domain (S Pelligrini, personal communication). All of these kinases, except Jak3, are expressed in many tissues, corresponding with the wide distribution of cytokine activity. Jak3 appears to be predominantly expressed in myeolocytic and lymphocytic lineages. Functionally, the JAK kinases have been shown to be responsible for mediating the activation of STAT proteins (29–31, 60, 61).

Receptor Signal Transduction

The ability of ligands to rapidly activate genes is a common feature of all major classes of receptors. Recent studies suggest that the tyrosine kinase receptor family and cytokine receptor family may share several features in signal transduction, including tyrosine kinase activation, mitogenesis, activation of the ras-MAP kinase pathway, and activation of PI-3 kinase (47–62). Detailed studies on the tyrosine kinase family of receptors have provided valuable insight into how receptors propagate signals through the membrane (63). The signal is initiated when a ligand interacts with two independent receptor chains, thereby promoting receptor dimerization (see Figure 1). Once dimerized, the tyrosine kinases (e.g., either those intrinsic to the receptor or those associated with the receptor) are in apposition and can transphosphorylate each other, leading to kinase activation. Subsequently, the receptor chains become tyrosine phosphorylated rendering them competent to interact with intracellular signaling components (e.g., STAT proteins).

Studies examining the interaction between the GH and its receptor (64) suggest that members of the cytokine family are also likely to transduce signals through receptor dimerization. However, the GH receptor, as well as all other members of the cytokine family of receptors, does not encode intrinsic tyrosine kinase activity. Instead, the multimerized receptor components rapidly activate associated tyrosine kinases (e.g. JAK kinases). The ability to associate with and activate tyrosine kinases appears to be essential for the ability to propogate many if not most intracellular signals (e.g. the proliferation, gene induction, and activation-signaling molecules like phosphatidylinositol 3 (PI3)-kinase, MAP kinase, and ras).

Although a number of tyrosine kinases have been suggested as possible mediators of cytokine receptor signaling to the nucleus, the most convincing data support a central role for the JAK kinases. The initial evidence implicating this family of kinases came from studies on IFN-α stimulated signaling (29). It was shown that the inability of a cell line to respond to IFN-α (and IFN-β), characterized by gene induction, could be restored by the Tyk2 kinase. Subsequently, the role of other members of this family in mediating signaling for other cytokines has been demonstrated. For several different cytokines, one or more members of this family of kinases have been shown

CYTOKINE   SIGNAL   TRANSDUCTION

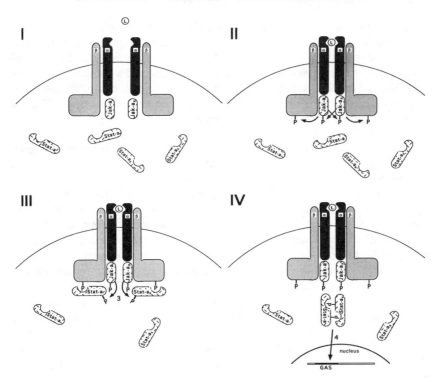

Figure 1 Cytokine Signal Transduction.

I:Signaling commences when the Ligand (L) induces receptor dimerization;

II: This brings the receptor associated Jak kinases (e.g., Jak-a and Jak-a$_x$) into apposition, enabling them to reciprocally transphosphorylate each other (1). The kinases, now activated, phosphorylate a distal tyrosine on the receptor (2);

III: This receptor phosphotyrosyl residue is subsequently recognized by the SH2 domain of a STAT protein (e.g., Stat-a and Stat-a$_x$), drawing them into the receptor complex, where they are activated by tyrosine phosphorylation (3);

IV:The activated STAT proteins are now rendered competent for hetero- or homo-dimerization, and nuclear translocation (4), and GAS binding. Jak-a$_x$ denotes a Jak kinase that can be either identical to or different from Jak-a. Stat-a$_x$ denotes a STAT protein that can be either identical to or different from Stat-a. The α-chain of the receptor binds ligand, while the β-chain transduces signals.

to be activated. Furthermore, discrete domains have been identified in a number of cytokine receptors that specifically interact with and mediate the activation of the Janus kinases. These and other types of studies have clearly implicated the JAK kinases in the activation of STATs. Although there is no formal evidence to date, many studies suggest that the JAK kinases may directly activate the STAT proteins [(35, 61); J Krolewski and C Schindler, unpublished observation].

IL-3 FAMILY OF RECEPTORS   The IL-3 family, which includes IL-3 (multi-colony stimulating factor), IL-5, and GM-CSF (granulocyte-macrophage stimulating factor), represents a set of functionally overlapping ligands. Each ligand binds a receptor consisting of a unique α component and a common β component. These three ligands play an important role in the maturation, proliferation, and activation of several myeloid lineages. In particular, IL-3 and GM-CSF have overlapping roles in the growth and maturation of many myeloid progenitor cells (65). The α components for IL-3, IL-5, and GM-CSF share unique amino acid motifs not found in the other cytokine receptors, which again highlights their unique relationship (66). The common β component is required for a number of intracellular signals, including: tyrosine phosphorylation of a set of cellular proteins, induction of immediate early genes, proliferation, and activation of components in the ras-related pathways (67–71). Of note, in the murine system, the IL-3 receptor can employ a second unique β component (72, 73, 73a). Detailed mutational studies of the β component have identified at least two domains that are important for signal transduction. A membrane-proximal region containing a conserved box 1/box 2 motif is responsible for the interaction with, and activation of, Jak2 (74, 75). The more distal domain is important for the activation of the ras related pathways. More detailed mutational studies of the Epo receptor have reached the same conclusions. Both a point mutation in the Epo receptor (39, 76), and a larger deletion of the β chain, lead to a loss in the ability to activate Jak2 or any downstream signals (77). These studies suggest that Jak2 activation is one of the earliest intracellular signaling events and assign a pivotal role to this family of tyrosine kinases.

As is the case for other cytokine families, activation of these receptor-kinase complexes leads to the stimulation of family specific signal transducing factors (STFs). However, unique to this family, IL-3 has been shown to activate two biochemically distinct STAT-containing STFs, depending on the differentiation state of the target cell. In immature myeloid cells STF-IL3a is activated, while in mature myeloid cells a biochemically distinct STF, STF-IL3b, is activated (96). STF-IL3a has been shown to consist of two tyrosine phosphorylated proteins of 77 and 80 kDa, whereas STF-IL3b appears to consist of a 94 and 96 kDa tyrosine phosphorylated protein. Purification of the 77 and 80 kDa proteins from murine myeloid cells has determined that they are two distinct murine homologs of prolactin activated ovine MGF/Stat5 (42, 77a). These studies also indicate that the 94 and 96 kDa proteins of STF-IL3b represent additional isoforms of Stat5.

IL-2 FAMILY OF RECEPTORS   The IL-2 receptor belongs to a family of receptors that have one unique chain, but all share the IL-2R γ chain (γC). This family includes IL-2, IL-4, IL-7, IL-9, IL-13, and IL-15 (47, 62, 78). The discovery that patients with X-Linked Severe Combined Immunodeficiency Syndrome (XSCID) have deletions in the cytoplasmic domain of γC demonstrates that this receptor component plays an important role in T-cell development (79). However, mice that are homozygous for the IL-2 null mutation have a normal number of lymphocytes, which again suggests a critical role for γC in T-cell growth-promoting signaling pathways (80). Studies from several laboratories have now confirmed that γC is also an essential component of the receptors for IL-4, IL-7 and IL-15 (78, 81–85). The role of γC in the receptors for IL-9 and IL-13 has been implicated, but not yet demonstrated (62).

In contrast to the IL-3 family of receptors, the IL-2 family employs two receptor-signaling components in addition to the shared chain. Two of the three components of the IL-2 receptor, β and γC, are important in signal transduction, while the α-component is required for high ligand affinity. As is the case for other cytokine receptors (see above), dimerization of receptor-signaling components (i.e. the β-chain and γC) is necessary for IL-2 signal transduction (86, 87). Furthermore, only the cytoplasmic domains of these two components have been shown to be required for normal signaling when chimeric receptor constructs are studied. More-detailed studies on the β-chain have identified two important domains. The more distal domain has been shown to interact with the tyrosine kinase lck, but this kinase is not essential for proliferation (47). The membrane-proximal domain has been shown to be essential for the proliferative response and induction of some immediate early genes (47). Interestingly, this domain appears to associate with and activate Jak1 (88). Recently, a new member of the Janus family, Jak3, has been shown to be activated, along with Jak-1, in response to IL-2 stimulation (43, 59). This kinase appears to interact with the γC (88, 89). IL-15 has also been shown to signal through the IL-2 β-chain and γC (78), and activate Jak1 and Jak3 (J O'Shea, personal communication). Studies on the IL-4 receptor have demonstrated that IL-4 also activates Jak-1 and Jak-3 (43, 59). However, unlike IL-2, the receptor for IL-4 (and IL-7) appears to consist of only two chains, γC and a ligand binding α-chain. Both of the IL-4 receptor chains have been shown to make important contributions to signal transduction (81, 82, 90).

The receptors for IL-2, IL-7, and IL-9 are expressed predominantly on lymphocytes, providing a physiological rational for an overlap in their signaling pathways (62). The receptors for IL-4 and IL-13, while also expressed on lymphocytes, are more widely distributed, which indicates that they might intrinsically differ from the other members of the IL-2 family. The demonstration that only IL-4 (and presumably IL-13; 91) requires the activation of an additional intracellular signaling molecule, 4PS, for a proliferative response, further supports this concept (92). 4PS (also referred to as IRS-2) is closely related to IRS-1 (Insulin Receptor Substrate 1), which is important in insulin receptor signaling. Furthermore, IRS-1 can functionally substitute for 4PS, and its activation (i.e., by tyrosine phosphorylation) is dependent on the IL-4 receptor α-chain (90). The distinction between the IL-2/IL-7 and IL-4/IL-13 signaling is further illustrated by the nature of the STAT-containing STFs activated by these ligands. IL-2 and IL-7 induce the same STF, which differs from the STF induced by IL-4 and IL-13 (93–96). A 94–95 kDa Stat5-related protein has been identified in STF-IL2 (and in STF-IL7, STF-IL9, and STF-IL15), whereas STF-IL4 (and presumably STF-IL13) appears to consist of a Stat6 homodimer (45a, 93, 94, Schindler et al., unpublished observation). These differences in signaling are consistent with the differing physiological roles of these cytokines.

IL-6 FAMILY OF RECEPTORS   The IL-6 receptor belongs to a family of receptors that share a common gp130 receptor component. Members of this family, which includes IL-6, IL-11, Oncostatin M (OnM), Leukemia Inhibitory Factor (LIF), and Ciliary Neurotrophic Factor (CNTF), are pleiotropic and stimulate responses in a wide range of tissues (47, 62). The prototypic IL-6 receptor consists of two chains, a ligand-binding α-chain (gp80), and a signal transducing β-chain (gp130). The α-chain is converted into a high-affinity IL-6 binding site only after it associates with gp130. Additionally, while the cytoplasmic domain of the α-chain appears to be dispensable, the cytoplasmic domain of gp130 is essential for signal transduction (62). Detailed studies have demonstrated that an active receptor complex requires gp130 dimerization (97, 98). These dimerized gp130 chains, which mediate signal propagation, have been shown to associate with JAK kinases and to activate Stat3 (99, 100).

The structures of the LIF and CNTF receptors are a variation on this theme (62). The LIF receptor consists of a gp130 chain and the closely related LIFβ chain. LIFβ and gp130 dimerization is required for both high-affinity ligand binding and signal transduction (97, 98). The CNTF receptor employs a ligand-binding chain (CNTFα), which is tethered to the membrane by a glycosyl-phosphatidylinositol linkage, in addition to a LIFβ and gp130 heterodimer (97, 98). The receptors for OnM and IL-11 have not been fully characterized, but both have been shown to require gp130 (101). OnM can bind and stimulate the LIF receptor, but also appears to bind a unique gp130-containing receptor (102).

As mentioned above, gp130 and LIFβ cytoplasmic domains associate with JAK kinases that undergo ligand-dependent tyrosine phosphorylation. Jak-1 is activated by all members of this family, although not to equivalent levels (99) and not in all cell types. Jak2 and Tyk2 have also been shown to be activated by members of this family in several some cell types (99, 100). More-detailed studies demonstrate that Jak1 and Jak2 associate with a membrane-proximal region of both gp130 and LIFβ in a ligand dependent manner (99, 100). Both

Stat1 and Stat3 are rapidly activated in response to this family of ligands, but the ratio appears to vary. Stat3 activation is more prominent in some tissues and in response to several of these ligands (6, 99, 100). Once activated, these STATs associate into three distinct DNA-binding complexes: Stat1 homodimers, Stat3 homodimers, and Stat1:Stat3 heterodimers (6, 96). The role of these individual complexes in mediating signal transduction has not yet been determined except in acute transfections where positive activation occurs (P Lamb, M Seidel, Z Zhong, JE Darnell, unpublished observations).

Studies on the IL-6 family of ligands and receptors have provided important new insights into the JAK-STAT pathway. They have afforded some of the best evidence to date that STATs receive their signals directly from the receptor complex. Detailed studies on the five tyrosine motifs in the cytoplasmic domains of gp130 (and LIFβ) demonstrate that a single tyrosine motif can mediate the interaction with Stat3. Furthermore, the addition of this tyrosine motif to the Epo receptor renders it capable of activating Stat3. Thus, the unique binding site(s) for Stat3 on gp130 and LIFβ serve to bring this molecule into the phosphorylated (i.e. active) receptor-kinase complex for subsequent activation (103). (This result was first obtained with the IFN-γ receptor, as we discuss below.)

The structural and functional similarities between the G-CSF, IL-12, and gp130 receptors, suggest that they may also transmit signals to the nucleus with a similar set of JAKs and STATs (52, 103, 104).

THE IFN FAMILY OF RECEPTORS   Although the IFN signaling pathways have provided significant insight into the components of the JAK-STAT pathway, their receptors remain to be fully elucidated. As previously alluded to, the first STAT proteins were purified from IFN-α–stimulated extracts, and the first Janus kinase implicated in signal transduction was Tyk2, which was shown to be essential for IFN-α receptor function and signal transduction. Subsequently, Jak1, and Jak2 were also implicated, not only in IFN-α and IFN-γ signal transduction but also in the signal transduction of many other cytokines (see Table 1). More-detailed studies taking advantage of kinase-defective cell lines demonstrated that two kinases are required for signaling, and that their activation is reciprocally dependent on the presence of the other kinase (30, 60, 105). These studies are consistent with observations that the signaling domains of the IL-2, IL-3, and IL-6 receptor families—many of which have been shown to associate with Janus kinases (see above)—require dimerization for activation. One function of this dimerization may be to bring two receptor-associated kinases together for appropriate activation.

Signaling in response to IFN-γ is initiated by a unique multicomponent ligand and a species-specific receptor. Two components of this receptor, IFNγR-α (106) and IFNγR-β (53, 54), have been identified and cloned in both the human and murine systems. Both of the human receptor chains appear to confer the normal constellation of responses to human IFN-γ when they are expressed in murine cells (54). However, the converse appears not to be true for the murine receptor components in human cells (53). IFNγR-β has only recently been cloned, but IFNγR-α has been studied extensively. Studies from several groups indicate that Jak1 specifically associates with an LPK (Leu-Pro-Lys) in the cytoplasmic domain of IFNγR-α in a ligand-independent manner [B Schreiber, personal communication, (107–109)]. One group has reported that Jak2, which is also activated by IFN-γ, associates with IFNγR-α, but in a ligand-dependent manner (109). Additionally, IFNγR-α is rapidly tyrosine-phosphorylated, in response to ligand, on tyrosine 440 (Y440). This tyrosine has been shown to be crucial for signal transduction (107, 108, 110). Recently, a tyrosine phosphorylated peptide from this region has been shown to specifically interact with Stat1, presumably with the Stat1 SH2 domain (110). The ability of an activated receptor to draw a ligand-specific STAT into the receptor-kinase complex provides an elegant explanation for how other ligands may signal in the JAK-STAT pathway, and is consistent with the studies on the tyrosine motifs in gp130.

The receptor for type I IFNs, which binds many closely related ligands (i.e. all of the 20 or so IFN-αs, IFN-β, and IFN-ω), may have a more complex structure than the type II IFN receptor (i.e. the IFNγR). Several cross-linking studies have identified three receptor components (53, 111–114). Two of the three receptor components, IFNARα (56) and IFNARβ (115), have been cloned. Studies on IFNARα suggest that its ectodomain may play a role in ligand binding, while the cytoplasmic domain associates with, and mediates the activation of, Jak1 (56). In contrast, the role of IFNARβ ligand binding is more controversial (112, 114). However, it appears to play an important role in signal transduction. Its expression correlates with signal transduction and not with ligand binding (114). Additionally, it is rapidly tyrosine-phosphorylated in response to IFN-α or IFN-β (114), and may associate with Tyk2 (116). Future studies on the type I receptor will be directed at demonstrating that the cloned components, and potentially a third component, can confer the normal constellation of responses to type I IFNs. It will be interesting to determine whether these receptors can discriminate between the various type I IFNs, since preliminary studies suggest that different type I INFs may trigger differing intracellular events [(29); J Yuan and C Schindler, unpublished observations].

Concluding Remarks

The data presented in this section support the model of signal tranduction presented in Figure 1. The signaling cascade is initiated when the ligand, by interacting with its specific receptor, promotes dimerization of the receptor-signal transducing component(s) (β chains). The JAK kinases, which physically associate with specific domains in the membrane-proximal region of the β chains, are drawn into apposition. This is consistent with experimental evidence suggesting that two kinases are required to activate each other by reciprocal transphosphorylation. The kinases then phosphorylate a tyrosine in a distal region of the receptor β chains. This active receptor is now capable of drawing a specific member of the STAT family into the receptor-kinase complex, through interaction with a STAT SH2 domain. The STAT proteins are subsequently activated (by tyrosine phosphorylation), most likely by the JAK kinases themselves, rendering them competent for dimerization (i.e. heterodimerization and/or homodimerization), nuclear translocation and DNA binding.

This model, which is supported by a substantial body of evidence derived from the study of numerous members of this family, describes the ability of cytokine receptors to activate unique sets of STAT proteins. These activated STATs can then, with high fidelity, directly transduce signals to their targets in the nucleus. The ability of receptors to determine which of the STAT proteins they will activate, by recognizing the unique features of the STAT SH2 domains (see below), provides additional support for this model, and casts the receptor as the central figure in regulating signal fidelity. Furthermore, this model accounts for the close relationship in the signal transducing pathways activated by the tyrosine kinase family of receptors and the cytokine family of receptors.

STAT PROTEINS

Definition

The STAT proteins are defined as a group by sequence similarities in blocks of amino acids scattered over a stretch of 700 amino acids that is virtually the entire length of the proteins (1) (Figure 2). The proteins were recognized for their dual functions in *s*ignal *t*ransduction in the cytoplasm and *a*ctivation of *t*ranscription of the nucleus, hence the name STAT. All known members of the STAT family have a single tyrosine residue in the region of residue 700 that becomes phosphorylated during cytoplasmic activation, and all of the proteins in their activated form can bind DNA in a sequence-specific manner.

Gene Number and Distribution

At present, the sequence of six genes from mammals that encode STAT family members has been reported (Table 2; Figure 2). These are numbered 1 to 6 in order of their discovery. Four of these genes (Stats1–4) have been located on chromosome maps: Stats1 and 4 are close together in the proximal region of mouse chromosome 1 (human 2q homology) while Stat2 is on mouse chromosome 10 and Stat3 is on chromosome 11 [(61); N Jenkins, N Copeland, C

```
hStat1    WNDGCIMGFI SKERERA.LL KDQQ.PGTFL LRFSESSREG AITFTWVERS    620
mStat1    WNDGCIMGFI SKERERA.LL KDQQ.PGTFL LRFSESSREG AITFTWVERS    620
mStat3    WNEGYIMGFI SKERERA.IL STKP.PGTFL LRFSESSKEG GVTFTWVEK.    630
mStat4    WIDGYIMGFV SKEKERL.LL KDKM.PGTFL LRFSES.HLG GITFTWVDQS    616
hStat2    WNDGRIMGFV SRSQERR.LL KKTM.SGTFL LRFSESS.EG GITCSWVEH.    618
sStat5    WNDGAILGFV NKQQAHD.LL INKP.DGTFL LRFSDSE.IG GITIAWKFDS    635


hStat1    QNGGEPDFHA VEPYTKKELS AVTFPDIIRN YKVMAAENIP ENPLKYLYPN    670
mStat1    QNGGEPDFHA VEPYTKKELS AVTFPDIIRN YKVMAAENIP ENPLKYLYPN    670
mStat3    DISGKTQIQS VEPYTKQQLN NMSFAEIIMG YKIMDATNIL VSPLVYLYPD    680
mStat4    EN.GEVRFHS VEPYNKGRLS ALAFADILRD YKVIMAENIP ENPLKYLYPD    665
hStat2    QDDDKVLIYS VQPYTKEVLQ SLPLTEIIRH YQLLTEENIP ENPLRFLYPR    667
sStat5    PDRN...LWN LKPFTTREGS IRSLADRLGD LNYLI..... .....YVFPD    672
```

Figure 2 Sequences of the STAT family members. Single letter amino acid code for sequences of reported STAT family members in August 1994. Mouse and human sequence are given for Stat1; human, Stat2; mouse Stat3 and Stat4; sheep, Stat5. Amino acid coordinates for individual proteins are labeled on the right. The shaded area shows identity.

Schindler, JE Darnell, unpublished observations]. The actual number of genes in the family is unknown. Since the nucleic acid sequence of the presently known genes has drifted so greatly, simple low-stringency hybridization or PCR amplification of related sequences has not readily uncovered new members. Genes for Stat1 and Stat2 were cloned after purification of IFN-α–activated DNA-binding proteins yielded peptide sequences. Degenerate oligonucleotides based on these sequences were used to clone the Stat1 and Stat2 cDNAs (23, 25). Stat3 and Stat4 were cloned by low stringency hybridization or PCR products using a Stat1 probe that encoded the SH2 domain (6, 40, 61, 116a) the most highly conserved amino acid stretch in the protein. Stat3 was also independently cloned through the purification of an IL-6 activated DNA binding factor, analogous to the method used to clone Stat1 and Stat2 (41). Stat5 was cloned by the purification of tyrosine phosphorylated DNA binding factors from PRL stimulated mammary tissue and IL-3 stimulated myeloid cells (42, 77a). Likewise, Stat6 was purified and cloned from IL-4 stimulated myeloid cells (45a). As will be discussed later, there are a number of tyrosine phosphorylated DNA binding factors that behave similarly to the known STATs; thus uncharacterized family members may still exist.

In addition to new genes in the STAT family, available evidence hints at the possibility of differential splicing, which could also greatly amplify the number of differently functional proteins in the family. Stat1α, the first protein to be described in the family, is known to exist as a 750 amino acid protein and as a -COOH truncated 712 amino acid protein (Stat1β, 84 kilodalton protein). These two proteins, as will be outlined below, have different func-

Table 2 Activation of STAT proteins by various ligands

Ligands	STATs						
	1	2	3	4	5	6	STF
IFN family							
IFN-α	+	+	+	−			
IFN-γ	+	−	−	−			
IL-10	+/−	−	+	−			
gp130 family*							
IL-6	+/−	−	+	−			
IL-11	+/−	−	+	−			
OnM	+/−	−	+	−			
LIF	+/−	−	+	−			
CNTF	+/−	−	+	−			
G-CSF	+/−	−	+	−			
IL-12			+	+			
γ-C family							
IL-2**	+/−		+		+		+[a]
IL-4						+	+[b]
IL-7					+		+[a]
IL-9					+		+[a]
IL-13						?	+[b]
IL-15					+		+[a]
gp140 family							
IL-3					+		+[c,d]
IL-5					+		+[c]
GM-CSF					+		+[c]
Growth hormone family							
EPO	−	−	−	−	?		+[e]
GH	+/−	−	+	−	+		
PRL	+/−	−	−	−	+		
Receptor tyrosine kinases							
EGF	+	−	+	−			
PDGF	+	−	+	−			
CSF-1	+		+				

The table indicates which STATs are main targets (+) of activation (tyrosine) phosphorylation and DNA binding) for ligands listed on left. Minor targets are indicated (+/−). STF (signal transduction factor) indicates an inducible DNA-binding activity for a DNA element containing the GAS consensus (TTNNNNNAA) and in most cases also shown to contain tyrosine phosphate (96); a–e indicate differences in gel shift mobility and/or different cross reactivity to different anti-Stat1 antisera and/or differences in competition with ligands. These factors do not contain Stats1–4. Some of the factors may be identical (c,a,b,b, etc.) because they have the same properties in response to the listed ligands (see text for details).

*The gp130 family is listed together because this group of ligands utilize the gp130 chain or a closely related molecule in their receptors.

**Some but not all cells that respond to IL-2 activate Stat1 and Stat3 but the major DNA binding complex does not contain Stats 1 or 3.

tional capacities and arise because of differential mRNA splicing (25). Since the genomic structure of the STAT proteins 1 and 2 reveals as many as 25 exons, some as short as seven amino acids, ample opportunity for differential splicing is afforded (116b).

Phylogenetic Distribution

The mammalian sequences that define the STAT family come from humans, mice and sheep. Other vertebrates might be expected to have similar proteins, and a cDNA virtually identical to Stat3 has been isolated from frogs (Z Zhong, JE Darnell, A Hemmati-Brivanlou, unpublished observations). In addition to vertebrates, a *Drosophila melanogaster* gene that is 20–30% identical in amino acid sequence to the mammalian STAT family members has been isolated (R Yan, N Dostatni, C Desplan and JE Darnell, unpublished observation). The earlier report that *hopscotch*, a *Drosophila* gene, encodes a kinase with similarity to the JAK family members (117) and the fact that the *hopscotch* protein will activate the new *Drosophila* STAT(R Yan & JE Darnell, unpublished observation) suggest that the JAK-STAT signaling pathway will also exist in invertebrates.

At present, no genes similar to either JAKs or STATs are known in single-cell organisms or plants. The pathway seems to serve the physiologic purpose of cell-cell or organ-organ communication over long distances through elaboration of polypeptide "cytokines" or "growth factors." Perhaps the pathway arose in evolution in parallel with multicellular animals that have a defined internal space. It will be interesting to see whether Coelenterates or Porifera have a functional JAK-STAT pathway.

Sequence Identities and Functional Domains

Studies to date on the known STAT proteins have been limited to identifying sequence similarities, to locating some of the important functional residues, and to studies on the nature of the interactions between family members. The six known mammalian proteins, as well as the *Drosophila* protein, have an overall sequence identity of 28 to 40% in their first 700 amino acids (Figure 2). Sequences distal to these 700 amino acids vary considerably. Because the STAT proteins have a long list of known or suggested functional requirements (1) (for example, receptor-kinase interaction, homo- and heterodimerization, nuclear translocation, possible association with nuclear proteins including other transcription factors and protein tyrosine phosphatases) the proteins would be expected to have many conserved domains.

Maximum sequence alignment between the family members shows strongest homology within amino acid residues 600–700. This region, which has over 50% amino acid identity between different STAT proteins, matches the sequence of the SH2 domains of other proteins in the amino-terminal portion quite well (118), but is more divergent in the carboxyl-terminal portion of the SH2 region (119). The SH2 domain in STAT proteins undoubtedly functions in binding phosphotyrosine, as is the case for other SH2 domains. Mutation of arginine 602 in Stat1 prevents phosphorylation of the STAT protein, which suggests a role for the SH2 domain in associating with the receptor-kinase complex (120, 121). This Stat1 SH2 residue (i.e. R_{602}) corresponds to the critical arginine located in the floor of the "SH2 pocket", which has been shown to bind phosphotyrosine (122, 123a). The most recent evidence on the amino acid sequences responsible for affinity of particular STAT proteins for particular activated receptors comes from exchanging the SH2 domains of Stat proteins 1 and 2, and introducing vectors encoding these chimeric proteins into cells (123a). Recall that although Stat1 and Stat2 are phosphorylated after IFN-α treatment only Stat1 is phosphorylated after IFN-γ treatment. However, when the SH2 of Stat1 is replaced by the Stat2 SH2, the resulting chimeric molecule fails to be phosphorylated in response to IFN-γ, but can still be phosphorylated in response to IFN-α. However, a different Stat1 chimeric molecule containing ~20 amino acids of Stat2, including the tyrosine site of Stat2, is phosphorylated after stimulation with IFN-γ. Finally, a chimeric Stat2 molecule with a Stat1 SH2 domain is activated by IFN-γ, whereas the wild type Stat2 is only activated by IFN-α. Thus the SH2 domain, previously shown to be required for phosphorylation, determines which receptor will bind a STAT substrate. The STAT substrate is then acted upon, regardless of which JAK kinases are associated with the receptor. Other SH2 interactions during dimerization of the STAT proteins will be discussed later.

Between amino acid 500 and 600 there is a distant but probably meaningful similarity in the STAT proteins to SH3 domains. SH3 groups in other proteins are also less well-conserved than the SH2 groups (118, 124). No critical residues in the potential SH3 region of the STAT proteins have yet been identified.

Through the use of Chou-Fasman rules (125a) for examining amino acid sequences for possible helical folding in protein, three regions (~135–165, ~195–240, and 255–285) of the STAT proteins have been identified as potentially helical in structure (23).

Aside from the mutations in the SH2 domain that establish its importance, only one other region of known function has been identified. Mutations of the single tyrosine, which is phosphorylated [residue 701 in Stat1 (28); and 690 in Stat2 (121)] in response to ligand, to phenylalanine block interferon-induced phosphorylation and the function of these proteins in either IFN-α– or IFN-γ–treated cells (5, 120). The same tyrosine residue in Stat1 is also phosphorylated in EGF-, PDGF-, and CSF-1–treated cells (35, 120) and in cells treated with a tyrosine phosphatase inhibitor (Z Zhong, C Schindler and JE Darnell, unpublished observations). A reasonable suggestion arises that only this residue is presented to and recognized by cellular tyrosine kinases. A single tyrosine residue in the homologous position in Stat5 also is phosphorylated in response to prolactin and IL-3 (77a, 125a). In the sequences of Stat1, Stat2, and Stat3, a lysine is present two amino acids downstream from the phosphorylation site, in the region that Cantley and colleagues assign the greatest importance for the SH2 binding of phosphotyrosine-containing peptides (126). However, the downstream amino acid sequence is not similar in Stat4 or Stat5, and the importance of residues surrounding the substrate tyrosine has not yet been fully elucidated for any of these molecules.

A number of mutations in the Stat1 protein are known to have an effect on the functioning of the Stat1 protein. For example, the remaining Stat1 protein is no longer phosphorylated in mutants lacking the amino-terminal 140 amino acids (121), even though the SH2 group and the phosphotyrosine are at a great distance from the deletion. Phosphorylation does occur in mutants of Stat1 missing the first 60 amino acids, but the protein is destabilized (K Shuai and JE Darnell, unpublished observations). Removal of the 22 amino acids (199–220) containing repeated leucines, in the second potential helical region of the Stat1 protein, also renders the 91 kD protein nonphosphorylated (121). Since phosphorylation is required for dimerization and DNA binding, it has not been possible to locate the portion of the molecule engaged in DNA binding. Mutants that allow phosphorylation and alter DNA binding are currently being analyzed (Z Wen, C Horvath and JE Darnell, unpublished observations).

Distribution of STATs in Cells and Tissues

The distribution of STATs in various cell types has been reported. Stat1 and Stat3 are present in many if not all cell types, though the concentration of the proteins varies considerably. Recent in situ hybridization to tissues during early mouse development shows high levels of Stat3 mRNA in the decidua and in the visceral endoderm at 7.5 to 9 days. Stat1 is also present in the decidua, but in an uneven, honeycomb-like appearance with only low levels in the fetus or extra-embryonic tissues (S Duncan, Z Zhong and JE Darnell, unpublished observations). Stat2 is also apparently present in a wide number of cell types, since IFN-α responses are widespread. Stat5 and particularly Stat4 are found in more limited distribution (40, 42). Stat4 is present in high concentration in testes and in moderate levels in thymus, but not in many other tissues or cultured cells. Stat5 was originally called mammary growth factor, but Stat5 mRNA has been found at higher concentrations in a number of other tissues and myeloid cell lines (42, 77a). The presence of Stat5 in cultured cells has not yet been carefully examined.

The concentration in cells of STAT proteins also appears to be regulated. In the first reported experiments with STAT mRNAs, a much higher level of

Stat1 and Stat2 mRNA was observed in cells after IFN-α or IFN-γ treatment than before treatment (23, 25). This induction occurs at the transcriptional level, since run-on experiments showed Stat1 and Stat2 genes to be transcriptionally activated by IFN-α and IFN-γ (127). This higher level of mRNA for Stat1 and 2 also corresponds to increased levels of STAT proteins as detected by immunoblots. The rise in protein in IFN-treated cells is paradoxical, because the activity of Stat1 and 2 as DNA-binding proteins disappears within two to four hours (20, 28), whereas the total STAT protein concentration has risen considerably during this time. This phenomenon remains unexplained. Additionally, high levels of unphosphorylated Stat4 proteins are present in myeloid precursor cells cultured in Epo or IL-3 (61). When these growth factors are removed, the cells stop growing and differentiate, with a commensurate sharp decline in the levels of Stat4. Thus, it is very clear that the STAT proteins change concentrations in cells, and that the Stat1 and 2 genes are transcriptionally activated by IFN-α and IFN-γ. However, the physiologic meaning of the changes and concentrations of proteins is not yet clear.

Ligands that Activate Various STAT Proteins

A matter of obvious physiological importance is the pattern of activation of STAT proteins by different ligands (see Table 2). This pattern, added to that of Table 1 showing which JAK kinases are activated by various polypeptides, provides at least a preliminary outline for the molecular pathways from the cell surface to particular genes whose transcription is stimulated by individual ligands.

What are the tests for STAT activation? DNA binding assays have led to the identification of STATs activated by a number of ligands. Stat1, Stat2, Stat3, Stat5, and Stat6 were all purified and cloned from ligand stimulated extracts using such an assay (22, 23, 25, 41, 42, 45a, 77a). Moreover, DNA binding sites based on the Stat1 binding element (i.e. GAS, see below) have proved invaluable in detecting the activation of STAT or STAT-like activities after stimulation with many ligands (1, 37, 93, 96). These DNA binding assays have become even more informative with the generation of STAT specific antisera, which can be used to identify individual STAT protein in DNA binding complexes. Furthermore, since these antisera precipitate tyrosine phosphorylated STAT proteins from ligand-treated cells, similar immunoprecipitation assays have been used extensively to detect specific STAT activation. Finally, the identification of several uncharacterized STAT-like components of STFs has been made possible by demonstrating reactivity to antibodies made against conserved regions of Stat1. A number of different STFs with STAT-like characteristics have been identified in this fashion (37, 38, 93–96, 128, 129).

Presently, at least 25 different ligands have been shown to activate Stats1–6 by the tests outlined above. Additionally, STAT-like proteins have been identified in STFs activated by IL-2, IL-7, IL-9, IL-10, IL-13, Epo, GH, and CSF-1 (93, 94, 96, 128, 129a, 129b, Schindler et al., unpublished observation).

The patterns of activation described so far show some STATs to be activated by many ligands and others to be highly specifically activated. For example, Stat1 and Stat3 are both activated by a large number of ligands, including those with receptors that have intrinsic tyrosine kinases as well as members of the cytokine receptor family. There are instances, for example IFN-γ treatment, where Stat1 is activated and not Stat3. The reverse is also true: IL-6 at low concentration has the capacity to activate Stat3 more strongly than Stat1. So some specificity is observed even though Stat1 and Stat3 both are activated by many identical ligands (e.g. EGF, PDGF, and CSF-1). It is also noteworthy that receptors comprised of similar protein chains (for example CNTF, LIF, IL-6, OnM, and IL-11, all of which share a gp130 receptor component) all activate the same set of STATs and, in general, favor Stat3 over Stat1, particularly at limiting ligand doses (6, 34, 38, 96, 99, 103, 130).

Stat2 is unique at the moment because as far as has been reported, this protein is activated only in cells treated with type I IFNs (27; J Yuan and C Schindler, unpublished observations). Additionally, Stat2 is the only family member that is not known to directly bind DNA as a homo- or heterodimer. Stat2 is able to bind DNA only after heterodimerization with Stat1 and association with a non-STAT protein of 48 kDa. This 48 kDa protein (p48) is a member of the IRF-1 family of ISRE binding proteins (24). It seems highly likely that other such multiprotein factors will emerge with further study.

Stat4 at the moment is unusual because it has been found to be expressed at high levels in the testes, yet does not appear to be phosphorylated in this tissue. Recently however, Stat4 has been found to be activated in T helper, class I, cells after treatment with IL-12 (K Murphy, personal communication). Stat5 has been shown to be activated by prolactin and the IL-3 family of ligands (42, 77a). Additionally, some of the IL-2 family of ligands appear to activate a Stat5-like protein (93, Schindler et al., unpublished observation). Lastly, Stat6 has recently been identified in the STF activated by IL-4 (45a, 94).

Much of the experimentation on STAT activation has been carried out in cultured cells, and the physiologic importance in the animal of various ligand-induced STAT activations will need to be retested in animals bearing null mutations—in particular STAT proteins. Nevertheless, current experiments show that the acute phase response (presumably IL-6–mediated) favors activation of Stat3 (36, 130), while EGF causes equal activation of Stats1 and 3 in mouse liver, just as is the case in cultured hepatoma cells (6, 96). Thus it seems highly likely that the STAT proteins will be shown to play important roles in the animal.

Protein:Protein Interactions within the STAT Family

The activation of more than one STAT protein in a cell by a particular ligand makes the study of the interactions between STAT proteins themselves or other possible DNA-binding proteins a pressing matter if the transcriptional specificity of response to individual ligands is to be understood.

The original purification of the IFN-α-induced DNA binding factor, ISGF3, yielded four potentially interacting proteins; a 91 kD, 84 kD and 113 kD protein that are now termed Stats1α,β; Stat2; and a 48 kD protein from the family of DNA binding proteins that includes IRF-1 (24). These studies initiated an interest in protein:protein interactions among the STATs. Initially, UV cross-linking data indicated that the 48 kD protein component made contact with DNA (22, 26). More extensive crosslinking studies have demonstrated that Stat1α or β (91 or 84 kD) and the Stat2 (113 kD) proteins also contact DNA (130a). Whether there is physical contact between the 48 kD molecule and the two STAT proteins has not been proven, but all of these proteins can be crosslinked to a 9 bp stretch of DNA.

The first evidence that a STAT-STAT interaction occurred came from coprecipitation of extracts from interferon-treated cells that were labeled with ^{35}S or ^{32}P (27). Antisera that only precipitated Stat2 from untreated cell extracts precipitated ^{35}S- or ^{32}P-labeled Stat2 and Stat1 after IFN-α treatment. This ligand-induced co-precipitation has subsequently been repeated with several additional anti-Stat1 and anti-Stat2 antisera. Physical evidence that Stat1 homodimerizes after treatment with IFN-γ includes zonal sedimentation analysis and comparison of migration of the phosphoprotein in various concentrations of gel matrices (Ferguson plots; 131). Furthermore, whereas the phosphorylated form of Stat1 is found in the dimeric form and DNA binding complex, the non-phosphorylated form is apparently free in the cytoplasm and sediments with an apparent molecular weight of 90 kD (close to its calculated size from amino acid sequence). Dimerization appears to involve phosphotyrosine-SH2 interactions, most likely intermolecular, because specific tyrosine phosphorylated peptides (e.g. TSFGY*DKPH representing the tyrosine 440 motif from IFNγR-α; LDGPKGTGY*IKTELI representing the phosphorylated sequence from Stat1; and KVNLQERRKY*LKHR representing the phosphorylated sequence from Stat2) disrupt the otherwise stable Stat1 homodimer or Stat1:Stat2 heterodimer (110, 131, S Gupta and C Schindler, unpublished observation). Tyrosine phosphorylated peptides that are recognized by other SH2 domains (e.g. EPQY*EEIPYL from *src*) do not disrupt these complexes. These results argue for a role of phosphotyrosine SH2 interaction in the activated Stat1 homodimers, but do not rule out the possibility that other interactions may strengthen or contribute to dimerization.

Apparent STAT heterodimerization has also been noted in cells treated with EGF or IL-6, where three distinct protein-DNA complexes are activated. The fastest moving complex is identical to Stat1 homodimers produced by IFN-γ in both mobility and antibody reactivity. The slowest moving complex reacts with antisera against Stat3 and the responsible protein was purified to homogeneity as Stat 3 (41, 45). Antisera against either Stat1 or Stat3 blocks the formation of the intermediate complex (6, 96, 130). In addition, Stat1- or Stat3-specific antisera can co-precipitate Stat1 and Stat3 from extracts of cells treated with CNTF, IL-6, or EGF. Similar observations were made on extracts from cultured cells or mouse liver after intraperitoneal injection of EGF or lipopolysaccharide (which stimulates the acute phase response in liver through secretion of IL-6) (132). All of these experiments suggest that Stat:Stat3 heterodimers can form in cells treated with EGF or IL-6. Together with evidence that heterodimers form between Stat1 and Stat2 in cells treated with IFN-α, these results strongly suggest that heterodimers between STAT proteins exist. However, recent evidence suggests that certain heterodimers will not form: IFN-α treatment leads to activation of Stats1, 2, and 3, but even in cells lacking Stat1 no evidence for a Stat2:Stat3 heterodimer or Stat2:Stat2 homodimer has been found [(34); C Horvath and JE Darnell, unpublished observations].

A word about the functionality of the STAT proteins during transcription is in order. Through somatic-cell genetic selection, mutant cell lines deficient in various STAT proteins and JAK kinases have been selected. By separately introducing into such cell lines cDNA constructs that express Stat1α, Stat1β, or Stat2, the following conclusions have been reached about the STAT proteins: In the cell line U3, which lacks Stat1 completely, either Stat1α or Stat1β, introduced separately, restores the ability to form ISGF3 and the transcriptional induction by IFN-α of endogenous genes (31); however, only Stat1α, and not Stat1β, functions in transcriptional activation of IFN-γ–triggered genes in those cells. The use of U2 cells that fail to make a normal p48 protein has also shown the necessity for that protein in the IFN-α pathway (D Levy, GR Stark, IM Kerr, personal communication).

More recently, the U6 cell line, which lacks the Stat2 mRNA and protein, has been characterized (133). The inability of these cells to respond to IFN-α can be restored by the introduction of a Stat2 cDNA expressed vector. As expected, the U6 cells respond normally to stimulation with IFN-γ.

Finally, recent results on the activation of STAT proteins at the receptor-kinase interface provide evidence for Stat1:Stat2 during activation. In U3 cells, Stat2 can be activated by IFN-α on the correct tyrosine (residue 690; 121). However, in U6 cells there is no phosphorylation of Stat1 after IFN-α (133). These results imply some interaction with, or at least some requirement for, the activation of the Stat1 protein in response to IFN-α. Such an interaction could take place on the receptor-kinase complex and thus could indicate the site of STAT dimer formation in the cell.

STAT Interactions with Other Proteins

Although p48 is the only non-STAT protein that has been shown to functionally interact with STATs, it is of course possible that other similar interactions will occur. For example, the IFN-γ inducible GRR element in the promoter of the FcγRI gene has been mapped to a 38 base pair region. The 3' most 22 bases in this site bind Stat1. However, the entire DNA element binds an additional larger IFN-γ induced Stat1 containing complex (134). Since the entire 38 base pair element directs a much stronger IFN-γ induction in transiently transfected cells, there may be a functional interaction between Stat1 and an additional protein, which would bind to the sequence flanking the GAS element in the GRR.

DNA SEQUENCE ELEMENTS THAT BIND STAT PROTEINS

The prototype sequence elements found in genes activated by IFN-α are termed interferon-stimulated response elements (ISREs). The importance of these elements was initially recognized by finding a highly conserved ~10–12 base pair stretch in the promoters of about ten different genes that were transcriptionally induced by IFN-α (1). The elements were capable of directing IFN-α–dependent transcription during acute transfection experiments, and mutations in the elements confirmed their necessity as binding sites. Many of these sites were also determined to be capable of binding the same protein complex, the interferon-stimulated gene factor 3, ISGF3. A similar set of experiments identified the IFN-γ inducible element, the GAS element, first in a single gene (the GBP gene) and later in a larger series of immediate early IFN-γ–induced genes (see Table 3).

EGF, PDGF and CSF-1 have been shown to induce binding of a factor to the SIE site (for *sis*, an oncogenic form of PDGF, inducible element) found in the promoter of the cfos gene (135). Cochran and colleagues have used a mutagenic analysis of the SIE to produce a "high affinity" version of this sequence (the M67 oligonucleotide). EGF, PDGF and CSF-1 have all been shown to cause phosphorylation of Stat1 and binding to the M67, which is now recognized to be a GAS element (see Table 3).

Subsequently, DNA binding sites in the promoters of a variety of other genes that bound ligand-activated proteins were recognized to be similar to the GAS consensus, TTNNNNNAA. For example, functionally identified blocks of sequence from the promoters of IL-6 responsive genes (136) were aligned in

Table 3 Functional GAS elements

Gene	Species	Element					
GBP	Human	ATT	AC	T	CT	AAA
FcgRI	Human	TTT	CC	C	AG	AAA
ICSBP	Human	TTT	CT	C	GG	AAA
ICAM-1	Human	TTT	CC	C	GG	AAA
c-fos$_{M67}$	Human	TTT	CC	C	GT	AAA
c-fos$_{wt}$	Human	GTT	CC	C	GT	CAA
IFP-53	Human	ATT	CT	C	AG	AAA
IRF-1	Human	TTT	CC	C	CG	AAA
IRF-1	Mouse	TTT	CC	C	CG	AAA
Ly6E	Mouse	ATT	CC	T	GT	AAG
CDC23$_b$	Human	TAC	CT	G	AG	AAA
IgHϵ	Mouse	TTC	AC	A	TG	AAC

an effort to determine the consensus sequence for the acute phase response element (APRE). This element bound an IL-6 induced transcription factor. The APRE was recognized to be a GAS-like element. Subsequently, GAS-like elements have been implicated in signaling mediated by a number of additional receptors (see Table 2). However, unlike the IFN and IL-6 families, for most of these ligands both the natural GAS elements and the genes they activate remain to be defined. For example, a set of genes activated by IL-2 has been reported, but the promoter elements responsible for IL-2 activation have not yet been determined (137). Even in the case of EGF and PDGF, aside from the cfos gene, little progress has been made in locating similar activation elements in other induced genes.

Optimum Binding Sites vs Natural Sites

Palindromic sequences, e.g. TTCCNGGAA, have been determined to be good Stat1 binding elements. In one study, this sequence has been selected from a random pool of degenerate 14 base pair oligonucleotides, by employing antibodies to Stat1 or Stat3, to select bound oligonucleotides. Repeated cycles of such selection followed by PCR amplification have identified individual "high-affinity" sites for both proteins with the sequence of TTCCNGGAA.

Although variations of the consensus sites appear in nature, it is not yet clear if there is a specific conserved sequence preference that will distinguish Stat1 and Stat3 homo- and heterodimers nor, of course, any of the other newly recognized STATs.

It is clear that different binding sites from different genes do not all bind equally to proteins activated by different ligands. For example, the Ly6E gene (mouse) has a site that binds Stat1 homodimers much better than Stat3 homodimers. The transcription factors activated by IL-3 and IL-4 don't bind LY6E sites well, but do bind well to the GAS site in the IFP gene (96). Furthermore, only the IL-4–activated factor binds to the IL-4–responsive elements in the promoters of the CDC-23 gene and the germ line immunoglobulin epsilon heavy chain (IgHε) (94, 138, 139). Yet the IL-4–activated factor has a good affinity for the palindromic element, suggesting that the sequence in this element may have been evolutionarily conserved to mediate induction by multiple cytokines (96, 127, 136, 140, 141).

Thus, while it is clear that a common palindromic sequence with TT and AA at its terminus will bind proteins activated by many ligands, the actual functional sequences in inducible genes need more attention.

CONCLUSION

In the past decade, as the number of known polypeptide effectors has climbed steadily into the hundreds, an important question recurs: How is specificity in response to a very large number of ligands achieved? A restricted pattern of expression of either a ligand or its receptor within an organism is recognized as one important means to effect a specific outcome (142), but many of the ligands that operate through the JAK-STAT pathway circulate in the blood, and cells can possess receptors simultaneously for many different ligands. Moreover, it is widely believed that many polypeptide ligands exert an early effect by stimulating (or, less often, inhibiting) transcription. Thus, for the JAK-STAT pathway, there is a great likelihood of specificity in the ligand-dependent activation or inactivation of immediate early genes. We direct attention to three possible loci in the cell that bear on this question: (a) ligand-receptor-kinase complexes, (b) activated STAT proteins and their interactions, and (c) DNA-binding sites—overlapping and restricted gene sets.

Receptor Kinase Specificity

The receptor kinase complex may have both a qualitative and quantitative effect on transcription. First, it is clear that the number of kinases and the number of STATs is far less than the number of receptors. So each receptor can neither have a dedicated kinase nor activate a unique STAT or STAT-containing transcription complex. Consistent with this idea, many ligands seem to activate overlapping sets of JAKs and STATs. The unique patterns of activation exhibited by the two types of IFNs may be more exceptional. Additionally, the observation that some JAKs and STATs are expressed and activated in a limited tissue distribution may add to specificity (40, 42–44, 77a). Moreover, the ability of receptors to interact with only a specific set of STATs indicates that it is the receptor itself that transmits the specificity of the ligand-receptor interaction into the cell (110, 143).

Second, with regard to the quantitative aspects of signal transduction for different receptors, several points need to be made. The activating potential of a given ligand depends on the number of receptors present, and on the time the signal continues after a ligand-receptor interaction, depending for example on the rate of internalization and the fate of a given receptor. So while two ligands may activate the same STAT, they might not do so at the same level, nor for the same period of time (96). Furthermore, the affinity of a particular STAT for a particular receptor-kinase complex may vary from one receptor to another, so that the combination of a ligand with its receptor will not necessarily generate the same number of active STAT molecules. Of additional importance, some receptors will activate other signals that, along with the JAK-STAT pathway, have an impact on the transcriptional outcome. For example, phosphotyrosine phosphatases are known to be activated by ligand binding (144), and such activation could easily affect the length or the strength of the STAT signal response emanating from a particular receptor. Activation of the IRF-1(ISGF-2) gene by IFN-α and IFN-γ serves to illustrate some of these points. IFN-α and IFN-γ only activate this gene through the GAS site to which the Stat1 homodimer (activated by either ligand) binds (127). The gene lacks any identifiable ISRE (145). In the IFN-α case the transcription of the IRF1 gene goes up and down within about two hours, as does the supply of phosphorylated Stat1. However, IFN-γ transcriptional stimulation lasts for at least 16 hours in spite of the equally rapid disappearance of the phosphorylated Stat1 homodimer. This continued transcription is characteristic for other IFN-γ–stimulated genes as well; that is, they become transcriptionally activated and remain active long past the time when the initiating Stat1 homodimer has disappeared (13, 20). Thus, some event(s) secondary to the initiation of transcription by these two ligands change the impact on the overall outcome. So while complete specificity of response from each ligand-receptor cannot be envisaged, it can be said that significant differences in the signal emanating from different receptors could be expected even though similar STAT proteins are activated.

STAT Protein Interactions

With respect to the STAT proteins themselves, the greatest contribution to raising the potentiality for specificity in ligand responses is the rising number of STATS (two discovered in 1992; six known by 1994) and the possibility for selective heterodimerization. As noted earlier, Stat1 and Stat3 can homodimerize, whereas Stat1 and 2, and Stat1 and 3, can heterodimerize. However, it has been difficult to find evidence for Stat2 homodimerization or Stat2 and 3 heterodimerization, even when both are activated. If the transcriptional activation (or inactivation) potential of homo- and heterodimers varies, then slight differences in the array of STATs activated by different ligands could result in a very different pattern of transcriptional activators. A further or independent increase in specificity could arise from interactions with non-STAT proteins, as is the case for ISGF-3 (i.e. Stat1:Stat2 + p48) or the FcγRI promoter (i.e. the ability of the GRR to mediate interactions between Stat1 homodimers and an additional transcription factor; 134). Hence, there is ample opportunity for maintaining signaling fidelity with a limited repertoire of STAT proteins.

Specificity at the Gene Level

The end point of this signaling cascade, and thus the final level of specificity, resides in the induction of genes. However, relatively little is known about the differences in the pattern of genes stimulated by different ligands. For example, most of the genes activated are unique to a given ligand or common to several ligands. Based on the limited data available, it appears that there will be at least some genes that are ligand specific, and some that are common to several ligands. Differences in the nature of the GAS elements in the promoters of genes may explain some of these differences. For example, studies have determined that the optimal DNA binding site for Stat1 and Stat3 homodimers are identical (C Horvath, Z Zhong, Z Wen, JE Darnell, unpublished observation) yet studies with natural GAS sites have demonstrated different affinities for STAT dimers (96). Additionally, STF-IL4 (likely to be a Stat6 homodimer; 45a) is able to bind a unique GAS element that other STATs are unable to recognize (94). Much more work remains to be done if we are to understand these important issues. First, the nature of the sets of genes activated by distinct ligands needs to be determined for many more ligands. Second, the promoters of these sets of genes need to be characterized for the presence of GAS elements, and the potential differences in these GAS elements determined.

In spite of the shortcomings in our understanding of specificity of ligand stimulated gene activation and our consequent lack of understanding for the differences in overall biologic effect mediated by different ligands, an optimistic final statement appears in order. After binding to their receptors, many ligands are thought to promote a change in the phenotype of the cell by initiating a specific gene activation program. The discovery of the JAK-STAT pathway and of the pervasive use of this pathway by many ligands, suggests that the pathway is crucial for initiating changes in cell phenotype in response to different ligands.

Literature Cited

1. Darnell JE Jr., Kerr IM, Stark GM. 1994. *Science* 264:1415–21
2. Wilks AF. 1989. *Proc. Natl. Acad. Sci. USA* 86:1603–7
3. Wilks AF, Harpur A, Kurlan RR, Ralph SJ, Zurcher G, et al. 1991. *Mol. Cell. Biol.* 11:2057–65
4. Firmbach-Kraft I, Byers M, Shows T, Dalla-Favera R, Krolewski JJ. 1990. *Oncogene* 5:1329–36
5. Shuai K, Stark GR, Kerr IM, Darnell JE Jr. 1993. *Science* 261:1744–46
6. Zhong Z, Wen Z, Darnell JE Jr. 1994. *Science* 264:95–98
7. Pellegrini S, Schrindler C. 1993. *TIBS* 18:338–42
8. Ihle JN, Witthuhn BA, Quelle FW, Yamamoto K, Thierfelder WE, et al. 1994. *TIBS* 19:222–27
9. Shuai K. 1994. *Current Opin. Cell Biol.* 6:253–59
10. Friedman RL, Manly SP, McMahon M, Kerr IM, Stark GR. 1984. *Cell* 38:745–55
11. Larner AC, Jonak G, Cheng Y-S, Korant B, Knight E Jr., et al. 1984. *Proc. Natl. Acad. Sci. USA* 81:6733–37
12. Larner AC, Chaudhuri A, Darnell JE Jr. 1986. *J. Biol. Chem.* 261:453–59
13. Decker T, Lew DJ, Cheng Y-S, Levy DE, Darnell JE Jr. 1989. *EMBO J.* 8:2009–14
14. Lew DJ, Decker T, Darnell JE Jr. 1989. *Mol. Cell. Biol.* 9:5404–11
15. Levy D, Larner A, Chaudhuri A, Babiss LE, Darnell JE Jr. 1986. *Proc. Natl. Acad. Sci. USA* 83:8929–33
16. Reich N, Evans B, Levy DE, Fahey D, Knight E Jr., et al. 1987. *Proc. Natl. Acad. Sci. USA* 84:6394–98
17. Decker T, Lew DJ, Mirkovitch J, Darnell JE Jr. 1991. *EMBO J.* 10:927–32
18. Lew D, Decker T, Strehlow I, Darnell JE Jr. 1991. *Mol. Cell. Biol.* 11:182–191
19. Levy DE, Kessler DS, Pine R, Reich N, Darnell JE Jr. 1988. *Genes & Devel.* 2:383–93
20. Levy DE, Kessler DS, Pine RI, Darnell JE Jr. 1989. *Genes & Devel.* 3:1362–72
21. Dale C, Iman AMA, Kerr IM, Stark GR. 1989. *Proc. Natl. Acad. Sci. USA* 86:1203–07
22. Fu X-Y, Kessler DS, Veals SA, Levy DE, Darnell JE Jr. 1990. *Proc. Natl. Acad. Sci. USA* 87:8555–59
23. Fu X-Y, Schindler C, Improta T, Aebersold R, Darnell JE Jr. 1992. *Proc. Natl. Acad. Sci. USA* 89:7840–43
24. Veals SA, Schindler C, Leonard D, Fu X-Y, Aebersold R, et al. 1992. *Mol. Cell. Biol.* 12:3315–24
25. Schindler C, Fu X-Y, Improta T, Aebersold R, Darnell JE Jr. 1992. *Proc. Natl. Acad. Sci. USA* 89:7836–39
26. Kessler DS, Veals SA, Fu X-Y, Levy DE. 1990. *Genes & Devel.* 4:1753–65
27. Schindler C, Shuai K, Prezioso VR, Darnell JE Jr. 1992. *Science* 257:809–15
28. Shuai K, Schindler C, Prezioso VR, Darnell JE Jr. 1992. *Science* 259:1808–12
29. Velazquez L, Fellous M, Stark GR, Pellegrini S. 1992. *Cell* 70:313–22
30. Watling D, Gushin D, Müller M, Silvennoinen O, Witthuhn BA, et al. 1993. *Nature* 366:166–70
31. Müller M, Briscoe J, Laxton C, Gushin D, Ziemiecki A, et al. 1993. *Nature* 366:129–35
32. Pellegrini S, John J, Shearer M, Kerr IM, S'ark GR. 1989. *Mol. Cell. Biol.* 9:460ɔ–12
33. McKendry R, John J, Flavell D, Muller M, Kerr IM, et al. 1991. *Proc. Natl. Acad. Sci. USA* 88:11455–459
34. Sadowski HB, Shuai K, Darnell JE Jr., Gilman MZ. 1993. *Science* 261:1739–44
35. Silvennoinen O, Schindler C, Schlessinger J, Levy DE. 1993. *Science* 261:1736–39
36. Ruff-Jamison S, Chen K, Cohen S. 1993. *Science* 261:1733–36
37. Larner AC, David M, Feldman GM, Igarashi K-I, Hackett RH, et al. 1993. *Science* 261:1730–33
38. Bonni A, Frank DA, Schindler C, Greenberg ME. 1993. *Science* 262:1575–79
39. Witthuhn BA, Quelle FW, Silvennoinen O, Yi T, Tang B, et al. 1993. *Cell* 74:227–36
40. Zhong Z, Wen Z, Darnell JE Jr. 1994. *Proc. Natl. Acad. Sci. USA* 91:4806–10
41. Akira S, Nishio Y, Inoue M, Wang X-J, Wei S, et al. 1994. *Cell* 77:63–71
42. Wakao H, Gouilleux F, Groner B. 1994. *EMBO J.* 13:2182–91
43. Witthuhn BA, Silvennoinen O, Miura O, Lai KS, Cwik C, et al. 1994. *Nature* 370:153–57
44. Kawamura M, McVicar DW, Johnston JA, Blake TB, Chen Y-Q, et al. 1994. *Proc. Natl. Acad. Sci. USA* 91:6374–78
45. Wegenka UM, Lütticken C, Buschmann J, Yuan J, Lottspeich F, et al. 1994. *Mol. Cell. Biol.* 14:3186–96

45a. Hou J, Schindler U, Henzel W, Ho T, Brasseur M, et al. 1994. *Science* 265: 1701–5
46. Bazan JF. 1990. *Proc. Natl. Acad. Sci., USA* 87:6934–38
47. Cosman D. 1993. *Cytokine* 5:95–106
48. Miyajima A, Kitamura T, Harada N, Yokota T, Arai K. 1992. *Annu. Rev. Immunol.* 10:295–331
49. Vignon I, Mornon J-P, Cocault L, Mitjavila M-T, Tambourin P, et al. 1992. *Proc. Natl. Acad. Sci. USA* 89:5640–44
50. Bartley TD, Bogenberger J, Hunt P, Li Y-S, Lu HS, et al. 1994. *Cell* 77:1117–24
51. deSauvage FJ, Hass PE, Spencer SD, Malloy BE, Gurney AL, et al. 1994. *Nature* 369:533–38
52. Chua A, Chizzonite R, Desai B, Truitt T, Nunes P, et al. 1994. *J. Immunol.* 153:128–36
53. Soh J, Donnelly RJ, Kotenko S, Mariano TM, Cook JR, et al. 1994. *Cell* 76:793–802
54. Hemmi S, Böhni R, Stark G, DiMarco F, Aguet M. 1994. *Cell* 76:803–10
55. Ho A, Liu Y, Khan T, Hsu D, Bazan J, et al. 1993. *Proc. Natl. Acad. Sci. USA* 90:11267–271
56. Novick D, Cohen B, Rubinstein M. 1994. *Cell* 77:391–400
57. Stahl N, Yancopoulos GD. 1993. *Cell* 74:587–90
58. Barbieri G, Velazquez L, Scrobogna M, Fellous M, Pellegrini S. 1994. *Eur. J. Biochem.* 223:427–35
59. Johnston J, Kawamura M, Kirken R, Chen Y-Q, Blake T, et al. 1994. *Nature* 370:151–153
60. Muller M, Laxton C, Briscoe J, Schindler C, Improta T, et al. 1993. *EMBO J.* 12:4221–28
61. Yamamoto K, Quell F, Thierfelder W, Kreider B, Gilbert D, et al. 1994. *Mol. Cell. Biol.* 14:4342–49
62. Kishimoto T, Taga T, Akira S. 1994. *Cell* 76:253–62
63. Ullrich A, Schlessinger J. 1990. *Cell* 61:203–12
64. deVos A, Ultsch M, Kossiakoff A. 1992. *Science* 255:306–12
65. Sachs L. 1993. *Blood Cells* 19:709–26
66. Goodall G, Bagley C, Vadas M, Lopez A. 1993. *Growth Factors* 8:87–97
67. Horie M, Broxmeyer H. 1993. *J. Biol. Chem.* 268:968–73
68. Kanakura Y, Drucker B, Cannistra S, Furukawa Y, Torimoto Y, et al. 1990. *Blood* 76:706–15
69. Okuda K, Sanghera JS, Pelech S, Kanakura Y, Hallek M, et al. 1992. *Blood* 79:2880–87
70. Gold M, Duronio V, Saxena S, Schrader J, Aebersold R. 1994. *J. Biol. Chem.* 269:5403–12
71. Satoh T, Nakafuku M, Miyajima A, Kaziro Y. 1991. *Proc. Natl. Acad. Sci. USA* 88:3314–18
72. Sakamaki K, Wang H-M, Miyajima I, Kitamura T, Todokoro K, et al. 1993. *J. Biol. Chem.* 268:15833–39
73. Hara T, Miyajima A. 1992. *EMBO J.* 11:1875–84
73a. Quelle F, Sato N, Witthuhn B, Inhorn R, Eder M, et al. 1994. *Mol. Cell. Biol.* 14:4335–41
74. Sato N, Sakamaki K, Terada N, Arai K, Miyajima A. 1993. *EMBO J.* 12: 4181–89
75. Silvennoinen O, Witthuhn BA, Quelle FW, Cleveland JL, Yi T, et al. 1993. *Proc. Natl. Acad. Sci. USA* 90:8429–33
76. Miura O, Cleveland J, Ihle J. 1993. *Molecular and Cellular Biology* 13: 1788–95
77. Quelle F, Quelle D, Wojchowski D. 1993. *J. Biol. Chem.* 268:17055–60
77a. Azam M, Erdjument-Bromage H, Kreider BL, Xia M, Quelle F, et al. 1995. *EMBO J.* In press
78. Giri J, Ahdieh M, Eisenman J, Shanebeck K, Grabstein K, et al. 1994. *EMBO J.* 13:2822–30
79. Noguchi M, Yi H, Rosenblatt H, Filipovich A, Adelstein S, et al. 1993. *Cell* 73:147–57
80. Schorle H, Holtschke T, Hünig T, Schimple T, Horak I. 1991. *Nature* 352: 621–24
81. Russell SM, Keegan AD, Harada N, Nakamura Y, Noguchi M, et al. 1993. *Science* 262:1880–83
82. Kondo M, Takeshita T, Ishii N, Nakamura M, Watanabe S, et al. 1993. *Science* 262:1874–77
83. Kondo M, Takeshita T, Higuchi M, Nakamura M, Sudo T, et al. 1994. *Science* 263:1453–54
84. Noguchi M, Nakamura Y, Russell SM, Ziegler SF, Tsang M, et al. 1993. *Science* 262:1877–80
85. Kawahara A, Minami Y, Taniguchi T. 1994. *Mol. Cell. Biol.* 14:5433–40
86. Nakamura Y, Russell S, Mess S, Friedmann M, Erdos M, et al. 1994. *Nature* 369:330–33
87. Nelson B, Lord J, Greenberg P. 1994. *Nature* 369:333–36
88. Tanaka N, Asao H, Ohbo K, Ishii N, Takeshita T, et al. 1994. *Proc. Natl. Acad. Sci. USA* 91:7271–75
89. Russell SM, Johnston JA, Noguchi M, Kawamura M, Bacon CM, et al. 1994. *Science* 266:1042–45
90. Keegan A, Nelms K, White M, Wang

L-M, Pierce J, et al. 1994. *Cell* 76:811–20

91. Punnonen J, Aversa G, Cocks B, McKenzie A, S. Menon, et al. 1993. *Proc. Natl. Acad. Sci. USA* 90:3730–3734.

92. Wang L-M, Myers M, Sun X-J, Aaronson S, White M, et al. 1993. *Science* 261:1591–94

93. Pernis A, Gupta S, Yopp J, Kashleva H, Schindler C, et al. 1994. Submitted

94. Schindler C, Kashleva H, Pernis A, Pine R, Rothman P. 1994. *EMBO J.* 13:1350–56

95. Kotanides H, Reich NC. 1993. *Science* 262:1265–67

96. Rothman P, Kreider B, Levy D, Wegenka U, Eilers A, et al. 1994. *Immunity* 1:457–68

97. Davis S, Aldrich TH, Stahl N, Pan L, Taga T, et al. 1993. *Science* 260:1805–08

98. Murakami M, Hibi M, Nakagawa N, Nakagawa T, Yasukawa K, et al. 1993. *Science* 260:1808–10

99. Lutticken C, Wgeneka UM, Yuan J, Buschmann J, Schindler C, et al. 1994. *Science* 263:89–62

100. Stahl N, Boulton TG, Farruggella T, Ip NY, Davis S, et al. 1994. *Science* 263:92–95

101. Yin T, Taga T, Tsang M, Yasukawa K, Kishimoto T, et al. 1993. *J. Immunol.* 151:2555–61

102. Thoma B, Bird T, Friend D, Gearing D, Dower S. 1994. *J. Biol. Chem.* 269:6215–22

103. Stahl N, Farruggella TJ, Boulton TG, Zhong Z, Darnell JE, et al. 1995. *Science*: In press

104. Fukunaga R, Ishizaka-Ikeda E, Pan C-X, Seto Y, Nagata S. 1991. *EMBO J.* 10:2855–65

105. Silvennoinen O, Ihle J, Schlessinger J, Levy DE. 1993b. *Nature* 366:583–85

106. Aguet M, Dembić Z, Merlin G. 1988. *Cell* 55:273–80

107. Cook JR, Jung V, Schwartz B, Wang P, Pestka S. 1992. *Proc. Natl. Acad. Sci. USA* 89:11317–321

108. Farrar MA, Campbell JD, Schreiber RD. 1992. *Proc. Natl. Acad. Sci. USA* 89:11706–10

109. Igarashi K, Garotta G, Ozmen L, Ziemiecki A, Wilks A, et al. 1994. *J. Biol. Chem.* 269:14333–36

110. Greenlund A, Farrar M, Viviano B, Schreiber R. 1994. *EMBO J.* 13:1591–1600

111. Colamonici O, Pfeffer L, D'Alessandro F, Platanias L, Gregory S, et al. 1992. *J. Immunol.* 148:2126–32

112. Benoit P, Maguire D, Plavec I, Kocher H, Tovey M, et al. 1993. *J. Immunol.* 150:707–16

113. Soh J, Mariano T, Lim J-K, Izotova L, Mirochnitchenko O, et al. 1994. *J. Biol. Chem.* In press

114. Constantinescu S, Croze E, Wang C, Murti A, Basu L, et al. 1994. *Proc. Natl. Acad. Sci. USA* 91:In press

115. Uzé G, Lutfalla G, Gresser I.1990. *Cell* 60:225–34

116. Colamonici O, Yan H, Domanski P, Handa R, Smalley D, et al. 1994. *Mol. Cell Biol.* 14:8133–42

116a. Raz R, Durbin JE, Levy DE. 1994. *J. Biol. Chem.* 269:24391–95

116b. Yan R, Quershi S, Zhong Z, Wen Z, Darnell JE. 1995. *Nuc. Acids Res.* In press

117. Binari R, Perriman N. 1994. *Genes & Devel.* 8:300–12

118. Koch CA, Anderson D, Morgan MF, Ellis C, Pawson T. 1991. *Science* 252:668–74

119. Fu X-Y. 1992. *Cell* 70:323–35

120. Shuai K, Ziemiecki A, Wilks AF, Harpur AG, Sadowski HB, et al. 1993. *Nature* 366:580–83

121. Improta T, Schindler C, Horvath CM, Kerr IM, Stark GR, et al. 1994. *Proc. Natl. Acad. Sci. USA* 91:4776–80

122. Overduin M, Rios CB, Mayer BJ, Baltimore D, Cowburn D. 1992. *Cell* 70:697–704

123a. Heim M, Kerr I, Stark G, Darnell JE. 1995. *Science*. In press

124. Pawson T, Schlessinger J. 1993. *Curr. Biol.* 3:434–41

125a. Gouilleux F, Wakao H, Mundt M, Groner B. 1994. *EMBO J.* 13:4361–69

126. Songyang Z, Shoelson SE, Chaudhuri M, Gish G, Pawson T, et al. 1993. *Cell* 72:767–78

127. Pine R, Canova A, Schindler C. 1994. *EMBO J.* 13:158–67

128. Sliva D, Wood T, Schindler C, Lobie P, Norstedt G. 1994. *J. Biol. Chem.* 269:26208–14

129. Finbloom DS, Petricon EF, Hackett RH, David M, Feldman GM, et al. 1994. *Mol. Cell. Biol.* 14:2113–18

129a. Lehmann J, Seegert D, Strehlow I, Schindler C, Lohmann-Matthes M, et al. 1994. *J. Immunol.* 153:165–72

129b. Eilers A, Baccarini M, Horn F, Hipskind RA, Schindler C, et al. 1994. *Mol. Cell. Biol.* 14:1364–73

130. Ruff-Jamison S, Zhong Z, Wen Z, Chen K, Darnell JE Jr., et al. 1994. *J. Biol. Chem.* 269:21933–21935

130a. Quershi SA, Salditt-Georgieff M, Darnell JE. 1995. *Proc. Natl. Acad. Sci. USA*. In press

131. Shuai K, Horvath CM, Tsai-Huang LH, Qureshi S, Cowburn D, et al. 1994. *Cell* 76:821–28

132. Wegenka UA, Buschmann J, Lutticken C, Heinrich P, Horn F. 1993. *Mol. Cell. Biol.* 13:276–88

133. Leung S, Qureshi SA, Kerr IM, Darnell JE Jr., Stark GR. 1994. *Mol. Cell. Biol.* In press

134. Pearse RN, Feinman R, Shuai K, Darnell JE, Ravetch JV. 1993. *Proc. Natl. Acad. Sci. USA* 90:4314–18

135. Wagner B, Hayes T, Hoban C, Cochran B. 1990. *EMBO J.* 9:4477–84

136. Yuan J, Wegenka UM, Hutticken C, Buschmann J, Decker T, et al. 1994. *Mol. Cell Biol.* 14:1657–68

137. Beadling C, Johnson KW, Smith KA. 1993. *Proc. Natl. Acad. Sci. USA* 90:2719–23

138. Köhler I, Rieber EP. 1993. *Eur. J. Immunol.* 23:3066–71

139. Rothman P, Li SC, Gorham B, Glimcher L, Alt F, et al. 1991. *Mol. Cell. Biol.* 11:5551–61

140. Gilmour KC, Reich NC. 1994. *Proc. Natl. Acad. Sci. USA* 91:6850–54

141. Caldenhoven E, Coffer P, Yuan J, Stolpe AVD, Horn F, et al. 1994. *J. Biol. Chem.* 269:21146–54

142. Johnston DS, Nusslein-Volhard C. 1992. *Cell* 68:201–19

143. Levy D, Darnell JE Jr. 1990. *New Biologist* 2:923–28

144. Boulton TG, Stahl NS, Yancopoulos GD. 1994. *J. Biol. Chem.* 269:11648–55

145. Sims SH, Cha Y, Romine MF, Gao P-Q, Gottlieb K, et al. 1993. *Mol. Cell. Biol.* 13:690–702

Annu. Rev. Biochem. 1995. 64:653–88

INTERFACIAL ENZYMOLOGY OF GLYCEROLIPID HYDROLASES: Lessons From Secreted Phospholipases A$_2$[1]

Michael H. Gelb

Departments of Chemistry and Biochemistry, University of Washington, Seattle, Washington 98195

Mahendra K. Jain

Department of Chemistry and Biochemistry, University of Delaware, Newark, Delaware 19716

Arthur M. Hanel

Departments of Chemistry and Biochemistry, University of Washington, Seattle, Washington 98195

Otto G. Berg

Department of Molecular Biology, Uppsala University Biomedical Center, Uppsala, Sweden

KEY WORDS: lipases, membranes, lipid-protein interactions, phospholipids, inhibitors, interfacial catalysis, interfacial equilibria, interfacial activation

CONTENTS

[1]We would like to dedicate this article to Professor GH de Haas.

0066-4154/95/0701-0653$05.00

ABSTRACT

Interfacial enzymes operate at an organized interface such as lipid aggregates in contact with the aqueous phase. The enzyme phospholipase A$_2$ is a well studied interfacial enzyme, and a discussion of its behavior at interfaces is the topic of this review. Knowledge gained from studies of phospholipases A$_2$ can be applied toward the quantitative analysis of other interfacial enzymes. The kinetic analysis of these enzymes is greatly simplified if one establishes certain experimental conditions that limit the exchange of enzyme and substrate between different substrate aggregates. With such constraints, the kinetics can be analyzed in terms of classical Michaelis-Menten theory adopted for the action of enzymes at interfaces. It is also possible to describe other enzyme properties such as inhibition and substrate preferences in a meaningful way using formalism that is well known in solution-phase enzymology.

PERSPECTIVES AND SUMMARY

What is interfacial enzymology? Operationally, it is defined as the action of an enzyme on substrates at an organized interface. The most prevalent biological interfaces are lipid aggregates in contact with an aqueous phase. This definition is aided by consideration of the following types of enzymes. Water-soluble enzymes that must associate with the interface to gain access to their substrates at the interface are interfacial enzymes. These enzymes are more difficult to characterize kinetically than are water-soluble enzymes, because interfacial catalysis is controlled not only by the usual variables such as the enzyme and substrate concentrations but also by the organization and dynamics of the interface where catalysis occurs. It is apparent that the reaction velocity of an interfacial enzyme is sensitive to the *two-dimensional surface concentration* of components at the interface, where the enzyme and substrate are concentrated into small regions of the total reaction volume. The fraction of enzyme bound at the interface depends, however, on the three-dimensional (bulk) concentration of the interface. Also, the residence time of the enzyme at the interface determines the *processivity* of the reaction, which refers to the ability of an enzyme to catalyze more than one turnover cycle without leaving the interface. Finally, interfacial enzymes can be described by steady-state rate equations (Michaelis-Menten approach) that are modified to take into account a reduction in dimensionality. In this review, such concepts are explored in detail.

Enzymes that permanently reside within or on the membrane and operate on substrates at the interface are also interfacial enzymes. Integral and peripheral membrane enzymes that catalyze the transformation of substrates that exist in the aqueous phase are not interfacial enzymes, although the possibility that the enzyme at the interface has different kinetic parameters than those in the aqueous phase should be considered. For example, CTP:phosphocholine cytidyltransferase, which acts on water-soluble substrates, is activated by interaction with membranes (1). Water-soluble enzymes that operate on substrates in the aqueous phase regardless of whether the substrate partitions between the two phases or exists only in the aqueous phase are not interfacial enzymes. Integral membrane enzymes that operate on substrates that are completely within the membrane phase appear to be rare, as most membrane-residing substrates are sufficiently amphiphilic to be present at the interface.

Interfacial catalysis is an important area of enzymology, as many interfacial enzymes have been described. One of the purposes of this review is to summarize how interfacial enzymes work, with a focus on the thinking that goes into the development of experimental concepts and strategies for the proper quantitative analysis of interfacial catalysis in general. Such studies form the basis for understanding the important features of interfacial catalysts, including interfacial activation, substrate preferences, and inhibition. The main focus in this review is on a family of 14-kDa secreted phospholipases A_2 (sPLA$_2$), because these are probably the best characterized interfacial enzymes.

THE KINETIC PARADIGM

Binding of substrate followed by catalysis is the key feature of catalytic turnover by all enzymes. This forms the basis for the well-established Michaelis-Menten formalism for the analysis of solution enzymology. Interpretation of kinetics of action of enzymes on one-dimensional polymeric substrates such

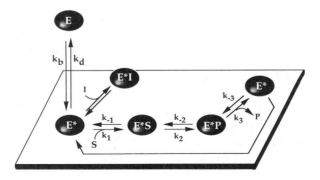

Figure 1 Michaelis-Menten kinetic scheme for interfacial catalysis. All enzyme species at the interface are shown in the box and are designated with an asterisk. *S, P,* and *I* designate substrate, products, and competitive inhibitor, respectively. The enzyme in the aqueous phase is designated *E.*

as DNA or RNA (e.g. see 2, 3) or on two-dimensional surfaces such as membranes or micelles (e.g. see 4) requires consideration of additional kinetic and structural data. The minimum modification of the Michaelis-Menten formalism that appears to be adequate for analyzing most of the features of interfacial catalysis is the incorporation of an additional equilibrium that defines the relative fraction of the enzyme in the aqueous phase versus at the interface (Figure 1). The enzyme in the aqueous phase (E) binds to the interface to give the enzyme at the interface (E^*). E^* can then bind a substrate molecule in its catalytic site to give the Michaelis complex E^*S, which undergoes chemical transformation to the enzyme/product(s) complex (E^*P). After product dissociation, either E^* can remain bound to the interface and catalyze another reaction or it can dissociate into the aqueous phase and then rebind in order to participate in further turnover. Thus, the values of the rate and equilibrium constants that define the E to E^* step determine the processivity of the enzyme at the interface. Two extreme modes are possible; the lowest degree of processivity is one in which the enzyme desorbs from the interface after each catalytic cycle (hopping mode), and the highest degree of processivity occurs when the enzyme leaves the interface—if at all—only after all of the substrate that the bound enzyme encounters at the interface becomes hydrolyzed (scooting mode). Based on this minimal kinetic picture, it is possible to make inferences about: (*a*) the structural motifs on the enzyme that recognize the interface; (*b*) the structural features of the enzyme that control binding and hydrolysis at the catalytic site; and (*c*) the effect of the change in the dimensionality on the kinetic parameters. Such factors are considered in this review.

The observed rate of most interfacial enzymes depends on the nature of the interface. The question of the biological relevance of such polymorphs as substrates is open ended. It may be emphasized at the outset, however, that although catalysis occurs at the interface, such effects are not necessarily due to a change in the intrinsic primary catalytic parameters of the enzyme (for which there is little evidence yet). Thus far, it appears that the organization and dynamics of the substrate interface (often called the "quality of the interface") control the E to E^* equilibrium, the size of the aggregate and the mole fraction of the substrate at the interface, and the effective exchange rates of the components (enzyme, substrate, product) between the coexisting interfaces. Complexities in the observed steady-state reaction progress appear if the E to E^* equilibrium and/or the exchange of substrate(s) and removal of product(s) between enzyme-containing aggregates and excess aggregates change with the reaction progress. Evaluation of the contributions of such factors requires a deeper understanding of the organization and dynamics of the interface (briefly reviewed below). A lack of appreciation of the role of such "physical" factors has caused considerable confusion in the interpretation of the kinetics of interfacial enzymes.

CRITERIA FOR INTERFACIAL CATALYSIS

There are probably at least several hundred interfacial enzymes, as they include most enzymes involved in lipid metabolism. Evidence for interfacial catalysis of the type discussed in the previous section has evolved largely from the study of lipolytic enzymes, and it is based on arguments along the following lines (reviewed in 4): (*a*) The concentration of solitary naturally occurring glycerolipids in the aqueous phase is very low (<100 pM), and their rate of desorption from the interface into the aqueous phase is very slow. Therefore, water-soluble interfacial enzymes must associate with aggregates to gain access to their substrates. (*b*) The apparent rate of hydrolysis of lipid dispersions is significantly altered by the presence of amphiphiles (other lipids, detergents, or organic solutes) at the interface (see e.g. 5, 6). (*c*) Other factors such as thermotropic phase properties of lipids that modulate the organization, packing, and dynamics of the interface also modulate the observed catalytic rate (7). For similar reasons, the surface pressure of lipid monolayers has a significant effect on the rate of interfacial catalysis (8). (*d*) A suggestive, but not unequivocal, demonstration of the interfacial nature of catalysis comes from the observation of an increase in the reaction rate when the substrate concentration exceeds its critical micelle concentration. For example, the rates of sPLA$_2$- and lipase-catalyzed hydrolysis of short-chain phospholipids and triglycerides, respectively, dispersed as solitary monomers in aqueous solution, are often quite low, and the rates can increase by orders of magnitude when the concentration of substrate is high enough to cause aggregation (9, 10). (e) Finally, as elaborated in this review, a proof for the interfacial nature of enzyme reactions comes from elements of two-dimensional processivity (11).

ORGANIZATION AND DYNAMICS OF GLYCEROLIPID AGGREGATES IN AQUEOUS DISPERSIONS

To understand interfacial catalysis, one needs to appreciate the organization and dynamics of lipid aggregates (summarized in Table 1) (12, 13). Properties of aqueous dispersions of lipids are best understood in terms of their amphiphilic character. Such dispersions exhibit polymorphisms determined by the effective relative sizes of the polar head group and the apolar hydrocarbon chains. The size, shape, and morphology of the aggregate and the dynamics of amphiphiles depend on not only the concentration, composition, and structures of the amphiphiles but also environmental factors such as temperature, pH, and ionic strength.

Aqueous dispersions of naturally occurring phospholipids (cylindrical-shaped) form bilayer-enclosed vesicles, and the properties of these aggregates do not change with their bulk concentration. Cone-shaped amphiphiles such as fatty acids, lysophospholipids, and short-chain synthetic phospholipids such as dioctanoyl phosphatidylcholine (PC) form micelles. Interestingly, bilayered vesicles are formed when fatty acids and lysophospholipids are added together in a 1:1 mixture (14). Addition of a micelle-forming detergent above a certain mole fraction to vesicles of long-chain phospholipids leads to the destruction of vesicles and the formation of mixed micelles. Other polymorphs have been described such as rods, discs, lipoprotein particles, reverse micelles, and emulsions (13). All of these aggregates are held together by noncovalent forces, and thus other amphiphiles can be incorporated with or without disrupting the overall morphology.

The important differences between micelles and vesicles are as follows (Table 1): Small vesicles contain from a few to several thousand phospholipids, whereas micelles are much smaller (typically 10–100 amphiphiles). The concentration of solitary phospholipids in the aqueous phase in equilibrium with vesicles is very low (<100 pM), compared to μM to mM amounts for micelles. The lateral diffusion coefficient for phospholipids in bilayers is about 10^{-9} cm^2 s^{-1}, which means that the rate of lateral diffusion–limited access of substrate to a membrane-localized enzyme would be >200,000 s^{-1} (surface fraction)$^{-1}$ (4), i.e. in order for the enzymatic turnover to be limited by lateral encounter with substrate, the turnover number must exceed this value. In vesicles the rates of the fusion without an inducing agent or intervesicle exchange and transbilayer flip-flop of phospholipids are negligibly small (half-times > 10 hours). With mixed micelles, transfer of long-chain phospholipids between

Table 1 Structures and dynamics of phospholipid aggregates[a]

Property	Bilayered vesicle	Micelle	Mixed micelle	Monolayer
Aggregation number	>6000	20–200	20–200	>10^{12}/cm^2
Solubility of monomers	<100 pM	μM–mM		<100 pM
Approximate number of carbon atoms in amphiphile chain	2 chains, >12 per chain	2 chains of 6–8 per chain, or 1 chain of 8		2 chains with total number >20, or 1 chain of >18
Diameter	>200 Å	~30 Å	~30 Å	Dimensions of trough
Half-times for exchange	>10 h for fusion (without fusogen), flip-flop, and intervesicle exchange	10^{-1}–10^{-4} sec for intermicelle exchange	10^{-1}–10^{-4} sec for intermicelle exchange of detergent, 0.1–10 sec for intermicelle exchange of phospholipid	

[a] References for listed properties are given in the text.

micelles occurs on a time-scale of seconds (15), even though half-times for intermicelle exchange of detergent and short-chain amphiphiles is of the order of 10^{-4} s (16). Since long-chain phospholipids do not desorb from micelles into the aqueous phase, intermicelle exchange requires collision-dependent fusion of micelles to give transient particles in which phospholipid exchange occurs followed by fission to re-form stable mixed micelles. The importance of such differences between vesicles and micelles to the analysis of interfacial enzymes is discussed below.

Most amphiphiles form monolayers at the air-water interface. Monolayers offer the ability to vary the surface pressure and thus the surface density of phospholipids. With medium chain–length amphiphiles, the rate of reactions catalyzed by interfacial enzyme can be monitored as the products leave the interface (8). The monolayer system is problematic in some respects, however. Not only is it difficult to establish the fraction of total enzyme at the interface, which is expected to be small because of the relatively large volume of the aqueous phase compared to the surface area of the interface, but the adsorption of the enzyme on the walls of the trough can be significant. Thus, the effective rate of catalytic turnover in monolayers is orders of magnitude smaller than that at other interfaces. Although the intrinsic rate of exchange of solute between the phospholipid monolayer and the aqueous solvent can be quite fast, the overall transfer of solute or enzyme added from the bulk aqueous phase to the monolayer requires minutes because of the presence of an unstirred solvent layer (about 1000 nm thick) in contact with the monolayer (4). Depending on the trough geometry, such factors can cause a lag of several minutes in the onset of lipolysis. It is also difficult to evaluate the effect of surface pressure on the partitioning and kinetics of the various processes that control the distribution of not only the enzyme at steady state but also the products.

CONCEPTS OF INTERFACIAL CATALYSIS

Interfacial Versus Catalytic Site Binding

According to Figure 1, the binding of the enzyme to the interface and the binding of a single substrate in the catalytic site are distinct processes. This implies that the enzyme has a surface that contacts the interface that is topologically distinct from the catalytic site. Experimental proof for this two-step binding is presented below in the case of sPLA$_2$. Such considerations lead to the concept of a *neutral diluent,* which is defined as an amphiphile that has no affinity for the catalytic site of the enzyme but forms an organized interface to which the enzyme can bind. Thus the enzyme bound to an interface of neutral diluent has its catalytic site filled with solvent. Neutral diluents are powerful tools for analyzing interfacial enzymes as will become apparent in discussions below.

Bulk Versus Surface Concentration of Substrate

It is apparent from Figure 1 that the bulk substrate concentration (moles per unit volume) controls the E to E^* equilibrium, whereas the surface concentration of substrate (moles per unit surface area of interface) controls the E^* to E^*S equilibrium. Equilibrium constants describing these two processes cannot be compared because they have different units; they may be normalized, however, by use of a thermodynamic box as is described later. For aggregated substrates, the bulk concentration can be expressed in terms of moles of substrate at the interface per unit volume or in terms of moles of aggregate per unit volume if the aggregation number of the particle is known. For the former, only those substrates at the interface that the enzyme "sees" should be considered. For example, in a bilayer vesicle, the enzyme may never see the substrate on the inner monolayer, and even less of the interface is seen by the enzyme if the vesicles are multilamellar.

In contrast to reactions in solution, interfacial reactions do not take place in dilute solution. As a consequence, non-ideal effects may play a role such that the dissociation constant for a complex of E^* with a catalytic-site ligand could change as the composition of the aggregate changes during the course of the reaction. What is a useful measure of concentration in the interface is determined primarily by the entropy of interface mixing. For an ideal solution this would imply that the mole fraction is the best choice for all concentrations. Although the interface is probably not an ideal solution, this seems to be the most useful alternative. When constituents of the interface are similar in size, the mole fraction in the interface is equal to the occupied surface fraction.

The Environment of All Enzymes Must Be Considered: Global Versus Local Steady States

An experimentalist is almost always forced to make kinetic measurements with a large number of enzyme molecules (ensemble) because of the limited sensitivities of the assays (there are rare exceptions such as the measured conductance of a single ion channel in a patch clamp experiment). In solution enzymology, it is axiomatic that all enzymes in the ensemble see the same average solution composition. This is a key assumption in all steady-state kinetic modeling that permits kinetic expressions in terms of the bulk substrate, product, inhibitor, and activator concentrations. This assumption is valid because in solution enzymology the mixing of components by Brownian motion throughout the bulk solution is usually fast on the time-scale over which a substantial change in the concentrations of components in the vicinity of an enzyme molecule occurs. If this were not the case, there would be regions of solution in which the concentrations of components were different, and enzymes in each region would respond to, for example, different extents of substrate depletion and product inhibition. Since only total concentrations of species in solution are measured, it would not be easy to extract the fundamental quantities that describe the interaction of the enzyme with substrates (K_M), products (K_P), or inhibitors (K_I), as well as the underlying rate constants.

In interfacial enzymology, all enzymes may not necessarily be in the same average environment. In situations in which the enzyme partitions between the interface and the aqueous phase, one must consider only the fraction of enzyme at the interface where catalysis occurs. Enzyme bound to a phospholipid vesicle may remain there for some time and then dissociate and rebind to a new interface. The compositions of vesicles that have had or have bound enzyme and those that have never made contact with enzyme will be different, and thus the vesicles provide different environments in which enzymes exist. Thus, it is necessary to consider the local steady state that describes the behavior of the enzyme acting in a single substrate aggregate. In contrast, the global steady state includes ensemble-averaged rate constants for the exchange of enzymes, substrates, and products between the different aggregates that exist in the entire reaction mixture. Interfacial enzymatic processes are too complex to analyze if the local and global steady states are convoluted with each other. In the next section strategies are described for constraining the system so that the primary rate and equilibrium parameters shown in the box in Figure 1 can be extracted. After all, these quantities describe the action of the enzyme at the interface where the lipolysis occurs and relate to the structural features of the substrate and enzyme.

ANALYSIS OF INTERFACIAL ENZYMES IN THE SCOOTING MODE

Coalescence of Local and Global Steady States

To extract the primary rate and equilibrium constants that describe the action of an interfacial enzyme at the interface, it is necessary to establish four experimental constraints on the system. The most important constraint is that the enzyme must remain bound to the vesicle interface during the time-course of the interfacial reaction. More precisely, the residence time of the enzyme at the interface must be longer than the period over which the reaction is analyzed. For example, if the entire reaction progress is to be monitored, the enzyme should not leave the interface of a vesicle before all of the available substrate (in the outer layer of the vesicle) has been hydrolyzed. Such virtually infinitely processive behavior has been termed interfacial catalysis in the *scooting mode* (11). The second requirement is that the substrates and products must not undergo intervesicle exchange during the analysis (see below, however). Thus, vesicles must remain intact after their outer-layer phospholipids have been hydrolyzed, which in fact occurs in sPLA$_2$-catalyzed reactions (11). This is probably the result of the fact that equimolar dispersions of the products of sPLA$_2$ hydrolysis, fatty acid and lysophospholipid, form bilayered vesicles (14). Even if this were not the case, however, one could use a system composed of an enzyme-resistant, vesicle-forming amphiphile, such as a diether phospholipid, containing a small mole fraction of hydrolyzable substrate. Thus, this required constraint on the vesicle system is achievable with virtually any interfacial enzyme. Two additional constraints are that the vesicles must have a narrow size dispersity and that the vesicle to enzyme ratio must be greater than 5 or so (see below, however); this latter requirement ensures that enzyme-containing vesicles contain at most one enzyme.

Consider the simplification of the system if all of the above constraints are imposed. As time progresses, all enzyme-containing vesicles will behave identically. When one such vesicle is say 50% hydrolyzed and the enzyme on this vesicle is experiencing a certain degree of substrate depletion and product inhibition, an enzyme in another vesicle is experiencing the same environment. This is true only if the vesicles have the same size, because for a given amount of lipolysis, the mole fraction of components depends on the number of substrates per vesicle. Experimentally, one determines the total product formed from all of the enzyme-containing vesicles, but this is equivalent to determining the product formed in each vesicle. In short, within these constraints in the scooting mode, the distinction between local and global steady states is no longer operative (18).

Not all of these requirements need to be met if only the initial reaction velocity is sought. Such information is useful in determining the substrate preferences of an interfacial enzyme or in the analysis of inhibitors as is discussed below. Thus, if the reaction is measured over a short time period such that the mole fraction of substrate in the vesicles remains close to its initial value (unity for a vesicle composed purely of substrate), it is not required that there be only one enzyme per vesicle or that the vesicles be of the same size. If there is adequate substrate in the aggregate such that intervesicle substrate exchange is not required to maintain the substrate mole fraction at a nearly constant value, then such exchange is of no consequence. If substrate and product exchange is required to maintain a constant substrate level, however, then the rate of such exchange must be faster than the rate of substrate depletion by the enzyme; otherwise the observed kinetics will reflect the substrate exchange rather than the catalytic properties of the enzyme. As is discussed later, such a concern about substrate replenishment is critical when studying the action of interfacial enzymes on small substrate aggregates such as micelles, where substrate depletion occurs rapidly.

Consider the complexity of the system if the enzyme hops from one vesicle to another. At some point in time an enzyme may leave a vesicle and bind to one that has not seen an enzyme previously. This enzyme experiences no substrate depletion or product inhibition; it is effectively at time zero. Thus intervesicle exchange of enzyme leads to time scrambling, which makes kinetic analyses difficult to say the least. Of course the same problem arises if either the substrates or products undergo inter-aggregate exchange, except as noted above. Additional complexities arise if the enzyme is not tightly bound to the interface. The fraction of enzyme at the interface may depend on the composition of the vesicles, and since the composition is altered by the action of the enzyme there may be a time-dependent shift in the E to E^* equilibrium with an accompanying change in the steady-state reaction velocity. This is never a concern, of course, in solution enzymology.

It is also possible to imagine a situation in which the enzyme displays low processivity and thus leaves the vesicle after a few turnovers. In this way all of the vesicles become hydrolyzed in a synchronous fashion, avoiding time scrambling. Such a mode is discernable if the reaction progress curve shows no sign of substrate depletion under conditions in which only the enzyme is undergoing intervesicle exchange. Analysis of sPLA$_2$s in this fast-hopping limit leads to other problems, however, as is discussed later.

The constraints on the system summarized in this section may seem a bit artificial. Does the enzyme scoot in an in vivo setting or only in the hands of the interfacial enzymologist? The point is really that the lipolysis reaction occurs at the interface by the enzyme at the interface. The primary rate and equilibrium parameters (steps within the box of Figure 1) obtained experimentally by constraining the system are just as applicable to the action of the enzyme under conditions in which it spends less time at the interface, as long as the enzyme remains at the interface long enough for E^*, E^*S, and E^*P to reach a steady state (which is typically msec for most enzymes). It will become apparent in this review that the quantitative analysis of, for example, substrate preferences and inhibitors of interfacial enzymes in the scooting mode yields quantities that are applicable to the action of the enzyme under any degree of processivity.

Interfacial Steady-State Rate Equations

The initial velocity per enzyme, v_0, for the hydrolysis of vesicles in the scooting mode is given by Equation 1 (19).

$$v_0 = \frac{k_{cat}}{(X_s + K_M^*)} \qquad\qquad 1.$$

This is the familiar Michaelis-Menten rate equation except that the Michaelis constant, K_M^*, and substrate concentration, X_S, are in units of mole fraction;

thus, K_M^* is the mole fraction of substrate at the interface that gives the half-maximal reaction velocity. Since X_S cannot exceed unity, it is possible to imagine an interfacial enzyme for which $K_M^* > 1$, and thus E^* is never saturated with substrate.

The entire reaction progress curve for the hydrolysis of a monodisperse population of vesicles under the scooting mode condition and with at most one enzyme per vesicle is given by the integrated Michaelis-Menten equation (19):

$$k_i = -\ln\left(1 - \frac{P_t}{P_{max}}\right) + \left(k_i \frac{N_s}{v_0} - 1\right)\frac{P_t}{P_{max}} \qquad 2.$$

Here, P_t is the product formed at time t, and P_{max} is the maximum possible product formed in the scooting mode, which is simply the number of enzymes times the number of substrates in the outer monolayer of a vesicle, N_S. The constant k_i is given by Equation 3:

$$k_i N_s = \frac{k_{cat}}{K_M^*\left(1 + \frac{1}{K_P^*}\right)} \qquad 3.$$

Here, K_P^* is the interfacial dissociation constant for the E^*P complex in units of mole fraction. Thus, when the mole fraction of P in the interface is K_P^*, the ratio E^*/E^*P is unity. The reaction progress curve as dictated by Equation 2 has a mixture of first-order and zero-order components. Depending on the kinetic parameters and the size of the vesicles (N_S), one of these components may dominate the observed kinetics. With small vesicles, the mole fraction of substrate drops more quickly than with larger vesicles, and the progress curve takes on a first-order appearance earlier in time (velocity decreasing at each point in time). With large vesicles, the reaction velocity remains approximately constant for several minutes (zero-order kinetics). This is analogous to solution enzymology, where the shape of the reaction progress curve depends on the total amount of substrate.

Neutral Diluents and Interfacial Equilibria

A complete analysis of an interfacial enzyme requires the determinations of a number of equilibrium constants that describe the interaction of the enzyme with ligands. Many of the constants describe processes that occur at the interface (for example $E^* + S \leftrightarrow E^*S$). Thus, one needs a neutral diluent (defined above) to provide a matrix in which E^* exists. Ligands (L) that bind to E^* can be added, and the amount of E^*L complex can be detected by spectroscopic or chemical means (examples given below for sPLA$_2$). From an analysis of the E^*/E^*L ratio versus the mole fraction of L at the interface of a neutral diluent, one can obtain the equilibrium dissociation constant for the

E^*L complex, K_L^*, in units of mole fraction. Since a neutral diluent, by definition, has no affinity for the catalytic site of E^*, L and neutral diluents do not compete with each other for the binding to E^*, thus K_L^* depends, to a first approximation, only on the structure of L. It should be mentioned, however, that K_L^* is not always completely independent of the structure of the neutral diluent; K_L^* depends not only on the free energy of interaction of the ligand with E^* but also on the free energy of interaction of the free L with its environment (in this case the neutral diluent aggregate), and this latter interaction can, in general, depend on the structure of the neutral diluent (20).

Competitive Inhibition of Interfacial Enzymes

Catalytic site-directed inhibitors of interfacial enzymes can function in ways analogous to those observed with soluble enzymes, i.e. by forming dead-end enzyme-inhibitor complexes. A competitive inhibitor is one that binds to the catalytic site of E^* and prevents substrate binding. Its effect on the initial reaction velocity in the scooting mode is described by Equation 4:

$$\frac{v_0^0}{v_0^I} = 1 + [(1 + 1/K_I^*)/(1 + 1/K_M^*)][X_I/(1 - X_I)] \qquad 4.$$

Here, v_0^0 and v_0^I are the initial velocities measured in the absence and presence of a competitive inhibitor, respectively, X_I is the mole fraction of inhibitor at the interface, and K_I^* is the equilibrium dissociation constant for the E^*I complex (in units of mole fraction). Equation 4 is the standard equation for competitive inhibition adapted for interfacial catalysis in the scooting mode.

LOW-MOLECULAR-WEIGHT SECRETED PLA₂s

For more than four decades sPLA$_2$s ($M_r \approx 14,000$) enzymes have served as prototypes for the study of interfacial catalysis (5, 8, 9, 18, 21). This class of water-soluble, calcium-dependent enzymes associates with organized interfaces and hydrolyzes the sn-2 ester of glycerophospholipids to produce a fatty acid and a lysophospholipid (22). More than 100 evolutionarily related enzymes (23) have been isolated from pancreatic juice, venoms (snake, insect, lizard), inflammatory exudates, specific tissues, and blood cells. Amino acid sequences are available for most of these (24, 25).

The roles of sPLA$_2$s in signal transduction and inflammatory processes have been extensively investigated (for reviews from 1990 onward, see 26–45). Previous reviews on the in vitro properties of purified sPLA$_2$s include those on structure/function (9, 21, 25, 27, 46–49), interfacial recognition (50, 51), interfacial kinetics (4, 18, 52–56), and inhibition (32, 57–61). Elements of these reviews, structural studies, and more recent data that relate to the fundamental concepts of interfacial catalysis described in the first part of the present review are summarized in this section.

Molecular Structure

High-resolution X-ray structures are available for several sPLA$_2$s, both in free form and complexed to catalytic-site inhibitors (62–72). The molecular structures of sPLA$_2$s give a clear picture of why binding of these enzymes to the interface and loading of their catalytic sites with a single phospholipid molecule are distinct processes. The enzyme contains a catalytic-site slot that extends all the way through the globular structure (Figure 2). The phospholipid substrate must travel about 15 Å into the slot to reach the catalytic network. The substrate's acyl chains contact hydrophobic amino acids that line the walls of the slot, some of which have been shown by mutagenesis to be important for substrate binding (73), and the polar head group protrudes out of the far end of the slot and into solvent. It is thought that the collar on the surface of the protein that surrounds the slot opening functions as the interface binding surface (i-face). Spectroscopic studies show that tryptophan-3 of porcine pancreatic sPLA$_2$ (pp-PLA$_2$), which lies on the i-face, makes direct contact with the interface, and is not in contact with bulk solvent when the enzyme is bound at the interface (74–79). Thus, the enzyme probably forms a water-tight seal with the interface, which presumably allows a single phospholipid molecule to dislodge from the plane of the bilayer and travel into the enzyme's catalytic-site slot. In this way the substrate alkyl chains never come in contact with bulk solvent, and this presumably facilitates the movement of substrate (50, 66). Also, the residues of the i-face are in an ensemble of conformations (80), and a conformational ensemble may be effective in accommodating diverse molecular structural and organizational features of the interfaces that sPLA$_2$s may encounter.

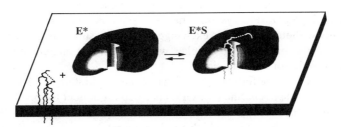

Figure 2 Schematic drawing of sPLA2 bound at the interface. The polar head group of a phospholipid substrate at the interface must break some of its interactions with neighboring phospholipids (dotted line) before dislodging from the interface to enter the catalytic slot of E^*. The polar head group of the enzyme-bound substrate may engage in interactions with the enzyme (dotted line).

These structures suggest that phospholipids that cannot dislodge from the bilayer will not be good substrates for sPLA$_2$s. Indeed, topologically restricted or polymerized phospholipids are poor sPLA$_2$ substrates (81–83).

Interestingly, the recently solved X-ray structure of prostaglandin H$_2$ synthase-1 reveals an architecture that is grossly reminiscent of that of sPLA$_2$s (84). The synthase has a membrane-binding appendage that is at the opening of a long channel leading to the catalytic metallo-porphoryin cofactor. The arachidonate substrate presumably dislodges from the membrane and travels down the enzyme's channel.

The X-ray structures of numerous sPLA$_2$s reveal that the catalytic machinery is conserved. The catalytic site contains an aspartate-histidine-water catalytic triad that is analogous to the triad found in serine proteases except that the serine is replaced by a water molecule hydrogen bonded to the imidazole ring. It has been proposed that this water is the attacking nucleophile and that the tetrahedral intermediate formed from the substrate's sn-2 ester is liganded directly to calcium (9, 66). The functions of the conserved catalytic-site residues have been probed by mutagenesis (78, 85–90).

Catalytic Parameters of pp-PLA$_2$ in the Scooting Mode

In 1986 it was shown that pp-PLA$_2$ catalyzes the hydrolysis of vesicles of the anionic phospholipid 1,2-dimyristoyl phosphatidylmethanol (DMPM) in the highly processive scooting mode (11). Thus, in the presence of excess vesicles over enzymes, only a fraction of vesicles are hydrolyzed because the enzyme cannot hop from one vesicle to another. Also, enzyme bound to vesicles of the anionic diether phospholipid 1,2-ditetradecyl phosphatidylmethanol does not hydrolyze vesicles of DMPM added subsequently. This work set the stage for a detailed kinetic analysis of interfacial catalysis according to the scheme in Figure 1 (19, 91–96). All of the constraints discussed above that are necessary for interpreting the entire reaction progress curve have been established with the pp-PLA$_2$/DMPM system. Thus, it is possible to describe the reaction analytically without intervesicle exchange of enzyme, substrates, and products, without collapse or fusion of vesicles, with at most one enzyme per vesicle, and with vesicles of uniform size.

It was shown by the following criteria that the amphiphile rac-2-hexadecyl-sn-glycero-3-phosphocholine (2H-GPC) is a neutral diluent for pp-PLA$_2$ (93). Fluorescence spectroscopy was used to show that the enzyme binds to 2H-GPC micelles. Phenacyl bromides, which alkylate the catalytic-site histidine residue of pp-PLA$_2$ (97), inactivate the enzyme in aqueous solution or bound to 2H-GPC micelles with the same kinetics, indicating that 2H-GPC does not occupy the catalytic site. The binding of catalytic-site ligands causes a perturbation of the UV spectrum of tryptophan-3 (9, 80), and such a spectral change is not seen with enzyme bound to 2H-GPC micelles. Thus, E, E^*, and E^*L can be characterized unequivocally (80).

The rate and equilibrium constants that describe the action of pp-PLA$_2$ on DMPM vesicles are listed in Figure 1. Before discussing the values of the interfacial constants, we summarize briefly how these were obtained, as many of the strategies differ from those used conventionally to analyze enzymes in aqueous solution (see 18, 19 for detailed descriptions). Fitting the entire reaction progress curve to the integrated Michaelis-Menten equation (Equation 2) gives $v_0 = 320$ s^{-1} and $k_i N_S = 35$ s^{-1}. The equilibrium dissociation constant for the complex E^*P, $K_P^* = 0.025$ mole fraction, was obtained by measuring the protection of the catalytic site from alkylation by phenacylbromide afforded by adding products (1:1 mixture of fatty acid and lysophospholipid) to enzyme bound to micelles of the neutral diluent 2H-GPC; when the mole fraction of products at the interface equals K_P^*, the half-time for enzyme alkylation is doubled. Through use of the same approach, K_I^* values for several catalytic-site competitive reversible inhibitors were determined.

Addition of inhibitors to DMPM vesicles decreases the initial reaction velocity, v_0, which depends on K_M^*, K_I^*, and X_I according to Equation 4. Since K_I^* and X_I are known, the values of $K_M^* = 0.26 - 0.68$ mole fraction were obtained by using several different inhibitors (93, 98). An independent evaluation of K_M^* comes from the addition of 2H-GPC to DMPM vesicles. The effect of this neutral diluent is only to lower the mole fraction of DMPM at the interface, and this causes a decrease in v_0 according to the Michaelis-Menten equation (Equation 1, surface dilution). Only concentrations of 2H-GPC up to 30 mole % can be used, since higher concentrations lead to the disruption of vesicles and the formation of mixed micelles (93). This analysis yields a value of $K_M^* = 0.3$ mole fraction, which is in good agreement with the value obtained with inhibitors. Still another independent estimate of K_M^* comes from the equation $k_i N_S / v_0 = (1 + 1/K_M^*)/(1 + 1/K_P^*)$ (19). The experimental values $k_i N_S = 35$ s^{-1}, $v_0 = 320$ s^{-1}, and $K_P^* = 0.025$ mole fraction yield a value of $K_M^* = 0.29$. With experimental values of $v_0 = 320$ s^{-1} and $K_M^* = 0.3$, $k_{cat} = 400$ s^{-1} according to Equation 1.

Hydrolysis of DMPM vesicles by pp-PLA$_2$ in oxygen-18 water gives myristate with only a single oxygen-18 (92). A lack of doubly labeled fatty acid is also seen in the hydrolysis of phospholipid/Triton X-100 mixed micelles by cobra venom sPLA$_2$ (99). Thus, once the sn-2 ester is hydrolyzed, the fatty acid and lysophospholipid products are released without resynthesis of the sn-2 ester. This establishes that $k_2 \gg k_{-2}$ and $k_3 \gg k_{-2}$.

Substitution of the sn-2 carbonyl carbon with ^{14}C results in a significant primary kinetic isotope effect of 1.12 ± 0.02 for the hydrolysis of phospholipid/Triton X-100 mixed micelles by cobra venom sPLA$_2$, whereas the isotope effect vanishes during the hydrolysis of radiolabeled DMPM vesicles by cobra venom and pp-PLA$_2$ (92). This result shows that the ratio of rate constants k_2/k_{-1}, also known as the forward commitment to catalysis (100, 101), is >> 1. Thus the general expression for $k_{cat}/K_M^* = k_1k_2/(k_{-1} + k_2)$ becomes approximately $k_{cat}/K_M^* = k_1$, which has a value of 1350 s^{-1} using the values of k_{cat} and K_M^* determined previously. The rate constant for the dissociation of substrate from the E^*S complex is not known accurately. An estimate comes from the value of the equilibrium dissociation constant for the complex of E^* with the diether phospholipid substrate analog 1,2-ditetradecyl phosphatidylmethanol, $K_S^* = 0.025$, obtained by the protection from alkylation method (93). Thus, since $K_S^* = k_{-1}/k_1$, an approximate value of $k_{-1} = 35$ s^{-1} is obtained.

According to Figure 1, $k_{cat} = k_2/(1 + k_2/k_3)$. Thus if the chemical step is rate limiting, k_{cat} approaches k_2, whereas if product release is rate limiting, k_{cat} approaches k_3. To investigate this issue, the hydrolysis of sn-2 thiolester phospholipid analogs were studied. It was anticipated that the substitution of sulfur for oxygen at the sn-2 position would alter k_2 more than k_3, and the effect of this element substitution on the overall value of k_{cat} was studied (96). For example, bovine pancreatic sPLA$_2$ hydrolyzes vesicles of the 1,2-dithiolester analog of DMPM in the scooting mode with a v_0 that is fivefold lower than the hydrolysis of the oxy-ester. Sulfur substitution does not affect the affinity of E^* for reaction products (K_P^*), however. Since it is unlikely that sulfur substitution changes the rate constants for the formation and dissociation of the E^*P complex by the same factor, the element effect is primarily due to a change in the rate of the chemical step (k_2). The fivefold element effect is also seen with various catalytic-site mutants that have greatly diminished values of v_0; this suggests that the chemical step is fully rate determining for maximal turnover by wild-type and mutant enzymes.

To the best of our knowledge this is the first determination of the interfacial Michaelis-Menten rate and equilibrium parameters for a lipolytic enzyme. Interestingly, the value obtained for the Michaelis constant, $K_M^* = 0.3$ mole fraction, is close to the substrate concentrations that the enzyme "sees" in the vesicle. It has often been stated that lipolytic enzymes are difficult to inhibit because they are immersed in an array of highly concentrated substrate molecules and thus will always operate at saturation (E^* close to 0). Clearly, there is no basis in fact for such concerns.

When pp-PLA$_2$ operates on pure vesicles of DMPM, most of the enzyme is in the E^*S form and turnover is limited by the esterolysis step and not by the dissociation of products or the diffusion of substrates into the catalytic site of the enzyme. When the substrate mole fraction drops below K_M^*, the turnover number is k_{cat}/K_M^*, and this is equal to the rate constant for the $E^* + S \rightarrow E^*S$ step ($k_1 = 1350$ s^{-1}). This step is not limited by two-dimensional diffusion of the substrate to the mouth of the catalytic-site slot, and it may be limited by the dislodgement of the phospholipid from the plane of the membrane.

It was found that 18 different sPLA$_2$s from pancreas and venoms all display scooting mode behavior on DMPM vesicles (95), suggesting that most, if not all, sPLA$_2$s are able to bind tightly to anionic vesicles. Bovine pancreatic sPLA$_2$ and its mutants (73, 78) and the human nonpancreatic sPLA$_2$ that plays a prominent role in inflammatory processes (102) have been subjected to the same type of analysis as that just described for the pp-PLA$_2$.

Role of Calcium

Ca^{2+} is required for the catalytic activity of sPLA$_2$s on all forms of substrates (9, 21). X-ray crystallographic studies place the Ca^{2+} surrounded by a loop of polypeptide chain in the catalytic site, where it promotes the binding of phospholipid analogs (66–69). Based on these structures, the following roles of the metal cofactor are suggested. The Ca^{2+} is directly bonded to one of the non-bridging oxygens of the sn-3 phosphate, and thus the metal forms part of the substrate-binding site. Ca^{2+} is also thought to interact directly with the carbonyl group of the sn-2 ester and the oxyanion of the putative tetrahedral intermediate that forms during catalysis.

The role of Ca^{2+} in catalysis by sPLA$_2$s has been studied for several years, but only recently has the problem been resolved (98). The difficulty lies in the fact that Ca^{2+} may promote interfacial binding of enzyme by altering the E to E^* equilibrium or by altering the E^* to E^*S equilibrium (which will alter the position of the E to E^* step because the equilibria are coupled), or by a combination of both. The use of a neutral diluent makes it possible to measure the effect of Ca^{2+} and other metal ions on the individual equilibria (98). In the absence and presence of a neutral diluent, respectively, the values of $K_{Ca} = 0.35$ mM and $K_{Ca}^* = 0.28$ mM are obtained. Thus, E and E^* display comparable affinities for Ca^{2+}, which is equivalent to saying that the E to E^* equilibrium is not perturbed by the metal. The corresponding values for Cd^{2+} are $K_{Cd} = 0.16$ mM and $K_{Cd}^* = 0.065$ mM. The enzymatic activity in the presence of Cd^{2+} is only 0.2% of that measured in the presence of Ca^{2+}, and thus Cd^{2+} is a partial inhibitor of pp-PLA$_2$.

Protection from alkylation studies show that the binding of substrate analogs to pp-PLA$_2$ at the interface requires calcium (98), and thus the general thermodynamic scheme for the binding of substrate and Ca^{2+} simplifies to the following two-step reaction:

$$E^* + Ca \underset{K_{Ca}^*}{\overset{\rightarrow}{\leftarrow}} E^*Ca \underset{K_L^{*Ca}}{\overset{\overset{\displaystyle L\searrow}{\rightarrow}}{\leftarrow}} E^*Ca{-}L$$

In other words, the species E^*S without bound calcium does not form. K_L^{*Ca} is the equilibrium constant for the dissociation of catalytic-site ligand from the E^*Ca-L complex, and the effective dissociation constant for calcium in the presence of L is $K_{Ca}^*(L) = K_{Ca}^*(1 + X_L/K_L^{*Ca})$. When $L = S$, the expression becomes $K_{Ca}^*(S) = K_{Ca}^*/(1 + X_S/K_M^*)$. $K_{Ca}^*(S)$ is the concentration of calcium that gives half the maximal value of the initial velocity for the hydrolysis of DMPM in the scooting mode. The experimental values of $K_{Ca}^*(S) = 0.10$ mM and $K_{Ca}^* = 0.35$ mM yield a value of $K_M^* = 0.40$, which is consistent with the value of the Michaelis constant determined as described above. In summary, calcium does not promote the binding of enzyme to the anionic interface, but it promotes the binding of substrate to the catalytic site of E^* and also serves as a catalytic cofactor.

Ca^{2+} is required for the hydrolysis of monodisperse 1,2-dibutyrl PC by the sPLA$_2$ from *Crotalus adamanteus* venom (103). The role of Ca^{2+} in promoting enzyme-substrate interaction has been studied for other snake venom sPLA$_2$s by measuring the apparent K_M values for phospholipid substrates as a function of the Ca^{2+} concentration (104). It was found that the apparent K_M values for the hydrolysis of monodisperse and micellar forms of short-chain PCs by sPLA$_2$s from *Agkistrodon halys blomhoffii* and *Trimeresurus flavoviridis* venoms are decreased about 10-fold in the presence of Ca^{2+}. In contrast, the K_M values for monodispersed PCs measured with the *Naja naja atra* venom and bovine pancreatic enzymes are Ca^{2+} independent. The apparent K_M for the hydrolysis of the monodispersed analog of dihexanoyl PC in which both esters are replaced with thiolesters by pp-PLA$_2$ is Ca^{2+} independent (105). In addition, the presence of Ca^{2+} does not significantly alter the affinity of sPLA$_2$s from *A. halys blomhoffii*, *T. flavoviridis*, and three different cobra venoms for monodispersed *n*-alkylphosphocholines (8–14-carbon *n*-alkyl chains) (106–109). These results, showing an apparent calcium-independent binding of substrate and substrate analogs to the catalytic site of sPLA$_2$s, seem paradoxical in light of the X-ray structures and studies with neutral diluent discussed above. Further studies are needed to understand these discrepancies fully.

The affinity of pp-PLA$_2$ for micelles of the substrate analogs *n*-alkylphosphorylcholine (12–16-carbon *n*-alkyl chains) requires Ca^{2+} at alkaline pH but not at neutral pH (74, 110); it is not clear, however, if the E to E^* or E^* to E^*S equilibrium is being affected. Apparently one effect of enzyme-bound Ca^{2+} is to increase the pK_a of the αNH_3^+ group of the N-terminal alanine. This group lies at the end of an α-helix that is a major portion of the i-face, and this helix is stabilized by interaction of the N-terminal αNH_3^+ group with the catalytic network involving Ca^{2+}. Thus, the enzyme binds to micelles in the absence of Ca^{2+} only when the pH is below the pK_a of the αNH_3^+ group (pK_a = 8.4) (74, 110). Micellar binding is seen at higher pHs only when Ca^{2+} is present, presumably because the pK_a of the N-terminal alanine is shifted to 9.3. The affinity of pp-PLA$_2$ for micellar n-alkylphosphorylcholines (14–16-carbon n-alkyl chains) is moderately increased (< fivefold) in the presence of 100 mM Ca^{2+} (111). Multiple effects may be occurring with such a high metal concentration. The problem is compounded by the fact these amphiphiles have a significant affinity for the catalytic site [$K_I^* = 0.65$ (80)].

Substrate Specificity

The literature is full of improper analyses of PLA$_2$ substrate specificity, and the problem is even worse with interfacial enzymes. Substrate specificity can only be interpreted in terms of Equation 5, which gives a measure of how good one substrate is versus another, or more precisely what is the relative velocity of two substrates (S_1 and S_2) when both are present in the reaction mixture at equal concentration.

$$v_1/v_2 = [(k_{cat}/K_M)_1/ (k_{cat}/K_M)_2] \times [S_1]/[S_2] \qquad 5.$$

According to Equation 5, such specificity is given by the relative k_{cat}/K_M values for the two substrates. Equation 5 transcribes to the analysis of interfacial enzymes as long as the enzyme operates in the scooting mode; in this case it is the interfacial k_{cat} and K_M^* values that are relevant.

Substrate specificity studies of interfacial enzymes are often carried out by comparing the velocities of hydrolysis of aggregates of different pure phospholipid species. Such an approach makes the results difficult to interpret because they reflect both the specificity of the enzyme at the interface (Equation 5) and the differences in affinities of the enzyme for different aggregates (E to E^*). The latter can be measured by direct binding studies, and the former can be measured by allowing the enzyme to operate in the scooting mode on vesicles containing competing substrates (91).

Analysis of the substrate specificity of pp-PLA$_2$s, *Naja naja naja* sPLA$_2$ and the sPLA$_2$ from human inflammatory exudate has been carried out in a competitive fashion in which both [3]H- and ^{14}C-labeled phospholipids are present in DMPM vesicles, and the enzyme is allowed to "choose" between them while operating in the scooting mode (91). The results show that sPLA$_2$s do not discriminate significantly (< fivefold) among phospholipids with different naturally occurring polar head groups or between saturated versus unsaturated fatty acyl chains (91, 102, 112). A completely different and erroneous picture would be obtained if absolute reaction velocities on vesicles of different pure phospholipid species were compared. The sPLA$_2$ from inflammatory exudates is reported to hydrolyze 1-palmitoyl-2-arachidonyl-*sn*-glycero-3-phosphoethanolamine in DMPM vesicles sevenfold faster than the analogous PC (113). Such studies were carried with only one substrate present at a time in the vesicles, and therefore the relative velocities obtained are not an accurate representation of the substrate specificity. Indeed when both substrates are present in DMPM vesicles and compete with each other for hydrolysis by the enzyme, no discrimination between them is observed (102). Likewise, the reported faster rates of hydrolysis of phosphatidylethanolamine (PE) versus PC vesicles by the same enzyme do not reflect the intrinsic specificity (114, 115).

Naja naja naja venom sPLA$_2$ hydrolyzes PC present in Triton X-100 mixed micelles 10-fold faster than PE in the same detergent, whereas PE is about 2-fold preferred when mixed micelles containing both substrates are hydrolyzed (116, 117). The relatively poor activity of the cobra venom enzyme on PE mixed micelles is not simply due to differences in the E to E^* equilibrium, since such specificity differences are seen in the presence of a sufficiently high bulk concentration of mixed micelles such that the velocity is maximal (118). Apparently, other factors contribute to the differences in rates.

Inhibitors

Competitive inhibitors of sPLA$_2$ bind to the catalytic site of E^* and block the binding of substrate. In evaluating reports of inhibitors of lipolytic enzymes, it is important to keep track of the mole fraction of inhibitor in the substrate aggregate. For example, an inhibitor with a reported IC$_{50}$ value of 1 μM suggests a high degree of potency, but if the substrate concentration in the assay is also in the μM range, the inhibitor may make up a significant fraction of the aggregate, and the E to E^* equilibrium could be altered in favor of E. In this case, the compound is a nonspecific inhibitor, since it does not function by selective binding to the enzyme. Examples of such inhibitors include mepacrine, indomethacin, aristolochic acid, and others (93). Thus, the mole fraction of inhibitor at the interface is the relevant quantity, since this is what the enzyme at the interface "sees." If the inhibitor exists mainly in the aqueous phase and binds to E^* in a way that prevents the binding of substrate, the degree of inhibition will also correlate with the bulk concentration of inhibitor. The use of the scooting mode analysis is advantageous in this context since the enzyme remains tightly bound to the interface even when solutes are added to the bilayer up to 20 mole % (93, 119, 120).

Tight-binding sPLA$_2$ inhibitors (IC$_{50}$ < 0.05 mole fraction) include phospholipid analogs in which the *sn*-2 ester has been replaced with an amide (121–126), a phosphonate (102, 120, 127–130), a difluoromethylene ketone (131), or a sulfonamide (132). Inhibitors whose structures show less resemblance to naturally occurring phospholipids include fatty acid amides (133), phosphate diesters (134), phospholipid analogs containing an *sn*-2 amide and an *sn*-3 carboxylate replacing the polar head group (135), substituted N,N-*bis*-carboxymethyl anilines (136), a triterpenoid from pink peppercorn berries (137), fungal secondary metabolites thielocin A1β and B3 (138, 139), and spiro-ketals (140).

Besides phenacylbromides, which alkylate the catalytic histidine, other inhibitors such as alkylbenzoylacrylic acids (141) and phospholipid analogs containing an sn-2 p-nitrophenyl glutaryl chain (142) cause covalent inactivation of sPLA$_2$s. The terpenoids monoalide and scalaradial isolated from sea sponges inactivate sPLA$_2$ by reacting with lysine residues (143–148). Manoalide-modified sPLA$_2$s retain their abilities to bind to vesicles but have diminished turnover numbers in the scooting mode, possibly because the i-face is not fully desolvated (149). It is generally accepted that lipocortins (anexins), once thought to be the PLA$_2$ inhibitory factor induced by anti-inflammatory glucocorticoids, inhibit the action of sPLA$_2$s in vitro by binding to the interface and blocking access of the enzyme (150).

It is also possible to imagine an inhibitor that binds to the catalytic site of enzyme in the aqueous phase. If the EI complex is still able to bind to the interface and if the E^*I complex is not stable such that inhibitor readily dissociates, perhaps by favorable inhibitor-interface interaction, such a compound will obviously be a poor inhibitor of interfacial catalysis. Thus, it makes the most sense to analyze competitive inhibitors under conditions in which the enzyme is operating at the interface (see 151 for a detailed discussion of these issues). Additional problems arise in the analysis of inhibitors of sPLA$_2$ in aqueous solution as is described in a later section. Agents that bind to the i-face of the enzyme in solution could effectively reduce the fraction of enzyme in the E^* form. With small molecules, this type of inhibition is expected to be rare, since the interface probably makes contact with a large number of residues on the i-face (50). Some of the sPLA$_2$ inhibitory protein isolated from plasma of certain snakes probably works by this mechanism (see e.g. 152, 153).

Interfacial Activation

The reaction rate for the action of sPLA$_2$s on water-soluble short-chain phospholipids often increases dramatically when the substrate aggregates as its concentration approaches the critical micelle concentration (9). Such modulation of catalytic behavior by the interface, called interfacial activation, could occur either by an increase in substrate affinity (K_S type) or the rate of the chemical step (k_{cat} type). The first of these possibilities has been investigated in the context of the thermodynamic box shown below (154).

Here, K_d and K_d^l are the equilibrium constants for the dissociation of E and EI, respectively, from the interface, K_I and K_I^* have their usual meanings, and K' is $[I]/X_I$. For a variety of catalytic site-directed competitive inhibitors, it was found that the ratio $K_d/K_d^l = 40$, i.e. inhibitor binding to the catalytic site stabilizes the binding of the pp-PLA$_2$ to the interface by 40-fold. By microscopic reversibility, interfacial binding of enzyme must also stabilize inhibitor binding by the same factor. Thus there is allosteric modulation of the

$$E + I \;\underset{}{\overset{K_I}{\rightleftharpoons}}\; EI$$

$$K_d \;\Bigg\updownarrow \qquad \Bigg\updownarrow K' \qquad\qquad \Bigg\updownarrow K_d^I$$

$$E^* + I^* \;\underset{K_I^*}{\overset{}{\rightleftharpoons}}\; E^*I$$

enzyme by the interface of the K_S type, and this contributes to the phenomenon of interfacial activation.

Insight into this allosteric modulation comes from recent high-resolution solution-state NMR of pp-PLA$_2$ in solution and bound to micelles of dodecyl-phosphocholine (155, 156). Overall the structures of E revealed by NMR and X-ray diffraction are similar; however, in solution the first few residues of the N terminus are flexible, whereas in the crystal they are part of a rigid α-helix. In addition, the surface loop containing tyrosine 69 is more flexible for the solution enzyme. Upon micelle binding, the N-terminal residues Ala-1, Leu-2, and Trp-3 move closer together, as do Thr-52, Tyr-73, and Tyr-69. NMR of the E^*I complex formed after adding an sn-2 amide phospholipid analog reveals that the N terminus now adopts an α-helix, and the N-terminal amino group of Ala-1 is now visible, which suggests that it is engaged in H bonding to the protein as seen in the X-ray structure. A conformational change in the N-terminal helix is also seen in the X-ray structure of the human nonpancreatic sPLA$_2$ when it binds an inhibitor (64).

The next obvious question is whether k_{cat} for E and E^* are the same or different. The NMR studies suggest that the catalytically most active state of E^* is formed only when a phospholipid is bound in the catalytic site (156); this is not unequivocal, however. A cursory look at the observed rates in the presence of micellar versus monomeric short-chain substrates would suggest that the k_{cat} for E^* is significantly higher than that for E (8, 9). There are several problems with this interpretation, however. Besides the obvious problems in measuring K_M^* at the micellar interface, there are some curious observations regarding the activity of the E form. For example, the apparent K_M values for water-soluble substrates often do not depend on the calcium concentration (104, 105), nor do they correlate with the observed IC_{50} and K_I values for inhibitors (157). Unless such anomalies are resolved, it is difficult to interpret the apparent hyperbolic velocity versus substrate concentration curves that are seen for the action of E on short-chain substrates at concentrations below their critical micelle concentrations.

An additional mode of interfacial activation may also be considered. The interface could modulate the substrate conformation so that it is more accessible to chemical action by the enzyme (158, 159). At least for sPLA$_2$s, this does not appear to offer any kinetic advantage, because the chemical step is rate limiting for interfacial catalysis in the scooting mode, as described earlier in this review. In addition, the conformations of the glycerol backbone in membrane phospholipids and of a phospholipid analog bound to the catalytic site of sPLA$_2$s are similar (48, 69).

KINETIC COMPLEXITIES AT INTERFACES

As will become apparent in the following section, the kinetic behavior of interfacial enzymes under conditions in which the constraints of the scooting mode are not satisfied is considerably more complex, often to the point that the primary rate and equilibrium parameters (Figure 1) cannot be resolved. Many of the problems can be appreciated now that interfacial catalysis has been described analytically in terms of Figure 1.

Inter-aggregate Substrate Exchange and Interfacial Catalysis on Mixed Micelles

The reaction progress curve for the action of sPLA$_2$ on small DMPM vesicles in the scooting mode is dominated by the first-order term in Equation 2, and thus the reaction rate continually decreases due to substrate depletion and product inhibition (19). Under such conditions, agents that increase the rate of inter-aggregate exchange of substrate can cause apparent activation even though the agent may not interact directly with the enzyme. An example of this type of nonspecific activator is the cyclic cationic peptide polymyxin B, which does not interact with sPLA$_2$s but causes rapid inter-aggregate substrate exchange (94).

Long-chain phospholipids codispersed in detergent have been used to characterize interfacial enzymes including sPLA$_2$s (6, 21). The surface concentration of phospholipid in detergent micelles can be altered by varying the detergent/phospholipid mole ratio, and the bulk concentration of substrate can be varied by changing the number of mixed micelles at a fixed detergent/phospholipid mole ratio. The effect of these changes on the kinetics of action of cobra venom sPLA$_2$ on phospholipid–Triton X-100 mixed micelles has been studied in detail (see e.g. 6, 118, 160). The binding of cobra venom sPLA$_2$ to Triton X-100 micelles occurs only if phospholipid is present, and the velocity decreases as the mole fraction of phospholipid in the detergent decreases (surface dilution).

Besides the problems associated with non-ideal mixing and polydispersity of mixed micelles (4), a major concern with these systems is whether the local and global steady states are the same. Micelles are much smaller than vesicles (Table 1), and thus enzyme-catalyzed depletion of substrate in enzyme-containing micelles is fast, and intermicelle exchange of enzyme and/or substrate is required to maintain the reaction progress. For example, a typical lipolytic enzyme catalyzes > 100 turnovers per second. If a mixed micelle has 10 phospholipids, the substrate is 50% consumed in about 50 msec unless there is intermicelle exchange of components that is fast on the 50 msec time-scale. Thus the kinetics in micelles may reflect the rates of intermicelle exchange events rather than the kinetic properties of the enzyme. The following recent evidence indicates that this may be the case, at least for some mixed micelles (161). Addition of more than 30 mole % detergents (bile salts, Triton X-100, lysophospholipids) to phospholipid vesicles leads to the disruption of bilayers and the formation of mixed micelles. Accompanying this change in morphology is a > 10-fold drop in the $sPLA_2$ reaction velocity. Spectroscopic studies show that this is not due to enzyme desorption from the mixed-micelle interface, and the sudden drop in velocity indicates that the decrease is not due to surface dilution of substrate. In addition, studies in mixed micelles with sn-2 thiolester–containing phospholipids, which have served as tools to establish that the chemical step is rate limiting for the $sPLA_2$-catalyzed hydrolysis of vesicles in the scooting mode (discussed above), when present in mixed micelles, are hydrolyzed at similar rates to those of oxy-esters. All of these results strongly suggest that lipolysis in mixed micelles is limited by a physical step such as the replenishment of substrate.

Mixed micelles have been used in the analysis of competitive inhibitors of $sPLA_2$ (see e.g. 123, 126). Although the enzyme may not be tightly bound to the interface, the likelihood that inhibition will result from a shift in the E to E^* equilibrium is minimized if the mole fraction of inhibitor is kept low and the bulk substrate concentration is high. One can take a guess about the effect of rate-limiting substrate replenishment on the analysis of inhibitors: It is probably the case that the relative potencies of a series of inhibitors can be obtained from kinetic studies employing mixed micelles, but it may be difficult to extract the absolute K_I^* values. This is because the local substrate concentration in enzyme-containing mixed micelles is lower than that calculated from the total amounts of components, and this leads to an overestimation of the potency of the inhibitor. Additionally, if the reaction velocity is partially controlled by intermicelle exchange of enzyme, the binding of inhibitor to enzyme will modulate the rate of depletion of substrate in the enzyme-containing mixed micelle.

Intervesicle Exchange of Enzyme

If $sPLA_2$ is operating in the hopping mode (low processivity), the rate of turnover can be much slower than that in the scooting mode (high processivity), even if there is a sufficiently high bulk concentration of phospholipid to ensure that all of the enzyme is at the interface. This is because fast kinetic studies have shown that pp-PLA$_2$ binds to vesicles in a two-step process consisting of a diffusion-limited collision with vesicles to give E' followed by a slow first-order process (half-time of 0.2 s), which generates E^* (162). The second step probably involves the desolvation of the region of the interface and enzyme that contact each other. Catalysis occurs only via E^*, possibly because in the absence of desolvation, it is difficult to dislodge the phospholipid from the plane of the bilayer into the catalytic site. Fast hopping implies that after the products are released from the catalytic site, E^* is converted to E' at a rate that is comparable to or faster than the binding of new substrate to form E^*S. Thus, the E' to E^* conversion becomes a part of every few turnover cycles. Since the turnover numbers of sPLA$_2$s in the scooting mode are fast (100–400 s^{-1}), fast hopping is a very slow mode of interfacial catalysis. Furthermore, the measured rates obtained in the fast-hopping regime will be a mixture of interfacial constants (steps within the box of Figure 1) with an indeterminate contribution from the rates for the E' to E^* step; in the scooting mode, this step occurs only in the pre–steady state. The pro-enzyme form of pp-PLA$_2$ contains a dysfunctional i-face and is a very slow lipase even if it is fully bound at the vesicle interface. This is probably because it is bound to vesicles in a form that resembles E' more than E^* and undergoes slow turnover in the hopping mode (4).

The phrase "quality of the interface" has been introduced to explain observations that the kinetics of hydrolysis of phospholipid aggregates by sPLA$_2$s depends on the form of the substrate and the conditions of the assay (8). Factors that alter the kinetics of sPLA$_2$s include change in the temperature near the phase transition of the vesicles (7, 163), the surface pressure of phospholipid monolayers (8, 164), incorporation of alkanols into the interface (5, 165), and the ionic strength of the reaction mixture (166, 167). As described below, the bulk of the evidence indicates that these effects are due to a change in the E to E^* equilibrium.

The reaction progress curves for the action of sPLA$_2$ on vesicles under conditions in which the enzyme is not fully bound to the vesicles are complex. The binding of most, and perhaps all, sPLA$_2$s to pure PC vesicles (zwitterionic interface) is orders of magnitude weaker than their binding to anionic vesicles, as shown by spectroscopic studies that directly monitor enzyme-interface interactions (4, 50, 75). For example, numerous sPLA$_2$s bind essentially irreversibly to vesicles of anionic phospholipids such as DMPM (95, 102). The binding of *Naja melanoleuca* sPLA$_2$ to PC vesicles occurs reversibly with a dissociation constant for the enzyme-vesicle complex of 5 μM (expressed as lipid monomer concentration), and the binding of pp-PLA$_2$ and human sPLA$_2$s to zwitterionic vesicles is hardly detectable even with mM amounts of lipid (75, 95, 102). Addition of anionic amphiphiles such as reaction products (1:1 mixture of fatty acid and lysophospholipid), anionic phospholipids, or sodium dodecylsulfate but not cationic amphiphiles to PC vesicles or mixed micelles leads to enhanced rates of lipolysis as well as a shift in the E to E^* equilibrium (91, 95, 102, 168–170).

Based on these results, it is apparent that the complexity of the reaction progress curves for the hydrolysis of zwitterionic vesicles is due in part to the fact that as the reaction proceeds, alteration in the composition of the interface by the reaction products will shift the E to E^* equilibrium with a concomitant increase in the reaction velocity. This hypothesis for apparent activation is supported by multiple lines of evidence (4, 75, 168). The initial velocity for the hydrolysis of dimyristoyl PC vesicles by pp-PLA$_2$ is slow, and the reaction accelerates over several minutes. This lag is completely eliminated if substrate vesicles initially contain 10 mole % reaction products, but not fatty acid alone. The fact that the reaction begins immediately after adding enzyme to product-containing vesicles implies that the kinetics of the lag are due not to slow binding of enzyme to the interface but rather to a slow buildup of products. The initial velocity in the presence of products and the velocity at the end of the lag are similar, which suggests that the only effect of added products is to eliminate the lag. The lag decreases as the amount of enzyme is increased, because the buildup of products in vesicles occurs more rapidly in the presence of more enzyme (168).

The lag increases with increasing amounts of substrate vesicles, which indicates that the mole fraction of products in the vesicles is what determines the lag (168). The lag in the hydrolysis of PC vesicles is minimal when the temperature is near the gel-to-liquid phase transition, and it increases as the temperature is lowered or raised (7, 163, 168, 171–174). This probably accounts for the apparent activation of sPLA$_2$s at the phase-transition temperature. It may also be noted that no anomalous change in kinetics is seen near the phase-transition temperature when pp-PLA$_2$ is operating on DMPM vesicles in the scooting mode (11). This suggests that the activation is related to the E to E^* equilibrium.

Spectroscopic studies of PC/products ternary mixtures suggest that the segregation of fatty acid occurs as the lag ends, and this also correlates with enhanced binding of enzyme to the interface (175, 176). If such segregation is maximal near the phase-transition temperature, this would account for the apparent activation of sPLA$_2$s at this temperature. Additional evidence for this comes from studies employing epifluorescence microscopy to visualize directly the action of sPLA$_2$ on phospholipid monolayers containing fluorescently labeled phospholipids (177, 178). With this technique, a monolayer of dipalmitoyl PC doped with small amounts of rhodamine-labeled amphiphile is formed, and the surface pressure is varied until fluid and liquid crystalline domains coexist; the former appears more fluorescent because the dye is excluded from the crystalline domains (179, 180). Addition of cobra venom sPLA$_2$ leads to a loss of the solid domains as seen by frayed indentations. As the solid domains become hydrolyzed, enzyme domains form in their place (the enzyme is tagged with fluorescein so its location on the monolayer can be viewed). Further studies were carried out with mixed monolayers of di-palmitoyl PC, 1-palmitoyl-lyso-PC, and palmitic acid (181, 182). Lateral compression of these ternary monolayers leads to the formation of domains that have a high anionic character, since they preferentially bind a fluorescent cationic dye or sPLA$_2$.

Enzyme Aggregation

There is vigorous debate about whether sPLA$_2$s form functional aggregates at the interface. The only unequivocal data that relies on direct measurement of the aggregation state of the functional enzyme is for the action of several sPLA$_2$s on DMPM vesicles in the scooting mode (95). The total moles of phospholipid hydrolyzed after the completion of the reaction divided by the number of phospholipids in the outer monolayer of DMPM vesicles (determined by a variety of techniques) gives the moles of catalytically active enzyme. When this number is divided into the mass of enzyme in the assay, a M_r within 10% of the monomeric M_r is obtained in all cases. In addition, spectroscopic studies using fluorescently tagged pp-PLA$_2$ reveal interprotein fluorescence energy transfer only when the enzyme is crowded on the interface of anionic or zwitterionic vesicles or micelles (95).

Most sPLA$_2$s have a tendency to aggregate both in solution or in the presence of amphiphiles (6, 18, 53, 169). It is not clear if these aggregates are functionally active or if they are formed under catalytic conditions; activation of cobra venom sPLA$_2$ by dimerization has been invoked, however (21). Based on such a possibility, attempts have been made to model events of the lag phase seen in the hydrolysis of zwitterionic vesicles after accumulation of a critical mole fraction of reaction products (introduced above). Such models are designed to simulate numerically the characteristics of the reaction progress during the lag phase with the predictions of the various modifications of the Michaelis-Menten formalism (171, 183–187). The key conclusion is that the one-step E to E^* equilibrium (Figure 1) does not account adequately for the events of the lag. A minimum model in which the enzyme first forms a dimer in solution (E_2) or at the interface (E_{2B}) and then converts to a catalytically active dimeric form (E^*_{2B}) nicely fits the data (187):

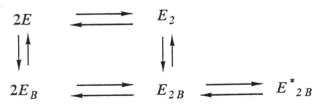

It is difficult to evaluate the full significance of this suggestion, because the values of the underlying rate and equilibrium constants have not been determined yet. Apparently, the need for the dimer formulation arises from the fact that although the binding of the enzyme to zwitterionic vesicles is poor, in the presence of a high vesicle concentration the maximum rate should reach the value observed in the presence of reaction products, but in fact it does not in the case of venom $sPLA_2$. Thus, according to this formalism, the dimeric enzyme is intrinsically a better catalyst, and the driving force for dimerization comes putatively from the lateral phase segregation of reaction products. Besides the lack of direct evidence for dimerization, it should be recalled that $sPLA_2$s are fully active as monomers on anionic vesicles, where there is no discernable lag and the kinetics are independent of the phase properties of the bilayer.

An alternative explanation for the events of the lag within the confines of Figure 1 may be found in stopped-flow studies of the binding of pp-PLA_2 to the interface (162) (discussed above). Perhaps on zwitterionic vesicles the enzyme is mainly in the E' state, and the transition to the E^* state, which is required for catalysis, is promoted by reaction products. In addition, different residence times of the E' and E^* forms could account for the differences in the kinetic behavior of $sPLA_2$ in the lag phase as well as in the steady state established after the formation of a critical mole fraction of products. Clearly, further studies are needed to understand fully the role of enzyme aggregation in the kinetics of action of $sPLA_2$ on PC vesicles.

pp-PLA_2 and *A. piscivorus piscivorus* $sPLA_2$ catalyze the hydrolysis of the water-soluble chromogenic compound 4-nitro-3-octanolyloxybenzoate (188, 189), and this leads to octanoylation of lysine 56 (porcine pancreatic) or lysines 7 and 10 (snake venom). Acylated pp-PLA_2 and venom $sPLA_2$ are 180- and 7-fold, respectively, more active than their non-acylated forms on the chromogenic substrate, and pp-PLA_2 is 100-fold more active on PC monolayers. Whereas the non-acylated enzymes are monomeric in solution, the acylated enzymes form dimers. Incubation of pp-PLA_2 and venom $sPLA_2$ with small unilamellar vesicles of PC radiolabeled in its *sn*-2 chain leads to dimeric, radiolabeled enzymes with enhanced activity (188, 189). The authors suggest that substrate-derived enzyme acylation is the basis for interfacial activation of $sPLA_2$s (189). Mutation of the acylation site of pancreatic $sPLA_2$ (lysine-56) or altering the structures of all of the lysines by mutation to arginines or by amidination, however, leads to enzymes that are fully active on substrate interfaces (95, 166, 190, 191). Also, interfacial activation on PC vesicles by reaction products is reversed when vesicles without product are added (186). Chemical fatty acylation of pp-PLA_2 and bee venom $sPLA_2$ with reactive esters leads to enzymes that no longer display a lag in the hydrolysis of PC monolayers (190, 192–195). Thus, there seems to be general consensus that acylation increases the affinity of enzyme for zwitterionic interfaces, but the proposal of substrate-based enzyme acylation as the basis for interfacial activation seems highly unlikely when all of the evidence is considered.

Hydrolysis of Water-Soluble Short-Chain Phospholipids

The analysis of sPLA$_2$-catalyzed hydrolysis of short-chain phospholipids dispersed as solitary monomers in aqueous solution has been clouded by observations that pre-micellar protein-lipid microaggregates form (see e.g. 127, 157, 169, 196–198). Such aggregation may be due to the segregation of short-chain phospholipids in contact with the i-face. The kinetic analysis of such behavior is difficult, because the reaction rate could be controlled by the replenishment of substrate in small enzyme-lipid microaggregates. Although it is possible that, in some cases, the enzyme-substrate complex is non-aggregated, caution should be exercised in interpreting the results with short-chain phospholipids.

CONCLUSIONS AND FUTURE PROSPECTS

Interfacial catalysis has come of age—to the point that the Michaelis-Menten formalism, with an additional step for binding of the enzyme to the interface, can be used to characterize many features of interfacial enzymes analytically. The concept of a neutral diluent has been developed, and this allows interfacial equilibria to be analyzed. There is a wealth of information on the interfacial enzymology of sPLA$_2$s, and such studies will pave the way toward detailed quantitative analyses of other interfacial enzymes such as triacylglycerol lipases, 85-kDa cytosolic phospholipase A$_2$, and phospholipases C (see e.g. 199–202). One area ripe for further development is the appreciation and understanding of factors that control the E to E^* equilibrium in physiological settings. Examples of agents that modulate the binding of sPLA$_2$s to cell membranes are bacterial permeability-inducing factor (203, 204), heparin-proteoglycan (205), and protein receptors (206, 207). In addition, segregated phospholipid domains have been observed in living cells (208), and these will almost certainly play a role in modulating the activity of interfacial enzymes.

ACKNOWLEDGMENTS

Research support is from the National Institutes of Health (HL36235, HL500400, and Research Career Development Award to M. H. G. and GM29702 to M. K. Jain). A. M. Hanel is the recipient of a postdoctoral fellowship from the American Cancer Society.

Literature Cited

1. Cornell RB. 1991. *Biochemistry* 30: 5873–80
2. Kornberg A, Baker T. 1992. *DNA Replication.* New York, NY: Freeman
3. Berg OG. 1985. *Biophys. J.* 47:1–14
4. Jain MK, Berg O. 1989. *Biochim. Biophys. Acta* 1002:127–56
5. Jain MK, Cordes EH. 1973. *J. Membr. Biol.* 14:101–18
6. Roberts MF, Deems RA, Dennis EA. 1977. *Proc. Natl. Acad. Sci. USA* 74: 1950–54
7. Upreti G, Jain MK. 1980. *J. Membr. Biol.* 55:113–23
8. Verger R, de Haas GH. 1976. *Annu. Rev. Biophys. Bioeng.* 5:77–117
9. Verheij HM, Slotboom AJ, de Haas GH. 1981. *Rev. Physiol. Biochem. Pharmacol.* 91:91–203
10. Sarda L, Desnuelle P. 1958. *Biochim. Biophys. Acta* 30:513–21
11. Jain MK, Rogers J, Jahagirdar DV, Marecek JF, Ramirez F. 1986. *Biochim. Biophys. Acta* 860:435–47
12. Cevc G, Marsh D. 1987. *Phospholipid Bilayers: Physical Principles and Models.* New York: Wiley
13. Jain MK. 1988. *Introduction to Biological Membranes.* New York: Wiley-Interscience
14. Jain MK, Van Echteld CJA, Ramirez F, Gier J, de Haas GH, Van Deenen LLM. 1980. *Nature* 284:486–87
15. Nichols JW. 1988. *Biochemistry* 27: 3925–31
16. Sloty CE, Roberts MF. 1994. *Biochemistry* 33:11608–17
17. Deleted in proof
18. Jain MK, Gelb MH, Rogers J, Berg OG. 1995. *Methods Enzymol.* In press
19. Berg OG, Yu B-Z, Rogers J, Jain MK. 1991. *Biochemistry* 30:7283–97
20. Lin H-K, Gelb MH. 1993. *J. Am. Chem. Soc.* 115:3932–42
21. Dennis EA. 1983. *The Enzymes* 16:307–53
22. Hanahan DJ. 1954. *J. Biol. Chem.* 207: 879–84
23. Davidson FF, Dennis EA. 1990. *J. Mol. Evol.* 31:228–38
24. van den Bergh CJ, Slotboom AJ, Verheij HM, de Haas GH. 1989. *J. Cell. Biochem.* 39:379–90
25. Waite M. 1987. *The Phospholipases.* New York: Plenum
26. Kudo I, Murakami M, Hara S. 1993. *Biochim. Biophys. Acta* 117:217–31
27. Dennis EA. 1994. *J. Biol. Chem.* 269: 13057–60
28. Anderson BO, Moore EE, Banerjee A. 1994. *J. Surg. Res.* 56:199–205
29. Vadas P, Pruzanski W. 1993. *Circ. Shock* 39:160–67
30. Vadas P, Browning J, Edelson J, Pruzanski W. 1993. *J. Lipid Mediat.* 8:1–30
31. Pruzanski W, Vadas P, Browning J. 1993. *J. Lipid Mediat.* 8:161–67
32. Glaser KB, Mobilio D, Chang JY, Senko N. 1993. *Trends Pharmacol. Sci.* 14:92–98
33. Mayer RJ, Marshall LA. 1993. *FASEB J.* 7:339–48
34. Bomalaski JS, Clark MA. 1993. *Arthritis Rheum.* 36:190–98
35. Mukherjee AB, Cordella-Miele E, Miele L. 1992. *DNA Cell Biol.* 11:233–43
36. Pruzanski W, Vadas P. 1991. *Immunol. Today* 12:143–46
37. Bomalaski JS, Clark MA. 1990. *Adv. Exp. Med. Biol.* 279:231–38
38. Kramer RM, Johansen B, Hession C, Pepinsky RB. 1990. *Adv. Prostaglandins Thromboxanes Leukot. Res.* 20:79–86
39. Dennis EA. 1990. *Adv. Prostaglandins Thromboxanes Leukot. Res.* 20:79–86
40. B'er'eziat G, Etienne J, Kokkindis M, Olivier JL, Pernas P. 1990. *J. Lipid Mediat.* 2:159–72
41. Vadas P, Pruzanski W. 1990. *J. Rheumatol.* 17:1386–91
42. Weiss J, Wright G. 1990. *Adv. Exp. Med. Biol.* 275:103–13
43. Pruzanski W, Vadas P. 1990. *Adv. Exp. Med. Biol.* 279:239–51
44. Kramer RM, Hession C, Johansen B, Hayes G, McGray P, et al. 1989. *J. Biol. Chem.* 264:5768–75
45. Bonventre JV. 1992. *J. Am. Soc. Nephrol.* 3:128–50
46. Dennis EA, Deems RA, Yu L. 1992. *Adv. Exp. Med. Biol.* 318:35–39
47. Kuipers OP, van den Bergh CJ, Verheij HM, de Haas GH. 1990. *Adv. Exp. Med. Biol.* 279:65–84
48. Scott DL, Sigler PB. 1994. *Adv. Protein Chem.* 45:53–88
49. Scott DL, Mandel AM, Sigler PB, Honig B. 1994. *Biophys. J.* 67:493–504
50. Ramirez F, Jain MK. 1991. *Proteins* 9:229–39
51. Biltonen RL, Lathrop BK, Bell JD. 1991. *Methods Enzymol.* 197:234–48
52. Vernon LP, Bell JD. 1992. *Pharmacol. Ther.* 54:269–95
53. Hazlett TL, Deems RA, Dennis EA. 1990. *Adv. Exp. Med. Biol.* 279:49–64
54. Jain MK, Gelb MH. 1991. *Methods Enzymol.* 197:112–25

55. Gelb MH, Jain MK, Berg O. 1992. *Bioorg. Med. Chem. Lett.* 2:1335–42
56. Bell JD, Biltonen RL. 1991. *Methods Enzymol.* 197:249–58
57. Nuhn P, Koch K. 1993. *Pharmazie* 48:494–508
58. Gelb MH, Berg O, Jain MK. 1991. *Curr. Opin. Struct. Biol.* 1:836–43
59. Gelb MH, Jain MK, Berg OG. 1994. *FASEB J.* 8:916–24
60. Mobilio D, Marshall LA. 1989. *Annu. Rep. Med. Chem.* 24:157–66
61. Blackwell GJ, Flower RJ. 1983. *Br. Med. Bull.* 39:260–64
62. Fremont DH, Anderson DH, Wilson IA, Dennis EA, Xuong NH. 1993. *Proc. Natl. Acad. Sci. USA* 90:342–46
63. Scott DL, Achari A, Vidal JC, Sigler PB. 1992. *J. Biol. Chem.* 267:22645–57
64. Scott DL, White SP, Browning JL, Rosa JJ, Gelb MH. 1991. *Science* 254:1007–10
65. Wery J-P, Schevitz RW, Clawson DK, Bobbitt JL, Dow ER, et al. 1991. *Nature* 352:79–82
66. Scott DL, White SP, Otwinowski Z, Yuan W, Gelb MH, Sigler PB. 1990. *Science* 250:1541–46
67. White SP, Scott DL, Otwinowski Z, Gelb MH, Sigler PB. 1990. *Science* 250:1560–63
68. Scott DL, Otwinowski Z, Gelb MH, Sigler PB. 1990. *Science* 250:1563–66
69. Thunnissen MMGM, Eiso AB, Kalk KH, Drenth J, Dijkstra BW, et al. 1990. *Nature* 347:689–91
70. Ward KB, Pattabiraman N. 1990. *Adv. Exp. Med. Biol.* 279:23–26
71. Renetseder R, Brunie S, Dijkstra BW, Drenth J, Sigler PB. 1985. *J. Biol. Chem.* 260:11627–34
72. Dijkstra BW, Kalk KH, Drenth J, de Haas GH, Egmond MR, Slotboom AJ. 1984. *Biochemistry* 23:2759–66
73. Dupureur CM, Yu B-Z, Mamone JA, Jain MK, Tsai M-D. 1992. *Biochemistry* 31:10576–83
74. van Dam-Mieras MCE, Slotboom AJ, Pieterson WA, de Haas GH. 1975. *Biochemistry* 14:5387–92
75. Jain MK, Egmond MR, Verheij HM, Apitz-Castro R, Dijkman R, de Haas GH. 1982. *Biochim. Biophys. Acta* 688:341–48
76. Jain MK, Maliwal BP. 1985. *Biochim. Biophys. Acta* 814:135–40
77. Jain MK, Vaz WLC. 1987. *Biochim. Biophys. Acta* 905:1–8
78. Dupureur CM, Yu B-Z, Jain MK, Noel JP, Deng T, et al. 1992. *Biochemistry* 31:6402–13
79. Ludescher RD, Johnson ID, Volwerk JJ, de Haas GH, Jost PC, Hudson BS. 1988. *Biochemistry* 27:6618–28
80. Jain MK, Maliwal BP. 1993. *Biochemistry* 32:11838–46
81. Soltys CE, Roberts MF. 1993. *Biochemistry* 32:9545–52
82. Jain MK, Rogers J, Marecek JF, Ramirez F, Eibl H. 1986. *Biochim. Biophys. Acta* 860:462–74
83. Cuccia L, Hebert N, Beck A, Just G, Lennox RB. 1994. *J. Am. Chem. Soc.* In press
84. Picot D, Loll PJ, Garavito RM. 1993. *Nature* 367:243–49
85. Li Y, Tsai M-D. 1993. *J. Am. Chem. Soc.* 115:8523–26
86. Kuipers OP, Dekker N, Verheij HM, de Haas GH. 1990. *Biochemistry* 29:6094–102
87. Kuipers OP, Franken PA, Hendriks R, Verheij HM, de Haas GH. 1990. *Protein Eng.* 4:199–204
88. Thunnissen MMGM, Franken PA, de Haas GH, Drenth J, Kalk KH, et al. 1992. *Protein Eng.* 5:597–603
89. Dupureur CM, Deng T, Kwak J-G, Noel JP, Tsai M-D. 1990. *J. Am. Chem. Soc.* 112:7074–76
90. Bekkers AC, Franken PA, Toxopeus E, Verheij HM, de Haas GH. 1991. *Biochim. Biophys. Acta* 1076:374–78
91. Ghomashchi F, Yu B-Z, Berg O, Jain MK, Gelb MH. 1991. *Biochemistry* 30:7318–29
92. Ghomashchi F, O'Hare T, Clary D, Gelb MH. 1991. *Biochemistry* 30:7298–305
93. Jain MK, Yu B-Z, Rogers J, Ranadive GN, Berg O. 1991. *Biochemistry* 30:7306–17
94. Jain MK, Rogers J, Berg O, Gelb MH. 1991. *Biochemistry* 30:7340–48
95. Jain MK, Ranadive G, Yu B-Z, Verheij HM. 1991. *Biochemistry* 30:7330–40
96. Jain MK, Yu BZ, Rogers J, Gelb MH, Tsai MD, et al. 1992. *Biochemistry* 31:7841–47
97. Volwerk JJ, Pieterson WA, de Haas GH. 1974. *Biochemistry* 13:1446–54
98. Yu B-Z, Berg OG, Jain MK. 1993. *Biochemistry* 32:6485–92
99. Lombardo D, Fanni T, Pluckthun A, Dennis EA. 1986. *J. Biol. Chem.* 261:11663–65
100. Northrop DB. 1977. *Isotope Effects on Enzyme-Catalyzed Reactions.* Baltimore: University Park Press
101. O'Leary MH. 1989. *Annu. Rev. Biochem.* 58:377–401
102. Bayburt T, Yu BZ, Lin HK, Browning J, Jain MK, Gelb MH. 1993. *Biochemistry* 32:573–82
103. Wells MA. 1972. *Biochemistry* 11:1030–41

104. Teshima K, Kitagawa Y, Samejima Y, Kawauchi S, Fujii S, et al. 1989. *J. Biochem.* 106:518–27
105. Volwerk JJ, Dedieu AGR, Verheij HM, Dijkman R, de Haas GH. 1979. *Rec. Trav. Chim. Pays-Bas.* 98:214–20
106. Ikeda K, Samejima Y. 1981. *J. Biochem.* 89:1175–84
107. Ikeda K, Samejima Y. 1981. *J. Biochem.* 90:799–804
108. Teshima K, Ikeda K, Imamura M, Miyake T, Inoue S, et al. 1989. *J. Biochem.* 105:1044–51
109. Teshima K, Ikeda K, Hamaguchi K, Hayashi K. 1981. *J. Biochem.* 89:1163–74
110. Slotboom AJ, Jansen EHJM, Vlijm H, Pattus F, Soares de Araujo P, de Haas GH. 1978. *Biochemistry* 17:4593–600
111. den Kelder GMD, Hille JDR, Dijkman R, de Haas GH, Egmond MR. 1981. *Biochemistry* 20:4074–78
112. Schalkwijk CG, Marki F, van den Bosch H. 1989. *Biochim. Biophys. Acta* 1044:139–46
113. Diez E, Louis-Flamberg P, Hall RH, Mayer RJ. 1992. *J. Biol. Chem.* 267:18342–48
114. Murakami M, Kudo I, Umeda M, Matsuzawa A, Takeda M, et al. 1992. *J. Biochem.* 111:175–81
115. Baker RR, Chang H-Y. 1991. *Biochem. Cell Biol.* 69:358–65
116. Roberts MF, Adamich M, Robson RJ, Dennis EA. 1979. *Biochemistry* 18:3301–14
117. Roberts MF, Otnaess A-B, Kensil CA, Dennis EA. 1978. *J. Biol. Chem.* 253:1252–57
118. Hendrickson HS, Dennis EA. 1984. *J. Biol. Chem.* 259:5740–44
119. Jain MK, Jahagirdar DV. 1985. *Biochim. Biophys. Acta* 814:319–26
120. Jain MK, Yuan W, Gelb MH. 1989. *Biochemistry* 28:4135–39
121. de Haas GH, Bonsen PPM, Pieterson WA, van Deenen LL. 1971. *Biochim. Biophys. Acta* 239:252
122. Magolda RL, Galbraith W. 1989. *J. Cell. Biochem.* 40:371–86
123. de Haas GH, Dijkman R, Lugtigheid RB, Dekker N, van den Berg L, et al. 1993. *Biochim. Biophys. Acta* 1167:281–88
124. Davidson FF, Hajdu J, Dennis EA. 1986. *Biochem. Biophys. Res. Commun.* 137:587–92
125. Yu L, Deems RA, Hajdu J, Dennis EA. 1990. *J. Biol. Chem.* 265:2657–64
126. Yu L, Dennis EA. 1992. *J. Am. Chem. Soc.* 114:8757–63
127. Yuan W, Quinn DM, Sigler PB, Gelb MH. 1990. *Biochemistry* 29:6082–94
128. Yuan W, Fearon K, Gelb MH. 1989. *J. Org. Chem.* 54:906–10
129. Yuan W, Fearon K, Gelb MH. 1989. *J. Cell. Biochem.* 39:351–66
130. Yuan W, Gelb MH. 1988. *J. Am. Chem. Soc.* 110:2665–66
131. Yuan W, Berman RJ, Gelb MH. 1987. *J. Am. Chem. Soc.* 109:8071–81
132. Pisabarro MT, Ortiz AR, Palomer A, Cabre F, Garcia L, et al. 1994. *J. Med. Chem.* 37:337–41
133. Jain MK, Ghomashchi F, Yu B-Z, Bayburt T, Murphy D, et al. 1992. *J. Med. Chem.* 35:3584–86
134. Jain MK, Tao W, Rogers J, Arenson C, Eibl H, Yu B-Z. 1991. *Biochemistry* 30:10256–68
135. Beaton HG, Bennion C, Connolly S, Cook AR, Gensmantel NP, et al. 1994. *J. Med. Chem.* 37:557–59
136. LeMahieu RA, Carson M, Han R-J, Madison VS, Hope WC, et al. 1993. *J. Med. Chem.* 36:3029–31
137. Jain MK, Rogers JM, Smith AE, Boger ETA, Ostender RL, Yu B-Z. 1994. *Phytochemistry* In press
138. Tanaka K, Matsutani S, Kanda A, Kato T, Yoshida T. 1994. *J. Antibiot.* 47:631–38
139. Tanaka K, Matsutani S, Matsumoto K, Yoshida T. 1992. *J. Antibiot.* 45:1071–78
140. Marki F, Hanulak V. 1993. *J. Biochem.* 113:734–37
141. Kohler T, Heinisch M, Kirchner M, Peinhardt G, Hirschelmann R. 1992. *Biochem. Pharmacol.* 44:805–13
142. Washburn WN, Dennis EA. 1991. *J. Biol. Chem.* 266:5042–48
143. Lombardo D, Dennis EA. 1985. *J. Biol. Chem.* 260:7234
144. Bennett CF, Mong S, Clarke MA, Kruse LI, Crooke ST. 1987. *Biochem. Pharmacol.* 36:733
145. Reynolds LJ, Mihelich ED, Dennis EA. 1991. *J. Biol. Chem.* 266:16512–17
146. Reynolds LJ, Morgan BP, Hite GA, Mihelich ED, Dennis EA. 1988. *J. Am. Chem. Soc.* 110:5172–77
147. Barnette MS, Rush J, Marshall LA, Foley JJ. 1994. *Biochem. Pharmacol.* 47:166–67
148. de Carvalho MS, Jacobs RS. 1991. *Biochem. Pharmacol.* 42:1621–26
149. Ghomashchi F, Yu BZ, Jain MK, Gelb MH. 1991. *Biochemistry* 30:9559–69
150. Dennis EA, Davidson FF. 1990. *Prog. Clin. Biol. Res.* 349:47–54
151. Jain MK, Yu B-Z, Gelb MH, Berg OG. 1992. *Med. Inflamm.* 1:85–100
152. Fortes-Dias CL, Lin Y, Ewell J, Diniz CR, Liu R-Y. 1994. *J. Biol. Chem.* 269:15646–51

153. Ohkura N, Inoue S, Ikeda K, Hayashi K. 1993. *J. Biochem.* 112:413–19
154. Jain MK, Yu BZ, Berg OG. 1993. *Biochemistry* 32:11319–29
155. Dekker N, Peters AR, Slotboom AJ, Boelens R, Kaptein R, de Haas G. 1991. *Biochemistry* 30:3135–47
156. Peters AR, Dekker N, van den Berg L, Boelens R, Kaptein R, et al. 1992. *Biochemistry* 31:10024–30
157. Rogers J, Yu BZ, Jain MK. 1992. *Biochemistry* 31:6056–62
158. Lin G, Noel J, Loffredo W, Stable HZ, Tsai MD. 1988. *J. Biol. Chem.* 263:13208–14
159. Barlow PN, Lister MD, Sigler PB, Dennis EA. 1988. *J. Biol. Chem.* 263:12954–58
160. Hendrickson HS, Dennis EA. 1984. *J. Biol. Chem.* 259:5734–39
161. Jain MK, Rogers J, Hendrickson HS, Berg OG. 1993. *Biochemistry* 32:8360–67
162. Jain MK, Rogers J, de Haas GH. 1988. *Biochim. Biophys. Acta* 940:51–62
163. Tinker DO, Wei J. 1979. *Can. J. Biochem.* 57:97–107
164. Pattus F, Slotboom AJ, de Haas GH. 1979. *Biochemistry* 18:2691–97
165. Upreti G, Rainer S, Jain MK. 1980. *J. Membrane Biol.* 55:97–112
166. Jain MK, Maliwal BP, de Haas GH, Slotboom AJ. 1986. *Biochim. Biophys. Acta* 860:448–61
167. Thuren T, Vainio P, Virtanen JA, Somerharju P, Blomqvist K, Kinnunen PKJ. 1984. *Biochemistry* 23:5129–34
168. Apitz-Castro R, Jain MK, de Haas GH. 1982. *Biochim. Biophys. Acta* 688:349–56
169. Pluckthun A, Dennis EA. 1985. *J. Biol. Chem.* 260:11099–106
170. Volwerk JJ, Jost PC, de Haas GH, Griffith OH. 1986. *Biochemistry* 25:1726–33
171. Bell JD, Biltonen RL. 1989. *J. Biol. Chem.* 264:12194–200
172. Kensil CR, Dennis EA. 1979. *J. Biol. Chem.* 254:5843–48
173. Op den Kamp J-A, Kauerz MT, van Deenen LL. 1975. *Biochim Biophys. Acta* 406:169–77
174. Op den Kamp J-A, de Gier J, van Deenen LL. 1974. *Biochim. Biophys. Acta* 345:253–56
175. Yu B-Z, Jain MK. 1989. *Biochim. Biophys. Acta* 980:15–22
176. Jain MK, Yu B-Z, Kozubek A. 1989. *Biochim. Biophys. Acta* 980:23–32
177. Grainger DW, Reichert A, Ringsdorf H, Salesse C. 1989. *FEBS Lett.* 252:73–82
178. Grainger DW, Reichert A, Ringsdorf H, Salesse C. 1990. *Biochim. Biophys. Acta* 1023:365–79
179. McConnell HM, Tamm LK, Weis RM. 1984. *Proc. Natl. Acad. Sci. USA* 81:3249–53
180. Albrecht O, Gruler H, Sackmann E. 1978. *J. Phys.* 39:301–13
181. Reichert A, Wagenknecht A, Ringsdorf H. 1992. *Biochim. Biophys. Acta* 1106:178–88
182. Maloney KM, Grainger DW. 1993. *Chem. Phys. Lipids* 65:31–42
183. Menashe M, Romero G, Biltonen RL, Lichtenberg D. 1986. *J. Biol. Chem.* 261:5328–33
184. Romero G, Thompson K, Biltonen RL. 1987. *J. Biol. Chem.* 262:13476–82
185. Bell JD, Biltonen RL. 1989. *J. Biol. Chem.* 264:225–30
186. Bell JD, Brown SD, Baker BL. 1992. *Biochim. Biophys. Acta* 1127:208–20
187. Bell JD, Biltonen RL. 1992. *J. Biol. Chem.* 267:11046–56
188. Tomasselli AG, Hui J, Fisher J, Zurcher-Neely H, Reardon IM, et al. 1989. *J. Biol. Chem.* 264:10041–47
189. Cho W, Tomasselli AG, Heinrikson RL, Kezdy FJ. 1988. *J. Biol. Chem.* 263:11237–41
190. Lugtigheid RB, Nicolaes GA, Veldhuizen EJ, Slotboom AJ, Verheij HM, de Haas GH. 1993. *Eur. J. Biochem.* 216:519–25
191. Noel JP, Bingman CA, Deng TL, Dupureur CM, Hamilton KJ, et al. 1991. *Biochemistry* 30:11801–11
192. Drainas D, Lawrence AJ. 1978. *Eur. J. Biochem.* 91:131–38
193. Diaz REC, Elansari O, Lawrence AJ, Lyall F, McLeod WA. 1985. *Biochim. Biophys. Acta* 830:52–58
194. van der Wiele FC, Atsma W, Dijkman R, Schreurs AMM, Slotboom AJ, de Haas GH. 1988. *Biochemistry* 27:1683–88
195. van der Wiele FC, Atsma W, Roelofsen B, van Linde M, Binsbergen JV, et al. 1988. *Biochemistry* 27:1688–94
196. van Oort MG, Dijkman R, Hille JD, de Haas GH. 1985. *Biochemistry* 24:7993–99
197. Bukowski T, Teller DC. 1986. *Biochemistry* 25:8024
198. van Eijk JH, Verheij HM, Dijkman R, de Haas GH. 1983. *Eur. J. Biochem.* 132:183
199. Rebecchi M, Peterson A, McLaughlin S. 1992. *Biochemistry* 31:12742–47
200. Volwerk JJ, Filthuth E, Griffith OH, Jain MK. 1994. *Biochemistry* 33:3464–74
201. Lewis KA, Garigapati VA, Zhou C,

Roberts MF. 1993. *Biochemistry* 32: 8836–41

202. Hanel AM, Schuttel S, Gelb MH. 1993. *Biochemistry* 32:5949–58

203. Wright GW, Ooi CE, Weiss J, Elsbach P. 1990. *J. Biol. Chem.* 265: 6675–81

204. Weiss J, Wright G, Bekkers AC, van den Bergh CJ, Verheij HM. 1991. *J. Biol. Chem.* 266:4162–67

205. Murakami M, Hara N, Kudo I, Inoue K. 1993. *J. Immunol.* 151:5675–84

206. Ishizaki J, Hanasaki K, Higashino K, Kishino J, Ohara O, et al. 1994. *J. Biol. Chem.* 269:5897–904

207. Lambeau G, Ancian P, Barhanin J, Lazdunski M. 1994. *J. Biol. Chem.* 269: 1575–78

208. Rodgers W, Glaser M. 1993. *Biochemistry* 32:12591–98

Annu. Rev. Biochem. 1995. 64:689–719

METABOLIC COUPLING FACTORS IN PANCREATIC β-CELL SIGNAL TRANSDUCTION

Christopher B. Newgard and J. Denis McGarry

Gifford Laboratories for Diabetes Research and Departments of Biochemistry and
Internal Medicine, University of Texas Southwestern Medical Center at Dallas,
5323 Harry Hines Boulevard, Dallas, Texas 75235

KEY WORDS: β-cell dysfunction, hyperlipidemia, glucose phosphorylation, insulin, glucose

CONTENTS

ABSTRACT

This chapter focuses on the biochemical mechanisms that mediate glucose-
stimulated insulin secretion (GSIS) from β-cells of the islets of Langerhans
and the potentiating role played by fatty acids. We summarize evidence supporting the idea that glucose metabolism is required for GSIS and that the
GLUT-2 facilitated glucose transporter and the glucose phosphorylating enzyme glucokinase play important roles in measuring changes in extracellular
glucose concentration. The idea that glucose metabolism is linked to insulin

689

0066-4154/95/0701-0689$05.00

secretion through a sequence of events involving changes in ATP:ADP ratio, inhibition of ATP-sensitive K^+ channels, and activation of voltage-gated Ca^{2+} channels is critically reviewed, and the relative importance of ATP generated from glycolytic versus mitochondrial metabolism is evaluated. We also present the growing concept that an important signal for insulin secretion may reside at the linkage between glucose and lipid metabolism, specifically the generation of the regulatory molecule malonyl CoA that promotes fatty acid esterification and inhibits oxidation. Finally, we show that in contrast to its short term potentiating effect on GSIS, long-term exposure of islets to high levels of fatty acids results in β-cell dysfunction, suggesting that hyperlipidemia associated with obesity may play a causal role in the diminished GSIS characteristic of non insulin-dependent diabetes mellitus (NIDDM).

PERSPECTIVES AND SUMMARY

In mammalian species, the islets of Langerhans synthesize polypeptide hormones that direct the storage and utilization of simple metabolic fuels. Fuel homeostasis is in large part controlled by the interplay of two products of the islets, glucagon and insulin. The two hormones exert opposing effects on cellular metabolism, and their secretion is controlled in a reciprocal manner by factors such as diet, nutritional status, and stress (1). Thus, a high insulin:glucagon ratio is a hallmark of the postprandial state, and activates anabolic pathways such as storage of glucose as glycogen and fatty acids as triglycerides. Conversely, a high glucagon:insulin ratio is typical of catabolic situations such as fasting or "fight or flight," and results in increased levels of simple fuels in the blood such as glucose arising from glycogenolysis and free fatty acids from lipolysis.

The mechanisms underlying the remarkable titration of glucagon and insulin levels are complex. In recent years, particular attention has been focused on understanding the biochemical events that couple fluctuations in the levels of metabolic fuels to changes in the rate of insulin secretion from the β-cells of the islets of Langerhans. Interest in this area naturally evolves from its relevance to diabetes. Insulin-dependent diabetes mellitus (IDDM), also known as Type I diabetes, occurs as a consequence of autoimmune destruction of the insulin-containing β-cells, while leaving the glucagon producing α-cells essentially untouched. The discovery of insulin in 1922 by Banting & Best resulted in implementation of insulin injection as a therapeutic strategy for IDDM, an approach that remains largely unchanged to the present day. While clearly life-saving, insulin injection fails to reproduce the exquisite control of fuel homeostasis afforded by the normal islets, which regulate delivery of insulin and glucagon on a moment-to-moment basis in response to changes in fuel levels. Thus, a number of alternative approaches involving delivery of insulin from transplanted cells capable of normal fuel-sensing responses have been contemplated (2, 3, 4). Clearly, a prerequisite for coherent development of such approaches is a detailed understanding of the biochemistry of fuel-mediated insulin release in the normal islet. Noninsulin dependent diabetes mellitus (NIDDM), also known as Type II diabetes, appears to be a consequence of two distinct cellular/metabolic lesions, insulin resistance and β-cell dysfunction, and is often linked with obesity. Insulin resistance is manifested by a markedly reduced effectiveness of the hormone for stimulation of glucose uptake and suppression of lipolysis in insulin-sensitive tissues such as muscle and fat. The β-cell defect has two components. First, in both diabetic and nondiabetic insulin-resistant states, insulin secretion in response to nonstimulatory concentrations of glucose (≤ 5 mM) is markedly elevated relative to normal controls (5, 6). The progression from obesity and insulin resistance to frank diabetes is associated with the appearance of the second lesion, the loss of the insulin secretory response to levels of glucose (generally 8 mM and above) that cause large increases in insulin release from normal islets. Since insulin resistance is rarely sufficient to cause diabetes, the β-cell lesion appears to be a crucial pathophysiological component of the disease. Understanding the mechanisms underlying the insulin secretory response to glucose and other fuels in normal islets is therefore critical for understanding the basis for β-cell dysfunction in NIDDM.

Insulin secretion from islet β-cells is affected by a wide array of factors. Perhaps the most important factor is glucose, since the response to many of the other effectors either requires or is potentiated by the presence of stimulatory levels of the sugar. The stimulatory effect of glucose has been linked to its metabolism in β-cells, and acceptance of this concept has led to the more general paradigm that metabolic fuels capable of affecting insulin release do so via entry into the β-cells and generation of metabolic signals, and are thus distinguishable from the mode of action of hormones or other signaling molecules that bind to membrane-localized receptors. In this article we review recent advances in understanding of metabolic coupling factors in fuel-mediated insulin release. We restrict ourselves to mechanisms involved in glucose and lipid-stimulated insulin secretion, and the interplay of glucose and lipid metabolism in the creation of relevant signals. We only mention here that the effects of glucose and lipids are often augmented by other agents. Several amino acids, most notably leucine and arginine, have powerful potentiating effects on glucose-stimulated insulin release. The effect of leucine appears to be mediated in part by allosteric activation of glutamate dehydrogenase and augmented glutamate metabolism (7). Leucine catabolism through the branched chain ketoacid dehydrogenase complex (BCKDH) may also contribute, since culturing of islets at high glucose sharply down-regulates BCKDH mRNA and abrogates leucine-induced insulin release (8). Several mechanisms have been suggested for arginine-induced insulin release (9). Schmidt, et al recently proposed that the arginine effect is mediated by its conversion into citrulline by nitric oxide synthetase (NOS) and consequent generation of nitric oxide (10). This model has been refuted by experiments showing that arginine analogs that are not NOS substrates are also able to stimulate insulin release (11). The effect of arginine may instead be mediated by a direct membrane-depolarizing effect of this charged amino acid (12, 13). Insulin secretion can also be mediated by peptide hormones such as glucagon or glucagon-like peptide-1 (GLP-1), or other pharmacologic agents that act through increases in cellular levels of cyclic AMP (14, 15, 16). The mechanisms by which these and other nonfuel secretagogues affect insulin secretion have been reviewed recently (14) and are discussed here only in the context of glucose and lipid-mediated effects.

REGULATION OF GLUCOSE-STIMULATED INSULIN RELEASE

Development of the Fuel Hypothesis

Early insights into the mechanism of glucose-stimulated insulin secretion were provided by Grodsky and co-workers, who showed that insulin release from the isolated perfused pancreas could be stimulated by glucose and, to a lesser extent, by other metabolizable sugars such as mannose and fructose, but not by poorly metabolized sugars such as galactose, xylose, or L-arabinose (17). The concept gained support with the demonstration that nonmetabolizable analogs of glucose such as 2-deoxyglucose or 3-O-methyl glucose were ineffective at promoting insulin release, and that inhibitors of glycolysis such as mannoheptulose or glucosamine blocked the response to D-glucose (18, 19). Equally compelling evidence was later provided when insulin secretion and the rate of glucose metabolism were found to increase in parallel in response to elevations in the extracellular glucose concentration (20, 21). These observations suggested that glucose and related metabolic fuels stimulate insulin release via entry into the islet β-cell and metabolism to generate signals that activate exocytosis of insulin-containing granules (the "fuel hypothesis"). An alternative model, whereby glucose was proposed to stimulate insulin release by interacting with "glucoreceptor" molecules in the plasma membrane, derived from experiments in which the metabolism of glucose and the insulin secretory effects of the sugar could be dissociated (22). Over time, the weight of evidence has accumulated in favor of the fuel hypothesis, which is now broadly accepted (14, 20, 23).

Several important questions are raised by a model in which glucose metabolism is responsible for triggering insulin release. First, what is the nature of the sensor or regulator that allows changes in extracellular glucose concentration to be translated into proportional increases in insulin secretion? Second, what is the identity of the glucose-derived or glucose-initiated signals that trigger insulin secretion? Third, how do metabolic signals triggered by glucose communicate with the secretory apparatus? Although these issues have been investigated during the past three decades, and have been extensively reviewed (14, 18, 23, 24, 25, 26, 27), the application of pharmacologic and molecular approaches has led very recently to dramatic new findings, the focus of this article.

The rate of glucose metabolism in islet β-cells increases in proportion to the extracellular concentration of the sugar over the range of 5–30 mM (20, 23). Hepatocytes and islet β-cells are similar in that they share expression of several specialized gene products involved in glucose metabolism such as the GLUT-2–facilitated glucose transporter (28, 29), glucokinase (30, 31), and glucose-6-phosphatase (32). Despite these similarities, glucose is handled differently by the two cell types. In liver, a large portion of glucose influx occurring at levels of the sugar characteristic of the fed state (7–8 mM) is diverted into glycogen synthesis and the pentose monophosphate shunt, with a relatively small glycolytic component. In islets, glycolysis is by far the predominant pathway. Glycogen synthesis accounts for approximately 1–7% of glucose flux, and glycogen levels do not approach those found in liver unless islets are maintained under chronic hyperglycemic conditions (33). In the presence of stimulatory glucose levels, pentose shunt activity accounts for only 2% of glucose metabolism (34). Despite their expression of glucose-6-phosphatase and pyruvate carboxylase, islets do not carry out gluconeogenesis, as they appear to lack phosphoenolpyruvate carboxykinase and fructose-1, 6-bisphosphatase activities (23). Glycolytic flux or "glucose usage" is usually determined in islets by measurement of ^3H_2O generation from 5-^3H glucose. Using this method, Matschinsky and colleagues demonstrated that islets contain both a high-affinity and low-affinity component of glucose usage (23, 35). The high-affinity component was defined as that operative over a range of glucose concentrations from 1–100 μM, with an apparent K_m of 215 μM and a V_{max} of 18 pmol/islet/hour. The low-affinity component was defined as that occurring over the range of glucose concentrations from 1–100 mM, with an apparent K_m of 11.3 mM and a V_{max} of 160 pmol/islet/h. Thus, within the range of physiologically relevant glucose concentrations (4–9 mM), the low-affinity component of glycolytic flux is clearly the predominant pathway for glucose metabolism in intact islets.

The Role of Glucose Phosphorylation

Since the rate of glucose metabolism in β-cells is proportional to the amount of insulin released, understanding of the factors controlling glucose flux should define the sensor or regulator for the response. Evidence has mounted over more than two decades that glucokinase is the rate-limiting or pace-setting step in islet β-cell glucose metabolism. Early indications of the importance of glucose phosphorylation that came from studies with inhibitors such as manno-heptulose (19, 36) were later supplemented by direct measurement of glucose uptake and phosphorylation in islet cells (21). Matschinsky & Ellerman showed in 1968 that intracellular concentrations of glucose closely approximated those of the extracellular environment, and that islet homogenates contain both a high K_m and low K_m glucose phosphorylating activity (21). Glucose is phosphorylated in mammalian tissues by members of the hexokinase gene family. Features that distinguish hexokinases I, II, and III from hexokinase IV or glucokinase include a lower K_m for glucose (10–100 μM vs approximately 8 mM), a larger protein mass (100 vs 52 kDa), and allosteric inhibition by glucose-6-phosphate (37). Biochemical and molecular approaches have now established that the high K_m glucose phosphorylating activity of islets is at-tributable to glucokinase (21, 23, 30, 31) and that hexokinase I probably contributes most of the low K_m activity (38). Interestingly, hexokinase activity actually slightly exceeds glucokinase activity in normal islet extracts (23, 39), despite the vast predominace of the low-affinity component of glucose usage. The proposed explanation for this finding is that hexokinase activity is inhib-ited in the intact β-cell, possibly by glucose-6-phosphate (35, 40). Measure-ment of the activities of other potential regulatory enzymes of glycolysis such as phosphofructokinase and pyruvate kinase revealed that these enzymes are present in islets at maximal activities exceeding that of glucokinase by at least an order of magnitude (35). Finally, the curve describing the glucose depen-dence of glucokinase enzyme activity in islet extracts is virtually superimpos-able upon that describing the low-affinity component of glucose usage (23). These and other observations have led to wide acceptance of the notion that glucokinase constitutes the β-cell glucose sensor.

Recent genetic studies have provided strong new evidence for an important role of glucokinase in regulation of glucose-stimulated insulin secretion (GSIS). Froguel and colleagues first reported linkage of mutations in the glucokinase gene with a subtype of noninsulin dependent diabetes known as maturity-onset diabetes of the young (MODY) (41). Subsequent studies have identified a large number of glucokinase mutations, mostly point mutations occurring in the protein-coding region of the gene, in MODY pedigrees (42). Patients are universally heterozygous, with one mutant and one normal allele of glucokinase. The importance of glucokinase for normal islet function is illustrated by the fact that most MODY patients with glucokinase mutations studied to date exhibit altered GSIS (43, 44). The major effect is on the threshold of the glucose response, such that higher glucose levels are required for triggering insulin secretion in MODY patients than in normal individuals

(43). The effect of the various mutations has been evaluated by expression of the mutant genes in bacteria and kinetic analyses of the purified, expressed proteins (45). The magnitude of this rightward shift in glucose dose response is generally proportional to the degree to which enzymatic activity is altered by the mutation (43), although there are exceptions to the general rule, such as the V203A mutant, which predicts a severely impaired enzyme (45) but which causes only a mild shift to the right in terms of the threshold for glucose response (43). Polonsky and co-workers have proposed that such patients may be able to compensate for severe mutations by increasing expression of the normal allele (43), but this remains to be demonstrated.

An important role for glucokinase has also been demonstrated in studies in which the insulin promoter was used to express an antisense glucokinase ribozyme in β-cells of transgenic mice, resulting in a 70% reduction in immunodetectable glucokinase protein and enzyme activity (46). Pancreas perfusion studies on these mice revealed a reduced and "right-shifted" GSIS response relative to controls. Unlike MODY patients, however, the transgenic mice were normoglycemic, possibly because of reduced hepatic glucokinase activity in the patients that could result in less effective glucose disposal. The precise relationship between β-cell glucokinase activity and insulin release has not yet emerged; in humans because of the inability to obtain islet samples for measurement of glucokinase activity, and in transgenic animals expressing anti-glucokinase ribozyme because only a single transgenic line with a fixed reduction of glucokinase has been examined. Furthermore, glucose usage has not been studied in either model.

Glucose phosphorylating enzymes have also been overexpressed in islets and insulinoma cell lines. Although these studies generally support the concept of an important role for the glucose phosphorylation step in dictating the concentration dependence and magnitude of the GSIS response, they also introduce a note of caution. Recombinant adenovirus has been used to deliver genes to isolated islets of Langherhans with high efficiency (47, 48). Becker et al constructed recombinant adenoviruses containing the islet (AdCMV-GKI) and liver (AdCMV-GKL) isoforms of glucokinase, and used them to achieve 20- to 30-fold overexpression of glucokinase enzyme activity in normal islets. Perifusion of islets overexpressing liver or islet glucokinase revealed only a 36–53% enhancement in insulin secretion in response to 20 mM glucose or 20 mM glucose + 30 mM arginine relative to normal islets (TC Becker, RJ Noel, JH Johnson, J Takeda, GI Bell, & CB Newgard, manuscript submitted for publication). Remarkably, no enhancement of 2- or 5-^{3}H glucose usage or lactate production was observed in these islets relative to untreated islets or islets treated with a control virus containing the β-galactosidase gene (AdCMV-βGAL) from *Escherichia coli*.

Low K_m hexokinases have also been overexpressed in normal islets by several groups. In one such study, electroporation was used to cotransfect fetal islets with a plasmid containing the hexokinase I cDNA and one containing an insulin promoter/chloramphenicol transferase (CAT) reporter gene. This manuever resulted in a shift in the threshold for glucose-induced insulin promoter activation from 5 mM in untransfected islets to ≤ 1 mM in the transfected cells (49). Epstein and co-workers used the rat insulin I promoter/enhancer to express yeast hexokinase in islets of transgenic mice. These animals showed increased insulin secretion and an enhanced capacity to dispose of a glucose load relative to normal mice (50). Becker et al overexpressed rat hexokinase I in isolated rat islets with a recombinant adenovirus (AdCMV-HKI) (47). Islets from transgenic animals expressing yeast hexokinase (51) or AdCMV-HKI-treated islets (47) cultured at 11 mM glucose exhibited an approximate ten–fold increase in hexokinase enzyme activity and a 2–4–fold enhancement of insulin secretion, glucose usage, and lactate production. The enhanced usage and insulin secretion were apparent only at basal (3 mM) but not stimulatory (20 mM) glucose levels. It is remarkable that such similar findings were reported by the two groups, since yeast hexokinase is not allosterically regulated by glucose-6-phosphate, while hexokinase I is inhibited by this ligand with a K_i of approximately 100 µM (37). Furthermore, in the adenovirus experiments, a large portion of the overexpressed hexokinase I was found to be associated with mitochondria (47), in which state the enzyme is thought to be more active and less inhibitable by glucose-6-phosphate (37). Thus the discrepancy between the high levels of enzyme activity and the modest changes in glucose usage and insulin secretion achieved in the two types of experiment does not appear to be explained by inhibition of the overexpressed hexokinase activity within the intact cell. The results of the glucokinase and hexokinase overexpression experiments are consistent with a model in which other steps in metabolism become rate-limiting for glycolytic flux and insulin release under conditions where these glucose phosphorylating enzymes are overexpressed. An interesting problem for future study is to understand the mechanisms underlying differential effects of overexpresed glucokinase versus other hexokinases on islet cell metabolism and insulin release.

While the magnitude of the insulin secretion response is not directly proportional to the extent of overexpression of glucose phosphorylating enzymes, the studies cited above show that overexpression of hexokinase alters the dose dependence of glycolytic flux and insulin secretion. A similar conclusion has been reached from studies with insulinoma cell lines. Most insulinoma cell lines are derived from islet tumors produced in X-irradiated rats (RIN cells), or more recently, by expression of SV40 T-antigen in islet β-cells via a construct containing the rat insulin promoter/enhancer. Some of these lines, such as RINm5F (53), RIN1046–38 (54), βTC-1 (55), or MIN-7 (56), either exhibit no GSIS or respond maximally to submillimolar concentrations of the sugar. Those lines responding to low glucose levels have an elevated high-affinity component of glucose metabolism and an increased hexokinase:glucokinase ratio relative to normal islet β-cells (53, 59). Other lines, such as MIN-6 (56), βTC-7 (57), or INS-1 (60), exhibit a normal dose-response curve for GSIS. When studied at low passage number, βTC-7 cells exhibit a glucokinase:hexokinase ratio similar to that of normal islets. With time in culture, these cells exhibit a left-shifted glucose dose-response curve that corresponds to a sixfold increase in hexokinase activity, with little change in the level of glucokinase or GLUT-2 expression (57). Similarly, overexpression of hexokinase I in MIN-6 cells causes a shift in the threshold for responsiveness from 5 mM in untransfected cells to less than 1 mM in transfected cells (61). Finally, RIN1046–38 cells that are engineered for GLUT-2 expression (see below) undergo a spontaneous increase in glucokinase activity but contain four times as much hexokinase as normal islets and are maximally responsive to glucose at micromolar concentrations of the sugar (62). Treatment of these cells with 2-deoxyglucose, which upon conversion to 2-deoxyglucose-6-phosphate inhibits hexokinase but not glucokinase activity, results in a partial normalization of the glucose dose-response curve, such that treated cells are maximally responsive at 5 mM rather than 50 mM glucose (62). All of these studies point to a critical role for the glucokinase:hexokinase ratio in dictating the dose threshold for GSIS. They also provide guidance for those investigators interested in engineering glucose-responsive cell lines that could be used for cell-based insulin delivery in IDDM (2, 4, 62, 63).

The Role of Glucose Transport

Early observations indicating that islets are capable of rapid equilibration of glucose across the plasma membrane coupled with the compelling data concerning glucose phosphorylation summarized above led to the assumption that glucose transport is not a regulatory or rate-determining step in GSIS (23). Further investigation of this concept in the past five years has led to the reemergence of the glucose transport step as a potentially important component of the glucose sensor. An important advance in this area came with the demonstration that antibodies and cDNA probes specific to the hepatic facilitated glucose transporter known as GLUT-2 cross-react with gene products expressed in islet β-cells (29). Subsequent kinetic studies revealed that islets transport 3-O-methyl glucose with kinetics similar to those measured in hepatocytes (28), and cloning of GLUT-2 cDNAs from human (65) and rat (28) islet libraries revealed their structures to be identical to their hepatic counterparts. The fact that GLUT-2, like glucokinase, has a higher K_m and V_{max} than other members of its gene family led to the proposal that the transporter may participate in controlling GSIS, perhaps by teaming with glucokinase to form a "glucose-sensing apparatus" (66, 67).

This concept gained support through a series of studies in which immunological methods were used to demonstrate sharply reduced GLUT-2 expression in islet β-cells in a wide array of rodent models of β-cell dysfunction (reviewed in ref. 67). For example, GLUT-2 expression was found to be unchanged in female Zucker diabetic fatty (ZDF) rats that retain normal GSIS and exhibit complete resistance to diabetes, but in male animals of the same strain that lack glucose responsiveness and develop diabetes with 100% incidence, GLUT-2 expression was found to be reduced in proportion to the severity of the hyperglycemia (5, 68). GLUT-2 expression was also suppressed in islets of db/db mice (69), GK rats (70), streptozotocin-injected neonatal rats (71), and Zucker or Wistar rats rendered diabetic by dexamethasone injection (72). While in all these cases loss of immunologically detectable GLUT-2 was associated with the loss of GSIS, depletion of GLUT-2 associated with overexpression of the H-ras oncogene (73), in response to culturing of islets in vitro (74), or as a result of hypoglycemic perfusion of the pancreas (75), did not serve to abrogate GSIS. Recently, quantitative histochemical methods were used to calculate the in situ glucose uptake capacity of islets from some of the models described above (76). Glucose uptake in response to an injection of the sugar was found to be reduced in islets of Zucker obese rats and db/db mice relative to the appropriate normal controls, but not in mice expressing the H-ras oncogene or partially pancreatectomized animals. Finally, a recent article describes a single NIDDM patient with a point mutation in the GLUT-2 gene (V197L) that renders the expressed mutant protein unable to transport glucose (77). This finding suggests that reduced expression of GLUT-2 may be sufficient to cause diabetes, although the contribution of diminished GLUT-2 in β-cells relative to other sites of expression of the gene, such as liver or gut, was not determined. In sum, many studies have shown a link between loss of GLUT-2 expression and β-cell dysfunction, but whether the reduced transporter expression is causal or secondary remains to be established.

In studies performed to date on NIDDM patients, mutations in the GLUT-2 gene are rare (77). Loss of GLUT-2 expression may instead be a byproduct of the diabetic environment since normal islets transplanted into db/db or streptozotocin-diabetic rats lose their GLUT-2 expression, while db/db islets transplanted into normal animals show an increase in immunodetectable GLUT-2 (69). The specific factor affecting GLUT-2 expression in these experiments remains to be identified. Logical candidates would include a dearth of insulin, reduced insulin action, or the effect of hyperglycemia per se. However, these ideas are inconsistent with experiments showing that insulin infusion sharply depresses GLUT-2 mRNA and activity (78), or that glucose infusion (78) or culturing of islets at high glucose (68) increases or maintains GLUT-2 levels, respectively. It is possible, of course, that expression of genes other than GLUT-2 are decreased in glucose-unresponsive islets, as evidenced by reduced glycerol phosphate shuttle activity in at least one model, the streptozotocin-injected neonatal rat (79). It is equally plausible that overexpression of certain other genes, including hexokinase (47, 80), glucose-6-phosphatase (81), or lactate dehydrogenase (82) might contribute to either the elevated basal insulin release or decreased glucose responsiveness characteristic of islets from animal models of NIDDM. Interestingly, no depletion in glucokinase enzymatic activity has been reported in any such models studied to date.

The correlative approaches described above, although resulting in some provocative observations, have not yet defined the precise role of GLUT-2 in normal islet β-cell glucose sensing and its importance in the pathogenesis of NIDDM relative to other gene products. Recent application of the tools of molecular biology have provided some new perspectives on these issues. Stable transfection of glucose-unresponsive insulinoma cells such as RIN1046–38 cells of intermediate passage number (62) or RINm5F cells (83), or an insulin-producing anterior pituitary cell line known as AtT-20ins (39, 84), with the GLUT-2 cDNA confers GSIS and increased insulin content. This effect appears to be specific to GLUT-2, since AtT-20ins cells transfected with GLUT-1 do not respond, even though such lines metabolize glucose at rates equal to those expressing GLUT-2 (84). While the enhancing property of GLUT-2 in this system thus appears to be distinct from any effect of the transporter on glucose flux, the continued requirement for glucose metabolism is indicated by the fact that nonmetabolizable analogs of glucose such as 3-O-methyl glucose or 2-deoxyglucose do not stimulate insulin release or potentiate the effect of forskolin in GLUT-2 transfected cells (84). These studies suggest that GLUT-2 expression is required for efficient transmission of signals derived from glucose metabolism, but they do not address the issue of whether the response is a direct consquence of the presence of GLUT-2 in the cell or is related to modulation of the expression of other mediators. In support of the latter idea, stable transfer of the GLUT-2 gene into insulinoma cells causes a fourfold increase in glucokinase enzymatic activity that is at least partially related to an increase in immunodetectable glucokinase protein (62). Using adenoviral vectors, it is possible to study cells expressing GLUT-2 prior to the induction of glucokinase activity, which occurs 48–72 hours after introduction of the GLUT-2 gene. Cells transduced with either AdCMV-GLUT2 or AdCMV-GKI viruses do not respond to glucose directly, although the GLUT-2–expressing cells do display glucose potentiation of forskolin-stimulated insulin release. Only when GLUT-2 and glucokinase are co-overexpressed by mixing of the two viruses do the RIN cells become glucose responsive (H BeltrandelRio & CB Newgard, manuscript in preparation). In sum, GLUT-2 expression appears to be a necessary but not sufficient element of GSIS; also required is the presence of substantial levels of glucokinase activity and possibly other factors. Issues for future investigation in this area include: (a) Does GLUT-1 overexpression increase glucokinase activity in a manner similar to that of GLUT-2 in RIN cells? (b) Does co-incubation of RIN cells with AdCMV-GLUT2 and AdCMV-GKI viruses alter the rate of glucose metabolism relative to that in untreated cells or cells treated with either virus alone? (c) What are the structural elements that distinguish GLUT-2 and GLUT-1 and confer the capacity for GSIS in insulin-secreting cell lines?

GLUCOSE-INITIATED SIGNALS FOR INSULIN RELEASE

The General Model

As the notion that the insulin secretion response is proportional to the rate of glucose metabolism has become generally accepted, the search for the identity of the glucose-induced or glucose-derived signals for insulin release has intensified. A major clue came with the discovery that pancreatic islets and insulinoma cell lines contain ATP-sensitive K^+ channels (85, 86). These channels are closed by incubation of islets with carbohydrate fuels that stimulate insulin release such as glucose, glyceraldehyde, or mannose, but are unaffected by poorly metabolized fuels such as 2-deoxyglucose or galactose (85, 87). Pancreatic islets also contain L-type voltage-gated Ca^{2+} channels that are opened in the presence of metabolizable fuels (88). Inhibition of L-type channels with dihydropyridines and other agents blocks insulin secretion in response to carbohydrate secretagogues. The importance of Ca^{2+} import in the GSIS reponse is further illustrated by the finding that islets cultured at low or absent extracellular Ca^{2+} levels respond poorly or not at all to fuel secretagogues (26, 27). These seminal experiments have led to a commonly accepted theory that glucose metabolism leads to an increase in the ATP:ADP ratio, resulting in inhibition of ATP-sensitive K^+ channels (K_{ATP} channels) and subsequent activation of voltage-gated Ca^{2+} channels (87). Increases in intracellular Ca^{2+} (Ca^{2+}_i) are thought to trigger insulin release through activation of protein kinases that interact with components of the microtubular/exocytotic machinery (26, 27). While essential features of this model are likely to be correct, a more detailed understanding of events that resolves longstanding questions is only beginning to emerge. One such question concerns the role of mitochondrial metabolism. The fact that inhibitors of electron transport or uncouplers of mitochondrial ATP production block GSIS suggests that glucose exerts its effect by stimulating mitochondrial metabolism (8, 23, 89). If correct, this model fails to explain why glyceraldehyde, a glycolytic substrate, is a powerful insulin secretagogue, while pyruvate and lactate, which are readily

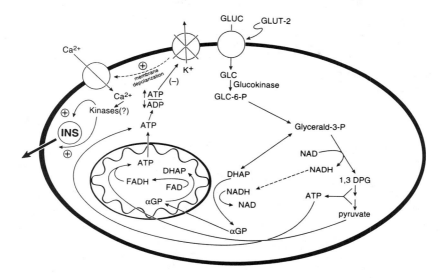

Figure 1   Schematic summary of signals for insulin release generated by glucose metabolism.

oxidized by islets, have little effect on insulin release (90, 91). A second question is whether changes in ATP:ADP ratio are of sufficient magnitude to explain inhibition of K_{ATP} channels (92). Also unclear is how factors that potentiate or enable GSIS interact with the basic pathway consisting of glucose → increased ATP:ADP ratio → inhibition of K_{ATP} channels → activation of voltage-gated Ca^{2+} channels. For example, glucagon-like peptide-1 (GLP-1), a hormone secreted from the gut in response to a carbohydrate load, enhances the GSIS response if given prior to a glucose challenge (15, 16, 93). It has also been suggested that certain lipid mediators such as diacylglycerol (DAG) or arachidonic acid are required for normal GSIS (26, 27, 94), but the precise role of these substances remains to be elucidated. Finally, changes in the levels of other nucleotides or nucleotide derivatives such as GTP (95) and cyclic ADP-ribose (96) have been proposed as mediators of GSIS. New information pertaining to these issues is summarized below and in Figure 1.

Relative Roles of Glycolytic vs Mitochondrial Metabolism

The fact that glyceraldehyde serves as a potent insulin secretagogue while pyruvate does not has never been satisfactorily explained. The dilemma is heightened by consideration of the following experimental data. First, pyruvate and lactate have mild potentiating effects in the presence of stimulatory glucose concentrations but fail to stimulate insulin release from islets in the absence of glucose, even at concentrations as high as 30 mM (90, 91). In contrast, other mitochondrial fuels such as 2-ketoisocaproate (the deamination product of leucine) (97) or methyl esters of succinate (98) are potent insulin secretagogues either in the presence or absence of glucose. Second, the inability of pyruvate or lactate to stimulate insulin release appears to be unrelated to the islets' capacity to oxidize these fuels. The rate of CO_2 generation from [U-^{14}C] lactate or pyruvate at a concentration of 20–30 mM is equivalent to the rate of CO_2 generation from [U-^{14}C] glucose at 10–15 mM (90, 91), yet only the latter stimulates insulin secretion. Third, fuels that stimulate insulin release (glucose, glyceraldehyde, and 2-ketoisocaproate) all increase the ATP:ADP ratio and the level of reduced pyridine nucleotides (NADH + NADPH) and also substantially increase intracellular Ca^{2+}, while pyruvate and lactate fail to alter any of these parameters when administered to islets in the absence of glucose (90, 91). In the presence of stimulatory glucose, pyruvate and lactate potentiate insulin secretion and enhance glucose-induced changes in pyridine nucleotides and Ca^{2+} influx without measurably affecting the ATP:ADP ratio (90, 91).

In recent years, a large number of pharmacologic agents have been employed to address the mysteries posed by the foregoing data and to better define the specific events that mediate fuel-mediated insulin release. Thus, glucose or methyl succinate-stimulated insulin release were found to be sharply reduced by a wide range of inhibitors of mitochondrial electron transport and ATP production, including antimycin A, rotenone, cyanide, and FCCP (89), suggesting that mitochondrial metabolism is an essential component of the response to these fuels. In contrast, a recent study by Dukes et al shows that incubation of β-cells or islets with 2α-methoxycynaocinnamate (αCHC), an inhibitor of mitochondrial pyruvate transport, or fluoroacetate, an inhibitor of the citric acid cycle, had no discernible effect on glucose or glyceraldehyde-induced closure of K_{ATP} channels (99). These same investigators reported that uncoupling of ATP production from glycolysis at the phosphoglycerate kinase step with arsenate, or at the pyruvate kinase step with fluoride, also had no effect on glucose-mediated K_{ATP} channel closure. In contrast, iodoacetate, which inhibits glyceraldehyde-3-phosphate dehydrogenase (G3PDH), was completely effective in preventing glucose-induced closure of K_{ATP} channels (99) and stimulation of insulin release (89). The conclusion advanced from these data (99) was that neither glycolytically derived ATP nor ATP produced directly from operation of the citric acid cycle participates significantly in closure of K_{ATP} channels that regulate fuel-mediated insulin release. Instead, generation of NADH in the G3PDH reaction was proposed as the critical signaling event. In support of this idea, treatment of islets with the glucose analog streptozotocin, which causes depletion of NAD^+, abrogated glucose-mediated closure of K_{ATP} channels and Ca^{2+} oscillations. Treatment of these same islets with nicotinamide to restore NAD^+ levels resulted in a partial recovery of glucose sensing, as evidenced by return of Ca^{2+} oscillations (99). Dukes and co-workers speculated further that NADH generated at the G3PDH step fuels mitochondrial ATP production via the malate-aspartate and glycerol phosphate-dihydroxyacetone phosphate shuttles, funneling reducing equivalents to sites 1 and 2 of the electron transport chain, respectively. The ATP produced in this way would then inhibit K_{ATP} channels and activate Ca^{2+} influx, as described earlier.

The notion of a critical role for NADH generated in the G3PDH reaction has several attractive features but also opens new questions. On the positive side, the model fits conceptually with the long-recognized fact that fuel-mediated insulin release is linked to the redox state in pancreatic islets (20, 23, 24, 25, 26, 27). It also identifies a candidate intermediate that lies between glyceraldehyde-3-phosphate and pyruvate that might explain why glucose and glyceraldehyde, but not pyruvate and lactate, are effective insulin secretagogues. Furthermore, the construct is consistent with the observations that activities of the mitochondrial FAD-linked glycerol phosphate dehydrogenase enzyme and the malate-aspartate shuttle in pancreatic islets greatly exceed those of other tissues such as liver, pancreatic acinar tissue, heart, or skeletal muscle (25, 100, 101). It is both supportive of the model and somewhat suprising that treatment of mouse islets with agents that cause even further activation of the glycerol phosphate (leucine, 2-amino[2, 2,1]heptane-2-carboxylate) or malate-aspartate (malate) shuttle activities results in enhanced Ca^{2+} oscillations in the presence of stimulatory glucose relative to islets incubated with glucose absent the shuttle activators (99).

Finally, recent experiments employing the tools of molecular biology provided support for a major role for signals generated between glyceraldehyde-3-phosphate and pyruvate. Similar to effects noted on acute insulin secretion, glucose but not pyruvate stimulates insulin promoter/enhancer activity, with the effect of glucose apparently requiring metabolism of the sugar (46). Given the apparent need for a metabolite generated between glyceraldehyde-3-phosphate and pyruvate, the lack of stimulatory effect of pyruvate might be ascribed to the absence of the key gluconeogenic enzyme phosphoenolpyruvate carboxykinase (PEPCK). Co-transfection of fetal islets with plasmids containing the PEPCK gene and an insulin promoter/enhancer-chloramphenicol acetyltransferase reporter gene resulted in a 4.6-fold enhancement of insulin promoter activity in the presence of 30 mM pyruvate relative to cells expressing the reporter gene alone (46). This effect was not apparent when the cells were grown at high glucose (16 mM), suggesting that the signals generated by pyruvate in PEPCK-expressing cells are not additive to those produced by stimulatory glucose concentrations. Since islets do express pyruvate carboxylase (102), the other enzyme required for conversion of pyruvate to phospho-enolpyruvate, the data are consistent with an important role for metabolism of triose phosphates in glucose sensing. This study points to intriguing similarities between glucose-stimulated insulin gene transcription and insulin secretion, but differences in such elemental features as the time frame of the two reponses (seconds to minutes for acute insulin release, minutes to hours for the transcriptional effect) suggest that extrapolation should be made cautiously. It will, for example, be important to determine whether overexpression of PEPCK will confer acute pyruvate-stimulated insulin release.

Other observations are more difficult to reconcile with a "special" role for ATP derived exclusively from shuttling of NADH from the cytosol to the mitochondria. First, 2-ketoisocaproate and methyl esters of succinate are potent insulin secretagogues despite the fact that they produce ATP and NADH by metabolism exclusively in the citric acid cycle. Second, the failure of pyruvate to stimulate insulin secretion in the absence of glucose could conceivably be ascribed to its preferential conversion to lactate and coincident oxidation of NADH. Indeed, incubation of islets with 30 mM pyruvate causes a 4.4-fold increase in lactate in the surrounding medium (91). This increment in lactate, however, represents only about 30% of the estimated total reducing equivalents (reduced flavoproteins + nicotinamide nucleotides) that can be derived from pyruvate oxidation. Further, pyruvate-driven oxidation of NADH does not explain the fact that lactate is ineffective as a secretagogue. Third, interpretation of pharmacologic studies must always be tempered by the potential for nonspecific effects. Thus, rotenone and antimycin A will block ATP production from NADH regardless of whether it is generated in the mitochondria or the cytosol. Likewise, iodoacetate blocks not only NADH formation at the G3PDH step but also flux through distal steps of glycolysis, and also has the potential to block other enzymes with reactive sulhydryl groups at their active sites (89). Finally, fluoride, an uncoupler of ATP production at the enolase/pyruvate kinase step of glycolysis was shown in one study to have no effect on glucose-induced closure of K_{ATP} channels (99), but was a strong inhibitor of GSIS in another (89). While further study of the relative importance of glycolytically derived NADH is clearly warranted, it seems unlikely that any single mediator will suffice to explain the process of fuel-mediated insulin release. As discussed further below, resolution of the remaining issues may require multifactorial models, possibly involving the interplay between carbohydrate and lipid metabolism. Figure 1 presents a schematic summary of the subset of metabolic signals that are directly related to carbohydrate metabolism per se.

K_{ATP} Channels and Intracellular Ca^{2+}

An early event in fuel-mediated insulin secretion is a decrease in K^+ efflux from islet β-cells (88, 103). The resting β-cell membrane is maintained at a very negative potential of −70 mV. As glucose is applied, the β-cell becomes progressively depolarized. Depolarization leads to increased electrical activity characterized by slow waves with superimposed oscillatory spikes that correlate with oscillations in intracellular Ca^{2+} and pulsatile insulin release (88, 104). ATP-sensitive K^+ channel activity was first demonstrated by Cook & Hales with patch-clamped β-cell membranes (86). Using whole-cell clamp techniques, Ashcroft et al demonstrated that β-cell K_{ATP} channels are closed by glucose, and that the glucose effect can be blocked by mannoheptulose, an inhibitor of glucose phosphorylation (85). Subsequent studies revealed that fuels that are well metabolized in β-cells such as glucose, glyceraldehyde, leucine, and mannose are capable of closing K_{ATP} channels, while poorly metabolized substrates such as 3-O-methyl glucose or galactose have no effect (87). Given the direct demonstration of inhibition of K_{ATP} channels by ATP in β-cell membranes, it was broadly assumed that glucose-induced closure of these channels was accomplished via changes in the concentrations of adenine nucleotides. Upon further investigation, reasons for questioning such a model emerged. First, patch-clamp studies with β-cell membranes demonstrated half maximal inhibition of K_{ATP} channels at ATP concentrations of approximately 12 μM, far below the physiological intracellular range of 2–5 mM (86, 105). A temporary resolution emerged from studies showing that ADP counteracts the inhibitory effect of ATP on K_{ATP} channels, preventing it completely when ADP is present in a 5:1 molar excess (87, 106). Such competition could only be demonstrated, however, for ADP levels greater than 300 μM, a concentration well above the estimated physiological levels of 30–80 μM, and clearly overshadowed by the intracellular ATP concentration. Second, data pertaining to the effect of glucose on islet β-cell adenine nucleotides, while somewhat confusing and contradictory, generally show only small increases in ATP levels and very modest decreases in either total or free ADP (23, 92). In one such recent study, free ADP levels fell from 43–46 μM in microdissected islets from pancreata perfused with 4.2 mM glucose to 28–34 μM in islets perfused with 8.3 mM glucose, while ATP levels were essentially unchanged. These changes occurred within 2–3 minutes of the switch in glucose concentration, at which time insulin release was stimulated by more than tenfold (92).

Recently, Hopkins et al proposed a two-site model that may explain how very small increments in ATP:ADP ratio occurring in fuel-stimulated β-cells are sufficient for regulation of K_{ATP} channel activity (107). The model holds that in addition to competing for the ATP inhibitory site, ADP binds to a second site that activates the channel. In support of this idea, ADP has been shown to activate K_{ATP} channels in the absence of ATP, proving that competition with ATP is not required to demonstrate a positive effect. Furthermore, in the absence of ATP, ADP is both an activator (over the range of 10–100 μM) and an inhibitor (at ≥ 100 μM), with the activating but not the inhibitory effect requiring magnesium. While low, physiologically relevant levels of ADP can activate K_{ATP} channels in patch-clamped membranes, they do not compete effectively with millimolar ATP for control of channel gating (107). Cook et al have proposed a "spare channel" hypothesis that assumes that most (99%) of K_{ATP} channels are maintained in a closed configuration by the high ATP: ADP ratio normally found in β-cells (108). Regulation of the few remaining channels might be possible with small, possibly localized changes in ATP:ADP ratio, resulting in exaggerated effects on membrane potential. There is precedent for ATP gradients within mammalian cells (109), and it has also been shown that glycolytically derived ATP is more effective at inhibiting K^+ channel activity in cardiac mycoctes than is ATP derived from mitochondrial metabolism (110). Whether localized changes in ATP:ADP ratio help to circumvent the seemingly unfavorable regulatory properties of K_{ATP} channels in islet β-cells remains to be determined.

Recently, Henquin and associates have demonstrated the existence of a pathway for fuel-mediated insulin release that is independent of changes in membrane potential (111, 112). Thus, islets incubated with diazoxide, which holds K_{ATP} channels open, are still sensitive to glucose stimulation as long as extracellular K^+ is kept at a high concentration (30 mM). This effect is achieved only with metabolized sugars, and is associated with an increase in ATP:ADP ratio, but not with changes in cAMP, inositol phosphate levels, or activation of protein kinase A or C (112). The effect also does not require changes in Ca^{2+}_i. The interpretation of these data is that an increase in ATP:ADP ratio may exert effects on insulin release at two sites, the ATP-sensitive K^+ channel, and a second more distal site, possibly relating to energetic requirements for granule exocytosis.

The increase in intracellular K^+ induced by K_{ATP} channel closure does not appear to signal to the secretory apparatus in its own right, but rather contributes to the critical event of membrane depolarization. Membrane depolarization is important because it causes activation of voltage-dependent, dihydropyridine-sensitive Ca^{2+} channels of the L-class that are expressed in the islets of Langerhans. Voltage-dependent Ca^{2+} channels exist as complexes comprised of α1, α2/δ, β, and γ subunits. The islets of Langherhans express two isoforms of α1 subunit, one of which appears identical to the major α1 isoform of heart, the second being more distantly related (41–68% identity) to other known α1 subunits, which has been dubbed the β-cell or neuroendocrine type (113). The α1 subunit is thought to contain both the calcium-conducting pore and the dihyropyridine binding activity (88). The relative contribution of the two α1 isoforms to nutrient-stimulated Ca^{2+} influx in islet β-cells remains to be established.

Regulation of intracellular Ca^{2+} in islet β-cells, as in all neuroendocrine cells, is highly complex. Since this area has been extensively reviewed (14, 26, 27), we restrict ourselves to a short summary of events. The fuel-mediated increase in $Ca^{2+}{}_i$ appears to derive from both extracellular and intracellular pools. A number of lines of evidence suggest that the influx of Ca^{2+} from the extracellular pool is essential for fuel-mediated insulin release. Thus, dihydropyridines block voltage-gated Ca^{2+} channels and also prevent fuel-mediated insulin release (14, 26, 27, 88). Stimulation of insulin secretion also can be prevented in perfused pancreas preparations or in cultured islets or β-cells in media with low or absent Ca^{2+}. Studies with ^{45}Ca show that glucose and other metabolized fuels stimulate Ca^{2+} influx and promote net retention of the ion. Furthermore, the amount of ^{45}Ca influx is proportional to the glucose concentration and the rate of glucose metabolism (20). Finally, the time course of Ca^{2+} uptake is consistent with its role in insulin release. Thus, stimulation of islets with glucose results first in increases in ATP:ADP ratio, reduced pyridine nucleotides, and inhibition of K_{ATP} channels (occurring within 40 seconds of the glucose stimulus), followed by an increase in intracellular Ca^{2+} within 2 minutes of the stimulus, coinciding roughly with the onset of insulin secretion (115). In addition to the effect of membrane depolarization, activation of voltage-gated calcium channels appears to be amplified by other metabolic signals. Whole cell Ca^{2+} currents in depolarized β-cell membrane patches are increased by the addition of glucose (43). Glucose may also lower the threshold electrical activity at which Ca^{2+} channels become active. These effects of glucose can be blocked by inhibitors of glycolytic and mitochondrial metabolism. A potential mechanism for the amplifying effect of glucose and other metabolic fuels has been suggested from the work of Turk, Gross, and co-workers (94). These investigators found that glucose induces activation of a recently identified ATP-regulated, Ca^{2+}-independent phospholipase (ASCI-PLA$_2$), which catalyzes formation of nonesterified arachidonic acid via membrane phospholipid hydrolysis. The enzyme was discovered in myocardium, but a protein with highly similar properties is also expressed in islets and β-cell lines (116). Nonesterified arachidonic acid generated by this enzyme activity is thought to accumulate in the plasma membrane, where it serves to amplify the effect of membrane depolarization on Ca^{2+} channel activity. Indeed, inhibition of ASCI-PLA$_2$ activity in isolated islets with a relatively specific haloenol lactone suicide substrate (HELSS) that resembles the normal plasmalogen substrate of the enzyme results in suppression of arachidonate production. HELSS also inhibits the glucose induced increase in intracellular Ca^{2+} and insulin secretion (94). The ASCI-PLA$_2$ enzyme has two properties that may allow its regulation by glucose metabolism: It binds to and is activated by ATP; and it binds to a complex consisting of four copies of the glycolytic enzyme phosphofructokinase (PFK). In myocardium, glycolytic flux and hydrolysis of membrane phospholipids are coordinately activated by ischemia.

Intriguingly, ischemic challenge also causes translocation of the ASCI-PLA$_2$/PFK complex from the cytosol to a membrane-associated compartment (58). There are several isoforms of PFK, and it remains to be determined which type is expressed in islets and whether the islet enzyme associates with ASCI-PLA$_2$ in a manner similar to the myocardial form. If so, this may represent an important new mechanism for coordinating glycolytic flux and phospholipid metabolism relevant to the regulation of insulin secretion.

Mobilization of Ca^{2+} from intracellular stores may also play an important role in fuel-stimulated insulin release. The major source of Ca^{2+}_i during stimulated insulin release appears to be the endoplasmic reticulum, although the mitochondria may also play a Ca^{2+} buffering role in β-cells (14, 26, 27). In addition to its activating effect on the recently discovered ASCI-PLA$_2$, it has long been known that glucose adminstration to islets results in activation of phospholipase C. Hydrolysis of phosphoinositides by phospholipase C leads to the production of two potentially important signaling molecules, inositol phosphates (particularly inositol-1,4,5-P3) and diacylglycerol (DAG), a potent activator of protein kinase C (26, 27). Inositol phosphates (IP$_3$) bind to the IP$_3$ receptors expressed in the endoplasmic reticulum of islet β-cells. The receptor functions as a homotetramer, with each monomer containing a cytoplasmic ligand binding domain and a membrane spanning region. Based on homology to the ryanodine receptor, a calcium release channel that is also expressed in islets, the transmembranous regions of the IP$_3$ receptor are thought to participate in Ca^{2+} efflux from the ER (117). Binding of IP$_3$ to the cytoplasmic domain of the receptor stimulates Ca^{2+} efflux from the ER and increases Ca^{2+}_i. The mechanism by which enhanced glucose flux activates phospholipase C is unknown. Interestingly, an alternative mechanism for glucose-induced mobilization of Ca^{2+}_i stores has recently been proposed. Takasawa et al (96) found that the NAD$^+$ metabolite cyclic adenosine diphosphate-ribose (cADP-ribose) induces Ca^{2+} release from cell-free islet microsomes. Further, cADP-ribose or Ca^{2+}, but not IP$_3$, were found to induce insulin secretion in digitonin-permeabilized islets (96). These observations have subsequently been challenged by others who failed to see effects of cADP-ribose (118). The discrepant results may be related to differences in the cell types studied by the two groups (normal rat islets vs ob/ob islets and HIT cells). Further study will be required to resolve this intriguing issue.

Mobilization of intracellular Ca^{2+} stores has traditionally been viewed as a secondary or amplifying event relative to the Ca^{2+} signal generated by voltage-dependent Ca^{2+} channels during glucose stimulation (14, 26). Recent studies indicate that the contribution of the intracellular pool may have been underestimated. Dukes and colleagues showed that glucose causes triphasic alterations in intracellular Ca^{2+} consisting of an initial brief decrease (phase 0), a transient increase (phase 1), and then regular oscillations (phase 2) (119).

Based on studies with pharmacologic agents such as caffeine and thapsigargin, these investigators suggest that phase 0 represents sequestration of Ca^{2+} in the ER and phase 1 represents depolarization-triggered release of ER calcium. In this model only the last phase is due to activation of cell surface Ca^{2+} channels. Although provocative, these data are subject to some of the same concerns surrounding the specificity of the pharmacologic agents that were raised earlier.

THE ROLE OF LIPIDS IN FUEL-MEDIATED INSULIN SECRETION

Is Malonyl-CoA Involved?

As noted above, inhibitor studies point to an important role for mitochondrial metabolism in GSIS. What has been perplexing is how metabolic events in the mitochondia are linked to the exocytosis of insulin granules. Although this question cannot be answered completely, very recent developments suggest that a key component of the linkage is vested in a principle already well established for liver but previously given only limited consideration as regards the β-cell, i.e. the intimate relationship between glucose metabolism and the fate of intracellular long-chain fatty acids.

The notion that fatty acid metabolism plays an important role in islet cell function is not new. Studies in the 1960s and early 1970s established that both in vivo and in vitro fatty acids are acutely stimulatory to pancreatic insulin secretion (120, 121, 122, 123). Subsequently, it was shown that isolated islets from genetically obese mice utilize palmitate efficiently for oxidation and phospholipid biosynthesis, the former process being suppressed in favor of the latter at glucose concentrations that were stimulatory to insulin secretion (124, 125). Similar findings were reported by Tamarit-Rodriguez et al using isolated islets from normal rats (126, 127). A particularly intriguing observation was that the starvation-induced blunting of glucose-stimulated insulin secretion (GSIS) in perfused rat islets (126) and in the perfused rat pancreas (128) could be largely offset by addition to the medium of 2-bromostearate (2-BrS), a known inhibitor of mitochondrial carnitine palmitoyltransferase I (CPT I). In a later series of studies, again with isolated rat islets, Vara & Tamarit-Rodriguez (129) noted that a common accompaniment of GSIS and of leucine-induced insulin secretion was the ability of both agents to divert fatty acids from the oxidative pathway into esterification products. Thus, by the mid-1980s it was beginning to appear that an important event associated with the enhancement of insulin secretion from the β-cell in response to glucose (and possibly some other secretagogues) is suppression of fatty acid oxidation; the attendant increase of fatty acid flux into esterified products such as diacylglycerols (DAG) and phospholipids was suggested to be instrumental in triggering insulin release (130).

Seeking to understand how glucose might inhibit fatty acid catabolism and stimulate lipid synthesis in the β-cell, Corkey et al (131) examined the acyl-CoA profile in a clonal pancreatic β-cell line (the HIT cell) before and after exposure to 10 mM glucose. The striking finding was that in response to glucose the cells exhibited a rapid rise in their content of malonyl-CoA that preceded the output of insulin. There was simultaneous suppression of $[1-^{14}C]$ palmitate oxidation together with increased formation of DAG and, presumably, of other esterified products. What emerged from these studies was that a central element of fuel "cross-talk," originally described in liver (132, 133) and subsequently implicated in tissues such as heart and skeletal muscle (134, 135), is probably also operative in the pancreatic β-cell, i.e. the ability of glucose-derived malonyl-CoA to inhibit CPT I, thus suppressing the flux of fatty acids through β-oxidation with concomitant elevation in the cytosolic content of long-chain acyl-CoAs. The latter, and/or some product(s) of their esterification to glycerol-3-phosphate, were proposed to act as proximal signals for insulin release (131). The fact that malonyl-CoA is a potent inhibitor of CPT I in rat islets (136), as is the case in all tissues studied to date (137), was consistent with this interpretation. The model received additional support when hamster insulinoma (HIT) cells were exposed to different combinations of palmitate, 2-bromopalmitate (2-BrP, another inhibitor of CPT I), glucose, and a variety of nonglucose insulin secretagogues (138). Invariably, hormone release was associated with increased malonyl-CoA production and (presumed) elevation of the cytosolic content of long-chain acyl-CoAs. In another series of experiments by Prentki and co-workers, this time using a more differentiated insulinoma cell line, the INS-1 cell (60), it was found that raising the glucose concentration from 5 to 20 mM elicited a 15-fold increase in the mRNA for acetyl-CoA carboxylase (ACC) (139). This effect of glucose was felt to be mediated by glucose-6-phosphate. It was suggested that since carbon flux through ACC controls malonyl-CoA levels, ACC, like glucokinase, might represent an important nutrient sensor in the β-cell.

These exciting new developments have evolved for the most part from studies with insulinoma cell lines. It was therefore necessary to establish that they have relevance to the functioning of a normal β-cell as it exists in its natural environment. Also to be proved was that the generation of malonyl-CoA from glucose was not simply a phenomenon that coincided with insulin secretion, but that the two events were causally related. Both issues were addressed by Chen et al (140) in experiments with the intact perfused rat pancreas. The rationale employed was that if the formation of malonyl-CoA from glucose is *essential* for GSIS, then blockade of the glucose malonyl-CoA conversion at a late stage in the pathway should impact negatively on insulin secretion. The agent chosen for this purpose was (-)-hydroxycitrate (HC), a known inhibitor of ATP-citrate lyase that catalyzes the penultimate step in the sequence leading from glucose to malonyl-CoA. It was found that HC markedly obtunded the rate of insulin secretion when the perfusate glucose concentration was raised from 3 to 20 mM. Citrate, used as a control, was without effect, ruling out the possibility that HC inhibited insulin secretion simply by virtue of its ability to chelate perfusate Ca^{2+}. In the same studies it was shown that inclusion of 0.5 mM palmitate in the control perfusion had the expected stimulatory effect on GSIS. Importantly, the fatty acid also completely offset the suppressive action of HC on insulin release.

A plausible explanation for these findings is that blockade of malonyl-CoA synthesis from glucose left CPT I in an active form, with the result that the cytosolic level of acyl-CoA did not rise appropriately; the latter deficit was corrected with the addition of exogenous palmitate. In keeping with this interpretation, HC was shown to antagonize the suppressive effect of a high glucose concentration on the oxidation of $[1-^{14}C]$ palmitate by isolated rat islets, presumably through its interference with malonyl-CoA synthesis (140). It thus seems reasonable to conclude that the malonyl-CoA-CPT I partnership constitutes an essential element in GSIS from the normal pancreas.

If an elevation in the cytosolic long-chain acyl-CoA concentration (glucose/malonyl-CoA-mediated) plays this postulated role in stimulus-secretion coupling, what might be the mechanism(s)? One possibility is that increased shunting of fatty acids into phospholipids provides essential building blocks for membrane turnover during the exocytosis of insulin granules (141). It is also conceivable that expansion of the DAG pool plays a contributory role, should activation of protein kinase C be involved in GSIS (see below). Alterations in Ca^{2+}_i dynamics, linked to stimulation of phosphoinositide synthesis and the generation of inositol phosphates, have also been considered as possible factors (138). Common to all of these scenarios would be a requirement of the β-cell for glycerol-3-phosphate with which to esterify the expanded acyl-CoA pool. It is attractive to speculate that the ability to generate both malonyl-CoA and glycerol-3-phosphate is what sets glucose and glyceraldehyde apart from other fuels as insulin secretagogues. Note that compounds such as pyruvate, lactate, and leucine, while capable of giving rise to malonyl-CoA, cannot support the formation of glycerol-3-phosphate in the β-cell because of the absence of phosphoenolpyruvate carboxykinase. Perhaps this is why these fuels require a threshold level of glucose in order to stimulate insulin secretion.

Apart from these potential effects of phosphatidic acid derivatives, long-chain acyl-CoA esters themselves might be more directly involved in the insulin secretory process. For example, they have been found to modulate Ca^{2+}_i handling in permeabilized HIT cells (142). They have also been implicated as regulatory factors for protein kinase C, K_{ATP} and/or Ca^{2+} channel activities (103, 138, 139, 140, 141), and in the budding of transport vesicles from Golgi cisternae (143). Moreover, the ability of fatty acyl-CoAs to interact with cellular proteins, thereby modifying their properties, is now a well established phenomenon. Of added interest in this regard is the mechanism by which compounds such as 2-BrS and 2-BrP cause enhancement of GSIS. As alluded to above, conventional thinking has been that these agents work through inhibition of CPT I, i.e. by subserving the role ordinarily played by glucose-derived malonyl-CoA. However, in order to inhibit CPT I, 2-BrS and 2-BrP must first be converted by intracellular long-chain acyl-CoA synthetase into their CoA ester form (144). There is evidence that in the pancreatic β-cell the CoA esters of these bromo-fatty acids might, in addition to causing suppression of CPT I, participate in the insulin secretory process by simulating the direct effects of natural acyl-CoAs (140).

In summary, elucidation of the key metabolic factors linking glucose metabolism to insulin secretion by the pancreatic β-cell has proved to be a formidable task. Although the picture is by no means complete, what seems to be emerging is that in the search for potential coupling factors the traditional emphasis on glycolysis and the tricarboxylic acid cycle, though highly informative, has provided limited information. Recognition of the fact that in the β-cell, just as in liver and other tissues, glucose metabolism is inextricably linked to the metabolism of lipids has brought us to a new and exciting threshold in the exploration of how the insulin secretory process in controlled. A schematic summary of the potential interactions of carbohydrate and lipid metabolism in the generation of signals for insulin release is presented in Figure 2.

Coupling of Signals to the Secretory Apparatus

As summarized above, a common signal generated by fuels that stimulate insulin release is an increase in Ca^{2+}_i. While significant progess has been made in understanding the mechanisms leading to increased Ca^{2+}_i, much less is known about how these changes, and changes in the concentration of bioactive phospholipids such as diacylglyerol (DAG), are mechanistically linked to secretory granule exocytosis. Ca^{2+} and DAG have been suggested to exert their effects through activation of protein kinases. Pancreatic islets express a number of kinases that require Ca^{2+} and calmodulin and that appear to be activated in response to glucose (26, 27). For example, myosin light-chain kinase purified from islets or β-cell lines phosphorylates skeletal muscle myosin light chain in vitro in a Ca^{2+}- and calmodulin-dependent manner (145, 146), leading to the suggestion that phosphorylation of myosin light chain might regulate the movement of insulin secretory granules. Similarly, Ca^{2+}- and calmodulin-dependent phosphorylation of several peptides in the 50–60 kD-range has been reported. Some of these peptides have identical electrophoretic mobility as α- and β-tubulin, respectively. Furthermore, diminished phosphorylation of the candidate peptides is observed by pretreatment of the sample with anti-tubulin

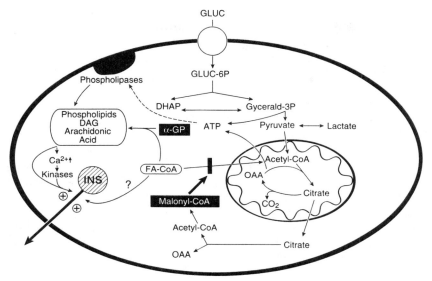

Figure 2 Schematic summary of potential interplay between carbohydrate and lipid metabolism in generation of signals for insulin release.

antibody, suggesting another mechanism whereby this class of kinases might impact secretory granule movement (147).

One of the most controversial areas in the field of insulin secretion concerns the role of protein kinase C. Recent reviews suggested that PKC is likely to be more important in mediating the effects of nonglucose secretagogues such as muscarinic agonists and phorbol esters than the effect of stimulatory glucose per se (94, 148, 149). This position has been recently challenged by the very convincing demonstration that glucose activates PKC translocation and resultant phosphorylation of myristoylated alanine-rich C kinase substrate (MARCKS), a protein phosphorylated in a relatively specific manner by PKC (150, 151). Furthermore, glucose activation is prevented by mannoheptulose or blockade of voltage-gated Ca^{2+} channels with nitrendipine (150). Calle et al suggested that the failure of other investigators to obtain similar results may be due to the use of islets cultured overnight vs freshly isolated islets, since the former are significantly less responsive to glucose (146). While this matter awaits further clarification, it is of interest to note that phosphorylation of MARCKS, a calmodulin-binding protein, results in a decrease in its binding affinity. Thus, PKC-induced phosphorylation of MARCKS may represent a means for generating "free" calmodulin for Ca^{2+} binding and cellular signaling. This may also be a mechanism for interaction of the PKC and Ca^{2+}/ calmodulin pathways in insulin release (150). In sum, understanding of the distal events in fuel-mediated insulin secretion is currently limited, but early results clearly suggest that this area deserves more attention.

The Role of Lipids in β-cell Dysfunction

In this section we focus on the most common form of diabetes, namely, obesity-related NIDDM. Available evidence indicates that this is a polygenic disease characterized, in its early stages, by insulin resistance and hyper-insulinemia, and, depending upon genetic and environmental factors (such as food availability and dietary habits), able to progress to impaired glucose tolerance and ultimately to full-blown diabetes (152). It is generally assumed that the primary inherited defect is insulin resistance in muscle and that this gives rise to compensatory hyperinsulinemia (152). However, an alternative view is that the converse sequence might be more correct, namely, that hyper-insulinemia is the earlier derangement and that this brings about the insulin resistance (153, 154). In this formulation, hypersecretion of insulin is seen as a driving force for hepatic very low density lipoprotein (VLDL) production, resulting in excessive delivery of lipid substrate to adipose tissues and muscle. The predicted effect would be fat accretion and interference with insulin-mediated glucose disposal in the muscle bed (153, 154). No doubt these contrasting positions will be debated for some time to come.

Regardless of the order of appearance of hyperinsulinemia and peripheral insulin resistance during the pre-NIDDM phase, the emergence of frank hyperglycemia is undoubtedly associated with a serious disruption of β-cell function such that the pancreas can no longer sustain the demand for enhanced insulin output. Both in human and rodent NIDDM this is associated with a major impairment of GSIS and, in all rat and mouse models studied to date, by a parallel reduction in the number of β-cells displaying the high K_m GLUT-2 transporter on their surface (28) (see above). Of particular interest is the study by Thorens et al (69), showing that transplantation of GLUT-2 deficient β-cells from diabetic *db/db* mice into nondiabetic recipients resulted in restoration of transporter expression, while, conversely, normal islets transplanted into diabetic animals lost their surface GLUT-2. These findings pointed to an extrinsic, presumably metabolic, factor(s) other than hyperglycemia as the cause of down-regulation of GLUT-2 in the transition from pre-NIDDM to NIDDM. In the search for alternative possibilities attention is again focusing on the role of abnormal lipid metabolism. Some potentially relevant observations are summarized below.

Reaven et al (155) conducted a detailed study of normal, mildly diabetic and severely diabetic subjects to examine the relationship between various hormonal and metabolic parameters over a period of 24 h during which three standard meals were consumed. One striking observation was that plasma-free fatty acid (FFA) levels were chronically elevated in diabetics, and in rough proportion to the degree of hyperglycemia. Also evident was that the patients with mild NIDDM displayed marked hyperinsulinemia, whereas insulin responses were clearly lower in the more severely affected group. Although the emphasis of the study was on the detrimental effects of excessively high concentrations of FFA on glucose disposal, the question might now be asked as to whether a relationship also existed between the level of circulating FFA and the status of β-cell function in these individuals. More recent experiments with the rat would be consistent with such a view. Thus, Sako & Grill (156) observed that after 3 h of infusion of Intralipid into normal rats (which resulted in acute elevation of the plasma FFA concentration) both basal and glucose-induced insulin secretion from the subsequently perfused pancreas were markedly enhanced. However, when the lipid infusion was continued for 48 h, basal secretion rates from the pancreas were the same as in controls while GSIS was suppressed by ~50%. In a parallel study using perfused rat islets, Elks (157) noted that acute exposure to 1 mM palmitate stimulated insulin release, as expected, but that as the exposure time was extended to 2, 3 or 4 h there was a time-dependent suppression of both first and second phase insulin secretion. No significant effects were seen when the concentration of palmitate was reduced to 0.35 mM. In addition, the inhibitory influence of 1 mM palmitate on GSIS was accompanied by a reduced rate of $[U-^{14}C]$ glucose oxidation. Both effects were prevented by inclusion in the perfusion fluid of 2-BrS or tetradecylglycidate to inhibit CPT I, and thus palmitate oxidation. Broadly similar results were reported by Zhou & Grill (158) from experiments in which isolated rat islets were exposed to palmitate, oleate, or octanoate for periods of up to 48 h. In all cases basal insulin secretion was enhanced while GSIS was diminished. The negative effect of the long-chain fatty acids, but not that of the medium-chain substrate (whose oxidation is carnitine-independent), was largely attenuated by the CPT I inhibitor, etomoxir.

From the studies cited above, the effect of elevated concentrations of fatty acids on β-cell function is clearly biphasic; acutely they augment GSIS, but over the longer term they cause elevation of basal insulin secretion while damping the normally robust response to high levels of glucose, i.e. they bring about a β-cell phenotype characteristic of NIDDM (159). The question arises as to whether β-cell failure in spontaneous NIDDM might be attributable, at least in part, to abnormal lipid dynamics. Evidence consistent with this notion has recently been obtained from studies in the obese Zucker drt rat (80, 160). In this strain the males begin to develop hyperglycemia at ~9 weeks of age and this becomes progressively more severe. Two weeks prior to hyperglycemia, plasma FFA levels began to rise, reaching almost 2 mM at the time that hyperglycemia appeared. As expected, diabetic animals exhibited enhanced basal insulin release and greatly reduced GSIS from the perfused pancreas.

The increase in basal insulin release may be related to an observed threefold increase in the rate of glucose usage at substimulatory glucose and an accompanying threefold enhancement in hexokinase activity (80). The enhanced basal glucose usage and β-cell hyperplasia that is also observed in obese Zucker rats could be reproduced by incubating normal islets with palmitate for 48 h (80). The loss of GSIS in the diabetic animals was correlated with a sharp decrease in β-cell GLUT-2 expression (160). Interestingly, their islet triglyceride content had risen to 10 times the value in lean control animals, and this correlated directly with their circulating FFA and glucose concentrations. Pairfeeding of the obese rats to lean littermates markedly reduced their plasma FFA levels and prevented all β-cell abnormalities, as well as the hyperglycemia (160). Although operative mechanisms remain to be worked out, the distinct possibility exists that excessive levels of circulating lipid substrate, either in the form of lipoprotein triglyceride or FFA, underlie both the insulin resistance and β-cell abnormalities that typify obesity/NIDDM syndromes (80, 153, 154, 160).

ACKNOWLEDGMENTS

Work from the authors' laboratories cited in this article was supported by grants from the U.S. Public Health Service (PO1-DK42582, RO1-DK46492, and R01-DK18573), BetaGene Inc. (to C.B.N.), the Chilton Foundation, and Sandoz Pharmaceuticals (to J.D.M.). We thank our colleagues in the Gifford Laboratories for Diabetes Research for their invaluable contributions along the way.

Literature Cited

1. Unger RH. 1981. *Diabetologia* 20:1–11
2. Beltran del Rio H, Schnedl W, Ferber S, Newgard CB. 1994. In *Pancreatic Islet Transplantation,* ed. W Chick, R Lanza. 1:169–183. Austin: Landes
3. Lacy PE, Scharp DW. 1986. *Annu. Rev. Med.* 37:33–46
4. Newgard CB. 1992. *Bio-Technology* 10:1112–20
5. Johnson JH, Ogawa A, Chen L, Orci L, Newgard CB, et al. 1990. *Science* 250:546–49
6. Pfeiffer MA, Halter JB, Porte D Jr. 1981. *Am. J. Med.* 70:579–88
7. Sener A, Malaisse-Lagae F, Malaisse WJ. 1981. *Proc. Natl. Acad. Sci. USA* 78:5460–64
8. MacDonald MJ, McKenzie DI, Kaysen JH, Walker TM, Moran SM, et al. 1991. *J. Biol. Chem.* 266:1335–40
9. Sener A, Blachier F, Rasschaert J, Malaisse WJ. 1990. *Endocrinology* 127:107–13
10. Schmidt HHW, Warner TD, Ishii K, Sheng H, Murad F. 1992. *Science* 255:721–23
11. Laychock SG, Modica ME, Cavanaugh CT. 1991. *Endocrinology* 129:3043–52
12. Charles S, Tamagawa T, Henquin JC. 1982. *Biochem. J.* 208:301–9
13. Henquin JC, Meissner HP. 1981. *Am. J. Physiol.* 240:E245–50
14. Liang Y, Matschinsky FM. 1994. *Annu. Rev. Nutr.* 14:59–81

15. Orskov C. 1992. *Diabetologia* 35:701–11
16. Thorens B, Waeber G. 1993. *Diabetes* 42:1219–25
17. Grodsky GM, Batts AA, Bennett LL, Vcella C, McWilliams NB, Smith DF. 1963. *Am. J. Physiol.* 205:638–44
18. Ashcroft SJH. 1981. In *The Islets of Langherhans,* ed. SJ Cooperstein, D Watkins, pp. 117–48. London: Academic
19. Coore HG, Randle PJ, Simon E, Kraicer PF, Shelesnyak MC. 1963. *Nature* 197:1264–66
20. Malaisse WJ, Sener A, Herchuelz A, Hutton JC. 1979. *Metabolism* 28:373–86
21. Matschinsky FM, Ellerman JE. 1968. *J. Biol. Chem.* 243:2730–36
22. Landgraf R, Kotler-Brajtburg J, Matschinsky FM. 1971. *Proc. Natl. Acad. Sci. USA* 68:546–40
23. Meglasson MD, Matschinsky FM. 1986. *Diabetes/Metab.Rev.* 2:163–214
24. Hedeskov CJ. 1980. *Physiol. Rev.* 60:442–508
25. MacDonald MJ. 1990. *Diabetes* 29:1461–66
26. Prentki M, Matschinsky, FM. 1987. *Physiol. Rev.* 67:1185–248
27. Turk J, Wolf BA, McDaniel ML. 1987. *Prog. Lipid Res.* 26:125–81
28. Johnson JH, Newgard CB, Milburn JL, Lodish HF, Thorens B. 1990. *J. Biol. Chem.* 265:6548–51
29. Thorens B, Sarkar HK, Kaback HR, Lodish HF. 1988. *Cell* 55:281–90
30. Iynedjian PB, Pilot P-R, Nouspikel T, Milburn JL, Quaade C, et al. 1989. *Proc. Natl. Acad. Sci. USA* 86:7838–42
31. Magnuson MA, Shelton KD. 1989. *J. Biol. Chem.* 264:15936–42
32. Waddell ID, Burchell A. 1988. *Biochem. J.* 255:471–76
33. Malaisse WJ, Sener A, Koser M, Ravazzola M, Malaisse-Lagae F. 1977. *Biochem. J.* 164:447–554
34. Ashcroft SJH, Weerasinghe LCC, Bassett JM, Randle PJ. 1972. *Biochem. J.* 126:525–32
35. Trus MD, Zawalich WS, Burch PT, Berner DK, Weill VA, et al. 1981. *Diabetes* 30:911–22
36. Coore HG, Randle PJ. 1964. *Biochem. J.* 91:56–59
37. Wilson JE. 1984. In *Regulation of Carbohydrate Metabolism,* pp. 45–85. Boca Raton: CRC
38. Meglasson MD, Burch PT, Hoenig M, Chick WL, Matschinsky FM. 1981. *Proc. Natl. Acad. Sci. USA* 80:85–89
39. Hughes SD, Johnson JH, Quaade C, Newgard CB. 1992. *Proc. Natl. Acad. Sci. USA* 89:688–92

40. Giroix M-H, Sener A, Pipeleers DG, Malaisse WJ. 1984. *Biochem. J.* 223:447–53
41. Froguel P, Vaxillaire M, Sun F, Vehlo G, Zouali H, et al. 1992. *Nature* 356:162–64
42. Froguel P, Zouali H, Vionnet N, Velho G, Vaxillaire M, et al. 1993. *N. Engl. J. Med.* 328:697–702
43. Byrne MM, Sturis J, Clement K, Vionnet N, Pueyo ME, et al. 1994. *J. Clin. Invest.* 93:1120–30
44. Vehlo G, Froguel P, Clement K, Pueyo ME, Rakotoambinina B, et al. 1992. *Lancet* 340:444–48
45. Gidh-Jain M, Takeda J, Xu LZ, Lange AJ, Vionnet N, et al. 1993. *Proc. Natl. Acad. Sci. USA* 90:1932–36
46. Efrat S, Leiser M, Wu Y-J, Fusco-DeMane D, Emran OA, et al. 1994. *Proc. Natl. Acad. Sci. USA* 91:2051–55
47. Becker TC, BeltrandelRio H, Noel RJ, Johnson JH, Newgard CB. 1994. *J. Biol. Chem.* 269:21234–38
48. Becker TC, Noel RJ, Coats WS, Gomez-Foix AM, Alam T, et al. 1994. *Methods Cell Biol.* 43:161–89
49. German MS. 1993. *Proc. Natl. Acad. Sci. USA* 90:1781–85
50. Epstein PN, Boschero AC, Atwater I, Cai XG, Overbeek PA. 1992. *Proc. Natl. Acad. Sci. USA* 89:12038–42
51. Voss-McCowan ME, Xu B, Epstein PN. 1994. *J. Biol. Chem.* 269:15814–18
52. Gazdar AF, Chick WL, Oie HK, Sims HL, King DL, et al. 1980. *Proc. Natl. Acad. Sci. USA* 77:3519–23
53. Halban PA, Praz GA, Wollheim CB. 1983. *Biochem. J.* 212:439–43
54. Clark SA, Burnham BL, Chick WL. 1990. *Endocrinology* 127:2779–88
55. Efrat S, Linde S, Kofod H, Spector D, Delannoy M, et al. 1988. *Proc. Natl. Acad. Sci. USA* 85:9037–41
56. Miyazaki J-I, Araki K, Yamato E, Ikegami H, Asano T, et al. 1990. *Endocrinology* 127:126–32
57. Efrat S, Leiser M, Surana M, Tal M, Fusco-Demane D, et al. 1993. *Diabetes* 42:901–7
58. Hazen SL, Wolf MJ, Ford DA, Gross RW. 1994. *FEBS Lett.* 339:213–16
59. Whitesell RR, Powers AC, Regen DM, Abumrad NA. 1991. *Biochemistry* 30:11560–66
60. Asfari M, Janjic D, Meda P, Li G, Halban PA, et al. 1992. *Endocrinology* 130:167–78
61. Ishihara H, Asano T, Katsunori T, Katagiri H, Inukai K, et al. 1994. *J. Biol. Chem.* 269:3081–87
62. Ferber S, BeltrandelRio H, Johnson JH,

Noel RJ, Cassidy LE, et al. 1994. *J. Biol. Chem.* 269:11523–29
63. Newgard CB. 1994. *Diabetes* 43:341–50
64. Newgard CB, Ferber S, Quaade C, Johnson JH, Hughes SD. 1994. See Ref. 161, pp. 119–53
65. Permutt MA, Koryani L, Keller K, Lacy PE, Scharp DW, et al. 1989. *Proc. Natl. Acad. Sci. USA* 86:8688–92
66. Newgard CB, Quaade C, Hughes SD, Milburn JL. 1990. *Biochem. Soc. Trans.* 18:851–53
67. Unger RH. 1991. *Science* 251:1200–5
68. Orci L, Ravazzola M, Baetens D, Inman L, Amherdt M. 1990. *Proc. Natl. Acad. Sci. USA* 87:9953–57
69. Thorens B, Wu Y-J, Leahy JL, Weir GC. 1992. *J. Clin. Invest.* 90:77–85
70. Ohneda M, Johnson JH, Inman LR, Chen L, Suzuki K, et al. 1993. *Diabetes* 42:1065–72
71. Thorens B, Weir GC, Leahy JL, Lodish HF, Bonner-Weir S. 1990. *J. Clin. Invest.* 86:1615–22
72. Ogawa A, Johnson JH, Ohneda M, McAllister CT, Inman L, et al. 1992. *J. Clin. Invest.* 90:497–504
73. Tal M, Wu Y-J, Leiser M, Surana M, Lodish H, et al. 1992. *Proc. Natl. Acad. Sci. USA* 89:5744–48
74. Tal M, Liang Y, Najafi H, Lodish HF, Matschinsky FM. 1992. *J. Biol. Chem.* 267:17241–47
75. Chen C, Thorens B, Bonner-Weir S, Weir GC, Leahy J. 1992. *Diabetes* 41:1320–27
76. Liang Y, Bonner-Weir S, Wu Y-J, Berdanier CD, Berner DK, et al. 1994. *J. Clin. Invest.* 93:2473–81
77. Mueckler M, Kruse M, Strube M, Riggs AC, Chiu KC, et al. 1994. *J. Biol. Chem.* 269:17232–35
78. Chen L, Alam T, Johnson JH, Hughes S, Newgard CB, Unger RH. 1990. *Proc. Natl. Acad. Sci. USA* 87:4088–92
79. Giroix M-H, Baetens D, Rasschaert J, Leclercq-Meyer V, Sener A, et al. 1992. *Endocrinology* 130:2634–40
80. Milburn JL Jr, Hirose H, Lee YH, Nagasawa Y, Ogawa A, et al. 1995. *J. Biol.* 270:1295–99
81. Khan A, Chandramouli V, Ostenson CG, Low H, Landau BR, et al. 1990. *Diabetes* 39:456–59
82. Sekine N, Cirulli V, Regazzi R, Brown LJ, Gine E, et al. 1994. *J. Biol. Chem.* 269:4895–902
83. Tiedge M, Hohne M, Lenzen S. 1993. *Biochem. J.* 296:113–18
84. Hughes SD, Quaade C, Johnson JH, Ferber S, Newgard CB. 1993. *J. Biol. Chem.* 268:15205–12

85. Ashcroft FM, Harrison DE, Ashcroft SJH. 1984. *Nature* 312:446–48
86. Cook DL, Hales CN. 1984. *Nature* 311:271–73
87. Misler S, Falke LC, Gillis K, McDaniel ML. 1986. *Proc. Natl. Acad. Sci. USA* 83:7119–23
88. Boyd AE III. 1992. *J. Cell. Biochem.* 48:234–41
89. MacDonald MJ, Fahien LA. 1990. *Arch. Biochem. Biophys.* 279:104–8
90. Malaisse WJ, Kawazu S, Herchuelz A, Hutton JC, Somers G, et al. 1979. *Arch. Biochem. Biophys.* 194:49–62
91. Sener A, Kawazu S, Hutton JC, Boschero AC, Devis G, et al. 1978. *Biochem. J.* 176:217–32
92. Ghosh A, Ronner P, Cheong E, Khalid P, Matschinsky FM. 1991. *J. Biol. Chem.* 266:22887–92
93. Holz GG, Kuhtreiber WM, Haebner JF. 1993. *Nature* 361:362–65
94. Turk J, Gross RW, Ramanadham S. 1993. *Diabetes* 42:367–74
95. Kowluru A, Metz SA. 1994. See Ref. 161, pp. 249–83
96. Takasawa S, Nata K, Yonekura H, Okamoto H. 1993. *Science* 259:370–73
97. Hutton JC, Sener A, Herchuelz A, Atwater I, Kawazu S, et al. 1980. *Endocrinology* 106:203–19
98. MacDonald MJ, Fahien LA. 1988. *Diabetes* 37:997–99
99. Dukes ID, McIntyre MS, Mertz RJ, Philipson LH, Roe MW, et al. 1994. *J. Biol. Chem.* 269:10979–82
100. MacDonald MJ. 1981. *J. Biol. Chem.* 256:8287–90
101. MacDonald MJ. 1982. *Arch. Biochem. Biophys.* 213:643–49
102. MacDonald MJ, Kaysen JH, Moran SM, Pomije CE. 1991. *J. Biol. Chem.* 266:22392–97
103. Dean PM, Matthews EK. 1970. *J. Physiol.* 210:255–64
104. Longo EA, Tornheim K, Deeney JT, Varnum BA, Tillotson D, et al. 1991. *J. Biol. Chem.* 266:9314–19
105. Rorsman P, Trube G. 1985. *Pfluegers Arch.* 405:305–9
106. Dunne MJ, Petersen OH. 1986. *FEBS Lett.* 208:59–62
107. Hopkins WF, Fatherazi S, Peter-Riesch B, Corkey BE, Cook DL. 1992. *J. Membr. Biol.* 129:287–95
108. Cook DL, Satin LS, Ashford MLJ, Hales CN. 1988. *Diabetes* 37:495–98
109. Nichols CG, Lederer WJ. 1990. *J. Physiol.* 423:91–110
110. Weiss JN, Lamp ST. 1987. *Science* 238:67–69
111. Gembal M, Gilon P, Henquin J-C. 1992. *J. Clin. Invest.* 89:1288–95

112. Gembal M, Detimary P, Gilon P, Gao Z-Y, Henquin J-C. 1993. *J. Clin. Invest.* 91:871–80
113. Seino S, Chen L, Seino M, Blondel O, Takeda J, et al. 1992. *Proc. Natl. Acad. Sci. USA* 89:584–88
114. Duchen MR, Smith PA, Ashcroft FM. 1993. *Biochem. J.* 294:35–42
115. Smith PA, Rorsman P, Ashcroft FM. 1989. *Nature* 342:550–53
116. Gross RW, Ramanadham S, Kruszka KK, Han X, Turk J. 1993. *Biochemistry* 32:327–36
117. Mignery GA, Sudhof TC. 1990. *EMBO J.* 9:3894–98
118. Islam MS, Larsson O, Berggren P-O. 1993. *Science* 262:584–85
119. Worley JF, McIntyre MS, Spencer B, Mertz RJ, Roe MW, et al. 1994. *J. Biol. Chem.* 269:14359–62
120. Goberna R, Tamarit J Jr, Osorio J, Fussganger R, Tamarit J, et al. 1974. *Horm. Metab. Res.* 6:256–60
121. Madison LL, Seyffert WA Jr, Unger RH, Barker B. 1968. *Metabolism* 17:301–4
122. Malaisse WJ, Malaisse-Lagae F. 1968. *J. Lab. Clin. Med.* 72:438–48
123. Greenough WB II, Crespin SR, Steinberg D. 1967. *The Lancet* 115:1334–36
124. Berne C. 1975. *Biochem. J.* 152:661–66
125. Berne C. 1975. *Biochem. J.* 152:667–73
126. Tamarit-Rodriguez J, Vara E, Tamarit J. 1984. *Horm. Metab. Res.* 16:115–19
127. Tamarit-Rodriguez J, Vara E, Tamarit J. 1984. *Biochem. J.* 221:317–24
128. Bedoya FJ, Ramirez R, Arilla E, Goberna R. 1984. *Diabetes* 33:858–63
129. Vara E, Tamarit-Rodriguez J. 1986. *Endocrinology* 119:404–7
130. Vara E, Tamarit-Rodriguez J. 1986. *Metabolism* 35:266–71
131. Corkey BE, Glennon MC, Chen KS, Deeney JT, Matschinsky FM, Prentki M. 1989. *J. Biol. Chem.* 264:21608–12
132. McGarry JD, Foster DW. 1980. *Annu. Rev. Biochem.* 49:395–420
133. McGarry JD, Mannaerts GP, Foster DW. 1977. *J. Clin. Invest.* 60:265–70
134. Duan C, Winder WW. 1992. *J. Appl. Physiol.* 72:901–4
135. Saddik M, Gamble J, Witters LA, Lopaschuk GD. 1993. *J. Biol. Chem.* 263:25836–45
136. Shrago E, MacDonald MJ, Woldegiorgis G, Bremer J, Schalinske K. 1985. In *Clinical Aspects of Human Carnitine Deficiency*, ed. P Borum, pp. 28–37. Oxford: Pergamon
137. McGarry JD, Woeltje KF, Kuwajima M, Foster DW. 1989. *Diabetes/Metab. Rev.* 5:271–84
138. Prentki M, Vischer S, Glennon MC, Regazzi R, Deeney JT, et al. 1992. *J. Biol. Chem.* 267:5802–10
139. Brun T, Roche E, Kim K, Prentki M. 1993. *J. Biol. Chem.* 268:18905–11
140. Chen S, Ogawa A, Ohneda M, Unger RH, Foster DW, McGarry JD. 1994. *Diabetes* 43:878–83
141. Vara E, Fernandez-Martin O, Garcia C, Tamarit-Rodriguez J. 1988. *Diabetologia* 31:687–93
142. Deeney JT, Tornheim K, Korchak HM, Prentki M, Corkey BE. 1992. *J. Biol. Chem.* 267:19840–45
143. Planner N, Orci L, Glick BS, Amherdt M, Arden SR, et al. 1989. *Cell* 59:95–102
144. Declercq PE, Falck JR, Kuwajima M, Tyminski H, Foster DW, et al. 1987. *J. Biol. Chem.* 262:9812–21
145. MacDonald MJ, Kowluru A. 1985. *Diabetes* 31:566–70
146. Penn EJ, Brocklehurst KW, Sopwith AM, Hales CN, Hutton JC. 1982. *FEBS Lett.* 139:4–8
147. Colca JR, Katagal N, Lacy PE, McDaniel ML. 1983. *Biochem. J.* 212:819–27
148. Metz SA. 1988. *Diabetes* 37:3–7
149. Wollheim CB, Regazzi R. 1991. *FEBS Lett.* 268:376–80
150. Calle, R, Ganesan S, Smallwood JI, Rasmussen H. 1992. *J. Biol. Chem.* 267:18723–27
151. Ganesan S, Calle R, Zawalich K, Greenawalt K, Zawalich W. 1992. *J. Cell Biol.* 119:313–24
152. DeFronzo RA, Bonadonna RC, Ferrannini E. 1992. *Diabetes Care* 15:318–68
153. McGarry JD. 1992. *Science* 258:766–70
154. McGarry JD. 1994. *J. Cell. Biochem.* 55S:29–38
155. Reaven GM, Hollenbeck C, Jeng C-Y, Wu MS, Chen Y-DI. 1988. *Diabetes* 37:1020–24
156. Sako Y, Grill VE. 1990. *Endocrinology* 1580–89
157. Elks ML. 1993. *Endocrinology* 133:208–14
158. Zhou Y-P, Grill VE. 1994. *J. Clin. Invest.* 93:870–76
159. Pieber TR, Stein DT, Ogawa A, Alam T, Ohneda M, et al. 1993. *Am. J. Physiol.* 265:E446–53
160. Lee Y, Hirose H, Ohneda M, Johnson JH, McGarry JD, et al. 1994. *Proc. Natl. Acad. Sci. USA.* 91:10878–82
161. Draznin B, LeRoith D, eds. 1994. *Molecular Biology of Diabetes.* Totowa, NJ: Humana

Annu. Rev. Biochem. 1995. 64:721–62

THE CATALYTIC MECHANISM AND STRUCTURE OF THYMIDYLATE SYNTHASE[1]

Christopher W. Carreras[2] and Daniel V. Santi

Departments of Pharmaceutical Chemistry and Biochemistry and Biophysics,
University of California San Francisco, San Francisco, California 94143-0448

KEY WORDS: methylase, enzyme mechanism, folate, enzyme catalysis, nucleic acid
modification, nucleotide metabolism

CONTENTS

---

[1]Abbreviations used: TS, thymidylate synthase; DHFR, dihydrofolate reductase; SHT, serine transhydroxymethylase; dUMP, 2'-deoxyuridine-5'-monophosphate; CH$_2$H$_4$folate, 5,10-methylene-5,6,7,8-tetrahydrofolate; dTMP, thymidine-5'-monophosphate; H$_2$folate, 7,8-dihydrofolate; CB3717, 10-propargyl-5,8-dideazafolate; FdUMP, 5-flouro-2'-deoxyuridine-5'-monophosphate; BrdUMP, 5-bromo-2'-deoxyuridine-5'-monophosphate; PABA, p-amino benzoic acid; PLP, pyridoxal-5'-phosphate; PAGE, polyacrylamide gel electrophoresis.

[2]Current Address: Department of Chemical Engineering Stanford University, Stanford, California 94305-5025

0066-4154/95/0701-0721$05.00

ABSTRACT

Thymidylate synthase (TS, EC 2.1.1.45) catalyzes the reductive methylation of dUMP by CH_2H_4folate to produce dTMP and H_2folate. Knowledge of the catalytic mechanism and structure of TS has increased substantially over recent years. Major advances were derived from crystal structures of TS bound to various ligands, the ability to overexpress TS in heterologous hosts, and the numerous mutants that have been prepared and analyzed. These advances, coupled with previous knowledge, have culminated in an in-depth understanding of many important molecular details of the reaction. We review aspects of TS catalysis that are most pertinent to understanding the current status of the structure and catalytic mechanism of the enzyme. Included is a discussion of available sources and assays for TS, a description of the enzyme's chemical mechanism and crystal structure, and a summary of data obtained from mutagenesis experiments.

INTRODUCTION

Thymidylate synthase (TS, EC 2.1.1.45) catalyzes the reductive methylation of dUMP by CH_2H_4folate to produce dTMP and H_2folate. The enzyme is unique among those that use folate cofactors in that CH_2H_4folate serves as both a one-carbon donor and a reductant, and H_2folate is a product. Thus, for each mole of dTMP produced, one mole of H_4folate is consumed. Replenishment of the H_4folate pool requires the sequential action of dihydrofolate reductase and serine transhydroxymethylase (Figure 1). TS has been extensively studied in terms of structure, function, and inhibition. Usually, TS is a dimer of identical subunits of 30–35 kDa each. In protozoa, TS and DHFR exist on a single polypeptide with the DHFR domain on the amino terminus and TS on the carboxy terminus (1). The native protein is a dimer of two such subunits and is in the size range of 110–140 kDa. Some, probably all, plants also contain the bifunctional DHFR-TS (2).

Several studies have made a significant contribution to our understanding of the catalytic mechanism of TS. A mechanism for TS proposed in 1959 (3) involved a bisubstrate adduct that closely resembled the currently accepted structure of a covalent intermediate in the TS reaction. In 1968, nucleophilic catalysis was recognized as a feature that could simplify an apparently complex chemical reaction to one amenable to enzyme catalysis (4, 5). Shortly afterwards, a chemical mechanism for TS catalysis was proposed that served as the platform upon which details of the reaction were formulated (6, 7). In subsequent years, the connectivity of the covalent TS-FdUMP-CH_2H_4folate complex (8, 9) and the complete amino acid sequence of *Lactobacillus casei* TS (10–12) were determined. The valuable molecular knowledge provided by the

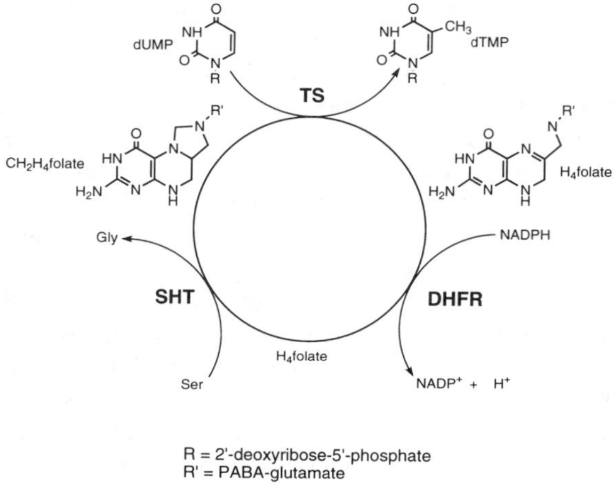

R = 2'-deoxyribose-5'-phosphate
R' = PABA-glutamate

Figure 1 The dTMP cycle.

aforementioned works enabled the in-depth studies of the mechanism of TS that occurred over the next decade. In 1987, the X-ray crystal structure of *L. casei* TS was solved (13), and this structure together with tools of recombinant DNA technology and cumulative knowledge of TS, provided the starting point for an understanding of structure-function relationships of TS. Over this period, it became apparent that TS is a paradigm for other enzymes that catalyze one-carbon transfers to the 5-position of pyrimidine heterocycles. These include the dCMP and dUMP hydroxymethylases, and RNA-uracil and DNA-cytosine methyl transferases.

Here, we review work on TS that is most pertinent to an understanding of the current status of the structure and catalytic mechanism of the enzyme. Because of space limitations, we have omitted many topics and citations covered in previous reviews (14–23). We make no attempt to review properties of TS that are specific to the bifunctional DHFR-TS found in protozoa and plants or aspects of TS related to the fields of pharmacology, chemotherapy, and molecular or cell biology.

SOURCES AND ASSAYS OF THYMIDYLATE SYNTHASE

Thymidylate Synthases thus far Expressed

Detailed studies of TS require large amounts of enzyme only obtainable in high expression systems. The first such systems available were methotrexate resistant *L. casei* cells that produced large quantities of both TS and DHFR (24–26). In the past decade, TS has been obtained from expression systems made by introduction of appropriate plasmids containing the desired coding, transcription, and translation sequences into *Escherchia coli* hosts. With optimal heterologous expression systems, TS is produced to a level that usually represents 5–30% of the soluble protein of extracts (cf 27–30).

Isolation of the TS gene is now straightforward. With prokaryotes, the TS gene can usually be obtained from a DNA library by positive selection in Thy⁻ strains of *E. coli*; DNA fragments containing the complete gene support cell growth in the absence of thymine, whereas others do not (see 27, 31, 32). Other approaches take advantage of the high sequence homology of TSs (Figure 2) (33–35). With over 25 TS gene sequences available, it is usually feasible to select one or several genes that have appropriate codon usage and to isolate the desired TS by hybridization to DNA libraries of the organism of interest (see 36, 37). Another approach is to use PCR to obtain a hybridization probe; degenerate oligonucleotide primers corresponding to two highly conserved regions of TS are synthesized and used to amplify a region of the TS gene (see 37, 38). The PCR product is then used as a hybridization probe to isolate TS sequences from appropriate DNA libraries.

Heterologous expression of TS in *E. coli* has usually been straightforward. Typically, an inducable promoter compatible with complementation of Thy⁻ *E. coli*, such as the *trp, tac,* or *lac* promoters, is used in the setting of commonly used vectors (see 27, 30, 36, 39). When the only objective is high expression of TS, vectors containing the T7 promoter have been effective (see 29, 40, 41). However, the T7 system is not compatible with bacterial-complementation systems. When expression was difficult in these vectors, success was achieved by incorporating a 5′ untranslated leader sequence from *L. casei* TS and modifying the codon usage of the 5′ part of the coding sequence (36).

Using a host cell that is deficient in TS (Thy⁻) is often advantageous. This permits complementation studies, and purification of expressed TS is not complicated by the presence of host TS. For studies of low activity mutants, Thy⁻ hosts are almost a necessity. Many Thy⁻ hosts are available from the American Type Culture Collection. A popular Thy⁻ host for TS expression is χ2913 (42). Recently, an *E. coli* strain was reported that lacks the entire chromosomal TS gene (43). In addition to the aforementioned advantages, use of this strain avoids possibilities of homologous recombination. Finally, for studies involving concomitant expression of multiple TS mutants, the use of a *recA⁻* strain of *E. coli* to avoid problems associated with lingering heterologous multimeric plasmids is advisable (44).

Using the above approaches, some 29 TS sequences from different organisms have been cloned and deposited into the DNA databases. By our count, over 11 monofunctional TSs have been overexpressed in heterologous systems.

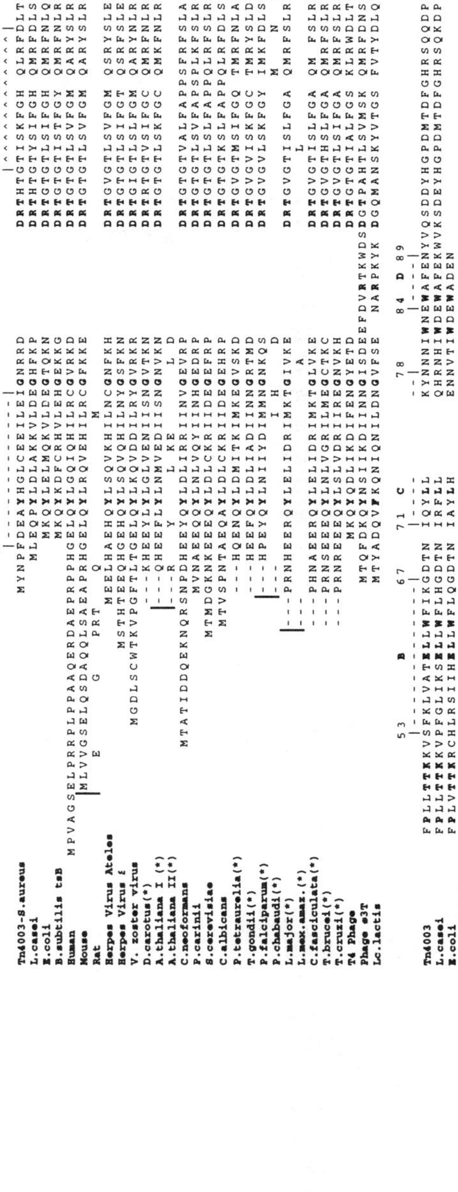

Figure 2 (See page 727 for caption.)

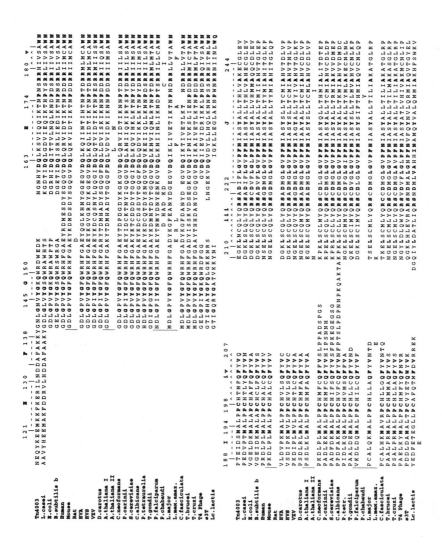

```
                250  ^^^^  11 <<<<  263  267  M 273             282        P*                316
                                                |-|-|          ^ ^ ^|                            |
Tm4003          GEFIHTFGDAHIYSNHMDIHTLQLSRDSYLPPQLINTDK        SIPDINYEDLEINYES  HPAIKAPIAV
L.casei         GDFIHTTGDAHIYVNLDDLYNLDQTHEQLSREPRPLPKLLINRKP   HDIDFDMKDILLNYDP  HPAIKAPLPAV
E.coli          GDFWTHGDAHIYSNHMDQTHIEQLNLSDPRPLPRLIIKRKP      ESIFPYRFHDFEIEGYDP HPHIKAPPVAI
B.subtilis b    GDFIHTLGDAHIYQNHIEPLKIQLERDVRREPRFARKV        DSINFAEDFIIEDYDP  HPHIKAVSV
Human           GDFIHTLGDAHIYLNHIEPLKIQLNREPRFPKKILRKV        EKIDDFKAEDFQIEGYNP HPTIKKEMAV
Mouse           GDFVHTLGDAHIYLNHIEPLKIQLQREPRFPKKILRKV        ETIDDFKVEDFQIEGYNP HPTIKKEMAV
Rat             GDFIHTLGDAHIYLNHIEPLKIQLQREPRFPKKILRKV                           HPTIKKEMAV
EVA             GDFIHTLGDAHIYVNHDALTEQLRETPRRFPTKFFARKV        ASIDDFKNDILENYNP  YSIKNFMMAV
BVB             GDLIHTLGDAHIYLNVDIDALKMQLRKFPFTLKFARNV        SCIDDFKADFPIDLNYNP HPIIKNHMAL
VZV             GDFIHVLGDAHIYSNLSDLQETSRMLRKTFFPTLKINGE       KKDIDSFEAAPFDLLITGYNP HOKIENHMAV
D.carotus       GDFIHVLGDAHIYKTNHVRPLQEQLL-NPRFPVKINPEK       KQIDSFFVAPFDLTGYDP HOKIENHMAV
A.thaliana I    HEFILQMGDAHIYRDHIEPLKTQLEEPRDPPKLKWARSEEI      GDIDGFKVPDFVVEGYKP WGKIDNKNMSA
A.thaliana II   GDFIHVSGDCIIYKDHIEALQQLRSPPRFPPTSLNRSI        TDIDDFTLDDFNIQNYHP YETIKKKMSI
C.neoformans    GEFIHTLGDAHIYKDNHIDALKKEQITRTNPRFPVIKEKNEI     KDIDDFKLDDFEINP   HPRIQNHMAV
S.cerevisiae    GEFIHTLGDAHIYLDHIDAKKEQFRIPRFPDFKNEIKGN        KSIDDFKKEDFEIVGYEP YPPIKKKMSV
C.albicans      GEFIHTLGDAHIYQNTHIVHNEALEALQIERERPLPRFPFPTLKLNPERI KSLFPDYTKEDFQINA  HDKIQNHMAV
P.aeri          KEFIHVLGNAHVYNNNIDSLKIQLNRIPPRFPFPTKLNPEFI     KNIEDFTISDFTQNTVH HORIKNHMAV
T.gondii        AEFIHVLGNAHVYNNNIDSLKIQLNIPPRFPFPTKLNPEFI                         
P.falciparum                                                                                    
P.chabaudi                                                                                      
L.major         GELVHTLGDAHVYRNVDALKAQLERVPHAFPPTIFKEER        QYLDYELTDMEVIDYVP HPAIKKEMAV
L.mmm.amai.                                                                                     
C.fasciculata   GELVHTGTAVVYSNHVEAAKEQLQKVPAFPVVFKKER          EFLDYESTDMEVVDDYVP YPPIKNEMAV
T.brucei        GELVHTLGDAHVYSNHVEAALKAQLKVPRFPPIKFKQDK        EFLEDFQESDIEIDYLP YPPISKNEMAV
T.cruzi         GELVHTLGDAHVYSNHVEPNEQLKRKVPAFPYVFRRRR         EFLDYEGCDMEVIDYAP YPPISKNKAV
T4 Phage        GDLIFSGDNTMYQNHVEQCKEILRERVDELIISGLPYKFRYLSTEQLKYK FRYLSTEQLKHGGKLLEVAV
φ3T Phage       KFFYFVNLHITDNQFEQANELMKRTASEAPRLLNVFDG         KDFHFTDDFLVLKH   GDKLLEVAV
Lc.lactis                                  KE                 TNFFIKFDDFELVDYFVKFQLKFDLAI
```

Figure 2 Aligned sequences of thymidylate synthases from various sources. The numbering scheme used is that of the *L. casei* enzyme. The position of α-helices and β-sheets in *E. coli* TS are designated with dashes and carets, respectively (35). Asterisks designate bifunctional enzymes where TS is the C-terminal domain and the DHFR domain is not shown. Sequences are from the GenBank or Dayhoff databases.

These include *L. casei* (28, 30), *E. coli* (27), *Lactobacillus lactis* (45), *Candida albicans* (29), *Pneumocystis carinii* (37, 46), *Cryptococcus neoformans* (47), T4 phage (32), Varicella *zoster* virus (42), human (36), mouse (48), and rat (J Ciésla & F Maley, submitted).

In addition, about six bifunctional protozoan DHFR-TSs have been cloned and expressed. These include those from *Leishmania major* (40), *Plasmodium faliciparum* (38, 49, 50), *Trypanosoma cruzi* (51), *Trypanosoma brucei* (P Gamarro & D Santi, submitted), and *Toxoplasma gondii* (M Trujillo & D Santi, in preparation; see Reference 1 for a review of protozoan bifunctional enzymes).

Finally, TSs have been cloned but not overexpressed from *Saccharomyces cerevisiae* (52), *Bacillus subtilis* (53), *Staphylococcus aureus* tn4003 (54), *Bacillus subtilis* phage φ-3T (55), saimirine herpes virus *1* (56), ateline herpes virus *2* (57), and bifunctional DHFR-TSs from *Paramecium tetraurelia* (I M Schlichtherle, direct GenBank submission), *Arabidopsis thaliana* (58), and carrot (59).

Assays of TS

In addition to catalysis of dTMP formation, TS participates in several binding and "partial" reactions that allow isolation of individual steps within the catalytic pathway. Below, we summarize the most widely used assays for TS activity and binding.

CATALYTIC ASSAYS   A qualitative assay for TS activity in vivo involves complementation of growth of Thy⁻ *E. coli* in the absence of thymine (see 39, 60). Although this method provides a rapid genetic method for the assessment of whether or not TS is active, it is sensitive to the level and activity of the expressed enzyme and is not quantitative. The minimum level of TS that can be detected corresponds to about ~0.002 U/mg in soluble extracts (60).

The simplest, most commonly used in vitro assay for TS is a spectrophotometric assay that measures the increase in absorbance at 340 nm ($A\varepsilon_{340} = 6400$ $M^{-1}cm^{-1}$) concomitant with H_2folate production (61). The sensitivity of this assay is about 10^{-4} U/mL TS activity. The spectrophotometric assay may be limited by interfering absorbance in crude extracts or added agents or by agents that oxidize the cofactor. When possible, it is clearly the assay of choice.

A more sensitive, but less convenient, assay of TS activity monitors the loss of tritium from [5-³H]dUMP into solvent upon dTMP formation (62, 63). Usually, the nucleotides are adsorbed on a solid medium such as activated charcoal, and the radioactivity in water is measured. This assay is highly sensitive and can be used to detect as little as 10^{-8} U/mg TS activity. A more convenient and accurate, albeit less sensitive, method of measuring tritium release monitors the decrease in ³H/¹⁴C ratio of [2-¹⁴C, ³H]dUMP after evaporation of solvent (64). Because TS also catalyzes a slow exchange of the

5-hydrogen of dUMP for solvent (see above), it may be important to verify that the loss of tritium is due to dTMP formation. This can be easily accomplished by HPLC separation of $[2\text{-}^{14}C, {}^{3}H]$dUMP and $[2\text{-}^{14}C]$dTMP and quantitation of the isotopes (see 65).

The cofactor-independent dehalogenation of BrdUMP or IdUMP to form dUMP provides a spectrophotometric method for measuring the enzyme's ability to bind nucleotides and activate C-5 for substitution (66). However, the reaction is slow compared to dTMP formation and suffers from low sensitivity. The assay is useful for determination of K_i values of compounds that compete for the dUMP binding site and in mechanistic studies of partial reactions of TS mutants.

DIRECT BINDING ASSAYS   TS forms a stable covalent complex with FdUMP and CH2H4folate ($K_d \approx 10^{-12}$ M) that may be used to titrate the active sites of the enzyme (67–69). The reaction has a wide range of applications that encompass the detection and quantitation of TS, FdUMP, or CH2H4folate. The reaction has also been used to assess the polyglutamate content of CH2H4folate cofactors (70, 71). When a radioactive label is present on either FdUMP or CH2H4folate, SDS/PAGE (68, 72), nitrocellulose filter binding (67), or trichloroacetic acid, precipitation (73) may be used to separate bound and free ligands. A spectrophotometric assay for formation of the TS-FdUMP-CH2H4folate complex follows the formation of a new chromophore at 330 nm (8, 68).

TS forms a tight binary complex with 5-NO$_2$dUMP ($K_d \approx 10^{-8}$ M) that may be quantitated by uv/vis spectroscopy ($A\varepsilon_{345nm}$ = 4810/mol of bound NO$_2$dUMP) (74). Unlike the TS-FdUMP-CH$_2$H$_4$folate complex, the interaction does not require folate and is not stable towards denaturants.

TS is effectively inhibited by pyridoxal 5'-phosphate (PLP), which binds in the dUMP site and forms a thiohemiacetal with the active site cysteine (75, 76). A characteristic difference spectrum accompanies complex formation and shows an increase in absorbance at 328 nm ($A\varepsilon_{328}$ = 4670 M^{-1}cm^{-1}). Displacement of PLP with nucleotides reverses this absorbance change, and provides a useful method for the direct determination of nucleotide K_d values (76, 77).

The formation of TS-dUMP binary complexes results in the quenching of intrinsic fluorescence of the protein and provides an extremely binding sensitive assay (68, 78, 79). However, the assay is susceptible to interfering absorbances and may fail with certain TS mutants (77).

CATALYTIC MECHANISM: MODEL AND ENZYME STUDIES

Chemical Mechanism

Figure 3 shows the salient features of the chemical mechanism of TS. After formation of a reversible ternary complex, nucleophilic attack by a thiol

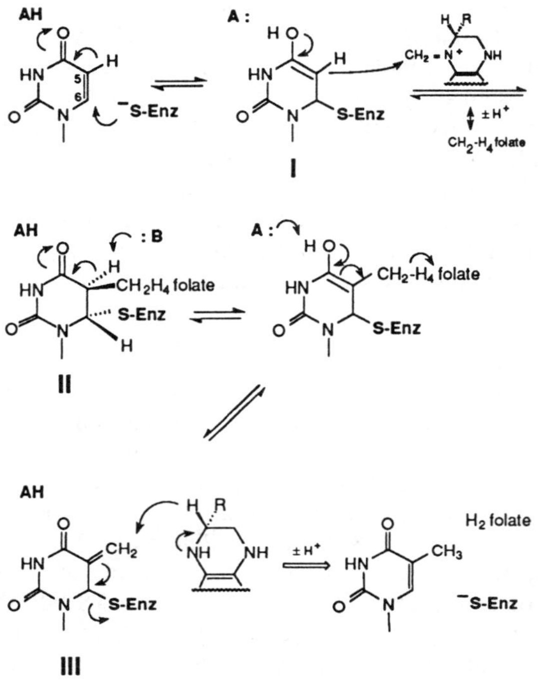

Figure 3 Salient features of the chemical mechanism of TS.

(Cys198 in *L. casei* TS) at the C-6 of dUMP converts the 5-carbon to a nucleophilic enol(ate), intermediate I. This is followed by covalent bond formation between C-5 of dUMP and the one-carbon unit (C-11) of CH_2H_4folate, which has been activated by formation of an iminium ion at N-5. The C-5 proton of intermediate II is abstracted, followed by β-elimination of H_4folate to give the exocyclic methylene intermediate III. Hydride transfer from non-covalently bound H_4folate to the exocyclic methylene intermediate III and β-elimination of the enzyme provide the products, H_2folate and dTMP, and the active enzyme.

For purposes of discussion, we have divided the mechanism of the TS reaction into four events: 1. Activation of C-5 of dUMP for substitution. 2. Activation of CH_2H_4folate. 3. Transfer of a methylene group from CH_2H_4folate to dUMP. 4. Reduction of the transferred methylene group.

ACTIVATION OF C-5 OF dUMP   Nucleophilic attack at the 6-position of dUMP converts the 5-carbon to a nucleophilic enol or enolate. This necessary step was first demonstrated in chemical models (see 14, 22, 23 for reviews). The evidence that this step is involved in the initial stages of the TS reaction is unequivocal.

Studies of the TS-FdUMP-CH$_2$H$_4$folate complex provided the earliest and strongest support for this mechanism of activation. The structure of the complex and mechanism of its formation was initially deduced from chemical considerations (8, 9, 80, 81). Later, researchers showed that proteolytic digestion of the complex yielded a peptide covalently bound to FdUMP and the cofactor through the thiol of Cys198 (10, 82). The structure was supported by ^{19}F NMR studies (83–85) and finally proven by solution of the crystal structure (86). Studies of α-hydrogen 2° isotope effects using [6-^{3}H]FdUMP suggested that rehybridization of C-6 occurs either as a late transition state in a rate-determining step of complex formation or in a post rate–determining step (87).

Other 5-substituted-dUMP analogues, which serve as mechanism-based inhibitors of TS, provided additional evidence for nucleophilic attack at C-6 of the pyrimidine. Inhibitors such as 5-trifluoromethyl- (88), ethynyl- (89, 90), bromovinyl- (91) and *trans*-5-(3,3,3-Trifluoro-1-propenyl)-dUMP (92) all have latent reactive functions at the 5-position of dUMP. After attack of Cys198 at the 6-position and formation of the enol or enolate at the 5-position, these functional groups are activated and react with a second nucleophile of the enzyme or with water. One inhibitor, 5-NO$_2$-dUMP, has an exceptionally electrophilic 6-carbon that simply reacts with the Cys198 to form a reversible, but highly stabilized, adduct (93).

The understanding of the early steps in catalysis also has benefited from studies of partial or alternate reactions of TS, such as the 5-hydrogen exchange of [5-^{3}H]dUMP for solvent protons and the dehalogenation of 5-Br or IdUMP (64, 66, 94). These "partial" reactions of TS do not require folates and have been useful in demonstrating transient 5,6-dihydropyrimidine adducts and in localizing lesions in the catalytic pathway caused by mutations (44, 72, 95, 96).

Nucleophiles such as thiolate or alkoxide catalyze the exchange of the 5-hydrogen of 1-substituted uracils for protons from the solvent (14, 23). In the absence of CH$_2$H$_4$folate, thymidylate synthase catalyzes exchange of the 5-hydrogen of dUMP approximately 10^5-fold slower than dTMP formation (64). As shown in Figure 4 for [5-^{3}H]dUMP, after nucleophilic attack at C-6, the activated 5-position is protonated to give the 5,6-dihydropyrimidine adduct, II. Reversal of these steps, with release of triton to solvent results in the observed exchange reaction. In nonenzymatic reactions, both the triton and the proton are equally accessible to solvent and (except for isotope effects) have an equal probability of being removed. However, the enzymatic reaction would be expected to proceed stereospecifically so that the same proton added to the 5-position would be released. Hence, if the exchange reaction was completely

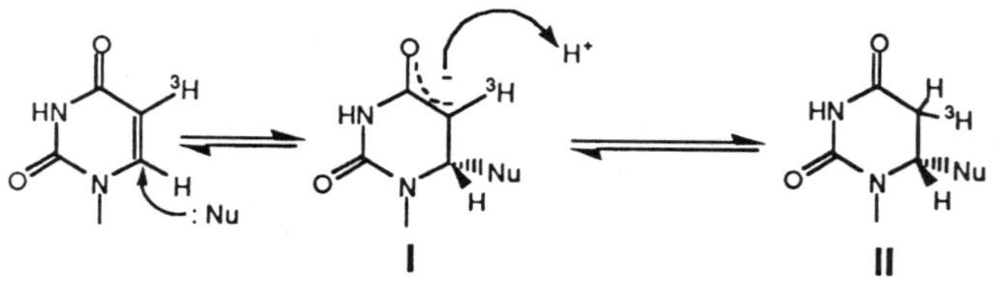

Figure 4 Mechanism of nucleophilic catalysis of 5-H exchange from [5-³H] dUMP.

stereospecific, no exchange would be observed. This apparent enigma was rationalized by invoking partial stereospecificity or a stereospecific *trans*-addition/*cis*-elimination of protons across the 5,6-double bond of dUMP analogous to the normal reaction (64).

In the presence of exogenous thiols, TS catalyzes the dehalogenation of BrdUMP (IdUMP); dUMP, Br⁻ (I⁻), and a disulfide are the products (Figure 5). The reaction also occurs chemically, albeit at a slower rate (66, 97). Although TS catalyzes debromination about 400-fold slower than dTMP formation, the reaction is substantially faster than the 5-hydrogen exchange reaction. In the favored (E1Hal) mechanism, Cys198 adds to C-6 to form the 5-Bromo-5,6-dihydropyrimidine adduct, intermediate I (Figure 5) (66, 98, 99). The activated halogen is then abstracted as a bromonium ion by exogenous thiol to yield a sulfenyl bromide and, intermediate II. The latter yields dUMP, while the sulfenyl bromide reacts with another molecule of thiol to yield the corresponding disulfide and Br⁻. C-5 of the 5-bromo-5,6-dihydropyrimidine intermediate also undergoes nucleophilic attack by exogenous thiol to displace Br⁻ and yield small amounts of 5-alkylthiopyrimidines.

The TS-catalyzed and nonenzymatic dehalogenation reactions proceed with

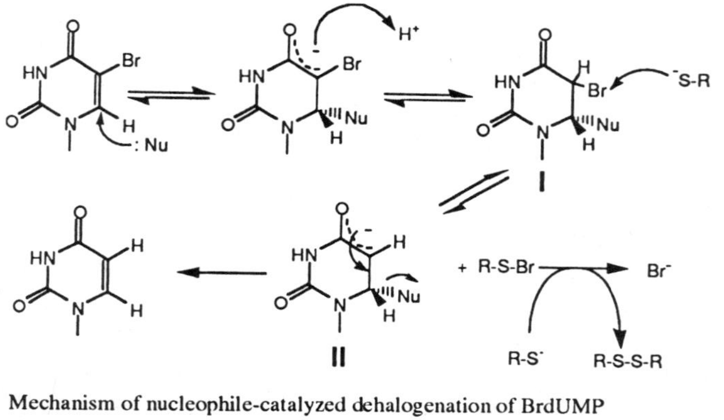

Figure 5 Mechanism of nucleophile-catalyzed dehalogenation of BrdUMP

theoretically maximal inverse 2° kinetic isotope effects when 6-^{3}H substrates are used (66, 99). Observation of a maximal isotope effect in these reactions demonstrates sp^2 to sp^3 rehybridization of C-6 before or during the rate-limiting step.

ACTIVATION OF CH$_2$H$_4$FOLATE BY IMINIUM ION FORMATION   The one-carbon unit of 5,10-CH$_2$H$_4$folate is transferred to the 5-position of dUMP in the TS reaction. In general, formaldehyde or adducts of formaldehyde, such as CH$_2$H$_4$folate, do not undergo direct reaction with nucleophiles as sp^3-hybridized carbon adducts but rather as the highly reactive sp^2-hybridized carbon formed by elimination (100). In the case of CH$_2$H$_4$folate, the reactive species could be the 5- or 10-iminium ions formed by opening of the 5-membered imidazolidine ring of CH$_2$H$_4$folate, which are both too unstable for detection (101). In kinetic studies of the hydrolysis of CH$_2$H$_4$folate the 5-iminium ion was proposed because N-5 is more basic than N-10, and N-10 is the better leaving group (102). Several other studies also inferred a 5- rather than a 10-iminium ion as the predominant reactive form of CH$_2$H$_4$folate. For example, reduction of CH$_2$H$_4$folate with KBH$_4$ resulted in 5-methyl-CH$_2$H$_4$folate, which must have arisen from the 5- rather than 10-iminium ion (103). In addition, a kinetic study of the condensation of formaldehyde with a series of tetrahydroquinoxaline analogues, in which the basicity of N-10 varied, indicated that these compounds react though an iminium ion formed at the tetrahydroquinoxaline equivalent of N-5 (104, 105).

Studies of the TS-FdUMP-CH$_2$H$_4$folate complex support the N-5 iminium ion of CH$_2$H$_4$folate as an intermediate in the TS reaction. A peptide to which FdUMP and CH$_2$H$_4$folate were covalently linked was isolated and found to be stable in the absence of anti-oxidants (82, 106). Because a 10-substituted H$_4$folate with a free 5-NH would not have withstood the aerobic conditions of isolation (107), the stability of this peptide provided evidence that the linkage of the one-carbon unit was to N-5 of the cofactor. Observations of kinetic and equilibrium secondary isotope effects using [11-^{3}H]CH$_2$H$_4$folate also provided evidence for an iminium ion intermediate in the TS reaction (87). In another experiment, N-5 was shown to be the site of attachment of the cofactor in the TS-FdUMP-CH$_2$H$_4$folate complex by treatment with reagents that specifically cleave the C-9–N-10 bond of either N-5- or N-10- substituted H$_4$folates (108). Finally, the crystal structure of the TS-FdUMP-CH$_2$H$_4$folate complex unequivocally showed covalent linkage between FdUMP and N-5 of the cofactor (86).

Because the iminium ion is too reactive to exist free in solution at significant concentrations (101), it must form on TS after initial binding of CH$_2$H$_4$folate. Structures of ternary complexes and modeling studies show that folates are not bound in a conformation that could easily accommodate the 5-membered imidazolidine ring of CH_2H_4folate (86, 109, 110). It follows that either (*a*) CH_2H_4folate initially binds to a conformation of the enzyme not yet observed crystallographically and is subsequently converted to the iminium ion, or (*b*) the 5-membered ring is opened concomitant with binding. Our understanding of this issue could be aided by studies with stable closed-ring analogues of CH_2H_4folate; one such analogue, 5-deaza-CH_2H_4folate, does not bind well to TS (111).

TS could assist opening of the imidazolidine ring of CH_2H_4folate in several ways. The enzyme could assist protonation of N-10 to enhance its leaving-group ability. Because imidazolidine-ring opening is general acid catalyzed (105, 112), a strategically placed general-acid on the enzyme could assist ring opening. In support of this, resonance Raman spectroscopic studies of the TS-FdUMP-CH_2H_4folate complex indicated the presence of hydrogen bonds to N-10 and/or the benzoyl carbonyl of the PABA moiety (113), each of which is observed crystallographically (86). In particular, Glu60 of *E. coli* TS coordinates to N-10 of the cofactor through a bridging water molecule in the TS-FdUMP-CH_2H_4folate crystal structure. The enzyme could also assist ring opening through perturbations of the imidazolidine ring (114). For example, rotation of the PABA moiety of CH_2H_4folate would disrupt conjugation between N-10 and the aromatic ring and make N-10 more basic. This would enhance protonation, which would in turn increase reactivity. Finally, if enzyme-cofactor interactions are incompatible with the constraints of the five-membered imidazolidine ring, strain could cause the N-10-C-11 bond to lengthen. In this regard, studies of heterolytic S_N1 C-O bond cleavages indicate that the ground-state lengths of bonds being broken correlate with their chemical reactivities: "...the longer the bond, the faster it breaks" (115–117). This last mechanism is an application of theories of strain and distortion in enzyme catalysis (118, 119).

METHYLENE TRANSFER FROM CH_2H_4FOLATE TO dUMP   The next step of the TS reaction requires joining of the activated forms of dUMP and CH_2H_4folate. There is ample chemical precedent for this type of substitution reaction (i.e. hydroxymethylation or Mannich reaction of uracils) (6, 120), which may be envisioned as nucleophilic attack of C-5 of dUMP on the N-5 iminium ion of CH_2H_4folate. This results in an intermediate in which C-5 of dUMP is connected to N-5 of H_4folate by a methylene bridge. This intermediate has been isolated by means of rapid quenching of ongoing reactions of wild-type TS (121), or SDS-quenching of Glu 60 mutants of *L. casei* TS (96).

Definitive evidence for a bridged intermediate comes from structural studies of the covalent TS-FdUMP-CH_2H_4folate complex, which is a close analogue of intermediate II (Figure 3) in dTMP synthesis. Both NMR and crystallographic studies show C-5 of FdUMP attached to a methylene carbon (83–86). The crystal structure of this complex also shows that C-6 of FdUMP is attached to Cys198, while C-5 is attached to N-5 of H_4folate through a methylene bridge (86).

REDUCTION OF THE TRANSFERRED METHYLENE   The final steps of the TS reaction are elimination of H_4folate from intermediate II, followed by reduction of the one-carbon unit. An early suggestion for this involved direct S_N2 displacement of the cofactor by a cofactor C-6 hydride (3). Although this mechanism would provide the correct products, precedent for direct nucleophilic displacement at the α-carbon of a tertiary amine is lacking. It is now generally accepted that removal of the C-5 proton from intermediate II (Figure 3) results in formation of H_4folate and the exocyclic methylene of intermediate III; the latter has high carbonium ion character at C-7 and would readily accept the hydride from C-6 of the noncovalently bound H_4folate.

Precedence for the exocyclic methylene intermediate III comes from model studies of reactions of esters and ethers of 5-hydroxymethyluracils. These are unusually reactive toward nucleophilic substitution at C-7 because they are activated by nucleophilic addition to C-6 and form an exocyclic methylene intermediate at C-5 (122, 123). Detailed discussions of model systems may be found in previous reviews (14, 19, 22, 23).

The enzyme could promote formation of the exocyclic methylene by assisting elimination of H_4folate from intermediate II (Figure 3). Removal of the dUMP C-5 hydrogen could be facilitated by general base catalysis. Several amino acid residues and ordered water molecules are proximal to C-5 and might assist in this process, although mutagenesis studies have thus far failed to implicate any side chain residue (see below). A conformational isomerization of the covalent ternary intermediate has been proposed that would decrease the pK_a of the C-5 proton by placing it in an axial position so its conjugate base would be resonance delocalized through the pyrimidine (see below). Finally, the leaving group ability of H_4folate could be enhanced by protonation of N-5. There is no acidic residue of TS appropriately placed to accomplish this, but intramolecular transfer from the 4-enol form of the pteridine has been proposed through water-assisted protonation of N-3 by Asp221 (86).

Following elimination of H_4folate, the exocyclic methylene intermediate III (Figure 3) is expected to be sufficiently reactive that transfer of hydride from C-6 of H_4folate would only require appropriate juxtapositioning of the reactive groups. The final β-elimination of the enzyme from C-6 of the pyrimidine likewise would probably not require enzyme assistance.

Kinetic Mechanism

STEADY-STATE KINETICS   Steady-state kinetic studies of TS indicate that when monoglutamylated folates are used, TS follows a sequential ordered mechanism in which dUMP binds first (124, 125). H2folate is the first product to dissociate from the enzyme, followed by dTMP. For *L. casei* TS, K_m values for dUMP and CH2H4folate are typically about 5 and 10 μM, respectively, and k_{cat} is ~5 s^{-1} (44). The ordered mechanism is supported by experiments that show that increasing CH2H4folate concentrations reduce the rate of dissociation of [^{3}H]FdUMP from covalent TS-[^{3}H]FdUMP-CH2H4folate complex (126). Crystallographic results also show that the pyrimidine of dUMP forms a significant part of the binding site for CH2H4folate (86, 109).

Studies utilizing γ-linked polyglutamyl CH2H4folates suggest that folylpolyglutamates may bind in the absence of dUMP and that the steady-state kinetic mechanism becomes random if phosphate buffer is used (127, 128). In support of these hypotheses, a crystal structure has been obtained in which the folate analogue CB3717 is bound to *E. coli* TS in the absence of nucleotide (129).

For the sequential ordered mechanism of TS, the k_{cat}/K_m value for dUMP reduces to the rate of association of dUMP with the enzyme under conditions of saturating CH_2H_4folate (130, 131). The use of this relationship to interpret TS mutants should be guarded since the order of substrate binding seems to be subject to change.

Although various steps of the TS reaction mechanism have been proposed to be rate determining (44, 87, 114), such proposals were based on studies with the TS-FdUMP-CH_2H_4folate complex or with TS mutants. This important aspect of the TS reaction has yet to be established.

STRUCTURAL STUDIES OF THYMIDYLATE SYNTHASE

The crystal structure of *L. casei* thymidylate synthase was solved in 1987 (13). Subsequently, crystal structures of *E. coli* (35, 86), T4 phage (132), *Leishmania major* (133), human (134) and *Pneumocystis carinii* (K Perry, R Stroud & D Santi, unpublished data) TSs were solved and confirmed the striking structural homology among these enzymes. Structures of binary complexes of TS with nucleotides (135), folate analogues (129), ternary complexes with nucleotides and folates (86, 109, 110, 136, 137), as well as complexes with novel inhibitors (138–142), have also been reported. Some of these structures represent snapshots of intermediates that occur during the catalytic cycle and have provided information about changes the enzyme undergoes during catalysis (see 143 for a review).

Structure of TS

A diagram of the TS monomer bound to dUMP and the folate analogue CB3717 is shown in Figure 6 (109). The native enzyme is a symmetric dimer of structurally similar subunits. The overall fold of the enzyme is common to all

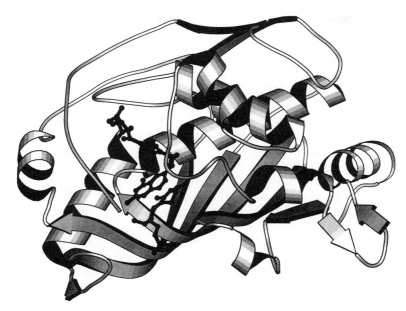

Figure 6 Structure of *E. coli* TS bound to dUMP and the folate analogue CB3717 [Brookehaven Protein Data Bank (pdb) file 2tsc] (109, 137).

TSs, and is defined by a series of 8 α-helices, 10 strands of β-sheet, and several segments of coil that connect the secondary structural elements. The subunit interface is formed by a 6-stranded twisted β-sheet composed of ~25 residues (~1200 Å^2) that packs against the same sheet formed by the other subunit of the dimer (13, see 144 for a detailed description of intersubunit contacts). Each subunit has a deep active site cavity that contains a bound inorganic phosphate, which has been observed in all TS structures solved in the absence of nucleotides (13, 35, 132, 134). Cys198g[3] is on the side of the cavity, and three arginine residues (Arg178′, Arg179′, and Arg218) and Ser219 are at the base of the cavity coordinated to the bound phosphate. Arginines 178′ and 179′ are not part of the same polypeptide chain as the other active-site residues and are donated from the other subunit of the dimer for the purpose of ligating the bound phosphate (13). The observation that each active site contains residues from both subunits provided a structural basis for TS being an obligate dimer.

[3]Residue numbers refer to those of *L. casei* TS. Primes are used to designate residues that are donated from the second subunit of the dimer. To convert to the *E. coli* numbering scheme, subtract 2 from residues numbered 1–91, and subtract 52 from residue numbers >143. For conversion to other numbering schemes, see Figure 2.

Structure of the TS•dUMP Complex

The structure of the TS•dUMP binary complex shows that the protein conformation in this complex is virtually identical to that observed in the phosphate-
bound structure (13, 135). A ligand-induced ordering of the polypeptide occurs
upon the binding of dUMP and is manifested by a decrease in the overall
B-factor of the molecule (135, 143). dUMP is bound in the active site cavity
in an extended conformation with its pyrimidine ring in the *anti* orientation (χ
$= -168°$), its deoxyribose ring in the C-4' *endo* conformation, and the C-4'-C-5'
bond *−sc* (*trans, gauche*).

dUMP BINDING   The phosphate of dUMP is coordinated by the same highly
conserved arginine residues and Ser219 that ligate inorganic phosphate in the
TS•P$_i$ structure, as well as by conserved Arg23 (135). Each of these residues
forms at least one hydrogen bond with the dUMP phosphate moiety, while
Arg178' forms an additional hydrogen bond with Tyr261, and Arg218 forms
additional hydrogen bonds with the peptidyl carbonyl oxygens of both Arg178'
and Pro196 (137). The guanidinium group of Arg218 is about 3.5 Å from the
Cys198 sulfur, too far to form a hydrogen bond but close enough to stabilize
the more reactive thiolate form of Cys198 via ion-pairing (13).

The largest difference in protein structure between the TS•P$_i$ and TS•dUMP
structures shows that Arg23 has moved to interact with the phosphate moiety
of dUMP and has become more ordered (135). This movement may be in
preparation for the conformational change where Arg23 forms a salt bridge
with the C-terminal carboxylate following folate binding (see below). Structures of Arg179 mutants bound to dUMP show Arg23 in either conformation
depending on the mutation (RM Stroud, JS Finer-Moore & E Fauman, in
preparation). Although the R179E TS•dUMP structure is similar to the wild-
type TS•dUMP structure in that Arg23 interacts with the dUMP phosphate,
the R179K mutant shows Arg23 pointing away from the dUMP phosphate,
similar to its position in the wild-type TS•P$_i$ structure. These observations are
consistent with two conformations of similar energy and sensitivity to small
changes in microenvironment, necessary features of a molecular switch that is
sensitive to dUMP binding.

The deoxyribose of dUMP interacts with Tyr261 and His259, each of which
form hydrogen bonds with the 3'-hydroxyl of dUMP (109, 135). The C-2'
carbon of dUMP lies in the back of the binding pocket and is 3.0 to 3.5 Å
from the backbone amide of Asp221 and the Ser219 side chain. These residues
would sterically hinder binding of nucleotides containing a 2'-OH and could
be responsible for the enzyme's decreased affinity for ribose nucleotides (19,
109, Liu, in preparation).

No bond is observed between Cys198 and C-6 of dUMP in the structure of

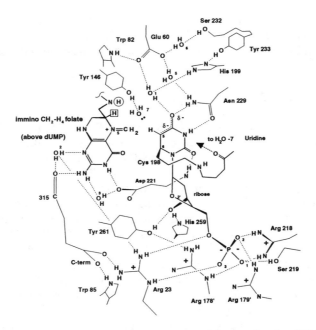

Figure 7   Schematic diagram of the hydrogen-bond network in the active site of TS (143).

the TS•dUMP complex, and the pyrimidine ring is planar (135). The pyrimidine moiety is involved in an extensive hydrogen bond network with Asp221, Asn229, Glu60, Trp82, His199, and several ordered water molecules (Figure 7) (86, 109, 135). The backbone NH of Asp221 donates a hydrogen bond to O-2 of dUMP. Asn229 forms the only direct contact between a TS side chain and the pyrimidine ring and assists in discrimination between dUMP and dCMP (146). The amide carbonyl oxygen of Asn229 accepts a hydrogen bond from NH-3 of dUMP, while the amide nitrogen donates a hydrogen bond to O-4 of dUMP. An ordered water molecule is hydrogen bonded to O-4 of dUMP, His199, and Glu 60, which is also hydrogen bonded with the indole nitrogen of Trp 82. Another ordered water molecule bridges Asn229 and Glu60. Researchers have proposed that these water molecules assist proton transfers to and from the pyrimidine during catalysis (86, 137, 143, 146, 147).

Other TS•nucleotide binary complexes show ligands bound in approximately the same site as observed for the TS•dUMP complex but in different conformations. For example, a crystal form of the TS•dUMP complex obtained under nonreducing conditions showed an altered conformation of the nucleotide (135). Oxidation of Cys198 was observed in this structure and blocked dUMP from occupying its usual position. Structures of V316Am TS complexed with dUMP or FdUMP also showed altered nucleotide binding modes relative to that observed for TS•dUMP (148). Although neither of these binding modes is productive, they illustrate the laxity in the conformations dUMP may adopt while bound to the enzyme.

Structures of Ternary Complexes

Several structures of ternary complexes between *E. coli* TS, nucleotides, and folates or folate analogues have been reported, including the TS-dUMP•CB3717 (109), TS-FdUMP•CB3717 (110), TS-FdUMP-CH$_2$H$_4$folate (86), and TS•dTMP•H$_2$folate (137) complexes. The TS-dUMP•CB3717 and TS•dTMP•H$_2$folate structures are the highest-resolution TS structures solved thus far, at 1.97-Å and 1.83-Å resolutions, respectively (109, 137). Ternary complex structures show nucleotides bound to the enzyme in positions and conformations similar to those observed in the structure of the TS•dUMP binary complex. Consistent with the ordered kinetic mechanism in which dUMP binds first, the nucleotide forms part of the binding site for folates and folate analogues, which are observed bound on top of the nucleotide (see Figure 2).

Ternary complexes are more ordered and have undergone an extensive conformational change compared to the TS•dUMP structure. Helices and sheets move as units, usually less than 1 Å, in response to ligand binding in a process that has been termed segmental accommodation (109). This mechanism avoids disruption of the favorable conformations and hydrogen-bond patterns found in these secondary structural elements. Upon ternary complex formation, the C-terminal tetrapeptide displays the largest change in position and is more ordered than in the TS•P$_i$ and TS•dUMP structures. The C-terminal residue (a Val in TS from most species) moves more than 4 Å to interact with the side chain of conserved Thr24 (109). The C-terminal carboxylate forms direct hydrogen bonds with Trp85 and Arg23, while the backbone carbonyl oxygen of Ala315 forms direct and water-mediated hydrogen bonds to the 2-NH$_2$ and 3-NH groups of the cofactor, respectively. The remainder of the C-terminal segment also interacts with conserved residues following the conformational change (Figure 8) .

Spectroscopic (68, 149), electrophoretic (68), protease protection (150), and gel-filtration (151) studies produced early indications of the aforementioned conformational changes. EPR spectroscopy of TS spin-labeled on its C-terminal residue indicates that in solution the C-terminus of TS is in an equilibrium between the two crystallographically observed conformations (152). The equilibrium is perturbed by ligand binding so that the relative amount of the conformation observed in ternary complex structures increases both upon formation of the TS•dUMP complex and upon its conversion to the TS-dUMP•CB3717 complex.

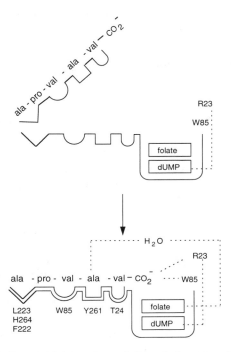

Figure 8 Schematic diagram of interactions of C-terminal residues before and after the ligand-induced conformational change. (Modified from Reference 44, with permission.)

NUCLEOTIDE BINDING   In general, nucleotides in ternary complexes are in conformations similar to that observed for dUMP in the TS•dUMP structure (86, 109, 110, 137). The TS-FdUMP•CB3717, TS-FdUMP-CH₂H₄folate, and TS-dUMP•CB3717 structures show a covalent bond between Cys198 and C-6 of the bound nucleotides. The pyrimidine rings are puckered in response to saturation of the 5,6 double bond. As in the TS•dUMP structure, nucleotides in ternary-complex structures show the phosphate moiety ligated by a quartet of arginine residues and Ser219, and the 3'-OH of the deoxyribose moiety forms hydrogen bonds with Tyr261 and His259.

The positions of Asp221, Asn229, Glu60, Trp82, His199, and two ordered water molecules that comprise the hydrogen bond network coordinating the pyrimidine in binary complexes are preserved. However, there is a difference in the structure of a third ordered water molecule near C-5 of the pyrimidine, hydrogen-bonded to the carbonyl oxygen of Ala196 and the hydroxyl of Tyr146 in the TS-dUMP•CB3717 structure (109, 110). This water molecule has also been proposed to assist proton transfers to and from the pyrimidine (137). In the TS•dTMP•H$_2$folate structure, the C-7 methyl group of dTMP shifts the location of this water by ~0.5 Å (137). Its occupancy is decreased two-fold and it is no longer hydrogen bonded to the protein. Disruption of these hydrogen bonds may contribute to the enzyme's differential affinity for dUMP vs dTMP (137).

FOLATE BINDING   Folate conformations are similar in each of the ternary complex structures (86, 109, 110, 137). Folates bound to TS adopt a folded conformation in which the pterin or quinazoline ring system is roughly perpendicular to the PABA moiety. 5,10-CH$_2$H$_4$folate can not be directly docked into this site with both its PABA and pterin moieties superimposed onto their crystallographically observed positions without perturbation of the 5-membered imidazolidine ring. Distortions of the cofactor necessary for accommodation of these moieties may be conducive to formation of the electrophilic N-5 iminium ion (86, 114). In addition, ternary complex structures show Glu60 interacting with N-10, either directly (109, 110) or through an ordered water molecule (86), suggesting that Glu60 could assist opening of the CH$_2$H$_4$folate imidazolidine ring via partial protonation (86, 109). Thus, CH$_2$H$_4$folate may initially encounter its binding site with an intact imidazolidine ring, and ring opening may then occur concomitantly with binding.

Interactions between TS and bound folates are similar in the various ternary complex structures (86, 109, 110, 137). For a detailed listing of protein-folate contacts, we refer the reader to tables (110), and summarize only the major features of the interaction here (See Figure 7). In the TS-FdUMP-CH$_2$H$_4$folate structure, a methylene group is covalently bonded to both N-5 of the folate and C-5 of FdUMP (86). The pterin (quinazoline) makes its largest contact with the bound pyrimidine of dUMP. The stacked ring systems of the two ligands are packed by extensive interactions with nonpolar residues, including conserved Trp82 and Trp85 (109, 110). The side chain of Asp221 is hydrogen bonded to N-3 of the pterin through a bridging water molecule (86). The main chain carbonyl of Ala315 accepts a hydrogen bond from the 2-amino group of the folate, and N-1 forms two water-mediated hydrogen bonds, one to the carbonyl of Ala315 and another to the side chain of Arg23. Although protonation of N-5 would provide an attractive mechanism for the β-elimination of H$_4$folate from the TS-dUMP-CH$_2$H$_4$folate intermediate II (Figure 3), no groups capable of this function are proximal to N-5.

The aromatic ring of the PABA moiety binds in a hydrophobic pocket lined with the side chains of Phe228, Leu224 and Ile81 (109, 110). As described above, Glu60 interacts with N-10. The PABA benzoyl carbonyl accepts a hydrogen bond from an ordered water molecule that is also hydrogen bonded to the main chain carbonyl of Pro313. The α-carboxyl of the glutamate is ligated by several water molecules, including two that are hydrogen bonded to Lys50.

In vivo, folates are modified with additional glutamate residues linked through peptide bonds to their γ-carboxyl groups. This modification may allow folates to bind to the free enzyme and allow a random kinetic mechanism (128). Because ternary-complex structures show the cofactor's pterin ring system bound on top of and trapping dUMP, the cofactor may be able to remain tethered to the enzyme through its PABA-polyglutamate moiety while the pterin moves to allow productive dUMP binding. Structures of the TS•CB3717 and TS•tetraglutamyl CB3717 binary complexes (129) confirm that folates can bind to the enzyme in the absence of nucleotides (153) and induce the C-terminal conformational change characteristic of ternary complexes. A ternary complex structure of *E. coli* TS complexed with dUMP and tetraglutamyl CB3717 shows the general position of the tetraglutamate moiety, which adopts an extended conformation to interact with a positively charged binding site that includes His53, Lys311, and Arg55 (136). This same positively-charged site may be responsible for reported interactions between TS and its mRNA (154, 155).

Stereochemistry of Methylation

The crystal structures of the TS-FdUMP-CH_2H_4folate and the TS-dUMP• CB3717 complexes allow assignment of the absolute stereochemistry of the covalent intermediate II (Figure 3) and insight into its stereochemistry of formation and conversion to dTMP (86, 143, 147). The structure and modeling studies confirmed and expanded previous proposals made from ^{19}F NMR studies of the complex (83–85). Consistent with ^{19}F NMR studies of the native TS-FdUMP-CH_2H_4folate complex (84), the crystal structure of this complex shows the thiol of Cys198 and one carbon unit of the cofactor situated across the 5,6-bond of FdUMP in the high energy *trans*-diaxial relationship (86). The pyrimidine ring exists in a half-chair conformation to accommodate the sp^3 hybridization of C-5 and C-6. Previously, ^{19}F NMR showed that, upon denaturation and proteolysis of the complex, the large enzyme and cofactor substituents relax to the more stable *trans*-diequatorial conformation (83). The absolute configurations of C-5 and C-6 of FdUMP in the complex are both *S*, and the configuration of C-6 of the cofactor is *R*. The latter agrees with the absolute stereochemistry of CH_2H_4folate determined independently (156). An additional observation relevant to the stereochemical pathway is that when (6*R*, 11*S*) [11-^{2}H,^{3}H]CH_2H_4folate is used in the TS reaction, (*S*)-methyl[*methyl*-^{1}H, ^{2}H,^{3}H]dTMP is produced (157, 158).

In the first step of the TS reaction, Cys198 must attack C-6 from a position approximately perpendicular (83) to the *si* face of C-6 of the pyrimidine ring, which is in accord with its position in the noncovalent TS•dUMP (135) and ternary-complex structures (86, 109, 110, 137). Earlier biochemical studies proposed that the antiperiplanar opening of the 5-membered imidazolidine ring of (6R, 11S) [11-^{2}H,^{3}H]CH$_2$H$_4$folate by cleavage of the C-11–N-10 bond gives the 5-iminium ion with the Z configuration that, to explain the formation of chiral dTMP, cannot undergo rotation about N-5 before it reacts with the activated pyrimidine (157). This arrangement of the activated pyrimidine and open-ring cofactor poised for reaction is mimicked in the TS-dUMP•CB3717 complex, where C-6 of dUMP is covalently bound to the enzyme, and CB3717 is positioned over the re face of the pyrimidine (147). The iminium ion then reacts with C-5 of the activated dUMP intermediate from a position approximately perpendicular to the re face of the pyrimidine to give the trans-diaxial product.

As modeled from the TS-FdUMP-CH$_2$H$_4$folate and TS-dUMP•CB3717 complexes, the structure of the initial ternary covalent intermediate has the N-5–C-11 bond syn or slightly gauche to the C-5 hydrogen, and the equatorially placed 5-H is not in a position where its abstraction can be facilitated by conjugation of the developing negative charge with the 4-carbonyl of dUMP (86, 147). It was proposed that abstraction of the C-5 proton is facilitated by a conformational isomerization in which the pyrimidine ring of the intermediate takes the opposite pucker, placing the 5,6-substituents in the diequatorial position (Figure 9) (83, 86). This arrangement places the C-5 proton in an axial position so it may be removed perpendicularly from the si face of the pyrimidine to yield a stable carbanion equivalent. From modeling the TS-FdUMP-CH$_2$H$_4$folate complex, Matthews et al have shown that this conformational isomerization results in a nearly 180° change in torsional angle about the C-5–C-11 bond (86). This alteration in turn produces an anti-periplanar arrangement of the N-5-C-11 and C-5–H bonds and a turning over of the methylene group (86, 143). This conformation also has a coplanar arrangement of atoms C-4, O-4, C-5, and C-11, which would promote acidity of the C-5 proton by allowing delocalization of the resulting carbanion equivalent. The modeled conformation could readily undergo β-elimination of H$_4$folate across the C-5-C-11 bond to give an exocyclic methylene with the configuration required to give a dTMP methyl group with the correct chirality upon hydride transfer (157). Finally, β-elimination of the enzyme from the methylated adduct would provide the products dTMP and H$_2$folate. The structure of the TS•dTMP•H$_2$folate complex provides a molecular picture of the enzyme-product complex (137).

Thus, the conformation of the native enzyme-bound covalent intermediate II (Figure 3), as determined by structural studies of the TS-FdUMP-CH$_2$H$_4$folate complex, does not appear to immediately precede the elimination of H$_4$folate in the pathway to dTMP. Chemical considerations (83) and molecular modeling from crystal structures of analogues (86, 143, 147) have led to the

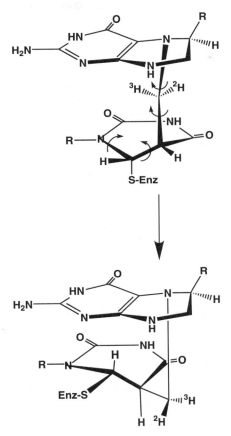

Figure 9 Proposed conformational isomerization of the TS-dUMP–CH₂H₄folate covalent intermediate preceding β-elimination.

conclusion that a conformational isomerization provides a later intermediate suitable for all of the subsequent reactions necessary to form dTMP.

Structural Differences Between the Subunits

There are reports that subunits of the TS dimer are nonequivalent in certain functions (41, 159–163). Usually, these have been observed as negative effects that binding of ligands to one subunit has on binding or reactivity to the other. In extreme cases, occupancy of one site completely prevents binding to the second site (161–163). Researchers have not established whether putative negative cooperative effects are preexisting or induced by ligand binding to the first site.

The potential sources for artifacts in experiments on negative cooperativity are manifold, and caution is warranted in evaluating experiments dealing with this phenomenon. For example, early reports vary considerably in the stoichiometry of the TS-FdUMP-CH_2H_4folate complex, which was suggestive of some form of negative cooperativity. However, the stoichiometry currently seems to be one mole of ligand per monomer, and the earlier discrepancies probably resulted from partially oxidized enzyme or an underestimation of the extinction coefficient of TS (144). The effect of C-terminal deletion on a single subunit of the TS dimer is an example of a discrepancy that remains unresolved. Studies of the carboxypeptidase inactivation have indicated that loss of TS activity occurs upon removal of a single C-terminal valine from the dimer (160, 164). However, other experiments have demonstrated TS activity in heterodimeric TSs that lack a single C-terminal residue from one subunit of the dimer (165). A clear-cut example of negative cooperativity is in the reaction of FdCMP with the N229D mutant of *L. casei* TS. Whereas TS reacts with FdUMP and CH_2H_4folate to form a ternary complex containing 2 mol FdUMP/mol TS dimer, the same enzyme preparation only binds 1 mol FdCMP/mol TS dimer (163).

One should be able to use crystallography to detect asymmetry in TS dimers when the dimer is the asymmetric unit of the crystal's space group. The dimer was the asymmetric unit in crystals used to solve four of the *E. coli* TS structures, TS-dUMP•CB3717 (109, 110), TS-FdUMP•CB3717 (86), TS•CB3717 (129), and TS•CB3717glu₄ (136). In addition, Finer-Moore et al also calculated structures of the TS•dUMP complex without averaging reflections related by the crystallographic two-fold axis (135). The small differences between the subunits in these structures could be attributed to nonequivalent crystal contacts, and no evidence could be found for asymmetry of the subunits. However, the equivalence of the subunits in these structures does not necessarily contradict reports of cooperativity. For example, preexisting or induced nonequivalence in unbound or partially liganded forms may not be manifested in the complexes occupied at both sites. Thus, although the TS•(dUMP)₂ structure has identical monomers, we do not know whether this is true for the free enzyme or a singly liganded enzyme, nor can we judge the effect of binding of a single folate to this complex. Likewise, ternary complex structures describe the final state of the protein and do not provide information about subunit nonequivalencies that may have occurred before reaching this state. What is still needed are appropriate crystal forms and structures of TS dimers in which single monomers are occupied.

The only structure that exhibited a clear structural difference between subunits of TS is of a TS-dUMP•CB3717 complex solved from crystals grown in the absence of reducing agents (109). In this structure, several cysteine residues appear to be oxidized, and dUMP is present in only one active site. The position of CB3717 is shifted by approximately 1 Å between the sites, and in both sites, CB3717 occupies a conformation different from that observed in the presence of reducing agents. The fact that the protein was oxidized and that the cofactor is in a nonproductive conformation discounts the relevance of this structure to the issue of cooperativity. However, this structure does show the existence of an alternate folate binding site that may or may not have relevance.

In our opinion, the issue on nonequivalent sites in relevant interactions of TS remains unresolved. On one side, combined biochemical experiments seem to indicate possible nonequivalence of sites. On the other, such experiments are prone to artifacts, and crystal structures thus far obtained have not provided conclusive evidence for or against structurally nonequivalent subunits.

Structures with Bound Inhibitors

Because TS is a target for drug design numerous crystal structures of TS with bound inhibitors have been solved. We have previously discussed structures of complexes with CB3717, a potent TS inhibitor that saw clinical trials and provided insight into a new generation of TS inhibitors of potential clinical relevance. A major structure-based drug design effort has been reported from Agouron Pharmaceuticals and has resulted in the synthesis of novel TS anti-folates and solution of their crystal structures in complexes with *E. coli* TS. Reviews of this work have appeared (139–141).

Another structure-based drug design effort has utilized a docking algorithm in combination with iterative crystallographic analysis to identify several novel TS inhibitors (138). Though none were particularly tight-binding, the study demonstrated the value of ligand-docking methods combined with structural analysis in obtaining novel solutions to the problem of identifying TS inhibitors.

A recently reported crystal structure of the TS-CF$_3$dUMP covalent complex (142) is of interest because CF$_3$dUMP is a mechanism-based inhibitor of TS (9, 166). Nucleophilic addition of Cys198 to C-6 of CF$_3$dUMP results in release of fluorine atoms from the CF$_3$ group and creation of an electrophilic difluoromethylene group that reacts with Tyr146. Although TS inhibition by CF$_3$dUMP has been previously studied, crystallographic analysis resulted in the identification of the nucleophile that covalently reacts with the CF$_3$ group.

MUTANT THYMIDYLATE SYNTHASES

Overview of Mutagenesis

Numerous mutant thymidylate synthases have been reported. Most TS mutants are of the *E. coli* (39, 167, 168) and *L. casei* (60) enzymes; however, T4 (169), mouse (170), and human (171) TSs have also been mutated. Some mutants have been made by conventional methods and rationale; a single residue was targeted for a single substitution for a specific reason. Other experiments have used multiple substitutions to provide so-called replacement sets of amino acids at a chosen position in the enzyme (39, 43, 60). In this semi empirical approach, no a priori judgment is made as to which substitution will be most informative. Since the latter approach has yielded a large number of mutants, a brief description of the methods used to produce and analyze multiple mutants is given below.

The *L. casei* TS gene was designed and chemically synthesized to contain 35 unique restriction sites and codons optimized for expression in *E. coli* (30). Replacement sets of mutant enzymes were created by the replacement of restriction fragments of the synthetic gene with fragments containing oligonucleotide mixtures coding for all 20 amino acids at targeted positions (60, 172). Plasmids encoding mutant enzymes were screened for their ability to complement Thy⁻ *E. coli* and then sequenced. Mutants were expressed in a Thy⁻ *E. coli* host at 5–20% of the soluble protein. An automated method of purification facilitated the rapid acquisition of large quantities of mutant enzymes for biochemical analyses (173).

A clever approach used for the generation of multiple mutants of *E. coli* TS involved the introduction of an amber stop codon at target sites in the gene (39, 43). The plasmid containing the mutant gene was then transformed into different Thy⁻ *E. coli* suppressor strains, each of which produced an altered tRNA that inserted a different amino acid in response to the amber codon. The ability of different amber mutants to complement the growth of these Thy⁻ strains in the absence of thymine was used to determine which amino acid substitutions allow biosynthesis of active TS. An advantage of this method is that it avoids the necessity of time-consuming DNA sequencing of mutants for identification; a disadvantage is that the system is not amenable to production of the quantities of enzyme needed for more intensive studies.

A compilation of the *E. coli* and *L. casei* TS replacement sets prepared and tested for activity by complementation of Thy⁻ *E. coli* is given in Table 1 (39, 43, 60). The complementation test only tells whether the enzyme is sufficiently active to sustain the growth of host cells (~ 0.002 U/mg TS activity in soluble extract). The screen depends on various factors, including expression level, and does not provide a quantitative assessment of activity. With few exceptions, the data obtained with the *E. coli* and *L. casei* enzymes by two different approaches are in agreement. There is also good agreement of the complementation data of *L. casei* with in vitro activities of mutants purified and examined thus far. The major differences are that the in vitro assays are quantitative and more sensitive, and some mutants that score negative in the complementation assay show measurable activity in vitro.

Table 1 Complementation of Thy⁻ *E. coli* by thymidylate synthase mutants

Residue[1]	Source	Non-Polar									Polar						Basic			Acidic		Reference
		Gly	Ala	Pro	Phe	Leu	Val	Ile	Met	Trp	Thr	Asn	Cys	Ser	Tyr	Gln	His	Lys	Arg	Glu	Asp	
Glu 16	*E. coli*	●	●	●	●	●						●	●	●	●	●	●	●	●	●		39
Asp 22	*E. coli*	○	○	○	○	○						●	●	○	○	○	○	○	○	●	●	43
Arg 23	*E. coli*	◐	○	◐	○	○						●	●	○	○	○	○	○	●	○		39
	L. casei	○	○	○	○	○	◐	◐			○	○	○	○	◐	○	○	○	○	○	○	60, D. Santi unpublished
Thr 24	*E. coli*	○	○	●	○	○			●		○	●	●	●	○	●	○	○				43
Gly 25[2]	*E. coli*	●	●	●	●	●						●	●	●	●	●	●	●	●	●		43
Thr 26	*E. coli*	○	●	●	●	○			●		●	●	○	○	●	○	●	○	●	●		43
Gly 27	*E. coli*	●	●	●	●	●			●		●	●	●	●	●	●	○	○	●	●		43
Thr 28	*E. coli*	●	●	●	◐	●			●		●	●	●	○	●	●	○	○	●	●		43
Leu 29[3]	*E. coli*	●	●	●	●	●						●	●	●	●	●	●	●	●	●		43
Ser 30	*E. coli*	●	●	○	◐	○						●	●	○	○	●	○	○	●	●		43
Ile 31	*E. coli*	○	○	○	●	●		●				●	●	●	●	○	○	○	○			43
Phe 32	*E. coli*	○	○	○	●	●					○	○	○	○	○	○	○	○	○			39
Gly 33	*E. coli*	◐	●	○	○	○					●	○	○	○	○	○	○	○	○			43
Gln 35	*E. coli*	●	●	●	●	●						●	●	●	●	●	●	●	●	●		39
Arg 37	*E. coli*	●	●	●	●	●						●	●	●	●	●	●	●	●	◐		39
Glu 60	*L. casei*	○	○	○	○	○	○	○	○	○	○	○	○	○	○	○	○	○	○	●	◐	96
Trp 82	*E. coli*	○	○	○	◐	○				●		○	○	○	○	○	○	○	○			39
	L. casei	○	○		●	○	○		○	●	○		○	●		○					○	J.Kealey & D. Santi, unpublished
Asp 83	*E. coli*	●	●	●	●	●						●	●	●	●	●	●	●	●	●		39
Tyr 146	*L. casei*	●	●	◐																		60
Asp 157[4]	*E. coli*	●	●	●	●	●						●	●	●	●	●	●	●	●	●		39
Asp 162	*E. coli*	○	○	○	○	○						●	◐	◐	○	○	○	○	○	●		39
Asn 173[5]	*E. coli*	●	●	●	●	●						●	●	●	●	●	●	●	●	●		39
Arg 178	*E. coli*	○	○	◐	○	○						○	◐	○	●	○	●	●	●	○		39
	L. casei		○	○	○	○	○						○					○	○		○	60, D. Santi unpublished
Arg 179	*E. coli*	●	●	●	●	●						●	●	●	●	●	●	●	●	●		39
	L. casei	●	●	○	●	●			●	●	●		●	●	●	●	●	●	●			60, D. Santi unpublished
Pro 196	*L. casei*		●	●	●	●		●	○			●	●	●						●		60
Pro 197	*L. casei*		●	◐	●	●	●	●	●		●	●	●	◐	◐		○					60
Cys 198	*E. coli*	○	○	○	○	○					●	●	○	○	○	○	○	○	○			39
	L. casei	○	○	○	○	○	○		○	○		●	○	○			○	○	○			60
His 199	*E. coli*	●	●	●	○	●						●		●	●	●	○	●	○			39
	L. casei	●	●	●		○	○		○	○	●	○	●		●		●	○	○			60, D. Santi, unpublished
Tyr 146	*L. casei*	○	○	○		○	○		●	○	○	○		○	●	○		○	○	○		60
Gln 203	*E. coli*	○	○	○	○	○						○	○	○	○	●	●	○	○	●		39
Arg 218	*E. coli*	○	○	○	○	○						○	○	○	○	○	○	●	○			39
	L. casei	○	○		○	○			○	○			○	○		○		●	○			60, D. Santi unpublished
Asp 221	*E. coli*	○	○	○	○	○						◐	○	○	○	○	○	○	○		●	39
	L. casei		○		○				○									○			●	D. Santi, unpublished
Asn 229	*E. coli*	◐	○	○	○	◐			●	●	●	●	●	◐	○	◐	○	○	○			39
	L. casei	○	○		●	●	●	●	●	●	●	●	○	○	○		●		○	○		180
Gly 256	*E. coli*	●	●	○	○	○						○	●	○	○	○	○	○	○			39
Tyr 261	*L. casei*	○	○	○		○	○		●	○	○	○		○	○	○		○	○	○		60
Glu 275[6]	*E. coli*	●	●	●	●	●						●	●	●	●	●	●	●	●	●		39
Val 316	*L. casei*	○	●	●	●	●	●	●	●	●	○	●	●	●	●	●	●	○	○	○		44

[1] The numbering system used is for *L. casei* TS; to convert to the *E. coli* numbering system, subtract 2 from *L. casei* residues 1 to 91, and subtract 52 from *L. casei* residue numbers >143. Dark circles represent mutants that complement Thy⁻ *E. coli*, open circles are mutants that do not allow complementation. Mutants showing weak complementation are shown as grey circles.

[2] Gly 25 of *E. coli* TS is a His in *L. casei* TS.

[3] Leu 29 of *E. coli* TS is a Tyr in *L. casei* TS.

[4] Asp 157 of *E. coli* TS is a Lys in *L. casei* TS.

[5] Asn 173 of *E. coli* TS is a Thr in *L. casei* TS.

[6] Glu 275 of *E. coli* TS is a Thr in *L. casei* TS.

TS is extremely permissive in the amino acid substitutions it tolerates, including many for residues that are highly conserved and/or important for structure or function. The observation that so many residues of TS are not essential raises the issue of why these residues are conserved. Currently, there is no satisfactory answer to this question, but speculations may be made. Because no mutant has been reported that catalyzes dTMP formation more efficiently than wild-type TS, conservation may simply result from optimization of TS activity. Also, some residues may be conserved for reasons other than catalysis, such as mRNA binding (154, 155) or participation in multienzyme complexes (174–176). The mechanism by which mutants of residues with important functions can be functional is also uncertain. The tolerance of such residues toward mutation has been explained as a structural plasticity of the enzyme. Here we assume that when two or more residues contribute to a function, the loss of function of one residue can be accommodated by a more efficient utilization of the other(s) (35, 60).

Mutation of Residues that Interact with dUMP

Residues that interact with dUMP are highly conserved throughout the TS sequences. Nevertheless, a large variety of substitutions are permitted at such residues without complete loss of activity.

A quartet of arginine residues (Arg23, Arg178', Arg179', and Arg218) forms a positively charged binding site for the 5'-phosphate of dUMP (109, 110, 135). Arg 178' and 179' are derived from a monomer different from the active site they occupy and function in. Substitutions for Arg178 that allow complementation of Thy⁻ E. coli include basic and other hydrophilic substitutions, though proline is also accepted in E. coli TS (43, 60). The purified L. casei R178T TS, which is active by complementation, shows a 5-fold reduction in k_{cat}, a 1000-fold increase in dUMP K_m, and a 100-fold increase in CH_2H_4folate K_m (S Kawase & D Santi, unpublished data). The apparent detrimental effect on binding of dUMP is expected from the role of Arg178, but the effect on K_m of the cofactor has no known structural basis. The purified R178F mutant of L. casei TS shows no detectable TS activity (>10^4-fold decrease) and has been used in subunit-complementation experiments (144, 177). In addition, experiments aimed at incorporating unnatural amino acids into proteins introduced azaleucine at position 178 of E. coli TS and created an active enzyme (178).

Most replacements for Arg179 of either E. coli or L. casei TS result in enzymes that complement Thy⁻ E. coli (39, 60). The purified R179A, E, K, and T mutants of L. casei TS show relatively small differences in kinetic parameters compared to wild-type TS (131). The most significant effect of all of these mutations was localized to a decrease in the bimolecular rates of association of dUMP with the enzyme, which decrease 10- to 20-fold relative to wild-type TS. Supporting the notion of TS plasticity, Arg179′ binds the phosphate of dUMP, yet can be replaced with residues with properties as dramatically different as glutamate without loss of activity. In the crystal structure of the R179E mutant, Arg178′ and Arg218 ligate the dUMP phosphate as in the wild-type TS•dUMP complex (J Finer-Moore, R Stroud, unpublished data), and Arg23 has moved slightly to form a hydrogen bond with a phosphate oxygen different from that in the wild-type TS•dUMP structure. Glu179 is shifted from the position normally occupied by Arg179; its negative charge is neutralized by an ion pair with Arg23, and it forms a hydrogen bond with a conserved water molecule that remains hydrogen-bonded to the dUMP phosphate (145).

In addition to dUMP binding, Arg23 forms a salt bridge with the C-terminal carboxylate following the conformational change that occurs upon folate binding (86, 109, 110, 137). Few members of the Arg23 replacement set complement Thy⁻ *E. coli* (39, 60); those members that do are structurally dissimilar. Studies with the purified Arg23 mutants (V, Q, E, N, I, G, F, A, P, L, D, H, S, Y, and K in single letter amino acid codes) of *L. casei* TS show that all are weakly active in vitro, with 10- to 200-fold reductions in k_{cat}, up to 10-fold increases in dUMP K_m values, and 4- to 30-fold increases in CH_2H_4folate K_m values relative to those of wild-type TS (D Santi, unpublished data). The inability of Arg23 mutants to complement Thy⁻ *E. coli* is best explained by their poor binding to substrates, which would reduce k_{cat}/K_m below the essential level needed for cell growth. The purified R23V mutant of mouse TS also shows a 100-fold reduction in specific activity (170). The effect of mutations of Arg23 on folate binding likely results from their resultant inability to form the salt bridge with the C-terminal carboxylate (86, 109, 110, 137).

Structural studies indicate that Arg218 may have a role in stabilizing the active-site thiolate of Cys198 by ion pair formation (13, 86, 143). None of the Arg218 mutants show activity by complementation, but no mutants have yet been studied in vitro.

Residues that interact with the deoxyribose moiety of dUMP include Tyr261 and His259, which both form hydrogen bonds with the 3′-OH of dUMP (109, 110, 135). The Y261M mutant of *L. casei* TS was the only one of 14 Tyr261 mutants isolated that complemented Thy⁻ *E. coli* (60); because this mutant is an outlier, the activity should be verified in vitro. Thus far, Tyr261 mutants have not been purified, and His 259 has not been mutated.

Cys198 provides an essential nucleophilic catalyst for the TS reaction. The C198S mutant of *E. coli* TS is the only Cys198 mutant of either *E. coli* or *L. casei* TS that allows complementation of Thy⁻ *E. coli* (39, 60). Studies with purified C198S *E. coli* (167), and T4 (169) TSs showed that these mutants catalyze dTMP formation with 5000- and 1500-fold reductions in k_{cat}, respectively, and only minor changes in substrate and cofactor K_m values relative to the wild-type enzymes. None of the Cys198 mutants of *L. casei* TS showed detectable activity in vitro (D Santi, unpublished data). Because alkoxides are capable nucleophilic catalysts in model reactions of TS (6, 122), the weakly acidic serine might serve as a nucleophile in the TS reaction; however, why this substitution is active in *E. coli* but not in *L. casei* TS is unknown.

Pro196 and Pro197 are thought to assist positioning of the active site thiolate of Cys198 (13). Almost all mutants of these residues in *L. casei* TS complemented Thy⁻ *E. coli* (60) and showed high activity in vitro (D Santi, unpublished data). Studies with purified Pro196 mutants (V, R, Y, T, F, H, K, N, C, D, and I) of *L. casei* TS showed that, relative to wild-type TS, these enzymes have up to 10-fold increases in K_m values for dUMP and CH_2H_4folate, and 2- to 30-fold reductions in k_{cat}. The P196D and P196N mutants were particularly active, having k_{cat} values only a few-fold lower than wild type TS. In the Pro196 replacement set, the effect of larger increases in K_m values were compensated for by smaller reductions in k_{cat}, so catalytic efficiencies (k_{cat}/K_m) were essentially constant. Purified P197 mutants (A, C, Q, H, I, L, M, S, T, W, and Y) of *L. casei* TS showed up to 10-fold increases in K_m values for dUMP and CH_2H_4folate, and 2- to 10-fold reductions in k_{cat} (D Santi, unpublished data).

As observed in crystal structures, His199, Tyr146, Trp82, Asn229, Glu60, and several highly ordered water molecules form a hydrogen-bond network around the pyrimidine (Figure 7). These residues have been mutated to establish their roles and to search for possible general acid/base catalysts to assist proton transfers at O-4 and C-5 of dUMP during catalysis.

Tyr146 forms a hydrogen bond with an ordered water molecule that is proximal to C-5 of dUMP in ternary complexes and may assist in removal of the C-5 proton (110, 137). Ala, Ser, and Pro replacements for Tyr146 of *L. casei* TS yield enzymes that complement Thy⁻ *E. coli*. Purified Y146F and Y146A mutants of *E. coli* TS show 200-fold decreases in k_{cat} relative to wild-type TS (F Maley, personal communication), indicating that Tyr146 is not essential for catalysis (60).

Trp82 forms hydrophobic contacts with both the pterin and the pyrimidine of the substrates and participates in the pyrimidine-coordinating hydrogen bond network through a hydrogen bond with Glu60 (109, 110, 137). Of some 13 Trp82 *E. coli* TS mutants tested for complementation, only the W82F mutant showed activity (39). The purified W82F and W82M mutants of *E. coli* TS showed 0.70 and 0.09% of the activity of wild-type TS, respectively (F Maley, personal communication). In *L. casei* TS, Trp82 was mutated to Tyr, Phe, and His without losing the ability to complement Thy⁻ *E. coli*. Studies with purified Trp82 mutants of *L. casei* TS (F, H, Y, S, N, M, L, A) show that these enzymes have up to 100-fold increases in their dUMP K_m, up to 20-fold increases in their CH_2H_4folate K_m, and 10- to 200-fold reductions in k_{cat} (J Kealey & D Santi, in preparation). The indole ring of Trp82 is therefore not essential for activity.

His199 was suggested as a candidate general base to assist removal of the C-5 proton (13). However, most mutations of His199 resulted in an active enzyme, indicating that this residue is not essential for catalysis (30, 39, 60, 169, 179). Purified *E. coli* His199 TS mutants G, N, V, and L (179), *L. casei* His199 TS mutants P, T, A, G, S, and V (D Santi, unpublished data), and the H199V mutant of T4 TS (169) showed k_{cat} decreases of only 10- to 100-fold and no significant changes in K_m values. The pH-rate profiles of the H199G, H199Q, and H199N mutants of *E. coli* TS are bell-shaped curves in contrast to the sigmoidal curve observed with wild-type TS. This difference, together with the decreased TS activity of these mutants at high pH values, was interpreted to suggest that His199 stabilizes the acidic form of an important catalytic group (179). If correct, this group may be part of the hydrogen bond network that coordinates the pyrimidine.

Asn229 makes the only direct contact between a protein side chain and the dUMP pyrimidine ring, forming a cyclic hydrogen bond network with O-4 and NH-3 of dUMP (109, 110, 135). Although about half of the Asn229 mutants isolated do not complement the Thy⁻ host, in vitro studies showed that Asn229 tolerates substitution by a variety of amino acids with only small losses of TS activity (39, 180).

Wild-type TS binds to dUMP about 1000-fold tighter than dCMP. With the exception of N229Q TS, N229 mutations decrease the enzyme's ability to discriminate between dUMP and dCMP (77, 146, 181). Surprisingly, Asn229 mutants of *L. casei* TS bind dUMP with near wild-type affinity and show an increased affinity for dCMP. Therefore Asn229 may provide specificity of TS binding for dUMP by excluding dCMP from the active site (77).

When TS sequences are compared to the sequence of T4 phage dCMP hydroxymethylase, the Asn229 of TS aligns with an aspartic acid residue of the hydroxymethylase (182, 183). The N229D mutation of TS reverses the enzyme's specificity and converts TS into a dCMP methylase (146, 181). Likewise, an analogous Asp-to-Asn mutation in dCMP hydroxymethylase changes its substrate specificity to a dUMP hydroxymethylase (184). The crystal structure of *Hha I* DNA-cytosine methylase shows that this enzyme also interacts with its target cytosine residue via an aspartate (185). The interactions of asparagine with uracil, and aspartate with cytosine as cited above, may represent a thus-far unrecognized general theme in protein-nucleic acid recognition. Such interactions may differentiate between uracil vs cytosine-containing sequences with up to 3 kcal/mol of binding energy.

It has been proposed that the carboxylic acid of Asp 229 of N229D TS forms a hydrogen bond bridge by donating a proton to N-3 and accepting a

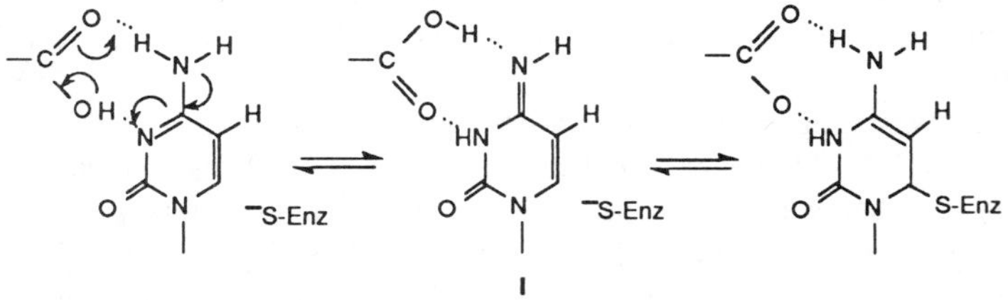

Figure 10 Proposed hydrogen bond network surrounding dCMP in the N229D TS•dCMP complex.

proton from the 4-NH$_2$ group of dCMP (Figure 10). Protonation of N-3 has been suggested to stabilize the normally unfavorable imino tautomer of cytosine (intermediate I), which would allow for a mechanism of methylation analogous to that of dUMP (146, 181). This proposal requires that the active-site environment cause a significant increase in the pK$_a$ of Asp 229 or the N-3 of dCMP, a hypothesis that remains to be tested.

The H199A/N229V and H199V/N229V double mutants were also constructed to test the possibility that plasticity of TS would allow one residue to accommodate for the lost function of the other (W Huang & D Santi, unpublished data). These mutants were active but had reduced specific activities in accord with values predicted by noninteractive effects of the individual mutants (186).

Glu60 seems to coordinate the hydrogen bond network surrounding the pyrimidine of dUMP, and also interacts with N-10 of the folate cofactor in ternary complexes. Glu60 is quite sensitive to mutagenesis; except for the Asp mutant, *L. casei* Glu60 TS mutants do not complement growth of Thy$^-$ cells and show low TS activity that can only be detected using sensitive assays (187). The only Glu60 replacement of *E. coli* TS reported is the glutamine mutant, which shows a large loss in activity (188). The purified *L. casei* E60L and E60A TS mutants showed an activity 25,000-fold lower than that of the wild-type enzyme (96). Furthermore, unlike wild-type TS, these mutants catalyzed tritium loss from [5-^{3}H]dUMP faster than dTMP formation, indicating the accumulation of a steady-state intermediate with an acidic 5-hydrogen and a change in partitioning of that intermediate towards reactants. Covalent TS-dUMP-CH$_2$H$_4$folate complexes formed and reacted slowly and could be isolated on SDS-PAGE. The complex was chemically and kinetically competent to form dTMP, thereby fulfilling criteria for a catalytic intermediate. This work complements a previous study that reported the isolation of the covalent intermediate of TS by rapid acid-quenching (121). Addition of acetate or formate greatly increased the rate of dTMP catalysis by Glu60 mutants, indicating that the function of Glu60 could be complemented by exogenous carboxylic acids (W Huang & D Santi, in preparation).

Every side-chain functional group that participates in the hydrogen-bond network surrounding the dUMP pyrimidine can be mutated without a complete loss of enzymatic activity. Although optimal catalysis requires wild-type residues, none of the residues of the network are completely essential. This leaves only water molecules as potential essential catalysts for proton-transfer reactions of thymidylate synthase. Indeed, the attractive, yet unproven hypothesis is that water molecules are the important proton transfer catalysts in the TS reaction.

Mutation of Residues that Interact with CH2H4folate

Residues involved in CH_2H_4folate binding that have been mutated include those of the C-terminus as well as Trp82, Trp85, Asp 221, Lys50, and Lys51.

Folate binding induces a conformational change in which C-terminal residues move to cover the bound substrates (109, 110). Experiments that utilized carboxypeptidase-cleaved TS were the first to demonstrate the importance of the C-terminal valine for folate binding and catalysis (150, 153, 160, 164). Carboxypeptidase cleavage of the C-terminal valine from a single subunit of TS completely inactivated the enzyme and decreased the enzyme's affinity for tetraglutamyl CH_2H_4folate. V316Am TS, a deletion mutant lacking the C-terminal valine from both subunits, was able to participate in partial reactions—catalysis of the dehalogenation of BrdUMP and formation of a covalent V316Am TS-FdUMP-CH_2H_4folate ternary complex—but not catalyze dTMP formation. The V316Am TS-FdUMP-CH_2H_4folate complex was impaired in its ability to isomerize from the noncovalent to covalent form, suggesting that the mutant is impaired in its ability to undergo the C-terminal conformational change (72). In support of this, the crystal structure of the V316Am TS-FdUMP-CH_2H_4folate complex showed that no covalent bonds were present between the enzyme and ligands and that the enzyme had not undergone the folate-dependent conformational change (189).

Studies of a complete replacement set of purified Val316 mutants of L. casei TS showed that Val316 tolerates almost all substitutions without a complete loss of activity; indeed, many were almost as active as wild-type TS (44). A quantitative structure-activity relationship correlated the side-chain hydrophobicity of the C-terminal residue with both k_{cat} and the CH_2H_4folate K_m of these mutants (44). This correlation was interpreted as reflecting the importance of hydrophobic interactions of the C-terminal side chain in folate binding, completion of the conformational change, and passage through the rate-limiting step of the TS reaction.

As with V316Am TS, deletion mutants lacking 4, 8, and 10-terminal residues lack TS activity but dehalogenate BrdUMP with wild-type efficacy (C Carreras & D Santi, in preparation). Mutants lacking eight or ten C-terminal residues could not form covalent complexes with FdUMP and CH_2H_4folate. Exogenous addition of an 8-residue synthetic peptide with the same sequence as the C-terminus did not revive activity of an A309Am (n-8) mutant, suggesting that the driving force for closure of the C-terminus emanates from the entropic advantage of its attachment to the enzyme, rather than from intrinsic binding interactions.

Trp82 and Trp85 are conserved residues that pack against the bound folate analogue in ternary complex structures (86, 109, 110, 137). Trp82 has been mutated as part of an effort to understand the hydrogen bond network surrounding the pyrimidine; some mutants, including W82F, were active (see above). Trp85 has been replaced only with Phe, and this enzyme is active. However, the W82F/W85F double mutant is inactive, suggesting that the presence of at least one of these tryptophans is essential (L Liu & D Santi, unpublished).

The side chain of Asp 221 forms a hydrogen bond with N-3 of the folate and may assist elimination of the cofactor from the TS-dUMP-CH_2H_4folate intermediate (86). Substitutions at this position generally give enzymes that do not complement Thy$^-$ E. coli , though the D221C mutant appears to be an exception (39). The purified D221A and D221N mutants are apparently inactive but bind nucleotides with wild-type affinity and dehalogenate BrdUMP more efficiently than wild-type TS (D Santi, unpublished data). Although further studies of the mutants are warranted, these results are consistent with a role for Asp221 in breakdown of the TS-dUMP-CH_2H_4folate catalytic intermediate.

Lys50, Lys51, and Lys58 are residues that have been identified as important for binding the polyglutamate moiety of the cofactor, first in cross-linking experiments (190, 191) and subsequently in crystallographic studies (136). Residues of T4 TS that are analogous to Lys50 and Lys51 were changed to Arg; each of these replacements yielded active enzyme, though moderate increases in CH_2H_4folate K_m values were noted (169).

Heterodimeric Thymidylate Synthase Mutants

The active site of TS has residues derived from both subunits (13). The demonstration that TS dimers can be reversibly unfolded and dissociated with only a small loss of activity (192) made it possible to produce heterodimeric enzymes with a single functional active site. Cys198 and Arg178′ are essential residues of TS active sites that emanate from different subunits. The C198A and R178′F single mutations each completely inactivate the enzyme. Refolding the C198A mutant in the presence of R178′F TS resulted in formation of heterodimers with one active subunit (R178′+C198), one inactive subunit (R178′F+C198A), half the activity of wild-type TS, and similar K_m values for substrate and cofactor (177). This method of heterodimer formation has been used to study the dimer interface through creation of cross-species TS heterodimers (144) and to create active, asymmetric enzymes that have a C-terminal deletion on only one subunit of the dimer (165).

Other Mutations

Conserved residues of TS that do not have apparent roles in binding or catalysis have been mutated, and most mutants were found to possess at least some TS activity. Kim et al mutated 10 highly conserved residues of the enzyme's N-terminus (43). Asp22, Gly25, Gly27, Thr28, Leu29, Ser30, Ile31, Phe32, Gln35, and Arg37 could all tolerate multiple substitutions and retain their ability to complement Thy⁻ E. coli. In addition, several mutants of the conserved Tyr6 in human TS have been isolated, including the Y6H mutant, which was isolated on the basis of its resistance to fluorouracil (171).

The small domain (residues 90–139) of *L. casei* TS is not present in TS from most other species. Its deletion from *L. casei* TS results in loss of the enzyme's ability to catalyze dTMP formation, although the cofactor-independent dehalogenation of BrdUMP proceeds normally (95). Analysis of the structure of *L. casei* TS showed that deletion of the small domain exposes to solvent a phenylalanine residue (Phe87) in the D helix that is a glutamate in the corresponding position of the *E. coli* enzyme. This knowledge was used to design a second site mutation, F87D, that partially restored TS activity in the deletion mutant, although cofactor binding was still impaired. The conclusion of this study was that the small domain assists in stabilization of helices C and D, which contain residues that are important for cofactor binding and catalysis.

TS has also been used as a model system for the designed introduction of disulfide bonds into proteins (193) and subsequent folding and thermal stability studies (S Agarwalla & P Balaram, submitted). The T155C/E188C mutant has two engineered cysteine residues; Cys155 is close enough to Cys188' of the opposing subunit so that an intersubunit disulfide bond can form. Similarly, a second, symmetry-related intersubunit disulfide is formed between Cys188 and Cys155'. The resulting covalent dimer remains soluble and retains secondary structure as high as 90°C, in contrast to the wild-type enzyme that precipitates at 52°C. The temperature optimum for dTMP formation was also increased from ~40 to 55 °C. The results demonstrate the efficacy of engineered disulfide bridges in thermostabilization of a multisubunit enzyme.

SUMMARY AND PERSPECTIVES

Knowledge of the catalytic mechanism and structure of TS has increased substantially over recent years. Major advances were derived from crystal structures of TS bound to various ligands, the ability to overexpress TS in heterologous hosts, and the numerous mutants that have been prepared and analyzed. These advances, coupled with previous knowledge, have culminated in an in-depth understanding of many important molecular details of the reaction. Steps up to and including the formation of the covalent ternary complex are particularly well understood. Also, important roles for water in the structure and function of TS have emerged, although knowledge of this aspect of catalysis is still rudimentary.

Several important issues specific to TS require resolution. First, although the steps in conversion of the covalent intermediate to products are in accord with data and rationale, they remain largely hypothetical. Second, the question of why TS is so highly conserved yet allows so many mutations deserves an answer. Third, the mechanism of general acid/base proton transfers at O-4 and C-5 of dUMP requires definition and understanding. Finally, TS structures and experimental results have revealed that conserved water molecules play important roles in binding and catalysis. Some creative approaches are needed to obtain experimental evidence for their function.

The knowledge base available for TS is sufficiently large that the enzyme should also be useful as a platform for studies of related enzymes and broader issues in protein chemistry. Such research has uncovered one mechanism by which enzymes discriminate between uracil and cytosine heterocycles and has accelerated work on enzymes such as dUMP and dCMP hydroxymethylases and the RNA uracil and DNA cytosine methyl transferases. Hence, TS should continue to serve as a useful paradigm for related enzymes. TS may also be an appropriate model to study aspects of protein folding, mutagenesis, and plasticity of function in protein-ligand interactions.

Finally, the structures of many TSs are now known, including those from most pathogens of humans. The enzyme is ideally positioned to explore how to best use structure-based inhibitor design of species-specific enzymes. In addition to its role as a model, success would have a high probability of providing useful chemotherapeutic agents.

ACKNOWLEDGMENTS

The authors thank Drs. Janet Finer-Moore, Robert Stroud, Frank Maley, Tom W Bruice, and Kathryn Ivanetich for careful reading of sections of the manuscript. We thank Ralph Reid for the sequence alignment in Figure 2, Robert Stroud for Figure 7, and David Yee for Figure 6. DVS acknowledges past and present personnel whose names appear in the cited references, and the National Institutes of Health (grants CA-14394, AI 19358-13, and AI 30261-3) for support.

Literature Cited

1. Ivanetich KM, Santi DV. 1990. *FASEB J.* 4:1591–97
2. Cella R, Parisi B. 1993. *Physiol. Plant.* 88:509–21
3. Friedkin M. 1959. In *The Kinetics of Cellular Proliferation*, ed. FJ Stohlman, p. 97. New York: Grune & Stratton
4. Santi DV, Pogolotti AL. 1968. *Tetrahedron Lett.* 59:6159–62
5. Santi DV, Brewer CF. 1968. *J. Am. Chem. Soc.* 90:6236–38
6. Santi DV, Brewer CF. 1973. *Biochemistry* 12:2416–24
7. Pogolotti AL, Santi DV. 1974. *Biochemistry* 13:456–66
8. Santi DV, McHenry CS, Sommer H. 1974. *Biochemistry* 13:471–81
9. Danenberg PV, Langenbach RJ, Heidelberger C. 1974. *Biochemistry* 13:926–33
10. Bellisario RL, Maley GF, Galivan JH, Maley F. 1976. *Proc. Natl. Acad. Sci. USA* 73:1848–52
11. Bellisario RL, Maley GF, Guarino DU, Maley F. 1979. *J. Biol. Chem.* 254:1296–300
12. Maley GF, Bellisario RL, Guarino DU, Maley F. 1979. *J. Biol. Chem.* 254:1301–4
13. Hardy LW, Finer-Moore JS, Montfort WR, Jones MO, Santi DV, Stroud RM. 1987. *Science* 235:448–55
14. Pogolotti AL, Santi DV. 1977. *Bioorg. Chem.* 1:277–311
15. Lewis CA, Dunlap RB. 1981. In *Topics in Molecular Pharmacology*, ed. ASV Burger, GCK Roberts, pp. 169–219. New York: Elsevier/North-Holland
16. Danenberg PV, Lockshin A. 1981. *Pharmacol. Ther.* 13:69–90
17. Meyers CE. 1981. *Pharmacol. Rev.* 33:1–15
18. Bruice TW, Garrett C, Wataya Y, Santi DV. 1980. *Methods Enzymol.* 64:125–35
19. Santi DV, Danenberg PV. 1984. In *Folates and Pterins*, ed. RL Blakley, SJ Benkovic, pp. 345–98. New York: Wiley
20. Cisneros RJ, Silks LA, Dunlap RB. 1989. *Drugs Future* 13:859–81
21. Maley F, Maley G. 1990. *Prog. Nucleic Acid Res. Mol. Biol.* 39:49–80
22. Bruice TW, Santi DV. 1991. In *Enzyme Mechanisms from Isotope Effects*, ed. P Cook, pp. 1–500. Boca Raton: CRC
23. Ivanetich KM, Santi DV. 1992. *Prog. Nucleic Acid Res. Mol. Biol.* 42:127–56
24. Crusberg TC, Leary R, Kisliuk RL. 1970. *J. Biol. Chem.* 245:5292–96
25. Dunlap RB, Harding NG, Huennekens FM. 1971. *Biochemistry* 10:88–97
26. Leary RP, Kisliuk RL. 1971. *Prep. Biochem.* 1:47–54
27. Belfort M, Maley GF, Maley F. 1983. *Proc. Natl. Acad. Sci. USA* 80:1858–61
28. Pinter K, Davisson VJ, Santi DV. 1988. *DNA* 7:235–41
29. Singer SC, Richards CA, Ferone R, Benedict D, Ray P. 1989. *J. Bacteriol.* 171:1372–78
30. Climie S, Santi DV. 1990. *Proc. Natl. Acad. Sci. USA* 87:633–37
31. Taylor GR, Barclay BJ, Storms RK, Friesen JD, Haynes RH. 1982. *Mol. Cell. Biol.* 2:437–42
32. Belfort M, Moelleken A, Maley GF, Maley F. 1983. *J. Biol. Chem.* 258:2045–51
33. Perryman SM, Rossana C, Deng TL, Vanin EF, Johnson LF. 1986. *Mol. Biol. Evol.* 3:313–21
34. Deng TL, Li DW, Jenh CH, Johnson LF. 1986. *J. Biol. Chem.* 261:16000–5
35. Perry KM, Fauman EB, Finer-Moore JS, Montfort WR, Maley GM, et al. 1990. *Proteins* 8:315–33
36. Davisson VJ, Sirawaraporn W, Santi DV. 1989. *J. Biol. Chem.* 264:9145–48; 1994. *J. Biol. Chem.* 269(48):30740
37. Edman U, Edman JC, Lundgren B, Santi DV. 1989. *Proc. Natl. Acad. Sci. USA* 86:6503–7
38. Cowman AF, Morry MJ, Biggs BA, Cross GA, Foote SJ. 1988. *Proc. Natl. Acad. Sci. USA* 85:9109–13
39. Michaels ML, Kim CW, Matthews DA, Miller JH. 1990. *Proc. Natl. Acad. Sci. USA* 87:3957–61
40. Grumont R, Sirawaraporn W, Santi DV. 1988. *Biochemistry* 27:3776–84
41. Dev IK, Dallas WS, Ferone R, Hanlon M, McKee DD, Yates BB. 1994. *J. Biol. Chem.* 269:1873–82
42. Thompson R, Honess RW, Taylor L, Morran J, Davison AJ. 1987. *J. Gen. Virol.* 68:1449–55
43. Kim CW, Michaels ML, Miller JH. 1992. *Proteins* 13:352–63
44. Climie SC, Carreras CW, Santi DV. 1992. *Biochemistry* 31:6032–38
45. Greene PJ, Yu PL, Zhao J, Schiffer CA, Santi DV. 1994. *Protein Sci.* 3:1114–16
46. Santi DV, Edman UE, Minkin S, Greene PJ. 1991. *Protein Expr. Purif.* 2:350–54
47. Livi L, Edman UE, Schneider GP, Greene PJ, Santi DV. 1994. *Gene* 150(20):221–26
48. DeWille JW, Jenh CH, Deng T, Harendza CJ, Johnson LF. 1988. *J. Biol. Chem.* 263:84–91

49. Bzik DJ, Li WB, Horii T, Inselburg J. 1987. *Proc. Natl. Acad. Sci. USA* 84: 8360–64
50. Sirawaraporn W, Sirawaraporn R, Cowman AF, Yuthavong Y, Santi DV. 1990. *Biochemistry* 29:10779–85
51. Reche P, Arrebola R, Olmo A, Santi DV, Gonzalez-Pacanowska D, Ruiz-Perez LM. 1994. *Mol. Biochem. Parasitol.* 65:247–48
52. Taylor GR, Lagosky PA, Storms RK, Haynes RH. 1987. *J. Biol. Chem.* 262: 5298–307
53. Iwakura M, Kawata M, Tsuda K, Tanaka T. 1988. *Gene* 64:9–20
54. Rouch DA, Messerotti LJ, Loo LS, Jackson CA, Skurray RA. 1989. *Mol. Microbiol.* 3:161–75
55. Kenny E, Atkinson T, Hartley BS. 1985. *Gene* 34:335–42
56. Honess RW, Bodemer W, Cameron KR, Niller HH, Fleckenstein B, Randall RE. 1986. *Proc. Natl. Acad. Sci. USA* 83: 3604–8
57. Richter J, Puchtler I, Fleckenstein B. 1988. *J. Virol.* 62:3530–35
58. Lazar G, Hong Z, Goodman HM. 1993. *Plant J.* 3:657–68
59. Luo M, Cella R. 1994. *Gene* 140:59–62
60. Climie S, Ruiz-Perez L, Gonzalez-Pacanowska D, Prapunwattana P, Cho SW, et al. 1990. *J. Biol. Chem.* 265:18776–79
61. Wahba AJ, Friedkin M. 1961. *J. Biol. Chem.* 236:PC11–12
62. Roberts D. 1966. *Biochemistry* 5:3546–48
63. Hashimoto Y, Shiotani T, Weber G. 1987. *Anal. Biochem.* 167:340–46
64. Pogolotti AL, Weill C, Santi DV. 1979. *Biochemistry* 18:2794–98
65. Kunitani MG, Santi DV. 1980. *Biochemistry* 19:1271–75
66. Garrett C, Wataya Y, Santi DV. 1979. *Biochemistry* 18:2798–804
67. Santi DV, McHenry CS, Perriard ER. 1974. *Biochemistry* 13:467–70
68. Donato HJ, Aull JL, Lyon JA, Reinsch JW, Dunlap RB. 1976. *J. Biol. Chem.* 251:1303–10
69. Cisneros RJ, Dunlap RB. 1990. *Biochim. Biophys. Acta* 1039:149–56
70. Priest DG, Happel KK, Doig MT. 1980. *J. Biochem. Biophys. Methods* 3:201–6
71. Priest DG, Doig MT. 1986. *Methods Enzymol.* 122:313–19
72. Carreras CW, Climie SC, Santi DV. 1992. *Biochemistry* 31:6038–44
73. Cisneros RJ, Dunlap RB. 1990. *Anal. Biochem.* 186:202–8
74. Wataya Y, Matsuda A, Santi DV. 1980. *J. Biol. Chem.* 255:5538–44
75. Chen SC, Daron HH, Aull JL. 1989. *Int. J. Biochem.* 21:1217–21
76. Santi DV, Ouyang TM, Tan AK, Gregry DF, Scanlon T, Carreras CW. 1993. *Biochemistry* 32:11819–24
77. Liu L, Santi DV. 1993. *Biochemistry* 32:9263–67
78. Sharma RK, Kisliuk RL. 1975. *Biochem. Biophys. Res. Commun.* 64:648–55
79. Mittelstaedt DM, Schimerlik MI. 1986. *Arch. Biochem. Biophys.* 245:417–25
80. Santi DV, McHenry CS. 1972. *Proc. Natl. Acad. Sci. USA* 69:1855–57
81. Langenbach RJ, Danenberg PV, Heidelberger C. 1972. *Biochem. Biophys. Res. Commun.* 48:1565–71
82. Pogolotti AL, Ivanetich KM, Sommer H, Santi DV. 1976. *Biochem. Biophys. Res. Commun.* 70:972–78
83. James TL, Pogolotti AL, Ivanetich KM, Wataya Y, Lam SSM, Santi DV. 1976. *Biochem. Biophys. Res. Commun.* 72: 404–10
84. Byrd RA, Dawson WH, Ellis PD, Dunlap RB. 1977. *J. Am. Chem. Soc.* 99: 6139–41
85. Lewis CAJ, Ellis PD, Dunlap RB. 1981. *Biochemistry* 20:2275–85
86. Matthews DA, Villafranca JE, Janson CA, Smith WW, Welsh K, Freer S. 1990. *J. Mol. Biol.* 214:937–48
87. Bruice TW, Santi DV. 1982. *Biochemistry* 21:6703–9
88. Santi DV, Pogolotti AL, James TL, Wataya Y, Ivanetich KM, Lam SSM. 1976. *ACS Symp. Ser.* 1976:57–76
89. Barr PJ, Nolan PA, Santi DV, Robins MJ. 1981. *J. Med. Chem.* 24:1385–88
90. Barr PJ, Robins MJ, Santi DV. 1983. *Biochemistry* 22:1696–703
91. Barr PJ, Oppenheimer NJ, Santi DV. 1983. *J. Biol. Chem.* 258:13627–31
92. Wataya Y, Matsuda A, Santi DV, Bergstrom DE, Ruth JL. 1979. *J. Med. Chem.* 22:339–40
93. Matsuda A, Wataya Y, Santi DV. 1978. *Biochem. Biophys. Res. Commun.* 84: 654–59
94. Wataya Y, Santi DV. 1975. *Biochem. Biophys. Res. Commun.* 67:818–23
95. Schellenberger U, Balaram P, Francis VSNK, Shoichet BK, Santi DV. 1994. *Biochemistry* 33:5623–29
96. Huang WD, Santi DV. 1994. *J. Biol. Chem.* 269(50):31327–29
97. Sander EG. 1977. In *Bioorganic Chemistry*, ed. EE van Tamelen, pp. 273–96. New York: Academic
98. Sedor FA, Sander EG. 1976. *J. Am. Chem. Soc.* 98:2314–19
99. Wataya Y, Santi DV. 1977. *J. Am. Chem. Soc.* 99:4534–36
100. Jencks WP. 1964. *Prog. Phys. Org. Chem.* 2:63

101. Kallen RG, Jencks WP. 1966. *J. Biol. Chem.* 241:5851–63
102. Kallen RG, Jencks WP. 1966. *J. Biol. Chem.* 241:5845–50
103. Gupta VS, Huennekens FM. 1967. *Arch. Biochem. Biophys.* 120:712–18
104. Benkovic SJ, Benkovic PA, Comfort DR. 1969. *J. Am. Chem. Soc.* 91:5270–79
105. Benkovic SJ, Benkovic PA, Chrzanowski R. 1970. *J. Am. Chem. Soc.* 92:523–28
106. Bellisario RL, Maley GF, Galivan JH, Maley F. 1976. *Proc. Natl. Acad. Sci. USA* 73:1848–52
107. Blakley RL. 1969. *The Biochemistry of Folic Acid and Related Pteridines.* New York: Am. Elsevier. 569 pp.
108. Pellino AM, Danenberg PV. 1985. *J. Biol. Chem.* 260:10996–1000
109. Montfort WR, Perry KM, Fauman EB, Finer-Moore JS, Maley GF, et al. 1990. *Biochemistry* 29:6964–77
110. Matthews DA, Appelt K, Oatley SJ, Xuong NH. 1990. *J. Mol. Biol.* 214:923–36
111. Gangjee A, Patel J, Kisliuk RL, Gaumont Y. 1992. *J. Med. Chem.* 35:3678–85
112. Fife TH, Pellino AM. 1981. *J. Am. Chem. Soc.* 103:1201–7
113. Fitzhugh AL, Fodor S, Kaufman S, Spiro TG. 1986. *J. Am. Chem. Soc.* 108:7422–24
114. Santi DV, McHenry CS, Raines RT, Ivanetich KM. 1987. *Biochemistry* 26:8606–13
115. Jones PG, Kirby AJ. 1984. *J. Am. Chem. Soc.* 106:6207–12
116. Jones PG, Kirby AJ. 1986. *J. Chem. Soc. Chem. Commun.,* pp. 444–45
117. Edwards MR, Jones PG, Kirby AJ. 1986. *J. Am. Chem. Soc.* 108:7067–73
118. Jencks WP. 1969. *Catalysis in Chemistry and Enzymology.* New York: McGraw-Hill. 836 pp.
119. Fersht A. 1985. *Enzyme Structure and Mechanism,* pp. 311–43. San Francisco: Freeman
120. Delia TJ, Scovill JP, Munslow WD, Burkhalter JH. 1976. *J. Med. Chem.* 19:344–46
121. Moore MA, Ahmed F, Dunlap RB. 1986. *Biochemistry* 25:3311–17
122. Santi DV, Pogolotti AL. 1971. *J. Heterocyclic Chem.* 8:265–72
123. Santi DV, Brewer CF. 1968. *J. Am. Chem. Soc.* 90:6236–38
124. Daron HH, Aull JL. 1978. *J. Biol. Chem.* 253:940–45
125. Lorenson MY, Maley GF, Maley F. 1967. *J. Biol. Chem.* 242:3332–44
126. Danenberg PV, Danenberg KD. 1978. *Biochemistry* 17:4018–24
127. Lu YZ, Aiello PD, Matthews RG. 1984. *Biochemistry* 23:6870–76
128. Ghose C, Oleinick R, Matthews RG, Dunlap RB. 1990. In *Chemistry and Biology of Pterins 1989,* ed. HC Curtius, S Ghisla, N Blau. pp. 1–414. Berlin: de Gruyter
129. Kamb A, Finer-Moore JS, Stroud RM. 1992. *Biochemistry* 31:12876–84
130. Segel IH. 1975. *Enzyme Kinetics: Behavior and Analysis of Rapid-Equilibrium and Steady-State Enzyme Systems.* New York: Wiley-Interscience. 957 pp.
131. Santi DV, Pinter K, Kealey J, Davisson VJ. 1990. *J. Biol. Chem.* 265:6770–75
132. Finer-Moore JS, Maley G, Maley F, Stroud RM. 1994. *Biochemistry* 33:15459–68
133. Knighton DR, Kan C-C, Howland E, Janson CA, Hostomsak Z, et al. 1994. *Nat. Struct. Biol.* 1:186–94
134. Schiffer CA, Davisson VJ, Santi DV, Stroud RM. 1991. *J. Mol. Biol.* 219:161–63
135. Finer-Moore JS, Fauman EB, Foster PC, Perry KM, Santi DV, Stroud RM. 1993. *J. Mol. Biol.* 232:1101–16
136. Kamb A, Finer-Moore JS, Calvert AH, Stroud RM. 1992. *Biochemistry* 31:9883–90
137. Fauman EB, Rutenber EE, Malye GF, Maley F, Stroud RM. 1994. *Biochemistry* 33:1502–11
138. Schoichet BK, Stroud RM, Santi DV, Kuntz ID, Perry KM. 1993. *Science* 259:1445–50
139. Webber SE, Bleckman TM, Attard J, Deal JG, Kathardekar V, et al. 1993. *J. Med. Chem.* 36:733–46
140. Reich SH, Fuhry MA, Nguyen D, Pino MJ, Welsh KM, et al. 1992. *J. Med. Chem.* 35:847–58
141. Appelt K, Bacquet RJ, Bartlett CA, Booth CLJ, Freer ST, et al. 1991. *J. Med. Chem.* 34:1925–34
142. Eckstein J, Foster P, Wataya Y, Santi DV, Stroud RM. 1994. *Biochemistry.* 33:15086–94
143. Stroud RM, Finer-Moore JS. 1993. *FASEB J.* 7:671–77
144. Greene PJ, Maley F, Pedersen-Lane J, Santi DV. 1993. *Biochemistry* 32:10283–88
145. Deleted in proof
146. Liu L, Santi DV. 1992. *Biochemistry* 31:5010–14
147. Finer-Moore JS, Montfort WR, Stroud RM. 1990. *Biochemistry* 29:6977–86
148. Perry KM, Carreras CW, Chang LC, Santi DV, Stroud RM. 1993. *Biochemistry* 32:7116–25

762   CARRERAS & SANTI

149. Galivan JH, Maley GF, Maley F. 1975. Biochemistry 14:3338–44
150. Galivan J, Maley F, Baugh CM. 1977. Arch. Biochem. Biophys. 184:346–54
151. Lockshin A, Mondal K, Danenberg PV. 1984. J. Biol. Chem. 259:11346–52
152. Carreras CW, Naber N, Cooke R, Santi DV. 1994. Biochemistry 33:2071–77
153. Galivan JH, Maley F, Baugh CM. 1976. Biochem. Biophys. Res. Commun. 71:527–34
154. Chu E, Koeller DM, Casey JL, Drake JC, Chabner BA, et al. 1991. Proc. Natl. Acad. Sci. USA 88:8977–81
155. Chu E, Voeller D, Koeller DM, Drake JC, Takimoto CH, et al. 1993. Proc. Natl. Acad. Sci. USA 90:517–21
156. Fontedilla-Camps JC, Bugg CE, Temple C, Rose JD, Montgomery JA, Kisliuk RL. 1979. J. Am. Chem. Soc. 101:6114–15
157. Sleiker LJ, Benkovic SJ. 1984. J. Am. Chem. Soc. 106:1833–38
158. Tatum C, Vedras J, Schleicher E, Benkovic SJ, Floss H. 1977. J. Chem. Sci. Chem. Commun. 7:218–20
159. Danenberg KD, Danenberg PV. 1979. J. Biol. Chem. 254:4345–48
160. Aull JL, Loeble RB, Dunlap RB. 1974. J. Biol. Chem. 249:1167–72
161. Galivan JH, Maley GF, Maley F. 1976. Biochemistry 15:356–62
162. Leary RP, Beaudette N, Kisliuk RL. 1975. J. Biol. Chem. 250:4864–68
163. Liu L, Santi DV. 1994. Biochem. Biophys. Acta. 1209:89–94
164. Cisneros RJ, Zapf JW, Dunlap RB. 1993. J. Biol. Chem. 268:10102–8
165. Carreras CW, Costi PM, Santi DV. 1994. J. Biol. Chem. 269:12444–46
166. Santi DV, Sakai TT. 1971. Biochemistry 10:3598–607
167. Dev IK, Yates BB, Leong J, Dallas WS. 1988. Proc. Natl. Acad. Sci. USA 85:1472–76
168. Maley GF, Maley F. 1988. J. Biol. Chem. 263:7620–27
169. LaPat-Polasko L, Maley G, Maley F. 1990. Biochemistry 29:9561–72
170. Zhang HC, Cisneros RJ, Deng WL, Johnson LF, Dunlap RB. 1990. Biochem. Biophys. Res. Commun. 167:869–75
171. Hughey CT, Barbour KW, Berger FG, Berger SH. 1993. Mol. Pharmacol. 44:316–23
172. Wells JA, Vasser M, Powers DB. 1985. Gene 34:315–23
173. Kealey JT, Santi DV. 1992. Protein Expr. Purif. 3:380–85
174. Plucinski TM, Fager RS, Reddy GPV. 1990. Mol. Pharmacol. 38:114–20
175. Reddy GPV, Pardee AB. 1980. Nature 304:86–88
176. Kozloff LM. 1983. Adv. Exp. Med. Biol. 163:359–74
177. Pookanjanatavip M, Yuthavong Y, Greene PJ, Santi DV. 1992. Biochemistry 31:10303–9
178. Lemeignan B, Sonigo P, Marlieré P. 1993. J. Mol. Biol. 231:161–66
179. Dev IK, Yates BB, Atashi J, Dallas WS. 1989. J. Biol. Chem. 264:19132–37
180. Liu L, Santi DV. 1993. Proc. Natl. Acad. Sci. USA 90:8604–8
181. Hardy LW, Nalivaika E. 1992. Proc. Natl. Acad. Sci. USA 89:9725–29
182. Lamm N, Wang Y, Mathews CK, Rüger W. 1988. Eur. J. Biochem. 172:553–63
183. Thylén C. 1988. J. Bacteriol. 170:1994–98
184. Graves KL, Butler MM, Hardy LW. 1992. Biochemistry 31:10315–21
185. Kumar S, Blumberg DL, Canas JA, Maddaiah VT. 1994. Cell 15:349–55
186. Wells JA. 1990. Biochemistry 29:8509–17
187. Huang W, Santi DV. 1994. Anal. Biochem. 218:454–57
188. Zapf JW, Weir MS, Emerick V, Villafranca JE, Dunlap RB. 1993. Biochemistry 32:9274–81
189. Perry KM, Carreras CW, Chang LC, Santi DV, Stroud RM. 1993. Biochemistry 32:7116–25
190. Maley F, Maley GF. 1983. Adv. Exp. Med. Biol. 163:45–63
191. Maley GF, Maley F, Baugh CM. 1982. Arch. Biochem. Biophys. 216:551–58
192. Perry KM, Pookanjanatavip M, Zhao J, Santi DV. 1992. Protein Sci. 1:796–800
193. Gokhale RS, Agarwall S, Francis VS, Santi DV, Balaram P. 1994. J. Mol. Biol. 235:89–94

Annu. Rev. Biochem. 1995. 64:763–97

DIVERSITY OF OLIGONUCLEOTIDE FUNCTIONS

Larry Gold

NeXstar Pharmaceuticals, Inc., 2860 Wilderness Drive, Boulder, Colorado 80301
and Molecular, Cellular, and Developmental Biology, University of Colorado,
Boulder, Colorado 80309

Barry Polisky

NeXstar Pharmaceuticals, Inc., 2860 Wilderness Drive, Boulder, Colorado 80301

Olke Uhlenbeck

Chemistry and Biochemistry, University of Colorado, Boulder, Colorado 80309 and
Member of the NeXstar Pharmaceuticals, Inc., Scientific Advisory Board

Michael Yarus

Member of the NeXstar Pharmaceuticals, Inc., Scientific Advisory Board and
Molecular, Cellular, and Developmental Biology, University of Colorado, Boulder,
Colorado 80309

KEY WORDS: SELEX, RNA ligands, RNA structure, drug discovery, combinatorial
libraries

CONTENTS

0066-4154/95/0701-0763$05.00

ABSTRACT

SELEX is a technology for the identification of high affinity oligonucleotide ligands. Large libraries of random sequence single-stranded oligonucleotides, whether RNA or DNA, can be thought of conformationally not as short strings but rather as sequence dependent folded structures with high degrees of molecular rigidity in solution. This conformational complexity means that such a library is a source of high affinity ligands for a surprising variety of molecular targets, including nucleic acid binding proteins such as polymerases and transcription factors, non-nucleic acid binding proteins such as cytokins and growth factors, as well as small organic molecules such as ATP and theophylline. The range of applications of this technology for new discovery extends from basic research reagents to the identification of novel diagnostic and therapeutic reagents. Examples of these applications are described along with a discussion of underlying principles and future developments expected to further the utility of SELEX.

INTRODUCTION

Historically, oligonucleotides have been thought of as tapes. DNA and RNA usually are read by polymerases and ribosomes, leading to proteins that have complex three-dimensional shapes. The three-dimensional structure of tRNA (1) was anything but tape-like, and that structure might have changed our collective thinking about oligonucleotides as tapes, but it did not—tRNA was an old molecule, a special case. Horace Freeland Judson notes that the Cambridge/MRC dogma of the 1960s was nucleic acids as information and proteins as conformation (2)—tapes versus shapes. So strongly entrenched was the idea of oligonucleotide tapes that the Nobel Prize committee found the discovery of RNA catalytic activity in 1982, a process dependent on RNA shape, to be revolutionary. Despite these developments, most nucleic acids continue to be understood as tapes. In fact, as we argue here, the reality could be just the opposite—basepairing and other intramolecular interactions may cause most oligonucleotides to have defined three-dimensional structures. Indeed, cells often counteract this tendency of single-stranded nucleic acids to adopt stable structures by building complex machines (such as replisomes, spliceosomes, and ribosomes, with their associated factors) that must denature structures during the reading of the tape. Although ribozyme (3, 4) and tRNA structures, as well as the possible role of ribosomal RNA as the catalytic component for transpeptidation (5), have contributed to the idea that other, undiscovered natural oligonucleotides can have profound shapes (beyond simple hairpins) and activities; the list of such oligonucleotides is modest at this time.

In prebiotic times oligonucleotides may have been tapes and shapes, information and conformation. This idea is central to the postulate of the RNA world (6, 7), in which RNA played an early, broad role in both genetics and enzymology. As early as 1976, without any relevant data, White (8–10) proposed an RNA world in which the early nucleotides were chemically more diverse than the present set; after translation was invented, the nucleotides became simpler as oligonucleotide functions were restricted almost entirely to information storage and transfer. More recently, Yarus (11) proposed that ancient catalytic RNAs might have functioned through readily available, stable, and functionally diverse bound metal ions that later were largely or entirely replaced by proteins with similar bound metals. Thus, even as ribozymes were discovered with new activities (12, 13), the scientific community accepted the structural and chemical limitations of the present set of mononucleotides and oligonucleotides and understood ribozymes to be unusual. Implicit in the RNA world is the idea that translation made available polymers with much better building blocks for catalysis. Amino acids are marvelously varied in their chemical potential, they are small compared to nucleotides, and they are polymerized into molecules with a relatively neutral backbone compared to natural oligonucleotides. However, see Herschlag et al (14) for an insightful discussion of how ribozymes accomplish their remarkable feats.

While the idea of an RNA world is focused on catalysis, prebiotic oligonucleotides also would have needed affinities for their substrates and other molecules involved in the process of cellularization. In the present biosphere, large molecular machines that recognize various reactants sometimes contain oligonucleotides but not usually. For example, the proteins that facilitate mitosis, neural connections and other cell-to-cell recognition, and immune surveillance and antibody synthesis, and the proteins that assemble into complex viruses all function as machines without oligonucleotides (to our knowledge). If antibodies had been made of oligonucleotides, the scientific community would have been as surprised as it was in 1982 when ribozymes were discovered. Yet many oligonucleotides with unexpected properties (including high binding affinity and specificity toward a variety of molecular targets, as well as new ribozymes) have been discovered recently through deep screening of oligonucleotide combinatorial libraries. Single-stranded oligonucleotides have the properties of both shapes and tapes—their unique shapes, like those of proteins, are dictated by their linear sequences and their intramolecular contacts. Rare sequences, present in large oligonucleotide libraries, can be identified. In this review we discuss the methods by which these fascinating oligonucleotides are found, their properties, and some of the applications of these novel oligonucleotides.

OLIGONUCLEOTIDES THAT BIND TO PROTEINS THAT RECOGNIZE NUCLEIC ACIDS AS THEIR NORMAL FUNCTION

Bacteriophage T4 DNA Polymerase

The methodology for exploration of oligonucleotide libraries is called SELEX (Systematic Evolution of Ligands by Exponential enrichment) (15). The ligands that emerge from SELEX have been called aptamers (16). The first target of SELEX was bacteriophage T4 DNA polymerase (15). For that DNA polymerase, a known translational repressor with a known target mRNA (17), eight nucleotides from the loop region of a hairpin within the translational operator were randomized to provide a library of 65,536 (4^8) sequences. The library was mixed with the protein and then partitioned by passage through nitrocellulose filters; those RNAs bound to the filter by the protein were eluted, reverse transcribed into DNA, and amplified with the polymerase chain reaction (PCR) to recreate a bacteriophage T7 promoter for subsequent transcription of the library for a second round of SELEX. After four such rounds the library complexity was reduced to essentially two sequences that bound with similar K_ds to the polymerase; one sequence was identical to the natural operator and the other contained four differences in the randomized domain (15). Site-directed mutagenesis of the wild-type sequence probably would not have uncovered the variant. For the eight loop nucleotides of interest, an expensive, labor-intensive, yet reasonable approach might have been to syn-

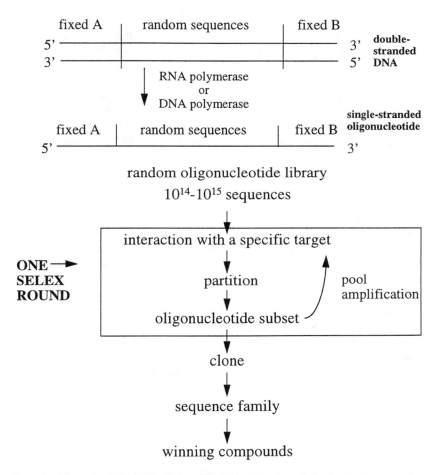

Figure 1 Schematic of SELEX procedure. Fixed A and B refer to defined sequences present on each member of the library that flank the random region. Fixed sequences permit amplification and transcription. The steps in the SELEX process are further described in the text.

thesize and test the 24 single nucleotide alterations. SELEX was informative because all possible combinations in the randomized region were tried.

The generalized SELEX protocol is shown in Figure 1. The protocol can be used for single-stranded DNA or for single-stranded RNA with minor alterations. Selections have been done with double-stranded nucleic acid libraries (18–20), but the significant and distinct thrust of SELEX is to explore the unexpected properties of single-stranded oligonucleotides.

Bacteriophage R17 Coat Protein

An important SELEX experiment (21) was done on the bacteriophage R17 coat protein, which binds a small hairpin in its genome tightly and specifically (22). The R17 coat protein was chosen as a target because the sequence requirements of its target were well understood (23); this experiment essentially was a reconstruction. An RNA library randomized over 32 positions (to provide $4^{32} = 1.8 \times 10^{19}$ different sequence possibilities, of which 3×10^{14} were tested) was used for SELEX. The sequences of individual oligonucleotides were determined after 11 rounds of the SELEX protocol. Those sequences showed features identical to those of the most avidly bound target RNA ever identified.

Analysis of the isolated sequences made clear upon what basis the ligands were identified. The family of isolated sequences has shared primary features. By inspection, sometimes aided by computer alignment and statistical analysis, one can deduce in many, but not all, SELEX experiments the minimal high-affinity ligand within the full-length oligonucleotides (see Figure 2). The shared domains often permit design of truncated oligonucleotides that are minimal high-affinity ligands.

The sequence GAGG of the natural site, a Shine-Dalgarno translational initiation site for the replicase gene of R17, tended toward the sequence CACC, because SELEX does not demand biological fitness as is demanded in genetics and evolution in vivo. The SELEX experiment yielded ligands with higher affinity than the natural site, probably because in general, biology optimizes for additional (often unknown) parameters rather than simply maximizes affinity.

SELEX Theory

In early SELEX experiments, differences were apparent in the nature of the oligonucleotides (DNA vs RNA), the size of the randomized regions, the number of rounds, and the details of partitioning and elution. An attempt was made to put SELEX on a firm theoretical foundation by simulating the process with simple assumptions regarding those variables, the background retention of the oligonucleotide pool during the partitioning event, and the affinities of the best oligonucleotides in the library relative to the affinities of other oligonucleotides in that library (24). The analysis assumes that the target and oligonucleotide library are brought to equilibrium, a state that may be difficult to achieve in some situations.

The lessons in the study are clear. If the target protein binds all oligonucleotides in the library, the highest-affinity ligands in the library must compete with the much more abundant weak binders. Similarly, if a class of intermediate-affinity ligands is present, less abundant than the bulk but more abundant

BINDING SEQUENCES FOR THE R17 COAT PROTEIN

GGGAGAUUCUUAGUA<u>CUC</u>**ACC**A**UCA**GGGGGCA

AAAUUAUCUUCGG<u>AUGC</u>**ACC**A**UCA**GGGCAUGG

UUGUCUUUCAUGUAGUAA<u>GC</u>**AC**G**AUCA**<u>CGGC</u>G

ACGAGAUUUAUUUAGAU<u>GUC</u>**AC**G**AUCA**<u>CGGGC</u>

GAGAUCAAUAGUA<u>AGG</u>**ACC**A**UCA**<u>GGCC</u>UGG

UGUAUAG**AGC**A**UCA**<u>GCCUAUACA</u>UUGCGUGGC

AUGAGAUAGAUCAUG<u>CUC</u>**AGG**A**UCA**<u>CCGGG</u>

AGAGGACU<u>C</u>AUUAG**AGG**A**UCA**<u>CCCUAGUG</u>CGG

AUUGCGUAA<u>UGUU</u>**ACC**A**UCA**<u>GGAACA</u>CCGCGU

<u>GAGUAAGAU</u>**AGC**A**UCA**<u>GCAUCUUGUUC</u>CCGCC

<pre>
U C
A   A
  N N'
  N N'   The shared structure
A
  N N'
  N N'
</pre>

Figure 2   Ten sequences, from Reference 21, showing the common primary sequences (A–AUCA) and proven basepairs.

than the highest-affinity ligands, then the highest-affinity ligands must compete toward the end of the SELEX process with those sequences as well. The number of SELEX rounds required to identify the highest-affinity ligands is larger than one might expect. To circumvent these problems, Irvine and colleagues (24) suggest using sufficient target protein in several early rounds to ensure the capture of the highest affinity ligands when far more abundant lower-affinity ligands are present. Later rounds would utilize lower target quantities to increase the stringency of SELEX at a time when the highest-affinity ligands are numerically sufficient to survive a competitive binding situation.

High background during the partitioning step increases the number of rounds and, in the worst situation, leads to the isolation of sequences that partition without facilitation by the protein target. For example, nitrocellulose filter-specific ligands, agarose-specific ligands, etc are often found, especially if they are more abundant in the library than the oligonucleotides that bind tightly to the intended target; obviously to counter-SELEX against the partitioning matrix or alternate partitioning protocols can be important.

The number of rounds required depends on the frequency of winners in the pool. In the T4 DNA polymerase SELEX two sequences out of 65,000 were found in 4 rounds; in the R17 coat protein SELEX roughly one sequence in 10^9 was selected in 11 rounds. From these data (and the many SELEX experiments reported below), a reasonable rule for predetermining the number of rounds until the highest affinity ligands are found might be one round for 10-fold enrichment. Rare high-affinity sequences usually will not be found after a very small number of rounds (see Table 1).

The length of the randomized domain in an oligonucleotide library is specified (although mutagenesis during amplification at each round may also be done). The arrangement and type of randomization are also variable. Total randomization of a block of contiguous nucleotides, partial randomization of a block, short blocks interspersed with fixed sequences, or linkage of blocks of random sequences to create a very large random region (>200 positions) are all viable strategies for SELEX. Twenty-five fully randomized nucleotides ($4^{25} = 10^{15}$) is near the practical limit of saturation in a SELEX experiment (e.g. all sequences of length 25 can readily be examined). More important is the likelihood of finding an oligonucleotide with the desired properties in a particular library: Some SELEX experiments have used libraries with randomized nucleotides within a predetermined structure, although most SELEX experiments have accepted whatever structure emerged during the experiment. Use of enough randomized nucleotides is critical to get a high affinity ligand and probably a stable structure (the simplest known oligonucleotide structure,

Table 1   High affinity sequences

Target	Library	Nts	Rds	Motif	K_d	Freq
T4 DNA Polymerase	RNA	8	4	hairpin*	5	4
R17 Coat Protein	RNA	32	11	hairpin	5	9
E. coli rho Factor	RNA	30	8	Cs, hairpin	1	10
E. coli S1 Protein	RNA	30	13	pseudoknot	4	12
E. coli 30S Particles + S1	RNA	30	13	pseudoknot	5	12
E. coli 30S Particles − S1	RNA	30	12	Shine-Dalgarno	7	11
QB Replicase	RNA	30	11	pseudoknot	5	12

E. coli metJ Protein	RNA	40	15	unknown	1	unknown
HIV-1 rev Protein	RNA	32	10	bulge	1	11
	5-iodoU-RNA	37	13	hairpin/bulge	1	
HIV-1 tat Protein	RNA	40	10	hairpin/bulge	5	11
HIV-1 int Protein	RNA	30	18	complex	20	12
HIV-1 Reverse Transcriptase	RNA	32	9	pseudoknot	5	10
	ssDNA	35	15	stem/bulge	4	12
MMLV Reverse Transcriptase	RNA	36	17	hairpin	10	11
AMV Reverse Transcriptase	RNA	36	12	hairpin	2	11
FIV Reverse Transcriptase	RNA	40	18	hairpin	2	13
U1A	RNA	10	3	hairpin*	5	4
	RNA	13	3	hairpin*	500	4
	RNA	25	3	none	500	3
Hel-N1	RNA	25	3	U-rich	unknown	3
Thrombin	RNA	30	12	hairpin	9	11
	modified RNA	30	15	hairpin	100	13
	ssDNA	60	5	G-quartet	25	4
	modified ssDNA	20	6	hairpin	400	6
Elastase	modified RNA	40	15	G-quartet	15	14
	ssDNA	40	17	G-quartet	10	14
sPLA2	RNA	30	12	complex	1	11
	modified RNA	50	11	hairpin	1	11
Nerve Growth Factor	RNA	30	10	pseudoknot	100	12
Basic Fibroblast Growth Factor	RNA	30	13	hairpin	0.20	11
	modified RNA	30	11	unknown	0.30	12
Vascular Endothelial Growth Factor	RNA	30	13	hairpin/bulge	0.20	12
	modified RNA	30	11	hairpin	0.03	9
Anti-gp10 Antibodies	RNA	10	3	hairpin*	unknown	6
SLE Monoclonal Antibody	RNA	30	13	hairpin	3	13
	ssDNA	40	8	unknown	1	10
Anti-Insulin Receptor	RNA	40	11	hairpin/bulge	2	11
IgE	modified RNA	40	9	G-quartet	35	9
Human Chorionic Gonadotropin	modified RNA	40	16	hairpin	60	14
Substance P	RNA	60	24	hairpin/bulge	190	14
Cyanocobalamin	RNA	72	8	pseudoknot	88	14
Theophylline	RNA	40	8	stem/bulge	110	14
Citrulline	RNA	74	7	stem/bulge	650	12
ATP	RNA	120	8	stem/bulge	700	10
Reactive Green	RNA/DNA	100,120	6	hairpin	60000	8,11
D-Tryptophan	RNA	120	7	unknown	18000	unknown
L-Arginine	RNA	25	7	stem/bulge	300000	9
L-Valine	RNA	25	9	stem/bulge	12000000	9

The Table shows features of SELEX experiments carried out against a variety of molecular targets described in the text. The type of library used is listed. Modified RNA refers to libraries consisting of 2'NH$_2$ or 2'F substituted pyrimidines instead of 2'OH pyrimidines. Modified ssDNA refers to the use of 5-(1-pentynyl)-2'-deoxyuridine replacing thymidine in the library (64). Nts refers to the length of the random region. Rds refers to the number of SELEX rounds. Motif refers to conserved secondary structural features of ligands that showed high affinity for the target. An asterisk indicates that the library was constrained, eg. the random region was flanked by a region of symmetry permitting only hairpins to be explored. K$_d$ is the equilibrium dissociation constant for the interaction of the ligand for the target expressed as nM. Freq refers to the frequency of the "winning" oligonucleotide in the starting random population. The numbers are the reciprocal of the exponent describing the abundance of the oligonucleotide. For example, 10 means that the frequency of the winning oligonucleotide was 10^{-10} of the starting population. These frequencies were estimated from the information content of the winning oligonucleotide as calculated according to Schneider et al (57), or were estimates based on the final published sequences in those cases where sequence family data were not available.

a hairpin with a short stem and a small loop, would require about 10 nucleotides). In some experiments the same target has been tried with libraries containing different lengths of randomized nucleotides, leading to identical winning ligands. However, this observation is not generic. Tuerk identified an oligonucleotide from a library randomized over 32 nucleotides that bound tightly to HIV reverse transcriptase (see below), but he found an even better oligonucleotide when he randomized 30 additional nucleotides adjacent to his first winner and performed additional SELEX rounds (C Tuerk, unpublished data).

A separate issue is the use of RNA versus DNA libraries (or chemically modified libraries). Again the data are sparse. For the few cases in which proteins or small molecules have been SELEX targets for both RNA and DNA libraries, the data suggest that binding affinity and specificity are similar for the winners from either library. The winning RNA or DNA oligonucleotide does not bind well when the same sequence is tested on both oligonucleotides.

Finally, target proteins probably have more than one oligonucleotide solution available through SELEX. In certain experiments several different oligonucleotide families are found; often those oligonucleotides compete with each other for binding as though a dominant target epitope can be contacted by more than one oligonucleotide shape, or by different sequences that present the same shape to the target, even though it may not be obvious that the ligands share a shape. In some cases after many SELEX rounds one finds oligonucleotides that can bind simultaneously to different epitopes on the target protein.

Other Bacterial, Nucleic Acid–Binding Proteins

Many bacterial proteins have been used as targets for SELEX, including the *E. coli* transcription termination factor (*rho*). While ligands that were rich in pyrimidines and apparently unstructured were identified—in agreement with the literature (25)—an additional ligand family was found that consisted of a small hairpin with a precise loop sequence and 5′ extension (26), features subsequently seen in some natural termination sites.

Ribosomal protein S1 was used as a target. The winning ligands were not pyrimidine-rich and unstructured, even though the literature predicted such ligands based on the natural sites (27). The highest-affinity ligands contained pseudoknots with several conserved loop sequences (28). These ligands also were identified when the target for SELEX was intact 30S ribosomal particles, which contain S1. Only when the 30S particles were deprived of S1 did the winning ligands contain the expected translational initiation (Shine-Dalgarno) sequences, as though the highest-affinity sites for 30S mRNA inspection during initiation could be somewhat more complicated than we previously thought they were (29, 30). The pseudoknot ligand and a Shine-Dalgarno sequence can bind to 30S particles at the same time. More recently, intact bacteriophage Q_β replicase was used as a target for SELEX (the replicase is a tetramer, one subunit of which is S1 ,31). The S1 pseudoknot was found repeatedly along with a small number of sequences that bind less tightly and that may interact with a different replicase subunit (D Brown, personal communication).

Recently the Uhlenbeck lab used SELEX to make an internal deletion in tRNA that still contained the structural information required for EF-Tu recognition (32). A randomized Ala-tRNA^{Ala}-minihelix library was screened for binding to Tu, cloned, and sequenced. Thus, SELEX can be used to identify features of a family of ligands for a protein when mere inspection of a set of known natural ligands (such as tRNAs) does not provide those clues (in this case because the natural set must have many features in common for the various other functions of tRNA).

Another SELEX experiment was aimed at understanding tRNA structure/function and binding to both phenylalanyl-tRNA synthetase and, again, Tu. The study (33) is generically informative for a SELEX issue not often discussed. Four individual ligands identified in this work as apparent synthetase binders were not aminoacylated by the synthetase. This problem was addressed in a way that has general power. These ligands were found to have a less abundant epitope-target than did the other ligands (by RNA-excess binding experiments) and to not bind to an independent preparation of synthetase purified by a different method, but to bind to a contaminant in the synthetase preparation. The ligands identified through SELEX might be used to stain native gels of the protein target as a routine method to verify that the evolved ligand binds to the intended protein.

The *E. coli* metJ repressor has been studied with both double-stranded DNA and single-stranded RNA (Yi-yuan H, Stockley PG, Gold L, Submitted; Yi-yuan H, Gold L, submitted). Double-stranded DNA "SELEX" (18, 19, 36) yields sequences that are very similar to the natural set of operators. The metJ apo-repressor selects RNAs with sequences that are unlike DNA operator sequences and that bind to the protein with comparable affinity to that seen between holo-repressor and operator (double-stranded) DNA.

Retroviral Proteins

The HIV-1 Rev protein has an RNA target, the *rev*-response element (RRE) (37–39). SELEX experiments (40, 41) revealed a simple structure that had most of the features of the natural RRE. The SELEX-derived oligonucleotides have been refined by further SELEX experiments; the best ligands observed thus far have K_ds some 10-fold lower than those of the RRE (42, 43).

The use of modified nucleotides in SELEX (Eaton & Pieken, 44) can provide dramatic improvement in the binding characteristics of the winners. Recently, the SELEX protocol has been modified to search for oligonucleotides that are covalently bound to the *rev* protein (Jensen KB, Atkinson BL, Willis MC, Koch TH, Gold L, submitted). Starting with the same pool of sequences used by Jensen et al (43), transcription reactions used 5-iodo-UTP in place of UTP (the 5 position on the pyrimidines is essentially free for substitution in the SELEX protocol, 44). The crosslinking-SELEX method included partitioning via gel electrophoresis of the protein-oligonucleotide complex under denaturing conditions. Ligands that were photocrosslinked to *rev* protein were found, presumably utilizing the 5-halo-uridine photochemistry described by Koch and his colleagues (46). Other ligands were found that specifically crosslinked to HIV-1 *rev* protein without photolysis. The crosslinking-SELEX protocol requires that the cDNA synthesis step (or the first strand synthesis in DNA-SELEX) be insensitive to the covalent adduct. For these particular experiments, cDNA synthesis was accomplished after protease degradation of the covalently bound protein. Any remaining peptide projecting from an adducted nucleotide did not slow cDNA synthesis.

The HIV-1 Tat protein has been studied through SELEX. SELEX yielded a winner that seemed to have little structure or sequence in common with the natural *tar* site (47, 48).

The reverse transcriptases (RTs) from HIV, MMLV (Moloney murine leukemia virus), AMV (avian myeloblastosis virus), and FIV (feline immunodeficiency virus) have been subjected to SELEX experiments. High affinity RNA pseudoknots were found for HIV RT; the K_ds for those RNAs were about 1 nM (49), as were the K_is for cDNA synthesis. However, the ligand for HIV RT did not inhibit either MMLV or AMV RTs. The RTs of MMLV, AMV, and FIV were used in subsequent SELEX experiments (50; Chen H, McBroom DG, Zhu Y-Q, North TW, GOLD LG, submitted); the ligands derived for those three proteins had high affinities (in the low nM range) and did not crossreact with each other's targets or with HIV RT. The consensus ligands for each RT do not share sequences with any of the other three consensus ligands. The lesson is that protein homologs, obviously quite similar at the sites that bind primer/template complexes and the SELEX winners, yield oligonucleotide ligands that easily distinguish one RT from another.

SELEX was performed with a single-stranded DNA library and HIV RT. The winning ligands bound with nM K_ds but did not resemble the RNA pseudoknots selected for the same enzyme (D Schneider & J Feigon, personal communication). Many of the selected DNA ligands inhibit HIV RT activity in vitro but have little inhibitory activity toward the same enzymes from AMV or MMLV.

HIV-1 *int* protein has been targeted by SELEX (Allen P, Worland S, Gold L, submitted; 53). An RNA ligand was identified that binds to integrase at the DNA binding site, has a K_d of about 10 nM, and does not obviously resemble the natural double-stranded DNA substrate.

Mammalian RNA-Binding Proteins

Two recent SELEX experiments (54, 55) were aimed at the mammalian RNA-binding proteins , Hel-N1 and U1snRNP-A (U1A) (reviewed recently in 56). With both targets the SELEX protocol was employed for just three rounds. In early rounds of SELEX, the most abundant sequences partitioned by any target are those that have low information content (57). Thus, early in SELEX (as noted above) one finds many sequences with improved (but not optimal) affinities if those sequences bind to the target with only slightly higher affinity than do the bulk of the sequences in the library. The Hel-N1 protein consensus and the U1A consensus were present in the initial library at high frequencies, about 10^{-3}, and thus it is not surprising that the consensus sequences had weak binding. It is unlikely that the highest-affinity ligands for either target protein have been identified; more rounds of SELEX are needed.

Nagai and colleagues have published the U1A structure without ligand (58), and they have also solved the co-crystal with the natural stem/loop RNA ligand recently (K Nagai, personal communication). The RNA used for the X-ray work is the natural site and is identical to the highest affinity winner obtained from SELEX. The structure of the co-crystal suggests that the loop undergoes induced fit as it binds to U1A because there are no intramolecular interactions between the loop nucleotides in the bound structure. Mere entropic fixation—by an RNA helix or a bulge—of the key loop nucleotides possibly provides a ligand suitable for high-affinity binding. However, we wonder if hypothetical very tight-binding ligands (found after more SELEX rounds with a library randomized over more nucleotides) would rigidify key single-stranded nucleotides with RNA structural motifs that allow a closer fit to the binding pocket of the U1A protein.

OLIGONUCLEOTIDES THAT BIND TO OTHER PROTEINS

Proteases

Thrombin was the first protease studied with SELEX using a library of single-stranded DNA (59). The winning sequence, $GGNTGGN_{2-5}GGNTGG$, would be expected to occur in approximately one per 20,000 molecules in the initial library, and thus five rounds of SELEX were enough to find these molecules. The ligands inhibited the thrombin catalyzed cleavage of fibrinogen to yield fibrin, and thus had anticlotting activity. The structure of the thrombin aptamer, determined by NMR (60) and X-ray crystallography (61), is an intramolecular G quartet, binding in the co-crystal in a sandwich between the exosite of one thrombin and the heparin-binding site of another. The ligand binds in solution with 1:1 stoichiometry (61), not 1:2, and the data suggest that the fibrinogen exosite is the location of the bound aptamer (62). The DNA ligand has promising properties as an anticoagulant (63). A second SELEX experiment from the same group identified a modified DNA ligand containing 5-(1-pentynyl)-2′deoxyuridine instead of thymidine that has a completely different sequence than the first, unmodified DNA ligand (64).

Kubik et al found an RNA that binds to thrombin but does so in the heparin-binding site (65). The winning RNA ligands are not G quartets, and there is no competition between the RNA and DNA ligands (65). These RNA ligands also had anticoagulant activity. The information content of the winning ligand family suggests that these sequences were in the library only once per 10^{11} sequences.

The important lessons from the two quite different SELEX experiments on the same target are (*a*) DNA and RNA ligands can recognize different target epitopes, (*b*) the ligands have completely different primary sequences and structures, and (*c*) both protein epitopes were suitable as targets for the generation of antagonists. What structural features of DNA and RNA (and the protein epitopes) directed such similar polyanions to different sites are not understood.

Human neutrophil elastase has been used to identify a DNA ligand that binds with a K_d of about 10 nM (Lin Y, Padmapriya A, Morden K, Jayasena S, submitted). This ligand binds specifically to elastase and not to other serine proteases; however, this specific ligand is not a potent inhibitor of elastase because it binds to the protein at some distance from the active site. By tethering a small peptide inhibitor to that oligonucleotide, a mM antagonist with poor specificity (the peptide) became a nM antagonist (as the oligonucleotide containing the peptide, now brought into proximity of the active site) with high specificity.

Blended-SELEX (from which one obtains oligonucleotides with components in addition to the standard nucleotides) can be used to make a drug more avid and more specific for its intended target, perhaps solving some of the toxicity problems associated with drugs that bind to unintended or unknown cellular components. The protocol for blended-SELEX has several variations, each intended to deliver an appropriate small molecule to a target, or to incorporate the small molecule into the ligand pool rather than into a finished SELEX winner (D Smith, personal communication). If each oligonucleotide in a library contains an additional component that is known to bind to a specific site on a target protein, blended-SELEX might yield oligonucleotide winners that bind to that protein epitope. For example, a library of oligonucleotides with glucose adducted might seek out the active site of hexokinase during SELEX. An experiment with elastase has altered a nonspecific suicide substrate for that protease into a highly specific compound through blending (D Smith, personal communication).

Secreted Phospholipase A2

sPLA$_2$ may be involved in inflammatory cascades (67). An RNA ligand and a modified RNA ligand (containing 2' amino replacements for the 2' hydroxyls of the pyrimidines in the RNA pool) have been made. The ligands bind with sub-nM K$_d$s and inhibit the activity of the enzyme (D Parma, personal communication). An unrelated oligonucleotide, found by accident, inhibits phospholipase A$_2$ with weaker affinity and less specificity (68). Most remarkably, the discovery accident involved testing several antisense oligonucleotides and realizing "that a subset of oligonucleotides directly inhibited human type II PLA$_2$ enzyme activity." (68) This suggests that antisense compounds may have unexpected effects on cells, because the shapes of these oligonucleotides allow them to interact with molecules other than the intended nucleic acid target.

Growth Factors

Nerve growth factor (NGF) was used for RNA SELEX (J Binkley, personal communication). The goal in this case was to find a winning ligand even though the nonspecific binding of the library oligonucleotides was extremely avid, because NGF is an extremely basic protein. Despite the higher nonspecific binding, enrichment was accomplished without difficulty, both under moderate and high salt conditions.

Oligonucleotides have been identified that bind to basic fibroblast growth factor (bFGF) (69) and to vascular endothelial growth factor (VEGF) (70), two proteins involved in angiogenesis (71, 72). Folkman postulated long ago (73) that angiogenic factors are required for tumor growth and thus these growth factors are sensible targets for anticancer therapies. The bFGF ligand has been tested with five members of the FGF family (74) and also with five other heparin-binding proteins. The K$_d$ of the bFGF ligand for these other target proteins is 100 to 100,000 times higher than for bFGF. The K$_d$ of the bFGF ligand for denatured bFGF is 100,000 times weaker as well (69). Similar specificity experiments have been done with the VEGF ligands, and again the oligonucleotides do not bind to other heparin-binding proteins or even to reduced VEGF (70). In direct assays for the binding of each radio-labeled growth factor to cell lines that express the cognate receptor, the SELEX-derived oligonucleotides are antagonists. The K$_d$s of the inhibitory ligands are around 1 nM, and the K$_i$s for receptor binding are very similar. Thus the SELEX ligands bind to the growth factors in a manner that competes with receptor binding. Some non-inhibitory ligands with pM K$_d$s have been isolated against both bFGF and VEGF.

Antibodies

An early antibody target for SELEX was serum prepared against a 13-amino-acid sequence (NH$_2$-MASMTGGQQMGRC-COOH) from the amino terminus of the bacteriophage T7 g10 protein (75). The RNA library contained only 10 randomized positions within the loop of a fixed hairpin. The serum, bound to protein A beads, was challenged for three SELEX rounds under very demanding binding conditions. After three SELEX rounds one winning sequence was found. The resultant ligand binds to the anti-g10 antibodies in the serum at the antigen recognition site (demonstrated through binding competition). Oligonucleotides can mimic a protein antigen even though that antigen is not a polyanion.

Systemic lupus erythematosis (SLE) is an autoimmune disease characterized by the production of antibodies against nuclear antigens, including DNA. One such antibody (76) was used as the target protein for SELEX. High affinity RNA and DNA ligands were found (J Ruckman, personal communication), with affinities at or below 1 nM.

SELEX has been used to target a mouse monoclonal antibody specific for an epitope of the human insulin receptor (Doudna JA, Cech TR, Sullenger BA, submitted). Patients with extreme insulin resistance type B often generate autoantibodies to this receptor. The RNA arising from the SELEX experiment specifically recognized the anti-insulin receptor antibody and could block its interaction with the insulin receptor in vitro. Most strikingly, the RNA crossreacted with auto-antibodies from other patients with extreme insulin resistance. Thus, the RNA is structurally similar to the protein epitope on the insulin receptor (a similar story about a protein epitope on bacteriophage T4 gene 32 protein that resembles nucleic acids is in 78). In principle, these RNA mimics could be used to inhibit the auto-antibody interaction selectively in vivo.

The IgE class of antibodies also has been targeted by SELEX (T Wiegand, personal communication). In these experiments the goal was to find antagonists that prevent binding of IgE to the IgE receptor. Thus the constant domain of IgE rather than the variable (antigen recognition) domain was the desired epitope for the oligonucleotides. This experiment yielded oligonucleotides that competitively inhibit the interaction of IgE with the FcεRI receptor.

Hormones

Oligonucleotides were sought that bound tightly to human chorionic gonadotropin (hCG) and not to the closely related glycoprotein luteinizing hormone (hLH). The protocol employed counterselection against hLH and a modified library containing 2'-fluoro pyrimidines (to achieve stability against nucleases in a diagnostic setting). Oligonucleotides were found that bind hCG in urine with a K_d of 50 nM (D Nieuwlandt, personal communication), roughly a 10^4-fold enhancement over the unselected random oligonucleotide library. These ligands bind hLH some 250-fold less tightly than they bind hCG; even though LH is nearly identical to CG for all but the carboxy terminus of one subunit.

Some Remarks Concerning SELEX with Generic Protein Targets

Thus far the data suggest that any protein target is suitable for SELEX, and that the K_ds obtained are comparable to those obtained with nucleic acid–binding proteins. We are aware of some 30 other target proteins that have been put through the SELEX drill. From those data and the data presented here, K_ds for average (non-nucleic acid–binding in vivo) proteins range between 50 nM and 50 pM. These K_ds are true affinities, measured in solution, under a small number of arbitrary solvent conditions (usually similar to the conditions used for SELEX itself). These affinities could sensibly be compared with the affinities toward various target antigens of Fab fragments of antibodies when measured in solution; the affinities of oligonucleotides identified through SELEX are more avid than those usually measured for Fab fragments (79).

OLIGONUCLEOTIDES THAT BIND TO PEPTIDES

Small peptides are good therapeutic targets, because many peptides are available in blood or in extracellular tissue spaces and could be titrated by an antagonist. If oligonucleotides could be found that bind tightly to a peptide, they would almost certainly antagonize the activity of the peptide by interfering with the binding of that peptide to its receptor (as in the case of the bFGF and VEGF ligands, described above). Interference of peptide activity could result from direct occlusion of the receptor-binding domain or from fixing an otherwise floppy peptide in a shape that is not appropriate for the receptor. We are aware of oligonucleotides that bind to two small peptides, bradykinin and substance P (M Wecker, personal communication; 80). The ligands for substance P had K_ds around 200 nM, less avid than ligands aimed at virtually every protein target tried so far. Even such modest affinity could provide antagonism in a functional substance P assay, especially when the affinities for the true receptor(s) are also modest. Whether extremely high-affinity oligonucleotides can be aimed at small floppy peptides is unclear, because the entropy loss of the target peptide upon binding may exceed the available binding energy from potential contact points. Large protein targets provide larger and more rigid surfaces for contact with an oligonucleotide.

OLIGONUCLEOTIDES THAT BIND TO SMALL MOLECULES

The second SELEX paper, historically, sought oligonucleotides that bound to organic dyes (16). The methodology includes randomization, partitioning, and amplification, the same critical steps utilized with bacteriophage T4 DNA polymerase (15). Selection experiments aimed at small molecules have been carried out in a similar manner regardless of the precise target. The oligonucleotide library is allowed to interact with the target, which is immobilized on a support. After a wash step to remove unbound material, bound oligonucleotides are removed by a molar excess of the free target (ideally), EDTA, or alteration in ionic strength. Eluted oligonucleotides are amplified, and the process is repeated until the sequence complexity of the pool is reduced. Removing oligonucleotide species that bind the resin instead of the resin-target complex is important. In general, in early rounds, very little binding (<1%) of the radiolabeled oligonucleotide pool is observed. After 7–10 rounds, typically 40–60% of the input RNA can be specifically eluted. This format provides a convenient platform for analysis of binding specificity. By using variants of the target molecule to elute RNAs from the column, ordering dissociation constants and providing information about the contribution of various target functionalities to the binding energy is possible.

Organic Dyes

Ellington and Szostak (16) described the isolation of RNA sequences that had affinity for anthraquinone-based dyes such as Cibacron Blue 3GA and Reactive Blue 4 linked to agarose. RNA was initially allowed to bind in high salt buffer (0.5M LiCl); bound RNA was eluted with water. After 4–5 rounds, more than 50% of the applied RNA was bound and eluted by water. RNAs were found that re-bound to the dye column on which they were selected and not to other dye columns, indicating that the selected RNAs could discriminate between functional groups on the dyes. The sequences of binding RNAs in different clones were heterogeneous, indicating that the initial random sequence population contained many sequences that could recognize dyes. The affinities of these RNAs for the dyes on the matrix were estimated to be in the 100–600 µM range.

Ellington and Szostak (81) also carried out SELEX experiments against the same targets using a random DNA library, in part to compare the RNA and DNA solutions. As has been observed with several other targets (thrombin, reverse transcriptase, lupus antibodies, bFGF, and VEGF), the RNA and DNA SELEX solutions differ in primary sequence, and neither the RNA version of the DNA winner nor the DNA version of the RNA winner binds to the target.

Nucleotides

RNA species that specifically recognize ATP were obtained (82). In these experiments, RNA retained on an ATP-agarose column was eluted with free ATP. After eight rounds of selection, clones were obtained that contained a well-defined 11-nucleotide consensus sequence flanked by two base paired stems of variable sequence. A 40 nucleotide truncate was synthesized that contained the consensus region and flanking regions and that bound to ATP with a K_d about the same as that of the full-length RNA (0.7 μM). To assess what parts of the target interact with the oligonucleotide, ATP analogues were used to elute bound RNA from the ATP-agarose column. The 2' OH moiety and five other positions on the base and sugar of ATP were clearly implicated in the recognition while the phosphates were not. Kethoxal sensitivity of guanosines in the consensus region was shown to be altered after ATP binding, suggesting that the ligand induces a conformational change in the RNA or directly occludes those guanosine residues.

GDP/GMP sites are also frequent in a pool of oligonucleotides randomized over 25 nucleotides. The high-affinity winners have relatively small structures, consisting mainly of an 11-nucleotide internal loop that provides a base-specific site (83).

Recently, the minimal ATP recognition domain has been used as a core structure to search for longer oligonucleotides that have catalytic activity (84, and see below). Randomized oligonucleotide regions were appended to a mutagenized version of the ATP-binding domain, and the resulting pool was challenged to catalyze thiophosphate transfer from ATP-γ–S. A ribozyme was discovered that autophosphorylates with a k_{cat} of about 0.3 min^{-1}, a rate that the authors estimate represents about 10^9-fold enhancement over the uncatalyzed reaction. This work is an elegant metaphor for a mechanism by which oligonucleotides capable of small molecule recognition could have acquired more sophisticated functions during evolution.

Amino Acids

The observation that arginine, uniquely among the standard amino acids, effectively inhibits self-splicing by the *Tetrahymena* pre-rRNA in vitro (85) provides an intellectual foothold to investigate RNA–amino acid interactions via SELEX. The recognition site for arginine in the RNA includes an arginine codon, and this site is conserved in all modern group I introns (86). These observations may point toward the pathway by which the genetic code arose. Arginine linked to Sepharose has been used as a target for SELEX (87) using free arginine as the eluant. Three arginine-binding stem-loop RNA motifs were isolated that had K_ds for arginine in the 200–400 μM range. These motifs had distinct recognition specificities for various arginine analogues used as eluants for RNA bound to the arginine column. An intronlike motif was not recovered, perhaps due to the requirement for a larger sequence context. A second notable aspect of the competition elution was the unselected affinity of the three motifs for guanosine 5' monophosphate, consistent with the idea that the guanidinium side chain of arginine is an important component of the recognition specificity. The K_ds of RNA containing the selected motifs for arginine were similar to that seen for the group I intron, while the RNA motifs are much smaller than the intron. This observation suggests that SELEX will yield a binding site in the context of the minimal supporting scaffold required to position that site properly.

Arginine-binding RNA species have also been obtained through an alternative two-step approach in which the initial selection was for L-citrulline affinity (88). Citrulline differs from arginine in that it contains a keto group rather than an amino group, and its urea group is neutral in contrast to the positively charged guanidinium group of arginine; thus citrulline may be a more challenging target for SELEX than is arginine. Members of the RNA family selected on citrulline-agarose each contained two separate consensus motifs of 10 and 6 nucleotides. All members could be folded into a similar secondary structure with the two conserved motifs comprising a bulged internal loop flanked by paired regions. A truncate designed from the predicted secondary structure had a K_d of about 65 μM for L-citrulline and showed slightly lower affinity for D-citrulline and minimal affinity for arginine. A pool of new RNA was prepared based on the sequence of the truncate but containing a 30% mutation rate per position. This pool was then applied to an arginine-agarose column to select for alterations permitting arginine recognition. After five rounds of selection, about half of the selected RNAs contained minor variants of the initial citrulline consensus motifs. The arginine-binding RNA is a triple mutant of the parent citrulline binder. However, this RNA bound to arginine (K_d of 56 μM) about 120-fold better than to citrulline and showed a 7-fold stereospecific preference for L-arginine. RNA structural solutions for ligands aimed at two chemically related but distinct targets can be connected by a few changes that dramatically alter recognition properties. A more direct experiment has been done in this regard (83); specificity of binding can be influenced by one nucleotide in a ligand, suggesting again that mutation and selection would suffice to allow substantial evolution of oligonucleotide recognition.

The potential of RNA to recognize selectively hydrophobic side chains represents another challenge to SELEX. RNA is essentially devoid of hydrophobic appendages (ignoring tRNA and its remarkable collection of modified nucleotides; 89). Hydrophobic side chains are involved in some DNA–protein interactions, such as sequence-specific recognition by the *engrailed* protein (90), but these involve the methyl group of thymine and so are not possible for unmodified RNA. Yarus and Majerfeld (91) selected RNAs capable of binding to valyl-agarose. RNA was eluted with L-valinamide. After nine rounds, RNA species were isolated whose predicted secondary structure contained a conserved internal bulge flanked by a sequence-conserved stem featuring two G-U pairs. This motif shows about 15-fold stereoselectivity preferring L- to D-valine, strongly prefers valine to threonine (isosteric to valine except for a polar oxygen in place of a methyl), and discriminates between other hydrophobic side chains both larger (leucine and isoleucine) and smaller (alanine) than valine. These results indicate that SELEX has produced an RNA pocket of definite size, shape, and polarity. Introduction of RNA libraries made from modified monomers with hydrophobic appendages (44) would, presumably, generate motifs with improved affinity characteristics.

SELEX has also been used to identify RNA species that recognize D-tryptophan-agarose with a K_d of 18 μM (92). These RNAs do not recognize L-tryptophan-agarose (K_d > 12 mM), indicating that recognition is stereospecific. These results again illustrate the principle that RNA can assume tertiary conformations that have analogous recognition properties to those of proteins.

Cyanocobalamin (Vitamin B12)

Extrapolation from existing synthetic and catabolic pathways back to the types of RNAs that might have functioned in a pre-DNA and protein world has been an intense area of speculation since the initial proposal of an RNA world (6). Benner et al (93) have proposed that modern reactions catalyzed by B12-dependent enzymes, such as the reduction of ribonucleotides into deoxyribonucleotides, may have been descended from B12-dependent RNA enzymes. With this same hypothesis, Lorsch and Szostak (94) carried out SELEX against cyanocobalamin coupled to agarose and identified a high-affinity RNA ligand with a K_d of 320 nM for B12 in solution. This sequence was subjected to mutagenesis and further selection, resulting in a pool of sequences that each contained a highly conserved 31-base sequence, of which 14 positions were absolutely invariant. The pattern of sequence conservation in the pool predicted a compacted pseudoknot secondary structure for the B12-binding domain. A truncate containing this motif bound to B12 with a K_d of 90 nM in solution. Chemical modification experiments of the RNA with dimethyl sulfate and diethyl pyrocarbonate in the presence and absence of B12 are consistent with a conformational change in the RNA upon binding to B12, although whether the footprint data represent direct interference with the probes is difficult to know. The size of the conserved region indicates that the design solution for a tight binding surface or pocket for B12 is more complex than that for dyes, ATP, or the amino acids studied to date. The frequency of these solutions in the original random population is correspondingly and perhaps remarkably low, about 10^{-14}. While the implications of these results for prebiotic biochemistry remain lost in the murky depths of evolution (no similar sequences are present in the existing RNA data base), they nonetheless provide another example of structural versatility of oligonucleotide libraries. Speculation abounds that oligonucleotides in vivo are contacted and altered by various small molecules, leading to interesting ideas about novel regulatory mechanisms (85; J Roth, personal communication; see below).

Theophylline

Theophylline (1,3 dimethylxanthine) is a naturally occurring alkaloid that has been recognized for decades as a potent bronchodilator for the relief of acute and chronic asthmatic symptoms (95). Theophylline can have serious adverse effects and serum levels must be monitored to prevent toxicity. Theophylline is closely related to other xanthine derivatives such as caffeine (1,3,7 methylxanthine) and theobromine (3,7 dimethylxanthine) that are commonly found in serum; diagnostic reagents for theophylline must show low cross-reactivity for these compounds. Monoclonal antibodies raised to theophylline show cross-reactivities to caffeine and theobromine of 0.2–0.3% of the theophylline signal (96). Theophylline and its cousins thus represent an interesting model system for investigating the potential of oligonucleotides to recognize therapeutically important targets with high affinity and specificity.

SELEX against immobilized theophylline was carried out with a pool of about 10^{13} RNA molecules containing a 40-nucleotide random region (97). Bound RNA was eluted with free theophylline. The majority of applied RNA was eluted by theophylline after eight rounds. In addition to simple theophylline elution, counterselection was done by a prior elution with caffeine to deplete the population of theophylline-binding RNAs with affinity for caffeine. The family of sequences that comprised the eighth round of SELEX was found to contain two conserved motifs six and nine nucleotides long, separated by a sequence of variable length. The 15 conserved nucleotides were absolutely conserved at 14 positions. Each of the various family members could be folded into a similar secondary structure in which the conserved motifs were brought together into a proposed theophylline binding pocket stabilized by flanking stems composed of nonconserved sequences. Again, the idea that SELEX generates a binding pocket that is "rigidified" by flanking sequences—much as the active site of an enzyme can be thought of as a floppy peptide that is immobilized by the rest of the polypeptide—is strengthened by these observations. RNA-theophylline binding was examined for both the full-length molecule and a 38-nucleotide truncate whose sequence was designed from the predicted secondary structure motif. The K_ds of the two RNAs for theophylline were 0.6 µM and 0.1 µM respectively. Caffeine is bound about 10^4-fold less tightly than theophylline. The RNA winner did better than the antibodies aimed at theophylline (96). Truncation improves binding sometimes, presumably because the unselected and fixed regions of a long RNA can interfere with optimal folding or presentation of the target to the RNA.

One of the experimental virtues of SELEX directed at small molecules such as theophylline (or ATP or others) is the multitude of related molecules that can be employed as competitors in binding experiments designed to probe the contribution of various moieties to the binding energy and to provide insight into the nature of the recognition site in RNA. Competition with a variety of xanthine derivatives permitted several conclusions (97) that are consistent with multiple models for recognition. One plausible scheme postulates hydrogen bonds from the N-7 of theophylline to an acceptor in RNA, possibly other H-bonds from RNA to the oxygen at C-2 and C-6, a steric boundary in the RNA near the C-8, and a hydrophobic region in the pocket to accommodate the N-3 methyl. ^{1}H NMR analysis of the 38 nucleotide RNA, which gives information about the overall number and type of base pairs, is consistent with the proposed secondary structure model. This conformation appears to change as theophylline is bound; the alteration in the spectrum upon theophylline binding has been used to determine that each RNA binds a single theophylline molecule. The details of the RNA conformational change upon theophylline binding are currently under investigation. Two-dimensional NOESY spectra are consistent with a model in which theophylline stabilizes an interaction between the conserved three-base asymmetric bulge (CCU) with some members of the conserved six-nucleotide symmetric bulge (A Pardi, personal communication).

In many ways specific sites for small molecules provide the most impressive evidence for the variety of oligonucleotide surfaces. Small molecules present fewer opportunities for specific interactions, and a smaller scope for a chelation effect to increase affinity (98). Thus, SELEX successfully aimed at small molecules dramatically confirms the versatility of the nucleotide building block.

On a deeper level, structural analysis is critical to the more general issue about the nature of the interactions involving SELEX-generated oligonucleotides and their target molecules. Researchers commonly assume that high affinities and specificities are correlated with molecular rigidity in the ligand. This is clearly not always true—certain peptide–receptor interactions involve stabilization of one of a number of peptide conformations. However, among the highest affinities known for peptides aimed at receptors, the conotoxins suggest that rigid peptides may do better than floppy ones (99, 100). Endothelin is another example of a relatively rigid peptide that binds to its receptors with high affinities (101). Those oligonucleotides that are most rigid bind most tightly to their targets, as long as no floppy oligonucleotide provides significantly more specific contacts with high binding energy. NMR and X-ray crystallography are the major tools for assessing stable higher-order structure in RNA, although the number of RNAs whose structure has been solved by these methods remains small (the 1994 Protein Data Base lists 2441 protein structures and 18 RNA structures, 9 of which are tRNAs). This situation is certain to change as small RNA ligands discovered by SELEX become the focus of structural efforts.

As pointed out by Ellington (102), SELEX directed against low-molecular-weight targets has revealed a few principles: Targets with heterocyclic rings are likely to be excellent SELEX targets that yield highly discriminatory oligonucleotide ligands, and tighter binding oligonucleotides tend to show better discrimination of their targets from their cousins, presumably due both to more contacts with the target and, perhaps, the increased molecular rigidity that limits interactions with related compounds. Finally, the chromatographic partitioning used against most small molecule targets is attractive because of the ease of applying extreme selective pressure using counterselective elution. However, SELEX with chromatographic partitioning is somewhat more difficult to drive toward the highest-affinity ligands because target density makes true equilibrium difficult to attain during the binding step.

RIBOZYMES VIA SELEX

Alterations of Natural Ribozymes

Natural ribozymes, including the self-splicing group I intron (103), have been subjected to genetic study. Initially, a collection of deleted forms of the *Tetrahymena* ribozyme were screened for the ability to catalyze cleavage-ligation reactions with single-stranded DNA substrates (104). The most successful ribozyme was identified by selective amplification; it contained a 93 nucleotide deletion near the 3′ terminus. Interestingly, the deleted form cleaved DNA more efficiently than the wild type cleaved RNA.

Green, Ellington, and Szostak (105) performed a far more extensive SELEX experiment on the *Tetrahymena* ribozyme. A region of nine nucleotides that were strongly conserved or covaried was randomized and functional variants selected. Later, other SELEX experiments were aimed at minimizing the size of a group I intron ribozyme (106) and at converting the RNA cleavage activity of a slightly different ribozyme to a DNA cleavage activity (107). Similar alterations have modified the Mg^{2+} requirement of the ribozyme to Ca^{2+} (108, 109). Selection experiments have been carried out on a hairpin ribozyme by Burke and his collaborators (110–112). These mutagenesis and selection experiments have focused on the transesterifications and hydrolytic reactions of known ribozymes.

New Ribozymes Through Direct Selection

There are but a few papers in which new ribozymes are uncovered by directly partitioning (from a randomized library) oligonucleotides able to catalyze a specific reaction. Pan and Uhlenbeck isolated RNAs that were cleaved site-specifically by Pb^{2+} to give a 2′3′ cyclic phosphate and subsequently selectively hydrolyzed to give 3′ phosphates (113); these cleavage reactions mimic the mechanisms by which degradative ribonucleases function. These new ribozymes were identified by partitioning those molecules that had been cleaved to linear form by the metal from circular RNA libraries.

Bartel and Szostak isolated new ribozymes that can ligate an oligonucleotide with a 3′ OH to the 5′ triphosphate end of the catalyst, displacing pyrophosphate (114). This ribozyme was found from a pool (more than 10^{15} molecules) of RNA randomized over more than 200 nucleotides. The experimental design placed, by annealing, the 3′ OH of an oligonucleotide adjacent to the 5′ triphosphate of the oligonucleotide library; thus the partitioning principle is identical to that seen earlier. The ligation event provides a new sequence, this time at the 5′ end of the presumptive ribozyme, for amplification. The new ribozymes have not yet been characterized structurally, but they will be fascinating. Recently Lorsch and Szostak (94) reported the selection of a new polynucleotide kinase activity; in this case the ATP-binding motif (see above) was surrounded by random sequences (a total of 100 nucleotides) and asked to bind ATP-γ-S in such a manner as to allow transfer of the terminal phosphate to the ribozyme. Partitioning was then based upon the selectability of the sulfur. Normal amplification and reiterative selections identified the new catalyst. Both of the new leadzymes, as well as the new ligase and the new kinase, catalyze chemistry at phosphorus centers.

An unpublished experiment (M Wecker & D Smith, personal communication) sought a new ribozyme that does something other than phosphorus chemistry. Transcription of an RNA library containing only 30 randomized positions was initiated with a reactive 5′ nucleotide, GMP-S (guanosine monophosphorothioate). The library was mixed with a small peptide, bradykinin, that had been made reactive toward GMP-S by bromoacetylating the peptide at the amino terminus. Specific sequences were found that accelerated the reaction (displacement of Br) between the activated RNA and the modified peptide by more than 6000-fold.

Illangesekare et al (115) have selected, by trapping the aminoacyl RNA product, an RNA that greatly accelerates the aminoacylation of its own 2′(3′) terminus. Aminoacylation utilized the universal biological substrate for aminoacyl-tRNA synthesis, aminoacyl-AMP. This suggests that RNA can catalyze substitution at an activated carbonyl carbon, and further suggests that an aminoacyl-tRNA synthetase composed of RNA is plausible.

SELEX Aimed at Transition-State Analogues

Antibodies can be found that catalyze reactions (116, 117). On occasion these abzymes evolve naturally, but most frequently they are provoked by raising antibodies to transition-state analogues (TSAs). Although many labs have sought oligonucleotides that bind tightly to TSAs, only one successful new ribozyme has been reported. We take the position that negative experiments are informative here, and thus evaluate the single success to date, a failure that is in press, and a second important failure that has yet to be analyzed to completion.

The published ribozyme based on SELEX-derived TSA binders is from the Schultz lab (118). A specific compound likely to resemble the transition state for an isomerization of substituted biphenyls was used to find binders; from a pool of about 10^{15} sequences of RNA randomized over 128 nucleotides, one sequence was identified that catalyzed the isomerization of a biphenyl substrate some 88-fold over the spontaneous rate. The reaction was inhibited by the TSA itself with a K_i of about 7 μM, predicting the observed rate enhancement. The authors suggest that more efficient catalysts "will require the isolation of RNAs with higher preferential affinities for the TSA." (118) This catalyst deserves further study, however modest its observed rate acceleration. The determination of the structure of one sequence 128 nucleotides in length is not easy; families of ligands and ligand truncation will simplify the analysis.

A negative experiment (119) was performed with a TSA for a Diels-Alder reaction. The RNA library was randomized over 80 nucleotides, and many similar sequences were found after 20 rounds of SELEX. The highest-affinity ligand bound the TSA with a K_d of about 0.35 mM, some 50-fold weaker than the ligand against the biphenyl TSA (118), and thereby not likely to enhance the catalytic rate appreciably. Nevertheless, nearly 200 different sequences were assayed for catalysis of the Diels-Alder reaction, and none were active. This failure may have been due to the inability to provide oligonucleotide sequences that bound the TSA more tightly than the substrates of the reaction, and to the modest affinity for the TSA.

Subsequently, K Morris and N Janjic (personal communication) have used a TSA for the cleavage of a trityl ester (120). In this case, RNA ligands were identified with 7 μM K_ds for the TSA, and these oligonucleotides had been counter-selected for diminished affinity for the substrate. The affinity for this TSA is similar to the affinity of the ligand aimed at the biphenyl isomerase TSA (above), but thus far the RNAs identified through SELEX have not catalyzed cleavage of the trityl ester substrate.

An exposition on the large potential catalytic potential of RNAs (11) suggests that some difficulties lie with the scientists rather than with the RNA libraries. Any polar polymer capable of complex higher order structure is capable of efficient and varied catalysis, because catalysis of many kinds is performed substantially by immobilized divalent metals. Esterase activity can be driven by metallocatalysts, and therefore the failure of the TSA approach is surprising. However, no K_ds in the nmolar and pmolar range were obtained, thus diminishing the potential rate enhancement by a TSA binder. However, perhaps the demands made during SELEX aimed at TSAs are too stringent for good catalysis. Since TSAs are only approximations of the true transition states, very specific binding to TSAs may somehow not allow binding to the true transition states. Such ideas demand a rationalization of the relative ease with which one finds monoclonal antibodies that bind to TSAs and catalyze reactions.

OLIGONUCLEOTIDES VERSUS OTHER COMBINATORIAL LIBRARIES

Why Oligonucleotides Provide Good Ligands

SELEX provides highly avid and specific ligands and at least some new catalysts, probably due to the tendency of rather short oligonucleotides to form structures. The oligonucleotides found through SELEX are often as small as 25 nucleotides, providing a contact region with the target of about 10–15 nucleotides (121; some ligands are even smaller and have even fewer nucleotides that contact the target, 122). The nucleotides that make contact with the target are almost always found within intramolecular structures: sometimes in hairpin loops, sometimes in bulges, sometimes in the loops of G quartets, and sometimes in pseudoknots. The average sizes of the contact domains are about 300–400 Å^2. This is roughly the size of the antigen recognition regions of Fab fragments of antibodies. This surface is responsible for the contacts that provide nM affinity (and better) for the SELEX-derived ligands. However, even pM K_ds require a very small fraction of the available free energy of binding implied by 300 Å^2 of contact. The surprise in SELEX is that the small number of monomers in oligonucleotides (compared to proteins) can nevertheless provide oligonucleotides with binding affinities and specificities that rival or exceed those of proteins, probably because the interactive forces used by standard nucleotides are complex.

Randomized peptide, protein, and small molecule libraries are being developed to find high-affinity ligands to use as diagnostic agents or drugs (123–128). Short peptide libraries have not been robust thus far. The structural freedom of peptides and the resulting entropic cost upon binding to a target (129) probably limit the use of peptide libraries if the demands on the peptide include high affinity and specificity. Amino acids are not interactive with each other. Proteins probably are large because only large proteins are structurally stable. Rarely do short peptides have such fixed structures in solution (99, 101). The failure of amino acid side chains to interact strongly with each other confounds the computational approach to the protein-folding problem.

In response to this entropic limitation, some effort has gone toward creating peptide libraries within the context of predetermined structural motifs, for example through the use of cyclic peptide libraries. Similarly, randomization has been created within the context of antibody structure to provide a library of about 6.5×10^{10} Fab fragments displayed on filamentous phage (127). This library has yielded higher-affinity and higher-specificity ligands than have other peptide and protein libraries, but even this library has not yet provided ligands with the qualities of oligonucleotide ligands. Neither random cyclic peptides nor randomized amino acids held in the lattice of the variable domain of an antibody structure are likely to be structurally robust.

Conversely, rather short oligonucleotides have stable, intramolecular structures. The bases are remarkably interactive with each other (130). Oligonucleotides pay the entropic cost of forming fixed structures prior to interaction with their target, allowing ligands to exist that require little in the way of induced fit upon binding. The winning SELEX answers have interesting structures (pseudoknots, bulges, G quartets, large hairpin loops that probably energy minimize to smaller structure) and all use many non-Watson-Crick basepairs. Because oligonucleotides form intramolecular structures easily, predetermining the structure of the oligonucleotides in the library prior to starting SELEX is neither necessary nor even useful. A simple way to reconstruct this entire discussion is to ask which random polymer—peptide, antibody, or oligonucleotide—will denature at the higher temperature.

Why Proteins Won in Nature

Essentially all catalysis, receptor-mediated signaling, and molecular machine functions in the present biosphere are handled by proteins. The early world was probably sequence sparse. It seems unlikely that 10^{15} different sequences often resided in a small volume of liquid; the SELEX experiment was not done often. The sparseness we imagine might have then confronted partitioning through the early membranes of the emergent biosphere; polyanions seem somehow inappropriate for facile movement from outside to inside those early compartments, especially when compared to less anionic proteins.

Finally—and most importantly—biology takes advantage of the tendency of proteins to ignore each other. Organisms in the present biosphere have enormously high intracellular protein concentrations, and yet proteins interact with each other specifically and appropriately. Imagine the difficulty of the protein-folding question, still unsolved in its intramolecular form, now complicated by the frequent collisions between the nascent, unfolded polypeptide and all the other protein surfaces in the cytoplasm. Proteins solve this problem through their intrinsic chemical dislike for each other, the very property that makes short peptides unstructured and the average protein large relative to its substrates and its active site. This is also the property that makes peptide and antibody combinatorial libraries a difficult source of high-affinity and high-specificity ligands. The biosphere probably evolved chaperonins (131) to help proteins fold if they could not efficiently fold on their own within a solvent loaded with all other proteins. This problem may be responsible for the formation of inclusion bodies, the insoluble aggregates formed when a foreign protein is overexpressed in a bacterium; the foreign protein did not have an evolutionary history in which to solve the folding problem caused by the presence of other proteins in the solvent.

Now imagine the same problem for oligonucleotides, in the contexts of both SELEX and biology. Oligonucleotides at 1 μM concentration—the conditions for a SELEX experiment corresponding to about 10 μg/ml, and just under 10^{15} sequences per ml—fold and behave essentially autonomously. Under these conditions, oligonucleotides ignore each other in spite of their capacities to anneal through short stretches of complementary bases. Were nature to utilize oligonucleotides at 100 mg/ml (the approximate protein concentration of cytoplasm) as a substitute for proteins, autonomy would be impossible, no oligonucleotide would be able to ignore its neighbors, and we would have a gel not an organism. The superiority of oligonucleotides as ligands and their defects as concentrated solutes spring from the same sources. Biochemical clarity is, perhaps, the major biological problem solved by the chemistry of proteins relative to oligonucleotides.

CONCLUSIONS AND SOME THOUGHTS ABOUT THE NEAR FUTURE

Studies of the Capacities of Oligonucleotides

We predict that White's ideas about plausibly ancient nucleotides will be tested (8–10), as will ideas about the capacities of the present oligonucleotides. The list of new ribozymes will grow, and they will include chemical activities that seem relevant to the early establishment of creatures. Redox reactions, energetics, partitioning of materials within protected structures (the early membranes), even oligonucleotides that can replicate themselves, and the earliest stages of translation, are all likely to progress in laboratories.

Ligand Improvement

Higher affinity and specificity will come from pre-SELEX modifications to the libraries to include chemical reactivities not found in standard RNA and DNA. For modifications to a library one must use mononucleotide triphosphates that are incorporated into oligonucleotides by RNA or DNA polymerase and those oligonucleotides must be templates for amplification, i.e. the enzymatic criteria for SELEX must be satisfied. Acceptable modifications at the phosphate, the bases, and the sugars include monothioates, pyrimidine adducts at the 5 position, purine adducts at the 8 position, and adducts at the 2′ position of the sugar (44). Similarly, it would be simple to modify a library chemically after enzymatic synthesis but prior to partitioning (to provide new functions within the body or at the ends of each molecule) and/or after partitioning (to allow amplification if the new functions block amplification), thus creating a library that is outside the range of the synthetic and amplification enzymes. Affinities will also be enhanced (and other properties selected) by post-SELEX oligonucleotide modifications, including adducting compounds on the ends of the oligonucleotides. Affinities and specificities will be further enhanced through technologies such as blended-SELEX and covalent-SELEX (see above), strategies in which reactive oligonucleotide substituents are brought close either to binding pockets on a target protein or to reactive atoms on the target for chemical attack. When oligonucleotides are intended for therapeutic and diagnostic uses in blood or urine, $2'$-NH_2 UTP and $2'$-NH_2 CTP can be used in the synthesis of the library so that all molecules in that library (and thus the winners) are resistant to potent human endonucleases (132).

Complex Targets

Most SELEX experiments have been aimed at purified targets, whether proteins, small molecules, or even other nucleic acids (133). In principle and in practice SELEX works against complex targets. Blood clots, wounded arteries, tumor cells, red blood cells, white blood cells, and protein mixtures are each suitable as targets for SELEX. The only requirement for sensible data interpretation is that the mixtures of winning oligonucleotides from any SELEX experiment be sorted into individual clones that can be classified with respect to the molecular target from within the mixture used during the selection.

Biological Regulation

SELEX can be used to answer three questions: 1. If one wants to uncover the role of an arbitrary protein in vivo, one might use SELEX-derived antagonists in much the way that people use antisense oligonucleotides, especially because introduction of an RNA antagonist into cells by gene therapy is similar to the introduction of antisense; 2. If one wants an oligonucleotide that binds tightly enough to the target protein to be informative during structural work, one should push SELEX until winners are found, and those winners will be suitable materials for structural work on the complex; 3. If one wants to know what natural sequences are recognized in vivo by a protein, one might do genomic-SELEX (i.e., create the library from the organism of interest and use SELEX to find the best binders from the precise set of sequences available to the protein in real life).

Genomic-SELEX may prove useful in studies of regulation. If SELEX is done on every new and old protein of interest, using genomic libraries, then complex regulatory pathways that control gene expression—which at present are invisible—may become known. One often purifies a presumptive regulatory protein—often a nucleic acid–binding protein—and wonders about the spectrum of biologically relevant nucleic acid targets. Genomic-SELEX can answer that question, perhaps in even the most complex biological systems in which saturation genetics are unavailable, and in situations in which it is not known if the target in vivo is DNA or RNA. Genomic-SELEX even allows scientists to ignore their bias against the regulatory role of housekeeping proteins such as aconitase (134). We think that every protein may have a high affinity oligonucleotide target in vivo to achieve a subtle regulatory function. This view differs from that of the recent review article on RNA-binding proteins (56). Professional RNA binding proteins contain a variety of conserved protein motifs. Yet, for high affinity, high specificity binding to specific oligonucleotides, we see no evidence that makes these "professionals" any better than the remaining proteins in the biosphere that have other primary functions. Every protein can, possibly, bind tightly and specifically to some oligonucleotide, and genomic-SELEX makes possible the discovery of natural and functionally interesting oligonucleotides.

The work on small molecule targets opens another arena of regulatory biology. Connell and Christian (135) have suggested that oligonucleotides selected for binding to cofactors might be informative regarding metabolism. SELEX has yielded oligonucleotides that bind to the flavin in FMN and FAD (136), as well as ATP and GTP (above). Probably much of metabolism revolves around constant monitoring of intracellular energy charge and redox state, and many RNAs in a cell could participate in that monitoring by binding to small molecules and being altered, for example, in their splicing, translation, or degradation rates. Genomic-SELEX will identify such cellular RNAs easily.

The ultimate arbiter of many SELEX applications is not binding in vitro but rather efficacy in vivo in a cell or an animal. Sequence randomization followed by screening and/or selection in vivo has a substantial history; recent examples include randomization of domains of the HIV genome to find the fittest viral genome (137) and randomization of yeast spliceosomal RNA components to find interactions between those RNAs (138). Randomization is within the framework of classic genetics except that the degree of mutation (and hence the number of mutants analyzed) is immense. SELEX in vitro can be followed by transformation/transfection and in vivo selection; rounds of SELEX may even be performed on oligonucleotides harvested from cells or animals after selection and/or growth.

Applications

SELEX technology will play a role in medical diagnostics, in medical imaging in vivo, in affinity chromatography, and in basic research. Departmental monoclonal antibody facilities might be replaced by an inexpensive, shared SELEX facility, alongside a facility that screens large libraries of antibodies by phage display. After a few years of data collection we will know a great deal about oligonucleotides and antibodies aimed at the same targets. Various reporter compounds (biotin, fluorescent compounds, reactive adducts) can be put on the ends of SELEX-generated oligonucleotides without affecting either their affinity or specificity; this property of oligonucleotides leads to affinity columns in which the compound added to the end of the oligonucleotide is a chromatography matrix. While SELEX usually yields winners that bind to structural epitopes on target proteins, it would be trivial to select oligonucleotides that bind specifically to, for example, SDS-denatured proteins so that protein purification could be followed. We and others are developing oligonucleotide antagonists and agonists to be used directly as drugs. Detection-measurement in vitro and imaging in vivo seem equally likely. Nuclease-resistant oligonucleotides with adducts will probably be useful for therapeutics; some adducts will alter the pharmacokinetic behavior of oligonucleotides, much the way that adducts to proteins have been used (e.g. polyethyleneglycol).

SUMMARY—SELEX TODAY

We see SELEX as a useful tool in the discovery of new drugs and diagnostic agents, as well as for basic research. Even today, after many SELEX experiments, we are still unsure of the limits to the technology. The emergent principle of SELEX that reflects the unexpected properties of oligonucleotides is that ligands with specific and tight binding activity toward a surprising variety of target molecules can be found in large oligonucleotide libraries, as predicted forcefully five years ago (15).

ACKNOWLEDGMENTS

We thank all of our collaborators and friends at NeXstar Pharmaceuticals, Inc. and the University of Colorado. Work at the University was supported by the National Institutes of Health, the National Science Foundation, and the Keck Foundation.

Literature Cited

1. Kim SH, Suddath FL, Quigley GJ, McPherson A, Sussman JL, et al. 1974. *Science* 185:435–40
2. Judson HF. 1979. *The Eighth Day of Creation.* New York: Schuster
3. Bass B, Cech TR. 1984. *Nature* 308: 820–26
4. Michel F, Westhof E. 1990. *J. Mol. Biol.* 216:581–606
5. Noller HF, Hoffarth V, Zimniak L. 1992. *Science* 256:1416–19
6. Gilbert W. 1986. *Nature* 319:618
7. Gold L, Allen P, Binkley J, Brown D, Schneider D, et al. 1993. In *RNA World.* Cold Spring Harbor, New York: Cold Spring Harbor Lab. Press:497–509
8. White HB 3d. 1976. *J. Mol. Evol.* 7:101–4

9. Senkbeil E, White HB 3d. 1978. *J. Mol. Evol.* 11:57–66
10. White HB 3rd. 1982. In *The Pyrimidine Nucleotide Coenzymes,* ed. J Everse, B Anderson, K You, pp. 1–17. New York: Academic
11. Yarus M. 1993. *FASEB J.* 7:31–39
12. Haseloff J, Gerlach NL. 1988. *Nature* 334:585–91
13. Symons RH. 1992. *Annu. Rev. Biochem.* 61:641–71
14. Herschlag D, Eckstein F, Cech TR. 1993. *Biochemistry* 32:8299–311
15. Tuerk C, Gold L. 1990. *Science* 249: 505–10
16. Ellington AD, Szostak JW. 1990. *Nature* 346:818–22
17. Andrake M, Guild N, Hsu T, Gold L, Tuerk C, Karam J. 1988. *Proc. Natl. Acad. Sci. USA* 85:7942–46
18. Oliphant AR, Brandl CJ, Struhl K. 1989. *Mol. Cell. Biol.* 9:2944–49
19. Kinzler KW, Vogelstein B. 1989. *Nucleic Acids Res.* 17:3645–53
20. Beutel BA, Gold L. 1992. J. Mol. Biol. 228:803–12
21. Schneider D, Tuerk C, Gold L. 1992. *J. Mol. Biol.* 228:862–69
22. Bernardi A, Spahr P-F. 1972. *Proc. Natl. Acad. Sci. USA* 69:3033–37
23. Romaniuk P, Lowary P, Wu H-N, Stormo G, Uhlenbeck O. 1987. *Biochemistry* 26:1563–68
24. Irvine D, Tuerk C, Gold L. 1991. *J. Mol. Biol.* 222:739–61
25. Platt T, Richardson JP. 1992. *Transcriptional Regulation,* ed. KR Yamamoto, S McKnight, pp. 365–88. Cold Spring Harbor, New York: Cold Spring Harbor Lab. Press
26. Schneider D, Gold L, Platt T. 1993. *FASEB J.* 7:201–7
27. Boni IV, Isaeva DM, Musychenko ML, Tzareva NV. 1991. *Nucleic Acids Res.* 19:155–62
28. Ringquist S, Jones T, Snyder EE, Gibson T, Bonni I, Gold L. 1995. *Biochemistry:*In press
29. Gold L. 1988. *Annu. Rev. Biochem.* 57:199–233
30. Gold L, Pribnow D, Schneider T, Shinedling S, Singer BS, Stormo G. 1981. *Annu. Rev. Microbiol.* 35:365–403
31. Blumenthal T, Carmichael GC. 1979. *Annu. Rev. Biochem.* 48:525–48
32. Nazarenko IA, Harrington KM, Uhlenbeck OC. 1994. *EMBO J.* 13: 2464–71
33. Peterson ET, Blank J, Sprinzl M, Uhlenbeck OC. 1993. *EMBO J.* 12:2959–67
34. Deleted in proof
35. Deleted in proof
36. Blackwell TK, Weintraub H. 1990. *Science* 250:1104–10
37. Malim MH, Hauber J, Le SY, Maizel JV, Cullen BR. 1989. *Nature* 338:254–57
38. Zapp ML, Green MR. 1989. *Nature* 342:714–16
39. Dayton ET, Powell DK, Dayton AF. 1989. *Science* 246:1625–29
40. Bartel DP, Zapp ML, Green MR, Szostak JW. 1991. *Cell* 67:529–36
41. Tuerk C, MacDougal S, Hertz G, Gold L. 1993. *The Polymerase Chain Reaction,* ed. R Ferre, K Mullis, R Gibbs, A Ross. New York: Birkhauser/Springer-Verlag
42. Giver L, Bartel D, Zapp M, Pawul A, Green M, Ellington AD. 1993. *Nucleic Acids Res.* 21:5509–16
43. Jensen KB, Green L, MacDougal-Waugh S, Tuerk C. 1994. *J. Mol. Biol.* 235:237–47
44. Eaton B, Pieken W. 1995. *Annu. Rev. Biochem.* 64:837–863
45. Deleted in proof
46. Willis MC, Hicke BJ, Uhlenbeck OC, Cech TR, Koch TH. 1993. *Science* 262: 1255–57
47. Dingwall C, Ernberg I, Gait MJ, Green SM, Heaphy S, et al. 1989. *Proc. Natl. Acad. Sci. USA* 86:6925–29
48. Tuerk C, MacDougal-Waugh S. 1993. *Gene* 137:33–39
49. Tuerk C, MacDougal S, Gold L. 1992. *Proc. Natl. Acad. Sci. USA* 89:6988–92
50. Chen H, Gold L. 1994. *Biochemistry* 33:8746–56
51. Deleted in proof
52. Deleted in proof
53. Goff SP. 1992. *Annu. Rev. Genet.* 26: 527–44
54. Levine TD, Gao F, King PH, Andrews LG, Keene JD. 1993. *Mol. Cell. Biol.* 13:3494–504
55. Tsai DE, Harper DS, Keene J. 1991. *Nucleic Acids Res.* 19:4931–36
56. Burd CG, Dreyfuss G. 1994. *Science* 265:615–21
57. Schneider TD, Stormo GD, Gold L, Ehrenfeucht A. 1986. *J. Mol. Biol.* 188: 415–31
58. Nagai K, Oubridge C, Jessen TH, Li J, Evans PR. 1990. *Nature* 348: 515–20
59. Bock LC, Griffin LC, Latham JA, Vermaas EH, Toole JJ. 1992. *Nature* 355: 564–66
60. Macaya RF, Schultze P, Smith FW, Roe JA, Feigon J. 1993. *Proc. Natl. Acad. Sci. USA* 90:3745–49
61. Padmanabhan K, Padmanabhan KP, Ferrara JD, Sadler JE, Tulinsky A. 1993.

Proc. Natl. Acad. Sci. USA 268:17651–54

62. Paborsky LR, McCurdy SN, Griffin LC, Toole JJ, Leung LLK. 1993. *J. Biol. Chem.* 268:20808–11
63. Griffin LC, Tidmarsh GF, Bock LC, Toole JJ, Leung LLK. 1993. *Blood* 81:3271–76
64. Latham JA, Johnson R, Toole JJ. 1994. *Nucleic Acids Res.* 22:2817–22
65. Kubik MF, Stephens AW, Schneider DA, Marlar R, Tassett D. 1994. *Nucleic Acids Res.* 22:2619–26
66. Deleted in proof
67. Vadas P, Waldemar P. 1990. *Phospholipase A2.* New York: Plenum
68. Bennett CF, Chiang M-Y, Wilson-Lingardo L, Wyatt JR. 1994. *Nucleic Acids Res.* 22:3202–9
69. Jellinek D, Lynott CK, Rifkin DB, Janjic N. 1993. *Proc. Natl. Acad. Sci. USA* 90:11227–31
70. Jellinek D, Green LS, Bell C, Janjic N. 1994. *Biochemistry* 33:10450–56
71. Basilico C, Moscatelli D. 1992. *Adv. Cancer Res.* 59:115–65
72. Klagsbrun M, Soker S. 1993. *Curr. Biol.* 3:699–702
73. Folkman J. 1971. *N. Engl. J. Med.* 285:1182–86
74. Mason IJ. 1994. *Cell* 78:547–52
75. Tsai DE, Kenan DJ, Keene JD. 1992. *Proc. Natl. Acad. Sci. USA* 89:8864–68
76. Tillman DM, Nainn-Tsyr J, Hill RJ, Marion T. 1992. *J. Exp. Med.* 176:761–69
77. Deleted in proof
78. Krassa KB, Green LS, Gold L. 1991. *Proc. Natl. Acad. Sci. USA* 88:4010–14
79. Griffiths AD, Williams SC, Hartley O, Tomlinson IM, Waterhouse P, Crosby WL, et al. 1994. *EMBO J.* 13:3245–60
80. Nieuwlandt D, Wecker M, Gold L. 1994. *Biochemistry.* In press
81. Ellington AD, Szostak JW. 1992. *Nature* 355:850–52
82. Sassanfar M, Szostak J. 1993. *Nature* 364:550–53
83. Connell GJ, Yarus M. 1994. *Science* 264:1137–41
84. Lorsch JR, Szostak JW. 1994. *Biochemistry* 33:973–82
85. Yarus M. 1988. *Science* 240:1751–58
86. Hicke BJ, Christian EL, Yarus M. 1989. *EMBO J.* 8:3843–51
87. Connell GJ, Illangesekare M, Yarus M. 1993. *Biochemistry* 32:5497–502
88. Famulok M. 1994. *J. Am. Chem. Soc.* 116:1698–706
89. Limbach PA, Crain PF, McCloskey

JA. 1994. *Nucleic Acids Res.* 22:2183–96

90. Kissinger CR, Liu BS, Martin-Blanco G, Kornberg TB, Pabo CO. 1990. *Cell* 63:579–90
91. Majerfield I, Yarus M. 1994. *Nat. Struct. Biol.* 1:287–92
92. Famulok M, Szostak J. 1992. *J. Am. Chem. Soc.* 114:3990–91
93. Benner SA, Ellington AD, Tauer A. 1989. *Proc. Natl. Acad. Sci. USA* 86:7054–58
94. Lorsch JR, Szostak JW. 1994. *Nature* 371:31–36
95. Hendeles L, Weinberger M. 1983. *Pharmacotherapy* 3:2–43
96. Poncelet SM, Limet JN, Noel JP, Kayaert MC, Galanti L, Collet-Cassart D. 1990. *J. Immunoassay* 11:77–88
97. Jenison RD, Gill SC, Pardi A, Polisky B. 1994. *Science* 263:1425–29
98. Page MI, Jencks WP. 1971. *Proc. Natl. Acad. Sci. USA* 68:1678–83
99. Olivera BM, River J, Clark C, Ramilo CA, Corpuz GP, Abogadie FC. 1990. *Science* 249:217–332
100. Skalicky JJ, Metzler WJ, Ciesla DJ, Galdes A, Pardi A. 1993. *Protein Sci.* 2:1591–603
101. Sakurai T, Yangisawa M, Masaki T. 1992. *Trends Pharmacol. Sci.* 131:103–8
102. Ellington AD. 1994. *Curr. Biol.* 4:427–29
103. Cech TR. 1990. *Annu. Rev. Biochem.* 59:543–68
104. Robertson DL, Joyce GF. 1990. *Nature* 344:467–68
105. Green R, Ellington AD, Szostak JW. 1990. *Nature* 347:406–8
106. Green R, Szostak JW. 1992. *Science* 258:1910–15
107. Beaudry AA, Joyce GF. 1992. *Science* 257:635–41
108. Lehman N, Joyce GF. 1992. *Nature* 361:182–85
109. Lehman N, Joyce GF. 1993. *Curr. Biol.* 3:723–34
110. Joseph S, Berzal-Herranz A, Chowrira BM, Butcher SE, Burke JM. 1993. *Genes Dev.* 7:130–38
111. Berzal-Herranz A, Joseph S, Chowrira BM, Butcher SE, Burke JM. 1993. *EMBO J.* 12:2567–73
112. Berzal-Herranz A, Joseph S, Burke JM. 1992. *Genes Dev.* 6:129–34
113. Pan T, Uhlenbeck OC. 1992. *Biochemistry* 31:3887–95
114. Bartel DP, Szostak JW. 1993. *Science* 261:1411–18
115. Illangesekare M, Sanchez G, Nickles T, Yarus M. 1994. *Science.* In press

116. Lerner RA, Benkovic SJ, Schultz PG. 1991. *Science* 252:659–67
117. Hilvert D. 1993. *Acc. Chem. Res.* 26: 552–58
118. Prudent JR, Uno T, Schultz P. 1994. *Science* 264:1924–27
119. Morris K, Tarasow T, Hilvert D, Gold L. 1994. *Proc. Natl. Acad. Sci. USA* 91:13028–32
120. Iverson BL, Cameron KE, Jahangiri GK, Pasternath DS. 1990. *J. Am. Chem. Soc.* 112:5320–23
121. Gold L, Allen P, Binkley J, Brown D, Schneider D, Eddy S, et al. 1993. See Ref. 7, pp. 497–509
122. Valegard K, Murray JB, Stockley PG, Stonehouse NJ, Liljas L. 1994. *Nature.* In press
123. Cwirla SE, Peters EA, Barrett RW, Dover WJ. 1990. *Proc. Natl. Acad. Sci. USA* 87:6378–82
124. Scott JK, Smith GP. 1990. *Science* 249: 386–90
125. Devlin JJ, Panganiban LC, Devlin PE. 1990. *Science* 249:404–6
126. Needels MC, Jones DG, Tate EH, Heinkel GL, Kochersperger LM, Dower WJ, et al. 1993. *Proc. Natl. Acad. Sci. USA* 90:10700–4
127. Griffiths AD, Williams SC, Hartley O, Tomlinson IM, Waterhouse P, Crosby WL. 1994. *EMBO J.* 13:3245–60
128. Ohlmeyer MHJ, Swanson RN, Dillard LW, Reader JC, Asouline G, Kobayashi R. 1993. *Proc. Natl. Acad. Sci. USA* 90:10922–26
129. Matthews BW, Craik CS, Neurath H. 1994. *Proc. Natl. Acad. Sci. USA* 91: 4103–5
130. Turner DH, Bevilacqua PC. 1993. See Ref. 7, pp. 447–64
131. Hendrick JP, Hartl F-U. 1993. *Annu. Rev. Biochem.* 62:349–84
132. Pieken WA, Olsen DB, Benseler F, Aurup H, Eckstein F. 1991. *Science* 253:314–17
133. Pei D, Ulrich HD, Schultz PG. 1991. *Science* 253:1408–11
134. Klausner RD, Roualt TA, Hartford JB. 1993. *Cell* 72:19–28
135. Connell GJ, Christian EL. 1993. *Origins Life Evol. Biosphere* 23:291–97
136. Burgstaller P, Famulok M. 1994. *Angew Chem. Int. Ed. Engl.* 33:1084–87
137. Berkhout B, Klaver B. 1993. *Nucleic Acids Res.* 21:5020–24
138. Madhani HD, Guthrie C. 1994. *Genes Dev.* 8:1071–86

Annu. Rev. Biochem. 1995. 64:799–835

6-PHOSPHOFRUCTO-2-KINASE/ FRUCTOSE-2,6-BISPHOSPHATASE: A METABOLIC SIGNALING ENZYME

Simon J. Pilkis

Department of Biochemistry, University of Minnesota, Minneapolis, Minnesota 55455

Thomas H. Claus

Institute for Metabolic Disorders, Miles, Inc., West Haven, Connecticut 06516

Irwin J. Kurland

Department of Physiology and Biophysics, State University of New York, Stony Brook, New York 11794-8661

Alex J. Lange

Department of Biochemistry, University of Minnesota, Minneapolis, Minnesota 55455

KEY WORDS: glycolytic/gluconeogenic pathway flux, gene therapy, diabetes, gene expression, phosphorylation

CONTENTS

799

0066-4154/95/0701-0799$05.00

PERSPECTIVE

Fructose 2,6-bisphosphate (Fru-2,6-P$_2$) was discovered during investigations into the mechanism whereby glucagon, via its second messenger cAMP, stimulates gluconeogenesis and inhibits glycolysis in liver. It is a potent activator of the glycolytic enzyme 6-phosphofructo-1-kinase and an inhibitor of the gluconeogenic enzyme fructose-1,6-bisphosphatase, and is thus an important regulatory/signal metabolite that provides a switching mechanism between these two opposing pathways of hepatic carbohydrate metabolism. Fru-2,6-P$_2$ is synthesized from Fru-6-P (fructose-6-phosphate) and ATP by 6PF-2-K (6-phosphofructose-2-kinase), and is degraded to Fru-6-P and inorganic phosphate by Fru-2,6-P$_2$ase. These two opposing enzyme reactions are catalyzed by a single unique protein, 6PF-2-K/Fru-2,6-P$_2$ase. It is one of only five known bifunctional enzymes that catalyze opposing reactions, and is unique because it is the only one whose target is a metabolite rather than a protein.

The general properties of 6PF-2-K/Fru-2,6-P$_2$ase have been described previously in the *Annual Review of Biochemistry* (1). The present review focuses on the 6PF-2-K/Fru-2,6-P$_2$ase gene family, on hormonal control of 6PF-2-K/Fru-2,6-P$_2$ase gene expression, and on structure-function relationships of both the kinase and bisphosphatase domains and the molecular mechanism(s) of their regulation by phosphorylation. At least five tissue-specific mammalian bifunctional enzyme isoforms, encoded by four different genes, have now been identified and sequenced. The properties and regulation of these enzyme forms as well as two yeast isoforms are reviewed. Recent work on the X-ray crystal structure of the Fru-2,6-P$_2$ase domain is also summarized. The reader is referred elsewhere (1–9) for more comprehensive discussions of the role of Fru-2,6-P$_2$ in the regulation of glycolysis and gluconeogenesis.

ISOFORMS OF 6PF-2-K/FRU-2,6-P$_2$ASE

Since the discovery of rat liver 6PF-2-K/Fru-2,6-P$_2$ase (10–12), four other mammalian isoforms have been identified in skeletal muscle (13, 14), heart (15, 16), testis (17), and brain (18). In all of the isoforms, there is a high degree of conservation of the core structure of both the kinase and bisphosphatase domains (Figure 1). For example, the primary sequences of the bovine heart and liver core domains are 86% identical, whereas the NH$_2$- and COOH-terminal regions are only 29% and 42% identical, respectively (19). The major difference between isoforms is the length and composition of the NH$_2$- and/or

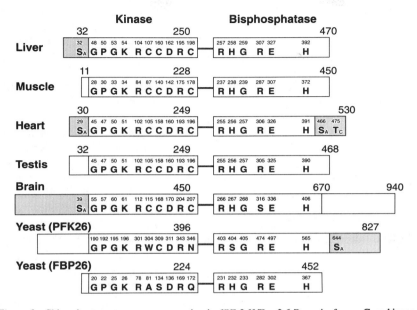

Figure 1 Chimeric core structure conservation in 6PF-2-K/Fru-2,6-P$_2$ase isoforms. Core kinase and bisphosphatase domains are depicted as unfilled boxes containing numbered active-site residues. Stippled boxes represent NH$_2$- or COOH-terminal regions that contain phosphorylation sites; the affected residues and their subscripts denote phosphorylation by cAMP-dependent protein kinase (subscript A) or protein kinase C (subscript C). NH$_2$- or COOH-terminal regions that do not contain phosphorylation sites are shown as unfilled boxes.

COOH-terminal regions and the presence or absence of various protein kinase phosphorylation sites in these regions. The rat liver isoform has a 32-amino-acid NH$_2$ terminus that contains a single cAMP-dependent protein kinase site at Ser32 (20), while the bovine heart isoform has a 30-amino-acid NH$_2$ terminus with a corresponding consensus cAMP phosphorylation site at Ser29 (21). The bovine heart isoform also contains two additional phosphorylation sites in the COOH-terminal region, which extends 60 amino acids beyond that of the liver isoform. The Ser466 site is a cAMP-dependent protein kinase site, while Thr475 is a protein kinase C site (21, 22). A second possible protein kinase C phosphorylation site may be at Ser84 (23). The brain enzyme has extended NH$_2$- and COOH-terminal regions, such that the enzyme is twice the size (110 kDa) of the liver form (55 kDa) (24). It also has a consensus cAMP-dependent protein kinase site just N-terminal to the core kinase domain. Neither the muscle nor the testis isoforms contain documented phosphorylation sites. Introduction of a serine at residue 30 of the testis enzyme by site-directed mutagenesis produced an enzyme that was a substrate for cAMP-dependent protein kinase, and phosphorylation caused reciprocal changes in its two activities (25). The testis isoform is similar in size to the liver enzyme (17), while the muscle isoform has 9 unique NH_2-terminal amino acids in place of the 32 amino acids of the liver form (13).

Two yeast forms of 6PF-2-K/Fru-2,6-P_2ase have also been identified (Figure 1). Yeast PFK26 is similar in size to the brain isoform, with extra coding sequences at both termini (26). It contains a cAMP-dependent protein kinase site just COOH-terminal to the core bisphosphatase domain (26). The other yeast isoform, FBP26, is similar to the muscle isoform, both in size and in lack of phosphorylation sites (27).

In addition to the highly conserved structures of 6PF-2-K/Fru-2,6-P_2ase across tissues, isoforms of tissue-specific 6PF-2-K/Fru-2,6-P_2ase have a high degree of identity across different species. For example, the identity at the amino acid level of the human (28, 29) and rat (30–32) liver enzymes to the bovine liver (19) enzyme is 98.3% and 97%, respectively. The chicken liver enzyme has 89.1%, 88.4%, and 88.6% identity to the human, rat, and bovine enzymes, respectively (33).

Even though their core structures are highly conserved, there are differences in kinase and bisphosphatase properties among the tissue-specific isoforms (Table 1). The rat liver, bovine testis, and bovine brain isoforms have similar kinase and bisphosphatase activities. The major difference among these isoforms is the affinity of the bisphosphatase for Fru-2,6-P_2, which is at least 1000-fold greater for the liver enzyme than for the other two enzymes. The skeletal muscle enzyme has very little kinase activity and high bisphosphatase activity. In addition, the affinity of the kinase for Fru-6-P is much less than that of the liver enzyme. At the other extreme is the bovine heart enzyme, which has little bisphosphatase activity at physiological pH. Also shown in Table 1 are the kinetic properties of a "catalytic core" bifunctional enzyme in which the first 22 NH_2-terminal and last 30 COOH-terminal amino acids were deleted from the liver enzyme, leaving intact the kinase and bisphosphatase catalytic domains (34). This form has low kinase and high bisphosphatase activity. The yeast isoform, PFK26, has only kinase activity because the active-site histidine of the bisphosphatase that corresponds to the rat liver phosphoacceptor, His258, has been replaced by Ser404 (Figure 1). When this Ser residue is replaced by His, bisphosphatase activity is restored (35). The other yeast isoform, FBP26, has high bisphosphatase activity and low kinase activity (27).

Yeast is the most primitive form to possess both 6PF-2-K and Fru-2,6-P_2ase activities (26, 27). Fungi and trypanosomes, but not bacteria, contain measurable levels of Fru-2,6-P_2 and its metabolizing enzymes (36). Bifunctional enzymes are present in amphibian muscle (37) and fish liver (38), and are

Table 1 Comparison of the kinetic properties of various 6PF-2-K/Fru-2.6-P$_2$ase isoforms

	Liver	Liver catalytic core	Skeletal Muscle rat	Heart bovine	Testes bovine	Brain bovine
Size (aa)	470	418	450	570	468	940
Kinase						
V_{max} mU/mg	113	16	5.3	180	90	90
K_m Fru-6-P (μM)	20	753	803	50	16	27
V_{max}/K_m	5.6	0.02	0.007	3.6	5	3.3
Bisphosphatase						
V_{max} mU/mg	45	194	112	2.2	22	29
K_m Fru-2,6-P$_2$ (μM)	0.005	0.01	0.01	40	>1	0.4
V_{max}/K_m	9000	19400	11200	0.055	22	0.4
E-P formation mol/mol	1.0	1.0	1.0	0.05	0.6	0.7
Kinase/ Bisphosphatase	2.5	0.08	0.05	80	4	3

Data from: Liver (196,203,205,206); Liver core (205,206); rat skeletal muscle (196); bovine heart (197,198,200); bovine testis (15); bovine brain (16).

similar to the rat skeletal muscle and liver isoforms, respectively. In plants, where Fru-2,6-P$_2$ plays an important role in carbon partitioning during sucrose synthesis (39), there is evidence for a bifunctional enzyme (40), but some plants may contain a separate Fru-2,6-P$_2$ase (40, 41). With the exception of the yeast enzymes, no amino acid or nucleotide sequences are available for these lower life forms, so their evolutionary relationship to mammalian forms cannot be evaluated.

6PF-2-K/FRU-2,6-P2ASE GENES

The discovery of Fru-2,6-P$_2$ and rat liver 6PF-2-K/Fru-2,6-P$_2$ase (10, 11) was followed by the isolation of the rat gene that encodes the liver enzyme (13, 42) and its cognate cDNA (30–32). In addition to this gene, distinct genes are now known to code for the heart (15, 16), testis (17), and brain (18) isoforms. There are at least two distinct genes that encode the different 6PF-2-K/Fru-2,6-P$_2$ase isoforms in yeast (26, 27). Based on the presence of Fru-2,6-P$_2$, it has been deduced that genes coding for Fru-2,6-P$_2$-metabolizing enzymes also exist in other primitive life forms (36).

The rat gene is approximately 60 kilobases (kb) and is located on the X-chromosome (13, 42, 43). A human gene also has been isolated, and the exon-intron structure of the rat and human genes is completely conserved (44). In addition to the liver mRNA, the rat gene generates both a skeletal muscle and a fetal mRNA by beginning transcription at three different promoters (13, 32, 42, 45). The muscle promoter lies 4.3 kb upstream of the liver promoter and the fetal promoter is located 1.5 kb upstream of the muscle promoter. The processed fetal mRNA differs from the muscle mRNA in that it splices two upstream exons to the muscle first exon to generate an mRNA that is 300 nucleotides longer than the muscle form. The remainder of the gene codes for 13 exons that are common to all three mRNAs produced. Initiation at either the skeletal muscle or fetal promoter generates the skeletal muscle isoform of the enzyme. This is the consequence of the liver first exon (which contains the N-terminal translated region) being spliced out during processing of the skeletal muscle or fetal primary transcripts. This gene is referred to as the liver/skeletal muscle gene.

Six *cis*-acting sequences, which contribute to transcriptional activity and tissue specificity, have been identified in the liver promoter of the rat liver/skeletal muscle $6PF-2-K/Fru-2,6-P_2ase$ gene (46, 47). Nuclear factor I (NF-I) binding to sites I (-43 to -66 bp), II (-78 to -102 bp), and V (-225 to -237 bp) has been demonstrated by DNase I footprinting. Sites III (-112 to -132 bp) and IV (-200 to -216 bp) account for approximately half the promoter activity. Site III binds the ubiquitous Oct-1 factor and HNF-3 in a mutually exclusive way. Site IV binds C/EBP-related factors as well as a liver-specific factor. Site VI (-267 to -283 bp) contains a poly(dG) sequence, which is thought to bind proteins involved in nucleosome positioning (48). Liver-specific DNase I hypersensitivity sites, presumably reflecting their involvement in regulation of transcriptional activity, have been identified in regions of the liver promoter at -200 bp and at -1000 bp (49).

Synthesis of fetal $6PF-2-K/Fru-2,6-P_2ase$ mRNA is controlled by *ets*-related transcription factors and Sp1 (45). These control sites for the fetal promoter were first described as an upstream enhancer region of the muscle promoter, before a distinct fetal mRNA was recognized (50). The muscle promoter was originally characterized as being composed of six binding sites for transacting factors that could stimulate a minimal muscle promoter, but when the promoter is in its native context it could stimulate transcription in L6 myoblasts, but not in myocytes (50).

A second distinct 22-kb gene, located on rat chromosome 10, encodes the heart form of $6PF-2-K/Fru-2,6-P_2ase$ (16, 51). The gene is composed of 16 exons. Although 12 successive exons are similar to exons of the liver/skeletal muscle gene, the heart gene contains dissimilar exons at both the 5′ and 3′ ends, which is consistent with the differences in the NH_2 and COOH termini of the heart, skeletal muscle, and liver forms of the enzyme. At least five mRNAs are encoded by the bovine heart $6PF-2-K/Fru-2,6-P_2ase$ gene: Three code for a 58-kDa protein with variation in the number of amino acids coded by exon 15; one codes for a 54-kDa protein by not using exon 15; and one codes for a truncated form of the enzyme (52). When expressed in *Escherichia coli,* all these forms, with the exception of the truncated form, exhibit 6PF-2-K activity (52).

Although the gene structures have not yet been elucidated, testis and brain isoforms of 6PF-2-K/Fru-2,6-P$_2$ase are also separate gene products. This was deduced from comparisons of the cDNAs with the cDNAs of the liver/skeletal muscle and the heart isoforms (17, 19, 24).

HORMONAL CONTROL OF 6PF-2-K/FRU-2,6-P$_2$ASE GENE EXPRESSION

Hormonal regulation of hepatic 6PF-2-K/Fru-2,6-P$_2$ase gene expression has been studied in vivo. In rats, the amount of liver 6PF-2-K/Fru-2,6-P$_2$ase protein decreased during starvation and diabetes and was restored by refeeding a high-carbohydrate diet or by insulin administration, respectively (32). The increase in enzyme protein after refeeding or insulin administration correlated with an increase in mRNA. Paradoxically, the mRNA abundance in livers of starved and diabetic rats was unchanged from that in liver from normally fed rats (32, 53), which suggests that insulin, in addition to its stimulation of gene transcription, may affect translation and/or processing of the mRNA or degradation of the protein. In diabetic rats, vanadate was shown to act similarly to insulin in restoring levels of 6PF-2-K/Fru-2,6-P$_2$ase content and mRNA, but recovery took 15 days, whereas insulin required only 60 hours (54). The mechanism of the insulin-mimetic effects of vanadate is not known.

Liver 6PF-2-K/Fru-2,6-P$_2$ase protein and its cognate mRNA were both reduced in livers of adrenalectomized rats (55). Subsequent administration of glucocorticoid increased mRNA 90-fold by affecting the transcription rate of the gene. The skeletal muscle isoform mRNA was also increased, albeit to a lesser degree than the liver mRNA (55). Gene expression of the liver enzyme is also depressed in thyroidectomized rats and restored by thyroid hormone administration (56). Intraperitoneal injection of glucagon into normally fed rats decreased the amount of 6PF-2-K/Fru-2,6-P$_2$ase protein and mRNA (57). The effect on mRNA was primarily due to inhibition of the transcription rate, but glucagon may also have a small effect to accelerate mRNA degradation. Changes in hormones also probably account for the decrease in 6PF-2-K/Fru-2,6-P$_2$ase and mRNA in maternal rat liver during the last days of gestation and the subsequent increase after delivery (58), but the mechanism of these effects is unknown.

Bifunctional enzyme gene transcription is also regulated during liver regeneration (59, 60). Immediately after partial hepatectomy, 6PF-2-K/Fru-2,6-P$_2$ase mRNA declined to nearly undetectable levels and then greatly increased during the proliferative phase of liver regeneration. The changes in mRNA abundance were paralleled by changes in gene transcription. There are other examples of possible linkage of 6PF-2-K/Fru-2,6-P$_2$ase gene expression and cell proliferation. The enzyme itself may be a product of a proto-oncogene (61). Furthermore, 6PF-2-K activity is increased upon transformation of chick embryo fibroblasts by retroviruses carrying the *v-src* or the *v-fps,* but the activation did not correlate with an increase in enzyme amount (62, 63). Mitogenic concentrations of insulin, epidermal growth factor, and phorbol esters have been shown to increase Fru-2,6-P$_2$ content and increase rates of glycolysis in a number of cancer as well as normal cells (64, 65), and the reader is referred to extensive reviews of this work (5, 66, 67). In most of these cases, however, bifunctional enzyme mRNA and gene transcription have not been measured during and/or correlated with cell proliferation. Their role, if any, in tumorogenesis and control of tumor cell glycolysis remains to be defined.

Hormonal control of bifunctional enzyme gene expression has also been studied in isolated and cultured cells. In primary cultures of rat hepatocytes, 6PF-2-K/Fru-2,6-P$_2$ase mRNA rapidly declined to undetectable levels in the absence of added hormones (68, 69). Dexamethasone increased mRNA levels, and this effect was blocked by actinomycin D and cycloheximide (68, 69). Insulin or thyroxine had no effect but potentiated the dexamethasone induction (68). Dibutyryl cAMP decreased the stimulatory effect of dexamethasone at early times, but increased mRNA levels, compared with dexamethasone alone, after 24 hours (68). A stimulatory effect of cAMP on 6PF-2-K/Fru-2,6-P$_2$ase mRNA abundance was also observed in isolated hepatocytes from fetal rats (70), whereas cAMP repressed 6PF-2-K/Fru-2,6-P$_2$ase mRNA in hepatocytes from adult rat liver (69, 70). The reason for this difference is unknown. Vanadate also elevates Fru-2,6-P$_2$ in primary cultures of rat hepatocytes (71), perhaps by inhibiting Fru-2,6-P$_2$ase (72).

In the rat hepatoma cell line, FTO-2B, dexamethasone or insulin increased the level of 6PF-2-K/Fru-2,6-P$_2$ase mRNA, and dibutyryl cAMP abolished this induction (73). The induction by insulin and dexamethasone was mediated by enhanced transcription of the gene. The insulin effect, but not the glucocorticoid effect, was abolished when glucose was removed from the media (73).

Based on Northern blot analysis with liver- and muscle-specific probes, the predominant mRNA in FTO-2B cells was shown to contain the muscle exon I but not the liver exon I, yet the size was that of the liver mRNA (73). This is now known to be the fetal mRNA form (45), and it is clear that the hormonal effects can be attributed to this form. The liver isoform mRNA, which is also expressed in these cells at very low levels (74), was induced by dexamethasone, confirming the results in primary hepatocytes (68, 69) and other rat hepatoma cells (73, 75, 76). Its induction by dexamethasone was suppressed by insulin, however (74). What is not clear is whether hormone response elements affected all three promoters of this gene in the same way and whether these effects are related in some way to tissue-specific expression.

The liver form of 6PF-2-K/Fru-2,6-P$_2$ase mRNA is the predominant form produced in another rat hepatoma cell line, FAO-1 (75). As was the case with FTO-2B cells, dexamethasone and insulin induced 6PF-2-K/Fru-2,6-P$_2$ase mRNA, and their effects were antagonized by cAMP. Since the mRNAs in FAO-1 cells and FTO-2B cells are initiated from two distinct promoters, but respond similarly to hormones, their control is probably mediated through common response elements.

In HII4E hepatoma cells, addition of either dexamethasone or insulin increased Fru-2,6-P$_2$ content and 6PF-2-K/Fru-2,6-P$_2$ase protein and mRNA in a time- and dose-dependent manner (76). Addition of both hormones had a synergistic effect on Fru-2,6-P$_2$ content that was due to stimulation of bifunctional enzyme gene expression as well as modulation of the phosphorylation state of the enzyme by insulin. None of these effects were mediated by changes in cAMP. Cyclic AMP–independent increases in 6PF-2-K/Fru-2,6-P$_2$ase protein by insulin have also been demonstrated in isolated hepatocytes, primary cultures of hepatocytes, and in rats subjected to euglycemic clamps (77–80). HTC cells, a more dedifferentiated hepatoma cell line, also express a bifunctional enzyme mRNA, but it has been reported not to be under hormonal regulation (81) even though glycolysis and Fru-2,6-P$_2$ metabolism in these cells have been reported to be regulated by glucocorticoids (82).

The only functional hormone response element identified to date in the rat liver/skeletal muscle 6PF-2-K/Fru-2,6-P$_2$ase gene is a glucocorticoid response element (GRE) located in the intron downstream of the liver first exon (42). This element has been shown to drive glucocorticoid-sensitive transcription in CAT reporter gene constructs using either the muscle or liver promoter of the 6PF-2-K/Fru-2,6-P$_2$ase gene or the weak heterologous thymidine kinase promoter. The element is composed of two tandem dimer hormone receptor–binding sites, each composed of 15 basepairs and separated by 12 basepairs. Deletion of the distal binding site abolished hormone-stimulated CAT activity, whereas deletion of the proximal site had only a slight effect (83). However, mutation of the critical G/C pairs in either the distal or the proximal binding site abolished the hormone-stimulated response. A DNase hypersensitivity site was also mapped to the liver/skeletal muscle bifunctional enzyme GRE site in intron I (49).

The finding that multiple bifunctional enzyme isoforms are encoded by at least four genes has revealed a complexity of isoform expression. While the brain isoform is expressed only in neural tissues (18, 24), there is growing evidence that several isoforms are expressed in a given tissue with one major isoform predominating. For example, liver, skeletal muscle, and heart isoforms and their mRNAs are present in rat liver, but the liver mRNA and isoform predominate (84). Skeletal muscle has also been shown to contain the liver isoform and its mRNA (75, 85). The heart isoform and its mRNA are expressed at a low level in brain (86). The nature of the signals that govern tissue-specific expression and/or promoter selection of the various bifunctional enzyme genes is not known, nor is it known whether there is any physiologic significance to the multiplicity of bifunctional enzyme expression in a given tissue.

STRUCTURE/FUNCTION STUDIES OF 6PF-2-K/FRU 2,6-P₂ASE

6-Phosphofructo-2-Kinase

GENERAL PROPERTIES   The kinase prefers ATP as a phosphate donor, but it can also utilize GTP (87) and (γ-S)ATP (88). The preferred phosphate group acceptor is D-Fru-6-P, although L-sorbose-6-phosphate can also serve as an acceptor (89). The pH optimum is between 8 and 9 (90, 91). The reaction is not inhibited by ATP (87, 92–94), in contrast to the case of mammalian 6PF-1-Ks. The kinase also catalyzes exchange reactions between ATP and ADP and between Fru-6-P and Fru-2,6-P₂ (90), suggesting a covalent mechanism of catalysis. However, analysis of the stereochemical course of the reaction was consistent with transfer of the phospho group between the two substrates without a phosphoenzyme intermediate (95), and no intermediate has been detected.

The kinase reaction is inhibited by both of its products, ADP and Fru-2,6-P₂ (94, 96). The former is competitive with ATP, while the latter is noncompetitive with either substrate. This pattern is consistent with a sequential-ordered mechanism where ATP binds first followed by Fru-6-P. Inorganic phosphate (91, 97, 98) and arsenate (97) increase the affinity of the kinase for Fru-6-P without affecting the K_m for ATP. Other effectors of the enzyme include citrate (91, 99), P-enolpyruvate (91, 99), and glycerol-3-P (94, 100) as inhibitors and AMP as a possible activator (91, 92).

Experiments with reagents that modify sulfhydryl groups have shown that cysteine residues are essential for kinase catalysis as well as for the Fru-6-P/Fru-2,6-P₂ exchange reaction (101, 102). Even reagents that target nucleotide-binding sites (e.g. FSBA) were found to interact with the reactive sulfhydryl groups (103). The enzyme subunit contains a total of 10 cysteine residues, but only four (Cys107, Cys160, Cys183, and Cys198) are essential for 6PF-2-K activity (102, 103).

STRUCTURAL ANALOGY TO 6PF-1-K: HOMOLOGY MODELING   Various structures of ATP-binding proteins were examined as models for the 6PF-2-K topology. The NH₂-terminal domain of the bacterial 6PF-1-K family (from *B. stearother-*

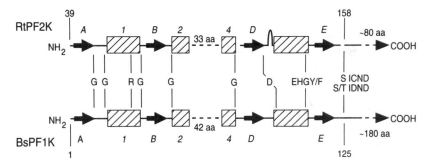

Figure 2 Partial sequence alignment of rat 6-phosphofructo-2-kinase (RtPF2K) domain with a fragment of *B. stearothermophilus* 6-phosphofructo-1 kinase (BsPF1K). Numbering schemes for both rat liver kinase (*top*) and BsPF1K (*bottom*) are used. Solid arrows, β strands labeled A–E: hatched boxes, helices labeled numerically. Conserved residues in both structures are shown. The residue in RtPF2K that corresponds to the catalytic residue Asp127 in BsPF1K is Cys160.

mophilus and *E. coli* (104–106) provided the most convincing framework for the rat liver kinase domain (107). Bacterial 6PF-1-Ks are homologous to the NH₂- and COOH-terminal halves of mammalian 6PF-1-Ks; gene duplication is postulated to have driven the evolution of the larger mammalian enzymes (108). Although no mammalian 6PF-1-K protein structures have been solved, several sequences from rabbit muscle and mouse and rat liver were available that aided in the modeling of rat liver 6PF-2-K (109–111).

When the nucleotide binding fold (nbf) pattern in rat liver 6PF-2-K was aligned with spatially equivalent 6PF-1-K residues from both bacterial and mammalian sequences, there was a similarity in the number of observed and predicted α helices and β strands (107) (Figure 2). Residues that were invariant, or closely conserved, in the alignment play important structural or functional roles in the 6PF-1-Ks (105, 106). Figure 2 also shows a partial sequence alignment of the two enzymes. Five absolutely conserved glycines mark the turns following strand A, helix 1, strand B, and helix 4 in the first half of the nbf. The second half of the nbf (past strand D, spatially paired with strand A) showed convincing sequence similarity between the two kinases: For example, 6PF-2-K residues [Glu148-His-Gly-Phe-(6 amino acids, or aa)-Ser-Ile-Cys-Asn-Asp162] matched the *B. stearothermophilus* 6PF-1-K residues [Glu114-His-Gly-Phe-(7aa)-Thr-Ile-Asp-Asn-Asp129] (105, 106). The latter motif, which marks the turn following strand E, is highly conserved and forms part of the active site of all 6PF-1-Ks (106). These residues are also predicted to form part of the active-site cavity of 6PF-2-K with Cys160 of 6PF-2-K placed in a catalytically sensitive position. The corresponding residue in *E. coli* 6PF-1-K, Asp-127, serves as a base catalyst in the reaction (106, 112).

Although not shown in Figure 2, the sequence and proposed structural similarity between 6PF-2-K and the 6PF-1-Ks extend past strand F in bacterial 6PF-1-K to approximately amino acid 201 in rat liver 6PF-2-K. Strand F of 6PF-1-K is the first strand of a second parallel β-sheet domain that is in spatial opposition to the NH$_2$-terminal, larger α/β fold (107). The smaller, four-β-strand domain of 6PF-1-K extends over approximately 100 residues before the protein chain rejoins the fold of the larger domain (105, 106). The larger domain of 6PF-1-K has been shown to contribute most of the residues that bind ATP, whereas the sugar phosphate is contacted by residues from both the large and small domains (106).

Although the last approximately 45 residues of 6PF-2-K show little overall sequence identity with the corresponding last 100 residues of the 6PF-1-K fold, there is a sequence-similar pattern (Asn-Arg230-Val-Gln-Asp-His-Val-Glu-Ser-Arg238) to a region marking the end of the small domain in 6PF-2-K (Thr-Arg243-Val-Thr-Leu-Val-Gly-His-Gln-Arg252). The Arg230 and Arg238 residues in this sequence are conserved in five 6PF-1-Ks (107).

MODEL-AIDED SITE-DIRECTED MUTAGENESIS STUDIES.

Catalytic residues   The alignment shown in Figure 2 suggested that the Cys160 in rat liver 6PF-2-K may correspond to the base catalyst, Asp127, of the bacterial 6PF-1-Ks. Mutation of Asp127 of bacterial 6PF-1-K to Ser resulted in a 10,000-fold decrease in V$_{max}$ (112). If Cys160 functions as a weak base catalyst, its mutation to the potentially strong base catalyst, Asp, might increase 6PF-2-K activity, but instead there was a 10-fold decrease (113; IJ Kurland, SJ Pilkis, unpublished results). Mutation of the corresponding cysteine residue (Cys138) in the muscle isoform of the 6PF-2-K domain to Asp or Ser did not decrease V$_{max}$ significantly (114). Crepin et al (114) concluded that this cysteinyl residue did not serve as a base catalyst and was not in the 6PF-2-K active site. Kurland et al (113) changed this same skeletal muscle residue to Ala and Gln as well as to Ser. The presence of Ser or Gln resulted in a threefold increase in V$_{max}$, supporting the notion that this residue is not functioning as a base catalyst. However, mutation to Ala resulted in an enzyme without significant kinase activity, which suggested that Cys138 plays a significant structural role in the active site (113).

In support of a critical structural role for this cysteinyl residue, alignment of the liver, heart, brain, testis, and yeast (PFK26) 6PF-2-K/Fru-2,6-P$_2$ases indicates that of all the cysteinyl residues shown to be important for kinase activity in mammalian 6PF-2-Ks, only Cys138 is conserved in the yeast PFK26 isoform, which is principally a kinase (26). Cys138 is not conserved in yeast FBP26 (Figure 1), which is principally a bisphosphatase (27). Furthermore, the skeletal muscle Cys138Gln mutant, in addition to having a threefold greater

V_{max}, had a fourfold increase in its K_m for ATP, which suggests that this residue may lie near the active site.

In addition to Asp127, several other residues from the active site of bacterial 6PF-1-K play a catalytic role (105, 106). The model predicts that Asp140, Asp108, and Arg82 of muscle 6PF-2-K may also be important in catalysis (107, 115), but mutations of these residues have not been reported. Crepin et al (114) did mutate other potential base catalysts in skeletal muscle 6PF-2-K, including Glu157 and Asp162 to Asp and Ser, respectively, but observed no effect on V_{max}. Asp162 corresponds to Asp129 in bacterial 6PF-1-K, which is essential for catalysis and might bind Mg^{2+} (105, 106).

A consideration of the relative turnover numbers of 6PF-2-K and 6PF-1-Ks also supports the hypothesis that the 6PF-2-K reaction may proceed without the intervention of a strong base catalyst. The specific activity of liver 6PF-2-K (50 mU/mg) is three orders of magnitude less than that of mammalian 6PF-1-K (100 U/mg), while the V_{max} of the bacterial 6PF-1-K Asp127Ser mutant is roughly equivalent to that of the native skeletal muscle and liver 6PF-2-Ks. Catalysis by 6PF-2-K may be mediated solely by preferential binding of the substrates in the transition state compared to the ground state, similar to the mechanism of stabilization of enzyme transition states by catalytic antibodies (116, 117).

Fructose-6-phosphate-binding site Based on homology modeling to bacterial 6PF-1-K, the arginine residues of 6PF-2-K—Arg195, Arg230, and Arg238— were predicted to be involved in Fru-6-P binding (107). Mutation of Arg195 to Ala or His had no effect on V_{max} of the kinase, but increased the K_m of Fru-6-P more than 3000-fold and the K_a of phosphate by 100-fold (118). Mutation of the adjacent basic residue, Lys194, to Ala had no effect on V_{max} or the K_m for Fru-6-P. Mutation of either Arg230 or Arg238 to Ala increased the K_m for Fru-6-P by 2–3-fold but, more significantly, also increased the K_m for ATP by 30–40-fold. These results indicate that Arg195 is a critical residue for the binding of Fru-6-P, and that this interaction is highly specific, as mutation of the adjacent Lys194 to Ala had no effect on Fru-6-P binding. This residue also apparently plays an important role in phosphate binding. The other two arginine residues, Arg230 and Arg238, which were shown to be important Fru-6-P-binding residues in bacterial 6PF-1-K (119–121), are involved principally in ATP binding in 6PF-2-K (118).

In addition to these site-directed mutagenesis studies, Rider et al (122) identified Arg225 as a residue that binds Fru-6-P, based on protection by Fru-6-P against inactivation by phenylglyoxal. The specificity of the chemical modification reaction is questionable, however. Along these same lines, Kitamura et al (123) isolated a peptide, after incubation with N-bromo [^{14}C] acetylethanolamine-P, which was labeled on Cys107. They postulated that this was a Fru-6-P-binding residue because Fru-6-P partially prevented kinase inactivation. It seems more likely that Cys107, like Cys160, helps maintain the proper conformation of the active site.

In the chicken liver enzyme, different residues may be used to bind Fru-6-P. Chemical modification and pH kinetic results indicate that lysine, rather than arginine, residues are essential for Fru-6-P binding in this isoform (33). One candidate is Lys200, which is a glutamic acid in the mammalian isoforms (19). The Fru-6-P concentration dependence of the chicken liver enzyme is hyperbolic in the absence of P_i (33, 124), whereas the rat liver isoform exhibits cooperativity (1, 4, 32, 97).

ATP-binding site   The observation that the amino acid sequence Gly48-Leu-Pro-Ala-Arg-Gly53-Lys-Thr in the kinase domain of 6PF-2-K was similar to that of other nbf signature sequences led to the model based on homology to bacterial 6PF-1-Ks (107). Site-directed mutagenesis of residues within the 6PF-2-K sequence confirmed that it was indeed a nbf signature sequence. For example, mutation of Gly48 to Ala in the rat liver enzyme completely eliminated kinase activity without affecting bisphosphatase activity (118), and the same results were obtained when Pro50 or Gly53 was mutated to Ala (125). Other residues, besides those of this sequence, must also be involved in nucleotide binding, because mutation of Arg230 and Arg238 to Ala reduced the affinity for ATP by 30–40-fold (118).

These two arginyl residues are conserved in the avian liver enzyme. ATP binding to this enzyme exhibited negative cooperativity (33, 124) however, whereas the rat liver enzyme did not (99, 126, 127). This difference may be due to a different ATP signature sequence (Gly46-Leu-Arg-Arg-Pro-Gly-Lys-Thr53) than is found in all the mammalian liver isoforms (Figure 2). Another difference between the avian and rat liver isoforms is the observation that Mg^{2+} ions inhibited the former but had no effect on the latter enzyme (33, 124). The presence of two aspartate residues in the avian liver enzyme (Asp160 and Asp161) may be responsible for this difference (33).

Phosphate-binding site   Bacterial 6PF-1-Ks are subject to regulation by many allosteric effectors (128–133) but have a single effector site that binds the activators ADP, GDP, and P_i, as well as the inhibitor, P-enolpyruvate (134, 135). Of these effectors, only P_i has a striking effect on 6PF-2-K (97, 98). Alignment of 6PF-2-K with the 6PF-1-Ks predicts that many of the effector-site residues that are necessary for binding the adenine, ribose, or α-phosphate moieties of ADP are absent in 6PF-2-K (107, 118). Also absent is the major intersubunit contact helping to shape the ADP effector site in 6PF-1-K (118). Indeed, analysis of the 6PF-2-K structure precludes the existence of allosteric sites for ATP, AMP, or ADP, a finding consistent with the lack of stimulatory effects of ADP on 6PF-2-K (94, 115, 126). The residues involved in P_i binding to the effector site in 6PF-1-K are the same as those that bind the β-phosphate of ADP, and these residues are conserved in 6PF-2-K (107, 118). Thus, it seems reasonable to postulate that P_i affects 6PF-2-K allosterically by its interaction with what remains of the bacterial effector site (118). The residues at this site have not yet been identified by functional studies, but have been predicted by homology modeling to be Leu43, Arg63, Asp96, Ala131, Thr132, and Lys187 (115, 118).

Inorganic phosphate binds to the Fru-6-P substrate site as well as the effector site in bacterial 6PF-1-Ks (134–136). The same appears to occur in 6PF-2-K, since mutation of a major Fru-6-P-binding-site residue, Arg195, to Ala increased the K_a for phosphate 100-fold (118). Thus, Fru-6-P may bind to one of the terminal nitrogen groups of Arg195 and P_i to the other to affect Fru-6-P binding and/or Fru-2,6-P_2 release (118). The mechanism of this effect is unclear, largely because mutation of Arg195 affects both Fru-6-P affinity and activation by P_i.

Summary Figure 3 illustrates the topology of bacterial 6PF-1-K and the proposed partial topology of the 6PF-2-K domain. Also shown are residues whose functional role in 6PF-2-K has been identified, along with the corresponding residues in 6PF-1-K. The identity of the 6PF-2-K residues was made possible by the use of homology modeling with bacterial 6PF-1-Ks and site-directed mutagenesis of key residues. These studies have provided a better understanding of the molecular mechanism whereby 6PF-2-K catalyzes its reaction, and have revealed both similarities and differences between 6PF-2-K and 6PF-1-K.

The greatest similarity between the two kinases is the presence of structurally analogous nbfs. The most important difference is the presence of a strong base catalyst (Asp127) in 6PF-1-K and the absence of one in 6PF-2-K, which probably accounts for the low turnover number of 6PF-2-K. The other major difference involves the small subdomain of the kinases. In 6PF-1-K, this domain contains three basic residues, Arg162, Arg243, and Arg252, that are essential for Fru-6-P binding (119–121). In 6PF-2-K, the corresponding residues are Arg195, Arg230, and Arg238. Arg195 is an essential residue for Fru-6-P binding, which confirms the model of Bazan et al (107). In fact, it appears to be more important for binding than is the corresponding Arg162 in 6PF-1-K, since mutation of Arg195 had a much greater effect on Fru-6-P affinity (118) than did mutation of Arg162 (119, 120). In contrast, mutation of Arg230 and Arg238 had little effect on Fru-6-P affinity, but instead dramatically reduced the affinity of 6PF-2-K for ATP. This result was surprising, because it was predicted that most of the ATP-binding residues would be located in the larger subdomain by analogy to 6PF-1-K. Furthermore, neither

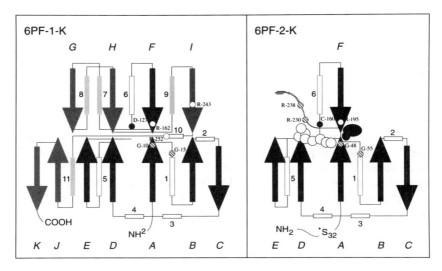

Figure 3 Comparison of the topology of 6PF-1-K based on the 6PF-1-K crystal structure, with the proposed (partial) topology for the 6PF-2-K domain. The β-strand and α-helical regions are shown as filled arrows and unfilled boxes, respectively, and they are labeled so that the structural elements in both enzymes have corresponding designations. The amino acid residues in the β strands and α helices are as reported (36). The structural elements not conserved in 6PF-2-K are stippled. Residues that bind ATP are shown as stippled circles and those that bind Fru-6-P are shown as unfilled circles. The base catalyst, Asp127, in 6PF-1-K and the corresponding Cys160 in 6PF-2-K are shown as filled circles.

mutation of Arg243 nor mutation of Arg252 in bacterial 6PF-1-Ks had any effect on ATP affinity (119–121). It appears that the COOH-terminal ~ 45 amino acids of the 6PF-2-K fold provide residues that are important for ATP binding, which is not true for the longer ~ 100 amino acids that constitute the four-β-strand, smaller domain of bacterial 6PF-1-Ks. Clearly, the 6PF-2-K domain forms a more economical—and structurally and functionally unrelated—counterpart to the 6PF-1-K small-domain excursion (Figure 3).

Fructose-2,6-Bisphosphatase

GENERAL PROPERTIES   Fru-2,6-P$_2$ase specifically hydrolyzes phosphate from the C-2 position of Fru-2,6-P$_2$; it does not hydrolyze Fru-1,6-P$_2$ or glucose-1,6-P$_2$, nor does it hydrolyze phosphate from the C-6 position (10, 137, 138). The reaction does not require divalent cations (137), has a pH optimum of 5.5–6.5 (90, 137, 139), and is inhibited by both the substrate, Fru-2,6-P$_2$, and the product of the reaction, Fru-6-P (137, 139). In the presence of a Fru-6-Pdepleting system, the K_m for Fru-2,6-P$_2$ is 5 nM, and micromolar substrate concentrations inhibit the reaction (139). Inorganic phosphate and glycerol-3-phosphate inhibit Fru-2,6-P$_2$ase activity at subsaturating substrate concentrations by antagonizing substrate binding, and thus increase the apparent K_m for Fru-2,6-P by 20-fold (94, 139). At saturating substrate concentrations, P$_i$ and glycerol-3-P activate Fru-2,6-P$_2$ase by antagonizing product binding and substrate inhibition and thus increase the K$_i$ for Fru-6-P (85, 139).

Catalysis of the bisphosphatase reaction involves the formation and breakdown of 3-phosphohistidine on the enzyme subunit (139, 140). The formation and breakdown of the phosphoenzyme are sufficiently fast for it to be a reaction intermediate, and the steady-state phosphoenzyme level correlates well with the hydrolytic rate (139). Recently, ^{31}P NMR confirmed the presence of 3-phosphohistidine in the actively turning over enzyme and suggested that this steady-state E-P intermediate exists primarily as the E-P•Fru-6-P complex (141). The reaction scheme for the Fru-2,6-P$_2$ase reaction is given below.

```
Fru-2,6-P₂                              Fru-6-P    H₂O                    Pᵢ
   ↓                                       ↑        ↓                     ↑
───────────────────────────────────────────────────────────────────────────
E        E•Fru-2,6-P₂        E-P•Fru-6-P       E-P      E-P•H₂O   E•Pᵢ   E
                             Fru-2,6-P₂ → ↓
                                    E-P•Fru-2,6-P₂
```

STRUCTURAL ANALOGY TO PHOSPHOGLYCERATE MUTASES: HOMOLOGY MODELING   The Fru-2,6-P$_2$ase domain of 6PF-2-K/Fru-2,6-P$_2$ase was modeled on the structure of yeast phosphoglycerate mutase (PGM), based on similarity of the sequences surrounding the phosphohistidine in both enzymes (107) (Figure 4). The yeast PGM structure is a typical three-layer α/β fold: a largely parallel β-sheet core sandwiched by α helices (142, 143). The active site lies in a crevice at the COOH-terminal end of the β sheet, a site topologically favored in structures similar to nbfs (144). PGM does not bind nucleotides, yet assumes an nbf-like α/β structure favorable for binding phosphoglycerates and for spatial grouping of catalytic residues far apart in sequence. In particular, His179 is juxtaposed with the phosphorylated His8, and the two histidine residues form a "clapping hands" structure. This approximately 170-amino-acid spacing between active-site histidines is closely reproduced in the related human muscle and brain PGMs (145–147), and the human (148), rabbit (149), and mouse (150) bisphosphoglycerate mutases (BPGM).

The phosphohistidine in Fru-2,6-P$_2$ase, His258, is located only eight residues from the presumed NH$_2$ terminus of the domain (amino acid 251). The companion to His258 is His392. This assignment was the result of a secondary structure study that located similar α and β secondary elements in Fru-2,6-P$_2$ase to those found in yeast PGM, or that were predicted to occur in other

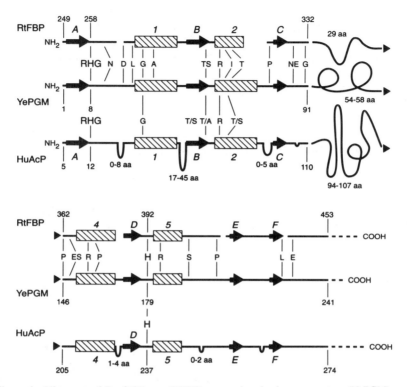

Figure 4 Alignment of Fru-2,6-P2ase (RtFBP), yeast phosphoglycerate mutase (YePGM), and human prostatic acid phosphatase (HuAcP). The yeast numbering system is followed. β-strand and α-helical regions are shown as arrows and boxes, respectively. Conserved residues in each structure are also shown. Different lengths of surface loops in each structure refer to varying lengths found in the PGM or AcPase family.

homologous mutases (107), aided by the known three-dimensional framework of the yeast enzyme (143) (Figure 4). Thus, the spacing between the two His residues is 134 amino acids rather than the 170 in yeast PGM. Complete structural alignment of these proteins required introduction of several gaps in the shorter Fru-2,6-P$_2$ase sequence that mapped to exposed surface loops in yeast PGM. The largest gap localized to a long meandering loop in the mutase between core strand C and helix 4 (Figure 4). Residues 333–361 mapped to a 25-amino-acid longer chain of yeast PGM (residues 92–145) that represents a large excursion from the core α/β mutase fold (143). This region of PGM is almost devoid of conventional secondary structure and forms a loosely packed lobe that overhangs the active-site crevice (143, 151). The analogous, but shorter, excursion in Fru-2,6-P_2ase probably adopts a different conformation than in yeast PGM and is therefore not in structural alignment with the mutase.

A crucial structural idiosyncrasy in helix 4 (which precedes His179) of the yeast PGM structure is conserved in rat Fru-2,6-P_2ase: The helix is "kinked" by Pro160, allowing packing of the long helix to the curled β-sheet core (143). The analogous helix 4 in Fru-2,6-P_2ase is bent by Pro376, an identity that determines the position of the helices of α and β strands in the COOH-terminal half of the bisphosphatase and corroborates the choice of His392 as the second active-site histidine.

MODEL-AIDED SITE-DIRECTED MUTAGENESIS STUDIES

Catalytic residues   The conservation of the Arg-His-Gly motif in the PGM and Fru-2,6-P_2ase families (152), the demonstration of a functional homology of Fru-2,6-P_2ase, PGM, and 2,3 BPGM (153), and the observation that His258 was labeled upon incubation with [2-^{32}P]Fru-2,6-P_2 (139, 140) all pointed to His258 being the phosphoacceptor in rat liver Fru-2,6-P_2ase. The importance of this residue in the rat liver bisphosphatase reaction was confirmed when it was mutated to Ala, and the mutant protein was devoid of Fru-2,6-P_2ase activity and did not form a phosphoenzyme intermediate (154). His392 also plays an important role in phosphoenzyme formation, since its mutation to Ala decreased activity by 50–100-fold and decreased the rate of phosphoenzyme formation 1500-fold (154). A third important residue, Glu327, which is conserved in all known Fru-2,6-P_2ase and PGM structures, is predicted to be in close proximity to His392. Mutation of Glu327 to Ala, Gln, or Asp also reduced activity by as much as 50-fold and decreased phosphoenzyme formation by at least 1000-fold (155).

Substrate- and product-binding sites   Residues 333–361, which were predicted to form a loosely packed loop overhanging the active site (107), contain eight glutamic, two aspartic, and four basic residues (Arg352, Lys356, Arg358, and Arg360) that are clustered at the COOH-terminal end. Mutation of Lys356 to Ala increased both the K_m for Fru-2,6-P_2 and the K_i for Fru-6-P 3000-fold without affecting the K_i for P_i (156). In addition, this mutant did not exhibit substrate inhibition. Mutation of Arg352 to Ala produced the same changes as the Lys356Ala mutant (157). These results indicate that this surface loop participates in forming the substrate/product-binding site by contacting the C-6 phosphoryl group of Fru-2,6-P_2/Fru-6-P, and that the Fru-6-P-binding site is, at least in part, responsible for substrate inhibition and P_i activation (157).

The third basic residue of this loop, Arg360, contributes to binding the C-2 phosphoryl group of Fru-2,6-P_2 to the free enzyme. Mutation of this residue to Ala increased the K_m for Fru-2,6-P_2 by 10-fold and the K_i for P_i by 12-fold without affecting V_{max} or the K_i of Fru-6-P (157). This residue also contacts the C-2 phosphoryl group of Fru-2,6-P_2 in the E-P•Fru-2,6-P_2 complex, since the Arg360Ala mutation greatly diminished substrate inhibition (157).

Mutation of the fourth basic residue in the surface loop, Arg358, had no effect on enzyme activity, which indicates that this residue plays no role in binding and/or catalysis (157). This result indicates that the dramatic changes in the kinetic properties of the other three basic residue mutants are highly specific, and that the spatial orientation of the side chains of these residues is critical for efficient contacts with either the C-2 or C-6 phosphoryl groups of Fru-2,6-P_2.

The yeast PGM model identified two additional basic residues as important sites for substrate binding in rat liver Fru-2,6-P_2ase (Figure 4). Both Arg257 and Arg307 are conserved in all known Fru-2,6-P_2ases and PGMs (107), and the corresponding residues in yeast PGM (Arg7 and Arg59) have been postulated to bind phosphoglycerate (158). Mutation of Arg257 and Arg307 to Ala increased both the K_m for Fru-2,6-P_2 and the K_i for P_i, but had no effect on the K_i for Fru-6-P (159). Therefore, these two residues contact the C-2 phosphoryl group of Fru-2,6-P_2. The two mutations had opposite effects on V_{max}, however; the V_{max} of Arg257Ala was 11-fold higher than that of the native enzyme, whereas V_{max} of Arg307Ala was 700-fold lower.

The only exception to conservation of the Arg residue corresponding to Arg307 of rat liver Fru-2,6-P_2ase is the substitution of this Arg by a Ser residue in the bovine brain isoform (24). The K_m for Fru-2,6-P_2 of this isoform is 70 μM (18), which is nearly 2000-fold higher than that for the rat liver enzyme. Mutation of Arg307 to Ala in the liver bisphosphatase also greatly increased the K_m for Fru-2,6-P_2 (159).

Nucleotide-binding site   Fru-2,6-P_2ase does not hydrolyze nucleoside triphosphates (160, 161), but the active site is predicted to reside in a crevice at the COOH-terminal end of a β sheet, a site topologically favored in nbf structures for binding nucleotides and other negatively charged ligands (107). The beginning of this region is marked by the sequence Gly274-Leu-Ser-Ala-Arg-Gly279-Lys-Gln, a pattern similar to nbf "signature" sequences. Lee et al (160) reported that the separately expressed bisphosphatase domain was activated by ATP and GTP at saturating substrate concentrations, but was inhibited competitively by these nucleoside triphosphates and by guanosine at subsaturating Fru-2,6-P_2 concentrations. These results suggest that the active site of the bisphosphatase domain can also bind nucleotides; the results demonstrate that the previously reported site of nucleoside triphosphate activation of the partially purified rat liver enzyme is the bisphosphatase domain itself and not the kinase domain (see 161 and 162 for review).

The pattern of inhibition/activation by nucleoside triphosphates is identical to that observed for P_i, which acts as a competitive inhibitor with Fru-2,6-P_2 and which also overcomes substrate inhibition by preventing the formation of an E-P•Fru-2,6-P_2 inhibitory complex (156, 157). Competitive inhibition by guanosine and the lack of activation by this nucleoside at saturating substrate concentrations support the notion that the nucleoside interacts with the enzyme's nbf, and that the effect of GTP/GDP/ATP to activate the enzyme is mediated by their phosphate group(s). The effects of GTP/ATP/GDP to overcome substrate inhibition suggested that their phosphate groups contact Lys356 and/or Arg360, residues shown to bind Fru-2,6-P_2 in the E-P•Fru-2,6-P_2 inhibitory complex (157). Site-directed mutagenesis experiments revealed that Arg360, but not Lys356, is involved in this interaction, and along with binding studies and ^{31}P NMR spectroscopy provide the first direct evidence for nucleoside triphosphate binding to the Fru-2,6-P_2ase active site (160).

Additional evidence for nucleotide binding to the Fru-2,6-P_2ase family comes from studies on a phosphatase involved in assembly of the nucleotide loop of cobalamin in *Salmonella typhimurium* (163). The phosphatase is a polypeptide of 26 kDa with homology to PGM and to eukaryotic Fru-2,6-P_2ase enzymes. The characteristic Arg-His-Gly motif is conserved, as is the second active-site histidine and other catalytic and substrate binding-site residues.

Summary   Studies on the structure/function relationships of the Fru-2,6-P2ase have revealed a number of functional analogies to catalysis by serine proteases. Both enzyme families catalyze their reactions via a catalytic triad. The serine proteases provide a general base, a His residue, that can accept the proton from the hydroxyl group of the reactive Ser, thus facilitating formation of the covalent tetrahedral transition state, which is stabilized by the negatively charged group of Asp (164, 165). Three residues, His258, His392, and Glu327, constitute the catalytic triad in the Fru-2,6-P2ase reaction mechanism. The hydrolysis reaction catalyzed can be divided into two half reactions. The first half reaction involves nucleophilic attack at the C-2 phosphoryl group of Fru-2,6-P2 by His258 to form a phosphoenzyme intermediate. The negatively charged carboxyl group of Glu327 stabilizes the protonated state of His392 during phosphoenzyme formation (Figure 5).

The second half reaction of Fru-2,6-P2ase is initiated by the nucleophilic attack of the phosphoenzyme intermediate by a water molecule (Figure 5). The leaving group (Fru-6-P) is highly basic, making it a nucleophile that is capable of reverse nucleophilic attack. In order to facilitate the forward reaction, it is necessary to convert the leaving group to a less basic form by donating a proton to it from His392. A similar mechanism occurs in the serine proteases.

Replacement of any one of the catalytic triad residues in the active site of Fru-2,6-P_2ase caused only a 50–100-fold reduction in V_{max} (154, 155), whereas it caused a 10^4–10^6-fold reduction in the protease, subtilisin (166). This dif-

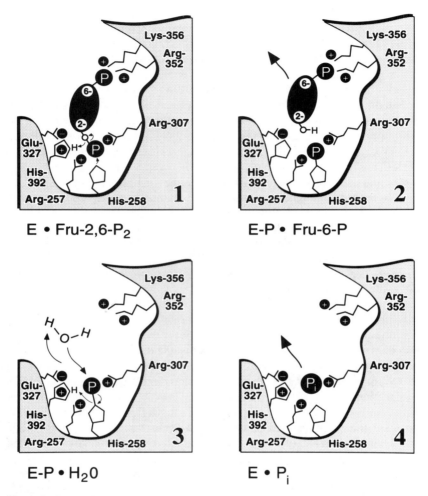

Figure 5 Schematic depiction of Fru-2,6-P2ase catalysis. Fru-2,6-P2 is represented in cartoon fashion with emphasis on the C-2 and C-6 phospho groups. *Step 1:* During E-P formation at His258, His392 acts as a proton donor to the oxygen on the C-2 phosphoester bond to enhance phosphoester bond cleavage and E-P formation (not shown). The C-2 phospho group is bound tightly by Arg257 and Arg307, while the C-6 phospho group is bound by Lys356. *Step 2:* After C-2 phospho group transfer to His258, the C-6 phospho group of Fru-6-P remains bound to Lys356 and Arg352. The negatively charged carboxylate group of Glu327 acts to maintain His392 in its protonated form. *Step 3:* During E-P hydrolysis, Glu327 may act as a base catalyst by affecting the ionization of H2O. *Step 4:* Phosphate is released from Arg257 and Arg307.

ference is not surprising, since Fru-2,6-P$_2$ase has a turnover number of only 6 min^{-1}, whereas that of subtilisin is several orders of magnitude higher. The turnover number of Fru-2,6-P$_2$ase has been shown to be equal to the rate constant for hydrolysis of the phosphohistidine intermediate, and this rate is only about 1000-fold greater than that of the uncatalyzed hydrolysis of phosphohistidine (167). Therefore, it is unlikely that a mutant that can still form an E-P intermediate can suffer more than a 1000-fold reduction in rate. Furthermore, the 50–100-fold reduction in V$_{max}$ of the Fru-2,6-P$_2$ase mutants is significant, since these mutants catalyze rates of phosphohistidine hydrolysis that are only 20–10-fold greater, respectively, than the uncatalyzed rate.

Even though Fru-2,6-P$_2$ase has a very low turnover number, its high affinity for Fru-2,6-P$_2$ makes it a highly efficient enzyme. The catalytic efficiency, defined as k$_{cat}$/K$_m$, of the enzyme is 0.2×10^8 M^{-1}sec^{-1}, which is close to the diffusion control limit (168). Why has mammalian Fru-2,6-P$_2$ase been chosen to optimize binding, rather than catalysis, during evolution? It is reasonable to argue that by using this strategy, it is possible to maintain Fru-2,6-P$_2$ at a low level, and hence to conserve energy by preventing a high degree of substrate cycling at this step. Since Fru-2,6-P$_2$ functions specifically as a regulator of glycolysis and gluconeogenesis, it would be advantageous to the cell, based purely on energy considerations, to maintain a low Fru-2,6-P$_2$ level, and yet still be able to regulate its level efficiently by covalent modification of the enzyme.

Arginine-257 and Arg307 may be residues that are functionally analogous to those in the "oxyanion hole" of serine proteases (164) that are responsible for tight binding and stabilization of the transition state. These two arginines are located close to the active-site histidines (107, 159), but they play different roles in this interaction. Arginine-257 binds substrate in the ground state more strongly than in the transition state, while the opposite is true for Arg307. This differential interaction can be attributed to differences in geometry and electron distribution of the substrate between the two states (169, 170). The formation of the E-P transition state may result in reduced stabilizing interaction between Arg257 and the phospho group, but the gain in stabilizing interaction between Arg307 and the phospho group is apparently enough to compensate for the loss.

The residues involved in the catalytic triad (His258, Glu327, and His392) and the oxyanion hole or tight binding-site residues (Arg257 and Arg307) of rat liver Fru-2,6-P$_2$ase are not unique to this enzyme, but are conserved in the two yeast Fru-2,6-P$_2$ases as well as the yeast PGMs and AcPases (Table 2). They are also conserved in the S. typhimurium phosphatase. On the other hand, the three residues involved in the substrate specificity pocket of the liver enzyme (Arg352, Lys356, and Arg360) are less highly conserved, being found in all mammalian Fru-2,6-P$_2$ases but not in the yeast PGM or AcPases or the

S. typhimurium phosphatase (Table 2). In the latter enzyme, the residues corresponding to Arg352, Arg356, and Arg360 are Trp134, Trp138, and Val142. These residues would be predicted to be involved in forming hydrogen bonds and/or van der Waal's interactions with the dimethylbenzimidazole group of α-ribazole-phosphate. The presence of these hydrophobic residues in the *S. typhimurium* phosphatase surface loop region provides indirect, but strong, support for the loop acting as a specificity pocket in this enzyme as well.

RELATIONSHIP TO ACID PHOSPHATASES   Acid phosphatases (AcPases) catalyze the phosphoryl transfer to water or to some alcohols (171, 172). Unlike alkaline phosphatases, which utilize a Ser residue (171–173), the AcPases proceed through the transient phosphorylation of a His residue (174–177). This functional similarity with the catalytic mechanism of PGMs and Fru-2,6-P_2ases (178, 179) suggested that there may be a similarity.

A gapped alignment of the five complete (and one partial) AcPase sequences showed that the yeast, *E. coli,* and human enzymes were distantly related (107, 180–184) (Figure 4). The regions surrounding the two histidine residues, typically about 200–260 amino acids apart in the AcPases, showed a significant degree of similarity to the active-site His residues in the mutase/bisphosphatase alignment. The NH_2-terminal His8/258 of the mutase/bisphosphatase fold, embedded in a conserved Arg-His-Gly-Glu/Gln motif, matched that of an analogous NH_2-terminal His residue in a similar Arg-His-Gly-Glu/Asp motif (Figure 4).

Analysis of AcPase sequences showed that predicted AcPase core α and β secondary structural elements matched those in yeast PGM. In addition, significant sequence similarity was observed in regions distinct from the active-site His residues. For example, the sequence pattern surrounding the conserved Arg59 in yeast PGM, Tyr-Thr-Ser-X_3-Arg-Val-X-Gln-Thr-Ala, is closely matched by the AcPase pattern, Phe/Tyr-Thr-Ser/Ala-X_3-Arg-Val-X-Asp-Thr/Ser-Ala (107). The difference in chain length between proposed active-site His residues in the AcPase and mutase/bisphosphatase sequences mapped to two excursions from the core α/β mutase fold (Figure 4). The first AcPase excursion is a loop, distal from the predicted active site, between helix 1 and strand B. The second segment mapped to the main region of variability in the mutase/bisphosphatase, the 54-residue excursion of yeast PGM between strand C and helix 5. These results suggest that the AcPases share a common protein fold with yeast PGM, and that they catalyze their reactions via a phosphoenzyme intermediate at the NH_2-terminal His residue. *E. coli* glucose-1-phosphatase may also be a member of the AcPase family (185). However, mammalian microsomal glucose-6-phosphatase, which catalyzes its reaction via a phosphoenzyme intermediate (186), does not belong to this family (187).

BIFUNCTIONAL ENZYME   823

Table 2 Conservation of key amino acids in various members of the Fru-2,6-P₂ase/PGM/acid phosphatase family

Attribute/Function	Rat liver Fru-2,6-P$_2$ase	Yeast PFK26	Yeast FBP26	PGM	Yeast AcPase	S. typhimurium Phosphatase
Molecular size (aas)	220	241	216	248	297	234
Catalytic triad						
Phosphoacceptor	His258	Ser404	His232	His8	His57	His40
Affects protonation state of proton donor His residue	Glu327	Glu497	Glu302	Glu86	Glu182	Glu110
Proton donor	His392	His565	His367	His179	His319	His152
Tight binding sites						
Stabilizes ground state; interacts with C-2 phospho group	Arg257	Arg403	Arg231	Arg7	Arg56	Arg39
Stabilizes transition state; interacts with C-2 phospho group	Arg307	Arg474	Arg282	Arg59	Arg157	Arg90
Surface loop specificity pocket						
Loop size (aas)	29	29	31	54–58	94–107	29
Interacts with C-6 phospho group	Arg352	Arg522	Arg327	Basic residues present	No alignment of basic residues	Trp134
Interacts with C-6 phospho group	Lys356	Lys526	Lys331			Trp138
Interacts with C-2 phospho group	Arg360	Arg530	Arg335			Val142

Recently, the X-ray crystal structure of rat AcPase was elucidated, and shown to contain a PGM-like protein fold (188). In addition, site-directed mutagenesis of the *E. coli* AcPase active-site His residues greatly diminished activity, as did mutation of the arginyl groups that correspond to Arg257 and Arg307 in rat liver Fru-2,6-P$_2$ase (189). No residue analogous to Glu327 of Fru-2,6-P$_2$ase was found in three of six AcPases, however, including the *E. coli* enzyme (190). In contrast, an alignment of the six sequences revealed a conserved Asp (Asp303 in the *E. coli* AcPase) (190). Mutation of Asp303 to alanine decreased V$_{max}$ by as much as 1300-fold. It was concluded that Asp303, rather than His304, is involved in proton donation to the substrate leaving group, while His304, mutation of which also caused a large decrease in V$_{max}$ (189), may be involved in the maintenance of the phosphohistidine protonated state. Similarly, Asp258 rather than His257 may be the proton donor in rat prostatic AcPase (191).

CRYSTAL STRUCTURE OF FRU-2,6-P2ASE   Crystallization of the rat liver Fru-2,6-P2ase domain was facilitated by the finding that a 30-amino-acid COOH-ter-

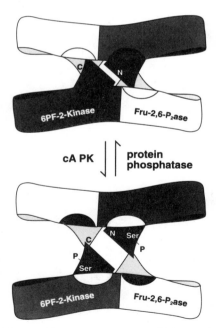

Figure 6   Schematic depiction of the postulated NH$_2$- and COOH-terminal interactions of hepatic 6PF-2-K/Fru-2,6-P2ase and their role in mediating the effect of cAMP-dependent phosphorylation on the enzyme. The active site of each domain is indicated by the semicircle and the activity by the shading: dark, high activity; and light, low activity.

minal truncation mutant (residues 251–440) could be expressed in high yield in *E. coli* and easily purified to homogeneity (160, 192). This mutant had kinetic properties essentially identical to the bisphosphatase of the intact enzyme, except that its V_{max} was 5–10-fold higher. This bisphosphatase domain was crystallized in the presence of inorganic phosphate and either polyethyleneglycol or isopropanol in two different space groups, $P2_1$ (193) and P_1 (192), respectively. A seleno-methionine-substituted bisphosphatase was crystallized under similar conditions, and the structure was solved by MAD phasing (194). The structure contained a PGM-fold. The active site had the characteristic pairing of two histidyl residues, with their imidazole rings forming a "clapping hands" structure about 4 Å apart. Glu327 was within 3 Å of His392. As was also predicted by the model, the surface loop region overhangs the active site. The crystal structure therefore confirms the model of the Fru-2,6-P_2ase domain proposed by Bazan et al (107) based on the PGM structure.

Regulation of 6PF-2-K/Fru-2,6-P₂ase by Phosphorylation/Dephosphorylation

EFFECT ON DIFFERENT ISOFORMS   The rat liver isoform of 6PF-2-K/Fru-2,6-P_2ase was the first form of the bifunctional enzyme shown to be regulated by cAMP-dependent protein kinase–catalyzed phosphorylation (10, 11, 20, 195, 196). The enzyme is phosphorylated on Ser32, and the phosphorylation site (Arg-Arg-Arg-Gly-Ser) is a classic cAMP-dependent consensus sequence, except that there are three basic residues NH_2-terminal to the phosphorylated serine (20). The effect of phosphorylation on 6PF-2-K is to increase the K_m for Fru-6-P by 20–30-fold (197). This effect is pH dependent; at physiological pH values (7.0–7.4), the major effect is to increase the K_m for Fru-6-P without affecting the K_m for ATP. There is also a decrease (50–65%) in the V_{max} (197). At higher pH values (8.0–8.5), phosphorylation has been reported either to lower the V_{max} dramatically (91), or to have little effect (197–199). Phosphorylation increases the V_{max} of Fru-2,6-P_2ase by 2–4-fold (20, 139) by enhancing the breakdown of the E-P intermediate, presumably by increasing the rate of dissociation of Fru-6-P from the E-P•Fru-6-P complex (139). Phosphorylation has no effect on the K_m for Fru-2,6-P_2.

The rat liver enzyme is the only one of the five tissue-specific mammalian isoforms in which both activities are affected by phosphorylation. This is presumably also the case for the chicken liver enzyme, which has the same phosphorylation site sequence as the rat liver enzyme (33), but activity changes have not yet been reported. With the exception of the heart isoform (21, 22), the other mammalian isoforms either lack phosphorylation sites [muscle (200, 201) and testis (202) isoforms] and/or their activities are unaffected by phosphorylation [brain (18) isoform].

The 58-kDa form of bovine heart 6PF-2-K/Fru-2,6-P$_2$ase contains two phosphorylation sites (Ser466 and Thr475) located just COOH-terminal to the Fru-2,6-P$_2$ase domain (21). Cyclic AMP–dependent phosphorylation at Ser466 activates the kinase by decreasing the K_m for Fru-6-P by 50% (21, 22). No effects of phosphorylation on Fru-2,6-P$_2$ase activity have been detected, perhaps because activity is very low at physiological pH (22, 203). No changes in 6PF-2-K activity were detected after phosphorylation of the enzyme by protein kinase C at Thr475 (22). The 54-kDa form does not contain the COOH-terminal phosphorylation sites, but both forms have a consensus cAMP-dependent phosphorylation site in the NH$_2$-terminal region at Ser29 (21, 22), which may be phosphorylated in vitro (22). Rat heart 6PF-2-K/Fru-2,6-P$_2$ase is a substrate for both cAMP-dependent protein kinase and the Ca^{2+}/calmodulin-dependent protein kinase (204). Both protein kinases catalyze phosphorylation of Ser466 and Ser483, which decreases the K_m for Fru-6-P by 50%.

The only other 6PF-2-K/Fru-2,6-P$_2$ase isoform that undergoes a cAMP-dependent phosphorylation-induced activity change is the yeast PFK26 isoform. Phosphorylation at the consensus phosphorylation site (RRYS, Ser644) at the COOH-terminal end of the Fru-2,6-P$_2$ase domain causes activation (205). This is analogous to what occurs with the heart enzyme.

STRUCTURAL BASIS FOR REGULATION OF 6PF-2-K/FRU-2,6-P$_2$ASE BY PHOSPHORYLATION   The NH$_2$- and COOH-terminal regions of 6PF-2-K/Fru-2,6-P$_2$ases are responsible for both the structural (Figure 1) and kinetic (Table 1) differences between isoforms. Since these regions are also the sites of phosphorylation, they must be responsible for any phosphorylation-induced kinetic changes as well. Phosphorylation of liver 6PF-2-K/Fru-2,6-P$_2$ase affects both kinase and bisphosphatase activity, even though the phosphorylation site is located prior to the beginning (residue 39) of the kinase domain. Therefore, the NH$_2$ and COOH termini must communicate with each other. In order to address the mechanism whereby cAMP-dependent phosphorylation reciprocally affects liver kinase and bisphosphatase activity, point mutations at Ser32 and deletion mutants at both the NH$_2$ and COOH termini of the enzyme have been engineered.

Mutation of Ser32 to Asp mimicked the effects of phosphorylation of the enzyme (197). The V$_{max}$ of the kinase was decreased, the K_m for Fru-6-P was increased, and the bisphosphatase V$_{max}$ increased to about the same extent as in the phosphorylated native enzyme. Thus, the phosphorylation effects are due to introduction of a negatively charged group. Phosphorylation of a Ser32Ala mutant on the adjacent Ser33 did not affect activity (197), which suggests that the spatial orientation of the phosphoserine group is important for transmission of the effect.

Deletion of the first 22 amino acids from the liver enzyme, to create a mutant equal in length to the native skeletal muscle isoform, but that still retained the cAMP-dependent phosphorylation site, increased the K_m of the kinase for Fru-6-P to a value (140 µM) that was closer to that of the muscle (335 µM) than the liver (7 µM) enzyme (199). The kinase V_{max} was only slightly reduced, but the V_{max} of the bisphosphatase was increased 4–5-fold. Phosphorylation of this mutant produced changes in kinase activity similar to the native liver enzyme: a decrease in V_{max} and an increase in the K_m for Fru-6-P, but the change in the latter was much less (3-fold) than in the native enzyme (12-fold) (199). Phosphorylation had no significant effect on the elevated bisphosphatase activity.

Deletion of the last 30 amino acids from the COOH terminus of rat liver 6PF-2-K/Fru-2,6-P_2ase reduced the kinase V_{max} by 50%, increased the K_m for Fru-6-P by 2-fold, and increased the V_{max} of the bisphosphatase 9-fold (206). Phosphorylation of this mutant further decreased the kinase V_{max} and further increased the K_m for Fru-6-P by 3-fold, but had no effect on the elevated bisphosphatase activity.

The extended COOH-terminal region of the heart isoform also appears to interact with the bisphosphatase domain to inhibit activity and/or prevent E-P formation. When the entire human heart Fru-2,6-P_2ase domain was expressed in bacteria, it produced an enzyme that was devoid of activity and did not form a phosphoenzyme intermediate. However, truncation of this form to a size that corresponds to the liver Fru-2,6-P_2ase domain resulted in an active enzyme that did form a phosphoenzyme intermediate (207).

Combining the NH_2- and COOH-terminal deletions of the rat liver enzyme in one mutant created a 418-amino-acid catalytic core, which still retained its cAMP-dependent phosphorylation site (208). This mutant had bisphosphatase kinetic properties and a decrease in kinase V_{max} similar to those of the COOH-terminal truncation mutant, as well as a 35-fold increase in K_m for Fru-6-P (Table 1). Phosphorylation of the NH_2- and COOH-truncated mutant had no further effect on either the kinase or bisphosphatase activity (208, 209), which strongly supports the hypothesis that the effects of phosphorylation are mediated by interactions of the NH_2 and COOH termini of the protein.

Based on these considerations, it has been postulated that the NH_2- and COOH-terminal regions of liver 6PF-2-K/Fru-2,6-P_2ase interact with the active sites of the kinase and bisphosphatase domains, respectively (206, 208, 209). But they also affect activity of the opposite domain. For example, deletion of the COOH-terminal region increased the K_m for Fru-6-P of the kinase and decreased its V_{max}, while deletion of the NH_2-terminal region increased bisphosphatase activity (199, 208, 209). In fact, deletion of either region activated the bisphosphatase more (7–9-fold) than did phosphorylation of the native enzyme (2–3-fold) (206, 208). Deletion of either 24 or 30 amino acids from the NH_2 terminus of the testis isoform also decreased the affinity for Fru-6-P, decreased kinase V_{max}, and increased bisphosphatase activity 2-fold (202).

Studies on the expression of the two domains of the enzyme in bacteria also provide support for interaction between the NH_2- and COOH-terminal regions. When the bisphosphatase domain was expressed, it behaved as a 26-kDa monomer and retained the same activity as found in the native bifunctional enzyme (210). When the kinase domain was expressed, it aggregated into a tetrameric structure with abnormal kinetic properties (211). These results suggest that subunit-subunit interactions are necessary for kinase activity, but not for the bisphosphatase. They are consistent with the NH_2- and COOH-terminal deletional studies, and suggest that NH_2- and COOH-terminal regions of the native enzyme are responsible for maintaining the dimeric structure, with the N-terminal region of one subunit interacting with the COOH-terminal region of the other subunit.

A model has been developed to explain the effect of phosphorylation on the liver bifunctional enzyme (Figure 6). In the dephosphorylated homodimer, the NH_2- and COOH-terminal regions are postulated to interact with their own respective active sites on the same subunit to produce intrasteric inhibition of the bisphosphatase and intrasteric activation of the kinase. In addition, these effects depend on NH_2- and COOH-terminal interactions between the two subunits, which are arranged in an antiparallel configuration. Deletion of either the NH_2- or COOH-terminal regions, or both regions, leads to disruption of these interactions, and results in activation of the bisphosphatase and, in the case of the NH_2terminal deletion, a decrease in kinase activity. The intrasteric inhibition is analogous to that documented for myosin light chain kinase and twitchin kinase (212–214). Phosphorylation by cAMP-dependent protein kinase, by introduction of a negative charge at Ser32, is postulated to partially disrupt these interactions, leading to inhibition of kinase activity and activation of bisphosphatase activity (Figure 6, page 824).

Dephosphorylation of the hepatic bifunctional enzyme is catalyzed by protein phosphatases 2A and 2C (215). Recent evidence suggests that insulin enhances the activity of a protein phosphatase, presumably phosphatase 2A, to dephosphorylate the liver bifunctional enzyme via a cAMP-independent mechanism (76, 77, 79).

Regulation of 6PF-2-K/Fru-2,6-P₂ase by ADP-Ribosylation

Hepatic 6PF-2-K/Fru-2,6-P_2ase is ADP-ribosylated by arginine-specific ADP-ribosyl transferase (216). All three arginine residues in the cAMP-dependent phosphorylation site sequence can undergo ADP-ribosylation, but it occurs preferentially at Arg29 and Arg30. ADP-ribosylation of 6PF-2-K/Fru-2,6-P_2ase blocked its phosphorylation by cAMP-dependent protein kinase and decreased its kinase activity and recognition by liver-specific anti-6PF-2-K/Fru-2,6-P$_2$ase antibodies, but did not alter Fru-2,6-P$_2$ase activity. ADP-ribosylation may constitute an additional posttranslational regulatory mechanism of hepatic 6PF-2-K/Fru-2,6-P$_2$ase, but its physiological relevance remains uncertain. An example of what occurs is the case of liver regeneration, during which the enzyme did not respond to glycerol-3-phosphate, was not affected by cAMP-dependent phosphorylation, and was not recognized by a liver-specific antibody (217).

FUTURE DIRECTIONS

Although much has been learned about 6PF-2-K/Fru-2,6-P$_2$ase in the past decade, there are still many unanswered questions. Determination of the crystal structure of the intact bifunctional enzyme, in both the phosphorylated and dephosphorylated states, will permit elucidation of the molecular details of how the kinase and bisphosphatase domains interact with each other and—most importantly—how phosphorylation at a single serine reciprocally affects the enzyme's activities. This mechanism may be even more complex than that described for phosphorylase, the only other mammalian enzyme for which the molecular details of the effect of phosphorylation on enzyme structure are known (218, 219).

Predictive structural studies of the rat liver bifunctional enzyme suggested an evolutionary lineage with two glycolytic enzymes, 6PF-1-K and PGM, which have been confirmed by site-directed mutagenesis studies. When it was proposed that the kinase domain was structurally analogous to that of 6PF-1-K, it was pointed out that the similarity was probably due to convergence at the active site of two enzymes that catalyze the same reaction and bind the same substrates (107). The relationships between the independent kinase and bisphosphatase domains of 6PF-2-K/Fru-2,6-P$_2$ase and other enzymes will likely result in a convincing phylogenetic branching for the divergence of PGM-fold proteins, which include AcPases and the bacterial phosphatase involved in nucleotide loop closure in cobalamin synthesis. Comparison of the crystallographic structure of Fru-2,6-P$_2$ase with yeast PGM, AcPases, and eventually the *S. typhimurium* phosphatase should yield another measure of evolutionary divergence. Identification of monofunctional Fru-2,6-P$_2$ases, predicted to be present in plants and yeast, will reveal new information on the evolution of the structure and function of 6PF-2-K/Fru-2,6-P$_2$ase. Even without such information, it is clear that the structural design of 6PF-2-K/Fru-2,6-P$_2$ase is a paradigm for bifunctional proteins.

The various forms of the bifunctional enzyme appear to have evolved by adaptation of their NH$_2$- and COOH-terminal sequences to meet the metabolic exigencies of a particular tissue. Different bifunctional enzyme genes probably arose by gene duplication, followed by the evolution of mechanisms for tissue-specific expression. An active area of future research will be to elucidate these mechanisms.

In mammalian cells, gene expression of the liver/skeletal muscle form is under multihormonal control. Glucocorticoid is the most important regulatory hormone, and its response element is located in the first intron of the gene. A fertile area of research in the future will be elucidation of the molecular mechanism whereby insulin and cAMP modulate bifunctional enzyme gene expression at the transcriptional level. Elucidation of transcription regulation will require identification of the positive insulin and negative cAMP response elements in the gene and the proteins that interact with them. Once these protein factors are identified, the important question of "regulation of the regulators" can then be addressed.

Finally, the use of the bifunctional enzyme as a model system for metabolic pathway engineering of the glycolytic and gluconeogenic pathways can be used to determine the relative importance of covalent modification and gene expression in the regulation of glycolytic/gluconeogenic pathway flux (220, 221). In addition, the ability to construct mammalian expression vectors for expression of this and other key regulatory enzymes will provide a foundation for future attempts at gene therapy for disease states, such as diabetes and metabolic enzyme deficiency states, in which glucose metabolism is deranged.

Any *Annual Review* chapter, as well as any article cited in an *Annual Review* chapter, may be purchased from the Annual Reviews Preprints and Reprints service.
1-800-347-8007; 415-259-5017; email: arpr@class.org

Literature Cited

1. Pilkis SJ, El-Maghrabi MR, Claus TH. 1988. *Annu. Rev. Biochem.* 57:755–83
2. Hers HG, Hue L. 1983. *Annu. Rev. Biochem.* 52:617–53
3. Claus TH, El-Maghrabi MR, Regen DM, Stewart HB, McGrane M, Kountz PD, et al. 1984. *Curr. Top. Cell Regul.* 23:57–86
4. Pilkis SJ, ed. 1990. *Fructose-2,6-Bisphosphate*. Boca Raton, FL: CRC Press. 272 pp.
5. Hue L, Rider H, Rousseau GG. 1990. See Ref. 4, pp. 173–92
6. Pilkis SJ, El-Maghrabi MR, Claus TH. 1990. *Diabetes Care* 13:582–99
7. Pilkis SJ, Claus TH. 1991. *Annu. Rev. Nutr.* 11:465–515
8. Pilkis SJ, Granner D. 1992. *Annu. Rev. Physiol.* 54:885–909
9. Rousseau GG, Hue L. 1993. *Prog. Nucleic Acid Res. Mol. Biol.* 45:93–127
10. El-Maghrabi MR, Pilkis J, Fox E, Claus TH, Pilkis SJ. 1982. *J. Biol. Chem.* 257:7603–7
11. El-Maghrabi MR, Fox E, Pilkis J, Pilkis SJ. 1982. *Biochem. Biophys. Res. Commun.* 106:794–802
12. Lively MO, El-Maghrabi MR, Pilkis J, D'Angelo G, Colosia AD, et al. 1988. *J. Biol. Chem.* 263:839–49
13. Darville MI, Crepin KM, Hue L, Rousseau GG. 1989. *Proc. Natl. Acad. Sci. USA* 86:6543–47
14. Kitamura K, Uyeda K, Kangawa K, Matsuo M. 1989. *J. Biol. Chem.* 264:9799–806
15. Kitamura K, Uyeda K. 1987. *J. Biol. Chem.* 262:679–81
16. Tsuchiya Y, Uyeda K. 1994. *Arch. Biochem. Biophys.* 310:467–74
17. Sakata J, Abe Y, Uyeda K. 1991. *J. Biol. Chem.* 266:15764–70

18. Ventura F, Rosa JL, Ambrosio S, Pilkis SJ, Bartrons R. 1992. *J. Biol. Chem.* 267:17939–43
19. Lange AJ, El-Maghrabi MR, Pilkis SJ. 1991. *Arch. Biochem. Biophys.* 290:258–63
20. Murray K, El-Maghrabi MR, Kountz P, Lukas T, Soderling TR, Pilkis SJ. 1984. *J. Biol. Chem.* 259:7673–81
21. Kitamura K, Kangawa K, Matuso H, Uyeda K. 1988. *J. Biol. Chem.* 263:16796–801
22. Rider MH, Van Damme J, Lebeau EL, Vertommen D, Vidal H, Rousseau GG, et al. 1992. *Biochem. J.* 285:405–11
23. Rider MH, Van Damme J, Vertommen D, Michael A, Vanderkerckhove J. 1992. *FEBS Lett.* 310:139–42
24. Ventura F, El-Maghrabi MR, Lange AJ, Pilkis SJ, Bartrons R. 1995. *Biochem. Biophys. Res. Commun.* In press
25. Abe Y, Uyeda K. 1994. *Biochemistry* 33:5766–71
26. Kretschmer M, Fraenkel DG. 1991. *Biochemistry* 30:10663–72
27. Kretschmer M, Paravicini G, Fraenkel DG. 1992. *Biochemistry* 31:7126–33
28. Lange AJ, Pilkis SJ. 1990. *Nucleic Acids Res.* 18:3652
29. Lange AJ, Li L, Vargas AM, Pilkis SJ. 1993. *J. Biol. Chem.* 268:9466–72
30. Colosia AD, Lively M, El-Maghrabi MR, Pilkis SJ. 1987. *Biochem. Biophys. Res. Commun.* 143:1092–98
31. Darville MI, Crepin KM, Vandekerckhove J, Van Damme J, Octave JN, et al. 1987. *FEBS Lett.* 224:317–21
32. Colosia AD, Marker A, Lange AJ, El-Maghrabi MR, Granner DK, et al. 1988. *J. Biol. Chem.* 263:18668–77
33. Li L, Lange AJ, Pilkis SJ. 1993. *Biochem. Biophys. Res. Commun.* 190:397–405
34. Kurland IK, Chapman B, El-Maghrabi MR, Pilkis SJ. 1995. *J. Biol. Chem.* In press
35. Kretschmer M, Langer C, Prinz W. 1993. *Biochemistry* 32:11143–48
36. Van Schaftingen E, Mertens E, Opperdoes FR. 1990. See Ref. 4, pp. 229–44
37. Pyko M, Rider MH, Hue L, Wegener G. 1993. *J. Comp. Physiol.* 163:89–98
38. DeFrutos PG, Baanante IV. 1994. *Arch. Biochem. Biophys.* 308:461–68
39. Stitt M. 1990. *Annu. Rev. Plant Physiol. Plant Mol. Biol.* 41:153–85
40. Larondelle Y, Mertens E, Van Shaftingen E, Hers H-G. 1986. *Eur. J. Biochem.* 161:351–57
41. MacDonald FD, Cseke C, Chou Q, Buchanan BB. 1987. *Proc. Natl. Acad. Sci. USA* 84:2742–46
42. Lange AJ, Espinet C, Hall R, El-Maghrabi MR, Vargas AM, et al. 1992. *J. Biol. Chem.* 267:15073–80
43. Olson S, Uyeda K, McBride OW. 1989. *Somat. Cell Mol. Genet.* 15:617–21
44. Lange AJ, Pilkis SJ. 1995. *Genomics.* In press
45. Dupriez VJ, Darville MI, Antoine IV, Gegonne A, Ghysdael J, Rousseau GG. 1993. *Proc. Natl. Acad. Sci. USA* 90:8224–28
46. Lemaigre FP, Durviaux SM, Rousseau GG. 1991. *Mol. Cell. Biol.* 11:1099–106
47. Lemaigre FP, Durviaux SM, Rousseau GG. 1993. *J. Biol. Chem.* 268:19896–905
48. Lewis CD, Clark SP, Felsenfeld G, Gould H. 1988. *Genes Dev.* 2:863–73
49. Zimmermann LN, Rousseau GG. 1994. *Eur. J. Biochem.* 220:183–91
50. Darville MI, Antoine IV, Rousseau GG. 1992. *Nucleic Acids Res.* 20:3575–83
51. Darville MI, Chikri M, Lebeau E, Hue L, Rousseau GG. 1991. *FEBS Lett.* 288:91–94
52. Vidal H, Crepin KM, Rider MH, Hue L, Rousseau GG. 1993. *FEBS Lett.* 330:329–33
53. Crepin KM, Darville MI, Hue L, Rousseau GG. 1988. *FEBS Lett.* 227:136–40
54. Mialpeix M, Bartrons R, Crepin K, Hue L, Rousseau GG. 1992. *Diabetologia* 33:243–48
55. Marker AJ, Colosia AD, Lange AJ, Tauler A, El-Maghrabi MR, Pilkis SJ. 1989. *J. Biol. Chem.* 264:7000–4
56. Wall SR, Von den Hove MF, Crepin KM, Hue L, Rousseau GG. 1989. *FEBS Lett.* 257:211–14
57. Rosa JL, Ventura F, Tauler A, Bartrons R. 1993. *J. Biol. Chem.* 268:22540–45
58. Casado M, Bosca L, Martin-Sanz P. 1993. *Endocrinology* 133:1044–50
59. Rosa JL, Tauler A, Lange AJ, Pilkis SJ, Bartrons R. 1992. *Proc. Natl. Acad. Sci. USA* 89:3746–50
60. Ventura F, Rosa JL, Ambrosio S, Gil J, Bartrons R. 1991. *Biochem. J.* 276:455–60
61. Hue L, Rousseau GG. 1993. *Adv. Enzyme Regul.* 33:97–110
62. Marchand M, Maisin I, Hue L, Rousseau GG. 1992. *Biochem. J.* 285:413–17
63. Bosca L, Majera M, Ghysdael J, Rousseau GG, Hue L. 1986. *Biochem. J.* 236:595–99
64. Bosca L, Rousseau GG, Hue L. 1985. *Proc. Natl. Acad. Sci. USA* 82:6440–44
65. Farnararo M, Vasta V, Bruni P, D'Alessandro A. 1984. *FEBS Lett.* 171:117–20
66. Rousseau GG, Fischer Y, Gueuning MA, Marchand MJ, Testar X, Hue L. 1988. *Prog. Cancer Res. Ther.* 35:188–98

67. Bosca L, Aragon JJ, Sols A. 1985. *Curr. Top. Cell Regul.* 27:411–18
68. Kummel L, Pilkis SJ. 1990. *Biochem. Biophys. Res. Commun.* 169:406–13
69. Lange AJ, Kummel L, El-Maghrabi MR, Tauler A, Colosia AD, Marker AJ, et al. 1989. *Biochem. Biophys. Res. Commun.* 163:753–60
70. Casado M, Bosca L, Martin-Sanz P. 1989. *Biochem. J.* 257:795–99
71. Mialpeix M, Katz NR, Bartrons R. 1990. *Cell Biochem. Funct.* 8:237–41
72. Rider MH, Bartrons R, Hue L. 1990. *Eur. J. Biochem.* 190:53–56
73. Cifuentes ME, Espinet C, Lange AJ, Pilkis SJ, Hod Y. 1991. *J. Biol. Chem.* 266:1557–63
74. Lemaigre FP, Lause P, Rousseau GG. 1994. *FEBS Lett.* 340:221–25
75. Espinet C, Vargas AM, El-Maghrabi MR, Lange AJ, Pilkis SJ. 1993. *Biochem. J.* 293:173–79
76. Vargas AV, Sola MM, Lange AJ, Poveda G, Pilkis SJ. 1994. *Diabetes* 43:792–99
77. Assimacopoulis JF, Renaud BJ. 1990. *J. Biol. Chem.* 265:7202–6
78. Probst L, Unthan-Fechner K. 1985. *Eur. J. Biochem.* 153:347–53
79. Muller A, Unthan-Fechner K, Probst I. 1988. *Eur. J. Biochem.* 176:415–20
80. Sanchez-Gutierrez JC, Sancho-Arias JA, Lechuga CG, Valle JC, Samper B, et al. 1994. *Endocrinology* 134:1868–73
81. Crepin KM, Darville MJ, Hue L, Rousseau GG. 1989. *Eur. J. Biochem.* 183:433–40
82. Loisseau AM, Rousseau GG, Hue L. 1985. *Cancer Res.* 45:4263–69
83. Lange AJ, Miksicek RJ, Pilkis SJ. 1994. *Mol. Endocrin.* In press
84. Taniyama M, Kitamura K, Thomas H, Lawson JWR, Uyeda K. 1988. *Biochem. Biophys. Res. Commun.* 157:949–54
85. Van Shaftingen E, Hers H-G. 1986. *Eur. J. Biochem.* 159:359–65
86. Watane F, Sakai A, Furuya E, Uyeda K. 1994. *Biochem. Biophys. Res. Commun.* 198:335–40
87. El-Maghrabi MR, Claus TH, Pilkis J, Pilkis SJ. 1981. *Biochem. Biophys. Res. Commun.* 101:1071–77
88. Rider MH, Kuntz DA, Hue L. 1988. *Biochem. J.* 253:597–601
89. Pilkis SJ, Pilkis J, El-Maghrabi MR. 1985. *J. Biol. Chem.* 260:7551–56
90. El-Maghrabi MR, Pate TM, Pilkis SJ. 1984. *Biochem. Biophys. Res. Commun.* 123:749–56
91. Van Schaftingen E, Davies DR, Hers H-G. 1981. *Biochem. Biophys. Res. Commun.* 103:362–69
92. Van Schaftingen E, Hers H-G. 1981. *Biochem. Biophys. Res. Commun.* 101:1078–86
93. Furuya E, Uyeda K. 1981. *J. Biol. Chem.* 256:7109–12
94. Pilkis SJ, Claus TH, Kountz PD, El-Maghrabi MR. 1987. *The Enzymes* 18:3–45
95. Kountz PD, Freeman S, Cook AG, El-Maghrabi MR, Knowles JR, Pilkis SJ. 1988. *J. Biol. Chem.* 263:16069–72
96. Kitajima S, Sakakibara R, Uyeda K. 1984. *J. Biol. Chem.* 259:6896–903
97. Kountz PD, McCain RW, El-Maghrabi MR, Pilkis SJ. 1986. *Arch. Biochem. Biophys.* 251:104–13
98. Laloux M, Van Schaftingen E, Francois J, Hers HG. 1985. *Eur. J. Biochem.* 148:155–59
99. Van Schaftingen E, Hers H-G. 1981. *Biochem. Biophys. Res. Commun.* 101:1078–86
100. Claus TH, Schlumpf JR, El-Maghrabi MR, Pilkis SJ. 1982. *J. Biol. Chem.* 257:7542–50
101. El-Maghrabi MR, Pate TM, Murray K, Pilkis SJ. 1984. *J. Biol. Chem.* 259:13096–102
102. El-Maghrabi MR, Pate TM, Pilkis J, Pilkis SJ. 1984. *J. Biol. Chem.* 259:13103–10
103. El-Maghrabi MR, Pate TM, D'Angelo G, Correia JJ, Lively MO, Pilkis SJ. 1987. *J. Biol. Chem.* 262:11714–20
104. Evans PR, Hudson PJ. 1979. *Nature* 279:500–4
105. Evans PR, Farrants GW, Hudson PJ. 1981. *Philos. Trans. R. Soc. Lond. Ser. B* 293:53–62
106. Shirakihara Y, Evans PR. 1988. *J. Mol. Biol.* 204:973–94
107. Bazan JF, Fletterick RJ, Pilkis SJ. 1989. *Proc. Natl. Acad. Sci. USA* 86:9642–46
108. Poorman RA, Randolph A, Kemp RG, Henrikson RL. 1984. *Nature* 309:467–69
109. Gehnrich SC, Gekakis N, Sul HS. 1988. *J. Biol. Chem.* 263:11755–59
110. Lee CP, Kao MC, French BA, Putney SD, Chang SH. 1987. *J. Biol. Chem.* 262:4195–99
111. French BA, Chang SH. 1987. *Gene* 54:65–71
112. Hellinga HW, Evans PR. 1985. *Eur. J. Biochem.* 149:363–73
113. Kurland IJ, El-Maghrabi MR, Pilkis SJ. 1993. *Biochem. Biophys. Res. Commun.* 195:228–36
114. Crepin KM, Vertammen D, Dom G, Hue L, Rider MH. 1993. *J. Biol. Chem.* 268:15277–84
115. El-Maghrabi MR, Colosia AD, Lange

AJ, Tauler A, Kurland IJ, Pilkis SJ. 1990. See Ref. 4, pp. 87–124

116. Gibbs RA, Taylor S, Benkovic S. 1992. *Science* 255:803–5

117. Benkovic S. 1992. *Annu. Rev. Biochem.* 61:29–54

118. Li L, Lin K, Kurland IJ, Correia JJ, Pilkis SJ. 1992. *J. Biol. Chem.* 267:4386–93

119. Schirmer T, Evans PR. 1990. *Nature* 343:140–45

120. Berger SA, Evans PR. 1990. *Nature* 343:575–76

121. Valdez BC, French BA, Younathan ES, Chang S. 1989. *J. Biol. Chem.* 264:131–35

122. Rider MH, Hue L. 1992. *Eur. J. Biochem.* 207:967–72

123. Kitamura K, Uyeda K, Hartman FC, Kangawa K, Matsuo H. 1989. *J. Biol. Chem.* 264:6344–48

124. Li L, Li P, Xu G-J. 1991. *Sci. China* 34:916–22

125. Kurland IJ, Pilkis SJ. 1995. *J. Biol. Chem.* In press

126. El-Maghrabi MR, Claus TH, Pilkis J, Pilkis SJ. 1981. *Biochem. Biophys. Res. Commun.* 101:1071–79

127. Furuya E, Uyeda K. 1981. *J. Biol. Chem.* 256:7109–17

128. Uyeda K. 1979. *Adv. Enzymol. Relat. Areas Mol. Biol.* 48:193–244

129. Sanwal BD. 1970. *Bacteriol. Rev.* 34:20–39

130. Abrahams SL, Younathan ES. 1971. *J. Biol. Chem.* 246:2464–67

131. Younathan ES, Voll RJ, Koerner TAW. 1981. In *The Regulation of Carbohydrate Formation and Utilization in Mammals*, ed. CM Veneziale, pp. 69–98. Baltimore: University Park Press

132. Evans PR, Farrants GW, Hudson PJ. 1981. *Philos. Trans. R. Soc. Lond. Ser. B* 293:53–62

133. Kemp RG, Foe LG. 1983. *Mol. Cell. Biochem.* 57:147–54

134. Blangy D, Buc H, Monod J. 1968. *J. Mol. Biol.* 31:13–35

135. Hengartner H, Harris JL. 1975. *FEBS Lett.* 55:282–85

136. Hellinga HW, Evans PR. 1987. *Nature* 327:437–39

137. Van Schaftingen E, Hers HG. 1982. *Eur. J. Biochem.* 124:143–49

138. Richards CS, Yokoyama M, Furuya E, Uyeda K. 1982. *Biochem. Biophys. Res. Commun.* 104:1073–80

139. Stewart HB, El-Maghrabi MR, Pilkis SJ. 1985. *J. Biol. Chem.* 260:12935–41

140. Pilkis SJ, Walderhaug M, Murray K, Beth A, Venkataramu SD, et al. 1983. *J. Biol. Chem.* 258:6135–41

141. Okar D, Kakalis JT, Narula SE, Armitage I, Pilkis SJ. 1994. *Biochem. J.* In press

142. Campbell JW, Watson HC, Hodgson GI. 1974. *Nature* 250:301–3

143. Winn SI, Watson HC, Harkins RN, Fothergill LA. 1981. *Philos. Trans. R. Soc. Lond. Ser. B* 293:121–30

144. Branden CI. 1980. *Q. Rev. Biophys.* 13:317–32

145. Shanske S, Sakoda S, Hermodson MA, DiMauro S, Schon EA. 1987. *J. Biol. Chem.* 262:14612–17

146. Blouquit Y, Calvin MC, Rosa R, Prome D, Prome JC, Pratbernou F, et al. 1988. *J. Biol. Chem.* 263:16906–10

147. Sakoda S, Shanske S, DiMauro S, Schon EA. 1988. *J. Biol. Chem.* 263:16899–905

148. Joulin V, Peduzzi J, Romeo PH, Rosa R, Valentin C, Dubart A, et al. 1986. *EMBO J.* 5:2275–83

149. Yanagawa SI, Hitomi K, Sasaki R, Chiba H. 1986. *Gene* 44:185–91

150. Patterson C. 1988. *Mol. Biol. Evol.* 5:603–25

151. Amato SV, Rose ZB, Liebman MN. 1984. *Biochem. Biophys. Res. Commun.* 121:826–33

152. Pilkis SJ, Lively MO, Pilkis J, El-Maghrabi MR. 1987. *J. Biol. Chem.* 262:12672–75

153. Tauler A, El-Maghrabi MR, Pilkis SJ. 1987. *J. Biol. Chem.* 262:16808–15

154. Tauler A, Lin K, Pilkis SJ. 1990. *J. Biol. Chem.* 265:15617–22

155. Lin K, Li L, Correia JJ, Pilkis SJ. 1992. *J. Biol. Chem.* 267:6556–62

156. Li L, Lin K, Correia JJ, Pilkis SJ. 1992. *J. Biol. Chem.* 267:16669–75

157. Li L, Lin K, Pilkis J, Correia JJ, Pilkis SJ. 1992. *J. Biol. Chem.* 267:21588–94

158. Fothergill-Gilmore JA, Watson HC. 1990. *Adv. Enzymol.* 62:227–313

159. Lin K, Li L, Correia J, Pilkis SJ. 1992. *J. Biol. Chem.* 267:19163–71

160. Lee YH, Okar D, Lin K, Pilkis SJ. 1994. *J. Biol. Chem.* 269:11002–10

161. Van Schaftingen E, Davies DR, Hers H-G. 1982. *Eur. J. Biochem.* 124:143–49

162. Van Schaftingen E. 1987. *Adv. Enzymol.* 59:315–33

163. O'Toole GA, Trzebiatowski JR, Esialante-Semerena JC. 1995. *J. Biol. Chem.* In press

164. Kraut J. 1977. *Annu. Rev. Biochem.* 46:331–58

165. Craik CS, Roczniak S, Langman C, Rutter W. 1987. *Science* 237:909–13

166. Carter P, Wells JA. 1988. *Nature* 332: 564–68
167. Hultquist DE, Moyer RW, Boyer PD. 1966. *Biochemistry* 5:322–31
168. Fersht A. 1985. In *Enzyme Structure and Mechanism*, pp. 147–53. New York: Freeman. 2nd ed.
169. Hockney DD. 1990. *The Enzymes* 19:1–36
170. Knowles JR. 1980. *Annu. Rev. Biochem.* 49:877–919
171. Coleman JE, Besman MJA. 1987. In *Hydrolytic Enzymes*, ed. A Neuberger, K Brocklehurst. Amsterdam: Elsevier
172. Taga EM, Moore DL, Van Etten RL. 1983. *Prostate* 4:141–50
173. Sowadski JM, Handschumacher MD, Murthy HMK, Foster BA, Wyckoff HW. 1985. *J. Mol. Biol.* 186:417–33
174. Igarashi M, Takahashi H, Tsuyama N. 1970. *Biochim. Biophys. Acta* 220:85–92
175. Ostrowski W. 1978. *Biochim. Biophys. Acta* 526:147–53
176. Van Etten RL, Hickey ME. 1977. *Arch. Biochem. Biophys.* 183:250–59
177. Van Etten RL. 1982. *Ann. N.Y. Acad. Sci.* 390:27–51
178. Han CH, Rose ZB. 1979. *J. Biol. Chem.* 254:8836–40
179. Hass LT, Place AR, Miller KB, Powers DA. 1980. *Biochem. Biophys. Res. Commun.* 95:1570–76
180. Bajwa W, Meyhack B, Rudolph H, Schweingruber A-M, Hinnen A. 1984. *Nucleic Acids Res.* 12:7721–39
181. Elliott S, Chang CW, Schweingruber ME, Schaller J, Rickli EE, Carbon J. 1986. *J. Biol. Chem.* 261:2936–41
182. Touati E, Danchin A. 1987. *Biochimie* 69:215–21
183. Pohlmann R, Krentler C, Schmidt B, Schroder W, Lorkowski G, Culley J, et al. 1988. *EMBO J.* 7:2343–50
184. Vihko P, Virkunnen P, Henttu P, Roiko K, Solin T, Huhtala ML. 1988. *FEBS Lett.* 236:275–81
185. Pradel E, Marck C, Boquet PL. 1990. *J. Bacteriol.* 172:802–7
186. Countaway JL, Waddell ID, Burchell A, Arion WJ. 1988. *J. Biol. Chem.* 263:2673–78
187. Lange AJ, Argaud D, El-Maghrabi MR, Pan N, Maitra S, Pilkis SJ. 1994. *Biochem. Biophys. Res. Commun.* 201:302–9
188. Schneider G, Lindquist Y, Vihko P. 1993. *EMBO J.* 12:2609–15
189. Ostanin KV, Harms EH, Sterus PE, Kucrel R, Zhou MM, Van Etten RL. 1992. *J. Biol. Chem.* 267:22830–36
190. Ostanin KV, Van Etter RL. 1993. *J. Biol. Chem.* 268:20778–84
191. Porvani KS, Heriala AM, Kurkela RM, Taavitsainen PA, Lindquist Y, Schneider G, et al. 1994. *J. Biol. Chem.* 269:22643–46
192. Lee YH, Lin K, Okar D, Alfano NL, Sarma R, Pflugrath JW, et al. 1994. *J. Mol. Biol.* 235:1147–51
193. Lee YH, Pflugrath J, Sarma R, Pilkis SJ. 1995. *J. Biol. Chem.* In press
194. Lee YH, Pflugrath J, Pilkis SJ, Ogata C. 1995. *J. Biol. Chem.* In press
195. Sakakibara R, Kitajima S, Uyeda K. 1984. *J. Biol. Chem.* 259:41–46
196. Rider MH, Foret D, Hue L. 1985. *Biochem. J.* 231:193–96
197. Kurland I, El-Maghrabi MR, Correia J, Pilkis SJ. 1992. *J. Biol. Chem.* 267: 4416–23
198. El-Maghrabi MR, Claus TH, Pilkis J, Pilkis SJ. 1982. *Proc. Natl. Acad. Sci. USA* 79:315–19
199. Kurland IJ, Li L, Lange A, Correia JJ, El-Maghrabi MR, Pilkis SJ. 1993. *J. Biol. Chem.* 268:14056–64
200. Sakata JS, Uyeda K. 1990. *Proc. Natl. Acad. Sci. USA* 87:4951–55
201. Sakata JS, Abe Y, Uyeda K. 1991. *J. Biol. Chem.* 266:15764–70
202. Tominaga N, Minami Y, Sakakibara R, Uyeda K. 1993. *J. Biol. Chem.* 268: 15951–57
203. El-Maghrabi MR, Correia JJ, Heil P, Pate T, Cobb C, Pilkis SJ. 1986. *Proc. Natl. Acad. Sci. USA* 83:5005–9
204. Depre C, Rider M, Veitch K, Hue L. 1993. *J. Biol. Chem.* 268:13274–79
205. Francois J, Van Schaftingen E, Hers H-G. 1988. *Eur. J. Biochem.* 171:599–608
206. Lin K, Kurland IJ, Li L, Lee YH, Okar D, Marecek JF, et al. 1994. *J. Biol. Chem.* 269:16953–60
207. Lange AJ, Pilkis SJ. 1995. *Arch. Biochem. Biophys.* In press
208. Kurland IJ, Chapman B, Pilkis SJ. 1995. *J. Biol. Chem.* In press
209. Kurland IJ, Pilkis SJ. 1995. *Protein Sci.* In press
210. Tauler A, Rosenberg AH, Colosia AD, Studier W, Pilkis SJ. 1988. *Proc. Natl. Acad. Sci. USA* 85:6642–46
211. Tauler A, Lange AJ, El-Maghrabi MR, Pilkis SJ. 1989. *Proc. Natl. Acad. Sci. USA* 86:7316–20
212. Hu S-H, Parker MW, La JY, Wilce HCJ, Benson GM, Kemp BE. 1994. *Nature* 369:581–84
213. Olah GA, Mitchell RD, Sosnick TR, Walsh DA, Trewhella J. 1993. *Biochemistry* 32:3649–57

214. Kemp BE, Pearsson RB. 1991. *Biochim. Biophys. Acta* 1094:67–76
215. Pelech S, Cohen P, Fisher MJ, Pogson CI, El-Maghrabi MR, Pilkis SJ. 1984. *Eur. J. Biochem.* 145:39–49
216. Rosa JL, Perez X, Ventura F, Tauler A, Gil J, Shimoyama M, et al. 1995. *Biochem. J.* In press
217. Rosa JL, Ventura F, Carreras J, Bartrons R. 1990. *Biochem. J.* 270:645–49
218. Sprang SR, Acharya KR, Goldsmith EJ, Stuart DI, Varvill K, Fletterick RJ, et al. 1988. *Nature* 336:215–21
219. Johnson L. 1994. *Nature: Struct. Biol.* 1:657–59
220. Scott P, Lange AJ, Pilkis SJ, Kruger NJ. 1994. *Plant J.* In press
221. Argaud D, Lange AJ, El-Maghrabi MR, Pilkis SJ. 1995. *J. Biol. Chem.* In press

Annu. Rev. Biochem. 1995. 64:837–63

RIBONUCLEOSIDES AND RNA

Bruce E. Eaton

Department of Chemistry, Washington State University, Pullman, Washington 99164-4630

Wolfgang A. Pieken

NeXagen Inc., Boulder, Colorado 80301

KEY WORDS: modified nucleosides, synthesis, crosslinking, RNA, SELEX, nucleotide
5′-triphosphates, nuclease stability, polymerase substrates

CONTENTS

ABSTRACT

Landmark discoveries such as the autocatalytic cleavage activity of certain RNA molecules, as well as small oligoribonucletide ribozymes and later the in vitro evolution of novel bioactive oligoribonucleotides (SELEX), have created entire new fields of biochemical research. The discovery of SELEX has provided a method for producing high-affinity nucleic acid ligands with high binding specificity to important medicinal targets. Including modified nucleotides into RNA ligands derived from SELEX may yield improved RNA thera-

837

0066-4154/95/0701-0837$05.00

peutics. The chemistry of oligoribonucleotides in comparison to oligodeoxyribonucleotides has led to resurgent attention on the role of modified nucleotides in RNA structure and function. Such modifications are also employed to impart stability towards endonuclease degradation on oligoribonucleotides.

SYNTHETIC METHODS FOR THE MODIFICATION OF RIBONUCLEOTIDE BASES

Uridines Modified at the 5-Position

Numerous examples of 5-position modified uridines are constituents of naturally occurring RNA (1). The structure and biological function of RNA is fine-tuned by modification of, mainly, the pyrimidine bases. Until recently, the biological role of RNA was thought to consist only of information storage and transfer (2). The discovery of in vitro evolution (SELEX) (3) has provided a method for producing high-affinity nucleic acid ligands with high binding specificity to various medicinally important targets, including HIV reverse transcriptase (4), HIV rev protein (5a,b), basic fibroblast growth factor (6), and thrombin (7). Posttranscriptional modification is important for the proper functioning of native RNA, and the inclusion of modified nucleotides in SELEX may improve RNA therapeutics. The bases altered at the 5-position of uridine are attractive because they are less likely to alter H-bonding interactions, the preferred pseudorotation of the ribose, or the glycosidic bond. In addition, the 5′-triphosphates of 5-position modified uridines are substrates for T7 polymerase, whose action represents an important step in SELEX.

This review describes only chemistries used to modify nucleosides, as opposed to base modification and glycosylation. A plethora of methods available for the modification of uridines entail synthesis of uracil followed by glycosylation, and these will not be discussed. The methods described here use the intact nucleoside, and many of these techniques are simple and convenient. One approach, beginning with β-uridine as a starting reagent, toward modifi-

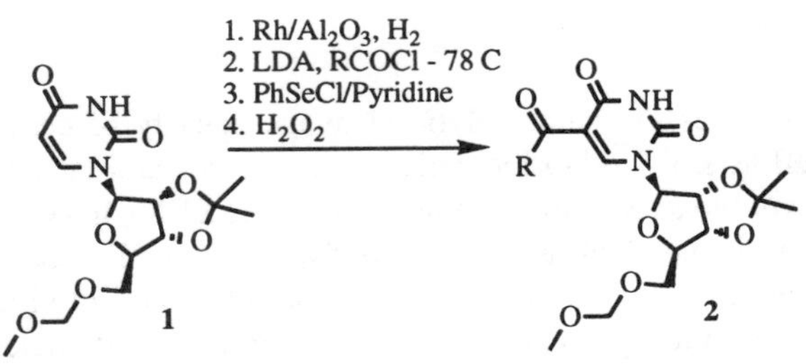

R	acylation yield(%)
Ph	79
Et	72
i-propyl	80
t-butyl	94
EtO	80

Figure 1 Synthesis of 5-acyluridines by conventional multi-step organic methods.

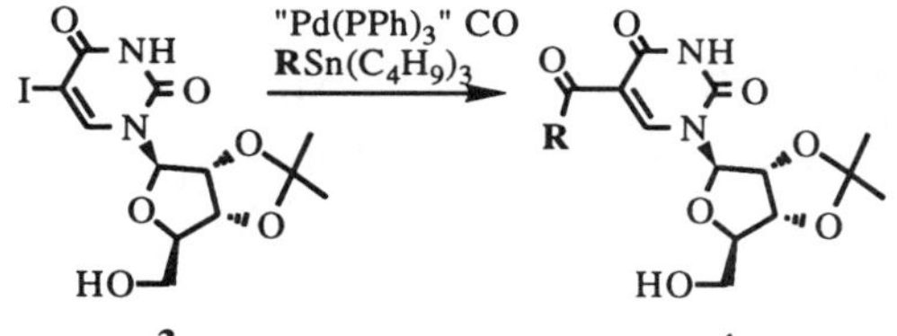

3a: R = CHCH$_2$ 89 %
3b: R = E-CHCHSi(CH$_3$)$_3$ 73 %
3c: R = C(OC$_2$H$_5$)CH$_2$ 81 %
3d: R = C$_6$H$_5$ 90 %

Figure 2 One-step palladium carbonylative coupling for the synthesis of 5-acyluridines.

cation appears in Figure 1 (8). Hydrogenation of analogue 1, treatment with LDA to form the enolate, acylation with an acyl chloride, and addition of PhSe followed by peroxide oxidation and elimination reforms the 5-6 double bond to give analogue 2. The group R could be a variety of alkyl groups and phenyl. The bulky R groups reportedly produced higher yields of acylation. The 2′,3′-isopropylidene and 5′-MOM protecting groups were crucial to the success of the reaction. It would be difficult or impossible to selectively remove the 5′-MOM group to prepare the 5′-triphosphate for SELEX. This method of preparation of 5-carbonyl uridine derivatives requires numerous steps.

Figure 2 presents a complimentary one-step synthesis of related modified uridines that does not require 5′-protection (GJ Crouch, MC Holmes, A Mundt & BE Eaton, unpublished data). The procedure used is essentially the same as that reported for 2′-deoxyuridine (9). Carbonylative C-C coupling of commercially available tin reagents to analogue 3 (10a,b,c) gave the α,β-unsaturated uridine (analogue 4). These reactions are typically performed under mild conditions (70°C, 50 psi CO, THF), and the products (analogue 4) are ready for triphosphate synthesis without additional protection and deprotection steps. These modified uridines are useful Michael acceptors and can be easily elaborated with other functional groups such as amino acids.

Other methods for preparing 5-carbonyl uridine analogues include hy-

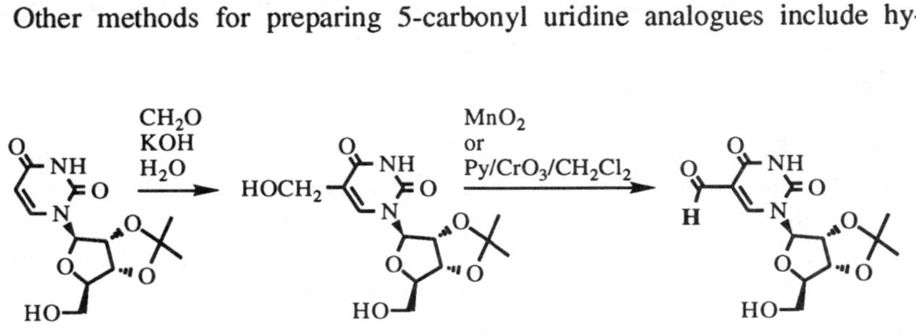

Figure 3 Hydroxy-methylation and subsequent oxidation in the synthesis of 5-formyluridine.

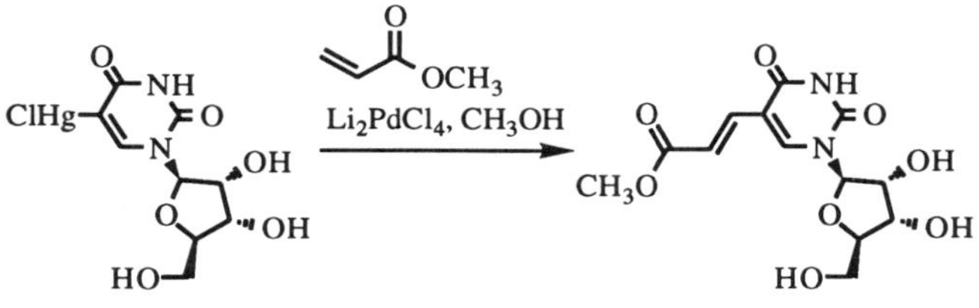

Figure 4 Palladium catalysis for the synthesis of 5-acrylateuridine.

droxymethylation of analogue 5 to give analogue 6 in good yield (Figure 3) (11). Oxidation of analogue 6 by either MnO_2 (12) or pyridinium dichromate/DMF/CH_2Cl_2 (13) gave analogue 7; the latter conditions resulted in a better yield. If we follow the procedure outlined in Figure 2 using Bu_3SnH, the combination of CO and analogue 3 give analogue 7 directly in good yield (GJ Crouch & BE Eaton, unpublished data).

Eckstein (12) has shown that, when β-5-formyluridine is treated with 4 N NaOH in methanol, anomerization occurs to give a mixture α:β of approximately 2:3, suggesting that acid labile protecting groups are most desirable. The crosslinking of two bases has been accomplished by reductive amination of 5-β-formyluridine with 1,3-diaminopropane and $NaCNBH_3$, although experimental details and yields were not given (14). The 5'-triphosphate-α-formyluridine has been used to affinity label *Escherichia coli* DNA–dependent RNA polymerase, and as a result, the formyluridine was found to be a noncompetitive inhibitor of this enzyme (15).

Bergstrom (16a–16e) reported the first use of palladium to form C-C bonds to the 5-position of uridine. One use of this modification procedure appears in Figure 4. This chemistry is not limited to acrylate but also works with allyl halides and acetates. In addition, ribose protection is not required. This procedure may also prove useful because uridine or other ribose-modified nucleosides can be activated for substitution by simple mercuration.

More recent work on palladium catalyzed C-C coupling uses analogue 10 as a substrate and affords modest to excellent yields of uridine products analogue 8 (Figure 5) (17). A variety of vinyl and aryl substituents could be attached to the 5-position of uridine by means of this method. Whether successful coupling required the ribose acetate protection is not clear. Unfortunately, in contrast to 5-iodouridine, analogue 10 is not commercially available. However, good to excellent yields were obtained through this method, which appears superior for the synthesis of these compounds. Similar results have been observed by using conditions analogous to those in Figure 2 but that lack the CO.

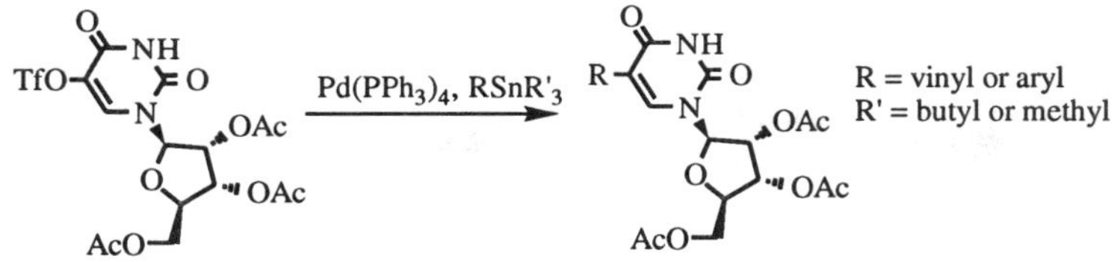

Figure 5 Palladium-catalyzed coupling of vinyl and aryl groups to the 5-position of uridine.

One of the most versatile methods for the attachment of alkyl groups to the 5-position of uridine involves palladium-catalyzed coupling of terminal acetylides (Figure 6) (18). Although this chemistry was originally used for modification of 2′-deoxyuridine (19), it appears to also work well for unprotected uridine (BR Rollman, BE Eaton & A Pardi, unpublished data). Surprisingly, the palladium coupling can accommodate both primary thioester and tosylate groups; hence making this a convergent synthesis for the preparation of uridine crosslinking reagents. The ability to perform the 5-position modification with 5′-DMT-protected analogue 12 makes this a streamlined approach to the automated synthesis of oligonucleotides.

A common modified base in tRNA is 5-methyluridine. Synthesis of 5-methyluridine is readily accomplished by palladium catalyzed cross coupling of analogue 14 and $(CH_3)_3Al$ (Figure 7) (20). No other aluminum reagents give preparatively useful reaction as compared to alternative synthetic methods. Other protection schemes and palladium catalysts gave slightly different yields making this synthetic method of limited scope.

Thiol-containing uridines have been of interest for many years because of their biological activity. Electrophillic substitution of the 5-position of uridine has been used to incorporate a thiocyanate, which is then reduced to the corresponding 5-thiol (Figure 8) (21). The reagent ClSCN is prepared by

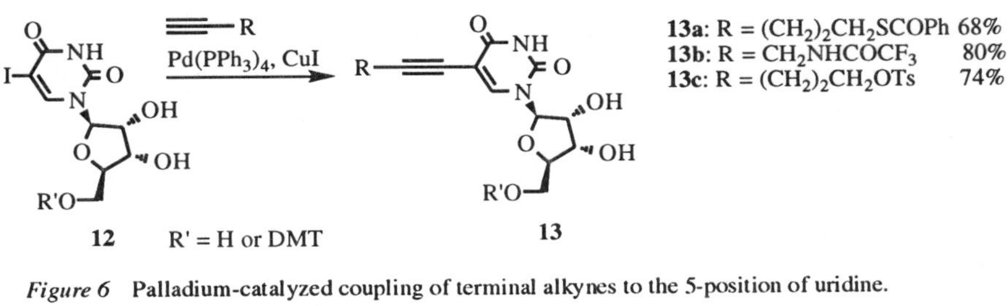

Figure 6 Palladium-catalyzed coupling of terminal alkynes to the 5-position of uridine.

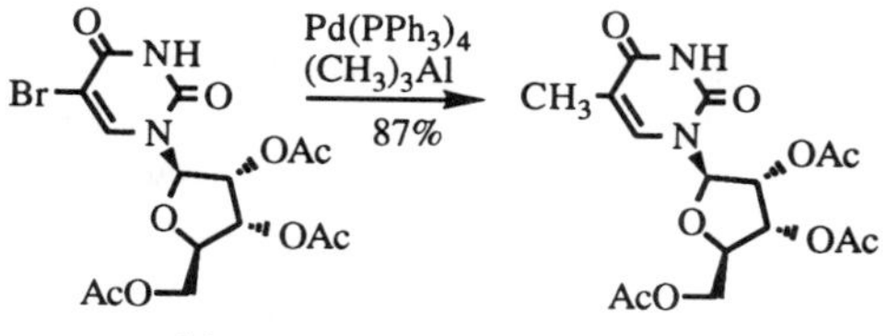

Figure 7 Synthesis of 5-methyluridine.

treating KSCN with Cl$_2$. Substitution of the 5-position is accomplished by treating analogue 16 with ClSCN in acetic acid. Dithiothreitol, mercaptoethanol, and glutathione could all be used as reducing agents. The biological activity of analogue 17 suggests that it may be reduced to analogue 18 in vivo.

Propargyl ethers have also been prepared at the 5-position of uridine. Alkylation of 5-hydroxyuridine with propargyl bromide in basic aqueous methanol gave 5-(2-propynyloxy)uridine in modest yield (22). Several other related uridine ethers have been prepared and tested for antiviral activity (23).

Adenosine and Guanosine Modified at the 8-Position

To enrich the oligonucleotide libraries of SELEX, purine as well as pyrimidine bases should be modified. The modification position least likely to interfere with H-bonding in purines is the 8-position. This section discusses chemistry used to modify intact adenosine and guanosine. We describe carbon-carbon bond forming reactions first, followed by transformations that attach heteroatoms.

Perhaps one of the most obvious routes to 8-alkyl-adenosine derivatives is as follows: protection by dimethylation of the 6-amino and 5'-hydroxyl as well as isopropylidene protection of the 2',3'-hydroxyls, followed by lithium halogen exchange of the 8-bromo group and subsequent alkylation with either methyl or ethyliodide (24). Because of the extensive protection chemistry, this method is of limited scope.

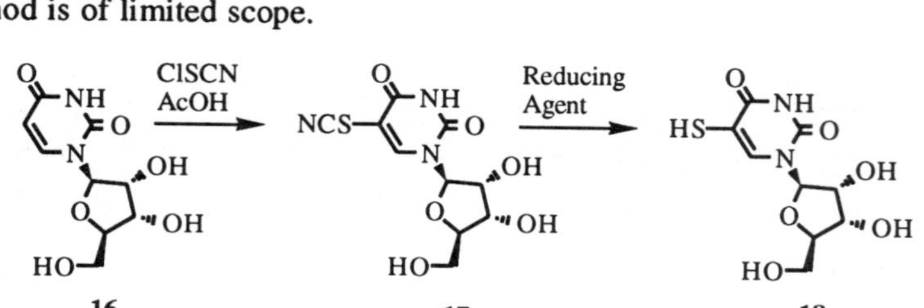

Figure 8 Electrophillic substitution of the 5-position of uridine with thiocyanate to form 5-thiol-uridine.

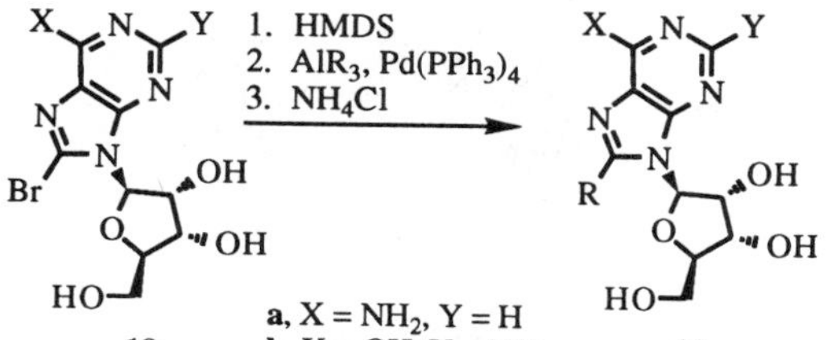

R	yield(%)
a Me	95
Et	42
i-Bu	72
b Me	24

a, X = NH₂, Y = H

b, X = OH, Y = NH₂

Figure 9 Attaching alkyl groups to the 8-position of adenosine and guanosine.

Figure 9 shows a more convenient approach to the alkylation of adenosine and guanosine (25). Silyl protection of the hydroxyl groups with HMDS followed by palladium-catalyzed coupling using trialkylaluminum reagents and either 8-bromoadenosine (analogue 19a) or 8-bromoguanosine (analogue 19b) gave modest to excellent yields. Why analogue 19b gave a significantly lower yield than analogue 19a is unclear. Only saturated aluminum alkyl reagents were reported.

One of the first, and potentially more versatile, examples of palladium catalyzed bond formation to the 8-position of adenosine was reported by Bergstrom (26) (Figure 10). This method of modifying adenosine requires the protection of all hydroxyl groups because of the Grignard reagent. After the coupling reaction is performed, the TMS groups can be removed by an aqueous work up.

All methods discussed to this point for the modification of purine nucleosides require the protection of the ribose hydroxyls. Palladium catalyzed carbon-carbon coupling of 8-bromo-adenosine with tin reagents avoids these protection steps. Figure 11 shows one of the first reported methods for 8-position modification of purines using tetraalkyltin reagents (27a,b). The reported yields are based on starting analogue 23 and not the equivalents of alkyl groups.

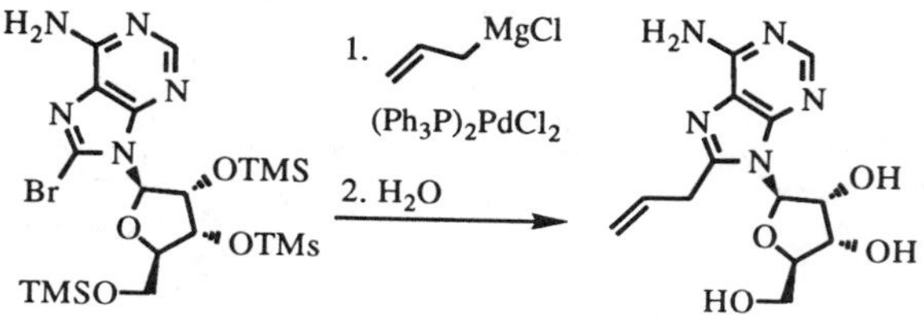

Figure 10 Synthesis of 8-allyladenosine by palladium catalysis.

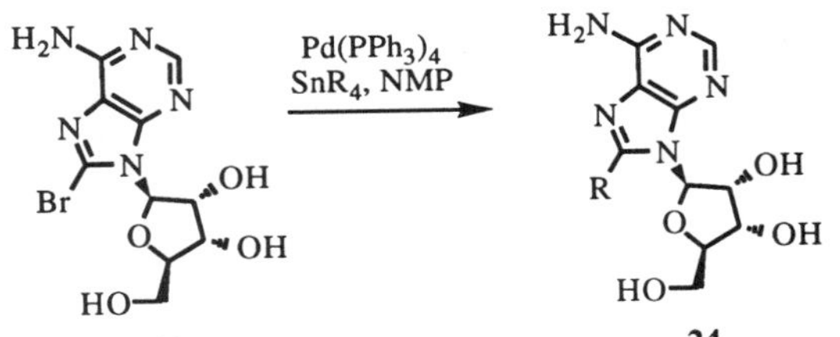

23 24

R	yield(%)
Me	74
Et	80
Vinyl	70

Figure 11 Palladium-catalyzed coupling of alkyl and vinyl groups to the 8-position of adenosine.

One limitation of this method is that few tetraalkyltin reagents are commercially available.

Guanosine has been modified at the 8-position using related tin/palladium chemistry (28). Nucleoside (analogue 25) was treated with the tin reagent *p*-formylphenyltributylstannane and $Pd(PPh_3)_4$ to give a good yield of analogue 26 (Figure 12). It is not clear whether successful coupling can occur without the ribose protecting groups.

Iron-mediated alkylation of the 8-position of guanosine and adenosine has also been accomplished. However, only guanosine can be specifically modified at the 8-position (Figure 13) (29). Adenosine resulted in modification at the 2- and 8-positions, and both mono and disubstituted adenosines were formed.

In addition to carbon bond–forming reactions, carbon-nitrogen and carbon-sulfur bonds, have been reported. Secondary amine functionality has been placed at the 8-position of both adenosine and guanosine (Figure 14) (30). Formation of the tertiary amine was only reported for guanosine. Presumably, amines with other functionality may also be attached to the 8-position of purines by this simple method. The mechanism of these transformations is unclear, but it undoubtedly is not direct synchronous displacement of bromide. Chemistry analogous to that shown in Figure 14 has resulted in conjugation of histamine to analogue 30a (31). Primary amines have also been appended

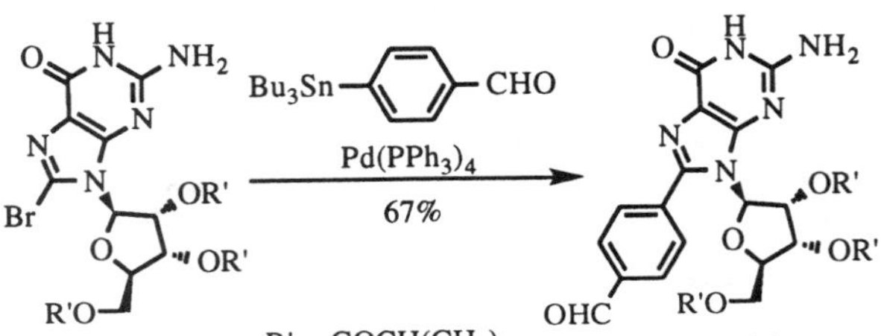

R' = $COCH(CH_3)_2$

25 26

Figure 12 Palladium catalyzed cross-coupling of an aryl stannane with 8-bromoguanosine.

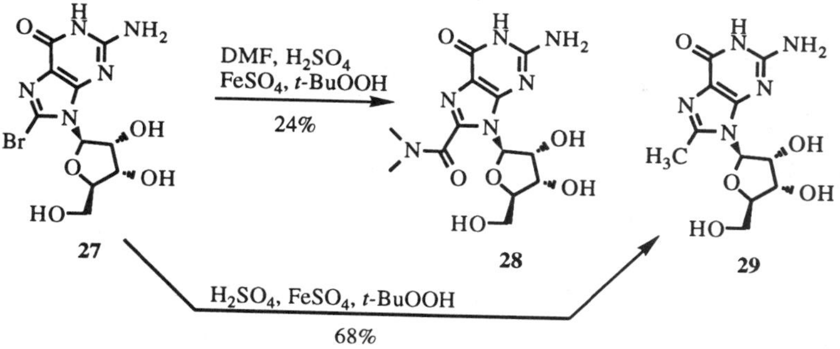

Figure 13 Iron mediated amidation and methylation of the 8-position of guanosine.

to the 8-position adenosine. Treating 8-azido-adenosine (32) with 1,3-propane dithiol resulted in disulfide formation and loss of N_2 (33). Preparation of 8-aminoguanosine was accomplished by treating adenosine with NH_2OSO_3H at pH 2–4 (34).

Some of the more interesting and biologically active purine nucleoside analogues contain sulfur and selenium. Both adenosine and guanosine have been modified at the 8-position with sulfur and selenium (35). Treating either analogue 30a or analogue 30b with sodium thiosulfate in 75% water/ethanol gave good yields of the sulfur-containing nucleosides (analogue 32A, analogue 32C) (Figure 15). The use of potassium selenosulfate as a nucleophile also gave good yields of nucleosides (analogue 32B, analogue 32D).

Guanosine derivative (analogue 33) makes stable complexes of methylmercury. Depending on the pH and the type of methylmercury used, different complexes were formed (Figure 16) (36). At low pH, cationic mono-, di-, and tri-methylmercury complexes were formed. When treated with methylmercury reagents at high pH, neutral complexes were isolated where substitution occurred at N1 and N7.

Other than the alkyl substituted purines, all of the modifications discussed

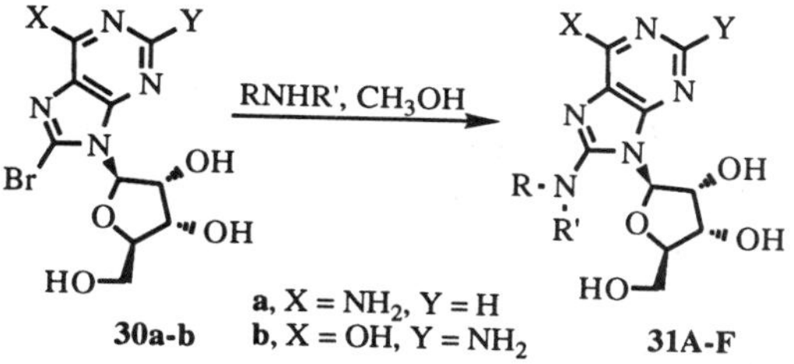

#	R	R'	X	Y	yield(%)
A	Me	H	NH$_2$	H	55
B	Et	H	NH$_2$	H	55
C	OH	H	NH$_2$	H	39
D	Me	H	OH	NH$_2$	63
E	Et	H	OH	NH$_2$	63
F	Me	Me	OH	NH$_2$	58

a, X = NH$_2$, Y = H
b, X = OH, Y = NH$_2$

30a-b   31A-F

Figure 14 Substitution of 8-bromoguanosine and 8-bromoadenosine with amines.

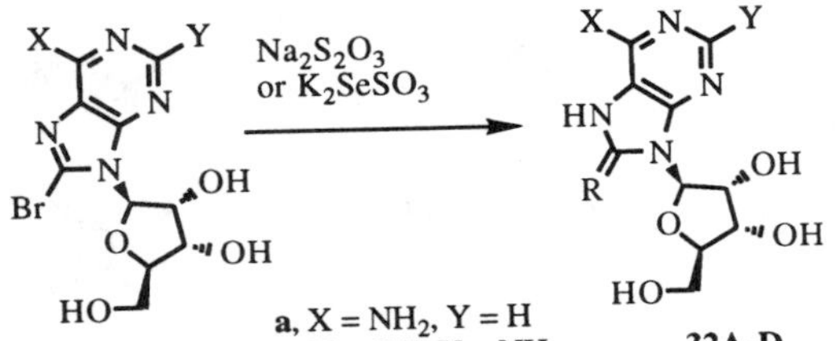

30a-b a, $X = NH_2$, $Y = H$; b, $X = OH$, $Y = NH_2$ **32A-D**

#	R	X	Y	yield(%)
A	S	NH_2	H	69
B	Se	NH_2	H	60
C	S	OH	NH_2	78
D	Se	OH	NH_2	65

Figure 15 Substitution of 8-bromoguanosine and 8-bromoadenosine by sulfur of selenium.

here could be used as sites of attachment for crosslinks. These functional groups could be used to form intermolecular crosslinks between RNA and proteins or intramolecular crosslinks to stabilize RNA structural motifs. Crosslinking has been used extensively to obtain information regarding the sites of contact on the RNA-protein interface. A large body of literature discusses the use of photochemical crosslinking. In the section below, we discuss methods for chemical crosslinking RNA to ribosomal proteins.

RNA CROSSLINKING

Several different types of crosslinking have been performed on RNA. The majority of work on crosslinking of RNA to proteins involves addition of a

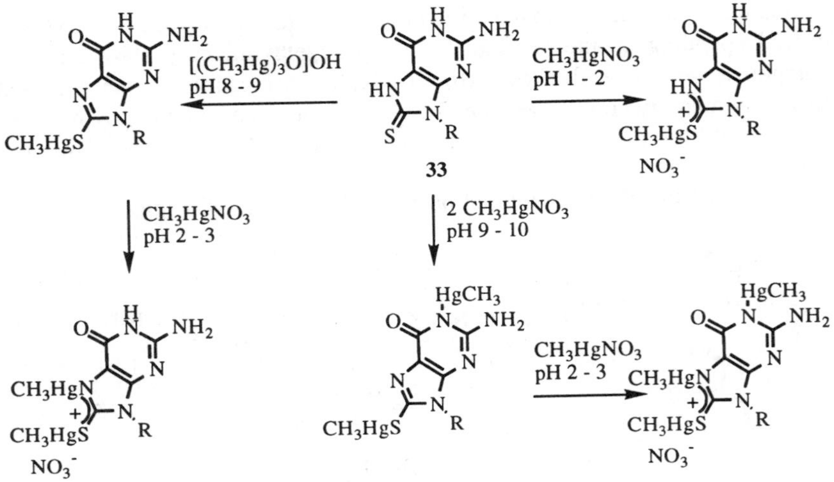

Figure 16 pH-dependent methylmercury complex formation with 8-thioguanosine.

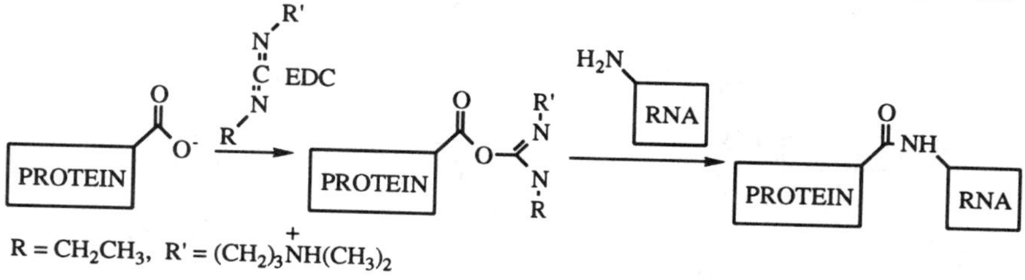

$R = CH_2CH_3, \quad R' = (CH_2)_3\overset{+}{N}H(CH_3)_2$

Figure 17 Carbodiimide crosslinking of the carboxylates of proteins to the amino groups of RNA.

bifunctional organic reagent that is either chemically or photochemically activated. Photochemically activated reagents have been used extensively and are not reviewed here. Some chemically activating nucleoside bases also crosslink to proteins, and some work has been done on RNA-RNA linkages. This section is divided into two topics: reagents used in crosslinking, and oligonucleotide modifications used for crosslinking both to proteins and within the same RNA.

Reagents for RNA Crosslinking

One of the reagents used most frequently in the bioconjugate chemistry of RNA-bonded proteins is 1-ethyl-3-dimethyl-aminopropylcarbodiimide (EDC). EDC crosslinking has been used for *E. coli* 16S rRNA and 30S ribosomal proteins. Protein S12 of 30S was found bound to the terminal G in the sequence CAACUCG at positions 1316–1322 in 16S rRNA (37). Rat liver rRNA-protein interactions studied with EDC crosslinking could be performed under conditions that did not change the sedimentation velocity and gave intact rRNA, implying that the conformation of the ribosomal protein-RNA complex was not changed dramatically (38).

Figure 17 presents the general mechanism for RNA-protein crosslinking. Presumably, EDC first reacts with a carboxylate in the protein followed by attack of an exocyclic amine base of the RNA. Alternatively, an amino or other nucleophillic group on the protein can react in the coupling step to give protein-protein crosslinking. As a result, purification procedures must be specific for isolation of the desired crosslinked adduct.

A bifunctional reagent, diepoxybutane, has been used in crosslinking experiments on bovine leukemia virus viral RNA and protein (39). Viral proteins p12 and pp15 were found to be crosslinked to the viral RNA. In contrast to EDC, good nucleophiles are required for a reaction to occur with diepoxybutane. Amino groups in both the RNA and protein are likely points of attachment, as well as protein thiols. In the absence of acid catalysis, hydroxyls probably do not react with this crosslinking reagent.

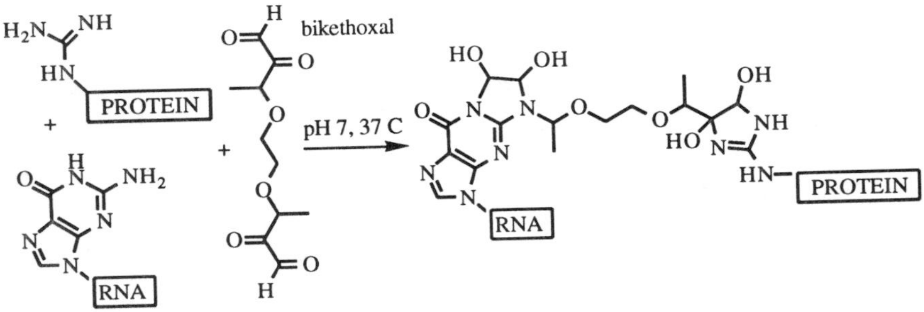

Figure 18 Chemoselective crosslinking of protein arginine side chains to RNA

The reaction of EDC and diepoxybutane is irreversible and may result in crosslinks that are kinetic rather than thermodynamic, which makes interpretation of RNA-protein contacts difficult. Reversible chemical crosslinking reagents have the advantage that they can equilibrate between reactive sites and can be removed subsequent to isolation of the RNA-protein complexes. Bikethoxal has been prepared on such a reagent and used to link *E. coli* RNA and protein in 30S, 50S, and the 5S RNA-L18 complex (40). The 1,2-dicarbonyl compounds react specifically with guanine in RNA (41a,b) and the guanidyl of arginine or the ε-amino of lysine in proteins (Figure 18) (42). The bikethoxal crosslinks could be reversed under mild basic conditions in the absence of borate. A bifunctional 1,2-dicarbonyl reagent 1,4-phenyl-diglyoxal (PDG) related to bikethoxal produces an intramolecular RNA-RNA crosslink between guanosine-2 and guanosine-112 in the stem region of *E. coli* 5S (43). The crosslink was formed after 5-6 h at 4°C and pH 7.2 and could be removed at pH 8.9. The crosslink rendered the stem resistant to RNase T_1, which allowed for identification of the linkage site.

One of the more recently discovered and interesting classes of RNA crosslinking reagents is transdiaminedichloroplatinum. This reagent has been used to crosslink S18 to 16S RNA in *E. coli* ribosomal 30S subunits (44), the tRNA[Val] to valyl-tRNA synthetase (45), and tRNA[Leu] to elongation factor Tu (46). Presumably, the coordination geometry in these crosslinked complexes is similar to that observed by X-ray crystal-structure analysis for the platinum complex in tRNA[Phe] (Figure 19) (47). The platinum complexes are formed under mild conditions (e.g. pH 7.2, 2.5 h, 25°C) in the dark. The crosslinks could be removed by treatment of the RNA-protein complex with thiourea (1 h, 37°C). The use of organometallic reagents in examining RNA-protein structure holds great promise.

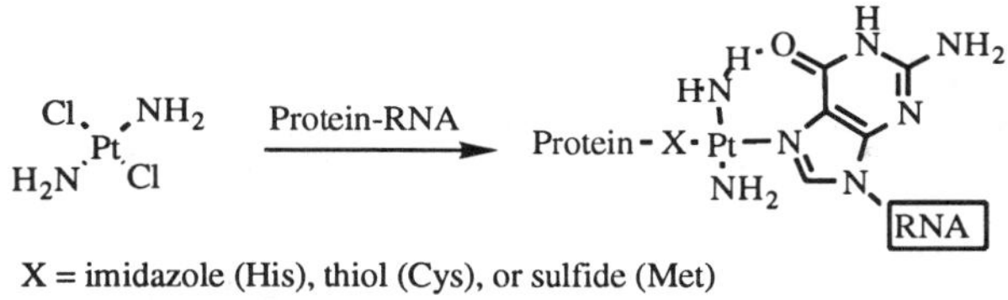

X = imidazole (His), thiol (Cys), or sulfide (Met)

Figure 19   Protein-RNA crosslinking under mild conditions with transplatinum.

Crosslinking Oligonucleotides: RNA-Protein Bond

Most of the examples of modified bases discussed in this review should be capable of forming reversible covalent crosslinks, either with a protein target or within the RNA folded structure. Irreversible crosslinking to 4-position modified cytidine has been used to link RNA and protein in the *E. coli* 30S ribosomal complex (Figure 20) (48). The first step in forming analogue 34 is the addition of HSO_3^- to the 6-position of cytidine. This reaction can also occur for uridine but is reversible. Once the bisulfite has added the base, the N4 nitrogen is converted to a hydrazino group with hydrazine. The hydrazine derivative reacts with the carbonyl groups of aldehydes and ketones. Treatment of analogue 34 with bromopyruvate gives analogue 35, which now contains an α-bromo substituent that is most reactive toward thiols. Cysteine thiols on the protein then substitute the bromide and form an irreversible RNA-protein crosslink. Significant crosslinking occurred under mild conditions (pH 6–8, 37°C, 1–2 h). There is no apparent reason why this chemistry should be limited to bromopyruvate as the linker, and linker molecules that contain the required functional groups are available in varying lengths.

Irreversible crosslinking of modified bases has been accomplished by means of photochemical crosslinking of 5-iodouridine to the R17 coat protein (49a). The observed extremely high specificity and large amount of crosslinking suggested that this technique may prove most useful for making specific

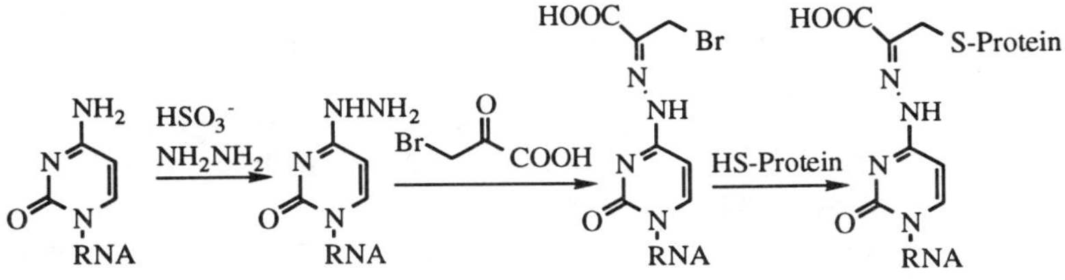

34   35

Figure 20   Chemical activation of the 4-amino group of cytidine in RNA and crosslinking to free thiol groups in a protein.

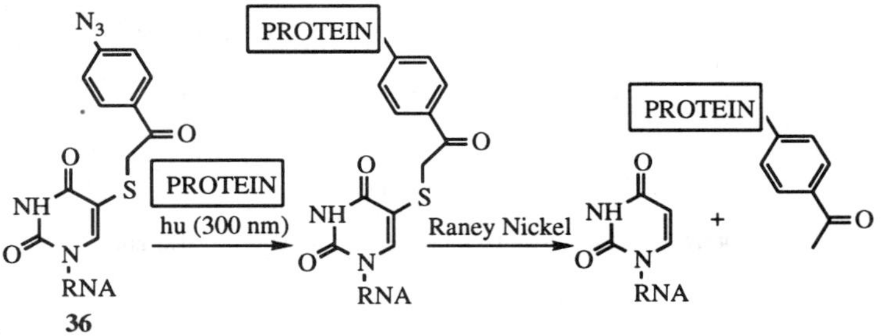

Figure 21 Photochemical RNA crosslinking to proteins by incorporation of a modified uridine containing an azido group.

RNA-protein crosslinks. Another approach to using a modified base to photo-chemically crosslink uses a group attached to the 5-position of uridine (Figure 21) (49b). The azidio group has a half-life of 10 s at pH 7.0 under the photochemical conditions (302 nm mercury vapor lamp) used. In contrast to photochemical crosslinking by 5-iodouridine containing nucleic acids, use of analogue 36 is advantageous in that the RNA may be chemically cleaved from the protein. This technique is disadvantageous in that the reactive species formed on irradiation will probably not be as specific in its reaction with the groups on the protein.

RNA-RNA Crosslinking with Disulfides

Recently, considerable effort (and money) has been put into developing methods for the introduction of disulfide crosslinks into DNA (50a–50f). In contrast, the crosslinking of RNA structural motifs has received less attention. Most of the work in this area has focused on the synthesis of uridine ana-logues with appended thiol groups. There appears no reason why the chem-istry developed for DNA cannot be applied to RNA structural studies. One example is the disulfide crosslinking of the 3′-5′ termini of an RNA hairpin loop (51). The sequence studied was 5′-U*GACUUCGGUCU,* where U* represents the N-3 ethyl-*t*-butyldisulfide uridine. This hairpin was prepared via automated phosphoramidite synthesis, and the modified disulfide phos-phoramidite reagent was prepared in nine-steps beginning with uridine. Once incorporated into the oligonucleotide, the disulfide was reduced to the thiol with dithiothreitol. Air oxidation (5 h, pH 8) gave the disulfide crosslinked hairpin.

SYNTHETIC METHODS FOR THE MODIFICATION OF RIBONUCLEOTIDE SUGARS

2'-Modified Nucleotides

An explosion of interest in understanding of the properties and function of ribonucleic acids occurred in the past decade. Landmark discoveries such as the autocatalytic cleavage activity of certain RNAs (52) and small oligoribonucleotide ribozymes (53), and later the in vitro evolution of novel bioactive oligoribonucleotides (3, 54), have founded entire new fields of research. The rich chemistry of oligoribonucleotides in comparison to oligodeoxyribonucleotides has renewed the attention on the role of the 2'-hydroxyl group for RNA structure and function. Specific modifications at the 2'-position of the ribose have been established as investigative tools for addressing this question (55). Such modifications are also employed to impart stability towards endonuclease degradation on oligoribonucleotides.

In the in vitro evolution technology, particularly the SELEX method introduced by Tuerk and Gold, the oligoribonucleotide library is regenerated at every iteration by transcription of a DNA library template with a DNA-dependent RNA polymerase, preferably from phage T7 (56). This step allows introduction of nucleotide analogues to the RNA pool at every iteration in the form of nucleoside 5'-triphosphates. The chemical modifications become an integral part of the activity of an RNA ligand generated by the SELEX process. In contrast, modification of native oligonucleotides, such as ribozymes or antisense molecules, must be limited to a few select positions in order to retain full activity.

Below, this review focuses on nucleotide analogues that show potential for combinatorial library enrichment processes. Therefore, the literature on modified oligoribonucleotides as antisense inhibitors of gene function is beyond the present scope but has been subject to recent reviews (57, 58). Until recently, nucleoside analogues modified at the furanose ring were chiefly employed as polymerase inhibitors by chain termination for inhibition of viral proliferation (59).

Chemical Synthesis of 2'-Modified Ribonucleotides

The modification of the 2'-position of pyrimidines has a long history and has been reviewed extensively (60, 61). Although the 2'-modified arabinonucleosides have enjoyed steady attention as potential antiviral compounds (62, 63), relatively few examples of 2'-modified ribonucleosides are known. Only very recently, in the wake of antisense research, has the interest in substitution of the 2'-hydroxyl group of ribonucleosides by novel substituents been rekin-

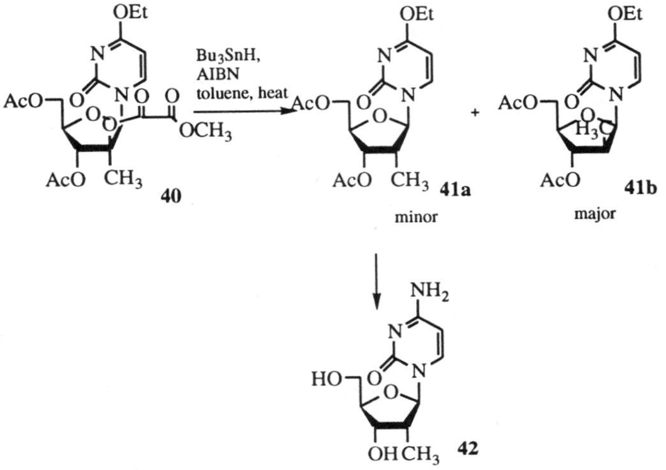

Figure 22   Synthesis of 2′-C-methyl-2′-deoxycytidine by methylation of the nucleoside.

dled. At present, only a few 2′-analogues have been converted to the 5′-triphosphate derivatives and studied for polymerase substrate activity.

The most prominent route for preparation of 2′-modified ribopyrimidines proceeds through introduction of nucleophiles to the corresponding 2,2′-anhydropyrimidine precursor. This reaction is limited to preparation of 2′-halides and introduction of a few other good nucleophiles such as azide and thiolates (60, 61). Introduction of azide, for example, gave the 2′-azidouridine (64). This was reduced to yield 2′-aminouridine or converted to the 2′-azidocytidine and reduced to the 2′-aminocytidine (65).

The 2′-O-methyl derivatives of ribonucleosides are accessible through methylation of the 3′-5′-protected precursor (66). Similarly, other 2′-O-alkylribonucleosides, and 2′-O-allyl derivatives (67), have been prepared (68). Several new 2′-modified ribonucleosides deserve attention as potential polymerase substrates. The synthesis of 2′-methyl cytidine (analogue 42) was recently described (Figure 22) (69). Radical deoxygenation of the methyl oxaloyl ester of the tertiary alcohol (analogue 40) gave an isomeric mixture of 2′-deoxy-2′-methyl derivatives (analogue 41) from which the desired ribo isomer was isolated as the minor component.

An alternate approach to the 2′-deoxy-2′-methylnucleosides elaborated the 2′-substituent prior to introduction of the base to the 1′-position (Figure 23) (70). Nucleosidation of the protected 2-methylribose (analogue 43) yielded a 7:1 mixture of desired β to α anomer from which the desired 2′-methylthymid-

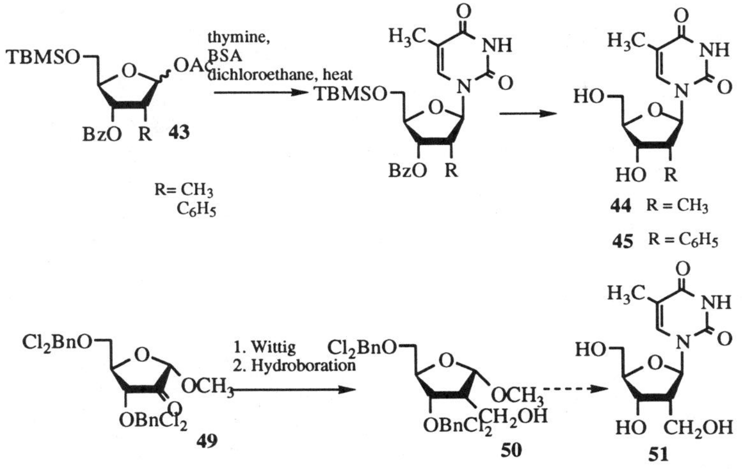

Figure 23   Synthesis of 2'-alkyl and 2'-aryl substituted nucleosides via substituted furanoses.

ine (analogue 44) was isolated after deprotection. The 2'-phenylthymidine (analogue 45) and the corresponding 2'-alkylcytidines (analogue 46, analogue 47) were obtained in an analogous manner. Functionalized alkyl substituents have also been introduced to the 2'-position. Wittig olefination of protected 2-dehydro-2-oxoribose analogue 49, followed by hydroboration, gave exclusively the desired α-hydroxymethyl isomer (analogue 50). The acetyl derivative of the 2'-hydroxymethylribose displayed the same β selectivity during glycosylation of the nucleobase as observed with other ribose derivatives bearing acetylated 2'-groups, yielding nucleoside (analogue 51) (71). The 2'-methoxymethylthymidine (analogue 52) was obtained in a similar manner.

The recently described 2'-fluoromethyl and 2'-trifluoromethylthymidine derivatives (analogue 53, analogue 54) promise an interesting comparison of chemical properties (72). The former was prepared from analogue 49 via the fluoroolefin analogue 55 (Figure 24), whereas nucleophilic addition of the trifluoromethyl group to analogue 49 with subsequent radical deoxygenation gave a 4:1 mixture of the desired α-trifluoromethylribose (analogue 56), which was converted to the 2'-trifluoronculeoside (analogue 54).

We recently realized that the problem of introduction of substituents to the 2'-position of 2,2'-anhydrouridine resembles the well-studied regiospecific intramolecular opening of α-hydroxy epoxides by nucleophiles tethered to the α-hydroxy group (73a-c). With this analogy in mind, we prepared the easily accessible 5'-protected 2'-carbonylimidazole intermediate analogue 57 (Figure

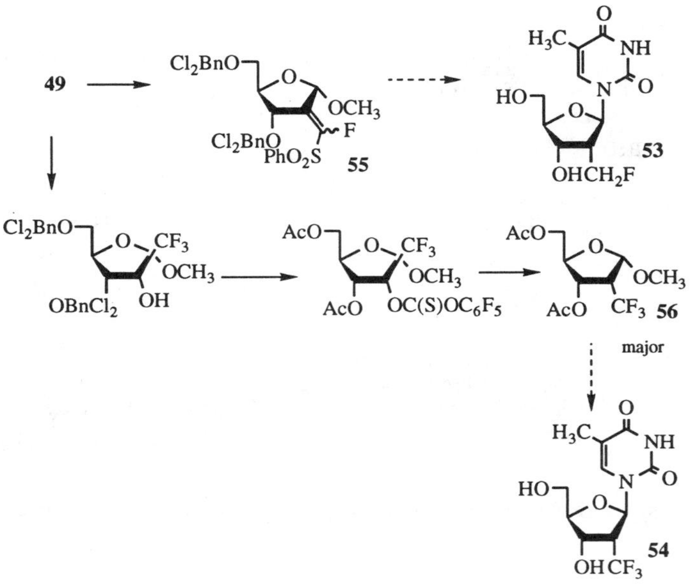

Figure 24 Synthesis of 2′-C-fluoromethyl-2′-deoxynucleosides.

25) that was quantitatively converted to the O-substituted carbamate derivative (analogue 58). The latter cyclizes and deprotects in a single reaction step under mild base conditions at room temperature. This technique allowed us to generate the novel O-benzyl-2′-hydroxylaminouridine analogue 59 (D Sebesta, S O'Rourke & W Pieken, unpublished data). We are currently exploiting this route to prepare a variety of novel 2′-modified pyrimidines.

We have similarly exploited the intramolecular nucleophilic opening of the 2′,2′-anhydrouridine to prepare 2′-aminouridine in a mild and efficient way (D McGee, Y Zhai & WA Pieken, unpublished data). Reaction of 5′-protected 2,2′-anhydrouridine with neat trichloroacetonitrile yields the oxazole derivative (analogue 60) that can be hydrolyzed by sodium hydroxide to the 5′-protected 2′-aminouridine (analogue 61).

Another interesting 2′-substituent is the 2′-mercapto group (76). The synthesis of both 2′-thiopyrimidines (77) and 2′-thioadenosine (78) has recently been reported. The 2′-thiouridine was obtained from reaction of 2,2′-anhydrouridine with 4-methoxybenzylmercaptane (79) and subsequent hydrolysis under acidic conditions. Under basic conditions, free 2′-thiopyrimidines rapidly oxidize to the disulfides. The crystalline hydrochloric salt of 2′-thiocytidine was obtained for the first time from the protected 2′-thiouridine derivative.

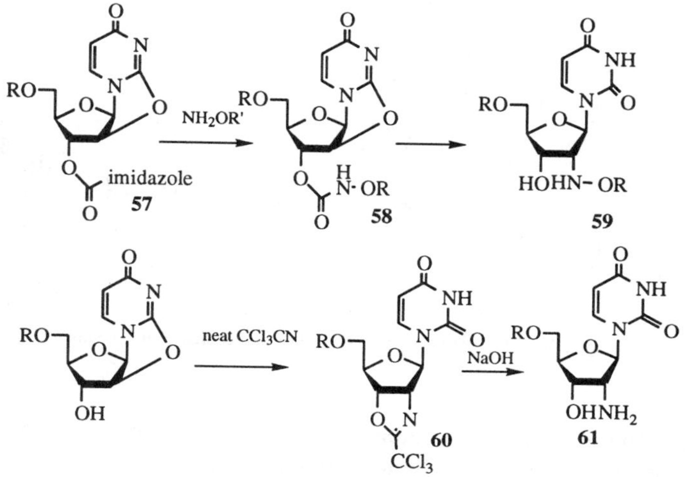

Figure 25 Intramolecular introduction of nucleophiles to the 2′-position of ribopyrimidines.

The 2′-thioadenosine was prepared by means of displacement of the triflate derivative of protected 2′-arabinoadenosine with 9-(4-methoxyphenyl)xanthene-9-thiol (80) and subsequent mild deprotection under acidic conditions.

Synthesis of Nucleoside 5′-Triphosphates

Polymerization of nucleotides with template-dependent RNA or DNA polymerases requires the nucleoside 5′-triphosphates as substrates. Early methods of preparation of these compounds relied on condensation of activated nucleoside 5′-monophosphate with phosphoric acid or pyrophosphate (81). These methods have been previously reviewed (82, 83).

A facile method for preparation of nucleoside 5′-triphosphates from the 2′- and 3′-protected nucleoside was recently introduced by Ludwig and Eckstein (84). Salicyl phosphorochloridate was used to introduce a phosphorous III to the 5′-position (Figure 26). Displacement of the salicyl group with pyrophosphate led to a cyclic 5′-phosphite diphosphate intermediate. Oxidation of this intermediate either with sulfur or iodine led to either the nucleoside 5′-α-thio triphosphate or the 5′-triphosphate after hydrolysis of the cyclic 5′-triphosphate. This method has been employed in the preparation of 2′-azido-, 2′-amino-, and 2′-fluoro 2′-deoxynucleoside 5′-triphosphates (85, 86). A facile one-step preparation of 5′-triphosphates was introduced recently. This method avoided protection of the 2′- and 3′-hydroxyl groups by using trimethylphosphate as the solvent (87). Condensation of the intermediate 5′-phosphorodi-

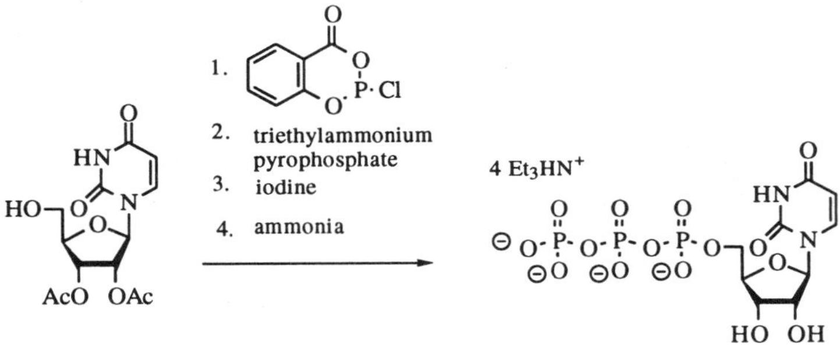

Figure 26 Ludwig-Eckstein reaction for preparation of nucleoside 5′-triphosphates.

chloridate with tributylammonium phosphate yielded the linear 5′-triphosphate, presumably via the cyclic metatriphosphate.

A solid-phase synthesis of 2′-*O*-methylribonucleoside 5′-triphosphate has also been reported (88). Commercial 5′-dimethoxytriyl-protected 2′-*O*-methyl-nucleoside, anchored to solid controlled-pore glass support via a 3′-succinate ester, was subjected to the Ludwig-Eckstein phosphorylation conditions subsequent to 5′-deprotection. Solid-phase anchoring of the nucleoside provided the advantage of facile separation from unwanted side products and allows potential automation of triphosphate synthesis. It is limited to availability or ease of preparation of anchored, protected nucleosides.

2′-MODIFIED TRIPHOSPHATES AS SUBSTRATES FOR POLYMERASES

Polymerization of 2′-modified nucleoside 5′-diphosphates with the template-independent polynucleotide phosphorylase has in the past been widely used for assembly of modified homopolymers. Among the analogues polymerized by this enzyme are those bearing 2′-*O*-methyl (89), 2′-fluoro (90), 2′-amino (91), 2′-azido (92), and 2′-chloro substituents (93). In contrast, the data are limited for the much more useful activity of 2′-modified nucleoside 5′-triphosphates as substrates for DNA-dependent RNA polymerases. For the purpose of this review, substrate activity is defined as incorporation and elongation of nucleotide analogues.

Table 1 summarizes the substrate activity of 2′-modified nucleoside 5′-triphosphates with DNA-dependent RNA polymerases. A comparative view of the data suggests that a clear understanding of the factors governing acceptance of nucleotide analogues as substrates by DNA-dependent RNA polymerases awaits further study. One can, however, clearly rule out the substituent-dependent preference of nucleoside conformation as a determinant for activity. The influence of the 2'-halo, 2'-O-methyl, 2'-azido, and 2'-amino substituent on nucleoside conformation has been previously reviewed (94). The extend of preference for C3'-endo conformation in solution correlates with the electronegativity of the substituent, with the exception of the 2'-amino group.

Of all RNA polymerases studied so far (95, 96; L Beebe, G Kirschenheuter & W Pieken, unpublished data), 2'-O-methyl analogues, which occur naturally as a posttranscriptional modification of tRNA (98), allow, at best, marginal chain extension, but these analogues may serve as effective inhibitors of native RNA synthesis. The E. coli RNA polymerase displayed differential activity with nucleotide analogues, dependent on the 2'-modification as well as on the nucleobase (99–102). Replacement of Mg^{2+} with Mn^{2+} as divalent metal cofactor increased the transcription yield for the 2'-fluoroUTP analogue to almost 50% of that observed with UTP.

In contrast to the E. coli enzyme, several nucleotide analogues are efficient substrates for the phage T7 DNA-dependent RNA polymerase (85). This observation does not extend to 2'-deoxynucleotides, which had an efficiency of only 70% full length transcript per template site (103). Furthermore, no substrate activity was found for 2'-O-methylUTP or araCTP with this enzyme (L Beebe & W Pieken, unpublished data).

Comparison of the different studies is complicated by the complexity of the transcription systems. In particular, the study of Eckstein and coworkers demonstrated the dependence of analogue substrate activity on template sequence (85). This phenomenon may also account for the different activity of 2'-fluoroUTP and 2'-fluoroATP with the pol(dAT) template with the E. coli polymerase. Several authors have remarked on the differential analogue activity during initiation versus elongation (85, 101). This variability further complicates comparison of results between different transcription systems. In transcriptions of very long templates, several authors have observed premature termination at a specific length for some nucleotide analogues but not for others (85, 100). The 10-fold difference of the K_m for 2'-fluoroUTP in the E. coli system compared to the T7 system is also noteworthy. Whether the recognition of the 2'-site is nonconserved between the two polymerases or whether the observed differences result from inherent factors in the investigated transcription systems remains unknown. However comparison of the crystal structure of T7 RNA polymerase with that of HIV-RT and Klenow fragment pointed to possible differences between T7 and the other two polymerases for recognition of 2'-deoxyNTP substrates (105).

Nonetheless, enzymatic polymerization is a viable method for generation of 2'-modified RNA, although yields are often lower. Generation of 2'-aminopyrimidine containing RNA pools by T7 RNA polymerase transcription has

Table 1 2'-modified substrates for DNA dependent RNA polymerases

Enzyme	Substrate	Template	Relative yield[a]	K_m/mM	Relative V_{max}[a]	K_i/mM	Reference
E. coli	2'-aminoUTP	poly(dAT)	0.021	—	—	0.73	100
E. coli	2'-chloroUTP	poly(dAT)	0.007	—	—	0.75	100
E. coli	2'-aminoATP	poly(dAT)	—	—	—	2.30	100
E. coli	2'-azidoATP	poly(dAT)	—	0.15–0.20	0.100–0.200	0.20	102
E. coli	2'-fluoroATP	poly(dAT)	—	0.25	0.007	0.20	102
E. coli	2'-fluoroUTP	poly(dAT)	0.200	—	—	—	100
E. coli	2'-deoxyTTP	poly(dAT)	0.050	—	—	—	102
E. coli	dTTP, dATP	poly(dAT)	0.005	—	—	—	102
Pseudomonas putida	2'-O-methylATP	P. putida gh–1 DNA	0.006	—	—	—	95
Qβ	2'-O-methylGTP	poly(dU$_2$C)	0.008	—	—	—	96
Qβ	2'-O-methylGTP	poly(C)	—	—	—	0.04	96
T7	2'-fluoroATP	E. coli tRNA^{ASP}	0.500	—	—	—	85
T7	2'-fluoroATP	yeast tRNA^{ASP}	0.400	—	—	—	85
T7	2'-fluoroUTP	E. coli tRNA^{ASP}	0.400	2.20[b]	0.092[b]	—	85
T7	2'-fluoroUTP	yeast tRNA^{ASP}	0.040	—	—	—	85
T7	2'-fluoroCTP	E. coli tRNA^{ASP}	0.100	2.80[b]	0.041[b]	—	85
T7	2'-aminoUTP	E. coli tRNA^{ASP}	0.600	0.20[b]	0.117[b]	—	85
T7	2'-azidoCTP	E. coli tRNA^{ASP}	none	—	—	—	85

[a] Values are relative to those determined with native ribonucleoside 5'-triphosphates.
[b] The template in this experiment was linearized pSPT19 plasmid DNA.

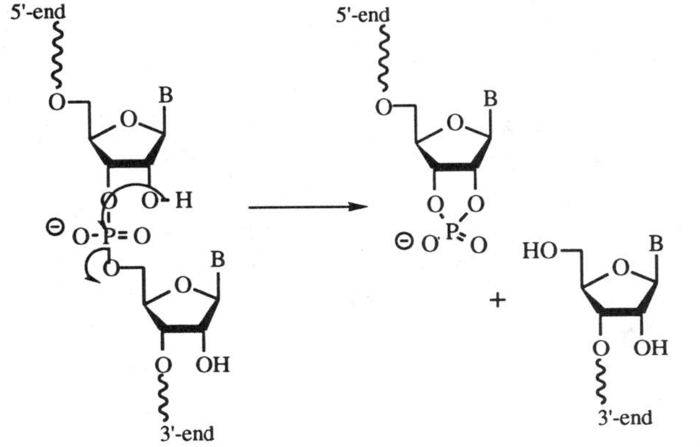

Figure 27   Mechanism of strand cleavage by endoribonucleases.

been successfully employed for enrichment of stabilized RNA ligands by the SELEX method (S Jayasena, unpublished data). More studies, particularly with novel 2′-analogues, are needed to shed light on the mechanism of enzymatic recognition of 2′-substituents.

NUCLEASE STABILITY OF 2′-MODIFIED OLIGONUCLEOTIDES

Endonucleolytic degradation of oligoribonucleotides, unlike that of oligodeoxyribonucleotides, proceeds through activation of the 2′-hydroxyl group for inline attack on the internucleotidic phosphorous (Figure 27). The resulting cyclic intermediate divides into the 5′-terminal oligonucleotide fragment bearing a 2′,3′-cyclic phosphodiester at its 3′-end and the 3′-terminal fragment bearing a free 5′-hydroxyl group (107). This mechanism of degradation is inhibited by introduction of 2′-substituents not activatable for nucleophilic attack on the adjacent phosphorous. Therefore, all 2′-modifications discussed above, with the exception of the 2′-mercapto group, necessarily constitute mechanism-based RNA endonuclease inhibitors.

Consequently, oligoribonucleotides bearing 2′-modifications have been tested for nuclease stability. Nearly complete stability towards RNase A endonucleolytic degradation has been observed for poly(2′-chloroU) and poly(2′-chloroC) homopolymers (94). Furthermore, degradation by snake-venom phosphodiesterase, as well as spleen phosphodiesterase, was markedly slowed down compared with native poly(rU) and poly(rC). These polymerases do not involve the 2'-hydroxyl group in chain cleavage. Both oligoribonucleotide analogues displayed physical characteristics similar to those of RNA. The finding that both homopolymers were completely resistant to degradation by DNase I further attests to these RNA-like characteristics.

A similar observation was made with the poly(2'-azidoC), poly(2'-azidoU), and the corresponding poly(2'-aminoU) and poly(2'-aminoC) homopolymers (92). All four were resistant to RNase A degradation. The poly(2'-azidoC) was only slowly hydrolyzed by snake venom phosphodiesterase, whereas the poly(2'aminoC) was degraded nearly as fast as native poly(rC). Whereas poly(2'-azidoC) was resistant to attack by spleen phosphodiesterase, poly(2'-azidoU) was broken down, albeit at a slower rate than the native poly(rU). The poly(2'-aminoU) did not hybridize to poly(rA). The low propensity of 2'-aminopyrimidine nucleotides for duplex formation was recently reconfirmed in heteropolymers containing different degrees of 2'-amino substitutions (108, 109).

The expected resistance toward endoribonucleolytic hydrolysis was also observed for heteropolymeric 2'-O-methyloligoribonucleotides (110) and for poly(2'-fluoroU) (111). Both types of oligonucleotides were susceptible to degradation by certain phosphodiesterases. The 2'-O-methyl oligoribonucleotides proved unstable towards snake venom phosphodiesterase hydrolysis. The poly(2'-fluoroU), on the other hand, remained stable towards DNase I attack, but it was rapidly degraded by DNase II from spleen.

In summary, any 2'-modification that disallows attack on the adjacent phosphorous inhibits degradation of RNA oligonucleotides by endonucleases. Although several modifications slow down nucleases that operate by a hydrolytic mechanism, none were completely resistant to all nucleases studied. The 2'-fluoro modification is also able to serve as substrate for certain DNA-recognizing endonucleases.

Oligoribonucleotides modified only at the pyrimidine sites with 2'-amino- or 2'-fluoronucleosides are stabilized by three orders of magnitude compared to their native RNA analogues when incubated with rabbit serum (112). In this medium, degradation by a pyrimidine-specific endoribonuclease activity appears to be the major path of RNA breakdown. Furthermore, when stabilization of the pyrimidine positions by a combination of 2'-fluoro and 2'-amino modifications in a ribozyme sequence was coupled with stabilization against exonuclease attack by introduction of three or two phosphorothioate internucleotide linkages at the 3'- and 5'-terminus, respectively, the oligonucleotide half-life in concentrated fetal calf serum was increased to greater than 24 h(113). These data indicate that considerable stabilization of oligoribonucleotides in biological media can be achieved only by partial substitution with nucleotide and backbone analogues. The type and degree of modification required for stabilizing oligoribonucleotides in vivo remain to be seen.

Literature Cited

1. Limbach PA, Crain PF, McCloskey JA. 1994. *Nucleic Acids Res.* 22:2183–96
2. Gold L, Tuerk C, Allen P, Binkley J, Brown D, et al. 1993. *The RNA World*, ed. A Gespeland, J Atkins, pp. 497–508. Plainview, NY: Cold Spring Harbor Lab.
3. Tuerk C, Gold L. 1990. *Science* 249:505–10
4. Tuerk C, MacDougal S, Gold L. 1992. *Proc. Natl. Acad. Sci. USA* 89:6988–92
5a. Tuerk C, MacDougal-Waugh S. 1993. *Gene* 137:33–39
5b. Jensen KB, Green L, MacDougal-Waugh S, Tuerk C. 1994. *J. Mol. Biol.* 235:237
6. Jellinek D, Lynott CK, Rifkin DB, Janjic N. 1993. *Proc. Natl. Acad. Sci. USA* 90:11227–31
7. Bock LC, Griffin LC, Latham JA, Vermaas EH, Toole JT. 1992. *Nature* 355:564–66
8. Hayakawa H, Tanaka H, Miyasaka T. 1985. *Tetrahedron* 41:1675–83
9. Crouch GJ, Eaton BE. 1994. *Nucleosides Nucleotides* 13:939–44
10a. Asakura J, Robins MJ. 1990. *J. Org. Chem.* 55:4928–33
10b. Sy WW. 1990. *Synth. Commun.* 20:3391–94
10c. Asakura J, Robins MJ. 1988. *Tetrahedron Lett.* 29:2855–58
11. Scheit KH. 1966. *Chem. Ber.* 99:3884
12. Armstrong VW, Witzel G, Eckstein F. 1986. *Nucleic Acid Chemistry*, ed. LR Townsend, RS Tipson, pp. 65–69. New York: Wiley Interscience
13. Whale RF, Coe PL, Walker RT. 1992. *Nucleosides Nucleotides* 11:1425–42
14. Kunst A, Gassen HG. 1981. *Nucleic Acids Res.* Symp. Ser. 9:165–68
15. Armstrong VW, Sternbach H, Eckstein F. 1976. *Biochemistry* 15:2086–91
16a. Bergstrom DE, Ruth JL. 1975. *J. Am. Chem. Soc.* 98:1587–89
16b. Bergstrom DE, Ogawa MK. 1978. *J. Am. Chem. Soc.* 100:8106–12
16c. Bergstrom DE, Ruth JL. 1978. *J. Org. Chem.* 43:2870
16d. Bergstrom DE, Ruth JL, Warwick P. 1981. *J. Org. Chem.* 46:1432–41
16e. Bergstrom DE. 1982. *Nucleosides Nucleotides* 1:1–34
17. Crisp GT, Flynn BL. 1990. *Tetrahedron Lett.* 31:1347–50
18. Hobbs FW Jr. 1989. *J. Org. Chem.* 54:3420–22
19. Robins MJ, Barr P. 1983. *J. Org. Chem.* 48:1854–62
20. Hirota K, Kitade Y, Kanbe Y, Isobe Y, Maki Y. 1993. *Synthesis*, pp. 213–15
21. Nagamachi T, Fourrey J-L, Torrence PF, Waters JA, Witkop B. 1974. *J. Med. Chem.* 17:403–6
22. Otter BA, Saluja SS, Fox JJ. 1972. *J. Org. Chem.* 37:2858–63
23. Torrence PF, Spencer JW, Bobst AM. 1978. *J. Med. Chem.* 21:228–31
24. Barton DHR, Hedgecock CJR, Lederer E, Motherwell WB. 1979. *Tetrahedron Lett.* 279–80
25. Hirota K, Kitade Y, Kanbe Y, Yoshifumi M. 1992. *J. Org. Chem.* 57:5268
26. Bergstrom DE. 1982. *Nucleosides Nucleotides* 1:1–34
27a. Mamos P, Van Aerschot AA, Weyns NJ, Herdewijn PA. 1992. *Tetrahedron Lett.* 33:2413–16
27b. Van Aerschot AA, Mamos P, Weyns NJ, Ikeda S, De Clercq E, Herdewijn PA. 1993. *J. Med. Chem.* 36:2938–42
28. Sessler JL, Wang B, Harriman A. 1993. *J. Am. Chem. Soc.* 115:10418–19
29. Maeda M, Nushi K, Kawazoe Y. 1974. *Tetrahedron* 30:2677–82
30. Long RA, Robbins RK, Townsend LB. 1967. *J. Org. Chem.* 32:2751–56
31. Prakash TP, Kumar RK, Ganesh KN. 1993. *Tetrahedron* 49:4035-50
32. Haley BE, Hoffman JF. 1974. *Proc. Natl. Acad. Sci. USA* 71:3367–71
33. Cartwright L, Hutchinson DW, Armstrong VW. 1976. *Nucleic Acids Res.* 3:2331–39
34. Kohda K, Baba K, Kawazoe Y. 1990. *Tetrahedron* 46:1531–40
35. Jankowski AJ, Wise DS Jr, Townsend LB. 1989. *Nucleosides Nucleotides* 8:339-48
36. Norris AR, Kumar R, Buncel E. 1984. *J. Inorg. Biochem.* 22:11–20
37. Chiaruttini C, Expert-Bezancon A, Hayes D, Ehresmann B. 1982. *Nucleic Acids Res.* 10:7657–76
38. Buisson M, Rebound AM. 1982. *FEBS Lett.* 148:247–50
39. Uckert W, Wunderlich V, Ghysdael J,

Portetelle D, Burny A. 1984. *Virology* 133:386–92

40. Brewer LA, Goelz S, Noller HF. 1983. *Biochemistry* 22:4303–9
41a. Shapiro R, Cohen BI, Shiuey SJ, Maurer H. 1969. *Biochemistry* 8:238–45
41b. Shapiro R, Hachmann J. 1966. *Biochemistry* 5:2799–2807
42. Glass JD, Pelzig M. *Biochem. Biophys. Res. Commun.* 81:527–31
43. Wagner R, Garrett RA. 1978. *Nucleic Acids Res.* 5:4065–74
44. Moine H, Bienaimé C, Mougel M, Reinbolt J, Ebel J-P, et al. 1988. *FEBS Lett.* 228:1–6
45. Tukalo MA, Kubler M-D, Kern D, Mougel M, Ehresmann C, et al. 1987. *Biochemistry* 26:5200–8
46. Rasmussen N-J, Wikman FP, Clark BFC. 1990. *Nucleic Acids Res.* 18:4883–90
47. Rubin JR, Sabat M, Sundaralingam M. 1983. *Nucleic Acids Res.* 11:6571–86
48. Nitta N, Kuge O, Yui, S, Tsugawa A, Negishi K, Hayatsu H. 1984. *FEBS Lett.* 166:194–98
49a. Willis MC, Hicke BJ, Uhlenbeck OC, Cech TR, Koch TH. 1993. *Science* 262:1255–57
49b. Hanna MM, Dissinger S, Williams BD, Colston JE. 1989. *Biochemistry* 28:5814–20
50a. Wang H, Osborne SE, Zuiderweg ERP, Glick GD. 1994. *J. Am. Chem. Soc.* 116:5021–22
50b. Ferentz AE, Keating TA, Verdine GL. 1993. *J. Am. Chem. Soc.* 115:9006–14
50c. Erlanson DA, Chen L, Verdine GL. 1993. *J. Am. Chem. Soc.* 115:12583–84
50d. Stevens SY, Swanson PC, Voss EW Jr, Glick GD. 1993. *J. Am. Chem. Soc.* 115:1585–86
50e. Ferentz AE, Verdine GL. 1992. *Nucleosides Nucleotides* 11:1749–63
50f. Glick GD. 1991. *J. Org. Chem.* 56:6746–47
51. Goodwin JT, Glick GD. 1994. *Tetrahedron Lett.* 35:1647–50
52. Zaug AJ, Cech TR. 1986. *Science* 231:470–75
53. Uhlenbeck OC. 1987. *Nature* 328:596–600
54. Ellington AD, Szostak JW. 1990. *Nature* 346:818–22
55. Heidenreich O, Pieken W, Eckstein F. 1993. *FASEB J.* 7:90–96
56. Gold L, Polisky B, Uhlenbeck O, Yarus M. 1995. *Annu. Rev. Biochem.* 64:763–97
57. Stein CA, Cheng Y-C. 1993. *Science* 261:1004–12
58. Milligan JF, Matteucci MD, Martin JC. 1993. *J. Med. Chem.* 36:1923–37

59. Huryn DM, Okabe M. 1992. *Chem. Rev.* 92:1745–68
60. Moffatt JG. 1979. In *Nucleoside Analogues*, ed. RT Walker, E De Clercq, F Eckstein, pp. 71–163. New York: Plenum
61. Townsend LB. 1988. *Chemistry of Nucleosides and Nucleotides*, pp. 59–67. New York: Plenum
62. Yoshimura Y, Iino T, Matsuda A. 1991. *Tetrahedron Lett.* 32:6003–6
63. Matsuda A, Nakajima Y, Azuma A, Tanaka M, Sasaki T. 1991. *J. Med. Chem.* 34:2919–22
64. Verheyden JPH, Wagner D, Moffatt JG. 1971. *J. Org. Chem.* 36:250–54
65. Wagner D, Verheyden JPH, Moffatt JG. 1972. *J. Org. Chem.* 37:1876–78
66. Sproat BS, Lamond AI. 1991. *Oligonucleoeotdes and Analogues: A Practical Approach*, ed. F Eckstein, pp. 49–86. New York: Oxford Univ. Press
67. Sproat BS, Iribarren AM, Garcia RG, Beijer B. 1991. *Nucleic Acids Res.* 19:733–38
68. Lesnik EA, Guinosso CJ, Kawasaki AM, Sasmor H, Zounes M, et al. 1993. *Biochemistry* 32:7832–38
69. Matsuda A, Takenuki K, Sasaki T, Ueda T. 1991. *J. Med. Chem.* 34:234–39
70. Schmit C. 1994. *Synlett* 238–40
71. Imazawa M, Eckstein F. 1979. *J. Org. Chem.* 44:2039–40
72. Schmit C. 1994. *Synlett* 241–42
73a. Roush WR, Adam MA. 1985. *J. Org. Chem.* 50:3752–57
74b. McCombie SW, Metz WA. 1987. *Tetrahedron Lett.* 28:383–86
74c. Jung ME, Young H. 1989. *Tetrahedron Lett.* 30:6637–40
74. Deleted in proof
75. Deleted in proof
76. Imazawa M, Ueda T, Ukita T. 1975. *Chem. Pharm. Bull.* 23:604–10
77. Divakar KJ, Mottoh A, Reese CB, Sanghvi YS. 1990. *J. Chem. Soc. Perkin Trans.* 1:969–74
78. Marriott JH, Mottahedeh M, Reese CB. 1991. *Carbohydr. Res.* 216:257–69
79. Divakar KJ, Reese CB. 1982. *J. Chem. Soc. Perkin Trans.* 1:1625–28
80. Marriott JH, Mottahedeh M, Reese CB. 1990. *Tetrahedron Lett.* 31:2646–57
81. Moffat JG. 1964. *Can. J. Chem.* 42:599–604
82. Townsend LB, eds. 1991. *Chemistry of Nucleosides and Nucleotides, 2:* 81–112, 143–46. New York: Plenum
83. Eckstein F. 1985. *Annu. Rev. Biochem.* 54:367–402
84. Ludwig J, Eckstein F. 1989. *J. Org. Chem.* 54:631–35

85. Aurup H, Williams DM, Eckstein F. 1992. *Biochemistry* 31:9636–41
86. McGee D, Vargeese C, Zhai Y, Kirschenheuter G, Siedem C, Pieken W. 1995. *Nucleosides Nucleotides.* In press
87. Misra NC, Broom AD. 1991. *J. Chem. Soc. Chem. Commun.*, pp. 1276–77
88. Gaur RK, Sproat BS, Krupp G. 1992. *Tetrahedron Lett.* 23:3301–4
89. Bobst AM, Rottman F, Cerutti PA. 1969. *J. Mol. Biol.* 46:221–34
90. Guschlbauer W, Blandin M, Drocourt JI, Thang MN. 1977. *Nucleic Acids Res.* 4:1433–43
91. Hobbs J, Sternbach H, Sprinzl M, Eckstein F. 1973. *Biochemistry* 12:5138–45
92. Ikehara M, Fukui T, Kakiuchi N. 1978. *Nucleic Acids Res.* 5:1877–87
93. Hobbs J, Sternbach H, Sprinzl M, Eckstein F. 1972. *Biochemistry* 11:4336–44
94. Saenger W. 1983. *Principles of Nucleic Acid Structure*, ed. CR Cantor, pp. 55–69. New York: Springer-Verlag
95. Gerard GF, Rottman F, Boezi JA. 1971. *Biochemistry* 10:1974–81
96. Brooks RR. 1979. *Biochem. J.* 183:65–71
97. Deleted in proof
98. Droogmans L, Haumont E, De Nenau S, Grosjean H. 1986. *EMBO J.* 5:1105–9
99. Armstrong VW, Eckstein F. 1976. *Eur. J. Biochem.* 70:33–38
100. Pinto D, Sarocchi-Landousy M-T,

Guschlbauer W. 1979. *Nucleic Acids Res.* 6:1041–48
101. Ishihama A, Enami M, Nishijima Y, Fukui T, Ohtsuka E, Ikehara M. 1980. *J. Biochem.* 87:825–30
102. Hurwitz J, Yarbrough L, Wickner S. 1972. *Biochem. Biophys. Res. Commun.* 48:628–31
103. Milligan JF, Uhlenbeck OC. 1989. *Methods Enzymol.* 180:51–62
104. Deleted in proof
105. Sousa R, Chung YE, Rose JP, Wang B-C. 1993. *Nature* 364:593–99
106. Deleted in proof
107. Saenger W. 1983. *Principles of Nucleic Acid Structure*, ed. CR Cantor, pp. 174–75. New York: Springer-Verlag
108. Aurup H, Tuschl T, Benseler F, Ludwig J, Eckstein F. 1994. *Nucleic Acids Res.* 22:20–24
109. Miller PS, Bhan P, Kan L-S. 1993. *Nucleosides Nucleotides* 12:785–92
110. Sproat BS, Lamond AI, Beijer B, Neuner P, Ryder U. 1989. *Nucleic Acids Res.* 9:3373–87
111. Janik B, Kotick MP, Kreiser TH, Reverman LF, Sommer RG, Wilson DP. 1972. *Biochem. Biophys. Res. Commun.* 46:1153–60
112. Pieken WA, Olsen DB, Benseler F, Aurup H, Eckstein F. 1991. *Science* 253:314–17
113. Heidenreich O, Benseler F, Fahrenholz A, Eckstein F. 1994 *J. Biol. Chem.* 269:2131–38

Annu. Rev. Biochem. 1995. 64:865–96

THE NUCLEAR PORE COMPLEX

Laura I. Davis

Department of Genetics, Howard Hughes Medical Institute, Duke University Medical Center, Durham, North Carolina 27710

KEY WORDS: nuclear pore complex, nucleocytoplasmic transport, NE, protein transport

CONTENTS

ABSTRACT

The nuclear pore complex (NPC) creates an aqueous channel across the nuclear envelope through which macromolecular transport between nucleus and cytoplasm occurs. Nucleocytoplasmic traffic is bidirectional and involves diverse substrates, including protein and RNA. It is unclear whether import and export are mechanistically similar, but evidence suggests that numerous pathways may be involved. The discovery of filaments that extend out from each side of the NPC suggests that the NPC may also have a structural role, perhaps providing a connection between cytoskeletal elements of the nucleus and

865

cytoplasm. If this suggestion is valid, it remains to be determined whether this aspect of NPC function is related to its role in nuclear transport. This review discusses recent developments regarding the structure of the NPC, characterization of its constituent proteins (nucleoporins), the mechanism by which transport occurs, the function of individual nucleoporins, and the pathway of NPC assembly and disassembly.

PERSPECTIVES AND SUMMARY

The nuclear pore complex (NPC) provides the sole avenue for macromolecular transport between nucleus and cytoplasm. This ~124 megadalton (mDa) complex perforates the double lipid bilayer of the nuclear envelope (NE) of all eukaryotes and creates an aqueous channel that allows passage of substrates with diameters as large as 230 Å. We know relatively little about the composition of the NPC or the mechanism by which it regulates nucleocytoplasmic traffic; any other functions remain virtually unexplored.

Transport across the NPC is bidirectional and involves diverse substrates, including protein and RNA. Nuclear proteins, long thought to be transported unidirectionally, have now been shown to move in both directions (shuttling). This is also true of most small nuclear RNAs (snRNAs), which are first exported to the cytoplasm where they bind to snRNP proteins prior to import back into the nucleus. Newly synthesized, processed mRNAs must also cross the NE before they engage in translation. Proteins larger than ~40 kDa require an exposed nuclear localization sequence (NLS) in order to gain access to the nucleus. Whether RNA has an analogous signal is not known. Although we know that protein import and RNA export occur concurrently and utilize the same NPCs, we do not know whether the two processes are mechanistically similar, or whether proteins enter and exit using the same mechanism. Because RNA appears to be bound to protein throughout its lifetime, its transport across the NPC could be mediated by NLSs that can engage the NPC from either side. Alternatively, exit from the nucleus may use a fundamentally different process, perhaps involving signals present on RNA. Finally, some movement of proteins and RNA may be mediated by diffusion. Surprisingly, evidence supporting each of these models has recently begun to accumulate, suggesting that there may be numerous pathways out of the nucleus.

The observation that proteins can shuttle necessitates revisiting an old question about the relative contributions of transport and intranuclear binding to the ultimate disposition of macromolecules between the nucleus and the cytoplasm. It has been widely accepted that proteins concentrate in the nucleus because they are actively transported across the NPC. However, few studies have directly addressed this issue. Whether one views protein import as active transport or as facilitated diffusion profoundly affects the way one thinks about both the mechanism by which translocation occurs and that by which soluble factors influence nuclear accumulation.

Although some have speculated that the NPC may also play a structural role within the cell, few have investigated this aspect of NPC function. A particularly exciting development has been the discovery of filamentous structures extending out from each side of the NPC. On the nuclear side these filaments form a basket which, at least in some organisms, appears to be connected at the distal end to a regular hexagonal lattice interconnecting the NPCs. These striking micrographs suggest a novel function for the NPC as an integral part of cellular infrastructure, perhaps providing a connection between cytoskeletal elements of the nucleus and cytoplasm. One challenge in the next few years will be to determine whether this suggestion is valid and, if so, whether this aspect of NPC function is related to its role in nuclear transport.

Owing largely to difficulty in purifying the intact NPC, few of the proteins that compose this structure were identified until recently. Over the past few years the genes encoding six metazoan and ten yeast NPC proteins have been isolated. Although sequence analysis has provided few clues about function, the isolation of genes encoding the nucleoporins has opened the problem to genetic analysis, which has already begun to be fruitful. As more components of the NPC are identified, simplified in vitro assays should allow us to isolate and characterize those nucleoporins that recognize and transport substrates.

NPC ARCHITECTURE

Much of our current understanding of NPC structure comes from the analysis of amphibian oocytes, particularly those of *Xenopus laevis*. NPCs are packed in a paracrystalline array within the NE of these cells, and the NE detaches easily when manually isolated nuclei are placed on a grid, alleviating the requirement for chemical extraction.

Four basic elements of structure have been observed (1–8; see Figure 1). The waist of the NPC consists of a spoke-ring assembly anchored within a specialized region of the NE (referred to as the pore membrane), where the lipid bilayers of the inner and outer NE are fused. A central plug lies within the aqueous channel formed by the spoke-ring assembly. Extending from the rings are cytoplasmic filaments on one side and a nuclear basket or "fish trap" on the other. At present, models that incorporate all four structures prevail, and each element is discussed below as a distinct entity. However, no single method of specimen preparation preserves all of these structures, and thus their relationship to one another is not entirely clear.

Spoke-Ring Assembly

Two- and three-dimensional density maps of the spoke-ring assembly have been generated from negatively stained preparations of detergent-released

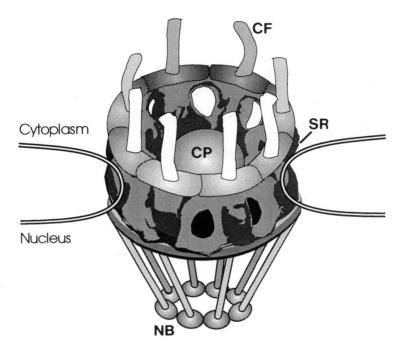

Figure 1 Structural elements of the NPC, including the spoke-ring complex (SR), central plug (CP), cytoplasmic filaments (CF), and nuclear basket (NB).

NPCs (2, 3) or frozen/hydrated NPCs in amorphous ice (1). Although different preservation conditions and image processing techniques were used in each study, the resulting maps are remarkably similar. This structure consists of nuclear and cytoplasmic rings connected to one another by two sets of eight spokes that form a waist at the pore membrane (see Figure 2). The spokes extend into the center of the pore, where they join at their ends to form an inner annulus surrounding a channel of ~40 nm.

Hinshaw et al (2) delineated four morphologically distinct subunits (ring, annular, columnar, and lumenal) that together make up a spoke-ring unit (see Figure 2). This unit is repeated eight times around the circumference of the pore and in mirror image across the nucleocytoplasmic plane. The two halves of the assembly are thus likely to be held together by homotypic interactions between the top and bottom subunits. The annular subunits are oriented diagonally, connecting the top of one spoke to the bottom of the adjacent one and giving the pore its handedness (1, 2). The column subunits form the arms of the spokes and connect the annular subunits to the rings. They also connect to

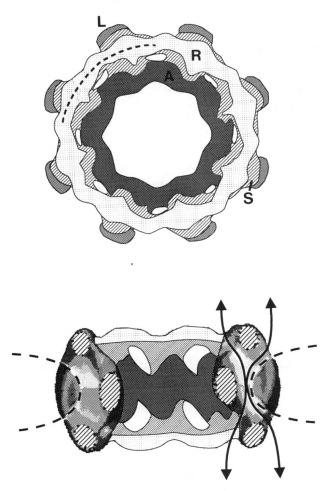

Figure 2 Surface views of a 3-dimensional electron density map of the NPC calculated from negatively stained samples: En face *(top)* and cut away side view *(bottom)*. The dotted line in the bottom panel represents the position of the nuclear envelope, while the arrows show the possible alternate routes for passive diffusion through the peripheral channels. The four subunits of the NPC are as follows: (A) annular subunits, (R) rings, (L) lumenal subunits, (S) column subunits. (Adapted from Ref. 2).

the lumenal subunits, the bulk of which lie in the lumenal space of the NE. This interaction probably anchors the spoke-ring complex in the pore membrane. There is a substantial space between the lipid bilayer and the peripheral subunits (Figure 2), suggesting that the NPC does not pose a barrier to diffusion of integral membrane proteins from the outer to the inner NE (1, 2).

In addition to the large central channel, both maps show separate smaller channels that could allow passive diffusion of small proteins and solutes (1, 2), but the maps differ about their position. Figure 2 shows eight peripheral channels about 10 nm in diameter formed between adjacent columnar subunits, the pore membrane, and the inner annulus in demembranated NPCs (2). However, analysis of membrane-associated NPCs instead revealed more internal channels lying between the central transporter and the inner annulus of the spoke-ring complex (1). Further analysis will be required to resolve these inconsistencies.

Based on mass estimates using quantitative scanning transmission electron microscopy (EM) (7), an average molecular mass of 1.75 mDa (about 20 proteins) was calculated for each subunit, assuming that the spoke-ring assembly makes up the entire mass of the intact NPC except for the central plug (2). However, this assembly lacks the distinct asymmetry between nuclear and cytoplasmic rings evident in the preparations used for mass estimates, which probably retained ring-associated structures that are lost upon detergent extraction (7). Therefore, the spoke-ring complex described by Hinshaw et al (2) may actually correspond to the "spoke" structure described by Reichelt et al (7), in which case one obtains a figure closer to 800 kDa per subunit.

Central Plug

Viewed as an aqueous conduit, the 420-Å-diameter channel created by the annular subunits is wide enough to accommodate the largest substrates able to cross the NPC, which have diameters of ~230 Å (9). However, the channel is often occluded with electron-dense material. Due to the variable presence and inconsistent structure of this material in negatively stained preparations, Hinshaw et al (2) refrained from including it in reconstructions. In contrast, about 80% of detergent-released NPCs viewed in amorphous ice retained central densities that showed good rotational symmetry, leading Akey & Radermacher (1) to model this structure as a cylindrical transporter closed at each end and tapered inwardly at the pore waist in an hourglass configuration. Previously, colloidal gold probes had been used to delineate two distinct substrate configurations with respect to the central transporter—one population at the rim and another over the center (10). The occluded ends of the transporter could thus reflect substrates docked over a central transport channel. Alternatively, the channel may be gated at each end (1). Because the internal diameter of the transporter channel at its narrowest point is only 10 nm, it would have to be able to expand significantly in order to accommodate large substrates (1).

The variable presence and inconsistent shape of the central plug raise the possibility that it is not a true NPC structure but instead consists of material caught in transit through the central channel at the time of preparation. If the shape of traversing particles is defined by the boundaries of the spoke-ring complex, then one might expect them to assume a fairly consistent hourglass configuration. The plug could also consist of peripheral structures that have retracted into the channel during preparation. In fact, specimens in which nuclear baskets are visible rarely show occlusion of the central channel (4, 5), suggesting that these two structures may be one and the same. Treatment with chelating agents, which dissociates the nuclear basket, led to loss of the central plug in negatively stained preparations (4).

Peripheral NPC Components

Although NPCs prepared as described above are largely free of attached material, fibrous structures extending from both rings were observed many years ago in negatively stained thin sections through *Xenopus* oocytes (8). Field emission in lens scanning EM (5, 6, 11, 12) and transmission EM analyses of quick-frozen/freeze-dried metal-shadowed preparations (4) have now elucidated these as integral components of the NPC, which may have a much more open and extended structure than previously appreciated.

Views of the cytoplasmic face of *Xenopus* or *Triturus* oocyte NPCs confirmed the presence of eight short cylindrical filaments that extend into the cytoplasm from the regions between each ring subunit (4, 5; see Figure 3). In some preparations, these filaments appear T-shaped (5), suggesting that they contain two subunits. Extraction or collapse of one or more of the subunits onto the cytoplasmic ring could explain why it is denser than the nuclear ring (7) and why the presence of particles associated with the ring varies from sample to sample (3). Occasionally, filaments are observed to extend between cytoplasmic rings of adjacent complexes, suggesting that they are interconnected (4, 5). However, the paucity of these connections argues that they may result from coincidental alignment of collapsed filaments. Nucleoplasmin adsorbed to colloidal gold has been observed to align on filaments extending off the cytoplasmic surface of the NPC (13). The cytoplasmic filaments may thus serve as docking sites for incoming protein substrates. They could also be involved in connecting the NPC to cytoskeletal elements.

The structure observed on the nuclear side is in the shape of a basket or "fish trap" and looks very different from the cytoplasmic filaments (4–6, 12; see Figure 3), showing clearly that the two sides of the NPC are asymmetric. The basket is composed of thin filaments that extend 50–100 nm out from the nuclear ring and terminate in a distal annulus (4–6, 12). While the annulus

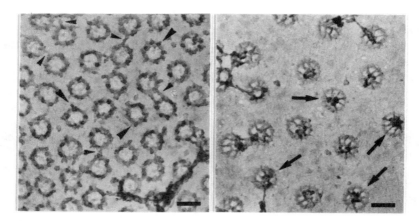

Figure 3 Electron micrographs of the cytoplasmic (*left*) and nuclear (*right*) faces of quick-frozen/freeze-dried/rotary metal–shadowed spread *Xenopus* NEs. The large arrowheads mark examples of collapsed cytoplasmic filaments protruding from the cytoplasmic rings of NPCs. Small arrowheads mark fibers that reach between two NPCs. Arrows at right mark NPCs with relatively well-preserved baskets. Scale bars, 100 nm (adapted from Ref. 4).

may be structurally distinct, it often appears to result from aggregation of terminal knobs present on each filament. Chelating agents cause the filaments to dissociate reversibly at the distal ends (4), raising the possibility that the conformation of the basket is regulated physiologically. Because the FG nucleoporin nup153 contains Zn^{2+} fingers and has been localized specifically to the distal ring of the baskets (14), this protein is an excellent candidate to be a target of such regulation.

In *Triturus* oocytes (and to a lesser extent in *Xenopus*), field emission scanning EM has revealed a regular hexagonal meshwork of 8–10-nm filaments (the NE lattice) in which the distal annuli of the baskets appear to be embedded (11). This structure is distinct from the nuclear lamina, which is located more closely apposed to the NE; its composition is not known. The lattice appears to be labile and is either absent or disordered over much of the surface, particularly in *Xenopus* (11). Its presence has not been demonstrated in somatic cells, and it is unclear whether this structure is peculiar to amphibian oocytes or whether the relative paucity of NPCs in somatic cells makes it less easy to identify. The lattice could also be an artifact, perhaps resulting from aggregation of NPC baskets that have pulled away from the nuclear rings. The relative lack of NPCs immediately under the lattice would tend to support this view but the highly ordered structure argues against it . Furthermore, the lattice is best visualized when nuclei are processed quickly, suggesting that swelling of the nucleus upon longer incubations causes tearing or dissociation (11).

COMPOSITION OF THE NPC

Molecular weight estimates from quantitative scanning transmission EM (7) suggest that the NPC contains at least 100 different polypeptides (nucleoporins). These estimates assume that the NPC is composed entirely of protein. Density measurements of enriched NPCs from yeast (15) agree with this assumption, and no NPC-specific RNA species have been identified. However, abundant cytoplasmic ribonucleoprotein (RNP) particles termed vaults (16) have an eight-fold symmetrical structure similar to that of the central plug of the NPC (17). Antibodies to the major vault protein decorate the NPC (18), although this finding should be interpreted with caution because of the abundance of vaults in the cytoplasm. Peripheral NPC structures probably do not contain a significant RNA component, because they are insensitive to RNAse treatment (4, 5).

As yet only a fraction of nucleoporins have been identified. Efforts to purify the NPC have been hampered by the lack of a functional assay to monitor purification. In addition, NPCs are much more labile than the structures in which they are embedded, so that extraction conditions required to release them from the nuclear lamina (and perhaps other nuclear and cytoplasmic structures) lead to the loss of morphologically identifiable complexes. A fraction highly enriched in NPCs has now been obtained from *Saccharomyces cerevisiae* (15). Approximately forty major polypeptides and an equal number of minor ones co-enrich in this fraction (15). This number is somewhat smaller than predicted by mass estimates of metazoan NPCs (7), and morphological analysis of isolated yeast NPCs suggests that they lack components present in metazoans (15). Isolated yeast NPCs are both narrower and thinner than those of *Xenopus*, and lack clear outer nuclear and cytoplasmic rings, although the inner spoke-ring assembly and central plug are apparent. However, negatively stained yeast nuclei showed peripheral structures not present in the enriched preparations (15), which resemble the nuclear "fish trap" and connecting lattice visible in amphibian oocytes (11). Thus, the smaller mass estimate may to some extent reflect loss of subunits during enrichment.

Pore Membrane Proteins (Poms)

Three integral proteins specific to the pore membrane have been identified and molecularly characterized (19–23; see Table 1). gp210 was first identified as the major ConA-binding protein of rat liver pore complex-lamina fractions (23). It appears to be evenly distributed around the waist of the NPC and is oriented with its 200-kDa amino terminal domain in the lumen of the NE (24). gp210 has two hydrophobic segments near the amino terminus (21), but only the proximal one serves as a transmembrane domain (24). The transmembrane domain is sufficient to target gp210 to the pore membrane, suggesting that it may associate with other poms in the context of the lipid bilayer (25). By analogy to viral fusogens, it has been suggested that conformational changes could trigger the second hydrophobic domain to interact with the lipid bilayer and facilitate fusion to form the pore membrane (21, 24). Microinjection of mRNA encoding antibody against the lumenal domain of gp210 caused inhibition of nuclear import (26), suggesting that conformational changes can be transmitted across the pore membrane. Whether such changes participate in physiological regulation of nuclear transport remains to be determined.

Highly enriched NPCs from *Saccharomyces cerevisiae* also contain a single ConA-binding protein of 152 kDa (20; see Table 2). This integral membrane protein, Pom152p, is novel and does not resemble mammalian gp210 (20). Pom152p was efficiently targeted to the NPC when expressed in mammalian cells, suggesting that the signals for pore membrane localization are conserved and that metazoans may contain a functional homologue of *POM152* (20). Notably, disruption of *POM152* produced no detectable phenotype.

The other known mammalian pom (pom121) also has a single transmembrane domain (22), but in this case the bulk of the protein is exposed to the spoke-ring complex (27). This domain contains XFXFG repeats characteristic of a the FG nucleoporins described below (22; see Figure 4). pom121 could target the peripheral FG nucleoporins to the NPC via interaction between the repetitive domains, although the repeats of individual nucleoporins do not appear to be necessary for localization to the NPC (28–31). However, if several nucleoporins form a complex, then a single repetitive domain may be sufficient to tether it to the membrane.

FG Nucleoporins

This family of nucleoporins consists of eight major (and numerous minor) proteins enriched in NE preparations. It was first identified in mammalian cells by virtue of reactivity with anti-NPC monoclonal antibodies and the lectin WGA (19, 32–35; see Table 1, Figure 4). The same antibodies recognize a similar set of NPC proteins in *Saccharomyces cerevisiae* (36–38; see Table 2, Figure 4). All except pom121 are peripheral proteins.

GLYCOSYLATION   All known metazoan FG nucleoporins are modified by the addition of single O-linked N-acetylglucosamine residues to serine and threonine (32, 33, 39, 40). This modification is catalyzed by a cytosolic transferase (41, 42) within 5 min of synthesis (33). It has been found on an increasing number of cytoplasmic, nuclear, and cytoskeletal proteins (for review see 43) and may be as ubiquitous as glycosylation of membrane and secretory proteins. So far, attempts to demonstrate O-linked GlcNAc addition in yeast have failed. It is not clear whether this modification is absent in lower eukaryotes or

Table 1 Metazoan nucleoporins

Protein	M_{app} (kDa)	XFXFG repeats	Localization	Complex	Comments	References
nup250	250	(yes)[a]	cytoplasmic rings and filaments	p75	may be identical to nup214(CAN)	58
nup214(CAN)	225	yes	cytoplasmic rings and filaments	—	may be identical to nup250. Fused to DEK or SET	54
nup180	180	(no)[b]	cytoplasmic rings and filaments			57
nup155	180	no	—			188
nup153	190	yes	distal annulus of nuclear basket	dimer	contains cys_2 cys_2 Zn^{++} fingers	14, 58, 60
p75	75	(no)[b]	cytoplasmic rings and filaments	p250		58
p62	62	yes	spoke-ring complex and trans-porter	p58, p54	p62/p58/p54 complex required for protein docking	34, 58, 59
p58	58	(no)[b]	spoke-ring complex and trans-porter	p62, p54	p62/p58/p54 complex required for protein docking	58, 59
p54	54	(no)	spoke-ring complex and trans-porter	p62, p58	p62/p58/p54 complex required for protein docking	58, 59
gp210	190	no	pore membrane	—	large amino terminal domain in lumen of NE. Antibodies to lumenal domain block transport in vivo.	21, 23–26
POM121	145	yes	pore membrane	—	large carboxyl terminal domain faces NPC	22, 27

[a] Tentative, based on the fact that protein is recognized by anti-FG nucleoporin antibodies.
[b] Tentative, based on the fact that the protein is not recognized by anti-FG nucleoporins.

Table 2 Yeast nucleoporins

gene	MW$_{app}$ (kDa)	FG repeat type	MAb reactivity	Mutant phenotypes	Synthetic lethality	Complex	Comments	References
NUP1	130	KPXF$^{S/}_{T}$FG FGXN	306 414	protein import, mRNA export, NE detachment	*NUP2* *SRP1*	Srp1p		31, 36, 50
NUP2 (*NSP2*)	95	KPXF$^{S/}_{T}$FG	306 414	not essential	*NUP1* *NSP1* *SRP1*	Srp1p		28, 50
NSP1	110	KPAFSFGAK FGXN	306 414	protein import, defective at docking step	*NUP2* *NUP49* *NUP116* *NUP145*	Nic96p Nup49p Nup54p	carboxyl terminal coiled coil domain with similarity to metazoan p62	30, 48, 53, 111, 112, 74
NUP49 (*NSP49*)	49	GLFG	192	protein import, defective at translocation step	*NSP1*	Nic96 NSP1p Nup54p		37, 52, 53, 74

Gene	MW	Motif	Function	Homologous genes	Interacting proteins	Description	References
NUP54	54	—	—	—	Nic96 Nup49p Nsp1p	—	53
NUP100	100	GLFG	not essential	*NUP116*	—	RNA binding motif within 190 amino acid region of similarity to *NUP116* and *NUP145*	37
NUP116 (*NSP116*)	118	GLFG FGXN	RNA export, membrane seal formed over NPC	*NSP1* *NUP100* *NUP145*	—	RNA binding motif within 190 amino acid region of similarity to *NUP100* and *NUP145*	29, 37, 52, 64
NUP145	65 80	GLFG	RNA export, (protein import), reticulation of NE and clustering of NPCs	*NSP1* *NUP116* *NUP100*	—	RNA binding motif within 190 amino acid region of similarity to *NUP100* and *NUP116*	29, 51
NIC96	90	none	—	—	Nsp1p Nup49p Nup54p	—	53
POM152	150	none	not essential	—	—	integral membrane protein	20

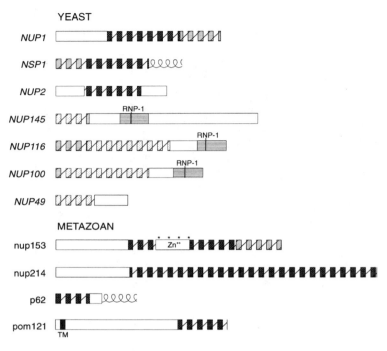

Figure 4 Domain structure of the FG nucleoporins from *Saccharomyces cerevisiae* (*top*) and metazoans (*bottom*). Repetitive domains are indicated by filled short boxes: XFXFG (darkly hatched), FGXN repeats (intermediate), and GLFG repeats (lightly hatched). Blank boxes indicate unique domains.

whether glycosidases released during sample preparation cleave the sugars too quickly to allow detection.

The function of O-linked GlcNAc is unknown. All of the sugar residues in p62 fall within a serine/threonine-rich linker between the repetitive domain and the carboxyl terminal alpha-helical region (44), leading to the suggestion that the sugars create a structurally rigid and extended linker between the two functional domains of p62. Another possibility is that glycosylation reversibly masks regulatory phosphorylation sites (43). The consensus sequence for GlcNAc addition is loose but appears to require a proline within three amino acids of the target serine or threonine (43, 44). This consensus sequence encompasses that of a number of proline-directed kinases. Kelly et al (45) have now shown that numerous YSPTSPS repeats within the CTD of RNA polymerase II are O-glycosylated and that phosphorylation and glycosylation appear to be mutually exclusive. Because phosphorylation of the CTD is thought to regulate promoter association (for review see 46), reversible glycosylation could affect this process. It should be noted, however, that a physiological effect of O-linked GlcNAc has yet to be demonstrated. Furthermore, addition of terminal galactosamine residues to the sugars on the FG nucleoporins prior to in vitro nuclear reconstitution had no apparent affect on either NPC formation or protein import capacity (47).

DOMAIN STRUCTURE   All of the FG nucleoporins share domains consisting of repeated short sequence motifs, either KPXFS/TFG, GLFG, or FGXN (see Figure 4). These repeats are separated by spacers that are either highly charged or rich in polar residues. The repetitive motifs are probably distantly related to one another and are predicted to assume a β-sheet conformation (36, 48, 49). The anti-FG monoclonal antibodies recognize the repeats (36, 37, 50), although some also appear to require O-linked N-acetylglucosamine (19, 39). It is likely that the subset of proteins recognized by each antibody reflects the type of repetitive domain it prefers (see Table 2).

The function of the repetitive domains is a mystery. The XFXFG and GLFG repeats are not necessary for localization or function of the yeast nucleoporins Nsp1p (48) or Nup145p (29), and they can be deleted with little affect on growth. As mentioned above, however, similar domains of different nucleoporins may have redundant functions, particularly if the proteins operate as a complex. Strains deleted for the XFXFG repeats of both *NUP2* and *NSP1* are viable (28), suggesting that at least two are dispensable. Although some double and triple truncations/deletions that include the GLFG repeats of *NUP116*, *NUP100*, and *NUP145* are synthetically lethal, these deletions also remove all or part of another domain that is conserved among all three proteins (29, 51; see Figure 4). Truncation of the carboxyl terminal domain of Nup1p, which contains a degenerate FGXN domain, leads to inviability (31).

In addition to the shared repeats, each of the FG nucleoporins has distinct domains that appear to mediate localization and function. The carboxyl terminal domains of Nsp1p and Nup145p are sufficient for viability (29, 48), and the synthetic lethality observed when deletions of *NUP2* are combined with *nup1* or *nsp1* mutants can be rescued by the unique amino terminal domain of Nup2p (28). All of the mutations/truncations affecting function have so far been localized to the unique domains (29, 31, 48, 52).

Nucleoporin Interactions and Localization

YEAST NUCLEOPORINS   The seven known yeast FG nucleoporins fall into two subfamilies: *NUP1*, *NUP2*, and *NSP1* encode proteins with XFXFG repeats whereas *NUP100*, *NUP116*, *NUP49*, and *NUP145* gene products have GLFG repeats (see Table 2, Figure 4). Nsp1p and Nup49p are physically associated in a complex that also contains the uncharacterized FG nucleoporin Nup54p, as well as a novel essential protein, Nic96p (53). Temperature-sensitive mutations within the region of *NSP1* encoding the heptad repeat domain are synthetically lethal with mutations of *NUP116, NUP49,* and *NUP145* (29, 52). In the case of *NSP1* and *NUP49,* the synthetic phenotype is most easily explained by destabilization of the complex. Other cases are more difficult to interpret, such as that of Nsp1p and Nup116p [which do not appear to be associated (53)], but could reflect destabilization of partially redundant complexes. This may also be the case for *NUP1* and *NUP2,* truncations of which are synthetically lethal (28). Both of the encoded proteins can associate with the Srp1 protein, but in separate complexes (50). As discussed below, a *Xenopus* homologue of *SRP1* has recently been shown to encode a cytosolic factor required for binding of nuclear proteins at the NPC (50a). Thus, Nup1p and Nup2p may function as partially redundant docks for Srp1p/substrate complexes.

Functional redundancy is clearest in the case of Nup145p, Nup100p, and Nup116p, which share a 190-amino-acid stretch that exhibits 78% similarity between Nup100p and Nup116p, and 55% similarity between these two Nups and Nup145p (29, 51). This domain contains a conserved octapeptide similar to the RNP-1 consensus found in single-stranded nucleic acid binding proteins (29). Both Nup145p and Nup116p bind poly(G) in vitro, but not ssDNA, dsDNA, or other homopolymers (29). Single or double deletions of this domain are viable, whereas removal of all three is lethal (29, 51), suggesting that the nucleic acid binding domain is essential but functionally redundant. These three nucleoporins cannot be completely redundant, however, because *NUP145* is essential (29) and strains disrupted for *NUP116* are lethal in combination with either a *NUP100* deletion or an amino terminal truncation of *NUP145* (51). So far, no partners for these proteins have been identified.

METAZOAN NUCLEOPORINS   Four of the eight major metazoan FG nucleoporins have been molecularly characterized (14, 22, 44, 54–56; see Table 1, Figure 4). All of these share XFXFG repeats, outside of which they are unique. As mentioned above, pom121 is an integral membrane protein whose repeat domain is exposed to the NPC (22, 27). All of the others are peripheral proteins that localize predominantly to the nuclear baskets and cytoplasmic filaments, as discussed below. The exception is p62, which may be part of the central plug or spoke-ring complex (10, 57, 58). Because p62 is part of a stable complex with two other proteins, p58 and p54 (58, 59), the latter are likely to have a similar localization.

Anti-nup153 antibodies stain the nuclear side of the NPC (14, 19, 58, 60). In formaldehyde-fixed, negatively stained preparations, the antibodies decorate filamentous material that extends up to 600 nm out from the nuclear face of the NPC (60), whereas in quick-frozen/metal-shadowed preparations, staining is specific for the distal ring of the nuclear baskets (58). Because the baskets were not well preserved in fixed preparations, the observed filaments may have reflected dissociated or "unraveled" baskets. On the other hand, more extended structures [for example the NE lattice (11)] may be missing from the frozen specimens, leaving only residual staining at the distal ring. Localization to the nuclear basket is particularly interesting in light of the fact that nup153 has four cys_2-cys_2 Zn^{++} fingers embedded within the XFXFG repeats, and exhibits Zn^{++}-dependent binding to *E. coli* DNA in vitro (14). Based on this observation, it was suggested that nup153 might bind to specific regions of chromatin and localize them to the vicinity of the NPC (14). The fact that EDTA causes reversible dissociation of the basket filaments at the distal end (4) raises the possibility that nup153 could mediate divalent cation-dependent aggregation of filament termini to form the distal ring, perhaps through homotypic protein interactions between neighboring filaments. Binding to chromatin (perhaps at repetitive sequence elements) or to an RNP transport substrate approaching the NPC could produce the same result. The ability of Nup153 to bind RNA has not been tested.

At least three proteins are localized exclusively on the cytoplasmic rings and filaments, including the FG nucleoporins nup214(CAN) (54) and nup250 (58), which may be the same protein. Upon gentle extraction, p250 remains associated with a 75-kDa protein in a complex of at least 1.5–2 mDa (58). Antibodies against a 180-kDa NPC protein also decorate the cytoplasmic filaments (57). Neither p75 nor p180 binds WGA or is recognized by the anti-FG repeat antibodies (57, 58).

nup214(CAN) has been identified as a fusion partner in two putative oncogenic chimeras (61, 62). In each case, it is the large carboxyl terminal XFXFG domain that is translocated onto foreign proteins. The fact that nup214(CAN) is present in both suggests that changes in the dynamics of nucleocytoplasmic transport could result in oncogenesis (54). However, it is also possible that the chimera has a novel function, perhaps conferred by mislocalization of nup214.

Although we are just beginning to determine the relationships among the nucleoporins, one concept has already emerged clearly. Individual FG nucleoporins are components of different complexes, which appear to be distributed throughout the subunits of the NPC. A crucial question is whether their functions are as diverse as their localization, or whether there is a continuity that we have yet to understand. It is somewhat puzzling that, aside from similarity between the coiled coil domains of p62 and Nsp1p (55), there is no significant homology between yeast and metazoan FG nucleoporins outside the repetitive domains. It is possible that the unique domains of the FG nucleoporins evolved late and confer quite different functions upon yeast and metazoan proteins. This possibility is difficult to reconcile with the apparent conservation in structure and function of the NPC.

Because all of these proteins are positioned within regions of the NPC likely to be in direct contact with transport substrates, one intriguing possibility is that the repetitive domains provide an array of substrate binding sites that spans the entire NPC. The repeats themselves, or the spacer sequences that separate them, could provide such sites. As discussed in the next section, substantial evidence indicates that the FG nucleoporins as a group are required for both the docking and translocation steps of protein import. An attractive feature of this model is that limited redundancy is easy to visualize, because each part of the NPC appears to have more than one FG nucleoporin. Furthermore, it suggests that the essential domains of these proteins might be required because they contribute to the structural integrity of individual NPC subunits, which may have evolved differently in metazoans and yeast. A structural role for the FG nucleoporins is clearly demonstrated by the morphological abnormalities observed in yeast strains carrying mutations or deletions in *NUP116* (64), *NUP145* (51), and *NUP1* (31).

NPC FUNCTION

Protein Import

Many nuclear proteins far exceed the diffusional limit of the NPC (65) and yet are efficiently transported (66–70). Even small proteins such as histones exhibit saturable import kinetics (71). These results suggest that proteins destined for the nucleus engage either a facilitated diffusion or an active transport pathway. Selectivity is governed by the presence of a nuclear localization sequence (NLS), which consists of one or two short, largely basic tracts of amino acids and is both necessary and sufficient to engage the import machinery (72).

Much of our current knowledge about the mechanism of protein import comes from in vitro assays that faithfully reconstitute this process. These assays measure the nuclear accumulation of a fluorescently labeled protein in either semipermeabilized cells (73, 74) or isolated nuclei (75, 76). Docking of substrates at the NPC can be separated from the subsequent translocation step by incubation at 4°C, or in the absence of ATP (13, 77). The fact that bound substrate can be chased into the nucleus by addition of factors required for translocation (78) provides strong evidence that binding is a true intermediate in the transport pathway.

DOCKING This step requires a functional NLS and added cytosol (13, 73, 77–79). One of the cytosolic factors was predicted to be an NLS-binding protein, based on in vivo studies suggesting that nuclear proteins bind to a saturable cytosolic factor prior to import (71). A candidate NLS receptor, originally identified by chemical cross-linking to synthetic NLS peptides (80), was subsequently shown to increase nuclear import when added to highly diluted cytosolic fraction (81), and to reconstitute binding completely when combined with a 97-kDa protein purified from the cytosolic fraction (82). The second factor is NEM sensitive (82) and is probably analogous to one of the NEM-sensitive factors previously identified in *Xenopus* extracts (78, 79).

Görlich et al (50a) have now cloned the gene encoding one of the cytosolic factors from *Xenopus* extracts. This factor (termed importin) is similar in molecular weight to the candidate NLS receptor described above and may be the same protein. Importin is 58% similar to yeast Srp1p. Srp1p binds to the FG nucleoporins Nup1p and Nup2p, and mutations in *SRP1* are synthetically lethal with *nup1* mutants (50). These results suggest that importin and Srp1p are functional homologues that target NLS-containing proteins to the NPC via association with the FG nucleoporins. Previous studies in metazoan systems support this model. Depletion of cytosol by preincubation with FG nucleoporins immobilized on WGA-Sepharose rendered it inactive, suggesting that at least one cytosolic factor interacts directly with one of the FG nucleoporins (83). Furthermore, depletion of WGA-binding proteins from *Xenopus* extracts prior to nuclear reconstitution resulted in the formation of NPCs that were incapable of docking (84). Although addition of WGA directly did not interfere with docking, either in reconstituted nuclei incubated with *Xenopus* extracts (77, 78) or upon injection into intact cells (85, 86), WGA completely blocked docking when added to semipermeabilized cells supplemented with purified factors (82).

Because depletion of the p62/p58/p54 complex is sufficient to prevent docking (59), this nucleoporin complex is a good candidate for the docking site in metazoan cells. However, anti-p62 antibodies and WGA decorate the center of the NPC, suggesting that this complex is localized at the central transporter or the spoke-ring complex (10, 57, 58). Such a location is somewhat difficult to reconcile with the observation that colloidal gold particles absorbed with nuclear proteins accumulated on filaments extending from the cytoplasmic side of the NPC (13). It is possible that binding to filaments serves to concentrate proteins at the NPC, but is not absolutely required for docking. Alternatively, depletion of proteins that form more internal ring subunits could also release the filaments. The filaments are composed of at least three proteins, nup180 (57), p75 (58), and nup214(CAN) (54), which is probably identical to p250 (58). nup214(CAN) binds WGA, and is therefore also a candidate for the docking protein (54). Antibodies to p180 do not inhibit docking either in vitro or in vivo.

In some assays hsp70 also appears to be required for protein import. Microinjection of anti-hsp70 antibodies inhibited transport of SV40 large T antigen (87, 88), as did depletion of hsp70 from semipermeabilized cells (89). These experiments are difficult to interpret because of the pleiotropic effects likely to result from depletion of hsp70. However, a direct role in presenting the NLS is suggested by experiments in which the expression of human hsp70 restored nuclear transport of mutant SV40 large T antigen in mouse cells, where the mutant protein otherwise remained cytoplasmic (90). These experiments suggest that heat shock proteins may be required to keep the NLS of some proteins exposed, such that they can bind to the NLS receptor.

A number of other NLS-binding proteins have been identified in both yeast and metazoan cells (91–100). Many of these are nucleolar proteins and have motifs typical of RNA-binding proteins (91, 98–101). NLS binding could be a fortuitous result of their extensive stretches of acidic residues. Alternatively, because many of these proteins shuttle between the nucleus and cytoplasm (91, 102), they could play ancillary roles as NLS receptors, perhaps targeting nuclear proteins after import.

TRANSLOCATION   Efficient translocation of nuclear proteins requires cytosolic factor(s) distinct from those involved in docking (78, 82). Fractionation of *Xenopus* cytosols identified a fraction (B) which could reconstitute import when combined with fraction A, which is required for binding (78). Furthermore, fraction B induced translocation of docked substrates when added to washed nuclei previously incubated with fraction A. The active component from fraction B was subsequently found to be the small G-protein Ran-TC4 (103). Ran-TC4 was also identified as a mediator of GTPγS inhibition of nuclear import in permeabilized cells, where it was further shown to be required for import (104). The guanine nucleotide exchange factor for Ran, RCC1, is an abundant protein that has been implicated in numerous nuclear functions, including control of DNA replication, RNA export, and transcription (reviewed in 105). All of these effects could stem from a requirement for a GTPase-mediated targeting pathway for nuclear import (103, 104).

The exact role of Ran remains to be elucidated. While it is required only for the translocation step (103), addition of nonhydrolyzable GTP analogues inhibited both binding and translocation (103, 104). This inhibition could be explained if Ran binds substrates in the cytoplasm and carries them to the nucleus (106). However, the fact that docking could be reconstituted in the absence of Ran (82) argues against such a model and suggests that GTP hydrolysis might be required to release proteins docked at the NPC, thus allowing the next step of translocation to take place and freeing docking sites for new binding. Whether Ran itself is responsible for this step is not clear. Ran alone was not sufficient to restore normal levels of import (103), and GTPγS inhibition was very sensitive to the concentration of extracts used (104).

These results suggest that there may be other GTPases and/or exchange factors involved that have yet to be identified. Several GTP-binding proteins cofractionate with NEs (107, 108), but their identity remains to be determined.

It is still unclear which NPC proteins directly mediate the translocation step, but again a number of studies have implicated one or more of the FG nucleoporins. This step is inhibited by WGA or anti-FG repeat antibodies (35, 74, 77, 101–103, 109, 110) neither of which interfere with passive diffusion of small dextrans (35, 101). Steric effects cannot be entirely ruled out however, because the diffusion channels may be spatially separate from the central transport channel (1, 2). Yeast strains carrying temperature-sensitive mutations in the FG nucleoporins *NUP1*, *NSP1*, or *NUP49* show inhibition of nuclear import after shift, both in vivo (31, 111, 112) and in vitro (74). Whether the defect is in docking or translocation is difficult to discern in vivo. However, in vitro assays suggest that *nsp1* mutants are defective specifically for translocation (74). Deletion or mutation of *NUP145* (51), *NUP116* (64), or *NUP1* (31) leads to profound changes in NPC and NE morphology. Thus, many of the transport phenotypes associated with the FG nucleoporin mutants may be secondary to gross structural defects in the NPC. The ultrastructure of the NE/NPC has not been investigated in either *nup49* or *nsp1* mutant strains.

The mechanism by which macromolecules are selectively transported across the NPC is unknown. If the central plug is a transporter that has gated openings at either end (1, 10, 113), translocation would require both opening and dilation of the transporter channel, because it is too narrow to accommodate large substrates (1). Reconstructed images of the transporter suggest that it exists in both an open and a closed state (113). The transporter model could explain the selectivity of transport, if the substrate-NLS receptor complex were required to trigger opening. Other studies suggest that the central channel has a much more open configuration, which could accommodate the largest macromolecules transported without dilating (2, 4). In this case, it is not clear how diffusion of large molecules could be prevented. Because translocation requires ATP (73, 75–78), it has been suggested that motor proteins might be involved. A myosin heavy chain–like polypeptide has been suggested as a component of the NE or NPC (114) but has so far not been shown to have a role in nuclear import.

The fact that proteins can concentrate in the nucleus has generally been taken as evidence that import occurs via active transport. However, facilitated diffusion followed by intranuclear binding would also result in nuclear accumulation (115), and there is currently no evidence that the NPC transports substrates up an effective concentration gradient. To determine the relative contributions of transport and intranuclear binding, Paine et al (116) used an oil-immersed nucleus coupled either to a latex bead or a bolus of cytosol, each loaded with labeled substrate. Surprisingly, cytosol was found not to be required for ATP-dependent equilibration of substrate between nucleus and latex bead but was absolutely required for concentration within the nucleus (116).

Taken at face value, these results suggest that movement across the NPC has properties of facilitated diffusion rather than active transport and argue against a unidirectional gated channel or motor. Furthermore, they suggest that intranuclear binding rather than import itself is mediated by the cytosolic factors that induce nuclear accumulation of docked substrates. While this conclusion would seem to be at odds with that reached from the in vitro data described above, in fact it is not. The concentration of substrate used in those systems is exceedingly low (73, 75, 78), so that only accumulated substrate can be detected. Therefore, simple equilibration of substrate would be scored as lack of import. It is therefore quite plausible that the role of Ran-TC4 (and possibly other G-proteins) in nuclear import (103, 104) is actually to target proteins to the appropriate intranuclear binding site, rather than to translocate them across the NPC. The nuclear NLS-binding proteins described above could also be part of this targeting machinery.

As opposed to active transport, facilitated diffusion is a reversible process. Therefore, a crucial question is whether movement of proteins is bidirectional, and whether export is mediated by the same machinery as import. A number of studies indicate that nuclear proteins do in fact "shuttle" between nucleus and cytoplasm (91, 102, 117–122; see 123 for review). Movement into the cytoplasm is correlated inversely with intranuclear binding capacity (122). However, there is conflicting evidence as to whether export is energy dependent or requires an NLS. Studies with nucleolin suggest that export is temperature dependent but does not require a specific signal (122). On the other hand, β-galactosidase cannot exit the nucleus unless fused to an NLS (120). The progesterone receptor was found to redistribute from nucleus to cytoplasm in the presence of sodium azide or upon shift to low temperature, arguing that ATP-dependent import is countered by passive outward diffusion (119). Finally, recent studies on the HIV rev protein (117) strongly suggest that rev export is dependent upon binding to mRNA. This result suggests that proteins exiting the nucleus as RNP particles may use a fundamentally different export pathway. Some of the conflicting results obtained so far may thus reflect the fact that proteins can exit the nucleus by numerous mechanisms. It remains to be determined whether NLS-dependent traffic is mechanistically reversible.

RNA Export

Our understanding of RNA export has lagged significantly behind that of protein import, in part because we lack in vitro assays. Furthermore, the appearance of RNA in the cytoplasm is dependent upon a number of factors, including processing and stability, that probably have nothing to do with its ability to cross the NPC. For example, the observation that unprocessed mRNA generally does not enter the cytoplasm may be largely attributable to the fact that newly synthesized messages are retained within the nucleus upon formation of a spliceosomal commitment complex and are thus not free to interact with the NPC until processing is complete (124, 125). The fate of mRNA between the completion of processing and its appearance in the cytoplasm is not at all clear. Three questions are of particular importance. How does RNA reach the NPC? Are there specific docking sites at the NPC for RNA or RNP proteins? How are RNAs translocated across the NPC?

MOVEMENT OF RNA WITHIN THE NUCLEUS   In situ hybridization has been used to visualize either poly(A) RNA or specific transcripts within the nucleus (126–131). In some cases, mRNA appears to be localized on "tracks" extending out from transcript domains (129–131). In other studies it appears to fill chromatin-free regions of the nucleus (126). These conflicting observations have led to two different models, one suggesting that simple diffusion accounts for movement of RNA to the NE (126) and the other proposing that RNA moves along a solid phase (130). Another alternative is that movement occurs by diffusion but that chromatin-free diffusion channels lead from transcript domains to the NPC and are thus not random. The latter two models both imply a structural connection between the NPC and transcript domains, an idea first proposed some years ago (132). As yet there is no evidence that such a connection exists. However, the area immediately under the NPC is virtually always devoid of heterochromatin, which otherwise coats the inner surface of the NE (133), raising the possibility that NPC components play a role in seeding channel formation.

EXPORT SIGNALS   It is not clear whether RNA or its associated proteins contain specific signals that are recognized by some component of the NPC. Colloidal gold coated with any of a number of RNA species, including poly(A) and poly (I), crossed the NPC upon injection into *Xenopus* oocyte nuclei (9). Accumulation around the NPC prior to export suggested the presence of a saturable receptor (9). These results argue that some general feature of polynucleotides serves as the signal for transport across the NPC. However, a requirement for RNA-associated proteins cannot be ruled out, because such proteins may have bound to RNA-gold particles prior to export. A role for cap binding proteins is suggested by experiments in which hypermethylated caps and cap analogues inhibited export (134), although such inhibition could occur well upstream of the NPC. As discussed above, some hnRNP proteins exhibit rapid shuttling, suggesting that they do not dissociate from messages until the hnRNP complex reaches the cytoplasm (118). This is also the case for a number of nucleolar proteins (91, 102). Such behavior is consistent with the idea that RNP proteins chaperone RNA out of the nucleus.

RNP particles in transit through the NPC were visualized in negatively stained sections of *Chironomus* salivary glands a number of years ago (135). More recently, tomographic reconstruction has yielded a number of interesting insights into this process (136). In a manner consistent with earlier observations (137–139), RNP particles initially appear to bind to nuclear filaments extending from the NPC—filaments that may represent the nuclear basket. Analysis of five different particles provided good evidence that the positions of various domains of the RNP particle are consistent with respect to each other and the NPC. Translocation proceeds through the central channel and always initiates with the 5' domain, which changes from a globular to a more linear conformation as it approaches the NPC. Consecutive domains are then transported in order, unwinding such that the material in transit has a diameter of ~25 nm. The 5' end appears to engage ribosomes very soon after entry into the cytoplasm. These observations suggest that RNP export across the NPC has at least three steps; docking, unwinding, and translocation.

MUTATIONAL ANALYSIS   The analysis of *Saccharomyces cerevisiae* mutants defective for RNA export has the potential to significantly increase our understanding of this process. The first such mutant to be identified (*rna1-1*) is pleiotropically defective for RNA processing (140–143) and export (143, 144). In situ analysis using oligo dT showed that poly(A) RNA accumulates in the nucleus after shift to the nonpermissive temperature (145, 146). Although Rna1p is largely cytoplasmic, localization studies of the *S. pombe* homologue showed a significant concentration of Rna1p at the NE, in a punctate pattern reminiscent of NPC staining (147). Interestingly, *rna1-1* is synthetically lethal with mutations in the yeast FG nucleoporin *NUP1* (31). Taken together, these results suggest that Rna1p could function by transiently associating with the NPC to facilitate egress of RNA. The processing defects associated with *rna1-1* would then be the consequence of "backing up " a linear pathway that ends in export.

Two extensive hunts for other RNA transport mutants have been carried out (146, 148, 149). In one case, in situ hybridization with oligo dT was used to screen a bank of temperature-sensitive mutants for those that exhibited increased nuclear fluorescence (146). A similar assay was used to screen survivors of a [³H]-amino acid suicide selection (148, 149). Like *rna1-1*, many of the mutants showed pleiotropic defects in RNA processing (148). One of the mutants (*mtr1*) is an allele of *prp20*, which had been shown by others to accumulate nuclear poly(A) RNA (145, 150). The apparent defect in mRNA export exhibited by the *prp20* mutant is of particular interest because *prp20* is the yeast homologue of RCC1 (151–155), the guanine nucleotide exchange factor for Ran-TC4. Ran may thus have effects on both protein import and RNA export, or it may be required for import of proteins that shuttle RNA out. Some of the *mtr* mutants exhibit striking changes in nuclear structure (148), including reticulation of the outer NE and NPC defects similar to those observed in some nucleoporin mutants (51, 64). Further analysis of these mutant collections is likely to reveal proteins that function at numerous points in the export pathway.

A number of mutations in known NPC proteins also show increased nuclear poly(A) RNA levels. These include *nup1* (31), *nup116* (64), and *nup145* (29). As discussed earlier, all of these mutants also show structural defects in the NE. Strains deleted for *NUP116* have NPCs that become "sealed" at 36°C by overgrowth and fusion of the outer NE membrane over the NPC, which probably completely blocks exit (64); *nup1* mutants do not have obviously defective NPCs but do have pleiotropic phenotypes, making it difficult to determine the primary defect (31). On the other hand, nuclear accumulation of poly(A) RNA is apparent within 3 hr of shutting off Nup145p synthesis, at which time protein import is unaffected (29). This suggests that Nup145p affects RNA transport specifically. Furthermore, Nup145p binds RNA in vitro (29). However, the gross perturbations in NPC structure observed in Nup145p truncations (51) suggest that caution should be exercised in interpreting these results, because the morphology of cells depleted of Nup145p has not been determined.

NUCLEAR ENVELOPE ASSEMBLY AND DISASSEMBLY

Postmitotic Assembly

In higher eukaryotes the NE undergoes dramatic reorganization during mitotic prophase, including disassembly of the nuclear lamina and NPCs, and vesiculation of the membrane. The extent of NPC disassembly is not completely understood but probably involves dissociation into individual subunits. This process is reversed during telophase, when the NE reassembles.

Several in vitro assays have been developed that faithfully reconstitute both NE disassembly (160–162) and reassembly (163–167). Most of these utilize amphibian egg extracts, which contain a stockpile of nuclear proteins and NE precursors sufficient to supply the rapid cell divisions that occur prior to the mid-blastula transition (168). Untreated extracts will form NEs around exogenously added DNA (163–165). These reconstituted nuclei are competent for nuclear transport (75, 76), chromatin decondensation, and DNA replication (164, 165, 169). Egg extracts supplemented with MPF will promote disassembly of the NE of exogenously added nuclei (161, 162, 170). These systems have been utilized to investigate the mechanisms that regulate three landmark events: binding and fusion of vesicles to form the nuclear membrane, assembly of the nuclear lamina, and appearance of NPCs. In whole extracts these are concerted events, but they can be isolated from one another to some extent by manipulation of extract fractions. Such manipulation allows one to investigate the order of assembly of various NE components and to determine whether these steps are interdependent.

THE NUCLEAR LAMINA   Lamina polymerization occurs in response to dephosphorylation triggered by MPF inactivation (reviewed in 171). There is conflicting evidence about whether lamina assembly is required for NE formation. Introduction of anti-lamin antibodies during metaphase blocks NE formation (166, 172). However depletion of lamin L_{III} (the major lamin isoform in meiotic cells) had no effect in vitro (173). The identification of a minor lamin L_{II} that may not have been depleted in the latter experiments could explain the discrepancy between these findings (174). The lamina is not required for NPC formation per se because annulate lamellae, which have no associated lamins, form spontaneously in vitro in the absence of chromatin (175). Furthermore, disassembly of the lamina can occur substantially before the NE vesiculates and thus does not itself trigger disassembly (161, 170), but this result leaves open the question of whether depolymerization is required for the NE to be released and vesiculation to proceed.

VESICLE TARGETING AND FUSION   Extracts can be fractionated into soluble and vesicular fractions. Neither of these alone can assemble the NE, but when mixed together they fully reconstitute assembly (164, 165, 169, reviewed in 176). Under conditions that allow some chromatin decondensation, vesicle binding occurs in the absence of cytosol (177, 178). Two distinct populations of vesicles have been identified: light (NEP-B) and heavy (NEP-A) (179). When incubated separately, only NEP-B vesicles bound to chromatin (179). Binding requires trypsin-sensitive components on both chromatin and vesicle surfaces (177, 180). In less highly fractionated systems, salt extraction of vesicles prior to incubation did not inhibit the binding step (180). However, NEP-B binding was abolished after salt extraction (179). The discrepancy between these results may be attributable to salt-extractable factors that are restored to stripped vesicles upon incubation in the less fractionated "soluble" fraction but not in the highly fractionated "cytosols" that were used to assay NEP-B activity. Taken together, these results suggest that NEP-B vesicles carry the targeting factor, which is likely to be composed of integral membrane proteins associated with peripheral components. This factor could be the lamin receptor with associated lamins, or an integral membrane protein of the NPC associated with one or more of the peripheral nucleoporins. NEP-B vesicles contain a number of FG nucleoporins (181), and the number of NPCs that ultimately form correlates with the relative amount of NEP-B fraction added (179). If the relative amount of vesicle fraction is lowered, "prepores" appear on the surface of chromatin (169). These structures are much thinner than mature NPCs and are not associated with membranes. These results are consistent with a model in which components of the NPC (which may be in equilibrium between soluble and vesicular fractions) can bind to both vesicles and chromatin. Association of these proteins could then be responsible for targeting NEP-B vesicles.

NPC FORMATION   NPCs are thought to form by association of nucleoporins across the membrane, followed by fusion of the lipid bilayers to produce the pore membrane (179, 182). The initial trigger for NPC formation remains unclear, but a key step may be aggregation of subunits within the plane of the membrane. The fact that cytoplasmic membranes containing mature NPCs (annulate lamellae) form readily upon incubation of extracts in the absence of chromatin (175) suggests that binding to chromatin is not required to initiate NPC formation. This important observation suggests that NPC assembly occurs spontaneously upon inactivation of MPF. An attractive possibility is that aggregation of NPC subunits occurs in response to changes in phosphorylation state. Once triggered, NPC formation could proceed by stepwise association and aggregation of NPC components beginning on one side of the NE, or by spontaneous association of components on both sides. Aggregation of NPC subunits in the membrane could trigger a conformational change in one or more of the poms, causing the lumenal segments to associate with one another across the cisternal space and induce fusion of lipid bilayers to generate the pore membrane (21).

The requirement for various nucleoporins in the process of NPC assembly has not been investigated in detail, and the experiments that have been performed have produced conflicting results. In both cases, WGA was used to deplete *Xenopus* extracts of FG nucleoporins prior to NE reassembly around template DNA (100, 183). When demembranated sperm nuclei were used as the template, somewhat fewer NPCs were observed relative to whole extract controls, but their morphology appeared normal (100). However no NPCs were observed when NEs were assembled around lambda DNA, even though vesicle binding and fusion occurred normally (183). While this discrepancy has not been resolved, it could be explained if, after demembranation, some FG nucleoporins remained associated with sperm nuclei and contributed to NPC formation (182).

Interphase NPC Assembly

Assembly of new NPCs must also occur during interphase. The absolute number of NPCs doubles as the NE increases its surface area during S phase (184), presumably because this maintains a constant surface area to volume ratio. In some lower eukaryotes the NE and NPCs remain intact throughout mitosis. Thus, interphase insertion is the only mechanism by which new NPCs are formed in these organisms.

Although no information is available about the mechanism by which NPCs are inserted into the NE during interphase, several models can be proposed. One view is that whole pieces of annulate lamellae are inserted by membrane fusion. This could occur at "gapped" regions of NE in a manner analogous to vesicle targeting and fusion during mitosis, except that NPCs are preformed. Alternatively, a single membrane fusion event could occur, but this would require NPCs to split at their nucleocytoplasmic face and the pore membrane to resolve. While this scenario seems unlikely, it should be noted that the intermediates thus generated would resemble quite remarkably the NE morphology of yeast *nup145* mutants (51), which could be limiting for a step in NPC assembly. While either of these models would neatly explain annulate lamellae function, the expected fusion intermediates have never been observed, in spite of numerous ultrastructural studies of annulate lamellae closely juxtaposed to the NE (reviewed in 185).

An alternative model is that interphase NPC assembly is mechanistically identical to postmitotic assembly, the only difference being that nuclear NPC components would have to be imported prior to assembly. The poms are probably free to diffuse across the pore membrane, as are other integral membrane proteins of the inner NE (186). In this model, one could view annulate lamellae as a reservoir of NPC components that can be disassembled and then reassembled at the NE when needed. As in mitosis, assembly/disassembly could be regulated by phosphorylation (presumably during S phase). Storing NPC components as annulate lamellae would allow NPC proteins to be synthesized throughout the cell cycle. Tight linkage between synthesis and assembly is evidenced by the fact that overexpression of many yeast nucleoporins is detrimental (28, 36, 64, 187). Assembly into stacked membrane sheets may also serve to isolate the NPCs and prevent interference with nuclear import.

We do not yet know enough about NPC assembly to determine the relationship between mitotic and interphase processes, let alone whether either or both are mechanistically conserved in fungi. The simplest view, however, is that they are all fundamentally the same. Furthermore, interphase NPC assembly may be the predominant mechanism by which new NPCs form in metazoan cells as well as in fungi. According to this view, mitotic assembly/disassembly is not required for NE growth but evolved primarily because it facilitates the orderly segregation of chromosomes.

CONCLUDING REMARKS

It is likely that most of the constituent proteins of the NPC will be molecularly characterized within the next few years. Over the next decade the combination of genetics, simplified in vitro assays, and increasingly refined EM analysis should provide a wealth of information regarding the structure of the NPC, its role in nuclear transport and cellular architecture, and the mechanism of NPC assembly. As other investigations begin to elucidate the contribution of nuclear structure to cellular regulation, one can hope that the two fields will merge, providing us with a much clearer picture of the overall organization of the nucleus.

Literature Cited

1. Akey CW, Radermacher M. 1993. *J. Cell Biol.* 122:1–19
2. Hinshaw JE, Carragher BO, Milligan RA. 1992. *Cell* 69:1133–41
3. Unwin PNT, Milligan RA. 1982. *J. Cell Biol.* 93:63–75
4. Jarnik M, Aebi U. 1991. *J. Struct. Biol.* 107:291–308
5. Goldberg MW, Allen TD. 1993. *J. Cell Sci.* 106:261–74
6. Ris H, Malecki M. 1993. *J. Struct. Biol.* 111:148–57
7. Reichelt R, Holzenburg A, Buhle EL, Jarnik M, Engel A, Aebi U. 1990. *J. Cell Biol.* 110:883–94
8. Franke WW, Scheer U. 1970. *J. Ultrastruct. Res.* 30:288–316
9. Dworetzky SI, Feldherr CM. 1988. *J. Cell Biol.* 106:575–84
10. Akey CW, Goldfarb DS. 1989. *J. Cell Biol.* 109:971–82
11. Goldberg MW, Allen TD. 1992. *J. Cell Biol.* 119:1429–40
12. Ris H. 1991. *EMSA Bull.* 21:54–56
13. Richardson WD, Mills AD, Dilworth SM, Laskey RA, Dingwall C. 1988. *Cell* 52:655–64
14. Sukegawa J, Blobel G. 1993. *Cell* 72:29–38
15. Rout MP, Blobel G. 1993. *J. Cell Biol.* 123:771–83
16. Rome L, Kedersha N, Chugani D. 1991. *Trends Cell Biol.* 1:47–50
17. Kedersha NL, Heuser JE, Chugani DC, Rome LH. 1991. *J. Cell Biol.* 112:225–35
18. Chugani DC, Rome LH, Kedersha NL. 1993. *J. Cell Sci.* 106:23–29
19. Snow CM, Senior A, Gerace L. 1987. *J. Cell Biol.* 104:1143–56
20. Wozniak RW, Blobel G, Rout MP. 1994. *J. Cell Biol.* 125:31–42
21. Wozniak RW, Bartnik E, Blobel G. 1989. *J. Cell Biol.* 108:2083–92
22. Hallberg E, Wozniak RW, Blobel G. 1993. *J. Cell Biol.* 122:513–21
23. Gerace L, Ottaviano Y, Kondor-Koch C. 1982. *J. Cell Biol.* 95:826–37
24. Greber UF, Senior A, Gerace L. 1990. *EMBO J.* 9:1495–1502
25. Wozniak RW, Blobel G. 1992. *J. Cell Biol.* 119:1441–49
26. Greber UF, Gerace L. 1992. *J. Cell Biol.* 116:15–30
27. Soderqvist H, Hallberg E. 1994. *Eur. J. Cell Biol.* 64:186–91
28. Loeb JDJ, Davis L, Fink GR. 1993. *Mol. Biol. Cell.* 4:209–22
29. Fabre E, Boelens WC, Wimmer C, Mattaj IW, Hurt EC. 1994. *Cell* 78:275–89
30. Hurt EC. 1990. *J. Cell Biol.* 111:2829–37
31. Bogerd AM, Hoffman JA, Amberg DC, Fink GR, Davis LI. 1994. *J. Cell Biol.* 127:319–32
32. Hanover JA, Cohen CK, Willingham MC, Park MK. 1987. *J. Biol. Chem.* 262:9887–94
33. Davis LI, Blobel G. 1987. *Proc. Natl. Acad. Sci. USA* 84:7552–56
34. Davis LI, Blobel G. 1986. *Cell* 45:699–709
35. Finlay DR, Newmeyer DD, Price TM, Forbes DJ. 1987. *J. Cell Biol.* 104:189–200
36. Davis LI, Fink GR. 1990. *Cell* 61:965–78
37. Wente SR, Rout MR, Blobel G. 1992. *J. Cell Biol.* 119:705–23

894   DAVIS

38. Aris JP, Blobel G. 1989. *J. Cell Biol.* 108:2059–67
39. Park MK, D'Onofrio M, Willingham MC, Hanover JA. 1987. *Proc. Natl. Acad. Sci. USA* 84:6462–66
40. Holt GD, Snow CM, Senior A, Haltiwanger RS, Gerace L, Hart GW. 1987. *J. Cell Biol.* 104:1157–64
41. Haltiwanger RS, Holt GD, Hart GW. 1990. *J. Biol. Chem.* 265:2563–68
42. Starr CM, Hanover JA. 1990. *J. Biol. Chem.* 265:6868–73
43. Haltiwanger RS, Kelly WG, Roquemore EP, Blomberg MA, Dong L-YD, et al. 1992. *Biochem. Soc. Trans.* 20:264–69
44. Cordes V, Waizenegger I, Krohne G. 1991. *Eur. J. Cell Biol.* 55:31–47
45. Kelly WG, Dahmus ME, Hart GW. 1993. *J. Biol. Chem.* 268:10416–24
46. Corden JL. 1990. *Trends Biochem. Sci.* 15:383–87
47. Miller MW, Hanover JA. 1994. *J. Biol. Chem.* 269:9289–97
48. Nehrbass U, Kern H, Mutvei A, Horstmann H, Marshallsay B, Hurt EC. 1990. *Cell* 61:979–89
49. Buss F, Kent H, Stewart M, Bailer SM, Hanover JA. 1994. *J. Cell Sci.* 107:631–38
50. Belanger KD, Kenna MA, Wei S, Davis LI. 1994. *J. Cell Biol.* 126:619–30
50a. Görlich D, Prehn S, Laskey RA, Hartmann E. 1994. *Cell* 79:767–78
51. Wente SR, Blobel G. 1994. *J. Cell Biol.* 125:955–69
52. Wimmer C, Doye V, Grandi P, Nehrbass U, Hurt EC. 1992. *EMBO J.* 11:5051–61
53. Grandi P, Doye V, Hurt EC. 1993. *EMBO J.* 12:3061–71
54. Kraemer D, Wozniak RW, Blobel G, Radu A. 1994. *Proc. Natl. Acad. Sci. USA* 91:1519–23
55. Carmo-Fonseca M, Kern H, Hurt EC. 1991. *Eur. J. Cell Biol.* 55:17–30
56. D'Onofrio M, Starr CM, Park MK, Holt GD, Haltiwanger RS, et al. 1988. *Proc. Natl. Acad. Sci. USA* 85:9595–99
57. Wilken N, Kossner U, Senecal J-L, Scheer U, Dabauvalle M-C. 1993. *J. Cell Biol.* 123:1345–54
58. Pante N, Bastos R, McMorrow I, Burke B, Aebi U. 1994. *J. Cell Biol.* 126:603–17
59. Finlay DR, Meier E, Bradley P, Hoercka J, Forbes DJ. 1991. *J. Cell Biol.* 114:169–83
60. Cordes VC, Reidenbach S, Kohler A, Stuurman N, van Driel R, Franke WW. 1993. *J. Cell Biol.* 123:1333–44
61. Von Lindern M, van Baal S, Wiegant J, Raap A, Hagemeijer A, Grosveld G. 1992. *Mol. Cell. Biol.* 12:3346–55
62. Von Lindern M, Fornerod M, van Baal

S, Jaegle M, de Wit T, et al. 1992. *Mol. Cell. Biol.* 12:1687–97
63. Deleted in proof
64. Wente SR, Blobel G. 1993. *J. Cell Biol.* 123:275–84
65. Paine PL, Moore LC, Horowitz SB. 1975. *Nature* 254:109–14
66. Feldherr CM, Pomerantz J. 1978. *J. Cell Biol.* 78:168–75
67. Feldherr CM, Cohen RJ, Ogburn JA. 1983. *J. Cell Biol.* 96:1486–90
68. Lang I, Scholz M, Peters R. 1986. *J. Cell Biol.* 102:1183–90
69. Peters R. 1984. *EMBO J.* 3:1831–36
70. Feldherr CM, Kallenbach E, Schultz N. 1984. *J. Cell Biol.* 99:2216–22
71. Breeuwer M, Goldfarb DS. 1990. *Cell* 60:999–1008
72. Dingwall C, Laskey RA. 1986. *Annu. Rev. Cell Biol.* 2:367–90
73. Adam SA, Marr RS, Gerace L. 1990. *J. Cell Biol.* 111:807–16
74. Schlenstedt G, Hurt E, Doye V, Silver PA. 1993. *J. Cell Biol.* 123:785–98
75. Newmeyer DD, Finlay DR, Forbes DJ. 1986. *J. Cell Biol.* 103:2091–2102
76. Newmeyer DD, Lucocq JM, Burglin TR, de Robertis EM. 1986. *EMBO J.* 5:501–10
77. Newmeyer DD, Forbes DJ. 1988. *Cell* 52:641–53
78. Moore MS, Blobel G. 1992. *Cell* 69:939–50
79. Newmeyer DD, Forbes DJ. 1990. *J. Cell Biol.* 110:547–57
80. Adam SA, Lobl TJ, Mitchell MA, Gerace L. 1989. *Nature* 337:276–79
81. Adam SA, Gerace L. 1991. *Cell* 66:837–47
82. Adam EJH, Adam SA. 1994. *J. Cell Biol.* 125:547–55
83. Sterne-Marr R, Blevitt JM, Gerace L. 1992. *J. Cell Biol.* 116:271–80
84. Finlay DR, Forbes DJ. 1990. *Cell* 60:17–29
85. Yoneda Y, Imamoto-Sonobe N, Yamaizumi M, Uchida T. 1987. *Exp. Cell Res.* 173:586–95
86. Dabauvalle M-C, Schultz B, Scheer U, Peters R. 1988. *Exp. Cell Res.* 174:291–96
87. Yoneda Y, Imamoto-Sonobe N, Matsuoka Y, Iwamoto R, Kiho Y, Uchida T. 1988. *Science* 242:275–78
88. Imamoto N, Matsuoka Y, Kurihara T, Kohno K, Miyagi M, et al. 1992. *J. Cell Biol.* 119:1047–61
89. Shi Y, Thomas JO. 1992. *Mol. Cell. Biol.* 12:2186–92
90. Jeoung o-I, Chen S, Windsor J, Pollack RE. 1991. *Genes Dev.* 5:2235–44
91. Meier UT, Blobel G. 1992. *Cell* 70:127–38

92. Benditt JO, Meyer C, Fasold H, Barnard FC, Riedel N. 1989. *Proc. Natl. Acad. Sci. USA* 86:9327–31
93. Lee W-C, Mélèse T. 1989. *Proc. Natl. Acad. Sci. USA* 86:8808–12
94. Li R, Thomas JO. 1989. *J. Cell Biol.* 109:2623–32
95. Silver P, Sadler I, Osborne MA. 1989. *J. Cell Biol.* 109:983–89
96. Yamasaki L, Kanda P, Lanford RE. 1989. *Mol. Cell. Biol.* 9:3028–36
97. Stochaj U, Silver PA. 1992. *J. Cell Biol.* 117:473–82
98. Xue Z, Shan X, Lapeyre B, Mélèse T. 1993. *Eur. J. Cell Biol.* 62:13–21
99. Shan X, Xue Z, Mélèse T. 1994. *J. Cell Biol.* 126:853–62
100. Goldfarb DS. 1988. *Cell Biol. Int. Rep.* 12:809–32
101. Lee W-C, Xue Z, Mélèse T. 1991. *J. Cell Biol.* 113:1–12
102. Borer RA, Lehner CF, Eppenberger HM, Nigg EA. 1989. *Cell* 56:379–90
103. Moore MS, Blobel G. 1993. *Nature* 365:661–63
104. Melchoir F, Paschal B, Evans J, Gerace L. 1993. *J. Cell Biol.* 123:1649–59
105. Dasso M. 1993. *Trends Biol. Sci.* 18:96–101
106. Goldfarb DS. 1994. *Curr. Biol.* 4:57–60
107. Rubins JB, Benditt JO, Dickey BF, Riedel N. 1990. *Proc. Natl. Acad. Sci. USA* 87:7080–84
108. Seydel U, Gerace L. 1991. *J. Biol. Chem.* 266:7602–8
109. Featherstone C, Darby MK, Gerace L. 1988. *J. Cell Biol.* 107:1289–97
110. Dabauvalle M-C, Benavente R, Chaly N. 1988. *Chromosoma* 97:193–97
111. Mutvei A, Dihlmann S, Herth W, Hurt EC. 1992. *Eur. J. Cell Biol.* 59:280–95
112. Nehrbass U, Fabre E, Dihlmann S, Herth W, Hurt EC. 1993. *Eur. J. Cell Biol.* 62:1–12
113. Akey CW. 1990. *Biophys. J.* 58:341–55
114. Berrios M, Fisher PA, Matz EC. 1991. *Proc. Natl. Acad. Sci. USA* 88:219–23
115. Paine PL. 1993. *Trends Cell Biol.* 3:325–29
116. Vancurova I, Lou W, Paine TM, Paine PL. 1993. *Eur. J. Cell Biol.* 62:22–33
117. Meyer BE, Malim MH. 1994. *Genes Dev.* 8:1538–47
118. Pinol-Roma S, Dreyfuss G. 1992. *Nature* 355:730–32
119. Guiochon-Mantel A, Lescop P, Christin-Maitre S, Loosfelt H, Perrot-Applanat M, Milgrom E. 1991. *EMBO J.* 10:3851–59
120. Guiochon-Mantel A, Delabre K, Lescop P, Milgrom E. 1994. *Proc. Natl. Acad. Sci. USA* 91:7179–83
121. Madan A, DeFranco DB. 1993. *Proc. Natl. Acad. Sci. USA* 90:3588–92
122. Schmidt-Zachmann MS, Dargemont C, Kuhn LC, Nigg EA. 1993. *Cell* 74:493–504
123. Laskey RA, Dingwall C. 1993. *Cell* 74:585–86
124. Chang DD, Sharp PA. 1989. *Cell* 59:789–95
125. Legrain P, Rosbash M. 1989. *Cell* 57:573–83
126. Zachar Z, Kramer J, Mims IP, Bingham PM. 1993. *J. Cell Biol.* 121:729–42
127. Carter KC, Taneja KL, Lawrence JB. 1991. *J. Cell Biol.* 115:1191–1202
128. Carter KC, Bowman D, Carrington W, Fogarty K, McNeil JA, et al. 1993. *Science* 259:1330–35
129. Huang S, Spector DL. 1991. *Genes Dev.* 5:2288–2302
130. Lawrence JB, Singer RH, Marselle LM. 1989. *Cell* 57:493–502
131. Xing Y, Lawrence JB. 1991. *J. Cell Biol.* 112:1055–63
132. Blobel G. 1985. *Proc. Natl. Acad. Sci. USA* 82:8527–29
133. Fawcett DW. 1981. *The Cell.* Philadelphia, PA: Saunders. 862 pp. 2nd ed.
134. Hamm J, Mattaj IW. 1990. *Cell* 63:109–18
135. Stevens BJ, Swift H. 1966. *J. Cell Biol.* 31:55–77
136. Mehlin H, Daneholt B, Skoglund U. 1992. *Cell* 69:605–13
137. Franke WW, Scheer U, Krohne G, Jarash E-D. 1981. *J. Cell. Biol.* 91:39s–50s
138. Scheer U, Dabauvalle M-C, Merkert H, Benavente R. 1988. *Cell Biol. Int. Rep.* 12:669–89
139. Schroder HC, Bachmann M, Diehl-Seifert B, Muller WEG. 1987. *Prog. Nucleic Acid Res. Mol. Biol.* 34:89–142
140. Hopper AK, Banks F, Evangelidis B. 1978. *Cell* 211–19
141. Knapp G, Beckmann JS, Johnson PF, Fuhrman SA, Abelson J. 1978. *Cell* 14:221–36
142. Piper PW, Aamand JL. 1989. *J. Mol. Biol.* 208:697–700
143. Hutchinson HT, Hartwell LH, McLaughlin CS. 1969. *J. Bacteriol.* 99:807–14
144. Shiokawa K, Pogo AO. 1974. *Proc. Natl. Acad. Sci. USA* 71:2658–62
145. Forrester W, Stutz F, Rosbash M, Wickens M. 1992. *Genes Dev.* 6:1914–26
146. Amberg DC, Goldstein AL, Cole CN. 1992. *Genes Dev.* 6:1173–89
147. Melchior F, Weber K, Gerke V. 1993. *Mol. Biol. Cell.* 4:569–81
148. Kadowaki T, Chen S, Hitomi M, Jacobs

E, Kumagai C, et al. 1994. *J. Cell Biol.* 126:649–59

149.   Kadowaki T, Zhao Y, Tartakoff AM. 1992. *Proc. Natl. Acad. Sci. USA* 89: 2312–16

150.   Amberg DC, Fleischmann M, Stagljar I, Cole CN, Aebi M. 1993. *EMBO J.* 12:233–41

151.   Clark K, Ohtsubo M, Nishimoto T, Goebl M, Sprague GF. 1991. *Cell Regul.* 2:781–92

152.   Fleischmann M, Clark MW, Forrester W, Wickens M, Nishimoto T, Aebi M. 1991. *Mol. Gen. Genet.* 227:417–23

153.   Ohtsubo M, Yoshida T, Seino H, Nishitani H, Clark KL, et al. 1991. *EMBO J.* 10:1265–73

154.   Matsumoto T, Beach D. 1991. *Cell* 66: 347–60

155.   Nishitani H, Ohtsubo M, Yamashita K, Iida H, Pines J, et al. 1991. *EMBO J.* 10:1555–64

156.   Deleted in proof

157.   Deleted in proof

158.   Deleted in proof

159.   Deleted in proof

160.   Suprynowicz FA, Gerace L. 1986. *J. Cell Biol.* 103:2073–81

161.   Miake-Lye R, Kirschner MW. 1985. *Cell* 41:165–75

162.   Lohka MJ, Maller JL. 1985. *J. Cell Biol.* 101:518–23

163.   Lohka MJ, Masui Y. 1983. *Science* 220: 719–21

164.   Lohka MJ, Masui Y. 1984. *J. Cell Biol.* 98:1222–30

165.   Newport J. 1987. *Cell* 48:205–17

166.   Burke B, Gerace L. 1986. *Cell* 44:639–52

167.   Forbes DJ, Kirschner MW, Newport JW. 1983. *Cell* 34:13–23

168.   Laskey RA, Kearsey SE, Mechali M. 1985. In *Genetic Engineering, Principles and Methods*, ed. JK Setlow, A Hollaender, pp. 135–48. New York: Plenum

169.   Sheehan MA, Mills AD, Sleeman AM, Laskey RA, Blow JJ. 1988. *J. Cell Biol.* 106:1–12

170.   Newport J, Spann T. 1987. *Cell* 48:219–30

171.   Dessev GN. 1992. *Curr. Opin. Cell Biol.* 4:430–35

172.   Benavente R, Krohne G. 1986. *J. Cell Biol.* 103:1847–54

173.   Newport JW, Wilson KL, Dunphy WG. 1990. *J. Cell Biol.* 111:2247–59

174.   Lourim D, Krohne G. 1993. *J. Cell Biol.* 123:501–12

175.   Dabauvalle M-C, Loos K, Merkert H, Scheer U. 1991. *J. Cell Biol.* 112:1073–82

176.   Wiese C, Wilson KL. 1993. *Curr. Opin. Cell Biol.* 5:387–94

177.   Newport J, Dunphy W. 1992. *J. Cell Biol.* 116:295–306

178.   Boman AL, Delannoy MR, Wilson KL. 1992. *J. Cell Biol.* 116:281–94

179.   Vigers GPA, Lohka MJ. 1991. *J. Cell Biol.* 112:545–56

180.   Wilson KL, Newport J. 1988. *J. Cell Biol.* 107:57–68

181.   Vigers GPA, Lohka MJ. 1992. *J. Cell Sci.* 102:273–84

182.   Dabauvalle M-C, Scheer U. 1991. *Bio Cell.* 72:25–29

183.   Dabauvalle M-C, Loos K, Scheer U. 1990. *Chromosoma* 100:56–66

184.   Maul GG, Maul HM, Scogna JE, Lieberman MW, Stein GS, et al. 1972. *J. Cell Biol.* 55:433–47

185.   Kessel RG. 1989. *Electron Microsc. Rev.* 2:257–348

186.   Powell L, Burke B. 1990. *J. Cell Biol.* 111:2225–34

187.   Hurt EC. 1989. *J. Cell Sci.* 12:243–52

188.   Radu A, Blobel G, Wozniak RW. 1993. *J. Cell Biol.* 121:1–9

Ann. Rev. Biochem. 1995. 35:897–934

THE SMALL NUCLEOLAR RNAs

ES Maxwell

Department of Biochemistry, North Carolina State University, Raleigh, NC 27695

MJ Fournier

Department of Biochemistry and Molecular Biology, University of Massachusetts, Amherst, MA 01003

KEY WORDS: small nucleolar RNAs, rRNA processing, ribosome synthesis, nucleolar
              function, intronic snoRNA

CONTENTS

ABSTRACT

The present review summarizes key progress made in characterizing the small nucleolar RNAs (snoRNAs) of eukaryotic cells. Recent studies have shown snoRNA populations to be substantially more complex than anticipated initially. Many newly discovered snoRNAs are synthesized by an intron-process-

897

ing pathway, which provides a potential mechanism for coordinating nuclear RNA synthesis. Several snoRNAs and snoRNP proteins are known to be needed for processing of ribosomal RNA, but precise functions remain to be defined. In principle, snoRNAs could have several roles in ribosome synthesis including: folding of pre-rRNA, formation of rRNP substrates, catalyzing RNA cleavages, base modification, assembly of pre-ribosomal subunits, and export of product rRNP particles.

PERSPECTIVES AND SUMMARY

Eukaryotic cells contain diverse populations of metabolically stable small nuclear RNAs, or snRNAs, which occur in both the nucleoplasm and the nucleolus. Intense study over the past 15 years has shown that several nucleoplasmic snRNAs play essential roles in tRNA and mRNA maturation, in particular in pre-mRNA splicing. The snRNAs associated with the nucleolus have received much less attention, but examination of these species has been rapidly expanding during the last several years. The location of the nucleolar snRNAs, referred to as snoRNAs, implies that these molecules function in ribosome biogenesis. Experimental evidence for this long-standing assumption has recently come with demonstrations showing that several snoRNAs are intimately involved in processing of ribosomal RNA.

SnoRNAs have been identified in a wide range of eukaryotes and are presumed to be ubiquitous. Most of the snoRNAs characterized are from yeast, in which nearly 25 species are known, or from vertebrates, in which more than 15 snoRNAs have been defined. The total number of snoRNAs is not known for any source, and final counts are expected to be substantially higher as new snoRNAs are steadily being discovered. Several structural homologs have been discovered in both simple and complex eukaryotes, arguing that at least some snoRNA functions are conserved in all eukaryotes.

Like the splicing snRNAs, the snoRNAs are complexed with specific proteins and exist as discrete ribonucleoprotein particles (snoRNPs). These snoRNPs are believed to combine with pre-rRNA, ribosomal proteins, and non-ribosomal proteins to form large complexes involved in rRNA maturation and ribosome subunit assembly. Depletion or mutation of specific snoRNAs or associated proteins has been shown to disrupt processing of the large rRNA transcripts and impair production of 18S, 5.8S, and 25S/28S rRNAs. The precise basis of these effects is not known, but it is likely to involve overlapping processes in ribosome synthesis. In principle, the snoRNAs could participate at several stages, including: (*a*) folding of pre-rRNA as RNA chaperones, (*b*) protein folding, (*c*) formation of rRNP substrates, (*d*) RNA processing and base modification reactions, (*e*) assembly of ribosomal subunits, and (*f*) export of assembled subunits.

The snoRNA genes exhibit an unusual and surprising dichotomy of organization. One class is organized in a classic way with the coding sequence flanked by promoter, terminator, and enhancer elements. The second class is quite unique, however, with the snoRNA coding sequence embedded within introns of protein genes. Synthesis of these intronic snoRNAs involves processing of the pre-mRNA, although the pathway and the precise mechanisms remain to be defined. Many of the proteins encoded by the host genes are tied to ribosome synthesis or function. This novel genetic arrangement suggests a potential means for regulating nuclear RNA metabolism and coordinating ribosome production with other cellular activities.

This review focuses on the rapidly growing class of snoRNAs and their potential roles in ribosome biogenesis. We explore the occurrence, diversity, and structures of the snoRNAs, the nature of the snoRNP complexes, the links between snoRNAs and ribosome biogenesis, and finally, snoRNA gene structure and synthesis. Our discussion incorporates published work and communications available as of September 1994. We encourage our readers to also consult earlier reviews on snoRNAs (1–8), the nucleolus (9, 10), and rRNA synthesis and ribosome biogenesis (11–15). Our aim is to provide a progress report on what promises to be an exciting and rapidly developing field of investigation in the coming years.

CLASSIFICATION AND PROPERTIES

History of snoRNA Research

Research on the nucleolar snRNAs can be divided roughly into two phases. The first phase commenced in the late 1960s and early 1970s, when the first snoRNAs were discovered in mammalian cells and shown to be associated with ribosomal RNA [reviewed in (7)]. Kinetic labeling and transcription inhibitors showed the nucleolar small RNAs to be metabolically stable products of RNA Polymerase II. Roles in ribosome synthesis were predicted, and this expectation was strengthened by demonstrations of chemical cross-linking of U3 snoRNA to rRNA. The focus then shifted to identifying individual snoRNAs (in particular U3 variants), to characterizing U3 gene structure and expression in mammals and *Xenopus*, and to defining the structural properties of vertebrate U3 RNA and U3 snoRNP complexes. Most of the biochemical interest in the nucleolus was directed at understanding rRNA gene structure and transcription and at defining the pathway of rRNA processing .

A new wave of snoRNA study commenced in the late 1980s as interest in ribosome synthesis, snoRNAs, and new RNA functions expanded. Novel strategies and technologies for analyzing snoRNAs emerged from studies of mRNA splicing. The development of antibodies for trimethylguanosine (TMG) and the nucleolar protein fibrillarin made it possible to detect snoRNAs and to distinguish these from nucleoplasmic RNAs. Simultaneously, antibody and nucleic acid probes were being used to assign ribosome synthesis functions to the sub-regions of the nucleolus and to map the location of the U3 snoRNA. Finally, the development of new or improved methods for analyzing rRNA synthesis led to the discovery that snoRNAs are intimately involved with rRNA processing. All aspects of snoRNAs are now under study, and the pace of research is quickening.

Occurrence and Diversity of snoRNAs

SnoRNAs are typically identified by the following criteria: (*a*) presence in isolated nucleoli, (*b*) nucleolar-like extraction properties (salt-resistant), (*c*) association with rRNA or nucleolar proteins (usually fibrillarin), and (*d*) the presence of sequence elements conserved in known snoRNAs (1, 2, 7, 8). Nearly 40 distinct snoRNA sequences (16) have been identified from a wide variety of organisms, with the majority identified in mammals, *Xenopus*, and yeast; major properties of the snoRNAs are summarized in Tables 1 and 2. The nucleolar and nucleoplasmic snRNAs exhibit similar size ranges within individual organisms, ranging from 67–280 nucleotides in metazoans and from 85–605 nucleotides in unicellular organisms.

No unifying system of nomenclature is in place for the snoRNAs. Fortunately, the situation is not very complicated, as most RNAs have been named by one of two conventions and the number of unique names is still manageable. The first mammalian snRNAs characterized were given "U" designations (U1, U2, etc.) because they were rich in uracil bases (7, 17). This nomenclature continues today, although uracil content is no longer a consideration. Newly isolated vertebrate snRNAs are now being awarded the next consecutive U-number and homologs found in other organisms (unicellular and metazoan) are given the same name. Yeast snRNAs were initially named by two systems, corresponding to map positions on 2-D polyacrylamide gels [snR1, snR2, etc.; (18, 19)] or by the estimated number of nucleotides [snR128, snR190, etc.; (20)]. The latter system has been abandoned and consecutively numbered snR-names are now used for new species. However, several gaps exist in the snR-series, and sequences for some named RNAs are not available.

SnoRNA abundance can vary greatly for different sources and within single organisms. The vertebrate RNAs are the most abundant and have the largest range of copy numbers. Mammalian and frog nuclei contain approximately 200,000 copies of U3, while minor species (U8, U13, U14, U15, etc.) are in the range of 10,000 and 20,000 copies (6, 7, 21). Copy numbers in yeast are much lower, ranging from a few hundred to somewhat more than 1000 per cell [(18, 19); A Balakin, M Fournier, unpublished observations].

The snoRNA populations are much larger than first thought, and all esti-

Table 1 Metazoan small nucleolar RNAs

snoRNA	Size (nts)	TMG Cap[a]	Box Elements[b]	Antibody Reactivity	rRNA Association[c]	References[d]
U3	206–228	+	A,B,C,C',D	fibrillarin	5'ETS: rat(x); human(x); mouse(x,p)	25,26,150,151; 32,142,187,188; 21,34,35,107,143,145; 72,108,146,147,165
U8	136–140	+	C,D	fibrillarin	ITS1-5.8S:frog(p)	8,39,118,119,149
U13	105	+	C,D	fibrillarin	5.8S/28S rRNA:frog(p)	21,24,64,189
U14	86–96	–	C,D	fibrillarin		21
U15	146–150	–	C,D	fibrillarin	18S rRNA:mouse(h)	44,105,190
U16	106	–	C,D	fibrillarin		31,185
U17(E1)	205–221	–	–	–	5'ETS/18S rRNA:HeLa(x)	181,182; 100,111,183,184
U18	67–70	–	C,D	fibrillarin		114
U19	200	–	–	–		2
U20	80	–	C,D	fibrillarin		115
U21	93	–	C,D	fibrillarin		116
U22	125	–	C,D	fibrillarin	18S rRNA:frog(p)	123
U23	147	–	–	–		e
U24	77	–	C,D	fibrillarin		e
E2	154	–	–	–	28S rRNA:HeLa(x)	111,183,184
E3	135	–	–	–	18S rRNA:HeLa(x)	111,183,184
7.2/MRP	260–280	–	–	Th/To autoimmune antigens		27,65,67,157–160

[a] TMG, trimethylguanosine; plant U3 species have a monomethyl phosphate cap

[b] Conserved nucleotide box designated C' in U3 corresponds to box C in other snoRNA species.

[c] ETS, external transcribed spacer; ITS, internal transcribed spacer; x, crosslinking; p, snoRNA loss impairs processing; h, in vitro hybridization.

[d] See reference 16 for compilation of snoRNA sequences.

[e] JP Bachellerie, unpublished observations

Table 2 Unicellular small nucleolar RNAs

snoRNA	Size (nts)	TMG Cap	Box Elements	Associated Proteins[a]	Null Phenotype	rRNA Association[b]	References
S. cerevisiae[c]							
U3	333	+	A,B,C,D	NOP1/SOF1	lethal	35S(h,x,p)	19,33,40,58,59,79,106,109,122,154,155
U14	125–28	–	C,D	NOP1	lethal	27S(h),35S(x,p)	29,37,38,45,68,110[d]
U18	102	–	C,D	NOP1			
MRP	340	–	–	POP1/SNM1	lethal	5.8S(p)	74,75,125,126,156
snR3	194	+	–	NOP1	viable	35S(p)	19,106,131
snR4	~192	+	–	NOP1	viable	35S(h)	19,106,131
snR5	198	+	–	NOP1	viable	20S(h)	19,106,131
snR8	189	+	–	NOP1	viable	20S(h)	19,106,131
snR9	187	+	–	NOP1	viable	35S(h),20S(h)	19,106,131
snR10	245	+	–	NOP1/GAR1/SSB1	cs	35S(h,p)	80,82,106,121,131
snR11	258	+	–	NOP1/SSB1	viable		19,82
snR13	124	+	C,D		viable		19,29
snR30	605	+	–	NOP1/GAR1	lethal	35S(x,p)	80,112,192
snR31	222	+	–	NOP1	viable	35S(p)	124,191
snR32	188	+	–	NOP1	viable		191

	Size		Box	Protein	Viability	Processing	Ref
snR33	183	+	—	NOP1	viable		191
snR34	203	+	—	NOP1	viable		e
snR35	204	+	—	NOP1	viable		e
snR36	182	—	—	NOP1	viable		e
snR37	386	+	—	NOP1	viable		d
snR38	~93		C,D	NOP1			f
snR39	~85		C,D	NOP1			f
snR40	95	—	C,D	NOP1			d
snR41	110	—	C,D	NOP1			d
snR189	192	+	—	NOP1	viable	20S(h)	20,106
snR190	191–94	—	C,D	NOP1	viable		68
D. discoideum							
U3(D2)	210	?	C				30
T. brucei							
U3(RNA B)	144	+	A,C	fibrillarin		pre-rRNAs(h)	41,103
T. thermophila							
U3	256	?	A,C,D				42,169

cs, cold sensitive; NOP1, homolog of fibrillarin; SOF1, suppressor of NOP1; GAR1, glycine-arginine rich protein-1; SSB1, single-strand binding protein-1; SNM1, MRP-specific protein; pre-rRNAs, pre-rRNAs, cosedimentation with several pre-rRNAs species; ?, not known

a All protein associations except POP1 (MRP) and SOF1 (U3) were identified by antibody precipitation assays. POP1 and SOF1 were identified genetically.

b 35S, 27S, 20S, yeast pre-rRNAs; h, hydrogen-bonding; x, crosslinking; p, depletion affects pre-rRNA processing.

c Homologs for U3 (167,168), U14, snR38., and snR39 exist in S. pombe (J Ni, D Samarsky, MJ Fournier, unpublished observations); U14 has been detected in K. lactis and C. albicans (D Samarsky, MJ Fournier, unpublished observations).

d A. Balakin, MJ Fournier, unpublished observations

e D Samarsky, A Balakin, MJ Fournier, unpublished observations

f J Ni, MJ Fournier, unpublished observations

mates must be considered tentative. The early electrophoresis patterns of mammalian small nuclear RNAs showed only the most abundant 8 to 10 species, which included the 5 splicing snRNAs, U3 snoRNA, and 7S RNA of the signal recognition particle. Many additional RNAs have since been revealed with sensitive post-labeling methods and TMG antibodies. Present patterns show many more snRNAs in mammals (21), and 17 snoRNA species have now been characterized (Table 1). Counts are expanding for other organisms, too. *Saccharomyces cerevisiae* was initially estimated to have 25 to 30 snRNAs of all types (18, 19), but is now believed to contain 50 or more non-splicing snRNAs [(1); A Balakin, MJ Fournier, unpublished observations]. Contributing to the growing number of snoRNAs in yeast is an increase in size range, with several small RNAs initially thought to be precursor tRNAs now known to be snoRNAs (A Balakin, M Fournier, unpublished observations). More than two dozen snoRNAs have been characterized from yeast (Table 2).

Primary Structure and Phylogeny

Three types of modified 5′ ends have been identified among the snoRNAs. About half of the RNAs from vertebrates and the vast majority from yeast have a 2,2,7-trimethylguanosine cap (Tables 1 and 2). These are believed to be formed at the ends of RNA Polymerase II transcripts. TMG cap formation has been demonstrated in *Xenopus* oocyte injection experiments using in vitro transcripts. Initial results showed that human U3 can be trimethylated in the cytoplasm, which suggests that it follows the path used by the splicing snRNAs (22). Interestingly, it has since been shown that *Xenopus* U3 can be hypermethylated in the nucleus and is retained in that compartment (23). This finding argues that maturation does not require a cytoplasmic phase. U8 can also be trimethylated in both the nucleus and the cytoplasm, and like U3, it normally remains in the nucleus (M Terns, C Grimm, J Dahlberg, unpublished observations). Studies with A-capped RNAs showed that a TMG cap is not required for nuclear retention, stability, or assembly into snoRNPs [(24); M Terns, C Grimm, J Dahlberg, unpublished observations] or for function (24), at least in the oocyte system. Functional mapping has shown that TMG synthesis depends on a signal in a 3′ terminal segment of human U3 (22) and conserved box D sequences in *Xenopus* U3 and U8 near the 3′ end (M Terns, C Grimm, J Dahlberg, unpublished observations). A second cap structure consisting of a γ-monomethyl phosphate has been observed for plant U3. This RNA is transcribed by RNA Polymerase III, and this modification has only been seen with Pol III products (U6, 7SK and B2) (25, 26). Information for this alteration is contained in a stem-loop structure at the 5′ end.

More than half of the mammalian snoRNAs characterized possess an unmodified 5′ phosphate at the 5′ terminus. This is characteristic of snoRNAs processed from pre-mRNA introns [(4); see below]. One RNA is known to have an unmodified 5' terminus consisting of a 5' triphosphate (7–2/MRP) (27, 28). Yeast has a few uncapped snoRNAs, including three intronic species (A Balakin, J Ni, MJ Fournier, unpublished observations) and two snoRNAs processed from a transcript not believed to be mRNA (snR190 and U14) (29). A simple 3'-OH occurs at the 3' ends of all snoRNAs examined to date.

Internal modified bases have been identified in a few mammalian and yeast snoRNAs and this may be a common property of snoRNAs (7). Unfortunately, information on this point is very limited owing to the difficulty of obtaining adequate amounts of pure material for analysis. Modified sugars and pseudo-uridine have been observed in rat U3 and U8 (7), and pseudouridine and ribose methylations in yeast (18).

A few snoRNAs are widely conserved, including four homologs in vertebrates and yeast (U3, U14, U18 and 7-2/MRP); three of these occur in plants as well (U3, U14 and 7-2/MRP). Conservation at a functional level is suggested by the fact that the three latter RNAs are all needed for rRNA processing in yeast and animal systems (see below). U3 has also been identified in insects, slime mold, and ciliated protozoa (16). Most tantalizing are recent results which show that the archaebacterium *Sulfolobus acidocaldarius* contains a small RNA with features reminiscent of U3 (S Potter, P Durovic, P Dennis, unpublished observations).

Conserved Elements and Secondary Structure

Primary sequences have been useful in identifying and classifying snoRNAs and in predicting secondary folding. The most abundant sequence data exist for U3 and U14. Approximately 20 sequences are available for U3 from 11 organisms and more than 30 sequences for U14 have been determined from 10 different sources [(16) and references therein; ES Maxwell, unpublished observations]. A more limited number of phylogenetic sequences are available for the other snoRNAs.

Several sequence and secondary structure elements are conserved among subsets of snoRNAs. The conserved sequences include four elements, known as boxes A through D, and several cases of impressive complementarity with rRNA. None of these elements is universal, but they are quite common. The box elements were first defined for homologs of U3 (8, 21, 30). Boxes A and B are limited to U3, whereas boxes C and D occur in several other RNAs as well, typically together (Tables 1 and 2). Recently, a second box-C-like element (C') has been defined in U3 based on recognition of a common helix-box C/D motif in other snoRNAs (31).

Box A contains a sequence complementary to rRNA that has been implicated in rRNA binding by cross-linking [(32, 33); see below]. Box B has been postulated to interact with rRNA as well; however, this has not been shown directly (34, 35). The box C, C′ and D elements have been analyzed by mutagenesis and implicated in snoRNA synthesis or snoRNP formation. Alterations in box C of human U3 interferes with formation of fibrillarin-containing complexes in vitro, although it is not known if direct binding is involved (36). In yeast, boxes C and D (both six nucleotides) are required for accumulation of U14 (37, 38). A base found to be essential in box C of U14 is also required for accumulation of a second yeast snoRNA (snR190), but not for U3 [(38); G Huang, MJ Fournier, unpublished observations]. U3 accumulation is impaired, however, by mutations in box C′, which suggests that it is functionally equivalent to box C of U14 and snR190, and perhaps to box C of other RNAs with this motif (D Samarsky, Q Wang, A Balakin, MJ Fournier, unpublished observations). These effects are most likely due to a defect in snoRNA maturation or snoRNP assembly rather than to a block in transcription, but this has not been shown. Boxes C and D influence *Xenopus* U8 stability too, though with effects that are much weaker, and box D is required for TMG cap formation of *Xenopus* U3 (24).

Several snoRNAs contain unusually long regions of complementarity to rRNA (see Figure 1), which raises the possibility of direct interaction at these sites (see below). The elements extend up to 21 contiguous nucleotides and are complementary to regions conserved in both 18S and 25/28S RNAs. Genetic analysis has shown two such elements in yeast U14 to be important for production of 18S RNA (see below). Computer screening of intron sequences for rRNA complementary elements led to the discovery of two intronic snoRNAs in *S. cerevisiae* (J Ni, MJ Fournier, unpublished observations) and three others in vertebrates (B Michot, JP Bachellerie, unpublished observations; N Watkins, J Ni, ES Maxwell, MJ Fournier, unpublished observations).

Secondary structure information is only available for a few snoRNAs. There is still a paucity of phylogenetic sequences and only three RNAs have been subjected to structural probing (U3, U14 and 7-2/MRP). Consensus models have been developed for the U3 variants, based on evolutionary sequences and biochemical probing of human, *Xenopus*, yeast, trypanosome, and *Tetrahymena* RNAs, although the yeast and trypanosome models lack some features conserved in metazoans [(34, 39–42); C Branlant, unpublished observations). Two models have been reported for U14 based on folding predictions and direct probing of mouse and yeast RNAs (29, 43). These are quite different, predicting a highly folded molecule for mouse (43) and one with little secondary structure in the case of yeast (29). U14 RNAs typically have complementary ends and flanking precursor segments. These terminal segments are predicted to pair in precursors and play a role either in U14 processing or snoRNP formation (44, 45). This view is supported by mutational results showing that

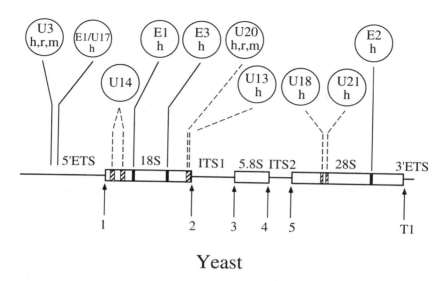

Vertebrate

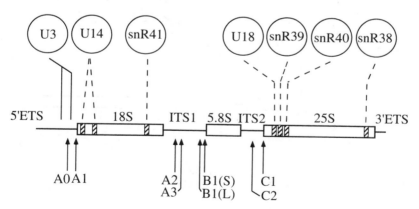

Yeast

Figure 1 Potential snoRNA binding sites on ribosomal RNA. Two classes of interactions in vertebrates and yeast are identified. One corresponds to sites where snoRNAs have been chemically cross-linked to rRNA (——); the second are sequence complementarities of 13 to 21 nucleotides (---). SnoRNAs are identified above the rRNA maps, and positions of known processing sites are shown below. Yeast U14 and snR30 have been cross-linked to rRNA, but the linkage sites are unknown.

Maps are not to scale. ETS, external transcribed spacer; ITS, internal transcribed spacer; solid bars identify sites of cross-linking within rRNA; hatched bars show sites of sequence complementarities; h, human; r, rat; m, mouse.

Vertebrate

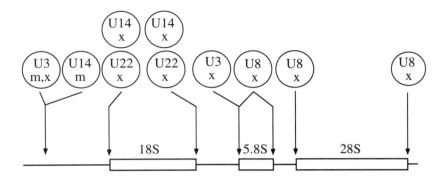

Yeast

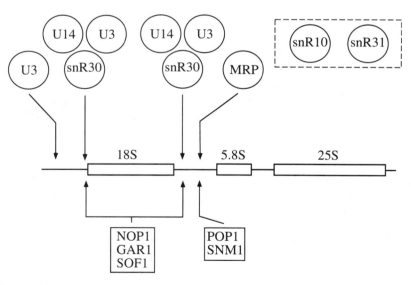

Figure 2 SnoRNAs and associated proteins involved in rRNA processing. Events disrupted by snoRNA or snoRNA-protein loss are identified for vertebrates and yeast. SnoRNAs are shown above the rRNA maps, and proteins are shown below. Arrows identify cleavage sites affected. Several cleavages are influenced by more than one snoRNA or protein. Processing is also affected by several other nucleolar and non-nucleolar proteins not known to be associated with snoRNAs (see text). SnR10 and snR31 influence processing of the primary transcript; snR10 also affects late cleavages. NOP1 mutations can affect processing at different points. m; mouse, x, *Xenopus*.

complementarity is essential for accumulation of yeast U14. The terminal stem adjoins the vital box C and D elements, and this Y-shaped motif is a likely recognition signal(s) (21, 37, 38). This motif appears at the ends of several snoRNAs, but near the 3′ end in the case of the box C′ variant in U3 (31). Not all box C– and box D–containing sequences conform to this motif.

Phylogenetic sequences and chemical modification data have also been used to develop secondary models for RNase MRP RNA [(46–48); T Kiss, W Filipowicz, unpublished observations]. These structures share interesting features with models of eukaryotic RNase P RNA and the phylogenetically conserved prokaryotic RNase P RNA. Folding predictions have been made for a number of other snoRNAs, as well, although generally with fewer distantly related sequences and without biochemical probing data. To date, detailed biochemical probing of snoRNP complexes has only been conducted for human, *Xenopus*, and yeast U3 [(34, 39); C Branlant, unpublished observations] and mouse MRP (46).

SnoRNA RIBONUCLEOPROTEIN STRUCTURE

Immunological and Biochemical Properties

The snoRNAs occur as ribonucleoprotein complexes rather than as free RNA. Early evidence of this organization came from immunological data using antibodies from patients with autoimmune diseases. Immunofluorescent staining of cells frequently revealed specific antigens within the nucleolus (49–51). This is in contrast to the "speckled" nucleoplasmic staining patterns obtained with Sm and RNP antibodies that recognize splicing complexes (5, 52). Most nucleolar antibodies are directed against fibrillarin, an evolutionarily conserved protein (53–59) of approximately 34 kDa, localized in the dense fibrillar region of the nucleolus (49, 50); early stages of rRNA processing and ribosome assembly occur here (see below). Fibrillarin has been characterized from a number of eukaryotes, including human (53, 54), mouse (55), *Xenopus laevis* (56), *Physarum polycephalum* (57), and *Saccharomyces cerevisiae* (designated NOP1) (58, 59). While typically associated with autoimmune disease, fibrillarin antibodies can also be induced in mice with mercuric chloride (60, 61) and silver nitrate (62).

Fibrillarin antibodies coprecipitate numerous vertebrate snoRNAs (although incompletely) including TMG-capped U3, U8, and U13 as well as the intron-encoded U14, U15, U16, U18, U20, U21, U22, and U24 snoRNAs [Table 1; (1, 2, 4) and references therein]. Consistent with its association with snoRNAs, fibrillarin is coordinately expressed with U3 (63) and U8 (64) during *Xenopus laevis* oogenesis. Most yeast snoRNAs are also precipitated with human fibrillarin antibodies, which demonstrates the evolutionary conservation of this protein and its association with snoRNAs (Table 2 and references therein). Another class of autoimmune antibodies designated Th or To has been shown to precipitate the 7-2/MRP RNA, another nucleolar species named 8-2, and the small cytoplasmic Ro RNAs (27, 65). Immunolocalization studies have shown the 7-2/MRP ribonucleoprotein in the granular component of the nucleolus (66) where the final stages of rRNA maturation appear to occur (see below).

Biochemical analysis has revealed two classes of snoRNA:protein complexes in both unicellular and multicellular organisms. Sedimentation analysis of nuclear extracts has demonstrated complexes of two sizes estimated at 10-15S and 70-90S (21, 67, 68). The smaller particles correspond to monomeric snoRNPs (34, 69), and the larger complexes are believed to be snoRNPs associated with rRNAs in putative pre-rRNA maturation complexes. The larger aggregrates may include: (a) complexes seen in EM analysis of nascent pre-rRNA transcripts undergoing 5′ ETS processing—corresponding to the terminal balls of "Miller Christmas Trees" (70), and/or (b) U3-containing ETS complexes detected when processing extracts are resolved on non-denaturing polyacrylamide gels (71, 72).

Current knowledge of snoRNP structure and protein composition is very limited, with most information derived from prefatory studies of the human U3 monoparticle. Early work revealed that fibrillarin antibodies coprecipitated about 6 proteins with U3 (34), but the nature of these associations is not clear. More recently, a human core U3 snoRNP has been purified using a combination of anti-TMG immunoaffinity selection and anion-exchange chromatography (69). The complex possesses 3 core polypeptides of 15, 50 and 55 kDa but lacks fibrillarin, which is lost during particle purification.

SnoRNA-Associated Proteins

Sedimentation size of the individual snoRNP complexes suggests a protein composition of a dozen or more polypeptides by analogy with splicing snRNPs. Only a few snoRNA-associated proteins have been identified thus far. Immunoprecipitation of numerous metazoan and almost all yeast snoRNAs with fibrillarin antibodies suggests that fibrillarin/NOP1 is a common component of many snoRNPs. Consistent with RNA binding, fibrillarin possesses both RNA recognition motifs (RRM) and glycine-arginine rich (GAR) sequences (53–59). The fibrillarin-snoRNP associations could involve indirect protein-protein interactions as well as direct binding to snoRNA or rRNA. Consistent with the notion that fibrillarin is not a core snoRNA-binding protein is the finding that it is easily washed from the human U3 snoRNP in low salt conditions (69).

Only two binding sites of snoRNA-associated proteins have been mapped thus far. Antibodies against the major 55 kDa core protein of the human U3 snoRNP have demonstrated that it is a U3-specific protein which binds to the central region of U3 RNA (69). Similarly, the vertebrate Th protein recognized by Th/To antibodies has been shown to bind to the central region of human 7-2/MRP RNA (73). Two MRP-associated proteins designated POP1 and SNM1 have been identified in yeast, but it is not known if these bind directly to snoRNA (74, 75). Th and POP1 antibodies both precipitate MRP RNA and the RNA of RNase P (74, 76–78). This shows that both complexes share common protein components, but the relationship of Th and POP1 is not clear. SNM1 is associated with MRP but not with RNase P RNA (75).

Three other nucleolar proteins have been implicated in snoRNA binding in yeast by immunological data (Table 2). The protein antigens and associated snoRNAs are: SOF1 (U3) (79), GAR1 (snR10 and snR30) (80, 81), and SSB1 (snR11) (82). Potentially important RNA-binding sequence elements identified in the proteins include: (a) G beta-like repeats found in beta-subunits of G proteins and the yeast splicing factor PRP4 (SOF1) (79); (b) GAR domains (GAR1 and SSB1) (80–82); and (c) RRM domains (SSB1) (83). Specific snoRNAs were absent in many of these immunoprecipitates, but screening was limited to a few snoRNAs: Other species may yet be found associated with these antigens. Immunoprecipitation and genetic complementation results indicate that SOF1 is associated with NOP1 (79).

Although RNA binding motifs are present in nucleolar proteins associated with snoRNAs, these elements may be utilized for binding other RNAs, in particular rRNA, and their presence cannot be interpreted as binding sites for snoRNAs. Other nucleolar proteins known to influence rRNA processing contain the same motifs but are not known to be associated with snoRNAs; these include nucleolin (84–86), NSR1 (87, 88), NOP77 (89), and NOP4 (90). Thus, the nature of the associations between the immunoprecipitated snoRNAs and proteins awaits more detailed characterization. Development of new antibody probes specific for individual snoRNP proteins and snoRNP complexes will be instrumental in defining these relationships and for structure-function studies of the snoRNP complexes.

SnoRNA FUNCTION: ROLES IN RIBOSOME BIOGENESIS

The Nucleolus and Ribosome Synthesis

There is a substantial body of evidence linking snoRNAs to ribosome synthesis and, specifically, to rRNA processing. This includes demonstrations of snoRNA interaction with rRNA and disruption of processing upon depletion of snoRNAs or associated proteins. Six snoRNAs have profound effects on pre-rRNA processing and several others are strong candidates for roles in this or overlapping processes. The precise basis of the snoRNA effects has not been defined in any case, and many could be indirect. Despite this uncertainty, the fact remains that snoRNAs are required for rRNA synthesis and ribosome production.

In principle, snoRNAs could participate in any aspect of ribosome synthesis associated with the nucleolus, from rDNA transcription to assembly and export of ribosomal subunits. Functions that are only peripherally related to ribosome synthesis are also formal possibilities, including formation of the nucleolus itself (91, 91a), although no evidence for such roles exists at this time. Discussion about snoRNA function is continued below, after reviewing what is known about their location, organization and involvement in rRNA synthesis.

Ribosome synthesis events known to occur in the nucleolus include: (*a*) transcription of the large rDNA units that encode the 18S, 5.8S and 25S/28S rRNAs; (*b*) processing and base modification of these transcripts, and; (*c*) assembly of ribosomal subunits. Many molecules join the nucleolus during this process, including RNA Polymerase I, topoisomerase I, the UBF transcription factor, RNA maturation enzymes, 5S rRNA, and ribosomal proteins (7, 14, 15, 92). Also recruited are the snoRNAs, snoRNP proteins and a variety of nonribosomal proteins.

The nucleolus forms around the clusters of rDNA, and both rDNA and RNA Polymerase I are required for its formation; disassembly occurs in periods of quiescence (9, 10, 93, 94). Three discrete morphological zones have been defined for the nucleoli of complex eukaryotes, and ribosome synthesis flows through these regions in vectoral fashion. The regions correspond to: (*a*) the fibrillar center(s) (FC); (*b*) an adjoining dense fibrillar component (DFC), and (*c*) an outer granular component (GC). The site of transcription is still disputed, but is generally believed to take place at the interface of the FC and DFC regions [reviewed in (9)]. It is accepted that transcripts accumulate in this latter zone. The early rRNA processing steps appear to occur in the DFC region where transcribed spacers have been detected; these were not present in the GC (95). This view is supported by cytological results showing early precursors in the DFC and mature rRNAs in the GC (8, 9, 14, 96, 97).

Fibrillarin is heavily concentrated in the DFC region (50) but has also been detected in the GC (10). Because most snoRNAs are associated with this protein, although incompletely (34), snoRNP complexes are expected to correlate with this distribution. U3, which is required for several early cleavages (11, 12), has in one study been detected primarily in the DFC and to a lesser extent in the GC; however, it has been observed in all three regions in other investigations (94, 98). Locations have not been determined for other fibrillarin-associated snoRNAs involved in early and late processing events. However, the 7-2/MRP RNP complex, which is required for an intermediate cleavage step, has been localized to the GC (66).

Despite the uncertainties about sub-nucleolar location, it seems clear that ribosome assembly overlaps rRNA maturation. Many ribosomal and non-ribosomal proteins are associated with the largest detectable rRNA precursors (96, 97, 99). These RNAs were initially thought to be full length but are now known to be somewhat shorter (12). Processing and subunit formation are influenced by both ribosomal and non-ribosomal proteins, thus any interaction between snoRNPs and precursor rRNAs involves complexes which contain these proteins as well.

SnoRNA Involvement in rRNA Synthesis

INTERACTIONS WITH rRNA. Early biochemical evidence tying snoRNAs to ribosome production came from co-purification and chemical cross-linking results showing snoRNAs associated with rRNP complexes and deproteinized rRNA [reviewed in (7, 12)]. Sedimentation with higher order rRNP complexes has been described for several vertebrate snoRNAs (U3, U8, U13, 7-2/MRP RNA, U17/E1, and U19) and two plant snoRNAs (U3 and MRP) [summarized in (2, 100)]. Co-migration of snoRNAs and rRNA in deproteinized extracts suggested that these RNAs interact through hydrogen bonding. Such associations have been reported for mammalian and trypanosome U3 (101–103), mammalian U14 (104, 105), and 10 yeast snoRNAs (68, 106).

Stronger evidence of snoRNA binding to rRNA has come from chemical cross-linking studies with specific sites of interaction identified in several cases [reviewed in (1, 2, 12)]. Cross-links have been obtained for four mammalian RNAs (U3, E1/U17, E2, and E3) and three species from yeast, U3, U14, and snR30 (Fig. 1). Three of these RNAs are known to be involved in rRNA processing (U3, U14, and snR30). U3 has been cross-linked to the 5' ETS region in mammalian and yeast studies, although with some differences. Human and rat U3 were cross-linked in vivo at unidentified sites that encompass or are adjacent to the U3-dependent ETS cleavage site (32, 107), while in prepared extracts mouse U3 was linked at a position 362 nucleotides downstream of this site (108). Mouse U3 has also been linked to a unique element near the ETS cleavage point through a variant of the conserved box A (box A'). Two ETS cross-linking sites have been identified for yeast U3, one near the corresponding vertebrate cleavage site and a second close to the 5' end of 18S RNA (33). The cross-linking sites in rat and yeast U3 include box A and, in yeast, a 10-nucleotide sequence complementary to the proximal ETS site. The differences observed in the various studies may reflect different locations of the U3-dependent cleavage reactions and multiple sites of contact.

The strongest evidence that binding of U3 to the ETS is important has come from phylogenetic and mutational analyses in yeast. Deletion of the complementary ETS sequence abolished production of 18S RNA with the same consequences as loss of U3 itself [see below; (33)]. The conclusion that impaired binding of U3 is the basis of these effects has been strengthened by recent results showing that: (*a*) complementarity is conserved in another yeast (*H. wingei*) but with different sequences (C Branlant, unpublished observations); and (*b*) processing in *S. cerevisiae* depends upon maintaining complementarity between U3 and the ETS (109). Deletion of the distal U3 cross-link site in the mouse ETS does not block processing, which suggests that this rRNA element has a non-essential role in ETS cleavage or may be involved in other events (108). A conserved U3-complementary sequence in the 5' region of 18S rRNA has also been proposed to interact with U3 (J Hughes, unpublished observations).

Yeast U14 has been cross-linked to rRNA but the precise sites of interaction are not yet known (J Morrisey, D Tollervey, unpublished observations). Base-pairing of these RNAs was first indicated by functional mapping of U14. Two sequence elements complementary to 18S RNA are essential or conditionally essential for processing (37, 45, 110). Mutations in these elements abolish activity but function can be restored with compensatory base changes in the rRNA sequence, which indicates that these RNAs interact directly through hydrogen-bonding (W Liang, MJ Fournier, unpublished observations). The target sequence for the essential U14 element is at the 5' end of 18S RNA 83 nucleotides from a cleavage site dependent upon U14 (A1). Crosslinking results obtained for the human E1/U17, E2, and E3 RNAs imply that these species also interact with rRNA at sites in the interior of 18S or 28S RNA (111), but functional information for these RNAs is not yet available. A functional requirement has been shown for yeast snR30; however, the sites for cross-linking have not yet been determined (112).

Direct binding to rRNA is implied for additional snoRNAs that contain striking sequence complementarities with rRNA (defined here as greater than 12 bps) (Figure 1; 68, 104, 105, 113–116). Many shorter complementarities also exist, including the U3 cross-linking sites noted above, but these are statistically less rare. Most of the longer complementarities are conserved in eukaryotic rRNAs. The complementarities in this group of snoRNAs all occur within the 18S or 25S/28S rRNA sequences. One snoRNA, U20, contains an extraordinary segment of 21 consecutive nucleotides complementary to 18S RNA! This RNA occurs in both mammals and yeast [(115); JP Bachellerie, unpublished observations]. The U20 complementary sequence in 18S is only 51 nucleotides from the 3' end, well positioned to participate in 18S rRNA processing or other events involving this region.

Interestingly, some of the candidate rRNA binding sites for this last set of snoRNAs are tightly clustered. Yeast 25S RNA contains three such elements in a segment of about 250 nucleotides (A Balakin, J N MJ Fournier, unpublished observations). Three vertebrate snoRNAs are complementary to this general region, one shared with yeast (U18) (JP Bachellerie, B. Michot, unpublished observations). This situation is very intriguing and suggests that snoRNAs might function as chaperones. Folding of precursor or mature rRNAs could involve close packing of snoRNPs working in cooperative fashion. Of course, other schemes, some involving low density binding, can be imagined as well. There appear to be enough snoRNAs in yeast for close packing over the full length of rRNA.

DEPLETION OF snoRNAs BLOCKS rRNA PROCESSING.   The strongest evidence of snoRNA involvement in rRNA processing has come from depletion analyses where removal of specific snoRNAs disrupts processing. These analyses have been carried out with four experimental systems: (*a*) a mouse cell extract to which in vitro synthesized rRNA is added (71, 72, 117); (*b*) an analogous system developed from *Xenopus laevis* (118); (*c*) *Xenopus* oocytes where materials are introduced by microinjection (64, 119, 120) and; (*d*) yeast, using genetic manipulations (121). Both extract systems mediate specific 5′ ETS cleavages with short model substrates. The mouse system was developed first and has been successfully exploited to characterize both cis- and trans-acting components involved in processing [reviewed in (11, 12)]. Depletion of snoRNAs in the extract and oocyte systems is achieved by targeted RNase H digestion and in the yeast system by repressing snoRNA gene expression.

Eight different snoRNA species have been shown to be important for processing (Tables 1, 2; Figure 2). Three of the RNAs are conserved in vertebrates, plants and yeast (U3, U14, 7-2/MRP). The interference effects fall into two classes that correspond to early and late steps in processing. The early events occur in the 5′ portion of the precursor and are associated with production of 18S RNA. The later cleavages involve processing of the 5.8S and 25/28S RNAs. Six snoRNAs affect early cleavages, whereas only two have thus far been linked to later steps.

SnoRNAs NEEDED FOR EARLY CLEAVAGES.   The snoRNAs linked with early processing steps include U3, U14, U22, snR10 and snR30. Depletion of U3 in the mouse extract system disrupted an internal cleavage in the ETS segment (72). This effect was not observed in the first depletion studies conducted with *Xenopus* oocytes; instead, a later cleavage site was altered (119, 120). *Xenopus* rRNA has since been shown to also undergo the U3-dependent cleavage in the 5′ ETS segment in both mouse and frog extract systems (117, 118). Very low processing efficiency has been proposed as the basis for the negative results seen in the earlier oocyte study (12). Depletion of U3 in yeast also disrupts processing at a 5′ ETS site (A0) and at two sites that define the 20S precursor to 18S RNA (A1,A) (122). A deficiency in 18S rRNA results presumably from rapid turnover of unusual intermediates. The 5′ ETS cleavages in mammals, frogs and yeast appear to be universal, although specific locations and target sequences differ (11, 12).

The first U14 depletion studies were carried out in yeast in which cleavages that produce 18S RNA were found to be disrupted, which led to a deficiency in 18S rRNA (110). Similar effects have been seen for *Xenopus* U14 in oocyte assays (B Peculis, J Steitz, unpublished observations). In the mouse, U14 is required for the same 5′ ETS cleavage reaction affected by U3 (C Enright, B Sollner-Webb, unpublished observations). *Xenopus* U22 and yeast snR30 have also been shown to be required for production of 18S RNA using the oocyte and yeast depletion strategies (112, 123). The fact that some cleavages are affected by more than one snoRNA suggests that these RNAs may function together as part of a multi-snoRNP complex (122) analogous to the spliceosome; we have described this as a "processome" (1). This view is consistent with the evidence for direct rRNA binding of the U3, U14, and snR30 RNAs, all of which are required for early events.

Early processing steps in yeast are also influenced by two non-essential snoRNAs (snR10 and snR31) but with effects different from those associated with U3, U14 and snR30. Cells lacking snR10 process the largest precursor less efficiently, and the efficiencies of the subsequent cleavages are also altered (121). SnR31 influences processing of the primary precursor, but the effect is mild and could be indirect (124).

SnoRNAs AFFECTING LATE CLEAVAGES.   SnoRNAs implicated in the later processing reactions include *Xenopus* U3 and U8, and yeast 7-2/MRP RNA. U3 and U8 both influence processing at the ITS1-5.8S boundary, and U8 is required for cuts on the 3′ side of 5.8S RNA and both ends of 28S RNA (64, 119, 120). The U3 effect is unique thus far to *Xenopus* (119) and could represent a species difference, related perhaps to alternate pathways. It is also possible that this downstream effect is an indirect consequence of disrupting the earlier events (12), a concern that applies to any snoRNA showing multiple effects. The effects seen for U8 are much stronger which implies that U3 and U8 act in different ways (12, 119).

Yeast MRP RNA is required for an ITS1 cleavage (125, 126) and the resulting 5.8S-28S RNA product is trimmed by exonuclease to define one of two 5′ ends of 5.8S RNA [major short form; (75, 127)]. At present the MRP RNA is the best candidate for a direct role in a cleavage reaction. Complexes containing this RNA have endonuclease activity upon RNA transcribed from mitochondrial DNA (128 and refs. therein). Furthermore, the MRP and ribonuclease RNase P RNAs exhibit considerable structural homology (46–48) and the particles share a common antigen (76–78).

Some snoRNAs are able to mediate their processing-related functions with truncated substrates, indicating that activity does not require full-length precursor rRNA in every case. This is true for snoRNAs linked to both early and late processing events. U3 and U14 are both active in the mouse 5' ETS cleavage assay with a 422 nucleotide fragment [(71, 72, 129); C Enright, B Sollner-Webb, unpublished observations)] and yeast 25S RNA can be produced in vivo from transcripts lacking portions of the 5' ETS and 18S rRNA coding sequence (130). Interestingly, a large deletion in the yeast 25S rRNA coding segment blocks accumulation of 18S RNA, which suggests that synthesis of this RNA involves interactions of widely separated elements (H Li, MJ Fournier, unpublished observations).

NON-ESSENTIAL snoRNAs.   Only 4 of 21 snoRNAs analyzed in *S. cerevisiae* are essential for growth, which indicates that ribosome synthesis does not require a full complement of snoRNAs (Table 2). It does not follow, however, that the dispensable RNAs have no function, only that such functions are not essential. These RNAs could affect processing quantitatively with alterations not obvious in simple growth assays. Functional redundancy is also possible. Synergistic effects have been examined in two instances with yeast strains lacking either six or three non-essential snoRNAs, respectively. No new growth phenotypes were observed in either study [(131); D Samarsky, A Balakin, MJ Fournier, unpublished observations].

There is no evidence that eubacterial ribosome synthesis involves snoRNA-like RNAs. This situation begs the question of why snoRNAs exist in eukaryotes and how snoRNA functions are mediated in non-eukaryotes, if they occur (12). From the perspective of ribosome structure, the challenge of building a prokaryotic ribosome appears to be nearly as complex as that of building a eukaryotic ribosome. There are several major differences in these two processes, of course, and snoRNA functions could relate to these distinguishing features. Insight into these fascinating issues may be forthcoming from processing studies with archaebacteria. A small RNA with some similarities to U3 has been discovered in *Sulfolobus acidocaldarius*, and shown to influence ETS processing in an homologous in vitro assay (P Dennis, unpublished observations). The presence of a fibrillarin-like protein in archaebacteria suggests that other similarities with eukaryotic ribosome synthesis may occur (132).

SnoRNA-Associated Proteins Involved in Processing

Genetic analyses in yeast have shown that rRNA processing depends on a large and growing number of proteins, some of which are associated with snoRNAs. The current list includes several ribosomal proteins and over a dozen non-ribosomal proteins [reviewed in (11, 90) and references therein]. The effects of the ribosomal proteins are not known but can be rationalized on several bases including rRNP substrate formation and subunit assembly (13, 133). The non-ribosomal proteins are a diverse population consisting of both nucleolar and non-nucleolar proteins.

Five of six nucleolar proteins associated with yeast snoRNAs affect processing and, as noted earlier, are candidate snoRNP proteins based on immunoprecipitation data; the sixth (SSB1) has not been tested for involvement in rRNA synthesis (Tables 1 and 2; Fig. 2). The snoRNA-associated proteins linked to processing are: NOP1 (134, 135), SOF1 (79), GAR1 (80, 81), POP1 (74), and SNM1 (75). NOP1, the yeast counterpart of fibrillarin, is essential and its function can be provided with *Xenopus* and human fibrillarin (54). NOP1, SOF1 and GAR1 are required for the same early processing steps associated with 18S RNA synthesis as seen for U3, U14, and snR30, but some mutant NOP1 alleles have other affects on processing (135); NOP1 also influences rRNA methylation. Interestingly, some NOP1 mutations disrupt both processing and methylation whereas one mutant form affects nucleolar methylation (134, 135) without strongly impairing processing. POP1 and SNM1 are necessary for the distal cleavage mediated by 7-2/MRP RNA and POP1 is also required for tRNA processing. Since fibrillarin is associated with most snoRNAs, its requirement in processing is not surprising. However, this very abundant protein may well have other nucleolar functions, consistent with the mix of phenotypes already observed (136). Fibrillarin is the only vertebrate snoRNA protein linked to ribosome synthesis thus far, based on genetic complementation in yeast (54).

The non-ribosomal, non-snoRNA proteins that influence processing are not all confined to the nucleolus [reviewed in (11, 90)]. Some have been detected in the nucleolus and nucleoplasm and others in the nucleus and cytoplasm. Both specific and general effects on processing have been seen and some proteins almost certainly act indirectly. Two genes proposed to have direct roles specify 5' to 3' exonucleases that influence processing of 5.8S rRNA (XNR1, RAT1) (127). Biochemical functions have not been defined for the other non-snoRNA proteins; however, a variety of interesting activities have been postulated on the basis of sequence motifs and early functional results. These include nuclear transport (NSR1), RNA binding (NSR1, NOP77), RNA helicase activity (SPB4, DRS1, CA9), ribosome subunit assembly (DRS1), and catabolite repression (RNA1) (87–89). Depletion of NSR1 disrupts processing with the same effects as loss of U3, U14, and snR30 (137, 138), consistent with an intimate relationship to snoRNAs but pleiotropic effects are possible here, too. A protein that may be more closely linked is NOP77, which was identified as a synthetic lethal mutant of NOP1 (89). As with the snoRNAs, sorting out the roles of the growing number of proteins promises to be a difficult challenge.

Finally, although processing is the only function implicated for the snoRNAs thus far, the nature of the interference effects indicate that several roles can be anticipated. In addition to functioning in RNase complexes, snoRNAs could serve as chaperones in rRNA folding [see early views in references (139–141)] or assembly of rRNP complexes. Maturation and assembly are likely to take place in an overlapping fashion in large, dynamic complexes with different mixes of snoRNAs and precursor rRNAs. These events could be mediated by an initial processome complex formed with the primary transcript as appears to be the case in 5′ ETS processing (70, 71). All snoRNPs could enter a single complex at this stage or, alternatively, bind to initial and subsequent complexes in differential fashion. The snoRNAs, snoRNP proteins, and other proteins needed could be processome components or could play roles in its formation. SnoRNAs shown to cross-link with more than one pre-rRNA region or to affect several processing events could either bind to multiple rRNA sites simultaneously or switch sites as processing proceeds. Based on present knowledge, the mRNA splicing paradigm appears to be a good model for rRNA maturation. The extent to which the snoRNPs mimic the action of the splicing snRNPs remains to be seen.

SnoRNA GENES AND SYNTHESIS

Dichotomy of snoRNA Gene Organization

The snoRNA genes have two distinct genomic organizations (4). Many exhibit an archetypical structure, with the coding region flanked by promoter, terminator, and associated enhancer elements. Independent transcripts are produced from an snoRNA-specific promoter and typically possess a modified 5′ terminus, usually a TMG cap. In the second arrangement, the snoRNA coding sequences are positioned within introns of protein genes. These intronic snoRNAs are synthesized as part of the pre-messenger RNA and undergo subsequent processing from the intron to produce a cap-minus snoRNA.

This dichotomy of gene organization has been observed in vertebrates and yeast. Early gene characterization of the major, TMG-capped, metazoan snRNAs resulted in the isolation of U3 genes which are exclusively independent transcription units (142–152). The intron coding arrangement has been revealed only recently with characterization of genes for the less-abundant snoRNAs. First noted for mouse U14 (44), the number of intron-encoded snoRNA genes in various vertebrates has grown to more than 10 during the last two years [Table 3; see reviews (2, 4, 153)]. The first 21 snoRNA genes characterized in yeast were non-intronic (Table 2) suggesting that the intron coding scheme might be limited to complex eukaryotes. However, three intronic snoRNAs have now been found in the yeasts *Saccharomyces cerevisiae*

Table 3 Intron-encoded small nucleolar RNAs

snoRNA	Parent Gene	Intron Location	Box C & D/ 5'-3' stem	Observed rRNA Com-plementarity	References
U14	hsc70 (mouse, rat, CHO, human)	5,6,8	+/+	18S	44,105,193,194,[c]
	hsc70 (trout)	2,4,5,6,7,8	+/+	18S	195
	hsc70 (X. laevis)	4,5,7	+/+	18S	[d]
	S13 (X. laevis)	3,4	+/+	18S	[d]
U15	S3 (HeLa cell)	1,5 or 6	+/+		31
	S1 (X. laevis)	3,5,6	+/+	28S	185
U16	L1 (X. tropicalis, X. laevis, human)	3	+/+	28S	181,182
U17 (E1)	RCC1 (HeLa cell)	1,2	−/−		100,184
(XS8)	S8 (X. laevis)	1,2,3,4,5,6	−/−	18S/ETS	113
U18	L1 (X. laevis)	2,4,7,8	+/−	28S	114
	L1 (X. tropicalis)	2,7,8	+/−	28S	114
	L1 (human)	4	+/−	28S	114
	EF-1β (S. cerevisiae)	1	+/−	25S	[e]
U19[b]	? (HeLa cell)	?	−/−		2

snoRNA	Parent gene	Copies	+/−	Target	Ref.
U20	Nucleolin (HeLa cell, mouse, rat, hamster, chicken, *X. laevis*)	11	+/+	18S	115,d
U21	L5 (human, chicken)	5	+/+	28S	116
	ARF-1 (*Drosophila*)	1	+/+	28S	f
U22[b]	? (HeLa cell)	?	+/+		123
U23	Nucleolin (mouse, rat, HeLa cell, hamster, chicken, *X. laevis*, fish)	12	−/−		f
U24	L7 (human, chicken)	2	+/+	28S	f
E3	eIF-4AII (human)	8	−/−		153,184
snR38	EF-1γ (*S. cerevisiae, S. pombe*)	1	+/−	25S	g
snR39	YL8 (*S. cerevisiae*)	2	+/−	25S	g
	L7 (*S. pombe*)	1	+/−	25S	g
E2[a]	? (human)	?	−/−		184

[a] = candidate intronic snoRNA
[b] = intronic snoRNAs of unknown parent gene
[c] A Laszlo, unpublished observations
[d] L Xia, J Liu, C Sage, B Trexler, M Andrews, ES Maxwell, unpublished observations
[f] A Balakin, MJ Fournier, unpublished results
[f] JP Bachellerie, unpublished observations
[g] J Ni, MJ Fournier, unpublished observations

and *Schizosaccharomyces pombe* (Table 3; A Balakin, J Ni, MJ Fournier, unpublished observations). The occurrence of snoRNA coding sequences within pre-mRNA introns has raised interesting questions about the functional implications of this arrangement as well as the evolution of introns and the snoRNA genes themselves.

Independent snoRNA Genes

Most of the unicellular snoRNA genes characterized are from *Saccharomyces cerevisiae* and the overwhelming majority of these are non-intronic. Nearly all exist as dispersed, single copy genes [(18, 19); Table 2 and references therein]. Exceptions are U3 snoRNA which is encoded by two genes (154, 155) and snR190 and U14 which are produced from coding units separated by only 66 nucleotides and apparently cotranscribed in a common precursor (29, 68). The sequences of the non-intronic genes suggest that they are transcribed by RNA Polymerase II. TATA-like boxes are typically located 80 to 120 nucleotides upstream of the putative transcriptional start sites (see Table 2 for references). Detailed transcriptional mapping has not yet been carried out for any of the yeast snoRNAs; however, the positions of the TATA sequences are consistent with that for protein genes transcribed by Pol II. Evidence of Pol II involvement is based upon the common occurrence of TMG caps.

Little is known about processing of non-intronic snoRNAs in yeast, but most are presumed to be unprocessed at the 5′ end because of the addition of the TMG cap; exceptions are snR190, U14 and 7-2/MRP RNA (29, 156). The lack of a TMG cap on snR190 and U14 is due to processing, and the mature RNAs have heterogenous 5′ and 3′ termini (29). Interestingly, cap trimethylation occurs with U14 produced from a Pol II promoter, which indicates that information for this process is within the snoRNA (29).

Only 4 of the approximately twelve characterized snoRNAs of metazoans are specified by non-intronic genes. Conventional genes for U3 (142–152) and 7-2/MRP RNA (157–160) have been isolated and characterized from higher eukaryotes, but the TMG cap structure on human U8 and U13 (21) suggests that these are non-intronic as well. The U3 genes of vertebrates are multicopy and dispersed in the genome [reviewed in (6)]. U3 pseudogenes also occur (142, 161–163), and this has complicated efforts to estimate the number of bona fide U3 genes in mammals [pseudogenes have been reported for human U13 also (164)]. The structures of the characterized U3 pseudogenes suggest that they have arisen by reverse transcription through self-priming by snoRNAs (162). Several authentic U3 genes have been successfully identified by injecting candidate DNA sequences into *Xenopus* oocyte nuclei and monitoring synthesis of U3 RNA. To date, bona fide U3 genes have been identified for human (143–145), mouse (146, 147, 165), rat (142, 148), and *Xenopus laevis*

(149). U3 genes have been characterized in both monocot (151, 166) and dicot (150, 152) plants as well. Finally, single copy genes for 7-2/MRP RNA have been reported for several diverse metazoan organisms, including human (158), mouse (157), cow (160), *Xenopus laevis* (159) and the plant *Arabidopsis thaliana* (67).

The most detailed studies on gene structure and transcription for non-intronic snoRNAs have been carried out on the U3 and 7-2/MRP RNAs. Genes for these RNAs have been characterized in a wide range of eukaryotes. U3 genes have been isolated from several unicellular organisms, including the yeasts *Saccharomyces cerevisiae* (154, 155) and *Schizosaccharomyces pombe* (167, 168), the slime mold *Dictyostelium discoideum* (30), and the protist *Tetrahymena thermophila* (169). The two U3 genes of *S. cerevisiae* are dispersed in the genome, and each possesses a single intron (154, 155). None of the other characterized snoRNA genes from this yeast have introns, and the U3 genes in *S. pombe* are also intron-less (167), although introns have been found in U3 genes from other yeast (C Branlant, unpublished observations). *Dictyostelium* is estimated to have about 5 U3 genes dispersed in the genome (30). A candidate TATA box is located 84 nucleotides upstream from the transcriptional start site in *S. cerevisiae* (154, 155), and a conserved sequence immediately upstream of both *S. cerevisiae* U3 genes may define a recognition signal involved in expression. Interestingly, the multicopy U3 genes of *Tetrahymena* possess upstream sequences similar to the TATA box and USE elements of plant U3 genes that are transcribed by RNA Polymerase III [see below; (169)].

Signals important for U3 transcription in higher eukaryotes have been defined with U3 genes from human, rat, mouse, and *Xenopus*. These genes are transcribed by Pol II and possess both proximal and distal sequence elements (PSE and DSE) located approximately −30 to −55 and −230 to −250 nucleotides from the + 1 start site (142–146, 148, 149). The PSE and DSE of U3 are similar to those found in the splicing snRNA genes (6) but the U3 genes also possess a conserved "U3 box" at the DSE (142, 145, 148, 149) which is critical for transcription (149). Additional 5′ flanking sequence is also conserved, which suggests that this region is important for expression or regulation (145, 149). Consistent with this notion is the observed developmental regulation of U3 synthesis during *Xenopus* oocyte development (63) and the varying levels of U3 expression in differentiating mouse (147) and myoblast (170) cells.

In contrast to vertebrates, the plant U3 genes of tomato (150), tobacco (152), wheat (151), maize (166) and *Arabidopsis* (152, 171) are transcribed by RNA Polymerase III. These genes possess both Pol II and Pol III promoter elements. However, the distance between the upstream TATA box and USE elements is critical, and a shorter spacing in the plant U3 genes makes them Pol III promoters (171, 172). Consistent with Pol III transcription, plant U3

RNAs lack a TMG cap (replaced by an O-methyl cap; 25, 26), and synthesis is resistant to α-amanitin treatment (151, 171, 172). Comparison of monocot and dicot U3 genes has revealed an additional transcriptional element in the 5' flanking region of the monocot genes (166). Pseudogenes for U3 have been found in tomato, which demonstrates their occurrence in plants as well as in vertebrates (150). The relatedness of the Pol II and Pol III promoters in plants suggests that the two may have evolved from a single ancestral element (166).

The 7-2/MRP genes in both vertebrates and plants are transcribed by RNA Polymerase III (67, 157–159) in apparent contrast with the corresponding gene in yeast (156). The single-copy genes of mouse (157), human (158), and *Xenopus* (159) possess a mixture of Pol II and Pol III promoter elements. These elements include an upstream TATA box and potential Sp1 transcription factor sites coupled with PSE and DSE elements. Consistent with Pol III transcription, however, the 7-2/MRP genes contain a conserved box A within the coding region; a T-rich region flanking the 3' end presumably functions as a terminator. The 7-2/MRP gene from *Arabidopsis* possesses the appropriately spaced upstream USE and TATA elements characteristic of plant snRNA Pol III genes, and synthesis of 7-2/MRP RNA is indeed resistant to α-amanitin treatment (67).

Intronic snoRNAs

GENE STRUCTURE AND DIVERSITY. The occurrence of snoRNA coding units within pre-mRNA introns was first recognized for U14 with 3 unique sequences identified in introns 5, 6, and 8 of the cognate hsc70 heat shock gene (44). Suggestions of such a peculiar coding arrangement are now evident in earlier studies of L1 ribosomal protein genes of *Xenopus laevis* and *Xenopus tropicalis* (173, 174) and the nucleolin gene of rodents (175). Strong sequence conservation was noted in four different introns of the L1 gene and these segments are now known to specify U16: complementarity to 28S rRNA was also noted for these sequences (173, 174). Splicing of intron 3 from the *Xenopus* L1 gene revealed it to be regulated by sequences within the intron and the intron itself undergoes processing to produce a small, stable RNA (U16) (176–179). Highly conserved intron sequences in the rodent nucleolin gene have since been shown to include U20 and U23.

The realization that snoRNA coding sequences can occur within pre-mRNA introns has led to a rapidly expanding list of eukaryotic intronic snoRNA species [see Table 3; (2, 31, 44, 113–116, 180–185)]. Interestingly, all the intronic snoRNA coding sequences are located in parent genes that specify proteins involved in nucleolar function, ribosome structure, or protein synthesis (see Table 3). This recurring theme may have important functional significance (see below). Early results suggested that the encoding of specific snoRNAs within given parent genes and in defined intron locations might be conserved in evolution. However, with the characterization of U14, U17, and U18 genes from different organisms, it has become clear that a specific snoRNA can be encoded in different parent genes in different organisms, and the number of snoRNA genes in the parent gene and intron location are not strictly conserved (Table 3). New sequence results for U14-containing genes in *Xenopus laevis* have revealed that intronic snoRNAs can also occur in two different parent genes within a single organism (L Xia, J Liu, C Sage, B Trexler, M Andrews, ES Maxwell, unpublished observations). Curiously, at least one snoRNA is encoded in both intronic and non-intronic arrangements. U14 is intron-encoded in mammals, amphibians, and fish (Table 3), but is non-intronic in *S. cerevisiae* (29, 68). In maize, the U14 genes are tandemly arranged (185a). This fascinating mix of coding patterns may have interesting implications for snoRNA gene evolution (see below).

SnoRNA PROCESSING.   Intronic snoRNAs are transcribed as part of the parent pre-mRNA and then processed from the intron to yield a mature snoRNA. Defining the mechanisms of processing and the relationship to mRNA splicing are key issues currently under study. Processing of intronic snoRNAs was first demonstrated in *Xenopus* oocytes with injection of mouse hsc70 (U14) (180) and frog ribosomal protein L1 (U16) (181) transcripts into nuclei. Processed snoRNAs were excised from their respective introns possessing correct 5′ and 3′ termini. Extracts from HeLa and mouse cells have also been used to demonstrate processing of human U15 (31) and U17 (100) snoRNAs. Most recently, U17 and U20 excision has been demonstrated in transfection assays of HeLa cells (T Kiss, W Filipowicz; J Cavaille, JP Bachellerie, unpublished observations).

Early results demonstrated that processing, at least in the systems utilized, is not dependent on pre-mRNA splicing. Removal of one or both flanking exons resulted in correct processing in both the *Xenopus* oocyte system (U14) (180) and cell extracts (U15, U17) (31, 100). Both systems produce similar processing intermediates consisting of snoRNA attached to either the upstream or the downstream portion of the adjoining pre-mRNA precursor (31, 100, 114, 181, 182). The occurrence of these two major intermediates demonstrates that the 5′ and 3′ termini are processed independently. Particularly evident with in vitro extracts is the appearance of metastable intermediates possessing 3 to 6 extra nucleotides at the 3′ end (31, 100). These observations indicate that exonucleolytic trimming is also involved in processing. Consistent with this conclusion are observations that processing appears to be inhibited by 5′ capping or circularization of the precursor transcript (T Kiss, W Filipowicz; K Tycowksi, JA Steitz, unpublished observations).

Single pre-mRNA transcripts are generally presumed to give rise to both spliced mRNA and processed snoRNA. However, this important point has not been demonstrated, and it represents a difficult experimental challenge. If the two-product model is correct, it follows that endonuclease activity is required for processing of intronic snoRNAs. Two possible pathways for endonucleolytic cleavage are currently under consideration. The first involves simple intron excision-lariat formation, followed by debranching and exonucleolytic trimming of the linear intron. Alternatively, endonucleolytic cleavage could occur within one or both flanking intron sequences, followed by trimming. Transfection results demonstrating that mutation of the donor and acceptor splice sites affects U17 production suggest that lariat structures are snoRNA precursors (T Kiss, W Filipowicz, unpublished observations). The extent of exonucleolytic trimming that would be required to complete the processing of this snoRNA is striking at more than 1000 nucleotides! In contrast, processing of U16 in both the in vivo and in vitro systems results in cleavage of the flanking 5' and 3' intron sequences about 20 nucleotides upstream and downstream from the snoRNA (114, 182). Unfortunately, current information is still too limited to make general conclusions about the endonucleolytic cleavage pathway. It is possible that both processing pathways are used in an organism-specific manner or even in single cell types, perhaps in a differential fashion.

Current evidence indicates that exonucleolytic trimming is required for snoRNA maturation with either endonucleolytic cleavage scheme. Ongoing experiments have begun to define the structural elements necessary for defining the boundaries for trimming. Analyses with truncated transcripts in the oocyte and extract systems showed that flanking exon sequences are not required (31, 100, 180). The ability to eliminate essentially all flanking intron sequences and still correctly process U14 snoRNA in *Xenopus* oocyte nuclei indicates that the RNA elements essential for maturation are contained within the snoRNA molecule itself (R Leverette, M Andrews, ES Maxwell, unpublished observations). Consistent with this, U14 or U17 snoRNAs inserted into an intron of a non-parental human globin pre-mRNA are correctly processed in *Xenopus* oocytes or transfected HeLa cells, respectively (R Leverette, M Andrews, ES Maxwell, unpublished observations; T Kiss, W Filipowicz, unpublished observations). Interestingly, the non-intronic yeast U14 precursor is properly processed in *Xenopus* oocyte nuclei, suggesting that processing elements are evolutionarily conserved within the structure of both intronic and non-intronic U14 (ES Maxwell, J Liu, MJ Fournier, unpublished observations). Recent deletion experiments have indicated that essential elements for U14 maturation are contained in the terminal stem and flanking boxes C and D (N Watkins, ES Maxwell, unpublished observations). Mutational analysis has indicated that both boxes C and D are also important for U16 snoRNA maturation (E Caffarelli, I Bozzoni, unpublished results).

A candidate element for processing of at least some snoRNAs appears to be the base-paired terminal stem flanked by boxes C and D (see Table 3). In some cases, terminal helix complementarity extends into the flanking intron sequences (44), but analyses with one mouse U14.6 variant has demonstrated the extended helix is not essential for processing (185b). Exonucleolytic trimming could halt at the terminal stem. Alternatively, this terminal helix and either or both neighboring box elements could serve as a structural motif for binding of a protein component(s) that modulates processing (31). Such an RNP complex could define the boundaries of trimming and perhaps could include signals recognized in this process. Fibrillarin antibodies immunoprecipitate both L1 pre-mRNA transcripts containing U16 as well as processed U16, which suggests that assembly of the snoRNA ribonucleoprotein complex may begin while the snoRNA is still part of the parent pre-mRNA intron (181). The formation of an RNP structure as an essential step in snoRNA processing is particularly attractive, because no conserved sequence elements or secondary structural motifs are common to all intronic snoRNAs. Finally, the site of snoRNA processing within the nucleus is unknown, but a prime candidate is the spliceosome apparatus. Processing at this site would provide a means for coordinating and regulating pre-mRNA splicing and snoRNA processing pathways. Indeed, recent analysis of U16 processing has indicated that a pre-mRNA binding protein (hnRNP protein C) can bind to flanking intron sequences and affect the efficiency of snoRNA processing with respect to pre-mRNA splicing (186).

BIOLOGICAL AND EVOLUTIONARY IMPLICATIONS.   The discovery of intronic snoRNAs has raised intriguing questions about the evolutionary origins and functional implications of this unique genetic arrangement. All the intronic snRNAs characterized to date are nucleolar species, and the parent genes all specify protein products associated with the function of the nucleolus or ribosome structure and function (Table 3). It seems unlikely that these pairings are coincidental, which suggests that important functional relationships may exist for these components. The positioning of snoRNAs needed for the processing of rRNA within pre-mRNA introns could metabolically link and coordinate expression of Pol I and II transcripts, balancing mRNA transcription with rRNA synthesis and ultimately with ribosome biogenesis (31, 180). Several intronic snoRNAs are encoded in parent genes that produce large amounts of protein product. This organization could ensure high-level expression of the resident snoRNAs as well (4). The occurrence of snoRNA coding units in several introns of a parent gene is also consistent with the achievement of high production levels. It has been suggested that the positioning of U17 within introns of the RCC1 gene, important in chromatin condensation and DNA replication, provides a basis for coordinating ribosome biogenesis with the cell cycle (100). Similarly, placement of snoRNA genes within numerous ribosomal protein genes could provide a means to coordinate the synthesis of rRNA and ribosomal proteins (31, 180). Such a feedback scheme has been noted in yeast, in which ribosomal proteins bind to parent mRNAs to modulate translation (13, 14). Coordination could also occur at the level of the gene products, with snoRNAs or snoRNPs binding to ribosomal and non-ribosomal proteins to help regulate and direct ribosome synthesis. Extending this theme, it is also interesting to consider the possibility that ribosomal and non-ribosomal proteins might bind with intronic snoRNAs at the precursor level and determine whether or not the snoRNA is processed from the pre-mRNA transcript. Defining the physical and functional relationships of the parent genes and their respective snoRNAs will be major goals for the future.

The genetic origin of the intronic snoRNAs is also unknown, and this too promises to be a subject of much interest. Intronic snoRNAs, and indeed all small nuclear RNAs, could be products of ancient introns which have acquired a biological function (4, 180). Some primordial snRNAs could have remained within introns because of functional advantages (intronic snoRNAs) while others became independent transcription units with independent genomic locations (non-intronic snoRNAs and spliceosomal snRNAs). In another scenario, intronic snoRNA coding units could have become incorporated into introns relatively recently in evolution (4, 180). Observations consistent with this hypothesis are: (a) the non-conserved positions of U14, U17, and U18 snoRNAs in different parent genes of different organisms (Table 3); (b) the alteration of snoRNA gene copy number and intron position within homologous parent genes in different organisms (Table 3); (c) the absence of E3 in one of two human eIF-4 gene copies (153); and (d) the dichotomy of U14 genomic organization as both an intronic and non-intronic snoRNA (44, 68). Terminal repeat sequences are not evident in the intronic coding units, which argues that integration did not arise by familiar homing processes. Because most vertebrate snoRNAs known are intronic while those of yeast are not, there is also the interesting question of whether yeast was late in acquiring this coding arrangement or is among the first organisms to evolve independent snoRNA genes. Characterization of additional snoRNA genes from evolutionarily diverse organisms should provide valuable insights into these interesting issues.

FUTURE RESEARCH DIRECTIONS

Exciting progress has been made in understanding the biology of the snoRNAs, and several goals for future research are clearly defined. The rapid discovery of new snoRNA species in both unicellular and metazoan organisms suggests that many additional RNAs remain to be identified and characterized. Similarly, cataloging of the snoRNP proteins has only begun, and complete listings of these components is essential as well. Understanding the structures of the snoRNP particles and how these assemble into larger complexes will be essential to understanding the mechanisms of snoRNA action, as will the synthesis and regulation of the RNP components themselves. Little is known about the expression and regulation of snoRNA genes and maturation of snoRNA precursors. Polymerase factors and transcription signals need to be defined, and the nature and mechanisms of the post-transcriptional reactions must be specified. Characterizing the synthesis and function of the intronic snoRNAs promises to be especially interesting, from the dual perspectives of defining this new biosynthetic scheme and determining if functional relationships exist between the snoRNAs and proteins encoded at the same loci. There is the potential for discovering important new regulatory links or, at least, fascinating novel schemes involved in genome evolution.

Strong impressions about snoRNA function have been developed, but the actual roles and mechanisms of action remain to be described; these issues will drive much of snoRNA research for the foreseeable future. SnoRNAs interact with rRNA and many are required for rRNA processing. While some are closely tied to endonucleolytic reactions, snoRNAs can, in principle, have roles in every stage of rRNA maturation and ribosome subunit assembly that occur in the nucleolus. These events almost certainly overlap, and they can be presumed to involve large assemblages of substrate rRNAs, snoRNPs and ribosomal and non-ribosomal proteins. Chaperone functions for snoRNAs is an especially interesting possibility, to mediate folding of rRNA and rRNP complexes and perhaps even protein components. Much important progress can come from the in vivo genetic and biochemical systems and approaches currently in use. However, there is an urgent need for new in vitro systems to study rRNA and ribosome synthesis. Achieving ribosome synthesis in a cell-free system seems an almost impossible feat at present, but establishing new partial systems for studying at least some snoRNA-dependent events should be feasible. This will open the way for fractionation and reconstitution studies, as well as for defining mechanisms of action. In the meantime, we can be confident that many exciting and important discoveries will emerge from current strategies and that a much clearer picture of snoRNA function will develop in the next few years. The snoRNAs now represent a major frontier in eukaryotic RNA research. Unveiling their secrets promises to be an exciting adventure!

ACKNOWLEDGMENTS

We thank W. Liang and J. Ni for preparing the figures, and all the members of our laboratories for their many contributions to the work described and their valuable help in preparing this review. We are grateful to the many colleagues who shared with us their manuscripts and other unpublished results. We especially acknowledge Tamas Kiss and Jean-Pierre Bachellerie for critical reading of this manuscript and Elizabeth Furter-Graves for editorial advice. Work in our laboratories was supported by grants from the National Institutes of Health (GM19351 to MJF) and the National Science Foundation (MCB-9304672 to ESM) and by an Alfred P. Sloan Fellowship (ESM).

Literature Cited

1. Fournier MJ, Maxwell ES. 1993. *Trends Biochem. Sci.* 18:131–35
2. Filipowicz W, Kiss T. 1993. *Mol. Biol. Rep.* 18:149–56
3. Mattaj IW, Tollervey D, Seraphin B. 1993. *FASEB J.* 7:47–53
4. Sollner-Webb B. 1993. *Cell* 75:403–5
5. Baserga SJ, Steitz JA. 1993. *The RNA World,* ed. R. Gesteland, JF Atkins, pp. 359–81. Cold Spring Harbor, NY: Cold Spring Harbor Press
6. Dahlberg J, Lund E. 1988. See Ref. 196, pp. 38–70
7. Reddy R, Busch H. 1988. See Ref. 196, pp. 1–37
8. Gerbi SA, Savino R, Stebbins-Boaz B, Jeppesen C, Rivera-Leon R. 1990. *The Ribosome—Structure, Function and Evolution,* ed. WE Hill, A Dahlberg, RA Garrett, PB Moore, D Schlessinger, JR Warner, pp. 452–69. Washington, DC: Am. Soc. Microbiol.
9. Scheer U, Thiry M, Goessens G. 1993. *Trends Cell Biol.* 3:236–41
10. Schwarzacher HG, Wachtler F. 1993. *Anat. Embryol.* 188:515–36
11. Eichler DC, Craig N. 1994. *Prog. Nucleic Acid Res. Mol. Biol.,* ed. K Moldave, W Cohen. New York: Academic. In press
12. Sollner-Webb B, Tycowski KT, Steitz JA. 1994. *Ribosomal RNA: Structure, Evolution, Gene Expression and Function in Protein Synthesis,* ed. RA Zimmermann, AE Dahlberg. New York: CRC. In press
13. Woolford JL. 1991. *Adv. Genet.* 29:63–118
14. Warner JR. 1990. *Curr. Opin. Cell. Biol.* 2:521–27
15. Raue HA, Planta RJ. 1991. *Prog. Nucleic Acid Res. Mol. Biol.* 41:91–129
16. Gu J, Reddy R. 1994. *Nucleic Acids Res.* 22:3481–82
17. Busch H, Reddy R, Rothblum L, Choi YC. 1982. *Annu. Rev. Biochem.* 51:617–54
18. Wise JA, Tollervey D, Maloney D, Swerdlow H, Dunn EJ, Guthrie C. 1983. *Cell* 35:743–51
19. Riedel N, Wise JA, Swerdlow H, Mak A, Guthrie C. 1986. *Proc. Natl. Acad. Sci. USA* 83:8097–8101
20. Thompson JR, Zagorski J, Woolford JL, Fournier MJ. 1988. *Nucleic Acids Res.* 16:5587–601
21. Tyc K, Steitz JA. 1989. *EMBO J.* 8:3113–19
22. Baserga SJ, Gilmore-Hebert M, Yang XW. 1992. *Genes Dev.* 6:1120–30
23. Terns MP, Dahlberg JE. 1994. *Science* 264:959–61
24. Peculis BA, Steitz JA. 1994. *Genes Dev.* 8:2241–55
25. Shimba S, Buckley B, Reddy R, Kiss T, Filipowicz W. 1992. *J. Biol. Chem.* 267:13772–77
26. Liu MH, Busch RK, Buckley B, Reddy R. 1992. *Nucleic Acids Res.* 20:4299–4304
27. Reddy R, Tan EM, Henning D, Nohga K, Busch H. 1983. *J. Biol. Chem.* 258:1383–86
28. Yuan Y, Singh R, Reddy R. 1989. *J. Biol. Chem.* 264:14835–39
29. Balakin AG, Lempicki RA, Huang GM, Fournier MJ. 1994. *J. Biol. Chem.* 269:739–46
30. Wise JA, Weiner AM. 1980. *Cell* 22:109–18
31. Tycowski KT, Shu MD, Steitz JA. 1993. *Genes Dev.* 7:1176–90
32. Stroke IL, Weiner AM. 1989. *J. Mol. Biol.* 210:497–512
33. Beltrame M, Tollervey D. 1992. *EMBO J.* 11:1531–42
34. Parker KA, Steitz JA. 1987. *Mol. Cell. Biol.* 7:2899–2913
35. Parker KA, Bruzik JP, Steitz JA. 1988. *Nucleic Acids Res.* 16:10493–509

36. Baserga SJ, Yang XW, Steitz JA. 1991. *EMBO J.* 10:2645–51
37. Jarmolowski A, Zagorski J, Li HV, Fournier MJ. 1990. *EMBO J.* 9: 4503–9
38. Huang GM, Jarmolowski A, Struck JC, Fournier MJ. 1992. *Mol. Cell. Biol.* 12: 4456–63
39. Jeppesen C, Stebbins-Boaz B, Gerbi SA. 1988. *Nucleic Acids Res.* 16:2127–48
40. Segault V, Mougin A, Gregoire A, Banroques J, Branlant C. 1992. *Nucleic Acids Res.* 20:3443–51
41. Hartshorne T, Agabian N. 1994. *Nucleic Acids Res.* 22:3354–64
42. Orum H, Nielsen H, Engberg J. 1993. *Nucleic Acids Res.* 21:2511
43. Shanab GM, Maxwell ES. 1991. *Nucleic Acids Res.* 19:4891–94
44. Liu J, Maxwell ES. 1990. *Nucleic Acids Res.* 18:6565–71
45. Li HV, Fournier MJ. 1992. *EMBO J.* 11:683–89
46. Topper JN, Clayton DA. 1987. *J. Biol. Chem.* 265:13254–62
47. Forster AC, Altman S. 1990. *Cell* 62: 407–9
48. Schmitt ME, Bennett JL, Dairaghi DJ, Clayton DA. 1993. *FASEB J.* 7:208–13
49. Lischwe MA, Ochs RL, Reddy R, Cook RG, Yeoman LC, et al. 1985. *J. Biol. Chem.* 260:14304–10
50. Ochs RL, Lischwe MA, Spohn WH, Busch H. 1985. *Biol. Cell* 54:123–34
51. Reimer G, Steen VD, Penning CA, Medsger TA, Tan EM. 1988. *Arthrit. Rheum.* 31:525–32
52. Luhrmann R. 1988. See Ref. 196, pp. 71–99
53. Aris JP, Blobel G. 1991. *Proc. Natl. Acad. Sci. USA* 88:931–35
54. Jansen RP, Hurt EC, Kern H, Lehtonen H, Carmo-Fonseca M, et al. 1991. *J. Cell Biol.* 113:715–29
55. Turley SJ, Tan EM, Pollard KM. 1993. *Biochim. Biophys. Acta* 1216:119–22
56. Lapeyre B, Mariottini P, Mathieu C, Ferrer P, Amaldi F, et al. 1990. *Mol Cell. Biol.* 10:430–34
57. Christensen ME, Fuxa KP. 1988. *Biochem. Biophys. Res. Commun.* 155: 1278–83
58. Schimmang T, Tollervey D, Kern H, Frank R, Hurt EC. 1989. *EMBO J.* 8:4015–24
59. Henriquez R, Blobel G, Aris JP. 1990. *J. Biol. Chem.* 265:2209–15
60. Reuter R, Tessars G, Vohr HW, Gleichmann E, Luhrmann R. 1989. *Proc. Natl. Acad. Sci. USA* 86:237–41
61. Monestier M, Losman JM, Novick KE, Aris JP. 1994. *J. Immunol.* 152: 667–75
62. Hultman P, Enestrom S, Turley SJ, Pollard KM. 1994. *Clin. Exp. Immunol.* 96:285–91
63. Caizergues-Ferrer M, Mathieu C, Mariottini P, Amalric F, Amaldi F. 1991. *Development* 112:317–26
64. Peculis BA, Steitz JA. 1993. *Cell* 73: 1233–45
65. Hashimoto C, Steitz JA. 1983. *J. Biol. Chem.* 258:1379–82
66. Reimer G, Raska I, Scheer U, Tan EM. 1988. *Exp. Cell Res.* 176:117–28
67. Kiss T, Marshallsay C, Filipowicz W. 1992. *EMBO J.* 11:3737–46
68. Zagorski J, Tollervey D, Fournier MJ. 1988. *Mol. Cell. Biol.* 8:3282–90
69. Lubben B, Marshallsay C, Rottmann N, Luhrmann R. 1993. *Nucleic Acids Res.* 21:5377–85
70. Mougey EB, O'Reilly M, Osheim Y, Miller O, Beyer A, Sollner-Webb B. 1993. *Genes Dev.* 7:1609–19
71. Kass S, Sollner-Webb B. 1990. *Mol. Cell. Biol.* 10:4920–31
72. Kass S, Tyc K, Steitz JA, Sollner-Webb B. 1990. *Cell* 60:897–908
73. Yuan Y, Tan E, Reddy R. 1991. *Mol. Cell. Biol.* 11:5266–74
74. Lygerou Z, Mitchell P, Petfalski E, Seraphin B, Tollervey D. 1994. *Genes Dev.* 8:1423–33
75. Schmitt ME, Clayton DA. 1994. *Genes Dev.* 8:2617–28
76. Gold HA, Craft J, Hardin JA, Bartkiewicz M, Altman S. 1988. *Proc. Natl. Acad. Sci. USA* 85:5483–87
77. Gold HA, Topper JN, Clayton DA, Craft J. 1989. *Science* 245:1377–80
78. Liu MH, Yuan Y, Reddy R. 1994. *Mol. Cell. Biochem.* 130:75–82
79. Jansen RP, Tollervey D, Hurt EC. 1993. *EMBO J.* 12:2549–58
80. Girard JP, Lehtonen H, Caizergues-Ferrer M, Amalric F, Tollervey D, Lapeyre B. 1992. *EMBO J.* 11:673–82
81. Girard JP, Caizergues-Ferrer M, Lapeyre B. 1993. *Nucleic Acids Res.* 21:2149–55
82. Clark MW, Yip MLR, Campbell J, Abelson J. 1990. *J. Cell Biol.* 111:1741–51
83. Jong AY, Clark MW, Gilbert M, Oehm A, Campbell JL. 1987. *Mol. Cell. Biol.* 7:2947–55
84. Heine MA, Rankin ML, DiMario PJ. 1993. *Mol. Biol. Cell* 4:1189–1204
85. Ghisolfi L, Kharrat A, Joseph G, Amalric F, Erard M. 1992. *Eur. J. Biochem.* 209:541–48
86. Lapeyre B, Bourbon H, Amalric F. 1987. *Proc. Natl. Acad. Sci. USA* 84: 1472–76
87. Kondo K, Inouye M. 1992. *J. Biol. Chem.* 267:16252–58

88. Lee WC, Xue Z, Melese T. 1991. *J. Cell Biol.* 113:1–12
89. Berges T, Petfalski E, Tollervey D, Hurt E. 1994. *EMBO J.* 13:3136–48
90. Sun C, Woolford JL. 1994. *EMBO J.* 13:3127–35
91. Azum-Gelade MC, Noaillac-Depeyre J, Caizergues-Ferrer M, Gas N. 1994. *J. Cell Sci.* 107:463–75
91a. Jimenez-Garcia LF, Segura-Valdez ML, Ochs RL, Rothblum LI, Hannan R, Spector D. 1994. *Mol. Biol. Cell* 5:955–66
92. Sollner-Webb B, Mougey EB. 1991. *Trends Biochem. Sci.* 16:58–62
93. Thiry M, Goessens G. 1992. *Exp. Cell Res.* 200:1–4
94. Fischer D, Weisenberger D, Scheer U. 1991. *Chromosoma* 101:133–40
95. Puvion-Dutilleul F, Bachellerie JP, Puvion E. 1991. *Chromosoma* 100:395–409
96. Hadjiolov AA. 1985. *The Nucleus and Ribosome Biogenesis.* New York: Springer-Verlag
97. Kumar A, Warner JR. 1972. *J. Mol. Biol.* 63:233–46
98. Puvion-Dutilleul F, Mazan S, Nicoloso M, Christensen ME, Bachellerie JP. 1991. *Eur. J. Cell Biol.* 56:178–86
99. Warner JR, Soeira R. 1967. *Proc. Natl. Acad. Sci. USA* 58:1984–90
100. Kiss T, Filipowicz W. 1993. *EMBO J.* 12:2913–20
101. Zieve GW, Penman S. 1976. *Cell* 8:19–31
102. Prestayko AW, Tonato M, Busch H. 1970. *J. Mol. Biol.* 47:505–15
103. Hartshorne T, Agabian N. 1993. *Mol. Cell. Biol.* 13:144–54
104. Maxwell ES, Martin TE. 1986. *Proc. Natl. Acad. Sci. USA* 83:7261–65
105. Trinh-Rohlik Q, Maxwell ES. 1988. *Nucleic Acids Res.* 16:6041–56
106. Tollervey D. 1987. *EMBO J.* 6:4169–75
107. Maser RL, Calvet JP. 1989. *Proc. Natl. Acad. Sci. USA* 86:6523–27
108. Tyc K, Steitz JA. 1992. *Nucleic Acids Res.* 20:5375–82
109. Beltrame M, Henry Y, Tollervey D. 1994. *Nucleic Acids Res.* 22:5139–47
110. Li HV, Zagorski J, Fournier MJ. 1990. *Mol. Cell. Biol.* 10:1145–52
111. Rimoldi OJ, Raghu B, Nag MK, Eliceiri GL. 1993. *Mol. Cell. Biol.* 13:4382–90
112. Morrissey JP, Tollervey D. 1993. *Mol. Cell. Biol.* 13:2469–77
113. Cecconi F, Mariottini P, Loreni F, Pierandrei-Amaldi P, Campioni N, Amaldi F. 1994. *Nucleic Acids Res.* 22:732–41
114. Prislei S, Michienzi A, Presutti C, Fragapane P, Bozzoni I. 1993. *Nucleic Acids Res.* 21:5824–30
115. Nicoloso M, Caizergues-Ferrer M, Michot B, Azum MC, Bachellerie JP. 1994. *Mol. Cell. Biol.* 14:5766–76
116. Qu LH, Nicoloso M, Michot B, Azum MC, Caizergues-Ferrer M, et al. 1994. *Nucleic Acids Res.* 22:4073–81
117. Kass S, Craig N, Sollner-Webb B. 1987. *Mol. Cell. Biol.* 7:2891–98
118. Mougey EB, Pape LK, Sollner-Webb B. 1993. *Mol. Cell. Biol.* 13:5990–98
119. Savino R, Gerbi SA. 1990. *EMBO J.* 9:2299–2308
120. Savino R, Gerbi SA. 1991. *Biochimie* 73:805–12
121. Tollervey D, Guthrie C. 1985. *EMBO J.* 4:3873–78
122. Hughes JM, Ares M. 1991. *EMBO J.* 10:4231–39
123. Tycowski KT, Shu MD, Steitz JA. 1994. *Science.* 266:1558–61
124. Dandekar T, Tollervey D. 1993. *Nucleic Acids Res.* 21:5386–90
125. Chu S, Archer RH, Zengel JM, Lindahl L. 1994. *Proc. Natl. Acad. Sci. USA* 91:659–63
126. Schmitt ME, Clayton DA. 1993. *Mol. Cell. Biol.* 13:7935–41
127. Henry Y, Wood H, Morrissey JP, Petfalski E, Kearsey S, Tollervey D. 1994. *EMBO J.* 13:2452–63
128. Clayton DA. 1994. *Proc. Natl. Acad. Sci. USA* 91:4615–17
129. Craig N, Kass S, Sollner-Webb B. 1987. *Proc. Natl. Acad. Sci. USA* 84:629–33
130. Musters W, Boun K, van der Sande CAFM, vanHeerikhuizen H, Planta RJ. 1990. *EMBO J.* 9:3989–96
131. Parker R, Simmons T, Shuster EO, Siliciano PG, Guthrie C. 1988. *Mol. Cell. Biol.* 8:3150–59
132. Amiri KA. 1994. *J. Bacteriol.* 176:2124–27
133. Woolford JL, Warner JR. 1991. *The Molecular and Cellular Biology of the Yeast Saccharomyces: Gene Dynamics, Protein Synthesis, and Energetics,* pp. 587–626. Cold Spring Harbor, NY: Cold Spring Harbor Lab. Press
134. Tollervey D, Lehtonen H. Carmo-Fonseca M, Hurt EC. 1991. *EMBO J.* 10:573–83
135. Tollervey D, Lehtonen H, Jansen R, Kern H, Hurt EC. 1993. *Cell* 72:443–57
136. Olson MOJ. 1990. *The Eukaryotic Nucleus,* ed. SH Wilson, P Strauss, 2:519–59. London: The Telford Press
137. Lee WC, Zabetakis D, Melese T. 1992. *Mol. Cell. Biol.* 12:3865–71
138. Kondo K, Kowalski LR, Inouye M. 1992. *J. Biol. Chem.* 267:16259–65

139. Crouch RJ, Kanaya S, Earl P. 1983. *Mol. Biol. Rep.* 9:75–78

140. Bachellerie JP, Michot B, Raynal F. 1983. *Mol. Biol. Rep.* 9:79–86

141. Kupriyanova NS, Timofeeva MY. 1988. *Mol. Biol. Rep.* 13:91–96

142. Stroke IL, Weiner AM. 1985. *J. Mol. Biol.* 184:183–93

143. Suh D, Busch H, Reddy R. 1986. *Biochem. Biophys. Res. Commun.* 137:1133–40

144. Suh D, Wright D, Reddy R. 1991. *Biochemistry* 30:5438–43

145. Yuan Y, Reddy R. 1989. *Biochim. Biophys. Acta* 1008:14–22

146. Mazan S, Bachellerie JP. 1988. *J. Biol. Chem.* 263:19461–67

147. Mazan S, Gulli MP, Joseph N, Bachellerie JP. 1992. *Eur. J. Biochem.* 205:1033–41

148. Ach RA, Weiner AM. 1991. *Nucleic Acids Res.* 19:4209–18

149. Savino R, Hitti Y, Gerbi SA. 1992. *Nucleic Acids Res.* 20:5435–42

150. Kiss T, Solymosy F. 1990. *Nucleic Acids Res.* 18:1941–49

151. Marshallsay C, Connelly S, Filipowicz W. 1992. *Plant Mol. Biol.* 19:973–83

152. Marshallsay C, Kiss T, Filipowicz W. 1990. *Nucleic Acids Res.* 18:3459–66

153. Seraphin B. 1993. *Trends Biochem. Sci.* 18:330–31

154. Hughes JM, Konings DAM, Cesareni G. 1987. *EMBO J.* 6:2145–55

155. Myslinski E, Segault V, Branlant C. 1990. *Science* 247:1213–16

156. Schmitt ME, Clayton DA. 1992. *Genes Dev.* 6:1975–85

157. Chang DD, Clayton DA. 1989. *Cell* 56:131–39

158. Topper JN, Clayton DA. 1990. *Nucleic Acids Res.* 18:793–99

159. Bennett JL, Jeong-Yu S, Clayton DA. 1992. *J. Biol. Chem.* 267:21765–72

160. Dairaghi DJ, Clayton DA. 1993. *J. Mol. Evol.* 37:338–46

161. Reddy R, Henning D, Chirala S, Rothblum L, Wright D, Busch H. 1985. *J. Biol. Chem.* 260:5715–19

162. Bernstein LB, Mount SM, Weiner AM. 1983. *Cell* 32:461–72

163. Mazan S, Michot B, Bachellerie JP. 1989. *Eur. J. Biochem.* 181:599–605

164. Baserga SJ, Yang XW, Steitz JA. 1991. *Gene* 107:347–48

165. Mazan S, Qu LH, Sri-Widada J, Nicoloso M, Bachellerie JP. 1990. *FEBS Lett.* 267:121–25

166. Connelly S, Marshallsay C, Leader D, Brown JW, Filipowicz W. 1994. *Mol. Cell. Biol.* 14:5910–19

167. Selinger DA, Porter GL, Brennwald PJ, Wise JA. 1992. *Mol. Biol. Evol.* 9:297–308

168. Porter GL, Brennwald PJ, Holm KA, Wise JA. 1988. *Nucleic Acids Res.* 16:10131–52

169. Orum H, Nielsen H, Engberg J. 1992. *J. Mol. Biol.* 227:114–21

170. Glibetic M, Larson DE, Sienna N, Bachellerie JP, Sells BH. 1992. *Exp. Cell Res.* 202:183–89

171. Waibel F, Filipowicz W. 1990. *Nature* 346:199–202

172. Kiss T, Marshallsay C, Filipowicz W. 1991. *Cell* 65:517–26

173. Loreni F, Ruberti I, Bozzoni I, Pierandrei-Amaldi P, Amaldi F. 1985. *EMBO J.* 4:3483–88

174. Cutruzzola F, Loreni F, Bozzoni I. 1986. *Gene* 49:371–76

175. Bourbon HM, Amalric F. 1990. *Gene* 88:187–96

176. Caffarelli E, Fragapane P, Gehring C, Bozzoni I. 1987. *EMBO J.* 6:3493–98

177. Caffarelli E, Fragapane P, Bozzoni I. 1992. *Biochem. Biophys. Res. Commun.* 183:680–87

178. Fragapane P, Caffarelli E, Lener M, Prislei S, Santoro B, Bozzoni I. 1992. *Mol. Cell. Biol.* 12:1117–25

179. Prislei S, Sperandio S, Fragapane P, Caffarelli E, Presutti C, Bozzoni I. 1992. *Nucleic Acids Res.* 20:4473–79

180. Leverette RD, Andrews MT, Maxwell ES. 1992. *Cell* 71:1215–21

181. Fragapane P, Prislei S, Michienzi A, Caffarelli E, Bozzoni I. 1993. *EMBO J.* 12:2921–28

182. Caffarelli E, Arese M, Santoro B, Fragapane P, Bozzoni I. 1994. *Mol. Cell. Biol.* 14:2966–74

183. Ruff EA, Rimoldi OJ, Raghu B, Eliceiri GL. 1993. *Proc. Natl. Acad. Sci. USA* 90:635–38

184. Nag MK, Thai TT, Ruff EA, Selvamurugan N, Kunnimalaiyaan M, Eliceiri GL. 1993. *Proc. Natl. Acad. Sci. USA* 90:9001–5

185. Pellizzoni L, Crosio C, Campioni N, Loreni F, Pierandrei-Amaldi P. 1994. *Nucleic Acids Res.* 22:4607–13

185a. Leader D, Sanders, JF, Waugh R, Shaw P, Brown JW. 1994. *Nuc. Acids Res.* 22:5196–203

185b. Barbhaiya H, Leverette R, Li J, Maxwell ES. 1994. *Eur. J. Biochem.* 226:765–71

186. Santoro B, De Gregorio E, Caffarelli E, Bozzoni I. 1994. *Mol. Cell. Biol.* 14:6975–82

187. Reddy R, Henning D, Busch H. 1979. *J. Biol. Chem.* 254:11097–105

188. Reddy R, Henning D, Busch H. 1980. *J. Biol. Chem.* 255:7029–33
189. Reddy R, Henning D, Busch H. 1985. *J. Biol. Chem.* 260:10930–35
190. Shanab GM, Maxwell ES. 1992. *Eur. J. Biochem.* 206:391–400
191. Balakin AG, Schneider GS, Corbett MS, Ni J, Fournier MJ. 1993. *Nucleic Acids Res.* 21:5391–97
192. Bally M, Hughes J, Cesareni G. 1988. *Nucleic Acids Res.* 16:5291–5303

193. Sorger PK, Pelham HRB. 1987. *EMBO J.* 6:993–98
194. Dworniczak B, Mirault ME. 1987. *Nucleic Acids Res.* 15:5181–97
195. Zafarullah M, Wisniewski J, Shworak NW, Schieman S, Misra S, Gedamu L. 1992. *Eur. J. Biochem.* 204:893–900
196. Birnstiel M, ed. 1988. *Structure and Function of Major and Minor Small Nuclear Ribonucleoprotein Particles.* New York: Springer-Verlag

AUTHOR INDEX

A

Aamand JL, 888
Aaronson S, 629
Abamyr SM, 548
Abbas S, 120
Abbott M, 423
Abbott MH, 423
Abdel-Monem M, 157, 158
Abdel Motal UM, 470
Abe N, 411
Abe T, 504
Abe Y, 437, 800, 802, 803, 805, 825
Abelson J, 457, 594, 601, 602, 604, 888, 902, 910
Abogadie FC, 785, 789
Abrahams SL, 812
Abrahamsson T, 101
Abrams JS, 52
Abramson A, 520
Abramson T, 518, 519
Abreu E, 237, 238
Abumrad NA, 348, 352, 353, ·355, 356, 366, 368, 697
Acampora D, 216
Ach RA, 919, 922
Achari A, 666
Acharya KR, 829
Achkar CC, 228
Achtman M, 162
Ackerman S, 544, 545
Acton S, 304, 305, 412
Adachi E, 416, 420
Adachi T, 101, 104
Adam A, 31, 37
Adam EJH, 883, 884
Adam MA, 853
Adam SA, 611, 882–86
Adamich M, 673
Adams BA, 502, 509
Adams DH, 134
Adams ME, 505, 507, 508
Adams SL, 408
Adari HY, 605
Adashi EY, 358
Adelman R, 99
Adelstein S, 629
Adler DA, 117
Admon A, 336, 536, 537, 552
Adzuma K, 89, 156
Aebersold R, 179, 191, 195, 329, 330, 539, 623, 628, 634, 636, 637, 639–41
Aebi M, 888
Aebi U, 867, 870, 871, 873, 875, 880, 881, 883, 885
Affolter M, 295
Agabian N, 616, 903, 906, 912
Agapite J, 553

Agarwall S, 757
Agata K, 318, 319
Agellon LB, 236–40, 248, 250, 253
Aggerbeck LP, 236, 414
Agnew WS, 495, 500, 501
Agnish ND, 206, 208
Agranoff BW, 321
Aguet M, 624, 626, 631, 632
Ahdieh M, 629
Ahern TJ, 130
Ahlijanian MK, 503, 505
Ahluwalia N, 483
Ahmad NN, 422
Ahmed F, 734, 754
Ahmed-Ansari A, 487
Ahne F, 449
Aiello PD, 736
Ailhaud G, 348–53, 355, 356, 366, 368
Airey J, 487
Aisen P, 99
Aitken JR, 236
Ajisaka M, 41
Akaike T, 99
Akamatsu Y, 45, 321–24, 332, 337, 338
Akanuma Y, 351
Akey CW, 867, 868, 870, 871, 880, 883, 885
Akimoto S, 159
Akira S, 358, 359, 623, 626, 629, 630, 634, 639, 642
Akiyama K-Y, 470
Ala-Kokko L, 416, 422, 424, 425
Alam T, 695, 698, 699, 715
Albers JJ, 237–40, 247, 253
Albertine KH, 132
Alberts BM, 182, 192, 193, 223
Albrecht O, 679
Albright RE Jr, 239
Alden CJ, 597
Aldred P, 382, 387, 388, 391
Aldrich C, 270, 272, 471, 474, 479
Aldrich RW, 525, 526
Aldrich TH, 565, 630
Alexander GJ, 282
Alexander J, 470, 474, 484–86
Alexander RS, 381, 385
Alfano NL, 825
Alicot EM, 299
Alksne LE, 616
Allan BD, 85
Allan D, 333
Allen H, 466
Allen P, 765, 789, 838
Allen TD, 867, 871
Allen W, 51
Alles AJ, 207, 226

Alliotte T, 100
Allison LA, 534
Allmaier M, 451, 452
Allmang C, 607
Almarsson O, 81
Almers W, 502
Almouzni G, 543
Alt F, 645
Altaba ARI, 210, 213, 216
Altman S, 437, 907, 910, 916
Altschul SF, 290
Altura R, 446
Alvares K, 366
Alzari PM, 85, 303
Amabile-Cuevas CF, 453
Amachi T, 101
Amaldi F, 908–10, 913, 920, 923, 924
Amalric F, 902, 909, 910, 917, 923
Amanuma T, 99
Amato SV, 816
Amberg DC, 874, 876, 879, 882, 885, 888, 889
Ambrosio S, 800, 803, 805, 807, 818, 825
Ames BN, 98, 99, 105
Amherdt M, 698, 699, 711
Amiguet P, 294
Amiguet-Barras F, 294
Amin S, 206
Amiri KA, 917
Amit AG, 303
Amor-Gueret M, 377, 389
Ampe C, 359, 598
Amri E-Z, 348, 350, 352, 353, 355–57, 366–68
Amthauer R, 395, 567, 569, 571, 572, 575, 576, 580, 581, 583, 584, 586, 587
Anant S, 85
Ancelin ML, 325
Ancian P, 682
Ander S, 101
Anderl C, 439
Anderson BO, 665
Anderson CF, 181
Anderson DC, 114, 132, 133, 164, 296, 637
Anderson DH, 666
Anderson DW, 422
Anderson F, 543
Anderson K, 470, 474, 482
Anderson KS, 474, 478, 484
Anderson L, 293
Anderson M, 135
Anderson P, 466
Anderson R, 227, 482
Anderson RA, 337
Anderson RW, 472

935

SUBJECT INDEX

A

A1 protein
DNA polymerase III holoenzyme and, 191
protein-RNA recognition and, 611–12

A309Am mutant
thymidylate synthase and, 756

AAV P5 promoter
eukaryotic transcription factors and, 546

ABC transporters
MHC class I-restricted peptides and, 475

AbdB-like genes
retinoids in vertebrate development and, 225–26

Acetazolamide
carbonic anhydrases and, 387

Acetylcholinesterase
collagens and, 412
glycosylphosphatidylinositol linkage and, 565, 567

N-Acetylglucosamine
selectin-carbohydrate interactions and, 120

Acid phosphatases
6-phosphofructo-2-kinase/fructose-2,6-biphosphatase and, 822, 824

Aconitase
SELEX technology and, 793
superoxide radical and, 97–98, 107

Aconitine
voltage-gated ion channels and, 495

AcPases
6-phosphofructo-2-kinase/fructose-2,6-biphosphatase and, 816, 821–24, 829

Acquired immunodeficiency syndrome (AIDS)
mycobacteria and, 30

5-Acrylateuridine
synthesis of, 840

Actin
protein domains and, 271, 291

Actinomycin D
hepatitis delta virus and, 273
6-phosphofructo-2-kinase/fructose-2,6-biphosphatase and, 806

Activator interference
eukaryotic transcription factors and, 553

Acyl-CoA synthase
adipocyte differentiation and, 352

5-Acyluridines
synthesis of, 838–39

ADA proteins
eukaryotic transcription factors and, 553

ADD1 gene
adipocyte differentiation and, 367

Adenosine diphosphate (ADP)
pancreatic β-cell signal transduction and, 690, 700- 2, 705–7
6-phosphofructo-2-kinase/fructose-2,6-biphosphatase and, 808, 812–13, 828–29

Adenosine monophosphate (AMP)
6-phosphofructo-2-kinase/fructose-2,6-biphosphatase and, 812

Adenosine triphosphate (ATP)
DNA polymerase III holoenzyme and, 172–74, 176–77, 179–81, 183
eukaryotic phospholipid biosynthesis and, 326
eukaryotic transcription factors and, 544–46, 549
glycosylphosphatidylinositol linkage and, 576–78, 580, 586
MHC class I-restricted peptides and, 475, 477–80, 486
nuclear pore complex and, 882, 886
pancreatic β-cell signal transduction and, 690, 700- 7, 711
6-phosphofructo-2-kinase/fructose-2,6-biphosphatase and, 800, 808, 811–14, 818–19, 825
protein-RNA recognition and, 597, 604
SELEX technology and, 764, 780–81, 783–84, 787

Adipocyte differentiation
ADD1 gene and, 367
cAMP and, 352
C/EBP family and, 358–65
cell/cell contact at confluence and, 355

cell culture models and, 348–49
coordinate gene expression and, 356
determination and, 354–55
differentiation-specific element and, 367
fatty acid activated receptor and, 366
fatty acids and, 352–53
442/aP2 adipose-specific enhancer and, 352–53, 356, 360-61, 365–66, 370
glucocorticoid and, 352
HNF3/forkhead and, 367–68
IGF-1 and, 350–52
inducers and, 349–53
mitotic clonal expansion and, 355–56
mPPARγ2 and, 353, 365–66
perspectives and summary, 346–48
preadipocyte repressor element and, 368
second messenger pathways and, 349–53
sequence of events in, 353–57
stage-specific gene expression and, 368–70
terminal differentiation and, 356–57
transcriptional control and, 357–68

Adipose response elements (AREs)
adipocyte differentiation and, 348, 365–66

A-DNA
triplex DNA structures and, 73

ADP-ribosylation
6-phosphofructo-2-kinase/fructose-2,6-biphosphatase and, 828–29

Adrenal corticosteroids
plasma cholesteryl ester transfer protein and, 241

Agarose gel electrophoresis
plasma cholesteryl ester transfer protein and, 245

ω-Agatoxin
voltage-gated ion channels and, 505

Agkistrodon halys blomhoffii
interfacial enzymology and, 672

CUMULATIVE INDEXES

CONTRIBUTING AUTHORS, VOLUMES 60–64

CHAPTER TITLES, VOLUMES 60–64

From:

Name _____

Address _____

_____ Zip _____

Place
Stamp
Here

ANNUAL REVIEWS SERIES *Volumes not listed are no longer in print*	Prices, postpaid, **per volume.** USA/other countries	Regular Order Please send Volume(s):	Standing Order Begin with Volume:

❏ *Annual Review of* MICROBIOLOGY
Vols. 21-24, 26-45	(1967-70, 72-91) .... $41 / $46	
Vols. 46-47	(1992-93) .... $45 / $50	
Vol. 48-49	(1994 and Oct. 1995) .... $48 / $53	Vol(s). _____ Vol. _____

❏ *Annual Review of* NEUROSCIENCE
Vols. 1-14	(1978-91) .... $40 / $45	
Vols. 15-16	(1992-93) .... $44 / $49	
Vol. 17-18	(1994 and Mar. 1995) .... $47 / $52	Vol(s). _____ Vol. _____

❏ *Annual Review of* NUCLEAR AND PARTICLE SCIENCE
Vols. 12-41	(1962-91) .... $55 / $60	
Vols. 42-43	(1992-93) .... $59 / $64	
Vol. 44-45	(1994 and Dec. 1995) .... $62 / $67	Vol(s). _____ Vol. _____

❏ *Annual Review of* NUTRITION
Vols. 1-11	(1981-91) .... $43 / $48	
Vols. 12-13	(1992-93)..... .... $45 /$50	
Vol. 14-15	(1994 and July 1995) .... $48 / $53	Vol(s). _____ Vol. _____

❏ *Annual Review of* PHARMACOLOGY AND TOXICOLOGY
Vols. 2-3, 5-31	(1962-63, 65-91) .... $40 / $45	
Vols. 32-33	(1992-93) .... $44 / $49	
Vol. 34-35	(1994 and April 1995) .... $47 / $52	Vol(s). _____ Vol. _____

❏ *Annual Review of* PHYSICAL CHEMISTRY
Vols. 13-21, 23-27	(1962-70, 72-76)	
29-42	(1978-91) .... $44 / $49	
Vols. 43-44	(1992-93) .... $48 / $53	
Vol. 45-46	(1994 and Nov. 1995) .... $51 / $56	Vol(s). _____ Vol. _____

❏ *Annual Review of* PHYSIOLOGY
Vols. 19-53	(1957-91) .... $42 / $47	
Vols. 54-55	(1992-93) .... $46 / $51	
Vol. 56-57	(1994 and Mar. 1995) .... $49 / $54	Vol(s). _____ Vol. _____

❏ *Annual Review of* PHYTOPATHOLOGY
Vols. 3-20, 22-29	(1965-82, 84-91) .... $42 / $47	
Vols. 30-31	(1992-93) .... $46 / $51	
Vol. 32-33	(1994 and Sept. 1995) .... $49 / $54	Vol(s). _____ Vol. _____

❏ *Annual Review of* PLANT PHYSIOLOGY & PLANT MOLECULAR BIOLOGY
Vols. 17-23, 26-29	(1966-72, 75-78)	
31-42	(1980-91) .... $40 / $45	
Vols. 43-44	(1992-93) .... $44 / $49	
Vol. 45-46	(1994 and June 1995) .... $47 / $52	Vol(s). _____ Vol. _____

❏ *Annual Review of* PSYCHOLOGY
Vols. 4, 5, 8, 10, 22-24	(1953, 54, 57, 59, 71-73)	
26-30, 33-37, 39-42	(75-79, 82-86, 88-91)....$40 / $45	
Vols. 43-44	(1992-93) .... $43 / $48	
Vol. 45-46	(1994 and Feb. 1995) .... $46 / $51	Vol(s). _____ Vol. _____

❏ *Annual Review of* PUBLIC HEALTH
Vols. 1-12	(1980-91) .... $45 / $50	
Vols. 13-14	(1992-93) .... $49 / $54	
Vol. 15-16	(1994 and May 1995) .... $52 / $57	Vol(s). _____ Vol. _____

<table>
<tr><td colspan="3">**ANNUAL REVIEWS SERIES**
Volumes not listed are no longer in print</td><td>**Prices, postpaid, per volume.**
USA/other countries</td><td>Regular
Order
Please send
Volume(s):</td><td>Standing
Order
Begin with
Volume:</td></tr>
</table>

☐ *Annual Review of*   SOCIOLOGY

Vols.	1-17	(1975-91) ....$45 / $50	
Vols.	18-19	(1992-93) ....$49 / $54	
Vol.	20-21	(1994 and Aug. 1995)...............$52 / $57	Vol(s). _____ Vol. _____

Student discount prices are $10, per volume, off the prices listed above.

<table>
<tr><td colspan="3">**ANNUAL REVIEWS SERIES**
Volumes not listed are no longer in print</td><td>**Prices, postpaid, per volume.**
USA/other countries</td><td>Regular Order
Please send:</td></tr>
</table>

☐ **THE EXCITEMENT AND FASCINATION OF SCIENCE:**

Vol.	1	(1965) softcover ....$25 / $29	_____ copy(ies)
Vol.	2	(1978) softcover ....$25 / $29	_____ copy(ies)
Vol.	3	(1990) hardcover ....$90 / $95	_____ copy(ies)

(Volume 3 is published in two parts with complete indexes for Volumes 1, 2 and both parts of Volume 3. **Sold as a two part set only.**)

....

☐ **INTELLIGENCE AND AFFECTIVITY:** Their Relationship During Child Development,
   by Jean Piaget   (1981) hardcover ....$8 / $9   _____ copy(ies)

....

☐ **ANNUAL REVIEWS INDEX** on Diskette for DOS
   single copy $15
   one year subscription (4 quarterly editions) $50

A complete, searchable listing of titles, authors, keywords, and more for all *Annual Review* articles published since 1984.

Send to: **ANNUAL REVIEWS INC.**
a nonprofit scientific publisher
4139 El Camino Way • P. O. Box 10139
Palo Alto, CA 94303-0139 • USA

For individual articles from any *Annual Review*, call **Annual Reviews Preprints and Reprints (ARPR)** toll free 1-800-347-8007 from USA or Canada. From elsewhere call 1-415-259-5017.

☐ Please enter my order for the volumes indicated above.

Please note: Advance orders for volumes not yet published will be charged to your account upon receipt. Volumes will not be shipped before the month of publication indicated.

Prices are subject to change without notice.

☐ Add applicable CA sales tax   ☐ Add 7% Canadian GST

☐ Check or money order enclosed or   ☐ Charge my:

☐ VISA   ☐ MasterCard   ☐ American Express

☐ Apply Student/Recent graduate discount. $10 off, per volume. Proof of status enclosed.

☐ Optional UPS ground service, $2 extra per volume in the 48 contiguous states only. UPS requires street address.

Optional air service to anywhere in the world. Charged at actual cost and added to invoice.

☐ UPS Next Day Air
☐ UPS Second Day Air
☐ US Airmail

Signature _____

Account Number _____ Exp. Date _____/_____

Name _____
<small>Please print</small>

Address_____
<small>Please print</small>

_____ Zip Code _____

Daytime telephone _____ Fax _____

☐ Send a free copy of the current *Prospectus* or catalog. Area(s) of interest: _____

ANNUAL REVIEWS

a nonprofit scientific publisher
4139 El Camino Way
P.O. Box 10139
Palo Alto, CA 94303-0139 • USA

Annual Reviews publications may be ordered directly from our office; through stockists, booksellers and subscription agents, worldwide; and through participating professional societies. **Prices are subject to change without notice. We do not ship on approval.**

- **Individuals:** Prepayment required on new accounts. in US dollars, checks drawn on a US bank.
- **Institutional Buyers:** Include purchase order. Calif. Corp. #161041 • ARI Fed. I.D. #94-1156476
- **Students / Recent Graduates:** $10.00 discount from retail price, per volume. *Requirements:* **1.** be a degree candidate at, or a graduate within the past three years from, an accredited institution; **2.** present proof of status (photocopy of your student I.D. or proof of date of graduation); **3.** Order direct from Annual Reviews; **4.** prepay. This discount **does not** apply to standing orders, *Index on Diskette*, Special Publications, ARPR, or institutional buyers.
- **Professional Society Members:** Many Societies offer *Annual Reviews* to members at reduced rates. Check with your society or contact our office for a list of participating societies.
- **California orders** add applicable sales tax. • **Canadian orders** add 7% GST. Registration #R 121 449-029.
- **Postage paid** by Annual Reviews (4th class bookrate/surface mail). UPS ground service is available at $2.00 extra per book within the contiguous 48 states only. UPS air service or US airmail is available to any location at actual cost. UPS requires a street address. P.O. Box, APO, FPO, not acceptable.
- **Standing Orders:** Set up a standing order and the new volume in series is sent automatically each year upon publication. Each year you can save 10% by prepayment of prerelease invoices sent 90 days prior to the publication date. Cancellation may be made at any time.
- **Prepublication Orders:** Advance orders may be placed for any volume and will be charged to your account upon receipt. Volumes not yet published will be shipped during month of publication indicated.

NOTE For copies of individual articles from any *Annual Review*, or copies of any article cited in an *Annual Review*, call **Annual Reviews Preprints and Reprints (ARPR)** toll free 1-800-347-8007 (fax toll free 1-800-347-8008) from the USA or Canada. From elsewhere call 1-415-259-5017.

ANNUAL REVIEWS SERIES *Volumes not listed are no longer in print*	**Prices, postpaid, per volume.** **USA/other countries**	Regular Order Please send Volume(s):	Standing Order Begin with Volume:
☐ *Annual Review of* **ANTHROPOLOGY**			
Vols. 1-20 (1972-91)	$41 / $46		
Vols. 21-22 (1992-93)	$44 / $49		
Vol. 23-24 (1994 and Oct. 1995)	$47 / $52	Vol(s). _____	Vol. _____
☐ *Annual Review of* **ASTRONOMY AND ASTROPHYSICS**			
Vols. 1, 5-14, 16-29 (1963, 67-76, 78-91)	$53 / $58		
Vols. 30-31 (1992-93)	$57 / $62		
Vol. 32-33 (1994 and Sept. 1995)	$60 / $65	Vol(s). _____	Vol. _____
☐ *Annual Review of* **BIOCHEMISTRY**			
Vols. 31-34, 36-60 (1962-65,67-91)	$41 / $47		
Vols. 61-62 (1992-93)	$46 / $52		
Vol. 63-64 (1994 and July 1995)	$49 / $55	Vol(s). _____	Vol. _____
☐ *Annual Review of* **BIOPHYSICS AND BIOMOLECULAR STRUCTURE**			
Vols. 1-20 (1972-91)	$55 / $60		
Vols. 21-22 (1992-93)	$59 / $64		
Vol. 23-24 (1994 and June 1995)	$62 / $67	Vol(s). _____	Vol. _____

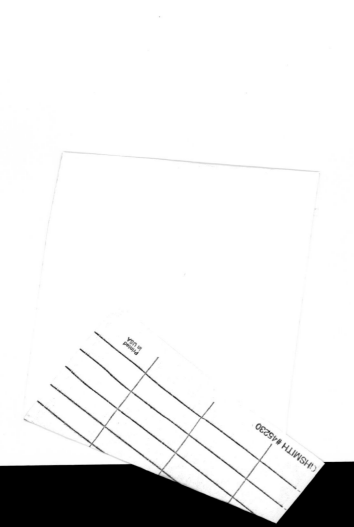